DOUGHERTY STATION

P9-CKV-579

Our study tools can help you master the material.

Maximize your study time and financial investment by using a variety of study tools that accompany this textbook.

Your purchase of a new copy of the text includes:

1. Free access to the text's Companion Website
2. A six-month subscription to **The Biology Place**
3. A six-month subscription to Evolution Lab (part of the *BiologyLabs On-Line* series) with **Evolution Lab** manual.

How to access the text's Companion Website:

1. Point your web browser to **www.awl.com/smith**
2. Click on the cover of Smith, *Ecology and Field Biology,* **Sixth Edition**

The website includes chapter objectives, self-testing programs, and message boards.

How to activate your subscription to **The Biology Place**:

1. Point your web browser to **www.awl.com/smith**
2. Click on the cover of Smith, *Ecology and Field Biology,* **Sixth Edition**
3. Click on **The Biology Place** icon
4. Select the "Register Here" link
5. Enter your pre-assigned Access Code exactly as it appears below:

Activation ID   BIELMQST12005131

Password   gullet

6. Select "Submit"
7. Complete the on-line registration form to establish your personal user ID and password
8. Once your personal ID and password are confirmed, go back to **The Biology Place** opening screen to enter the site with your new ID and password.

Your activation ID and password can used only once to establish your subscription. Once you register, the activation ID and password cease to function and are not transferable. *Write down your new ID and password. You will need them to access the site.*

If you did not purchase a new copy of this textbook, the Access Code may no longer be valid. However, if your instructor is recommending or requiring use of **The Biology Place**, you can find more information on purchasing a subscription directly on-line at **www.biology.com**, or you can purchase a subscription at your college bookstore if your professor has specifically requested it.

The Biology Place contains several tools to help you prepare for exams, including tutorial animations, simulations, chapter quizzes, articles, web links, and case studies.

Turn the page to learn more...

How to use your subscription to Evolution Lab:

1. Point your web browser to **biologylab.awlonline.com**
2. Click on "Evolution Lab". Enter the site using the same user name and password you created to access The Biology Place.
3. Use the lab manual that came bundled with your text to explore the many experiments possible on this site.

How to contact Technical Support:

For technical support, please visit the technical support website at **www.awlonline.com/techsupport**, or send an e-mail to **online.support@pearsoned.com**, with a detailed description of your computer system and the technical problem. You can also call our tech support hotline at 800-677-6337, Monday - Friday, 8:00 a.m. - 5:00 p.m. EST.

How you can contact us:

We invite your comments and suggestions via e-mail at **question@awl.com**.

We wish you great fun and success in your ecology course!

ECOLOGY & FIELD BIOLOGY

SIXTH EDITION

Robert Leo Smith
West Virginia University

Thomas M. Smith
University of Virginia

Illustrated by
Robert Leo Smith, Jr.

CONTRA COSTA COUNTY LIBRARY

Benjamin
Cummings

An imprint of Addison Wesley Longman, Inc.

3 1901 03717 0950

Capeto   San Francisco   Boston   New York   City
Montreal   Munich   Paris   Singapore   Sydney   Tokyo   Toronto

Acquisitions Editor: Elizabeth Fogarty
Project Editor: Heather Dutton
Assistant Editor: Chriscelle Merquillo
Managing Editor: Wendy Earl
Design Manager: Bradley Burch
Production Editor: Scott Hitchcock
Text Designers: Yvo Riezebos and Electronic Publishing Services Inc., N.Y.C.
Cover Designer: Yvo Riezebos
Cover Photo: © VCG 1999/FPG
Art Studio: Robert L. Smith, Jr. and Electronic Publishing Services Inc., N.Y.C.
Photo Researcher: Electronic Publishing Services Inc., N.Y.C.
Manufacturing Coordinator: Vivian McDougal
Project Coordination and Electronic Page Makeup: Electronic Publishing Services Inc., N.Y.C.
Marketing Manager: Josh Frost
Printer and Binder: Quebecor World
Cover Printer: Lehigh Press, Inc.

Library of Congress Cataloging-in-Publication Data
Smith, Robert Leo
    Ecology and field biology / Robert Leo Smith, Thomas M. Smith.—6th ed.
        p. cm.
    Includes bibliographical references.
    ISBN 0-321-04290-5
    1. Ecology.   2. Biology—Field work.   I. Smith, T. M. (Thomas
Michael), 1955-   . II.Title.
QH541.S6 2001
577—dc21   00-043121

ISBN 0-321-04290-5

2 3 4 5 6 7 8 9 10 - QWV - 04 03 02 01 00

Copyright © 2001, Benjamin Cummings, an imprint of Addison Wesley Longman, Inc.

All rights reserved.

Printed in the United States of America. This publication is protected by Copyright and permission
should be obtained from the publisher prior to any prohibited reproduction, storage in a retrieval
system, or transmission in any form or by any means, electronic, mechanical, photocopying,
recording, likewise. For information regarding permission(s), write to: Permissions Department.

The text paper of this book is recycled paper containing 10% post-consumer fiber by fiber weight.
The paper is produced from harvested forests managed under the tenets of the Sustainable Forestry
Initiative under third party verification.

Benjamin
Cummings

About the cover: Desert cacti at dawn in
Joshua Tree National Park, California.

To Carrie

The last of the F2 generation

BRIEF CONTENTS

CONTENTS

PART III The Organism In Its Environment   79

CHAPTER 5
ADAPTATION   80

CHAPTER 6
PLANT ADAPTATIONS TO THE ENVIRONMENT I: PHOTOSYNTHESIS, AND THE LIGHT ENVIRONMENT   85

CHAPTER 7
PLANT ADAPTATIONS TO THE ENVIRONMENT II: THERMAL, MOISTURE, AND NUTRIENT ENVIRONMENTS   98

CHAPTER 8
ANIMAL ADAPTATIONS TO THE ENVIRONMENT   120

PART VI Ecological Genetics   355

CHAPTER 19

POPULATION GENETICS   356

PART VII The Community   381

CHAPTER 20

COMMUNITY STRUCTURE   382

CHAPTER 21

COMMUNITY DYNAMICS   403

CHAPTER 22

PROCESSES CONTROLLING COMMUNITY DYNAMICS   427

CHAPTER 23
LANDSCAPE ECOLOGY   449

PART VIII
The Ecosystem   477

CHAPTER 24
ECOSYSTEM PRODUCTIVITY   478

CHAPTER 25
BIOGEOCHEMISTRY I:
NUTRIENT CYCLING   504

CHAPTER 26
BIOGEOCHEMISTRY II:
GLOBAL CYCLES
AND HUMAN IMPACTS   524

CHAPTER 27
BIOGEOGRAPHY OF ECOSYSTEMS 543

PART IX Comparative Ecosystem Ecology 561

CHAPTER 28
GRASSLAND TO TUNDRA 562

CHAPTER 29
FORESTS 601

CHAPTER 30
FRESHWATER ECOSYSTEMS 629

PREFACE

More readers, we suspect, flip past prefaces than read them; yet they remain important. In textbooks, the preface is where authors can give us a real glimpse of themselves and the reasons for writing the book, how they think it should be used, and what they hope readers will gain from it.

Ecology and Field Biology first appeared in 1966. That first edition had straightforward objectives: to present a balanced introduction to ecology—plant and animal, theoretical and applied, physiological and behavioral, population and ecosystem. These remain the objectives of the sixth edition, but the field of ecology has changed and grown. Today, ecology is an essential curricular component in a wide variety of departments, including biology, environmental sciences, forestry and agriculture, wildlife ecology, and anthropology as well as being the cornerstone of curriculum in ecology and evolutionary biology. In each of these programs, ecology is a component of quite different courses of study. For example, students within a biology or ecology program have a background in physiology and genetics, whereas those in an environmental sciences department are grounded in physical sciences such as atmospheric sciences or geology. These differing contexts within which ecology is taught cause instructors to emphasize different facets of the larger field of ecology.

Our book has grown with the field of ecology. This growth reflects a philosophy that this book functions not only as a classroom text, but as a reference that provides a full introduction to the ever-broadening field of ecology. Many curricula restrict ecology to a one-semester or even a quarter course; it is difficult, if not impossible, to adequately cover general ecology in that amount of time. To help the instructor, we have compartmentalized our presentation of topics to make the sixth edition a flexible text that still retains a logical flow.

REVISIONS IN THE TEXT

Since the first edition appeared in 1966, ecology has experienced major growth and development. Revisions of the text have reflected those changes. Each revision has been thorough, retaining the best of the old and incorporating new developments and approaches.

The present revision has resulted in major shifts in the presentation of certain topics:

- The material in Part 2 of the fifth edition has been restructured and expanded into Parts 2 and 3 of the sixth. Part 2 examines the physical environment, exploring climate (Chapter 2), the hydrologic cycle (Chapter 3), solar radiation and the light environment (Chapter 3), and soils (Chapter 4).

- New chapters in Part 3 provide an introduction to ecophysiology that emphasizes the evolutionary tradeoffs involved in adaptations allowing organisms to successfully survive, grow, and reproduce under varying environmental conditions.

- The discussion of community ecology in Part 7 has been extensively revised and reorganized to focus on structure, pattern in time and space, and processes.

- The topics of island biogeography, disturbance, patch dynamics, and edge and ecotone environments have been reorganized as a single unit and expanded to provide an introduction to landscape ecology.

- The discussions of ecosystem structure and energetics in Chapters 10 and 11 of the fifth edition have been combined into a single discussion of ecosystem productivity.

- Chapter 12 of the fifth edition, Biogeochemical Cycles, has been expanded and separated into two chapters in Part 8 of the sixth edition. The different constraints on the cycling of nutrients in aquatic and terrestrial ecosystems are contrasted and highlighted.

- New to the sixth edition is a chapter exploring global environmental change, specifically the potential impact of rising atmospheric concentrations of carbon dioxide and other greenhouse gases on the global climate system.

- A common theme weaving through the book concerns the tradeoffs involving the acquisition of carbon and other essential resources by organisms and the consequences of these tradeoffs on higher levels of organization.

ORGANIZATION

The text is divided into nine parts and 32 chapters.

Part 1 is introductory and remains much the same as in the fifth edition. Now condensed to one chapter, it provides a brief overview of ecology—what it is, how it developed, and its importance and relevance to current environmental issues—and introduces ecology as an experimental science. Although students may come into an ecology course with some grasp of the scientific method, they bring it from the context of laboratory experimentation, not from a field approach. This chapter introduces hypothesis testing and the use and place of predictive models in ecology.

Part 2 deals with the physical environment in which life exists. Chapter 2 discusses climate, which greatly influences the environmental conditions under which organisms live. Chapter 3 introduces the characteristics of light, temperature, water, and nutrients. Chapter 4 explores soil, its development, characteristics, and role as a substrate for life in terrestrial environments.

Part 3 emphasizes the adaptations of organisms to variations in their physical environment introduced in Part 2. These new chapters provide an introduction to ecophysiology, examining the evolutionary tradeoffs involved in adaptations that allow organisms to successfully survive, grow, and reproduce under varying environmental conditions. Chapter 5 introduces adaptation and homeostasis, concepts basic to an appreciation of physiological ecology and natural selection. Chapters 6, 7, 8, and 9 include much of the material found in Chapters 4–8 of the fifth edition. However, this material is now organized around trophic groups, dividing autotrophs and heterotrophs, and consumers and decomposers within the latter category. Organizing the chapters by trophic groups emphasizes the fundamentally differing constraints imposed on different organisms relating to how they acquire carbon and other essential nutrients. Such a division also provides a direct link with later discussions of processes controlling community dynamics in Part 7, and ecosystem productivity and the flow of energy and matter through ecosystems presented in Part 8. Organized in this way, these new chapters also introduce ecophysiology within the context of understanding the structure and dynamics of communities and ecosystems.

Chapter 6 emphasizes the photosynthetic process as the means of carbon acquisition for life, and its dependence on light. Chapter 7 covers the response of plants to temperature, moisture regimes, and nutrient availability. Chapter 8 deals with the means of animal acquisition of carbon and the response of animal life to temperature, moisture, and nutrients. Chapter 9 examines decomposers, their role in converting organic matter into inorganic nutrients, and the ultimate release of carbon fixed in photosynthesis, completing the circle started with photosynthesis in Chapter 6.

Parts 4 and 5 examine topics within the general area of population ecology. We have divided Part 5 (Population Ecology) from the fifth edition into two parts. Part 4 examines Interspecific Population Ecology; Part 5 explores Population Interactions. In Part 4, Chapter 10 covers demography and considers both unitary and modular populations. Chapter 11 explores population growth. Chapter 12 considers density-dependent and density-independent influences, emphasizing the role of intraspecific competition in population regulation, with examples from both plant and animal populations. Topics include dispersal, and that part of social behavior involving social dominance and territoriality as expressions of competitive interactions. As such, much of Chapter 12 deals

with behavioral ecology, a topic Chapter 13 expands on, with discussions of mating systems, sexual selection, reproductive effort, and parental care.

Part 5 concerns interspecific relationships. Chapter 14 examines interspecific competition, and Chapter 15 is an overview of predation theory, including optimal foraging theory. Chapter 16 expands on the general model of predation introduced in Chapter 15 to include discussions of two major predator-prey systems: plant-herbivore and herbivore-carnivore, including cannibalism, intraguild predation, and the reciprocal responses of prey and predator. Also discussed are predator-prey cycles. The first section of Chapter 17 explores parasitism and disease, as well as social parasitism, including population dynamics and evolutionary responses. The second section deals with mutualism, the coevolution of the various facets of mutualism, and its possible origins and effects on interacting populations. Part 5 ends with a consideration of human interactions with natural populations, including the effects of exploitation, pest control, population restoration, and extinction.

Chapter 19, the former Chapter 21 (Population Genetics) of the fifth edition, has been separated to form its own section, Part 6, to emphasize the role of population genetics in ecology, especially conservation biology and landscape ecology. The single chapter considers the basics of population genetics, the genetics of small and fragmented populations and its relationship to effective population size, and viable populations.

Part 7, The Community, covers much of what is considered community ecology, although no distinct line can be drawn between community ecology and the discussion of population interactions that form Part 5. Chapter 20 examines community structure, focusing on the descriptive study of the physical and biological structure of communities. Chapter 21 introduces community dynamics, examining how the structure of communities varies in time and space. This chapter includes much of the material in Chapter 30 of the fifth edition. Chapter 22 examines processes that control community dynamics. There we examine current theories regarding the underlying mechanisms that structure communities. The topics in Chapter 22 draw on the discussion of species adaptations to environmental variation presented in Part 2, linking pattern and process across spatial and temporal scales. In addition, the topics of island biogeography, disturbance, patch dynamics, and edge and ecotone environments, previously dispersed among a number of chapters, have now been reorganized as a single unit, Chapter 23, and expanded to provide an introduction to landscape ecology. These topics relate back to the material in Chapter 19 on the effects of habitat fragmentation and isolation on the genetics of small populations.

Part 8 examines the ecosystem concept. Building on the discussion of community ecology in Part 7, the topics of ecosystem structure and energetics (Chapters 10 and 11 of the fifth edition) have been combined into Chapter 24, an integrated discussion of ecosystem productivity. This chapter examines trophic structure and the flow of energy through the ecosystem. Discussion of the environmental constraints on primary and secondary productivity link directly with Part

2, which covered constraints and trade-offs in species adaptations to varying environmental conditions. Chapter 25 provides a general discussion of nutrient cycling within ecosystems, examining how biotic and abiotic factors interact to regulate the flow of essential nutrients through the ecosystem. The chapter makes the direct link between primary productivity (Chapter 6) and decomposition (Chapter 9) in regulating the flow of energy and matter through ecosystems. The discussion in this chapter draws heavily on Chapter 9 as it relates to the constraints of climate and litter quality on the rate of nutrient cycling in ecosystems. Particular emphasis is given to contrasting constraints on the cycling of nutrients in aquatic and terrestrial ecosystems. In addition to presenting an overview of nutrient cycling and biogeochemistry, the chapter discusses specific biogeochemical cycles from an ecosystem perspective and how the cycles of essential nutrient species are directly linked as components of organic matter. Chapter 26 expands the more localized view of biogeochemical cycles within an ecosystem to provide a global perspective, also examining the impacts of human populations on regional and global biogeochemical cycles. The final chapter in this part, Chapter 27, considers the biogeography of ecosystems and its relationship to biodiversity. This chapter leads into Part 9.

Comparative Ecosystem Ecology, Part 9, has been a hallmark of this text. These chapters provide a comparative study of the structure and function of major ecosystems, including differences in structure, energy flow, nutrient cycling, and adaptations of organisms to those environments. The common thread is the nitrogen cycle, which serves to emphasize the functional differences among the various major ecosystems.

Part 9 introduces concepts unique to certain ecosystems, such as nutrient spiraling in streams and microbial loops in marine ecosystems. In addition, it provides students with the background they need to understand the major problems receiving so much attention in ecology and environmental conservation—the destruction of old-growth forests, tropical deforestation, wetland values and losses, habitat fragmentation, acidification, and the like.

All of these chapters contain a good deal of ecophysiology and evolutionary adaptations as well as ecosystem structure and function. This content provides the instructor with additional material that can be incorporated with topics in other chapters. Chapter 28 provides an overview of grasslands, tropical savannas, shrublands, deserts, and tundra. Chapter 29 is devoted to various forest ecosystems, from the taiga to the tropical rain forest. Chapter 30 provides a survey of lentic and lotic ecosystems, with emphasis on the differences in the pathways of energy flow, energy flow between the two types of ecosystems, and nutrient cycling in flowing water. Also considered are the delineation of wetlands and the effects of dams on lotic systems. Chapter 31 deals with the marine environment from the intertidal zones to the open sea. It includes discussions of hydrothermal vents, roles of planktonic microbes in marine food webs, mangrove tidal swamps, and nutrient cycling in coral reef ecosystems.

New is Chapter 32, which explores the topic of global environmental change, specifically the potential impact of rising atmospheric concentrations of carbon dioxide and other greenhouse gases on the global climate system. The chapter discusses how forest clearing and fossil fuel consumption are altering the global carbon cycle, exploring key processes involved in the exchange of carbon among the atmosphere, oceans, and terrestrial ecosystems. The chapter examines how scientists are investigating the potential influence of rising atmospheric concentrations of CO_2 on the global climate system, and how changes in the Earth's climate system will influence sea level, natural ecosystems, agricultural production, and human health.

The text also features a set of essays, Ecological Applications, from *Elements of Ecology,* updated edition. These illustrate how the principles of ecology intimately relate to everyday life in ways that most of us never realize.

PEDAGOGY

The sixth edition features:

- Six new Ecological Applications essays
- Chapter-opening list of concepts
- Topical outline of each chapter
- Summary of each chapter
- Review questions, including critical-thinking questions
- Important terms boldfaced when they are first introduced
- Glossary of important terms at the back of the text
- Cross references to related topics found in other chapters.

ILLUSTRATION PROGRAM

Back in 1966 the first edition featured the use of spot drawing of organisms in graphs. This feature has now appeared in most other ecology texts.

The illustration program, an amalgam of the old and the new, adds a new visual appeal to the text. We have retained much of the superb pen and ink work from the first edition by the late Ned Smith. All of the remaining illustrations have been redesigned, redrawn, and generated on the computer by Robert Leo Smith, Jr. All the photographs, some retained from the fifth edition and some new, have been carefully selected to supplement the text, not simply to add color to it. Many new illustrations and graphs have been added. The maps of vegetative distribution found on the inside covers of the fifth edition now make up Appendix C.

APPENDIX

Appendixes A and B, both features of all previous editions, have been retained in the text. The list of more than 170 Journals of Interest to Ecologists, that in one way or another relate to ecology, is now found on the web.

Appendixes A and B provide a manual of methodology for:

- Sampling terrestrial vegetation
- Sampling aquatic vegetation
- Dendrochronology
- Palynology
- Estimating animal population size
- Finding population dispersion and interspecific association
- Community similarity
- Community ordination
- Species diversity
- Population structure
- Life and fecundity tables
- Reproductive values
- Rate of increase.

Appendix C consists of the Terrestrial Ecoregions of the World Map and the Holdridge Life Zone Map.

Some reviewers have suggested that the Appendixes incorporate computer programs. We did not for several reasons. One is space. The second is that several quantitative ecology texts that include computer programs are available. We highly recommend four: *Ecological Methods* by C. Krebs (1999 Benjamin Cummings), *Statistical Ecology* by Ludwig and Reynolds (Wiley 1989), *Applied Population Ecology using* RAMAS® *EcoLab,* 2nd ed., by Akcakaya et al. (Sinauer Associates 1999), and *Conservation Biology with* RAMAS® *EcoLab* by Shultz et al. (Sinauer Associates 1999).

One of us (RLS) has used basic computer programs in courses, with some reservations. Students are too prone to toss data into a computer program without understanding the underlying principles of the models. To get around this, I have had the students work out diversity indices, ordinations, and population problems without computers first. Only then do they understand what is involved in their calculations.

The most difficult and time-consuming part of the text to prepare has been the Bibliography, which contains several thousand titles. Citations of older references have been retained where appropriate; truth does not change over the years. A great number of new titles have been cited, necessary to thoroughly update the text.

This sixth edition has received input from a number of reviewers listed in the acknowledgments who pointed out inconsistencies, inaccuracies, and weakness and strengths, and suggested changes and the incorporation of new material. Although the manuscript has been reviewed, copyedited, checked, and rechecked for typographical and conceptual errors, some will slip through. If you find errors, please call them to my attention.

ANCILLARIES

A complete set of supplementary materials is available to support the use of *Ecology and Field Biology*, sixth edition, for both students and instructors.

Instructor's Art CD-ROM: The art from the text in Power-Point. *Ecology and Field Biology* is one of the few—perhaps the only—text to offer this resource for instructors.

Instructor's Manual: The manual contains new outlines, topics for discussion, and updated references. 0-321-04285-9

Test Bank: Completely new test questions: multiple choice, true/false, short answer (about 20 of each per chapter). These questions also test students' ability to read and understand graphs and figures. 0-321-04293-X

Transparency Acetates: 0-321-04292-1

MEDIA SUPPLEMENTS

Companion Website
http://www.awl.com/smith
For students—multiple choice questions, essay tests, and web links. For instructors—all the art pieces from the book for use in PowerPoint presentations.

Biology Labs Online: Evolution Lab
http://biologylab.awlonline.com
Part of the Biology Labs Online series, Evolution Lab presents evolution in action by showing adaptation by natural selection over centuries. A lab manual and password comes with every new student copy of the text.

The Biology Place
http://biology.com
A comprehensive collection of biology resources and information. Every new book purchase includes a subscription.

ACKNOWLEDGMENTS

No textbook is the product of the author alone. Although the author writes the text, the material of which it is composed represents the work of hundreds of ecological researchers who have spent lifetimes in the field and laboratory. Their published works on experimental results, observations, and conceptual thinking provide the raw material out of which a textbook is fashioned.

Revisions of a textbook depend heavily on the input of users. The sixth edition has received input from instructors and students who pointed out inconsistencies and inaccuracies, and suggested changes and the incorporation of new material. We took their suggestions seriously, and incorporated many of them; the book is much improved from their input. We are deeply grateful to the following reviewers who provided detailed critiques and helpful suggestions: Judy E. Bluemer, Morton College, Cicero, IL; Michelle A. Briggs, Lycoming College, Williamsport, PA; Dan Binkley, Colorado State University, Fort Collins, CO; Richard D. Brown, Northern Virginia Community College, Stirling, VA; Young D. Choi, Purdue University, Calumet Hammond, IN; Patricia J. Clark, Cumberland College, Williamsburg, KY; John Cruzan, Genera College, Beaver Falls, PA; Diane Dudzinski, Washington State Community College, Marietta, OH; William L. Hallahan, Nazareth College, Rochester, NY; Shannon Kuchel, Colorado Christian University, Lakewood, CO; John C. Maerz, Binghamton University, Binghamton, NY; Vicky Meretsky, Indiana University, Bloomington, IN; Juliana Mulroy, Denison University, Granville, OH; Howard S. Neufeld, Appalachian State University, Boone, NC; T. J. Sarro, Mount Saint Mary College, Newburgh, NY; Robert E. Stockhouse II, Pacific University, Grove, OR; Morrel H. Sweet II, Texas A & M University, College Station, TX.

In his role as fact checker, Scott Elliott, University of Washington, picked up a number of inaccuracies easily overlooked by authors and copy editors.

John Sencindiver of the Soil Science Department, West Virginia University, critically reviewed Chapter 4, Soils, and provided many helpful suggestions for its improvement.

Jay Zieman, Department of Environmental Sciences, University of Virginia, provided photos and artwork resources for the examples of seagrass dynamics in Chapter 21.

Barbara J. Bentz, Entomologists and Project Leader at the Rocky Mountain Research Station, U.S. Forest Service at Logan, UT, was extremely helpful in providing photo resources for examples of herbivory in the Rocky Mountain region.

Dan Binkley, Colorado State University, thoroughly reviewed Chapter 29, Forests, resulting in its reorganization and improvement in content.

TMS thanks the students of EVSC 320, Fundamentals of Ecology, at the University of Virginia for acting as subjects in a neverending experiment in teaching ecology to undergraduates. Their responses, critiques, and suggestions are reflected in the reorganization and content of this text.

The essays, which appeared in *Elements of Ecology*, fourth edition, are a collaborative effort among Tom Smith, Todd Dennis, a postdoctoral student in Environmental Science, and Elizabeth Zayatz, who was the technical editor for the essays.

As in the past, this text is sort of a family industry, but even more so with this edition. My first son, Robert Leo, Jr., a graphic artist, rendered all of the color graphics under con-

siderable time pressure while working on two other books as well. His familiarity with the text, artistic ability, and skill at computer graphics allowed a close collaboration in the development of illustrations, not possible otherwise. My second son, Thomas Michael, Associate Professor, Environmental Science Department, University of Virginia, who had input into the previous two editions and developed the Holdridge Life Zone Map in the fifth edition, has joined as coauthor. He brings to this text a fresh perspective, new ideas, a global approach, and an understanding of the needs and problems of both major and nonmajor students in an ecology course.

The book could not have arrived at its present stage without the help, encouragement, and especially, patience of the staff at Benjamin Cummings and Electronic Publishing Services Inc. Elizabeth Fogarty, Acquisition Editor, kept the faith that the project would be completed within the deadline in spite of the time-consuming major changes in the text and unanticipated interruptions in the flow of work experienced by the authors. Erika Buck, who traded the publishing world for graduate school, initially served as Project Editor and got the revision for the sixth edition under way. Ginnie Simione-Jutson, Senior Project Editor, took over the revision project and set up the ground rules and schedules, and handed the project over to Heather Dutton as Project Editor, who had the task of keeping the project running smoothly and gently prodding the three of us to keep moving. Chriscelle Merquillo secured and corresponded with the reviewers, and both she and Anne Hikido reviewed the incoming chapters and made sure text, captions, figures, and tables were complete before sending the material on to production.

Scott Hitchcock of Electronic Publishing Services Inc. had the task of pulling manuscript and art through final stages of the book, coordinating copyediting, page proofs, and art, and keeping the three of us on a very tight schedule. Francis Hogan of Electronic Publishing Services did an excellent job of selecting many new photos needed for color illustrations, skillfully choosing photos to match our specific requirements for each subject.

Through it all our spouses, wives Alice and Nancy, endured the time demand imposed on us by book production. Alice took care (and still does) of all the problems of living, while I devoted full time including weekend and evenings working on this book. She has patiently endured book widowhood for years. To her, I am very grateful.

Robert Leo Smith

PART ONE

Introduction

C H A P T E R 1

Ecology: Meaning and Scope

CONCEPTS

1. Ecology is the study of the structure and function of nature.

2. Ecology developed from many roots, but its beginnings trace back to natural history and plant geography.

3. Ecology has branched into many subdivisions, many of them specialized.

4. The principles of ecology provide the scientific basis for the solution to many of our environmental problems.

5. Modern ecology is an empirical, experimental science.

6. Science involves the testing of hypotheses.

7. A hypothesis is a statement that can be tested.

8. Testing hypotheses involves the collection of data by observation and experimentation, and the analysis of that data.

9. Useful in ecological studies are models, explicit sets of hypotheses relating pattern and process.

ECOLOGY DEFINED

What is ecology? Ask nonecologists and they will probably answer that ecology has something to do with the environment or with saving it. Before the 1960s, few of them could have given you any answer. If you had asked a biologist in the same time period, you would probably have gotten some vague answer implying that ecology was "quantified natural history." Ecology only became a household word in the 1960s, through the environmental movement, and then its popular meaning became confounded with environmentalism.

The origin of the word *ecology* is the Greek *oikos,* meaning "household," "home," or "a place to live." It is derived from the same root word as *economics,* "management of the household." Ecology, then, could be considered as the economics of nature. Clearly, ecology deals with the organism and its place to live, its environment. Ecologists continue to work toward a deeper definition. Here is a sampling of their attempts:

> "the study of structure and function of nature." (Odum 1971:3)
>> "the scientific study of the distribution and abundance of animals." (Andrewartha 1961:10)
>> "the scientific study of the interactions that determine the distribution and abundance of organisms." (Krebs 1985:4)
>> "the scientific study of the relationships between organisms and their environments." (McNaughton and Wolfe 1979:1)
>> "the study of the relationships between organisms and the totality of the physical and biological factors affecting them or influenced by them." (Pianka 1988:4)
>> "the study of the adaptation of organisms to their environment." (Emlen 1973:1)
>> "the study of the principles which govern temporal and spatial patterns for assemblages of organisms." (Fenchel 1987:12)
>> "the study of the patterns of nature and how those patterns came to be, and how they change in space and time." (Kingsland 1985:1)
>> "the study of organisms and their environment—and the interrelationships between the two." (Putman and Wratten 1984:13)
>> "the study of the relationship between organisms and their physical and biological environments." (Ehrlich and Roughgarden 1987:3)

The author of the term, Ernst Haeckel, defined ecology as "the body of knowledge concerning the economy of nature—the investigation of the total relationships of the animal both to its inorganic and its organic environment; including, above all, its friendly and inimical relations with those animals and plants with which it comes directly or indirectly into contact—in a word ecology is the study of all those complex interrelations referred to by Darwin as the conditions for the struggle for existence."

None of these definitions is fully satisfactory. They are either too restrictive or too vague, and the original definition relates only to animal ecology. Most definitions apply to population ecology and overlook ecosystem function. The subject has outgrown them.

Instead of trying to define ecology in a manner that attempts to encompass all that it involves, we might better to consider a functional definition of ecology. Ecology works at characterizing the patterns seen in nature, studying the complex interactions among organisms and their environments, and understanding the mechanisms involved in biological diversity.

THE DEVELOPMENT OF ECOLOGY

Just as there is no consensus on the definition of ecology, so there is no agreement on its beginnings. It is more like a multistemmed bush than a tree with a single trunk. Some historians trace the beginnings of ecology to Darwin, Thoreau, and Haeckel; others to the Greek scholar Theophrastus, a friend and associate of Aristotle, who wrote about the interrelationships between organisms and the environment.

Plant Ecology

The modern impetus to ecology came from the plant geographers. They discovered that, although plants differed in various parts of the world, certain similarities and differences demanded explanation. They looked to climate as a possible answer, because similar climates supported similar vegetation. Carl Ludwig Willdenow (1765–1812), one of the early influential plant geographers, championed this explanation. His ideas caught the attention of a wealthy young Prussian naturalist, Friedrich Heinrich Alexander von Humboldt (1769–1859) (Figure 1.1a). He sailed in 1799 with the French botanist Aimé Bonpland for a five-year expedition through tropical Spanish America. They traveled though Mexico, Cuba, Venezuela, and Peru and explored the Orinoco and Amazon rivers. Humboldt described these travels and tropical America's plant and animal life in a 30-volume work, *Voyage to the Equatorial Regions.* In these books Humboldt described vegetation in terms of physiognomy, correlated vegetation types with environmental characteristics, and coined the term *association.*

Travel to the tropics continued to stimulate plant ecologists. A Danish botanist, Johannes Warming (1841–1924) (Figure 1.1b), spent three years studying tropical vegetation in Brazil, which he described in a book 30 years later. His major contribution was the book *Plantesamfund: Grundtrak af den okologiske Plantegeografu,* published in 1895, in which he emphasized the importance of moisture, temperature, and soil in patterns of vegetation. The book greatly influenced the development of ecology. Following Warming, another tropical traveler, the German botanist Andreas Schimper (1856–1901), published *Pflanzengeographie auf physiologischer Grundlage* (1898). In it he explained

(a)   (b)   (c)

FIGURE 1.1   Plant geographers. (a) Friedrich Heinrich Alexander von Humboldt. (b) Johannes Warming. (c) F. E. Clements.

regional differences in vegetation as a function of moisture and temperature.

Other plant geographers stayed closer to home. Nevertheless they made major advances in plant ecology, especially the emerging concept of vegetation change over time. Anton Kerner (1831–1898) was one of these geographers. Commissioned to survey the vegetation of eastern Hungary and Transylvania, he described plant development in *Plant Life of the Danube Basin* (1863). He pioneered the use of experimental transplant gardens at various elevations in the Tyrolean Alps to study the growth and behavior of plants taken from alpine and lowland sites. Later, the Polish botanist Jozef Paczoski (1864–1941) described how plants modify their environment by creating microenvironments. Because his book was published in Slavic, it was belatedly discovered by ecologists outside the Slavic world. He is now recognized as the father of plant sociology or **phytosociology.**

The study of plant communities developed along separate paths in Europe and America. In Europe plant ecologists concentrated on describing the plant community. Christen Raunkiaer (1860–1938) of Denmark contributed a scheme of life form classification and quantitative methods of sampling vegetation, the data from which could be treated statistically. Later Josias Braun-Blanquet developed methods of community sampling, data reduction, and the classification and nomenclature of plant communities. A. G. Tansley, however, urged a more experimental approach to plant ecology. His views on ecology and research anticipated by years the type of ecological studies that emerged in the 1970s.

In America plant ecologists were far more interested in how plant communities develop. Human settlements were destroying the original forests and grasslands, creating pro-

found changes in vegetation. Interest in these disturbances led to pioneer studies of the dynamics of vegetation and vegetation changes over time, or plant succession. A doctoral study on the succession of plant life on Indiana sand dunes by H. E. Cowles (1897) established the study of plant succession as one of the central concepts of modern ecology. A major leader in these studies and the development of American plant ecology was F. E. Clements (1874–1945) (Figure 1.1c). Dogmatic and convincing, Clements quickly became the major theorist of plant ecology in America. He promoted the idea of the community as a superorganism and gave ecology a hierarchical framework still reflected in modern views of the field.

Animal Ecology

Animal ecology developed later than plant ecology and along lines divorced from it. The beginnings of animal ecology can be traced to two Europeans, R. Hesse of Germany and Charles Elton of England. Elton's *Animal Ecology* (1927) and Hesse's *Tiergeographie auf logischer grundlage* (1924), translated into English as *Ecological Animal Geography,* strongly influenced the development of animal ecology in the United States. Charles Adams and Victor Shelford were two pioneering U.S. animal ecologists. Adams published the first textbook on animal ecology, *A Guide to the Study of Animal Ecology* (1913). Shelford wrote *Animal Communities in Temperate America* (1913).

Shelford gave a new direction to ecology by stressing the interrelationship of plants and animals. Ecology became a science of communities. Some earlier European ecologists, particularly the marine biologist Karl Mobius, had developed the general concept of the community. In his essay "An Oyster Bank Is a Biocenose" (1877), Mobius explained that the oyster bank, although dominated by one animal, was really a complex community of many interdependent organisms. He proposed the word *biocenose* for such a community. The word comes from the Greek, meaning "life having something in common."

The appearance in 1949 of the encyclopedic *Principles of Animal Ecology* by five second-generation ecologists from the University of Chicago—W. C. Allee, A. E. Emerson, Thomas Park, Orlando Park, and K. P. Schmidt—pointed the direction modern ecology was to take. It emphasized feeding relationships and energy budgets, population dynamics, and natural selection and evolution.

Still another area of biology, animal behavior, grafted its branch onto ecology. Although Darwin, Wallace, and others described activities of animals, the formal study of animal behavior began with George John Romanes (1848–1894), who introduced the comparative method of studying nonhuman animals to gain insights into human behavior. His approach depended largely on inferences; in contrast C. Lloyd Morgan (1852–1936), an English behaviorist, emphasized the use of direct observation and experiment.

After the early 1900s, animal behavior study developed along four major lines. One was the study of behavioral mechanisms, perceptual and physiological. It became known as **behaviorism.** A second, more relevant to ecology, was the study of the function and evolution of behavior, including comparative physiology, pioneered by J. B. Watson in his book *Behavior: An Introduction to Comparative Psychology* (1914). Comparative psychology evolved into **ethology.** The three major founders of ethology were Konrad Lorenz, noted for his studies of genetically programmed behavior; Niko Tinbergen, who developed the scheme of four areas of inquiry (causation, development, evolution, and function); and Karl von Frisch, who pioneered studies of bee communication and behavior. After World War II a third field of animal behavior, wedded to ecology, appeared. It was **behavioral ecology,** which investigates the way animals interact with their living and nonliving environments, with a special emphasis on how that behavior is influenced by natural selection. Behavioral ecology begot a controversial offspring, **sociobiology,** pioneered by E. O. Wilson in *Sociobiology: The New Synthesis* (1975). Sociobiology, concentrating on field observations of social groups of animals, applies the principles of evolutionary biology to the study of social behavior in animals. It became controversial when some writers attempted to apply it to humans.

Physiological Ecology

Physiological ecology, or **ecophysiology,** is concerned with the responses of individual organisms to temperature, moisture, light, nutrients, and other factors of the environment. Early plant physiologists studied photosynthesis and plant growth, including the influence of environment on growth. Justus Leibig (1840) investigated the role of limited supplies of nutrients on the growth and development of plants and came up with the "law of the minimum." F. F. Blackman

(1905) extended this idea of limiting factors to include a maximum—a plant could get too much of a good thing. Soon both plant physiologists and plant ecologists began to work out the physiological relationships among plants, climate, atmosphere, and soil. As plant physiologists deciphered the mechanisms of photosynthesis and water relations in plants, ecophysiologists related these functions to plant distributions and adaptations. After World War II, the field grew rapidly. New instrumentation, modern experimental techniques, and rapidly advancing knowledge allowed ecophysiologists to study the interactions of plant physiology and environmental responses in the field and in laboratory growth chambers.

Animal ecophysiology developed out of animal physiology, concerned at first with the functioning of the human body. V. E. Shelford (1911) stimulated the study of animal ecophysiology when he applied the concept of limiting factors to animals in a "law of tolerance." This law linked the physiology of an organism to its environment. He suggested that organisms have both a negative and an optimal response to environmental conditions, and that these responses influence their distribution. This idea stimulated investigations, notably by Kurt Schmidt-Nielson (1975), into how such physiological responses as thermoregulation, energy metabolism, and water balance relate to the environmental conditions in which animals live.

Observers noted that certain plants and animals use chemical substances for defense, and that some plant exudates inhibit the growth of associated species. They began to investigate chemical substances in the natural world. There were studies of the role of chemicals in species recognition, courtship, and defense as well as studies of the chemicals themselves. Such work has grown into the specialized field of **chemical ecology.**

Population Ecology

As plant ecology was arising out of plant geography, other developments were under way. One was the voyage of Charles Darwin on the *Beagle,* during which he collected numerous biological specimens, made detailed notes, and mentally framed his view of life on Earth (Figure 1.2). Darwin (1809–1882) observed the relationships between organisms and environment. He attributed the similarities and dissimilarities of organisms within continental land masses and among continents to geographical barriers separating the inhabitants. He noted from his collection of fossils how successive groups of plants and animals, distinct yet obviously related, replaced one another over geological time.

In developing his theory, Darwin was influenced by the writings of Thomas Malthus (1766–1834). An economist, Malthus (1798) advanced the principle that populations grew in geometric fashion, doubling after some period of time. Experiencing such rapid growth, a population would outstrip its food supply. Ultimately the population would be restrained by a "strong, constantly operating force—among plants and animals the waste of seeds, sickness, and premature death. Among mankind, misery and vice." From this concept Darwin

FIGURE 1.2   Charles Darwin.

mentally in the Soviet Union by G. F. Gause (1934) with laboratory populations of protozoans, and in the United States by Thomas Park (1954) with flour beetles. Many of the concepts of population genetics have been combined with ideas from population ecology to make up the field of **evolutionary ecology,** concerned with the interactions of population dynamics, genetics, natural selection, and evolution.

Closely associated with population and evolutionary ecology—and often difficult to separate from them clearly and completely—is **community ecology.** More narrowly, community ecology is concerned with interactions among species—competition, predation, parasitism, and mutualism—and their influences on species abundance and distribution. More broadly community ecology also deals with the physical and biological structure of communities, community dynamics, and processes such as succession.

Ecologists now have a body of theory relating to competition, population growth, life history strategies, resource utilization, niches, coevolution, community structure, food webs, and the like. Theoretical ecologists take theories and equations developed in pure mathematics, physics, and even economics and apply them to ecological questions. They attempt to provide a substantial mathematical foundation for ecological concepts, upon which predictions can be based. Theoretical ecologists have stimulated new insights into relationships among species, utilization of resources, and life history patterns. Critics of theoretical ecology argue that it suffers from too many hypotheses that are untested or untestable in the field.

Ecosystem Ecology

Early ecologists, particularly plant ecologists, were concerned with observing the patterns of organisms in nature, attempting to understand how patterns were formed and maintained by interactions with the physical environment. Some, notably F. E. Clements, sought some system of organizing nature. He proposed that the plant community behaves as a complex organism or superorganism that grows and develops through stages to a mature or climax stage. His idea was accepted and advanced by other ecologists. A few ecologists, however, notably A. G. Tansley (1871–1955), did not share this view (see page 8). In its place he advanced a holistic and integrated ecological concept that combined living organisms and their physical environment into a system, which he called an ecosystem. A system is a complex in which the parts interact to produce the behavior of the whole.

Although Tansley formally advanced the concept of the ecosystem, limnologists in both Europe and America were using a holistic ecosystem type approach to the study of ecology of lakes that predated Tansley. Prominent among these scientists were A. Thienemann and F. A. Forel (1841–1912). Thienemann developed an ecological approach to freshwater biology. He introduced the ideas of organic nutrient cycling and feeding levels, using the terms *producers* and *consumers.* Forel was more interested in the physical parameters of fresh-

developed the idea of "the survival of the fittest" as a mechanism of natural selection and evolution.

Meanwhile, unknown to Darwin, an Austrian monk, Gregor Mendel (1822–1884), was studying in his garden the transmission of inheritable characters from one generation of pea plants to another. The work of Mendel would have answered a number of Darwin's questions on the mechanisms of inheritance and provided for his theory of natural selection the firm base it needed. Belatedly, Darwin's theory of evolution and Mendelian genetics were combined to form the study of evolution and adaptation, two central themes in ecology. Two papers by G. H. Hardy (1908) and W. Weinberg (1908) on genetic equilibrium in populations mark the beginning of **population genetics.** The theoretical basis of population genetics was advanced by Sewell Wright (1931), Sir R. S. Fisher (1930), and J. P. Haldane (1932, 1954).

The Malthusian concept of population growth and limitations stimulated the study of population dynamics. P. F. Verhulst (1838) of Italy formulated the mathematical basis for population growth under limiting conditions. Verhulst's work, expanded by R. Pearl and L. J. Reed (1929), was the basis for the contributions of A. Lotka and V. Volterra (1926) to the study of population growth, predation, and interspecific competition. Their work established the foundations of **population ecology,** concerned with population growth, regulation, and intraspecific and interspecific competition. The mathematical models of Lotka and Volterra were tested experi-

water habitats, especially lakes. In his monograph on Lake Leman, he introduced term **limnology** for the study of freshwater life.

Concepts of lake dynamics were further developed by S. A. Forbes (1844–1930), an entomologist at the University of Illinois and the Illinois State laboratory of Natural History (Illinois Natural History Survey), which he founded in 1878. He wrote a classic paper of ecology, "The Lake as a Microcosm" (1887). It concerned the interrelations of life in a lake, particularly through food chains, and the role of natural selection in the regulation of numbers of predators and prey. Thus limnological approaches to the study of lakes contributed the development of ecosystem ecology.

Unrelated to limnology, but destined to have an important influence on its future and that of ecology, was the work of Edgar Transeau in an Illinois cornfield. Transeau was not an ecologist, much less a limnologist. He was interested in improving farm production by understanding the photosynthetic efficiency of the corn plant. His landmark paper, "The Accumulation of Energy in Plants" (1926), marked the beginning of the study of primary production and energy budgets.

Thienemann's and Transeau's work stimulated the study of lakes by E. A. Birge and by C. Juday of the Wisconsin Natural History Survey. In a classic paper, "The Annual Energy Budget of an Inland Lake" (1940), Juday summarized the accumulation of energy by aquatic plants over a year and also its movement through various feeding groups, including the decomposers.

The work of Juday and Birge influenced R. A. Lindeman, a young limnologist at the University of Minnesota. Lindeman was interested in exploring plant succession in terms of energy. He turned his attention to Cedar Bog Lake in Minnesota. In a 1942 paper, "The Trophic-Dynamic Aspect of Ecology," Lindeman described succession in terms of energy flow through the lake ecosystem. He showed how short-term processes of feeding, or trophic relationships, affected the long-term changes in the lake. This paper, a significant advance in ecology, marked the beginning of **ecosystem ecology.**

Preceding Lindeman's contribution was the theoretical work of a physical chemist, A. J. Lotka. In his book *Elements of Physical Biology* (1925) he introduced thermodynamic principles of energy transformations in biology along the lines of physical chemistry. He considered food webs and the cycles of carbon dioxide, phosphorus, nitrogen, and water; and he viewed Earth as a single energy-transforming system. Most ecologists overlooked Lotka's contribution. However, his ideas and Lindeman's study stimulated further pioneering work on energy flow and nutrient budgets by G. E. Hutchinson (1957, 1969) and H. T. and E. P. Odum in the 1950s. J. Ovington (1962) in England and Rodin and Bazilevic (1967) in the Soviet Union investigated nutrient cycling in forests. The increased ability to measure energy flows and nutrient cycling by means of radioactive tracers and to analyze large amounts of data with computers permitted the development of **systems ecology,** the application of general systems theory and methods to ecology.

Long-Term Ecological Research

For many ecological processes, long periods of observation are required to detect changes. For such processes long-term research is needed. Many patterns and processes do not vary through time in a constant fashion. Series of observations must be collected for any trend or pattern to emerge.

Unfortunately long-term studies are uncommon, despite repeated evidence of the misleading nature of short-term research. Short-term funding lies at the root of the problem. Demands of both academic institutions and research funding agencies for scientists to obtain results and publish papers quickly precludes any projects running three to five years; most are for one to two years. Further complicating the situation are the processes that control tenure and advancement at academic institutions, which operate on time scales much shorter than most ecological processes.

To address the lack of organized research efforts over long time scales, the National Science Foundation (NFS) initiated a program in Long Term Ecological Research (LTER) in 1980. Researchers at a network of 21 sites across the United States now design studies on time scales of years, to decades, to a century (see *Bioscience* 1990, 40:495–523). They also address a wide range of spatial scales—meters to kilometers to cross-continent comparisons among research sites.

Sites in the LTER system currently extend from Puerto Rico to northern Alaska and represent a broad diversity of environments and ecosystems, including temperate and tropical forests, prairie, desert, alpine and arctic tundra, agricultural fields, lakes, rivers, coastal wetlands, and urban sites. All sites are large enough to incorporate moderate- to large-scale landscape mosaics, and a majority of these site include human-manipulated as well as natural ecosystems. In 1997, the LTER network expanded to include two new sites, Baltimore, MD, and Phoenix, AZ. The objective of adding urban sites is to examine the interactions between ecological and socioeconomic components of the urban environment, a growing area of concern since urbanization continues to increase.

The LTER sites are as varied in investigative design and objectives as they are in the range of ecosystems they encompass. To provide a focus for the individual projects, core research objectives were established early in the program. All LTER studies must at heart support these core objectives. These objectives are to understand the patterns and controls of primary production, food webs, population abundance and distribution, organic matter accumulation, and biogeochemical cycling as well as to answer questions related to disturbance frequency and effect. Research approaches include observation, experimentation, comparative analysis, retrospective studies, and dynamic modeling.

Our current understanding of how key ecological processes, such as primary productivity and decomposition, vary among ecosystems comes largely from piecing together disparate studies. Consequently, the development of comparable data sets and standardization in analytical methods and equipment were concerns addressed at the beginning of the LTER program. This approach permits comparative studies to take place within the LTER network and to extend to non-LTER projects. Coordinated and cooperative research across diverse ecosystems take many forms, including the installation of standard experimental designs across many sites. Such integrated intersite comparisons are essential to addressing questions concerning, for example, climate change and its potential influence on key ecological processes.

TENSIONS WITHIN ECOLOGY

The complexity of ecology's past has led some scientists into opposing camps. Often these fundamental divisions have been, or can be made, productive.

Plant Versus Animal Ecology

The first major split in ecology was the failure of plant ecology and animal ecology to meet on common ground. In England plant ecology was influenced strongly by A. E. Tansley and animal ecology by Charles Elton. At that time one journal, *The Journal of Ecology,* sponsored by the British Ecological Society, covered the field of ecology. In a few years Elton started *The Journal of Animal Ecology.* The two, plant ecology and animal ecology, went their separate ways.

In the United States the split was less amicable. Early on, a controversy developed over the term *ecology.* Botanists decided at the Madison (Wisconsin) Botanical Congress in 1893 to drop the *o* from *oecology* and adopt an anglicized spelling. Zoologists refused to recognize the term at all. The entomologist William Morton Wheeler complained that botanists had usurped the word and had distorted the science. He urged zoologists to drop the term and adopt the word *ethology.*

The schism was widened by a more fundamental difference in approach. Plant ecologists ignored any interaction between plants and animals. In effect, they viewed plants as growing in a world without parasitic insects and grazing herbivores. For years plant and animal ecologists kept to their separate ways. F. E. Clements and V. E. Shelford began to bring the two sides together with *Bioecology* (1939), in which they suggested that plants and animals be considered as interacting components of broad biotic communities or biomes.

Organismal Versus Individualistic Ecology

Although the division between plant and animal ecology narrowed, a new division was to plague ecology. It had its roots in the ideas of Clements, who strongly influenced philosophical ideas in ecology. Clements viewed the plant community as an organism. Like an individual organism, vegetation moved through several stages of development, from youthful colonization of bare ground to a mature, self-reproducing climax in balance with its climate-determined environment. The climax was the end or goal toward which

all vegetation progressed. If disturbed, vegetation responded by retracing its developmental stages to the climax again.

Clements's organismal approach was not lost on animal ecologists. In the United States the zoologist and animal behaviorist William Morton Wheeler, an international authority on ants and termites, advanced the idea that ant colonies behave as organisms. They carry out such functions as food gathering, nutrition, self-defense, and reproduction. Basing his ideas on those of C. Lloyd Morgan, a biological philosopher, Wheeler applied **emergence theory** to ecology. He proposed that natural associations have certain emergent properties as aggregations of organisms—predators and prey, parasites and hosts—that arose from lower levels of organization. All levels occurred together in an ecological community or **biocenosis.** The biocenosis modified its component species through behavioral changes and new levels of integration. Everything in the biocenosis was related to everything else. His view of a tight but orderly nature contrasted with the chaotic impacts imposed on nature by humans.

This organismic, levels-of-hierarchy view of nature advanced by Clements and Wheeler captured the thinking of that influential group of ecologists at the University of Chicago, the authors of *The Principles of Animal Ecology* (1949), Allee, Park, Park, Emerson, and Schmidt. In that book they stated that the organismic concept of ecology was "one of the fruitful ideas contributed by biological science to modern civilization."

Although the organismal concept dominated ecology until the early 1960s, many ecologists refused to accept it. Clement's organismic concept had its critics, notably H. A. Gleason and A. G. Tansley. In 1926 Gleason published "The Individualistic Concept of the Plant Association." In it he argued that the plant association was hardly an organism capable of self-reproduction. Instead, he argued, each community is unique. It arises randomly through environmental selection of seeds, spores, and other reproductive parts of plants that enter a particular area. Thus no sharp boundaries exist between plant communities. Plants exist along a continuum dictated by changing environmental conditions. The English ecologist A. G. Tansley, once enamored with the organismic concept, ultimately rejected it too. Vegetation, he allowed, might be called a quasi-organism, but certainly not an organism or a complex organism. In fact, Tansley rejected the whole idea of a biotic community as anthropomorphic. No social relationship exists among plants or between plants and animals as the term connotes, he argued. In its place Tansley substituted the term **ecosystem.** He viewed plants and animals as components of a system that also included physical factors.

Holism Versus Reductionism

By the mid-1960s the individualistic concept of Gleason had supplanted the organismic concept—almost. Many of its philosophical and functional attributes lived on in the "new ecology" of the 1960s. The new ecology, as defined by E. P. Odum (1964, 1971), is a "systems ecology," an "integrative discipline that deals with supraindividual levels of organization."

According to this concept ecosystems develop from youth to maturity. Each stage of development exhibits some of its own unique characteristics. Interactions among populations and between plants and animals result in a hierarchical organization. This organization involves interacting components that produce large functional wholes. The outcome is the emergence of new system properties that are not evident at the level below (Odum 1971, 1982). These emergent properties account for most of the changes in species and growth that take place over time. The approach is **holistic** (studying the total behavior or attributes of a complex system) because systems are considered too complex to study in bits. Because the whole is greater than the sum of its parts, ecosystems can be studied only as functional units.

This holistic approach has critics who take a **reductionist** approach. They consider that the ecosystem is the sum of its parts. By understanding how each part—the species, their numbers, and characteristics—functions, we can discover how the whole system operates. Rather than guiding the evolution of species, the nature of ecosystems results from the evolution of species.

Fenchel (1987:17) puts the reductionist's point of view well: "I find the entire argument as nonsensical as stating that an alarm clock is qualitatively different from its constituent wheel, bolts, and springs. A holist approach to an alarm clock … is to observe that when wound it will run. To arrive at a real understanding of the device one must take it apart in order to see how it works … to take a reductionist's approach."

The holist would counter that studying the wheels, bolts, and springs tells nothing about the way the whole system functions, what the clock really does. You could study a few separate components, but they are outside the context of the whole clock. Only when all parts of the system are functioning as a unit can the clock function. Then its emergent property, telling time, becomes apparent.

Is the sum of the parts of the clock greater than the whole or not? Allen and Starr in their book *Hierarchy* (1982) argue that the whole problem of emergent properties is a matter of scale and assert that some properties of the whole are emergent and cannot be derived from the behavior of the parts alone. They also point out that ecosystem models of holists are simply large-scale reductionism. Ecosystem ecologists cannot possibly study a model of an entire ecosystem. For one, boundaries of most ecosystems are arbitrary. Further, ecosystems are not closed. One ecosystem feeds into another. Thus ecologists can only study pieces of it. The only major difference between a reductionist and a holist is that the holist studies larger pieces, made up of assembled parts studied by the reductionists.

What keeps ecosystem ecologists (holists) and population and evolutionary ecologists (reductionists) apart is their approach to ecology. Population ecologists focus on species' interactions with their environment in the broadest sense. They are interested in the historical or ultimate reasons why natural selection favored different adaptive responses among species over evolutionary time. Ecosystem ecologists are more interested in the how of current or proximate outcomes of the functional interactions at the population, community, and ecosystem levels. These differences in approach may not be as great as they appear.

What can bring the two groups together? Population ecologists could approach population growth and population interactions such as mutualism, parasitism, predation, and competition as interacting systems (Berry 1981) and as components of a hierarchy of systems. Systems ecologists could integrate some evolutionary theory into system models, particularly in the area of ecosystem development and organization (Loehle and Peckmann 1988). Food web theory, for example, crosses the line into both evolutionary and systems ecology, involving both species interactions and the transfer of energy and nutrients through a hierarchy. Ecosystem functioning ultimately depends on species adaptations, which are the outcomes of evolution. For example, efficiency of water use by certain ecosystems such as grasslands and deserts results from the water use efficiency of the individual plants. The natural assemblage of plants and animals that make up the living component of an ecosystem is not a random collection of species; rather it is one that has been determined by the competitive abilities and other attributes of the component species (H. Odum 1983).

APPLIED ECOLOGY

Ecological theories and models help us understand the human impact on environments. They provide a basis for ecosystem and natural resource management, preservation, and restoration. All these activities make up **applied ecology.** For years theoretical and academic ecologists viewed applied ecology as an intellectual lightweight. Applied ecologists, for their part, often ignored theory, even when it could be of practical use. Fortunately ecologists of both persuasions now recognize that solutions applied to environmental problems must be based on sound theory developed through research.

Applied ecology began to take shape in the 1930s. In 1932 Herbert Stoddard pointed out the role of fire in the control of plant succession in his book *The Bobwhite Quail.* This topic was ignored by academic plant ecologists. Aldo Leopold pioneered the application of ecological principles to the management of wildlife in his classic *Game Management* (1933). In *Forest Soils* (1954), H. L. Lutz and R. F. Chandler discussed nutrient cycles and their role in the forest ecosystem. J. Kittredge pointed out the impact of forests on the environment in *Forest Influences* (1948).

Although applied ecology has been around since the early 1930s, it did not gain visibility until the 1970s, when ecology became involved in social, political, and economic issues. This involvement grew out of public awareness of the problems of pollution, toxic wastes, overpopulation, and a degraded environment. Although the public treated these issues as if they were new, ecologists had grappled with environmental problems for years. The ecological movement had its roots in Europe, especially Germany. An early founder of political ecology was Ernst Haeckel. From Germany it moved to northern Europe, Great Britain, and the United States. In England the animal ecologist Charles Elton helped found the Nature Conservancy. The plant ecologist A. G. Tansley founded the British Ecological Society and was active in the conservation movement.

In the United States George Perkin Marsh called attention to the effects of poor land use on the human environment in his dramatic book *Man and Nature* (1885). In the 1930s F. E. Clements urged that the Great Plains be managed as grazing land and not be broken by the plow. The plant ecologist Paul Sears wrote *Deserts on the March* (1935) in response to the Great Plains Dust Bowl of the 1930s. William Vogt's *Road to Survival* (1948) and Fairfield Osborn's *Our Plundered Planet* (1948) called attention to the growing population-resource problem. Aldo Leopold's *A Sand County Almanac* (1949), which called for an ecological land ethic, was read largely by those interested in wildlife management until the 1970s, when it became the bible of the environmental movement.

Rachel Carson did more than anyone else to bring environmental problems to the attention of the public (Figure 1.3). Since the publication of her book *Silent Spring* (1962), people have become more aware that chemical poisons and other

FIGURE 1.3 Rachel Carson.

pollutants are recycled through the environment. Once castigated as more fiction than fact, Carson's predictions came only too true as carnivorous birds fell victim to toxic chemicals. With a ban on DDT in the United States, some eagles, hawks, and osprey began a gradual comeback. Carson made people quick to recognize other continuing chemical dangers, such as dioxin and PCBs.

The alarming decline in species populations, the threatened extinction of many forms of life, the fragmentation and loss of habitat, and the burgeoning impact of human population growth on Earth's natural resources have impelled otherwise academic ecologists into applied ecology to become involved in resource management. Their entrance into applied ecology has resulted in a growing interest in and effort to apply ecological theory to environmental problems. This interest is reflected in the development of three new fields of ecology—conservation biology, restoration ecology, and landscape ecology—and the establishment of new journals in the field—*Conservation Biology, Landscape Ecology, Restoration Ecology, Journal of Applied Ecology, Ecological Applications, Environmental Management,* and *Ecosystems.*

Conservation biology is a new synthetic field that applies principles of many disparate fields—ecology, biogeography, population genetics, economics, sociology, anthropology, philosophy, and other fields—to the maintenance of biological diversity (Meffe and Carroll 1997). It aims to generate new scientific approaches by melding population biology with applied fields of ecology to solve the problems of protection and maintenance of biodiversity in the face of homogenizing and destructive forces affecting many ecosystems. Although conservation biology in some ways stands apart from other applied fields, its ultimate success will require its joining in part with forestry, range management, wildlife and fishery management, and resource economics. In some ways it has its roots in wildlife biology, a field in which biological knowledge and ecological principles were first applied systematically to conservation of organisms and their natural habitats.

Restoration ecology involves the application of principles of ecosystem development and function to the restoration and management of disturbed lands (see Jordan et al. 1987, Baldwin et al. 1994, Hobbs and Norton 1996, Meffe and Carroll 1997). Its goal is to return a particular habitat or ecosystem to conditions as similar as possible to the predegraded state. Restoration ecology covers a continuum from the reclamation of highly disturbed local sites and the re-creation of a particular ecosystem to restoration and maintenance of entire landscapes for sustainable production and conservation values. The restoration of degraded land means rebuilding functional ecosystems, but not necessarily restoring the site to resemble the original in all its aspects. Restoration ecology offers research ecologists the opportunity to learn about ecosystem structure and function by putting them back together, to develop guiding principles for restoration, and to test ecological ideas. Learning what does and what does not work provides insights into how ecosystems function. Although restoration ecology has been concerned primarily with small on-site reclamation, restoration ecologists are moving in the direction of a much larger scale—that of the landscape. Restoration ecology is also concerned with reestablishing landscape patterns and heterogeneity that have been altered by disturbance and increasing conservation values in fragmented or modified landscapes. This expanded interest is bringing restoration ecology into a close relationship with landscape ecology.

Landscape ecology is the ecology of landscapes—land areas composed of clusters of local ecosystems repeated in a similar manner across kilometers-wide areas, such as parcels of forest existing in an agricultural or urbanized landscape. Thus landscape ecology is concerned with spatial patterns in the landscape and how they develop, with an emphasis on the role of disturbance, including human impacts (Forman and Godron 1986, Forman 1995, Turner 1989, Farina 1998). It explores how a heterogeneous combination of ecosystems is structured, how it functions and changes over time. It studies the distribution of landscape elements, such as fields, forest, grassland, highway rights of way, and suburban subdivisions, and their spatial and temporal interactions and exchanges across ecosystem boundaries—movement and flows of plants, animals, energy, mineral nutrients, and water.

Beginning as a somewhat undefined subunit of ecology, landscape ecology has expanded greatly. Few ecologists would undertake a community or ecosystem study without considering landscape-level implications. It has become an integral part of forestry, wildlife management, ecosystem management, and conservation biology, all dealing with landscape level interactions. Landscape ecology employs the recently developed techniques of geographic information systems (GIS). These are computerized systems that generate maps containing spatial information about a portion of a landscape, such as topography, vegetation patterns, and other physical features that that can be displayed from remote sensing data. The information is stored digitally and can be presented visually or graphically, making it available for comparing various landscape features at different locations or at the same location over time. Such maps can be linked to population simulation models that enable ecologists to predict potential responses of different organisms to specific landscape changes (Pulliam et al. 1992).

We have begun to apply ecosystem and theoretical ecology more intensively to resource management in the past decade, even though economics too often takes precedence over sustainability. A recent approach is **ecosystem management,** grounded in landscape ecology. Like *ecology,* the term *ecosystem management* has a diversity of definitions. A simple but perhaps inadequate definition is the following: Ecosystem management considers ecological systems as functional units and stresses their long-term sustainability. It emphasizes a shift from short-term yields and economic gains to long-term management that sustains or restores a natural or modified ecosystem at a level that allows human use but does not result in long-term ecosystem degradation, including the loss of species. However defined, most definitions envision ecosystem management as requiring the integration of ecological, socioeconomic, and institutional perspectives.

Although a few features of ecosystem management are being implemented by some governmental agencies on publicly owned lands and by some nongovernmental institutions, ecosystem management has its vocal and often bitter critics. At the heart of the controversy is the fear among developers, many foresters, range managers, and others that ecosystem management with its emphasis on maintaining full array of ecosystem functions threatens short-term economic gains. They see it as marking the end of traditional, commodity-oriented monocultural use of land, threatening takeover of the management of private forests and ranges, and placing controls on development. Critics have cultivated in the public a mistaken belief that ecosystem management is a program designed as a way for federal government to expand authority over state and local land use. This, of course, is wrong because federal and state agencies require input from the public, and management of resources falls under the aegis of federal, state, and many local governmental agencies. For good discussions of ecosystem management, see Grumbine (1994) and Meffe and Carroll (1997).

In the past quarter century since people became concerned about growing environmental degradation, how has the situation changed? We started off well enough with environmental legislation: the National Environmental Policy Act (1969), designed to protect the environment from overzealous development and to mitigate losses, the Endangered Species Act (1973, amended 1982 and 1994), the Clean Water Act (1977, amended 1981, 1987, 1994), and the Clean Air Act of 1970 and 1977, among others. The early enthusiasm for a quality environment is still strong with the public. It has weakened, however, among many politicians; government is less sensitive about environmental issues. During the 1980s there was even an environmental backlash at the federal level, as the Reagan administration attempted but failed to undo all the environmental progress made during the previous two decades.

There has, of course, been progress. Water quality has improved considerably, and the air above some of our cities is cleaner. However, we have discovered that our environmental problems are not only more difficult to solve than once believed; many are growing worse. Toxic wastes pollute groundwater and land. Air is becoming more polluted worldwide. Haze has cut visibility in the eastern United States by more than 50 percent in the past 40 years. Acid rain affects lakes and streams. Increased concentrations of carbon dioxide and ozone threaten climatic stability. Roads cut into open country, and suburban expansions eat away at the hinterlands and farmlands. Continued deforestation in both temperate and tropical regions is fragmenting wildlife habitat, increasing the rate of extinction. A rapidly growing urban and suburban population with increasing interest in outdoor recreation is placing intolerable pressures on state and national parks that threaten their ecological integrity.

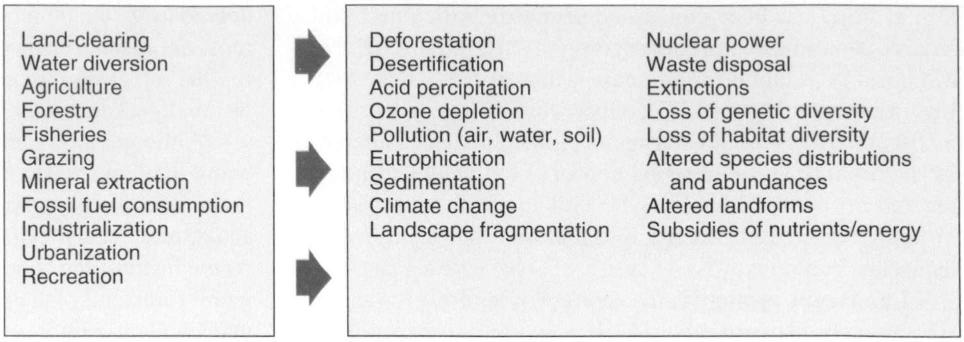

Land-clearing		Deforestation	Nuclear power
Water diversion		Desertification	Waste disposal
Agriculture		Acid percipitation	Extinctions
Forestry		Ozone depletion	Loss of genetic diversity
Fisheries		Pollution (air, water, soil)	Loss of habitat diversity
Grazing		Eutrophication	Altered species distributions
Mineral extraction		Sedimentation	and abundances
Fossil fuel consumption		Climate change	Altered landforms
Industrialization		Landscape fragmentation	Subsidies of nutrients/energy
Urbanization			
Recreation			

FIGURE 1.4 Human activities lessen the sustainability of the biosphere.

Even the oceans have not escaped, as human debris and chemicals have been deadening the seas and destroying marine life. In spite of surplus agricultural production, wetlands are still being drained for more cropland at an alarming rate, threatening the very existence of already dangerously declining wetland wildlife. All these activities have affected regional and global ecological processes and have lessened Earth's ability to support a diversity of life, including humans (Figure 1.4). With the human population growing at the rate of 1.8 percent annually, its pressures on Earth's resources will accelerate.

Among the many environmental problems facing humanity, four broad areas are critical: global climate change, biological diversity, sustainability, and rapidly growing human populations. We are causing dramatic changes in the distribution, abundance, and number of species. The loss of diversity can affect the stability of communities and populations on which our economy depends, as exemplified by the loss of commercially important fish species. This rapid diminution of Earth's resources affects our ability to sustain both natural and managed ecosystems and human life itself.

The basis and solution of our environmental problems are ecological in nature. To this end The Ecological Society of America has developed a three-pronged Sustainable Biosphere Initiative, involving research, education, and environmental decision making (Lubchenco et al. 1991). Research priorities focus on the critical areas of global change, biological diversity, and sustainable ecological systems. Researchers seek answers to such problems as the responses of ecological systems to stress, development and application of ecological theory to the management of ecological systems, and an ecological understanding of the effects of introduced species, pests, and pathogens.

Unfortunately, attempts to apply sound ecological principles to environmental problems often run headlong into economic, political, and social opposition, as witnessed by the debates over old-growth forests, regulation of fishing, land zoning, and wetlands preservation. Successful application requires a citizenry that understands ecology and its importance. We need ecological education at all levels. Ecological principles need to be clearly understood by economists, engineers, lawyers, businesspeople, and politicians, all of them decision makers who can hurt or improve the environment. Most decision makers are unaware of the facts, do not understand basic ecological concepts, or are even hostile to environmental considerations for political, economic, or special reasons. An educated public can cause decision makers to be more responsible.

The future of human life on Earth depends on far more ecological knowledge than we now possess, even though we are not applying all we know. For the first time in the history of Earth, *Homo sapiens* has become the completely dominant organism, changing Earth and its diversity of life at will with little regard for the consequences. It is little wonder, then, that some of the most intellectually challenging problems in ecology lie in that transition zone between theoretical and applied ecology.

ECOLOGY: AN EMPIRICAL AND EXPERIMENTAL SCIENCE

Early on, ecology was largely observational and descriptive. In North America so little was known about the nature and species composition of natural communities that ecologists concentrated on describing them and collecting factual information. They had little incentive or basis to undertake experimental studies. In Europe most ecologists concentrated on vegetation classification. Early steps at empiricism in North American ecology began with F. E. Clements. Studying development of vegetation in the midwestern grasslands, Clements recognized the inadequacy of simply observing vegetational changes. If he was to make progress in the study of vegetation dynamics, he had to devise some means of quantifying and recording changes over time and the differences between various areas. To do so, Clements devised the quadrat method (see Appendix A), which is still the sampling unit for many ecological studies today (Tobey 1981). His quadrats were square plots of varying sizes in which all plants could be listed, counted, and even mapped on grids within the quadrats. The quadrat method provided a distinct advantage in study of vegetation. The establishment of permanent plots allowed the study of the development of vegetation or succession on denuded plots, the changes in plants in response to environmental changes, and the changes in plant communities across a gradient. In spite of his quantitative approach to the study of vegetational development, Clements had little time for statistics, which he considered as simply expressing known facts without uncovering anything new. And, unlike A. E. Tansley, he opposed mathematical models in ecology.

A small group of population ecologists in England and United States, however, were developing mathematical models based on the Lotka-Volterra equations (see Chapters 10, 13, and 14) to study population growth, predator-prey relationships, and competition. Much of the impetus for mathematical models was directed toward applied ecology: pest control and fisheries. Empirical methods came into their own in the 1960s. Faced with the rapid emergence of molecular biology, the growth of physics, and the expressed opinions that ecology was a soft science, some ecologists—notably George E. Hutchinson, Robert MacArthur, Richard Levins, and Robert May—attempted to revolutionize and transform ecology into a highly predictive science with its own fundamental laws like those of physics (see Kingsland 1995). The result was a flurry of building mathematical models, many of them little more than mathematical exercises. Ecologists ignored the fact that they were dealing not with objects as in physics but living organisms, highly variable in their physiology, genetics, evolutionary history, and environmental interactions. This variability precluded ecology from becoming an exact science. After several decades of divisiveness much of the furor has died. It did bring to ecology, however, a strong emphasis on experimentation and hypothesis testing, and on development of models that would provide insights into and suggest solutions to ecological problems.

The specific goal of ecology is to understand the patterns and processes related to life on Earth—how the variety of plants and animals have adapted to the various environments and how they interact among themselves and with their abiotic environment. To gain some insight into these observed processes and patterns, ecologists, like all scientists, develop possible explanations about the causes of observed phenomena. These possible explanations are called hypotheses. A **hypothesis** is a statement about an observation that can be tested.

There are two approaches to the development and testing of hypotheses: inductive and deductive. In the **inductive** method, the scientist gathers empirical data and from it arrives at a generalization. The inductive method proceeds from specific observations to a general conclusion. Using a **deductive** method, a scientist develops a general idea about a phenomenon, performs experiments, and from them makes specific predictions that can be tested again. The experimenters go from a general idea to a specific prediction. Although the distinction between the two approaches may seem subtle, they are fundamentally different approaches. The inductive approach is the formation of general principles from specific observations; the deductive approach is the prediction of specific events from general principles. Each method has its advantages, and ecologists employ both.

Experimental Approach

Testing hypotheses entails first the collection of data by direct observation or experimentation. Ecologists employ direct observation and comparison in "natural experiments" (Diamond 1986). For example, ecologists may be interested in studying the relationship between the abundance of standing dead trees (snags) and the population density of woodpeckers in various forest stands. Because woodpeckers use snags as nest sites, the hypothesis is the abundance of snags represents a limiting resource to population size. By censusing the woodpecker populations and the number of dead trees in various forest stands, the ecologist could compare the woodpecker population with the number of snags. If the woodpecker density is higher on sites with more snags, then the observations support the hypothesis.

The observed patterns, however, only suggest a correlation (co-related); they do not address the question of cause and effect. There may be other reasons for the variation in woodpecker population density, which may also be related to the density of snags. If the density of large trees on which to feed limits the population, you might expect a relationship between woodpecker population density and stand age. In this case, the factor causing the variation in population density among stands is the availability of suitable foraging sites and not nesting sites. It just so happens that the abundance of nesting sites (standing dead trees) is correlated to the causal variable of food abundance (stand age). To help determine cause and effect, an ecologist may employ an experimental approach in which one or more variables are directly manipulated. This rigorous establishment of causation is what separates science from unwarranted assumptions about reasons for observed phenomena.

The experimental approach, unlike the collection of observations from unmanipulated systems, directly determines the response of one variable—the **dependent variable**—to variation in some other variable(s)—**independent variable(s)**. The investigators control the independent variable. They manipulate the variable in a predetermined way, called a **treatment,** and monitor the response of the dependent variable. For example, in a laboratory experiment to test a plant's growth response to increased CO_2, the concentration of CO_2 is the independent variable and plant growth is the dependent variable (Figure 1.5).

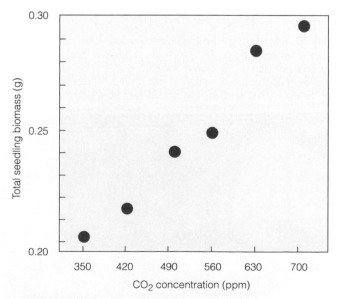

FIGURE 1.5 The effect of atmospheric CO_2 concentration on the growth of yellow birch seedlings in the greenhouse. Growth is measured as the accumulated biomass over the period of the experiment. Each point represents the biomass of a seedling grown under the corresponding atmospheric CO_2 concentration.

One important constraint in experiments is the need to control the variation in other independent variables that may influence the response of the dependent variable. For example, in an experiment testing the response of plant growth to elevated CO_2, the investigator must be sure that all other factors, such as moisture availability, temperature, and available light, that may influence plant growth are the same for all treatments. Otherwise the researcher cannot definitively interpret the plant response as a direct function of CO_2 alone.

Even though the investigator tries to reduce all possible differences among individuals or units used in the experiment and to maintain a rigid control over environmental conditions, there will always be some amount of variation among individual plants. One source of variation very difficult to control is the inherent genetic variation among individual organisms in a population. Because genetic and many other variations often cannot be controlled, the investigator must examine the response of a number of individuals rather than relying on a single observation. This approach is known as **replication;** the individuals receiving the treatment are the **replicates.** The investigator must determine the number of replicates required for each experiment to account for this uncontrolled variation. The decision will be related to the degree of variation exhibited by individuals within the population being examined.

Because the purpose of the experiment is to examine the response of the individuals or experimental units to the treatment (variation in the dependent variable), a group of individuals must be used as a control. The **control** is a group of individuals (replicates) that do not receive the treatment, but otherwise are handled exactly like the treated plant. The control forms the basis for comparison with the individuals receiving the treatment. In the case of the experiment to examine the response of plants to elevated CO_2, the investigator would grow the control individuals under current nor-mal (ambient) concentration of CO_2, thereby providing an estimate of (expected) growth for individuals in the population under baseline conditions.

Field Experiments

Ecologists conduct experiments in the laboratory and in the field (Figure 1.6). Laboratory experiments usually involve synthetic communities of one or a few species. On the basis of these simplified systems, such experiments can reveal a wide range of possible outcomes or relationships that the investigator can evaluate in the field under more natural or complex conditions.

Field experiments involve the manipulation of one or more independent variables in a natural system. The investigator accomplishes this manipulation by adding or removing a species, by erecting exclosures to prevent access to a space or resource, or by adding or withholding essential resources. In contrast to the laboratory experiment, the investigator often finds it impossible to control the many independent variables that may influence the response of the dependent variable. This circumstance limits the investigator's ability to assign a causal relationship between the response of the dependent variable and the treatments. To improve this situation, the investigator must include a set of experimental units or plots as controls. These control units must be similar to those receiving the treatment. The investigator uses this set of field controls for comparison with the set of units on which the treatments are applied. This comparison will form the basis for testing the hypothesis regarding the influence of the independent variable on the dependent variable.

Field experiments are generally on a larger scale than laboratory experiments, and in some ways they are more "realistic" in that the objects under study have not been isolated and controlled in a laboratory environment. However, there are limitations to field studies. One such limitation is

FIGURE 1.6 Although ecologists carry on experimental studies in the laboratory, much of their work is done in the field. (a) A grid established in an oak forest to study the effects of thinning and insect defoliation on the growth and spread of *Armillaria mellea.* The fungus *Armillaria mellea* (honey mushroom) is the cause of root rot in oaks, resulting in the death of trees, especially those of low vigor. (b) Collection of fungal samples from soil blocks removed along the grids.

(a)   (b)

the problem of adequately choosing the sample or replicates of the objects under study. Experiments require the selection of experimental units (such as plots or individuals), some of which will receive the treatment (i.e., manipulation of the independent variable) and some of which will act as the control (no manipulation). The control then serves to monitor the natural variation in the dependent variable.

Hypothesis Testing

Assume the investigator has developed a hypothesis that elevated levels of CO_2 result in increased plant growth. Plant growth response to elevated CO_2 is the dependent variable and the concentration of CO_2 is the independent variable. The investigator wants to know the effect of varying the independent variable on the dependent variable. The next step is to collect the data necessary to test the hypothesis and reach a conclusion. The experimental data would consist of growth measures (e.g., biomass gain over a period of time) for plants grown under ambient (the control samples) and elevated levels of CO_2 (treatment samples).

We test the hypothesis statistically by creating a null hypothesis (H_0) and an alternative hypothesis (H_1). The null hypothesis is the statement of no difference between control and treatment units—that is, the independent variable has no significant effect on the dependent variable. The alternative hypothesis is a statement of significant difference between control and treatment units. In the present experiment the null hypothesis would state that the elevated levels of CO_2 have no effect on growth, because generally in an experiment it is easier to disprove a null hypothesis than to prove a hypothesis. If the results of these experiments showed no difference in growth between the control and the treatment, the investigator would fail to reject the null hypothesis. If there was a difference, the investigator would not accept the null hypothesis. The testing of the null hypothesis involves statistical procedures, some of which are presented in Appendix A.

Models and Predictions

After hypothesis testing, we can formulate a relationship between a dependent and independent variable (for example, plant growth and the level of CO_2) to develop a model of how this two-variable system functions. A **model** is an abstract representation of the real system. It is an explicit set of assumptions, typically formulated in a mathematical way. When we construct and test models (examine their predictions), we are in fact testing the underlying assumptions on which they are structured. Failure of a model may well provide as much insight as its success by contributing information about the validity of the assumptions on which the model is based.

We can view hypotheses as models. The hypothesis that plant growth will increase under elevated CO_2 is an explicit assumption based on the physiological understanding of photosynthesis. We can test the model experimentally and either accept or reject it. Further, we can test the generality of this model with other species, or examine the interaction with

other independent variables such as temperature. This model is qualitative because it predicts only a direction of variation (increased growth) in the dependent variable rather than the amount of variation. In most cases, once a qualitative relationship has been established, a quantitative model is sought. The new model can be statistical or nonstatistical.

Statistical Models   Statistical models are mathematical descriptions of data. They predict the value of the dependent variable based on mathematical functions. An example of a statistical model is the simple linear regression model: $y = a + bx$, where y is the dependent variable, x is the independent variable, a is the parameter describing the y intercept, and b is the parameter describing the slope of the line. The y intercept is the value of y when x is equal to zero. The slope of the line quantitatively relates the change in y per unit change in x. The parameters, a and b, are solved for mathematically based on the paired observations of x and y.

For example, a long-standing theory is that a close relationship exists between the density of small trees and shrubs and the density of snowshoe hares (*Lepus americana*). Litvaitis et al. (1985) established two 49-hectare (ha) study sites in regenerating spruce-fir and hardwood stands. They determined snowshoe hare densities in spring and fall by mark capture-recapture methods (see Appendix A), quantified habitat use with fecal pellet counts, measured intensity of twig-clipping by hares, and sampled the density of understory vegetation. The results of their study, summarized in Figure 1.7, showed a linear relationship between hare density and stem density. The hares increased as understory stem density increased. The regression equation ($y = 0.000046x - 1.06$) is a statistical model for predicting the density of hares (y) given a value of stem density for the plot (x).

One important constraint of such statistical models, because they are a mathematical description of data, is their limited value when extrapolated beyond the bounds of observations. The model can be used legitimately to predict hare density for plots that have a stem density of 20 to 60×10^3 stems per hectare. Because no observations were made and included beyond these bounds, we have no information as to how hare density might vary with increasing density of

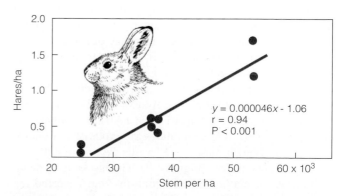

FIGURE 1.7   Relationship between stem density and estimated snowshoe hare density in Maine, 1981–1983. (From Litvaitis et al. 1985.)

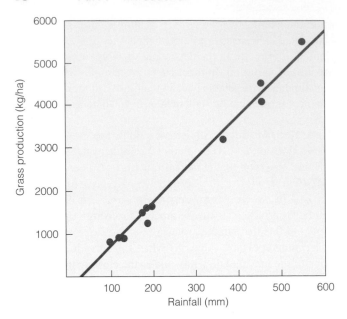

FIGURE 1.8   Production of the grassland of Namibia in relation to annual rainfall. (From H. Walter 1973.)

stems. With increasing density of stems, hare density may continue to increase, level off because some other factor now limits population density, or even decline because the stem density impedes mobility. To predict the effect of stem densities beyond the bounds of these observations, additional plots with stem densities above 50 and below 20 would have to be included in the analysis. Another way of looking at it is that the model (the regression equation) is a description of the correlation between the two variables, not a statement of causation.

Nonstatistical Models   In contrast to statistical models, nonstatistical models are much less well defined. In a way they make up an aggregate of all other mathematical models. The main difference between these two categories is that many nonstatistical mathematical models go beyond a description of the relationship between the dependent and independent variables and assign a mechanism to the parameters. For example, ecologists have used data on grassland productivity from arid and semiarid grasslands to develop a statistical model relating grassland productivity to annual rainfall (Figure 1.8) (H. Walter 1973). Although the model provides a good estimate of expected productivity (dependent variable) within the range of rainfall (independent variable) examined, it does not address the mechanism by which this relationship comes to exist.

To accomplish this goal, we need a much more complex model of plant processes. An example is a model for the grasslands of the Serengeti in East Africa (Coughenour et al. 1985). This model simulates the processes of photosynthesis and respiration as well as the cycling of nutrients through the ecosystem (Figure 1.9). The model is made up of a series of mathematical equations that describe the processes of CO_2 uptake, respiration, and loss of water from leaves and other plant functions. It can predict patterns of net primary productivity, evapotranspiration, and nutrient

cycling through grassland ecosystems as a function of soil, rainfall, and herbivory. Although the model addresses the specific mechanisms responsible for the relationship between rainfall and grassland productivity, it requires much data to define the parameters relating each of these processes to plant growth. Care must be taken that the model can be understood and interpreted. Although this approach is a valuable contribution to understanding the grassland ecosystem, the statistical model may, within the bounds of observation, prove to be a better predictor for the purposes of management. Both types of models are essential in the development and application of science.

Analytical Models   Within the large category of nonstatistical models are two contrasting types: analytical and simulation. Analytical models can be solved mathematically. The investigator can analyze the behavior of the model with respect to the parameters. Because only one solution exists for a given set of parameters, the model is deterministic. An example is the logistic population model (see Chapter 9)

$$dN/dt = rN(K - N)/K$$

where N is some measure of the population size such as the number of individuals or biomass, t is time, r is the instantaneous growth rate expressed in per capita units of N, and K is the carrying capacity or the maximum sustainable value of N for the given environment. The logistic model is analytical because the equation can be solved. For example, consider a hypothetical population of an initial size of 30 with the parameters of $r = 0.186$ and $K = 50$. If we insert these parameters into the logistic model and solve, we arrive at a predicted population size of 38 in 6 years and 49 in 20 years. Thus the behavior of the model can be analyzed with respect to the parameters. Only one solution exists for a given set of parameters (r and K) and initial population size (N).

Simulation Models   In contrast, simulation models cannot be solved analytically. An example is the group of models known as individual-based population models. In contrast to the logistic population growth model, individual-based population models simulate each individual organism within the population rather than an aggregated parameter describing the population as a whole. For example, the parameter r in the logistic equation is a population estimate of the average net fecundity rate per individual and K represents the carrying capacity of the habitat in terms of maximum sustainable population size. Individual-based models, however, simulate the establishment, growth, reproduction, and mortality of each individual in the population, with these processes responding directly to availability of resources like water, light, and nutrients. Fecundity rates may vary among individuals based on their size and age, and the carrying capacity of the habitat may vary as a function of resource availability. The population response is viewed as the composite of the individuals. One advantage of this approach is that it is not necessary to make two implicit assumptions associated with the population-level approach typified by the logistic model: (1) all individuals within the population are

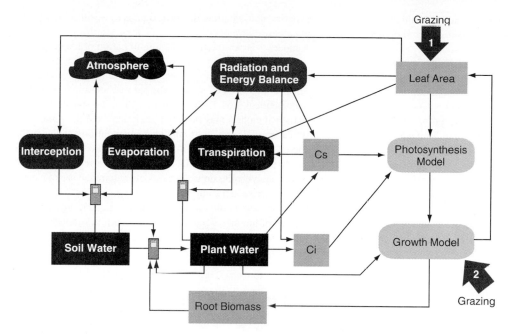

FIGURE 1.9 Nonstatistical model of the Serengeti grassland ecosystem. The diagram depicts the components of the system—soil water, plant water, leaf area, and root biomass (designated by rectangles)—and the processes, flows of water, and nutrient transfers among the compartments. All components are affected by grazing, which reduces the leaf area of plants and removes the growing points at the tips of roots and stems. The squares C_s and C_i are respectively stomatal and internal conductances of water. Doors indicate flow regulation. (Adapted from Coughenour et al. 1985.)

identical—the unique features of each are unimportant; (2) the population is perfectly mixed—there are no local spatial interactions of any important magnitude (Huston et al. 1988).

An example of individual-based models is the forest gap model (Botkin et al. 1971, Shugart and West 1977). Forest gap models simulate the establishment, growth, reproduction, and mortality of individual trees on a forest stand. Each individual responds to the environment of the forest stand (temperature and available light, water, and nutrients) and in return influences the environment of the stand (for instance, taller trees shade smaller ones). These models are much more complex than the population-level models in that each individual in the population must be simulated through time, requiring the use of complex computer programs. The models require a great amount of information on the life history of species to relate the processes of growth, reproduction, and mortality to environmental conditions. In fact the models must simulate (1) the environmental conditions of the forest stand; (2) the response of plants to that environment; and (3) how plants modify the environment of the plot, influencing the other individuals in the stand.

Both modeling approaches have their advantages and disadvantages. The more appropriate approach depends on the objectives of the investigator and the question being addressed.

Validation

Once a model has been developed, it must be validated. **Validation** is the test of how much confidence can be placed in the model. Does the model really agree with the behavior of the real-life system it is to mimic? Can it produce empirically correct predictions? To validate the model, the investigators must collect an independent set of data in the field, incorporate it into the model, and determine how well the predictions of the model based on the new set of data compare with data used to frame the model (Jeffers 1988). This step is important; ecology has many models, but few are validated.

Summary

1. Ecology is difficult to define precisely. A simple, but not inclusive, definition is the study of the interrelations of organisms with their physical and biological environments.

2. Ecology's origins are diverse, but a main root goes back to early natural history and plant geography. Those evolved into the study of plant communities. European plant ecologists concentrated on describing plant communities. American plant ecologists concentrated on studying the development and dynamics of plant communities.

3. Animal ecology developed later than plant ecology. It evolved into the study of natural selection and evolution, beginning with the major contribution of Darwin. It gave rise to behavioral ecology, which is concerned with the way animals interact with their living and nonliving environment as influenced by natural selection.

4. Studies of physiological responses of animals and plants to temperature, moisture, light, nutrients, and other environmental factors grew into physiological ecology. Studies of certain chemical reactions of organisms to their environment stimulated the development of chemical ecology. Chemical ecology

is the study of the uses of chemicals by plants and animals as attractants, repellents, and defensive mechanisms, and of their evolution and chemical structure.

5. Darwin's theory of natural selection and Mendel's work in genetics gave rise to the field of population genetics. The Malthusian concept of population growth and limitations gave birth to population ecology, concerned with population growth, regulation, competition, and predation. Concepts of population genetics and population ecology combined to form evolutionary ecology and theoretical ecology concerned with interactions of population, genetics, natural selection, and evolution. Closely allied with population ecology is community ecology, which in part deals with population interactions that affect community structure and in part with the physical structure and the dynamics and processes associated with community change.

6. As its various disciplines expand, ecology is becoming fragmented into specialties, often with a deepening lack of communication among them. Two major divisions are holistic ecosystem ecology, with emphasis on emergent properties, and reductionist evolutionary and population ecology, which views ecosystems as the sum of parts that can be studied separately to discover ecosystem functions.

7. Applied ecology is concerned with the application of ecological principles to major environmental and resource management problems. Traditionally, applied ecology meant forest, range, wildlife, and fishery management. Recently, applied ecology has spawned the new fields of conservation biology, restoration ecology, and landscape ecology. The future of the quality of human life in all its aspects and the sustainability of Earth depend on our ability to recognize and apply ecological principles to our use of natural resources. One approach is ecosystem management that attempts to integrate ecological, socioeconomic, and institutional perspectives to allow use in a manner that will not result in long-term ecosystem degradation.

8. Ecology has evolved from a descriptive to an empirical and experimental approach emphasizing hypothesis testing, statistical analysis, and development of models to provide new insights and develop new hypotheses. A hypothesis is any statement that can be tested.

9. Two approaches to testing hypotheses are the inductive method, which goes from the specific to the general, and the deductive method, which goes from the general to the specific. The inductive approach is useful for investigating correlations between classes of facts. In the deductive method, the investigator develops a research hypothesis, collects data to support or refute it, develops a mathematical model, attempts to fit the model to the data, and then tests and, if necessary, modifies the model.

10. Testing hypotheses entails the collection of data by direct observation or by experimentation. Experimentation involves simplification by manipulating one or a few variables while holding others constant. Both laboratory and field experiments involve manipulations of one or a few independent variables in.

11. Experimentation involves the determination of the response of one variable, the dependent variable, to variations in an independent variable or variables, manipulated by treatments. The association of the dependent to independent variables assesses the nature of the response of the dependent variable and tells something about the relationship between the two. The experimental approach involves replicates of treatments and controls that do not receive the treatment. Replicates allow the scientist to account for uncontrolled variations among experimental units; controls form the basis for comparison. The results of

experimentation allow the investigator to reject or accept the hypotheses developed.

12. A model is an abstraction and simplification of a natural phenomenon developed to predict a new phenomenon or to provide insights into existing ones. A verbal or graphic model may serve as the basis of a more formal mathematical model. A mathematical model may be statistical or nonstatistical. Nonstatistical models may be either analytical or simulation. Analytical models are mathematical formulations that can be solved directly. Simulation models may take on a variety of forms, including differential equations. Because they cannot be solved analytically, simulation models require the use of a computer to arrive at a solution.

13. Once developed, the model should be validated. Validation is the test of the model's ability to do what it is supposed to do. It measures quantitatively the extent to which the output of the model agrees with the behavior of the real-life system.

Review Questions

1. Define ecology.
2. Why was plant geography the apparent stimulus for the development of modern ecology?
3. What differences separate the organismal concept of ecology from the individualistic concept?
4. How do the two concepts in question 3 relate to holism and reductionism in ecology?
5. What is applied ecology and how does it relate to theoretical and ecosystem ecology?
6. The following statement appeared in *Sustainable Long-Term Forest Health and Productivity* by the Society of American Foresters: "... is the strategy by which in the aggregate, the full array of forest values and functions is maintained at the landscape level. Coordinated management at the landscape level, including cross ownership, is essential component." This statement caused serious divisiveness among foresters. Why?
7. Refer to the document, "The Sustainable Biosphere Initiative" (Lubchenco et al. 1991). Select one of the research topics and discuss how the results from such research would relate to our environmental problems.
8. Why is ecology not taken as seriously as it should be by political decision makers?
9. What makes ecology an empirical science?
10. What is a hypothesis?
11. How do the inductive and deductive approaches to testing hypotheses differ? What are advantages and weaknesses of each?
12. What distinguishes a dependent variable from an independent variable?
13. What is the importance of replicates and controls in ecological studies?
14. What is a model? Why are models useful in ecological research?
15. What is the difference between statistical and nonstatistical models?
16. Comment on the statement made by C. J. Krebs: "Hypotheses without data are not very useful, and data without hypotheses are wasted." What does Krebs mean?
17. Examine some papers in such ecological journals as *Ecology, Journal of Ecology,* and *Journal of Animal Ecology,* both in early and recent years. How do they differ in experimental approach? Are the hypotheses clearly stated? Are the data collected adequate for testing the hypotheses?

PART TWO

The Physical Environment

CHAPTER 2

Solar Radiation and Climate

CONCEPTS

1. Climate is a product of weather over time.

2. Solar radiation is a major determinant of climate.

3. Temperature change in rising and descending air masses is an adiabatic process.

4. All circulation systems within the atmosphere and oceans are driven by solar radiation.

5. A persistent physical effect from Earth's rotation is the Coriolis force.

6. Humidity specifies the amount of water vapor in the atmosphere.

7. Regional climate is influenced by landscape features.

8. Microclimate defines the climate in which organisms live.

Today's temperature, the TV weather reporter informs us, is so many degrees above or below normal; and the rainfall for the month is below normal. The reporter refers to the means or averages of temperature and precipitation over a certain period of time. These averages provide the norm about which meteorologists can describe daily weather fluctuations. A combination of temperature, moisture, precipitation, and winds for a given place expressed as means or averages describes its **climate.** Because of the daily, monthly, and yearly fluctuations, descriptions of climate never exactly match from one period to another, nor do they give us an exact picture. What might have been typical climate for a region 50 years ago may not be the climate there today.

Climate is a product of **weather,** which is air in motion, driven by unequal heating. Heating of the atmosphere involves an exchange of heat between the air and the surface area over which it flows. Loss of heat by an air mass over one region is balanced by the gain of heat elsewhere, but only over a long period of time. The short-term imbalance of heat gains and losses is responsible for most disturbances in the atmosphere. Masses of cold air and masses of warm air often clash along a moving front (the boundary separating two air masses with different properties) to produce major storms.

SOLAR RADIATION: THE KEY TO CLIMATE

Looking at a photograph of Earth taken from space, you can easily get the impression that all parts of the planet equally share the solar radiation reaching it. They do not. The polar regions receive much less solar radiation than the tropical and temperate regions. Variations in Earth's surface, water, rock, sand, soil, snow cover, and vegetation, along with Earth's daily rotation and yearly revolution about the sun, influence how the heat of the sun is absorbed by Earth and distributed over the globe. These variations and inequalities create different patterns of heating and cooling that influence Earth's climates.

Fate of Solar Radiation

Solar radiation travels more or less unimpeded toward Earth until it contacts Earth's atmosphere. The amount of solar radiation that reaches that point (a height of 83 km) as measured on a surface held perpendicular to the sun's rays is about 2 calories per cm^2 per minute (1376 W/m^2), a value known as the **solar constant.** Although called a constant, it actually fluctuates because of variations in ultraviolet outputs of the sun and solar flares.

However bright the sun may seem, only 50 percent of the solar energy traveling to Earth makes it through the atmosphere to Earth's surface. What happens to all the incoming energy? If you take the amount of solar radiation that reaches the atmosphere as 100 percent, 25 percent is reflected from clouds and atmosphere back to space along with another 5 percent reflected by Earth's surface (Figure 2.1) Another 25 percent is absorbed by dust, water vapor, and carbon dioxide

in the atmosphere. Thus reflection and absorption remove 55 percent of the solar radiation as it arrives. Earth's surface absorbs the remaining 45 percent as short-wave radiation. This energy is radiated back as long-wave radiation (electromagnetic wavelengths longer than 3 μm). Out of this 45 percent, 29 percent is radiated back to the atmosphere by the way of evaporation and thermals. The remainder becomes part of the Earth's own surface radiation (104 percent), of which 4 percent is lost directly to space. Of the 100 percent absorbed by water vapor and carbon dioxide in the atmosphere (that act much like a blanket over Earth), 12 percent escapes to outer space and 88 percent is reradiated back to Earth. This reradiation produces the greenhouse effect, which provides some stability to surface temperatures.

In passing through the atmosphere, the energy of certain wavelengths is absorbed, so that a limited spectrum of energy reaches Earth's surface (Figure 2.2). The atmosphere removes nearly all of the ultraviolet radiation. Atmospheric gases scatter shorter wavelengths, giving a bluish color to the sky and causing Earth to shine out in space, as evidenced by photographs taken from the moon. Water vapor scatters radiation of all wavelengths, so an atmosphere with much water vapor is whitish—thus the grayish appearance of a cloudy day. Dust scatters long wavelengths to produce reds and yellows in the atmosphere. Because of the scattering of solar radiation by dust and water vapor, part of it reaches Earth as diffuse light from the sky, called **skylight.** This skylight enables us to see in shaded areas and in twilight. Infrared radiation that reaches Earth and is sensed as heat (sensible heat) is absorbed and a portion is reradiated back as far infrared (4 to 100 μm). What we see as light is visible radiation that can be separated into a spectrum that ranges from violet to red. This part of solar radiation has little to do with Earth's heat budget, but it is of inestimable importance in energy fixation through photosynthesis.

Albedo

Earth does not absorb all the solar radiation impinging on it. Earth's surface reflects back a percentage of the solar radiation, called **albedo.** The amount reflected back determines how fast and to what degree the surface is heated. Water's surface has a low albedo for direct rays, approximately 2 percent, and a very high albedo for low-angle rays, so that there is glare on the water in late afternoons of summer. The albedo for snow and ice is high, 45 to 90 percent; hence the problem of snow blindness on sunny days in a snowy landscape. For forests and grassland, the albedo ranges from 5 to 30 percent, and for clouds overhead it is about 90 percent.

Because of the reflectance of the surface and the angles of the sun's rays, Earth's albedo varies from region to region throughout the year. Albedos for land masses do not vary much from one another during the summer. From tundra to desert in North America, summer albedos are about 16 percent. Greens and browns of the summer landscape absorb far more solar radiation than they reflect. During winter, the story

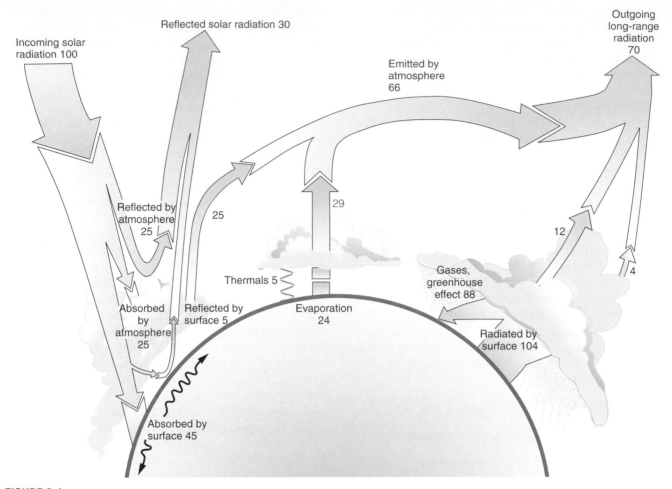

FIGURE 2.1   Disposition of solar energy reaching Earth's atmosphere.

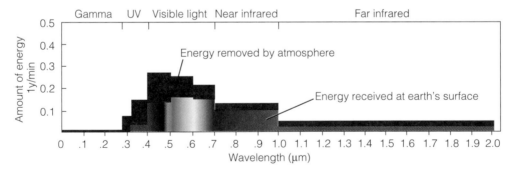

FIGURE 2.2   Energy in the solar spectrum before and after depletion by the atmosphere from a solar altitude of 30°. (From Reifsnyder and Lull 1965.)

is different. The treeless, snow-covered tundra has a winter albedo of about 85 percent, compared with 46 to 50 percent for forest and grassland at 45 to 55°N latitude, and 18 to 19 percent for western desert and shrubland and eastern croplands and woodlands at 25 to 35°N latitude.

Global albedos measured from outer space beyond the atmosphere range from high values of 50 to 60 percent at the polar regions to lows from 20 to 30 percent in tropical and equatorial latitudes. Although the Southern Hemisphere has less land mass and more ocean than the Northern Hemisphere, their albedos are similar. This similarity suggests that global albedos are influenced more by cloud cover than by surfaces of land and sea. However, the albedo of the Northern Hemisphere is more irregular than that of the Southern Hemisphere.

Humidity

When weather reporters tell us that the relative humidity is 50 percent, they are referring to the water vapor content in the air. Water vapor, water in the gaseous state, gets into the air by evaporation. This transformation of water from a liquid to a gaseous state requires energy, referred to as the **latent heat of evaporation.** In the air, water vapor acts as an independent gas that, like air, has weight and exerts pressure. The amount of pressure water vapor exerts independent of the dry air called **vapor pressure.** The water content the air is typically defined in terms of vapor pressure in units of megapascals (MPa). The pressure that water vapor exerts when the air is saturated (can hold no more water vapor) is called the **saturation vapor pressure.**

The maximum amount of water that a given volume of air can hold (the saturation vapor pressure) is a function of its temperature (Figure 2.3). Warm air can hold more water than cold air. The difference between saturation vapor pressure and the actual (ambient) vapor pressure at any given temperature is called the **vapor pressure deficit.** The amount of water in a given volume of air is its absolute humidity. Relative humidity, the measure of humidity with which the most of us are familiar, is the amount of water vapor in the air expressed a percentage of the saturation vapor pressure (the maximum) At saturation vapor pressure, the relative humidity is 100 percent.

If the air cools while the amount of moisture it holds (water vapor pressure) remains constant, the relative humidity increases, because cold air cannot hold as much water as warm air can (lower saturation vapor pressure). If the air cools beyond the saturation pressure (100 percent relative humidity), the moisture condenses into clouds. When the particles of water or ice become too heavy to remain suspended in the air, precipitation falls. For a given water content of the air (vapor pressure), the temperature at which saturation vapor pressure is achieved is called the **dew point temperature.** Think of finding dew or frost on a cool morning. As nightfall approaches, temperatures drop and relative humidity rises. If cool night air temperatures reach the dew point, water condenses and dew forms, lowering the amount of water in the air. As the sun rises, air temperatures warms and the amount of moisture that the air can hold increases. The dew evaporates, increasing vapor pressure in the air.

In any one area relative humidity varies widely from one place to another, depending on terrain. Variations in humidity are most pronounced in mountainous country. Low elevations warm up and dry out earlier in the spring than high elevations, and soil moisture becomes depleted in summer. Because of daytime heating, relative humidity is lower in the afternoon in the valley and on slopes directly exposed to the sun; it is highest on the mountain tops. At night the valley bottoms are most humid. However, the daily range of humidity is greatest in the valleys and lowest at higher elevations.

In all these situations the movement and transfer of water is a diffusion process. The diffusion gradient is the difference in the absolute humidity between any two levels. In the case of the atmosphere, the relative humidity or vapor pressure deficit is the measure of how dry the air is—its ability to receive water via evaporation. This concept is critical to understanding a wide array of processes important in ecology—

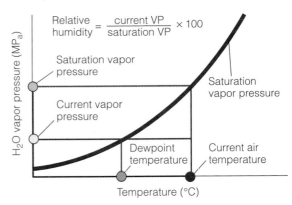

FIGURE 2.3 Saturation vapor pressure as a function of air temperature. For a given air temperature, the relative humidity is the ratio of actual vapor pressure to saturation vapor pressure. For a given vapor pressure, the temperature at which saturation vapor pressure occurs is called the dew point.

from the energy balance of the planet to the energy and water balance of plants and animals.

The Adiabatic Process

Earth is a giant heat machine driven by the heating and cooling of its atmosphere in accord with the law of gases. This law, highly simplified, states that pressure is proportional to temperature times density. At a constant temperature, a gas at high pressure is more dense than at low pressure. If pressure remains constant, a gas is less dense as its temperature increases. These two relationships figure prominently in the behavior of the atmosphere.

Molecules of a gas do not form a well-defined surface but spread out and move freely, colliding and rebounding like plastic balls, pushing and exerting pressure. When a volume of gas is compressed into a small space, density increases. The molecules are close together and collide frequently, producing **sensible heat,** heat that can be felt. When the same volume of gas is allowed to expand into a larger space, the molecules are less dense. More widely separated, the molecules collide less frequently, resulting in a drop in sensible heat (i.e., the gas is cooler). If a parcel of air compresses and warms, or expands and cools, with no interchange of energy (heat) with its surroundings, the situation is called an **adiabatic process.**

Now consider a column of air up through Earth's atmosphere. The pressure at any chosen height in the column equals the weight of the air column above it (Figure 2.4). Because of the pull of gravity and the weight of air above, the pressure is the greatest and the air the warmest near the Earth's surface. As you move up the column, the air continuously becomes less dense and thus cooler. The rate at which the air temperature changes with height in the air column (elevation) is the **lapse rate.** As you move up through the air column at any given time, you experience the prevailing decrease in temperature, the **environmental lapse rate.**

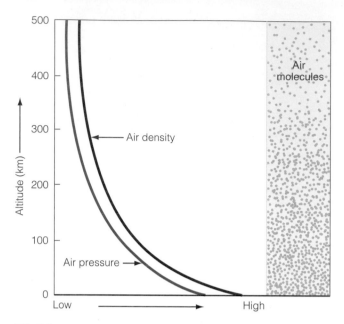

FIGURE 2.4   Both air pressure and air density decrease with increasing altitude.

If a parcel of air is unsaturated air (relative humidity less than 100 percent), the rate of adiabatic cooling or warming with change in elevation remains constant. The rate of heating or cooling is approximately 10°C for every 1000 meters. The rate of temperature change is called the **dry adiabatic lapse rate.**

For example, if a parcel of air 30°C at ground levels rises in elevation, its temperature will drop to 10°C by the time it reaches 2000 m (Figure 2.5). As it begins its descent, the density once again increases and the temperature rises.

Because cooler air holds less moisture, the relative humidity of a rising air mass increases as it cools. If the air cools to its dew point temperature, the relative humidity reaches 100 percent. Any further cooling and condensation will result in cloud formation. The process of condensation, the transformation of water from a vapor to a liquid, releases energy in the form of **latent heat** (heat that is either released or absorbed by a unit mass of a substance when it undergoes a change of state) to the rising air mass. The heat released offsets some of the cooling due to the expansion of the air. In this case, the air no longer cools at the dry adiabatic rate, but at a lesser rate called the **moist adiabatic lapse rate.** The moist adiabatic lapse rate has no single value, because it varies with the temperature of and the moisture in the air; but the rate averages about 6°C per 1000 m.

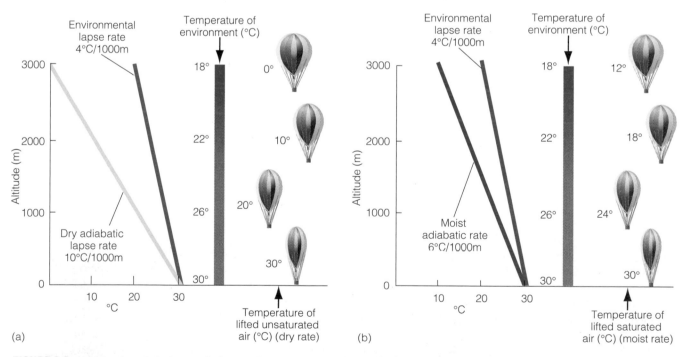

FIGURE 2.5   (a) The dry adiabatic rate. As long as the air parcel remains unsaturated, it expands and cools by 10°C per 1000 meters. The sinking parcel compresses and warms by 10°C per 1000 meters. (b) A rising parcel of saturated air under the same environmental temperature conditions and environmental lapse rate as in (a) will cool at the moist adiabatic rate. Note the slower rate of decline in the temperature of the rising air parcel because of the release of energy in the form of latent heat. In each example the lifted parcel of air is colder and heavier than the surrounding air and under certain conditions could sink back to its original position. These examples are characteristic of a stable atmosphere. (Adapted from Ahrens 1991.)

ROTATIONAL EFFECTS

The Coriolis Force

The movement of warming and cooling air masses about Earth is strongly affected by the rotation of Earth itself. Earth spins on its axis from west to east. This momentum causes an object or an air mass to move in the direction of Earth's rotation. Its path, however, is deflected by an apparent force produced by Earth's rotation, the Coriolis force (Figure 2.6). To understand how it works, consider first a nonrotating platform. If you are positioned at A and throw a ball to a catcher at point B, the ball will travel in a straight line (Figure 2.6a). Now allow the platform to rotate counterclockwise, the direction Earth spins when viewed from the North Pole, and then throw the ball from point A to the catcher at B. The ball will travel in a straight line to the expected point B, but by the time the ball arrives there the catcher will have moved to point B′ and you will have moved to point A′. You and the catcher expected the ball to follow the path A′B′ (Figure 2.6b). In following its actual path AB, the ball appears to be following the curve A′B. In landing at B the ball appears to have deviated to the right from the expected path. Thus while landing at B, the ball lands to the right of the catcher, giving you and the catcher the impression that the ball itself deflected to the right. If the platform is rotated clockwise, the direction Earth appears to rotate when viewed from the South Pole, the object will be deflected to the left. The Coriolis force, then, prevents a direct simple flow from the equator to the poles. In the Northern Hemisphere winds are directed to the right or clockwise, and in the Southern Hemisphere to the left or counterclockwise (Figure 2.6c).

Movement of Air Masses

Sunlight does not strike Earth uniformly (Figure 2.7). Because of Earth's shape, the Sun's rays strike more directly on the equator than on the polar regions, so that lower latitudes get more heat. The inclination of Earth's axis at an angle of 23½° and the rotation of Earth about the sun further increase the inequality of heat distribution. These factors combine to determine the amount of solar radiation reaching any point of Earth at any time. Earth's surface at all times lies half in the Sun's rays and half in shadow, marked by a dividing line, the circle of illumination. The equator bisects the circle of illumination at a right angle to the sun's rays. Thus at the equator, Earth experiences roughly 12 hours of daylight and 12 hours of night throughout the year, whereas the polar regions experience 24 hours of night in winter and 24 hours of daylight in summer. Only two times a year, at the vernal and autumnal equinoxes (March 20 or 21, September 22 or 23), does the circle of illumination pass through the poles, at which time they experience 12 hours of daylight and 12 hours of darkness (Figure 2.7a). At the time of summer and winter solstices, the circle of illumination is tangent to the Antarctic and Arctic circles (Figure 2.7b). Thus at the time of the winter solstice, the sun remains below the horizon at the North Pole and above the horizon at the South Pole. During the summer solstice, the reverse situation is true. Thus the lower latitudes, especially the equatorial regions, have constantly warm temperatures compared with the variable temperatures of the higher latitudes.

Air, heated close to the surface at the equatorial regions, rises until it reaches the stratosphere. In the stratosphere the

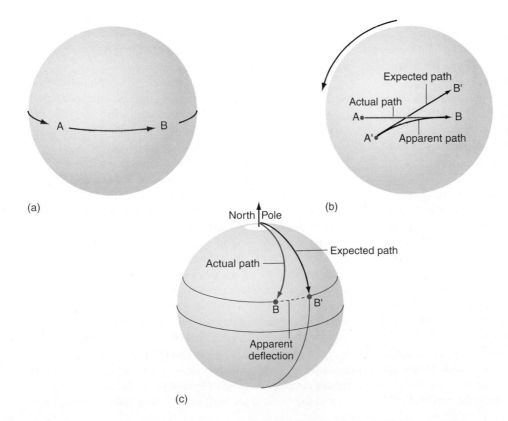

FIGURE 2.6   (a) Ball thrown from A to B on a nonrotating platform. (b) Deflection of the ball thrown from A to B on a rotating platform, showing actual, expected, and apparent paths. (c) Deflection of a projectile shot equator-ward from the North Pole. When viewed from the North Pole, the projectile appears to be deflected to the right. (After Neiburger et al. 1982:187, 188.)

(a)

Expected path
Actual path
A•
A′•   Apparent path   B′   B

(b)

North Pole
Expected path
Actual path
B   B′
Apparent deflection

(c)

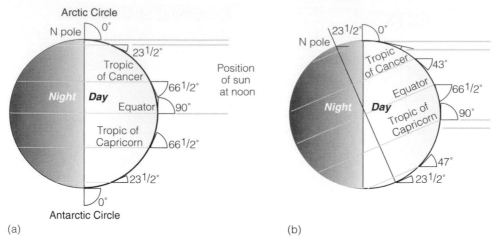

FIGURE 2.7   Altitude of the sun (a) at the equinoxes and (b) at the solstices. Latitude, the inclination of Earth's axis at an angle of 23½°, and the revolution of Earth about the sun determine the amount of solar radiation reaching any point on Earth at any time.

temperature no longer decreases with altitude. There the air masses that now possess a temperature equal to or lower than that of the stratosphere are blocked from any further upward movement. With more air rising, air masses are forced to spread north and south toward the poles. As the air masses approach the poles, they cool, become heavier, and sink. This heavier cold air then is pulled southward and northward by the void created as bottom air heated at the equator is forced to rise (Figure 2.8).

If Earth were stationary and without any irregular land masses and oceans, the atmosphere would flow in this unmodified circulatory pattern (Figure 2.8a). But Earth's surface is irregular; there is differential heating, and the Earth, as we already know, spins, creating the Coriolis effect. On the rotating Earth, air currents in the Northern Hemisphere start as south winds and move north, but not for long. The Coriolis force deflects the winds to the right, changing their direction to the northeast or east. Frictional drag slows the northward movement and causes the air masses to pile up at about 30°N latitude, where in the process they lose heat by radiation (Figure 2.8b). This combination of piling and heat loss forces the cooling air to descend, producing cells of semipermanent high pressure, at about 60° north and south latitude. Air that has descended at 60° latitudes flows both northward toward the pole and southward toward the equator. The northward-flowing air currents become the prevailing westerlies, and the southward flowing air, also deflected to the right, becomes the northeast trade winds of the lower latitudes. (Note: These terms refer to the originating direction of the wind, not the direction in which they are blowing.)

The air aloft gradually moves northward, losing heat. It descends at the polar region, where it loses additional heat at a surface already cooled by insufficient solar radiation and the albedo of the ice packs. This dense mass of cold air flows southward toward the equator. On its way it is deflected to the right and becomes the polar easterlies. Similar flows take place in the Southern Hemisphere, but the air flow is deflected to the left by rotating Earth. Thus the cold winds of

winter in the Southern Hemisphere are south winds, and the warm winds of spring flow down from the north.

This pattern of rising and descending air forms tubes about the Earth. The direct circulation cells near the equator are called Hadley cells, and the indirect middle latitude cells are called Ferrel cells after the meteorologists who described them.

The interaction of wind and heating produces more or less permanent high-pressure cells known as subtropical highs in the Atlantic and Pacific oceans; winds and cooling produce low-pressure cells such as the Aleutian and Icelandic lows. The highs are more pronounced during the summer months, the lows during the winter months. Also produced are monsoon winds, dry winds that blow from continental interiors to the oceans in summer, and winds heavy with moisture that blow from the oceans to the interior in winter, bringing with them heavy rains.

The northeast and southeast trade winds blow between the subtropical highs (horse latitudes) and the equatorial lows (doldrums). The region where the tradewinds meet is the **intertropical convergence zone (ITCZ)** (Figure 2.9). Located in the rising branches of the Hadley cells, it is a region of rising, unstable air, cloudiness, and rain. Although tropical regions about the equator are always exposed to warm temperatures, the sun is directly over the geographical equator only two times at year, at the spring and fall equinoxes. At the northern summer solstice, the sun is directly over the Tropic of Cancer; at the winter solstice (which is summer in the Southern Hemisphere), it is directly over the Tropic of Capricorn. As a result the ITCZ moves poleward and invades the subtropical highs in northern summer; in the winter it moves southward, leaving clear dry weather behind. As it migrates southward it brings rain to the southern summer. Thus as the ITCZ shifts north and south, it brings on the wet and dry seasons in the tropics (Figure 2.9). Because air and water heat slowly, a time lag of about one month develops between the change in the vertical orientation of the sun and the shift in the intertropical convergence.

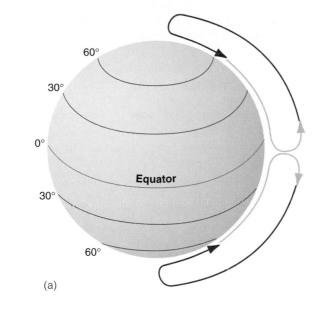

(a)

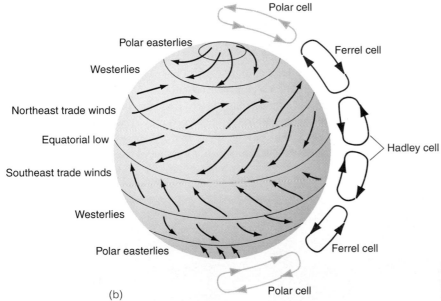

(b)

FIGURE 2.8 Circulation of air cells and prevailing winds (a) on an imaginary, nonrotating Earth and (b) on the rotating Earth.

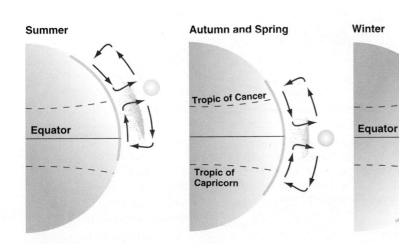

FIGURE 2.9 Shifts of the intertropical convergence, producing rainy seasons and dry seasons. Note that as the distance from the equator increases, the dry season is longer and the rainfall is less. These oscillations result from changes in the altitude of the sun between the equinoxes and the solstices as diagrammed in Figure 2.7. (From H. Walter 1977.)

For this reason the intertropical convergence does not move as far north and south as does the sun.

Lastly, there are moving air masses with their cyclonic and anticyclonic frontal systems. In the Northern Hemisphere, winds in high-pressure cells, called **anticyclones,** flow clockwise and move outward and downward from the center of the system. Because air moves downward from high altitudes, highs are characterized by minimum cloudiness and little precipitation. A high-pressure area is surrounded on all sides by low pressure. Low pressure cells, called **cyclones,** flow counterclockwise. The air moves inward and upward, resulting in cooling, increased relative humidity, and precipitation. In the Southern Hemisphere winds move counterclockwise around a high-pressure system and clockwise around a low-pressure system.

Ocean Currents

The upper or surface waters of the ocean are constantly in motion. This motion, reflected in waves and currents, is caused largely by winds blowing across the surface. These wind-driven ocean currents, modified by Coriolis forces, transport enormous quantities of water across vast distances. In doing so, ocean currents also become a mechanism for the transport of sensible heat from equatorial regions to arctic regions.

In the absence of any land masses, ocean currents could circulate unimpeded around the globe, as does the flow of ocean waters around Antarctica. Continental land masses, however, divide the ocean into two main bodies, the Atlantic and the Pacific. Both oceans are unbroken from high latitudes north and south to the equator; and both are bordered by land masses on either side that deflect ocean currents (Figure 2.10).

Each ocean is dominated by two great circular water motions or **gyres,** each centered on a subtropical high-pressure area north and south of the equator. Within each gyre the current, in response to the Coriolis force, moves clockwise in the Northern Hemisphere and counterclockwise in the Southern Hemisphere. The movements of the currents are also influenced by the prevailing winds, the trades or tropical easterlies on the equator side and the prevailing westerlies on the pole side. The eastward-flowing equatorial countercurrent separates the two gyres, north and south. This current results from the return of lighter (less dense) surface water piled up on the western side of the ocean basin by the equatorial currents.

As currents flow westward they become narrower and increase their speed. Deflected by the continental basin and pushed by the Coriolis force, they turn poleward, carrying warm water with them. The two major currents in the Northern Hemisphere are the Gulf Stream in the Atlantic and the Kuroshio (or Japanese) Current in the Pacific. The Gulf Stream flows north from the Caribbean, presses close to Florida, swings along the southeast Atlantic Coast of North America, and divides (Figure 2.10). One part becomes the Norwegian Current, carrying warm water past Scotland, warming and dampening the climate of Great Britain (which lies in about the same latitude range as central Canada). The other part of the current swings south as the Canary Current, completing the gyre. The Kuroshio Current gives rise to the south-flowing

FIGURE 2.10 Ocean currents of the world. (1) Antarctic West Wind Drift; (2) Peru Current; (3) South Equatorial Current; (4) Equatorial Countercurrent; (5) North Equatorial Current; (6) Kuroshio Current; (7) California Current; (8) Brazil Current; (9) Benguela Current; (10) South Equatorial Current; (11) Guinea Current; (12) North Equatorial Current; (13) Gulf Stream; (14) Norway Current; (15) Labrador Current; (16) Canary Current; (17)North Atlantic Current; (18) Sargasso Sea; (19) Monsoon Drift (summer: east; winter: west); (20) Mozambique Current; (21) West Australian Current; (22) East Australian Current; (23) Alaskan Current. Red arrows represent warm currents; blue arrows, cold water. (From Coker 1947.)

warm California Current, which bathes the Pacific Northwest in rain and fog, responsible for the growth of the magnificent Pacific Northwest coniferous rain forest. Counterparts in the Southern Hemisphere are the East Australian Current and West Australian Current in the Pacific and the Brazil Current in the Atlantic. The counterpart of the Canary Current in the South Atlantic is the Benguela Current, which flows along the African Coast; in the Pacific it is the Peru.

As the California and Peru currents swing westward to complete the gyres, they drag along with them cool water along coasts of California and Peru. To replace the displaced water, deep cold water rises to surface, bringing with it a load of nutrients from the bottom. The surfacing of this cold, nutrient-rich water into the sunlit area of the ocean is termed **upwelling.** Heavy phytoplankton production in the upwellings supports an abundance of marine life, especially fishes and birds. Because these waters are colder than the adjacent land, the currents supply little moisture to coastal lands, resulting in a string of deserts along the coast of Mexico and South America and chaparral along the coastal regions of California and Chile.

The combination of sea surface temperature, ocean currents, and the atmosphere determine climate. Changes in east–west zonal circulation of the atmosphere, as well as north–south movements, influence easterly and westerly flows of winds. These movements, coupled with displacements of major ocean currents, can have pronounced effects on climate over the short term. An example is the El Niño effect.

Every several years, strong southeast trade winds blow westward from regions of high pressure over the eastern Pacific to regions of low pressure near Indonesia. These westward-driven surface waters, heated by the equatorial sun, become warmer than average and pile up in the Pacific, west of the international date line. The sea level and barometric pressure rise in the western Pacific and drop in the eastern Pacific. As a result of the strong differential in air pressure, west winds replace the eastern trade winds, sending warm water and winds eastward toward South America. Because this warming of the surface waters of the ocean off the South American coast occurs around Christmas time, the phenomenon is called El Niño, meaning Little Boy (referring to the Christ child). Toward the end of El Niño, the atmospheric pressure reverses. It falls over the western Pacific and rises over the eastern Pacific. The surface waters cool, the trade winds strengthen, and cool waters move toward and pile up on the east coast of the Pacific, a phenomenon called La Niña (Little Girl).

El Niño represents one extreme and La Niña the other extreme of an irregular interannual oscillation in atmospheric pressure that extends across the Pacific, known as the Southern Oscillation (Figure 2.11). Climatologically, El Niño is characterized by low pressure and storms in the eastern Pacific and high pressure and drought in the western Pacific, wet winters in the southeastern and southwestern United States, relatively warm winters in Canada and the northern United States, and fewer Atlantic Ocean hurricanes. La Niña, in contrast, is characterized by high pressure and drought in the eastern Pacific and low pressure and

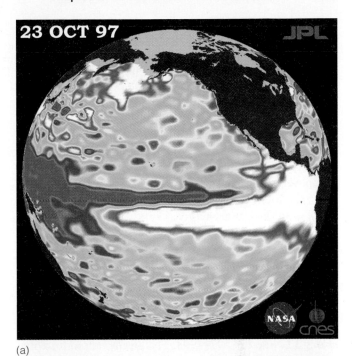

(a)

(b)

FIGURE 2.11 The Southern Oscillation. (a) Ocean temperatures and associated phenomena during a La Niña, 23 October 1997. (b) Ocean temperatures and associated phenomena during an El Niño, 27 February 1999. The red and white colors of the oceanic waters represent warm water. The dark blue colors represent cool water.

storms in the western Pacific, cold winters in Canada, wetter winters in the Pacific Northwest, warmer, drier winters in the southeastern and southwestern United States, and more Atlantic Ocean hurricanes.

In some years El Niño becomes exceptionally strong and the trade winds collapse. The warm ocean waters move eastward toward South America, and travel north and south along the coast. They block the upwelling of nutrient-rich cold water along the South American coast with catastrophic effects on marine life.

An exceptionally strong El Niño event occurred in 1982–1983, greatly reducing the phytoplankton population supported by upwelling nutrients. The decline in phytoplankton, the base of the marine food chain, caused a sharp decline in the schooling fish—notably anchovies and sardines, a major food source for marine birds and mammals. The loss of food caused reproductive failure, disruption of migratory behavior, and heavy mortality among seabirds. It brought about nearly 100 percent mortality in the Galapagos fur seals. It disrupted Pacific Coast fisheries and destroyed much of the California kelp.

In addition to affecting marine ecosystems off the western South American coast, El Niño also affected inland ecosystems. It brought drought to agricultural areas of South Africa, United States, and Canada and much of Australia. It caused torrential rains in usually dry areas and dislocated the rainy season in the tropics.

The El Niño of 1982–1983 was followed by another very strong El Niño in 1997–1998. That event caused torrential rains in California, cold temperatures in the southern United States, and record high temperatures in the northern United States. The torrential rains brought about devastating floods and mudslides in Peru, Ecuador, and Mexico and landslides and blackouts in California. Temperature-sensitive fish moved to cooler water, depleting the food source for starving sea lions. Winter was relatively warm and dry in the northern United States, and temperatures 2 to 4 °C above normal brought on early blooms to spring flowers. Winter was wet across the south. Florida experienced tornadoes and flooding damage to temperature-sensitive corals. The northwest United States experienced a warmer and drier than normal winter. In the western Pacific, Indonesia and Malaysia suffered drought. As El Niño faded it was replaced by La Niña in 1999, bringing on drought in the northeastern United states and a devastating hurricane season in the Atlantic.

Together the patterns of temperature, winds, and ocean currents influence the global patterns of precipitation (Figure 2.12). As easterly winds move across the tropical oceans, they gather moisture. The warm air cools as it rises. When the rising air reaches dew point, clouds form and precipitation falls. This pattern accounts for the high precipitation in the tropical regions of eastern Asia, South America, and Africa, as well as the relatively high precipitation in southeastern North America.

As the winds move northward and southward, they cool. In the horse latitudes, where the cool air descends, two belts of dry climate encircle the globe. The descending air warms and therefore holds more moisture. The dry air draws water from the surface, causing arid conditions. In these belts the world's deserts have formed (see Chapter 29).

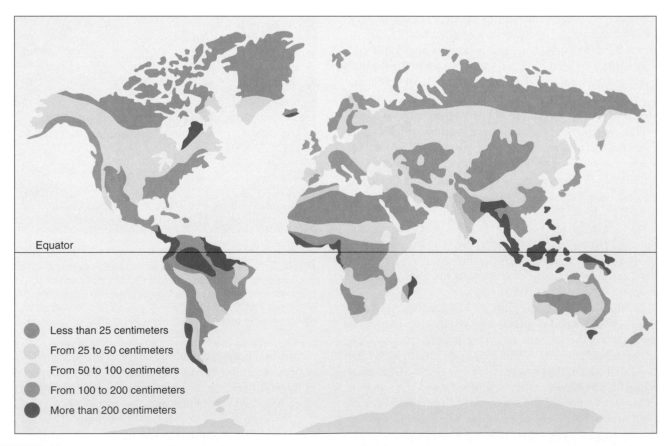

FIGURE 2.12 Annual world precipitation. The wettest and driest areas are related to mountain ranges, ocean currents, and winds.

REGIONAL CLIMATES

The massive circulation patterns of atmosphere and ocean currents determine the climatic pattern of relatively large regions—the macroclimate. Within these regions, however, continental location, nearness to large bodies of water or seas, and topographic features such as mountains and valleys influence the climate of a region and its seasonal changes.

There are several ways of depicting seasonal climate changes in a region by plotting the annual progression of temperature. One method is the **climograph,** which is a plot of mean monthly temperatures against mean monthly precipitation (Figure 2.13). Connecting the points for each month forms an irregular polygon, which can be compared with polygons for other areas. Climographs are highly useful in predicting responses of organisms to changes in physical environment and in assessing the suitability of a regional climate for the transplanting or reestablishing of a species population.

A different approach is the **climate diagram** developed by Walter and Leith (1966) and Walter (1985) that graphs the seasonal variations in temperature and precipitation (Figure 2.14). Months are plotted along the horizontal axis of the diagram: January to December for the Northern Hemisphere; July to June for the Southern Hemisphere. The left vertical axis plots the temperature, the right vertical axis the precipitation. The vertical axes are so scaled that 10°C equals 20 mm of precipitation, so that the relative position of the temperature and precipitation graphs reflects the

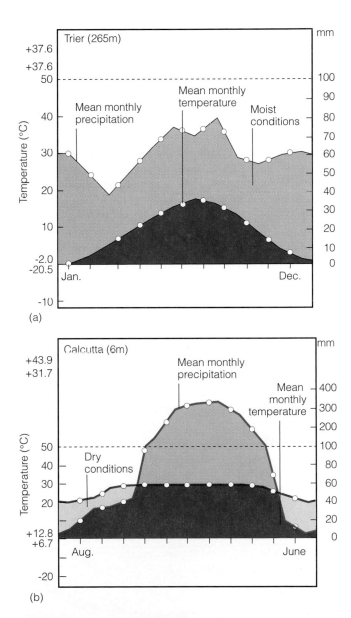

(a)

(b)

FIGURE 2.14. The structure of climate diagrams. The horizontal axis plots the months: January to December for the Northern Hemisphere; July to June in the Southern Hemisphere. The left vertical axis is temperature in °C. The right vertical axis is precipitation in mm. Temperature is plotted as a red line; precipitation is plotted as a blue line. Blue shading indicates that the precipitation line is above temperature and conditions are wet or moist. The gold shading shows that the temperature line is above precipitation, indicating dry conditions. (a) Climate diagram typical of a temperate region. (b) Climate diagram for a tropical region where rainfall exceeds 100 mm, requiring a change in scale above 100 mm. Rainfall above 100 mm is indicated by the dark blue shading.

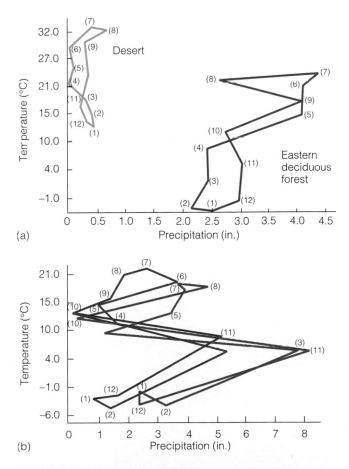

(a)

(b)

FIGURE 2.13 Temperature-moisture climographs. (a) The hot, dry desert climate differs graphically from the cool, temperate, moist climate of the East. Data for the graph are mean temperature and precipitation for Yuma, Arizona, and Albany, New York. (b) Conditions on the rain shadow side and the high-rainfall side in the Appalachian Mountains in West Virginia. Numbers in parentheses indicate the month.

availability of water. When the precipitation line lies above the temperature line, precipitation is adequate. When the temperature line lies above the precipitation line, evaporation exceeds precipitation. For regions where precipitation exceeds 100 mm, the precipitation scale for convenience is compressed so that a 10°C change is equivalent to 200 mm of precipitation.

Topographical Influences

Mountains influence regional climates in two ways: by modifying the patterns of precipitation and by creating climatic differences with altitude. As an air mass is intercepted by a mountain range, it ascends and cools at the dry adiabatic rate. When the moisture reaches its condensation point, it cools at a moist adiabatic rate and drops its moisture on the windward side. The air, no longer saturated, descends on the leeward slope, warms at the dry adiabatic rate, and picks up moisture from the land (Figure 2.15). As a result, the windward side of a mountain range supports more moisture-loving vegetation than the leeward side, where drier conditions exist in the rain shadow. Thus in North America the westerly winds that flow over the Sierra Nevada drop their moisture on the west-facing slopes, which support excellent forest growth, whereas the leeward sides grade into the hot deserts of the Southwest and the cool deserts of sagebrush in the Northern Great Basin. In the Allegheny Mountains of the Appalachian chain, the high ridges intercept much of the moisture in the air masses moving from the west. On this windward side, the average precipitation is 168 cm/yr and the slopes support northern hardwood forests. As a result the leeward side, the Ridge and Valley section, receives only about half as much precipitation, 77 cm/yr, and oak-pine forests dominate the vegetation (see Figure 2.13b).

In mountainous country climate changes on an altitudinal gradient. The change in climate going upslope mimics the broad climatic changes experienced by going to higher latitudes. As a rule of thumb, temperatures drop about 1°C for each 100 m rise in elevation, so that climates of higher altitudes bear little resemblance to those of the lowlands about their bases. Above the base region is the montane level, which has declining temperatures and increasing relative humidity as the altitude increases. At the highest levels the climate is very cold, and on top of the highest mountains is a land of perpetual ice and snow. This broad rule holds for mountains from tropics to boreal regions, although individual mountains possess their own particular weather conditions.

North-Facing and South-Facing Slopes

The greatest climatic differences exist between north-facing and south-facing slopes (Figure 2.16). South-facing slopes in the Northern Hemisphere receive the most solar energy. North-facing slopes receive the least energy.

At latitude 41°N (about central New Jersey and southern Pennsylvania) midday insolation on a 20° slope is, on the average, 40 percent greater on the south-facing slopes than on the north-facing slopes during all seasons. This difference has a marked effect on the moisture and heat budget of the two sites. High temperatures and associated low vapor pressures induce evapotranspiration of moisture from soil and plants. The evaporation rate often is 50 percent higher, the average temperature higher, the soil moisture lower, and the extremes of all of these are more variable on south-facing slopes. Thus the microclimate ranges from warm and dry (xeric) conditions with wide extremes on south-facing slopes to cool and moist (mesic), less variable conditions on north-facing slopes. Xeric conditions are most highly developed on the top of south-facing slopes, where air movement and soil drainage are greatest, whereas the most mesic conditions are at the bottom of the north-facing slopes.

The whole north-facing and south-facing slope complex is the result of a long chain of interactions: Solar radiation influences moisture regimes; the moisture regime influences

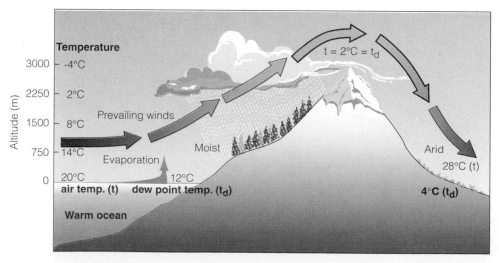

FIGURE 2.15 Formation of a rain shadow. Air is forced to go over a mountain. As it rises, the air mass cools and loses its moisture as precipitation on the windward side. The descending air, already dry, picks up moisture from the leeward side.

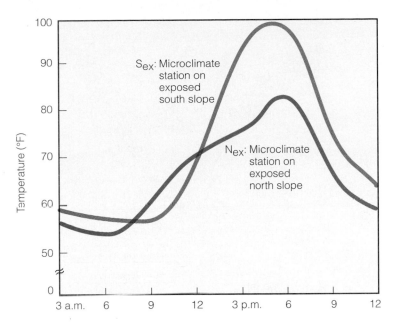

FIGURE 2.16 Changing microclimatic temperatures of north-facing and south-facing slopes on a sunny day in August. Note the differences in microclimate between the exposed north-facing and south-facing slopes. (Location: Greer, West Virginia; data courtesy Dr. W. A. van Eck.)

the species of trees and other plants occupying the slopes (Figure 2.17); the species of trees influence mineral recycling, which is reflected in the nature and chemistry of the surface soil and the makeup of the herbaceous ground cover.

Being mobile, few if any animals are typical of only north-facing or south-facing slopes, as far as we now know. However, their movements may be limited to some extent by the differences in conditions and food supplies on the slopes. Mule deer (*Odocoileus hemionus*) tend to use south-facing slopes more heavily in winter and early spring, and north-facing slopes in summer (Taber and Dasmann 1958). In the central Appalachians the red-backed vole (*Clethrionomys gapperi*), normally an inhabitant of cool fir (*Abies*), spruce (*Picea*), aspen (*Populus*), and northern hardwood forests throughout its range, is restricted in its local distribution to forested, mesic, north-facing slopes. Those species of soil invertebrates intolerant of humidity, such as some mites, can exist only in a dry habitat and therefore are confined to the south-facing slopes. In contrast, terrestrial salamanders inhabit the cool, moist north-facing slopes (Spotila 1972).

Inversions

Solar energy heats Earth's surface and the air above it by day (Figure 2.18a). As night approaches, the surface air above the ground loses more long-wave radiant energy than it receives (Figure 2.18b). This surface layer of cooler air deepens (Figure 2.19a) as the night progresses and forms a nighttime surface inversion in which the temperature increases with elevation. By morning, surface heating gradually eliminates this nighttime inversion (Figure 2.19b).

Such inversions are particularly pronounced in hilly and mountainous country in summer when the air mass is stable and the weather is calm and clear (Figure 2.20). At night, air in the valley cools next to the ground, forming a weak surface inversion. At the same time, cold dense air flows down slopes from the hill or mountaintop. Together they cause the

inversion to become deeper and stronger, and the cold dense air is trapped beneath a layer of warm air. In mountainous areas the top of the night inversion is usually below the main ridge. If air is sufficiently cool and moist, fog may form in the valley. Smoke from industry and other heated pollutants released in such an inversion will rise only until their temperature equals that of the surrounding air, a point called the thermal belt or "warm slope zone." Then smoke flattens out and spreads horizontally just below the thermal belt. These inversions break up when surface air warms during the day to create vertical convections and turbulence, or when a new air mass moves in.

Similar but more widespread inversions occur when a high-pressure area stagnates over a region. In a high-pressure area the air flow is clockwise and spreads outward. The air flowing away from the high must be replaced, and the only source for replacement air is from above. Thus surface high-pressure areas are regions of sinking air movements from aloft, called **subsidence.**

When high-level winds slow down, heavy cold air at high levels in the atmosphere tends to sink. As the parcel of air sinks, it is compressed, heats, and becomes drier. A layer of warm air then develops at a higher level in the atmosphere (Figure 2.21, p. 36) with no chance to descend. It hangs several hundred to several thousand feet above the earth, forming a **subsidence inversion.** Such inversions tend to prolong the period of stagnation and increase the intensity of air pollution. Subsidence inversions that bring about our highest concentrations of pollution are often accompanied by lower-level radiation inversions.

Along the west coast of the United States, and occasionally along the east coast, the warm seasons often produce a coastal or **marine inversion** (Figure 2.22, p.36). In this case cool, moist air from the ocean spreads over low land. This layer of cool air, which may vary in depth from a hundred to several thousand meters, is topped by warmer, drier air, which also traps pollutants in the lower layers.

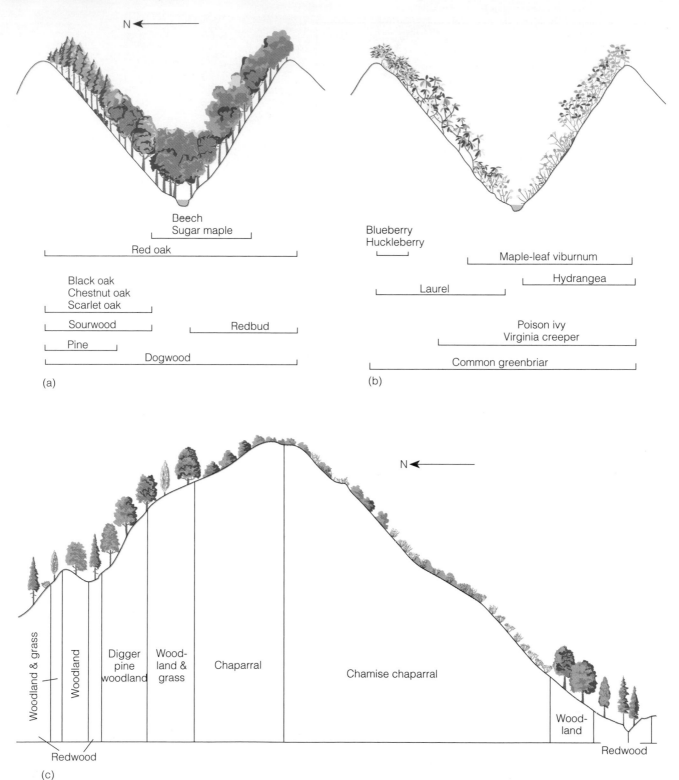

N ←

Beech
Sugar maple
Red oak

Black oak
Chestnut oak
Scarlet oak
Sourwood
Pine
Dogwood
Redbud

(a)

Blueberry
Huckleberry
Maple-leaf viburnum
Laurel
Hydrangea
Poison ivy
Virginia creeper
Common greenbriar

(b)

N ←

Woodland & grass | Woodland | Digger pine woodland | Wood-land & grass | Chaparral | Chamise chaparral | Wood-land | Redwood

Redwood

(c)

FIGURE 2.17   Influence of microclimate on the type and distribution of vegetation on north-facing and south-facing slopes. (a) Trees and (b) shrubs in the hill country of southwestern West Virginia. (c) The Point Sur area in California. In each case note the similarity of vegetation on the lower parts of north-facing and south-facing slopes, in contrast to vegetation change going upslope.

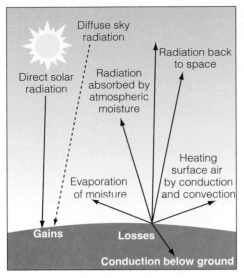

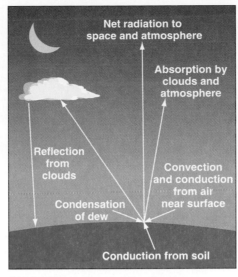

(a) Daytime surface heat exchange

(b) Nighttime surface heat exchange

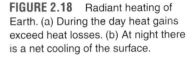

FIGURE 2.18 Radiant heating of Earth. (a) During the day heat gains exceed heat losses. (b) At night there is a net cooling of the surface.

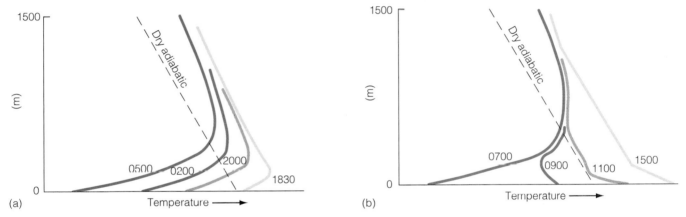

FIGURE 2.19 Formation and elimination of a nighttime surface inversion typical on clear, cool nights. (a) As the ground cools rapidly after sundown, a shallow surface inversion forms (1830). As cooling continues during the night, the inversion deepens from the surface upward, reaching its maximum depth just before dawn (0500). (b) After sunrise, the surface begins to warm and the night surface inversion (0700) is gradually eliminated during the forenoon of a clear summer day. A dry adiabatic layer above the soil deepens until it reaches its maximum depth about midafternoon (1500). (After Schroeder and Buck 1970.)

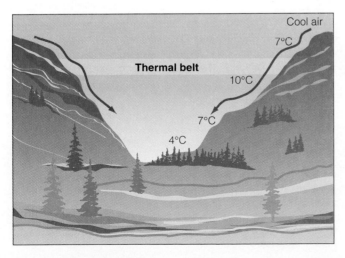

FIGURE 2.20 Topography can produce daily microclimatic extremes in valleys and depressions in the ground. At night air cools next to the ground, forming a weak surface inversion in which the temperature increases, rather than decreases, with height. At the same time, cool air moves downslope, deepening the inversion. When air is sufficiently cool and moist, fog forms in the valley. Smoke or air pollution released in such a situation will rise only until its temperature equals that of the surrounding air. Then it will flatten out just below the layer of warm air.

Surface air must flow out as subsidence progresses

Cool air from aloft begins to settle

Warm, very dry air approaches the surface

FIGURE 2.21 Descent of a subsidence inversion.

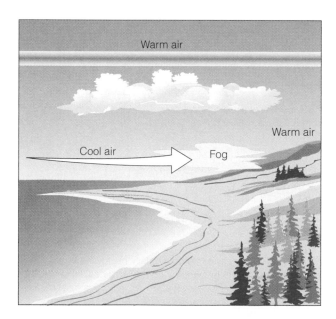

FIGURE 2.22 During a marine inversion, cool air from the ocean moves in beneath the heated layer above. (After Schroeder and Buck 1970.)

Urban Climate

The building of cities has altered not only the slope, soil, and vegetation of the land they occupy, but also the atmosphere, creating a distinctive urban microclimate.

The urban climate is a product of the morphology of the city. In the urban complex, stone, asphalt, and concrete pave-

ment and buildings with a high capacity for absorbing and reradiating heat replace natural vegetation having low conductivity of heat. Rainfall on impervious surfaces is drained away as fast as possible, reducing evaporation. Metabolic heat from masses of people and waste heat from buildings, industrial combustion, and vehicles raise the temperature of the surrounding air. Industrial activities, power production, and vehicles pour water vapor, gases, and particulate matter into the atmosphere in great quantities. The effect of this storage and reradiation of heat is the formation of a heat island about cities (Figure 2.23) in which the temperature may be 6 to 8°C higher than the surrounding countryside (Landsberg 1970, SMIC 1971).

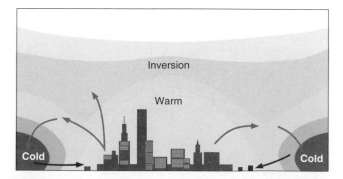

FIGURE 2.23 Idealized scheme of nighttime circulation above a city in clear, calm weather. A heat island develops over the city. At the same time, a surface inversion develops in the country. As a result cool air flows toward the city, producing a country breeze. Lines are temperature isotherms; arrows represent wind. (Adapted from H. Landsberg 1970.)

Heat islands are characterized by high temperature gradients about the city. Highest temperatures are associated with areas of highest density and activity; temperatures decline markedly toward the periphery of the city (Figure 2.24). Although detectable throughout the year, heat islands are most pronounced during summer and early winter and are most noticeable at night when heat stored by pavements and buildings is reradiated to the air. The magnitude of the heat island is influenced strongly by local climatic conditions such as wind and cloud cover. If the wind speed, for example, is above some varying critical value, a heat island cannot be detected.

During the summer the buildings and pavement of the inner city absorb and store considerably more heat than does the vegetation of the countryside. In cities with narrow streets and tall buildings, the walls radiate heat toward each other instead of toward the sky. At night these structures slowly give off heat stored during the day. Although daytime differences in temperature between the city and the country may not differ noticeably, nighttime differences become pronounced shortly after sunset and persist through the night. The nighttime heating of the air from below counteracts radiative cooling and produces a positive temperature lapse rate while an inversion is forming over the countryside. This, along with the surface temperature gradient, sets the air in motion, producing country breezes to flow into the city.

In winter solar radiation is considerably less because of the low angle of the sun, but heat accumulates from human and animal metabolism and from home heating, power generation, industry, and transportation. In fact, heat from these sources is 2.5 times that contributed by solar radiation. This energy reaches and warms the atmosphere directly or indirectly, producing more moderate winters in the city than in the country.

Urban centers influence the flow of wind. Buildings act as obstacles, reducing the velocity of the wind by as much as 20 percent compared with that of the surrounding countryside, increasing its turbulence, robbing the urban area of the ventilation it needs, and inhibiting the inward movement of cool air from the outside. Strong regional winds, however, can produce thermal and pollution plumes, transporting both heat and particulate matter out of the city and modifying the rural radiation balance a few miles downwind (Clarke 1969, Oke and East 1971).

Throughout the year urban areas are blanketed with particulate matter, carbon dioxide, and water vapor. The haze reduces solar radiation reaching the city by as much as 10 to 20 percent. At the same time, the blanket of haze absorbs part of the heat radiating upward and reflects it back, warming both the air and the ground. The higher the concentration of pollutants, the more intense the heat island is.

The particulate matter has other microclimatic effects. Because of the low evaporation rate and the lack of vegetation, relative humidity is lower in the city than in surrounding rural areas, but the particulate matter acts as condensation nuclei for water vapor in the air, producing fog and haze. Fogs are much more frequent in urban areas than in the country, especially in winter (Table 2.1).

Another consequence of the heat island is increased convection over the city. Updrafts, together with particulate matter and large amounts of water vapor from combustion processes and steam power, lead to increased cloudiness over cities and increased local rainfall both over cities and over regions downwind. An example of weather modification by pollution is the increase in precipitation and stormy weather about La Porte, Indiana, downwind from the heavily polluted

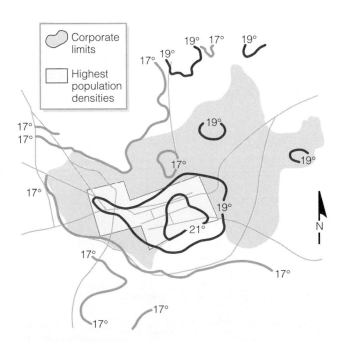

FIGURE 2.24 Thermal pattern of night air in a small city, Chapel Hill, North Carolina. The highest temperatures are inside the corporate limits, where the population and activity are the greatest. (From Kopec 1970.)

TABLE 2.1 Climate of the City Compared with the Country

Element	Comparison
Condensation nuclei and particles	10 times more
Gaseous mixtures	5–25 times more
Cloud cover	5–10 percent more
Winter fog	100 percent more
Summer fog	30 percent more
Total precipitation	5–10 percent more
Relative humidity, winter	2 percent less
Relative humidity, summer	8 percent less
Radiation, global	15–20 percent less
Duration of sunshine	5–15 percent less
Annual mean temperature	0.5–1.0° C more
Annual mean wind speed	20–30 percent less
Calms	5–20 percent more

Source: Adapted from H. E. Landsberg 1970.

areas of Chicago, Illinois, and Gary, Indiana, and close to moisture-laden air over Lake Michigan. Since 1925 there has been a 31 percent increase in precipitation, a 35 percent increase in thunderstorms, and a 240 percent increase in the occurrence of hail (Changnon 1968).

MICROCLIMATES

The regional climate describes the general climatic conditions of the locality that organisms inhabit. These conditions, however, do not describe the actual climate in which organisms live. Within the area they inhabit organisms encounter a wide range of little climates, or microclimates, in which heat, moisture, and air movements, influenced by soil, vegetation, ground cover, and other factors, all vary greatly from one microsite to another.

Climate near the Ground

On a summer afternoon the temperature under a calm, clear sky may be 28°C at 1.83 m (6 ft), the standard level of temperature recording. On or near the ground—at the 5 cm level—the temperature may be 5°C higher; and at sunrise, when the temperature for the 24-hour period is the lowest, the temperature may be 3°C lower at ground level (Biel 1961). Thus in an open field in the mideastern United States, the afternoon temperature near the ground may be equivalent to the temperature at 1.83 m in Florida, 700 miles to the south; and at sunrise the temperature may be equivalent to the 1.83 m temperature in southern Canada. Even greater extremes occur above and below the ground surface. In New Jersey, March daytime temperatures about the stolons of clover plants 1.5 cm above the surface of the ground may be 21°C, while 7.5 cm below the surface the temperature about the roots is − 1°C (Biel 1961). The temperature range for a vertical distance of 9 cm is 22°C. Under such climatic extremes most organisms exist.

The chief reason for the great differences between temperature at ground level and at 1.83 m is solar radiation. During the day in an open field, the soil, the **active surface** (where most solar energy is absorbed), absorbs solar radiation, which comes in short waves, and radiates it back as long waves to heat a thin layer of air above it. Because air flow at ground level is almost nonexistent, the heat radiated from the surface remains close to the ground. Temperatures decrease sharply in the air above this layer and in the soil below. The heat absorbed by the ground during the day is reradiated by the ground at night. This heat is partly absorbed by the water vapor in the air above. The drier the air, the greater is the outgoing heat and the stronger is the cooling of the surface of the ground and the vegetation. Eventually the ground and the vegetation are cooled to the dew point, and water vapor in the air may condense as dew on the vegetation and ground. After a heavy dew a thin layer of chilled air lies over the surface, the result of rapid absorption of heat in the evaporation of dew.

Influences of Vegetation and Soil

By altering wind movement, evaporation, moisture, and soil temperatures, vegetation influences the microclimate of an area, especially near the ground. Temperatures at ground level in the shade are lower than those in places exposed to the sun and wind.

Vegetation also reduces the steepness of the temperature gradient and influences the height of the active surface. With absent or very thin vegetation, temperature increases sharply near the soil; but as plant cover increases in height and density, the leaves of the plants intercept more solar radiation (Figure 2.25). Plant crowns then become the active surface, or more precisely the active layer. As a result temperatures are highest just above the dense crown surface and lowest at the surface of the ground. Maximum absorption of solar radiation in tall grass occurs just below the upper surface of the vegetation, whereas in short grass maximum temperatures are at ground level (Waterhouse 1955). (Among broad-leafed herbaceous plants, daily maximums occur on the upper leaf surfaces.) At night minimum temperatures are some distance above the ground, because the air is cooled above the tops of plants and the dense stalks prevent the chilled air from settling to the ground.

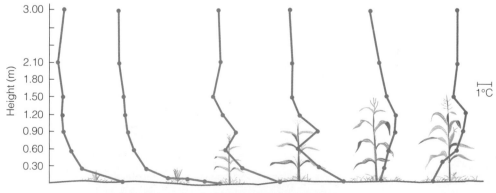

FIGURE 2.25 Vertical temperature gradients at midday in a cornfield, from seedling stage to harvest. Note the increasing height of the active surface. (Adapted from Wolfe et al. 1949.)

Within dense vegetation such as a forest, air movements are reduced (Figure 2.26) and calm exists at ground level. This calm is an outstanding feature because it influences both temperature and humidity and creates a favorable environment for insects and other animals.

Vegetation deflects wind flow up and over its top. If the vegetation is narrow, such as a windbreak or a hedgerow, the microclimate on the leeward side may be greatly affected. Deflection of wind produces an area of eddies immediately behind the vegetation, in which the wind speed is low and small particles such as seeds are deposited (Figure 2.27). Beyond is an area of turbulence, in which the climate tends to be colder and drier than normal. If some wind passes through the barrier and some goes over it, no turbulence develops, but the mean temperature behind the barrier is high in the morning and lower in the afternoon.

Humidity changes greatly from the ground up. Because evaporation takes place at the surface of the soil or at the active surface of plant cover, the vapor content (absolute humidity) decreases rapidly from a maximum at the bottom to atmospheric equilibrium above. During the night little difference exists above and on the ground. Within growing vegetation, however, relative humidity is much higher within and below plant cover than above it. In fact, near-saturation conditions may exist in part because of transpiration.

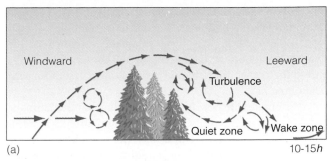

(a)   10-15*h*

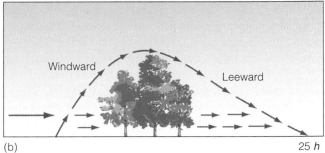

(b)   25 *h*

FIGURE 2.27 The influence of a dense windbreak (a) and a permeable windbreak on wind flow (b) in the shelter and in the open. Note the turbulence produced by wind and the quiet zone associated with the dense windbreak. The quiet zone experiences higher temperatures and a higher relative humidity, making the quiet zone an important microclimate and shelter for animal life living in the open and in windbreak. The permeable windbreak reduces wind speed but does not create a strong microclimate.

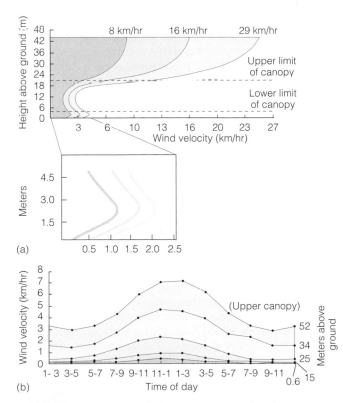

FIGURE 2.26 (a) Distribution of wind velocities with height as affected by the timber canopy of coniferous forests for wind velocities of 8, 16, and 24 km/hr at 43 m above the ground. (b) Average wind velocity during a June day inside a coniferous forest with a cedar understory in northern Idaho. Note the decrease in velocity near the ground. The inset is a detail of wind movement from ground level to 4.5 m. (From Gisborne 1941.)

Soil properties also influence microclimates. Dark-colored soils are better absorbers of heat than light colored soils, because light-colored soils reflect heat. Dry soils are poorer conductors of heat than moist soils. Poor conductivity and reflectivity are two reasons that temperatures are cool at relatively shallow depths in dry sandy soils. In a soil that conducts heat well, considerable heat energy will be transferred into the soil, from which it radiates to the surface at night.

Microclimates and Habitats

Organisms, animals especially, exploit their available range of microclimates to seek most favorable environments in which to live: in crannies and pockets, in tree cavities and logs, in and beneath vegetation, beneath stones and leaf litter, where temperature and moisture are most favorable.

Seeking thermal cover, a place to escape from the extremes of weather, is a major activity of animals from insects to large mammals. Thermal cover influences their movements, feeding, nesting, and resting sites. Large mammals, such as deer and elk (*Cervus elaphus*), seek thickets of conifers and evergreen shrubs, where wind flow is reduced and where it is warmer in winter and cooler in summer (Ozoga and Gysel 1972, Edgerton and McConnell 1976). Cavity-nesting and cavity-roosting birds seek hollows of trees, where wood is an excellent insulator against heat and cold (Kendeigh 1961). Salamanders seek leaf litter and decayed logs in summer, where the microclimate is cool and humid. Tropical salamanders inhabit bromeliads and banana plants, whose leaves afford cooler and more stable temperatures than their surroundings (Feder 1982). Tropical log-dwelling salamanders move to the warmer, sunlit side of the log during the cooler times of day and to the cooler, more humid portion of the log during the heat of the day.

Vertebrates are not alone in responding to microclimatic conditions. The microdistribution of many insects is influenced by microclimatic conditions. Some seek cool, moist places; others seek dry, warm conditions. Slugs (Limacidae) stay hidden beneath moist litter and rotting logs by day and move about only during the humid nights. Bark beetles (Scolytidae) inhabiting fallen logs avoid the driest, warmest parts of the log and seek the dampest section of the log near the ground to lay their eggs. Nectarivorous hummingbird flower mites (Gamasidae: Ascidae) inhabiting *Heliconia wagneriana* flowers seek the most shaded locations within the blooms and avoid poorly shaded parts and flowers (Dobkin 1985).

Plants, too, respond to microclimatic conditions. Unable to move, they depend on colonization of the most favorable microclimatic sites. Herbaceous growth in a dense forest, for example, may be limited to microsites that are flooded with large flecks of sunlight. Some of the most pronounced microclimatic effects on plants occur in areas of convex slopes and low concave surfaces. These places have much lower temperatures at night, especially in winter, much higher temperatures during the day, especially in summer, and a higher relative humidity. The concave surfaces radiate heat rapidly on still, cold nights, and cold air flows in from surrounding higher levels. On such sites the air temperature may be 8°C lower than in surrounding terrain, causing a temperature inversion on a small scale. Because low ground temperatures in these areas tend to result in late spring frosts, early fall frosts, and a subsequent short growing season, these depressions are called **frost pockets** (Geiger 1965).

The pockets need not be deep. Spurr (1957) found that the minimum temperatures in small depressions only 1 to 1.2 m deep were equivalent to those of a nearby valley 60 m below the general level of the land. Such variations in temperature due to local microrelief can strongly influence the distribution and growth of plants. Tree growth is inhibited; and because the low surfaces more often than not accumulate water as well as cold air, such sites may contain plants of a more northern distribution. Frost pockets may also develop in small forest clearings. The surface of the tree crowns channels cold air into the clearings as terrestrial radiation cools the layer of air just above.

Summary

1. Weather is air in motion driven by unequal heating. Climate is the summation of weather over time.
2. Solar radiation is the major source of thermal energy for Earth. Only 50 percent of solar radiation in the form of short-wave radiation reaches Earth. Considering the amount of incoming radiation as 100 percent, 30 percent is reflected back, and 25 percent is absorbed by dust, water, and carbon dioxide in the atmosphere. Earth's surface absorbs the remaining 45 percent as short-wave radiation, which is radiated back as long-wave radiation. About 12 percent of this reradiated long-wave radiation escapes to outer space; 88 percent is absorbed by water vapor and carbon dioxide in the atmosphere and in turn is radiated back to Earth.
3. Earth's surface reflects back a percentage of solar radiation impinging on it, called albedo. Because of the nature of Earth's surface, this albedo varies from region to region.
4. Water vapor in the atmosphere acts as an independent gas that has weight and pressure. The amount of pressure water vapor exerts is vapor pressure. The pressure it exerts when the air is saturated—that is, when the number of water molecules leaving the air equals the number returning to the air—is saturation vapor pressure. The amount of water vapor a given volume or weight of air contains is expressed as relative humidity, the amount of water vapor actually in the air relative to the amount it could hold if saturated. The amount of water vapor air can hold varies with its temperature.
5. Behavior of air masses involves adiabatic processes, in which heat is neither lost or gained from the outside. Daily heating and cooling cause air masses to rise and sink. Rising unsaturated air cools at the dry adiabatic rate. Because it releases latent heat, rising saturated air cools at a moist adiabatic rate. In a stable atmosphere the lifted parcel of air will be cooler and thus heavier than the surrounding air and will drop back to its original position. In unstable air, the lifted parcel of air will be warmer and thus lighter than the surrounding air and will rise upward from its original position. The behavior of air masses has a significant influence on climate.
6. Under certain conditions, the temperature of air masses increases with height rather than decreasing. Such an air mass is very stable, creating a temperature inversion that can trap atmospheric pollutants and hold them close to the ground. Inversions break up when air close to the ground gains heat, causing it to circulate and rise through the inversion, or when a new air mass moves into the area.
7. The movement of air masses about Earth is influenced by the Coriolis effect generated by the spinning Earth. Spinning deflects air masses to the right in the Northern Hemisphere and to the left in the Southern Hemisphere.
8. Solar radiation is the major determinant of climate. The heat budget is influenced by the position of incoming solar radiation on an essentially spherical Earth, the distribution of land and water, and Earth's daily rotation. Because of the Earth's shape, the planet experiences unequal heating. The prevailing winds, ocean currents, and rainfall patterns over the planet are all affected by variations in heat budgets.
9. Wind-driven ocean currents modified by the Coriolis force transport sensible heat and water from equatorial regions to arctic and antarctic regions. Each ocean is dominated by two great circular masses or gyres, which are centered on subtropical high-pressure areas north and south of the equator.
10. The climate of any given region is a combination of patterns of temperature and moisture, which are influenced by latitude and location in the continental land mass. These variations are influenced by differences in topography, aspect, and height. Mountain ranges influence regional climates by intercepting air flow, causing its moisture to drop out in the windward side and creating an area of dry air on the leeward side. Most pronounced are environmental differences among slope positions and aspects.

11. Climates of urban areas are characterized by the presence of a heat island. Compared with surrounding rural areas, a city has a higher average temperature, particularly at night, more cloudy days, more fog, more precipitation, a lower rate of evaporation, and lower humidity.

12. Although regional climates determine conditions over an area, the actual climatic conditions under which organisms live is the microclimate, the climate near the ground where conditions are variable and even extreme relative to the conditions above. Litter, logs, rocks, and vegetative cover provide microclimatic habitat for organisms.

Review Questions

1. What is the major determinant of climate?
2. Contrast the roles short-wave and long-wave radiation play in the heat budget of Earth.
3. Discuss what happens to solar radiation passing through the atmosphere and to the surface of Earth.
4. What effect does the greater heating of lower latitudes have on the air circulation on Earth?
5. Describe the relationship between air temperature and saturated vapor pressure.
6. What does relative humidity represent?
7. When air is pumped into a tire, the tire gets hot. Why? How is this phenomenon related to the change in temperature of a parcel of air as its approaches the ground?
8. Explain the difference between the environmental lapse rate and the dry adiabatic rate.
9. Why are moist and dry adiabatic rates of cooling different?
10. What is the Coriolis effect? How does it influence global air circulation? Ocean currents?
11. What is an inversion? How do nighttime valley inversions and subsidence inversions occur?
12. What is the significance of microclimates to local plant and animal distributions?
13. What are some climatological reasons that in many humid middle latitudinal regions, forest growth is greater on north-facing slopes than on south-facing slopes?
14. What effect would cloudiness and wind speed have on the degree and depth of an air temperature inversion near the active surface at night?
15. Contrast the microclimate beneath a forest canopy with the microclimate in a grass field.

Cross References

Plant responses to moisture, 104–111; animal responses to moisture, 99–100/127–128; thermal energy exchange, 48–50; plant responses to temperature, 100–104; seasonality, 95; nature of energy, 480–481; global climate change, 707–719; biomes, 547; life zones, 547–548; ecoregions, 549–552; grasslands, 563–568; savannas, 574; shrublands, 578–584; deserts, 584–587; forests, 602–613; tundra, 589–595.

CHAPTER 3

The Physical Environment

CONCEPTS

1. Solar radiation with wavelengths between 400 and 740 nanometers is visible light, also known as photosynthetically active radiation.

2. Light has spectral quality, intensity, duration, and directionality. Light impinging on physical matter may be reflected, absorbed, or transmitted.

3. The thermal environment of organisms involves radiation, convection, conduction, and evaporation.

4. The unique structure and properties of water determine its role and behavior in an organism's environment.

5. Precipitation and evaporation are the basic processes that move water between Earth and the atmosphere, forming the water cycle.

6. Certain elements are essential for the growth and reproduction of all organisms.

The physical environments of organisms are characterized by light, temperature, moisture, and nutrients. Solar radiation, directly and indirectly, determines the nature of those physical environments. Solar radiation, as outlined in Chapter 2, involves both light energy and thermal energy. The visible portion of solar radiation, with its daily and seasonal periodicity, is the energy source of photosynthesis, the basic process supporting life on Earth. The infrared portion of solar radiation, the primary source of Earth's heat budget, influences the thermal environment in which organisms live. Solar energy drives the movement of water between Earth and atmosphere and its presence, distribution, behavior, and availability in the environment.

LIGHT

Day and night, sunlight and shade, cloudy days, moonlit nights—the play of light has pronounced effects on life. It is the energy source for photosynthesis; it influences the distribution of plants and animals on land and in water; and it affects their daily activities and seasonal changes.

Nature of Light

Light is that part of solar radiation in the visible range, comprising wavelengths of 400 to 740 nanometers (nm; a nanometer is one billionth of a meter) (Figure 3.1). Collectively, these wavelengths are known as **photosynthetically active radiation (PAR)** because they contain the wavelengths used in photosynthesis. Light with wavelengths longer than the visible range is near infrared with wavelengths of 740 to 5000 nm, and far infrared or thermal radiation from 5000 to 100,000 nm. Light with shorter wavelengths than the visible range is ultraviolet light—UV-A with wavelengths of 315 to 380 nm, and UV-B with wavelengths of 280 to 320 nm.

Most of the ultraviolet radiation is absorbed by the stratospheric ozone layer. Solar UV-B (280–320 nm) radiation decreases on a latitudinal gradient from the tropics, where the ozone layer is the thinnest, to the poles, where the layer is the thickest. Altitudinally, solar UV-B radiation increases about 14 to 18 percent per 1000 m elevation. But in recent years this stratospheric ozone layer is being diminished by increasing concentrations of human-made ozone destroying chemicals such as chlorofluorocarbons (CFCs). This depletion, most pronounced over the poles and the tropics (Kerr 1992), allows an increased amount of ultraviolet radiation to reach the ground, mostly UV-B radiation, especially the wavelengths between 290 and 320 nm. (Caldwell et al. 1989). This increase can have specific pronounced effects on organisms, discussed in later chapters.

In addition to its spectral quality, light has three other characteristics: intensity, duration, and directionality. The intensity of light varies daily and seasonally. It is influenced by its angle of incidence, the angle at which it strikes a surface. The angle of incidence of solar radiation depends on the altitude of the sun—its height above the horizon. During a clear day the highest intensity of light occurs at noon, when the sun is overhead, and lowest intensity occurs at dawn and twilight, when the low angle of incidence means that light has a long path through the atmosphere. The seasonal intensity of light changes with latitude (Figure 3.2). In the Northern Hemisphere sunlight is most intense in June at 80°N latitude (near the North Pole) and at 40°N latitude (New York City); in winter at 80°N latitude the intensity is zero. At low latitudes (the Tropics) there is little seasonal fluctuation in direct light intensity, but note that direct sunlight is greatest at 0° latitude during the wintertime of the Northern Hemisphere.

Duration of light is also seasonal. It, too, is a function of latitude and season. On the first day of summer in the Northern Hemisphere, June 21, daylength at 80°N latitude is nearly 24 hours; at 40°N latitude, 17 hours; and at 0° latitude, about 12 hours. On the first day of winter at 40° the length of day is 9 hours. At the spring and fall equinoxes in March and September, the length of day is 12 hours everywhere in the world.

The directionality of light shifts daily and seasonally. The sun is nearly overhead in early summer; but as winter approaches, the altitude of the sun drops toward the horizon. In mountainous country south-facing and west-facing slopes receive the most light; deep narrow valleys may experience shorter days because of shading by opposite slopes. On a microscale many organisms find themselves always shaded on one side.

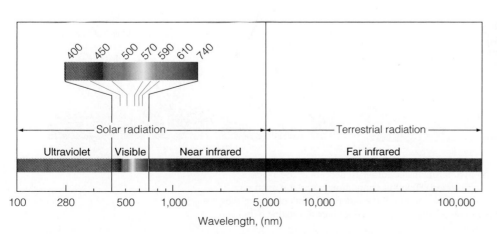

FIGURE 3.1 A portion of the electromagnetic spectrum, separated into solar and terrestrial radiation. (After Halverson and Smith 1979:6.)

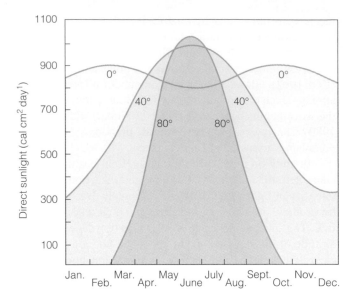

FIGURE 3.2 Amount of direct sunlight incident on a horizontal surface at the outer extremity of Earth's atmosphere as a function of time of year for north latitudes of 0°, 40°, and 80°. (From Gates 1972:28.)

Fate of Light

Light impinging on an object may be absorbed, reflected, or transmitted through it. For example, when light hits a leaf, plants reflect about 70 percent of infrared wavelengths striking the leaves perpendicularly. They reflect only 6 to 12 percent in the visible range. The degree of reflection varies with the nature of the leaf. Leaves with whitish hairs and cuticles reflect more light than deep green leaves. Plants reflect green light more strongly (10 to 20 percent) than red and orange light (3 to 10 percent). Because they reflect green light most strongly, leaves appear green. Only about 3 percent of UV light is reflected. Much of the visible radiation is absorbed. About 70 percent is used by green pigments in plants for photosynthesis. A remaining fraction of light is transmitted through the leaf, the amount depending on its thickness and structure. Thin leaves transmit more light than thick ones. Transmission is greatest for those wavelengths that are also reflected. Thus plants transmit mostly green and far infrared wavelengths.

In both grasslands and forest the leafy canopy intercepts most of the sunlight that floods open spaces (Figure 3.3). As light penetrates the canopy, different wavelengths are filtered out—the light becomes attenuated. The degree of attenuation is influenced by the density of the canopy, the optical properties of the leaves, and the number and sizes of gaps in the canopy. Blue and red are the most highly attenuated wavelengths, whereas green and far infrared wavelengths pass through relatively unaltered. Far infrared wavelengths are less strongly attenuated than visible light. These changes in spectral quality vary among forests. For example, beneath a canopy of a deciduous forest on a clear day, far infrared radiation dominates with lesser amounts of blue, green, and red (Vezina and Boulter 1966), giving an observer the impression of standing in reddish-green shade. In the depths of the forest, only far infrared wavelengths prevail. In the more open pine forest, little difference exists in the transmission of various wavelengths. On cloudy days selective absorption of spectral wavelengths is lower, and transmission is higher in both forests.

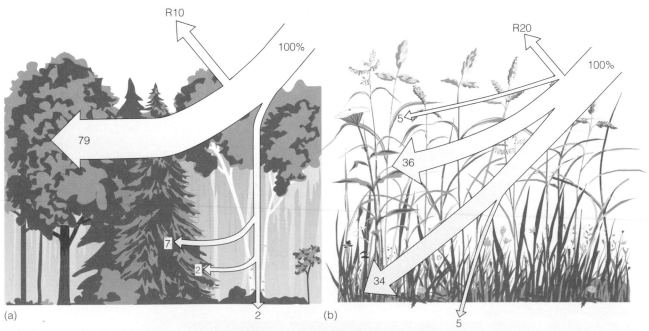

(a)   (b)

FIGURE 3.3 Attenuation of radiation. (a) In a boreal mixed forest 10 percent of the incident photosynthetically active radiation (PAR) is reflected (R) in the upper crown, and the greatest absorption occurs in the crown. (b) In a meadow 20 percent of the radiation is reflected from the upper surface, whereas the greatest absorption occurs in the middle and lower regions of the canopy. (Adapted from Larcher 1980.)

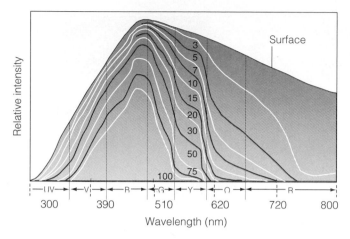

FIGURE 3.4 The spectral distribution of solar energy at Earth's surface and after it has been modified by passage through varying depths, measured in meters, of pure water. Note how rapidly red wavelengths are attenuated. At approximately 10 m, red light is depleted; but at 100 m, blue wavelengths still retain nearly one-half their intensity.

Light that strikes the surface of water is also attenuated (Figure 3.4). Depending on the angle at which light strikes the water, a greater or lesser amount is reflected from the surface. For example, only 6 percent of the light striking the surface of water at an angle of incidence of 60° from the perpendicular to the surface of the water is reflected; at 80°, 35 percent is reflected. The higher the sun is from the horizon, the more light will enter the water.

Light that enters the water is absorbed rapidly. Only about 40 percent reaches a meter deep in clear lake water. Moreover, water absorbs some wavelengths more than others. First to go is visible red light and infrared light in wavelengths greater than 750 nm. This absorption reduces solar energy by one-half. In clear water yellow goes next, followed by green and violet, leaving only blue wavelengths to penetrate deeper water. A fraction of blue light is lost with increasing depth. In the clearest of seawater, only about 10 percent of the blue wavelengths reaches more than 100 m.

Leaf Area Index and Extinction Coefficient

Light is transported in discrete bundles called **photons.** The rate at which photosynthetically active radiation impinges on a given area is called the photon flux density and is measured in units of μmols photons m^{-2} s^{-1}. As already discussed, the amount of PAR reaching Earth's surface varies as a function of the diurnal and seasonal cycles, as well as of latitude. These variations have a direct influence on rates of net photosynthesis and plant growth. However, a major influence on the availability of PAR to a given leaf is the reflectance and absorption of solar radiation by other leaves. The absorption and reflectance of PAR by leaves results in a dramatic variation in the vertical profile of light within a single plant or stand of plants.

The amount of light at any depth in a plant canopy is a function of the number of leaves above. As you move down through the canopy, this number increases and consequently the amount of light decreases. However, because leaves vary

in size and shape, the number of leaves is not the best measure of quantity.

The quantity of leaves, or foliage density, is expressed in terms of leaf area. Because most leaves are flat, the leaf area is the surface area of one or both sides of the leaf. When the leaves are not flat (e.g., pine needles), the entire surface area is sometimes measured. To quantify the changes in light environment with increasing area of leaves, a measure is used that defines the area of leaves per unit of ground area. This measure is called the **leaf area index (LAI):**

$$\text{LAI} = \text{m}^2 \text{ leaf area/m}^2 \text{ ground area}$$

A leaf area index of 3, for example, would mean that there are 3 square meters of leaf area over each meter of ground area. The greater the leaf area above any surface, the lower the quantity of light reaching that surface. This relationship can be seen in Figure 3.5. As you move from the top of the plant canopy to the ground level, the cumulative leaf area and leaf area index increase. Correspondingly there is a decrease in light (PAR).

The general relationship between available light and leaf area index is described by Beer's Law:

$$\text{AL}_i = e^{-\text{LAI}_i \cdot k}$$

The subscript i refers to the vertical height in the canopy. For example, a value of 20 refers to 20 m above the ground. The value AL_i is the light reaching any vertical position i in the canopy. The value is expressed as a proportion of light at the top of the canopy (a value ranging from 0.0 to 1.0). The number e is the base of the natural logarithm ($e - 2.718\ldots$); LAI_i is the cumulative leaf area index above height i; k is the **light extinction coefficient,** a measure of the degree to which leaves absorb and reflect light.

Assuming the light extinction coefficient to be $k = 0.6$, we can calculate changes in the vertical profile of light for the example presented in Figure 3.5. Moving from the top to the bottom of the canopy, the quantity of light declines exponentially (Figure 3.6).

The decrease, or attenuation, of light with increasing leaf area is also influenced by the arrangement of leaves on the plant. Leaves that are positioned horizontally will intercept more light than leaves that are positioned at an angle (Figure 3.7). Leaf angle influences the vertical distribution of light through the canopy as well as the total amount of light intercepted (absorbed and reflected) by the plant canopy.

Although light decreases downward through the plant canopy, some direct sunlight does penetrate openings in the crown and reaches the ground as sunflecks. Sunflecks can account for 70 to 80 percent of solar energy reaching a forest floor. Sunflecks and indirect light (skylight) enable many plants inhabiting the forest floor to endure the otherwise shaded conditions (Chazdon and Pearcy 1991).

Only about 1 to 5 percent of the light that strikes the canopy of a typical temperate deciduous forest (LAI = 3–5) in the summer reaches the forest floor. More light travels through a stand of pine trees (LAI = 2–4)—about 10 to 15 percent. In a tropical rain forest (LAI = 6–10), only 0.25 to

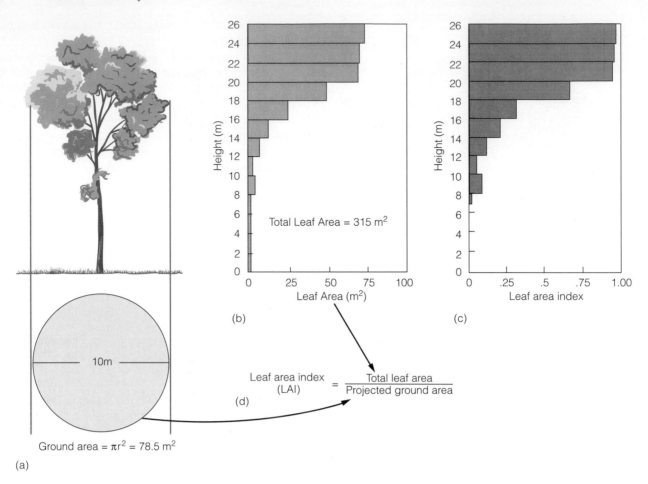

(b)

(c)

Leaf area index
(LAI) $=$ $\dfrac{\text{Total leaf area}}{\text{Projected ground area}}$

(d)

Ground area = πr^2 = 78.5 m²

(a)

FIGURE 3.5   The concept of leaf area index. (a) The leaf area index is the total area of leaves divided by the projected ground area. (b) By dividing the surface area of leaves in each vertical layer of the canopy by the projected ground area, (c) a vertical profile of the leaf area index can be established. (d) This profile can then be used to calculate the vertical distribution of light (see text for explanation).

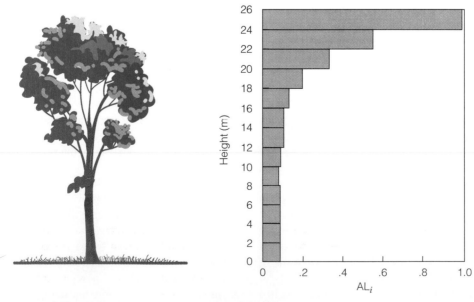

FIGURE 3.6   The vertical profile of light resulting from the distribution of the leaf area index shown in Figure 3.5. Values for available light at each height in the canopy (AL_i) are expressed as the proportion of light (PAR) at the top of the canopy (expressed as the value of 1.0).

FIGURE 3.7 Although horizontal leaves (right) capture the most sunlight per unit leaf area, the upper layers shade the lower, reducing the overall interception of light. An upright orientation of leaves (left) is more efficient in capturing the sun's energy. That arrangement is typical of plant stands in which individuals are growing closely, such as grasses.

2 percent gets through. Relatively open woodlands of trees, such as birches and oaks, allow light to filter through. There light attenuates gradually throughout the canopy, as it does in grasslands. In grassland, where the top of the canopy is relatively open, the middle and lower layers intercept the most light (see Figure 3.3b).

In many environments seasonal changes strongly influence the total leaf area. For example, in the temperate regions of the world many forest tree species are deciduous, shedding their leaves during the winter months. In these cases, the amount of light that penetrates a stand of vegetation varies with the season (Figure 3.8). In early spring in temperate regions, when leaves are just expanding, 20 to 50 percent of the incoming light may reach the forest floor. Spring flowering plants may use this flood of light by completing the reproductive phase of their life cycle before the canopy closes. When less than 10 percent of the light reaches the forest floor, flowering is over. In fall, when the leaves begin to drop, increased light again reaches the forest floor and another surge of flowering, involving goldenrods and asters, takes place.

Changes in the light in the aquatic environment also follow Beer's Law. However, other environmental conditions are also involved. Freshwater lakes, streams, and ponds and

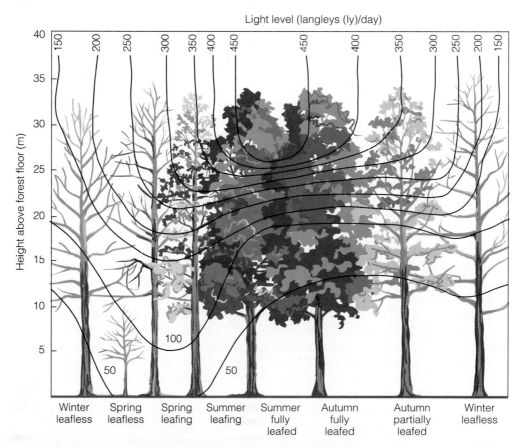

FIGURE 3.8 Light levels within and above a yellow-poplar (*Liriodendron tulipifera*) stand over a year. The greatest intensity of solar radiation occurs in summer, but the canopy attenuates most of the light, so that little reaches the forest floor. The most illumination reaches the forest floor in the spring, when trees are still leafless. The forest receives the least radiation in winter, with low sun angles and shorter daylengths. As a result, the amount of solar radiation reaching the forest floor during the winter months is little more than that during the months of midsummer, when the canopy is fully developed.

coastal waters have a certain degree of turbidity, which greatly affects light penetration. Because of the high level of yellow substances such as clay colloids and fine detrital material washed into water from terrestrial ecosystems, blue wavelengths are the most strongly attenuated and are removed at very shallow levels. Green penetrates most deeply, and where the concentration of yellow material is high, red wavelengths may penetrate as far as the green. In very yellow water, red is the last wavelength to be extinguished. Waters with organic stains, such as bog lakes and ponds, appear dark, and nutrient-poor lakes and ponds, low in phytoplankton, are clear and bluish. Waters supporting a dense growth of phytoplankton take on a decidedly greenish appearance. Heavy growth of phytoplankton and cyanobacteria can shut out light to deeper waters just as effectively, if not more so, than the dense canopy of a deciduous or tropical rain forest.

TEMPERATURE

All organisms live in a thermal environment characterized by heat and temperature. There is a difference between the two. **Heat** is a form of energy possessed by all substances that results from the random motion of molecules within the substance. The quantity of heat possessed by a substance depends on the kinetic energy of its molecules and its size. A large substance holds more heat than a small one. **Temperature** is the immediate direct measure of the average kinetic energy possessed by individual molecules of a substance. Temperature expresses the substance's intensity of hotness and provides a measure of its tendency to give up heat.

The environmental temperatures experienced by most organisms result, directly or indirectly, from solar radiation. The amount of solar radiation reaching any point on Earth at any time varies with the time of year, slope, aspect, cloud cover, time of day, and other factors. Seasonal fluctuations can be extreme from one point to another. In North Dakota, for example, where the annual mean temperature is between 3 and 9°C, air temperatures fluctuate from a low of −43°C in winter to 49°C in summer. In West Virginia, where the mean annual temperature is 12°C, temperatures range from −37 to 44°C.

Environmental temperatures in any area differ between sunlight and shade and between daylight and dark. In temperate regions surface temperatures of soil may be 30°C higher in the sunlight than in the shade. Daytime temperatures are often 17°C higher than nighttime temperatures; on deserts this spread may be as high as 40°C. Temperatures on tidal flats may rise to 38°C when exposed to direct sunlight and drop to 10°C within a few hours when the flats are covered by water.

All organisms require a certain temperature or range of temperatures to carry on their metabolic processes. Low temperatures slow metabolic processes because they slow chemical reactions. Very high temperatures denature enzymes and destroy their activity. Somewhere in between these extremes, organisms find their optimal temperatures for metabolic and other activities. Organisms capable of withstanding wide temperature regimes are called **eurytherms.** Those restricted to a more narrow range are called **stenotherms.**

The ability to withstand extremes in temperatures varies widely among living organisms, but there are temperatures above and below which no life can exist. A temperature of 52°C is about the highest at which any animal and protozoan can still grow and multiply. The observed thermal limit for the survival of metabolically active vascular plants ranges from about + 60 to − 60°C in different species. Some hot-spring algae can live in water as warm as 73°C (Brock 1967); and some arctic algae can complete their life cycles in places where temperatures barely rise above 0°C. Nonphotosynthetic bacteria inhabiting hot springs can grow actively at temperatures greater than 90°C (Bott and Brock 1969, Brock 1979).

THERMAL ENERGY EXCHANGE

Living organisms must maintain a balance between heat energy gained and heat energy lost (Figure 3.9) In effect, thermal energy absorbed from the environment plus metabolic energy produced must equal thermal energy lost from the body and energy stored.

A major source of heat transfer is **radiation.** This radiation may be direct sunlight, which can be intense. At noon in summer on a clear day, direct solar radiation can amount to 976.9 W/m^2/min (1.4 cal/cm^2/min). Another source is skylight or diffuse radiation, sunlight scattered by moisture and dust in the atmosphere. This heat may amount to 139.5 W/m^2/min (0.2 cal/cm^2/min). A third source is reflected sunlight bounced off objects in the environment. This can range from 69.78 to 209.3 W/m^2/min (0.1 to 0.3 cal/cm^2/min), depending on the type of surface reflecting the sunlight. In addition, thermal radiation, long-wave infrared radiation, is emitted from the surfaces of soil, rocks, organisms, and all other objects in the environment at their ambient temperatures (Figure 3.10). The rate at which a surface emits radiation relates to its surface temperature and its ability to give off thermal radiation. Still another source is the heat given off as a product of metabolism. In metabolic oxidation one mole of glucose gives off 686 kcal of energy, only 40 kcal of which is trapped as ATP; the rest goes off as heat. Just as the organism gains heat by radiation from the environment, so it loses heat to the environment by infrared or long-wave radiation.

Another source of heat transfer is **conduction,** the direct transfer of heat from one substance to another. It involves contact between bodies. The amount of heat lost (or gained) by conduction varies with the surface area exposed, the separation and temperature difference of the two surfaces, the thickness of insulation, such as fur, and conductivity. Air, for example, has low conductivity; water has high conductivity.

Convection is transfer of heat by the circulation of fluid (liquid or gas). Convection may occur naturally in the fluid surrounding an object, such as swirling air above a hot stove; or it may be forced, with pressures from fluids passing by or over an object, such as air flow generated by a moving fan. The amount of sensible heat (heat that can be felt) transferred by convection depends on the shape and area of the organism,

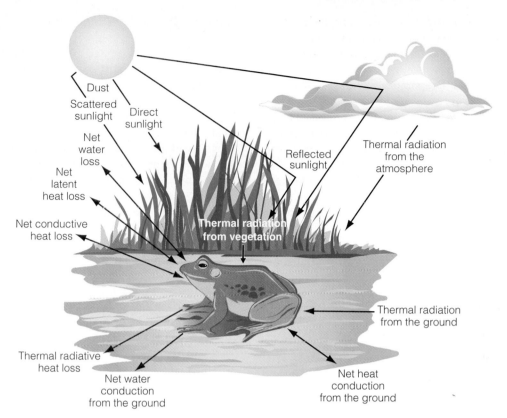

FIGURE 3.9 Exchange of energy between a frog and its environment. The frog receives short-wave radiation directly from the sun or scattered by clouds and objects; it receives long-wave radiation from the substrate, vegetation, and atmosphere. The frog loses energy by emitting long-wave infrared radiation; by exchanging energy with ambient air; and by convection, evaporation, and condensation. It will also lose or gain energy by conduction of heat to and from its body core and by conduction and thermal radiation to the ground. (After Tracy 1976:295.)

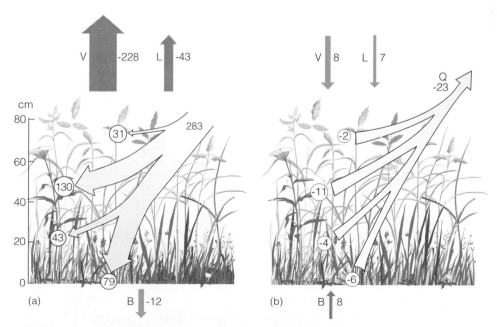

FIGURE 3.10 Energy exchange—absorption and emission—in a meadow. Q=net radiation; V=evaporation; L=sensible heat convection; B=soil heat flux; figures are in cal/cm2. (a) By day, the active layer lies between 35 and 55 cm; it absorbs 45 percent of net radiation. The lowermost active layer absorbs 28 percent. During input 80 percent of the radiant energy is used for the evaporation of water, 15 percent for sensible heat, and 5 percent to raise soil temperature. (b) At night, radiation as well as heat exchange is reversed. The vegetation is reradiating heat to the atmosphere and gaining some energy from the atmosphere through the evaporative process. (After Cernusa 1976.)

the velocity of the fluid, and the physical properties of the fluid. Forced convection can speed up another process, evaporation. Evaporation cools an organism because in changing from a liquid to a vaporous state, water must absorb a substantial amount of heat. Evaporation is influenced by the vapor pressure gradient between air and the object.

Influencing all forms of heat gain and loss by radiation, conduction, convection, and evaporation is the boundary layer. Every organism, in fact every object, has a boundary layer of still air adhering to the surface (see Gates 1972). The resistance this layer offers to absorption and loss of heat is influenced by the size, shape, texture, and orientation of the organism, wind speed, and temperature differences between the surface of the organism and the air.

The relationship tying all these components together is net energy exchange. Net heat gain by solar radiation, infrared radiation, conduction, convection, and metabolism plus increases in energy storage must equal the total heat lost by radiation, conduction, convection, evaporation, and decreases in energy storage.

One aspect is important: Heat produced continuously by organisms is lost passively to the environment. Such loss can take place only when the ambient or surrounding temperature is lower than the core body temperature. When ambient temperature equals core temperature, the route for passing heat off to the environment is lost. When ambient temperature exceeds body temperature, the flow is reversed and heat moves from the environment to the organism.

WATER

Leonardo da Vinci wrote, "Water is the driver of nature." Although most ecologists would now argue that solar radiation is really the driver of nature, da Vinci did not overstate the importance of water to life on Earth. Without the cycling of water, nutrients such as calcium and phosphorus could not make their endless odyssey through the environment, ecosystems could not function, and life could not be maintained. Strategies for obtaining and conserving water have shaped the nature of terrestrial organisms and communities. Adaptations to living in a watery environment have influenced the nature of aquatic life. The distribution of plants and animals is influenced by when, where, and how much precipitation, runoff, and evaporation occur. Thus water is essential to life on Earth.

The Structure of Water

Water can vanish as a gas in the atmosphere, reappear as dew, fog, clouds, rain, and snow, and in the cold change into a solid. Water can change into all these forms because of the physical arrangement of its hydrogen atoms and hydrogen bonds.

A molecule of water consists of one atom of oxygen (atomic number = 8) and two atoms of hydrogen (atomic number = 1), bonded together covalently (characterized by sharing valence electrons) (Figure 3.11a) The outer orbitals of the oxygen atom have six electrons, but are able to accommodate eight. The orbital of hydrogen has one electron, but is able to accommodate two. By sharing, the two hydrogen

FIGURE 3.11   (a) A molecule of water. An atom of oxygen is covalently bonded to two atoms of hydrogen. The angle between the two hydrogen bonds is 105°. (b) Hydrogen bonds between water molecules. The partial positive charge of the hydrogen atoms of one molecule attracts a negatively charged oxygen atom of another molecule to form a hydrogen bond between them. (c) In ice, below 0°C, hydrogen bonds hold water molecules rigidly in place to form a crystalline lattice.

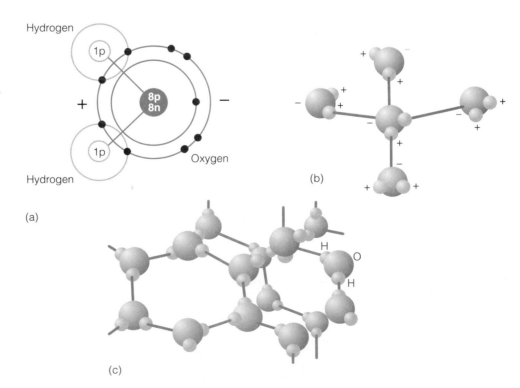

atoms become covalently bonded to the oxygen atom to form a molecule of water. The remaining four, or two pairs, of unshared electrons on the oxygen's outer shell tend to concentrate in a direction away from the O—H bond. If you draw a tetrahedron to enclose the water molecule, the two hydrogen nuclei occupy two corners of the tetrahedron and the unshared pairs of electrons occupy the other two corners. Because oxygen's outer electrons form this tetrahedron, the hydrogen atoms are attached to oxygen in a V-shaped arrangement at an angle of 105°. This arrangement yields an asymmetrical molecule in which the shared electrons are nearer the oxygen atom than the hydrogen atoms. The side on which the hydrogen atoms are located tends to be electropositive, and the opposite side with the oxygen atom tends to be electronegative. Consequently the water molecule exhibits polarity.

Because of this polarity, the positively charged region (hydrogen) of one water molecule is attracted to the negatively charged region (oxygen) of another. This electrostatic attraction forms a hydrogen bond that acts as a bridge between water molecules. The 105° angle of association between the hydrogen atoms encourages an open tetrahedral-like arrangement of water molecules, with hydrogen atoms acting as connecting links between them (Figure 3.11b). This association accounts for the latticelike arrangement of clusters of water molecules.

The lattice is at its most highly organized state when water is frozen (Figure 3.11c). Each oxygen atom is hydrogen-bonded to two hydrogen atoms, and each hydrogen is covalently bonded to one oxygen atom. Ultimately each water molecule is bound to four other water molecules by means of hydrogen atoms. One such unit built upon another gives rise to a lattice with large open spaces. Water molecules so structured occupy more space than they would in the liquid form. As a result water expands upon freezing and ice floats.

When heat is applied to ice, the molecules have a higher kinetic energy (energy of motion) and the weakened bonds are broken, causing the lattice to partially collapse. Although water molecules now move more freely and form a liquid, they can pack more tightly together. The volume occupied by the water molecules decreases and the density increases, until water achieves its greatest density at 3.98°C. At this temperature contraction of the molecules brought about by the partial collapse of the lattice balances the thermal expansion of the warming molecules. As water is heated above 3.98°C, more hydrogen bonds break and the distance between the molecules increases, resulting in a lower density. Given enough kinetic energy, individual water molecules can leave ice or liquid water and enter the atmosphere as gaseous water vapor. The vaporization of ice is called **sublimation** and the vaporization of a liquid is **evaporation.** Both processes occur more rapidly as the temperature increases.

Seawater behaves somewhat differently. Seawater is defined as water with a minimum salinity of 24.7 ‰. (‰ = parts per thousand). Its density, or rather its specific gravity relative to that of an equal volume of pure water (sp. gra. = 1) at atmospheric pressure, is correlated with salinity. At 0°C the density of seawater with a salinity of 35‰ is 1.028. The density of seawater increases at a lower temperature and decreases at higher temperatures. No definite freezing point exists for seawater. Ice crystals begin to form at a temperature that varies with salinity. As pure water freezes out, the remaining unfrozen water becomes saltier, which further lowers its freezing point. If the temperature decreases enough, a solid block of ice crystals and salt ultimately forms.

Physical Properties

Water can store tremendous quantities of heat with a relatively small rise in temperature. It can do so because of the large number of hydrogen bonds present. Energy applied to water goes into breaking the bonds and not into raising the kinetic energy (temperature) of the molecules. Thus water has a high **specific heat,** the number of calories necessary to raise 1 g of a substance 1°C. The specific heat of water is by definition unity or 1. The specific heat of water is exceeded by only a few other liquids, such as ammonia.

Not only does water have a high specific heat, it also possesses the highest heat of fusion and heat of evaporation of all known substances that are liquid between temperatures of 0 to 100°C. Water must lose considerable quantities of heat before it can change from a liquid to a solid form, ice; and conversely, ice must absorb considerable heat before it can convert to a liquid. Approximately 80 calories of heat are needed to convert 1 g of ice at 0°C to a liquid state at 0°C. This amount of heat is equivalent to that needed to raise the same quantity of water from 0 to 80°C.

Evaporation involves substantial amounts of heat; 536 cal are needed to overcome the attraction between water molecules and to convert 1 g of water at 100°C into vapor. Heat lost at the point of evaporation returns at the point of condensation (the conversion from vapor to liquid). Such phenomena play a major role in meteorological cycles and the evaporative cooling of organisms.

Hydrogen bonding and polarity give water two other properties, cohesion and adhesion. **Cohesion** is the holding together of like substances; the hydrogen bonds account for the cohesiveness of water. **Adhesion** is the holding together of unlike substances. The polarity of water molecules allows them to form hydrogen bonds with other polar molecules and ions. These properties account for the **capillarity** of water, its ability to move through the pores of paper, the fine pores of soil, and the conducting ducts of leaves. This capillarity is due in part to the adhesion of water to an electrostatic surface and the cohesion of water molecules. Water molecules are repelled by nonpolar molecules, such as oil, which explains why oil and water do not mix.

Molecular cohesion of water accounts for **surface tension.** Water molecules below the surface are symmetrically surrounded by other molecules of water. The forces of attraction are the same on all sides. At the water's surface, the molecules exist under a different set of conditions. Below is a hemisphere of strongly attractive water molecules; above is a much smaller attractive force of the air. The molecules on the surface therefore tend to contract and become taut.

In aquatic habitats surface tension is a barrier to some organisms and a support for others. It is a force that enables water to creep upward into fine tubes such as the pores of soil and the conducting network of plants. Aquatic insects and plants have evolved structural adaptations that prevent the penetration of water into the tracheal systems of the former and the stomata and internal air spaces of the latter.

Hydrogen bonding gives water another property, **viscosity.** Compared with other liquids, the viscosity of water is high. Observe a liquid flowing through a clear glass tube. The liquid moving through the tube behaves as if it were parallel concentric layers flowing one over the other. The rate of flow is greatest at the center; because of friction between the layers, the flow decreases toward the sides of the tube. This type of resistance between layers is called **lateral** or **laminar viscosity.** You can observe this phenomenon along the side of any stream or river with uniform banks. The water at streamside is nearly still, whereas the current in the center may be swift.

Viscosity of flowing water is complicated by another type of resistance, **eddy viscosity.** Water masses pass from one layer to another, creating turbulence both horizontally and vertically. Eddy viscosity, which is many times greater than laminar viscosity, is biologically important, especially for planktonic organisms. It prevents them from drifting down to the deep and brings nutrients to them.

Viscosity is the source of frictional resistance to objects moving through water. Because this resistance is 100 times that of air, animals must expend considerable muscular energy to move through the water. This resistance has been an important selection pressure in the evolution of a streamlined fusiform body shape in fish and such marine mammals as whales and dolphins and of rodlike shapes, protuberances, and other features of phytoplankton and zooplankton that modify their sinking rates.

The Water Cycle

Distribution of Water   Although water in a stream or a spring rain appears separate, it is part of a worldwide whole distributed in land, sea, and atmosphere and unified by the water cycle. It is influenced by solar energy, by the currents of the air and oceans, by heat budgets, and by water balances of land and sea. Through historical time the amount of free water has remained relatively stable, although the balance between land and sea has fluctuated.

Water occurs in six general forms in the biosphere: vapor, fresh surface water (streams and lakes), groundwater, snow and ice, salt water (oceans), and water inside the bodies of organisms. Oceans cover 71 percent of Earth's surface. With a mean depth of 3.8 km (2.36 mi), they hold 97.25 percent of all Earth's water. Fresh water represents only 3 percent of the planet's water supply (Figure 3.12). Two of that 3 percent consists of ice; and groundwater takes up another 0.70 percent. That leaves only 0.05 percent of the total water supply on Earth as fresh liquid. Considering the fresh liquid supply as 100 percent, lakes contain 60 percent of fresh water on Earth; flowing rivers and streams, 1 percent; soil moisture, 33 percent; and the atmosphere, 6 percent. If this atmospheric water were to condense on Earth, it would form a layer of water no more than 1 inch thick; yet it is the atmosphere, and its relation to land and ocean, that keep water circulating over Earth.

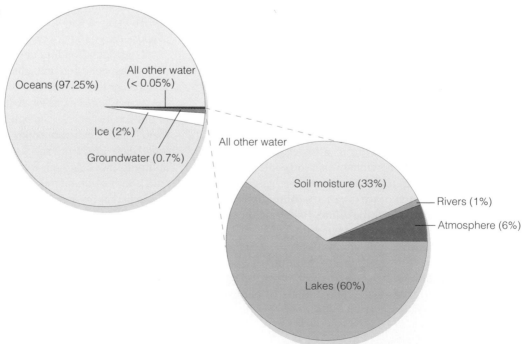

FIGURE 3.12   The distribution of Earth's water supply.

Of major importance is the fresh water found in groundwater. Groundwater fills the pores and hollows within the Earth just as water fills pockets and depressions on the surface. Estimates, necessarily rough and inaccurate, place renewable and cyclic groundwater at 7×10^6 km^3 (Nace 1969), or approximately 11 percent of the freshwater supply. Some of this groundwater is "inherited," as in aquifers in desert regions, where the water is thousands of years old. Because inherited water is not rechargeable, heavy use of these aquifers for irrigation and other purposes is depleting the supply. Loss is occurring in the aquifers of the midwestern United States and Saudi Arabia. In the foreseeable future the supply could be exhausted. A portion of the groundwater, approximately 14 percent, lies below 1000 m. Known as fossil water, it is often saline and does not participate in the hydrological cycle.

The atmosphere, for all its clouds and obvious close association with the water cycle, contains only 0.035 percent fresh water. If this atmospheric water were to condense on Earth, it would form a layer of water no more than one inch thick. Yet it is the atmosphere, and its relation to land and ocean, that keeps the water circulating over Earth.

The Local Water Cycle   Outside it is raining. The rain strikes a house and runs down the windows and walls into the ground. It disappears into the grass, drips from the leaves of trees and shrubs, and trickles down the trunks. When the rain stops, the windows and walls dry. The entire episode—the spring shower, the infiltration into the ground,

the throughfall in the trees and bushes, the runoff from the walks, the evaporation—epitomizes the water cycle on a local scale (Figure 3.13).

Precipitation is the driving force of the water cycle. Whatever its form, precipitation begins as water vapor in the atmosphere. When air rises it is cooled adiabatically, and when it rises beyond the temperature level at which condensation takes place, clouds form (Chapter 2). The condensing moisture coalesces into droplets 1 to 100 μm in diameter and then into rain droplets with a diameter of approximately 1000 μm (1 mm). Where temperatures are cold enough, ice crystals may form instead. Particulates smaller than 10 μm in the atmosphere act as nuclei on which water vapor condenses. At some point the droplets or ice crystals fall as some form of precipitation. As the precipitation reaches the Earth, some of the water reaches the ground directly, and some is intercepted by vegetation, litter on the ground, and urban structures and streets. It may be stored in lakes and ponds, run off, or in time infiltrate the soil.

Because of **interception,** various amounts of water evaporate into the atmosphere without ever reaching the soil surface. Grass in the Great Plains may intercept 5 to 13 percent of the annual precipitation (average loss, 8 percent), and litter beneath a stand of grass may intercept 3 to 8 percent of annual precipitation (average loss, 4 percent) (Corbett and Crouse 1968). Precipitation striking forest trees or forest canopy must penetrate the crowns before it reaches the ground. A forest in full leaf in the summer can intercept a significant portion of a light summer rain.

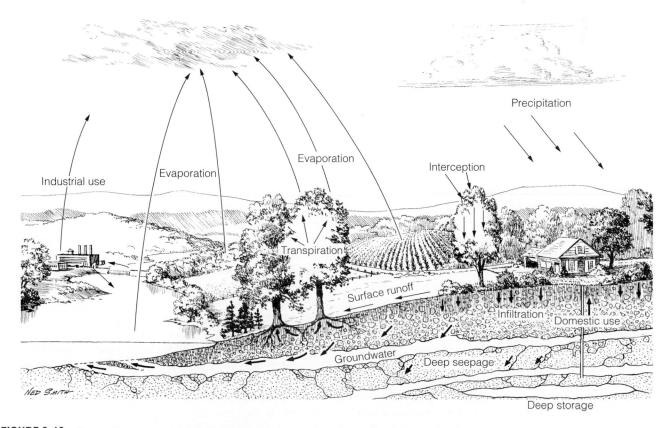

FIGURE 3.13   The water cycle on a local scale, showing the major pathways of water movement.

In a deciduous forest in summer a greater proportion of rainfall is intercepted during a light shower than during a heavy rain. During a light shower the rain does not exceed the storage capacity of the canopy, and the water held by the leaves subsequently evaporates. Water exceeding the storage capacity of the canopy either drips off the leaves as **throughfall** or runs down the stem, twigs, and trunk as **stemflow.** Water then enters the soil in a narrow band around the base of the tree. In winter deciduous trees intercept very little precipitation.

The amount of rainfall intercepted depends on the type and the age of the forest. In general, conifers intercept more rainfall on an annual basis than hardwoods. Mature pine stands, for example, will intercept 20 percent of a summer rainfall and 14 to 18 percent of the annual precipitation, whereas an oak forest will intercept 24 percent of a summer rain, but only 11 percent of the total annual precipitation (Lull 1967). Interception in urban areas is about 16 percent of total annual precipitation, approximately the same amount intercepted by deciduous forests in summer. Residential areas intercept about 13 percent (Lull and Sopper 1969).

The precipitation that reaches the soil moves into the ground by **infiltration.** The rate of infiltration is governed by soil, type of vegetation, slope, and the characteristics of the precipitation itself. In general, the more intense the rains, the greater the rate of infiltration, until the infiltration capacity of the soil, determined by soil porosity, is reached.

Long wet spells and heavy storms may saturate the soil, and intense rainfall or rapid melting of snow can exceed the infiltration capacity of the soil. When the soil can no longer absorb the precipitation, the water becomes **overland flow.** Sheet flow over the surface is changed to channelized flow as the water becomes concentrated in depressions and rills. This process can be observed even on city streets as water moves in sheets over the pavement and becomes concentrated in streetside gutters. Again the amount of runoff and associated erosion of soil depends on slope, texture of the soil, soil moisture conditions, and the type and condition of vegetation.

In the undisturbed forest infiltration rates usually are greater than intensity of rainfall, and surface runoff does not occur. In urban areas infiltration rates range from zero to a value exceeding the intensity of rainfall where soil surface is open and uncompacted. Because of low infiltration, surface runoff from urban areas might be as much as 85 percent of the precipitation (Lull and Sopper 1969). Because they are so compacted by frequent tramping and mowing, lawns have a low infiltration rate. Thus urban areas provide little or no recharge to groundwater.

Water entering the soil will **percolate,** or seep down to an impervious layer of clay or rock to collect as groundwater. From here the water finds its way into springs, streams, and eventually to rivers and seas. A great portion of this water is utilized by humans for domestic and industrial purposes, after which it reenters the water cycle by discharge into streams or into the atmosphere, as gravitational water.

Water remaining on the surface of the ground, in the upper layers of the soil, and collected on the surface of veg-etation, as well as water in the surface layers of streams, lakes, and oceans, returns to the atmosphere by evaporation. Evaporation is the movement of water molecules from the surface into the atmosphere at a rate governed by how much moisture the air contains (vapor-pressure deficit).

As the surface layers of soil dry out, a barrier through which little soil water moves develops and evaporation ceases. Further water losses from the soil take place through plants. Plants take in water from the soil through the roots and lose it through the leaves and other organs in a process called **transpiration.** Transpiration is the evaporation of water from internal surfaces of leaves, stems, and other living parts and its diffusion from the plant. The total flux of evaporating water—from the surfaces of the ground and vegetation—is called **evapotranspiration.**

The temperate deciduous forest and the urbanized areas represent two environmental extremes in water cycling. In comparison with the forest, urbanized areas are characterized by reduced interception, less infiltration, much less soil moisture storage, less evapotranspiration, and reduced water quality. Urban areas also exhibit increased overland or surface flow, increased runoff, and increased peak flows of streams and rivers.

ELEMENTAL NUTRIENTS

Living organisms require some 30 to 40 chemical elements for growth, development, and metabolism (Table 3.1). Organisms need some of these elements in large amounts. Known as **macronutrients,** such elements include oxygen, hydrogen, nitrogen, carbon, calcium, phosphorus, potassium, magnesium, sulfur, sodium, and chlorine. Other elements, needed in much smaller quantities, are **micronutrients.** They include, among others, cooper, zinc, boron, manganese, molybdenum, cobalt, iodine, selenium, silica, and iron. Some are essential to all organisms; others appear to be essential only for some organisms. If micronutrients are lacking, organisms fail as completely as if they lack nitrogen, calcium, or any other major element.

What are the original sources of nutrients that support life? The major sources in terrestrial environments are weathering of mineral soil, decomposition of organic matter, nitrogen fixation, atmospheric gases, deposition of atmospheric particles, precipitation, and ocean salt spray, all discussed in chapters that follow. The major sources of nutrients for aquatic life are inputs from the surrounding land in the form of drainage water, organic matter, sediment, and precipitation. The maintenance of inorganic nutrients in terrestrial and aquatic environments involves considerable recycling of nutrients between the abiotic environment and the organisms themselves (Chapter 25).

Being stationary, plants must obtain their nutrients from a highly localized area of soil, water, and air. Thus terrestrial plants depend heavily on the nature of the soil (Chapter 4). Because most animals are mobile, they can be more selective, but they still depend on the success of plants in sequestering nutrients in their tissues.

TABLE 3.1 Some Essential Elements

MACRONUTRIENTS	
Element	*Role*
Carbon (C) Oxygen (O) Hydrogen (H)	Basic constituents of all organic matter.
Nitrogen (N)	Utilized only in fixed form: nitrates, nitrites, ammonium. Component of chlorophyll and enzymes; building block of proteins.
Calcium (Ca)	In animals needed for acid–base balance, clotting of blood, and contraction and relaxation of heart muscles; controls movement of fluid through cells; gives rigidity to skeletons of vertebrates; forms shells of mollusks, arthropods, and one-celled Foraminifera. In plants combines with pectin to give rigidity to cell walls; essential to root growth.
Phosphorus (P)	Necessary for energy transfer in living organisms; major component of nuclear material of cells. Animals require a proper ratio of Ca:P, usually 2:1 in the presence of vitamin D, wrong ratio in vertebrates causes rickets. Deficiency in plants arrests growth, stunts roots, and delays maturity.
Magnesium (Mg)	In living organisms essential for maximum rates of enzymatic reactions involving transfer of phosphates from ATP to ADP. Integral part of chlorophyll and middle lamella of plants; involved in protein synthesis in plants. In animals activates more than 100 enzymes. Deficiency in ruminants causes a serious disease, grass tetany.
Potassium (K)	In plants, involved in the formation of sugars and starches. In animals, involved in synthesis of protein, growth, and carbohydrate metabolism.
Sulfur (S)	Basic constituent of protein. Plants utilize as much sulfur as they do phosphorus. Excessive sulfur is toxic to plants.
Sodium (Na)	Maintenance of acid–base balance, osmotic homeostasis, formation and flow of gastric and intestinal secretions; nerve transmission; lactation; growth and maintenance of body weight. Toxic to plants along roadsides when used to treat highways.
Chlorine (Cl)	Enhances electron transfer from H_2O to chlorophyll in plants. Role in animals similar to that of sodium, with which it is associated in salt (NaCl).
MICRONUTRIENTS	
Iron (Fe)	In plants involved in the production of chlorophyll; part of complex protein compounds that serve as activators and carriers of oxygen and as transporters of electrons in mitochondria and chloroplasts. In iron-rich respiratory pigment hemoglobin in blood of vertebrates and hemolymph of insects; thus essential for the functioning of every organ and tissue of the animal body. Synthesis into hemoglobin and hemolymph occurs throughout life. Deficiency results in anemia.
Manganese (Mn)	In plants enhances electron transfer from H_2O to chlorophyll and activates enzymes in fatty acid synthesis. In animals necessary for reproduction and growth.
Boron (B)	Fifteen functions ascribed to boron in plants, including cell division, pollen germination, carbohydrate metabolism, water metabolism, maintenance of conductive tissue, translocation of sugar. Deficiency in plants causes stunted growth in leaves and roots and yellowing leaves.
Cobalt (Co)	Required by ruminants for the synthesis of vitamin B_{12} by bacteria in the rumen. Deficiency results in anemia.
Copper (Cu)	In plants concentrated in chloroplasts; influences photosynthetic rates; involved in oxidation-reduction reactions; enzyme activor. Excess interferes with phosphorus uptake, depresses iron concentration in leaves, reduces growth. Deficiency in vertebrates causes poor utilization of iron, resulting in anemia and decreased calcification of bones.
Molybdenum (Mo)	In free-living nitrogen-fixing bacteria and blue-green algae acts as catalyst for conversion of gaseous nitrogen into usable form. High concentration in ruminants causes "teart" disease characterized by diarrhea, debilitation, and permanent fading of hair color.
Zinc (Zn)	In plants needed in the formation of growth substances (auxins); associated with water relations; component of several enzyme systems. In animals functions in several enzyme systems, especially the respiratory enzyme carbonic anhydrase in red blood cells. Deficiency in animals causes parakeratosis, a dermatitis.
Iodine (I)	Involved in thyroid metabolism. Deficiency results in goiter, hairlessness, and poor reproduction.
Selenium (Se)	In function closely related to vitamin E; prevents white muscle disease in ruminants. Borderline between requirement level and toxicity is narrow. Excess results in loss of hair, sloughing of hooves, liver injury, and death.

Acidity influences the availability and uptake of nutrients or elements and restricts the environment of organisms sensitive to acid situations. The measurement of acidity is pH, calibrated as the negative logarithms of the concentrations of hydrogen and hydroxyl ions in solution. The scale is based on the pH of pure water. Water is a weak electrolyte, a small fraction of which dissociates into ions: $H_2O \longrightarrow H^+ + OH^-$. In pure water the ratio of H^+ ions to OH^- ions is 1. Because both occur in a concentration of 10^{-7} moles per liter, a neutral solution has a pH of 7 $[-\log(10^{-7}) = 7]$. A solution departs from neutral when one ion increases and the other decreases. We use the concentration of one of the ions to describe a solution as an acid or a base. Customarily, we use the negative logarithm of the hydrogen ion. Thus a gain of hydrogen ions to 10^{-6} moles per liter means a decrease of OH^- ions to 10^{-8} moles/l and the pH of the solution is 6. The negative logarithmic pH scale goes from 0 to 14. A pH greater than 7 denotes an alkaline solution, a pH of less than 7 an acidic solution. A solution with a pH of 5 has 10 times the hydrogen concentration of one of pH 6; and a solution with a pH of 4 has 10 times as many hydrogen ions as one of pH 5, and 100 times the hydrogen concentration of a solution of pH 6.

Summary

1. Light has a profound effect on the ecology of almost all individuals, species, and communities. Visible light, that part of the electromagnetic spectrum between the wavelengths of 400 to 740 nm, is known as photosynthetically active radiation (PAR). Short wavelengths between 280 and 380 nm are ultraviolet light; wavelengths longer than 740 nm are infrared. In addition to its spectral qualities, light also possesses intensity, duration, and directionality, all of which vary diurnally and seasonally.

2. Light impinging on an object may be reflected, absorbed, or transmitted through it. Plants reflect green light most strongly and absorb red wavelengths used in photosynthesis.

3. Light passing through a canopy of vegetation or through water becomes attenuated. Certain wavelengths drop out before others. In a forest, green and far red wavelengths pass through relatively unaltered. In pure water, red and infrared light are absorbed first, followed by yellow, green, and violet; blue penetrates the deepest. Natural water is rarely clear, and attenuation of wavelengths is strongly affected by turbidity, organic stains, and phytoplankton.

4. The measure of the decrease in intensity between the surface and a given depth is the extinction coefficient. The extinction coefficient is the ratio of the intensity of light at the surface to that at a given depth in water or vegetative cover. In aquatic environments the extinction coefficient is influenced by turbidity and density of phytoplankton. In terrestrial vegetation the total surface area of leaves above a given area of ground, measured by a leaf index, influences the depletion of light. The higher the leaf area index, the greater is the attenuation of light.

5. All organisms live in a thermal environment characterized by heat and temperature. Heat is a form of energy that results from random motion of molecules within a substance. Temperature is a measure of a substance's tendency to give up heat.

6. Organisms maintain a balance between heat energy gained from and lost to the environment. Heat gains come from direct and reflected sunlight, diffuse radiation, infrared radiation, and metabolism. Heat is lost to the environment by infrared radiation, conduction, convection, and evaporation.

7. A water molecule consists of one atom of oxygen and two atoms of hydrogen. The water molecule is V-shaped with an unequal distribution of positive and negative charges, resulting in the polarity of the water molecule. This polarity allows hydrogen bonding between water molecules. The electrostatic attractive forces of the hydrogen bonds between individual molecules causes clusters of water molecules to be arranged in a lattice. The lattice is most highly organized when frozen. Heat weakens the bonds between molecules, yielding liquid and gaseous states.

8. The structure of water relates to its unique physical properties of specific heat, cohesion, adhesion, capillarity, surface tension, and viscosity, all of which influence the external and internal environment of organisms.

9. Water becomes available to organisms through the local water cycle. In the local water cycle, water moves through cloud formation in the atmosphere, precipitation, interception, and infiltration into the ground, reaching groundwater, streams, and springs, from which evaporation takes place, bringing water back to the atmosphere in the form of clouds.

10. All living organisms must obtain from their environment certain elements essential to reproduction, growth, and survival. These elements include nitrogen, calcium, phosphorus, potassium, and sodium in relatively large quantities. These elements are termed macronutrients. Organisms need other elements, such as copper, zinc, iron, and boron, in lesser and often minute quantities. These are known as micronutrients.

Review Questions

1. What is photosynthetically active radiation (PAR)?
2. What are the four key characteristics of light for ecology?
3. What are photons?
4. Why does the intensity of light vary daily and seasonally?
5. What is the fate of light impinging on an object?
6. Contrast the fate of light striking a forest canopy with that of light impinging on the surface of a lake.
7. What is the extinction coefficient? Leaf area index? How does the leaf area index relate to the extinction coefficient?
8. Distinguish between heat and temperature.
9. Describe radiation, conduction, convection, and evaporation.
10. What is the role of each of the processes in question 9 in the energy exchange between an organism and its environment?
11. What are the structural properties of water? What is the ecological importance of these properties?
12. What are the specific heat, latent heat, viscosity, and surface tension of water? How do they relate to the environment of an organism?
13. Follow the fate of a rain shower from cloud to groundwater and identify the various processes involved.
14. Distinguish between a macronutrient and a micronutrient.
15. What is the importance of the following elements for plants and animals: nitrogen, calcium, magnesium, sodium, iron, and boron?

CHAPTER 4

Soils

CONCEPTS

1. Soil is the thin mantle of weathered mineral and organic matter that supports terrestrial life.

2. A soil profile depicts layers or horizons that reflect that soil's development.

3. Soil has a number of key physical properties, including texture and moisture availability, and chemical properties that influence the nature of life that soil supports.

4. Soil formation involves interactions of environmental conditions, weathering, topography, and organisms.

5. Major soil development processes involve additions to and loss from the soil body and translocation and transformation of materials within the soil body.

6. Soil development processes result in unique soils, which can be broadly classified into 12 major soil orders.

Soil is the foundation of terrestrial communities. It is the major site of decomposition of organic matter and the return of mineral elements to the nutrient cycle (see Chapters 3). From it plants obtain the minerals and water they need for photosynthesis. Vegetation, in turn, influences the development of soil, its chemical and physical properties, and its organic matter content. Controlling much of the decomposition of organic matter and soil-forming processes are an abundance and diversity of soil organisms, including bacteria, fungi, and soil invertebrates. Thus soil acts as a pathway between the organic and mineral worlds.

DEFINITIONS OF SOIL

As familiar as it is, soil is difficult to define. Indeed, one eminent soil scientist, a pioneer of modern soil studies, Hans Jenny (1980), will not give an exact definition of soil. In his book *The Soil Resource* (1980:364), he writes:

> Popularly, soil is the stratum below the vegetation and above hard rock, but questions come quickly to mind. Many soils are bare of plants, temporarily or permanently; or they may be at the bottom of a pond growing cattails. Soil may be shallow or deep, but how deep? Soil may be stony, but surveyors (soil) exclude the larger stones. Most analyses pertain to fine earth only. Some pretend that soil in a flower pot is not soil, but soil material. It is embarrassing not to be able to agree on what soil is. In this pedologists are not alone. Biologists

cannot agree on the definition of life and philosophers on philosophy.

However, soils have been variously defined. One definition is that soil is a natural product formed from weathered rock by the action of climate and living organisms. Another common definition states that soil is a collection of natural bodies of Earth that is composed of mineral and organic matter and is capable of supporting plant growth. The term *natural bodies* emphasizes that soils are different. Further, if soils are considered natural bodies, that concept becomes the basis for soil classification.

In a more extensive definition, soil is a natural body composed of solids (mineral and organic matter), liquid, and gases that occurs on the land surface and occupies space. It is characterized by one or both of the following: horizons, or layers, that are distinguishable from the initial material as the result of additions, transfers, and transformation of energy and matter *or* the ability to support rooted plants in a natural environment.

Soil, however, is not just an abiotic environment capable of supporting some form of plant life. It is teeming with life—billions of minute animals, bacteria, and fungi. The interactions between the abiotic and biotic make soil a living system.

Soil is also recognized as a three-dimensional unit or body called an individual soil body, possessing length, width, and depth. A three-dimensional soil body large enough to study all physical and chemical properties and all horizons is called a **pedon** (Figure 4.1). Pedons, the basic unit in the study of soils, range in area from 1 to 100 m².

FIGURE 4.1   (a) Schematic diagram illustrating the concept of the pedon and its associated soil horizon. (Boul et al. 1998.) (b) A generalized profile of the soil. *O,* loose leaves, organic debris, and partially decomposed or matted inorganic material. *A,* dark-colored horizon mixed with mineral matter. *E,* horizon of maximum leaching; prominent in Spodosols, it may be faintly developed in other soils. *B,* zone of maximum accumulation of clay minerals or of iron and organic matter; *BE* represents a transitional area or continuum between the two horizons. *C,* the weathered material, either like or unlike the material from which the soil presumably formed. *BC* represents a transitional area or continuum between the *B* and *C* horizons. *R,* consolidated bedrock.

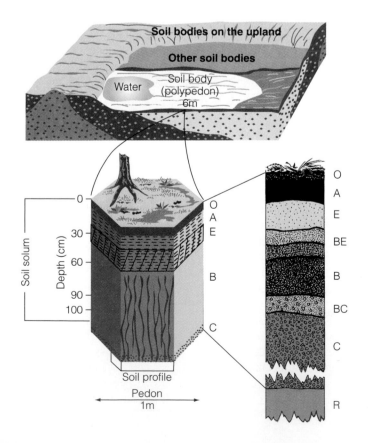

THE SOIL PROFILE

Soil, Hans Jenny (1980:6) wrote, "is a body of nature that has its own internal organization and history of genesis." Let us begin with its internal organization. A fresh cut along a road-bank or an excavation tells something about a soil. A close-up and even cursory look reveals bands and blotches of color from the surface downward. Closer examination, even handling the material, reveals changes in texture and structure. Any vertical cut through a body of soil or pedon is the **soil profile.** The apparent layers are called the **horizons.** Each horizon has a characteristic set of features, particularly color, that distinguishes it from other horizons. Each horizon has its own thickness, texture, structure, consistency, porosity, chemistry, and composition.

In general soils have six major horizons: *O, A, E, B, C, R* (Figure 4.1). In some soils the horizons are quite distinct. In other soils the horizons form a continuum with no clear-cut boundary between one horizon and another.

The *O* **horizon** is the surface layer, formed or forming above the mineral layer and composed of fresh or partially decomposed organic material that has not been mixed into mineral soil. It is usually absent in cultivated soils. This layer and the upper part of the next horizon, *A,* constitute the zone of maximum biological activity. Both are subject to the greatest changes in soil temperatures and moisture conditions, contain the most organic carbon, and are the sites where most or all decomposition takes place. The *O* layer fluctuates seasonally. In temperate regions it is thickest in the fall, when new litter is added, and thinnest in the summer after decomposition has taken place.

The *A* **horizon** (once called A_1) is the upper layer of mineral soil with a high content of organic matter. It is characterized by an accumulation of organic matter and by the loss of some clay, inorganic minerals, and soluble matter. The *E* **horizon** (once labeled A_2) is the zone of maximum leaching (eluviation, thus the label *E*). Its chemistry and structure have been altered by the downward movement of suspended and dissolved material through weathering and leaching. The *E* horizon is further characterized by the development of granular, platelike, or crumblike structure. Some soils possess both an *A* and an *E* horizon. Other soils lack one or the other, depending on the nature of weathering and the incorporation of organic matter with mineral soil.

The *B* **horizon** is the zone of illuviation, collection of leached material. It accumulates silicates, clay, iron, aluminum, and humus from the *E* horizon. It develops a characteristic physical structure involving blocky, columnar, or prismatic shapes (see page 58). Below the *B* horizon in some soils may be compact, slowly permeable layers, clay pans or fragipans. A **clay pan** possesses much more clay than the horizon above it. It is very hard when dry, and stiff when wet. A **fragipan** is a brittle, seemingly cemented, subsurface horizon low in organic matter and clay but high in silt or very fine sand. When dry a fragipan is very hard, and when moist it tends to rupture suddenly if pressure is applied. Both clay pans and fragipans interfere with root and water penetration in the soil.

The *C* **horizon** contains weathered material, either like or unlike the material from which the soil is presumed to have developed. Some active weathering takes place in this horizon, but it is little affected by soil formation. Below the *C* horizon is unweathered bedrock, the *R* **horizon.**

PROPERTIES OF SOILS
Physical Properties

Soils are distinguished by differences in their physical and chemical properties. Physical properties include color, texture, structure, depth, and moisture. All are highly variable from one soil to another.

Color Color has little direct influence on the function of a soil, but considered with other properties, it can tell a good deal about the soil. Color is one of the most useful and important characteristics for the identification of soil (see page 74). In temperate regions, brownish-black and dark brown colors, especially in the *A* horizon, generally indicate considerable organic matter. The *B* horizon of well-drained soils may range anywhere from very pale brown to reddish and yellowish color. However, it does not always follow that dark-colored soils are high in organic matter. Soils of volcanic origin, for example, are dark in color acquired from their parent material of basaltic rocks. In warm temperate and tropical regions, dark clays may have less than 3 percent organic matter.

Red and yellow soils derive their colors from the presence of iron oxides, the bright colors indicating good drainage and good aeration. Red and yellow colors increase from cool regions to the equator. Other red soils obtain their color from parent material such as red lava rock rather than from soil-forming processes. Well-drained yellowish sands are white sands containing a small amount of organic matter and such coloring material as iron oxide. Quartz, kaolin, carbonates of calcium and magnesium, gypsum, and various compounds of ferrous iron give whitish and grayish colors to the soils. Grayish colors indicate permanently saturated soils in which iron is in the ferrous form. Imperfectly and poorly drained soils are mottled with blotches of various shades of yellow-brown and gray. The colors of soils are determined by the use of standardized color charts, notably the Munsell color charts.

Texture The texture of a soil is determined by the proportion of particle size classes (Figure 4.2). Texture is partly inherited from parent material and partly a result of the soil-forming process.

Particles are classified on the basis of size into rock fragments, sand, silt, and clay. Rock fragments consists of particles larger than 2.0 mm in diameter and are not considered part of the fine fraction of soil. Sand ranges from 0.05 to 2.0 mm, is easily seen, and feels gritty. Silt consists of particles from 0.002 to 0.05 mm in diameter, which can scarcely be seen by the naked eye and feel and look like flour. Clay particles, less than

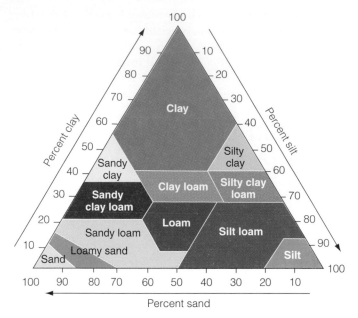

FIGURE 4.2 A soil texture chart showing the percentages of clay (below 0.002 mm), silt (0.002 to 0.05 mm), and sand (0.05 to 2.0 mm) in the basic soil textural classes. For example, a soil with 60 percent sand, 30 percent silt, and 10 percent clay would be considered a sandy loam.

0.002 mm, are too small to be seen under an ordinary microscope. Clay controls the most important properties of soils, including plasticity—the ability to change shape when pressure is applied and retain that shape when the pressure is removed—and the exchange of ions between soil particles and soil solution. The nature of a soil's texture is determined by the percentage (by weight) of sand, silt, and clay. Based on these proportions, soils are divided into textural classes (Figure 4.2).

Texture relates to pore space in the soil and therefore plays a major role in the movement of air and water in the soil, penetration by roots, and water storage capacity. In an ideal soil, particles make up 50 percent of the volume; the other 50 percent is pore space. Pore space includes spaces within and between soil particles, as well as old root channels and animal burrows. Coarse-textured soils possess large pore spaces that favor rapid water infiltration and rapid drainage. Up to a point, the finer the texture, the smaller the pores, and the greater the available active surface for water adherence and chemical activity. Very fine textured soils, such as heavy clays, easily become compacted if plowed, stirred, or walked upon. They are poorly aerated and difficult for roots to penetrate. The condition can be alleviated by the addition of organic matter to improve soil structure and sand to change texture.

Soil particles are held together in clusters or shapes of various sizes, called **aggregates** or **peds.** The arrangement of these aggregates is called **soil structure.** As with texture, there are many types of soil structure. Soil aggregates are classified as granular, crumblike, platelike, subangular blocky, angular blocky, prismatic, and columnar (Figure 4.3).

Structure is influenced by texture, plants growing on the soil, other soil organisms, and the soil's chemical status.

Depth   The layer of soil on which life depends is not spread evenly across the landscape. Its depth varies from place to place depending on slope, weathering, parent material, and vegetation. Soils developed under native grassland tend to be several meters deep, whereas soils developed under forests are relatively shallow with an *A* horizon of about 15 cm and a *B* horizon of about 60 cm. Soils at the bottom of slopes, on level ground, and on alluvial plains tend to be deep, whereas soils on ridgetops and steep slopes tend to be shallow, with bedrock close to the surface. Natural fertility and water-holding capacity of such soils are low.

Moisture   Dig into the surface layer of a soil about a day after a soaking rain and note the depth of water penetration. Unless the soil is clay, you should discover that the transition between wet surface soil and dry soil is sharp. The water that fell on the ground infiltrated the soil, filling the pore spaces and draining into the dry soil below. Depending on the amount of water, the downward flow or percolation halts within two to three days and the water hangs in the soil capillaries or pores.

The amount of water per unit volume a soil can hold is one of its most important characteristics. When a soil holds the maximum amount of water it can retain, it is at **field capacity (FC).** Typically field capacity can be estimated by saturating a given volume of soil and the applying a suction (negative pressure) of −0.05 MPa (megapascals). The soil is weighed, dried, and weighed again. The difference in weight is then expressed as gram water per volume of soil. If the process is repeated with a suction of −1.5 MPa, the resulting water content is called the **wilting point (WP)** or permanent wilting percentage. The term *wilting point* refers to the inability of plants to extract water from the soil. As discussed later, the ability to extract water as the water content declines varies from species to species. As a result, the negative pressure used to define wilting point of varies, ranging from −1 to −3.5 MPa. The amount of water retained by the soil between field capacity and wilting point (FC − WP) is the **available water capacity (AWC)** (Figure 4.4). The AWC provides an estimate of the water available for uptake by plants.

Both the field capacity and wilting point of a soil are heavily influenced by soil texture, and to a lesser extent by clay minerals, organic content, stoniness, and soil structure. Particle size of the soil directly influences the pore space and surface area onto which water adheres. Sand has 30 to 40 percent of its volume in pore space, clays and loams 40 to 60 percent. As a result, fine-textured soils have a higher field capacity than sandy soil, but the increased area results in a higher value of the wilting point as well (Figure 4.4). Conversely, coarse-textured soils (sands) have a low field capacity and a low wilting point. Thus AWC is highest in intermediate clay loam soils.

The topographic position of a soil affects the movement of water both on and in the soil. Water tends to drain downslope,

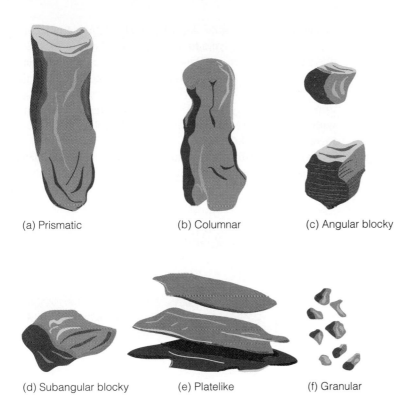

(a) Prismatic   (b) Columnar   (c) Angular blocky

(d) Subangular blocky   (e) Platelike   (f) Granular

FIGURE 4.3 Some types of soil structure. (a) Prismatic, (b) columnar, (c) angular blocky, (d) subangular blocky, (e) platelike, (f) granular.

leaving soils on higher slopes and ridgetops relatively dry, and creating a moisture gradient from ridgetops to streams. However, after a dry period more moisture may be stored from rain on the upper slope than on the lower slope.

In all there are seven drainage classes: (1) *very excessively drained* soils and (2) *excessively drained* soils, in both of which plant roots are restricted to upper layer of soil because of water deficiencies; (3) *well-drained* soils, in which plant roots can grow to a depth of 90 cm without restriction due to excess water; (4) *moderately well drained* soils, in which plant roots can grow to a depth of 50 cm without restriction; (5) *somewhat poorly drained* soils, which restrict the growth of plant roots beyond a depth of 36 cm; (6) *poorly drained* soils, which are wet most of the time and are usually characterized by alders, willows, and sedges; and (7) *very poorly drained* soils, in which water stands on or near the surface most of the year.

Poorly drained areas support hydric soils (see Chapter 31). Hydric soils develop where flooding or ponding occurs long enough during the growing season to inhibit the diffusion of oxygen into the soil. Under these anaerobic conditions iron in the soil reduces to the ferrous state, which gives a dull gray or bluish color to the horizon. Being soluble some of this iron moves into more aerobic sites and precipitates out as oxidized iron, giving a mottled appearance to some hydric horizons. Some hydric soils possess a thick, dark surface layer because anaerobic conditions inhibit the decomposition of organic matter. Hydric soils are important indicators of wetlands (see Chapter 31) (Hart et al. 1996).

The delineation of drainage classes of soils is important ecologically and economically. Somewhat poorly drained or

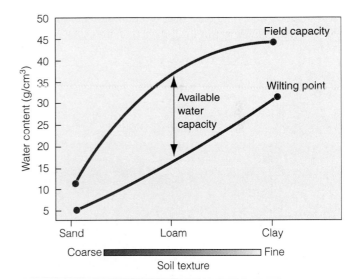

FIGURE 4.4 Variations in field capacity and wilting point as a function of soil texture. Field capacity and wilting point are expressed as water content of soil in g/cm³. The difference between field capacity and wilting point (as shown by the arrow for a loam soil) is called available water capacity. This value represents the soil water available for uptake by plants. The actual available water at any time will be the difference between the water content (value on *y* axis) and the wilting point.

hydric soils mark the beginning of wetland ecosystems. Historically, poorly drained soils were (and many still are) "improved" for agricultural and other uses by ditching and installing drainage tiles. Such drainage has been slowed by (often controversial) wetland laws. Drainage classes are also important considerations for the selection of construction and highway sites.

Chemical Properties

Soils also have different chemical properties. These properties are influenced by parent material, vegetation, moisture, clay organic matter content, and other factors. Chemical elements in the soil are dissolved in solution, are a constituent of organic matter, and are adsorbed on particles. These ions move from soil to plant, from plant to animals, and into the biogeochemical cycle (see Chapter 26). In soils, ions are limited in their mobility because they are closely held to particles of clay and humus.

Because of its unique structure, clay controls many of the important properties of soil. The basic clay mineral consists of three elements: aluminum (Al^{3+}), silica (Si^{4+}), and oxygen (O^{2-}). The base of the clay mineral is silica, which attracts four oxygens to form a tetrahedron (Figure 4.5a). These tetrahedrons are bound together by oxygen to form unit cells, $(Si_2O_5)_n$. These cells repeat in an orderly fashion to form lengthy sheets (Figure 4.5b). The other element, aluminum, reacts with oxygen in a similar manner to form sheets of octahedrons. The silica sheets bond to the aluminum sheets in layers to form platelike particles known in colloidal chemistry as **micelles** (Figure 4.6). Some clay particles consist of one sheet of silica and one sheet of aluminum held together by hydrogen bonds to form 1:1 clays (kandites). An example

of such a clay is kaolinite, $Al_4Si_4O_{10}(OH)_8$ (Figure 4.6a). Others bond two silica layers to one aluminum layer to form 2:1 clays (smectites), an example of which is the common clay mineral, montmorillonite (Figure 4.6b). A general formula for smectites is $Al_4Si_8O_{20}(OH)_4 \cdot nH_2O$. A third type of clay mineral has two layers of silica and one layer of aluminum with magnesium (Mg^+) or potassium (K^+) holding the layers together (Figure 4.6c).

There are differences among all these clays. In 1:1 clays the associated hydroxyl atoms balance the structure electro-

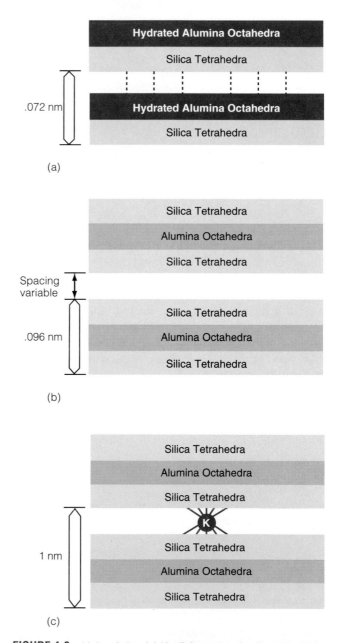

FIGURE 4.6   Units of clay. (a) Kaolinite, a 1:1 clay, is composed of pairs of silica and alumina sheets held together by hydrogen bonds. (b) A unit of montmorillonite, a 2:1 clay, consists of a silica sheet on each side of an alumina sheet. The interlattice space varies with the amount of water present. (c) A unit of illite consists of silica sheets on either side of an alumina sheet. Adjacent micelles are held together by potassium (K) bridges.

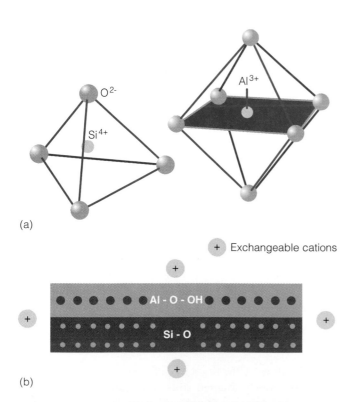

FIGURE 4.5   (a) Structure of clay colloids or micelles. Basic building units of aluminosilicate crystals are silica, Si^{4+}, a tetrahedron, and aluminum, Al^{3+}, an octahedron. (b) Sheets of aluminum and silica held together with shared bonds.

statically. In kaolinite the oxygen atoms on the surface of the silicate layer face the hydroxyl atoms of the aluminum layer. The oxygen and hydroxyl atoms form hydrogen bonds that hold the layers tightly together, so that water and cations cannot enter the interlayer spaces. This situation does not exist in the 2:1 clays. They have twice as many Si atoms as Al atoms in a unit cell, twice as many oxygen atoms, half the number of hydroxyls, and a variable number of water molecules. Because oxygen atoms at the surface of unit cells face each other, the bonding is weak, and water and its associated cations can enter the clay particles. For this reason the unit cells vary in size.

An important feature of the micelles, especially of the 2:1 clays, is that one element can substitute for another without changing the structure. For example, Al may substitute for Si in the tetrahedral sheet. Since the Si ion has a charge of 4+ and the Al ion only 3+, the unit cell would have a net negative charge of −1. Similarly, iron (Fe^{2+} and Fe^{3+}) or magnesium (Mg^{2+}) may substitute for Al^{3+} in the octahedral sheet, changing the electrical charge. Substituting one ion for another without changing the arrangement of the ion or the morphology of the mineral is called **isomorphous substitution.**

One of the most important features of clay minerals, this substitution results in net negative charges that must be balanced by positively charged cations. **Exchangeable cations** are loosely held on the surface of the micelles and can be replaced easily by others. The edges and sides, negatively charged, act as highly charged anions. They attract cations, water molecules, and organic substances. The total number of negatively charged exchange sites on clay and humus particles that attract positively charged cations is called the **cation exchange capacity (CEC)** (Figure 4.7) The CEC represents the net negative charges possessed by the soil. These negative charges enable a soil to prevent the leaching of its positively charged nutrient cations.

Exchange sites are occupied by such ions as calcium (Ca^{2+}), magnesium (Mg^{2+}), potassium (K^+), sodium (Na^+),

and hydrogen (H^+). Some of these ions, especially Al^{3+} and H^+, cling more tenaciously to micelles than do others. Less tenacious, in descending order, are Ca^{2+}, Mg^{2+}, K^+, NH^+, and Na^+. These latter ions are more easily displaced from the exchange site, and their place may be taken by aluminum or hydrogen ions. The percentage of sites occupied by basic cations, primarily Ca^{2+}, Mg^{2+}, Na^+, and K^+, is called **percent base saturation.** Acidic soils have a low percent base saturation because they have a high number of exchangeable hydrogen ions. Soils with a high CEC are potentially fertile. Soils high in both CEC and base saturation are fertile unless they are saline or contain toxic heavy metals.

Positively charged cations and negatively charged anions are dissolved in soil solution and occupy the exchange sites on the clay and humus particles. Cations occupying the exchange sites are in a state of dynamic equilibrium with similar cations in solution. Cations in soil solution are continuously being replaced by or exchanged with cations on the clay and humus particles, in response to changes in the concentration in the soil solution. For example, the removal of certain cations from soil solution by the roots of plants reduces the concentration of those cations and enhances the release of cations from the micelles.

Hydrogen ions added by rainwater, by acids from organic matter, and by metabolic acids from roots and microorganisms increase the concentration of hydrogen ions in the soil solution and displace other cations, such as Ca^{2+}, on the micelles. As more and more hydrogen ions replace other cations, the soil becomes increasingly acidic. Acidity is one of the most familiar of all chemical conditions in the soil. Typically soils range from a pH of 3 (extremely acid) to a pH of 9 (strongly alkaline). Soils just over a pH of 7 (neutral) are considered basic, and those of pH 5.6 or below are acid. As soil acidity increases, the proportion of exchangeable Al^{3+} increases and Ca^{2+}, Na^+, and other cations decrease. Such changes bring about nutrient deprivation in plants and microorganisms and also aluminum toxicity.

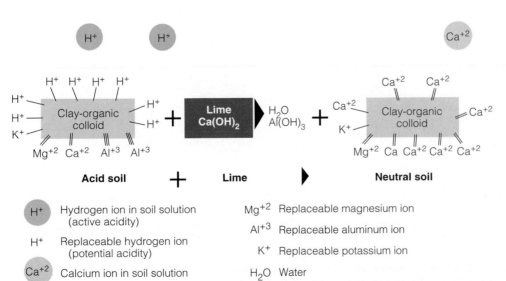

FIGURE 4.7 An example of cation exchange reactions when $Ca(OH)_2$ (lime) is added to an acidic soil. The aluminum ions in acid soil have been replaced by calcium ions, and the replaced aluminum ions react with soil water to form insoluble hydroxides and oxides, which will remain so as long as the soil stays neutral.

H^+	Hydrogen ion in soil solution (active acidity)
H^+	Replaceable hydrogen ion (potential acidity)
Ca^{+2}	Calcium ion in soil solution
Ca^{+2}	Replaceable calcium ion
Mg^{+2}	Replaceable magnesium ion
Al^{+3}	Replaceable aluminum ion
K^+	Replaceable potassium ion
H_2O	Water
$Al(OH)_3$	Aluminum hydroxide

THE LIVING SOIL

Soil possesses several outstanding characteristics as a medium for life. It is relatively stable structurally and chemically. The underground climate is far less variable than above-surface conditions. The atmosphere remains saturated or nearly so, until soil moisture drops below a critical point. Soil affords a refuge from high and low extremes in temperature, wind, evaporation, light, and dryness. These conditions allow soil fauna to make easy adjustments to unfavorable conditions. On the other hand, soil hampers movement. Except to such channeling species as earthworms, pore space is important. It determines living space, humidity, and gas exchange.

Only a part of the upper soil layer is available to most soil animals as living space. Spaces within the surface litter, cavities walled off by soil aggregates, pore spaces between individual soil particles, root channels, and fissures—all are potential habitats. Most soil animals are limited to pores and cavities larger than themselves.

Water in the pore spaces is essential: the majority of soil life is active only in water. Soil water is usually present as a thin film lining the surfaces of soil particles. This film contains, among other things, bacteria, unicellular algae, protozoa, rotifers, and nematodes. Most of the organisms are restricted in their movements by the thickness and shape of the water film in which they live. Some soil animals, such as

millipedes and centipedes, are highly susceptible to desiccation and avoid it by burrowing deeper.

When excessive water floods pore spaces, typically after heavy rains, conditions are disastrous for some soil inhabitants. If earthworms cannot evade flooding by digging deeper, they come to the surface, where they often die. Many small species and immature stages of larger centipedes and millipedes are immobilized by a film of water and cannot overcome the surface tension imprisoning them.

A wide diversity of life is found in the soil (Figure 4.8). The number of species of bacteria, fungi, protists, and representatives of nearly every invertebrate phylum found in the soil is enormous. More than 250 species of Protozoa live in English soils (Sandon 1927). In the soil of beech woods in Austria live at least 110 species of beetles, 229 species of mites, and 46 species of snails and slugs (Franz 1950).

Dominant among the soil organisms are bacteria, fungi, protozoans, and nematodes. Protozoans range in density from 10^4 to 10^5 per gram of soil. Nematodes number from 10 to 10^2 per gram of soil. These organisms obtain their nourishment from the roots of living plants and from organic matter. Some protozoans and free-living nematodes feed selectively on bacteria, which number from 10^8 to 10^9 per gram of soil.

Living within the pore spaces of the soil are the most abundant and widely distributed of all forest soil animals, the mites (Acarina) and springtails (Collembola), which make up

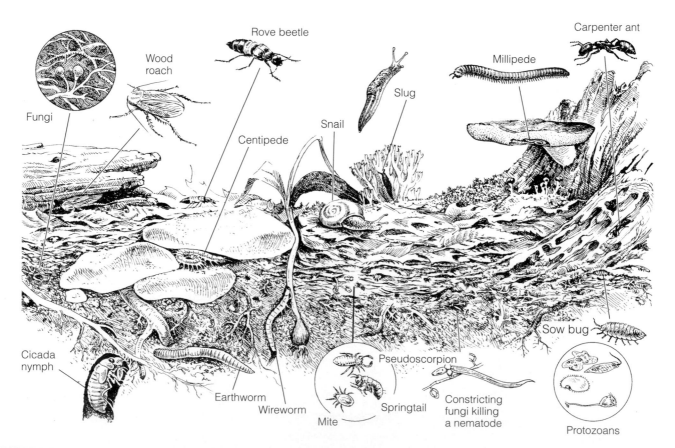

FIGURE 4.8 Life in the soil. This drawing shows only a small fraction of the kinds of organisms that inhabit the soil and litter. Note the fruiting bodies of fungi, which are consumed by animals, vertebrate and invertebrate.

over 80 percent of the total number of animals in the soil (Salt et al. 1948). Flattened dorsoventrally, they can wiggle, squeeze, and even digest their way through tiny caverns in the soil. They feed on fungi or search for prey in the dark interstices and pores of the organic mass.

The more numerous of the two, both in species and numbers, are the mites, tiny eight-legged arthropods from 0.1 to 2.0 mm in size. The most common mites in the soil and litter are the Orbatei. They live mostly on fungal hyphae that attack dead vegetation as well as on the sugars digested by the microflora of evergreen needles.

The Collembolae are the most generally distributed of all insects. Their common name, springtail, is descriptive of the remarkable springing organ at the posterior end, which enables them to leap comparatively great distances. The springtails are small, from 0.3 to 1 mm in size. They consume decomposing plant materials, largely for the fungal hyphae they contain.

Prominent among the larger soil fauna are the earthworms (Lumbricidae) (see Edwards 1998, Hendrix 1995). Earthworm activity consists of burrowing through the soil. Burrowing involves ingestion of soil, the ingestion and partial digestion of fresh litter, and the subsequent egestion of both mixed with intestinal secretions. Egested matter is defecated as aggregated castings on or near the surface of the soil or as a semiliquid in intersoil spaces along the burrow. These aggregates produce a more open structure in clayey soil and bind sandy soil together. In this manner earthworms improve the soil environment for other soil organisms by creating larger pore spaces and by mixing organic matter with the mineral soil.

Feeding on the surface litter are millipedes. They eat leaves, particularly those in which some fungal decomposition has taken place. Lacking the enzymes necessary for the breakdown of cellulose, millipedes live on the fungi contained within the litter. The millipedes' chief contribution is the mechanical breakdown of litter, making it more vulnerable to microbial attack, especially by saprophytic fungi.

Accompanying the millipedes are snails and slugs. Among the soil invertebrates they possess the widest range of enzymes to hydrolyze cellulose and other plant polysaccharides, possibly even the highly indigestible lignins.

Not to be ignored are termites (Isoptera), white wingless (except for sexual forms), social insects. Termites, together with some dipteran and beetle larvae, are the only larger soil inhabitants that can break down the cellulose of wood. They do so with the aid of symbiotic protozoans living in their gut. Termites dominate the tropical soil fauna. In the tropics termites are responsible for the rapid removal of wood, dry grass, and other materials from the soil surface. In constructing their huge and complex mounds, termites move considerable amounts of soil. The detrital organisms support the predaceous soil organisms, dominated in the deciduous forest by terrestrial salamanders that feed on a wide range of soil invertebrate fauna. Salamander densities are large enough to affect the invertebrate soil community (Wyman 1998, Burton 1976). Spiders, beetles, pseudoscorpions, centipedes, and predaceous mites feed on small arthropods. Protozoans, rotifers, myxobacteria, and nematodes fed on bacteria and algae. Various predaceous fungi live on bacteria-feeders and algal-feeders.

SOIL FORMATION

Soil formation is a dynamic process that takes place in different environments. It is strongly influenced by the parent material, climate—largely temperature and water exchanges—topography, and time.

Factors in Soil Formation

Parent Material   The unconsolidated mass on which soil formation takes place is parent material. This material may or may not be derived from the on-site geological substrate or bedrock on which it rests. Parent materials can be transported by wind, water, glaciers, and gravity and deposited on top of bedrock (Table 4.1). Because of the diversity of materials involved, transported soils are commonly more fertile than soils derived in place from parent materials.

Whatever the parent material, whether derived in place from bedrock or from transported material, it ultimately comes from geological materials: igneous, sedimentary, and metamorphic rocks. **Igneous** rocks result when a hot mixture of elements, called magma, cools. If the magma cools at Earth's surface, it forms extrusive igneous rocks; if it cools

TABLE 4.1 Types of Parent Material

Parent Material	Characteristics
Residual	In place.
Sedimentary rock	Deposition and compaction; shale, limestone, sandstone.
Igneous rock	Cooling of magma; granite and basalt.
Metamorphic rock	Changed igneous and sedimentary; slate and marble.
Transported	Carried by wind, water, ice, gravity.
Glacial ice	
Till	Unsorted loose material of particles, from clay to boulders, carried in, ahead of, or under glaciers.
Outwash	Water-deposited material from glacial melt.
Water Deposition	
Alluvium	Eroded soil material carried and deposited by rivers and streams.
Wind Deposition	
Aeolian	Shifting sand.
Loess	Nonstratified silt mixed with scattered particles of sand and clay.
Gravity	
Colluvium	Erosional deposits on hillsides.
Organic soils	Derived from plant material; peat.

below the surface, it forms intrusive igneous rocks. **Sedimentary** rocks form from material deposited in lakes and oceans. Over time and under the pressure of overlying materials, layers of these accumulated sediments consolidate into rocks. Over geological time, these rocks have been exposed by uplifting and wearing away of mountains. **Metamorphic** rocks are formed when either igneous rocks or sedimentary rocks are heated or subjected to intense pressure at considerable depths within the Earth. The original minerals melt and form new minerals. The composition of all these rocks largely determines the chemical composition of the soil.

Climate: Radiant Energy and Water   Climate is most influential in determining the nature and intensity of weathering and the type of vegetation that further affects soil formation. The soil material experiences daily and seasonal variations in heating and cooling (see Chapter 2). Open surfaces exposed to thermal radiation undergo the greatest daily fluctuations in heating and cooling; soils covered with vegetation the least. Hill slopes facing the sun absorb more heat than those facing away from the sun (see Chapter 2). Radiant energy has a pronounced effect on the moisture regime, especially the evaporative process (see Chapter 3) and dryness. Temperature can stimulate or inhibit biogeochemical reactions in soil material.

Water is involved in all biogeochemical reactions in the soil, for it is the carrier of the acids that influence the weathering process. Water enters the soil material as a liquid and leaves it as a liquid by percolation and as a gas through evaporation (see Chapter 3). The water regime in soil material is sporadic, and in many parts of the Earth highly seasonal. Water that enters the soil during heavy rainfall and snowmelt moves down through the soil. As it moves, it leaves behind suspended material and may carry away mineral matter in solution, a process called **leaching.** On sloping land, water distributes material laterally through the soil.

Topography   Topography is a major factor in soil development. It influences both the intensity of radiant energy impinging on the soil material and the amount of water that enters the soil; both relate directly to the weathering process. More water runs off and less enters the soil on steep slopes than on relatively level land. Water draining from slopes enters the soil on low and flat land. Thus soils and soil material tend to be dry on slopes and moist to wet on the low land. Steep slopes are subject to surface erosion and soil creep—the downslope movement of soil material, which accumulates on lower slopes and lowlands.

Biota   Vegetation, animals, bacteria, and fungi all contribute to the formation of soil. Vegetation, in particular, is responsible for organic material in the soil and influences its nutrient content. For example, forests store most of their organic matter on the surface, whereas in grasslands most of the organic matter added to the soil comes from the deep fibrous root systems. Organic acids produced by vegetation accelerate the weathering process. The role of biota in soil development is discussed later.

Time   The weathering of rock material; the accumulation, decomposition, and mineralization of organic material; the loss of minerals from the upper surface; gains in minerals and clay in the lower horizons; and horizon differentiation—all require considerable time. Well-developed soils in equilibrium with weathering, erosion, and biotic influences may require 2000 to 20,000 years for their formation; but soil differentiation from parent material may take place in as short a time as 30 years (Crocker and Major 1955, Thurman and Sencindiver 1986). Certain acid soils in humid regions develop in 2000 years because the leaching process is speeded by acidic materials. Parent materials heavy in texture require a much longer time to develop into soils, because of an impeded downward flow of water. Soils develop more slowly in dry regions than in humid ones. Soils on steep slopes often remain poorly developed regardless of geological age because rapid erosion removes soil nearly as fast as it is formed. Floodplain soils age little through time, because of the continuous accumulation of new materials. Such soils are not deeply weathered and are more fertile than geologically old soils, because they have not been exposed to the leaching process as long. The latter soils tend to be infertile because of long-time leaching of nutrients without replacement from fresh material.

Weathering

Weathering involves the physical disintegration and the chemical decomposition of parent materials. Physical disintegration breaks down parent material. Chemical decomposition releases soluble materials and synthesizes new minerals.

Physical Weathering   The formation of soil begins with the weathering of bedrocks and rock material. Physical or mechanical weathering comes about through the combined action of water, wind, and changing temperature. Rock surfaces flake and peel away. Water seeps into crevices, freezes, expands, and cracks the rock into smaller pieces. Water flowing along the surface carries loosened material away, exposing fresh surfaces to the physical weathering.

Chemical Weathering   Accompanying mechanical weathering and continuing long afterward is chemical weathering. Chemical weathering involves among other reactions, oxidation, reduction, oxidation-reduction, and hydrolysis. Oxidation, a chemical process by which an element loses electrons, occurs under the conditions of high oxygen supply and low biological oxygen demand. Reduction, the gain of electrons, occurs in a geochemical environment in which the material is water saturated, the oxygen supply is low, and biological oxygen demand is high. Hydrolysis involves H^+ ions that attack silicates. They are derived from the ionization of carbonic acid in water. The myriad chemical processes transform primary minerals, such as aluminosilicates and feldspars, into secondary minerals, particularly clays such as kaolinite and montmorillonite. These secondary minerals may weather into still other kinds of clay minerals. As iron is especially reactive with water and oxygen, iron-bearing minerals are prone to rapid decomposition. Iron remains oxidized in the red ferric state under aerobic conditions, or it may reduce to the gray ferrous state under wet anaerobic conditions. The clay particles produced are shifted and rearranged vertically and laterally within the mass by percolating water and on the surface by runoff, wind, or ice.

Rain falling on and filtering through the accumulating organic matter picks up acids and minerals in solution, reaches the mineral soil, and sets up a chain of complex chemical reactions that continue in the developing soil material. Water and carbon dioxide in the soil combine to form carbonic acid, which reacts with hydroxides of potassium, sodium, calcium, and magnesium to produce carbonates and bicarbonates. These products and other mineral elements either accumulate deeper in the developing soil material or are leached away, depending on the amount of water passing through. The greater the rainfall, the more water moves down through the soil and the less moves upward. Thus high precipitation results in heavy leaching and chemical weathering, particularly in regions of high temperatures.

Role of the Biota

Living organisms in and on the soil material have a pronounced influence on soil development. Lichens colonize bare rock and produce oxalic and other acids. These acids attack silicates by forming chemical complexes with the cations that the rocks contain. The activity of lichens encourages the colonization of rocks by mosses, producing a slowly developing layer of organic matter. In time other plants colonize the weathered material. Plant roots penetrate and further break down the parent material. The roots pump nutrients up from its depths and add them to the surface. In doing so,

plants translocate minerals deep in the parent material as well as recapture minerals carried deep into the soil by weathering processes. Through photosynthesis, plants capture the sun's energy and add a portion of it to the soil in the form of organic carbon. This energy source of plant debris enables bacteria, fungi, earthworms, ants, termites, and other soil organisms to colonize the area.

Higher organisms in the soil—millipedes, centipedes, earthworms, mites, springtails, grasshoppers, and others—consume fresh material and leave partially decomposed products in their excreta. Earthworms, ants, termites and burrowing rodents churn lower and upper layers of soil material. Microorganisms, the bacteria and fungi, further reduce this material into various carbohydrates, proteins, lignins, fats, waxes, resins, and ash, and eventually into inorganic products (see Chapter 9).

The fraction of organic matter that remains is called **humus,** dark-colored, noncellular, chemically complex organic material. Its characteristic constituents are humin, a group of unchanged plant chemicals, and other organic compounds such as fulvic acid and humic acid. Decomposition proceeds so slowly that the amount of organic material changes little each year. The formation of new humus balances annual loses by decomposition. The ratio of the formation of new humus to the destruction of old determines the amount of humus in the soil.

Soil organisms are intimately involved in the development of the *O* and *A* horizons of the soil, especially in temperate forest regions. Three types of humus formation exist in temperate forests, **mor, moder,** and **mull.** They result from an interaction of physical, chemical, and especially biological mechanisms (Figure 4.9).

Mor   Characteristic of dry or moist acid habitats, especially heathland and coniferous forest, mor is a well-defined, unincorporated, and matted or compacted organic

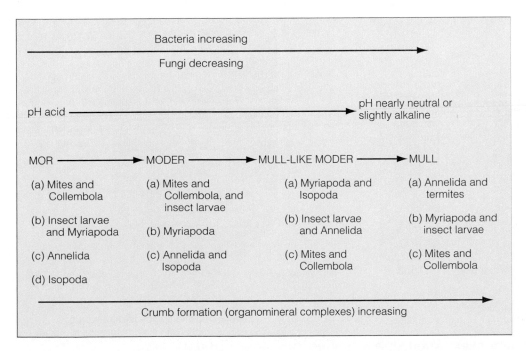

FIGURE 4.9 The sequence of humus types and related processes. Note the inverse relationship between bacteria and fungi as the humus sequence goes from mor to mull, as well as the pronounced changes in invertebrate life. (From J. A. Wallwork 1973:53.)

deposit resting on mineral soil (Figure 4.10). It results from an accumulation of litter that is slowly mineralized and remains unmixed with mineral soil. Thus a sharp break exists between the *O* and *A* horizons.

The formation process distinguishes mor from other humus types. The main decomposing agents are fungi, both free-living and mycorrhizal, which tend to depress soil animal activity and produce acids; nitrifying bacteria may be absent. Fungi consume mostly the vascular cells of leaves and pass on a residue of mesophyll tissue. Proteins within the leaf litter are stabilized by protein-precipitating material, making them resistant to decomposition in some cases. Because of limited volume and pore space, acidity, the type of litter involved, and the nature of its breakdown, mor is inhabited by a small biomass of soil animals. They have little mechanical influence on the mineral soil. Instead, these organisms live in an environment of organic material cut off from mineral soil beneath.

Mull Mull results from a different process. Characteristic of mixed and deciduous woods on fresh and moist soils with a reasonable supply of calcium, mull possesses only a thin scattering of litter on the surface, and the mineral soil is high in organic matter (Figure 4.10). All organic materials convert to true humic substances. Because of animal activity, these are inseparably bound to the mineral fraction, which absorbs them like a dye. There is no sharp break between the *O* and *A* horizons. Because of less acidity and a more equitable base status, bacteria tend to replace fungi as the chief decomposers, and conversion of the ammonium produced in the early breakdown of litter to nitrates by soil bacteria is rapid. Soil animals are more diverse and possess a greater biomass, reflecting a more equitable distribution of living space, oxygen, food, and moisture and a smaller fungal component. This faunal diversity is one of mull's greatest assets, because the humification process flows through a wide variety of organisms with differing metabolisms. These soil animals fragment plant debris, mix it with mineral particles, thus enhancing microbial and fungal

activity, and incorporate humified material with mineral soil. Plants extract nutrients from the soil and deposit them on the surface. Then the soil flora and fauna reverse the process.

Moder On the continuum from mull to mor lies moder (Figure 4.10). In this humus type, plant residues are transformed into the droppings of small arthropods, particularly Collembola and mites. Residues not consumed by the fauna are reduced to small fragments, little humified and still showing cell structure. The droppings, plant fragments, and mineral particles all form a loose, netlike structure held together by chains of small droppings. Under heavy precipitation humus leached from the droppings acts as a binding substance to form a dense, matted litter approaching a mor. On the continuum between moder and mull, we often see the droppings of large arthropods, which can take in considerable quantities of mineral matter with food. However, moder differs from mull in its higher organic content, restricted nitrification, and a more or less mechanical mixture of the organic components with the mineral, the two being held together by humic substances, yet separable. In other words the organic crumbs are deficient in mineral matter, in contrast to mull, in which mineral and organic parts are inseparably bound together.

Soil Development Processes

Soil is a dynamic and evolving entity, an open system continuously receiving inputs and contributing continuously to the environment (Figure 4.11). These interactions change par-

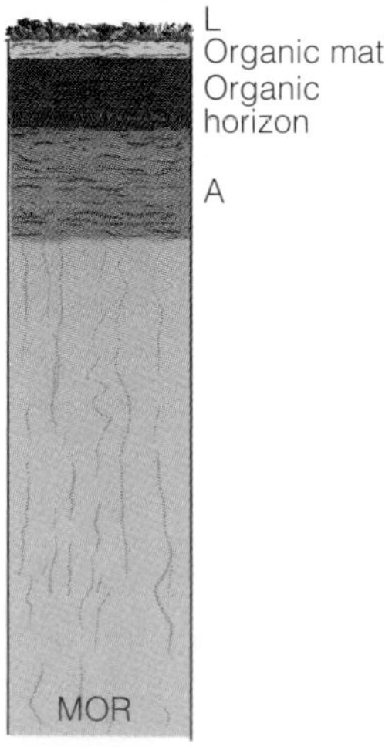

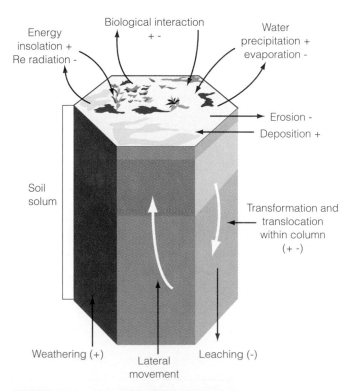

FIGURE 4.11 Schematic illustration of additions, losses, translocations, and transformations in the process of soil profile development. (Based on Boul et al. 1997.)

FIGURE 4.10 Comparison of layers in mor, moder, and mull humus.

TABLE 4.2 Some Processes of Soil Formation

Process	Description
Eluviation	Movement of material out of a portion of a soil profile.
Illuviation	Movement of material into a portion of a soil profile.
Calcification	Processes including accumulation of calcium carbonate in the *B* horizon and possible other horizons in the soil.
Decalcification	Reactions that remove calcium carbonate from one of more soil horizons.
Salinization	The accumulation of soluble salts, such as chlorides, sulfates, and bicarbonates of sodium, calcium, and magnesium, and potassium in salic horizons.
Desalinization	Removal of soluble salts from salic and soil horizons.
Alkalization	The accumulation of sodium ions on the exchange sites in a soil.
Lessivage	The mechanical migration of small mineral particles from the *A* to the *B* horizon of soil, producing *B* horizons relatively enriched in clay.
Podzolization	The chemical migration of aluminum and iron and/or organic matter, resulting in the concentration of silica in the illuviated layer.
Desilication	The chemical migration of silica out of the soil solum and thus the concentrations of sesquioxides in the solum, with or without the formation of ironstone (laterite and hardened plinthite) and concretions.
Melanization	The darkening of light-colored, unconsolidated, initial materials by an admixture of organic matter (as in a dark-colored *A* horizon).
Gleization	The reduction of iron under anaerobic soil conditions, with the production of bluish to greenish-gray matrix colors, with or without yellowish-brown, brown, and black mottles, and ferrous and manganiferous concretions.

Source: Adapted from Buol et al. 1997 *Soil Genesis and Classification* 4th ed. Ames, Iowa State University Press.

ent material and bring about **genesis** (formation) of a soil from parent material, resulting in the development of the distinctive horizons of a soil profile. Soil genesis results from four broad processes (Boul et al. 1997):

1. **Additions** of organic or inorganic material to the soil body. One depth or part of the soil receives more material than another.

2. **Losses** of material from the soil through surface erosion and leaching.

3. **Translocation**, vertically and laterally, of material from one point within the soil to another.

4. **Transformation** of mineral and organic substances and the rearrangement of soil material into structural units or peds.

Involved in these processes are a number of subprocesses, some of which are described briefly in Table 4.2. Vertical removal or translocation of soil material from one horizon is **eluviation.** The accumulation of soil material in another horizon is **illuviation.** Removal or eluviation of calcium carbonates (translocation), if present, from a soil body is **decalcification,** common to soils in humid regions. The accumulation of calcium carbonate in the subsoil (translocation), common to soils of arid regions, is **calcification.** The washing in suspension of fine clays and lesser amounts of coarse clays and silt into cracks and voids of soils is **lessivage** (translocation). The clays involved may be the product of weathering in the *A* horizon, present in the parent material, or carried in and deposited by wind. Involving both the addition to the soil body and subsequent downward transport of organic material is **melanization.** It results in the darkening of the upper horizons of the soil.

Incorporating both the transport and transformation of materials in the soil body is **gleization.** It involves the reduction of iron in the soil to a ferrous state under poor drainage conditions (see the discussion hydric soils on page 61), giving a dull gray or bluish color to the horizons. Gleization is a characteristic soil-forming process in cold wet situations, especially in the tundra; but it also common in other regions where the water table remains above the *B* and *C* horizons (perched condition).

SOIL CLASSIFICATION
Soil Orders

To apply what we know about soils, we must classify them. Soil scientists in the United States use *Soil Taxonomy,* developed by the Soil Conservation Service, now called the Natural Resources Conservation Service. There are other, similar approaches to classification including the Canadian Soil Classification System and the FAO-UNESCO system (see Landon 1984), which borrowed heavily from the Russian and U.S. soil classification systems. *Soil Taxonomy* recognizes 12 soil orders (see Table 4.3 and Figure 4.12). Their worldwide distribution (Figure 4.13, p.72) is determined largely by climate and vegetation, as indicated by the presence or absence of diagnostic horizons (see Figure 4.12). A brief description of soil orders follows.

Spodosols Highly acidic, Spodosols (wood ash) are characterized by a dark colored *B* horizon, called spodic, that forms under a gray to white colored *E* horizon. They develop under coniferous forests in cool humid regions to tropical rain forests in areas with sandy-textured, acid soil material. The

TABLE 4.3 The 12 Major Soil Orders

Order	Derivation and Meaning	Description	Location
Entisol	Coined from *recent*	Dominance of mineral soil materials; absence of distinct horizons.	All climates, any vegetation; young soils; floodplains, glaciated areas, mountains.
Inceptisol	L. *inceptum,* "beginning"	Texture finer than loamy; little translocation of clay; often shallow; little development of horizons.	All climates and vegetation; developing soil.
Vertisol	L. *verto,* "invert"	Dark clay soils; wide, deep cracks when dry.	Areas with clay-textured parent material; climate can dry and crack soil.
Aridosol	L. *aridus,* "arid"	Dry for extended periods; low in humus; high in base content; may have clay, carbonate, gypsum horizons.	Desert and semiarid regions.
Mollisol	L. *mollis,* "soft"	Surface horizons dark brown to black with soft consistency; rich in bases.	Semihumid regions; native grassland vegetation.
Spodosol	Gr. *spodos,* "ashy"	Light gray, whitish *A* horizon on top of a black and reddish *B* horizon; high in extractable iron and aluminum.	Cool humid regions; coniferous forests.
Alfisol	Coined from *Al* and *Fe*	Shallow penetration of humus; translocation of clay; well-developed horizons; not strongly leached.	Humid, temperate regions; deciduous and coniferous-deciduous forests.
Ultisol	L. *ultimus,* "last"	Intensely leached; strong clay translocation; low base content; red or yellow soils.	Warm humid to tropical climate; forest vegetation.
Oxisol	Fr. *oxydé,* "oxidized"	Highly weathered soils; red, yellow, or gray; rich in kaolinite, iron oxides, and often humus.	Tropical and subtropical climates and vegetation.
Histosol	Gr. *histos,* "organic"	High content of organic matter.	Bogs, marshes, muck.
Andisol	Jpn. *ando,* black," and *do,* "soil"	Developed from volcanic ejecta; not highly weathered; upper layers dark-colored.	Volcanic regions, especially tropical.
Gelisol	Gr. *gelid,* "very cold"	Presence of permafrost or soil temperatures of 0°C or less within 2 m of surface.	Two circumpolar regions and at high elevations in mountains.

vegetation supplies the persistent organic compounds that are essential to the formation of a spodic horizon. The main soil-forming processes involve first the combination of iron and aluminum oxides in complex ways with organic acids and humus of the *O, A,* and *E* horizons. These oxides and clays are carried downward by rainwater into the *B* horizon, where they accumulate and become immobilized in complexes of organic matter and oxides of iron and aluminum. Left behind in the upper horizons is the light- or ash-colored *E* horizon. The organic horizon is a mor with a layer of fermented litter on top of a layer of humus unmixed with mineral soil. Collectively these processes are called **podzolization** from the Russian *pod* (beneath) and *zol* (ash). Such soils are covered mostly with coniferous and coniferous-deciduous forests; in spite of their acidity, they can support pastures, hay, and such crops as potatoes, oats, and rye.

Alfisols Occupying about 13 percent of the land area of Earth, Alfisols are associated with humid temperate decidu-ous-coniferous forest, deciduous forest, and broadleaf ever-green forest. The *B* horizon of these soils has accumulations of clay, derived from eluviation of clay from the *A* and *E* horizons and subsequent illuviation in the *B* horizon, together with the formation of clay by in-place weathering of feldspars and other minerals. Unlike the Spodosols, Alfisols have comparatively little accumulation of organic matter on the surface. Instead organic matter tends to be well mixed by soil animals to form an darkened, organically enriched *A* horizon (melanization). In some soils, the division between the *A* and *B* horizons may be hard to distinguish and the *E* horizon may be missing. Because of their high base saturation and native fertility, Alfisols are very important agricultural soils.

Ultisols Formed on parent materials with few basic cations, the Ultisols have a lower base saturation than Alfisols. Characteristic of warm humid climates and old parent material, Ultisols are more intensely weathered than Alfisols, and they are acidic. They have a reddish or strong yellow color because

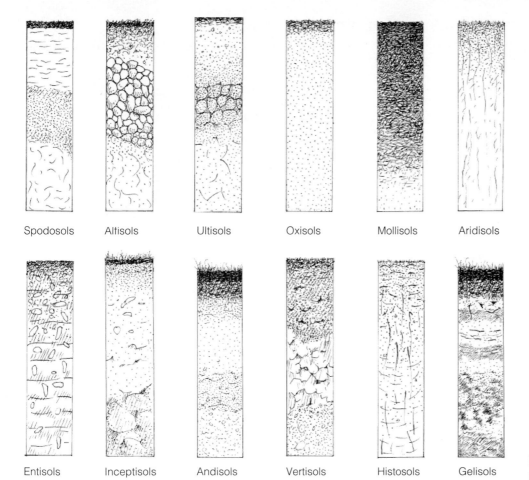

| Spodosols | Alfisols | Ultisols | Oxisols | Mollisols | Aridisols |

| Entisols | Inceptisols | Andisols | Vertisols | Histosols | Gelisols |

FIGURE 4.12 Profiles of the 12 major soil orders of the world.

of the release of iron in the form of secondary iron oxides from the silicates. Such red soils, characteristic of the southeastern United States and tropical regions, are a product of extensive leaching, lessivage, and some podzolization, resulting in clay accumulation in the *B* horizon. Added to this accumulation are clays formed by weathering in place in the *B* horizon. In contrast to Alfisols, the low nutrient content and high subsoil acidity of Ultisols limit agricultural productivity on them and historically have led to land abandonment. Ultisols can be productive agriculturally if fertilized and with proper management they are valuable for timber production, especially pines.

Oxisols Oxisols are associated with humid subtropical and tropical forested regions of the world, where rainfall is heavy and temperatures high. They make up about 22 percent of intertropical soils; the rest are mostly Ultisols and Alfisols. Uniformly high temperatures result in more intense soil development processes. Weathering of the geologically very old substrate in these regions is almost entirely chemical, brought about by water and its dissolved substances. Because precipitation usually exceeds evaporation, the water movement is almost continuously downward. With only a small quantity of electrolytes present in the soil water due to continual leaching by water, silica and aluminosilicates are carried downward,

while sesquioxides (clays lacking a silica base) of aluminum and iron remain behind. The sesquioxides are relatively insoluble in pure rainwater, but the silicates tend to precipitate as a gel in solutions containing humic substances and electrolytes. If humic substances are present, they act as protective colloids about iron and aluminum oxides and prevent their precipitation by electrolytes. The end product of such a process—desilication, the removal of silica by leaching—is a soil composed of hydrous oxides, clays, and residual quartz, deficient in bases, low in plant nutrients, and intensely weathered to great depths. The large amount of residual iron and aluminum left after the depletion of silica and bases becomes enriched as hydrous oxides, forming a variety of often brilliant reddish colors deep into the soil. For this reason these soils may lack distinct horizons. In some soils the subsoil has the ability to harden irreversibly when exposed to sun and air into plinthite (formerly called laterite). Traditionally, Oxisols have supported shifting tropical cultivation, pineapples, bananas, and coffee. Recently vast areas of tropical forests have been cleared for agriculture. With proper management Oxisols support soybeans, wheat, corn, and pasture grasses.

Mollisols The subhumid-to-arid and temperate-to-tropical regions of the world—the plains and prairies of North America, the steppes of Russia, the veldts and savannas of Africa,

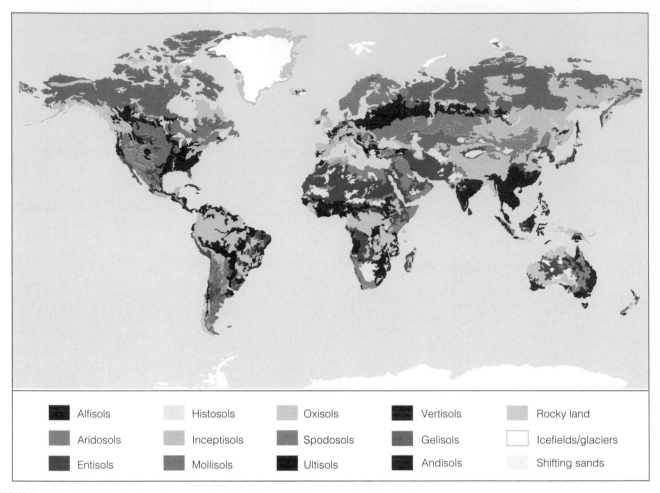

FIGURE 4.13   World distribution of major soil orders. (From Soil Conservation Service.)

Legend:
- Alfisols
- Histosols
- Oxisols
- Vertisols
- Rocky land
- Aridosols
- Inceptisols
- Spodosols
- Gelisols
- Icefields/glaciers
- Entisols
- Mollisols
- Ultisols
- Andisols
- Shifting sands

and the pampas of South America—support grassland vegetation. Dense root systems may extend many feet below the surface. Each year nearly all of the vegetative material above ground and a part of the root system are turned back to the soil as organic matter. Although the material decomposes rapidly the following spring, it is not completely gone before the next cycle of death and decay begins. Soil inhabitants, notably earthworms, ants, and burrowing rodents, mix organic matter with mineral soil, developing a soil high in organic matter and darkening the upper horizons. There is a deep, dark, base-rich *A* horizon that is granular and soft and an indistinct *B* horizon often high in calcium carbonate. This soil-forming process is **melanization.** Mollisols support the major grain-producing areas of the world. Agricultural exploitation has greatly diminished the organic and other features of these soils.

Aridosols   Aridosols are soils associated with arid and semiarid regions with relatively sparse growths of shrubs and grass (see Chapter 29). Because plant growth is limited, little organic matter and nitrogen accumulate in the soil. These soils are low in humus, high in base content, and often have gypsum, clay, or carbonate horizons. Scant precipitation results in slightly weathered and calcified soils high in plant

nutrients. The horizons are usually faint and thin. Depending on the geological history and nature of the parent material, soil-forming processes involve calcification and salinization. Because the amount of rainfall in semiarid grassland regions generally is insufficient to remove calcium and magnesium carbonates from the profile, they are carried down only to the average depth reached by the percolating waters. Grass maintains a high calcium content in the surface soil by absorbing large quantities from the lower horizons and redepositing it on the surface. Little clay is lost from the surface. Soils developed by this process, called **calcification,** primarily develop a calcic *B* horizon. The light-colored *A* horizon is low in organic matter, and the *B* horizon is characterized by an accumulation of calcium carbonate.

Within these regions are areas where soils contain excessive amounts of soluble salts, either from the parent material or from the evaporation of water that drains in from adjoining land. Infrequent rains penetrate the soil and carry the material downward to a certain depth. The water evaporates, leaving more or less cemented horizons of sodium, calcium, or magnesium salts below the surface in a soil-forming process called **salinization.** Under certain conditions the carbonates may cement soil particles and rock fragments together below the surface to produce hard layers known as

petrocalcic horizons or **caliche.** Caliche is impervious to plant roots and is difficult to excavate.

Entisols Entisols show no evidence of any horizon development. They occupy about 20 percent of the land area of Earth and are often found where other soil groups dominate. Entisols occur in areas where soil-forming processes have been interrupted or curtailed by flooding, erosion, and large-scale disturbance by human settlements and activities. They are associated with wetlands and alluvial deposits along rivers as well as areas subject to erosion and sedimentation that continually create new Entisols. If protected from erosion and other disturbances, Entisols may slowly show signs of soil development. Fertile alluvial lands renewed by depositions from flooding are important agricultural lands. In steep mountainous and hilly regions, erosion and mass wasting and in lowlands, flooding present major engineering problems for road building and construction.

Inceptisols Inceptisols are mineral soils rich in weatherable materials with some subsoil development. They are associated with steeply sloping mountain lands whose slopes, subject to surface erosion, add to the colluvium (accumulated materials) at the base of the slopes. Other areas of Inceptisols are glacial deposits and recent valley and delta depositions of rivers. The major pedogenic process associated with Inceptisols is leaching. Some Inceptisols support agriculture. Others are best devoted to forests and pastures.

Andisols Andisols form from volcanic ash, pumice, cinders, and related materials. Their occurrence relates to the global distribution of volcanic activity both recent and in the geological past, especially during the Holocene. Volcanic ejecta is mineralogically different from the parent materials of other soils. Its rapid cooling produces easily weathered material called volcanic glass. It is characterized by materials with a wide range of chemical composition and mineralogy that support a diversity of vegetation from desert shrub to tropical forests. Andisols are dominated by organic and inorganic compounds. The least weathered part of the soil is the surface layer. Because of the periodic additions of ejecta, the horizons are variously colored and do not necessarily relate to horizon development. Easy to till and locally fertile, depending on the nature of the volcanic ejecta, they have a stable, high water-holding capacity and thus are resistant to water erosion, but they are subject to wind erosion.

Vertisols Vertisols are soils with a very high (30 percent) content of sticky, swelling, and shrinking-type clays. They are dark, often blackish in color to a depth of 1 meter or more, and typically develop from calcium-rich parent material in regions whose climate features a dry period of several months. Vertisols occur in every continent except Antarctica. Extensive areas occur in Australia, India, Sudan, Ethiopia, Egypt, and the western United States, especially Texas and California. During the dry season, the clay shrinks and the soil cracks to the surface. Materials on the surface of the soil fall into the cracks. In the wet season, water entering the cracks cause the clays to expand.

The cracks close, embedding the material in the profile and causing some microbuckling on the soil surface. Such soils have limited agricultural use, mostly pasture, and create major engineering problems for road building and pipelines.

Histosols Histosols occur from the arctic tundra to the tropics and include all organic soils. The main soil-forming process is the accumulation of partially decomposed organic parent material produced by plants and animals under saturated anaerobic conditions. Such conditions result from high water tables, poor drainage, and climatic conditions of extremely high precipitation and low evaporation. Histosols are associated with wetlands, bogs, and other peatlands. Many Histosols have been drained for vegetable production and building sites, and mined for commercial horticultural peat, with the resulting destruction of peatland ecosystems. Sensitive to drainage, the organic matter decomposes steadily under aeration and gradually disappears, resulting in land subsidence and loss of organic matter.

Gelisols Gelisols are the newest recognized soil order. Gelisols are characterized by permafrost—permanently frozen material underlying the solum (see Chapter 29)—within 2 meters of the surface, rather than horizons. Soils included in this new order were previously scattered among other orders, notably Inceptisols, Entisols, and Histosols. Gelisols occur in the two circumpolar regions and at high elevations in mountains at lower latitudes. They have a deep active layer of accumulated organic matter supplied by tundra plants and animals, and mineral soil on top of the permafrost. During the warm season freezing and thawing mixes the material and incorporates organic material throughout, disrupting any strong horizon development. Such activity results in patterned ground forms, and **solification**—the slow flow of saturated soil and other unconsolidated material (see Chapter 29). Gelisols are highly fragile. Disturbance results in melting of permafrost. Because of cold temperatures, these soils cannot handle liquid, solid, or gaseous wastes (see Chapter 29).

Lower Soil Categories

Soil orders are further subdivided into suborders, great groups, families, and series. Suborders are based on soil properties that reflect environmental controls on current soil-forming processes. Great groups are a subdivision of suborders. They are based on the presence or absence of diagnostic horizons. Subgroups are a subdivision of great groups They include subgroups that show central properties of the great groups, subgroups of soils that intergrade, and soils with atypical properties that are not characteristic of any great group. Families are defined largely on physical and mineralogical properties of importance to plant growth. Families are subdivided into soil series. A soil series is made up all soils formed from a particular type of parent material that are similar in all characteristics and arrangement of the soil profile except for the texture of the *A* horizon. The soil series is the lowest taxonomic level in the soil classification scheme.

In the *Soil Taxonomy* system soil series are named after the locality in which they were first described. For example, the Ovid series was named after the town of Ovid and the Miami series after the Miami River in western Ohio (not to be confused with Florida's Miami).

Locally every soil series has its neighboring soil series with unlike properties (Figure 4.14), reflecting changes in slope, drainage patterns, and soil materials. These soil series may have abrupt boundaries or form a continuum. If several related soils found side by side have developed from the same parent material but differ mainly in natural drainage and slope, they form a **catena,** from the Latin for *chain*. In Figure 4.14 the groups of soils from C through G, all derived from loess, make up a catena.

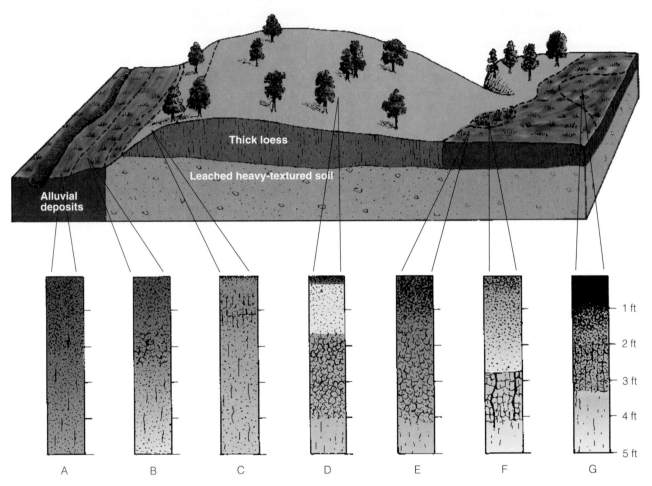

FIGURE 4.14 Topography and vegetation acting together produce a toposequence of soil series. This diagram shows the normal sequence of seven soil series from the Mississippi to the uplands in Illinois. It also illustrates how bodies of soil types fit together in the landscape. Boundaries between adjacent bodies are gradations or continuums, rather than sharp lines.

The lower part of the diagram pictures the profiles of the seven soils, showing the color and thickness of the surface horizon and the structure of the subsoil. Note how the natural vegetation has influenced surface color and how topographic position and distance from the bluff have influenced subsoil development.

Profile A (Sawmill) is an Entisol, a bottomland soil formed from recent sediments and not subject to much weathering. Profile B (Worthen) at the foot of the slope is also an Entisol. Developed from recent alluvial material, it shows little structure. Profile C (Hooper) on the slope developed from a thick loess on top of leached till, while the soil on the bottom of the slope developed directly from the till. Profile D (Seaton) is an upland soil formerly covered with timber. It possesses a light surface color and lacks structure, the result of rapid deposition of loess during early soil formation, holding soil weathering to a minimum. These two soils probably should be considered Entisols because they lack any well-formed horizons, which are the product of strong soil weathering. Profile E (Joy) represents upland soil developed under grass. Note the dark surface and the lack of structure, again the result of a rapid deposition of loess. Profile F (Edgington) is a depressional wet spot. Extra water flowing in from adjacent fields increased the rate of weathering, resulting in a light grayish surface and subsurface and a blocky structure to the subsoil. This indicates strongly developed gley soil. The depth of subsoil suggests that considerable sediment has been washed in from the surrounding area. Profile G (Sable) represents a depressional upland prairie soil. The deep, dark surface and the coarse blocky structure mark it as a Mollisol. Abundant grass growth produced the dark color. (After Veale and Wascher 1956.)

Although the major soil development process defines broad soil patterns, local patterns are controlled by topography. As the topography of an area develops over time through the action of physical and chemical weathering, a group of soils develops along with it. This group of soils, which may include several catenas, is known as a **toposequence.** In Figure 4.14 the groups of soils from A through G represent a toposequence, ranging from well-drained ridgetop to very poorly drained soils and alluvial deposits. If all the soil-forming factors remain constant except for vegetation, then vegetation leaves its imprint on soil formation. Soils so developed form a **biosequence.** In Figure 4.14 soils D (Seaton) formed under forest and E (Joy) formed under grass make up a biosequence. Groups of soils of different ages derived from the same parent materials but subject to soil-forming processes over different lengths of time make up **chronosequences.** Examples occur in areas of retreating glaciers, volcanic flows, and sand dunes.

MAPPING SOILS

The distribution of specific kinds of soils across the landscape is delineated on a soil map (Figure 4.15). The kinds of soil mapped may cover from a few to several hundred acres. These mapped units may include small areas of other soils too small to map separately.

Most soil map units carry one soil name and that soil dominates the unit. A unit with one soil name is called a consociation. Unlike soils may be grouped together into units called associations, complexes, or undifferentiated groups for reason of scale or practicality (Soil Survey Division Staff 1993). Phases of mapping units are commonly used in detailed soil mapping. Phases indicate some additional property about the soil or soils named in the mapping unit. For example, Berks silt loam, 15 to 25 percent slopes, has two phases: a surface texture phase (silt loam) and a slope phase.

Soil maps are developed through soil surveys, which date back to 1900. Modern soil mapping is done by the Soil Conservation Service in cooperation with other governmental agencies and soil scientists at Agricultural Experiment Stations of land grant universities.

As their first step in mapping soils, soil scientists collect all available information relevant to local soils: geology, climate, vegetation, topographic maps, and aerial photographs. They begin mapping by broadly delineating boundaries of soil types on aerial photographs. Then they visit the various sites, sampling the soil by spade and soil auger to determine the characteristics of the soil profile. They identify the soil in the field, refine the plotting of the soil boundaries, and place the identifying symbols on the map. Finally, they describe the soils, interpret the survey, and publish the results in a Soil Survey Report.

The Soil Survey Report describes the soil map units and the typical pedons for each soil series. It assesses the management of the soil for crops and pasture, its susceptibility to erosion, and its usefulness for crop production, woodland

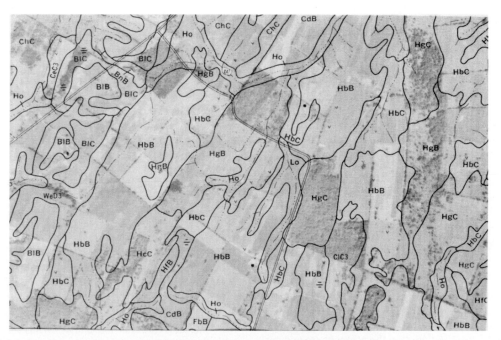

FIGURE 4.15   Soil map of a section of Jefferson County, West Virginia, indicating soil associations and soil series. The symbol H indicates the Hagerstown series and C the Chilhowie series. These soils are deep, well drained, and on fairly level land. The symbols indicate the characteristics of the soils. For example, HbB is Hagerstown silt loam, 2 to 6 percent slope; HgB is very rocky silt loam, 2 to 6 percent loam; HgC is very rocky silt loam, 6 to 12 percent slope; and CdB is Chilhowie silt clay, 2 to 6 percent slope. (Courtesy of Soil Conservation Service.)

management, and wildlife habitat. It also provides information on the suitability of various soils for road construction, housing development, building sites, sanitary facilities, water management, and recreation. Unfortunately, many site development engineers and real estate developers do not make use of this information. Their failure to do so results in damage and loss from soil instability.

MISMANAGED SOILS

Over much of the landscape, soils have been highly disturbed. Soils have been buried under fill, overturned and moved about by excavations, surface mining, and road construction, and exposed to erosion by wind and water. Upper horizons have been mixed by agricultural plowing and tillage and compacted by heavy machinery and trampling.

Soil compaction occurs when any weight pushes the soil particles together and reduces the size of the pores. The greatest compaction occurs under wet and moist conditions. Moist soil particles easily slide over one another. Heavy machinery from large tractors to construction equipment, trampling on lawns and playing fields, concentrated use of pathways and hiking and riding trails, and off-road use of all-terrain vehicles all compact the soil. Compacted soil cannot absorb water, so that the water flows across the surface, resulting in **soil erosion** —the carrying away of soil particles by wind and water.

Disturbance of topsoil by agriculture, construction, and road building exposes it to erosion. Stripped of its protective vegetation, soil is removed by wind and water faster than it can be formed. Loss of the upper layers of humus-charged, granular, highly absorptive topsoil exposes the humus-deficient, less stable, less absorptive, and highly erodible layers beneath. If the subsoil is clay, it absorbs water so slowly that heavy rains produce a highly abrasive and rapid runoff. Soil of the uplands ends up in muddied rivers, where it affects wildlife, settles out in dams, shortening their useful life, and builds up river deltas.

The intensity of water erosion is influenced by slope, the kind and condition of soil, and rainfall. The least conspicuous type of water erosion is **sheet erosion,** a more or less even removal of soil over a field. When runoff tends to concentrate in streamlets instead of moving evenly over a slope, the cutting force is increased and **rill erosion** produces small channels downslope. Where concentrated water cuts the same rill long enough, or where runoff in sufficient volume cuts deeply into the soil, highly destructive gullies result. **Gully erosion** often begins in wheel ruts made by off-road vehicles in fields and forest, logging roads and skid trails, and livestock and hiking trails.

Bare soil, finely divided, loose, and dry, as it often is after tillage, is ripe for wind erosion. Very fine particles of dust are picked up by the wind and carried as dust clouds. Often dust particles are lifted high in the atmosphere and carried for hundreds and even thousands of miles. This problem occurred in the Great Plains during the droughty Dust Bowl days of the 1930s, and it is happening on an increasing scale worldwide today.

Summary

1. Soil is the foundation of terrestrial ecosystems. It is the site of decomposition of organic matter and of the return of mineral elements to the nutrient cycle. It is the habitat of animal life, the anchoring medium for plants, and their source of water and nutrients.

2. Soil is the weathered outer layer of Earth's crust that supports life. Individual soils are three-dimensional bodies with width, length, and depth. The smallest basic unit of study is the pedon, characterized by a distinctive soil profile, a vertical section of the layers or horizons.

3. The soil profile, in general, posses five commonly recognized horizons, although all are not necessarily present in any one soil: the *O,* or organic, layer; the *A* horizon, characterized by accumulation of organic matter; *E,* the zone of leaching of clay and mineral matter; the *B* horizon, in which mineral matter accumulates; and *C,* the underlying material, exclusive of bedrock.

4. Differences between soils and between horizons within soils are reflected by variations in texture, structure, and color. Texture of the soil is determined by the proportion of soil particles of different sizes—sand, silt, and clay. Texture is important in the movement and retention of water in the soil.

5. Soil particles, particularly the clay-humus complex, are the key to nutrient availability and the cation exchange capacity of the soil—the number of negatively charged sites on soil particles that can attract positively charged ions. Cation exchange capacity relates to percent base saturation, the percentage of sites occupied by basic cations. Soils with a high cation exchange capacity are potentially fertile.

6. Soil is more than its physical characteristics. It is a living system embracing a wide array of organisms from bacteria and fungi to earthworms. Larger organisms live in pore spaces and channels in the soil. Bacteria, protozoans, and other microorganisms live in the thin film of water surrounding soil particles. Most microorganisms obtain their nourishment from organic matter. Larger soil inhabitants feed on fresh litter and other detrital material, and on bacteria and fungi. These detrital feeders support an array of predacious soil inhabitants from mites to spiders. These organisms through their activities strongly influence soil development.

7. Soil formation involves the interactions of the external environmental factors including radiant energy and water; topography, which influences thermal and water regimes; and geological substrate or parent material.

8. Soil begins with the physical and chemical weathering of geological bedrock or deposited material. Physical weathering involves the mechanical breakdown of the substrate. Chemical weathering consists of the transformation of primary minerals into secondary minerals and the translocation and leaching of those materials.

9. Strongly influencing weathering are organisms within and on the surface of the developing soil. Plants rooted in weathering material further break down the substratum, pump nutrients up from its depths, and add all-important organic material. Through decomposition and mineralization this dead organic matter is converted into humus, an unstable product that is continuously being formed and destroyed by mineralization. Organic matter or humus is usually grouped into three types, all influenced by the activity of soil organisms: mor, characteristic of acid habitats whose chief decomposing agents are fungi; mull, characteristic of deciduous and mixed woodlands,

whose chief decomposing agents are bacteria; and finally, moder, highly modified by the action of soil animals. Moder is characteristic of many deciduous forests

10. Soil profile development is the result of four processes: additions to the soil body, losses from the soil body, translocation of materials within the soil body, and transformations of materials within the soil body, such as mineral weathering and organic matter breakdown. Within this framework are a number of subprocesses. In grassland regions the development process involves melanization, which is the darkening of the upper horizons by accumulation and mixing of organic matter. In semiarid regions a development process is calcification, in which calcium carbonate accumulates at the average depth reached by percolating water. In forest regions, a major developmental process is podzolization—involving the leaching of calcium, magnesium, iron, aluminum, and organic matter from the upper horizon, and the retention of silica. Gleization takes place in poorly drained soils; organic matter decomposes slowly, and iron is reduced to the ferrous state.

11. The processes of soil formation result in distinctive local individual soil bodies. Individual soils with the same profile characteristics and derived from the same parent material make up a soil series. Various soil series are further classified into families, subgroups, great groups, suborders, and 12 recognized orders.

12. Humans have extensively modified soils across Earth. Plowing and earth moving have altered the upper horizons. Disturbance of soil surfaces has resulted in wind and water erosion, depleting the topsoil, exposing lower horizons, and reducing the ability of soil to support humankind.

Review Questions

1. What is soil?
2. What characterizes a soil profile?

3. What do texture, color, depth, and moisture tell you about a soil?
4. What is cation exchange capacity? How does it affect nutrient availability and soil acidity?
5. What do the following contribute to soil formation: physical and chemical weathering, plant life, soil organisms, exchanges of radiant energy and water with the environment, topography?
6. Why would you expect soils developed on a slope to be different from a lowland soil?
7. Explain the differences among mull, moder, and mor organic layers. In what way do they contribute to soil development?
8. Compare the following forms of soil developmental processes: melanization, lessivage, podzolization, calcification, salinization, gleization.
9. Name and provide several diagnostic characteristics of each of the 12 major soil orders.
10. From your local or state office of the Natural Resources Conservation Service office, obtain a soil survey for your area. Note the pattern of soil distribution. What are the local soil series? Relate local soils to agricultural development, urban development, soil erosion problems, and forest distribution.
11. Do local developers consider soil characteristics of their sites? What is the danger of not doing so?
12. This chapter relates mostly to natural, undisturbed soils. Over large areas these profiles have been highly disturbed by agriculture and other activities. How does soil tillage affect the soil profile? The organic horizon and soil organic matter content?
13. What effect might the application of soil pesticides have on soil organisms and soil development processes? (For some answers, see Tate 1987.)

Cross References

Local water cycle, 53; soil-plant-atmosphere continuum, 104–107; nutrient sources and cycling, 113–114; calcicoles and calcifuges, 117; nutrients and consumers, 123–126, decomposition, 146–156.

How a Lack of Mushrooms Helped Power the Industrial Revolution

With names like Earth Star, Satyr's Beard, Big Laughing Gym, Witch's Hat, Velvety Earth Tongue, Green-headed Jelly Club, and Angel Wings, mushrooms are a diverse and mysterious lot (Figure II-A). Reminiscent of plants, but actually fungi, mushrooms are—depending on species—a chef's delight, a poisoner's potion, or a recycler's dream. Some, such as truffles, chanterelles, and morels, are gustatory treasures. Others, such as the *amanitas* Death Cap and (the ironically named but strikingly beautiful) Destroying Angel, have served as instruments of murder. Some, such as Corpse Finder, have unveiled evil deeds by their habit of growing atop soil harboring *corpora delecti*. Some, such as White Rot Fungus, have changed the course of Earth's environment—and human history—with their extraordinary ability to decompose the toughest parts of plants.

Although the legacy of mushrooms is fascinating, perhaps their major impact on our lives stems not from their presence but from their absence, specifically during the first few hundred million years after some plants abandoned water for a new life on land. Indeed, a lack of mushrooms during a portion of Earth's history is one major reason that we now have fuel for our cars, oil for our furnaces, and electricity for our lights.

Our modern way of life harks back to the the Industrial Revolution in the mid-eighteenth century. It was a time that saw the development of the steam engine and a shift in labor from small workshops to factories, from humans to machines. The Industrial Revolution required huge amounts of energy, and so the source of energy shifted from the scarce resource of wood to the more abundant coal.

Coal, petroleum oil, and natural gas are referred to as fossil fuels because they are derived from the fossilized remains of plants. Fossil fuel formation is an exceedingly slow process that requires special conditions. Occasionally, when plants died, their partially decomposed tissues came to rest in rock pores, fissures between rock layers, or sand beds.

Later, impervious materials covered the organic matter, preventing its dispersal. Over millions of years, geological processes buried the remains of these plants deep in the earth's crust. The weight of the overlying rock subjected the plant remains to intense pressures and temperatures until, eventually, they became fossilized, or carbonized, into coal, crude oil, or natural gas.

Most of the organic carbon that gave rise to the fossil fuel reserves was deposited during the Carboniferous period some 360 to 285 million years ago (Figure II-B), a period termed the first Coal Age. This was a time when Earth's climate was warmer and wetter than today's and primitive land plants flourished around the margins of shallow inland seas. These primitive forests dominated the North American and Eurasian land-scapes until Earth's climate turned colder and drier and the vast forests began to disappear. A second Coal Age extended from 200 to 1.6 million years ago, but this period saw far less deposition of organic carbon than the Carboniferous.

A key to fossil fuel formation lies in the evolution of the primitive forest plants from their aquatic ancestors. The earliest land plant was most likely a multicellular green algae that became established on land over 425 million years ago. The move from an aquatic to a terrestrial environment posed a new problem for plants: how to avoid desiccation. Water that evaporates to the air from a plant's surface must be replaced with water from the surrounding environment. The bryophytes (mosses)—some of the earliest land plants—absorbed water directly through aboveground tissues. The development of roots and an internal vascular system was still to come.

The earliest known vascular plants appear in the fossil record during the Silurian period, some 400 million years ago. By the early Carboniferous period, forests appeared. The trees that inhabited these forests depended on barklike tissues for support

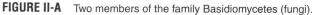

FIGURE II-A   Two members of the family Basidiomycetes (fungi).

rather than the woody tissues that support modern trees. This distinction is important to fossil fuel formation because bark differs from wood in a very important way: it is much higher in carbon compounds called lignins.

Although approximately 50 percent of the dry weight of plant tissues is composed of the element carbon, these carbon molecules exist in a wide variety of compounds. The major carbon compounds of plant tissues are lipids, proteins, and polysaccharides (such as cellulose and hemicellulose), and other large and complex structural molecules such as lignins. Decomposer organisms use the carbon in dead plant tissues as a source of energy. However, not all carbon compounds are of equal value as an energy resource. Lipids and proteins are the easiest to break down and yield the greatest amount of energy to decomposers. Therefore, these carbon compounds are the first to disappear during decomposition. Cellulose and hemicellulose likewise have relatively weak chemical bonds and open molecular structures that permit fairly rapid degradation, making them intermediate in energy quality to decomposer organisms. Lignins are complex carbon molecules that are extremely difficult to break down and yield little energy. Herbivores cannot digest lignins, termites are ineffective in breaking them down, and even bacteria do not significantly digest them. The decomposition of lignin compounds is carried out by the mushrooms and related fungi known collectively as the Basidiomycetes. The breakdown of lignins involves a bleaching process, known as "white rot," which degrades lignins and gives mushrooms access to nutrients bound up in these compounds. The process actually yields little or no net gain of metabolic energy to the mushrooms.

The difference in the quality of these various carbon compounds as an energy source for decomposers results in a distinctive pattern of decomposition (Figure II-C): the higher-quality carbon compounds disappear first, until the remaining carbon in the organic matter is largely composed of

lignin compounds. In fact, the dark brown organic matter called humus found on the surface of forest soils is largely made up of lignins. These lignin compounds also constitute the majority of the organic carbon that forms coal deposits.

The chemical fossils of the earliest known vascular plants contain lignin-like compounds. This places the evolution of lignins back at least to the Silurian, 400 million years ago. Because lignins are energetically expensive to synthesize, as land plants evolved they switched to using less costly carbon-based compounds, such as cellulose (the major constituent of wood), to form their supportive tissues. The majority of lignin compounds in modern plants are in the vascular tissues (veins of the leaves and the xylem and phloem of the woody tissues). This evolutionary progression toward the use of chemically simpler compounds required the passage of enormous amounts of time. Not until the Mesozoic, which began some 240 million years ago, did trees evolve the use of woody tissues for support, reducing their dependence on the more lignin-rich barklike tissues.

The dependence of Carboniferous trees on barklike tissues rather than wood for support means that the proportion of total plant carbon bound up in lignin compounds of those trees may well have been twice that of modern trees. The much higher lignin content of these early forests suggests a much slower rate of decomposition than that of today's forests. This rate was further depressed by the lack of organisms capable of degrading lignins. Although the first positive identification of *Basidiomycetes* is in fossil material from the Pennsylvanian epoch in the mid-Carboniferous period some 300 million years ago, it is believed that these early fungi were not decomposer organisms. The examination of thousands of coal samples has led scientists to conclude that there is no convincing evidence of the presence of fungal remains in coal deposits before the mid-Cretaceous period (130 million years ago). The presence of mushrooms and other fungi in the fossil

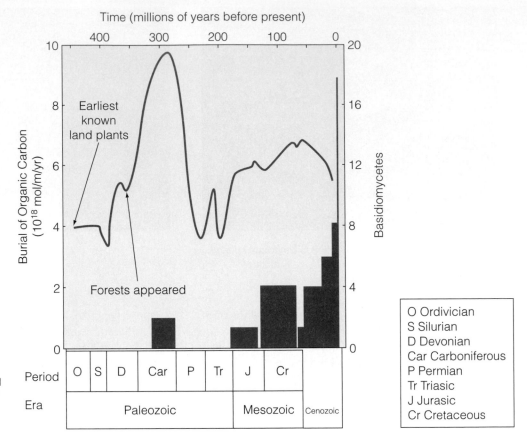

FIGURE II-B   Changes in the burial rate of organic carbon and the evolution of Basidiomycetes' diversity over geological time. (Adapted from Robinson 1990.)

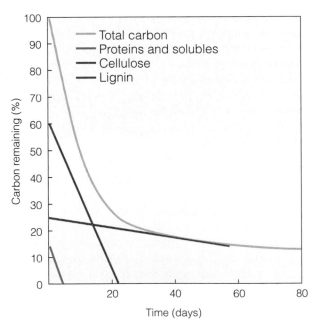

FIGURE II-C   Changes in carbon content of decomposing leaves. Total carbon is partitioned into various classes of carbon compounds. Note that lignins decay much slower than proteins and cellulose.

record increases from this period onward (Figure II-B). This suggests that effective lignin degradation did not appear until some 200 million years after lignins evolved, or about 100 million years after the peak of organic carbon deposition. Because of the time lag between the period when plant organic matter was highest in lignin content and the evolution of organisms (mushrooms) capable of degrading lignins, a vast reserve of organic matter built up during the

Carboniferous period. This reserve was eventually converted into fossil fuels and forms the lion's share of today's coal, crude petroleum, and natural gas resources.

So, next time you turn on a lamp or open the refrigerator, think for a second about how the early forests of the Carboniferous period and a 100-million-year absence of mushrooms made it all possible.

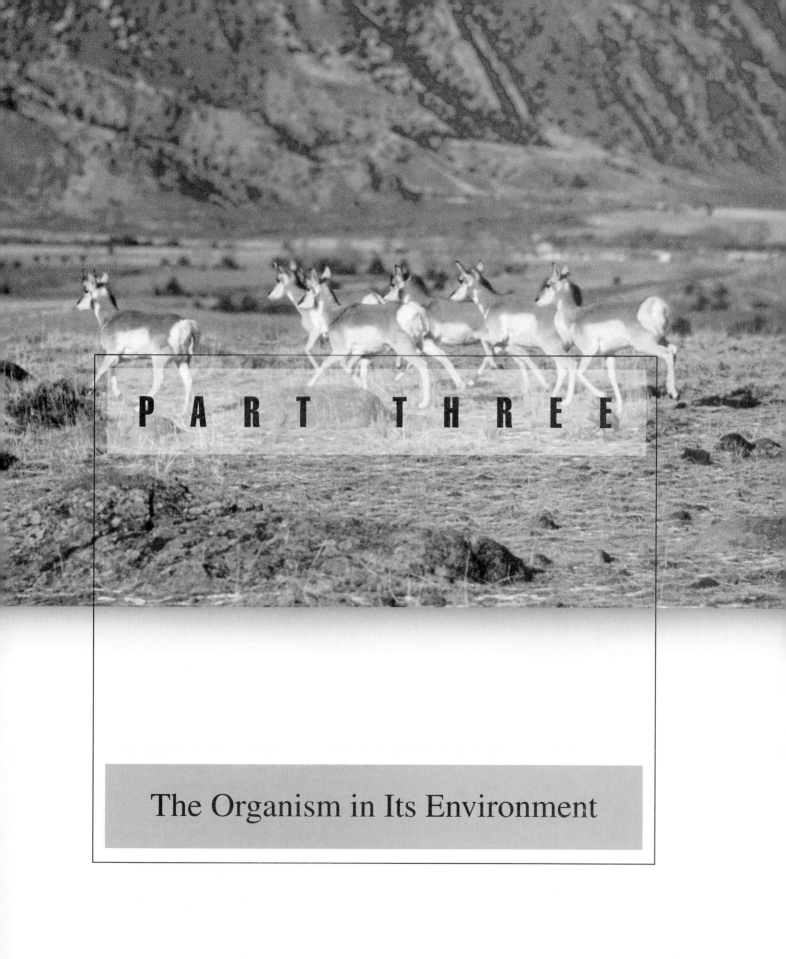

PART THREE

The Organism in Its Environment

CHAPTER 5

Adaptation

CONCEPTS

1. Adaptation is any heritable behavioral, morphological, or physiological trait that maintains or increases the fitness of an organism to live under a given set of environmental conditions. Adaptation, measured as fitness of the organism, is the result of natural selection.

2. The adaptiveness of an organism involves its ability to function between some upper and lower limits of environmental conditions. This adaptiveness is expressed as the law of tolerance.

3. Organisms must maintain a rather constant internal environment relative to external environments. They maintain this conformity through homeostatic mechanisms.

4. No single set of characteristics will enable an organism to function equally well under all environmental conditions.

Most of us can recall our first childhood visit to a zoo. We were amazed by the diversity of strange and wonderful animals: the giraffe with its long neck, the snow-white coat of the polar bear, and the orangutan with its exceedingly long arms. These animals no doubt seemed as if they were from another world, so unlike the animals that inhabit the environment we know.

However, in the savannas of Africa with its widely dispersed, umbrella-shaped trees, the ice flows of the Arctic, and the canopy of the tropical rain forest in Borneo, these animals look as natural as the birds at our backyard feeders, or the deer from the edges of forest and field at dusk. What appear to be peculiarities in the context of one environment appear as advantages—characteristics that enable the organisms to thrive in another environment. The long neck of the giraffe allows it to feed in areas of the tree inaccessible to other browsing animals in the savanna. The white coat of the polar bear makes it virtually invisible to potential prey on the snowy landscape of the Arctic. The long arms of the orangutan are essential for life in the canopy, where balance requires more than a sure step. These characteristics that enable an organism to thrive in a given environment are called **adaptations.**

THE MEANING OF ADAPTATION

Adaptation is one of those terms in ecology that has been burdened with different meanings. Perhaps the most common one is any behavioral, morphological, or physiological trait that is assumed to be the result of natural selection. A second meaning is any physiological or morphological feature or form of behavior used to explain the ability of an organism to live where it does. A third definition is an inherited characteristic that enhances an organism's ability to survive and reproduce in a given environment. A fourth definition is a change in physical, physiological, or behavioral traits that results from some current environmental pressure, such as "adapting" or "adjusting" to a change in temperature.

To understand adaptation we must return to some basics. Consider a local population of any organism living in a given environment. It consists of a number of individuals of different ages that possess heritable variations. Some of these individuals die early in life; others survive to maturity but fail to reproduce; and still others reproduce and leave behind various numbers of offspring. The fates of individuals hinge on genetically determined characteristics that enable them to cope with the physical and biological environment.

Among the reproducing individuals, some will leave more offspring than others. These individuals are considered more fit than the others because they contribute the most to the population gene pool. Organisms that leave few or no offspring contribute little or nothing to the gene pool and so are considered less fit.

Fitness, however, is more than a numbers game. Assume that individual A leaves four offspring and individual B leaves two offspring. By our general definition of fitness, we would initially say that A is more fit than B. However, suppose that of the four individuals produced by A only one survives to reproduce, whereas both offspring of B survive to reproduce. Although A left more offspring, B actually contributes proportionately more to future generations, making B more fit. It is not necessarily the number of offspring an individual leaves behind that measures fitness; rather it is the number of its descendants that influence the heritable characteristics of the population. The **fitness** of an individual is measured by the proportionate contribution it makes to future generations.

The differential reproductive success or fitness of individual organisms comes about through the process of **natural selection.** Under a given set of environmental conditions, those individuals most able to cope with the environmental situation are selected for, and those unable to do so are selected against. Thus natural selection selects for any heritable structural or behavioral characteristic that increases fitness.

The heritable characteristics an organism possesses, it owes to past generations. Its ancestors, in effect, experienced the selective screening of various combinations of genetic characteristics that produced the heritable characteristics owned by present individuals. The possession of these characteristics enables an organism to match the features of its environment. As long as environmental conditions under which individuals of the current generation exist are similar to those experienced by past generations, the organism is adapted to the environment. If environmental conditions change significantly, then the fitness and even the survival of individuals will be in jeopardy. **Adaptation**, then, is any heritable behavioral, morphological, or physiological trait that maintains or increases the fitness of an organism under a given set of environmental conditions.

TOLERANCE

The adaptation of an organism to its environment is exhibited by its ability to function between some upper and lower limits in a range of environmental conditions. The role of inherent limitations in the response of organisms to their environment was recognized as early as 1840 by the German organic chemist Justus von Liebig. In his book *Organic Chemistry and Its Application to Agriculture and Physiology,* von Liebig described his analysis of chemistry and plant growth. He set forth his conclusions in a simple statement, revolutionary for his day: "The crops of a field diminish or increase in exact proportion to the diminution or increase of the mineral substance conveyed to it in manure."

What Liebig was saying is that plant growth responds to the nutrient that is most limiting. If one of these mineral substances is present in minimal quantities only, the growth of the plant will be minimal. This statement became known as **the law of the minimum.**

The law of the minimum, strictly applied, carries with it some restrictions. It applies only under steady-state conditions, with all other resources being in excess of needs. If the quantity of the limiting resources increases, so that the rate of growth of the plant increases, then that substance is no longer limiting, but another might become so, as the increased growth places more demand on other resources. For example, if availability of phosphorus is low, molybdenum, another essential plant nutrient needed in very small amounts, may be sufficient for the plant's needs. However, if the availability of phosphorus increases, stimulating additional plant growth, molybdenum might become the limiting element. This concept applies not only to nutrients but also to other factors of the environment.

Too little of a nutrient or other resource, such as light, may be harmful; but too much of a good thing can be just as bad, and thus limiting. F. F. Blackman (1905) advanced this concept. He pointed out that a maximum quantity of a resource tolerated by organism would limit response as well. Blackman's observations became known as **the law of limiting factors.**

The law of the minimum and the law of limiting factors emphasize environmental conditions. The reproduction, growth, and functioning of individual organisms depend on the amount of the essential environmental requirements presented to them in minimal quantities during the most critical season of the year. Thus, for many birds and mammals, the availability of food is critical not during the winter but between the end of winter and early spring, when food resources have been exhausted. Organisms, then, live within a range of too much and too little, the limits of tolerance. V. E. Shelford in 1913 incorporated the concept of environmental limitations to organisms' ability to reproduce, grow, and survive into **the law of tolerance.**

The law of tolerance can be illustrated as a bell-shaped tolerance curve (Figure 5.1). The ordinate or *x* axis represents the gradient or range of a particular environmental factor. The abscissa or *y* axis represents the response of the species or individuals of a species on that gradient. The upper or middle part of the curve embraces the optimal state for reproduction or fitness. The lower parts of the graph are those portions of the gradient representing conditions under which individuals grow but possess lower fitness, survive but not reproduce, or fail to survive.

The tolerance curves are broad for many organisms. They are able to exist within a wide range of values for a particular environmental factor, such as salinity, temperature, or humidity. For example, the adult blue crab (*Callinecters sapidus*) of the American East and Gulf Coasts can live in salinities that range from seawater (about 34 ppt) to nearly fresh water. Other organisms, even life stages within the same species, have a narrow range of tolerance, often concentrated at either end of the total tolerance curve. Eggs and larva of the blue crab can survive only in salinities above 23 ppt.

The range of tolerance is not fixed. As seasons and conditions change, individuals may acclimate to them and shift their tolerance curves to the right or to the left (Figure 5.2). Consider, for example, fish inhabiting a pond in which the water temperature changes from summer through winter. As the water warms in spring, the tolerance of fish for warmer temperatures gradually increases; at the same time, the tolerance level for lower temperatures decreases. Similarly, as the water cools in fall and winter, the tolerance for low temperatures increases, while the tolerance for high temperatures decreases. Thus a temperature that would be lethal for a fish in winter, if the fish were suddenly exposed to it, can be tolerated in summer. Although the tolerance ranges shift with the season, all the shifting takes place within the adaptive physiological limits of the organism. This plastic, relatively short-term response of an individual to exposures to different or changing natural environments is **acclimatization.**

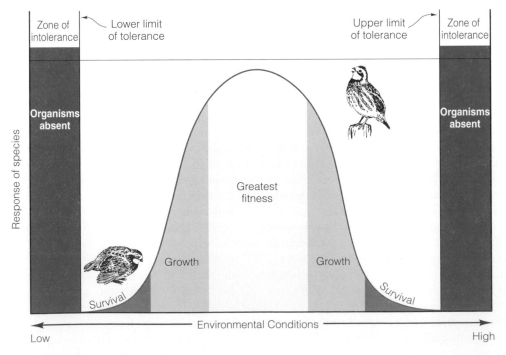

FIGURE 5.1   The law of tolerance

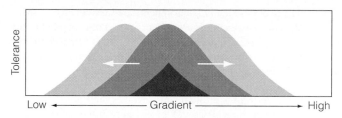

FIGURE 5.2 Seasonal shifts in tolerance ranges

The law of tolerance explains much about the local and geographical distribution of species. Most are restricted by a condition or conditions beyond the tolerable limits of the most sensitive stage of their life cycle. An organism may have a wide range of tolerance for some substance or condition, but a narrow range for another, and will thus be limited by the condition for which it has a narrow range. A mature tree, for example, may survive and grow under environmental conditions outside its natural range, but be unable to reproduce because its seedlings are outside their tolerance range. In some cases, a less than optimal state in one condition may lower the limits of tolerance for another; or the organism may achieve its best fitness at some intermediate point between interacting factors.

These variations in tolerance relate directly to differences in basic biochemical, physiological, morphological, and behavioral characteristics. The variation among species in these basic characteristics arises because no single set of characteristics enables an organism to function equally well under all environmental conditions. This constraint can be restated as the following premise:

> The set of characteristics (biochemical, physiological, morphological, and behavioral) that enables an organism to maximize its fitness under one set of environmental conditions limits its ability to do equally well (maximize fitness) under differing environmental conditions.

Hence as environmental conditions change, the set of characteristics that allow an organism to survive, grow, and reproduce will vary. Therefore, as the physical environment varies—for example, as conditions go from wet to dry—the characteristics that enable a plant to gain carbon, acquire essential resources for growth, and successfully reproduce will vary. We will use this premise as a framework for understanding the adaptations of organisms to the various features of the physical environment discussed in the coming chapters. In later parts of the book, we will examine the consequences of these patterns for interactions among species, the biological environment, and ultimately, species distribution and abundance.

HOMEOSTASIS AND FEEDBACK SYSTEMS

Faced with changing external environment, animals maintain a fairly constant internal environment They need some means of regulating internal conditions such as body temperature, water balance, pH, and amounts of salts in fluids and tissues,

relative to external conditions. The maintenance of a relatively constant internal environment in a varying external environment is called **homeostasis.**

Homeostasis involves the flow of external environmental information into a system with biological homeostatic devices or mechanisms that respond to changes. We are surrounded by all sorts of chemical and mechanical homeostatic devices. Consider the thermostat that controls the furnace. If we wish the temperature of the room to be 20°C (68°F), we set that point on the thermostat. When the temperature of the air falls below that point, a temperature-sensitive device within the thermostat trips the switch that turns on the furnace. When the temperature reaches the set point, the temperature-sensitive device responds by shutting off the furnace heating element. This type of response, halting or reversing the movement away from and returning it to the set point, is called **negative feedback.** The response of the system and its inputs are inversely related.

If the thermostat fails to function properly and does not shut the furnace off, then the furnace continues to burn and the temperature continues to rise, and eventually the furnace overheats. This continued movement way from the set point is **positive feedback.** The measure of this feedback is directly related to input. Examples of positive feedback are heat stroke, growth of cancerous cells, compound interest, and geometric growth of populations.

A key difference between mechanical and living systems is that in living systems the set point is not firmly fixed as it often is in mechanical systems. Instead organisms have a limited range of tolerances, called **homeostatic plateaus.** Homeostatic systems work within the maximum and minimum values by using negative feedback to regulate activity above the set point. If the system deviates from the set point, a negative feedback response ensues—a control mechanism inhibits any strong movement away from the set point. If it fails to do so, positive feedback drives the response away from the set point. Among animals, the control of homeostasis is both physiological and behavioral.

Among the homeotherms the maintenance of homeostasis is largely physiological, aided by behavioral and morphological mechanisms. To avoid summer heat, birds and mammals may seek shade; to keep warm in winter, they add an insulating layer of fur, feathers, or fat through physiological responses to changing daylengths. Poikilotherms maintain temperature homeostasis by gaining heat from or losing heat to the environment. Their control of homeostasis is largely behavioral. For example, poikilotherms may seek shade or warmth to maintain body temperatures. Many of their other homeostatic responses, however, are purely physiological, such as those of fish maintaining their internal osmotic pressure in an aquatic environment. Plants, too, employ homeostatic processes. These range from near instantaneous physical control of the uptake of carbon and the loss of water under different environmental conditions to acclimation to seasonal and annual variations of essential resources. The role of homeostasis will become quite evident in the material that follows.

AUTOTROPHS AND HETEROTROPHS

This general pattern of constraints and trade-offs on the adaptation of organisms to differing environmental conditions applies to a wide variety of essential processes related to survival, growth, and reproduction. But perhaps no process is more fundamental than the acquisition of that essential element, carbon. All life on Earth is composed of carbon-based compounds. To assimilate new tissues, meet the metabolic costs of maintenance, and ultimately produce offspring requires the acquisition of carbon.

Not all organisms derive this essential element from the same source. One fundamental classification of organisms in ecology is the dichotomy between autotrophs and heterotrophs. **Autotrophs** are organisms that utilize inorganic sources of carbon, primarily carbon dioxide. Comprising plants, algae, and certain bacteria, autotrophs use energy derived from solar radiation to fix carbon dioxide into organic compounds. In contrast, **heterotrophs** are organisms such as animals, bacteria, and fungi that use organic sources of carbon by consuming other organisms or their by-products. Autotrophs are often referred to as **primary producers,** and heterotrophs are termed **secondary producers** (see Chapter 24). Heterotrophic organisms are further classified into consumers and decomposers. **Consumers** are heterotrophic organisms that feed on other organisms, living or dead. In contrast, **decomposers** feed on dead organic matter or waste products.

The different ways in which organisms derive their source of carbon, particularly the division between autotrophs and heterotrophs, impose different sets of constraints that directly influence how these organisms interact with the environment. In the following chapters we will use this classification as a framework for examining the responses and adaptations of organisms to variations in the physical environment. Later in the text we will expand this framework to explore the consequences of these patterns for the distribution and abundance of organisms, and the cycling of energy and matter through the ecosystem.

Summary

1. To survive, function, and leave reproducing offspring, organisms must be adapted to their environment. Although the term *adaptation* has several meanings, ecologically it refers to any heritable behavioral, morphological, or physiological trait that maintains or increases the fitness of an organism under a given set of environmental conditions.

2. Organisms acquire their adaptations through the selective forces of the environment. The most adapted or most fit contribute to the next generation; the less fit do not. The key to adaptation is the genetic variability of local populations that exist under particular environments and have evolved genetic adaptations to them. Genetic diversity and phenotypic plasticity, the physical expression of the interaction between genotype and the environment, within a population enable individuals to respond to short-term or long-term changes in the environment.

3. The adaptiveness of organisms involves their ability to function within some upper and lower limits of the range of environmental conditions. The response of organisms to limiting environmental conditions is expressed in the law of tolerance as measured by fitness. The ranges of tolerance are not fixed but vary seasonally, within the total physiological tolerances of the organism. The ranges of tolerance influence distribution of organisms. Those organisms possessing a wide range of tolerances will be the most widely distributed.

4. To confront daily and seasonal environmental changes, organisms must maintain some equilibrium of their internal environment with the external one. The maintenance of a relatively constant internal environment in a varying external environment is called homeostasis. Homeostasis involves negative feedback responses. Through various sensory mechanisms, an organism responds physiologically or behaviorally to maintain an optimal internal environment relative to its external environment.

5. Characteristics that enable an organism to survive, grow, and reproduce under one set of environmental conditions inhibit its ability to do so under a different set of environmental conditions.

6. A fundamental adaptation of all organisms is the acquisition of carbon, because all life on Earth is composed of carbon-based compounds. Life can be divided into two broad groups based on their means of carbon acquisition: autotrophs that use inorganic sources of carbon and solar energy to fix carbon into organic compounds; and heterotrophs that acquire organic carbon by feeding on other organisms, living or dead.

Review Questions

1. Define adaptation.
2. How do organisms become adapted to their environment? How does adaptation relate to natural selection?
3. Contrast the law of the minimum with the law of tolerance.
4. What are ecotypes?
5. Define homeostasis. How does homeostasis function through negative feedback?
6. Diagram and explain a simple homeostatic mechanism involving a sensor (input), control center, and effector (output) in a vertebrate animal. Explain how this mechanism helps the organism to cope with its environment.
7. Discuss how adaptations might influence the distribution of plants and animals.

Cross References

Plant responses to moisture, 104–112; animal responses to moisture, 137–139; plant responses to temperature, 99–104; animal responses to temperature, 127–137; plant adaptations to light intensity, 90–94; photoperiodism, 94; seasonality, 95; nutrients and plants, 112–118.

C H A P T E R   6

Plant Adaptations to the Environment I: Photosynthesis and the Light Environment

CONCEPTS

1. Photosynthesis, the process by which the energy of the sun is used in the fixation of CO_2 into carbohydrates, consists of light and dark reactions.

2. Carbon dioxide enters the plant through the stomata or openings in the leaf; water escapes through transpiration by the same route.

3. Plants employ three types of photosynthesis with varying methods of carbon fixation: C_3, C_4, and CAM.

4. How plants allocate carbon to the synthesis of new tissue is important to their survival, growth, and reproduction.

5. Rates of photosynthesis respond to increases and decreases in photosynthetically active radiation (PAR).

6. Plant species exhibit adaptations to high-light or low-light conditions.

7. Plant response to changing daylengths influences the seasonal patterns of growth and reproduction.

The ultimate source of carbon for all living organisms is carbon dioxide (CO_2) in the atmosphere. However, not all living organisms can utilize this abundant form of carbon directly. Only one process can transform carbon in the form of carbon dioxide into organic molecules and living tissue. That process, carried out by autotrophic organisms—namely green plants, algae, and photosynthetic bacteria—is photosynthesis. It is essential for the maintenance of life on Earth. In this chapter we will examine the basic processes of photosynthesis and respiration in autotrophic organisms, and how the physical environment influences these processes. We will then integrate these basic processes to the whole plant, and examine the array of adaptations and responses that allow plants to survive, grow, and reproduce under differing environmental conditions. These patterns will form the framework for later discussions of population interactions and community and ecosystem dynamics.

PHOTOSYNTHESIS

Photosynthesis is the process by which energy from the sun—in the form of short-wave radiation—is harnessed to drive a series of chemical reactions that results in the fixation of CO_2 into carbohydrates (simple sugars) and the release of oxygen (O_2) as a by-product.

The process can be expressed in the simplified form

$$6CO_2 + 12H_2O \longrightarrow C_6H_{12}O_6 + 6O_2 + 6H_2O$$

The net effect of this reaction is the utilization of six moles of water (H_2O) and the production of six moles of oxygen (O_2) for every six moles of CO_2 that is transformed into one mole of sugar ($C_6H_{12}O_6$). The synthesis of various other carbon-based compounds—such as proteins, fatty acids, and enzymes—from these initial products occurs both in the leaves and in other parts of the plant.

Light and Dark Reactions

Photosynthesis, a complex sequence of metabolic reactions, can be separated two processes, often referred to as the light and dark reactions. The light reaction is the initial photochemical reaction where light energy is trapped in absorbing pigments called chlorophyll within the chloroplast—hence the name *light reaction*. The absorption of a photon of light raises the energy level of the chlorophyll molecule. The excited molecule is not stable; the electrons return rapidly to their ground state, releasing the absorbed photon energy in several ways. It can be released as heat in a process called thermal dissipation. It can be emitted as a photon of lower energy content (higher wavelength), a phenomenon called fluorescence. Or the electron excitation energy can be transferred to another acceptor molecule, resulting in a process called photosynthetic electron transport. The latter process results in the synthesis of adenosine triphosphate (ATP) from adenosine diphosphate (ADP) and NADPH from nicotinamide adenine dinucleotide phosphate ($NADP^+$). The high-

energy substance ATP and the strong reductant NADPH produced in the light reactions are essential for the second step in photosynthesis—the dark reactions.

In the dark reactions, CO_2 is biochemically incorporated into carbohydrates—simple sugars. The reactions are called dark because these processes do not directly require the presence of sunlight—they can occur in the dark. They depend, however, on the products of the light reactions and, therefore, on the essential resource driving photosynthesis—sunlight.

The process by which CO_2 is incorporated into simple sugars begins in most plants when the five-carbon molecule ribulose biphosphate (RuBP) combines with CO_2 to form two molecules of a three-carbon compound called 3-PGA (phosphoglyceric acid) (Figure 6.1); this initial reaction is called carboxylation. It is catalyzed by the enzyme **rubisco,** ribulose biphosphate carboxylase-oxygenase. (Rubisco is the most abundant enzyme on the planet.) The plant quickly converts the PGA formed in this process into an energy-rich sugar molecule, glyceraldehyde-3-phosphate (G3P). The synthesis of G3P from PGA requires both ATP and NADPH, the high-energy molecule and reductant that are formed in the light reactions. Some of the G3P is utilized to produce simple sugars [$(CH_2O)_6$], starches, and other products required for the growth and maintenance of the plant; the remainder is used to synthesize new RuBP to continue the process. The synthesis of new RuBP from G3P requires additional ATP. Hence the availability of light energy (solar radiation) can limit the dark reactions of photosynthesis through its control on the production of ATP and NADPH required for the synthesis of G3P and the regeneration of RuBP. This photosynthetic pathway involving the initial fixation of CO_2 into the three-carbon PGA molecules is called the Calvin-Benson cycle or the C_3 cycle (Figure 6.1), and plants employing it are known as C_3 plants.

The C_3 pathway has one major drawback. The enzyme rubisco that drives the process of carboxylation also acts as an oxygenase—it can catalyze the reaction between O_2 and RuBP. This reaction forms a two-carbon molecule, phosphoglycolate, and one molecule of 3-PGA, in contrast to the two molecules of 3-PGA formed in the reaction with CO_2. Further metabolism of phosphoglycolate into 3-PGA results in the release of CO_2. The oxygenation of RuBP and the eventual release of CO_2 are called **photorespiration.** It is a competitive reaction to the carboxylation process, and it thus reduces the efficiency of C_3 photosynthesis.

Uptake of Carbon Dioxide

The chemical reactions outlined in the preceding discussion assume the presence of CO_2 in the leaf, at the physical site where photosynthesis occurs. The plant's uptake of CO_2 from the atmosphere is one of the key processes in understanding the interactions between plants and the physical environment. In terrestrial plants, the CO_2 in the atmosphere (which we will refer to as C_a) diffuses or enters the plant through openings on the surface of the leaf called **stomata.** The CO_2 concentration inside the leaf (at the site where photosynthesis occurs) will be denoted C_i. The diffusion or movement of CO_2 from

Step 1
Carbon fixation. An enzyme called rubisco combines three molecules of CO_2 with three molecules of a five-carbon sugar called ribulose bisphosphate (abbreviated RuBP). Six molecules of the three-carbon organic acid 3–phosphoglyceric acid (3-PGA) result.

Step 2
Energy consumption and redox.
Two chemical reactions (indicated by the two arrows) consume energy from six molecules of ATP and oxidize six molecules of NADPH. Six molecules of 3-PGA are reduced, producing six molecules of the energy-rich three-carbon G3P.

Step 3
Release of one molecule of G3P.
Five of the G3P from step 2 remain in the cycle. The single molecule of G3P you see leaving the cycle is the net product of photosynthesis. A plant cell uses two G3P molecules to make one molecule of glucose, which has six carbons. Since the Calvin cycle incorporates only one molecule of CO_2 — and thus only one carbon — at a time, it takes six complete turns of the cycle to make two molecules of G3P that go into one glucose molecule.

Step 4
Regeneration of RuBP.
A series of chemical reactions uses energy from ATP to rearrange the atoms in the five G3P molecules, forming three RuBP molecules. These can start another turn of the cycle.

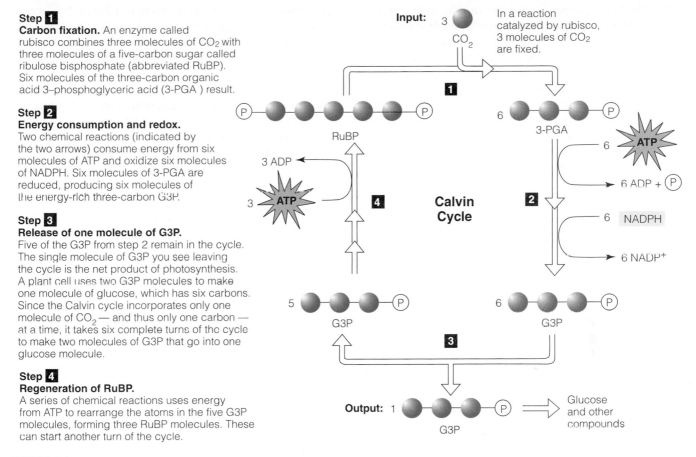

FIGURE 6.1 Calvin-Bensen cycle. (From Campbell et al. 1997.)

the air outside the leaf into the leaf is driven by the concentration or diffusion gradient $(C_a - C_i)$. Substances flow from areas of high concentration to areas of low concentration until they achieve an equilibrium (that is, the concentrations in the two areas are equal). As photosynthesis occurs, and CO_2 is combined with RuBP to form PGA, the concentration of CO_2 inside the leaf decreases, increasing the diffusion gradient, and CO_2 moves through the stomata into the leaf. As long as photosynthesis occurs, the CO_2 concentration inside the leaf draws down, maintaining the gradient $(C_a - C_i)$.

There are several sources of resistance to the flow of CO_2 from the external environment into the leaf and to the site of carboxylation within the mesophyll. The most important of these is stomatal resistance (r_s), which is a function of both the density and aperture of stomata on the leaf surface. The reciprocal of stomatal resistance is stomatal conductance (g_s). Conductance expresses the rate of flow or exchange and is typically described in units of mol/m²/s. The rate of CO_2 uptake, referred to as the photosynthetic or assimilation rate (A), can now be expressed as

$$A = (C_a - C_i)/r_s = (C_a - C_i) \times g_s$$

where the uptake of CO_2 (A) is a function of the diffusion gradient divided by the resistance to flow from the external environment into the leaf. Assimilation rate is most often

expressed in units of CO_2 uptake per unit leaf area per unit time (e.g., μmol/m²/s).

If photosynthesis were to stop and the stomata were to remain open, then CO_2 from the outside air would diffuse into the leaf until the internal CO_2, C_i, was equal to the outside concentration, C_a. Then no further movement of CO_2 into the leaf would occur. But this is not how plants function. As photosynthesis and the demand for CO_2 is reduced (for whatever reason), the stomata will tend to close, reducing the rate of flow of CO_2 into the leaf. The stomata close not only because terrestrial plants take up CO_2 through the stomata, but also because they lose water through the stomata to the surrounding atmosphere.

Transpiration

The loss of water through the stomata is called **transpiration.** It represents a major constraint on the uptake of CO_2 by terrestrial plants. The rate at which water moves from the leaf into the surrounding air is controlled by physical processes similar to those involved in the diffusion of CO_2 into the leaf. The transpiration rate is related to the diffusion gradient of water vapor from inside to outside the leaf. As with CO_2, water vapor will diffuse or move from areas of high concentration to those of low concentration (from wet to dry). For all practical purposes, the air inside the leaf is saturated with water. The flow of water from inside the leaf to the outside air thus will be a function of the amount of water vapor in the air—the humidity (see Chapter 2). Because the air inside the leaf is saturated, the vapor pressure deficit represents the water gradient from the leaf interior to the atmosphere; thus the vapor pressure deficit drives transpiration.

The diffusion gradient from the interior of the leaf to the outside air is approximately two orders of magnitude greater for water than for CO_2, and this creates a major problem for plants. For the leaf to function, it must replace water lost through the stomata with water taken up through the root system and transported to the leaf. To reduce the loss of water through transpiration, the stomata tend to close as the vapor pressure deficit rises (Figure 6.2). The increased stomatal resistance (decreased conductance) acts to reduce water loss, but it also reduces the uptake of CO_2.

We can now appreciate the trade-off faced by terrestrial plants. To carry out photosynthesis the plant must open the stomata to take up CO_2, but at the same time the plant will lose water through the stomata to the outside air, water that must be replaced through the plant's roots. If the availability of water to the plant is limited, the plant is faced with balancing the opening and closing of the stomata to allow for the uptake of CO_2 while minimizing the loss of water through transpiration. This balance between photosynthesis and transpiration is an extremely important constraint that influences the characteristics of plants and the responses of ecosystems under differing environmental conditions.

The major difference between terrestrial and aquatic plants in CO_2 uptake and assimilation is the lack of stomata in aquatic plants. CO_2 diffuses from the atmosphere into the surface waters, followed by mixing into the water column. Once dissolved, CO_2 reacts with the water to form bicarbonate (HCO_3^-). This reaction is reversible, and the concentrations of CO_2 and bicarbonate tend toward a dynamic equilibrium. In aquatic plants there is a direct diffusion of CO_2 from the boundary layer (waters adjacent to the leaf) across the cell membrane. Once inside, the process of photosynthesis proceeds in much the same manner as that outlined for terrestrial plants.

One difference is that some aquatic plants can also use bicarbonate as a carbon source. However, the plants must first convert it to CO_2 using the enzyme carbonic anhydrase. This conversion can occur in two ways: (1) active transport of bicarbonate into the leaf followed by conversion to CO_2; or (2) excretion of the enzyme into adjacent waters (boundary layer) and subsequent uptake of converted CO_2 across the membrane. Because the diffusion of CO_2 in water is 10^4 times slower than in the air, the major limitation on diffusion is the boundary layer resistance. This constraint can be particularly important in quiescent waters such as dense seagrass beds or rocky intertidal pools.

Dark Respiration

Photorespiration in C_3 plants reduces net photosynthesis by 30 to 50 percent and occurs only in photosynthetic cells in the light. Photorespiration is often referred to as light respiration, because it involves the utilization of oxygen and the release of carbon dioxide. However, this process is different from what physiologists refer to as true or dark respiration. Dark respiration involves the oxidation of carbohydrates (e.g., glucose) to generate energy in the form of ATP:

$$(CH_2O)_6 + 6O_2 \longrightarrow 6CO_2 + 6H_2O + ATP$$

Dark respiration takes place exclusively in the mitochondria.

Dark respiration can occur in all living cells, and it was believed to take place only in the dark. However, more recent work has shown that the process continues in photosynthetic cells in the light, at approximately 5 to 15 percent the rate of photosynthesis.

Dark respiration is often partitioned into two components: (1) growth and synthesis (R_g), and (2) maintenance (R_m). Respiration associated with growth and synthesis depends directly on the rate of photosynthesis. Maintenance respiration is proportional to the dry weight of the living tissue and is temperature sensitive, as discussed in the following section.

Alternative Photosynthetic Pathways: C_4 and CAM Plants

Some plant species that inhabit warmer and drier climates have an alternative photosynthetic pathway to that described for C_3 plants. The process, called C_4 photosynthesis, is facilitated by an internal leaf anatomy different from that of C_3 plants (Figure 6.3). C_4 plants possess two distinct types of photosynthetic cells: the mesophyll cells and the bundle sheath cells. The bundle sheath cells surround the veins or vascular bundles.

In C_4 plants, CO_2 reacts with the acceptor molecule phosphoenolpyruvate (PEP), a three-carbon compound, within the mesophyll cells (Figure 6.5). This reaction is in

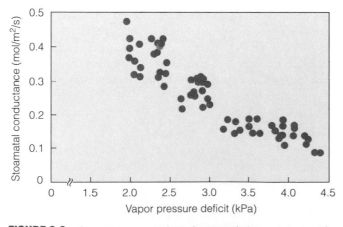

FIGURE 6.2 Change in stomatal conductance (g_s) as a function of vapor pressure deficit (VPD) for red maple (*Acer rubrum*). Increasing vapor pressure deficit indicates declining relative humidity of the atmosphere. As the air becomes drier, stomata close to prevent water loss (decreasing stomatal conductance).

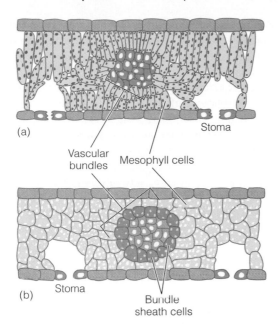

(a)

Stoma

Vascular bundles   Mesophyll cells

(b)   Stoma

Bundle sheath cells

FIGURE 6.3   Cross-sectional view of leaf of (a) a C_3 plant and (b) a C_4 plant. Note the bundle sheath cells surrounding the vascular bundle in the C_4 leaf.

contrast to the initial carboxylation involving RuBP in C_3 plants. The carboxylation in C_4 plants is catalyzed by the enzyme PEP carboxylase, producing oxaloacetate (OAA) as the first product of fixation. The OAA is then very rapidly transformed into malic and aspartic acids, which are four-carbon molecules. These products then move to the bundle sheath cells. Here the plant enzymatically breaks down malic and aspartic acids to release CO_2, a process called decarboxylation. The CO_2 is then fixed again within the bundle sheath cells using the C_3 pathway involving RuBP and rubisco. The remaining pyruvic acid diffuses back to the mesophyll cells, where it is used to regenerate PEP.

The initial fixation of CO_2 by PEP carboxylase acts to concentrate CO_2 within the bundle sheath cells, where carboxylation via the C_3 pathway occurs. As a result the concentration of CO_2 within the bundle sheath cells can reach 8 to 10 times as high as that in either the mesophyll cells or the surrounding atmosphere.

The extra step in the fixation of CO_2 gives C_4 plants certain advantages. PEP carboxylase has a much higher affinity for CO_2 than does rubisco, which has a tendency to catalyze O_2. This eliminates the process of photorespiration in the mesophyll cells. The PEP carboxylase in the mesophyll cells of C_4 plants fixes CO_2 in the outer leaf tissue and releases CO_2 to the bundle sheath cells in the inner tissues. This process draws down the CO_2 in the mesophyll to very low concentrations and increases the concentration of CO_2 in the bundle sheath cells. The high CO_2 concentrations in the bundle sheath cells reduce photorespiration (by favoring carboxylation over oxygenation), while the low concentrations within the mesophyll create a steep diffusion gradient for CO_2, increasing the fixation of CO_2 for a given stomatal conductance. The result is a higher rate of net photosynthesis than in C_3 plants.

The C_4 photosynthetic pathway is not found in algae, bryophytes, ferns, gymnosperms, or the more primitive angiosperms. C_4 species are mostly grasses native to tropical and subtropical regions, and some shrubs and dicot herbaceous plants of arid and saline environments. Grasses make up more than half of the known C_4 species. Most North American species of C_4 grasses have a subtropical distribution (Terri and Stowe 1976, Terri 1979). No known C_4 grasses grow on the tundra. The C_4 pathway occurs in unrelated taxa, so that it evolved several times (Ehleringer and Monson 1993), probably in response to past periods of lower atmospheric CO_2 concentrations.

Yet another photosynthetic process, known as the crassulacean acid metabolism (CAM), is found among a number of succulent, semidesert plants in some 15 families, including Cactaceae (Figure 6.5), Euphorbiaceae, and Crassulaceae (from which the method of carbon fixation receives its name). CAM plants resemble the C_4 pathway in that CO_2 is initially fixed into four-carbon compounds using PEP and is subsequently decarboxylated and refixed using RuBP and rubisco (see Figure 6.5). However, CAM plants do not have the specialized bundle sheath cells. In contrast, CAM plants open their stomata in the cooler temperatures of night, when the vapor pressure deficit is at its minimum. Such behavior functions to reduce water loss through transpiration. During the night, PEP carboxylase fixes CO_2 and accumulates large quantities of malic acid in the vacuoles of the mesophyll cells (see Figure 6.5). During the day, the plants close their stomata to reduce water loss. Malic acid is decarboxylated, and the resulting CO_2 is fixed using the conventional C_3 pathway.

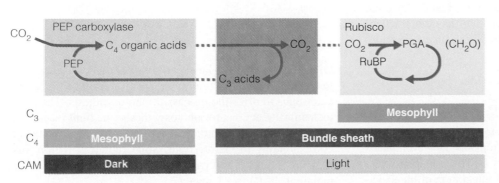

CO_2 PEP carboxylase C_4 organic acids CO_2 Rubisco CO_2 PGA   (CH_2O)

PEP   RuBP

C_3 acids

C_3

C_4 Mesophyll   Bundle sheath

CAM   Dark   Light

Mesophyll

FIGURE 6.4   Comparison of the basic features of C_3, C_4, and CAM photosynthetic pathways. Carboxylation reactions in C_4 plants take place in different cells; in CAM plants, they take place in the same cells but at different times of day. (Adapted from Jones 1992.)

FIGURE 6.5   Cactus, like this saguaro (*Cereus*), use the CAM cycle of photosynthesis and store available water in their tissues.

The CO_2 concentration in the leaf remains high during the day because it cannot diffuse through the closed stomata. As a result, rubisco operates in a high CO_2 environment, virtually eliminating photorespiration.

Although appearing to be similar in function to C_4 plants, CAM plants differ in that they separate the initial fixation of CO_2 from the conventional C_3 pathway at different times, rather than in different locations as in C_4 plants (Osmond 1978). The CAM pathway is slow and inefficient in the fixation of CO_2, but it acts to dramatically reduce water loss through transpiration.

CARBON ALLOCATION

Thus far we have focused on the processes of photosynthesis and respiration in the leaves of green plants. However, plants are not composed only of leaves; they also have supportive tissues such as stems and roots. In the integration of photosyn-

thesis and respiration within the whole plant, the balance of carbon uptake and loss is a direct function of the relative contribution of these different tissues to the total mass of the plant.

The total carbon uptake per unit time will be a function of the average rate of assimilation (net photosynthesis) per unit leaf area (photosynthetic surface) multiplied by the total surface area of photosynthetic tissue (leaf area). The total loss of carbon in respiration per unit time will be a function of the total mass of living tissue—that is, the sum of leaf, stem, and root tissues. The net carbon gain (photosynthesis − respiration) is then allocated to a variety of processes. Some of the carbon will be used in maintenance respiration, and the rest in the synthesis of new tissues in the process of plant growth.

How the carbon is allocated will have a major influence on the survival, growth, and reproduction of the plant (see Chapter 13). Different plant tissues are involved in the acquisition of essential resources necessary to support photosynthesis and growth. Leaf tissue is the photosynthetic surface, providing access to the essential resources of solar radiation and CO_2. Stem tissue provides vertical support, elevating leaves above the substrate and increasing access to solar radiation in dense stands of plants. It also provides the conductive tissue necessary to mobilize water and nutrients to the leaves. The root tissue provides access to below-ground resources such as water and nutrients in the underlying substrate. As we shall discuss in the following sections, allocation of carbon to the production of these tissues will be influenced by the availability of these essential resources for plant growth.

Under ideal conditions, the allocation of carbon to the further production of leaf tissue will promote the fastest growth. Increased allocation to leaf tissue increases the photosynthetic surface, which increases the rate of carbon uptake as well as carbon loss due to respiration. Allocation to nonphotosynthetic tissue such as stem and root increases the respiration rate but does not directly increase the capacity for carbon uptake. But the allocation of carbon to the production of stem and root tissue is essential for the acquisition of key resources necessary for carbon uptake. As these resources become scarce, it becomes increasingly necessary to allocate carbon to the production of these tissues at the cost of the production of leaves and the associated increase in photosynthetic capacity. The implications of these shifts in patterns of carbon allocation will be addressed in the following chapter.

LIGHT

Of the energy in the solar spectrum that reaches Earth's surface (see Figure 3.1), only a fraction can be used by autotrophs to drive the process of photosynthesis. The photochemical reactions for photosynthesis (the light reactions) use light energy within the range of 400–740 mm, the range corresponding to visible light. Solar radiation within this range of values is referred to as photosynthetically active radiation (PAR) (see Chapter 3).

Response of Net Photosynthesis to Variation in PAR

The actual quantity or flux of PAR is called the photon flux density and is typically measured in units of $\mu mol/m^2/s$. The function relating the net exchange of CO_2 (net photosynthesis) for a plant with variation in PAR is referred to as the **light response curve** (Figure 6.6). In the absence of PAR, photosynthesis ceases and (dark) respiration continues, resulting in a net release of CO_2 by the plant. As light becomes available, photosynthesis begins, offsetting the loss of CO_2 via respiration. Photosynthesis will continue to increase with increasing PAR as a direct function of the augmented production of ATP and NADPH in the light reactions (both compounds required in the regeneration of RuBP). At some value of PAR, the rate of CO_2 uptake in photosynthesis exactly offsets the loss of CO_2 in respiration. The resulting rate of net photosynthesis (net CO_2 uptake) is zero. This value of PAR is called the **light compensation point**. As the value of PAR increases above the light compensation point, the rate of photosynthesis increases relative to respiration and therefore the rate of net photosynthesis increases. The rate of net photosynthesis will continue to increase with increasing PAR until some other factor, such as the diffusion of CO_2 into the leaf, begins to limit photosynthesis. At this point, the rate of increase in photosynthesis with increasing PAR will begin to decline. Eventually a point is reached at which any further increase in PAR results in no further increase in photosynthesis. This value of PAR is called the **light saturation point,** and the corresponding value of net CO_2 uptake is called the light-saturated rate of net photosynthesis. Because of their high potential rates of carboxylation, many C_4 plant species do not exhibit light saturation, even under conditions of full sunlight.

In some plants, values of PAR above the light saturation point may actually result in a decline in net photosynthesis. This decline is referred to as **photoinhibition.** Photoinhibition may be in part due to an accumulation of electrons over and above those that can be transferred to NADP in the light reactions of photosynthesis (Hall and Rao 1994). Generally the effects of photoinhibition are reversible, with no perma-

nent damage resulting if values of PAR are reduced quickly. However, long-term exposure of these species to high values of PAR can result in permanent damage.

Plant Response to Reduced PAR

The immediate response of a leaf to reduced light levels (PAR below light saturation) is a reduction in the rate of net photosynthesis per unit of leaf area (see Figure 6.7). However, plants respond to a long-term reduction in PAR in a number of different ways, involving shifts in their biochemistry, physiology, and morphology. When leaves are grown under reduced light conditions, rates of photosynthesis are limited primarily by the photochemical reactions and their direct influence on the rates of RuBP regeneration. As a result, lower quantities of the enzyme rubisco, which catalyzes the fixation of CO_2, are required. The lower costs of producing rubisco, together with other compounds necessary to maintain the processes of photosynthesis and plant growth, reduce overall rates of respiration. Recall that the light compensation point is defined as the value of PAR at which the loss of CO_2 in respiration exactly offsets the rate of CO_2 uptake in photosynthesis. The result of reduced rates of respiration in leaves grown under shaded conditions is a reduction in the light compensation point. That is, leaves grown under shaded conditions typically can continue to maintain positive rates of CO_2 uptake under lower values of PAR. However, the processes resulting in the reduction in the light compensation point come with a cost. As a result of the lower levels of rubisco, the light saturation point is also reduced, reflecting the lower rates of carboxylation that can be maintained. In addition, the light-saturated rate of net photosynthesis is much reduced (Figure 6.7).

In addition to the changes in photosynthetic physiology, leaves grown under reduced light conditions often exhibit a different morphology. In general, leaves grown under reduced light conditions are larger (in surface area) and thinner than those grown under high light levels. The increased surface area facilitates the capture of the limiting resource—light.

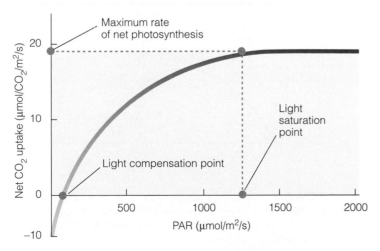

FIGURE 6.6 Response of net CO_2 uptake rate to photosynthetically active radiation (PAR). Net photosynthesis increases with increasing PAR up to a maximum, referred to as the light saturation point. A further increase in PAR results in no further increase in net photosynthesis; in some cases, it can result in a decline in net photosynthesis, a process called photoinhibition. The light compensation point is the value of PAR at which photosynthesis equals respiration and net photosynthesis is zero. For values of PAR below the light compensation point, respiration dominates and there is a net loss of CO_2 by the leaf.

The changes in leaf-level physiology and leaf morphology described above can occur on different leaves of the same plant, as with leaves from the top and bottom of the plant canopy. However, when a plant is grown under shaded conditions, a number of changes in carbon allocation also occur. In addition to producing broader, thinner leaves, plants grown under shaded conditions allocate more of the fixed carbon to leaf production and less to the roots. This shift in allocation is reflected in the decrease in the ratio of root mass (g) to leaf area (cm²) as light decreases (Figure 6.8). Just as with the changes in leaf morphology (thinner, broader leaves), this shift in allocation from roots to leaves increases the surface area for the capture of the limiting resource of light.

Plant Adaptations to Variation in the PAR Environment

Although the shifts in physiology, morphology, and carbon allocation outlined above can occur among individuals of the same species when grown under high- versus low-light conditions, the most marked differences in these characteristics are among different species that are adapted to either sun or shade environments (Figure 6.9). Species that are adapted to high-light environments are often referred to as **shade intolerant**, whereas species adapted to shaded environments are called **shade tolerant**. These names refer directly to the observed difference in the light compensation points for these two types of plants. The difference between shade tolerant and shade intolerant species in their ability to continue photosynthesis and

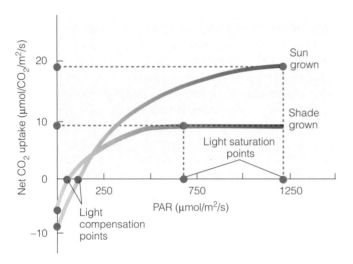

FIGURE 6.7   Patterns of photosynthetic response to light availability for leaves growing in sun and shade conditions. Shade leaves have a lower light compensation point and a lower light saturation point than sun-grown leaves.

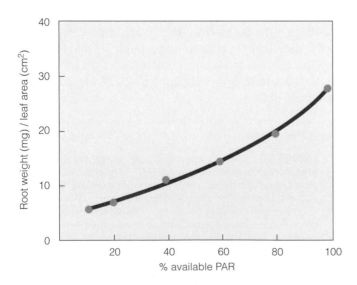

FIGURE 6.8   Changes in the ratio of root mass (mg) to leaf area (cm²) for individuals of *Eucalyptus dives* grown under different levels of photosynthetically active radiation (PAR). As light levels are reduced, the allocation to root tissue declines relative to the production of leaves. The increased allocation to leaves, together with the production of thinner leaves, acts to increase the photosynthetic surface area for the capture of light. (Smith et al. 2000)

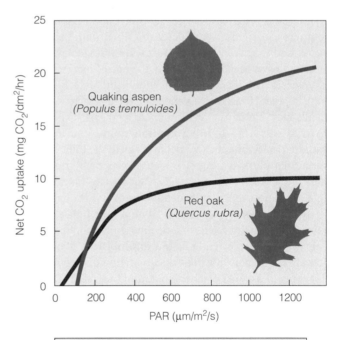

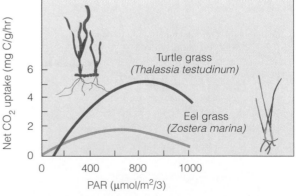

FIGURE 6.9   Photosynthetic light response curves for shade-adapted (shade tolerant) and sun-adapted (shade intolerant) species from (a) terrestrial (trees) and (b) aquatic habitats (seagrasses).

growth under reduced light levels has a direct impact on their ability to survive under shaded conditions.

Augspurger (1982) examined the influence of light availability on seedling survival and growth for a variety of tree species that grow in the tropical rain forests of Panama. Her findings show a marked difference between shade tolerant and shade intolerant species in their patterns of seedling survival and growth when grown under high- and low-light conditions (Figure 6.10). Shade tolerant species show little difference in survival and growth rates under sun and shade conditions. In contrast, both survival and growth rates of shade intolerant species were dramatically reduced under shaded conditions. These observed differences are a direct result of the difference in light compensation point and whole-plant carbon balance resulting from the differences in patterns of carbon allocation discussed earlier.

The dichotomy in adaptations between shade tolerant and shade intolerant species reflects a trade-off between characteristics that enable a species to maintain high rates of CO_2 uptake and growth under high-light conditions and the ability to continue survival and growth under low-light conditions. The changes in biochemistry, physiology, leaf morphology, and carbon allocation exhibited by shade tolerant species enable them to reduce the amount of photosynthetically active radiation required to maintain a positive carbon balance. However, these same characteristics limit their ability to maintain high rates of CO_2 uptake and growth when light levels are high. In contrast, plants adapted to high-light environments can maintain high rates of photosynthesis and growth under high-light conditions, but at the expense of continuing photosynthesis, growth, and survival under reduced light conditions.

Because of the rapid attenuation of light with depth, most aquatic plants live in the equivalent of a shaded environment, but with a major difference. Far infrared wavelengths are strongly absorbed by water, but not by terrestrial foliage. The red-to-far red ratio in water is between 3.6 and 7 to 1, compared with 1.15 to 1 for the forest interior.

Aquatic plants can so modify the light environment for plants in the lower water column that they greatly reduce PAR. Because phytoplankton species can become photoinhibited at high light intensities at the surface, especially in sunny weather, they have higher photosynthesis rates at greater depths. Some species of phytoplankton move up and down through the water column to escape the inhibitory effects of high light and to reach the depth at which light intensity is most favorable. Photosynthetic ability is further influenced by seasonal variations in water temperature and light, and by the adaptation of phytoplankton and macrophytes to different wavelengths of visible light. Most macrophytes grow at the depth at which light intensity is most favorable.

Response to Ultraviolet Radiation

Photosynthetically active radiation is not the only portion of the solar spectrum with a direct influence on plant adaptations. In recent years the stratospheric ozone layer has become diminished by increasing concentrations of human-made, ozone-destroying chemicals (see Chapter 26). This depletion of stratospheric ozone allows increased levels of ultraviolet radiation, especially UV-B radiation, to reach the Earth's surface. Laboratory and greenhouse experiments show that UV-B radiation can damage DNA, partially inhibit photosynthesis, alter the growth form of plants, and reduce yield (Caldwell et al. 1989, Tevini and Teramura 1989,

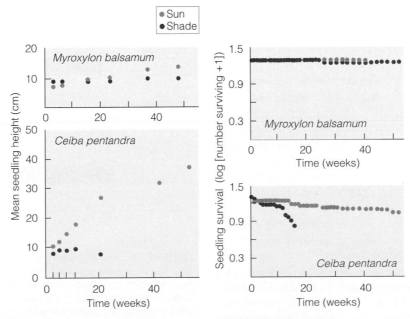

FIGURE 6.10 Seedling survival and growth over a period of one year for seedlings of two tree species on Barro-Colorado Island, Panama, grown under sun and shade conditions. *Ceiba pentandra* is a shade intolerant species; *Myroxylon balsamum* is shade tolerant. (From Augspurger 1982.)

Teramura 1990). These damaging effects, however, have not been clearly demonstrated in field-grown plants.

Plants have evolved defenses against UV-B radiation reaching the interiors of their leaves. One line of defense is leaf reflectance. The major barriers to UV-B penetration are the epidermal cells, containing anthocyanins, colorless flavonoids, and other phenols that absorb UV-B radiation yet transmit PAR to the interior of the leaf.

Plants exhibit a wide range in their ability to screen ultraviolet radiation. Tropical and alpine plants, naturally exposed to high levels of ultraviolet radiation, more effectively block UV-B radiation than temperate species. However, even within these plants, some attenuate UV-B better than others. Day et al. (1992) compared the screening ability of a group of Rocky Mountain plants growing at high elevations. The leaf epidermis of herbaceous dicotyledonous plants was the least effective. Between 18 and 41 percent of UV-B radiation reached the interior of the leaf. The needles of conifers were the most effective, attenuating essentially all incident UV-B radiation. This effectiveness is partly because UV-B absorbing pigments are found not only in cell vacuoles but also in epidermal cell walls. Between the two extremes are grasses and woody dicots. These researchers attributed the differences to the cost-effectiveness of such defenses. Because their longer-lived needles are exposed to larger lifetime doses of UV-B radiation, conifers may have evolved highly effective screening mechanisms. On the other hand, grasses and herbaceous and woody dicots with short-lived seasonal leaves would not have evolved such screening properties.

PERIODICITY AND PLANT PROCESSES

Although we have been emphasizing the role of light as an energy source for plants, light has another major role in the life of plants. It serves as a timing mechanism to keep plant activity in tune with the daily and seasonal changes in their environment.

In the northern and southern latitudes, the daily periods of light and dark lengthen and shorten with the seasons. The activities of organisms are geared to the changing seasonal rhythms of night and day. Most organisms of temperate regions have reproductive periods that closely follow changing daylengths of the seasons. Trilliums (*Trillium* spp.) and violets bloom in the lengthening days of spring before the forest leaves are out and an abundance of sunlight reaches the forest floor. Asters (*Aster* spp.) and goldenrods (*Solidago* spp.) flower in the shortening days of fall.

The signal for these responses is **critical daylength.** When the duration of light (or dark) reaches a certain portion of the 24-hour day, it inhibits or promotes a photoperiodic response of an organism to changing daylength (Figure 6.11). Critical daylength varies among organisms, but it usually falls somewhere between 10 and 14 hours. Through the year plants and animals compare that time scale with the actual length of day or night. When the actual daylength or nightlength is greater or lesser than the critical value, the organism responds

appropriately. Some organisms can be classified as **day-neutral,** affected not by daylength but rather by some other influence such as rainfall or temperature. Others are **short-day** or **long-day** organisms. Short-day organisms are those whose reproductive or other seasonal activity is stimulated by daylengths shorter than their critical daylength. Long-day organisms are those whose seasonal responses such as flowering and reproduction are stimulated by daylengths longer than their critical daylength.

Horticulturists exploit these short-day and long-day responses of plants to force them to come in bloom through the year. When they hold plants under short-day and long-night conditions, short-day plants are stimulated to flower and long-day plants are inhibited from flowering. When they increase daylength, short-day plants do not flower and long-day plants come into bloom. If the dark period (the subjective night in the circadian rhythm; see Chapter 8) of a short-day and a long-day plant is interrupted, each plant responds as if it had been exposed to a long day. The long-day plant flowers and the short-day plant does not. In reality short-day and long-day plants respond not to the length of light but to the

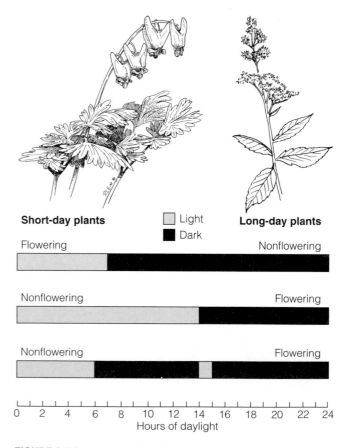

FIGURE 6.11   The influence of photoperiod on the time of flowering in long-day and short-day plants. If exposure to light is experimentally controlled, short-day plants are stimulated to flower under short-day conditions, are inhibited from flowering under long-day conditions, and respond to an interruption of a long dark period as though they had been exposed to long-day conditions. Long-day plants do not flower under short-day conditions, only under long-day conditions or interrupted short-day conditions.

length of darkness. The two might more accurately be called long-night and short-night plants (Figure 6.11).

Just as chlorophyll is involved in the fixation of light energy in the process of photosynthesis, another colored pigment, phytochrome, is the key to the plant's ability to detect changes in light and dark and measure the seasons. Phytochrome comes in two forms: P_r, which absorbs red light; and P_{fr}, which absorbs far red light. When P_r absorbs red light, it converts to P_{fr}; and when P_{fr} absorbs far red light, it converts to P_r. The plant synthesizes phytochrome P_r. If the plant is kept in the dark, P_r remains in that form and any P_{fr} remaining reverts to P_r in the dark. Thus after sunset each day any P_{fr} remaining converts to P_r. At sunrise P_r rapidly converts to P_{fr}. The conversion comes about because sunlight is richer in red than in far red light. The time that elapses between the conversion of P_{fr} to P_r and the conversion of P_r to P_{fr} measures the photoperiod and synchronizes plant activity to daily cycles of light and dark.

Responses to changing daylength are also seasonal, reflected in the seasonal activities of plants. Who is not aware of the seasonal changes in plants—the unfolding of leaves in spring and the dropping of leaves in fall, the blooming of flowers and the ripening of seeds? These and other biological events recurring with the passage of seasons and influenced by the attendant interaction of light, temperature, and moisture show the phenomenon of **seasonality.** The study of the causes of the timing of these events, the biotic and abiotic forces affecting them, and the interrelations among phases of the same or different species is called **phenology** (Leith 1974).

Seasonality in temperate and arctic regions results largely from changes in light and temperature. Seasonality in tropical regions is keyed to rainfall. In a very broad way seasonal changes in temperature and light regimes result in alternate warm and cold periods. However, the progression is gradual, and in temperate zones seasons can be identified as early or late spring, early or late fall, and so on. In the tropics the seasons are alternately wet and dry and their onset may be gradual or abrupt. The beginning of the rainy season is a dependable environmental cue by which plants and animals of the tropics can become synchronized to seasonal changes. In some tropics areas the division between the dry season and the wet season is marked by 100 mm in rainfall. The dry season in tropical rain forest has less than 100 mm rainfall per month. In extreme conditions, of course, no rain falls. The wet season has more than 100 mm rainfall per month. The onset of the rainy season, which may last up to six months, varies with the movement of the intertropical convergence. For this reason the wet season and the dry season are predictable.

Phenological responses reflect altitudinal and latitudinal changes in light, temperature, and moisture. The advance of spring in the temperate regions is marked by progressively later flowering of the same species of trees and herbs across a region, depending on elevation. Although these progressive changes are most pronounced across broad geographical areas (Figure 6.12), distinct variations, easily observed especially in mountainous country, exist within a given region. Leaf emergence and flowering of trees advance upslope in spring, and fall coloration moves downslope in fall. These variations reflect local microclimates that act as selection pressures on local populations, resulting in ecotypic variations in environmental response.

In tropical regions flowering and fruiting and leafy growth of plants reflect the alternation of wet and dry seasons

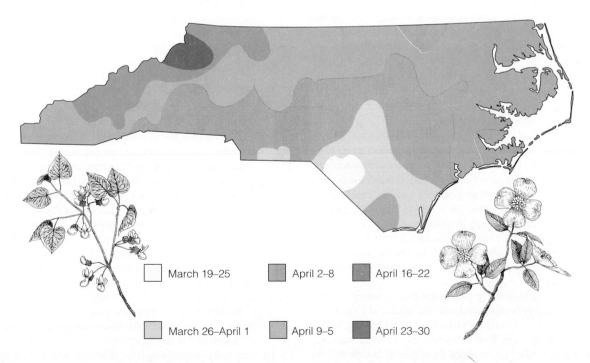

March 19–25   April 2–8   April 16–22

March 26–April 1   April 9–5   April 23–30

FIGURE 6.12 The arrival of spring 1970 across North Carolina, as indicated by the opening of the flowers of dogwood (*Cornus florida*) and redbud (*Cercis canadensis*). (From Leith 1974.)

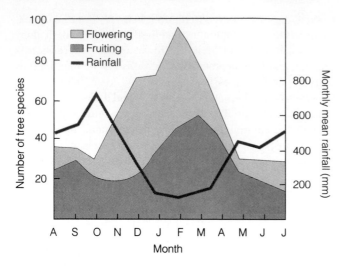

FIGURE 6.13   Synchronization of flowering and fruiting in rain forest tree species in Golfito, Costa Rica, with mean monthly rainfall. Note that flowering and fruiting reach their highest levels during the dry season, the months of January, February, and March. (Adapted from Janzen 1967.)

(Figure 6.13). The coming of the rainy season is marked by a flush of vegetative growth just as warming spring temperatures trigger leafy growth in temperate regions. Over much of the seasonal tropics, flowering and fruiting coincide with the dry season (Janzen 1967). Some species flower at the end of the rainy season, when soil moisture is still high. Other species flower at the end of the dry season.

Summary

1. Photosynthesis is the process by which autotrophs use the energy of the sun to convert carbon dioxide and water into carbohydrates. It is a two-step process involving the light and dark reactions. In the light reaction, light energy is trapped by absorbing pigments. In the dark reaction, CO_2 is biochemically incorporated into simple sugars.

2. Most plants use the Calvin or C_3 cycle. It involves the formation of three-carbon phosphoglyceric acid, 3-PGA, used in subsequent reactions, from a five-carbon molecule, RuBP. Involved is the enzyme rubisco, the most abundant enzyme on the planet. The C_3 cycle has one major drawback. Rubisco, acting as an oxygenase, catalyzes a reaction between O_2 and RuBP to form a two-carbon molecule and one molecule of PGA. This reaction is called photorespiration and competes with C_3 respiration, reducing its efficiency.

3. Terrestrial plants take up CO_2 through openings in leaves called stomata. The movement or diffusion of CO_2 from the outside air into the leaf is driven by a concentration or diffusion gradient.

4. The opening of stomata results in the loss of water through transpiration, a cost associated with the uptake of CO_2. However, transpiration also serves to dissipate excess energy, avoiding excessively high leaf temperatures. Lacking stomata, aquatic plants take up CO_2 from the water. Although they do not experience transpiration, aquatic plants experience a very slow diffusion of CO_2 from the water into the leaf.

5. Other plants, called C_4, use a four-carbon process in which carbon dioxide taken into the leaf reacts to form malic or aspartic acid, stored in the mesophyll cells. The CO_2 fixed in these compounds is then released and fixed once again using the Calvin cycle in the bundle sheath cells.

6. C_3 and C_4 plants possess structural and physiological differences that are important ecologically. C_4 plants carry on photosynthesis at higher leaf temperatures, higher light intensities, and lower CO_2 concentrations than C_3 plants.

7. Succulent plants of semiarid deserts use the crassulacean acid metabolism (CAM). These plants open their stomata and fix CO_2 as malic acid by night. By day they close their stomata and use both fixed and respiratory photosynthesis in the Calvin cycle.

8. Some of the carbon fixed by the plant is used in respiration. The rest goes to the synthesis of new tissue (plant growth). How much carbon is allocated to leaf tissue, stem, roots, flowers, fruit, and seed has a major influence on the plant's survival, growth, and reproduction.

9. The reduction of light directly reduces rates of photosynthesis through its impact on light reactions. To compensate for reduced photosynthetic rate per unit leaf area, plants allocate more carbon to the production of thinner, broader leaves. This shift in allocation and leaf morphology acts to compensate for the reduced photosynthesis per unit leaf area by increasing the photosynthetic surface area.

10. Plants that function best under different light intensities may be classified as shade intolerant (sun plants) or shade tolerant (shade plants). Each group is adapted to certain light regimes. Shade tolerant plants and leaves have low photosynthetic, respiratory, and metabolic growth rates. Sun plants and leaves have a high respiration rate and are adapted to high light intensity.

11. Plants are exposed to ultraviolet radiation as well as photosynthetically active radiation. Depletion of the ozone layer in the stratosphere by human-made pollution is allowing increased penetration of ultraviolet light. Many plants, especially those of the tropics and alpine regions and conifers, possess defenses against ultraviolet light. Flavonoids and phenolics in the epidermal cells of leaves absorb UV-B wavelengths before they reach and damage the interior of the leaf.

12. Daily and seasonal activities of plants are influenced by changing daylength. The signal for these responses is the critical daylength, the temporal point of reference that organisms use to compare the actual length of day or night. A colored pigment, phytochrome, is the key to the plant's ability to detect changes in light and dark and measure the seasons; this pigment is sensitive to red and far red light. Critical daylength, which varies among organisms, usually falls between 10 and 14 hours. Daily periods of light and dark gradually change with the seasons. When the actual daylength or nightlength is greater or lesser than the critical value, the organism responds accordingly. Organisms whose reproductive or other seasonal activity is stimulated by daylengths shorter than their critical daylength are short-day organisms. Long-day organisms are those whose seasonal responses such as flowering and reproduction are stimulated by daylengths longer than the critical daylength. Organisms not influenced by daylength are day-neutral.

13. Seasonal changes in light together with accompanying changes in temperature and precipitation bring about seasonality (seasonal periodicity) in recurring biological events, the study of

which is phenology. In temperate regions seasonality is influenced by changes in light and temperature; in tropical regions, by changes in rainfall. Such periodicities are reflected in leaf growth, flowering, and fruiting of plants.

Review Questions

1. What is rubisco?
2. What is photorespiration? How does it involve the enzyme rubisco? How does it influence the net uptake of CO_2 (assimilation rate)?
3. What process controls the flow of CO_2 from the outside air into the leaf?
4. What trade-off do terrestrial plants face when opening stomata to take in CO_2?
5. How do submerged aquatic plants and terrestrial plants differ in their uptake of carbon?
6. What is PEP carboxylase? How do C_4 and C_3 plants differ in their leaf anatomy?
7. Why do CAM plants open their stomata at night and close them during the day?
8. What is (dark) respiration?
9. How does light influence/limit the rate of photosynthesis?
10. What is the light compensation point? Light saturation point?
11. Why does the rate of photosynthesis increase with increasing PAR above the light compensation point?
12. How do the light compensation and light saturation points differ between shade tolerant and shade intolerant species?
13. How do patterns of carbon allocation differ for plants grown under high- versus low-light conditions? Why?
14. What are some defenses against UV radiation evolved by plants?
15. If UV-B radiation increases, what might the effects be on tropical plants? Temperate plants? Alpine and arctic plants?
16. Would you expect shade tolerance to be constant for a given species? Why or why not?
17. Many garden catalogs classify perennial plants according to their ability to grow in sun, partial shade, and shade. What problems do gardeners encounter if they ignore the indicated light preferences and plant a sun-loving (shade intolerant) plant in the shade of trees? (Consider the growth response of the plant.)
18. What is critical daylength and what is its significance for the flowering of plants?
19. Contrast the roles of the pigments chlorophyll and phytochrome in plants. How does the sensitivity to the light spectrum differ between these pigments?
20. What is a short-day response? A long-day response?
21. What is the relationship of seasonality to photoperiod in plants? How do you suppose this relationship is influenced by climatic gradient?
22. What is phenology?

Cross References

Solar radiation, 21; light, 43; adaptation, 81; intertropical convergence, 26; circadian rhythms 140; biological clocks, 140; photosynthesis and temperature, 100.

CHAPTER 7

Plant Adaptations to the Environment II: Thermal, Moisture, and Nutrient Environments

CONCEPTS

1. Plants maintain a thermal balance with the environment through the processes of evaporation and convection, influenced by leaf size and shape and stomatal control.

2. Photosynthesis and respiration respond directly to variations in temperature.

3. Plant growth reflects variations in temperature over the course of a year or growing season.

4. Plants respond physiologically to the stresses of heat and cold.

5. Plants gain water from the soil through a water potential gradient from the soil to the atmosphere maintained by transpiration.

6. Plants respond physiologically to both deficits and excesses in moisture.

7. The close relationship between nutrient availability and plant growth is reflected in net photosynthesis and carbon allocation.

8. Mutualistic relationships between bacteria and mycorrhizal fungi aid plants in their acquisition of nutrients.

PLANTS AND THE THERMAL ENVIRONMENT

Plants live in a thermal environment. As pointed out in Chapters 2 and 3, the thermal environment is dynamic in both time and space. At any given location, temperatures vary both diurnally and seasonally as a function of the input of solar (or short-wave) radiation. However, air temperatures do not necessarily reflect the temperatures experienced by plant tissues. Plants are constantly absorbing short-wave and long-wave radiation from the surrounding environment as well.

Thermal Energy Balance

To understand the thermal environment experienced by plants, one must examine the balance between the inputs of energy and the means by which this energy is utilized and dissipated to the surrounding environment. Biophysicists and ecologists use two different models to examine the energy balance of plants and plant canopies—radiative and turbulent transport. In our discussion, we will use the turbulent transport model because of its explicit consideration of leaf morphology and the process of transpiration, two topics that feature prominently in the subsequent discussion of plant adaptations to varying environmental conditions.

The net energy absorbed by a plant (per unit time) is referred to as the plant's net radiation balance (R_n). Plants both absorb and reflect solar radiation and absorb and emit long-wave radiation. The resulting net radiation balance can be calculated as the difference between the inputs and outputs of solar (short-wave) and long-wave radiation. Of the net radiation absorbed by a plant only a small proportion, usually less than 5 percent, is used in photosynthesis and stored as energy in chemical bonds (M). The remainder goes into physical storage (S), which includes energy used in heating the plant tissues as well as that used to raise the temperature of the air within the canopy (boundary layer). To prevent temperatures from rising to critical levels, excess energy must be dissipated to the surrounding environment as thermal energy. The two modes by which plants dissipate heat energy are convection (C) and evaporation (E) (see Chapter 3). The net energy balance of the plant can therefore be partitioned as

$$R_n = M + S + (C + \lambda E)$$

where λ is the latent heat of vaporization, the energy required to transform a unit of water (e.g., a mole) from liquid to vapor phase.

For plants, evaporation (E) includes both transpiration and direct evaporation of water from the leaf surface (evapotranspiration), although transpiration is by far the dominant mode. The ability to dissipate heat via evaporation is influenced by the same factors presented earlier in our discussion of transpiration: stomatal conductance and vapor pressure deficit. As stomata open to allow for the uptake of CO_2, water is lost via transpiration. As a result, significant heat energy is dissipated from stomatal opening associated with photosynthesis.

Convective heat loss (C) is a function of the temperature difference between the object and the surrounding fluid (air or water). Heat energy is transferred from areas of high to areas of low energy. For heat to be dissipated through convective means, the leaf temperature must be higher than that of the surrounding air. An additional factor controlling convective heat loss is the object's conductance to thermal exchange, which is highly influenced by the boundary layer.

As you will recall from Chapter 3, the boundary layer is the layer of air adjacent to the surface of an organism or object. As an object loses heat, the air within the boundary layer warms and transfers the heat to the surrounding atmosphere. How much heat is lost or gained is a function of the rate of exchange of the air between these two environments influenced by the boundary layer conductance.

Two major factors influence the boundary layer conductance to heat exchange in plants: the physical environment and the size and shape of the leaves. The boundary layer conductance to heat exchange increases with wind speed. As air flows over a leaf, mixing occurs between the boundary layer and the adjacent air, causing the temperature of the two air masses to converge. If the air is still, the air within the boundary layer warms as a result of heat being dissipated from the leaf, resulting in a divergence between the temperatures of the two air masses. The effect of wind speed on the sensible temperature within the boundary layer is what is commonly referred to as the wind-chill factor. The dissipation of heat to the surrounding air through convection is a function of the difference in temperature between an object (leaf) and the surrounding air—the air within the boundary layer. As heat is dissipated it warms the boundary layer and reduces the temperature differential, therefore reducing further heat loss. As wind speed increases, the air within the boundary layer mixes with surrounding air. Two things occur: (1) the temperature that the organism (leaf) senses within the boundary layer is the same as the surrounding air, rather than warmer; and (2) the drop in temperature within the boundary layer results in a further dissipation of heat. This cooling of the boundary layer and additional loss of body heat acts to cool the organism (leaf).

The second important factor influencing boundary layer conductance and the dissipation of heat through convection is the size and shape of the organism (leaf). The easiest example of the influence of morphology on convective heat exchange is the radiator. Radiators are designed to increase the ratio of surface area to volume. This functions to increase the surface area for heat exchange as air passes over the surface. Smaller, more lobed leaves are more effective at heat exchange than are larger, less lobed leaves (Figure 7.1)—they exhibit a higher boundary layer conductance to heat exchange. This is one reason that plants in hot dry environments have on average smaller leaves than species that occupy cooler and wetter environments.

The dissipation of excess energy through these two processes, transpiration (evaporation) and convection, enables plants to keep leaf temperatures from rising to critical values. Plants have a direct influence on the maintenance of their energy balance and leaf temperatures—in the short term

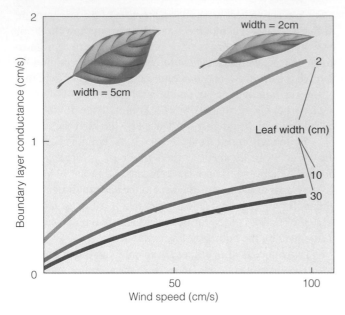

FIGURE 7.1   Variations in leaf morphology that have a direct influence on the leaf boundary layer conductance and the dissipation of heat via convection. Smaller and more lobed leaves increase the surface area (per unit leaf volume or mass) for heat exchange and help in the development of turbulence and mixing in the boundary layer.

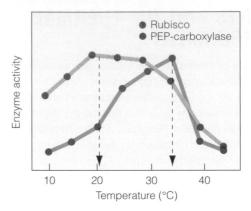

FIGURE 7.2   Comparison of the activity of the photosynthetic enzymes rubisco and PEP carboxylase as a function of leaf temperature. Rubisco activity has a direct influence on rates of carbon fixation and CO_2 uptake for C_3, C_4, and CAM plants, whereas PEP carboxylase is found in C_4 and CAM plants only. Temperature optima for the two enzymes are shown by the vertical arrows.

through stomatal control on transpiration, and in the long term (development and evolution) through variations in leaf morphology. However, the physical environment imposes constraints on the ability of plants to dissipate excess heat energy via both transpiration and convection. Air temperatures will directly influence heat loss in both transpiration and convection, while wind speed has a major influence on the boundary layer environment. Water, which is lost via transpiration, must be replaced by uptake of water from the soil; therefore, patterns of precipitation and soil moisture will directly influence the plant's energy balance. We shall discuss these constraints in more detail in a later section addressing plant water balance.

Thermal Effects on Photosynthesis and Respiration

Both photosynthesis and respiration respond directly to variations in the thermal environment. The rate of carbon fixation—carboxylation—is a direct function of rubisco activity, which is sensitive to temperature (Figure 7.2). Likewise, both photorespiration and dark respiration are temperature sensitive. Increased temperature favors oxygenation over carboxylation, thus increasing rates of photorespiration relative to photosynthesis. In addition, dark respiration rates increase as a function of temperature (Figure 7.3). The resulting patterns of photosynthesis and respiration as a function of temperature can be seen in Figure 7.4. As temperatures rise above freezing, both photosynthesis and respiration rates increase. Initially, photosynthesis increases faster than respiration. As temperatures continue to rise, the photosynthetic rate reaches

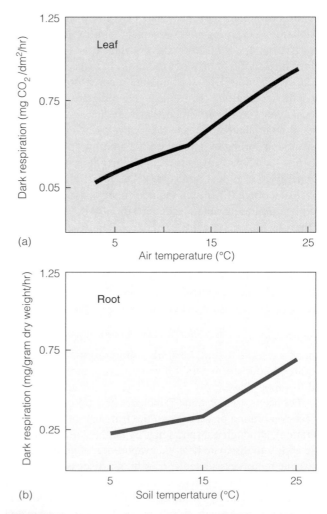

FIGURE 7.3   Variations in dark respiration for (a) leaf and root (Lawrence and Oechel, 1983) (b) tissues of quaking aspen (*Populus tremuloides*) as a function of air and soil temperature, respectively.

a maximum related to the temperature response of rubisco. As temperatures continue to rise, the photosynthetic rate declines and the respiration rate continues to increase. As temperatures rise further, even respiration declines as temperatures reach critical levels. The temperature response of net carbon gain, or net photosynthetic rate, can be calculated as the difference between the rate of carbon uptake in photosynthesis and the rate of carbon loss in respiration. Three cardinal values describe the temperature response curve: T_{min}, T_{opt}, and T_{max}. T_{min} and T_{max} are respectively the minimum and maximum temperatures at which net photosynthesis approaches zero—no net carbon uptake. T_{opt} is the temperature or range of temperatures over which net carbon uptake is at its optimum (i.e., maximum).

The photosynthetic temperature response curves for a number of terrestrial plant species are shown in Figure 7.5. Note that the species vary in the range of temperatures over which net photosynthesis is at its maximum, T_{opt}. In fact, the differences in T_{opt} for the species correspond to differences in the thermal environments that the species inhabit. Species found in cooler environments typically have lower T_{min}, T_{opt}, and T_{max} than species that inhabit warmer climates. These differences in the temperature response of carbon uptake are directly related to a variety of biochemical and physiological adaptations that act to shift the temperature responses of photosynthesis and respiration toward the prevailing temperatures in the environment.

C_3 and C_4 plants exhibit consistent differences in their photosynthetic responses to temperature (Figure 7.6). C_4 plants typically have a higher range of T_{opt} than that of C_3 plants. The higher temperature optima for photosynthesis in C_4 plants is directly related to the differences between PEP carboxylase and rubisco, the two enzymes that catalyze carboxylation (see Chapter 6). As temperatures rise, rubisco favors oxy-

genation over carboxylation, increasing photorespiration and reducing net photosynthesis. The lack of photorespiration in the initial fixation of CO_2 by PEP carboxylase in C_4 plants eliminates this problem. Second, the temperature optima for PEP carboxylase activity is higher than that for rubisco. As a result, the temperature optima for the C_3 pathway corresponds to that of rubisco, whereas the temperature optima for the C_4 pathway corresponds to the range of temperatures in which the activity of both enzymes is relatively high (see Figure 7.2).

Although species from different thermal habitats exhibit different temperature responses for photosynthesis and respiration, these responses are not fixed. When two groups of individuals of the same species are grown under different thermal conditions in the laboratory or greenhouse, divergence in the temperature response of net photosynthesis is often observed (Figure 7.7). In general, the range of temperatures over which net photosynthesis is at its maximum shifts in the direction of the thermal conditions under

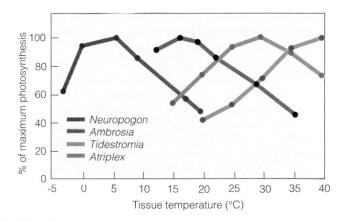

FIGURE 7.5   Relationship between net photosynthesis and temperature for a variety of terrestrial plant species from dissimilar thermal habitats: *Neuropogon acromelanus* (Arctic lichen), *Ambrosia chamissonis* (cool, coastal dune plant), *Atriplex hymenelytra* (evergreen desert shrub), and *Tidestromia oblongifolia* (summer-active desert perennial). (From Mooney et al. 1976.)

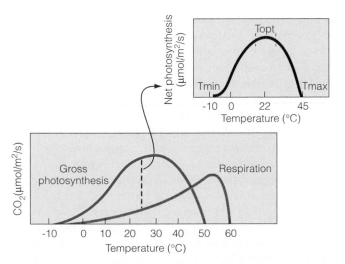

FIGURE 7.4   Generalized functions describing the relationship between leaf temperature and the processes of photosynthesis and respiration. The difference between photosynthesis and respiration at any given temperature defines the rate of net CO_2 uptake, net photosynthesis (inset).

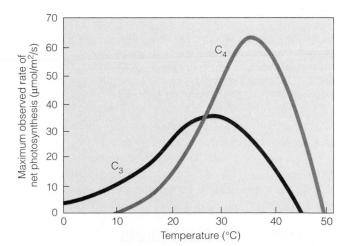

FIGURE 7.6   Temperature sensitivities of the maximum rates of net photosynthesis for C_3 and C_4 photosynthesis. (From Woodward and Smith 1994.)

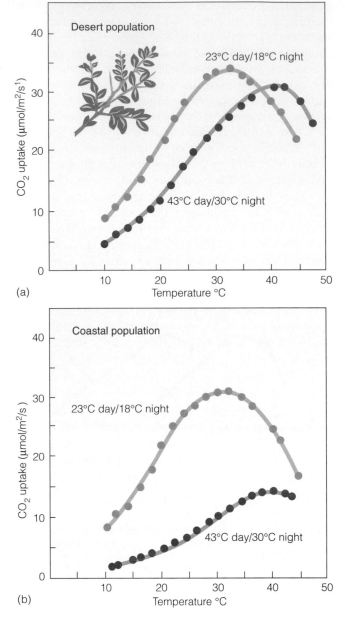

(a)

(b)

FIGURE 7.7   The relationship between temperature and net photosynthetic rate for cloned plants of big saltbush (*Atriplex lentiformis*) grown under two different day/night temperature regimes. Plants were grown under high light (1700–1800 μmol/m²/s) and normal atmospheric concentrations of CO_2 and O_2. Note the shift in T_{opt} corresponds to the temperature conditions under which the plants were grown. (After Pearcy 1977:485.)

which the plant is grown—that is to say, individuals grown under cooler temperatures exhibit a lowering of T_{opt}, whereas those individuals grown under warmer conditions exhibit an increase in T_{opt}. This same process of acclimation can be observed as seasonal shifts in the temperature response of net photosynthesis (Figure 7.8).

Temperature and Plant Growth

As temperatures vary over the course of the day, and from day to day with the changing of the seasons, the relative rates of photosynthesis and respiration likewise vary. The accumula-

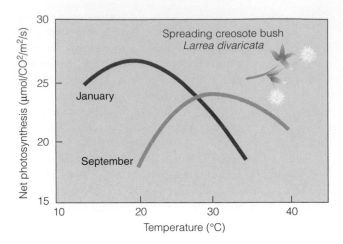

FIGURE 7.8   Seasonal shift in the relationship between net photosynthesis and temperature for spreading creosote bush (*Larrea divacarta*) shrubs growing in the field. Note that T_{opt} shifts to match the prevailing air temperatures. (After Mooney et al. 1978.)

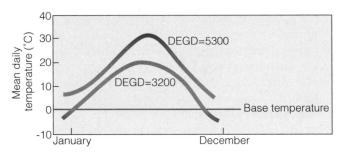

FIGURE 7.9   Mean daily temperature profiles and growing degree-days for two locations using a base temperature of 0°C. Growing degree-days (DEGD) is the sum of the departures of temperatures above some minimum or base temperature.

tion of carbon and the growth of individual plants will reflect these variations in temperature over the course of the year or growing season. A measure that is commonly used to relate variations in temperature over a single growing season to plant growth is the index of **degree-days**. The index of degree-days is calculated as the sum of the departures in temperatures above some **minimum** *or* **base temperature** (Figure 7.9). The base temperature typically reflects T_{min}, the temperature at which net photosynthesis is at or approaching zero. The integration of mean daily temperatures above the base temperature therefore reflects photosynthetic activity and carbon accumulation as it relates to temperature. Values of degree-days have been correlated with growth rate for species in both managed (agricultural crops and forest plantations) and natural ecosystems, as well as related to the geographic distribution of species (Figure 7.10).

Extreme Temperatures and Plant Survival

A number of critical temperatures have a direct effect on plant cell structure and metabolic processes. However, many studies

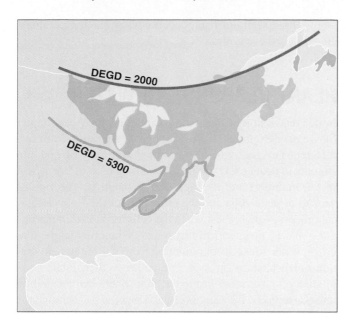

FIGURE 7.10 The geographic range of (*Betula alleghaniensis)* and the growing degree-day (variable DEGD) values of 2000 and 5300, which closely approximate the north and south boundaries of this species. Range map from Fowells (1965); isotherms used to compute the DEGD lines from the U.S. Department of Commerce publication (1968). (Figure from Botkin, et al [1972]. © 1972 by International Business Machines Corporation. Reprinted with permission.)

show that the annual minimum temperature may effectively limit plant distribution by exceeding their threshold for survival. Low-temperature thresholds for mortality for 220 arboreal species are shown in Figure 7.11, featuring three critical temperatures that may be crucial in controlling the distribution of plants. When the minimum temperature falls in the region of 0 to 10°C, mortality occurs in plants that are chill sensitive. These species are typically broadleaf evergreen.

Freezing temperatures can result in the formation of intra- and extracellular ice as well a phase change in membrane lipids. Various plant factors have been associated with the ability to tolerate freezing temperatures, including increased solute concentrations, lipid composition, and small cell size. The plant growth regulator, abscisic acid, has also been implicated in the development of freezing resistance.

The next significant threshold is at −15°C, the lowest temperature at which the majority of broadleaf evergreen species can survive. No specific process appears to coincide with this limit, although an increased probability of intracellular ice formation may be critical.

In areas where the minimum temperature falls in the range of −15 to −40°C, the dominant plant life form is broadleaf, winter-deciduous. These species depend on the ability to avoid intracellular ice formation by supercooling. The equilibrium state is for ice to form when the temperature falls below the freezing point depression appropriate for the solute concentration. Typically, cells can lower the freezing point by no more than 3°C by increasing solute concentrations. However, contents rarely freeze, in part because of the absence of suitable ice nucleation sites within the cells, allowing the water to remain in the unstable supercooled condition. Similarly, the cell wall water may remain liquid far below the theoretical freezing point, though in most plant tissues only a few degrees of supercooling can be achieved. Ice nucleation usually starts in the extracellular water, either because of low solute concentrations or the presence of ice-nucleating bacteria. Some hardwood species, including various species of oak, elm, maple, and dogwood, can supercool to the homogeneous nucleation temperature at which ice forms without a requirement for nucleation sites. This temperature occurs between −41 and −47°C for plant tissue. A number of species can survive temperatures below −40°C, often with no observable limit for survival. The majority of these species are conifers that occur in the far northern latitudes.

Tolerance to freezing is not uniformly distributed through a plant. Roots, bulbs, and rhizomes are the most sensitive to

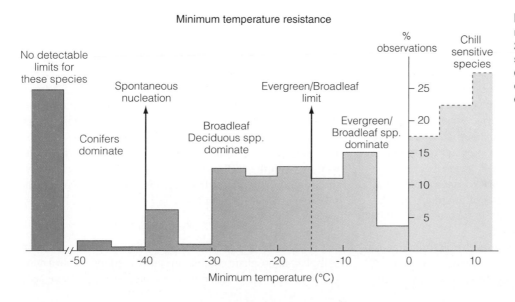

Minimum temperature resistance

FIGURE 7.11 Observations on the minimum temperature resistance of 220 arboreal species. Dashed lines show the projected response to chilling temperatures; arrows refer to observable limits (see text for discussion). (Woodward 1986.)

freezing, succumbing to temperatures between −10 and −30°C. Buds of woody plants, too, vary in their tolerance. Terminal buds of trees are less resistant to cold than lateral buds; most resistant are the basal reserve buds on twigs that will replace new spring growth killed by a late frost. Woody stems are more cold resistant than leaves and buds.

Further resistance to chilling and frost damage is obtained by insulation. Some species of arctic and alpine flowers, and some early spring flowers of temperate regions, possess hairs that act as heat traps and prevent cold injury. The interior temperature of cushion-type and rosette plants may be 20°C higher than the surrounding air.

Although species differ in their high-temperature tolerance, in general leaf temperatures of 45°C result in the disruption of cell metabolism. The underlying basis for acclimation to high temperatures is not well understood. Although heat shock proteins are likely to play a role, the details are not clear. Some species of plants, notably cacti, can acclimate to high temperatures. They can carry on protein synthesis at a sufficiently high rate in the face of rising temperatures to equal the protein breakdown rate and thus avoid ammonia poisoning (Steponkus 1981). Under heat stress some plants orient their leaves parallel with rather than horizontal to the sun's rays or fold their leaves at midday, thus reducing the surface area exposed to solar radiation (Mooney et al. 1977, Ehleringer et al. 1976, Ehleringer 1980, Ehleringer and Forseth 1980). Many desert plants accomplish the same result by possessing small leaves or no leaves at all, carrying on photosynthesis through their stems. Some desert plants possess a dense coating of white plant hairs that reflect visible light (Ehleringer et al. 1976).

Processes Other than Survival and Growth

We have focused our discussion of temperature response on survival and growth. Temperature, though, also affects other processes such as germination and reproduction. Understanding the role of temperature on the distribution and abundance of plant species is complicated by the fact that the temperature responses and tolerances for germination and survival may differ from those exhibited for growth.

Certain temperature thresholds induce flower formation, and still different temperatures are effective at bringing about the development and unfolding of the flowers. Winter annuals and biennials, as well as the buds of certain woody plants, require a cold winter season in order to flower normally in the spring ("chilling requirement"). They do not become ready to flower until they have been exposed for weeks to temperatures between −3 and +13°C.

Fruits and seeds require more heat to ripen than necessary for growth of the vegetative parts of the plant. For this reason vegetative reproduction may be favored in habitats with short or cooler growing seasons.

The range of temperatures required for germination of spores and seeds must correspond to external temperatures that guarantee sufficiently rapid development for the young

plant. Often seed dormancy occurs when the temperatures are outside these limits.

PLANT RESPONSE TO WATER

In our discussion of terrestrial plants thus far (see Chapter 6), we have examined the process of transpiration from two different perspectives: (1) the inevitable cost of opening stomata to take up CO_2, and (2) the dissipation of latent heat in the maintenance of plant energy balance. The rate of growth of plant cells and the efficiency of their physiological processes are highest when the cells are at maximum turgor—that is, they are fully hydrated. When turgor pressure drops, water stress occurs, ranging from wilting to dehydration and mechanical stresses. For the leaves to maintain maximum turgor, the water lost to the atmosphere in transpiration must be replaced by water taken up from the soil through the root system of the plant and transported to the leaves. The dual role of stomata in CO_2 uptake and water loss (transpiration) leads to the inevitable conflict between the maintenance of turgor and high photosynthetic rates as water availability declines.

Water Uptake and the Soil-Plant-Atmosphere Continuum

As the leaf loses water via transpiration, the turgor of the leaf cells drops, setting up a pressure gradient from the leaf to the root-soil surface resulting in water movement from the soil into the root, and from the root through the conductive tissue to the leaf (Figure 7.12). The movement of water through the soil-plant-atmosphere continuum is passive, a response to the pressure gradient. In the case of transpiration, the exchange of water from the leaf interior to the atmosphere is a function of the hydrostatic pressure gradient represented by the difference in the absolute water vapor pressure between the inside of the leaf and the outside air (see discussion of transpiration in Chapter 6). However, in the various transfers of water from the soil to the root, and from the root to the leaf, forces other than hydrostatic pressure are involved. For this reason, the concept of water potential is used to describe the pressure gradient. The symbol used to represent water potential is ψ, and it is defined in pressure units, generally megapascals (MPa). As with hydrostatic pressure, water flows from areas of high to areas of low water potential. In nature values of water potential typically range from zero, when water is freely available, to increasingly negative values. However, positive values are possible. Therefore, for water to flow from the soil into the roots, through the conductive tissues into the leaf, and from the leaf to the atmosphere, the following condition must hold:

$$\psi_{atmosphere} < \psi_{leaf} < \psi_{root} < \psi_{soil}$$

The total water potential (ψ_w) of any compartment of the soil-plant-atmosphere continuum can be partitioned into a number of components, with one being hydrostatic pressure (ψ_p),

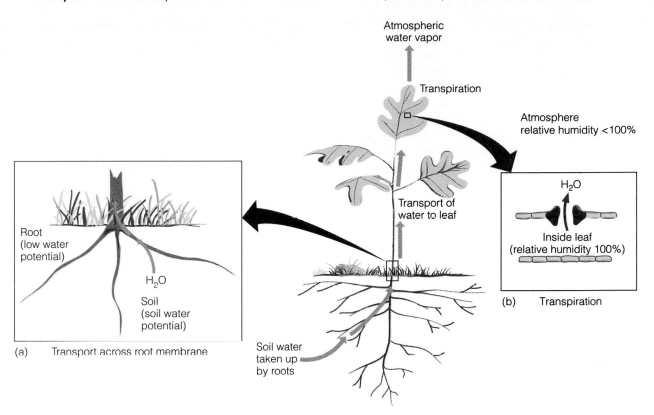

FIGURE 7.12 Transport of water along a pressure gradient from soil to leaves to atmosphere.

or physical potential. The other components are osmotic potential (ψ_π), matric potential (ψ_m), and gravitational potential (ψ_g):

$$\psi_w = \psi_p + \psi_\pi + \psi_m + \psi_g$$

We shall discuss each component in turn as it relates to the maintenance of the flow of water through the continuum.

For water to move or diffuse from soil solution into the roots, it must pass through the cell membranes of the root. Some membranes are permeable; they do not impede the movement of substances through them. Others are selectively permeable; they allow some substances to pass through them but not others. Such membranes are fairly permeable to water and are more or less permeable to other substances.

The general tendency of molecules is to move from a region of high concentration to one of low concentration. This movement or diffusion accounts for the spread of a solute (dissolved substance) throughout a solvent (the medium that dissolves the solute). To understand the movement of water and solutes into the root, consider this example. Suppose we enclose a solute such as salt (sodium chloride) in high concentration (and water in low concentration) in a funnel sealed with a semipermeable membrane and lower it into a beaker of distilled water. The membrane is permeable to water but not the salt. The volume of the fluid within the funnel will increase and move up the tube as water moves across the membrane into the solution by a process called **osmosis** (Figure 7.13). Water continues to move across the membrane until the **osmotic pressure** of the solute, decreasing as the solute

becomes more diluted by the pure water, is balanced by the physical pressure exerted by the fluid on the tube.

Osmotic pressure accounts for the internal pressure (called turgor) plants cells achieve when the water supply is adequate. As plants lose water through transpiration, the concentration of water molecules in the cells decreases, and

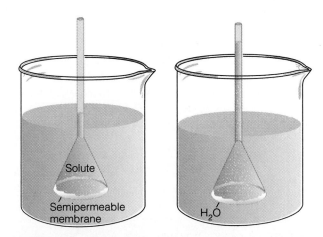

FIGURE 7.13 Demonstration of osmotic potential. Fluid within the funnel increases in volume and moves up the tube as water passes through the membrane into the solution by osmosis. Water continues to cross the membrane until the osmotic pressure of the solute, decreasing with dilution, is balanced by the gravitational pressure exerted by the fluid in the tube.

water, when available, moves from the soil solution into the plant. The tendency of solutes in a solution to cause water molecules to move from areas of high to low concentration is called its **osmotic potential.** The osmotic potential of a solution depends on its concentration. The higher the concentration of solutes, the lower its osmotic potential and the greater its tendency to gain water. Osmotic potential is a major component of the total leaf and root water potentials.

Water in the soil moves along the potential gradient into the roots with their higher solute concentration and lower water potential. What prevents the roots from achieving equilibrium with the osmotic energy of soil water, and pulls the water up through the plant, is a water potential gradient from the soil to the atmosphere. Plants pull water from the soil, where the water potential is the highest, to the atmosphere, where water potential is the lowest (most negative). When the relative humidity of the atmosphere drops below 100 percent, the water potential of the atmosphere rapidly becomes negative (Figure 7.14), increasing the capacity of the air to evaporate water and drive transpiration. Attracted by the low water potential of the atmosphere, moisture from the surface of and between the mesophyll cells within the leaf evaporates and escapes through the stomata. To replace the water lost, more water is pulled from the xylem (hollow conducting tubes throughout the plant) in the leaf veins. Transpiration and replacement of water at the cell's surface make the surface water potential more negative. The negative water potential extends down to the fine rootlets in contact with soil particles and pores. This tension pulls water from the root and up

through the stem to the leaf. The root water potential declines, so that more water moves from the soil into the root.

The loss of water through transpiration continues as long as the amount of energy striking the leaf is enough to supply the necessary latent heat of evaporation, moisture is available for roots in the soil, and the roots are capable of removing water from the soil. At field capacity (see Chapter 4), water is freely available and soil water potential is at or near zero. As water is drawn from the soil, the water content of the soil declines and the soil water potential (ψ_{soil}) becomes more negative.

The tendency for water to adhere to surfaces is called **matric potential.** As the water content of the soil declines, the remaining water adheres more tightly to the surfaces of the soil particles and the matric potential becomes more negative. For a given water content, the matric potential of a soil is influenced strongly by its texture (see Chapter 4). Soils composed of fine particles, such as clays, have a higher surface area (per soil volume) for water to adhere to than do sandy soils, and therefore maintain more negative matric potentials for the same water content.

As soil water potential becomes more negative, the root and leaf water potentials must decline to maintain the potential gradient. If precipitation does not recharge soil water and soil potentials continue to decline, eventually the plant will not be able to maintain the potential gradient. At this point the stomata will close to stop further loss of water through transpiration. However, this closure also prevents further uptake of CO_2. The soil water potential at which stomatal closure occurs is a function of the ability of the plant to further reduce leaf water potentials without disrupting basic physiological processes. The value of leaf water potential at which stomata close and net photosynthesis ceases varies among plant species (Figure 7.15) and reflects basic differences in their biochemistry, physiology, and morphology.

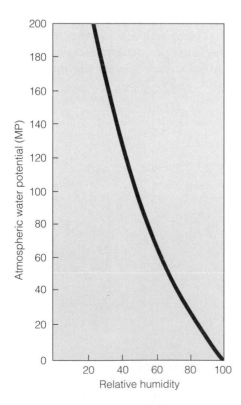

FIGURE 7.14 Relationship between relative humidity of the air and atmospheric water potential.

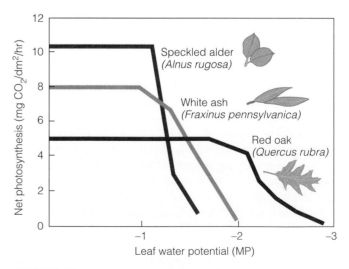

FIGURE 7.15 Changes in net photosynthesis as a function of leaf water potential for three tree species from the northeastern United States. The decline in net photosynthesis with declining water potentials (more negative values) results primarily from stomatal closure.

One additional component that can have a significant influence on plant water potential is **gravitational potential.** For water to move up the conductive tissue of a plant, it must overcome the force of gravity. The gravitational component (ψ_g) reflects differences in potential energy due to differences in height from the reference level. It is positive above the reference level, negative below it. ψ_g increases by 0.01 MPa m^{-1} above the ground. The gravitational potential can be quite important when considering the movement of water in tall trees (for a detailed review of water transpiration in plants, see Canny 1998).

Responses to Short-Term Moisture Stress

Although stomatal closure reduces the loss of water when the potential gradient can no longer be maintained, the leaves continue to intercept radiation. The leaves having lost the ability to dissipate excess energy through transpiration, their internal temperature will rise. The resulting heat stress can affect protein synthesis (Walbot and Cullis 1985). Some plant species, such as evergreen rhododendrons, respond to moisture stress by an inward curling of the leaves. Others show it in a wilted appearance caused by a lack of turgor in the leaves. Both leaf curling and wilting allow leaves to reduce water loss and heat gain by reducing the surface area exposed to solar radiation. Prolonged moisture stress inhibits the production of chlorophyll, causing the leaves to turn yellow or, later in the summer, to exhibit premature autumn coloration. As conditions worsen, deciduous trees may prematurely shed their leaves, the oldest ones dying first. Such premature shedding can result in dieback of twigs and branches.

Other physiological changes take place during severe water shortage. Some water-stressed plants reduce their osmotic potential by accumulating in the leaves such inorganic ions as calcium, magnesium, potassium, and sodium (Mattson and Haack 1987) and amino acids, sugars, and sugar alcohols (Kramer 1983). The reduced osmotic potential aids in reducing leaf water potentials, allowing for the potential gradient from plant to soil to be maintained.

Conifers and other evergreens in temperate climates can experience winter drought. When winter temperatures are warm enough to thaw the ice in the water ducts of woody plants, the trees lose this water by transpiration through the leaves. Because the soil is frozen, making water unavailable, the plants cannot replace the water lost by transpiration. The result is browning of needles and dieback of twigs from dehydration.

Plant Response to Long-Term Variations in Water Availability

In addition to the short-term responses to declining soil water potential, plants growing under conditions of low water availability develop a number of physiological and morphological responses that both conserve water and increase its uptake. These patterns can be observed among individuals of the same species when grown under different moisture conditions.

Individuals grown under low water availability tend to have smaller and thicker leaves than individuals of the same species grown under higher moisture conditions. The thicker leaves found under reduced water conditions tend to have a greater number of layers of mesophyll cells per unit surface area. This increase allows for an increased photosynthetic capacity and diffusion rate of CO_2 per unit leaf area, while reducing the surface area that absorbs radiation and loses water through transpiration. In addition, the proportion of carbon allocated to the production of leaves declines, further reducing the total surface area of leaves.

The reduced allocation to the production of leaves under reduced water availability is accompanied by an increased allocation of carbon to the production of roots. The increased surface area of roots allows the plant to expand the volume of soil that can be exploited for water. This overall shift in the allocation of carbon from the production of leaves to the production of roots with declining water availability can be seen in Figure 7.16.

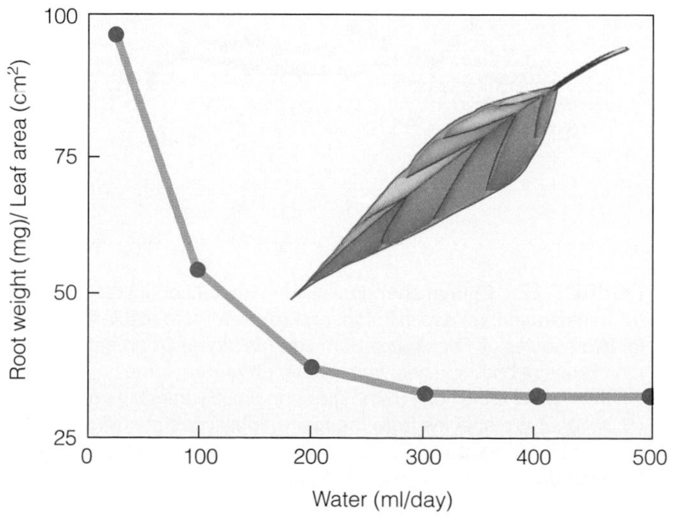

FIGURE 7.16 Relationship between plant water availability and the ratio of root mass (mg) to leaf area (cm^2) for broadleaved peppermint (*Eucalyptus dives*) seedlings grown in the greenhouse. Each point on the graph represents the average value for plants grown under the corresponding water treatment. As water availability decreases, plants allocate more carbon to the production of roots relative to leaves. This increased allocation to roots increases the surface area of roots for the uptake of water, while the decline in leaf area decreases water loss through transpiration.

Interspecific Variation in Adaptations to Mesic and Xeric Environments

The differences in physiology and morphology that yield contrasting plant responses to variations in water availability are most pronounced among species adapted to wet (mesic) versus dry (xeric) environments. This contrast can be seen in a comparison between two species of eucalyptus that inhabit different environments in southeastern Australia (Smith et al. 2000). *Eucalyptus saligna* inhabit the coastal regions where the mean annual rainfall is approximately 1500 mm. In contrast, *Eucalyptus dives* inhabit the drier interior regions, where rainfall is approximately 500 mm yr^{-1}. In an experiment designed to examine the response of these two species to variations in water availability, individuals of each species were grown in the greenhouse under controlled environmental conditions. A number of individuals of each species were grown under one of six water treatments, and a variety of variables relating to physiology, carbon allocation, and growth were monitored.

Diurnal variations in stomatal conductance, photosynthesis, and transpiration for individuals of the two species grown under high-water conditions are contrasted in Figure 7.17. In the case of both species, stomatal conductance declines over the course of the day (Figure 7.17a). This decline is initially a result of increasing vapor pressure deficit as air temperatures rise during the morning hours. The further decline in the latter part of the day is a result of reduced water availability as water is transpired (Figure 7.17b). Patterns of net photosynthesis track these changes in stomatal conductance (Figure 7.17c). Note that the values of net photosynthesis are consistently higher for the mesic species (*E. saligna*) over the course of the day. These higher values of CO_2 exchange are being maintained by the higher values of stomatal conductance. However, the higher values of net photosynthesis come at the cost of higher rates of transpiration. Recall from an earlier discussion (see Chapter 6, page 87) that the diffusion gradient of water from the inside to the outside of the leaf is approximately two orders of magnitude greater than the corresponding diffusion gradient for CO_2. As a result the higher stomatal conductance for *E. saligna* allows for a greater uptake of CO_2, but the water loss per unit of CO_2 taken up is higher than that for *E. dives*. Even though the mesic species, *E. saligna,* maintains higher rates of carbon gain, the CO_2 uptake per unit of water lost is consistently higher for the xeric species, *E. dives* (Figure 7.17d).

The ratio of carbon uptake per unit water transpired (net photosynthesis/transpiration) is called the **water use efficiency.** Water use efficiency is important under xeric conditions, where the ability to replenish water lost through transpiration is limited by soil water availability. One advantage of the C_4 photosynthetic pathway discussed earlier is that as a group, C_4 plants exhibit higher water use efficiency than C_3 plants. The increased water use efficiency is a function of the higher rates of carboxylation and the ability of C_4 plants to maintain a very low CO_2 concentration within the mesophyll cells. The low internal concentration of CO_2 in the mesophyll cells develops

a steep diffusion gradient for CO_2 from the inside of the leaf to the outside air. The steeper gradient allows C_4 plants to maintain a higher rate of photosynthesis than C_3 plants for a given stomatal conductance. Since both CO_2 uptake and water loss are a direct function of stomatal conductance, this allows for a higher CO_2 uptake per unit of water loss—that is, greater water use efficiency.

Patterns of carbon allocation for the two species of *Eucalyptus* follow those outlined for individuals of a given species under conditions of high and low water availability (Figure 7.18a). The ratio of root mass (g) to leaf area (cm^2)

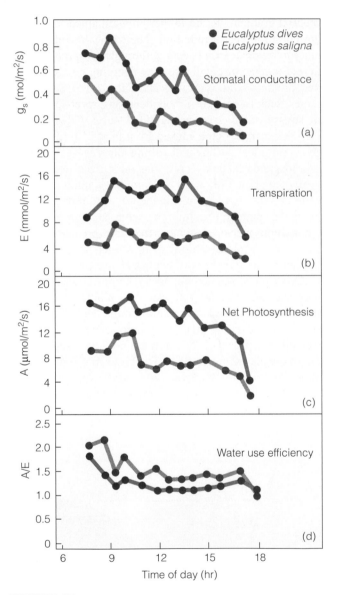

FIGURE 7.17 Diurnal changes in (a) stomatal conductance, (b) transpiration, (c) assimilation, and (d) water use efficiency for two species of *Eucalyptus* from contrasting environments. Individuals of both species were grown under the same environmental conditions in the greenhouse. Sydney blue gum (*E. saligna*) is a species from the high-rainfall coastal environment of southeastern Australia, and broadleaved peppermint (*E. dives*) is from the drier interior regions of the continent.

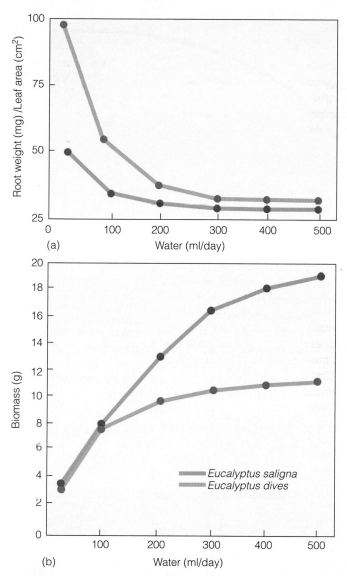

FIGURE 7.18 Comparison of patterns of carbon allocation and growth rate for two species of *Eucalyptus* along an experimental gradient of water availability. (a) Although both species exhibit the same patterns of response to declining water availability, the xeric species, *E. dives*, exhibits a consistently higher ratio of root mass (mg) to leaf area (cm²) than the mesic species, *E. saligna*, across the water gradient. (b) *E. saligna*'s growth rate (biomass gain over period of experiment) continues to increase with increasing water availability. *E. dives* reaches maximum growth rate at intermediate water treatments. (After Smith et al. 2000)

lower stomatal conductance, and consequently lower rates of net photosynthesis, than species adapted to mesic environments. However, the lower stomatal conductance results in a higher water use efficiency. A lower allocation of carbon to the production of leaves reduces water loss, while the increased allocation of carbon to root production increases the plant's access to soil water. However, the reduced rates of photosynthesis, together with the reduced surface area of leaves (photosynthetic surface), reduces overall carbon gain and plant growth (Figure 7.18b). As with the light resource, these adaptations can be seen as a trade-off between maintaining high rates of photosynthesis and growth when water is available, and the ability to continue survival, growth, and ultimately reproduction when water is consistently in short supply.

Link Between Plant Water and Energy Balance

One of the major means of thermal regulation in plants is the dissipation of latent heat through transpiration (see Chapter 6). However, if the plant is to maintain turgor, the water lost in transpiration must be replaced through the uptake of water from the soil. As soil water potentials decline, the ability to maintain the uptake of water from the soil decreases. The excess energy must be dissipated via other means or leaf temperatures can rise to critical levels. The other major means of heat dissipation in plants is convection. As we have discussed previously, the ability to dissipate heat through convection is influenced by leaf morphology. For the same unit of leaf area, smaller leaves increase the surface area for heat exchange as air passes over the canopy. Given the constraints on heat dissipation via transpiration, you might expect that in moving from mesic to xeric environments, the dominant mechanisms of maintaining plant energy balance would shift from transpiration to convection. In fact, the evolutionary consequences of such a shift appear in the results of a study by T. J. Givnish (1986), which expands on the earlier work of Parkhurst and Loucks (1972). Givnish examined changes in average leaf size for plant communities along a gradient of rainfall in the tropics. As precipitation decreases, the average leaf size decreases (Figure 7.19). Under mesic conditions large leaves can be supported, because excess energy can be dissipated via transpiration. However, as precipitation declines, the ability to dissipate excess energy as latent heat decreases. In this case, small leaves can be viewed as an adaptation to aid in the dissipation of excess energy via convection.

Plant Responses to Flooding

Too much water can place as much stress on plants as too little water. Symptoms of flooding are similar to those of drought, including stomatal closure, yellowing and premature loss of leaves, wilting, and rapid reduction in photosynthesis (Jackson and Drew 1984, Kozlowski and Pallardy 1984). The causes, however, are different.

Excess water around the roots reduces oxygen levels, causing anaerobiosis and death of the root tips. The remaining

increases with decreasing water availability. Note however, that the dry-adapted species (*E. dives*) has a consistently higher ratio of root mass to leaf area than the mesic species (*E. saliga*) across the entire gradient. This response is the same pattern seen in plant adaptations to variations in the light environment. Although individuals of the same species grown under contrasting environmental conditions exhibit a suite of responses that help to compensate for the shortage of an essential resource, these same responses are magnified in the adaptations of species to high and low resource availability. In the current case, species adapted to xeric environments tend to have

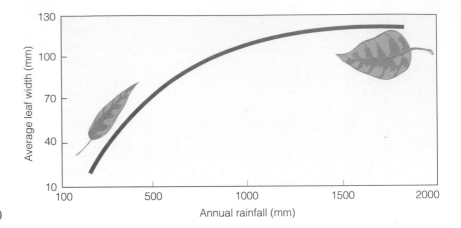

FIGURE 7.19   Average leaf width at low elevations as a function of average annual rainfall in tropical regions of Australia, Central America, and South America. (Adapted from Givnish 1986.)

root system is less permeable to water uptake, and the plant wilts. In addition, root death often adds detrital (dead plant) material to the xylem stream, which clogs the xylem, further reducing the uptake of water (Talboys 1978). Flooding also brings about morphological changes in the plant. Deprived of oxygen because water displaces air in the soil pores, adventitious roots grow horizontally along the oxygenated zone of the soil surface, where anaerobically generated toxins are absent. These roots emerge from the submerged part of the stem when original roots succumb, to replace them. Perennially high water tables force some plants found in poorly drained soils—for example, red maple (*Acer rubrum*) and

white pine (*Pinus strobus*)—to develop shallow, horizontal root systems, which make them highly susceptible to drought and windthrow.

Prolonged flooding often results in dieback or death of plants, particularly woody species. The response of trees to flooding depends on its duration, the movement of water, and the species of tree. Standing in stagnant water injures trees more than standing in oxygenated flowing water. If a tree's roots are flooded for more than half the growing season, death often results (Kozlowski 1984). Look for standing dead trees about beaver dams (Figure 7.20) and along highways where road construction has impeded soil drainage.

FIGURE 7.20   Trees flooded by a beaver dam will eventually die.

Plant Adaptations to Flooding

Growing plants need both sufficient water and a rapid gas exchange with their environment. Much of this exchange takes place between the roots and soil atmosphere. When soil pores are filled with water, no gas exchange can occur. The plants experience depressed O_2 levels and are asphyxiated; in effect, they drown. Roots depend on gaseous O_2 to carry on aerobic respiration. Without sufficient oxygen, the roots cannot respire aerobically and are forced to shift into anaerobic metabolism. In addition, anaerobic conditions in the soil inhibit uptake and transport of ions, and they depress the concentration of nitrogen, phosphorus, and potassium in the shoot. These conditions also cause the plant to accumulate ethylene. Ethylene gas, a growth hormone, is produced especially under flooded conditions (Jackson 1985) and is highly insoluble in water. Under flooded conditions ethylene diffusion from the roots is slowed and oxygen diffusion into the roots is inhibited. Ethylene, therefore, accumulates to high levels in the root. This stimulates adjacent cells in the cortex of the root to lyse and separate to form interconnected gas-filled chambers (Drew et al. 1979, Jackson 1982) called **aerenchyma.** These chambers, typical of hydrophilic plants, allow some exchange of gases between submerged and better aerated roots.

Hydrophilic plants have evolved certain adaptations that allow them to cope with a waterlogged environment. Most have aerenchyma through which oxygen diffuses from the shoots to the roots. Aerenchyma throughout the leaves and roots occupy nearly 50 percent of the plant tissue. Some plants of cold, wet, alpine tundras have similar spaces in leaf, stem, and root to carry oxygen to the roots (Keeley et al. 1984). In contrast to nonwetland species, wetland plants lose less oxygen from the roots, thus maintaining oxygen concentrations (Jackson and Drew 1984).

A few woody species can grow on permanently flooded sites, such as bald cypress (*Taxodium distichum*), mangroves, willows, and water tupelo (*Nyssa aquatica*). Cypresses that experience fluctuating water tables develop knees or pneumatophores, specialized growths of the root system, which have been suggested as an aid to their survival by oxygenating their root systems (Figure 7.21). Pneumatophores on mangroves also provide oxygen to roots during tidal cycles. These plants may also modify their metabolism to adapt to short-term hypoxia and reduce ethanol toxicity (Hook 1984).

Plant Adaptations to Salinity

Saline environments pose two dangers to plants: problems of osmotic regulation and direct toxicity. The immediate effect of an increase in salinity is osmotic. If the osmotic potential of the surrounding water or soil is lower than that of the root cells, water is drawn out of the cells and they will dehydrate. Although the absorption of inorganic ions may relieve the osmotic gradient, high concentrations of inorganic ions can be toxic, posing a second threat to survival.

Halophytes, plants adapted to saline environments, occupy both terrestrial and aquatic habitats. Terrestrial halophytes grow in soil with more than a 0.2 percent salt content, notably in desert regions, whereas aquatic halophytes are

FIGURE 7.21   Knees or pneumatophores rise from the lateral roots of swamp bald cypresses. These knees may be beneficial as aeration organs and help anchor the trees to the waterlogged soil.

FIGURE 7.22 *Salicornia,* a succulent annual or perennial, depending on the species, is a halophyte, able to exist only in saline environments.

found in salt-water environments. These plants use two mechanisms, salt exclusion and salt secretion, to protect the shoot and leaf cells of plants from high salt concentrations. It is generally thought that the air space of the root cortex is freely accessible from the rhizosphere, so that the internal endodermis must form the first effective barrier to the upward movement of solutes (salts) from the soil. This hypothesis is supported by the observation that the roots of plants in saline habitats have a much higher concentration of salts in their roots than in their leaves. Although the high osmotic potential of the roots allows the plant to maintain the potential gradient from the soil to the root, the leaf water potential must be at least equal to that of the root if water is to be retained. Although sodium chloride ions make up a significant portion of the total osmotic concentration, the remainder is presumably organic.

Some plants that do not exclude salt at the root, or are "leaky" to salt, secrete salts from specialized organs. The leaves of many salt marsh grasses are characteristically covered with salt particles that have been excreted through specialized glands embedded in the leaves. These salt-excreting glands selectively remove certain ions from the vascular tissues of the leaf. For example, in *Spartina* the excretion is high in sodium relative to potassium. Still others species accumulate ions in tissues away from metabolic sites and shed the leaves and their accumulated salts (Greenways and Murrs 1980).

Few halophytes are obligates, requiring a saline habitat (Barbour 1970). Halophytes are adapted to saline habitats through their ability to handle excess salts. Most halophytes of coastal marshes and swamps, such as salt marsh grass and mangroves, grow best at low salinity; their growth decreases as salinity increases. Others, such as glassworts (*Salicornia*) (Figure 7.22) grow best at moderate salinity; their growth declines at low and high salinity.

Halophytes are salt-resistant. Within limits, they can maintain homeostasis and carry on metabolic functions in the presence of excess salt. Some nonhalophytes growing along seashores, such as salt-spray rose (*Rosa rugosa*) and beach plum (*Prunus maritima*), also exhibit salt resistance. Salt resistance is an important consideration in the selection of trees and crop plants for growing in saline soils of semiarid regions. Some species of pines, Douglas fir, balsam fir, and black cherry (*Prunus serotina*) are particularly vulnerable to direct salt damage, whereas larch (*Larix*), paper birch (*Betula papyrifera*), and aspen are relatively salt-resistant.

PLANTS AND NUTRIENTS

Plants require some 16 elements for growth (see Table 3.1). These 16 elements are classified into two categories: macro- and micronutrients (see Chapter 3). The prefixes *macro* and *micro* refer only to the quantities of the element required for plant growth. They do not signify importance as relates to key plant processes. Micronutrients are only limiting on unusual geological formations, very old and weathered soils, or areas of extreme human disturbance. Of the macronutrients, C, H, and O form the majority of plant biomass. These elements are derived from CO_2 and H_2O and are made available to the plant as simple sugars through photosynthesis.

The remaining six macronutrients—nitrogen (N), phosphorus (P), potassium (K), calcium (Ca), magnesium (Mg), and sulfur (S)—exist in a variety of states in the soil and their availability to plants is affected by several important and different processes. A generalized diagram of the primary sources of nutrients and the processes controlling their availability to plants is shown in Figure 7.23. In the case of terrestrial environments, plants take up nutrients from the atmosphere and soil. In aquatic environments, plants take up nutrients from the substrate or directly from the water column. The weathering of rocks and minerals and the process of ion exchange in controlling the availability of nutrients to plants have been discussed in Chapter 4. As mineral nutrients released through these processes are taken up by the plant, they are incorporated into plant tissues. With the senescence of individuals or plant parts (such as the shedding of leaves in deciduous plants), plants return these nutrients to the soil surface, substrate, or water column.

Plants require nutrients in an inorganic or mineral form. Therefore, the nutrients that have been incorporated into living tissues (organic nutrients) and returned to the soil must first be transformed into an inorganic form before they are available for plant uptake. This transformation occurs during the process of decomposition, a topic to be discussed in Chapter 9. Once this transformation has occurred, the nutrients are once again available for uptake and incorporation into plant tissues during growth. This cycling of nutrients—from the soil or water to the plant and back to the soil, where it is then transformed into inorganic form through decomposition—is called nutrient cycling. We shall present this topic in much more detail in Chapter 25.

Nutrient Uptake and Plant Processes

Two major factors influence the rate of nutrient uptake by plants: availability and demand. The uptake of nutrients is typically expressed in terms of the Michaelis-Menten equation relating the uptake rate (V) as a function of the external concentration of the nutrient (C_{ext}):

$$V = (V_{max} \times C_{ext})/(K_m + C_{ext})$$

Here V is the rate of nutrient uptake, V_{max} is the saturation uptake rate, C_{ext} is the external concentration, and K_m is the value of C_{ext} at which V is half of V_{max} (Figure 7.24). As nutrient concentration (C_{ext}) rises above some minimum, the rate of uptake (V) increases. As nutrient concentration continues to rise, the rate of increase in uptake per unit increase in concentration declines. Eventually the plant reaches a maximum uptake rate (V_{max}), at which point any further increase in concentration results in no further increase in the rate of uptake.

The most important variable controlling the nutrient content of plant tissues is the rate of uptake. The relationship between nitrogen uptake rate per unit of root mass and leaf nitrogen concentration for a variety of plant species is shown in Figure 7.25a. The nutrient concentrations of plant tissues have a direct relationship to key processes related to plant growth, survival, and reproduction. An example is the relationship between leaf nitrogen concentration and maximum observed rates of photosynthesis (Figure 7.25b). The basis for this relationship is that over 50 percent of the total nitrogen in leaf tissues is associated with the maintenance of photosynthesis, including the synthesis of rubisco and chlorophyll (Evans 1989).

Plant Adaptations to Variations in Nutrient Availability

Root Growth Versus Shoot Growth As with plant response to water availability, one way in which plants

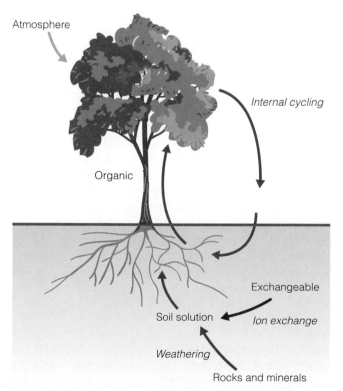

FIGURE 7.23 Diagram showing the major sources of mineral and organic nutrients and the processes (in italics) influencing their release and availability to terrestrial plants.

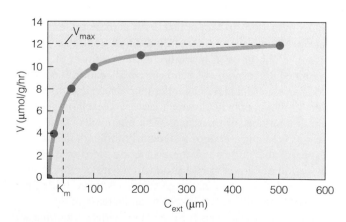

FIGURE 7.24 Plant uptake rate (V) of potassium as a function of availability (C_{ext}).

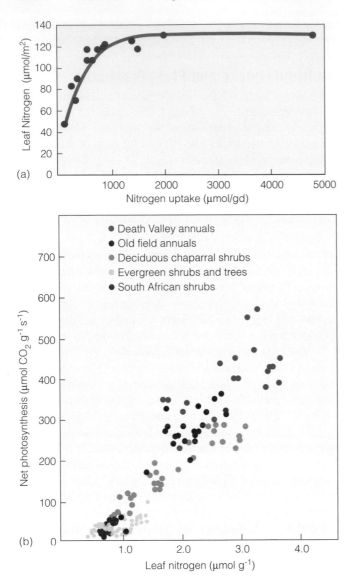

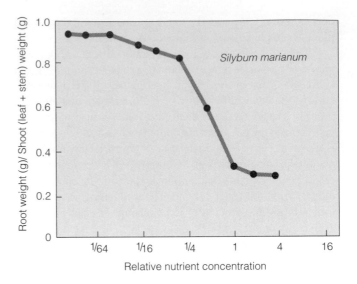

FIGURE 7.26 Changes in the allocation to root and shoot (stem+leaf) tissues for *Silybum marianum*, a species of thistle inhabiting south-eastern Australia, grown under varying concentrations of nitrogen. Nitrogen availability is expressed as relative concentration. (Adapted from Austin et al. 1985.)

FIGURE 7.25 Influence of (a) root nitrogen uptake on leaf nitrogen concentrations (adapted from Woodward and Smith 1994) and (b) leaf nitrogen concentrations on maximum observed rates of net photosynthesis for a variety of species from differing habitats (adapted from Field and Mooney 1983).

compensate for the low availability of nutrients and the associated reductions in root uptake rates is to increase the allocation of carbon to the production of root tissue. This response can be seen in Figure 7.26. Austin et al. (1985) examined the response of plant carbon allocation and growth along an experimental gradient of nutrient availability. Plants were grown in pots in the greenhouse, and each pot received one of a number of treatments. The treatments consisted of various concentrations of a standard nutrient solution. The results of the study show a consistent pattern of increased allocation of carbon to root production with declining nutrient availability across the range of species. This shifting pattern of carbon allocation is seen as an increase in the ratio of root mass to shoot mass for the plants, where shoot mass is the combined weight of leaves and stems.

The pattern of increased allocation to root production at the expense of above-ground tissues is most pronounced when a comparison is made between species adapted to nutrient-poor versus nutrient-rich environments. Bradshaw et al. (1964) conducted a number of experiments examining variation among plant species in their response to nutrient availability. A comparison of the growth response (biomass accumulation) of three grass species from differing habitats to variation in nitrogen availability is presented in Figure 7.27. Nitrogen availability (parts per million) was controlled by varying the concentration of the solution used to fertilize the plants. Growth response was measured as the accumulation of dry weight over the period of the experiment. Each data point on the graph is the average response of individuals grown under the defined nitrogen treatment. Note that the growth rate of individuals of *Agrostis stolonifera* continues to rise with increasing nitrogen availability across the range of nitrogen concentrations observed. In contrast, the growth rate of *Agrostis canina* approaches an asymptote at intermediate nitrogen concentrations, and the growth rate of *Nardus stricta* actually declines above concentrations of 27 ppm. These three species inhabit different environments. *Agrostis stolonifera* inhabits high-nutrient environments, *N. stricta* is found in extremely infertile environments, and *A. canina* occurs in areas of intermediate nutrient availability.

Adaptations such as shifts in carbon allocation to the production of roots rather than leaves are no doubt an important factor controlling the observed differences in maximum observed growth rates under high nutrient availability. J. P. Grime (1979), however, has suggested that low growth rate itself might be an adaptation to low-resource environments. Species with high potential growth rates under nutrient-rich conditions would undergo reduced metabolic function (stress), whereas plants with inherently low growth rates and associated low nutrient demand/uptake rates could function at their maximum metabolic rates.

Leaf Longevity   Another adaptation that has been associated with low-nutrient environments is increased leaf longevity. Reich et al. (1992) examined the relationship among leaf

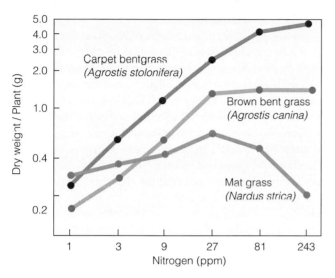

FIGURE 7.27   Growth response of three grass species from differing habitats to nitrogen availability. *Agrostis stolonifera* grows in high-nutrient environments, *Nardus stricta* is found in extremely infertile environments, and *Agrostis canina* occurs in areas of intermediate nutrient availability. Nitrogen availability (ppm) is controlled by varying the concentration of the solution used to fertilize the plants. Growth response is measured as the accumulated dry weight over the period of the experiment. Each data point is the average response of individuals grown under the defined nitrogen treatment. (Adapted from Bradshaw et al. 1964.)

longevity (life span), leaf nitrogen concentration, and net photosynthetic rate for a wide array of plant species from different habitats. The authors found a significant inverse relationship between leaf nitrogen concentration and leaf life span (Figure 7.28). This relationship suggests that species with short-lived leaves tend to have higher leaf nitrogen concentrations than those species having longer-lived leaves. A possible mechanism for this observed pattern becomes apparent when it is combined with the relationship between leaf nitrogen concentration and maximum observed photosynthetic rates discussed earlier (see Figure 7.25b). The result is an inverse relationship between leaf life span and maximum observed rates of photosynthesis. That is, species with short-lived leaves tend to have higher rates of photosynthesis than species with long-lived leaves.

One possible way of interpreting this relationship is to view the adaptation of leaf longevity in the context of a simple economic model (Mooney and Gulman 1982). There is a "cost" of producing a leaf, in both carbon and other essential nutrients. This cost must be "paid back" if a positive carbon balance is to be maintained. The lower the rate of photosynthesis, the longer the time period necessary for this payback to occur. This relationship in part explains the observed correlation between leaf longevity and maximum observed rates of photosynthesis. In habitats where nitrogen availability is low, the rate of nitrogen uptake and consequently the leaf nitrogen concentration will be low (see Figure 7.25). The low nitrogen concentration results in low rates of net carbon gain. Under this set of conditions, the habit of longer-lived leaves is an advantage in that it extends the period over which carbon uptake can occur. This relationship in part explains the dominance of evergreen needle-leaf tree species (pines) in the coastal Piedmont regions of the southeastern United States, which are characterized by low-nutrient, sandy soils.

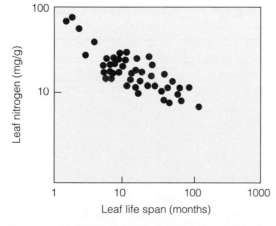

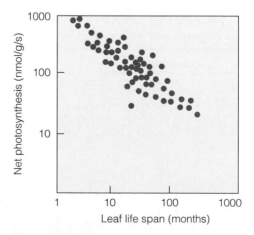

FIGURE 7.28   Relationship between (a) leaf longevity (life span) and leaf nitrogen concentration and (b) leaf longevity and net photosynthetic rate for a wide variety of plants from different habitats. Each data point represents a single species. Species having longer-lived leaves tend to have lower leaf nitrogen concentrations, and subsequently, lower rates of photosynthesis. (Adapted from Reich et al. 1996.)

Influence on Nutrient Availability Availability of nutrients is largely influenced by the processes outlined in Figure 7.23. Plants, however, also have a number of direct and indirect influences on the local availability of nutrients. Nutrients in soil solution are taken up as water is absorbed across the root membrane. In addition, active transport occurs; this is the dominant process in aquatic plants. As plants absorb nutrients from the surrounding environment, zones of nutrient depletion become established. Nutrients are then replenished as they diffuse into these depleted zones from the surrounding soil or water as a result of the diffusion gradients established through plant uptake.

The return of nutrients in senescent plant tissues to the soil or water has a major influence on the rates of nutrient availability to plants through nutrient cycling (see Chapter 25). One of the largest and most consistent inputs of dead plant material to decomposers in terrestrial environments is the natural process of leaf senescence. For deciduous tree species in the temperate zone, this senescence occurs in the autumn. As chlorophyll and other compounds within the leaves break down, leaves change from green to a variety of colors. During this process, plants transport a significant percentage of the nutrients from the leaves to the perennial parts of the tree prior to leaf-fall. This process is called **nutrient retranslocation,** which also occurs in evergreen species. The difference in nitrogen concentration between green and senescent leaves for a number of deciduous and evergreen tree species inhabiting the forests of central Virginia is shown in Table 7.1.

One possible adaptation to low-nutrient environments is an increased rate of nutrient retranslocation. This hypothesized relationship does not occur for the species included in Table 7.1; rather, the percentage of nitrogen reabsorbed prior to leaf senescence is relatively consistent across all species observed. However, other studies that examined species from widely differing nutrient habitats have found an inverse relationship between leaf nutrient concentrations and percent nutrient retranslocation (Chapin 1980).

Another very important type of plant adaptation to low nutrient availability is the evolution of mutualistic relationships with microorganisms, especially nitrogen-fixing bacteria and mycorrhizae, that increase the plant's access to nutrients. A **mutualism** is an interaction between two species that is beneficial to both (see Chapter 16).

Although nitrogen in the form of N_2 gas is a major constituent of the atmosphere, this abundant form of nitrogen cannot be used directly by plants; it must first be transformed into a usable form, a process called fixation (see detailed discussion of N cycle in Chapter 25). This task is accomplished by specific groups of bacteria in both terrestrial and aquatic environments. In aquatic environments, cyanobacteria carry out this transformation. In terrestrial environments, rhizobium bacteria are the nitrogen-fixing organisms. Cyanobacteria are free-living organisms, whereas the rhizobium bacteria are associated with (grow in and on) the root system of certain plant species. Unlike the free-living cyanobacteria, rhizobium bacteria depend on carbon provided by the plants on which they live as a source of food energy. In return, they provide the plant with nitrogen.

Another group of microorganisms associated with the root systems of terrestrial plants are mycorrhizal fungi. Mycorrhizal fungi are attached to the roots, extending from the roots out into the soil. The fungi assist the plant in the uptake of nitrogen and phosphorus, and in return the plant provides the fungi with carbon, a source of food energy. As with nitrogen-fixing rhizobium bacteria, the relationship between plant and fungi is mutually beneficial (for a detailed discussion of mychorrizal fungi, see Chapter 16).

TABLE 7.1 Nitrogen, total carbon, and lignin concentrations of green and senescent (litter) leaves for nine tree species found in Central Virginia. All values are expressed as percentage of total dry weight. Percent nitrogen reabsorption is the difference between the nitrogen content of green and senescent leaves expressed as a percentage of green leaf nitrogen content.

| | Green Leaf | Litter | | | | | |
Species	% N	% N	% C	% Lignin	C:N	Lignin:N	% N Reabsorption
White Oak	2.08	0.82	47.46	17.70	57.88	21.59	60.58
Scarlet Oak	2.14	0.85	48.74	21.80	57.34	25.65	60.28
Southern Red Oak	1.88	0.60	49.18	25.90	81.97	43.17	68.09
Red Maple	1.96	0.76	47.98	11.70	63.13	15.39	61.22
Tulip Poplar	2.55	0.90	51.27	18.20	56.97	20.22	64.71
Virginia Pine	1.62	0.54	52.41	27.00	97.06	50.00	66.67
American Hornbeam	2.20	1.16	46.62	22.50	40.19	19.40	47.27
Sweetgum	1.90	0.59	46.83	20.50	79.38	34.75	68.95
Sycamore	2.10	0.90	47.33	36.40	52.69	40.44	57.14
							Avg. = 61.66

In the mutualistic relationships between plants and both rhizobium bacteria and mychorrizal fungi, carbon fixed in photosynthesis is allocated to the process of nutrient uptake. In this manner, these mutualistic relationships are similar to the observed pattern of increased allocation of carbon to root production under low nutrient availability (see Figure 7.26). As with the latter phenomenon, the allocation of carbon to support mutualistic associations with microorganisms aiding the uptake of nutrients is an advantage only under reduced nutrient conditions. Under higher nutrient availability, the association would represent a cost (in terms of carbon that could otherwise be allocated to growth or reproduction) with little benefit.

In summary, characteristics of plant species adapted to low-nutrient environments include lower tissue nutrient concentrations, lower maximum rates of net photosynthesis, and increased allocation of carbon to root production (or mutualistic associations). These patterns result in overall lower maximum rates of growth, even under high nutrient availability. In addition, increased leaf longevity and nutrient retranslocation prior to leaf senescence have been suggested as mechanisms of nutrient conservation. These adaptations, too, can be viewed as a trade-off between the ability of a species to survive, grow, and reproduce under reduced nutrient conditions and its ability to maintain high rates of production under conditions of high nutrient availability.

Calcicoles and Calcifuges

Soil acidity affects the nutrient uptake of plants and indirectly affects their distribution. A close relationship exists between available calcium and the degree of soil acidity as measured by pH (see Chapter 4). Many plants grow well in mildly acidic or acid soils; other grow best in high-calcium (often called high-lime) soils. Plants have been broadly classified as **calcicole** (lime-loving), **calcifuge** (lime-hating), and **neutrophilus** (tolerant of either condition) (Larcher 1980). This relationship, however, is more than a simple one between plants and calcium. Calcium availability is only one of a number of pH-related soil conditions, including the toxicity of iron, aluminum, and other heavy metal ions, and deficiencies of nitrogen, phosphorus, and magnesium. Calcium deficiency and low pH are closely associated only because calcium is the most abundant cation in the soil.

True calcifuges, such as rhododendrons and azaleas, have a low lime requirement and can live in soils with a pH of 4.0 and less. Highly acidic soils associated with calcifuges invariably have a high concentration of aluminum and iron ions, toxic to most plants. Calcifuge plants are tolerant of aluminum and have a high demand for iron. They either possess specific sites within the cytoplasm where aluminum may accumulate harmlessly, chelate aluminum, or precipitate it at the cell surface. Calcifuge plants are especially sensitive to calcium. If grown on calcareous soils, they suffer from lime chlorosis, a disease in which roots and leaves become stunted and the leaves yellowed. This sensitivity may be due in part to the plant's mechanism for chelating aluminum, which has an affinity for iron at higher pH. This deficiency can be corrected by the foliar application of ferrous iron salts. It is the ability to grow in the presence of toxic ions, especially aluminum, that sets calcifuge plants apart from calcicoles.

This difference can create a problem for gardeners where calcifuge plantings of azaleas and rhododendrons grow adjacent to lawns of calcicole grasses such as fescue and bluegrass (*Poa pratensis*). Keeping the lawn grasses healthy requires the application of lime, but maintaining the evergreen shrubs requires an acid fertilizer.

True calcicolous plants, such as alfalfa (*Medicago sativa*), blazing star (*Chamaelirium luteum*), and southern red cedar (*Juniperus silicicola*) are restricted to soils of high pH, not because they have any particular demand for calcium, but because they are susceptible to aluminum toxicity, acidity, and other factors influenced by calcium. Aluminum toxicity begins to appear at a pH of 4.5. Free aluminum accumulates on the surface of the root and in the root cortex. It interacts with phosphorus to form highly insoluble compounds. Calcareous soils tend to be porous and well drained, with lower moisture and higher temperatures than noncalcareous soils. These conditions restrict certain plants, less competitive on more mesic sites, to calcium-rich soils.

Plants of Serpentine and Toxic Soils

Heavy metals, such as iron, nickel, chromium, cesium, zinc, and cobalt, are toxic to plants, causing chlorosis and stunted growth. They interfere with the uptake of other nutrients and inhibit root growth and penetration. For these reasons high concentrations of heavy metals can be a powerful force of natural selection, resulting in flora specialized to endure their toxic effects. This effect is most evident on mine slag heaps and the chemically distinct serpentine soils. Serpentine soils are derived from ultrabasic magnesium silicate rocks, usually greenish in color. These soils are very high in magnesium and iron, high in nickel, chromium, and cobalt, and low in calcium, phosphorus, sodium, and aluminum. Serpentine soils possess a distinctive flora tolerant of those conditions. Plant life consists of rare and endemic species and ecotypes of nonserpentine species that have developed a special tolerance to serpentine soils (Kruckeberg 1954, Franklin and Dyrness 1973). **Endemics** are species restricted to certain specialized habitats; **ecotypes** are ecological races that are well adapted to a local set of conditions. These endemics and ecotypes appear to substitute magnesium for other bases and to tolerate nickel and chromium in their tissues at levels highly toxic to other plants (Walker 1954).

The transition in vegetation between serpentine and adjacent soils is often sharp, marked by plants stunted in appearance. In Oregon, for example, open, stunted stands of pine replace Douglas fir; in California chaparral vegetation replaces oak woodland; and in New Zealand tussock grassland replaces southern beech (*Nothofagus*) (Whittaker 1954).

The evolution of tolerance to heavy metals has produced certain species that can minimize toxic effects. Their roots' uptake mechanism may exclude heavy metals from the plant; or the plant may carry the metals unchanged in its tissues. A few species are site-specific for heavy metals (Antonovics et al. 1971), indicating the presence of those metals in the soil. For example, in England and Europe the pansy (*Viola calaminaria*) and the pennycress (*Thlaspi calaminare*) grow only on zinc carbonate and silicate calamine soils (their scientific species names derive from the soils on which they grow). In central Africa a group of copper tolerant plants, especially *Becium homblei,* are used as indicator plants for the possible location of commercial ore deposits (Howard-Williams 1970).

The industrial age has produced large areas of soils contaminated with heavy metals, including mine tailings, coal spoils, strip-mined lands, and landfills. These areas have favored the rapid evolution within 30 to 100 years of ecotypes of certain plants, particularly grasses, tolerant of heavy metals (Law et al. 1977, McNeilly 1987). The evolution of such species depends on three factors: (1) the selection of tolerant seedlings from a normal surrounding population; (2) continued natural selection for metal tolerance and against susceptible genotypes, despite gene flow from the surrounding population; and (3) selection for the ability to survive in physically harsh, largely xeric, nutrient-poor environments. Such selection usually favors a tolerance for a specific heavy metal such as zinc or nickel associated with the site, not a broad range. These ecotypes generally are poor competitors in adjacent normal habitats. Races of tolerant grasses are now being used to revegetate toxic soils (Smith and Bradshaw 1979, Gemmel and Goodman 1980).

Summary

1. Both photosynthesis and respiration respond directly to variations in temperature. As temperatures rise above freezing, photosynthesis and respiration rates increase. At higher temperatures, photosynthetic rates decline and respiration rates increase. C_4 plants carry on photosynthesis at higher temperatures ranges than C_3 plants. The reason is that the enzyme PEP associated with C_4 plants functions at a higher temperature than rubisco.

2. Plant survival in extreme temperatures depends on resistance to freezing and tolerance to heat. Cold-resistant plants survive through the formation of intracellular ice and certain solutes that bring about supercooling. High temperatures result in the disruption of cell metabolism.

3. Certain temperature thresholds induce flowering; seeds of many plants require prior exposure to cold to stimulate germination.

4. Plants are constantly absorbing short-wave and long-wave radiation from the environment. Net energy absorbed is the plant's net radiation balance. To prevent temperatures from rising to critical levels, plants dissipate excess energy as heat to the surrounding environment through convection and evaporation. Influencing the dissipation of heat is the boundary layer, the thin layer of air adjacent to the surface of an organism or object. As an object loses heat, the air within the boundary layer warms and transfers heat to the surrounding air. Influencing the conductance of heat by the boundary layer in plants and the dissipation of heat through convection is the size and shape of the leaves. The dissipation of thermal energy through transpiration and convection enables plants to keep leaves from reaching critical temperatures.

5. Most plants maintain a stable water balance independent of environmental fluctuations by affecting osmotic potential and turgor pressure, collectively known as water potential. Under normal moisture conditions, water potential decreases from soil to leaf, pulling up water.

6. Plants respond to moisture deficits first by closing the stomata, reducing transpiration. Severe drought decreases photosynthesis, causes leaves to turn yellow and shed, and even kills. In the case of water limitation, the plant allocates more carbon to the production of roots at the expense of leaves. Under conditions of low water availability, these two shifts in carbon allocation, more roots and less leaves, function together to benefit the plant. The increased allocation to roots increases the access to the limiting resource, whereas the decreased surface area of leaves reduces water loss.

7. Too much water also creates stress. In extreme conditions, anaerobic respiration in the soil asphyxiates plants. Flooding accumulates toxic material in a plant, depresses the concentrations of ions and oxygen, and may stimulate the growth of adventitious roots along the oxygenated soil surface. External symptoms are similar to those of drought stress. Prolonged flooding can result in death.

8. Wetland plants possess adaptations to flooding. Notable are aerenchyma tissues that form interconnected gas spaces through the leaves to the roots, carrying water to waterlogged roots.

9. Plants vary in their requirements and tolerances for different elements. Each species has specific abilities or requirements to exploit a nutrient supply in its own manner. This fact is often reflected in the competitive abilities of different plants.

10. Low nutrient availability will reduce plant tissue concentrations, which can directly affect basic plant processes such as photosynthesis. In addition, under conditions of low nutrient availability, plants increase the allocation of carbon to root production. These two factors combine to decrease overall net carbon uptake and growth rate.

11. Some plants grow well on acid soil, are tolerant of aluminum, and are sensitive to calcium. These plants are called calcifuges. Other plants sensitive to aluminum do best in soils high in calcium. They are called calcicoles.

12. A few plants have evolved tolerances to toxic heavy metals, such as copper, nickel, and zinc. A special group of plants, the halophytes, is tolerant of high salinity. Plants in saline environments may accumulate and excrete salt through the leaves or store water in them.

Review Questions

1. What two physical mechanisms are involved in a plant's maintenance of heat balance?

2. When a plant dies from a rapid drop in temperature, what probably happened at the cellular level?

3. What characteristics of a plant, particularly of its leaves, influence the amount of heat it absorbs and loses? Name two and explain how they function.

4. What is the role of the boundary layer in the dissipation of heat by a plant?
5. What is supercooling and how does it function?
6. How does a plant draw water from the soil to the leaves?
7. How is osmosis involved in the movement of water from soil to the plant?
8. Contrast osmotic potential with matric potential.
9. What is the cost to a plant if it closes its stomata to reduce water loss?
10. What is the relationship between leaf morphology and available moisture?
11. What is meant by water use efficiency?
12. How do C_3 and C_4 plants differ in their water use efficiency?
13. What similarities exist between the responses of a plant to the stresses of drought and flooding?
14. As you are traveling along an interstate, you notice that some trees standing in a pool of water created by a road fill are dead. Explain why these trees died.
15. What is serpentine soil? What are the effects of heavy metals on plant tolerance and distribution?
16. What are halophytes? Where would you find them?

17. Why do plants allocate more carbon to root production in the face of declining availability of nutrients, especially nitrogen?
18. What is the possible relationship between leaf longevity and nitrogen availability?
19. What is the relationship between leaf senescence and nutrient translocation?
20. Distinguish between calcifuges and calcicoles. What are the outstanding characteristics of each?
21. A home owner has a lawn bordered with rhododendrons and azaleas. Of what should the home owner be aware in the maintenance of both lawn and shrubbery? Why? What fertilization program is needed to keep both the lawn and the border growing?
22. A horticultural book warns not to plant a variety of Japanese holly as a border plant along a concrete walkway. Why should the shrub not be planted there?

Cross References

Water, 50; nutrients, 54; thermal environment, 48; soil chemistry, 62; photosynthesis, 86; transpiration, 87.

CHAPTER 8

Animal Adaptations to the Environment

CONCEPTS

1. All animals directly or indirectly depend on plants for food.

2. Plant-consuming animals derive nutrients from plants with the aid of bacteria living in the digestive tract and thus convert plant tissue into animal tissue.

3. Ectothermic organisms depend on the heat of the environment to maintain their body temperature; endothermic ones maintain their body temperature by metabolic heat.

4. Organisms whose body temperatures follow ambient temperatures are poikilotherms; organisms that maintain a fairly constant body temperature in spite of variable environmental temperatures are homeotherms.

5. Animals regulate internal water by balancing intake of water with moisture losses.

6. Organisms possess a circadian rhythm of about 24 hours that becomes synchronized to a 24-hour environmental rhythm of light and dark.

7. An organism's biological clock is set by changes in daylength.

Carbon fixation by and the growth of plants, directly or indirectly, provide the nutritional resources of animals and influence the structure of their physical environment. By creating shelter and interacting in other ways with the environment, vegetational structure influences animal responses to environmental variables, especially temperature and moisture.

ANIMALS AND NUTRIENT ACQUISITION

The need for animals to derive their energy from organic carbon compounds presents this class of organisms with a wide array of potential food items. The ultimate source of these organic compounds is plants. The problem facing consumers is the conversion of this plant tissue to animal tissue. Plants and animals have different chemical compositions. Animals are high in fat and proteins, which they use as structural building blocks. Plants are low in proteins and high in carbohydrates, much of it in the form of cellulose and lignins in cell walls. Nitrogen is major constituent of protein. In plants the carbon to nitrogen ratio is about 40 to 1. In mammals the ratio is about 14 to 1. The task of converting cellulose and a limited supply of plant protein into animal falls to the plant eaters. The diversity of potential sources of energy in the form of plant and animal tissues requires an equally diverse array of physiological, morphological, and behavioral characteristics enabling animals to acquire and assimilate these resources.

Means of Acquiring Nutrients

There are many ways to classify organisms based on the resources they utilize and the means by which they exploit these resources. The most general of these classifications is the division based on how animals use plant and animal tissues as a source of food. Those organisms that rely on plant tissues as food are called **herbivores.** Those organisms that feed on animal tissues are classified as **carnivores.** Organisms that feed on both plant and animal tissues are called **omnivores.** And organisms that feed on dead plant and animal matter, called detritus, are detrital feeders or **detritivores.** Each of these four feeding groups has characteristic adaptations that allow them to exploit these different diets.

Herbivory Herbivory involves the consumption of plant material by animals, which can be categorized further by the type of plant material they eat. **Grazers** feed on leafy material, especially grasses. **Browsers** feed on mostly woody material. **Granivores** feed on seeds and **frugivores** eat fruit. Others types of herbivorous animals such as avian sapsuckers (*Sphyrapicus* spp.) and sucking insects such as aphids feed on plant sap, and hummingbirds, butterflies, moths, and ants feed on plant nectar.

Grazing and browsing herbivores, with some exceptions, live on diets high in cellulose. In doing so, they face several dietary problems. Their diets are rich in carbon, but low in protein. Most of the carbohydrates are locked in indigestible

cellulose and the proteins exist in chemical compounds. Lacking the enzymes needed to digest cellulose, herbivores depend on specialized bacteria and protozoa living mutually (see Chapter 16) in their digestive tracts. These bacteria and protozoans digest cellulose and proteins, and they synthesize fatty acids, amino acids, proteins, and vitamins. For most vertebrates, bacteria and protozoans are concentrated in the foregut or hindgut, where they carry on anaerobic fermentation. In herbivorous insects, bacteria and protozoans inhabit the hindguts. Some species of cellulose-consuming wood beetles and wasps depend on fungi. These insects carry fungal spores with them externally when they invade new wood.

Ruminants, such as cattle and deer, are exemplary cases of herbivores anatomically specialized for the digestion of cellulose. They possess a highly complex digestive system consisting of a four-compartment stomach—the rumen (from which the group gets its name), reticulum, omasum, and abomasum or true stomach (Figure 8.1) and a long intestine. The rumen and reticulum, inhabited by anaerobic bacteria and protozoans, function as fermentation vats. The ruminants provide the microbes with cellulose. The ruminants also regulate

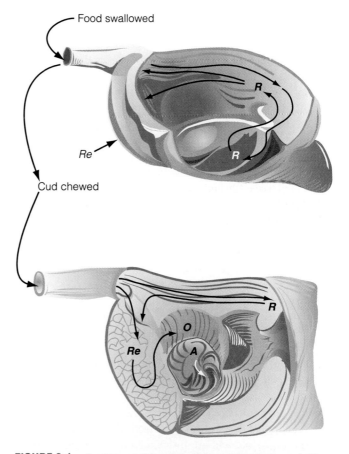

FIGURE 8.1 Anatomy and function of the ruminant stomach. The four-compartment stomach consists of the rumen (*R*), reticulum (*Re*), omasum (*O*), and abomasum (*A*). As indicated by the arrows, the consumed food enters the rumen. The ruminant regurgitates boluses of the fermented material (cud) and rechews it. Finer material enters the reticulum, then the omasum and abomasum. The coarser material reenters the rumen for further fermentation.

the pH and chemistry in the rumen by adding bicarbonates and phosphates from highly developed salivary glands (Hungate 1975, Hofmann 1989).

As ruminants graze, they chew their food hurriedly. The consumed material descends to the rumen and reticulum, where it is softened to a pulp by the addition of water, kneaded by muscular action, and fermented by bacteria. At leisure the animals regurgitate the food, chew it more thoroughly to reduce plant particle size, making it more accessible to microbes, and swallow it again. The mass again enters the rumen. Finer material is pulled into the reticulum, and from there it is forced by contraction into the third compartment, or omasum. The opening into the omasum dictates the size and amount of material entering it. In the omasum the material is further digested and finally forced into the abomasum, or true glandular stomach.

The digestive process carried on by the microorganisms produces short-chain volatile fatty acids. These acids are rapidly absorbed through the wall of the rumen into the bloodstream and are oxidized, providing the ruminant with a major source of energy. Part of the material in the rumen is converted to methane and lost to the animal, and part remains as fermentation products. Many of the microbial cells involved are digested in the abomasum to recapture still more of the energy and nutrients. Further bacterial action breaks down complex carbohydrates into sugars. In addition to carrying on fermentation, the bacteria also synthesize B-complex vitamins, the enzyme cellulase necessary for carbohydrate metabolism, and amino acids.

Variations exist among the digestive processes of ruminants and the kinds of forage they use. In general, three feeding types have evolved among ruminants, none of which is restricted to any one taxonomic group (Hofmann 1989). Grass and roughage eaters, such as cattle, sheep, and water buffalo (*Bubalus bubalus*), which make up about 25 percent of the 150 ruminant species, including six domestic species, are adapted to eat fibrous forage. They graze for long periods of time and then spend an almost equally long period resting and ruminating. About 40 percent of the ruminant species are concentrate selectors. These include the giraffe (*Giraffa* spp.), greater kudu (*Tragelaphus strepsiceros*), dikdiks (*Madoqua* spp.), and roe deer (*Capreolus capreolus*). Their digestive systems are far less suited to digest plant fibers. They seek high-quality, easily digested forage rich in soluble cell contents, such as soft, juicy dicot plants. They graze for short periods of time, followed by short periods of rumination. The plant material stays for only a short period of time in the relatively small rumen. The rest of the ruminants are intermediate between the other two types. These ruminants include red deer (*Cervus elaphus*), reindeer (*Rangifer tarandus*), impala (*Aepyceros melampus*), and gazelles (*Gazella* spp.). They eat a mixed diet, but forage selectively. They avoid fiber as long as possible and are influenced by seasonal fluctuations in forage quality. The white-tailed deer (*Odocoileus virginianus*), for example, eats low-fiber forages such as dicot herbs and leaves, young woody growth, and acorns, as long as they are available. During the winter season their only available food may be mostly fibrous woody plant material. Like concentrate feeders, these intermediates cannot digest

fibrous forage efficiently, because they lack the appropriate bacteria to produce the enzymes needed to break down the complex carbohydrates.

Grazing marsupials, such as the kangaroos, also possess a ruminant type of digestive system, which has arisen by parallel evolution. Although the stomach is not divided into four compartments, it has regions analogous to the rumen, reticulum, abomasum, and omasum.

Most of the digestion in ruminants takes place in the foregut. Among nonruminants, such as rabbits and horses, digestion takes place less efficiently in the hindgut. Nonruminant vertebrate herbivores, such as horses, have simple stomachs, long intestinal tracts that slow the passage of food through the gut, and a well-developed caecum, a blind pouch attached to the colon of the intestine where fermentation takes place. Among such herbivores fermentation follows rather than proceeds the small intestine. In many wood-eating insects a region of the hindgut, the ileum, is dilated to form a fermentation pouch housing bacteria or protozoa that digest wood particles. Such hindgut fermentation usually results in incomplete digestion. Such herbivores, however, can process large quantities of forage rapidly, even if inefficiently. Lagomorphs—rabbits, hares, and pikas— resort to a form of **coprophagy,** the ingestion of fecal material for further extraction of nutrients. Part of ingested plant material enters the caecum and part enters the intestine to form dry pellets. In the caecum the ingested material is processed by microorganisms and is expelled into the large intestine as soft, green, moist pellets surrounded by a proteinaceous membrane. The lagomorphs reingest the soft pellets, which are much higher in protein than and lower in crude fiber than the hard fecal pellets. The amount of feces recycled by coprophagy ranges from 50 to 80 percent. The reingestion is important because the pellets, functioning as "external rumen," provide bacterially synthesized B vitamins and ensure a more complete digestion of dry material and a better utilization of protein (McBee 1971). Coprophagy is widespread among the detritus-feeding animals, such as wood-eating beetles, isopods, and millipedes, that ingest pellets after they have been enriched by microbial activity.

Among marine fish, herbivores are small, found in restricted sets of families, and inhabit coral reefs, where characterized by high diversity they make up about 25–40 percent of the fish biomass. These herbivorous fish feed on algal growth that, unlike the food of terrestrial herbivores, lacks cellulose; but in their feeding these fish ingest large amounts of inorganic material and indigestible organic material. They gain access to the nutrients inside the algal cells by means of one or more of four basic types of digestive mechanisms. In some fish with a thin-walled stomach, low stomach pH weakens algal cell walls and allows digestive enzymes to come in contact with algal nutrients. Fish that possess gizzardlike stomachs ingest along with algae inorganic material that mechanically breaks down algal cells to release nutrients. Some reef fish possess specialized pharyngeal jaws that shred or grind algal material before it reaches the intestine. Other fish depend on microbial fermentation in the hindgut to assist in the breakdown of algal cells. These four types are not mutually exclusive. Some marine herbivores may combine low stomach pH or grinding and shredding with microbial fermentation in the hindgut (Choat and Clements 1998).

Carnivory Herbivores are the energy source for carnivores, the flesh eaters. Organisms that feed directly on the herbivores are termed **first-level carnivores** or second-level consumers. First-level carnivores represent an energy source for the **second-level carnivores.** Still higher categories of carnivorous animals feeding on secondary carnivores exist in some communities. As the feeding level of carnivores increases, their numbers decrease because of limited energy sources.

Unlike herbivores, carnivores are not faced with the digestion of cellulose or the quality of food. Because little difference exists in the chemical composition between the flesh of prey and the flesh of predators, there is no problem in digestion and assimilation of nutrients from their prey. Carnivores regurgitate indigestible bones, hair, feathers, and chitin. Their major problem is obtaining a sufficient quantity of food.

Lacking the need to digest complex cellulose compounds, carnivores have a short intestines and simple stomachs. The stomach is little more than an expanded hollow tube with muscular walls. It stores and mixes foods, and its intrinsic glands add mucus, enzymes, and hydrochloric acid to speed digestion.

Omnivory Omnivory relates to feeding on more than one feeding or trophic level. Most commonly omnivory involves animals that feed on both plants and animals. The red fox (*Vulpes vulpes*) feeds on berries, apples, cherries, acorns, grasses, grasshoppers, crickets, beetles, and small rodents. The black bear (*Ursus americanus*) feeds heavily on vegetation—buds, leaves, nuts, berries, tree bark—supplemented with bees, beetles, crickets, ants, fish, and small to medium-sized mammals. The food habits of many omnivores vary with the seasons, stages in the life cycle, and their size and growth (Figure 8.2).

Detritius Feeding Animals that feed on fallen leaves, twigs, and other dead organic material, collectively called **detritus,** are **detritivores.** Mostly invertebrates, they are so closely associated with the breakdown of organic matter and decomposition that the detritivores are considered in the following chapter on decomposition. Like herbivores, detritivores depend heavily on mutualistic relations (see Chapter 16) with microorganisms to aid in the breakdown of cellulose and lignins.

Nutritional Needs

Animals require mineral elements and 20 amino acids, of which 14 are essential. Amino acids make up proteins and vitamins. These needs differ little among vertebrates and invertebrates. Insects, for example, have the same dietary requirements as vertebrates, although they need more potassium, phosphorus, and magnesium than vertebrates and less

calcium, sodium, and chlorine (Dodd 1973). The ultimate source of most of these nutrients is plants. For this reason, the quantity and quality of plants affect the nutrition of consumers. When the amount of food is insufficient, consumers may suffer from acute malnutrition, leave the area, or starve. When food is of low quality, it reduces reproductive success, health, and longevity.

The highest quality plant food for herbivores, vertebrate and invertebrate, is high in nitrogen in the form of protein (Mattson 1980, Strong et al. 1984). As the nitrogen content of their food increases, assimilation of plant material improves, increasing growth, reproductive success, and survival. Nitrogen is concentrated in the growing tips, new leaves, and buds of plants. Its content declines as leaves and twigs mature and become senescent. Herbivores have adapted to this period of new growth. Herbivorous insect larvae are most abundant early in the growing season and complete their growth before the leaves mature. Many vertebrate herbivores, such as deer, give birth to their young at the start of the growing season, when the most protein-rich plant foods will be available for their growing young.

Although food selection is strongly influenced by availability and season, herbivores, both vertebrate and invertebrate, do show some preference for the most nitrogen-rich plants, which they probably detect by taste and odor (Verme and Ullrey 1984). For example, beavers show a strong preference for willows and aspen, two species that have a very low C:N ratio. Chemoreceptors in the nose and mouth of deer encourage or discourage consumption of certain foods. During drought, nitrogenous compounds are concentrated in certain plants, making them more attractive and vulnerable to herbivorous insects (Mattson and Haack 1987).

The need for quality foods differs among herbivores. Ruminant animals, as already pointed out, can subsist on rougher or lower quality forage because bacteria in the rumen can synthesize such requirements as vitamin B_1 and certain amino acids from simple nitrogenous compounds. Therefore the caloric content and the nutrients in a certain food might not reflect its real nutritive value for the ruminant. Nonruminant herbivores require more complex proteins. Seed-eating herbivores exploit the concentration of nutrients in the seeds. Such animals are not likely to have dietary problems. Among the carnivores, quantity is more important than quality. Carnivores rarely have a dietary problem because they consume animals that have resynthesized and stored protein and other nutrients from plants in their tissues.

Mineral availability and deficiencies appear to influence the abundance and relative fitness of some animals. One essential nutrient that has received attention is sodium, the most variable nutrient in forest and arctic ecosystems (Jordan et al. 1972). In areas of sodium deficiency in the soil, herbivorous animals face an inadequate supply of sodium in their diets. The problem has been noted in Australian herbivores (Blair-West et al. 1968), in African elephants (*Loxodonta africana*) (Weir 1972), in rodents (Aumann and Emlen 1965, Weeks and Kirkpatrick 1978), in white-tailed deer (Weeks and Kirkpatrick 1976), and in moose (*Alces alces*)(Belovsky 1981b, Belovsky and Jordan 1981).

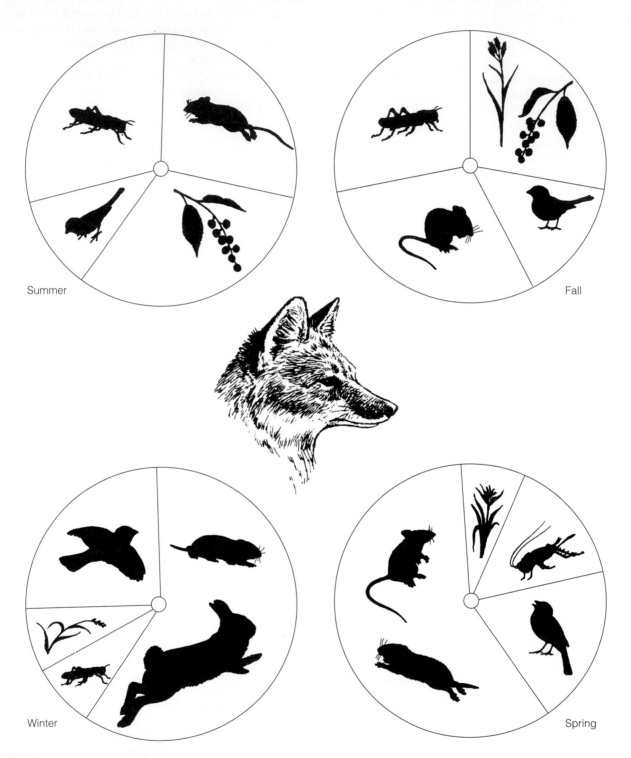

FIGURE 8.2   The red fox (*Vulpes vulpes*) is an example of an omnivore, an animal that feeds on more than one trophic or feeding level. The food habits of the red fox are seasonal. Note the prominence of fruits and insects in summer and of rodents and rabbits in spring and fall.

Sodium deficiency can influence the distribution, behavior, and physiology of mammals, especially the herbivores. The spatial distribution of elephants within Wankie National Park in central Africa appears to be closely correlated with the sodium content of drinking water, with the greatest number of elephants found at waterholes with the highest sodium content. In sodium-deficient areas in southwestern Australia, the European rabbit builds up reserves of sodium in its tissues during the nonbreeding season. These reserves appear to be exhausted near the end of the breeding season, which concludes abruptly. During the breeding season, the rabbits selectively graze on sodium-rich plants to the point of depletion, exhausting sodium availability.

Ruminants face severe mineral deficiency problems in spring. Attracted by the flush of new growth, deer, bighorn sheep (*Ovis canadensis*), mountain goats (*Oreamnos americanus*), elk (*Cervus elaphus*), and domestic cattle and sheep feed on new succulent grass, but with high physiological costs. Spring vegetation is much higher in potassium relative to calcium and magnesium than during the rest of the year (Figure 8.3). This high intake of potassium stimulates the adrenal gland to increase the excretion of aldosterone (Weeks and Kirkpatrick 1976). Aldosterone is the principle mineral corticoid that promotes retention of sodium by the kidney. While aldosterone stimulates the retention of sodium, it also facilitates the excretion of potassium and magnesium. Because concentrations of magnesium in soft tissues and skeletal stores are low in herbivores, these animals experience magnesium deficiency. This deficiency results in a rapid onset of diarrhea and often a neuromuscular derangement or tetany. The deficiency comes late in gestation for females and at the beginning of antler growth for male deer and elk, times when mineral demands are high.

To counteract this mineral imbalance in the spring, the large herbivores seek mineral licks, places in the landscape where animals concentrate to satisfy their mineral needs by eating mineral-rich soil (Figure 8.4). Although sodium chloride is associated with mineral licks, animal physiologists hypothesize that it is not sodium the animals seek but magnesium, and in the case of bighorn sheep, mountain goats, and elk, calcium as well (Jones and Hanson 1985).

Moose of the northern forest ecosystem face a particular problem obtaining sufficient sodium (Belovsky and Jordan 1981). Northern terrestrial vegetation contains only 3 to 28

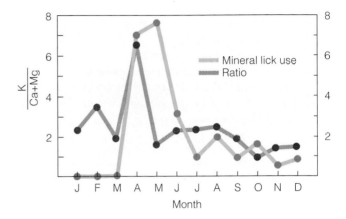

FIGURE 8.3 Monthly ratio of K to Ca+Mg in the diet compared with an index of the intensity of use of mineral licks by white-tailed deer (*Odocoileus viginianus*) in Indiana. Use of mineral licks by deer was most intense when the ratio of K to Ca+Mg in the diet was the lowest in the spring and the mineral needs the greatest. (From Jones and Weeks 1985:132.)

FIGURE 8.4 A mineral lick used by white-tailed deer.

ppm of sodium, far below the nutritional requirements of ruminants. To circumvent sodium stress, moose feed in summer on submerged aquatic plants, which contain between 2 and 9400 ppm of sodium. Moose may consume from 63 to 95 percent of the production of submerged aquatics, particularly pond lilies. Such heavy consumption appears to trigger declines in aquatic vegetation. The resulting decreased availability of sodium may be one reason for the periodic decline of moose populations in parts of their range (Belovsky 1981b).

The size of deer, their antler development, and their reproductive success all relate to nutrition (Moen and Severinghaus 1983). Other factors being equal, only deer obtaining high-quality foods grow large antlers. Deer on diets low in calcium, phosphorus, and protein show stunted growth, and bucks develop only thin spiky antlers (French et al. 1955). Reproductive success of does is highest where food is abundant and nutritious (Figure 8.5a). On the best range in New York State, 1.71 fawns on the average were born each year for each reproductive doe (Figure 8.5b). On a poor range,

however, average fawn production per doe was only 1.06 per year (Cheatum and Severinghaus 1950).

Other studies show a more general relationship between the level of soil fertility and the health of mammals. Studies of the relationship between soil and cottontail rabbits in Missouri showed no significant differences in body weights of rabbits collected from areas with soils of contrasting fertility (Williams 1965), but there was a positive correlation between soil fertility and fecundity both in Missouri (Williams and Cashey 1965) and in Alabama (Hill 1972) (Figure 8.6).

Link Between Food Source, Species Morphology, and Behavior

Consumers represent a diverse group of organisms that utilize a wide array of potential food sources. Each type of food used by animals presents a unique set of constraints related to the ability of the organisms to acquire and assimilate the food item. These constraints directly influence the evolution of characteristics relating to the physiology, morphology, and behavior of the species. The morphology of each species allows it to exploit a given food resource; however, it also functions to restrict its ability to exploit other, different food resources. The morphology required to extract nectar from a flower is quite different from that required to feed on insects visiting the flower or to feed on its seeds.

An example of the link between food resource, foraging behavior, and morphology is presented in Figure 8.7. Three species of bird that exploit different food resources are shown in Figure 8.7a: the ruby-throated hummingbird (*Archilochus colubris*), Figure 8.7b: the ovenbird (*Seiurus aurocapillus*), and Figure 8.7c: the black-backed woodpecker (*Picoides articus*). The hummingbird is a nectivore, feeding on the nectar

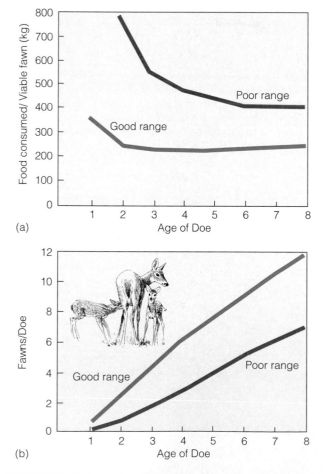

FIGURE 8.5   Differences in the reproductive success of female white-tailed deer on good and poor ranges in New York State. (a) Food consumed per viable fawn on poor range (Adirondack Mountains) was much greater than food consumed on good range (western New York). (b) Reproductive success was considerably greater on good range than poor range. (Data from Cheatum and Severinghaus 1950.)

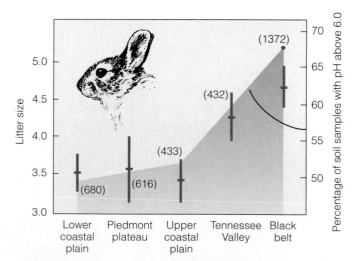

FIGURE 8.6   The mean size of second cottontail rabbit (*Sylvilagus floridanus*) litters from five Alabama soil regions plotted against the percentage of soil samples of pH above 6.0, indicative of high-fertility soils. Vertical lines give the range of two standard deviations. The numbers in parentheses are the numbers of soil samples analyzed for each soil region. Note the strong relationship between soil fertility and litter size. (From Hill 1972:1201.)

FIGURE 8.7 Three bird species: (a) the ruby-throated hummingbird (*Archilochus colubris*), (b) the ovenbird (*Seiurus aurocapillus*), and (c) the black-backed woodpecker (*Picoides articus*), and the substrates on which these three species forage. Patterns of habitat selection reflect the ability of the vegetation to provide suitable resources and foraging substrates given the resource needs of the species.

of flowers. Its morphology is designed for hovering over flowers, extracting nectar using its long thin beak. The ovenbird is an insectivore, feeding on macroarthropods in the leaf litter of the forest floor. It forages by walking along the forest floor, overturning the leaf litter, and capturing small insects. The woodpecker is also an insectivore, but it feeds by gleaning insects from the bark and boring into the trunks of trees to feed on insect larvae. Its short tail, and long tarsus and toes allow it to move vertically along the tree trunk.

The morphology of each of these bird species reflects the constraints of the food resource and the associated foraging behaviors required for its acquisition. In addition, these food resources are associated with a distinct substrate. For the hummingbird, it is the flowers from which nectar is extracted. In the case of the ovenbird, it is the broadleaf deciduous leaf litter on the forest floor. For the woodpecker, it is the trunk and larger branches of large coniferous trees. Each of these substrates used for foraging is associated with a distinctive vegetation type. The flowers are found in abundance in old fields, the leaf litter in mature deciduous forests, and the large trunks of coniferous trees in mature pine forests. The link among food resources, foraging substrate, and vegetation type shown in Figure 8.7 is the basis in part for the selection of differing vegetation types as habitat by these three species (see discussion of habitat selection in Chapter 12).

ANIMAL ADAPTATIONS TO THE THERMAL ENVIRONMENT

The physical aspects of the habitat, dominated in terrestrial environments by vegetation, directly or indirectly, influence the thermal and moisture environment experienced by animals. The structure of vegetation ameliorates the environment, provides refuge from environmental extremes, and modifies thermal radiation and associated moisture relations within the organisms' habitat. These conditions influence the physiological and behavioral responses of animals to their changing environment.

Thermal Balance

For an organism to maintain a somewhat constant body temperature, heat gained by the body must be balanced by heat losses. Heat exchange with the environment takes place by four major means, discussed in Chapter 3: conduction, convection, radiation, and evaporation. Added to this is metabolic heat production, part of which remains in the body and part of which is lost to the environment. The heat balance of an organism can be described by (Schmidt-Nielsen 1997)

$$H_{tot} = H_c \pm H_{cd} \pm H_r \pm H_e \pm H_m$$

where H_{tot} is the rate of metabolic heat production; H_c, the rate of heat gained or lost through convection; H_{cd}, the rate of heat gained or lost through conduction; H_r, the rate of heat gained or lost through radiation; H_e, the rate of heat lost through evaporation; H_m, the rate of heat storage in body through metabolic processes. The total heat gain of an organism is the sum of heat gain from metabolism plus heat gain from the environment. The relationship among all these factors depends heavily on temperature.

Thermal relations between an animal and its environment can be appreciated best by considering a thermal model of the animal body (Figure 8.8). The core of the animal body, its deep interior, is assumed to be at a uniform temperature, T_b. The ambient temperature of the environment is T_a. The temperature of the body's surface, T_s, because of the boundary layer and other factors, will differ from both T_b and T_a. Separating the body core from the body surface are layers (L) of muscle tissue and fat, across which the temperature gradually changes from T_b to T_a.

To maintain core body temperature, the animal must balance losses and gains. Thus thermal balance in the core of the animal is influenced by heat produced by metabolism; heat stored; heat flow to skin as affected by the thickness and conductivity of fat, fur, hair, feathers, and scales; heat flow to the ground; and heat lost by evaporation. A general formula for heat balance in animals is

$$M = K_o(T_b - T_a)$$

where M is effective net metabolic heat production (which in ectotherms would be minimal or nonexistent), K_o is overall thermal conductance of the organism, T_b is body core temperature, and T_a is environmental temperature. K_o is a constant for a given set of conditions, variable among species. It is a function affected by surface area, conduction, convection, and radiation (see Chapter 3). The larger the value of K_o, the greater is the loss of heat.

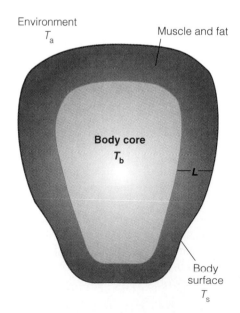

FIGURE 8.8 Temperatures in an animal body. Body core temperature is T_b, the ambient temperature is T_a, the surface temperature is T_s, and L is the thickness of the outer layer of the body.

How animals confront thermal stress is influenced heavily by their physiology and environment. Because air has a lower specific heat than water and absorbs less solar radiation, terrestrial animals are subject to more radical and potentially dangerous changes in their thermal environment than are aquatic organisms. Incoming solar radiation can produce lethal high temperatures; and radiational loss of heat to the air, especially at night, can result in lethal low temperatures. Aquatic animals live in a more stable energy environment, but generally they have a lower tolerance of temperature change.

Physiologically, animals can be divided into three groups: (1) those that maintain a fairly constant internal temperature regardless of external temperature, such as birds and mammals; (2) those that allow their body temperature to vary with ambient temperature, such as invertebrates, fish, amphibians, and reptiles; and (3) those that sometimes regulate their body temperature and sometimes do not, such as bees and bats. Animals belonging to the first group are **homeotherms** (Greek *homeo,* "the same"); to the second group, **poikilotherms** (Greek *poikilos,* "manifold" or "variegated"); and to the third group, **heterotherms** (Greek *hetero,* "different").

Homeotherms maintain a rather constant body temperature in the face of changing environmental temperatures by means of their own oxidative metabolic production of heat, called **endothermy.** Poikilotherms maintain their body temperature by using sources of heat energy such as solar radiation and reradiation rather than metabolism. This mechanism of maintaining body temperature is called **ectothermy.** Heterotherms utilize both endothermy and ectothermy, depending on environmental situations and metabolic needs. The terms *homeotherm* and *endotherm,* and *poikilotherm* and *ectotherm,* are often used synonymously, but there is a difference. The terms *ectotherm* and *endotherm* emphasize the mechanisms by which body temperatures are determined; the other terms emphasize the nature of the variations in body temperature.

Poikilotherms

Poikilotherms, such as amphibians and insects, have a high thermal conductance between the body and the environment and a low metabolic rate. For this reason body temperature and thus tissue temperatures change with environmental temperatures. Among such organisms the rate of metabolism at rest increases exponentially with increasing temperature. This relationship between a rise in environmental temperature and the rate of metabolism is described by van't Hoff's rule: for every 10°C rise in temperature, oxygen consumption rates and, therefore, metabolic rates double. This factorial increase in metabolism over each 10°C increment of temperature is called the temperature coefficient or Q_{10}:

$$Q_{10} = R_T/R_{T-10}$$

Here R_T is the rate at any given body temperature T and R_{T-10} is the rate at body temperature $T - 10$°C.

For example, suppose an organism has a metabolic rate of 1.8 cal/hr at a body temperature of 20°C and a rate of 1 cal/hr at 10°C. Q_{10} would be 1.8. For every 10°C rise in temperature, the metabolic rate would increase by a factor of 1.8. The Q_{10} resting metabolism for most organisms whose body temperatures follow environmental temperatures is, by van't Hoff's rule, around 2. However, metabolic rates are not truly exponential, and so Q_{10} is not constant over all temperatures.

Within the range of temperatures that poikilothermic animals can tolerate, the rate of metabolism, and therefore oxygen consumption, changes (Figure 8.9). Rising temperatures increase the rate of enzymatic activity according to van't Hoff's law, which controls metabolism and oxidation of carbohydrates. Lacking any physiological homeostatic mechanisms, terrestrial poikilotherms must depend on behavioral control over body temperatures. Lizards may vary their body temperatures no more than 4 to 5°C when active, amphibians 10°C. The range of body temperatures over which ectotherms carry out their daily activities is called the **active temperature range (ACT)**. By limiting their ACT, poikilotherms can adjust their physiological and developmental processes within a limited range at a low metabolic cost. Within this ACT, many poikilotherms have a low thermal sensitivity to a certain range of acute temperature changes. For example, barnacles of the intertidal zone may have a body temperature of 30°C at low tide when exposed to sun and air. Immersed in ocean water of 14°C at high tide, their body temperature drops to 16°C, yet their resting rate of metabolism shows little thermal sensitivity within that temperature range (Newell and Northcroft 1965). The evolution of such low thermal sensitivity adds some stability to their metabolic processes in an environment of recurring sharp changes in temperature.

Being ectothermic has advantages and disadvantages. Prisoners of environmental temperatures, poikilotherms of temperate regions, such as snakes, can become highly active only when the temperature is sufficiently warm. Because metabolic activity declines with decreasing temperature, these animals become sluggish in the cool of morning and evening. Similarly, they have to restrict their active life to the late spring, summer, and early fall. During periods of intense physical activity, when energy consumption is high, poikilotherms depend on the anaerobic breakdown of glycogen for energy. This breakdown results in an accumulation of lactic acid in the tissues that can be oxidized only after activity ceases. Anaerobic metabolism severely limits bursts of poikilothermic activity to a few minutes because of physical exhaustion. This tendency to exhaustion is one reason that so many predatory terrestrial poikilotherms such as snakes and alligators secure prey by ambush rather than by chase.

Because they do not depend on internally generated body heat, poikilotherms can reduce metabolic activity during periods of temperature extremes and of food or water shortage. Low energy demands enable poikilotherms to colonize areas of limited food and water, such as deserts. Because they do not have the problem of metabolic heat loss, poikilotherms are not limited to any minimum size or definite shape. Many are single celled, such as paramecia and amoebas; others, such as earthworms, millipedes, and snakes, have cylindrical bodies. Such characteristics enable poikilotherms to exploit resources and habitats unavailable to homeotherms. On the other hand, the same metabolic restrictions impose an upper size limit: ectotherms would not be able to absorb enough heat to warm a very large body. For this reason some paleontologists argue that the large dinosaurs had to be endothermic. A counterargument is that large ectotherms could develop and maintain body temperatures above air temperatures in a tropical environment because their low surface-to-volume ratio would limit cooling.

Adaptations to High Temperature Although poikilotherms as a group do not have a lower limit to body size, size is nevertheless an important aspect of their life (Stevenson 1985). The rate of heating and cooling in a poikilotherm decreases as size increases. That fact determines the behavioral options open to a poikilotherm for controlling T_b. Consider a beetle about 10 mg in weight. Because of its small weight, it heats and cools quickly, but rarely can it raise its T_b above ambient. However, it does have the ability to control its T_b by moving in and out of the sun. If the temperature of a rock or a log on which it rests becomes too hot, it can easily find a shaded crevice. This ability allows very small poikilotherms to make maximum use of favorable environmental temperatures.

Associated with the beetle on a rock may be a small lizard of about 0.2 kg. It, too, is small enough to heat up fast to an active temperature that allows it to make maximum use of favorable thermal conditions during the day. If the temperature becomes too hot, the lizard can quickly retreat to a cooler site. By changing locations, the lizard can maintain some level of stability in its body temperature.

For large terrestrial poikilotherms, those over 10 kg, such as very large lizards and tortoises, life is considerably different. Because they heat slowly, large poikilotherms need a much higher temperature than smaller ones to reach the

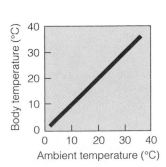
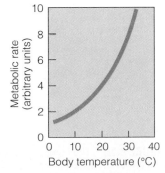

FIGURE 8.9 Relationship among body temperature, resting metabolic rate, and ambient temperatures in poikilotherms. (a) Body temperature is a function of ambient temperature. (b) Resting metabolism is a function of body temperature. (After Hill and Wyse 9189:83.)

same active T_b. Further, because large poikilotherms have difficulty locating shelter from environmental extremes, they are more or less restricted to environments with small seasonal fluctuations in temperature. Large poikilotherms, however, have the advantage of moving freely about in space and time during the day because their large body mass buffers their body temperatures.

Most aquatic poikilotherms (fish and aquatic invertebrates) generally encounter temperature fluctuations of lesser magnitude and usually have more poorly developed behavioral and physiological thermoregulatory capabilities than terrestrial forms. Immersed in a watery environment, most of these animals do not maintain any appreciable difference between their body temperature and that of the surrounding water. Any heat produced in the muscles is transferred to blood flowing through them, and carried to the gills and skin, where it is lost to the water by convection and conduction. Because of the close relationship between body temperature and environmental temperature, fish are readily victimized by rapid changes in water.

Aquatic poikilotherms live with limits of **thermal tolerance,** the range of temperature bounded by upper and lower limits within which an aquatic poikilotherm, particularly fish, have the highest survival. Within this range fish may seek certain preferred temperatures, which will vary seasonally.

Responses of poikilothermic animals to changes in thermal environments and the limits of thermal tolerance can be simulated in the laboratory experiments. Adaptations or adjustments that take place under experimental rather than natural conditions are called acclimation. Groups of fish are held at various constant temperatures for a sufficient length of time to ensure they are acclimated to that temperature. Then they are tested to determine their upper and lower limits of thermal tolerances at that temperature. Data obtained for a series of different temperatures can be plotted on a single graph (Schmidt-Nielsen 1997). An example is Figure 8.10. The horizontal axis is the acclimation temperature; the vertical axis is lethal temperature. The diagonal broken line is temperature line. If the fish is kept at some temperature on the diagonal line until it is fully acclimated, upper and lower tolerances can be determined for that temperature. A series of such plots establishes a complete range of thermal tolerances. The upper and lower limits for thermal tolerance for any given temperature located on the broken line can be determined.

As the graph suggests, the limits of thermal tolerances are not fixed. Their values are affected by the previous thermal history of the organism. Within limits, poikilotherms can adjust or acclimatize to higher or lower temperatures. If the organism lives at the higher end of the tolerance range, it acclimatizes so that both the lethal high and lethal low temperatures are higher than if it were living within the cooler end of the range. Aquatic poikilotherms can adjust slowly to seasonal temperatures.

The preceding discussion relates to a gradual acclimation to a given temperature. But what happens when this thermal environment changes suddenly? Consider a fish living in warm water of 20°C that is suddenly immersed in water of 4°C. The fish experiences thermal shock. It cannot respond physiologically to such a big temperature change over a short period of time.

Amphibians present a somewhat different situation. Permanently aquatic forms maintain body temperatures in the same manner as fish: they seek preferred temperatures within their habitat. For semiterrestrial and terrestrial frogs and salamanders, adjusting to temperature change is more complex. These salamanders exhibit seasonal variations in body temperature and possess little in the way of behavioral thermoregulation. Generally they are restricted to moist, shaded environments. Semiterrestrial frogs, such as bullfrogs and green frogs, can exert considerable control over their body temperature, which does not, as is often assumed, simply follow air temperature. By basking in the sun, called **heliothermism,** frogs can raise their body temperature as much as 10°C above ambient temperature. Because of associated evaporative water losses, such amphibians must either be near water or partially submerged (Figure 8.11). Forms that live near water also use evaporative cooling through the skin to reduce body heat loads. By changing position or location or by seeking a warmer or cooler substrate, amphibians can maintain body temperatures within a narrow range of variation (Lillywhite 1970).

Reptiles have their own response to temperature. Most are terrestrial and lack the buffering effects of water. Exposed to widely fluctuating temperatures of the terrestrial environment, reptiles must possess more refined means of temperature regulation.

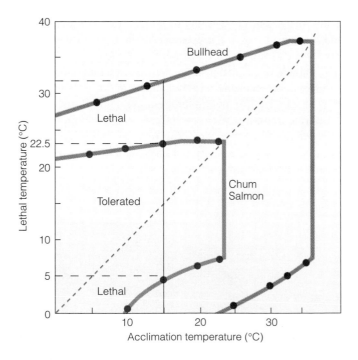

FIGURE 8.10   The complete ranges of thermal tolerance of chum salmon (*Oncorhynchus keta*), a cold-water fish, and the bullhead (*Ictalurus nebulosus*), a warm-water fish, are shown by the solid lines. If a fish is kept at some temperature along the diagonal broken line until fully acclimated, its tolerance to high temperature can be read on the upper solid line. Its tolerance to cold can be read on the lower solid line. For example, a salmon acclimated to 15°C has an upper lethal limit of 22°C and a lower limit of 5°C. (From Schmidt-Nielsen 1997, based on Brett 1968.)

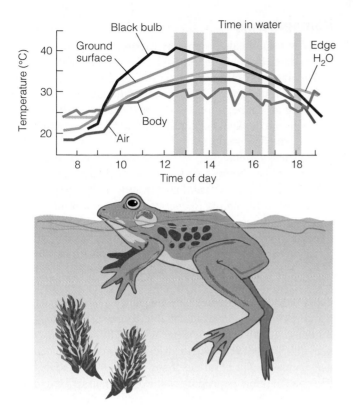

FIGURE 8.11 The body temperature of a bullfrog measured telemetrically. Dips in the black bulb temperature (maximum possible emittance, as measured by a true black body) indicate the effects of cloud cover, convection, or both. Water temperature around the pond's edge varies from one location to another by as much as 2 to 3°C. While in shallow water a frog may show a higher body temperature than that recorded for the edge water. Note how uniform a temperature the bullfrog maintains by moving in and out of the water. (From Lillywhite 1970:164.)

Reptiles exhibit little relationship between their core body temperature and ambient temperature, even though they are poikilothermic. Evaporative cooling by panting and by water loss through the skin keeps the body temperature from reaching the **critical thermal maximum (CTM),** the temperature at which an animal's capacity to move is so reduced that it cannot escape from thermal conditions that will lead to its death (Cowles and Bogert 1944). Thus reptiles possess some of the basic physiological mechanisms so highly developed in endotherms.

The simplest way for a reptile to regulate body temperature is heliothermism; but a more elaborate behavior common to many lizards is **proportional control.** If ambient temperature is lower than preferred, the lizard can spread its ribs, flatten its body, and orient itself so that its body is at right angles to the sun to gain the maximum amount of heat. If the temperature is too high, the lizard can pull its ribs together and orient its body parallel to the sun, decreasing the surface area exposed. Other behavioral means include burrowing into the soil and possibly changing color. Thus reptiles have at their disposal a variety of behavioral mechanisms useful for temperature regulation (Figure 8.12).

Adaptations to Low Temperatures Many poikilothermic animals of temperate and arctic regions must endure long periods of below-freezing temperatures in winter. They escape the cold through supercooling and resistance to freezing.

Supercooling (see Chapter 7) occurs when the body temperature falls below freezing without freezing body fluids. The amount of supercooling that can take place is influenced by the presence of certain solutes, such as glycerol, sorbitol, and mannitol, all polyhydric alcohols, in the body. Some arctic

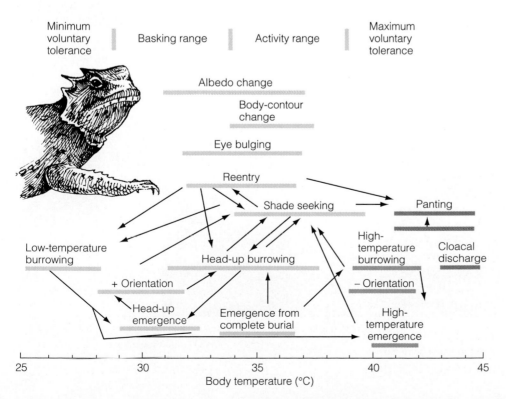

FIGURE 8.12 Behavioral mechanisms in the regulation of body temperature by the horned lizard (*Phrynosoma coronatum*). (From Heath 1965.)

fish, certain insects of temperate and cold climates, and reptiles exposed to occasional cold nights employ supercooling.

Some intertidal invertebrates of high latitudes and certain aquatic insects survive the cold by actually freezing and then thawing when the temperature moderates. In some, more than 90 percent of the body fluids may become frozen, and the remaining fluids contain highly concentrated solutes, including polypeptides and glycopeptides. Ice forms outside shrunken cells, and muscles and organs are distorted. After thawing, they quickly resume normal shape. Other animals, particularly arctic and antarctic fish and many insects, resist freezing because of the presence of glycerol in body fluids. Glycerol protects against freezing damage and lowers the freezing point, increasing the degree of supercooling. Wood frogs (*Rana sylvatica*), spring peepers (*Hyla crucifer*), and gray tree frogs (*H. versicolor*) can overwinter just beneath the leaf litter because they break down the high reserves of liver glycogen into glycerol. They transport it by the way of the bloodstream to other organs, providing protection against freezing. At the same time, these amphibians undergo carefully controlled extracellular ice formation in the body by moving water out of organs and tissues and sequestering it as ice in body spaces about the organs (Storey and Storey 1995). Dormant poikilotherms also experience such physiological changes as lower blood sugar, increased liver glycogen, and increased carbon dioxide levels in the blood, and darkened skin.

In addition to supercooling, many insects exhibiting frost hardiness enter a resting stage called **diapause,** characterized by a cessation of feeding, growth, mobility, and reproduction. Among many insects diapause is a genetically determined, obligatory resting stage before development can proceed. It is timed mostly through photoperiod and is associated with falling temperatures. Diapause prevents the appearance of a sensitive stage of development at a time when low temperatures would kill. Diapause ends with the lengthening of days and the return of warm temperatures.

Homeotherms

Birds and mammals escape the thermal constraints of the environment by being endothermic. They maintain their thermal metabolic optimum by oxidizing glucose and other energy-rich molecules. They regulate the gradient between body and air temperatures by seasonal changes in insulation (the type and thickness of fur, structure of feathers, and layer of fat), which poikilotherms do not possess, by evaporative cooling, and by increasing or decreasing metabolic heat production. Homeothermy allows these animals to remain active regardless of environmental temperatures, although at high energy costs. The ability to operate at high temperatures provides homeotherms with great endurance and a means of remaining active at low temperatures.

However, homeothermy has placed a size restraint on animals possessing it. A close relationship exists between body size and basal metabolic rate of a resting, fasting animal. For each species there is a range of environmental temperatures within which the metabolic rates are minimal, the **thermoneutral zone** (Figure 8.13). Outside this zone, marked by upper and lower **critical temperatures,** metabolism increases. The basal metabolic rate, as measured by oxygen consumption, is proportional to body mass raised to the $\frac{3}{4}$ (0.75) power. As body weight increases, the weight-specific metabolic rate decreases. Conversely, as body mass decreases, basal metabolism increases (exponentially with very small body size) (Figure 8.14).

Within any taxonomic group of endotherms, small animals have a higher metabolic rate per unit of body weight than large ones. Part of the reason lies in the ratio of surface area to body mass. An animal loses heat to the environment in proportion to the surface area exposed relative to volume of body mass. Given the same environmental conditions, a large animal of a similar shape loses proportionately less heat to the environment than a small one. To maintain a constant internal temperature, small

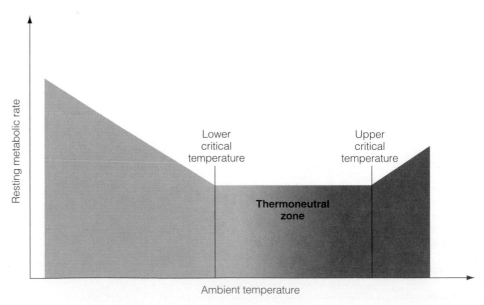

FIGURE 8.13   General resting metabolic response of homeotherms to changes in ambient temperature.

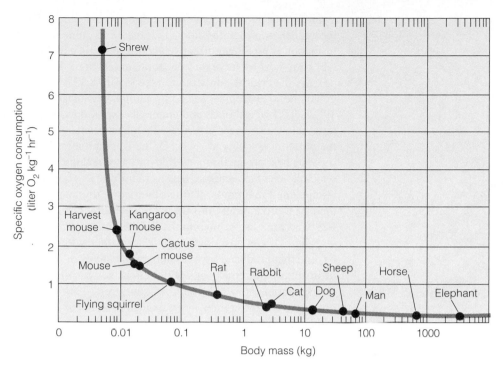

FIGURE 8.14 Observed rates of specific oxygen consumption per unit body mass by various mammals. Oxygen consumption increases rapidly with decreasing body mass. Note that the abscissa has a logarithmic scale and the ordinate an arithmetic scale. (After Schmidt-Nielson 1979:1885.)

endothermic animals have to burn energy rapidly. In fact, weight-specific rates of small endotherms rise so rapidly that below a certain size they could not meet energy demands. In general, 5 grams is about as small as an endotherm can be and still maintain a metabolic heat balance. Few endotherms ever get smaller than that. Two who do are the world's smallest mammals: the pygmy white-toothed shrew (*Suncus etruscus*) of Africa and Kitti's hog-nosed bat (*Craseonycteus thonglongyai*) of Thailand, both of which weigh 2 grams as adults.

To regulate heat exchange between themselves and the environment, homeotherms utilize a number of physiological and morphological mechanisms. One is a countercurrent heat exchanger (Figure 8.15). Some mammals, particularly those of the Arctic, such as porpoises and whales, have extensive areas in the extremities where the veins are closely juxtaposed to the arteries. Much of the heat lost by outgoing arterial blood is picked up or exchanged to the returning venous blood. Thus the venous blood is warmed on its return and reenters the body core only slightly cooler than the outgoing arterial blood. Such vascular arrangements are common in the legs of mammals and birds, and in the tails of rodents, especially the beaver (*Castor canadensis*).

The arteries and veins of many endotherms are divided into a large number of small, parallel, intermingling vessels that form a discrete vascular bundle or net known as a **rete**. Within the rete the blood flows in two directions and a heat exchange takes place. Such a rete in the head functions to cool the highly heat-sensitive brain of the oryx, an African antelope exposed to high daytime temperatures (Figure 8.16). Among some Arctic mammals, such as the arctic fox (*Alopex lagopus*), countercurrent circulation prevents excessive heat loss through the extremities. The footpads of this fox as well as wolves and sled dogs routinely cool to near 0°C (Irving 1972, Hill 1976).

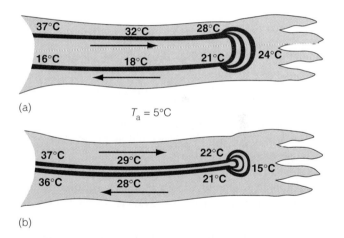

FIGURE 8.15 A model of countercurrent flow in the limb of a mammal, showing hypothetical temperature changes in the blood (a) in the absence and (b) in the presence of countercurrent heat exchange.

Physiologically it is more difficult for homeotherms to adapt to high temperatures than to cold. To maintain T_b they must lose heat to the environment. If the ambient temperature is the same or exceeds the core body temperature, then heat will flow from the environment into the animal and induce heat stress. To counteract this stress many animals employ evaporative cooling, largely through sweating and panting, but at the risk of reducing their water balance. **Panting** is very rapid shallow breathing that increases evaporation from the upper respiratory tract.

Birds do not sweat, and an insulating covering of feathers inhibits water loss through the skin. Birds lose body heat largely through radiation, conduction, and convection. When conditions

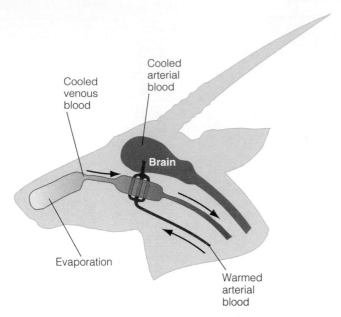

FIGURE 8.16   A desert gazelle can keep a cool head in spite of a high body core temperature by means of a rete. Arterial blood passes in small arteries through a pool of venous blood cooled by evaporation as it drains from the nasal region.

demand it, birds can decrease their heat load by evaporative cooling through panting. However, panting requires work, and work only adds more metabolic heat. Some groups of birds, particularly the goatsuckers (Caprimulgidae), owls (Strigidae), pelicans (Pelecanidae), boobies (Sulidae), doves (Columbidae), and gallinaceous (chickenlike) birds (Phasianidae), avoid this dilemma by **gular fluttering,** the rapid vibration of the gular area while keeping the mouth open. Supplementing panting, it increases air flow over highly vascular oral membrane. Evaporative cooling by gular fluttering uses less energy than panting.

Parts of the body may serve as thermal windows in the radiation of heat. The large ears of such desert mammals as the kit fox (*Vulpes macrotis*) and the antelope jackrabbit (*Lepus alleni*) function as efficient radiators to the cooler desert sky (Schmidt-Nielsen 1964). By seeking shade, where the ground temperatures are low and solar radiation is screened out, or by sitting in depressions, where radiation from the hot ground surface is obstructed, the jackrabbit can radiate 5 kcal/day through its two large ears (400 cm²). This loss is equal to one-third of the metabolic heat produced in a 3 kg rabbit. Such a radiation loss alone reduces enough heat without much loss of water.

Some animals of arid regions simply avoid heat by adopting nocturnal habits and remaining underground or in the shade during the day. Some desert rodents that are active by day periodically seek burrows and passively lose heat through conduction by pressing their bodies against burrow walls.

Birds and mammals can maintain their core body temperatures within a narrow range by changes in the insulating thickness of hair, fur, feathers, and fat. Prior to winter many mammals acquire a heavier coat of hair, which thins out with the coming of warm weather. Birds can add to their thermal

stability in the cold of winter by fluffing their feathers. During sudden or prolonged cold spells there is a point at which insulation is no longer effective and the animals must maintain body heat by increased metabolism (Figure 8.17). This point is the critical temperature (see Figure 8.13). It varies greatly between tropical and arctic animals (Scholander et al. 1950a, b). Tropical birds and mammals exposed to temperatures below 23.5 to 29°C increase their heat production. If air temperature is lowered to 10°C, the tropical animal must triple its heat production; and if lowered to freezing, the animal no longer can generate heat as rapidly as it is being lost. Arctic small mammals, on the other hand, do not increase their heat production until the air temperature has fallen to −29°C. Large arctic mammals, like the musk ox (*Ovibos moschatus*), can sustain the coldest weather without heat beyond that produced by normal basal metabolism. Because of insulation, Eskimo dogs and arctic foxes can sleep outdoors at temperatures of −40°C without stress. This ability is not due to any difference in metabolism itself but to effective insulation and cold acclimation.

When body temperatures fall below the critical level, endotherms increase metabolism by shivering. **Shivering,** the uncoordinated, involuntary, high-frequency contraction of skeletal muscles, converts chemical energy to thermal energy just as voluntary muscular contractions do. Among birds

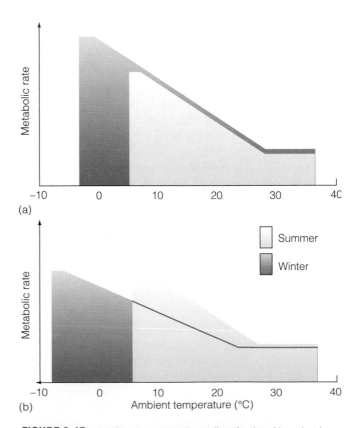

FIGURE 8.17   (a) Simple metabolic acclimatization. Note that in winter the metabolic rate increases as the ambient temperature declines. The plateau on the left indicates maximal metabolic rates. (b) Simple insulatory acclimatization. Note how insulation reduces the metabolic rate in winter and permits tolerance of much lower temperatures.

exposed to cold, shivering and voluntary muscular activity are primary sources of extra heat.

Mammals acclimated to cold temperatures can increase their heat production by **nonshivering thermogenesis** (Hill and Wyse 1988). Nonshivering thermogenesis is the generation of heat from the metabolism of brown fat, a highly vascular brown adipose tissue around the head, neck, thorax, and major blood vessels. Brown fat occurs in the young of most species and in mammals that hibernate; it increases in mass when animals are chronically exposed to low temperatures (Smith and Horwitz 1969, Chaffee and Roberts 1971). Heat generated from the metabolism of this fat is transported to the heart and brain. Nonshivering thermogenesis allows mammals to be active at lower temperatures than if they had to depend on shivering alone. This form of heat production and exercise are additive; the two together can be important in maintaining body temperatures.

Because of the high cost of endothermy, many birds and mammals employ behavioral mechanisms to minimize the need to expel heat or generate it when cold. Animals seek shade (if available) in the heat and will huddle together when cold.

Heterotherms

Among both poikilotherms and homeotherms are species that sometimes regulate their body temperature and sometimes do not, the **heterotherms.** These animals at different stages of their daily and seasonal cycle or under certain environmental situations take on the characteristics of both endotherms and ectotherms. They can undergo rapid, drastic, repeated changes in body temperature, but they also can generate significant amounts of metabolic heat.

Insects are ectothermic and poikilothermic; yet in the adult stage most species of flying insects are heterothermic. When flying, insects have high rates of metabolism with heat production as great as or greater than homeotherms, and they show the same general relationship between body mass and energy metabolism (Figure 8.18). They reach this high metabolic state in a simpler fashion than homeotherms because they are not constrained by pulmonary and cardiovascular pumping systems. Insects take in oxygen by demand through openings or spiracles on the body wall and transport it throughout the body by a tracheal system.

Temperature is critical to the flight of insects. Most cannot fly if the temperature of the thoracic muscles is below 30°C; nor can they fly if the muscle temperature is over 44°C. This constraint means the insect has to warm up before it takes off, and it has to get rid of excess heat in flight. With wings beating up to 200 times per second, flying insects can produce a prodigious amount of heat (Heinrich and Esch 1994).

Some insects, such as butterflies and dragonflies, can warm up by orienting their bodies and spreading their wings to the sun. They warm up by shivering the flight muscles in the thorax. Moths and butterflies may vibrate their wings to raise thoracic temperatures above ambient (Pivnick and McNeil 1986, Rawlins 1980). Bumblebees do the same through abdominal pumping without any external wing movements. They use their flight muscles uncoupled from the

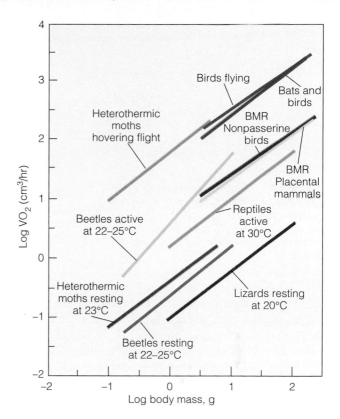

FIGURE 8.18 The regression of energy metabolism on body mass of heterothermic insects and some selected terrestrial vertebrates at rest, during terrestrial activity, and during flight. (From Bartholomew 1981:60.)

wings (Heinrich 1976, 1979). They must maintain a constant thoracic temperature during flight and not lose too much heat through the abdomen. To accomplish both, the bee must maximize its rate of heat loss at a high ambient temperature and minimize its heat loss at a low one.

One heat retention mechanism is insulation. In the bumblebee, the sphinx moth and other moths, and butterflies, the thorax and dorsal surface of the abdomen are insulated with pile (which appears as fine hair). The ventral surface of the abdomen, free of pile, acts as a thermal window. The bumblebee has abdominal air sacs that act as another insulating device to retard heat flow from the thorax to the abdomen. The blood flow between the cooler abdomen and the warm thorax is heated and cooled by a countercurrent mechanism. Cool blood pumped forward by the heart is warmed in the aorta by passing through the flight muscles and, if necessary, can be cooled on the way back to the abdomen (Figure 8.19). Without such heat retention mechanisms cool blood circulating into the thorax would chill the muscles, interfering with their function. In flight on a hot summer day, the insect may need to reduce the thoracic temperatures to prevent overheating. By increasing blood flow to the central abdomen, bumblebees in particular lose excess heat to the air. In very warm situations, bumblebees can employ evaporative cooling by regurgitating fluids, wetting the proboscis, and moving it about in the air.

When crawling about flowers securing nectar, flying insects except the bumblebee suddenly become ectothermic,

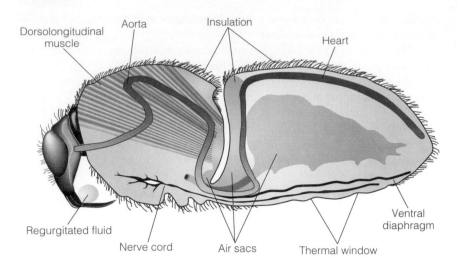

FIGURE 8.19 Temperature regulation in the bumblebee involves insulation on the thorax and the dorsal side of the abdomen. The heart pumps cool blood (hemolymph) to the thorax, where it can be warmed or cooled. (After Heinrich 1976:564.)

and thoracic temperatures may drop to ambient. Bumblebees can maintain a high thoracic temperature when crawling about flowers, permitting immediate flight without another warm-up.

Certain vertebrate poikilotherms that exhibit some degree of thermoregulation employ countercurrent heat exchangers. The swift, highly predaceous tuna and mackerel sharks possess a rete in the band of dark muscle tissue that aids a sustained swimming effort. Metabolic heat produced in the muscle warms up the venous blood, which gives up heat to the adjoining, newly oxygenated blood returning from the gills. Such a countercurrent heat exchange increases the power of the muscles, because warm muscles can contract and relax more rapidly.

To escape extreme temperatures, a few homeotherms become heterothermic, allowing their core body temperatures to fall to near ambient temperature. The simplest response is **torpor,** experienced by a number of birds, such as hummingbirds (Trochilidae) and poorwills (*Phalaenoptilus nuttallii*), and by small mammals such as bats, pocket mice (*Perognathus* spp.), kangaroo mice (*Microdipodops pallidus*), and white-footed mice. Daily torpor is not necessarily associated with any scarcity of food or water. Rather it seems to have evolved as a means of reducing energy demands over that part of the day or night during which the animals are inactive. Arousal returns body temperature rapidly to normal as the animal renews its metabolic heat.

A deeper state of torpor with a substantial drop in metabolic rate is **hibernation,** in which torpor is used make stored energy supplies—body fat or food cache—last through the period of seasonal dormancy. It is characterized by a cessation of coordinated locomotory movements; a reduction of heart rate, respiration, and total metabolism; and a body temperature below 10°C. The animal, however, retains the ability to warm up spontaneously and to emerge periodically from the hibernating state using only endogenously generated heat (French 1988, 1992).

Most animals that hibernate are small and possess high metabolic rates that require a high food intake. During winter or other periods of food scarcity or adverse environmental conditions, these animals could not find sufficient food to survive without hibernation. Hibernation occurs among rodents, bats, some insectivores such as the European hedgehog (*Erinaceus europaeus*), some Australian marsupials such as the eastern pygmy-possum (*Cercartetus nanus*), and some birds such as hummingbirds, swifts (Apodidae), and mousebirds (*Colias* spp.) of Africa.

Entrance into hibernation is a highly regulated sequence of events that coordinates the shutdown of cellular activity (Boyer and Barnes 1999). This shutdown stabilizes physiological systems in the face of profound changes in tissue temperature and availability of oxygen and energy. To ensure a store of energy to draw upon during the hibernation period, some hibernators, such as the woodchuck (*Marmota monax*), feed heavily in late summer to build up large fat reserves. Others, like the chipmunk (*Tamias striatus*), lay up a store of food in underground chambers instead. All hibernators, however, have to acquire a metabolic regulatory mechanism different from that of the active state. Tissues of deeply hibernating mammals must be able to function adequately at both of the temperature ranges at which they exist during the year.

During hibernation, heartbeat is lowered or even intermittent, respiration is reduced, and carbon dioxide levels in the blood are increased, building up respiratory acidosis (Lyman et al. 1982). Acidosis affects cellular processes, inhibits glycolysis, lowers the threshold for shivering, and reduces the metabolic rate. Body temperature decreases during entry into torpor. The heat-generating mechanism for thermoregulation is not activated when the body temperature drops, because the hypothalamic set point for body temperature control apparently is reprogrammed (Boyer and Barnes 1999). Few hibernators remain torpid continuously. Most awaken spontaneously from time to time and then drop back into torpor. The chipmunk with its large store of seeds spends much less time in torpor than the larger fat-storing hibernators. Many chipmunks may not enter torpor at all, even though they remain underground for several months.

As the animal arouses from hibernation it hyperventilates without any change in breathing intervals, CO_2 in the body decreases, and blood pH rises. These changes are followed by shivering in the muscles, resulting in high lactic acid production and a rise in metabolism (Figure 8.20).

Hibernation in the black bear (*Ursus americanus*) and grizzly bear (*Ursus arctos*) is unique, if indeed bears can be considered true hibernators or even heterotherms. Bears do not undergo profound hypothermia. It would take much too long to warm up a large body. Their large size allows them to build sufficient fat reserves to last them through the period of winter dormancy even though metabolism is near normal. Bears move into their hibernating state gradually in the fall when acorns and berries are abundant. They gorge themselves, consuming up to 20,000 calories a day. They seek a den in a cave, beneath an overhanging rock, or in a large hollow tree. Their temperature drops only slightly, from 37 to 35°C, within normal limits for an active bear. Their heartbeat decreases from a summer sleeping rate of 40–50 beats per minute to 8–10 beats per minute. Their metabolism drops to 50–60 percent of normal, requiring the burning of some 4000 calories of energy a day. This high metabolic rate ultimately results in a 20–25 percent loss in body weight. During win-

ter dormancy bears do not drink, eat, defecate, or urinate, although urine forms and enters the urinary bladder (Nelson 1980, Barbosa et al. 1997). Even after bears emerge from hibernation five to six months later, they may not eat or drink for several weeks.

Bears can endure a long period of food deprivation and maintain a high rate of metabolism because they possess a physiological mechanism that changes urea metabolism, so uremia and dehydration do not occur. Urea, water, and other constituents in urine are absorbed by the bladder wall and reenter the bloodstream. There, urea is degraded into amino acids, which are reincorporated into plasma proteins (Nelson et al. 1983, Nelson and Beck 1984).

For the female bear hibernation is also the time of giving birth to two to four young. The blind, almost naked cubs are only eight inches long and weigh 220 to 280 g. By the time the female comes out of hibernation, the young suckling on the sleeping sow will weigh 1.8 to 2.5 kg.

A dormancy similar to that of hibernation is employed by some desert mammals, ground squirrels, and birds during the hottest and driest parts of the year. Such summer dormancy is called **estivation.** There is little evidence to show that estivation is triggered by either heat or drought and no clear physiological distinction seems to exist between hibernation and estivation. J. W. Hudson (1973) considers it synonymous with hibernation. L. R. Walker et al. (1979) suggest that estivation be termed *shallow torpor* instead.

ANIMAL ADAPTATIONS TO THE MOISTURE ENVIRONMENT

The responses of animals to moisture changes are more complex than those of plants. The mechanisms involved to rid the body of excess water and solutes or to conserve them range from the contractile vacuoles of protozoans to the complex kidney and urinary systems of birds and mammals. Further, animals possess a protective outer covering that protects against passive water loss, and they can move to avoid unfavorable situations or to find new water.

Maintenance of Water Balance

If an organism is not to dehydrate, its input of water must equal its losses. Moisture is a constant problem in two environments: salt water and the desert. Many organisms of these two environments use the similar strategies to overcome the problem.

Vertebrates depend heavily on kidneys as the mechanism for maintaining an osmotic balance. The kidney incorporates a countercurrent multiplier system (suggestive of the countercurrent heat exchanger) in the nephron that maintains water and solute balances in the body. It employs an active tubular transfer system, with a hairpin turn as a part of each tubule, known as the loop of Henle. This loop enables filtrate to flow in opposite directions in the descending and ascending parts, creating a countercurrent flow across a steep osmotic gradient. This system concentrates the urine and

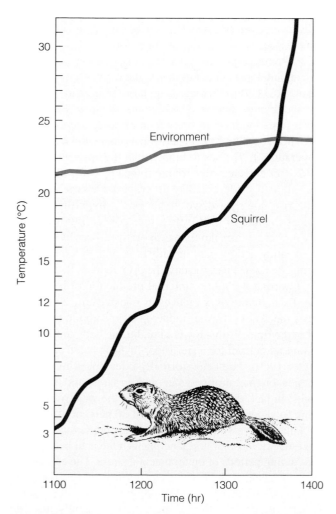

FIGURE 8.20 Rise in the body temperature of the arctic ground squirrel upon arousal from hibernation, stimulated by shivering. With a second rapid rise, at 24°C, the animal opens its eyes and sits up. (After Mayer 1960.)

returns water to the body. The final concentration of urine depends on the length of the loop; the longer the loop, the more concentrated the urine. Efficient water recovery is essential for all animals living in an environment where fresh water is limiting.

Animals of marine and brackish environments have cells that are more dilute than seawater. These organisms are **hypoosmotic.** To survive in this situation, these animals must inhibit the loss of water by osmosis through the body wall and prevent an accumulation of salts in the system. Marine teleost (bony) fish absorb salty water into the gut. They secrete magnesium and calcium through the kidneys and pass these ions off through urinary ducts as a partially crystalline paste. The fish excrete sodium and chlorine by pumping the ions across membranes of special cells in the gills. This pumping process is called **active transport.** It involves the movement of salts against a concentration gradient at the cost of metabolic energy. Sharks and rays retain urea to maintain a slightly higher concentration of salt in the body than that of surrounding seawater.

Marine invertebrates, such as clams and sponges, have no problem with osmotic imbalance because they are **isoosmotic.** They have the same osmotic pressure as seawater.

These two approaches to maintaining osmotic balance contrast with freshwater organisms. Possessing body fluids that are osmotically more concentrated than the surrounding medium, the **hyperosmotic** aquatic organisms need to prevent osmotic inflow. In freshwater fish, intake of water is mainly through the gills and excess water is eliminated through urine. In expelling excess water, the fish also lose solutes that must be replaced. The main replacement of solutes is active uptake in the gills.

Birds of the open sea can use seawater because they possess a special salt-secreting gland located on the surface of the cranium (Schmidt-Nielsen 1960). Gulls (Laridae), petrels (Hydrobatidae), and other seabirds excrete from these glands fluids in excess of 5 percent salt. Petrels and other tube-nosed swimmers forcibly eject the fluid through the nostrils; others drip the fluid out of the internal or external nares.

Some desert birds, like marine birds, utilize a salt gland to maintain a water balance. Otherwise marine, desert, and salt marsh birds have the same basic adaptations for conservation of water and elimination of sodium and chlorine ions. Because birds normally secrete semisolid uric acid as an adaptation related to weight reduction for flying, they are well fitted for water conservation in arid environments.

Among the reptiles the diamondback terrapin (*Malaclemys terrapin*), an inhabitant of the salt marsh, lives in a variably saline environment. The terrapin maintains a relatively high osmotic pressure in its cells when the water is dilute; yet it possesses the ability to accumulate substantial amounts of urea in the blood through the functioning of the salt glands located near the eyes when it finds itself in water more concentrated than 50 percent seawater.

Marine mammals that have to drink seawater possess large kidneys with long loops of Henle that enable them to get rid of excess salt by forming highly concentrated urine. The kidneys of marine mammals are so efficient that their urine is hyperosmotic, possessing a greater osmotic pressure than blood and seawater.

Terrestrial mammals living in an arid environment must conserve water, which they can accomplish through a highly efficient kidney. The kangaroo rat of the southwestern North American desert and its ecological counterparts, the jerboas (*Jerboa* spp.) and gerbils (*Gerbillus* spp.) of Africa and the Middle East, and the marsupial kultarr (*Antechinomys laniger*) of Australia feed on dry seeds and dry plant material even when succulent green plants are available. They rarely drink. Instead, these mammals obtain water from their own metabolic processes: the oxidation of carbohydrates and fats that yields water. To aid in its retention, these desert rodents, like marine mammals, also have exceptionally long loops of Henle that produce highly concentrated urine. To conserve water further these rodents remain by day in sealed burrows. They possess no sweat glands, and their feces are dry. In addition, some desert mammals can tolerate a certain degree of dehydration. Desert rabbits may withstand water losses up to 50 percent of their body weight; camels can tolerate a 27 percent loss. Some desert rodents have evolved specialized incisors or hairs that enable them to scrape high saline tissues off leaves before ingestion.

Extreme adaptations to aridity exist among some of the African antelopes. Outstanding is the oryx (*Oryx leucoryx*) (Figure 8.21). Many African ungulates migrate to escape the heat and dryness, but the oryx remains. During the day it permits a large increase in body temperature, which in effect stores heat, causing **hyperthermia,** a substantial rise in body temperature. If T_b rises to higher than the ambient temperature T_a, this enhances the driving force ($T_b - T_a$) for dry heat loss and reduces dependence on evaporative heat loss. Even if the rise does not drive T_b above T_a, hyperthermia still reduces the rate of heat influx, thus conserving water. The oryx further reduces daytime evaporative losses by suppressing sweating and by panting only at very high temperatures. During the hottest part the day, the oryx reduces its metabolic rate, lowering the rate of internal production of heat at that time. By night the oryx reduces its nonsweating evaporation across the skin by further reducing its metabolic rate below that of daytime. With a lowered nighttime body temperature, the saturation level for water vapor in the exhaled air is lower (Taylor 1969). The oryx normally does not drink water. It depends on metabolic water and grasses and succulent shrubs. In fact, the oryx can obtain all the water it needs by eating food containing an average of 30 percent water.

Other animals in arid environments possess adaptations suggestive of those of arid land plants. One is to become dormant during periods of environmental stress. The animal may adopt an annual cycle of active and dormant stages. For example, Couch's spadefoot toad (*Scaphiopus couchi*) of the southwestern desert of the United States spends the dry period—lasting for eight to nine months—in an underground burrow lined with a gelatinous substance that reduces evaporative losses through the skin. It emerges when rainfall saturates

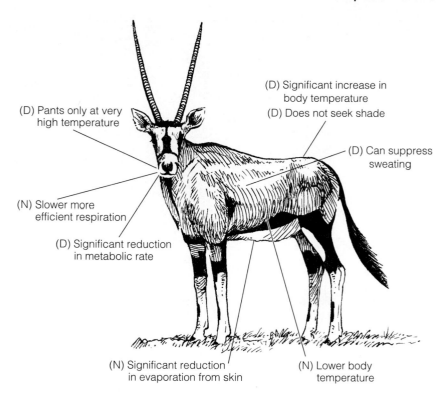

(D) Pants only at very high temperature

(D) Significant increase in body temperature

(D) Does not seek shade

(D) Can suppress sweating

(N) Slower more efficient respiration

(D) Significant reduction in metabolic rate

(N) Significant reduction in evaporation from skin

(N) Lower body temperature

FIGURE 8.21 The physiological adaptations to aridity and heat of an African ungulate, the oryx (*Oryx leucoryx*). D=day; N=night.

the ground, moves to the nearest puddle, mates, and lays eggs. Young tadpoles hatch in a day or two, mature rapidly, and metamorphose into functioning adults. They dig their own retreats in which to estivate until the next rainy period.

Response to Drought and Flooding

Animals move to more favorable environments when moisture conditions become critical. Of course, under extreme conditions, animals can die from dehydration or drown. Drying of marshes, for example, causes nesting losses among waterfowl and muskrats and makes them highly vulnerable to predation. Flooding and heavy rains can drown the young of ground-nesting birds and mammals. The anaerobic conditions of waterlogged soils force earthworms to the surface, where many of them die.

Drought is a period during which rainfall is totally absent or substantially lower than usual for the area in question. Seasonal in some climates and periodic in others, drought can have pervasive effects on animal abundance and distribution. During the wet season on the African savanna, the African buffalo (*Syncerus caffer*) finds food abundant; but during the dry season, the quality of food declines as the grasses dry. Buffalo become more selective, seeking green leaves, moving to the moist riverine habitat, breaking into smaller units, and utilizing different areas. As the dry season progresses, buffalo become less selective, consuming dry stems and leaves they otherwise would have rejected (Sinclair 1977).

Eventually, available protein drops below the levels needed by the body, and the animals use up their fat reserves. Undernourished and lacking the protein intake necessary to maintain the immune system, old animals become most vulnerable to the diseases and parasites they normally harbor. The number that die depends on the rapidity with which the adults use up their energy reserves before the coming of the rainy season. If the next season sees more rainfall and that rainfall extends sporadically into the dry season, the mortality of adults the following year is reduced. Rainfall influences the quantity and quality of dry season forage, which in turn influences mortality of adult buffalo and thus may influence the density of buffalo populations (Sinclair 1977).

Drought stress promotes outbreaks of leaf-eating insects by influencing thermal and nutritional conditions that favor insect growth (Mattson and Haack 1987). High temperatures raise the rate and efficiency of enzymatic reactions in insects. They enhance the insects' detoxification systems, allowing them to override plant defenses. High temperatures may also promote genetic changes favoring insect population growth. The increased contents of nitrogen, minerals, and sugars in the leaves of drought-stressed plants provide a rich food for the leaf-eating insects.

Moisture influences the speed of development and fecundity of some insects. If the air is too dry, the eggs of some locusts and other insects become quiescent. There is an optimum humidity at which the nymphs of some insects develop the fastest. Others, such as the nondiapause migratory grasshopper *Melanoplus sanguinipes,* require a minimum soil moisture before eggs will develop (Mukerji and Gage 1978). Excessive moisture may be directly lethal to some insects, but more important, it encourages the development and spread of pathogenic microorganisms, especially fungi, among insect populations (Ferro 1987, Martinat 1987).

ANIMAL ADAPTATIONS TO THE LIGHT ENVIRONMENT

We tend to associate light as an environmental factor with plants because it is the energy source of photosynthesis. Most of us may not consider light as an important environmental influence on animals. Yet the daily and seasonal changes in the light environment trigger daily and seasonal responses in the activities of animals.

Cicardian Rhythms

At dusk in the forests of North America, a small squirrel with silky fur and large black eyes emerges from a tree hole. With a leap the squirrel sails downward in a long sloping glide, maintaining itself in flight with broad membranes stretched between its outspread legs. Using its tail as a rudder and brake, it makes a short, graceful upward swoop that lands it on the trunk of another tree. This is *Glaucomys volans,* the flying squirrel, perhaps the most common of all our tree squirrels. Because of its nocturnal habits, this mammal is seldom seen. Unless disturbed, it does not come out by day. It emerges into the forest world with the arrival of darkness; it retires to its nest before the first light of dawn.

The squirrel's day-to-day activities form a 24-hour cycle. The correlation of the onset of activity of the flying squirrel with the time of sunset suggests that light has some regulatory effect on the activity of the squirrel. If the flying squirrel is brought indoors and confined under artificial conditions of night and day, it will restrict its periods of activity to darkness and its periods of inactivity to light. Whether the conditions under which the squirrel lives are 12 hours of darkness and 12 hours of light or 8 hours of darkness and 16 hours of light, the onset of activity always begins shortly after dark.

Such behavior in itself does not mean that the squirrel has any special timekeeping mechanisms. It could easily be responding behaviorally to nightfall and daybreak. However, if the same squirrel is kept in constant darkness, it still maintains its pattern of activity and inactivity from day to day in the absence of all time cues. Under these conditions the squirrel's activity rhythm deviates from the 24-hour periodicity (Figure 8.22). Its daily cycle of activity and inactivity under the conditions of constant darkness (when it normally is active) varies from 22 hours and 58 minutes to 24 hours and 21 minutes, the average being less than 24 hours (DeCoursey 1961). Because the cycle length deviates from 24 hours, the squirrel gradually drifts out of phase with the periods of daylight and night of the external world.

This innate rhythm of activity and inactivity covering approximately 24 hours is characteristic of all living organisms except bacteria. Because these rhythms approximate, but seldom match, the periods of Earth's rotation, they are called **circadian rhythms** (from the Latin *circa,* "about," and *dies,* "day"). The period of the circadian rhythm, the number of hours from the beginning of a period of activity one day to the beginning of activity on the next, is called the **free-running cycle.** In other words, the rhythm of activity exhibits a self-sustained oscillation under constant conditions of dark or light.

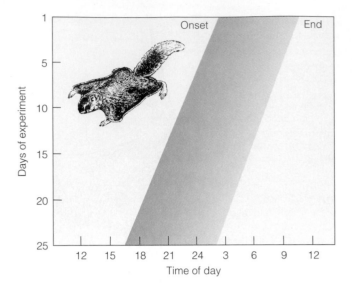

FIGURE 8.22   Drift in the phases of activity of a flying squirrel held in continuous darkness at 20°C for 25 days. Note that the onset of activity becomes gradually later each day. (After DeCoursey 1960:51.)

Circadian rhythms have a strong genetic component and are transmitted from one generation to another. They are affected little by temperature changes, are insensitive to a great variety of chemical inhibitors, are not learned, and are not imprinted on the organism by the environment. They influence not only the time of physical activity and inactivity but also physiological processes and metabolic rates. They do not have any special adaptation to specific local or regional environmental conditions. What circadian rhythms do is provide a mechanism by which organisms can maintain synchrony with their environment.

Thus plants and animals are influenced by two daily periodicities, the external rhythm of 24 hours and the internal circadian rhythm of approximately 24 hours. If the two rhythms are to be in phase, some external environmental "time-setter" must adjust the endogenous rhythm to the exogenous. The most obvious cues are temperature and light. Of the two the master time-setter is light. It brings the circadian rhythms of organisms into phase with their external environment.

The Biological Clock

The circadian rhythms and their sensitivity to light are the mechanisms underlying the biological clock, the timekeeper of physical and physiological activity in living things. In one-celled organisms and plants, the clock appears to be located in individual cells. In multicellular animals the clock is associated with the brain.

Skillful surgical procedures have allowed circadian physiologists to discover the location of the physiological clock in some mammals, birds, and insects. In most insects studied, the clock, including the photoreceptors, is located either in the optic lobes or in the tissue between the optic lobes and the brain. In the cockroach and cricket the receptors for the entrainment of circadian rhythms of locomotion and stridulation are located in the compound eye, but the controlling clock is in the brain (Beck 1980, Saunders 1982). In birds the clock evidently is located in the pineal gland, in the lower central part of the brain (Farner and Lewis 1971, Gwinner 1978). In mammals the clock appears to be located in a part of the hypothalamus just above the optic chasm, the place where the optic nerves from the eyes intersect, although the pineal gland also plays a part.

To function as a timekeeper, the clock has to have an internal mechanism with a natural rhythm of approximately 24 hours, which can be reset by recurring environmental signals, such as the changes in the time of dawn and dusk. The clock has to be able to run continuously in the absence of any environmental time-setter; and it has to be able to run the same at all temperatures. Cold temperatures must not slow it down, nor warm temperatures speed it up. Operation of the clock in mammals involves a special hormone, melatonin, produced by the pineal gland, that serves to measure time. More melatonin is produced in the dark than in the light. The amount produced is a measure of changing daylength.

Models of Biological Clocks

Two basic models of biological clocks have been proposed. One is an oscillating circadian rhythm sensitive to light (Bunning 1964) (Figure 8.23). The cycle or time-measuring process begins with the onset of light or dawn. The first half is light sensitive, and the second half requires darkness (Figure 8.23a). Short-day effects are produced when light does not extend into the dark period (Figure 8.23b). Long-day effects are produced when light does extend into the dark period (Figure 8.23c). Because this simple model does not explain all photoperiodic responses, a variation is a two-oscillator model in which one oscillation is regulated by dawn and the other by dusk. There are more complex models of the clock, but this basic model underlies most of them.

Is there a population of clocks? Does one master clock drive all other clocks? It appears that the biological clock is organized as a hierarchy of clocks. The various overt rhythms influencing physiological and behavioral phenomena are controlled by "slave" or subservient clocks coupled to a master clock. These clocks or pacemakers may be integrated groups of cells within organs such as the central nervous system, where they have specific timekeeping functions. When the master clock is reset by a light signal, it in turn resets the other clocks. One slave clock may take longer to resynchronize than another with a different period. This resetting of the master clock results in transient disturbances of the phase relationships among the subservient clocks. This desynchronization of the overt physiological rhythms of the body causes, for example, the discomfort of jet lag.

How biological clocks function is the domain of the physiologist. Ecologists are more interested in their adaptive value. One adaptive value is that the biological clock lets organisms anticipate environmental cues.

Certain circadian activities, for example, relate to aspects of the environment that are ecologically more important than light or dark per se. The transition from night to day is accompanied by such environmental changes as a rise in humidity

and a drop in temperature. Woodlice, centipedes, and millipedes, which lose water rapidly in dry air, spend the day in the dark and damp under stones, logs, and leaves. They emerge at dusk when the humidity of the air is more favorable. These animals show an increased tendency to escape from light as the length of time they spend in darkness increases. On the other hand, their intensity of response to low humidity decreases with darkness. Thus these invertebrates come out at night into places too dry for them during the day; and they quickly retreat to their dark hiding places as light comes (Cloudsley-Thompson 1956, 1960).

The circadian rhythms of many organisms relate more to the biotic than the physical aspects of their environment. Predators, such as insectivorous bats, must relate their feeding activity to the activity rhythm of their prey. Moths and bees must visit flowers when they are open to obtain nectar. The flowers must have a rhythm of opening and closing that coincides with the time when insects that pollinate them are flying. The clock is the most energy-efficient way to adapt to the periodicities of the environment.

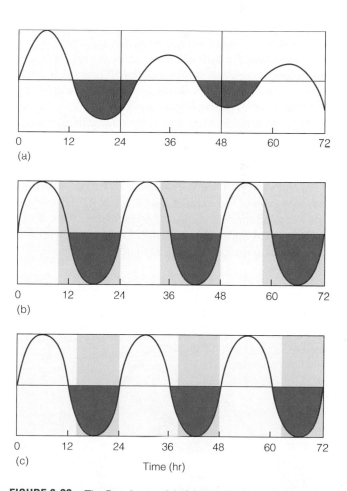

FIGURE 8.23   The Bunning model of the entrainment of circadian rhythms to daylength. (a) Oscillations of the clock cause an alternation of half-cycles with quantitatively different sensitivities to light (light vs. solid). The free-running clock in continuous light or continuous darkness tends to drift out of phase with the 24-hour photoperiod. (b) Short-day conditions allow the dark to fall into the light half-cycle. (c) In the long day, the light falls into the dark half-cycle. (From Bunning 1960:253.)

Critical Daylengths

In the northern and southern latitudes the daily periods of light and dark lengthen and shorten with the seasons. The activities of plants and animals are geared to the changing seasonal rhythms of night and day. The flying squirrel, for example, starts its daily activity with nightfall, regardless of the season. As the short days of winter turn into the longer days of spring, the squirrel begins its activity a little later each day (Figure 8.24). Most animals of temperate regions have reproductive periods that closely follow changing daylengths of the seasons. For most birds the height of the breeding season is lengthening days of spring; for deer the mating season is the shortening days of fall. The signal for these responses is **critical daylength.**

Many organisms possess both long-day and short-day responses. Because the same duration of dark and light occurs two times a year, as the days lengthen in spring and shorten in fall, the organisms could get their signals mixed. For them the distinguishing cue is the direction from which the critical daylength is approached. One critical daylength is reached as long days move into short, another as short days move into long.

For example, diapause is controlled by photoperiod. The time measurement in such insects is precise, usually falling in a light phase somewhere between 12 and 13 hours. A quarter-hour difference in the light period can determine whether an insect goes into diapause or not (Saunders 1982). Adkisson (1966) found that the cotton boll-worm (*Heliothis zea*) failed to enter diapause if the larvae were exposed to a light period of 13.25 hours; its critical day length is 13 hours. Experimentally, the boll-worm terminated diapause most rapidly when exposed to photoperiods of 14 hours. Thus to the cotton boll-worm the shortening days of late summer and fall forecast the coming of winter and call for diapause; and the lengthening days of late winter and early spring are the signals for the insect to resume development, pupate, emerge as an adult, and reproduce.

Gonadal development and spring migratory behavior in birds increase with daylength. This was experimentally demonstrated over 60 years ago when Rowan (1925) forced juncos (*Junco hyemalis*) into the reproductive stage out-of-season by artificial increases in daylength. Since then numerous experimental studies involving many species of birds have demonstrated that the reproductive cycle in birds is controlled by physiological response to daylength (Farner 1959, 1964a, b; Wolfsson 1959). After the breeding season, the gonads of birds regress spontaneously. During this time, light cannot induce gonadal activity. The short days of early fall hasten the termination of this period. Progressively shorter days of winter begin to stimulate the birds again. The lengthening days of early spring then bring the birds into reproductive stage.

In mammals photoperiod influences activities such as food storage (Muul 1969) and reproduction. Consider, for example, such seasonal breeders as sheep and deer, whose reproductive cycle is initiated in the shortening days of fall, activated by the hormone melatonin. Because more melatonin is produced during the dark cycle of the circadian rhythm, these animals receive a high concentration of melatonin for a longer period of the 24-hour day as the daylength shortens and nightlength increases. This increase in melatonin reduces the sensitivity of the hypothalamus to the negative feedback of steroids from the ovaries and testes, allowing the anterior pituitary to release pulses of luteinizing hormone (LH), increasing follicular growth in the ovaries and sperm production in the testes.

In the male deer, development of antlers as well as the reproductive cycle relate to photoperiod. The lengthening days of spring stimulate an increase in growth hormones and prolactin. (Figure 8.25). These hormones stimulate the growth of antlers in the spring and early summer. During the shortening days of late summer, growth hormones and prolactin decrease. Under the influence of melatonin, testosterone and sexual resurgence increase. In the presence of testosterone, antler growth ceases, velvet sheds, and antlers harden by the onset of the season of rut. The decline of testosterone in winter results in the shedding of antlers. Normally the deer is in velvet about one-third of the year. When the duration of that year is changed artificially by altering the frequency of daylength, deer replace antlers as often as two, three, or four times a year, or only once every other year, depending on how much the light cycle has been altered. Thus growth, development, loss of antlers, and the reproductive cycle are controlled by hormones, testosterone and prolactin, whose production is mediated by the photoperiodic responses of the hypothalamus.

TIDAL AND LUNAR CYCLES

For some organisms, tidal and lunar rhythms are of greater ecological importance than light-dark cycles. Animals that inhabit the intertidal zones of the sea show rhythms in their behavior that coincide with the cycles of high and low tides. Internal timing processes entrained to tidal cycles have been demonstrated

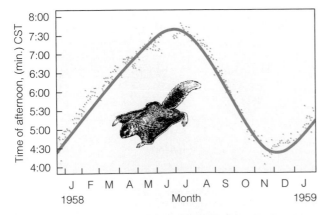

FIGURE 8.24   Onset of running wheel activity for one flying squirrel in natural light conditions throughout the year. The graph is the time of local sunset through the year. (From DeCoursey 1960:50.)

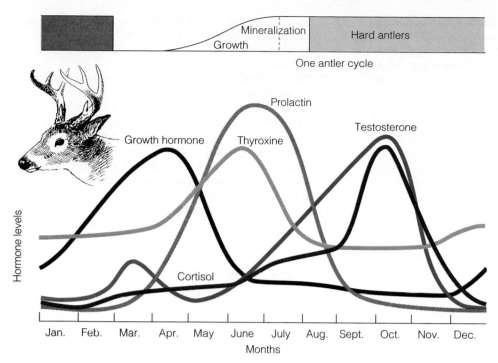

FIGURE 8.25 The seasonal course of hormonal levels during the annual cycle of the white-tailed deer and its relationship to antler growth. Note the response of growth hormones and prolactin to lengthening days and the decline in prolactin and the increase in production of testosterone during the shortening days of fall. (After Gross et al. 1974.)

in a number of intertidal animals, such as the European shore crab (*Carcinus maenas*) and fiddle crabs (*Uca minas* and *U. crenulata*). When these crabs were brought into the laboratory and held under constant temperature and light, devoid of tidal cues, they exhibited tidal rhythmicity in their activity as they would back in the marsh. Their rhythm mimics the ebb and flow of tides every 12.4 hours, one-half of the lunar day of 24.8 lunar hours, the interval between successive moonrises. Under the same constant conditions, fiddler crabs exhibit circadian rhythm of color changes, dark by day and light by night.

Is the clock involved unimodal with a 12.4-hour cycle, or bimodal with a 24.8-hour cycle, close to the period of the circadian clock? Does one clock operate on a solar-day rhythm of approximately 24 hours and a lunar clock operate on a tidal one of 24.8 hours? Based on their experimental evidence, J. D. Palmer and his associates at the University of Massachusetts (1996) suggest that daily and tidal rhythms in intertidal organisms involve one independent clock to synchronize daily activities and two strongly coupled circalunidian clocks to synchronize tidal activity. Each lunar day clock drives its own tidal peak. If one clock quits ticking in the absence of environmental cues, the other one still runs. This feature enables tidal organisms to synchronize their activities in a variable tidal environment. Day-night cycles reset solar-day rhythms, and tidal changes reset tidal rhythms. (For reviews, see Palmer 1995, 1990, and 1976, Enright 1975.)

Reproduction in some marine organisms is restricted to a period that bears some relationship to tides. These rhythmic phenomena occur every lunar cycle of 28 days, or in some instances every semilunar cycle of 14 to 15 days. Among these species is the grunion (*Leusethes tenuis*), a small California fish that swarms in from the sea to lay eggs on sandy beaches, and the intertidal midge (*Clunio marinus*). These periodicities are so exact that activities of these animals can be predicted ahead of time. Laboratory studies confirm the entrainment of activity cycles to moonlight (Enright 1975).

Summary

1. Major groups of animals are herbivores that convert plant biomass into animal biomass; carnivores that feed on herbivores and thus indirectly obtain energy fixed by plants; omnivores that feed on both plant and animal biomass; and detritivores that feed on dead organic matter.
2. The ultimate source of nutrients for animals, directly or indirectly, is plants. Herbivores, with the aid of bacteria inhabiting the digestive tract, convert cellulose and a limited supply of protein into animal tissue. Low concentrations of nutrients in plants can have an adverse effect on the growth, development, and fitness of herbivores.
3. Three essential nutrients that influence the distribution, behavior, and fitness of grazing herbivores are sodium, calcium, and magnesium. Among the herbivores, the quality of food, especially its protein content and digestibility, is of critical importance. Among the carnivores, quantity often is more limiting than quality.
4. Plant and animal tissues provide animals with a wide array of potential food resources. This assortment requires a variety of physiological, morphological, and behavioral characteristics that enable species to acquire and assimilate these very different food resources.
5. Most organisms (invertebrates, fish, amphibians, and reptiles) are poikilothermic. They gain their heat energy from the environment. Their rate of metabolism, as measured by oxygen consumption per unit of body weight, approximately doubles for every 10°C rise in temperature in accordance with van't Hoff's rule. Birds and mammals are homeotherms, maintaining their body temperatures by metabolically generating body heat. Heterotherms are animals (bees and bats) that regulate their body temperatures and sometimes allow their temperatures to drop to ambient.

6. Within limits poikilotherms acclimatize to higher or lower temperatures by avoiding temperature extremes, by adjusting tolerances to given temperature ranges, by heliothermism, and in some instances, by evaporative cooling. Some reptiles change their body temperatures by changing their rates of heartbeat and metabolism. Some cold tolerant poikilotherms utilize supercooling and the synthesis of glycerol in body fluids to resist freezing in winter.

7. Animals depending on the environment as a source of heat are ectothermic. Their body temperatures tend to follow ambient temperatures. Because of their variable body temperature, these animals are termed poikilothermic. They regulate their body temperatures behaviorally by moving in and out of warmer and cooler areas.

8. Animals that depend on internally produced heat to maintain body temperatures are endothermic. They maintain a rather constant body temperature independent of the environment.

9. Many animals are heterothermic. Depending on the environmental and physiological conditions, they function either as endotherms or ectotherms. They regulate body temperature metabolically or allow it to drop to ambient temperature, increasing or conserving energy.

10. Some homeotherms and heterotherms acclimatize by seasonal changes in body insulation, vascularization, evaporative heat loss, and nonshivering thermogenesis. Most homeotherms and heterotherms employ a well-developed countercurrent circulation, which involves the exchange of body heat between arterial and venous blood and their equivalent in insects. This exchange retains body heat by reducing heat loss through the body parts, or cools or heats blood flowing to such vital organs as the brain.

11. Some animals enter a state of dormancy during environmental extremes to reduce the high energy costs of staying warm or cool. They slow their metabolism, including heartbeat, respiration, and body temperature. Birds, such as hummingbirds, and mammals, such as bats, undergo daily torpor, the equivalent of deep sleep without the extensive metabolic changes of the seasonal torpor of hibernation. Hibernation involves a whole rearrangement of metabolic activity to run at a very low level.

12. Like plants, animals maintain a water balance with their environment, which most accomplish by means of an excretory system, dominated in many groups of organisms by a kidney. The system is most complex in mammals whose countercurrent multiplier system in the nephron of the kidney reabsorbs water from the urine against an osmotic gradient.

13. Many animals inhabiting a saline environment have salt-excreting glands. Animals of arid regions may reduce water loss by becoming nocturnal, producing highly concentrated urine, using only metabolic water, and tolerating a certain degree of dehydration.

14. Drought stress can lower food quality, and cause death from lack of water. Drought also promotes outbreaks of leaf-eating insects by influencing thermal and nutritional conditions in plants that favor insect growth.

15. Daily periodicities of organisms are under the influence of day-night cycles. Daily activities follow circadian rhythms, free-running under constant conditions with an oscillation that has its own inherent frequency. For most organisms the inherent clock deviates slightly from 24 hours. These circadian rhythms are synchronized with the 24-hour cycles of light and dark by a biological clock set by the major external time-setter, light.

16. The onset and cessation of activities of most organisms are usually synchronized with dusk and dawn. The response depends on whether the organisms are diurnal (light-active) or nocturnal (dark-active).

17. The biological clock is useful for synchronizing activities of organisms not only with night and day, but also with the seasons of the year. The possession of a self-sustained rhythm with approximately the same frequency as that of environmental rhythms enables organisms to "predict" such advance situations as the coming of spring. It brings plants and animals into a reproductive state at a time of year when probability of survival is highest. It synchronizes within populations such activities as mating and migration.

18. Other periodicities, particularly among marine organisms, are strongly influenced by lunar cycles and the tidal cycles associated with them.

Review Questions

1. How does the nutrient level in plants affect the well-being of animals that eat them?
2. Contrast poikilothermy, homeothermy, and heterothermy. How do they relate to ectothermy and endothermy?
3. Why are small endotherms so rare?
4. Why should the camel, a desert mammal, have such a heavy coat of hair?
5. Bats are small hibernating mammals. They do not store food nor do they put on a heavy layer of fat. How do bats, then, survive the long winter hibernation?
6. Compare the oxygen consumption per kilogram of body weight of three mammals: one 10 kg, the second 100 kg, and the third 200 kg. Use the formula on page 128.
7. Consider a population of fish acclimatized to the warm-water outflow into a river from a power-generating station. Discuss the physiological stresses experienced by the fish and the consequences if the power plant shut down for one week.
8. How do the mammalian kidney and related structures in other organisms enable them to maintain a water and solute balance with their environment?
9. If marine mammals and desert rodents can survive in a physiologically arid environment without fresh water, why can't humans?
10. Using the graph in Figure 8.10, determine and compare the thermal tolerances for the chum salmon and bullhead at an acclimation temperature of 20°C.
11. Prepare a graph of the awaking times of selected local birds, such as robins and cardinals, marked by the first morning songs, against the time of local sunrise. (For the dedicated only; this project requires early rising, especially as spring wears on.) What conclusions can you draw from the graph? How do your daily observations relate to seasonal periodicity?
12. Relate circadian rhythms and biological clocks to jet lag. Compare a change in activities (sleep, meals, and so on) during a flight from New York to Frankfort, from New York to Tokyo, and for each return flight. What happens to time-setting of the biological clock?

Cross References

Adaptation, 81; evaporation, 51, humidity, 22; microclimate, 38; nutrients, 54; soil, 59–65; thermal environment, 48; salt marsh, 689–695; desert, 548–589.

Decomposition

CONCEPTS

1. Decomposition is a sequential process involving leaching, fragmentation, anabolism, immobilization, and mineralization.

2. Decomposition involves all consumers; the true decomposers, however, are bacteria and fungi, and a wide diversity of detritivores and microbivores.

3. The nature and rates of terrestrial decomposition are influenced by litter quality and the physical environment, especially temperature and moisture.

4. In aquatic environments, phytoplankton and zooplankton have important roles in the decomposition process.

Decomposition is the breakdown of chemical bonds formed during the construction of plant and animal tissue. It is the end product of the consumer pathway from photosynthesis. Whereas photosynthesis involves the incorporation of solar energy, carbon dioxide, water, and inorganic nutrients into organic biomass (living matter), decomposition involves respiration: the release of energy originally fixed by photosynthesis, carbon dioxide, and water and ultimately the conversion of organic compounds into inorganic nutrients. Decomposition is a complex of many processes, including fragmentation, change in physical structure, ingestion, egestion, and concentration. These processes are accomplished by a variety of decomposer organisms. All heterotrophs to some degree function as decomposers. As they digest food, they break down organic matter, alter it structurally and chemically, and release it partially in the form of waste products. However, what we typically refer to as decomposers are organisms that feed on dead organic matter or detritus. This group is composed of bacteria, fungi, and detritivores, animals that feed on dead material.

CLASSIFICATION OF DECOMPOSERS

The innumerable organisms involved in decomposition are categorized into several major functional groups. Organisms most commonly associated with the process of decomposition are the **microflora,** comprising the bacteria and fungi. Bacteria may be aerobic, requiring oxygen for metabolism; or they may be anaerobic, able to carry on their metabolic functions without oxygen, by using inorganic compounds such as sulfates as the oxidant. This type of respiration by anaerobic bacteria, commonly found in the mud and sediments of aquatic habitats and in the rumen of ungulate herbivores (see Chapter 8), is fermentation. **Fermentation,** which converts sugars to organic acids and alcohols, is a less efficient means of breaking down organic matter than aerobic respiration. The process tends to lower the pH of the substrate, making it more acidic and favoring fungal activity. Many decomposer bacteria are **facultative anaerobes.** These bacteria use oxygen when it is available in respiration; but in its absence they can shift to fermentation, utilizing inorganic compounds as an oxidant. Other bacteria are **obligate anaerobes,** meaning they cannot survive in the presence of oxygen.

Bacteria are the dominant decomposers of dead animal matter, whereas fungi are the major decomposers of plant material. Fungi extend their hyphae into the organic material to withdraw nutrients. Fungi range in type from species that feed on highly soluble, organic compounds, such as glucose, to more complex hyphal fungi that invade tissues with their hyphae (Figure 9.1a).

Bacteria and fungi secrete enzymes into plant and animal tissues to break down the complex organic compounds. Some of the resulting products are then absorbed as food. After one group has exploited the material to the extent of its

ability, a different group of bacteria and fungi able to utilize the remaining material moves in. Thus a succession of microflora occurs in the decomposition of organic matter until the material is finally reduced to inorganic nutrients.

Decomposition is aided by the fragmentation of leaves, twigs, and other dead organic matter (detritus) by invertebrate detritivores. These organisms fall into four major groups as classified by body width (Figure 9.2): (1) microfauna and microflora include protozoans and nematodes inhabiting the water film in soil pores; (2) mesofauna, whose body width falls between 100 μm and 2 mm, includes mites (Figure 9.1b), potworms, and springtails that live in air-filled soil air spaces; and (3) macrofauna and (4) megafauna, represented by millipedes, earthworms (Figure 9.1c), and snails in terrestrial habitats, and by mollusks and crabs in aquatic habitats. Earthworms and snails dominate the megafauna over 20 mm. The macrofauna and megafauna can burrow into the soil or substrate to create their own space, and megafauna, such as earthworms, have major influences on soil structure. These detritivores feed on plant and animal remains and on fecal material.

Energy and nutrients incorporated into bacterial and fungal biomass do not go unexploited in the decomposer world. Feeding on bacteria and fungi are the **microbivores.** Making up this group are protozoans such as amoebas, springtails (Collembola), nematodes, larval forms of beetles (Coleoptera), and mites (Acari). Smaller forms feed only on bacteria and fungal hyphae. Because larger forms feed on both microflora and detritus, members of this group are often difficult to separate from detritivores.

Microbivores act as regulators of decomposition in two important ways: by controlling the abundance and distribution of fungi and by stimulating microbial activity (Lusenhop 1992, Seastedt 1984). By selectively feeding on certain fungal species, microarthropods—for example, Collembola—influence fungal distribution. By selecting certain species over others, microbivores can also influence the interactions among bacteria and fungi. In extreme cases, microbivores act to so reduce microbial populations that they delay ordinary decomposition. In contrast, microbivores can act to stimulate fungal and bacterial growth by the direct return of mineral nutrients through their feces, which microflora consume. In addition microbivores disperse fungal spores through the soil.

STAGES OF DECOMPOSITION

Decomposition of dead plant and animal matter moves through several stages, from deposition to the final breakdown into inorganic nutrients (Figure 9.3). Early stages of decomposition involve initial **leaching,** the loss of soluble sugars and other compounds that are dissolved and carried away by water, and **fragmentation,** the reduction of organic matter into smaller particles, either physically or chemically. Both of these abiotic processes result in the loss of mass and changes in the chemical composition of the detritus. As decomposition proceeds, further fragmentation and leaching take place.

FIGURE 9.1   (a) Fungi are major decomposers of leaf litter in the forest. (b) Mites are among the most abundant small detritivores. Some species are predacious. (c) Earthworms and millipedes are large detritivores important in the fragmentation of leaf litter.

Detritivores oxidate organic compounds, releasing energy through respiration. They then degrade them into smaller and simpler products. The release of organically bound nutrients into a inorganic form available for plants and microbes is called **mineralization.** At the same time, decomposer organisms use these nutrients for their own growth, incorporating them into microbial biomass. The incorporation of mineral nutrients into microbial biomass is known as **nutrient immobilization.** Only those nutrients not taken up by microbes are available to plants. Immobilization of nutrients—particularly potassium, calcium, and nitrogen—can affect plant growth.

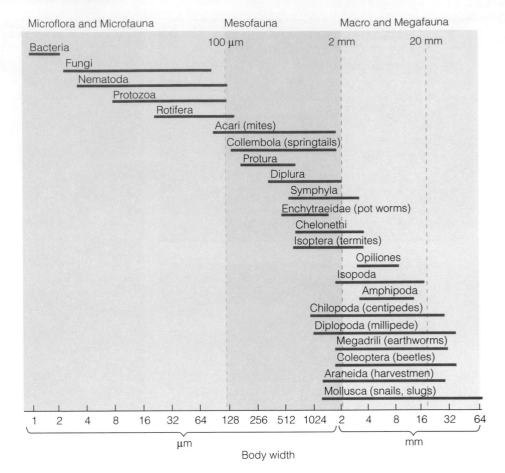

FIGURE 9.2 A general classification of soil fauna based on body width or diameters is a more functional classification relative to litter breakdown and decomposition. Bacteria and fungi fall into this scheme as microflora. Protozoans and nematodes fall into category of microfauna. (From Swift et al. 1979.)

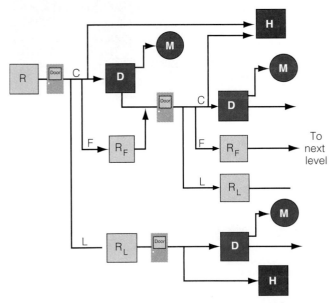

FIGURE 9.3 A model of pathways in the decomposition of dead plant and animal matter (R). Initial decomposition process involves fragmentation (F), the reduction of particle size of chemically unchanged litter; leaching (L) of soluble materials in an unchanged chemical form to another site; and breakdown of complex molecules (C) into smaller ones. Part of the resultant compounds enters humus (H), and part is resynthesized into decomposer tissue (D). In time, part of the decomposer tissue is mineralized (M) and part becomes new decomposer tissue. Similar pathways continue as the products from each stage of decomposition undergo further changes. Ultimately the original or primary material is completely mineralized. (Adapted from Swift et al. 1979:51, 53, 54.)

THE PROCESS OF DECOMPOSITION

Microbial decomposition of plant leaves can begin while the green leaves are still on the plant. Living plant leaves produce varying quantities of exudates that support an abundance of surface microflora (Ruinen 1962). These organisms feed on the exudates and on any cellular material sloughed off. The same exudates account for the nutrients leached from the leaves during a rainfall. In tropical rain forests leaves are heavily colonized by bacteria, actinomycetes, and fungi.

When leaves become senescent, the tissues are invaded by both bacteria and fungi, and decomposition accelerates. Once on the ground, the now dead plant tissues are subject to attack by different bacteria and fungi. Detritivores break the litter into smaller parts, making the litter more accessible to the microflora. The action of such litter feeders as millipedes and earthworms can increase exposed leaf area up to 15-fold (Ghilarov 1970). Because the net assimilation of plant detritus by detritivores is low, on average less than 10 percent, a great deal of material passes unchanged through the gut of these organisms. This fecal material is readily colonized by other microbes. Some detritivores, particularly earthworms, enrich the soil with vitamin B_{12}. In addition, they bury surface litter and mix organic matter with the soil, bringing the material into contact with other microbes (Hartenstein 1986). James (1991) found that in a Kansas tallgrass prairie, earthworms annually processed approximately 4 to 10 percent of the *A* soil horizon (see Chapter 4). In doing so, earthworms processed approximately 1 percent of the total organic matter in the top 15 cm of soil.

The activities of detritivores are important to the decomposition process (van der Drift 1971, Luxton 1982). Experiments by Witkamp and Olson (1963), Edwards and Heath (1963), Witkamp and Crossley (1966), and others have shown that suppression of activities of detritivores results in a marked slowdown in microbial decomposition. There is experimental evidence to indicate that predation by terrestrial salamanders on leaf-fragmenting detritivores reduces decomposition rates of forest floor litter (Wyman 1998). In the absence of detritivores, the decomposition of wood is slowed by half.

Meanwhile, other organisms are using organic material from the roots of living plants. The soil region immediately surrounding the roots, known as the **rhizosphere,** and the root surface itself, called the **rhizoplane,** support a host of microbial organisms that feed on senescent root material and root exudates. Root exudates may consist of simple sugars, fatty acids, and amino acids. In fact, some 10 sugars, 21 amino acids, 10 vitamins, 11 organic acids, 4 nucleotides, and 11 miscellaneous compounds have been identified in the rhizosphere (Curl and Truelove 1986).

Decomposition of animal matter is more direct than the decomposition of plant material. The chemical breakdown of flesh does not require all of the specialized enzymes needed to digest plant matter. That flesh not consumed by scavengers such as crows, vultures, and foxes is decomposed largely by bacteria rather than fungi and by certain arthropods such as blowflies (Calliphoridae).

In summer and fall, when temperatures are high, microbial activity and colonization of dead animal tissues by blowflies is intense. Blowfly maggots emerging from eggs laid in the tissues can consume a small mammal carcass in seven to eight days (Putnam 1978a, b). Between bacteria and maggots, 70 percent of the organic material of the small mammal carcass is consumed, leaving the remaining 30 percent, largely hair and bone, behind. Due to low temperatures and reduced microbial activity, decomposition during the winter and spring is restricted. As a result, the carcass can become mummified through reduced pH and other chemical changes in the tissues and is eventually fragmented and scattered.

Unlike the bodies of dead animals, most fecal matter represents an already highly decomposed substrate. However, the dung of large herbivores still contains an abundance of partially digested organic matter that provides a rich resource for specialized detritivores in addition to earthworms, bacteria, and fungi. Among these specialized detritivores are species of flies that lay their eggs in dung, upon which the larva will feed. The most notable of the coprophagus detritivores are the dung beetles (Figure 9.4) (Scarabaeinae, Aphodiinae, and Geotrupinae) (Hanski and Carubefort 1991). The eggs, larvae, and pupae of many species of the genus *Aphodius* develop within the dung pat. Other dung beetles, known as tumblebugs, form a mass of dung into a ball in which they lay their eggs, roll it a distance, dig a hole, and bury it as a food supply for the larvae. Aphodiinae dung beetles tunnel and form a dung ball underground. The earth-boring dung beetles, Geotrupinae, spend most of their lives in deep burrows, usually beneath carrion or dung. The female lays her eggs in a plug of dung at the end of the burrow.

FIGURE 9.4 The dung beetle is exceedingly abundant in both tropical and temperate regions. Thousands of individuals and dozens of species may be attracted to a single dropping.

ENVIRONMENTAL CONTROLS ON DECOMPOSERS

Two major factors influence the rate of decomposition through their direct effect on decomposer populations: (1) the quality of the dead organic matter as a food source, and (2) the physical environment, particularly temperature and moisture (climate).

Litter Quality

Litter provides both an energy (carbon) and nutrient (e.g., nitrogen) source for microbial decomposers. Characteristics that influence the quality of plant litter as an energy source relate directly to the types and quantities of carbon compounds present—namely the types of chemical bonds present, and the size and three-dimensional structure of the molecules in which these bonds are formed. Not all carbon compounds are of equal quality as an energy source to microbial decomposers.

Glucose and other simple sugars (or carbohydrates), the first products of photosynthesis, are very high quality sources of carbon for microbial decomposers. These molecules are physically small. The breakage of their chemical bonds yields much more energy than required to synthesize the enzymes needed to break them. Cellulose and hemicellulose are the main constituent of cell walls. These compounds are more complex in structure and therefore more difficult to decompose. They are of moderate quality as a substrate for microbial decay. The much larger lignin molecules are among the most complex and variable carbon compounds in nature. There is no precise chemical description of lignin; rather, it represents a class of compounds. These compounds possess very large molecules intricately folded into complex three-dimensional structures that effectively shield much of the internal structure from attack by enzyme systems. As such, lignins, major components of wood, are one of the slowest components of plant tissue to decompose.

The variation in rates of decomposition of different carbon compounds are revealed in an experiment that examined the rate at which carbon was consumed during the decomposition of straw placed on a soil surface (Figure 9.5). The total carbon content of the straw, expressed as a percentage of the original mass, declined exponentially over the period of the 80-day study. However, when the total carbon was partitioned into various classes of carbon compounds, the rates at which these compounds were decomposed varied widely. Proteins, simple sugars, and other soluble compounds made up some 15 percent of the original total carbon content. These compounds decomposed very quickly, disappearing completely within the first few days of the experiment. Cellulose and hemicellulose made up some 60 percent of the original carbon content. Although these compounds decompose more slowly than the proteins and simple sugars, by three weeks into the experiment these compounds had been completely broken down. The third category of carbon compounds examined, the lignins, made up some 20 percent of the total original carbon. These compounds were broken down very slowly

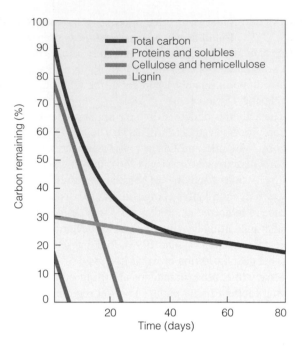

FIGURE 9.5 Variation in the rates of decay (mass loss) of different classes of carbon compounds in an experiment examining the decomposition of straw.

over the course of the experiment, with the vast majority of lignins remaining intact by day 80. As decomposition proceeded during the experiment, the quality of the carbon resource declined. High- and intermediate-quality carbon compounds declined at a relatively rapid rate. Thus the proportion of total carbon remaining as lignin compounds continually increased with time. The increasing component of lignin lowered the overall quality of the remaining litter as an energy source for microbial decomposers.

The influence of carbon quality on rates of decomposition is reflected in the decomposition of leaves of various forest trees. Easily decomposed and highly palatable leaves from such species as redbud, mulberry, and aspen support higher populations of decomposers than litter from oak and pine, which is high in lignin. Earthworms have a pronounced preference for such species as aspen, white ash, and basswood, do not entirely consume sugar maple and red maple leaves, and do not eat red oak leaves at all (Johnston 1936). In a European study (Lindquist 1942) earthworms preferred the dead leaves of elm, ash, and birch, ate sparingly of oak and beech, high in lignin, and did not touch pine or spruce needles. Millipedes likewise show a species preference (van der Drift 1951). Thus decomposition of litter from certain species proceeds more slowly than litter from others.

Measuring Litter Breakdown

One of the primary approaches that ecologists use to study the decomposition of plant tissues is the litterbag experiment. In this experiment, they place a known quantity of plant litter (usually leaves) in a fine-mesh bags made of synthetic materials (Figure 9.6). The bags often hinder access to larger detritivores, but they allow free access to microbial decomposers. Researchers place the bags on the surface of the soil, just beneath the litter layer, or in water in aquatic and wetland situations. After a period of time, the experimenters collect a number of the bags, take them back to the laboratory, and dry the contents. By comparing the mass of the remaining litter material with the original mass placed into the bags, they can calculate the mass loss over the intervening period. By repeating this process of sampling over an extended period of time, ecologists can follow the loss of litter mass and changes in the chemistry of the remaining litter materials.

Figure 9.7 shows data from two litterbag experiments examining the rates of decomposition for leaves of various tree species at two locations in eastern North America, New Hampshire and Virginia. In both graphs the x axis is time and the y axis is percentage of the original mass remaining. The loss of mass through time is generally expressed as a negative exponential function:

$$\text{Percent original mass remaining} = e^{-kt}$$

Here e is the natural logarithm, t is the time unit used (e.g., years, months, days), and k is the decomposition coefficient, which defines the slope of the negative exponential curve. The

FIGURE 9.6 Litterbag experiments involve the placement of a known mass of test litter in mesh bags placed just below the litter layer. After a period of time the weight of the remaining litter material is compared to the original mass placed in the bag, revealing the loss over the invervening time.

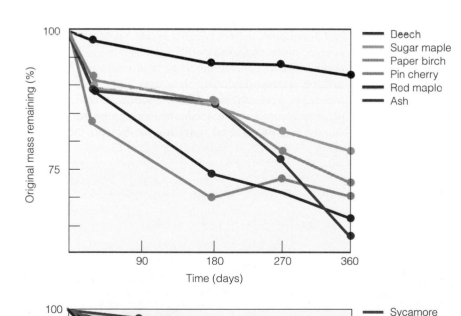

FIGURE 9.7 Patterns of mass loss through time for two litterbag experiments examining the decomposition of leaves from a variety of tree species in (a) New Hampshire (data from Melillo et al. 1982) and (b) central Virginia (data from Smith 2000). Mass loss (y axis) is expressed as percentage of the original mass (placed in litterbags) remaining as a function of time. (See text for description of litterbag experiments.)

larger the value of *k,* the faster the rate of decomposition (Figure 9.8). The species included in Figure 9.7 show a wide range of decay rates (*k*), with the percentage of original mass remaining at the end of one year ranging from 45 to 75 percent.

Studies in various ecosystems have examined the characteristics of the plant litter materials that are correlated with rates of mass loss (Cromack 1973, Melillo et al. 1982, Berg and McClaygertry 1989, Aber et al. 1982, Cornelisser 1996). Although a number of factors relating to carbon and nutrient quality of the litter materials have been identified, the most consistent litter characteristic accounting for variations in the observed rates of decay is the percentage of the total carbon content that is made up of lignin compounds (Melillo et al. 1982, Berg et al. 1993, Meentemeyer 1978). The inverse correlation between initial lignin content and rate of decay (*k*) for leaf litter from the nine tree species involved in Figure 9.7 is shown in Figure 9.9. As initial percentage lignin content increases, the rate of decay decreases. This relationship is easy to understand in light of the discussion of carbon quality (see Figure 9.5). Given the extremely low quality of lignin as an energy source, the greater the amount of carbon tied up in lignin, the lower the consumption rate by decomposers.

In addition to being the source of energy for decomposer organisms, the litter material is also the primary source of other essential nutrients. Litterbag experiments can also be an approach to the study of the nutrient dynamics of decomposing litter materials. For example, the same experiments used to examine patterns of mass loss with time as shown in Figure 9.7 also provided data on changes in the nitrogen content of the decomposing litter materials. Figure 9.10 shows the changes in nitrogen content, expressed as a percentage of the original nitrogen content of the litter material. For the six species examined in the

New Hampshire study (Figure 9.7a), the amount of nitrogen, in absolute terms, is increasing even though the mass of litter material is declining,

To understand why the nitrogen content of the litter materials shown in Figure 9.7a is increasing, we need to note that the nitrogen content reflects the nitrogen remaining in the plant tissues as well as in the biomass of the microdecomposers, bacteria and fungi, that are feeding on the litter. As they consume the litter materials, microbes incorporate nitrogen and other essential nutrients into new biomass or excrete it in waste products. Bacteria and fungi break down proteins in the dead plant tissues into amino acids. The amino acids are oxidized to carbon dioxide, water, and ammonia. Other groups of bacteria transform the ammonium to nitrates. This mineralization process transforms nutrients in organic matter into inorganic minerals available for uptake by plants, bacteria, and fungi.

The availability of any particular nutrient to decomposers depends on the ratio of energy supply to nutrient supply, expressed as the carbon:nutrient ratio, C:X. In the case of nitrogen, the ratio is expressed as C:N. The typical C:N ratio in plant litter is between 30:1 and 100:1, whereas the C:N ratio for bacteria and fungi is 10–15:1. When the C:X ratio of the litter material is high, the supply of that nutrient relative to the energy source is low. If the litter material cannot meet the demand for nutrients by the growing microbial population, the microbes will take up previously mineralized nutrients from external sources (e.g., soil and other organic matter) and immobilize it in microbial biomass, making it unavailable for plant use. Since both of these processes, mineralization and immobilization, are taking place as decomposer organisms are consuming the litter, the supply rate of mineral nutrients to the soil during the

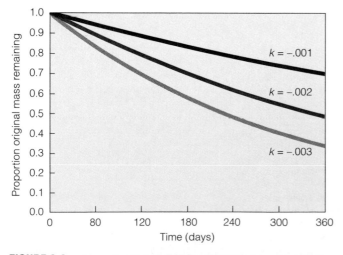

FIGURE 9.8   General model of decomposition where mass loss is expressed as an exponential decay function e^{-kt}. The three curves represent differing rates of decay as described by the decomposition coefficient *k*. The larger the value of *k*, the faster the rate of decomposition and the steeper the curve. Values of *k* shown are based on a time scale (*t*) of days.

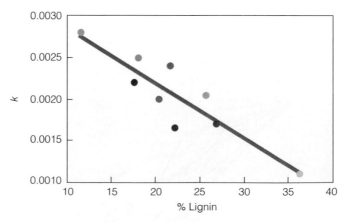

FIGURE 9.9   (a) Relationship between initial percent lignin and rate of decomposition (*k*) for nine species of leaf litter, shown in Figure 9.7b. Values of *k* were estimated using the general model outlined in Figure 9.8. The inverse correlation suggests that decomposition rate declines with increasing lignin content of the leaf material.

process of decomposition, the **net mineralization rate,** is the difference between the rates of mineralization and immobilization (Figure 9.11).

The increasing nitrogen content of the decomposing leaf litter seen in Figure 9.7a results from of immobilization of external nitrogen sources by the growing population of microbial decomposers. Although the data presented in Figure 9.10a show a steady increase in nitrogen over the first year, this pattern will not continue indefinitely. Upon colonization of the new litter material, the microbial population

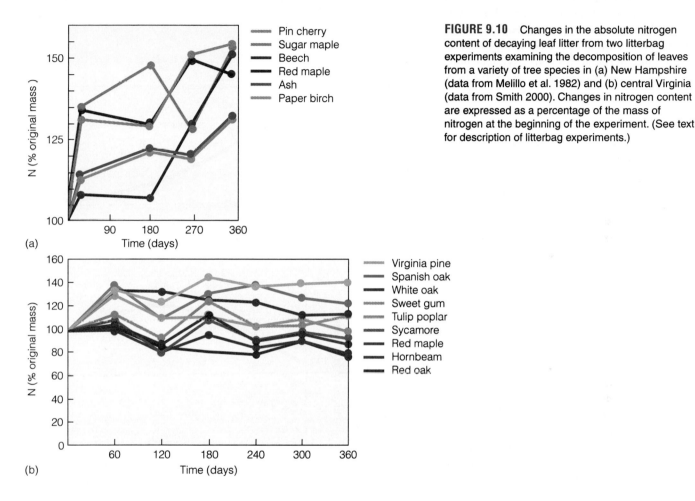

FIGURE 9.10 Changes in the absolute nitrogen content of decaying leaf litter from two litterbag experiments examining the decomposition of leaves from a variety of tree species in (a) New Hampshire (data from Melillo et al. 1982) and (b) central Virginia (data from Smith 2000). Changes in nitrogen content are expressed as a percentage of the mass of nitrogen at the beginning of the experiment. (See text for description of litterbag experiments.)

FIGURE 9.11 The relationship between immobilization rate and mineralization rate during the decomposition of plant material. Net mineralization rate (net release or uptake of a nutrient) is the difference between rates of mineralization and immobilization. When the initial ratio of carbon to a nutrient (C:X) is high—in this case, nitrogen (C:N)—the immobilization rate is high and the mineralization rate is low, resulting in the uptake of nitrogen by decomposers during the initial stages of decomposition.

increases (Figure 9.12a). As the population consumes the dead organic matter, the total carbon content declines, and because of the preferential consumption of nonlignin compounds, the percentage of the remaining carbon that is composed of lignins increases (Figure 9.12b). Initially, the nitrogen content of the litter is insufficient to meet microbial demand and immobilization increases the nitrogen content of the remaining litter mass (Figure 9.12c). The declining carbon and increasing nitrogen content of the litter results in a continuous decline in the C:N ratio throughout the decomposition process (Figure 9.12d). As the quantity and quality of the remaining carbon declines and the nitrogen content increases, nitrogen is no longer the limiting factor for microbial production. At this point, generally characterized by a "critical C:N," mineralization exceeds immobilization and there is a net release or loss of nitrogen from the decomposing litter (Figure 9.12c).

Although the processes and patterns outlined in Figure 9.12 describe the decomposition of all plant litters, the actual rates of mineralization and immobilization, and thus the net release of mineral nutrients to the soil (net mineralization rate), are influenced by the initial litter chemistry. Note that unlike the six species from the study in New Hampshire (Figure 9.10a), not all of the nine species from the study in Virginia (Figure 9.10b) show a continuous increase in nitrogen content throughout the experiment. All nine species initially increase in nitrogen content (values greater than 100%). Following this initial increase, three species continue to show an increase, one remains fairly constant, and a few even decline. An analysis of factors related to the observed variations in nitrogen dynamics during decomposition shows that the changes in nitrogen content (percent increase) during the first 90 days of the experiment are positively correlated with the C:N of the litter materials. The major factor influencing the observed variation in C:N among species is not differences in the carbon content, but differences among species in the initial nitrogen content. Those species with higher initial nitrogen concentrations show less immobilization and, therefore, less of an increase in nitrogen content during the initial stages of decomposition. The rate of immobilization is directly influenced by the nutrient concentration of the litter material and its ability to meet the nutrient demands of the microbial decomposer populations (Figure 9.13). Although the discussion and examples have focused on nitrogen, the same pattern of immobilization and mineralization as a function of litter nutrient content applies to all essential nutrients (Figure 9.14).

Influence of the Physical Environment

In addition to the quality of the dead plant and animal material as a food source, the physical environment also has a direct effect on both macro- and microdecomposers and, therefore, on the rate of decomposition. Both temperature and moisture greatly influence microbial activity. Low temperatures reduce or inhibit microbial activity; so do dry condi-

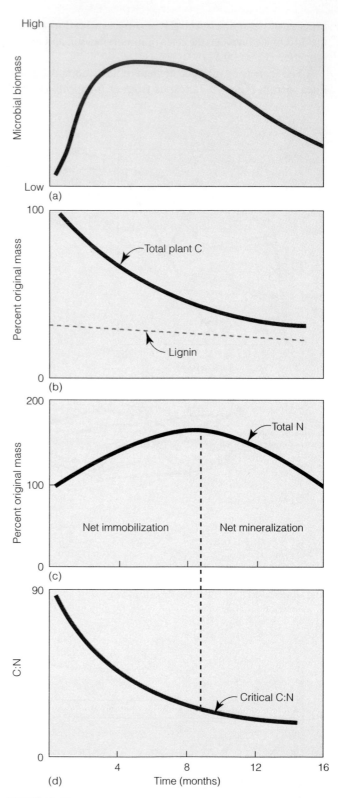

FIGURE 9.12 Hypothetical dynamics of (a) microbial biomass, (b) carbon content, (c) nitrogen content, and (d) ratio of carbon to nitrogen (C:N) during the process of decomposition.

tions. The optimum environment for microbes is a warm, moist one. Alternate wetting and drying and continuous dry spells tend to reduce both the activity and populations of microflora (Swift et al. 1979).

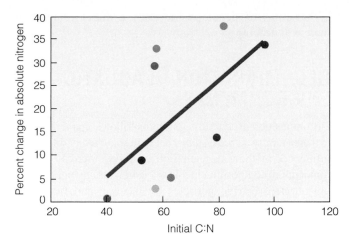

FIGURE 9.13 Relationship between nitrogen immobilization in decaying leaf litter and the initial ratio of carbon to nitrogen (C:N) for the nine species of leaf litter shown in Figure 9.10b. The rate of immobilization is measured as the increase in absolute nitrogen content of the litter over the first 90 days of the experiment. As the initial C:N of the leaf litter increases, the rate of immobilization increases.

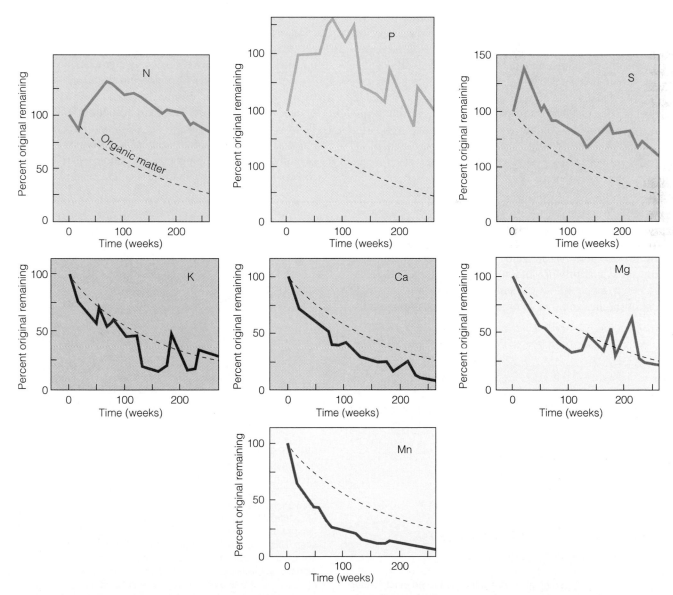

FIGURE 9.14 Patterns of immobilization and mineralization for a variety of nutrients. Data from the decomposition of Scots pine litter over a five-year period. (Data from Staff and Berg 1982; figure adapted from Aber and Melillo 1991.)

The data presented in Figure 9.15 show the effects of climate on microbial decomposer activity and subsequent rates of decomposition. This effect of climate shows up in the pattern of decomposition of red maple (*Acer rubrum*) leaves at study sites in New Hampshire and Virginia. Although the reported values for lignin content were the same at both sites, the decomposition rate is lower in New Hampshire. These observed differences can be attributed directly to the differences in the climate at the two sites. The mean daily temperature at the New Hampshire site is 7.2°C and the mean potential evaporation is 621 mm. The mean daily temperature at the Virginia site is 14.4°C and the mean potential evaporation is 806 mm.

The interaction between litter quality and climate on rates of decomposition was examined by Meentemeyer (1978) in a study of variations in decomposition rates at a number of sites in Europe and North America. Meentemeyer found that at each site there was a significant inverse relationship between rate of decomposition (k) and the initial lignin content of the litter species. This is the same relationship shown in Figure 9.9. However, the slope of the regression describing the relationship between k and lignin content varied among the sites (Figure 9.16). These differences were directly related to the climate at the sites, as represented by the value of actual evapotranspiration (AET). **Actual evapotranspiration** is the combined value of evaporation and transpiration at a site, here measured in mm/yr. As such it is directly influenced by both the availability of water (precipitation) and the evaporative demand of the atmosphere, which is a direct function of temperature (see discussion of relative humidity and evaporation in Chapter 2). Actual evapotranspiration is highest in areas of high precipitation and temperature, declining with either decreasing temperature (reduced evaporative demand) or decreasing precipitation (reduced availability of water to meet evaporative demand).

For a given lignin content, litter decomposed more quickly (higher value of k) in warmer, wetter climates. The higher rates of decomposition are a direct function of increased decomposer activity.

DECOMPOSITION IN AQUATIC ENVIRONMENTS

Decomposition in aquatic ecosystems follows a pattern similar to that in terrestrial ecosystems, but with some major differences influenced by the watery environment. As in terrestrial environments, decomposition involves leaching, fragmentation, colonization of detrital particles by bacteria and fungi, and consumption by detritivores and microbivores. In flowing water ecosystems, leaves, twigs, and other particulate matter are colonized by aquatic fungi. One group of aquatic arthropods, called shredders, fragment the organic particles and in the process also eat bacteria and fungi on the surface of the litter (see Chapter 31; downstream collectors filter from the water fine particles and fecal material left by the shredders. Grazers and scrapers feed on algae, bacteria, fungi, and organic matter collected on rocks and large debris (all discussed in Chapter 31). Algae take up nutrients and dissolved organic matter from the water.

In still, open water of ponds and lakes, and in sea, dead organisms and other organic material, called particulate organic matter (POM), drift toward the bottom. On its way this POM is constantly ingested, digested, and mineralized until much of the organic matter settles on the bottom in the form of humic compounds. How much depends in part on the depth of the water through which the particulate matter falls. In shal-

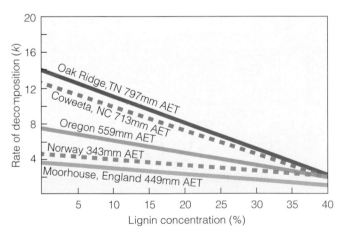

FIGURE 9.16 Relationship between lignin content and rate of decomposition (*k*) for different litter types at five study sites in Europe and North America. Values of *k* shown are based on an annual time scale. Although there is a significant inverse relationship between lignin content and rate of decomposition at each site, the slope of the regression describing change in decomposition rate as a function of lignin content varies among the sites. This variation is a function of the differences in climate at the five sites. The slope of the regression increases with increasing actual evapotranspiration. That is, for a given lignin content, the rate of decomposition increases with increasing actual evapotranspiration. This variation in decomposition reflects the faster rates of decay in warmer, wetter climates. (From Meentemeyer 1978.)

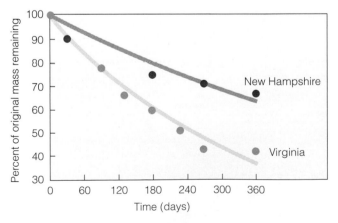

FIGURE 9.15 Patterns of mass loss with time for red maple (*Acer rubrum*) leaf litter at two sites: New Hampshire and Virginia. Data are from Figure 9.7. Differences in the rates of decomposition are a result of differences in the climate (temperature) at the two sites.

low water much of it may arrive in relatively large packages to be further fragmented and digested by bottom-dwelling detritivores such as crabs, snails, and mollusks (Newell 1965).

Bacteria work on the bottom or benthic organic matter. Bacteria living on the surface can carry on aerobic respiration, but within a few centimeters below the surface of the sediment, the oxygen supply is exhausted. Under this anoxic condition, a variety of bacteria capable of anaerobic respiration take over decomposition. With oxygen depleted, these bacteria employ other inorganic electron acceptors, in particular NO_3 near the top of the bottom mud, resulting in **denitrification,** the conversion of NO_3 to N_2. Bacteria use Fe^{3+} and SO_4 in the middle layers of mud, resulting in sulfate and iron reduction; they use HCO_3 in the deep mud, resulting in methane production.

Aerobic and anaerobic decomposition by benthic bacteria form only a part of the decomposition process. Dissolved organic matter (DOM) in the water column also provides a source of fixed carbon for decomposition. Major sources of this DOM are the macroalgae, phytoplankton, and zooplankton inhabiting the open water. Phytoplankton and other algae excrete quantities of organic matter at certain stages of their life cycle, particularly during rapid growth and reproduction. During photosynthesis the marine alga, *Fucus vesiculosus,* produces an exudate containing on average 42 mg C/100 g dry weight of algae/hr. Total exudate accounts for nearly 40 percent of the net carbon fixed (Sieburth and Jensen 1970). Cellular breakdown (autolysis) of phytoplankton and zooplankton accounts for 25 to 75 percent of the regeneration of available nitrogen and phosphorus. The exuded matter goes into solution rather than to bacterial decomposers (Johannes 1968). In fact 30 percent of the nitrogen contained in the bodies of zooplankton is lost by autolysis within 15 to 30 minutes after death, too rapidly for any bacterial action to occur.

Although phytoplankton can take up organic and inorganic compounds, only bacteria in the water can assimilate the many kinds of dissolved organic molecules that occur in low concentrations in the water, convert them into bacterial biomass, and make them available for other feeding groups, notably zooplankton. Dissolved organic matter then becomes a substrate for the growth of bacteria. Both dissolved and colloidal matter condense on the surface of air bubbles in the water, forming organic particles on which bacteria flourish (Riley 1963, Wright 1970, Cole 1982). By incorporating the nutrients in these particles into their own biomass, bacteria concentrate them. As in terrestrial ecosystems the utilization of these organic nutrients by bacteria results in both increase and immobilization of nutrients. As in terrestrial systems, bacterial consumption can reduce the supply of available nutrients to phytoplankton, thus reducing algal blooms.

Ciliates and zooplankton eat bacteria and in turn excrete nutrients in the form of exudates and fecal pellets in the water. Zooplankton, too, in the presence of an abundance of food, consume more than it needs and will excrete half or more of the ingested material as fecal pellets. These pellets make up a significant fraction of the suspended material. Bacteria attack these pellets to obtain the nutrients and growth substances they contain. Thus the cycle starts again.

Summary

1. Decomposition is the breakdown of chemical bonds formed during the construction of plant and animal tissue, thereby converting organic into inorganic nutrients.
2. True decomposers are bacteria, fungi, and other associated organisms. Involved in decomposition are an array of organisms known as detritivores, from microfauna such as protozoans to megafauna such as earthworms. They fragment and digest large detrital material into smaller pieces more easily attacked by microflora. Feeding on the microflora are microbivores.
3. Decomposition of plant and animal matter goes through several stages from deposition to final breakdown into inorganic substances. Early stages are leaching, which involves the loss of soluble sugars and other compounds, and fragmentation, the reduction of organic matter into smaller particles that are acted on by bacteria and fungi.
4. Decomposition involves mineralization, the conversion of material from organic to inorganic forms. Resynthesis (incorporation) of these inorganic nutrients into decomposer tissue is a process called nutrient immobilization.
5. The decomposition of animal matter involves the breakdown of animal tissue by bacteria and certain arthropods. Fecal matter of consumers is colonized by such specialized detritivores as flies and dung beetles.
6. Litter quality is influenced by the quantities and ratios of the types of carbon compounds present in the litter. Three general types of carbon compound are high-quality, easily digested simple sugars; the more complex cellulose and hemicellulose, relatively difficult to decompose; and lignin, with much larger molecules that are very difficult to decompose. Differences in the ratios of these types of carbon in plants affect the rates of decomposition among the different kinds of litter.
7. Studies of litter decomposition involve litterbag experiments. These experiments enable ecologists to study the rates of decomposition of various types of litter by calculating the mass loss over a period of time.
8. In addition to being an energy source for decomposers, litter is also a source of nutrients, influenced by the carbon:nutrient (C:X) ratio. During decomposition both mineralization and immobilization are taking place. If decomposers deplete the immediately available nutrients in the litter, they need to draw nutrients from the soil. Because decomposers are taking up more nutrients than are being replaced by mineralization, plants may experience nutrient depletion. The supply rate of mineral nutrients to the soil during decomposition is the net mineralization rate.
9. As decomposition proceeds, carbon in the litter mass declines through decomposer respiration, and other nutrients increase, exceeding the needs of the decomposers. When mineralization rate exceeds immobilization rate, nutrients are released from the decomposing litter to the soil.
10. Environmental conditions directly affect decomposer organisms and thus decomposition. Rate of decomposition is strongly influenced by moisture and temperature, which directly affect microbial activity. Alternate wetting and drying and continuous dry spells affect both activity and populations of microflora. Differences in temperature, both locally and regionally, result in spatial differences in decomposition rates.
11. Decomposition in aquatic systems follows much the same pattern as in terrestrial ecosystems, involving leaching, fragmentation, colonization of detrital particles by bacteria and fungi,

and consumption by microbivores. A major difference involves dissolved organic matter in the water column, excreted by phytoplankton and zooplankton. Bacteria take up this dissolved organic matter, which is not broken down into inorganic forms, and convert it into bacterial biomass that in turn becomes food for zooplankton. Thus, in contrast to terrestrial ecosystems, bacteria in aquatic systems act more as converters rather than as decomposers, whereas phytoplankton and zooplankton play a major part in the release of nutrients available for uptake.

Review Questions

1. What is decomposition? Why is it important ecologically?
2. What is the major role of each of the following in the decomposition process: detritivores; microbial grazers; bacteria and fungi; bacteria in aquatic systems; zooplankton in aquatic systems?
3. What are the sequential stages of decomposition?
4. Distinguish between nutrient mineralization and nutrient immobilization.
5. What is the importance of detritivores in decomposition?
6. Describe the differences between the decomposition process of plant and animal matter.
7. Why do some plant materials decompose more rapidly than others under the same environmental conditions?
8. What is the significance of the C:X ratio?
9. Contrast aquatic decomposition with terrestrial decomposition.
10. Distinguish between decomposition under aerobic and anaerobic conditions.
11. (a) In autumn straw and dead grass were plowed into the soil, where they were colonized by bacterial and fungal decomposers. The following spring the new plant growth was yellowed and stunted in growth, evidence of nitrogen deficiency. If decomposers are supposed to mineralize nutrients, why were the plants experiencing a deficiency of nitrogen? (b) By what process will inorganic nitrogen eventually be returned to the soil?

St. Patrick and the Absence of Snakes in Ireland

There are no snakes in Ireland. Popular legend has it that in the fifth century Patrick, a monk who immigrated from Britain and later became his adopted country's patron saint, drove the snakes of Ireland into the sea to their destruction (Figure III-A). Although Patrick, whose feast day we celebrate each March 17, contributed much to the cultural development of the Irish, he is not responsible for that country's lack of snakes. The reasons for their absence lie elsewhere and include the interplay organisms have with their physical environment.

Snakes, like all reptiles and amphibians, are ectotherms (cold-blooded). As is true of all animals, snakes need to maintain certain body temperatures for the natural processes of muscle and nerve activity, as well as digestion, to take place. Being ectotherms, snakes are more or less "thermal prisoners" of their surroundings: their body temperature is dictated by that of their immediate external environment. The range of environmental temperatures within which snakes are active is narrow, from about 10 to 40° C, with an optimum "operating" temperature of about 30° C. Compared to the range of atmospheric temperatures that exist on the surface of the earth (−110° C to 60° C), snakes' temperature tolerances are quite restricted. Although snakes depend on the external environment to provide the needed energy to maintain body temperatures within the range necessary for life processes, they are not passive prisoners of the constant variations in their thermal environments. Snakes and other terrestrial ectotherms can manipulate heat exchange between their bodies and their environments by a combination of behavioral and physiological processes. For instance, a snake can control its absorption of solar radiation—and thereby its body temperature—by altering the color of its absorptive surface (the skin) or changing the orientation of its body relative to the sun.

Many reptiles can change their color by dispersion or contraction of dark pigments, such as melanin, in their skin. Because dark skin substantially increases the amount of solar energy that is absorbed, many individuals living in the cooler parts of a species' range are darker than their same-species counterparts in warmer climates (Figure III-B). Additionally, many snakes of the temperate regions can change their color to accommodate seasonal changes in the amount of solar radiation. Some snakes capitalize on this mechanism for increasing solar radiation by having dark skin on their heads that they expose to the sun before other parts.

FIGURE III-A St. Patrick, who, legend has it, drove the snakes from Ireland.

(a)

(b)

FIGURE III-B   Light and dark phases of the Mexican moccasin (Agkistrodon bilineatus).

Warming the brain and the sensory organs such as the eyes and the tongue first enhances a snake's ability to detect both danger and food. Finally, pregnant females of some species are darker than males and nonpregnant females, presumably to maintain warmer-than-normal body temperatures that are thought to accelerate embryonic development.

As noted, a second way that snakes control their absorption of solar radiation is by increasing or decreasing the amount of body area exposed to radiation. By orienting its body to lie at right angles to the direction of the sun, and by spreading and flattening to increase its body's surface area, a snake can attain temperatures much higher than the surrounding air. When a snake's body has reached a suitable temperature, it avoids further heating by lightening its skin color, changing its posture to one more parallel to the sun's rays, and eventually moving into shade or underground where heat absorption is reduced. Furthermore, the temperature of the substrate that the snake is in contact with is also important because a cool snake can crawl on a warm rock or other surface and absorb its heat.

The behavioral and physiological means by which snakes regulate body temperature permit them to occupy a surprisingly large part of Earth's land surface. Snakes occur in all continents except Antarctica. However, the vast majority of snake species are found within the tropical and subtropical regions (at least three-fourths of the snake species existing on Earth are found between 22° N and 22° S latitude). Species diversity of reptiles decreases rapidly from the equator toward the higher latitudes. In North America, the number of lizard species is highest in the warm desert regions of the southwest and declines continuously as you move northward (Figure III-C). The same pattern of species diversity is evident in Europe, where the number of reptile species declines markedly as you move from the warmer Mediterranean coast toward northern Europe and the British Isles. Although one species, the European adder (*Vipera berus*), is found above the Arctic Circle in Scandinavia, its unusually northern distribution comes at a cost. The adder has a very limited period of activity, often only three to four months a year. In addition, the species may take up to four years to attain sexual maturity, and females may breed only once every three or four years, using the intervening period to build up the fat reserves necessary to produce offspring.

The progressive decline in solar radiation and temperatures from the tropics to the poles not only reduces the abundance and diversity of reptiles but also has a direct influence on their body size. The reason for these patterns is that heat exchange occurs across the surface of a body but warming usually occurs throughout the entire body's mass or volume. Large bodies, because of their low surface area to volume ratio, take longer to warm than smaller ones. This physical reality results in upper limits to the size of snakes (and other reptiles) depending on the distance of the snake's habitat from the tropics. All of the large snakes, such as the anaconda and python, are found within the tropic and subtropical regions. Other large reptiles—such as the iguanas, monitor lizards (which include the goanna of Australia and the Komodo dragon of Indonesia), and the crocodilians (alligators, caimans, and crocodiles)—are likewise limited in their distribution to the warm, aseasonal environments of the subtropics and tropics. The maximum body size for ectotherms declines as you move north and south from the equator. This pattern is the exact opposite of that observed for endotherms (warm-blooded animals), where average body size increases from the tropics to the poles,* a pattern referred to as Bergman's rule. The environmental constraint on the upper limit of body size for ectotherms is at the very heart of the current debate over whether the dinosaurs were cold- or warm-blooded. The fossil record places the dinosaurs well into the northern latitudes, with specimens found in Alaska and Siberia.

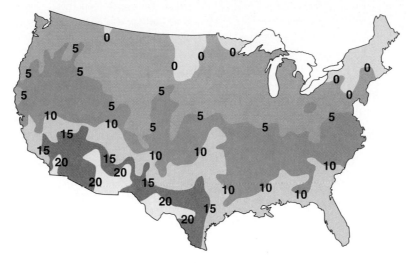

FIGURE III-C Patterns of lizard species diversity (number of species) in the contiguous United States.

As for snakes, the fossil record reveals that the diversity of reptiles and amphibians (terrestrial ectotherms) has always been low in northern Europe. During the Pleistocene era (from 1.6 million to 10,000 years ago), a time when the temperature of Earth was considerably lower than that now observed, much of the Northern Hemisphere was covered by glaciers. Although certain refugia (small pockets of land not covered by ice) existed, the massive ice sheet virtually obliterated all life in northern Europe, including Ireland and Britain. As temperatures rose and the glaciers retreated during the Holocene (10,000 years to present), the region was recolonized by plants and animals, but the cool temperatures of northern Europe proved inhospitable to most snakes. Nevertheless, the island of Britain, similar in climate to Ireland, ended up with three species of snakes that survive there to this day. Two enjoy an extensive distribution, but the third is found only in a small area of southern England.

Three species of snakes is a paltry sum, but nonetheless it's more than Ireland has. Why the difference? After the retreat of the continental glaciers at the end of the last ice age, Britain had a land bridge connecting it with the mainland of Europe, whereas Ireland did not. The limited diversity of snakes on the mainland meant that the pool of species available to recolonize both Britain and Ireland was small in the first place. However, Ireland's lack of a land bridge with continental Europe coupled with the limited dispersal abilities of most snake species have to date prohibited their successful recolonization there.

In fact, although devoid of snakes, Ireland does have three species of terrestrial ectotherms. Two species of amphibian (smooth newt, *Triturus vulgaris*; common frog, *Rana temporaria*), and one reptile, the small viviparous lizard *Lacerta vivipara*, were able to reach Ireland's shores despite the absence of a land bridge. So, given the physiological constraints imposed by temperature on the distribution of snakes together with the recent (in geological terms) climate history of the region, even if St. Pat were to have run the snakes out of Ireland, it would have been a rather small task.

* The reason for this is large bodies lose less heat than small bodies to the environment because they expose less surface area per unit volume. This is an important consideration for animals that generate their own internal heat. (Note that ectotherms tend to be limited by how much heat they can extract from their environment and endotherms by how much heat they lose to the environment.)

PART FOUR

Intraspecific Population Ecology

CHAPTER 10

Properties of Populations

CONCEPTS

1. A population is a group of individuals of the same species occupying a given space at a given time.

2. Populations are characterized by density, age structure, a birth rate, and a death rate.

3. Populations may be unitary or modular.

4. Individuals within a population may be distributed randomly, uniformly, or in clumps.

5. Dispersal of individuals influences the density and distribution of populations.

6. Age structure reflects the history of a population.

7. Mortality patterns are basic to understanding population dynamics.

8. A life table summarizes survivorship, mortality, and life expectancy of individuals in a population.

9. Fecundity may be expressed as a net reproductive rate that combines survivorship with an age-specific schedule of births.

A **population** is a group of organisms of the same species occupying a particular space at the same time. Both space and time are defined by some situational criteria. For example, the population of a town is defined by the space enclosed by political boundaries. If the town is a resort area or supports an academic institution, the population within the defined space will differ depending on the defined time period. A population possesses certain emergent properties not associated with individuals. Populations are characterized by density, the number of organisms occupying a defined unit of space. They have an age structure, the ratio of one age class to another, and a sex ratio, the proportion of one sex to another. Populations have a birth rate and an immigration rate that add to the population, and a death rate and an emigration rate that shrink the population. The difference between gains and losses becomes the growth rate of a population.

UNITARY AND MODULAR POPULATIONS

To study a population, we need to determine the number of individuals present in a given area. If you are studying animals, typically there is no major problem in defining individuals. You can recognize a deer as an individual. It consists of morphological parts: head, four legs, a tail, hair, and other sharply defined characteristics. Other deer, regardless of age, have a similar appearance. Thus individual deer are easy to identify and count.

But what about the trees among which the deer are standing (Figure 10.1)? Each stem appears as an individual tree, but is it? If you give the trees some careful thought, you will appreciate the fact that each is a vertical structure made up of modules of smaller units: leaves, buds, twigs, branches, and seasonally, flowers and fruits. If you exposed the roots, you may discover that some of the trees are connected by the same root structure. Each stem is a part of a larger unit, a clump of closely associated trees. Each stem is a **module,** a part of the whole.

An individual, strictly speaking, is derived from a zygote. The deer is such an individual. The trees can also be traced back to sexual reproduction, arising from seeds produced when wind-carried pollen fertilized ovules in the female flowers. Once established, some species of trees, such as aspen (*Populus* spp.), and many species of herbaceous plants, such as strawberries (*Fragaria*) and mayapples (*Podophyllum peltatum*), produce new "individuals" or modules asexually by means of buds on shallow horizontal roots or by stems that touch the ground (Figure 10.2) (resulting in two levels of population structure). Individual plants produced by sexual reproduction are termed **genets** or genetic individuals. The asexually produced individuals derived from the genetic parent are called **ramets.** They may remain physically linked to the genet or exist independently of the parent plant. In the latter case, the connections between the original stem and its ramets die, resulting in populations of ramets that can produce their own lateral extensions or ramets. Although

FIGURE 10.1 Populations consist of either unitary or modular organisms. The white-tailed deer is a unitary organism with a discrete growth form. The surrounding trees appear as individuals, but some of them are probably modules of the original parent tree, arising from buds on the horizontal roots to form clones. In addition, the tree is made up of iterated growth forms of buds, leaves, stems, and twigs.

living independently, ramets all possess the same genetic constitution as the parent genet. By producing ramets a genet can extend over a considerable area.

A group of ramets arising from the same genet make up **clone.** Because all individuals of the clone have arisen from the same zygote and have the same genotype, we can view them collectively as a single organism or genet, no matter how many "individuals" the clone may possess. Because they are a part of the original growth, iterations from the parent plant, the genet may be very old, although the "individual" trees or plants, the ramets, may be very young. Technically, to describe such a population we should separate and count the genets from ramets, but this of course is difficult if not impracticable. For this reason, in plant population studies individual ramets are counted as and function as individual members of the population.

This feature is not restricted to plants. Some animal groups are also modular, including sponges, corals, and hydroids. All of them are sessile organisms, except during the dispersal stage.

Ramets are horizontal modules. Woody plants, in particular, exhibit vertical growth of modules. These modules give rise to buds, leaves, and stems and other structures, one above

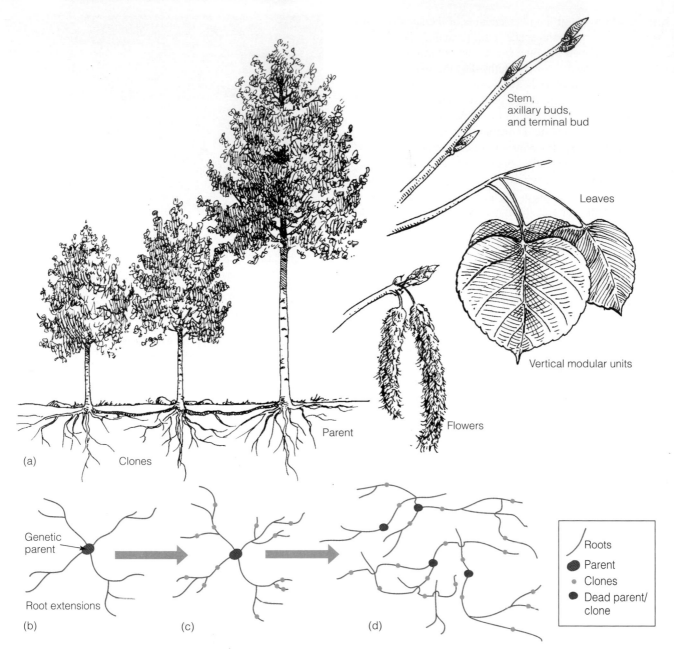

FIGURE 10.2 Horizontal and vertical modular growth in an aspen tree *(Populus tremuloides)*. Vertical growth involves modules of leaves, buds, and associated stems. Horizontal growth involves modules of roots and root buds, which give rise to clones. These clones are various ages, with the youngest individuals forming the leading edge of growth away from the parent. Connections between the original parent and its clones are lost in time, and some clones lead independent existences. (a) Genetic parent. (b) Clonal growth arising from root extensions. (c) Genetic parent is dead. Clonal growth arising from root extensions of the parent forms groups of new individual stems. Roots of some clones have separated from the parental and some clonal root extensions to establish new independent groups. However, all clones of the parent, independent or not, are identical genetically.

the other, while maintaining strong woody connections to older modules. Stems eventually develop into branches, limbs, and vertical extensions of trunks (Figure 10.2). Some modules develop into flowers, producing potential new genets. Birth, life, and death of vertical modules and competition among them for light and nutrients influence the nature of the vertical structure.

Such vertical modular growth in itself is a demographic process (see Harper 1977, White 1979, Harper and Bell 1979). Leaves, buds, stems, and roots are also populations in their own right. They compete with neighboring modules of the same type for moisture, light, and nutrients, and respond in their own way to environmental conditions, yet may supply carbon (energy) to neighboring modules or compete for it. They experience their own birth rates, death rates, and changes in age structure. The distribution of birth rates and death rates of the modules influence the growth form the plant will take, including the ordering of branches in trees.

METAPOPULATIONS

Few species populations are continuous. They exist as separate or spatially disjunct populations distributed in patches across a heterogeneous landscape. If the patches are sufficiently close that individuals move freely enough among them to stabilize local population fluctuations, then such populations, though separate, behave as a single continuous population. However, in many landscapes the habitat is so fragmented that species exist as distinct, partially isolated subpopulations, each possessing its own population dynamics. Such subpopulations exist in widely separated discrete patches of breeding habitat in a matrix of unsuitable habitats. Linking these populations is dispersal among them. Such separated populations interconnected by immigration are called **metapopulations.**

The classic concept of metapopulation was formulated mathematically by R. Levin in 1969, but received little attention until around 1990 when ecologists realized that continuous populations were being highly fragmented into discrete subpopulations. The Levin concept of metapopulation carried certain assumptions. Each subpopulation has its own birth rates and death rates independent of other subpopulations; limited dispersal links the subpopulations; and individual populations have a finite lifetime. All habitat patches are similar in the size and shape, continuously favorable, and equally accessible. Some suitable habitat patches have to remain empty to allow colonization (an important conservation implication to be discussed later). The pattern of habitat occupancy changes with local extinctions and recolonizations.

The concept of the metapopulation has evolved and continues to evolve to embrace much less restrictive assumptions (Hanski and Simberloff 1997). In real situations a continuum of metapopulations exist that diverge from the classic concept (Harrison and Taylor 1997) (Figure 10.3). At one end of the continuum is the mainland-island populations (see Chapter 23) in which island populations are maintained by continuous immigration from large, extinction-resistant mainland populations. At the other end of the continuum are totally isolated populations that rarely receive immigrants and are doomed to eventual extinction. In between are isolated populations linked together by dispersal. Successful movement from one patch to another is influenced by the distance and nature of the surrounding habitats. If one habitat becomes vacant through local extinction, it can be colonized by immigrants from another. Such local extinctions and recolonizations allow metapopulations to persist. The metapopulation concept is highly applicable to problems of habitat fragmentation and the conservation of species (see Chapters 23 and 18).

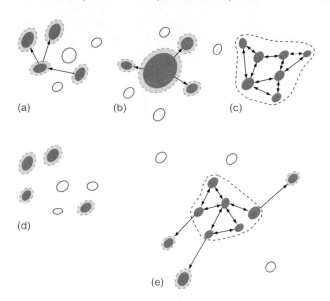

FIGURE 10.3 Different types of metapopulation. Filled circles, occupied habitat patches; empty circles, vacant habitat patches; dotted lines, boundaries of local populations; arrows, dispersal. (a) Classic (Levins); (b) mainland-island; (c) patchy population; (d) nonequilibrium metapopulation; (e) intermediate case combining features of a–d. (Harrison and Taylor 1997.)

POPULATIONS AS GENETIC UNITS

A population is a genetic unit. Each individual carries a certain combination of genes, a sample of the population's total genetic information. The sum of all genetic information carried by all individuals of an interbreeding population is the **gene pool.** Gene flow, the exchange of genetic information between populations, comes about through immigration and emigration.

Populations may also be considered as evolutionary units. Evolution results from changes in gene frequency in a given gene pool over a period of generations. An outcome may be changes in the physical expression of organisms in the population that reflect the altered genetic constitution. These changes in the genetic constitution often result from selective pressures brought to bear by the environment on individuals of the population.

DENSITY AND DISPERSION
Crude Versus Ecological Density

The size of a population in relation to a definite unit of space is its **density.** Every ten years the Census Bureau counts the number of people living in the United States and often presents the data as the number of individuals per square mile. Wildlife biologists estimate the number of game animals in a particular area; a forester determines the number and volume of trees in a timber stand. The measure of the number of individuals per unit area is called **crude density.**

Populations do not occupy all the space within a unit, because it is not all a suitable habitat. A biologist might estimate the number of deer per square mile, but the deer might avoid half the area because of human habitation and land use practices, lack of cover, or lack of food. A soil sample may contain 2 million arthropods per square meter, but these arthropods inhabit only the pore spaces in the soil. Goldenrods inhabiting old fields grow in scattered groups or clumps because of soil conditions and competition from other old field plants. No matter how uniform a habitat may appear, it is not uniformly habitable because of microdifferences in light, moisture, temperature, or exposure, to mention a few conditions. Each organism occupies only areas that can adequately meet its requirements, resulting in patchy distribution. Density measured in terms of the amount of area available as living space is **ecological density.**

Attempts have been made to make such ecologically realistic measurements. For example, one study in Wisconsin expressed the density of bobwhite quail as the number of birds per kilometer of hedgerow rather than per hectare (Kabat and Thompson 1963). However, ecological densities are rarely estimated, because it is difficult to determine what portion of a habitat represents living space, especially if organisms require different habitats during the year or during developmental change.

The density of organisms in any one area varies with the seasons, weather conditions, and food supply, to name only a few factors. However, an upper limit to the density is imposed by the size of the organism and its trophic level. Generally the smaller the organism, the greater its abundance per unit area. A 40 ha forest will support more woodland mice than deer and more trees 5.1 to 7.6 cm (2 to 3 inches) dbh (diameter breast height) than trees 30 to 35 cm (12 to 14 inches) dbh.

Density is one of the more important parameters of populations. It both determines in part and is determined by energy flow, resource availability and utilization, physiological stress, dispersal, and productivity of a population. The density of human populations, for example, relates to economic growth and the expansion and management of towns, cities, regions, states, and nations. Dense or sparse populations can place strains on economic and social institutions. The distribution of humans in a given region affects land use and pollution problems. Wildlife biologists need to know about densities of game populations to regulate hunting and manage habitats. Foresters base timber management and evaluation of site quality in part on the density of trees.

Patterns of Dispersion

Crude density also tells us nothing about how evenly individuals within the population are distributed over space and time within the area surveyed. Determining this dispersion is a major field problem.

Spatial Dispersion Individuals may be distributed randomly, uniformly, or in clumps (Figure 10.4). Distribution is considered random if the position of each individual is independent of the others or if the occupation of each spot is equally likely. Random distribution is rare, for it can occur

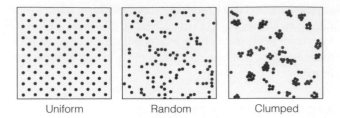

Uniform   Random   Clumped

FIGURE 10.4 Patterns of distribution.

only where the environment is uniform, resources are equally available throughout the year, and interaction among members of the population produces no patterns of attraction or avoidance. Some invertebrates of the forest floor, particularly spiders (Cole 1946, Kuenzler 1958), the clam *Mulinia lateralis* of the intertidal mudflats of the northeastern coast of North America, and certain forest trees (Pielou 1974) appear to be randomly distributed.

Uniform or regular distribution is the spacing of individuals more evenly than would occur by chance. Regular patterns of distribution result from intraspecific competition among members of a population. For example, territoriality under homogeneous environmental conditions can produce uniform distribution (Figure 10.5). In plants it may result from severe competition for crown and root space among forest trees (see Gill 1975), and for moisture among desert plants (Beals 1968) and savanna trees (Smith and Goodman 1986).

The most common type of distribution is clumped, also called clustered, contagious, and aggregated. This pattern of dispersion results from responses by organisms to habitat differences, daily and seasonal weather and environmental changes, reproductive patterns, and social behavior. The distribution of human beings is clumped because of social behavior, economics, and geography.

Clumped distributions among plants are often influenced by the nature of propagation and specific environmental requirements. Poorly dispersed seeds, such as those of oaks and cedar, are clumped near the parent plant or where they are placed by animals. Well-dispersed seeds are more widely distributed, but even they tend to be more abundant near the parent plant. Seed germination and survival of seedlings also influence the degree and type of aggregation.

Some animal aggregations represent individual responses to environmental conditions. Individuals may be drawn together by a common source of food, water, or shelter. Moths attracted to light, earthworms congregated in a moist pasture field, barnacles clustered on a rock—all have little or no social interaction. The individuals do not aid one another and only passively prevent other members of the same species from sharing the condition that brought each of them to the same location.

Aggregations on a higher social level reflect some degree of interaction. Prairie chickens congregate for communal courtship; elk band together in herds with some social organization, usually with a cow as the head (Altmann 1952); birds congregate on feeding grounds away from territorial

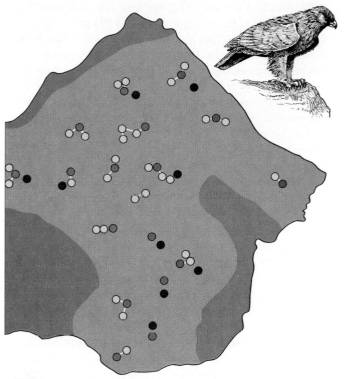

Group of sites belonging to one pair

Single site

Marginal site not regularly occupied

Breeding, year of survey 1967

Low ground unsuited to breeding eagles

FIGURE 10.5 Golden eagle territories in Scotland, showing the even dispersion of breeding sites. Territoriality in birds and other animals, a function of intraspecific competition, usually results in a fairly uniform distribution over an area of suitable habitat. (From Brown 1976:109.)

sites, yet show intolerance for each other near the nest. Aggregations of the highest social structure are found among insect societies, such as ants and termites, and in the naked mole rat. Here individual members are organized into social castes according to the work they perform.

Temporal Dispersion Organisms in populations are distributed in time as well as in space. Temporal distribution can be circadian, relating to daily changes in light and dark. The environmental rhythm of daylight and dark (see Chapter 8) is responsible for the daily movement of some animal populations, such as nectar-feeding insects seeking patches of open flowers, the daily movement of plankton from deeper to upper layers of water, and the withdrawal and emergence of nocturnal and diurnal animals.

Other temporal distributions relate to changes in humidity and temperature, seasons, lunar cycles, and tidal cycles. Seasonal changes are reflected in the sequential blooming of wildflowers in forest and field and in the return and departure of migrant animals. The populations of forests and fields are quite different in spring, summer, fall, and winter. Distributions may

also be related to longer periods of time, which encompass annual cycles, successional stages, and evolutionary changes.

Dispersal Movements Dispersal movements may be one way out of one habitat into another, or with no return trip. The former movement is termed **emigration,** the latter **immigration.** Both involve the same movements; the distinction is one of viewpoint. Emigrants from one area become immigrants to another. Dispersal with a return to the place of origin is termed **migration.**

Many organisms, especially plants, depend on passive means of dispersal involving gravity, wind, water, coats of mammals, the feathers of birds, or the guts of both. The distance these organisms travel depends on the quality of the dispersal agent. Seeds of most plants fall near the parent and their density falls off quickly with distance (Figure 10.6). Heavier seeds, such as those of oaks, have a much shorter dispersal range than the lighter wind-carried seeds of maples, birch, milkweed, and dandelions. Some plants, such as cherries and viburnums, depend on active carriers such as particular birds and mammals to disperse their seeds by eating the fruits and carrying the seeds to some distant point in their guts. Other plants possess seeds

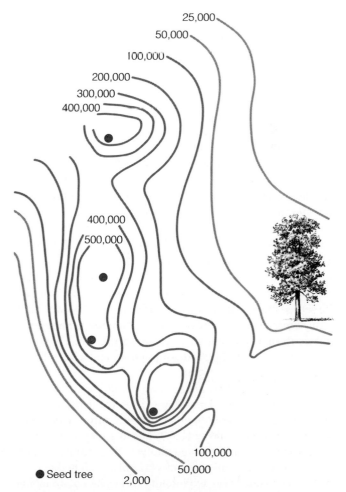

● Seed tree

FIGURE 10.6 Pattern of annual seedfall of yellow-poplar (*Liriodendron tulipifera*). Lines show equal seeding density. With this wind-dispersed species, seedfall drops off rapidly away from the parent seed trees. (After Engle 1960.)

armed with spines and hooks that catch on the fur of mammals, the feathers of birds, and the clothing of humans.

For mobile animals, dispersal is active, but many depend on a passive means of transport, such as wind and moving water. Wind carries the young of some species of spiders, larval gypsy moths, and cysts of brine shrimp. In streams the larval forms of some invertebrates disperse downstream in the current to suitable microhabitats. The larval offspring of sessile marine organisms such as barnacles become active planktonic swimmers in the marine environment and determine where they will settle.

Dispersal among mobile animals may involve both young and adults, males and females, but there is no hard and fast rule about who disperses. Young are the major dispersers among birds. Greenwood (1980) suggests their leaving be called **natal dispersal** and that the dispersal of adults from poor to better reproductive sites be called **breeding dispersal** (see also Greenwood and Harvey 1982). Among rodents subadult males and females make up most of the dispersing individuals (Tamarin 1978, Krebs et al. 1976, Beecham 1980). At times dispersing females outnumber males, and at other times the opposite is true; or there may be no difference at all (Johnson and Gaines 1987). Many dispersers, up to 40 percent, are in breeding condition. Among many groups of insects, dispersal takes place among a polymorphic component of normally flightless insects that acquire wings, notably aphids (Aphidoidae), leafhoppers (Cicadellidae), and long-winged (versus short-winged) water striders (Gerridae). Crowding, temperature change, quality and abundance of food, and photoperiod all have been implicated in stimulating the development of winged individuals. These winged forms have the option of flying to a new habitat or not flying, depending on environmental cues. Within insect populations in which all individuals are capable of flight, a portion may disperse and colonize new habitats; others may not. (For review, see Harrison 1980.)

The prereproductive period is the usual time of dispersal. Many species of birds disperse in spring, following the return of migrants or breakup of winter flocks among nonmigratory resident species. In some resident species, such as the ruffed grouse, young disperse in late fall and are settled on breeding areas by early winter. Rodents leave their home place mostly when the population is increasing and at the peak phase of population growth. Dispersal decreases during the periods of population decline (Krebs et al. 1976, Beecham 1980).

Because dispersers are seeking vacant habitat, the distance they travel will depend in part on the density of surrounding populations and the availability of suitable unoccupied areas. Murray (1967) stated a rule of dispersal: move to the first uncontested site you find and no further. Animals generally will either disperse in a straight line from their natal area or make exploratory forays into surrounding areas before leaving the natal site. They should settle in the first empty site.

Larsen and Boutin (1994) monitored the individual fates and documented the movements of 205 (94 percent) of 219 young red squirrels (*Tamiasciurus hudsonicus*) at Fort Assiniboine, Alberta, Canada. Prior to settlement, the young squirrels made exploratory forays of up to 900 m from their natal territory (Figure 10.7), which they did not abandon until they settled on their own territory. Just under one-half of the 73 offspring that acquired territories settled either on or adjacent to their mother's territory. The farthest distance a young squirrel settled was only 323 m from the natal territory, or about the distance of three territory widths.

In developing a model of dispersion based on straight-line movements, P. M. Waser (1985) measured dispersal distance in terms of the number of home ranges (based on average size of home range for the species) the disperser has to travel from its natal site (Figure 10.8). His model predicts the distribution of dispersal as a function of turnover of home ranges or territories vacated by the death of previous owners, and of the distance the disperser has to travel to find them.

Migratory Movements   Migration is a two-way movement in response to an evolutionary or environmental adaptation or pressure. Repeated return trips may be daily or seasonal. Zooplankton in the oceans move down to lower depths by day and move up to the surface by night. Their movement appears to be a response to light intensity. Bats leave their daytime roosting places in caves and trees, travel their feeding grounds, and return by daybreak. Other migrations are seasonal, either short range or long range. Annually earthworms make a vertical migration deeper into the soil to spend the winter below the freezing depths and move back to the upper soil when it warms in spring. Elk move down from their high mountain summer ranges to lowland winter ranges. On a larger scale, caribou move from summer calving range in the taiga to the arctic tundra for the winter, where lichens comprise their major food source (Figure 10.9). Gray whales (*Eschrichtius robustus*) move down from the food-rich arctic waters in summer to their warm wintering waters of the California coast, where they give birth to young. Similarly humpbacked whales (*Megaptera novaeangliae*) migrate from northern oceans to the central Pacific off the Hawaiian Islands. The most familiar of all are long-range and short-range migration of waterfowl, shorebirds, and neotropical migrants in spring to their nesting grounds and in fall to their wintering grounds.

Another type of migration involves only one return trip Such migrations occur among Pacific salmon (*Oncorhynchus* spp.) spawned in freshwater streams. The young hatch and grow in the headwaters of coastal streams and rivers. The young travel downstream and out to sea, where they reach sexual maturity. At this stage they return to the home stream to spawn and then die.

A third type of migration, exemplified by the monarch butterfly, is unusual because the fall migrants do not return north but their offspring do. About 70 percent of the last generation of monarch butterflies in summer moves south in

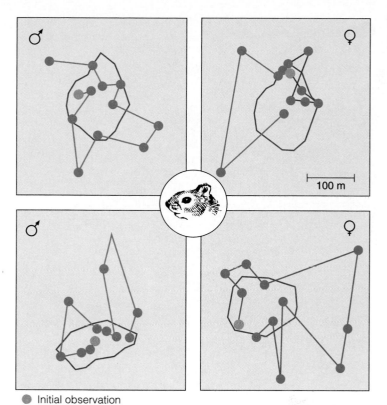

FIGURE 10.7 Direction dispersal forays made by young male radio-collared red squirrels. (After Larson and Boutin 1994: 218.)

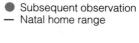

● Initial observation
● Subsequent observation
— Natal home range

FIGURE 10.8 Observed and expected dispersal distances (a) in the deer mouse (*Peromyscus maniculatus*) and (b) in the white-crowned sparrow (*Zonotrichia leucophrys*). The bar graphs indicate the observed dispersal distances. The reddish line indicates the dispersal distance expected if the search were along a radius away from the natal site. The yellow line in (b) is the expected pattern if the search were to the nearest empty site. (From Waser 1985:1173.)

FIGURE 10.9 Migratory pathways of four vertebrates. (1) Ring-necked ducks (*Aythya collaris*) breeding in the northeast migrate in a corridor along the coast to wintering grounds in South Carolina and Florida. (2) The canvasback duck (*Aythya valisinera*), whose major breeding areas are in the prairie pothole region, has a number of corridors; but most canvasbacks diverge to the Atlantic or Pacific Coast. (3) The gray whale (*Eschrichtius robustus*) summers in the Arctic and Bering seas; it winters in the Gulf of California and the waters off Baja California. (4) The barren-ground caribou (*Rangifer tarandus*) winters in the taiga and spends the summer on the arctic tundra.

Winter breeding grounds

noticeable flights to their wintering grounds in the highlands of Mexico, a trip that covers about 14,000 km. From the wintering grounds monarchs undertake a northward movement in January and arrive in the deep southern United States in early spring, where they start a new generation. The first spring generation continues the northward trek, following the milkweed north with the spring, thus restocking northern populations killed by frost. One generation succeeds another until the monarchs that finally arrive on the northern breeding ground are several generations removed from the ancestors that migrated south the previous fall. Such migrations are not confined to the monarch butterfly. Similar but less extensive migrations are undertaken by other insects, including two leafhoppers (*Macrosteles fascifrons* and *Empoasca fabae*), the harlequin bug (*Murgantia histrionica*), and the milkweed bug (*Oncopeltus fasciatus*).

AGE STRUCTURE

When you attend events, you may notice the differences in age of those attending. For example, rock concerts are dominated by younger people, symphonic concerts by older ones.

Parades and fairs have a greater diversity of age classes. Once aware of these differences, you can appreciate how the nature and character of a population are influenced by its age structure. Age distribution, termed **age structure**—structure of plant, animal, and human populations—determines in part population reproductive rates, death rates, vigor, survival, and other demographic attributes.

The age structure of populations can be analyzed using specific age categories, such as years or months. We may use other categories: (1) life-history stages, such as prereproductive, reproductive, and postreproductive stages in birds and mammals and eggs, pupae, larvae, and instars in insects; and (2) size classes in plants, such as heights of herbaceous plants or seedlings and diameters of trees, because reproduction is better predicted by size than by age in plants.

Age Structure in Animals

Theoretically, all continuously breeding populations should tend toward a **stable age distribution.** The *ratio* of each age group in a growing population remains the same if the age-specific birth rate and the age-specific death rate do not change. If the stable age distribution is disrupted by any cause, such as natural catastrophe, disease, starvation, or emigration, the age composition will tend to restore itself upon return to the previous conditions, provided, of course, the rates of birth and death are still the same.

When deaths balance births and the population is closed—that is, it experiences no movements into or out of the population—the population has reached a constant size. It assumes a special form of stable age distribution, known as a **stationary age distribution.** This distribution requires that the population is not growing. Under these conditions its age structure remains the same. Thus all stationary age distributions are stable, but only some stable age distributions are stationary.

A population can increase, decrease, or remain stable. The ratio of young to adults in a relatively stable population of most mammals and birds is approximately 2:1 (Figure 10.10). A normally increasing population should have an increasing proportion of young, whereas a decreasing population should have a decreasing proportion of young. Within this framework are a number of variations. A population, for example, may be decreasing yet show an increasing ratio of young to adults because of high adult mortality from various causes, such as overexploitation (see Chapter18). Whatever the relationships, the number in one age class that enters the next age class influences the age structure of a population from year to year. A deficiency of young can lead to an aging population, but a high proportion of old individuals may not be a cause of a deficiency of young. By combining information on population density, age ratios, and reproduction, a biologist can correlate changes in population structure with habitat changes and ecological and human influences.

The history of a population can be detected in a series of age distributions. Grant and Grant (1992) provide the age distribution of the cactus ground finch (*Geospiza scandens*) on Isla Daphne Major, Galapagos, for the years 1983 and 1984 (Figure 10.11). Reproductive activity and success are governed by rainfall. In 1982 rainfall was negligible, and reproduction failed because of failure of seed crop on which the finches depend. No young entered year class 1 in 1983. Heavy rainfall marked 1983, a year of El Niño (see Chapter 2), and so reproduction was highly successful because of an abundance of seeds and an abundance of caterpillars for feeding nestlings. This success was marked by the presence of year class 1 in 1984, but no young were available to enter year class 2. The very small age class 3 in 1983 results in a very small age class 4 in 1984. Note the shrinkage of other age classes as they move up the age ladder. Males live longer and breed later than females.

The loss of age classes can have a profound influence on a population's future. Consider an exploited fish population in which the older reproductive age classes are removed. If the population experiences reproductive failure for one or two years (as in the ground finch age pyramids), there will be no young fish to move into the reproductive age class to replace the fish removed, and the population can collapse.

Although useful in looking at a population's history and trends, age pyramids, especially for wild populations subject to exploitation, rarely indicate whether the rate of increase is positive, negative, or zero. When age distributions are not

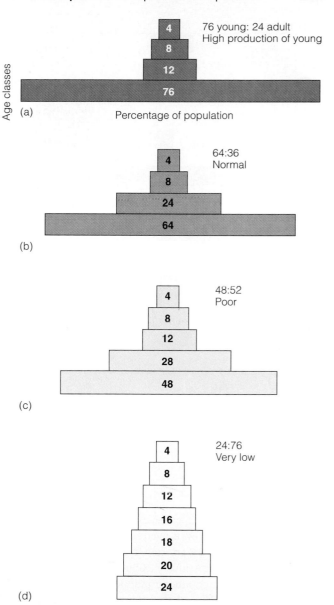

FIGURE 10.10 Theoretical age pyramids, especially applicable to mammals and birds. (a, b) Growing populations in general are characterized by a large number of young, giving the pyramid a broad base. (c) A population with a high proportion of individuals in the older age classes. Such a population is aging. (d) If the ratio of young to adults declines, the population is dominated by older individuals and the production of young is low. The pyramid of such a population is a narrow one with a small base of young. (After Alexander 1958.)

stable, changes in age distribution do not imply changes in survivorship or fecundity (see Caughley 1977) and cannot be safely used alone to predict population trends.

Age Structure in Plants

Modular structure and asexual reproduction in plants impose difficulties on research on age structure in plants. Determining age in plants involves following a cohort of marked individuals over time, or determining the ages of individuals by growth rings, bud scars, or other indicators.

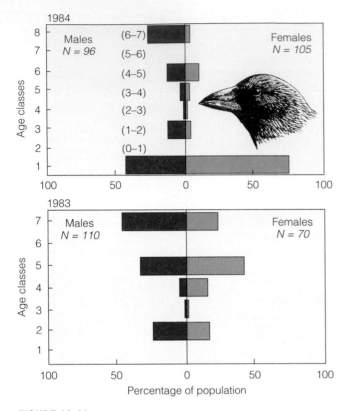

FIGURE 10.11 Age structure of the cactus ground finch (*Geospiza scandens*) on Isla Daphne Major for 1983 and 1984. Note the missing year classes and the shrinkage of a particular age class from one year to the next. (From Grant and Grant 1992:775.)

Among many plant species, especially woody ones, age may not tell enough. In even-aged stands of trees, the bulk of individuals falls into a very few age classes because they dominate the site, competitively excluding young age classes. Because of competition among individuals, some trees will achieve greater size than others, giving the impression of age differences, when in fact the trees are the same age. Age structure in such stands is less useful than some size criterion, such as diameter classes (Figure 10.12).

A similar approach is necessary for uneven-aged stands, because older individuals are not necessarily the larger dominant ones. What appear to be smaller young trees are in fact often the same age as the dominant individuals. One or two age classes dominate and hold onto the site until they die or are removed. Then new young age classes develop. Size classes and numbers within each class change as the population ages and mortality increases. In such a population, age alone is a poor criterion of population dynamics. In Table 10.1, for example, as the basal area (or size) of the individual trees increases, the number of trees that can occupy a given space decreases.

Plant ecologists have also applied analysis of age and size to modular units of plants, particularly leaves, twigs, and stems. Age distributions affected by growth and senescence of leaves have an important influence on the structure, growth form, and productivity of plant populations. For example,

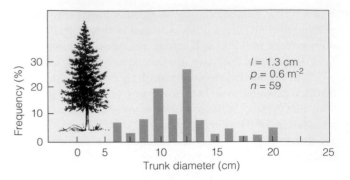

FIGURE 10.12 Size classes in an even-aged 54-year-old stand of balsam fir (*Abies balsamea*). Although all trees are the same age, note the difference in the size classes, as measured by diameter at breast height. The smaller trees give the impression of being younger individuals, when instead they are suppressed individuals. (l=interval of trunk diameter between successive bars, determined by dividing the range of observed diameters into 12 equal intervals; p=stand density; n=number of trees in sample.) (From Mohler et al. 1978.)

TABLE 10.1 Yield Table for Douglas Fir on Fully Stocked Hectare

Age (Years)	Trees	Av. dbh (cm)	Basal Area (m²)
20	1427	14.5	9.4
30	875	29.9	14.3
40	600	31.0	18.1
50	440	38.8	20.8
60	345	46.2	23.1
70	283	53.0	24.9
80	243	59.1	26.5
90	210	65.0	27.8
100	188	70.1	29.0
110	172	74.7	30.7
120	158	79.0	30.9
130	148	83.0	31.7
140	138	87.1	32.6
150	128	90.9	33.2
160	120	94.5	33.8

Source: Derived from McArdle et al. 1949.
dbh = diameter breast height
basal area = πr^2 when r is radius (1/2 bh)

Dickerman and Wetzel (1985) studied the clonal growth and ramet demography of the cattail *Typha latifolia* in south-central Michigan, providing a detailed picture of the shoot emergence, mortality, density, height growth, and effects of intraspecific competition on the population. They followed the fate of marked individual shoots over two years and were able to develop both age and height classifications of the population. Shoots emerged in three main pulses each year, resulting in three major groups. The first group emerged in early spring and was gone by late autumn. The second group appeared in midsummer, and about 80 percent of these shoots died by autumn. The remainder resumed growth the following spring.

The third group emerged in late summer, and about 90 percent of its shoots resumed growth the following spring.

A major problem in the study of age structure in plants is the seed bank in the soil. Seeds can remain in the soil for one to many years before they germinate. Thus seeds of a particular age are not the same as plants of the same age. New plants of a particular year may be from seeds that existed in the soil for one to many years. What, then, is the age of the individuals in the above-ground population derived from seeds of different ages?

SEX RATIOS

Populations of most sexually reproducing organisms tend toward a 1:1 sex ratio (the proportion of males to females). The **primary sex ratio** (the ratio at conception) also tends to be 1:1. (This may not be universally true and is, of course, difficult to determine.)

The **secondary sex ratio** (the ratio at birth) among mammals is often weighted toward males, but the population shifts toward females in older age groups. For example, the ratio among fetuses of elk in western Canadian national parks was 113 males to 100 females (Flook 1970). Between ages $1\frac{1}{2}$ and $2\frac{1}{2}$ the ratio of males to females dropped abruptly, until it remained at about 85:100, although in certain areas it dropped as low as 37:100. The greatest decline in the number of males occurred between the ages of 7 and 14. The decline of females was much less rapid. The difference in the rate of decline has an additional effect on the higher ratio of females to males. The loss of males allows more food and space for the females and young, further increasing the female rate of survival.

Among humans, too, males exceed females at birth, but as age increases the ratio swings in favor of females. In 1965 the ratio of males to females based on a stable age distribution in the United States was age 0 to 4, 104:100; age 40 to 44, 100:100; age 60 to 64, 88:100; age 80 to 84, 54:100. Lower mortality among females is characteristic of advanced and underdeveloped countries alike.

Among birds the sex ratios tend to remain weighted toward the males (Bellrose 1961). Fall and winter sex ratios of prairie chickens show a preponderance of males in adult groups; similar ratios are characteristic of other gallinaceous birds and some passerine birds.

It is not easy to explain why sex ratios should shift from an equal ratio at birth to an unequal one later in life. Perhaps a partial answer lies in the factors related to the genetic determination of sex and the physiology and behavior of the two sexes. Sex in organisms is determined by the X and Y chromosomes. The XY combination produces males in mammals and females in birds and some insects, notably butterflies. Perhaps the chromosome combination itself may be partially responsible for the biased losses of females in birds and males in mammals. Each gene in the X and Y chromosome is expressed in the XY combination, whereas in the XX combination the heterozygous pairs of alleles can mask the harm-

ful effects of single recessive genes (see Chapter 19). Therefore, XY adults may be more susceptible to disease, physiological stress, and aging than XX organisms.

Physiological and behavioral patterns affect mortality of the sexes differently. For example, during the breeding season the male elk battles other males for dominance of a harem, defends his harem from rivals, and mates with the females. These activities not only consume considerable energy, but leave little time for feeding, and the male often ends the breeding season in poor physical condition. Among birds the female may help the male defend territory, builds the nest, lays and incubates the eggs, broods the young, and often with the help of the male, feeds the young. While incubating the eggs and brooding the young, the female is much more vulnerable to predation and other dangers than is the male. Thus adult males in mammals and adult females in birds apparently expend more energy, are subject to greater stress, and are more vulnerable to predation and other dangers. Their higher vulnerability and mortality is consistent with the imbalance in the sex ratio in the older age cohorts (see Chapter 13).

MORTALITY AND NATALITY

The age structure of a population reflects two continuous processes in a populations: mortality (deaths) and birth. Mortality typically concentrates on the very young, reducing a potential abundance of new individuals entering perhaps an already crowded population, and on the old, removing senescent individuals to make room for more vigorous young.

Mortality is often expressed as a **crude death rate,** usually the number of deaths per 1000 in a given period. However, it is more precise to express mortality in a population either as the probability of dying or as a death rate. The **death rate** is the number of deaths during a given time interval divided by the average population; it is an instantaneous rate. For example, if the population size at the beginning of the period is 1000 and the number alive at the end of the period is 600, the average size of the population for the period would be 1600/2 or 800. The number of deaths is 400, and so the death rate is 400/800 or 0.50. The **probability of dying** is the number that died during a given time interval divided by the number alive at the beginning of the period, 400/1000 or 0.40. The complement of the probability of dying is the **probability of surviving,** the number of survivors divided by the number alive at the beginning of the period. Because the number of survivors is more important to a population than the number dying, mortality is best expressed in terms of **life expectancy,** the average number of years to be lived in the future by members of a given age in the population.

The greatest influence on population increase is the birth of new individuals, called **natality.** Natality may be described as maximum or physiological natality and as realized natality. **Physiological natality** represents the maximum possible number of births under ideal environmental conditions, the biological limit per individual. Because this number is rarely achieved in wild populations, its measure is of little value to a field biologist, but it is a useful yardstick against which to compare realized natality. **Realized natality** is the amount of successful reproduction that actually occurs over a period of time. It reflects the type of breeding season (continuous, discontinuous, or strongly seasonal), the number of litters or broods per year, the length of gestation or incubation, and so on. It is influenced by environmental conditions, nutrition, and density of the population.

Natality, measured as a rate, may be expressed either as crude birth rate or specific birth rate. **Crude birth rate** is expressed in terms of population size—for example, 50 births per 1000 population per year. **Specific birth rate,** a more accurate measure, is expressed relative to a specific criterion, such as age. It is usually expressed as an **age-specific schedule of births,** the number of offspring produced per unit time by females in different age classes.

The Life Table

A clear and systematic picture of mortality and survival in a population is best provided by a life table. It is a useful device to analyze probabilities of survivorship of individuals in a population, to determine ages most vulnerable to mortality, and to predict population growth. This device was developed by students of human populations and is widely used by life insurance companies. For them it is a critical tool in determining probabilities of survivorship upon which to base life insurance premiums. Population ecologists have adopted the life table for the study of natural populations.

The life table consists of a series of columns, each of which describes an aspect of mortality statistics for members of a population according to age. (For details of construction, see Appendix B.) Figures are presented in terms of a standard number of individuals all born at the same time, called a **cohort.** By convention, the initial number of individuals in a cohort is set at 1000, N_x. The columns include x, the unit of age or age level; $N_x l_x$, the number of individuals in a cohort that survive to that particular age level; d_x, the number of a cohort that die in an age interval x to $x + 1$ (from one age level listed to the next); and q_x, the probability of dying, or age-specific mortality rate. This rate is determined by dividing the number of individuals that died during the age interval by the number alive at the beginning of the age interval. Another column, 9, the survival rate, may be added. It can be calculated p from $1 - q$. To calculate life expectancy, given in column e_x, two additional statistics are needed: L_x and T_x. L_x is the average years lived by all individuals in each category in the population. It is obtained by summing the number alive at age interval x and the number at age $x + 1$, and dividing the sum by 2. T_x is the number of time units left for all individuals to live from age x onward. It is calculated by summing all the values of L_x from the bottom of the table upward to the age interval of interest. Life expectancy, e_x is obtained by dividing T_x for the particular age class x by the survivors for that age, as given in the l_x column.

Data for life tables are easy to obtain for laboratory animals and for humans. Laboratory animals are confined populations of known ages. For humans data on births and deaths are reported to and recorded by state governments and life insurance companies. Data on age, mortality, and survivorship for organisms in the wild are much more difficult to obtain. Mortality (d_x) can be estimated by determining the ages at death of a large number of animals born at the same time, if they were marked or banded. Ideally we would like to follow a group born during a given season in a particular year. However, to obtain a sufficiently large sample we may need to combine the fates of individuals born in different seasons or different years.

Another approach to gathering data is to determine the age at death of a representative sample of carcasses of the species under study collected in the same year. Age can be determined by examining wear and replacement of teeth in deer, growth rings in the cementum of teeth of ungulates and carnivores, annual rings in the horns of mountain sheep, and weight of the lens of the eye of rabbits and hares. This information also goes into the d_x column, providing a mortality schedule. Recording ages at death of a sample of a population wiped out by some catastrophe provides an excellent opportunity to determine age structure data for the l_x series, because all animals were alive just prior to sampling and experienced death in the same short time period. Similarly, determining the age of animals shot during a hunting season provides information for the l_x column, because the sample is from a living population.

There are three types of life tables: horizontal, vertical, and dynamic-composite. **Horizontal life tables,** also called **cohort** or **dynamic life tables,** are constructed by following a cohort of individuals, a group all born within a single short span of time, from birth to the death of the last member. Construction of such life tables is most easily accomplished when the species is short-lived (one to two years), so that the generations are discrete.

To develop a cohort life table for organisms with overlapping generations, we have to mark or in some manner follow all the individuals born in one year through the death of the last survivors. Because many animals get thoroughly mixed with individuals of other age classes, following them becomes a real problem. Table 10.2 is an example of a cohort life table for the red deer (*Cervus elaphus*). Lowe (1969) followed the 1957 cohort of red deer on the Isle of Rhum by making a total count of calves in 1957. From that year on, Lowe aged all deer shot under rigorously controlled conditions and by that means was able to sort out members of the 1957 cohort from all the rest. By 1966, 92 percent of the cohort was dead.

A variation of the horizontal or dynamic life table is the **dynamic-composite life table.** It records information in a manner similar to the horizontal or dynamic life table, but it considers as a cohort a composite of a number of animals marked over a period of years rather than at just one birth period. For example, wildlife biologists may mark or tag a number of newly hatched young birds or newly born mammals

TABLE 10.2 Cohort Life Table for
Red Deer on Isle of Rhum, 1957

x	N_x	l_x	d_x	q_x	e_x
			STAGS ♂		
*1	1000	1.000	84	.0840	4.76
2	916	0.916	19	.0207	5.14
3	897	0.897	0	0.0	3.25
4	897	0.897	150	0.1672	2.23
5	747	0.747	321	0.4300	1.58
6	426	0.426	218	0.5120	1.38
7	208	0.208	58	0.2788	1.31
8	150	0.150	130	0.8665	0.63
9	20	0.020	20	1.0000	0.50
			HINDS ♀		
*1		1.000	0	0	4.35
2	1000	1.000	61	0.0610	3.35
3	939	0.939	185	0.1970	2.53
4	754	0.754	249	0.3302	2.03
5	505	0.505	200	0.3960	1.79
6	305	0.305	119	0.3901	1.63
7	186	0.186	54	0.2903	1.35
8	132	0.132	107	0.8105	0.70
9	25	0.025	25	1.0000	0.50

Source: V. P. W. Lowe 1969.
*No data for first age class

TABLE 10.3 Time-Specific Life Table for
Red Deer on Isle of Rhum, 1957

x	N_x	l_x	d_x	q_x	e_x
			STAGS ♂		
1	1000	1.000	282	0.2820	5.81
2	718	0.718	7	0.0098	6.89
3	711	0.711	7	0.0098	5.95
4	704	0.704	7	0.0099	5.01
5	697	0.697	7	0.010	4.05
6	690	0.690	7	0.010	3.09
7	684	0.684	182	0.2660	2.11
8	502	0.502	253	0.5040	1.70
9	249	0.249	157	0.6306	1.91
10	92	0.092	14	0.1521	3.31
11	78	0.078	14	0.1794	2.81
12	64	0.064	14	0.2187	2.31
13	50	0.050	14	0.2799	1.82
14	36	0.036	14	0.3889	1.33
15	22	0.022	14	0.6363	0.86
16	8	0.008	8	1.0	0.50
			HINDS ♀		
1	1000	1.000	137	0.370	5.19
2	863	0.863	85	0.973	4.94
3	778	0.778	84	0.107	4.42
4	694	0.694	84	0.120	3.89
5	610	0.610	84	0.137	3.36
6	526	0.526	84	0.159	2.82
7	442	0.442	85	0.189	2.26
8	357	0.357	176	0.501	1.67
9	181	0.181	122	0.672	1.82
10	59	0.059	8	0.141	3.54
11	51	0.051	9	0.164	3.00
12	42	0.042	8	0.197	2.55
13	34	0.034	9	0.246	2.03
14	25	0.025	8	0.328	1.56
15	17	0.017	8	0.492	1.06
16	9	0.009	9	1.000	0.50

Source: V. P. W. Lowe 1969.

each year over a period of years. After following the fate of each year's group, they pool the data and treat all of the marked animals as one cohort (see Barkalow et al. 1970, Houston 1982). Or biologists may record the ages at death of animals found over a series of years and pool those data to construct a life table.

The third type of life table is the **vertical, time-specific,** or **static life table** (Table 10.3). It is constructed by sampling the population in some manner (such as hunting takes, or core samples of trees) and aging the organisms to obtain a distribution of age classes during a single time period. This life table involves the assumptions that each age class is sampled in proportion to its numbers in the population and ages at death, that the birth rate and death rate are constant, and that the population is neither increasing nor decreasing. It assumes, for example, that survivors of one year class were survivors from the year before and so on, the same as they would have been if they were a single cohort. Such assumptions, of course, are false because when data are so collected, the number of survivors in one year may be greater than the year previous. In this case the data have to be adjusted or smoothed to get rid of this anomaly (see Caughley 1977).

A basic difference exists between the static or vertical life table and the cohort or horizontal life table. The cohort follows the actual survivors from birth to death and reflects the life history of that particular group. The static life table is a cross section of a population at a specific time. It shows what survivorship would be if the population continued to survive at the observed rate. Thus the cohort life table shows the actual survivorship of the population and the static life table shows what survivorship might be. Note that in Table 10.2 the red deer cohort survived for 9 years, whereas in Table 10.3 the red deer population might survive to 16 years. The differences also appear in the survivorship curves presented later.

Both the static (time-specific) and dynamic-composite life tables are inaccurate. Mortality and reproduction vary from year to year over which the data are collected. The data reflect standing age distributions, which give the number of animals relative to the number of newborn animals in each age class at a particular time. The life tables, however, are based on a stable age distribution. If age distributions are unstable, populations are changing continuously and the data from which life tables are constructed do not reflect the true nature of the population. In spite of these shortcomings, such life tables may present a reasonable assessment of average conditions in the population, useful for comparing life-history trends within and between populations of the same species.

The life tables described are typical of long-lived species in which generations overlap and in which different age groups are alive at the same time. However, vast numbers of organisms have one annual breeding season and their generations do not overlap. All individuals belong to the same cohort or age class. Many insects and annual and biennial plants follow this pattern.

Some organisms, notably insects, have several stages to their life cycle. In this case, the life table is best divided into developmental stages rather than discrete time intervals. The l_x values are obtained by observing a natural population over the annual season and by estimating the size of the surviving population at each stage of development from eggs, larvae, pupae, and adults. If we keep records of weather conditions, abundance of predators, parasites, and diseases, we can also estimate how many die from various causes.

An example is a life table for a sparse stable population of gypsy moths in Connecticut (Table 10.4). A brief description of the life cycle will give the table more meaning. In April and May gypsy moth larvae emerge from eggs deposited by females the previous summer. In response to light, the larvae climb to the tops of trees and spin down from unfolding leaves by a silken thread. Because the thread is easily broken, larvae may be carried by the wind for long distances, which often results in a mass redistribution of the population.

The larvae go through a number of developmental stages called instars. Male larvae pass through five instars, females through six. Each instar, particularly the last three, is characterized by different feeding behaviors. Larvae reach full size by late June or July, stop feeding, and spin a few threads that will hold the pupae on the tree. The insects spend about two weeks as pupae and then emerge as adult moths. Males emerge first, several days ahead of females. Females, who do not fly, attract strong flying males by means of a powerful sex attractant. After mating, the females lay an egg mass containing a variable number of eggs (the sparser the population, the greater the number of eggs).

This life table represents the fate of a cohort from a single egg mass. The age interval or x column indicates the life-history stages, which are of unequal duration. The l_x column indicates the number of survivors at each stage. The d_x col-

TABLE 10.4 Life Table Typical of Sparse Gypsy Moth Populations in Northeastern Connecticut

x	l_x	d_{xf}*	d_x	$100\,q_x$
Eggs	550.0	Parasites	82.5	15
		Other	82.5	15
		Total	165.0	30
Instars I–III	385.0	Dispersion, etc.	142.4	37
Instars IV–VI	242.5	Deer mice	48.5	20
		Parasites and disease	12.1	5
		Other	167.3	69
		Total	227.9	94
Prepupae	14.6	Predators, etc.	2.9	20
Pupae	11.7	Vertebrate predators	9.8	84
		Other	0.5	4
		Total	10.3	88
Adults	1.4	Sex (SR = 30:70)	1.0	70
Adult, female	0.4	—	—	—
Generation	—	—	549.6	99.93

Source: R. W. Campbell 1969.
*d_{xf} = factor of mortality

umn gives the breakdown of deaths by causes in each stage. Note that life expectancy is not calculated because it is not meaningful. All the adult population will die in late summer.

Applying the life table concept to plants is an active area in the study of plant demography. One example is the life table (Table 10.5) for the 1978–1979 generation of the annual plant *Phlox drummondii* at Nixon, Texas, developed by Leverich and Levin (1979). On a study plot they located, mapped, and censused at intervals a cohort of individuals in the population from their emergence as seedlings in November 1978 to their death in 1979. In this table the N_x column records the actual number of individuals, and l_x is the probability of survivorship.

Plant ecologists have adapted the life table to study the population dynamics of plant modules, such as leaves, buds, and stems, that exhibit age-specific mortality. Life tables are a useful tool in studying the role of plant modular growth in primary production and life-history traits of plants.

McGraw (1989) constructed a life table for shoots of *Rhododendron maximum* (Table 10.6). A shoot is an apical meristem and the live leaves produced by that meristem. McGraw was interested in whether shoot life histories were patterned more clearly by age or size and whether interactions between age and size determined shoot fate (death, growth, flowering, and branching). He was able to use a life table approach because shoots are analogs of individuals in traditional populations. They grow, reproduce (branch), and die. They are easily aged by counting annual bud scars along the stem; new leaves and stems are produced annually. McGraw sampled the entire shoot population on a large rhododendron, judged median for the rhododendron population based on previous studies.

TABLE 10.5 Life Table for *Phlox drummondii* at Nixon, Texas

Age Interval (days) $x - x'$	No. Surviving to Day x N_x	Survivor-ship l_x	No. Dying During Interval d_x	Average Mortality Rate per Day q_x	Mean Expectation of Life (days) e_x
0–63	996	1.0000	328	0.0052	122.87
63–124	668	0.6707	373	0.0092	104.73
124–184	295	0.2962	105	0.0059	137.59
184–215	190	0.1908	14	0.0024	137.05
215–231	176	0.1767	2	0.0007	115.72
231–247	174	0.1747	1	0.0004	100.96
247–264	173	0.1737	1	0.0003	85.49
264–271	172	0.1727	2	0.0017	68.94
271–278	170	0.1707	3	0.0025	62.71
278–285	167	0.1677	2	0.0017	56.78
285–292	165	0.1657	6	0.0052	50.42
292–299	159	0.1596	1	0.0009	45.19
299–306	158	0.1586	4	0.0036	38.46
306–313	154	0.1546	3	0.0028	32.36
313–320	151	0.1516	4	0.0038	25.94
320–327	147	0.1476	11	0.0107	19.55
327–334	136	0.1365	31	0.0325	13.85
334–341	105	0.1054	31	0.0422	9.90
341–348	74	0.0743	52	0.1004	5.58
348–355	22	0.0221	22	0.1428	3.50
355–362	0	0.0000			

Source: Leverich and Levin 1979.
Note: dx expressed as proportion

McGraw divided the shoots into age classes, noted shoot death by counting those that no longer supported any live green leaves, and determined fertility by the number of daughter branches produced over a period of one year.

Survivorship and Mortality Curves

The life table is a key tool in the analysis of population dynamics. From it we can derive survivorship curves based on the l_x column and mortality curves based on the q_x column. These curves enable us to determine the ages at which a particular organism most often dies. This information provides some leads in determining the causes of death and ultimately the processes that affect the population dynamics of a given species. They enable biologists to compare survival between the sexes, between cohorts arising in different years, between populations, and between species.

Survivorship Curves The survivorship curve depicts age-specific mortality through survivorship. It is obtained by plotting the number of individuals of a particular age cohort against time. The usual form is to plot the logarithms of the numbers of survivors (usually on semilogarithmic graph paper) against age. The semilog scale translates absolute numerical change in a population to per capita rate of change.

TABLE 10.6 Life Table for Shoots of *Rhododendron maxiumum*

Age	l_x	d_x	q_x	e_x
0	1.000	0	0	5.60
1	1.000	0.016	0.016	4.60
2	0.984	0	0	3.67
3	0.984	0.075	0.077	2.67
4	0.909	0.185	0.024	1.85
5	0.724	0.346	0.477	1.19
6	0.378	0.270	0.714	0.82
7	0.108	0.095	0.882	0.62
8	0.013	0.013	1.000	0.50

Source: Adapted from McGraw 1989.
Note: d_x expressed as proportion

Survivorship curves may be classified into at least three hypothetical types (Figure 10.13) (Deevey 1947). These curves are conceptual models only, against which real-life survivorship curves can be compared. The Type I curve is convex. It is typical of populations whose individuals tend to live out their physiological life span; they exhibit a high degree of survival throughout life and experience heavy mortality in old age. Such a curve is typical of some plants, such as *Phlox drummondii* (Figure 10.14), and many mammals, such as red deer. Note the difference in the two curves generated by the cohort and static life tables (Figure 10.15). The Type II curve is linear and is typical of organisms with constant mortality rates. Such a curve is characteristic of the adult stages of many birds (Figure 10.16a),

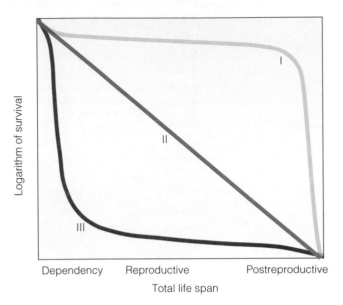

FIGURE 10.13 Three basic types of survivorship curves. The vertical axis may be scaled arithmetically or logarithmically. If it is logarithmic, the slope of the lines will show the following rates of change for the three types: Type I, curve for populations in which the survival of juveniles is high and mortality is concentrated among old individuals; Type II, curve for populations in which the rate of mortality is fairly constant at all age levels, so that there is more or less a uniform percentage decrease in survivorship over time; Type III, curve for a population in which high mortality is concentrated among juveniles.

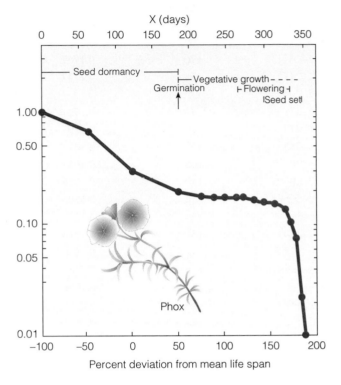

FIGURE 10.14 Type I survivorship curve for the annual *Phlox drummondii* at Nixon, Texas. (From Leverich and Levin 1979:885.)

FIGURE 10.15 A comparison of survivorship of the 1957 cohort of both male (stags) and female (hinds) red deer, based on a static life table. The cohort survivorship is based on the 1957 calves and applies to the post-1957 population. The static survivorship is based on the 1957 population and applies to the pre-1957 population. (Lowe 1969.)

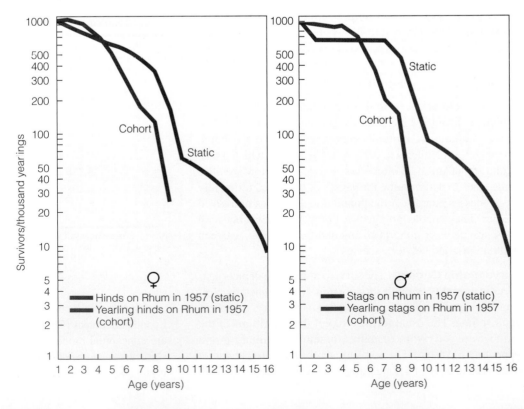

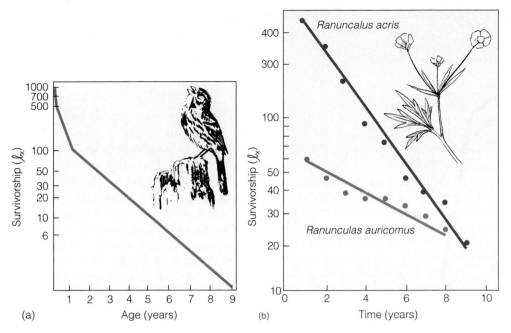

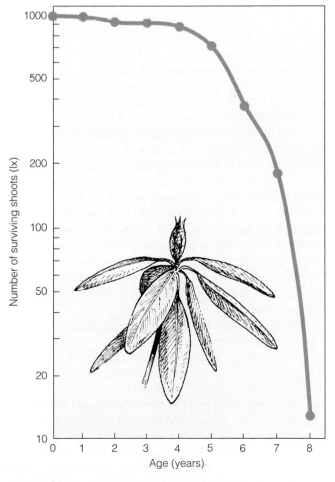

FIGURE 10.16 (a) Survivorship curve for the song sparrow (*Melospiza melodia*). The curve is typical of birds. After a period of high juvenile mortality, characterized by a concave or Type III curve, the survivorship curve becomes linear, suggestive of Type II. (From Johnson 1956.) (b) Type II survivorship curves of buttercups *Ranuculus acris* and *R. auricomus*. (From Sarukhan and Harper 1974.)

rodents, and some plants (Figure 10.16b). Type III is concave and is typical of organisms with extremely high mortality rates in early life, such as many species of invertebrates (Figure 10.17) and fish, and some plants.

Among plants, life table attributes and survivorship apply not only to individuals but also to the modular units: the populations of leaves, buds, flowers, fruits, and seeds, as well as clones (Figure 10.18). Mortality and survivorship of these subpopulations influence the individual growth form of the plant, its ability to compete, and its photosynthetic and reproductive performance.

Survivorship curves are useful in comparing the survival characteristics of the population of one area, time, sex, or

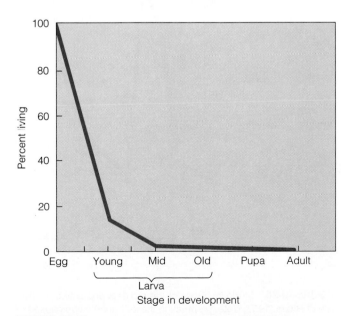

FIGURE 10.17 A Type III survivorship curve exhibited by the oystershell scale insect (*Lepidosaphes ulma*). (Adapted from Price 1975.)

FIGURE 10.18 Survivorship curve of plant modules: shoots of *Rhododendron maximum*. The curve is Type I. (Based on data from McGraw 1989.)

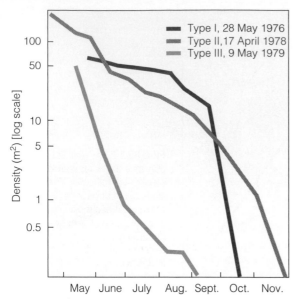

FIGURE 10.19 Variation in survivorship for three cohorts of the summer annual garden rocket (*Erucastrum gallicum*). Type I (a), Type II (b), and Type III (c) survivorship curves reflect responses to a year of abundant rainfall, a year of intermittent rainfall and drought, and a year with a spring drought, respectively. (Adapted from Klemow and Raynal 1983:697.)

species with another or in studying the influence of environmental conditions on survival of populations. For example, Klemow and Raynal (1983) followed the fate of a population of the summer annual garden rocket (*Erucastrum gallicum*) for five years. Survival of plants that emerged in the spring varied markedly with rainfall (Figure 10.19). Cohorts in a year with abundant rainfall exhibited a Type I survivorship; cohorts in a year of intermittent rainfall and drought had a Type II survivorship; and cohorts experiencing a spring drought had a Type III survivorship curve.

Mortality Curves By plotting the data in the q_x or mortality rate column of the life table against age, you get a mortality curve. It consists of two parts: (1) the juvenile phase, in which the rate of mortality is high; and (2) the postjuvenile phase, in which the rate first decreases as age increases, and then increases with age after a low point in mortality (Figure 10.20). Most populations have a roughly J-shaped curve.

Because q_x is the ratio of the number dying during an age interval to the number alive at the beginning of the period (or surviving the previous age period), that parameter is independent of the frequency of each of the previous age classes. Therefore the parameter is free of the biases inherent in the l_x column and survivorship curves. Most life tables of wild populations are subject to bias because the first-year age class is not adequately represented. This error distorts all succeeding l_x and d_x values—if the first values are inaccurate, all succeeding ones are inaccurate. But if the first values of q_x are wrong, the error does not affect the other values. For this reason mortality curves, which indicate the rate of mortality indirectly by the slope of the line, are more informative than survivorship curves.

Animal Natality

As pointed out earlier, natality is measured as a birth rate. In animals, especially, it is expressed as an age-specific schedule of births, the number of offspring per unit time by females in different age classes. Because the number of young produced is a function of the number of females in the population, the age-specific birth schedule usually counts only females giving rise to females. The age-specific schedule is obtained by determining the mean number of females born in each group of females, designated as m_x (for maternity; some population ecologists use b_x for births). If we take the l_x or survivorship column from the life table and the age-specific m_x values, we can construct a fecundity or fertility table, showing the number of offspring produced per unit time. The fecundity table enables us to calculate the net reproductive rate, R_0, the number of female offspring left during a lifetime by a newborn female. When reproductive success of known males can be determined, a fecundity table can be constructed for that sex also.

Table 10.7 presents a fecundity table for red deer based on the static or time-specific life table (see Table 10.3). The

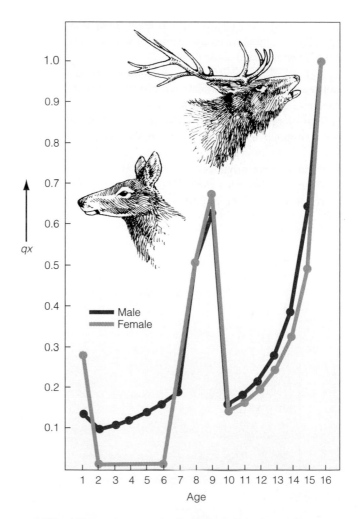

FIGURE 10.20 Mortality curve for the red deer population on the Isle of Rhum, 1957. Note the break in the J-shaped curve caused by a sharp rise in mortality for deer between the sixth and tenth years. Compare these curves with the survivorship curves for the same population in Figure 10.14. (Lowe 1969.)

TABLE 10.7 Fecundity Table for Red Deer

x	l_x	m_x	$l_x m_x$	$x l_x m_x$
1	1.000	0	0	0
2	0.863	0	0	0
3	0.778	0.311	0.242	0.726
4	0.694	0.278	0.193	0.772
5	0.610	0.308	0.134	0.667
6	0.526	0.400	0.210	1.26
7	0.442	0.476	0.210	1.47
8	0.357	0.358	0.128	1.024
9	0.181	0.447	0.081	0.729
10	0.059	0.289	0.017	0.170
11	0.051	0.283	0.014	0.154
12	0.042	0.285	0.012	0.144
13	0.034	0.283	0.010	0.130
14	0.025	0.282	0.007	0.098
15	0.017	0.285	0.005	0.075
16	0.009	0.284	0.003	0.048

$$R_0 = 1.316$$

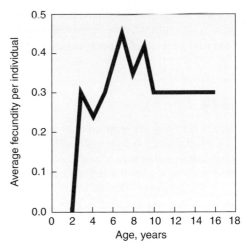

FIGURE 10.21 A fecundity curve for the red deer. Note that the highest fecundity is in the middle age groups of females. (Lowe 1969.)

average number of daughters produced by mothers of a given age is indicated as m_x. The females produced no young during years 1 and 2; therefore the m_x value for those years is 0. The m_x value roughly increase with age, then remains the same from age 10 on. To adjust for mortality in each age group, the m_x value for each age class is multiplied by the corresponding l_x or survivorship value. The resulting $l_x m_x$ value gives the mean number of females born in each age group adjusted for survivorship. For example, in female age class 3, the m_x value is 0.311, but when adjusted for survivorship, the value drops to 0.242,; and in age class 12, the m_x value is 0.285, but drops to 0.012 when adjusted for survivorship. The sum of these $l_x m_x$ values over all ages in which reproduction occurs represents the number of females a newborn female will leave during her lifetime, or R_0, the net reproductive rate. The R_0 value or female red deer on Rhum is 1.316. The female deer replace themselves when R_0 equals or is greater than 1. This point will be discussed in greater detail in Chapters 11 and 13. The m_x column of the fecundity tables also provides the data needed to plot a fecundity curve, which depicts the number of offspring per individual per age class (Figure 10.21). It suggests that individuals of certain age classes contribute more to population growth than those of other age cohorts. The average contribution to the next generation that members of a given age group make between their current age and death is their **reproductive value,** considered in Chapter 13.

Plant Natality

The distinct demographic differences between plants and most animals create problems in determining natality in plants. Individual plants vary widely in seed production from year to year

and from age class to age class. Moreover, seeds usually undergo a period of dormancy, which in some species may last for a number of years until conditions are right for germination. Germination is the formal equivalent of birth in plants.

The simplest situation exists with annual and biennial plants, because both produce seeds in one reproductive effort at the end of their individual life spans. By censusing the number of seeds produced by the different individuals, we can arrive at some idea of fecundity. Leverich and Levin (1979) did this for the annual Drummond's phlox, arriving at a fecundity schedule for both ovules and seeds (Table 10.8, seeds only). From this fecundity schedule, they determined R_0, the net rate of increase. In this case, involving an annual species with no overlapping generations, R_0 gives us not only the

TABLE 10.8 Fecundity Schedule for *Phlox drummondii* at Nixon, Texas, Based on Seed Production

x – x′	N_x	l_x	m_x^{seed}	$l_x m_x$
0–299	996	1.0000	0.0000	0.0000
299–306	158	0.1586	0.3394	0.0532
306–313	154	0.1546	0.7963	0.1231
313–320	151	0.1516	2.3995	0.3638
320–327	147	0.1476	3.1904	0.4589
327–334	136	0.1365	2.5411	0.3470
334–341	105	0.1054	3.1589	0.3330
341–348	74	0.0743	8.6625	0.6436
348–355	22	0.0221	4.3072	0.0951
355–362	0	0.0000	0.0000	0.0000

$$\Sigma = 2.4177$$

$$R_0 = \Sigma\, l_x m_x = 2.42 \text{ (per capita)}$$

$$R = \frac{\ln R_0}{365} = 0.0024 \text{ (per capita per day)}$$

Note: $x – x'$ = age interval; N_x = no. surviving to day x; m_x^{seed} = average no. of seeds per individual during interval; l_x = survivorship. $l_x m_x$ = contribution to net reproductive rate during interval.
Source: From Leverich and Levin 1979.

average number of offspring produced per individual over a lifetime, but also the multiplicative factor with which to obtain the size of the next generation, a point to be considered later.

Summary

1. A population is a group of organisms of the same species occupying a particular space at a given time. Populations may be made up of unitary individuals or modular units in which each individual or clone is part of a larger group. The concept of modular population can be extended to populations of modules that make up individual plants: leaves, buds, flowers, and stems.

2. Many populations of a species exist as separate populations or spatially disjunct subpopulations linked together by dispersal or immigration among them. Populations so structured are known as metapopulations.

3. Density is the size of a population in relation to a defined unit of space. The number of individuals per unit area is crude density. The number of individuals per unit of occupied habitat is ecological density.

4. Individuals in a populations are distributed in some kind of pattern over an area. Some are uniformly distributed, a very few are randomly distributed, and most exhibit a clumped distribution. Many clumped species are reflecting intraspecific interactions, either cooperative or competitive.

5. Movements within and among populations are emigration, the permanent movement of individuals out of a population; immigration, the movement of individuals into a population or vacant habitat; and migration, a daily or seasonal round-trip movement between different areas.

6. Dispersal distances of individuals are influenced by the mode of dispersal—active or passive—and among animals by age, sex, and reproductive state.

7. Age and sex ratios affect the structure of a population and influence the rates of births and deaths. Reproduction is limited to certain age classes; mortality is most prominent in others. Changes in age class distribution bring about changes in natality and mortality.

8. The sex ratio tends to be balanced between males and females, but populations can have manifestly unbalanced sex ratios. Some unbalanced sex ratios may be normal for the species; others may be produced by unusual events.

9. Population size depends on the number of individuals added to the group and the number leaving—the difference between the birth rate and the death rate, and the balance between immigration and emigration. Birth rate is more important than immigration rate in the addition of new individuals. Mortality reduces population more than emigration. Mortality is greatest among the very young and the old age groups.

10. Mortality and survivorship are best analyzed by means of a life table, which is an age-specific summary of deaths in a population. From a life table we can derive mortality and survivorship curves, which are useful in comparing demographic trends within and between populations living under different environmental conditions.

11. Reproductive rates of populations can be analyzed by means of a fecundity table that relates age-specific reproduction to age structure and survivorship.

Review Questions

1. What is a population?
2. What distinguishes a modular population from a unitary population? Give an example of each.
3. Distinguish among a genet, a ramet, and a clone. How do they relate to a modular population?
4. What is population density? Crude density? Ecological density?
5. What are the three basic types of spatial distribution of populations? Which one is most common and why?
6. How are populations dispersed in time?
7. What are three basic types of dispersal movements? How do they differ?
8. What is the significance of age structure in a population?
9. What is an age pyramid? What does it tell us about a population?
10. What is the difference between a stationary age distribution and a stable age distribution?
11. What are some of the problems in applying the concept of age structure to a plant population? What is the usefulness of applying the age structure concept to a plant population?
12. What is the significance of sex ratios to a population?
13. Distinguish between death rate, mortality rate, probability of dying, and life expectancy.
14. What is natality? Crude birth rate? Specific birth rate?
15. What is a life table? What are the differences among a cohort or dynamic life table, a static or time-specific life table, and a dynamic-composite life table?
16. What are the major columns of a life table? Explain each.
17. What is a survivorship curve? What are the three basic types?
18. What is a mortality curve? How does it differ from a survivorship curve?
19. What is the net reproductive rate? What are some problems in determining the net reproductive rate of plants?
20. What is a fecundity table? How is it used to obtain the net reproductive rate?
21. Using the following data for a population of ground squirrels, construct a life table and a fecundity table (first convert the raw population data N_x to a proportionality to get the l_x column):

$$x = 0, 1, 2, 3, 4, 5, 6, 7, 8, 9$$
$$N_x = 410, 164, 85, 44, 21, 12, 8, 4, 2, 1$$
$$m_x = 0.0, 1.0, 2.2, 2.4, 3.2, 1.8, 1.7, 2.0, 1.4, 1.4$$

22. Determine the net reproductive rate of the ground squirrel population discussed in question 21.

Cross References

Adaptation, 81; population growth, 184–187; mating systems, 216; reproductive costs 222; reproductive value, 225.

C H A P T E R 11

Population Growth

CONCEPTS

1. The difference between the instantaneous birth rate and instantaneous death rate is the rate of population increase.

2. Geometric increase of a population is characterized by a constant schedule of births and deaths.

3. No real-life population increases without limit.

4. As population density increases, population growth usually slows and fluctuates about some point of equilibrium with the environment.

5. Population fluctuations are influenced by density-dependent mechanisms.

6. When populations become very small, chance events alone can lead to extinction.

Populations are dynamic. Depending on the nature of the organism, population numbers may change from hour to hour, day to day, season to season, year to year. In some years certain organism are abundant; in other years they are scarce. Local populations appear, expand, decline, or become extinct. Eventually the area may be recolonized by individuals moving in from other populations. Such changes come about as the interaction of organisms and their environment affects birth rates, death rates, and the movement of individuals.

RATE OF INCREASE

Possessing a sufficiently large sample of individuals of all ages in the population to construct a life table and knowing the age-specific fecundity (discussed in Chapter 10), we can determine some characteristics of population growth. We will begin by constructing a life/fecundity table for the hypothetical squirrel population in Table 11.1. It provides the survivorship schedule, l_x, fecundity schedule, m_x, and an additional parameter p_x, the proportion of animals surviving in each age class; the parameter p_x is the complement of q_x and is equal to $1 - q_x$. With this data we can chart the growth of a population by constructing a **population projection table.**

We will illustrate the construction of a population projection table by using a hypothetical population of 10 female squirrels, all age 1, introduced into an empty oak forest. The year of establishment will be designated as year 0. For this population we will use a hypothetical life table (Table 11.1) to provide schedules of survivorship, fecundity, and the probability of survival. These 10 females in the year of introduction (year 0) will give birth to 40 young, half of which, 20, will be female. Because females form the reproductive units of the population, we follow only the females in the construction of the table (Table 11.2). The total population of the female squirrels for the year 0 stands at 30 squirrels, 10 of which are age 1 and 20 of which are age 0. Not all of these squirrels will survive into the next year. The survival of these two age groups is obtained by multiplying the number of each by the p_x value. Because the p_x of the females of age 1 is 0.5, we know that 5 individuals ($10 \times 0.5 = 5$) survive to year 1 (age 2). The p_x value of age 0 is 0.3, so that only 6 of the 20 in this age class in year 0 survive ($20 \times 0.3 = 6$) to year 1 (age 1). In year 1 we now have 6 one-year-olds and the 5 two-year-olds, with both age classes reproducing. The m_x value of the 6 one-year-olds is 2.0, and so they produce 12 offspring. The 5 two-year-olds have an m_x value of 3.0, and so they produce 15 offspring. Together the two age classes produce 27 young, which now make up age class 0. The total population for year 1 is 38. Survivorship and fecundity are determined in a similar manner for each succeeding year. Survivorship is tabulated year by year diagonally down the table to the right through the years, while new individuals are being adding each year to age class 0.

The steps for determining the number of offspring in year t (N_t) are given by the equation

$$N_{tx} = \sum_{x=1}^{\infty} N_{tx} m_x$$

where N_{t0} is the number in age class 0 at the given year t and N_{tx} is the number of age x at year t. For year 0 the calculation is

$$N_0 = (10)(2) = 20$$

and for succeeding years

$$N_1 = (6)(2) + (5)(3) = 27$$
$$N_2 = (8.1)(2) + (3)(3) + (3)(3) = 34.2$$

TABLE 11.1 **Life Table for a Hypothetical Squirrel Population**

x	l_x	q_x	p_x	e_x	m_x	$l_x m_x$	$x l_x m_x$
0	1.0	0.7	0.3	1.09	0	0	
1	0.3	0.5	0.5	1.47	2.0	0.60	0.60
2	0.15	0.4	0.6	1.43	3.0	0.45	0.90
3	0.09	0.55	0.45	1.05	3.0	0.27	0.81
4	0.04	0.75	0.25	0.75	2.0	0.08	0.32
5	0.01	1.0	0.00	0.5	0.00	0.00	0.00
Σ						1.40	2.63

TABLE 11.2 **Population Projection Table, Hypothetical Squirrel Population**

Age	YEAR										
	0	1	2	3	4	5	6	7	8	9	10
0	20	27	34.1	40.71	48.21	58.37	70.31	84.8	101.86	122.88	148.06
1	10	6	8.1	10.23	12.05	14.46	17.51	21.0	25.44	30.56	36.86
2	0	5	3.0	4.05	5.1	6.03	7.23	8.7	10.50	12.72	15.28
3	0	0	3.0	1.8	2.43	3.06	3.62	4.4	5.22	6.30	7.63
4	0	0	0	1.35	0.81	1.09	1.38	1.6	1.94	2.35	2.83
5	0	0	0	0	0.33	0.20	0.27	0.35	0.40	0.49	0.59
Total	30	38	48.2	58.14	68.93	83.21	100.32	120.85	145.36	175.30	211.25
Lambda	λ	1.27	1.27	1.21	1.19	1.21	1.20	1.20	1.20	1.20	1.20

and so on. Thus *the number of offspring added to age 0 each year is obtained by multiplying the number in each age group by the m_x value for that age and summing these values over all ages.*

From such a population projection table we can calculate **age distribution** (see Chapter 10), the proportion of individuals in the various age classes for any one year, by dividing the number in each age group by the total population size for that year. The general equation is

$$C_{tx} = \frac{N_{tx}}{\sum_{y=0}^{\infty} N_{ty}}$$

where C_{tx} is the proportion of age group x at year t, N_{tx} is the number in each age group x at year t, and N_{ty} is the number in age group y at year t.

Comparing the age distribution of the hypothetical squirrel population in year 3 with that of the population in year 7, we observe that the population attains a stable or unchanging age distribution by year 7 (Table 11.3). From that year on, the proportions of each age group in the population and the rate of growth remain the same year after year, even though the population is steadily increasing.

The population projection table demonstrates an important concept of population growth. The constant rate of increase of the population from year to year and stable age distribution depend on survivorship (l_x) from each age class to the next and on fecundities of each age class (m_x). Both factors were used in the development of the population projection table. By dividing the total number of individuals in year $x + 1$ by the total number of individuals in the previous year x, one can arrive at the finite multiplication rate, λ (lambda), for each time period.

Lambda can be used as a multiplier to project but not predict population size some time in the future:

$$N_t = N_0 \lambda^t$$

For our hypothetical population, we can multiply the population size, 30, at time (year) 0 by $\lambda = 1.20$, the value derived from the population projection table, to obtain a population size of 36 for year 1. If we multiply 36 again by 1.20, or the initial population size 30 by λ^2 (1.20^2), we get a population

size of 43 for year 2; and if we multiply the population at $N_0 = 30$ by λ^{10}, we arrive at a projected population size of 186 for year 10. These population sizes do not correspond to the population sizes early in the population projection table, because in the lower years λ is higher and the population has not reached a stable age distribution. Only after the population achieves a stable age distribution does the λ value of 1.20 project future population size. For example, if population size N in year 7 is used as N_0, then 121×1.20^3 projects a population size of 209 in the year 10. This example emphasizes that population projections using λ assume a stable age distribution.

We will use the information in Table 11.1 to obtain another measure of population increase. By multiplying age-specific survivorship (l_x) by age-specific fecundity (m_x) and summing all the $l_x m_x$ values for the entire lifetime, as we did in Table 10.3, we obtain the **net reproductive rate** R_0, defined as

$$\sum_{x=0}^{\infty} l_x m_x$$

If R_0 equals 1, the birth rate equals the death rate; individuals are replacing themselves and the population remains stable. If the value is greater than 1, the population is increasing; if it is less than 1, it is decreasing. R_0 for our hypothetical population is 1.40, and so our population is increasing.

In contrast to λ, which projects population growth from one point in time to another and is always positive, R_0 measures the finite rate of increase in terms of discrete generation time and can be negative. If generations are discrete, as they are in many insects, then the unit of time (t) and generation time (T) are the same. But what about populations with overlapping generation times for which growth does not take place in discrete generation time? To calculate growth in populations with overlapping generations, such as our squirrel population, the value of T has to be changed to a **mean cohort generation time (T_c)**. T_c is the mean period of time elapsing between the birth of the parents and the birth of the offspring. Mean cohort generation time is computed from the fecundity schedule by multiplying each $l_x m_x$ by its corresponding age x. All values are summed and divided by the sum of the $l_x m_x$ or R_0 to obtain T_c.

This parameter enables us to calculate r, the **per capita rate of increase** per unit time. r is a measure of the instantaneous rate of change of population size per individual. It is

TABLE 11.3 Approximation of Stable Age Distribution, Hypothetical Squirrel Population

| Age | PROPORTION IN EACH AGE CLASS FOR YEAR | | | | | | | | | | |
	0	1	2	3	4	5	6	7	8	9	10
0	.67	.71	.71	.71	.69	.70	.70	.70	.70	.70	.70
1	.33	.16	.17	.17	.20	.17	.17	.18	.18	.18	.18
2		.13	.06	.07	.06	.07	.07	.07	.07	.07	.07
3			.06	.03	.03	.04	.04	.03	.03	.03	.03
4				.02	.01	.01	.01	.01	.01	.01	.01
5					.01	.01	.01	.01	.01	.01	.01

the difference between instantaneous birth rates, b, and instantaneous death rate, d, or $r = b - d$. In a closed population, one in which no individual enters or leaves the population, r is known as the intrinsic rate of increase.

The relationship between the per capita growth rate r and per generation growth rate (R_0) is approximately

$$r = \frac{\ln R_0}{T_c}$$

For the squirrel population in Table 11.1 the cohort generation time is $\Sigma x l_x m_x / R_0 = 2.63/1.40 = 1.87$. The per capita rate of increase can be approximated by $\log_e 1.40/1.87 = 0.336/1.87 = 0.1797$.

The per capita rate of increase (or decrease) can be converted to the finite rate of increase by the formula $\lambda = e^r$, where e is the base of natural logarithms ($e = 2.71828...$). By determining e^r with an r value of 0.1797 or 0.180 we obtain $\lambda = 1.198$ or 1.20. This value agrees with that obtained from the population projection after stable age distribution was achieved.

The per capita rate of increase can be obtained more precisely from the Euler equation:

$$\sum_{\lambda}^{\infty} e^{-rx} l_x m_x = 1$$

This equation can be solved only by using a trial and error method (iteration, see Appendix B). **Iteration** is accomplished by introducing different values of r and rx above and below the estimated value of r until the sum on the left-hand side approximates the given value of 1 on the right. The iterated value of r for the squirrel population is 0.186, which gives λ a value of 1.203.

POPULATION GROWTH

Exponential Growth

If a population were suddenly presented with an unlimited environment, as can happen when a small number of bacteria, plants, or animals are introduced into a suitable but unoccupied habitat, it would tend to expand exponentially. Knowing λ, you can project population growth by the equation

$$N_t = N_0 \lambda^t$$

where N_0 is the population size at some given time, N_t is the projected population at some later time t, and λ^t is value of λ raised to the power t, the number of projected years. Such a calculation has been done earlier in this chapter. Calculating change in population size in this manner provides an assessment of geometric rate of increase involving discrete time units or discrete generations.

Assuming there were no movement in or out of the population and no mortality, then the excess of births would account for changes in population numbers. Under this condition population would increase like compound interest, a continual increase called **exponential growth.** But the growth rate of populations is tempered by death, and so a death rate is factored in with the birth rate in values for λ, R_0, and r.

Changes in population size can be projected by using the exponential equation

$$N_t = N_0 e^{rt}$$

where e is the base of natural logarithms $2.71828...$, r is the per capita rate of increase, and t is the unit of time. This equation is similar to the geometric growth projected by $N_t = N_0 \lambda^t$. The term e^{rt} takes the place of λ^t. Provided that the geometric growth rate and the exponential growth rate are the same, species with discrete and overlapping generations will produce similar growth curves, except that geometric growth rate of species with discrete generations will have discrete points (Figure 11.1), whereas species with continuous growth will exhibit continuous or instantaneous change. Comparisons of changes in population sizes projected by the two equations are:

| | Population Size | |
| | Exponential | Geometric |
Year	($r = 0.180$)	($\lambda = 1.20$)
0	10	10
1	12	12
3	17	17
5	25	25
7	35	36
9	52	52
12	87	89

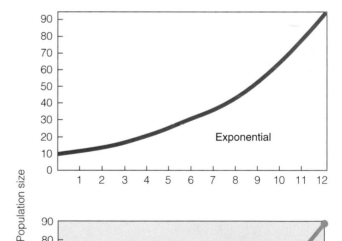

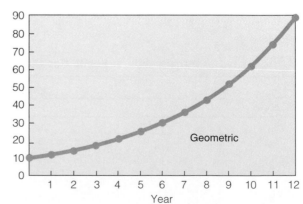

FIGURE 11.1   Geometric growth curve plotted for $\lambda = 1.20$ and exponential growth curve plotted for $r=0.186$. The curves are similar except that the geometric curve has discrete points.

The population growth described and graphed is for a specific value of $r = 0.187$. Different values of r describe different exponential curves, as illustrated in Figure 11.2.

The equations just discussed project the number of individuals added to the population. Exponential growth can also be expressed as the rate of change in numbers over time:

$$dN/dt = (b - d)N$$

or

$$dN/dt = rN$$

This equation states that the rate of change dN/dt is directly proportional to the size of the population (N); dN/dt is equal to N times the per capita rate of increase, r. With instantaneous change in population growth, the time interval increments approach zero (Figure 11.3).

An example of exponential growth is the rise of a reindeer herd on St. Paul, one of the Pribilof Islands in Alaska (Figure 11.4). Introduced in 1910, reindeer expanded from 4 males and 22 females to a herd of 2000 in only 30 years. Exceeding the ability of the range to support it, the population crashed in an exponential fashion. Such curves are typical of populations that grow rapidly and exceed available resources.

Knowing the value of r for a population allows the calculation of doubling time of that population's growth. The doubling of a population occurs at the time t when $N_t/N_0 = 2$. Therefore $N_0 e^{rt}/N_0 = 2$. Thus $e^{rt} = 2$, and $rt = \log_e 2 = 0.693$; then $t = 0.693/r$. The doubling time of the hypothetical squirrel population in Table 11.1 is $0.693/0.180 = 3.85$ years.

Logistic Growth

Exponential growth is not biologically realistic: no population can grow indefinitely. The environment is not constant, and resources such as food and space are limited. As population density increases, competition for available resources among its members also increases. Eventually the detrimental effects of increased density—increased mortality from disease, starvation, and predation; decreased fecundity; and

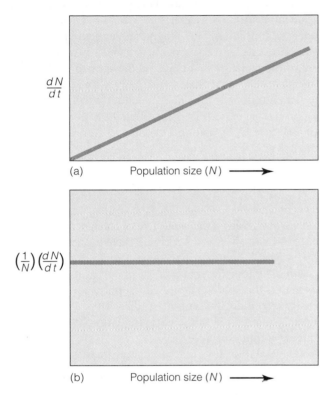

FIGURE 11.3 (a) Exponential population growth rate (dN/dt) as a function of population size. (b) Per capita exponential growth rate r, $(1/N)(dN/dt)$, as a function of population size. (After Gotelli 1995.)

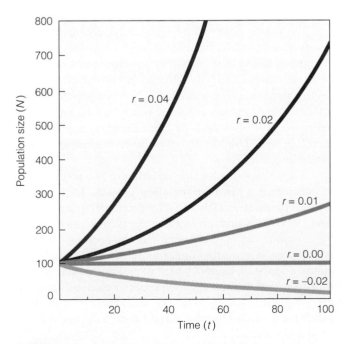

FIGURE 11.2 Trajectories of exponential population growth calculated from a starting population of 100 individuals for several values of r.

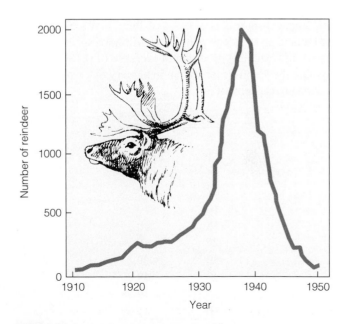

FIGURE 11.4 Exponential growth of the St. Paul reindeer herd is illustrated by the left side of the graph. When the herd outstripped its resources, the population crashed. (From Scheffer 1951.)

emigration—begin to slow population growth until it ceases, the population reaching a level at which theoretically it is in equilibrium with its environment.

This concept of the decline of the exponential growth rate as the size of the population increases was described by the mathematician Pierre-Francois Verhulst in 1838. He called this equation logistic because of its logarithmic-exponential form. His paper, however, was buried in a seldom-read journal. In 1920 Raymond Pearl and I. J. Reed of Johns Hopkins University published a nearly identical version of the equation in a paper on the growth of the population of the United States since 1790. They predicted a regular pattern of decline in exponential growth as the population size increased. They modified the exponential equation $dN/dt = rN$ by adding a variable to describe the effect of density:

$$\frac{dN}{dt} = rN\left(\frac{K-N}{K}\right)$$

Here dN/dt is the instantaneous rate of change, N is the number of individuals, K is the carrying capacity, and $(K - N)/K$ or $1 - N/K$ is the unutilized opportunity for population growth. The logistic equation for population growth supposes that the source supply can support an equilibrium density of K individuals of the species. Thus K is often biologically interpreted as the **carrying capacity** of the environment for this species.

The equation says that the rate of increase of a population over a unit of time is equal to the potential increase of a population times the unutilized portion of the resources. When N is low, $(K - N)/N \approx 1$ (thus most of the resources are unutilized). When $N \approx K$, $(K - N)/K \approx 0$ (thus most of the resources are utilized) and the population grows little or none at all. If $N > K$, then dN/dt is negative and N declines toward K (Figure 11.5).

Let us use the rate of increase of 0.180 derived from the hypothetical squirrel population of Table 11.1 as an example of logistic growth. The population will have a starting value of 10 individuals with a rate of increase of 0.180. We will assume the squirrels inhabit an imaginary oak woodlot with a carrying capacity of 50:

Year	Size
0	10
1	11
5	19
8	26
10	30
15	40
25	48
30	49.2
40	49.8
50	49.98
70	49.99

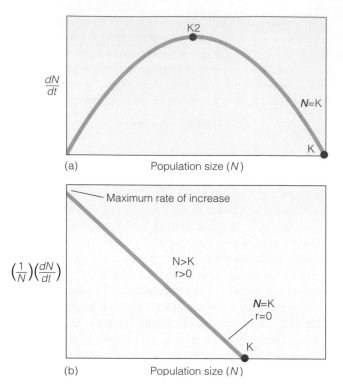

FIGURE 11.5   (a) Logistic population growth (dN/dt) as a function of population size. (b) Per capita logistic growth rate, r [$(1/N)(dN/dt)$], as a function of population size. (After Gotelli 1995.)

Note how the population grows slowly at first, accelerates, and then slows (Figure 11.6). The point in the logistic growth curve where population growth is maximal, known as the **inflection point,** is $K/2$ (in this example, 27). From this point on, population growth slows as each individual added to the population causes an incremental decrease in the per capita rate of increase.

The logistic equation involves several assumptions. Age distribution is stable, initially at least. No immigration or emigration takes place. Increasing density depresses the rate of growth instantaneously without any time lags in reproductive responses. The relationship between population size and the rate of growth is linear. When plotted logarithmically, the rate of growth declines directly as population increases (see Figure 11.5b). We also have to assume a predetermined level for K.

We can view the logistic equation in two ways. One is to consider it a law of population growth. That was the original proposition elaborated by Pearl (1927) (see Kingsland 1985). A second, preferable view is to regard the logistic equation as a mathematical description of how populations might grow under most favorable conditions and as a yardstick against which to measure actual population growth. Although some natural populations appear to grow logistically, rarely, if ever, does their growth curve match the predicted logistic curve, because the assumptions are too simplistic.

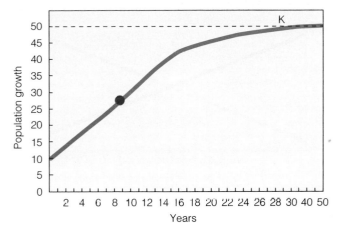

FIGURE 11.6 The roughly logistic growth curve for a hypothetical squirrel population, from Table 11.1, with an *r* value of 0.180. The curve is sigmoidal. Its upper limit, *K*, is termed the carrying capacity. The inflection point, at which density begins to slow population growth, is located at *K*/2. In this example the inflection point stands at population size 27.

For example, consider the growth of the human population in Monroe County, West Virginia, whose economic base has been agriculture and small industry. It was settled by Europeans in the early 1700s, and the population was well established in 1800, the year for which U.S. Census data are first available. The population reached 13,200 in 1900 and has fluctuated about that number since then. The population grew most rapidly from 1800 to 1850, so growth during that period provides the data to estimate *r*, the rate of increase, as 0.074. *K*, based on the population levels since the year 1900, is 13,200. With these values of *r* and *K* the logistic equation predicts that the population would reach the asymptote around 1870, 30 years before it actually did. The predicted growth curve rises much more steeply than the actual growth curve (Figure 11.7). The human growth curve approaches but does not conform to the logistic curve. The reasons for nonconformity are obvious. The age structure was not stable; birth rates and death rates varied from census period to census period; and immigration and emigration were common to the population.

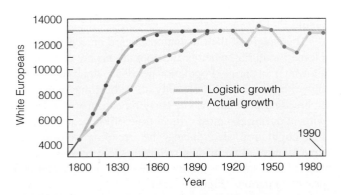

FIGURE 11.7 Actual and logistically predicted population growth of the white European population in Monroe County, West Virginia. (Data from U.S. Census Bureau.)

Time Lags

The logistic equation suggests that populations function as systems, regulated by positive and negative feedback. Growth is stimulated by positive feedback (as illustrated by the exponential growth curve), then slowed by the negative feedback of competition and dwindling resources. As *N* approaches *K*, the population theoretically responds instantaneously as density-dependent reactions set in.

Rarely does such feedback work as smoothly in real life as the equation suggests. Often adjustments lag, and available resources may allow the population to overshoot equilibrium (Figure 11.8). Unable to sustain itself on the remaining resources, the population declines to some point below carrying capacity, but not before it has altered resource availability for future generations. Population recovery as determined by reproductive rates is influenced by the density of the previous generation and the recovery of the resources, especially food supply. These factors build a time lag into population recovery.

To make the logistic equation more realistic, we need to factor into the equation a **reaction time lag (*w*),** a lag between environmental change and corresponding change in the rate of population growth:

$$\frac{dN}{dt} = rN\left(\frac{K - N_{t-w}}{K}\right)$$

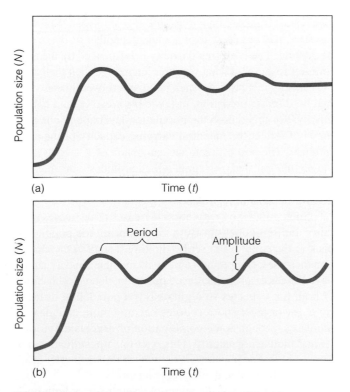

FIGURE 11.8 Logistic growth curves with a time lag. (a) A medium time lag generates damped oscillations and a convergence on *K*. (b) A large time lag generates cycles that rise and fall but do not converge on *K*. (After Gotelli 1995: 36, Figure 2.5).

Another factor is a **reproductive time lag (g),** a lag between environmental change and change in the length of gestation or its equivalent:

$$\frac{dN}{dt} = rN_{t-g}\left(\frac{K - N_{t-w}}{K}\right)$$

Means of incorporating time lags into the logistic equation are detailed by Krebs (1985) and Berryman (1981).

Time lags result in population fluctuations (Figure 11.8). The population may fluctuate widely without any reference to equilibrium size. Such populations may be influenced by some powerful outside, or extrinsic, force such as weather. A population may fluctuate about the equilibrium level, *K,* rising and falling between some upper and lower limits.

DENSITY-DEPENDENT RESPONSES TO GROWTH

Sooner or later all populations have to confront the carrying capacity of the environment in which they live. Although explicitly defined in the logistic equation, a constant equilibrium level rarely exists. Because the environment is variable, *K,* the number of individuals the environment can support at any one time, is also variable. Carrying capacity is the level at which available resources can sustain individuals in a population at survival level. The availability of resources, however, varies from season to season and year to year. While one resource may be adequate or abundant, another resource may be lacking. Thus carrying capacity is influenced by the most limiting resource. If that resource increases, then the carrying capacity increases. Further, populations over time are limited by disease, predation, unfavorable weather, and habitat quality. For this reason the population level at any one time may not reflect the potential carrying capacity of the environment. The best estimate we can obtain of *K* is to average the population size over time, which provides a mean population size. Populations tend toward this equilibrium through density-dependent regulation.

Such population regulation increases or decreases mortality and reproduction. Both tend to bring the population back to the equilibrium set by limiting factors. As a population increases, competition among its members and a scarcity of resources result in increased mortality, decreased natality, or both for a species so regulated. If a population drops to some lower level and resources become more abundant, it increases through some combination of decreasing mortality and increasing natality. This relationship is illustrated in Figure 11.9. For population regulation the birth rate, death rate, or both must be density-dependent.

Through most of the sigmoid growth curve, both positive and negative feedback operate, with a change in the relative importance of each. In the early stages of population growth, positive feedback dominates. As population reaches

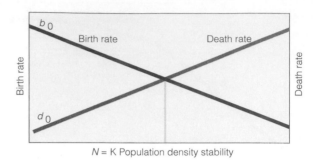

FIGURE 11.9 Density-dependent per capita birth rates and death rates change as a function of density. The population reaches stability at *N=K.*

K/2, negative feedback overcomes positive feedback. How rapidly these two responses function relates to the population's impact on resources and future growth. If individuals remove a resource faster than it is replaced, then the present population impoverishes the environment for the next generation, slowing population growth.

POPULATION FLUCTUATIONS AND CYCLES

As you may have observed, the abundance of some species of insects, birds, and mammals seems to remain about the same from one year to the next. Other species may be noticeably abundant some years and noticeably scarce in others. Why the difference?

Population fluctuations are mostly local phenomena. A species may show stability over the whole of its range, but be highly variable locally. Populations fluctuate because time lags in density-dependent mechanisms, particularly birth rates and death rates, tend to either undercompensate or overcompensate for population size. In addition, populations are affected by changes in carrying capacity and other extrinsic influences, especially weather, predation, and competition (Figure 11.10).

The pattern of the fluctuations, based on census data, is colored by timing, whether the populations are censused in fall, winter, or spring. Population trends in winter may reflect neither the true carrying capacity of the breeding habitat nor the breeding population. The carrying capacity of the environment during the most critical time of the year can influence population size during the breeding season. Fall populations of some species, reflecting the success of recruitment, may be larger or more variable from year to year than spring populations (Figure 11.11).

The nature of the fluctuation reflects the population's **resilience.** Resilience is the rate at which a population returns to equilibrium after a disturbance takes it away from balance

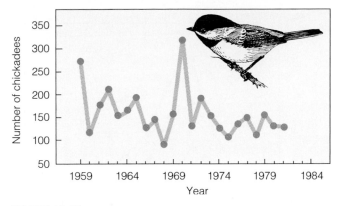

FIGURE 11.10 Fluctuation in a wintering population of black-capped chickadees (*Parus atrieapillos*) in northwestern Connecticut about a mean long-term density of 160 birds. The short-term decline in the population in 1968–1969 was attributed to an influx of competitive tufted titmice (*Parus bicolor*). In spite of fluctuations the population exhibited no sustained increase or decrease. (Data from Loery and Nichols 1985.)

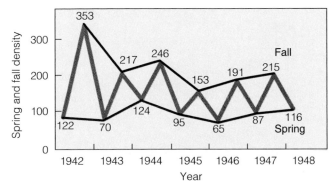

FIGURE 11.11 Trends of spring and fall populations in the northern bobwhite (*Colinus virginianus*) in Wisconsin. Note the seasonal fluctuations of population density. Fall densities fluctuate more than spring densities. (After Kabal and Thompson 1963:78.)

with its environment (Pimm 1991). In other words, resilience is a measure of how fast the population declines from above and how quickly it increases from below equilibrium. The resilience of a population is strongly influenced, if not determined, by its reproductive rate.

Size provides a clue. Small-bodied animals, such as meadow mice, fluctuate more widely than large-bodied ones, such as deer. Small animals have shorter lives, die more quickly, and thus decrease more dramatically from year to year. However, they reproduce fast, and can recover from their losses quickly. Such species have a high resilience. Large-bodied animals possess more stability about an equilibrium level because they live longer and are less subject to environmental vagaries. Very long-lived animals reproduce slowly and may require a substantial period of time to return to equilibrium. Consider, for example, the slow population growth of whales. Such species possess low resilience. The

measure of resilience, then, is the time required for the population to return toward equilibrium. The longer the time, the less the resilience.

The return time can be influenced by interactions with other species. A population of a given species does not live alone. If you disturb one species, you affect others as well. If species A depends on species B for food, both have experienced disturbance. Species A cannot return to equilibrium until species B has done so (for a good discussion, see Pimm 1991). This point will be emphasized throughout later chapters.

Population fluctuations that are more regular than we would expect by chance are called **oscillations** or **cycles.** In natural populations the two most common intervals between peaks are three to four years, typified by voles (Figure 11.12), and nine to ten years, typified by the snowshoe hare. These cyclic fluctuations are largely confined to simpler ecosystems, such as the boreal forest and tundra. Usually only local or regional populations are affected, although there is some evidence to suggest broader synchronies.

A number of theories have been advanced to explain cycles. A general theory among biologists holds that something in the physical environment, in the ecosystem, or in the population itself causes cycles. Predation has been singled out as a cause (see Chapter 15) (Keith et al. 1984, Korpimäki and Krebs 1996). Malfunction of the animal's endocrine system has been cited, as well as changes in the frequencies of genes that make much of the population less resistant to environmental changes, aggressive behavior (Krebs 1985), dispersal (Stenseth 1983), parasites (Hodson et al. 1985, 1992), and food shortages. Food shortages have been implicated in snowshoe hare decline (Keith 1974) (see Chapter 12) and in cycles of lemmings in the arctic tundra (Pitelka 1973, Batzli et al. 1980).

Experimental studies by Moss et al. (1996) suggest that causes of cycles may rest within the population itself, particularly for the red grouse (*Lagopus lagopus scoticus*). Their studies, which refute the prevailing theories, at least for the territorial and monogamous red grouse, point to changes in age structure and associated behavior. The social structure of red grouse is kin-oriented (see Chapter 13) with fathers and sons occupying clusters of territories, characterized by less aggressive territorial behavior (see Chapter 12) toward kin than non-kin. The population attains peak density when it reaches minimum territorial size. High density results in smaller clutches and limits the recruitment of young cocks into the spring population. The lack of recruitment reduces the number of kin cocks as well as the ratio of young cocks to old in the population. In the spring for five successive years during the increase phase of the six- to seven-year cycle, the experimenters removed enough territorial cocks to prevent the population from attaining peak population densities (Figure 11.13). The removal of the cocks resulted in a similar loss of hens from the breeding population, induced by the lack of territorial males. The combined loss of cocks and hens held the population below peak densities that trigger a cycling decline. Because of the

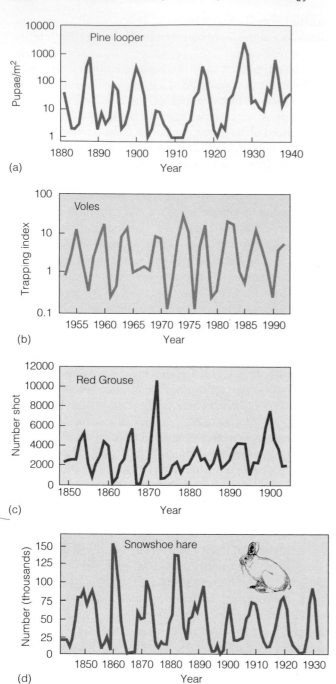

FIGURE 11.12 Examples of cyclic populations. (a) Pine looper (*Bupalus piniarius*) in Germany (Kendall et al. 1999 after Schwerdtfeger 1941). (b) Voles (*Microtus* and *Cleithrionomys*) at Kilpisjarvi, northern Finland (Kendall et al. 1999 after Hanski et al. 1993). (c) Red grouse (*Lagopus lagopus scoticus*) in Scotland (Middleton 1934). (d) Snowshoe hare (*Lepus americanus*). (From MacCulich 1937.)

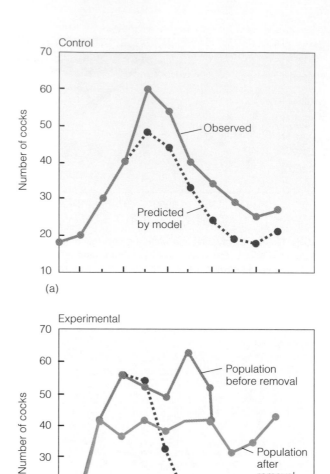

FIGURE 11.13 Spring numbers of red grouse cocks in the control (a) and experimental removal (b) areas. See text for description of the experiment. The upper points and graph in the experimental area show the number of cocks in spring prior to removal. The lower points and graph show the number of cocks in the spring population after removal. Vertical lines indicate the number of cocks removed. Hens followed a similar pattern in both areas. The control populations exhibited cyclic behavior. Removal of cocks damped the cycle. (Moss et al. 1996:1517).

resulting low density and large territories, recruitment of young becomes easier and population density increases. By reducing population density over the five years, the experimenters were able to manipulate the age structure so as to favor recruitment of young cocks and thus prevent the population cycle from occurring.

Cycling populations exhibit certain demographic characteristics. L. Keith and his associates (Meslow and Keith 1968; Keith and Windberg 1978; Keith et al. 1984) followed snowshoe hare populations through two periods of increase and three of decline in the Rochester district of central Alberta. These studies provide insights into the demographic features of the ten-year snowshoe hare cycle. The decline, which set in prior to the peak winter populations, was characterized by a high winter-to-spring weight loss, decrease in juvenile growth rate, decreased juvenile overwinter survival, reduction of adult survival beginning one year after the population peak and continuing to the low, and decreased reproduction (characterized by reductions in ovulation rates, third- and fourth-litter pregnancy rates, and length of breeding season).

EXTINCTION

When deaths exceed births and emigration exceeds immigration, populations decline. R becomes less than 1; r becomes negative. Unless the population can reverse the trend, it at worst faces extinction or at best an increased probability of becoming extinct.

Vulnerability to extinction varies widely among species (Table 11.4). Some species are common—widely distributed across their range and occupying a variety of habitats. Most species are relatively rare. Some occupy a wide natural range, but within it they are restricted to certain habitats. Others have a narrow range within which they are restricted to a very narrow habitat. They are much more vulnerable to extinction than the common ones; but even common species are not immune, as exemplified by the extinct passenger pigeon (*Ectopistes migratorius*) and the endangered black rhinoceros (*Diceros bicornus*).

There are several causes of declines in small populations. When only a few individuals are present, females of reproductive age may have little chance of meeting a fertile male. Many females remain unfertilized, reducing average fecundity. A small population suffers more from predation and sudden environmental changes, because there are fewer individuals to survive. They are also vulnerable to hybridization with related species. An example is the hybridization of sharp-tailed grouse (*Pedioecetes phasianellus*) with prairie chickens (*Tympanuchus cupido*). Losses feed upon losses until the population disappears.

Extinction is a natural process. Through millions of years of Earth's history, species have appeared and disappeared, leaving a record of their existence as fossils and trails in sedimentary rock. Some species could not adapt to geological and climatic changes. Others diverged into new species while the parent stock disappeared. Massive extinctions have occurred at several points in Earth's history: the late Ordovician; the late Devonian; the late Permian, which witnessed the extinction of up to 96 percent of species; and the Cretaceous-Tertiary, which saw the end of the dinosaurs (Figure 11.14).

Mass extinction is happening today at an accelerated pace. Some estimates place current extinction rates at 100 species a day, many of which are not yet known to science. The greatest number of extinctions has taken place since A.D. 1600. Well over 75 percent of modern-day extinctions have been caused by humans through the alteration and destruction of habitat,

TABLE 11.4 A Classification of Rare Species Based on Three Characteristics: Geographic Range, Habitat Specificity, and Local Population Size

Geographic Range	Habitat Specificity	Local Population Size	Examples
Extensive	Wide	Large	Commonly abundant in range of habitats (robin, *Turdus migratorius*; pigweed, *Chenopodium album*)
Large	Wide	Small	Constantly sparse over a large range and in several habitats (bald eagle, *Haliaeetus leucocephalus*; bristle grass, *Setaria geniculata*)
Large	Narrow	Large; dominant somewhere	Locally abundant over a large range in a specific habitat (California sea lion, *Zalophus californianus*; red mangrove *Rhizophora mangle*)
Large	Narrow	Small, nondominant	Constantly sparse in a specific habitat but over a large range (red-cockaded woodpecker, *Picoides borealis*; Pacific yew, *Taxus brevifolia*)
Small	Wide	Large; dominant somewhere	Locally abundant in several habitats but restricted geographically (orangebelly darter, *etheostoma radiosum*; Mountain pine, *Pinus mugo*)
Small	Wide	Small; nondominant	Constantly sparse and geographically restricted in several habitats (pigmy hippo, *Choeropsis liberiensis*; yellowwood, *Cladrastis kentukea*)
Small	Narrow	Large; dominant somewhere	Locally abundant in a specific habitat but restricted geographically (Kirtlandt's warbler, *Dendroica kirtlandii*; shale barren rockcress *Arabis serotina*)
Small	Narrow	Small; dominant	Constantly sparse and geographically restricted in a specific habitat) gaur, *Bos gauris*; Furbish's lousewort, *Pedicularis burbishiae*)

With the exception of the first set of characteristics, which describes a common species, combinations of the reduced states of any two of the three characteristics describe a form of rarity.
Source: After Rabinowitz 1981.

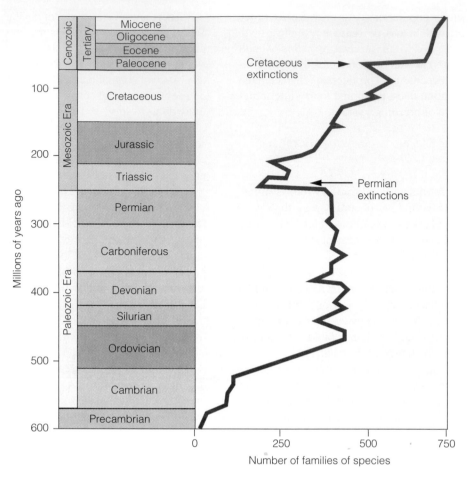

FIGURE 11.14   The geological time scale and mass extinctions in the history of life. The fossil record profiles mass extinctions during geological times. The first mass extinction occurred during the Cretaceous, which wiped out more than half of all species, including all the dinosaurs. The Permian mass extinction event resulted in the loss of 96 percent of all marine species, and perhaps as many as 50 percent of the total species on Earth.

introduced predators and parasites, predator and pest control, competition for resources, and hunting of various types.

Despite popular impressions, extinction does not take place simultaneously over the full range of a species. It begins with isolated local extinctions when environmental conditions deteriorate or the population is unable to replace itself. Local extinctions often begin when habitats are destroyed and the dispossessed find remaining habitats filled. Restricted to marginal habitats, the individuals may persist for a while as nonreproducing members of a population or succumb to predation and starvation. As the habitat becomes more and more fragmented, the species is broken down into small isolated or "island" populations out of contact with other populations of its species. As a result the population is subject to inbreeding and genetic drift, reducing the ability of the small population to withstand environmental changes (see Chapter 19).

The maintenance of local populations often depends heavily on the immigration of new individuals of the same species. As the distance between local populations of

"islands" (metapopulations) increases and as the size of the local population declines, their continued existence becomes more precarious. As the number falls below some minimum level (see Chapter 18), the local population may become extinct simply through random fluctuations. Although we equate such situations mostly with rarer species, even the more common species experience local extinctions. These often go unnoticed because the loss is masked by the influx of immigrants from surrounding areas. One study of a suburban population of robins showed that because of the losses of nests and young through predation by cats and interference by humans, the robins were not replacing themselves. Robins sang each spring only because new birds moved into the area. Thus suburbia became a population sink rather than a population source.

In fact, most local populations do not thrive for long. They are revived fast enough by new immigrants that replace the losses and keep the population going. As one local population slides down the slope toward extinction, a local population somewhere else is experiencing overcrowding and supplies new recruits for depleted habitats (see Chapters 10 and 12).

Extinctions are of two sorts: deterministic and stochastic. **Deterministic extinction** comes about through some force or change from which there is no escape. The Cretaceous-Tertiary extinctions are an example. So is destruction of habitat on a local or regional scale. Habitat destruction rarely causes the extinction of a species except when a species is endemic or it is already on the verge of extinction. Examples are the recently extinct dusky seaside sparrow (*Ammodramus maritimus nigrescens*) of the Merritt Island and St. Johns River marshes of southern Florida and the thicktail chub (*Gila crassicauda*) of the Central Valley of California, whose habitat was destroyed by drainage, dam building, and water diversion.

Small localized populations of a species are more often subject to **stochastic extinction,** which comes about from normal random changes within the population or environment. Such changes normally do not destroy a population but merely thin it out; a smaller population, though, faces an increased risk of extinction from some decimating event.

Stochastic events may be demographic or environmental. Demographic stochasticity is chance variations in individual births and deaths. Demographic stochasticity results from habitat deterioration and loss through normal successional processes, reducing population size and restricting individuals to local patches of habitat. In a small population, a high death rate or low birth rate can lead to a random or accidental extinction. When a population falls below a minimal viable size, it faces very great risk of going extinct. Environmental stochasticity is a random series of adverse environmental changes, which comes about mostly through deterioration in environmental quality. If all members of a local population are affected equally by an adverse environmental change, the population may be reduced to a level at which demographic stochasticity takes over. How long a population can exist at a low level depends on the size of individuals, longevity, mode of reproduction, and seed banks in plants.

A classic example of an extinction is that of the heath hen (*Tympanuchus cupido cupido*). Formerly abundant in New England, the heath hen, an eastern form of the prairie chicken, was driven by excessive hunting and habitat destruction to the island of Martha's Vineyard off the Massachusetts coast and to the pine barrens of New Jersey. By 1880 it was restricted to Martha's Vineyard. At this point the population was subjected to deterministic extinction over most of its range. The small population, confined to a small island, was highly vulnerable to stochastic events. At first the population prospered, growing from a population of 200 birds in 1890 to more than 2000 in 1920. Then a major stochastic event, a combination of fire, winter gales, and cold weather, reduced the population to 50. The heath hen never recovered, and the last bird died in 1932.

Summary

1. Populations increase when births and immigration exceed deaths and emigration. The difference between the two (when measured as an instantaneous rate) is the population's per capita rate of increase, r.

2. In an unlimited environment a population expands geometrically, a phenomenon that may occur when a small population is released in a unfilled habitat. Geometric increase is characterized by a constant schedule of birth and death rates, an increase in numbers equal to the per capita rate of increase, and the assumption of a fixed or stable age distribution, which is maintained indefinitely.

3. Because the environment is limited, geometric growth is not maintained indefinitely. Population growth eventually slows and arrives at a point of equilibrium, K, the number of the species the environment can support. Biologically this number, which varies with environmental conditions, is often interpreted as the environmental carrying capacity.

4. However, natural populations rarely achieve such equilibrium levels; instead a population fluctuates in numbers, above and below K, because of time lags in density-dependent mechanisms, particularly birth and death rates and environmental influences.

5. Population fluctuations more regular than expected by chance are called cycles. The two most common intervals between peaks are three to four years, typified by lemmings, and nine to ten years, typified by the snowshoe hare. Many theories have been advanced to explain cycles, but the true nature of cycles still needs to be solved.

6. When populations become quite small, they become vulnerable to extinction. Extinctions are of two sorts: deterministic and stochastic. Deterministic extinctions come about through some force or change from which there is no escape, such as destruction of habitat or changes in environment. Stochastic extinctions result from random changes within the population or the environment, to which small populations are very vulnerable.

Review Questions

1. What is the difference between lambda λ, R_0, and r? What is the relationship among them?

2. Distinguish between exponential growth and logistic growth. Give the equations for each.

3. What are the weaknesses and limitations of the logistic equation? Why is the equation useful?

4. What is carrying capacity (K)? Why do populations fluctuate about some estimated value of K?

5. What are some of the hypotheses regarding the causes of cycles? Consult the cited papers and discuss the evidence and arguments for each of the hypotheses.

6. Why do species go extinct?

7. Distinguish between deterministic and stochastic causes of extinction.

8. Has the human population in some parts of the world reached or exceeded carrying capacity (economic as well as ecological)? Consider some of the African countries and India. What evidence do you have?

9. What are some signs of stress in human populations? (See Brown et al. 1998.)

Intraspecific Competition: Population Regulation

CONCEPTS

1. Major density-dependent regulating mechanisms involve intraspecific competition.

2. A high population density can cause increased social stress, resulting in increased mortality and decreased fecundity.

3. Dispersal has its costs and benefits for the individual and the population of a species.

4. Social dominance, expressed as territoriality and social hierarchy, can function in population regulation.

5. Density-independent influences are not regulatory, but they can greatly influence both density-dependent mechanisms and population growth

6. With *k*-factor analysis, ecologists recognize the key factors determining population trends.

As discussed in the Chapter 11, density dependence is implicit in the concept of population regulation. Density-dependent effects influence a population in proportion to its size. At some low density no interaction occurs. Above that point the larger the population becomes, the greater is the proportion of individuals affected. Density-dependent mechanisms act largely through shortages and competition for resources. If the effects of a particular influence do not change proportionately with population density, or if the proportion of individuals affected is the same at any density, the influence is density-independent.

INTRASPECIFIC COMPETITION DEFINED

Population regulation implies a homeostatic feedback that functions with density. In part this feedback involves competition among individuals of the same species for environmental resources. **Competition** results only when a needed resource is in short supply relative to the number seeking it. As long as resources are abundant enough to allow each individual a sufficient amount for survival and reproduction, no competition exists. When resources are insufficient to satisfy adequately the needs of all individuals, the means by which they are allocated has a marked influence on the welfare of the population.

Nicholson (1954) demonstrated how intraspecific competition for a limited resource, such as food, affects population numbers. Although his long-term experiments on sheep blowflies (*Lucilia cuprina*) lacked all the complex interactions we would expect to find in nature, his work does show what might happen and established some basic concepts in the study of competition.

In one of his many experiments Nicholson fed to a culture of blowflies containing both adults and larvae a daily quantity of beef liver for the larvae and an ample supply of dry sugar and water for the adults. The number of adults in the experimental cages varied, with pronounced oscillations. When the population of adults was high, the flies laid such a great number of eggs that the resulting larvae ate all the food before they were large enough to pupate. As a result, no adult offspring came from the eggs laid during that period. The number of adults progressively declined through natural mortality, and fewer eggs were laid. Eventually the intensity of larval competition was so reduced that a fraction of the larvae obtained sufficient food to grow to a size large enough to pupate. These larvae in turn gave rise to egg-laying adults. Because of the developmental time lag between the survival of larvae and an increase in egg-laying adults, the population continued to decline, further reducing the intensity of larval competition and permitting an increasing number of larvae to survive. Eventually the adult population again rose to a very high level and the whole process started again.

Competition for limited food held this blowfly population in a stage of stability and prevented a continuing increase and decrease. The time lag involved in the addition of egg-laying adults to the declining population resulted in an alternate overshooting and undershooting of the equilibrium position, causing an oscillating population (Figure 12.1).

In this and the other experimental competitive situations the larvae and adults were seeking food, the rate of supply of which was not influenced by the activity of the flies. In effect the resource, the available food, was subdivided into many small parts to which the competitors, the larvae and adult flies, had general access. The individuals "scrambled" for their food. Gross crowding resulted in wastage because each competitor got such a small fraction of the food that it was unable to survive.

Nicholson called this type of competition **scramble.** In its purest form, all competing individuals garner such a small share of the resources that none survive. Among some populations outcomes are less severe; competition is scramblelike rather than pure scramble. Such competition, in which each individual is affected by the amount of shared resource remaining, can be called **exploitative competition.** Competing individuals do not necessarily react to each other, only to the level of resources. Scramble competition tends to produce sharp fluctuations in a population over time. It limits the average density of the population below that which the resources could support if an adequate amount of resources were obtained by only a part of the population.

That is exactly what takes place with **contest competition.** The deleterious effects of limited resources are confined to a fraction of the population, and members of the population interact directly. For that reason contest competition can also be called **interference competition.** Once a population characterized by contest competition passes the point at which resources become limiting, a fraction of the individuals obtains all the resources it needs. The remaining individuals get less and produce no offspring or die.

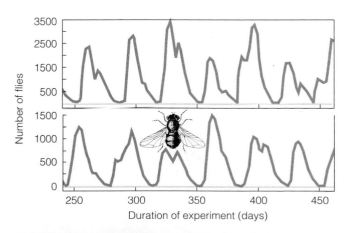

FIGURE 12.1 Fluctuations in the number of adult blowflies (*Lucilia cuprina*) in two cultures subjected to the same constant conditions, but restricted to different daily quotas of food: (a) 50 g; (b) 25 g. Although a greater food supply permitted a greater increase in density, the final outcome in each situation was similar. The adults, experiencing scramble competition, increased rapidly and then declined sharply as they reached the limits that the food could support. (From Nicholson 1957.)

Intraspecific competition influences births, deaths, and growth of individuals in a density-dependent manner. Its effects come slowly, involving at first the general welfare rather than survival of individuals. Later, as its impacts become accentuated, intraspecific competition affects individual fitness.

Effects on Growth and Fecundity

When a population reaches a point at which the resources, particularly food, are insufficient to meet the needs of individuals, something has to give. In populations characterized by scramble or exploitative competition, individuals respond to lowered level of food by reducing growth. Examples of this inverse relationship between density and rate of body growth are found among poikilothermic vertebrates. Dash and Hota (1980) discovered that frog (*Rana tigrina*) larvae reared experimentally at high densities grew slower, decreasing their chances of successfully transforming from tadpoles to frogs, and took longer to reach the minimum threshold size for metamorphosis to frogs, 0.75 g. Tadpoles held at lower densities grew more rapidly and larger and transformed at an average size of 0.889 g (Figure 12.2). In this situation intraspecific competition had minimal influence on population size, and total biomass remained approximately the same, but it was apportioned among many small individuals at high densities.

Relationships among density, growth, and fecundity extend to other vertebrate groups. Harp seals (*Phoca groenlandica*) become sexually mature when they reach 87 percent of their mature body weight of about 120 kg. At low population densities young animals attain this weight at a much faster rate than when population densities are high (Figure 12.3). Fertility in harp seals, as measured by the number of females giving birth to young, is density-dependent, too (Figure 12.4). Some birds exhibit a similar density-dependent relationship to fecundity. Roseberry and Klimstra (1984) found that the higher the breeding population of northern bobwhite (*Colinus virginianus*) relative to carrying capacity on their southern Illinois study area, the lower was the rate of summer population gain (Figure 12.5).

These examples suggest that vertebrate populations do respond to increasing numbers in a density-dependent fashion through intraspecific competition. The timing of the response

depends on the nature of the population. Fowler (1981) has hypothesized that among large mammals with a long life span and low reproduction, regulating mechanisms do not function until the population approaches carrying capacity. In other words, mortality and natality are more or less in balance, births compensating for deaths, until the popula-

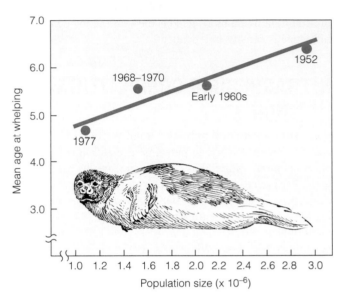

FIGURE 12.3   The mean age of maturity of harp seals (*Phoca groenlandica*)(and other marine and terrestrial animals) is related not so much to age as to weight. Seals arrive at sexual maturity when they reach 87 percent of average adult body weight. Seals attain this weight at an earlier age when population density is low. (From Lett et al. 1981:144.)

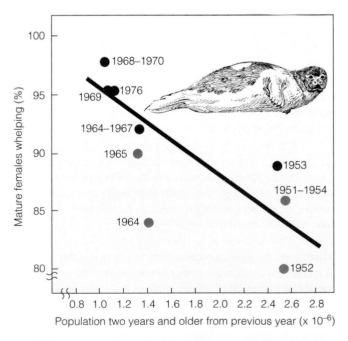

FIGURE 12.4   Fertility, too, is density-dependent in harp seals. As the population of seals (measured by including only animals two years and older from the previous year) increases, the percentage of females giving birth to young decreases markedly. (From Lett et al. 1981:146.)

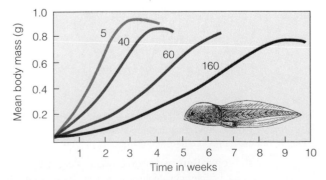

FIGURE 12.2   Influence of density on the growth rate of the tadpole *Rana tigrina*. Note how rapidly the growth rate declines as density increases from 5 to 160 individuals confined in the same space. (After Dash and Hota 1980:1027.)

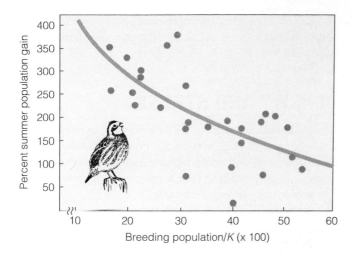

FIGURE 12.5 Population growth rate of northern bobwhite quail on the Southern Illinois Carbondale Research area (1954–1979) as a function of breeding population size, expressed as percent of carrying capacity (k). The breeding population shown here was adjusted to account for changes in land use and thus carrying capacity of the area. (From Roseberry and Klimstra 1984:96.)

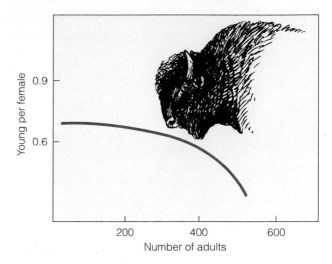

FIGURE 12.6 Nonlinear density-dependent change in the population of a large mammal, the American bison. The birth rate of the bison, expressed as young per female, is independent of density until the population reaches a certain density. Then birth rates decline with increasing density. (From Fowler 1981:607.)

tion is close to *K*, at which point density-dependent mechanisms set in and tend to overcompensate. The birth rate of bison (*Bos bison*) (Figure 12.6) shows such a response; but birth rates of other large mammals, such as elk (*Cervus elaphus*) (Figure 12.7), appear to be linear, but negatively density-dependent.

Effects on Plant Biomass

Consider an old field colonized by thousands of pine seedlings and follow its history through time. Initially all of the seedlings grow rapidly at the about same rate, accumulating biomass in a density-independent fashion. After 5 to 15 years, the canopy closes. At this point competition for canopy space, root space, light, water, and nutrients intensifies. Some individuals gain more of the available resources than others and thus grow faster. Eventually some individuals lack the ability to compete and die. The result is the gradual thinning of a stand and a developing hierarchy of size with few large and many small individuals.

As the number of individuals declines, both the mean weight of individual plants and the total yield (as measured by biomass) increase. This progressive decline in density in a population of growing individuals in a single species populations is known as **self-thinning.** How soon plant populations experience competition and depressing effects on the increase in the mean weight of plants depends on the initial density. If the density of seedlings is high, that population will experience the effects of competition, such as reduction in growth and increased mortality, much sooner than less dense populations. Regardless of their starting densities, populations eventually converge upon a common density that will decrease through time (Figure 12.8). If we plot the logarithm of mean weight or volume against the logarithm of plant density, the slope of the line averages around –³⁄₂ or

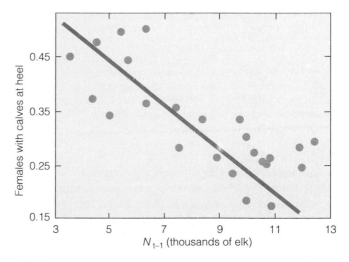

FIGURE 12.7 Linear relationship between calf recruitment and previous year population size in Yellowstone elk. (From Houston 1982:45.)

–1.5. In other words average plant volume or weight increases about 1.5 units for each unit decrease in density. Plant ecologists call this relationship the $-\frac{3}{2}$ **power law** of self-thinning (Yoda et al. 1963, White and Harper 1970).

As is evident in Figure 12.8 showing the relationship between tree volume and tree density for a series of loblolly pine (*Pinus taeda*) over a period of 50 years, populations with the highest density experienced increased mortality much sooner than the low-density populations (Peet and Christensen 1987). All populations moved along the same straight line described by the logarithm of average plant weight plotted against the logarithm of density of survivors, the slope of which is approximately –³⁄₂. Each line depicts the trajectory of the size versus density relationship for each stand of various starting densities. The points on the trajectory indicate a different point

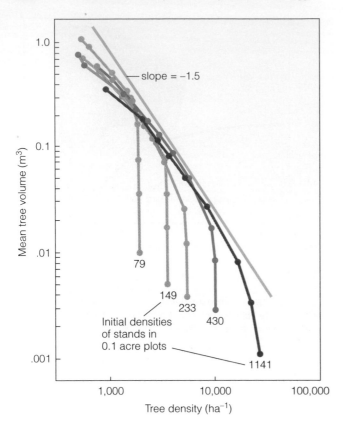

FIGURE 12.8 Self-thinning in a series of loblolly pine (*Pinus taeda*) stands over a 50-year period. The straight line has a slope of :1.5 as predicted by the law of thinning. All stands converge on the same density and tree size along the slope of the line. (From Peet and Christensen 1987.)

in time for each stand. In time, all populations, regardless of initial density, reached the same point, as indicated on the graph, where density declined as plant volume increased.

The $-\frac{3}{2}$ power law is based on the geometry of space occupied by a plant. The law assumes that all plants of any given species grow to a similar shape, regardless of size and growing conditions, and that the combined action of crown growth and self-thinning maintains a complete crown closure. Mortality in the population occurs when the percentage of canopy cover relative to ground area approaches 100 percent, blocking light to plants below. Data from a number of even-aged one-species stands suggest that plants from herbaceous to semiwoody chaparral (Guo and Rundel 1998) and coniferous and woody species obey it (Osawa and Allen 1993). For this reason, the $-\frac{3}{2}$ power law has been accepted as an ecological law (Weller 1987a, b, 1989). However, not all plants necessarily conform to the law, woody plants in particular. Plant shapes are variable, even within a species. Because thinning exponents vary according to plant geometry, they are not always near the idealized value of $-\frac{3}{2}$.

The thinning law applies to even-aged, monospecific stands. How, then, does the law apply to multispecies and uneven-aged stands in which competitive interactions are among individuals of species and ages with differing competitive abilities for resources? Of the few studies of such stands that do exist, the mixtures appear to conform broadly to the

−1.5 thinning law when the component species are considered collectively as a single population (Bazzaz and Harper 1976, Juloka-Sulonen 1983).

DENSITY AND STRESS

How do individuals respond as their population increases, as their living space becomes more crowded, as their food, shared by more and more hungry members of the group, becomes less available? One response is increased social stress, especially among vertebrates.

Stress in vertebrates, evidence suggests, can act on the individual through a physiological feedback involving the endocrine system. This feedback is most closely associated with the functioning of the pituitary and adrenal glands (Christian 1963, 1978, Christian and Davis 1964, Davis 1978). Stress triggers hyperactivation of the hypothalmus-pituitary-adrenocorticular system, which in turn alters the secretion of growth and sex (gonadotropic) hormones. Profound hormonal changes suppress growth, curtail reproductive functions, and delay sexual activity. Further, these hormonal changes may suppress the immune system and cause breakdowns in white blood cells, increasing an individual's vulnerability to disease (Sinclair 1977). Social stress among pregnant females may increase intrauterine mortality (spontaneous abortion) and cause inadequate lactation and subsequent stunting of nurslings. Thus, stress can result in decreased births and increased mortality (Lloyd et al. 1964, Myers et al. 1967, 1971)

Pheromones (chemicals released by animals that serve as communication among individuals of the same species) present in the urine of adult rodents may inhibit reproduction among members of a population. Such a function is suggested in a study involving wild female house mice (*Mus musculus*) living in high-density and low-density populations confined to grassy areas within a highway cloverleaf. Urine from females of a high-density population was absorbed onto filter paper. The paper was placed with juvenile wild female mice held individually in laboratory cages. Similarly, urine from females in low-density and sparse populations was placed with other juvenile test females. Juvenile females exposed to urine from high-density populations experienced delayed puberty, whereas females exposed to urine from low-density populations did not. The results suggest that pheromones present in the urine of adult females in high-density populations may delay puberty in juveniles and help slow population growth (Massey and Vandenberg 1980). Juvenile female house mice from low-density populations exposed to urine of dominant adult males accelerate the onset the puberty (Lombardi and Vandenbergh 1977).

Plants growing under crowded conditions, in which competition is high, respond in various ways to conditions of low nutrients, low moisture, and other environmental stress, such as too little or too much light (see Chapter 6). Individual plants react to increased density with decreased growth, as expressed by reduction in mean plant weight, loss of leaves and branches, and changes in growth form. In spite of this reduction, yield per unit area is constant over a wide range of densities. Genets may respond to high density by reducing the number of ramets, as perennial ryegrass (*Lolium perenne*) does, resulting in a lower density of tillers than genets in the population (Kays and Harper 1974). Plants also modify their morphology by reducing the number of nodes per stem, internode length, number of flowers and seeds, leaves per stem, and branches (for examples, see White 1984, Sarukhan et al. 1984). Such reductions, especially in leaf area, can increase mortality (Fowler and Antonovics 1981, Antonovics and Primack 1982).

Ruderal plants, those adapted to persistent and severe disturbance, such as ragweed and other annual weeds, respond to low moisture and nutrient stress by producing seeds at the expense of vegetative development. Individual plants are small and poorly developed, yet the number of seeds relative to individual plant biomass is high. Seeds of such plants can survive buried in soil for long periods of time. They can germinate quickly when disturbance exposes the seeds to light and fluctuating daily temperatures.

When subjected to competitive stress, biennials, such as fetterbush (*Pieris hieracioides*) and bluethistle (*Echium vulgare*), may delay reproduction for three to five years, even though they can reproduce in two years under ideal conditions. Under conditions imposed by dense populations, a lower proportion of plants survives to maturity, reproduction is delayed, and fewer seeds are produced by mature plants (Klemow and Raynal 1985). Some plants respond to stress by flowering sooner, others by flowering later. Plants growing on infertile, drought-prone sites exhibit a similar behavior, responding to unfavorable conditions rather than to high densities.

Such responses to stress influence individual fitness among plants and the maintenance and expansion of populations. However, there is still insufficient evidence that density-dependent stress in plants acts in any sort of regulatory way. (For detailed review of density-dependent regulation in plants, see Antonovics and Levin 1980.)

MECHANISMS OF POPULATION REGULATION

By what means does intraspecific competition regulate populations? Stress, dispersal, and social interactions all may play a part.

Dispersal

Instead of coping with stress (to state the situation somewhat anthropomorphically), some animals run away from a bad situation. They seek vacant habitats (see Chapter 10). Although dispersal is most apparent when population density is high, it is a relatively constant phenomenon. Some individuals leave the parent population whether it is crowded or not.

When a lack of resources forces out individuals, usually subadults driven out by adult aggression, the odds are they will perish, although a few may arrive at and settle in some suitable area. Such dispersal results from overpopulation; it has little influence on population regulation.

More important to population regulation is the dispersal that takes place when population density is low or increasing, but well before the population reaches a level at which resources are overexploited. Individuals who disperse are not a random selection of the population, but ones who are in good condition, belong to any sex or age group, have a good chance of survival, and show a high probability of settling in a new area (see Chapter 10). Some evidence exists that such individuals are genetically predisposed to disperse.

Dispersal carries with it certain costs and benefits (Table 12.1). A few dominant juveniles may be able to establish themselves reproductively when the natal area is vacated through mortality of the adults. Because there is little probability they will reproduce on their natal area, most juveniles or subadults can maximize their fitness only if they leave their

TABLE 12.1 Potential Costs and Benefits of Dispersal Choices (G = Genetic, S = Somatic)

STAY AT HOME: PHILOPATRY	
Costs	**Benefits**
Inbreeding depression (G)	Optimal inbreeding: maintain locally adapted genes (G)
Reduced fitness because of resource shortage (S)	Reduced physical risks: increased survivorship (S)
Reduced indirect fitness: competition with kin (S)	Familiarity with local terrain: security (S)
	Familiar social environment (S)
	Adaptive local traditions (S)
	Maintain kin association (S)
DISPERSE	
Costs	**Benefits**
Outbreeding depression: disrupt coadapted genes (G)	Outbreeding enhancement (G)
Hybrid young not well adapted (G)	Avoid overcrowding (S)
Alleles less suited to the environment (G)	Avoid competing with kin (S)
Greater risk in movement: predators, local diseases, unfamiliarity with terrain (S)	Improve fecundity (S)

Source: Adapted from Shields 1987.

birthplace, in spite of the risks. Where intraspecific competition is intense, dispersers can locate habitats where resources are more accessible, breeding sites are available, and competition is less. Dispersers also increase the probability of encountering more new individuals with which to mate, reducing the probability of inbreeding and increasing fitness of offspring, because of heterozygosity (see Chapter 19).

Intraspecific competition of some sort may be a driving force behind some dispersals. Christian (1971) hypothesized that increased population density in rodents increased levels of aggression. Aggressive individuals force subdominant ones to disperse. Under these conditions maximum dispersal would be at peak densities and involve social subordinates, mostly males.

Evidence obtained from fenced populations indicates that the highest rates of dispersal occur before peak densities are reached. Not all dispersers are necessarily subdominants, or the less aggressive, and a large fraction of dispersing animals are females. Some animals may be born to disperse. W. E. Howard (1960) suggested a genetic basis for dispersal, and Myers and Krebs (1971) and C. J. Krebs et al. (1976) found that dispersers were not a random subsample from their control populations of voles, *Microtus pennsylvanicus* and *M. ochrogaster*. Certain genotypes were more prone to dispersal than others when populations were increasing. However, there is little evidence on the heritability of tendencies to disperse.

Bekoff (1977) suggested that social interaction prior to dispersal rather than aggressiveness is the mechanism behind dispersal. Asocial individuals, either dominant individuals avoided by their sibs or subdominants avoiding their sibs, are the most likely to disperse because they fail to develop social ties. Because aggressive behavior is not an adequate stimulus for dispersal, no relationship exists between population density and dispersal.

Lidicker (1975) has hypothesized two types of dispersal, presaturation and saturation. **Presaturation dispersal** takes place during the increase phase of population growth before population reaches a peak or carrying capacity and before resources are depleted. The dispersers are in good condition, consist of any sex or age group, have a good chance of survival, and have a high probability of settling in a new area. Such dispersals seem to be density-independent. **Saturation dispersal** occurs when carrying capacity has been exceeded. The individuals, mostly juveniles and subdominants, have two options: to stay and either perish or not breed, or to leave the area. The mass movements of squirrels in the 1800s (Allen 1962) and muskrat dispersal described by Errington (1963) are examples of saturation dispersal. Such dispersal is density-dependent.

Dispersal requires a source and a **sink,** an empty or unfilled habitat, or even marginal or unsuitable habitat in which the animals can survive for a time. The dispersal sink must permanently remove animals from the resident or source population. Many dispersers die during their travels. Some dispersers discover and settle into patches of optimal habitat. Others move into areas where conditions (predation, poor nesting sites, lack of protective cover) preclude successful reproduction (Figure 12.9). In such habitats reproduction does not balance mortality. Even though sink habitats often

support very large populations, many would disappear without the immigration of surplus individuals from more productive source areas (Pulliam 1988). Because a species may actually be more abundant in a sink habitat than in the source habitat, the sink appears to be optimal. In reality, the habitat may be luring dispersers into areas subject to high predation or other causes of reproductive failure.

Such population sinks can be distinguished from optimal habitats only with some knowledge of the demographics of the species in each. Such knowledge is critical in the conservation of species. Because of population abundance, we may mistakenly select a sink habitat as an optimal source habitat when selecting refuges and unintentionally lead both sink and source populations to local extinction.

Does Dispersal Regulate Populations?

Does dispersal play any role in population regulation? This question has especially intrigued microtine biologists and ecologists, who have been seeking explanations for population cycles of voles and other rodents ever since they were described by Elton in 1926. Much of what we have learned about dispersal and population interactions is based on comparisons of fenced rodent populations in which dispersal could occur with those in which dispersal was prevented. The magnitude of dispersal was estimated by com-

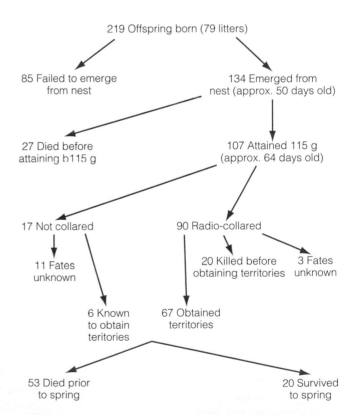

FIGURE 12.9   The fate, starting at birth, of 219 red squirrel (*Tamiasciurus hudsonicus*) offspring (79 litters) from the 1988, 1989, and 1990 cohorts at Fort Assiniboine, Alberta, Canada. Of the 90 radio-collared young dispersers, only 20 survived on new territories. The rest either died during dispersal or were lost to various dispersal sinks. (From Larsen and Boutin 1994:217.)

paring the numbers of marked animals disappearing from a control grid with the number of those animals reappearing in grids from which all animals are continuously removed (for reviews see Gaines and Johnson 1987, Gaines and McClenaghan 1980). For example, in an early experiment C. J. Krebs et al. (1969) enclosed three populations of meadow voles (*Microtus pennsylvanicus*) in southern Indiana in such a way that individuals could not immigrate or emigrate, but predators had access. They compared these populations with a control population whose members were able to disperse. The enclosed populations increased in size; in one grid the population was three times as high as in the control. Overpopulation in the enclosures resulted in overgrazing, habitat deterioration, and starvation. Physiological stress was relatively unimportant because *Microtus* can exist at densities several times higher than those normally experienced by other voles. This experiment led to a long series of studies involving fenced populations that investigated emigration from growing populations.

One of the first of these studies was done by Krebs and his associates (1976) in British Columbia. They set up control and experimental grids populated with *Microtus townsendii*. They established control populations whose members were allowed to emigrate through exit tubes. They cleared experimental grids of voles, continually kept the areas vacant, and monitored recolonization. Colonization of experimental areas was most rapid when populations in the control areas were increasing. In declining populations on the control grids, in which losses were due to death and not emigration, very little dispersal occurred.

The outcomes of these experiments, reviewed by Gaines and McClenaghan (1980) and Gaines and Johnson (1987), show that dispersal in fluctuating populations of voles is density-independent. Most dispersals take place during times of population increase. Although dispersal is positively correlated with population density and with the rate of population increase, there is no association between the proportions of the population leaving the area and the rate of population increase and decrease. The dispersers are not a random subset of the set of the resident population, but are mostly younger males.

Although dispersal may not function as a regulatory mechanism in a traditional sense, it can expand populations, aid in the persistence of local populations (rescue effect), and function as a form of natural selection by sorting out phenotypes and genotypes.

SOCIAL INTERACTIONS

Intraspecific competition expresses itself in social behavior as the degree of tolerance between individuals of the same species. Social behavior appears to be a mechanism that limits the number of animals that can live in a particular habitat, have access to food supply, and engage in reproductive activities.

Aggressive and submissive interactions (called agonistic behavior) are the basis of social organization, which takes two forms—dominance and territoriality. The difference between the two involves individual interactions and also the utilization of space. Social organization based on interindividual distance (distance from another individual that provokes aggressive or avoidance behavior) and dominance relationships among members of a social unit that share space is **social dominance.** Social organization involving the division and exclusive occupation of space by a social unit or individual with a defended boundary is **territoriality.**

Social Dominance

Social dominance is based on intraspecific aggressiveness and intolerance and on the dominance of one individual over another. Two opposing forces are at work simultaneously: mutual attraction versus social intolerance, a negative reaction against crowding. Each individual occupies a position in the group or local population based on dominance and submissiveness.

In its simplest form an alpha individual is dominant over all others, a beta individual is dominant over all but the alpha, and so on to the omega, which is totally subordinate. This relationship was first described by Schjelderup-Ebbe (1922) for the domestic chicken. It is a straight-line or linear peck order, so called because pecking follows dominance—that is, birds peck at others of lower rank (Figure 12.10a). Even within a peck order complexities may exist,

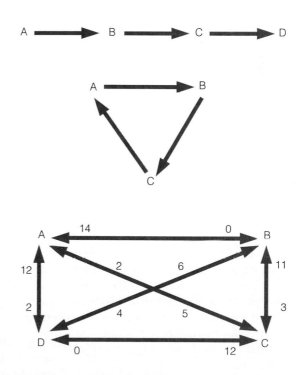

FIGURE 12.10 Examples of peck orders. (a) A straight-line peck order in which one animal is dominant over the animal below it. A is the alpha individual; D is the omega. (b) Triangular peck order in which A is dominant over B, B is dominant over C, yet C is dominant over A. (c) A more complex triangular peck order. The double arrows indicate encounters between individuals. The numbers represent the number of wins of one individual over another. For example, A is clearly dominant over B with 14 wins; B never dominated A. D is the omega individual.

such as triangular or nonlinear hierarchies. In these triplets, the first individual is dominant over the second, the second is dominant over the third, and the third is dominant over the first (Figure 12.10b). In such a situation, an individual of a lower rank can peck an individual of a higher rank.

Among some animals, birds in particular, peck order is replaced by peck dominance, in which social rank is not absolutely fixed. Threats and pecks are dealt by both members during encounters, and the individual that pecks the most is regarded as dominant (Figure 12.10c). The position of the individual in the social hierarchy may be influenced by levels of male hormones, strength, size, weight, maturity, previous fighting experience, previous social rank, injury, fatigue, close associates, and environmental conditions.

In mixed groups, males and females may have separate hierarchies with males dominant over females; or females may be equal to the males or dominant over them. In other species, dominance is unrelated to sex. Once social hierarchies are well established within a group, newcomers and subdominant individuals rise in rank with great difficulty. Strangers attempting to join the group are either rejected or relegated to the bottom of the social order.

Rise in hierarchy often is related to sexual activity and hormones. This is particularly true among those species that remain in flocks throughout the year. Individuals, male or female, that come into breeding condition early, even though subdominant in the winter group, rise in hierarchy through increased aggressiveness.

To the dominant individuals in such a social structure go most of the resources. Among many species the dominant male secures the most mates, thereby ensuring greater fitness at the expense of subdominant males. Dominant individuals have first choice of food, shelter, and space, and subdominant individuals obtain less. When shortages are severe, low-ranking individuals may be forced to wait until all others have fed, to take the leavings, face starvation, or disperse.

Under these circumstances, social interactions cannot be called regulatory. Rather, the outcome of this type of contest competition simply ensures that the dominant animals in the population will continue to reproduce successfully. The fitness of such individuals is secured at the expense of subdominant individuals.

Social dominance, however, can influence population regulation if it affects reproduction and survival in a density-dependent manner. An example is the wolf. Wolves live in small groups of 6 to 12 or more individuals called packs. The pack is an extended kin group consisting of a mated pair, one or more juveniles from the previous year who do not become sexually mature until the second year, and several nonbreeding related adults.

The pack has two social hierarchies, one headed by an alpha female and one headed by an alpha male, the leader of the pack to whom all other members defer. Below the alpha male is the beta male, closely related, often a full brother, who has to defend his position against pressures from other males below.

Mating within the pack is rigidly controlled. The alpha male (occasionally the beta male) mates with the alpha female.

She prevents lower ranking females from mating with the alpha and other males, while the alpha male inhibits mating attempts by other males. Thus each pack has one reproducing pair and one litter of pups each year. These pups are reared cooperatively by all members of the pack. At low wolf densities, some packs may rear two litters of pups a year (Ballard et al. 1987).

The size of a wolf population in a region is governed by the size of the packs, which hold exclusive areas. Regulation of pack size is achieved by events within the pack that influence the amount of food available to each wolf. The food supply itself does not affect births and deaths, but the social structure that leads to an unequal distribution of food does. The reproducing pair, the alpha female and the alpha male, has priority for food; they, in effect, are independent of the food supply. The subdominant animals, male and female, with little reproductive potential, are affected most seriously. At high densities the alpha female will expel other adult females from the pack. Other individuals may leave voluntarily. Unless these animals have an opportunity to settle successfully in a new territory and form a new pack, they fail to survive.

The social pack, then, becomes important in population regulation. As the number of wolves increases, the size of the pack increases. Individuals are expelled or leave, and the birth rate relative to the population declines because most sexually mature females do not reproduce. Overall the percentage of reproducing females declines. When the population of wolves is low in a region, sexually mature females and males leave the pack, settle in unoccupied habitat, and establish their own packs with one reproducing female. More rarely, the pack may produce two litters instead of one litter in a year (Ballard et al. 1987, Van Ballenberghe 1983). But at very low densities, females may have difficulty locating males to establish a pack and so fail to reproduce or even survive. (For details on social regulation of population size in wolves see Mech 1970, Zimen 1978, Fritts and Mech 1981, Ballard et al. 1987.)

Territoriality

A more complex of grouping is **territoriality,** a situation in which an individual animal or a social group defends an exclusive area, a **territory,** that is not shared with rivals (Figure 12.11). As a result territorial individuals tend to occur in more or less regular patterns of spatial distribution. Territoriality and social hierarchy represent degrees of manifestation of the same basic pattern of dominance. Thus it is often difficult to draw a sharp distinction between them. Depending on the season, the degree of crowding, and the distribution of resources, territoriality can grade into social hierarchy and vice versa.

For example, the dragonfly *Leucorrhini arbuicunda,* like other dragonflies, maintains a territory along the edge of a pond. When its population density is low, individuals are spaced 3 to 7 m apart and aggressive interaction is frequent. As density increases, the level of territorial defense and attachment to the site decreases. Thus territoriality can be considered a spatial organization of dominance hierarchies in which the individual holds the highest rank in the hierarchy in its own territory or center of activity.

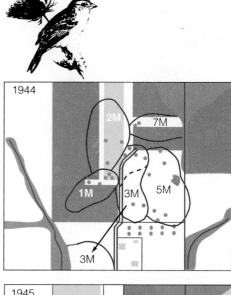

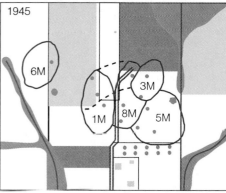

FIGURE 12.11 Mating territories of six banded male grasshopper sparrows (*Ammodramus savannarum*). Dots indicate song perches. Note how they are distributed near the territorial boundaries. The shaded areas represent crop fields, the white areas hayfields. Dashed lines indicate boundary shifts in territory prior to second nesting. Males returned to nearly the same territorial area the second year. Such behavior is philopatry. (From Smith 1963:160.)

Types of Territory Types of territories vary according to the needs of the animals that defend them (Nice 1941). One type is a general purpose territory, established during the breeding season. This type of territory is common among songbirds and some mammals, such as muskrats. Within it all activities from feeding to mating and rearing the young take place. Late in the breeding season or soon after, territorial defense breaks down.

A second type, common among hawks, is a mating and nesting territory with feeding done elsewhere. Mating territories are exemplified by leks of prairie grouse and the singing grounds of woodcock (*Philohela minor*). Swallows and many colonial birds defend only a nesting territory, the size of which is often determined by the distance the bird can strike from its nest. Other animals, such as hummingbirds and some squirrels, defend only a food resource, a feeding territory. Some birds defend a winter roosting territory when adequate roosting sites are scarce.

Territorial Defense Once an animal has established a territory, the owner must defend it against intruders. At first conflicts may be numerous. Birds, frogs, and insects usually defend their claims vocally by singing from some conspicuous spot. Songs and calls advertise the fact that the area is already occupied. They are a long-distance warning that potential trespassers should not waste their energies trying to settle there. Birds may shift their song perches throughout the territory (see Figure 12.11) or vary their song patterns, perhaps in an effort to suggest that more than one male is in the area.

Song can be effective in maintaining space between individuals. If a bird is removed from its territory, the space is quickly claimed by another deprived of a territory outright or forced to settle in some suboptimal area. If a territorial male is removed but his song has been recorded and is played in the territory, other males will stay away (Carrick 1963).

If songs and calls fail and another individual does move into the area, the owner may confront the trespasser with visual display. That display may involve raising the crest, fluffing the body feathers, spreading the wings and tail, and waving wings among birds, erecting the ears and baring the fangs among mammals (Figure 12.12). Such displays usually intimidate the invader, encouraging it to leave.

If intimidation displays fail, then the territory owner is forced to attack and chase the intruder, an activity easily observed among many birds in spring and among dragonflies about a pond's edge. The gall-forming aphid *Pemphigus betae* (about which more will be said later) defends her territory by engaging in end-to-end kicking-and-shoving contests with the intruder, which may last more than two days and may result in the death or one or both aphids (Whitham 1987).

Some animals defend a territory by the use of scent markers. Wolf packs and coyotes mark territories with well-placed scent posts, frequently renewed by urine (Peters and Mech 1975, Barrette and Messier 1980). These scent marks warn neighboring members of their species about boundary rights. Just as important, these scent posts tell members of a wolf pack that they are within their own territory and prevent accidental straying into hostile territory.

The use of scent and other chemical releasers, or **pheromones,** is widespread among animals, especially mammals and insects. They are secreted from endocrine glands, transmitted as a liquid or a gas, and smelled or tasted by others. They are important not only to mark territory and trails but also to convey such information as the identity of an individual, its sex and social rank, the location of food, and the presence of danger or a potential mate (see Wilson 1971, Whittaker and Feeney 1971, Free, 1987).

Why Defend a Territory? Why should an insect, bird, or mammal defend a territory? The reasons vary among animals. For some it is the acquisition and protection of a needed resource such as food, or a reduction in the risk of predation. For others it is the attraction of a mate. The basic benefit is always an increased probability of survival and improved reproductive success—in short, increased fitness.

Consider the territorial behavior of the gall-forming aphid *Pemphigus betae* on the leaves of its host plant, narrowleaf

FIGURE 12.12 Displays to defend territory or assert social dominance. (a) Agonistic displays among birds. (Pectoral sandpiper, *Calidris melanotos,* after Hamilton 1959; herring gull, *Larus argentatus,* based on photographs in Tinbergen 1953; hermit thrush, *Catharus guttatus,* after Dilger 1956; redpoll, *Carduelis fiammea,* after Dilger 1960; ring-necked pheasant, *Phasianus colchicus,* after Collias and Taber 1951.) (b) Aggressive (upper) and submissive (lower) expressions in the American wolf (*Canis lupus*), typical of canids. (c) Aggressive display in the black-tailed deer (*Odocoileus hemionus sitkensis*). The illustration shows details of the head during the snort that occurs when the buck is circling in a crouch position. Note the widely opened preorbital gland near the eye, curled upper lip, and bulged neck muscles. The snort is a sibilant expulsion of air through the closed nostrils, causing them to vibrate. (After Cowan and Geist 1961.)

cottonwood (*Populus augustifolia*), which grows along streams in western North America. The female sexual aphid deposits a single overwintering egg in the deeply fissured bark of the main branches and trunk of the tree. From the egg a wingless stem mother, who will give rise to new offspring,

emerges in spring at the time leaves are unfolding. The stem mothers move en masse from the trunk and branches and establish themselves on immature leaves. They probe the expanding tissues to form a small depression. In a short time leaf tissue will envelop the aphid, forming a hollow gall in which she will parthenogenetically produce up to 300 progeny. The number of aphids invading the tree may be high, up to 850 per 1000 leaves.

The aphids have a preferred site, the base of the largest leaves (Whitham 1987). These leaves have the lowest concentration of phenolics, complex molecules that inhibit feeding activities of herbivorous insects (see Chapter 16), and are the least susceptible to leaf fall. Such leaves are rare, representing only about 1.6 percent of available leaves, and so there is a great deal of competition for them. Kicking-and-shoving contests between prospective colonists settle disputes for microterritories 3–5 mm in length. The winners usually are the largest stem mothers, who may or may not share a leaf with another. Stem mothers who have a leaf to themselves have the highest reproductive success. When two aphids share a leaf, the stem mothers that secure the basal portion of the leaf produce on the average 56 percent more progeny than those displaced to the distal portion of the leaf. These differences in reproductive success result from microhabitat variations in the quality of the leaf. Only a few millimeters of difference in leaf position affects aphid reproduction (Figure 12.13).

To determine what would happen to reproductive success, Whitham removed one member of a competing pair. Immediately the remaining aphid crossed the former territorial boundary and enlarged her own. If the remaining member was the distal stem mother, she increased the number of her progeny 48 percent on average. If she was the basal stem mother, she improved her reproductive success by only 18 percent.

His study points out several aspects of territoriality. The territorial animal that claims the best territory has the highest reproductive success. The reproductive cost of losing is lower fecundity, if the losers reproduce at all. The winners are the most dominant individuals. Among aphids dominant individuals usually are the largest. Solitary stem mothers achieved the highest reproductive success, followed by the basal stem mother of a competing pair and then by the distal stem mother. Smaller stem mothers are forced to share leaves or have to settle on inferior smaller leaves. Thus the cost of losing a competitive interaction is reduced fitness.

Defending a territory can be a costly business, especially if optimal resources are limited. When the stakes of winning or losing between aphids are high, contests over a position on a leaf may last two days and result in the death of one or both contestants. Even on a less intense scale, territorial defense uses energy, consumes time, and interferes with feeding, courtship, mating, and rearing of young.

Like all economic endeavors, territorial ownership has costs and benefits, and the owner has to balance the two (Figure 12.14) (Davies and Houston 1984). Some territories are economically defendable and some are not. A general prerequisite is a predictable resource somewhat dispersed. Then acquisition of an area ensures its owner of resources, reduces foraging costs, and allows time for other activities. If resources are unpredictable and patchy, it may be advantageous for individuals to belong to a group and cooperatively seek needed resources without being restricted to one area. Spotted hyenas (*Crocuta crocuta*), for example, live in clan territories in the Ngorongora Crater of Kenya, where resources are predictable; whereas on the Serengeti Plains, where food is seasonal, they range over wide areas and do not defend a territory (Kruuk 1972).

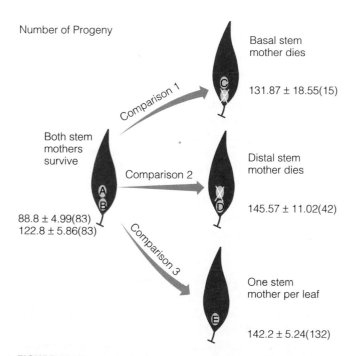

FIGURE 12.13 Reproductive success and fitness of territorial stem mothers of the gall-forming aphid *Pemphigus betae*. When either stem mother of a competing pair suffered an early death, the remaining stem mother produced significantly more progeny (A<C and B<D). Stem mothers occupying leaves singly from the beginning of gall formation produced more progeny than either member of a competing pair (A<B<E). When released from competition early in development, the surviving stem mothers on the average produced the same number of progeny as stem mothers solitary from the beginning of gall formation (C=D=E). Mean number of progeny from leaves of same quality and size ± SE (*n*) is indicated for 355 surviving stem mothers. (From Whitham 1986:140.)

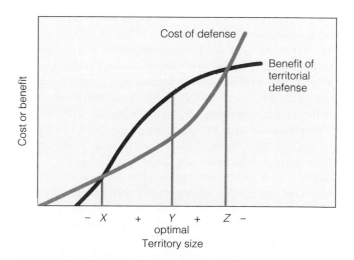

FIGURE 12.14 Model of a cost-benefit ratio defined by a cost curve and a benefit curve as it might apply to territorial defense. Costs of territorial defense increase as the size of territory increases. Defense is profitable only between *X* and *Z*. A maximum cost-benefit ratio is at *Y* in each case.

Territorial Size Closely associated with the ratio of costs of territorial ownership to benefits derived is territorial size. As the size of a territory increases, the cost of territorial defense increases. Many male birds in spring attempt to claim more ground than they can economically defend. They are forced to draw in the boundaries to make the area more manageable. However, there is a minimum size below which they cannot go, because it will be too small to meet their needs. (For models see Stamps and Krishnan, 1990, Adams 1994). The number of territory owners an area can hold is the total area divided by the minimum size of the territory. Minimum and maximum sizes, however, can change from year to year. Size is influenced by resource availability, habitat changes, adult mortality, and settlement patterns. Somewhere along the gradient of too small and too large is an optimal size for a territory, one from which the owner gains maximum benefits for the costs incurred (see Figure 12.14).

For example, the territories of the golden-winged sunbird (*Nectarina reichenowi*) vary greatly in size and in floral composition, but each territory contains just enough of a nectar supply to meet an individual's daily energy requirement (Gill and Wolf 1975). Such flexibility in the size of territories has been likened by J. Huxley (1945) to an elastic disk compressible to a certain size. Territory size decreases as density increases, but when the territory compresses to a certain size, the resident resists further compression and denies access to additional settlers (see Getty 1981, Stamps and Krishnan 1990). Because aggressive behavior varies among individuals, the most aggressive have the advantage and the less aggressive are forced to settle elsewhere.

This settlement pattern relates to the theory of ideal free distribution (Fretwell and Lucas 1970). It holds that at the outset all individuals of equal competitive abilities can move freely between habitats and settle where they can gain the greatest benefits. As the best habitats fill some individuals would do better by moving to suboptimal habitats. As suggested, this situation rarely occurs. All individuals do not receive equal rewards nor do they experience free distribution. In a given landscape a range of habitats exists from the best to the poorest (Figure 12.15). If habitat patch *A* is optimal for foraging and reproduction, individuals will settle there first. At low population densities all individuals will be able to settle in this optimum habitat. When population density increases, dominant and highly competitive individuals will exclude others from this optimum habitat. Subdominant individuals will have to occupy the poorer habitat patches *B*, where they may successfully breed. In turn, as *B* becomes filled surplus individuals are forced to settle in the poorest habitat *C*, which supports survival but not successful breeding. Thus in real-life situations, competitors are not equal and habitats vary in quality. Thus the distribution is more an ideal preemptive distribution rather than ideal free distribution (Pulliam and Danielson 1991).

Floaters As a result of contest competition among males for space, some individuals are denied territory. Thus a portion of the population does not reproduce because they are excluded from the suitable breeding sites by territorial individuals. They

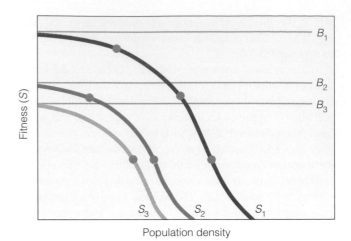

FIGURE 12.15 Relationship among habitat suitability, population density, and fitness. The suitability of habitat is ranked from B_1, the best, to B_3, the poorest. The curves from S_1, the highest fitness, to S_3, the lowest, show the fitness of individuals in habitats at different population densities. At low population density all individuals will be able to settle in optimum habitat B_1. At intermediate density some individuals will be forced to settle in habitat B_2. At high density all three habitats will be settled. The model implies that as density increases, habitat suitability for part of the population decreases. (After Fretwell and Lucas 1969:24.)

make up a surplus population, a **floating reserve** that would be able to reproduce if a territory became available to them.

Such a floating reserve of potentially breeding adults has been described for a number of species. The number of floaters may be great. Studies of a banded white-crowned sparrow population (*Zonotrichia leucophrys nuttalli*) in California indicated a nonbreeding surplus of potentially breeding birds (Petrinovich and Patterson 1982). In fact, 24 percent of the territory holders entered the breeding population two to five years after banding, and 25 percent of the nestlings that acquired territories did so two to five years after their birth. Territory holders that disappeared during the breeding season were quickly replaced.

Although the existence of floating reserve populations is acknowledged, few data exist on the social organization and behavior of surplus birds. Floaters may live singly off the territories as white-crowned sparrows do; or the floaters may form flocks with a dominance hierarchy on areas not occupied by territory holders, as the floating populations of red grouse do on the heather-dominated moors of Scotland (Watson and Moss 1971, 1972).

Floaters in some species live on the breeding territories of other individuals. An example of this strategy is provided by the detailed studies by S. M. Smith (1978) of the rufous-collared sparrow (*Zonotrichia capensis*) in Costa Rica (Figure 12.16). By observing banded birds, both territorial and nonterritorial, and by selectively removing certain individuals, Smith determined the role of the floater or "underworld" bird. Territorial sparrows on her study area occupied small territories ranging from 0.05 to 0.40 ha and made up 50 percent of the total population. The other 50 percent, underworld birds consisting of both males and females, lived in well-defined restricted home ranges

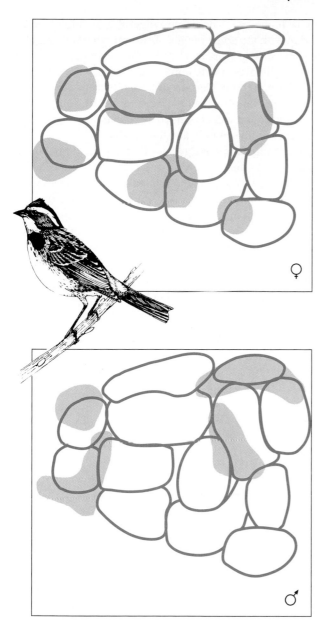

FIGURE 12.16   Territorial boundaries of rufous-collared sparrows (*Zonotrichia capensis*). Home ranges (shaded) of "underworld" females and males are superimposed on occupied territories. Eventually some of the females occupied a territory by replacing a missing bird. (Adapted from Smith 1978:570.)

within other birds' territories. Male home ranges, often disjoined, embraced three or four territories. Female home ranges were usually restricted to a single territory. Because home range boundaries of both sexes coincided with territorial boundaries, each territory held two single-sex dominance hierarchies of floaters, one male and one female. When a territory owner, male or female, disappeared, it was quickly replaced by a local underworld bird of appropriate sex on the territory. These floaters usually entered the territories as young birds hatched some distance away and were tolerated by the owners.

A similar situation exists among some territorial spiders (Reichert 1981) and the aphid *Pemphigus* (Whitham

1987). They, too, have a floating reserve of individuals who quickly claim vacated sites. Among the spiders, the floaters live in cracks and crevices within occupied territories. Periodically, floaters will unsuccessfully attempt to take over an occupied habitat.

Population Regulation   A consequence of territoriality can be population regulation. If no limit to territorial size exists and all pairs that settle on an area get a territory, then territoriality results only in spacing out the population. No regulation of the population results. If territories have a lower limit in size, then the number of pairs that can settle on an area is limited. Those that fail to get a territory have to leave. Thus territoriality might limit population breeding density over the short term, but it cannot regulate densities over a period of years (Newton 1998).

For example, in the arctic ground squirrel (*Spermophilus undulatus*) all females are allowed to nest. However, territorial polygamous males drive excess males from the colony into submarginal habitat, where they exist as a nonbreeding floating population. The number of breeding males remains constant because losses are continually replaced from the floating population. This floating population decreases drastically over the year, largely from predation and weather. In this species, territoriality stabilizes the number of breeding males but does not regulate the population, because only males appear to be surplus.

Krebs (1971) removed breeding pairs of great tits (*Parus major*) from their territories in an English oak woodland. The pairs were replaced by new birds, largely first-year individuals, that moved in from territories in hedgerows, considered suboptimal habitat (Figure 12.17). The vacated hedgerow territories, however, were not filled, suggesting that a floating reserve of nonterritorial birds did not exist. In this case territorial behavior limits the density of breeding birds in optimal

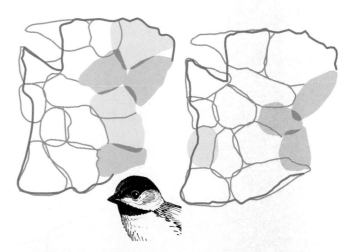

FIGURE 12.17   Replacement of removed individuals and settlement of vacated territory. Six pairs of great tits (*Parus major*) were removed between March 19 and March 24, 1969 (left shaded area). Within three days, four new pairs had taken up territories (right shaded area) and some residents had expanded their territories, so that territories again formed a complete mosaic over the woods. (From Krebs 1971:7.)

habitat, but does not regulate the population, because some birds are breeding in suboptimal habitat.

To demonstrate that territorial behavior limits a population, we have to show that a portion of a population, including males and females, does not breed because it is excluded from suitable breeding sites by dominant or territorial individuals. These surplus individuals should be able to breed if territories became available to them. Such is the case with the white-crowned sparrows, rufous-collared sparrows, and red grouse. The size of the reproductive population is limited by territoriality, and density-dependent population regulation results.

HOME RANGE

Territorial or not, all animals possess a **home range**—an area over which an animal lives, seasonally or throughout the year. Part of a home range may be defended, for example, about a

nest or home site. The home range of one individual may overlap with the home ranges of several other individuals of the same species (Figure 12.18). In these situations individuals living in overlapping home ranges experience some form of social hierarchy. Certain individuals are dominant and exert some control over resources in the areas of overlap. Subdominant individuals tend to avoid contact with dominant individuals both spatially and temporally.

Generally, home ranges do not have fixed boundaries. Seldom is a home range rigid in its use, size, and establishment. It may be compact, continuous, or broken into two or more discontinuous parts reached by trails and runways. Irregularities in distribution of food and cover produce corresponding irregularities in home range and in frequency of animal visitation. The animal does not necessarily visit every part daily. Its movements may be restricted to trails and most of its activities may be concentrated in a smaller core area used more intensively than other parts.

FIGURE 12.18 (a) Home ranges of paired female and male kit foxes (*Vulpes velox macrotis*) in western Arizona, determined by observations of marked animals. Each pair of foxes shares much the same home range, which suggests a degree of territoriality. (b) Home ranges of two unpaired kit foxes overlap the home ranges of two paired males. Much of the overlap resulted from the movement of nonpaired males during the breeding season to dens used by females of the other pairs. (From Zoellick and Smith 1992:86.)

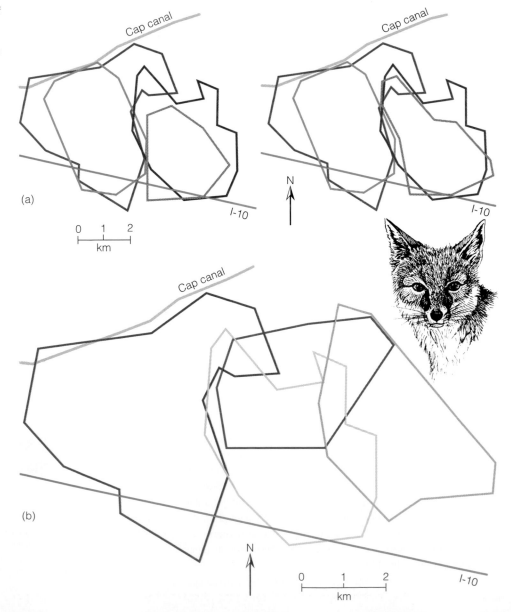

The size of a home range among mammals relates to body size (McNab 1963, Harestad and Bunnell 1979). Large mammals have larger home ranges than smaller ones, and carnivores generally have larger home ranges than herbivores and omnivores of similar size. Males and adults have larger home ranges than females and subadults. Difference in weight alone is sufficient to account for this within a species without invoking any competitive interactions. The home range of herbivores and omnivores increases at a nearly constant rate as body weight increases; among carnivores the home range increases at a greater rate as body weight increases (Figure 12.19).

Like territoriality, possession of a home range confers certain advantages. The animal becomes familiar with the local area, knowing where to find food, shelter, and cover from enemies with a minimum expenditure of energy. It can define a series of escape routes to cover and travel routes to food sources throughout the year.

KEY FACTOR ANALYSIS

How do we determine what density-dependent influences are at work in a given population? One method is **key factor analysis.** A key factor is a biological or environmental condition associated with mortality that causes major fluctuations in population size. Key factor analysis is based on a k value derived from the life table. Related to the mortality rate q, k (sometimes called killing power) is defined as $\log_{10} l_x - \log_{10} l_{x+1}$. The k value, unlike q, has the advantage of being additive. The summation of k values over age classes provides K, the total killing power, which reflects the rate or intensity of mortality. Like l_x, it is comparable between populations.

Key factor analysis has been useful in the study of insect populations that have discrete generations and life stages to which mortality can be assigned. It is more difficult to use in populations with overlapping generations. However, with modifications it can be used to detect when regulation may be occurring in the life cycle and to aid in the search for the causes of mortality.

To demonstrate how key factor analysis works, we can use the data from the life table of the gypsy moth. The first figure in the k factor table (Table 12.2) is the maximum potential natality for each generation. It is determined by multiplying the number of females of reproductive age by the maximum number of eggs per female. For our example we will simply consider the maximum fecundity for one female gypsy moth in a sparse population as 800 eggs. The l_x value of each successive stage of the life cycle as it appears in the life table is entered, and the values are converted to logarithms. Each logarithm is subtracted from the previous one to give a k or mortality value for each age class. These values are added to give a total generation mortality K. This is done for a number of successive generations. To identify the key

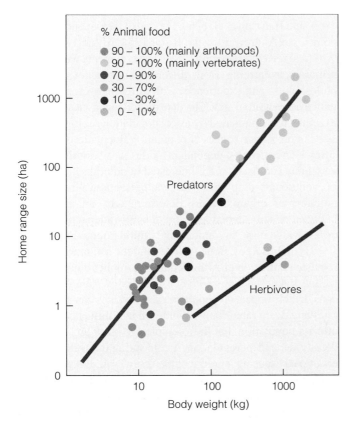

FIGURE 12.19 Relationship between the size of home range and body weight of North American mammals. (Adapted from Harestad and Bunnell 1979:390.)

TABLE 12.2 *k* Values for a Gypsy Moth Population Based on Life Table for a Sparse Population in Connecticut

Age	l_x	Logarithm of l_x	k Value	Cause of Morality
Maximum natality	800.0	2.903		
Eggs laid	550.0	2.740	0.163	Parasites, other
Instars (larvae) I–III	385.9	2.585	0.155	Dispersion
Instars IV–VI	242.0	2.384	0.201	Parasites, disease
Prepupae	14.6	1.164	0.122	Dessication
Pupae	11.7	1.068	0.096	Vertebrate predation
Adult	12.4	0.146	0.922	Natural mortality
			$K = 2.757$	

factor that influences trends in adult populations, the k values for each successive generation are plotted along with K. The plot shows whether the mortality rate for one particular stage or age class consistently displays over the generations a strong correlation with total mortality. Such graphs are shown for the great tit and winter moth in Figure 12.20. If

there is some correlation between the k value of a particular stage and total K, then the analysis can be carried further to determine the k factor within that stage. For details on the use of key factor analysis, see Dempster 1975.

DENSITY-INDEPENDENT INFLUENCES

During the late 1950s and early 1960s not all ecologists accepted the idea that populations were regulated by density-dependent influences. Many argued that density-independent mechanisms were more important. (For density-dependent viewpoints, see Nicholson 1954, 1956, Solomon 1957, Lack 1966, and Hairston et al. 1960. For density-independent viewpoints, see Andrewartha and Birch 1954, Milne 1957. For a historical review, see Krebs 1985.) The arguments were largely semantic, stemming from different approaches, different philosophies, and experiences with different taxa, all infused with a dose of advocacy. The density-independent school, for example, was dominated by insect population ecologists interested in proximate causes of population fluctuations and densities. The density-dependent side was dominated by vertebrate ecologists more interested in natural selection and in evolutionary problems. Most ecologists now agree that the numbers of organisms are determined by an interaction between density-dependent and density-independent influences, which may vary among and within populations.

By themselves, density-independent influences do not regulate population growth. Regulation implies a homeostatic feedback that functions with density. However, they can have considerable impact on population size and they can affect birth rates and death rates. Density-independent influences may so affect a population that they completely mask any effects of density-dependent regulation. A cold spring may kill the flowers of oaks, causing a failure of the acorn crop. Because of the failure, squirrels may experience widespread starvation the following winter. Although the proximate cause of starvation is the density of squirrels and the meager food supply, weather is the ultimate cause. In general, population fluctuations influenced by annual and seasonal changes in the environment tend to be irregular and correlated with variations in temperature and moisture. Conditions beyond the organisms' limits of tolerance can have a disastrous impact, affecting growth, maturation, reproduction, survival, movements, and dispersal of individuals within a population and even eliminating local populations.

In general the influence of weather is stochastic: it is irregular and unpredictable and it functions largely by influencing the availability of food. Pronounced changes in population growth often can be correlated directly with variations in moisture and temperature. For example, outbreaks of spruce budworm (*Choristoneura fumiferana*) are usually preceded by five or six years of anticyclonic weather characterized by low rainfall and high evaporation; the outbreaks end when wet weather returns. Such density-independent effects can take place on a

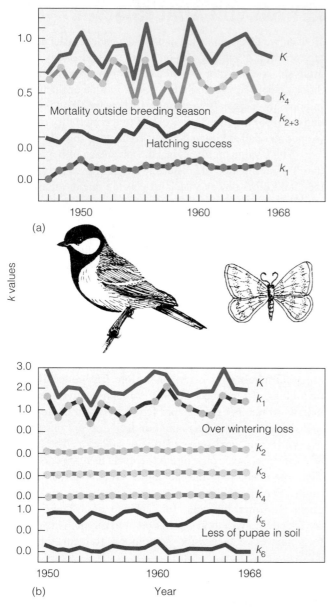

FIGURE 12.20 Key factor analysis applied to the great tit (*Parus major*) and the winter moth (*Operophtera brumata*). (a) The key factor in the life cycle of the great tit in Marley Woods, Oxford, is k_4, mortality outside of the breeding season. In the rest of the annual life cycle, variations in clutch size (k_1) and hatching success (k_{2+3}) are density-dependent and are sufficient to regulate the population. (b) Key factors in the life cycle of the winter moth are k_1, overwintering loss of winter moth eggs and larvae before the first larval census in spring, and k_5, a density-dependent loss of the pupae in the soil due to predation in the spring. (From Podoler and Rogers 1975:97–101.)

local scale where topography and microclimatic conditions influence the fortunes of local populations.

Consider the San Francisco Bay checkerspot butterfly (*Euphydryas editha bayensis*). Adults emerge from pupae from mid-March to early May and lay their eggs at the base of the plantain (*Plantago erectus*). Within two weeks the eggs hatch, and for three weeks the larvae feed on the plantain and owl's clover (*Orthocarpus densiflorus*), until they reach the third instar. Now the dry season sets in and the larvae go into diapause until December. Diapause is broken by the winter rains, which also stimulate the germination and growth of the food plants. The larvae go through three more instars before they pupate and emerge as adults. During a two-year period, 1983 and 1984, record rainfalls and cool weather retarded the development of the larvae, delaying the flights of the adults and the deposition of eggs. Meanwhile the food plants *Plantago* and *Orthocarpus* were developing normally. By the time the eggs hatched the host plants were already senescent or dying. Lacking food, the prediapause larvae faced heavy mortality from starvation. Normally the south-facing slope experiences earlier plant growth and faster development of larvae because of the warmer microclimate, while the north-facing slope experiences about a two-week delay in growth of food plants. Because of the delay in their development, the larvae were confronted with senescent plants on the south-facing slope, while the host

plants on the north-facing slope were not available for another week. If they had been available, the plants would have died before the larvae reached the third instar. The following year only 18 adults were captured (Dobkin et al. 1987). In such a manner microclimatic changes can lead to sharp declines or extinctions in local populations.

Deer in the northern part of their range are sensitive to severe winters. In Minnesota, Mech et al. (1987) found a significant relationship between snow accumulation over the previous three years and viability of offspring, as indicated by fawn-doe ratios and percent change in the deer population (Figure 12.21). The data show a winter-to-winter carryover. Snow limits food, affecting prenatal nutrition over winter and in utero development in two-year-old deer. When the sum of a three-year average of snow accumulation exceeds 10.2 m and an average single year accumulation exceeds 340 cm, one can expect fewer fawns and a decline in deer population compared with winters with less snow.

In desert regions a direct relationship exists between precipitation and rate of increase in certain rodents and birds. Merriam's kangaroo rat (*Dipodomys merriami*) occupies lower elevations of the Mojave Desert. The kangaroo rat has the physiological capacity to conserve water and survive long periods of aridity. However, it does require in its environment a level of moisture sufficient to stimulate the growth of herbaceous desert plants in fall and winter. The kangaroo rat

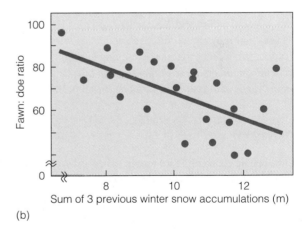
(b)

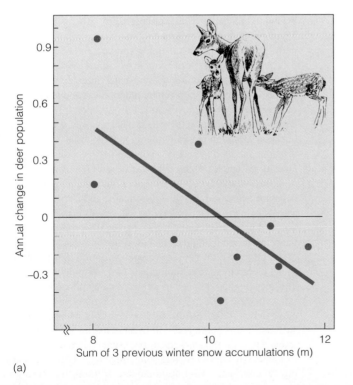
(a)

FIGURE 12.21 The relationship between the sum of the previous three winter monthly snow accumulations in northeastern Minnesota and the population of white-tailed deer. (a) Fecundity (fawn:doe ratio). (b) Percent annual change in next winter's population. (From Mech et al. 1987:619, 622.)

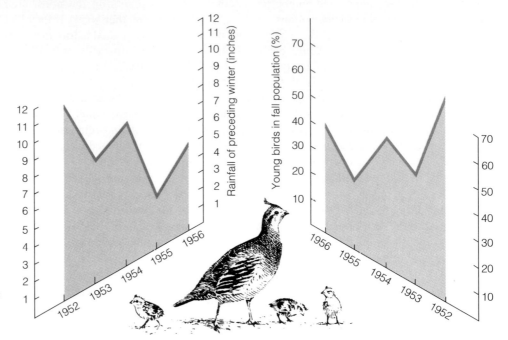

FIGURE 12.22 Relationship of winter rainfall to the percentage of young the following year in a Gambel's quail population in southern Arizona. Note how the production of young follows rainfall. (After Sowls 1960:187.)

becomes reproductively active in January and February when plant growth, stimulated by fall rains, is green and succulent. Herbaceous plants provide a source of water, vitamins, and food for pregnant and lactating females. If rainfall is scanty, annual forbs fail to develop and the production of kangaroo rats is low (Beatley 1969, Bradley and Mauer 1971). This close relationship to seasonal rainfall and relative success of winter annuals is also apparent in other rodents occupying similar desert habitats and in Gambel's quail (*Callipepla gambelli*) (Figure 12.22) and scaled quail (*Callipepla squamata*) (Francis 1970).

Summary

1. Intraspecific competition for resources in short supply is a density-dependent mechanism in the regulation of population numbers. There are two basic types of competition. In scramble competition for resources in short supply, all individuals have equal access to the resource and each attempts to get a part of it. In extreme cases scramble competition results in each individual obtaining insufficient amounts to survive and reproduce. In contest competition successful individuals usurp the resource and the unsuccessful are denied access to it. Contest competition is characteristic of species whose individuals can defend a resource from others.

2. Among animals, intraspecific competition can result in density-dependent reduction in growth and fecundity, delayed maturity, and an increase in mortality, especially of the young.

3. Among plants, response to density results in the declining number of individuals and an increase in the mean weight of surviving individuals. Regardless of starting densities, monospecific, even-aged populations of a plant species will arrive at a point at which growth compensates for loss through mortality. This relationship is expressed in the −1.5 self-thinning law.

4. Among animals, some mechanisms of intraspecific competition involve physiological and behavioral interactions among members of the population. Increased density may affect population growth through physiological responses of individuals. Crowding may produce stresses that can result in abnormal behavior, reduced growth, and infertility.

5. Social pressure and crowding may also induce emigration or dispersal. However, dispersal appears to be density-independent except at the highest population levels; most dispersal appears to occur when populations are increasing rather than at the peak population levels. Dispersal may function in population regulation by encouraging mostly subadults and possibly some subdominant individuals to leave their natal area and occupy vacant habitats. Although risks in moving may be great, successfully dispersing individuals improve their fitness by doing so.

6. Density influences social behavior of individuals in a population. Basic social behavior involves aggressive and submissive behavior expressed as social dominance and territoriality that act as spacing mechanisms among individuals in the population.

7. Social dominance results in the sharing of space and maintenance of interindividual distances. Some individuals are dominant; other exhibit degrees of subordination that result in a social hierarchy among individuals in the population.

8. Territoriality, the defense of a fixed area against others of the same species, divides space among individuals and excludes other individuals. Territoriality occurs only when a resource such as food is contained within an economically defensible space. Once established, territorial owners defend their space by aggressive signals such as song, display, or chemical marking. In contrast, social dominance results in the sharing of space and maintenance of interindividual distances, but with some individuals dominant. Territoriality functions in population regulation only if it creates a surplus population consisting of sexually mature individuals prevented from breeding by territory holders.

9. Nonterritorial animals occupy home ranges, the areas they occupy seasonally or throughout the year. Individuals may have overlapping home ranges, the use of which at any time may be influenced by a social hierarchy. Subordinate individuals avoid contact with dominant ones by using different parts of a shared area or home range and by using the area at different times.

10. One approach to the study of density-dependent mortality in animal populations is key factor analysis. A key factor is a biological or environmental condition associated with mortality that causes a major fluctuation in population size.

11. Density-independent influences affect but do not regulate populations. They can reduce local populations, even to the point of extinction, but their effects do not vary with density.

Review Questions

1. Distinguish between density-dependent regulation and density-independent influences.
2. Define competition. What is scramble competition? Contest competition?
3. How can density-independent influences affect density-dependent mechanisms?
4. How does population density influence individual and population growth in animals? In plants?
5. How might stress influence population growth in mammals? In plants? How do plants respond to stress?
6. What are some of the costs and benefits of dispersal?
7. Under what conditions might dispersal aid in population regulation?

8. Comment on the remark: "If this woods is cut, the animals will just move elsewhere." What is the probable fate of the displaced individuals and why?
9. Distinguish among social dominance, territoriality, and home range.
10. How is dominance expressed in social hierarchy?
11. How can social hierarchy and territoriality function in population regulation?
12. What is territoriality? How does it relate to social hierarchy?
13. How can territorial behavior increase the fitness of the male holder of the territory? The female? Speculate on the idea that the female may choose the territory and take the male that goes with it rather than choose the male.
14. What are the benefits and costs of territoriality?
15. What is a floating reserve? How does it function?
16. Look up the home range sizes of such large mammals as the elephant, grizzly bear, mountain lion, and others and then discuss the problem of maintaining viable populations of these wide-ranging mammals on such limited areas as national parks.
17. Explain key factor analysis. Why is it useful?

Cross References

Plant responses to moisture, 104–112; animal responses to moisture, 137–139; genetic drift, 373–378; plant-herbivore interactions, 286–287; habitat fragmentation, 450, 458; chemical defense in plants, 286–287; chemical defense in animals, 293; mating systems, 216–287.

C H A P T E R   1 3

Life History Patterns

CONCEPTS

1. Reproduction may be asexual or sexual. Sexual reproduction involves the production of gametes by males and females, either as separate individuals or as hermaphrodites.

2. Mating systems are broadly defined as monogamous and polygamous.

3. Sexual selection involves intrasexual and intersexual components.

4. Parental allocation of reproductive costs balances selection for fecundity with selection for survival of parents and offspring.

5. Equal investment in sons and daughters can result in a biased sex ratio in favor of the less expensive sex.

6. *r*-traits and *K*-traits describe differences in life history patterns.

7. Animals use environmental cues to select habitat.

Reproduction is the major vocation of all living things. By transmitting genetic characteristics from one generation to another, individuals maximize the numbers of their descendents. All their activities, their life history patterns, and traits are directed toward that one end. It colors the way in which individuals select and acquire a mate, their mode of reproduction, the number of young produced, and how often. It is the underlying reason for bright colors and fragrances of flowers, the songs of birds, the flashing of fireflies by night, the antlers of deer.

This concentration on activities and morphologies, which most of us rarely relate to sexual activity and reproduction, is metabolically expensive to individuals. Parents have to make "decisions" about the amount of energy to allocate to growth, defense, and reproduction. How organisms budget these energy expenditures has intrigued behavioral ecologists. How do individuals meet these costs and allocate them to reproduction and survival? How has sexual selection and the drive to contribute to a new generation affected the fitness of individuals? These are questions explored in this chapter.

PATTERNS OF REPRODUCTION

How organisms form new individuals falls into two categories: asexual and sexual reproduction. Asexual reproduction, which takes many forms, results in the formation of new individuals genetically the same as the parents (see Chapter 19). Paramecia reproduce by dividing in two. Hydras reproduce by budding; strawberry plants spread by runners. Aphids produce eggs by normal cell division or mitosis that develop

into female adults without fertilization (parthenogenesis). However, organisms that rely heavily on asexual reproduction turn on occasion to sexual reproduction. Hydras at some time in their life cycle produce egg and sperm. Plants produce flowers. At the end of the summer, aphids resort to sexual reproduction, halving the chromosomes and producing males.

Sexual reproduction is common to eukaryotic organisms. It involves the production of haploid gametes, egg and sperm that combine to form a diploid cell or zygote. This halving and recombination of genes allows the gene pool to become mixed. Mixing supplies the genetic variability necessary to meet changing selective pressures, and to prevent an accumulation of harmful mutations (see Chapter 19). For the individual, however, sexual reproduction is expensive. Each individual can contribute only one-half of its genes to the next generation. Because the success of that contribution depends on the contribution gained from a member of the opposite sex, it follows that each individual should acquire the best possible mate.

Sexual reproduction can take a variety of forms. The most familiar involves separate male and female individuals. It is common to most animals. Plants with that characteristic are called **dioecious;** examples are holly trees and stinging nettle (*Urtica* spp.). In such plants, fertilization takes place with the fusion of a large nonmobile female egg and a small mobile male gamete. Some individual organisms possess both male and female organs. They may be **monoecious** or **hermaphroditic.** In plants, individuals can be hermaphroditic either by having bisexual flowers with both male (stamens) and female (ovaries) organs, as in lilies and buttercups, or by being monoecious (Figure 13.1). In monoecy,

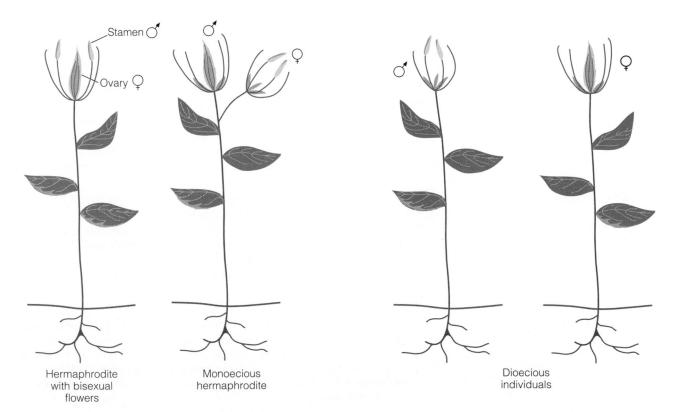

Hermaphrodite with bisexual flowers

Monoecious hermaphrodite

Dioecious individuals

FIGURE 13.1 Floral structure in dioecious and monoecious (hermaphroditic) plants.

the individual plant possesses separate male and female flowers, as do birch (*Betula*) and hemlock (*Tsuga* spp.) trees. Among animals, hermaphroditic individuals possess both testes and ovaries, a condition common to some invertebrates, such as earthworms, and some fish. Some hermaphrodites are simultaneous, as earthworms are; others are sequential. The latter type are one sex when young or small and develop into the opposite sex when mature or large, as snails and jack-in-the-pulpit (*Arisaema* spp.) do.

MATING SYSTEMS

On a brushy rise of ground at the edge of a woods, a pair of red foxes has a deeply dug burrow within which are the female and her litter of pups. Outside at the burrow entrance are scattered bits of fur and bones, the leftovers of meals carried to the den by the male for his mate and pups. Back in the woods, a doe has a young dappled fawn hidden in a patch of ferns on the forest floor. The fawn's father is nowhere near and has no knowledge of this offspring, sired during a short encounter of a few days the previous fall, along with several others.

The monogamous fox and the promiscuous deer represent two approaches to a system of mating. A **mating system** includes such aspects as the number of mates males and females acquire, the manner in which they are acquired, the nature of the pair bond, and the pattern of parental care provided by each sex. The structure of mating systems ranges from monogamy through many variations of polygamy. Mating systems may even vary within a species, involving different strengths of pair bonds (Table 13.1).

The nature and evolution of male-female relationships are influenced by ecological conditions, especially the availability and distribution of resources and the ability of individuals to control access to mates or resources. If the male has no role in the feeding and protection of young, and defends no resources available to them, the female gains no advantage by remaining with the male; and it makes no sense to the male to remain with her. If the habitat is sufficiently uniform so that little difference in territorial quality exists, the number of young raised in the poorer habitat is only slightly less than the number reared in the best. Selection would favor monogamy because female fitness in both would be nearly the same. However, if the habitat is diverse, with some parts more productive than others, competition may be intense and some males will settle on poorer territories. Under such conditions, a female may find it more advantageous to join another female in the territory of the male defending a rich resource than to settle alone with a male on a poorer territory. Selection under those conditions will favor a bigamous situation, even though the male may provide little aid in feeding the young.

TABLE 13.1 **An Ecological Classification of Mating Systems**

Monogamy	Neither sex has opportunity of monopolizing additional members of the opposite sex. Fitness often maximized through shared parental care. (Fox, beaver.)
Polygyny	Individual males frequently control or gain access to multiple females.
Resource defense polygyny	Males control access to females *indirectly,* by monopolizing critical resources. (Hummingbirds, redwinged blackbird.)
Female (or harem) defense polygyny	Males control access to females *directly,* usually by virtue of female gregariousness. (Elk, wild horses, seals.)
Male dominance polygyny	Mates or critical resources are *not economically monopolizable.* Males aggregate on leks during the breeding season and *females select mates* from these aggregations. (Sage grouse, prairie chicken.)
Explosive breeding assemblages	Both sexes converge for a short-lived, highly synchronized mating period. The operational sex ratio is close to unity and sexual selection is minimal. (Frogs, insects.)
Rapid multiple-clutch polygamy	Both sexes have substantial but relatively *equal* opportunity for increasing fitness through multiple breedings in rapid succession. Males and females each incubate separate clutches of eggs. (Some shorebirds.)
Polyandry	Individual females frequently control or gain access to multiple males.
Resource defense polyandry	Females control access to males *indirectly,* by monopolizing critical resources. (Spotted sandpiper, jacana.)
Female access polyandry	Females do not defend resources essential to males but, through interactions among themselves, may limit access to males. Among phalaropes, both sexes converge repeatedly at ephemeral feeding areas where courtship and mating occur. The mating system most closely resembles an explosive breeding assemblage in which the operational sex ratio may become skewed with an excess of females. (Phalaropes, small arctic shorebirds.)

Source: Emlen and Oring 1977.

Monogamy

Monogamy is the formation of a pair bond between one male and one female. It is most prevalent among birds and rare among mammals, except several carnivores, such as the fox and mustelids, and a few herbivores, such as the beaver.

Monogamy occurs mostly among those species in which cooperation by parents is needed to rear young successfully. Nearly 90 percent of all species of birds are seasonally monogamous, because in most birds the young are helpless at hatching and need food, warmth, and protection. The avian mother is no better suited to provide these needs than the father. Instead of wandering off to other females and seeking other mates, the male can increase his fitness more by continuing his investment in the young. Without him, the young carrying his genes may not survive. Among mammals the situation is different. The females lactate, providing food for the young. Males often can contribute little or nothing to the survival of the young, so it is to their advantage to mate with as many females as possible. Among the exceptions are the canids, among which the male provides for the female and young and defends the territory. Both male and female are able to regurgitate food for the weaning young.

Monogamy, however, has another side, **extra-pair copulations.** Among some 100 species of monogamous birds, both male and female cheat while each usually maintains a primary relationship (Mock and Fujioka 1990; Davies 1986; Davies and Houston, 1986). By engaging in extra-pair relationships, the female may increase her fitness by rearing young sired by two or more males. The male increases his fitness by having his offspring produced by several females.

Polygamy

Polygamy means acquiring two or more mates, none of which is mated to other individuals. It can involve one male and several females or one female and several males. A pair bond exists between the individual and each mate. When one member of the pair is freed from parental duty, partly or wholly, the emancipated member of the pair can devote more time and energy to intrasexual competition for more mates and resources. The more unevenly such critical resources as food or quality habitat are distributed, the greater is the opportunity for a successful individual to control the resource and thus available mates.

The number of members of the other sex an individual can monopolize depends on the degree of synchrony in sexual receptivity. For example, if females in the population are sexually active for only a brief period, as with the white-tailed deer, the number a male can monopolize is limited. However, if females are receptive over a long period of time, as with elk, the number a male can control depends on the availability of females and the number of mates the male can energetically defend.

Such variability in environmental and behavioral conditions results in various types of polygamy. There are two basic forms: **polygyny,** in which an individual male gains control of or access to two or more females; and **polyandry,** in which an individual female gains control of or access to two or more males. A special form of polygamy is **promiscuity,** in which males and females copulate with one or many of the opposite sex and form no pair bonds. Emlen and Oring (1977) have classified these types of polygamous relationships according to the means by which individuals gain access to the limiting sex (see Table 13.1).

In polyandry the female rather than the male is the competitive individual. In resource defense polyandry, best developed in two orders of shorebirds, Gruiformes (cranes and rails) and Charadriiformes (shorebirds), the female competes for and defends resources essential for the male (Oring and Louk 1986). As in polygyny, this mating system depends on the distribution and defensibility of resources, especially quality habitat. The production by females of multiple clutches, all of which require the brooding services of a male, leads to competition among females for access to available males. After the female lays a clutch, the male begins incubation. Then he becomes sexually inactive and is effectively removed from the male pool, resulting in a scarcity of available males.

SEXUAL SELECTION

Choosing a proper mate is essential if a plant or animal is to contribute genetically to the next generation. The tumult of bird song in spring, the frenzy of frogs calling in ponds, and the clashes among stags in the season of rut—all focus on sex. All this activity seems to involve males attempting to attract females; but at the same time, more quietly perhaps, females are also seeking out males. In any population there are just so many males and females. Because males are not as selective with whom they mate, females usually have no trouble finding a partner. Females, however, are selective, and males must prove their fitness; those who cannot go mateless. The result is intense rivalry among males for female attention (Figure 13.2). The outcome of male rivalry, called **intrasexual competition,** may indeed give the victors access to females. However, success depends not so much on who wins as on how the competition affects their acceptance by choosy females. In the end the female determines which male will sire her offspring.

Supposedly, males (and in some situations females) that win the intrasexual competition are the fittest. By selecting the fittest mate from among competing males, based on some specific characteristic during courtship, females attempt to assure their own fitness. Making such a choice is known as **sexual selection.** Ultimately sexual selection results in the evolution of morphological and behavioral traits that influence both competitive mating and mating systems.

Models of Sexual Selection

How does a female select a mate with the greatest fitness? That question still intrigues behavioral ecologists. Darwin (1871) attempted to answer it first. Because the elaborate and often outlandish plumages of birds and the horns, antlers, and large size of polygamous males seemed incompatible with natural selection, Darwin developed the concept of sexual

FIGURE 13.2   This bull elk is bugling a challenge to other males in a contest for control of a harem.

selection. He hypothesized that competition among males selected for weapons, body size, and striking plumage patterns. Females then selected males from among the winners, based on their appearance and behavior. Supposedly these traits reflected the highest fitness.

How does sexual selection square with natural selection when the traits selected by females seem to expose males to predation? The geneticist R. A. Fisher (1930) attempted to provide an explanation. He hypothesized that some novel genetic characteristic, such as a plumage pattern, that indirectly indicates maleness becomes the object of a female mating preference. Females who select males with this trait produce fitter progeny. The original selective advantage of the preferred characteristic paired with female preference for the characteristic increases the advantage accrued to the male possessing the trait. Positive feedback results in an accelerated increase or "runaway" both in the male trait and in the female preference for the trait, which may produce tracts favorable for mating but unfavorable the rest of the time.

Both of these hypotheses assume that mate competition is a male trait and that mate choice is a female trait. Trivers (1972) and, earlier, J. Maynard Smith (1956) suggested a more balanced view of sexual selection, based on the reproductive interests of the female (see Small 1992). The basic strategy for both male and female is to ensure their own maximum fitness. What increases male fitness is not necessarily what improves female fitness. Sperm is cheap. With little investment, males should mate with as many females as possible to achieve maximum fitness. By contrast, females invest considerably more in reproduction. It is to their advantage to be selective in choosing a mate, one who will pass on the best genes to the next generation.

This concept led to another explanation of sexual selection, advanced by the Israeli behavioral ecologist A. Zahavi (1975, 1977), the **handicap hypothesis.** It postulates the evolution of three characteristics: a male handicap, a female mat-

ing preference for the handicap, and a general viability trait. The handicap is a secondary male characteristic, such as bright plumage, that could reduce the male's survival. The viability trait is one that affects an individual's ability to survive, as by escaping a predator. If a male can carry handicaps and survive, it is proof of a superior genotype. Females showing preference for the handicapped male receive a selective advantage because their offspring carry genes for high viability. The handicap principle has sparked a good deal of controversy and research in sexual selection (Kodric-Brown and Brown 1984, Kirkpatrick 1986, 1987, Andersson 1994, Johnstone 1995).

Processes of Selection

J. Maynard Smith (1991) has nicely classified the processes of sexual selection into two major types: intrasexual selection and intersexual selection. Male-to-male (or in some cases female-to-female, as in some shorebirds) competition for the opportunity to mate with females is **intrasexual selection.** It leads to exaggerated secondary sexual characteristics, such as bright or elaborate plumage in male birds and antlers in deer. **Intersexual selection** is mostly female choice of a mate. Intersexual selection breaks down into parts, the female choice of a conspecific mate and the choice between conspecifics. Choice between conspecifics may be one of two types. In one the female makes a choice based on resources such as territory or access to food, which will improve her fitness. The other choice may involve genes only, as with many polygamous species. By selecting a male with exaggerated traits, the female will also be acquiring genes for highest fitness. The problem facing behavioral ecologists is finding out what females look for in a mate in each of these situations, and how they make their final choice.

Resource-Based Selection   For monogamous females the criterion for mate selection appears to be acquisition of a resource, usually a high-quality territory (for example, see Zimmerman 1971, Howard 1978b, Searcy 1979). The question is whether the female selects the male and accepts the territory that goes with him or whether she selects the territory and accepts the male that goes with it (see Petrie 1986). There is some evidence from the laboratory and the field that female songbirds base their choice, in part, on the complexity of the male's song. This evidence is based on studies in which hormone-implanted females (the hormone maintains the female in a sexually receptive condition during the experimental period) are exposed to the playback of actual and simulated conspecific bird song of varying complexity of males of their own species. These females responded with stronger solicitation displays to the more complex song types (Searcy et al. 1985, Searcy and Andersson 1986, Baker et al. 1986, Catchpole 1987) (Figure 13.3). In the field, males with more complex song repertoires appeared to hold the higher quality territories, so that the more complex song may convey that fact to the female (Figure 13.4). None of these studies, however, determined the fitness of females attracted to these males.

Another criterion of fitness in birds may be plumage. Female large cactus finches (*Geospiza conirostric*) on Isla Genovesa, Galapagos, select a mate on the basis of courtship

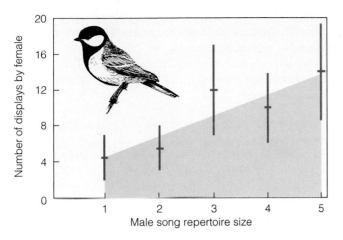

FIGURE 13.3 Mean number of copulation-solicitation displays given by 11 female great tits as a function of male song presented in repertoire sizes ranging from one to five song types. (From Baker et al. 1986:495.)

FIGURE 13.4 Male sedge warblers (*Acrocephalus schoenobaenus*) with more complex songs (larger repertoire of syllable types) attracted females for pairing before their rivals with simpler songs. Songs were recorded from a marked population. (From Catchpole 1980:150.)

performance and black plumage of males. Both black plumage and courtship behavior indicate older males with past breeding experience and proven viability (Grant and Grant 1987). By mating with experienced males, females gain a fitness advantage in production of young and their recruitment into the breeding population.

Such behavior relates to the bright plumage hypothesis advanced by Hamilton and Zuk (1982). This hypothesis proposes that only healthy males can develop bright plumage. Females use differences in the brightness of male plumage as a criterion of health. There is some evidence, based on the many studies stimulated by this model, that a relationship exists between infection of the male bird by coevolved endoparasites and the brightness of his plumage (Borgia and Collis 1989, Moller 1991, Zuk 1991, Saino et al. 1977, 1999). Males with the lowest parasitic infection have the brightest plumage. Females pick these males because they are the most disease-resistant. This hypothesis, controversial to say the least, still awaits an accumulation of data, intraspecific tests, and tests of alternative hypotheses (Zuk 1991).

Among polygamous species, the question becomes more complex. In those cases in which females acquire a resource along with the male, the situation is similar to that of monogamous relationships. Among birds, for example, the females show strong preference for males with high-quality territories (Plesczynska and Hansell 1980, Lenington 1980). On such territories with superior nesting cover and an abundance of food, females can attain reproductive success, even if this choice involves sharing the territory and the male with other females.

Among those fish species in which parental care of the eggs is an exclusive male function, females prefer to spawn with males already caring for another female's eggs, although the males give less care to adopted eggs (Sargent 1989). Males of some fish, such as the fantail darter (*Etheostoma fla-bellare*), capitalize on such female preferences by resorting to egg-mimicry. Territorial males excavate nests beneath flat rocks and care for eggs attached by the female to the ceiling

of the nest. Male darters have each of the seven to eight spines of the distal end of the first dorsal fin tipped with fleshy knobs that appear as eggs when the males are resting beneath the rocks. Female darters preferred males with eggs over males without eggs, and males with egg-mimics over males without them (Knapp and Sargent 1989).

Genes-Only Selection That females select males on the basis of resources they offer requires little explanation. What puzzles behavioral ecologists most is the basis of female choice in polygamy situations in which the males offer only genes and no resources. How does the female select the male with greatest fitness? This question has stimulated a body of growing research.

It would seem that the female has limited information with which to select a mate. She might select a winner from among males that best others in combat, as in bighorn sheep (*Ovis canadensis*), elk, and seals. She might select mates based on intensity of courtship display or some morphological feature that may reflect a male's genetic superiority and vitality. Whatever the situation, the selection process comes down to salesmanship on the part of the male and sales resistance on the part of the female (Williams 1975).

Among some polygamous species, such as elk and seals, a dominant male does control a harem of females. Even in this situation the females are not placid and do have some choice. Protestations by female elephant seals over the attention of a dominant male may attract other large males nearby, who may attempt to dislodge the male from the group. Such behavior ensures that females will mate only with the highest ranking male (Cox and LeBoeuf 1977).

In most cases females seem to control mate choice. Consider the long-tailed widowbird (*Euplectes progne*) of the African savanna. It is one of several species of widowbirds and whydahs (*Steganura* spp.) whose males possess highly elongated tails in their breeding plumage; but none can equal that of the long-tailed widowbird. The males are black with a red shoulder patch and a tail 50 cm long (Figure 13.5), which they display in courtship flight. A male may attract as many as six females (mottled brown, short-tailed birds the size of a sparrow) to their territories. A behavioral ecologist, M. Andersson (1982), investigated female choice in this species in an experiment involving males. From some males he cut the tail feathers to shorten them and glued these feathers onto the tails of other males to greatly lengthen them. On some males he clipped the feathers and glued them back in place, and others he left unclipped as controls. He quickly discovered that the females had a decided preference for the males with the artificially exaggerated tail feathers. These males attracted more mates and had a higher reproductive success (Figure 13.6) than both short-tailed and normal birds. Although the long tails did not seem to impair the males in territorial defense, Andersson did not determine whether the extra-long tails reduced the males' survival over time.

Such advertising can be costly for courting polygamous males, however. In the breeding season male three-spined sticklebacks (*Gasterostens aculeatus*), a small fish of marine and freshwater habitats in North America, Europe, and Pacific Asia, are polymorphic in the coloration of their throats. Some males have red throats and some do not. Females seem to prefer to mate with red-throated males. Semler (1971) presented females a choice between two genetically non-red males, one of which had been given an artificial red throat painted with nail polish. Females selected artificially colored males, even though they did not differ genetically from the plain-colored males. The polymorphism is probably maintained because red-throated males are subject to intense predation by rainbow trout. Thus a cost accrues to the red-throated males that acts against their reproductive advantage.

Lek Behavior   Extreme examples of genes-only female choice appear in the lek species. These animals aggregate into groups on communal courtship grounds called **leks** (or arenas) (see Payne 1984, Gosling 1986, Johnsqard 1994). Males on the lek defend very small territories that hold no resources and advertise their presence by colorful vocal and visual displays. Females visit the leks of displaying males, select a male, mate, and move on. Although few species engage in this type of mating system, it is widespread in the animal world, from insects and frogs to birds and mammals. The lek mating system is characterized by male defense of small clustered mating territories in an area where females have large overlapping ranges that the males cannot economically defend.

At least three hypotheses have been advanced to explain lek behavior (Davies 1991). One hypothesis, **female choice,** is that females show preference for a courtship arena because it is the safest place to mate or it forces males to cluster. Leks provide an unusual opportunity for females to choose a mate among the displaying males (Bradbury 1981). This advantage holds especially among lekking insects, such as certain tropical Drosophila, in which females are widely dispersed.

A related hypothesis is the **"hotspot" model** (Bradbury and Gibson 1983), which says males cluster in places where encounters with females are potentially high. If females have large overlapping ranges, males can ensure encounters with them by aggregating in overlap zones. For example, sage grouse (*Centrocercus urophasianus*) males display in leks near the females' wintering ranges (Figure 13.7). As females abandon the winter range for nesting areas, the males establish new lek areas near them (Bradbury et al. 1989).

A third hypothesis is the **"hotshot" model** (Beehler and Foster 1988). This model emphasizes an inequality of mating success. It envisions a strong hierarchy among the males with the dominant male, in smaller leks at least, displacing all others and leaving no opportunity for female choice. In other words, the visiting females select an arena and mate with the dominant males. Indeed, among some species this situation may be true, as in the case of the tropical manakins (Piprinae) and birds of paradise (*Paradisaea* spp.). Among these birds one dominant male on the lek may perform over 90 percent of successful copulations (Beehler 1983). In spite of the odds of not mating, males congregate on the lek because by displaying together they draw in females from a larger area. By associating with and congregating about "hotshot" males with the most effective displays, subdominant and satellite males may be able to steal mating opportunities. A majority of matings on the lek, however, are done by a small percentage of the males in the male dominance hierarchy formed in the absence of females.

The hotshot and hotspot models may not describe the real situation for many species. Consider the fallow deer (*Dama dama*), a lekking ungulate of Europe, whose repro-

FIGURE 13.5   The long-tailed widowbird (*Euplectes progne*), unmistakable with its extra long tail and bright red shoulder bordered with white, inhabits the open grasslands of eastern South Africa.

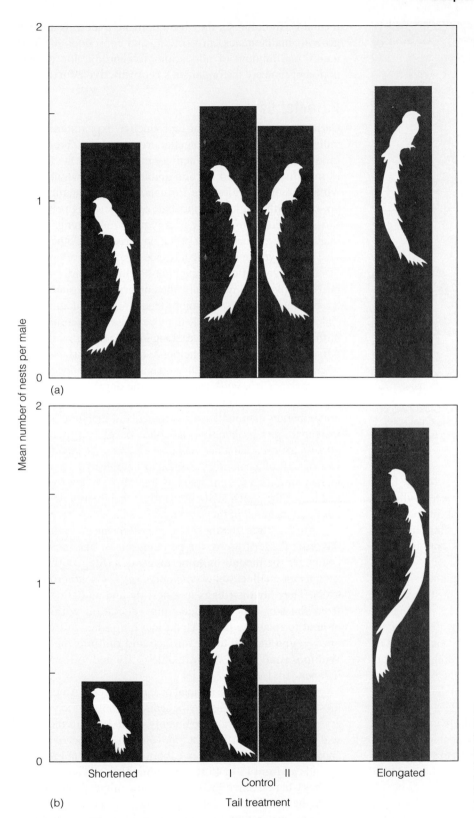

FIGURE 13.6 Mating success in male long-tailed widowbirds subjected to different tail treatments. (a) Mean number of active nests for the nine males of the four treatment categories before the experiment. (b) Number of new active nests in each territory after treatment of males. (From Andersson 1982:819.)

ductive behavior has been followed by Clutton-Brock and his associates (1988, 1989). The fallow deer may defend and mate in resource territories, especially at low population densities. At high densities the males commonly defend small territories about 15 to 20 m in diameter on traditional mating grounds. Females visit and move

through several territories before settling in one. The buck collects a harem of females, although each doe moves frequently among territories, rarely staying in one for more than several hours. The investigators found that bucks holding central territories possessed larger harems and had a higher mating success than those with holding edge

FIGURE 13.7 Male sage grouse (*Centrocercus urophasianus*) on a strutting ground or lek strut inflate the air sacs of the breast and spread their pointed tail feathers in courtship displays before the females.

territories. On the leks studied, the most successful male mated with 15 percent of 200 does and about 50 percent of the bucks on the lek did not mate at all. Thus it would appear that positions or hotspots on the lek could explain females' choices.

To test that hypothesis, the investigators pinned sheets of black polyethylene on the territories of the most successful bucks. Females refused to visit such territories, forcing the males to shift their territories and set up new display sites. The females moved with them, suggesting a strong choice of these particular males by the females. According to the hotshot model, males in territories adjacent to the original display site should shift their territories also, but they remained on the old lek. The results of these studies suggest that female fallow deer do choose their mates on the basis of male appearance and behavior and that territory position may not influence their choice. It is difficult, however, to separate territory position and mate choice, if the best males happen also to occupy the best sites.

REPRODUCTIVE EFFORT

Natural selection favors individuals that produce the maximum number of reproducing offspring in a lifetime. The size and number of offspring vary greatly among plants and animals. Whether the offspring are large or small, few or many, parents allocate a certain amount of resources and time to fecundity. In doing so, they have to make trade-offs among growth, maintenance, protection, and reproduction. The nature and amount of allocations to reproduction over a period of time are the organism's **reproductive effort.**

Parental Care

Caring for young is a major reproductive expenditure: providing food, shelter, and protection from predators, brooding, grooming, and any related activity that increases the fitness of the offspring. The kind and amount of care, which vary widely among species, are influenced by the maturity of young at birth. Some birds incubate eggs for a long period of time, investing considerable energy prior to hatching. Some mammals have much longer gestation periods than others and give birth to young in a more advanced stage of development.

The degree of development at hatching or birth among birds and placental mammals, the eutherians, translates into the two major types of young produced: precocial and altricial. **Precocial** young are able to move about at or shortly after birth, although they may have a long infancy and grow slowly. **Altricial** animals are born helpless, naked or nearly so, and often blind. They grow rapidly and mature early. Between the two extremes is a wide variation in the degree of development (Nice 1962, Derrickson 1992) (Table 13.2). Among noneutherian mammals (monotremes and marsupials), the marsupials give birth to tiny and poorly developed young that possess the minimal anatomical development needed to survive outside of the uterus. Attached to a nipple in the pouch or marsupium, the young marsupial continues its development and leaves the pouch at about the same weight as a newborn placental mammal of the same relative size.

Among birds, the most altricial offspring occur among the order Passeriformes, the perching birds. The most precocial are the mound-building megapods (Megapodiidae), such as the malleefowl (*Leipoa ocellata*) of southern Australia. They lay their eggs in a nest mound and allow heat from the sun and from decaying vegetation within the mound to incubate them. Upon hatching, the well-developed young burrow to the surface and run into the bush. Within several hours they are able to fly and to lead fully independent lives.

Most mammals are altricial or semiprecocial. Among the most precocial mammals are the seals. Not only is their delivery extremely rapid, approximately 45 seconds in the gray seal (Bartholomew 1959), but movements and vocalizations appear shortly after birth. Newly born fur seals are able to rise up and call from 15 to 45 seconds after birth and are capable of shaky but effective locomotion a few minutes later. Even while the umbilical cord is still attached, the pups are able to shake off water, nip at each other, and scratch dog-fashion with the hind flippers. The cows are protective and attentive only between parturition and estrus, about a week's time, although they nurse the young much longer.

Many species of fish, such as the cod (*Gadus morhua*) lay millions of floating eggs that drift free in the ocean with no parental care. Other species, such as bass, lay eggs in the hundreds and provide some degree of parental care. The qual-

TABLE 13.2 **Population Projection Table, Hypothetical Squirrel Population**

TYPE	CHARACTERISTICS	EXAMPLE
Feed Selves		
Precocials	Eyes open, down-covered, leave nest first day or two	
Precocials 1	Independent of parents	Megapods
Precocials 2	Follow parents but find own food	Ducks, shorebirds
Precocials 3	Follow parents and are shown food	Quail, chickens
	Fed by Parents	
Precocials 4	Follow parents and are fed by them	Grebes, rails
Semiprecocials	Eyes open, down-covered, stay at nest though able to walk	Gulls, terns
Semialtricials	Down-covered, unable to leave nest	
Semialtricials 1	Eyes open	Herons, hawks
Semialtricials 2	Eyes closed	Owls
Altricials	Eyes closed, little or no down, unable to leave nest	Passerines

Source: Adapted from Nice 1962.

ity of parental care appears to be related to egg size (Sargent et al. 1987). The larger the egg size relative to the size of the fish, the greater the amount of parental care. This extra care apparently relates to the longer time offspring take to develop and absorb the yolk sac and become juveniles. Juveniles that hatch from larger eggs experience lower mortality, faster growth, and earlier sexual maturity.

Among amphibians, parental care is most prevalent among tropical anurans (frogs and toads) and is poorly developed in salamanders. Females of some species of salamanders, such as the dusky salamander (*Desmogathus fucus*) of the eastern United States, remain with the eggs until they hatch. Males of some tropical anurans carry tadpoles in their vocal pouches or on their backs. In other species the eggs develop on the backs of females or in a brood pouch formed in the dorsal skin (Figure 13.8). Male midwife toads (*Alytes obstetricans*) of Europe carry eggs and developing young on their backs. All of these species eventually place them in a suitable environment for further growth.

Internal fertilization and terrestrial reproduction are fully developed among reptiles. Few eggs well supplied with yolk may be carried inside the mother's body until they hatch; or they may be placed in nests buried in the ground and given little subsequent care. Crocodiles, however, are the exception in reptiles. They actively defend the nest and young for a considerable period of time.

Except for the social insects, parental care is not well developed among invertebrates. Some retain eggs within the body until they hatch; others, such as crayfish, carry eggs externally. Invertebrate parental care is most highly developed in social ants, bees, and wasps. Social insects provide all five functions of parental care: food, defense, heat, sanitation, and guidance. (See Tallamy 1984 on insect parental care.)

We may be stretching the imagination a bit to consider maternal care in plants; yet the plant does have an influence on its offspring beyond its genetic contribution (Schall 1984). The seed-producing parent has a nurturing role on the devel-

FIGURE 13.8 The South American hylid frog *Gastrotheca cornuta* carries the developing eggs in a brood pouch of skin on her back.

oping seeds, which influences the offspring's fitness. This relationship was demonstrated in intensive greenhouse studies of the seed size and germination and seedling development in the Texas bluebonnet (*Lupinus texensis*). When released from the parent plant, the larger seeds, supplied with greater nutrient resources, had a higher germination rate and higher seedling survivorship and seedling biomass than did small seeds and seedlings.

Parental Investment

Costs of reproduction are high in terms of energy expenditure and survival (Clutton-Brock and Godfray 1991). The physical drains and energy costs of brooding, gestation, birth, feeding the young, lactation, and defense of the females are great. For polygamous males, in particular, the strain of defending females, fighting off male competitors, and sexual activity shortens life expectancy and future reproductive activity.

Parents have only a limited amount of energy. From this amount they have to allocate a certain amount to their own maintenance and survival. To make its maximum contribution to future generations, an organism must balance the profits of immediate reproductive investments of time and energy against the costs to future prospects, including fecundity and its own survival.

Suppose you were given $1000 to spend. You could spend the entire $1000 at once on some item. You would have made a one-time investment of your money; you would have nothing left for future expenditure. As another alternative, you might spend part of the money on one item and place the balance in a savings account. Some time later you might withdraw some of the money for another expenditure, which would reduce the amount you have to spend in the future. At each withdrawal, you further reduce the amount for future expenditure, until the whole $1000 is gone. How long the money lasts depends on the amount and number of withdrawals for present expenditure.

In a similar manner organisms work under a budget. Because the organism has to allocated its energy budget to different expenditures, it will have to make some trade-off. The organism can expend all its energy in one major reproductive effort in its lifetime, or it can opt for repeated reproductive bouts, producing fewer young each time and saving some of its energy for future reproduction.

Those organisms that go for one major reproductive effort in a lifetime are termed **semelparous**. They invest all their energy in growth, development, and energy storage, and then expend that energy in one massive suicidal reproductive effort. Such a reproductive strategy is employed by most insects and other invertebrates, by some species of fish, notably salmon, and by many plants—annuals, biennials, short-lived perennials such as thistles and century plants, and some bamboos. Many semelparous plants, such as annual ragweed, are small, short-lived, and occupy ephemeral or disturbed habitats. For them it would not pay to hold out for future reproduction, for their chances of future reproduction are slim. They gain their maximum fitness by expending all their energies in one bout of reproduction.

Other semelparous organisms, however, are long-lived and delay reproduction. Mayflies (Ephemeroptera) may spend several years as larvae before emerging from the surface of the water for an adult life of several days devoted to reproduction. Periodical cicadas spend 13 to 17 years below ground before they emerge as adults to stage an outstanding exhibition of single-term reproduction. The Hawaiian silverswords (*Argyroxiphium*) live 7 to 30 years before flowering and dying. The Haleakala silversword (*A. sandwicense*) (Figure 13.9), found on the summit of Haleakala volcano on Maui in Hawaii, lives under environmental conditions that are extreme: intense light, periods of strong heat and cold, and little moisture. The plant accumulates energy and moisture within its leaves, which it expends in a very short period of flowering once in its life-

FIGURE 13.9 The Haleakala silversword (*Argyroxiphium sandwicense*) stays in a rosette stage for an unknown number of years before forming a massive flowering stalk up to 2 m tall covered with maroon-colored flowers in large heads. After the seeds form, the plant dies.

time. Environmental conditions preclude repeated reproduction. In general, for a species to evolve semelparity, that mode of reproduction has to increase fitness enough to compensate for the loss of repeated reproduction.

Organisms that produce fewer young at one time and repeat reproduction throughout their lifetime are termed **iteroparous.** Iteroparous organisms include most vertebrates, many invertebrates, perennial herbaceous plants, shrubs, and trees. For an iteroparous organism the problem is timing reproduction—early in life or at a later age. Whatever the choice, it involves trade-offs. Early reproduction means less growth, earlier maturity, reduced survivorship, and reduced potential for future reproduction. Later reproduction allows for increased growth, later maturity, and increased survivorship, but decreased fecundity after each bout of reproduction. In effect, if an organism is to make a maximum contribution to future generations, it has to balance the profits of immediate reproductive investments against costs to future prospects, including fecundity and its own survival. It has to make a trade-off between present progeny and future offspring

(Williams 1966). This trade-off relates to the reproductive values of individuals in a population, a concept related to fecundity introduced at the end of Chapter 10. Reproductive value is the average contribution to the next generation that members of a given age group make between their current age and death. Examples of reproductive values are given in the V_x column of the fecundity table for the cactus ground finch in Table 13.3.

The concept of reproductive value was developed by R. A. Fisher (1930). In a population of constant size, he defined the reproductive value as the age-specific expectation of future offspring as

$$V_x = \sum_{y=x}^{\infty} \frac{l_y}{l_x} m_y$$

where l_x = the probability of surviving to age x and consequently l_y/l_x = probability of living from age x to age y, m_y = average reproductive success of an individual at age y, and V_x = reproductive value of an individual of age x.

For newborn individuals in a population that is neither increasing nor decreasing, the reproductive value is the same as the net reproduction rate, $R_0 = 1$. Populations change in size, however, and so the definition of reproductive value becomes the present value of future offspring. In other words, the reproductive value is the sum of the number of offspring the individual is expected to produce each subsequent year of its life; for age y, the term in this sum is the product of the number of offspring an individual of age y produces times the probability of the age x individual surviving to age y. This value represents the number of offspring an individual dying

TABLE 13.3 Fecundity and Reproductive Values for the Cactus Ground Finch

Age Class (x)	Probability of Surviving to x (l_x)	Fledglings Per Season (m_x)	Product of Survival and Reproduction (l_x m_x)	Reproductive Value (V_x)
		MALES		
0	1.000	0.0	0.0	0.762
1	0.512	0.0	0.0	1.561
2	0.317	0.0	0.0	2.404
3	0.293	0.104	0.030	2.601
4	0.268	0.250	0.067	2.730
5	0.268	0.250	0.067	2.480
6	0.244	0.475	0.116	2.450
7	0.244	0.125	0.030	1.975
8	0.244	1.350	0.330	1.850
9	0.122	0.600	0.073	1.000
10	0.073	0.0	0.0	0.667
11	0.073	0.0	0.0	0.667
12	0.073	0.667	0.049	0.667
13	0.049	0.0	0.0	0.0
14	0.049	0.0	0.0	0.0
15	0.024	0.0	0.0	0.0
16	0	. . .	. . .	. . .
			$R_0 = \Sigma l_x m_x = 0.762$	$\Sigma = 21.814$
		FEMALES		
0	1.000	0.0	0.0	2.101
1	0.512	0.364	0.186	4.294
2	0.279	0.187	0.052	6.873
3	0.279	1.438	0.401	6.686
4	0.209	0.833	0.174	7.006
5	0.209	0.500	0.104	6.173
6	0.209	0.833	0.174	5.673
7	0.209	0.250	0.052	4.839
8	0.209	3.333	0.696	4.588
9	0.139	0.125	0.017	1.888
10	0.070	0.0	0.0	3.500
11	0.070	0.0	0.0	3.500
12	0.070	3.500	0.245	3.500
13	0	. . .	. . .	. . .
14	. . .	. . .	. . .	. . .
15	. . .	. . .	. . .	. . .
			$R_0 = \Sigma l_x m_x = 2.101$	$\Sigma = 60.621$

Source: Grant and Grant 1992:773.

at age $x + 1$ would have to produce at age x to leave behind as many offspring as another individual of age x who did not die early and produced offspring at a normal rate for the rest of its life. The general equation for reproductive value is

$$\frac{V_x}{V_0} = \frac{e^{rx}}{l_x} \sum_{y=x}^{\infty} e^{-ry} l_y m_y$$

(For calculation of reproductive value see Appendix B)

The reproductive value of a newborn individual is influenced by state of the population. In an expanding population the reproductive value of young is low for two reasons. First, in an increasing population the probability of death before reproduction may also increase. Second, because the future breeding population of which the young will be a part will be larger, they will contribute less to the overall gene pool than current progeny. Conversely, young born into a declining population will contribute more to the future than present progeny and so are worth more.

Reproductive values are usually calculated for females because reproductive values for males are difficult to obtain, but calculations can be done if we know the number and fate of the offspring sired by the male. Grant and Grant have done this for the cactus ground finch (Table 13.3 and Figure 13.10). Reproductive values for females are higher those of males. Generally reproductive values rise to a peak at an early age and then decline. However, the reproductive values of the cactus ground finch vary, depending on annual fluctuations in rainfall.

In the polygamous red deer (*Cervus elapus*) on the Isle of Rhum, the reproductive value of the hinds (as plotted in Figure 13.11) declined gradually from four years of age. This decline reflected the physiological drain of reproduction on the females. The reproductive value of the stag remained constant from three to seven years, then declined sharply, indicating the loss of vigor after several years of intense reproductive activity. Unlike old hinds, stags apparently do not invest heavily in breeding activities during the declining years.

If an organism is to make the maximum contribution to future generations, it has to balance the profits of immediate reproductive investments against costs to future prospects, including fecundity and its own survival. The situation is defined by partitioning the equation for reproductive value into two parts, one for present reproduction and one for residual future reproduction:

$$V_x = m_x + \sum_{t=x+1}^{\infty} \frac{l_t}{l_x} t$$

The first component of the equation, m_x, is contemporary reproductive output; the second component is the residual reproductive value or expected future progeny (future investment). Energy expended in repeated reproduction weighs against future prospects, as measured by declining fecundity and reduced potential for survival (Figure 13.12). Thus a parent that has invested heavily in early reproduction will have less to invest in later fecundity and survival. Increased early

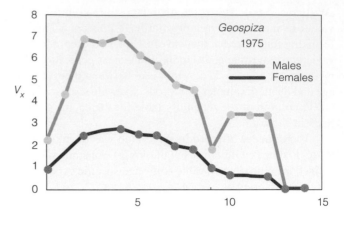

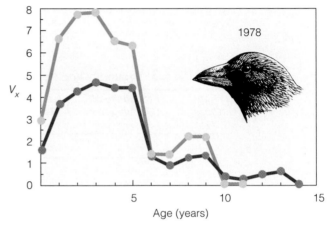

FIGURE 13.10   Reproductive value as a function of age in two cohorts of the cactus ground finch (*Geospiza scandens*) in 1975 and 1978. (From Grant and Grant 1992:772.)

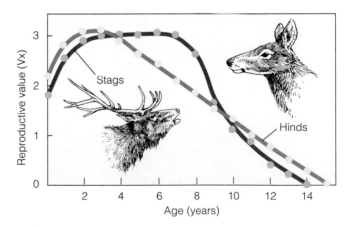

FIGURE 13.11   Plot of reproductive values for red deer (*Cervus elapus*) hinds and stags. Reproductive value for both sexes was calculated in terms of the number of female offspring surviving to one year old that parents of different ages can be expected to produce in the future. (From Clutton-Brock 1984:154.)

investments will require that the organism increase its level of present investment in reproduction, because of decreased probability of survival. Somewhere in between lies the best strategy for optimal lifetime reproductive success, the sum of present and future reproductive success.

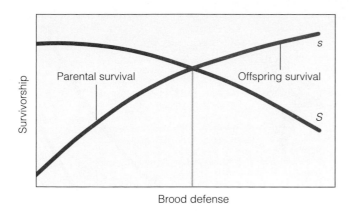

FIGURE 13.12 A graphic analysis of C. B. Williams' (1966) trade-off model between fecundity and adult survival. Parental investment in offspring is measured by brood defense. As investment increases, offspring survival (s) increases and parental survival (S) decreases. The point of intersection represents optimal parent investment for survivorship and future fecundity. Beyond this point parents are sacrificing future investment for present investment. (After Sargent and Gross 1985:44.)

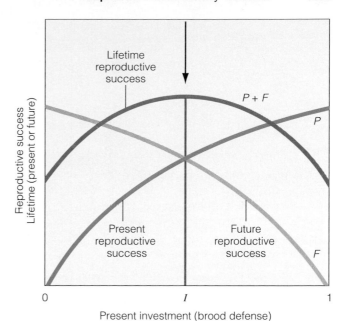

FIGURE 13.13 Effects of present investment on reproductive success. Present reproductive success (P) is assumed to increase with diminishing returns (reproductive success of adult offspring) on present investment, whereas future reproductive success declines with present investment. Lifetime reproductive success is the sum of present and future reproductive success ($P + F$). The optimal level of present investment (I) is that which brings about maximum lifetime reproductive success ($P + F$). Although theoretical, these models and that of Figure 13.12 provide some insights on why birds abandon a nest after some effort in defense or why grizzly bears abandon young as a reproductive tactic (see Tait 1980). (After Coleman et al. 1985:60.)

Semelparity Versus Iteroparity Why some plants are annuals or semelparous and others perennial or iteroparous has intrigued ecologists. What are the evolutionary advantages of semelparity and iteroparity, of reproducing and dying in a single season or reproducing over a number of seasons? What would an annual plant gain by becoming a perennial? L. C. Cole (1954) assessed this question mathematically. He calculated that the gain would be equivalent to adding one more offspring to an annual's output. Because it is easier to make one more sperm, egg, or seed than to survive for an additional year, Cole wondered why an organism should be a perennial. The answer, Charnov and Schaffer (1973) proposed, is that survivorship varies from one life cycle phase to another, and the young have lower survivorship than older individuals. Iteroparity is advantageous if juvenile mortality is high, adult mortality is low, and the population growth rate is slow. Semelparity is favored if the population growth rate is high, juvenile survival is high, and adult survival and the probability of surviving to reproduce a second time are low.

For an iteroparous organism to achieve high fitness, it has to settle for neither maximum present reproduction nor maximum future reproduction. An optimal level of parental investment balances the rate of return from investment in the present with the rate of return from investment in the future. Present investment in fitness should continue to grow through reproductive success of offspring while future investments in fitness continue to decline (see Sargent and Gross 1985, Coleman et al. 1986) (Figure 13.13).

Empirical evidence from natural populations comparing growth, survival, and future reproduction of reproductive and nonreproductive individuals is weak and confusing. There have been manipulative experiments with plants to increase and decrease the costs of reproduction in plants by removing leaves (which decreases the leaf area of the plant) or pollinating the flowers (which increases reproductive output). In some of these studies, plants manipulated for one to two years

show lower survival, growth, and reproduction the following year; others do not. For four years Primack and Hall (1990) followed the costs of reproduction for randomly selected hand-pollinated pink lady's slipper orchids (*Cypripedium acaule*) compared with insect-pollinated plants. Because the lady's slipper depends on bees for pollination, plants growing in the wild rarely develop fruit. Costs of annual reproduction in hand-pollinated plants did not show up until the third and fourth years, when the fruiting plants had lower rates of growth and flowering than the control plants. An average-size fruiting plant experienced a 10 to 13 percent decrease in leaf area and a 5 to 16 percent decrease in the probability of flowering the following year.

Demonstrating the costs of reproduction is no easier in animals. In some mammals it appears that energy expended in repeated reproduction weighs against future prospects as measured by declining fecundity and reduced potential for survival. Red deer females debilitated by intense reproduction in one year perform poorly the next year. Milk hinds, ones who suckled a calf the previous year, have higher reproductive costs and higher mortality than yeld hinds, ones that did not produce a calf the previous year (Figure 13.14) (Clutton-Brock 1984). In the American bison (*Bison bison*) rearing sons is more debilitating to the mother than rearing daughters, because the sons suckle longer—up to 15 months (Wolff 1988). Cows that have produced sons breed later than

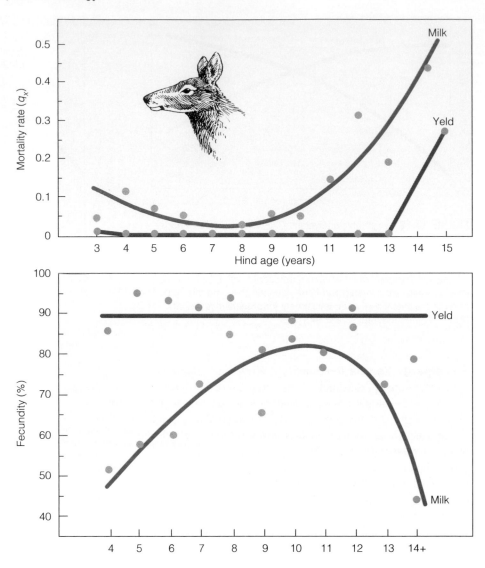

FIGURE 13.14   (a) Age-specific mortality of red deer milk hinds (who have reared a calf to weaning age) and yeld hinds (who have not). Milk hinds have a higher reproductive cost, which increases mortality. (b) The effect of that mortality and past investment in young is decreased future fecundity. (From Clutton-Brook 1984:224, 216.)

other cows and are more likely to be barren than cows that have produced daughters the previous year.

Parental Energy Budgets

How organisms allocate energy to growth and reproduction is central to their reproductive strategy. If an organism directs more energy to reproduction, it has less energy to allocate to growth and maintenance. As a result the individual may grow more slowly to the next age or fail to reproduce or to survive. For example, Lawler (1976) found that reproductive females of the terrestrial isopod *Armadillidium vulgare* had a lower rate of growth than nonreproductive females. The nonreproductive females devoted as much energy to growth as the reproductive females devoted to both growth and reproduction. Tinkle (1969) demonstrated that among lizards, species with larger fecundity have poorer individual survival. They channeled more energy into egg production than to growth over a short period, resulting in heavy physiological stress. Many North American freshwater fish experience a considerable loss of size or growth from reproduction, up to 25 percent or more in some trout (*Salvelinus*) (Bell 1980). Energy costs are more dif-

ficult to determine in homeotherms, although the costs show up in survival and future reproduction. Birds expend about four times the basal metabolic rate in feeding their young (Robbins 1983), and lactating mammals use up 2.5 to 5 times basal metabolic energy (Gittleman and Thompson 1988).

Plants, too, must allocate a portion of their net annual assimilation to reproductive effort. The amount plants allocate to vegetative reproduction can influence seed production and the growth of the genet (Fitter 1986). Annual species of plantain (*Plantago*) had higher reproductive costs, based on milligrams of seeds produced per square centimeter of leaf area, than perennial species (Primack 1979). Perennials, however, can allocate a high amount of yearly biomass to reproduction. For example, the meadow buttercup (*Ranuculus acris*) allocates up to 60 percent of its biomass to reproduction, and the showy goldenrod (*Solidago speciosa*) of old fields and woods allocates up to 35 percent (Abrahamson and Gadgil 1973). Many plants that grow under a range of environmental conditions are flexible in adjusting energy allocation to reproduction. Plants that grow in more adverse environments allocate more energy to reproduction than those in more moderate habitats.

The shortcomings of such estimates are many. In plants the estimates fail to take into account below-ground energy allocations, including the production of bulbs, corms, and roots. Simply comparing above-ground biomass of vegetative parts to biomass of reproductive output may not be an accurate assessment of the energy costs of reproduction, as Reekie and Bazzaz (1987) found in intensive studies of reproductive costs in quack grass (*Agropyron repens*). They point out that resource allocation is not necessarily a reliable indicator of costs of reproduction, because plants experience different costs. A plant with a low cost of reproduction could afford to allocate a larger proportion of its resources to reproduction.

Clutch Size   To produce viable offspring, organisms need a certain minimal amount of energy to be apportioned among a certain number of young. As parents divide available energy among an increasing number of offspring, the fitness of individual offspring declines. Eventually parents reach a point at which decreasing expenditure or investment per offspring from birth to independence results in declining payoff in the fitness of their offspring (Smith and Fretwell 1974).

One method by which parents can optimally apportion investment in offspring is adjustment of brood or clutch size. Two basic choices are to apportion parental investment among many small young or to concentrate that investment in a few larger ones. Within that range of choices the parents have to adjust the number of young they can rear with the resources available without significantly reducing their own fitness and survival, or the fitness of their offspring.

David Lack (1954) originally hypothesized that among birds clutch size evolved through natural selection to correspond to the average largest number of young the parents can feed. There is indeed evidence that the optimal clutch size should be lower than that which would produce the maximum number of young each season. This hypothesis can and has been tested experimentally in birds simply by manipulating clutch size, especially by adding additional eggs, usually one, to the nest. In some studies enlarged broods suffered greater mortality and yielded fewer offspring than unmanipulated natural broods, acting as controls for comparison. In others increased broods produced more fledglings on the average than control broods (Lessels 1991). In one such study Gustafsson and Sunderland (1988) experimentally modified the clutch size of altricial young of the collared flycatcher (*Muscicapa latirostris*) by increasing or reducing the natural clutch size by one or two eggs. The unmanipulated natural clutch size had no effect on adult survival, juvenile survival, or future reproduction. The large-sized clutches produced more young but at the cost of lower juvenile survival, subsequent reduced fecundity of the parents, and lower fecundity in the offspring. Reduced clutches produced fewer young. Natural clutch size produced more recruits than the enlarged or reduced clutches. Such studies suggest that physiological and ecological factors probably set clutch size among birds at an optimum level that results in maximum lifetime reproduction, as Lack postulated.

A similar relationship holds among some insect parasitoids, but because no parental care is involved, the constraints involve hosts in which the insects lay their eggs. Two models are proposed (Parker and Courtney 1984): (1) individuals should adjust their clutch size so that large clutches are laid on hosts that can support the growth of more offspring; and (2) as time between bouts of egg-laying increases, usually when hosts are rare, large clutches should be laid in higher quality hosts. The swallowtail butterfly that lays its eggs on the leaves of the pipevine (*Aristolochia reticulata*) follows this pattern (Pilson and Rausher 1988). The females lay larger clutches on younger plants with more edible leaves and terminal buds and consistently discriminate between high- and low-quality plants. Hardy and his associates (1992) manipulated the clutch size of an ectoparasitoid wasp (*Goniozus nephantidis*), which lays a clutch of up to 20 eggs on small caterpillars. They artificially created clutches of different sizes on hosts of approximately the same weight. Increasing clutch size had little effect on juvenile survival, but it did result in smaller adult females. In nature clutch size of such parasitic wasps is strongly influenced by host size. The larger hosts support larger clutches.

Brood Reduction   When parents cannot predict available food resources at the time of nesting (which they usually cannot), they may have to adjust to the situation by brood reduction. The decision is when to withdraw investment from one or more offspring. Theoretically parents should attempt to rear all of the offspring until they arrive at the point where the potential fitness of the discontinued offspring (if cared for) is less than the augmented fitness of the remaining offspring.

Decisions in brood reduction, of course, are not consciously made by the parents. They come about largely through asynchronous hatching and siblicide. In asynchronous hatching the young are of several ages. The older siblings may beg more vigorously for food, forcing the harried parents to ignore the calls of the younger, smaller sib, who perishes, or the older or more vigorous young may simply kill their weaker sib. For example, the common grackle (*Quiscalus quiscula*) begins incubation before its clutch of five eggs is complete, a pattern of hatching ensuring survival of some young under adverse conditions (Howe 1976). The eggs laid last are heavier and the young from them grow fast. However, if food is scarce, the parents fail to feed these late offspring because of more vigorous begging by the siblings. The last-hatched young then die of starvation. Thus asynchronous hatching favors the early-hatched young at the expense of those hatched later. A similar situation exists in western gulls (*Larus occidentalis*) (Sydeman and Emslie 1992). Although parents attempt to ensure the survival of all young, their investment in older young is protected by starvation of late-hatched birds if a food shortage arises.

A number of birds, including raptors, herons, egrets, gannets, boobies, and skuas, practice siblicide. The parents normally lay two eggs, possibly to ensure against infertility of a single egg. The larger of the two hatchlings kills the smaller sibling or runt, and the parents redirect all resources to the surviving chick (Godfray and Harper 1990). These birds are not alone in siblicidal tendencies. The females of some parasitic wasps lay two or more eggs in a host and the larvae fight each other until only one survives (Godfray 1987).

Latitudinal Variation in Clutch Size It has been commonly observed that clutch size in birds is higher in high latitudes than low. Birds in temperate regions have larger clutch sizes than birds in the tropics (Figure 13.15), and mammals at higher latitudes have larger litters than those at lower latitudes (Lord 1960). Lizards exhibit a similar pattern. Those living at lower latitudes have smaller clutches, have higher reproductive success, reproduce at an earlier age, and experience higher adult mortality than those living at higher latitudes (Tinkle and Ballinger 1972, Andres and Rand 1974).

Lack (1954) proposed that such variations in clutch size are adaptations to food supply. According to Lack's hypothesis, temperate species have larger clutches because increasing daylength allows parents a longer time to forage for food to support larger broods. In the tropics, where daylength does not change, food becomes a limiting resource. Cody (1966) modified this concept by employing the principle of allocation of energy. He proposed a second hypothesis, that clutch size results from different allocations of energy to egg production, avoidance of predators, and competition.

Another hypothesis, proposed by Ashmole (Ashmole 1963, Ricklefs 1980), states that clutch size varies in direct proportion to the seasonal variation in resources, largely food supply, used by a population. Population density is regulated primarily by mortality in winter, when resources are scarce. Available resources during the breeding season relative to breeding population density influence clutch size and number of young reared. Greater winter mortality makes more food available to the survivors during the breeding season. This resource availability is reflected in larger clutches. Thus geographical variation in mean clutch size is more strongly correlated with winter food availability and production, which directly influences the size of the breeding population, than with summer food production (for a discussion on this topic see Ricklefs 2000).

Insects, too, such as the milkweed beetle (*Oncopeltus*) support the hypotheses (Landahl and Root 1969). Temperate and tropical milkweed beetles have a similar duration of the egg stage, egg survivorship, developmental rate, and age at sexual maturity. Although the clutch sizes are the same, the temperate species lay a larger number of eggs because they lay more clutches. Total egg production of the tropical species is only 60 percent that of the temperate species.

Plants follow the general principle of allocation on a latitudinal basis. McNaughton (1975) investigated the allocation of resources to reproduction in a series of greenhouse studies of cattail (*Typha*). He measured reproductive effort by the growth of rhizomes, the cattails' principal means of expansion within a given habitat. Cattails shed their very small seeds to the wind as means of colonizing a new habitat. McNaughton considered three species of cattails on a climatic gradient: the common cattail (*Typha latifolia*), a climatic generalist that grows from the Arctic Circle to the equator; the narrowleaf cattail (*T. augustifolia*), restricted to the northern latitudes of North America; and *T. domingensis*, restricted to the southern

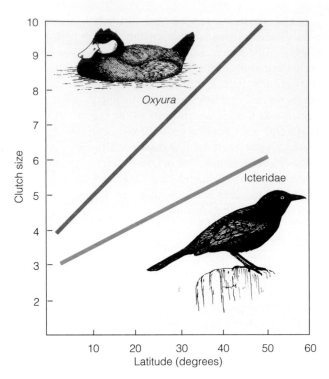

FIGURE 13.15 Relationship between clutch size and latitude in birds. Represented as examples are the subfamily (Icteridae) (blackbirds, orioles, and meadowlarks) in North and South America and the worldwide genus *Oxyura* (ruddy and masked ducks) of the subfamily Anatinae. (Adapted from Cody 1966.)

latitudes of North America. The cattails exhibited a climatic gradient in the allocation of energy to reproduction. *Typha augustifolia* and northern populations of *T. latifolia* grew earlier and faster and produced a greater number of rhizomes than the *T. domingensis* and southern populations of *T. latifolia,* although the southern cattail produced larger rhizomes. Faced with stronger intra- and interspecific competition, the southern cattail invested more energy in vegetative growth. Likewise, tropical members of the sunflower tribe Heliantheac in the aster family (Asteraceae) have significantly smaller clutch sizes (ovules per flowering head) than those of temperate and alpine regions, with an average of 51.2, 82.7, and 89.6 ovules respectively (Levin and Turner 1977).

Although reproductive output among organisms does appear to be greater at higher latitudes, the number of comparable species for which we have data is few, too few to consider the hypotheses as a general law. Many more studies along a latitudinal gradient from the tropics to arctic regions are needed.

Hermaphroditism and Gender Change The discussion of reproductive costs relates mostly to species among which males and females are separate individuals. Cost allocation among hermaphrodites, particular sequential hermaphrodites, has interesting differences. Hermaphroditism, especially common in plants, probably evolved among those organisms in which the female could take on the male function by relinquishing only a small amount of female function. Similarly, the male could add a female function without seriously compromising sperm production (Charnov 1982). In hermaphroditism the contribution of combined male and female exceeds the contributions that would be made by separate sexes.

Mating in hermaphroditic organisms can involve either outcrossing or self-fertilization. Apparently most hermaphroditic animals are not self-fertilized. Earthworms mate with other individuals, as do hermaphroditic coral reef fish. Some animal hermaphrodites are completely self-sterile, and only a few, like certain land snails, are self-fertilized.

Among animal hermaphrodites, how is reproductive effort allocated between egg and sperm and between male and female behaviors? The male side of behavior in a hermaphrodite attempts to fertilize the eggs of other hermaphrodites, increasing the reproductive success of the male. The female side attempts to secure fertilization of her eggs to enhance her reproductive success.

Once the eggs are laid, should the hermaphrodite male defend the eggs of its partner or those of its female aspect? Fischer (1981) investigated this problem in the hermaphroditic coral reef fish *Hypoplectrus nigracans,* the black hamlet. He found that the reproductive success of the hamlet as a male depends on its ability to reproduce as a female, because spawning partners "trade eggs," giving up eggs to be fertilized in exchange for the opportunity to fertilize those of another individual. Courtship is largely a female function rather than a male function; it advertises that an individual has eggs. Each fish parcels out its daily clutch in four to five

spawns, and it alternates sex roles with each spawn, taking turns giving up parcels to be fertilized.

Many hermaphroditic plants are self-compatible but have evolved different means to prevent self-fertilization. Anthers and pistils may mature at different times, or the pistil may extend well above the stamens. Other plants have evolved more effective means. One is a genetic mechanism that prevents the growth of a pollen tube down the style of the same individual. Some hermaphroditic species are divided into two or three morphologically different types between which pollination takes place.

Although many hermaphroditic organisms possess mechanisms to reduce self-fertilization, the capacity for self-fertilization does carry advantages, especially among plants. A single self-fertilized individual is able to colonize a new habitat and then reproduce itself, establishing a new population. Other hermaphrodites produce self-fertilized flowers under stressed conditions, ensuring a new generation. Jewelweed (*Impatiens* spp.), for example, under normal and optimal environmental conditions produces cross-pollinated (chasmogamous) flowers. Under adverse environmental conditions or after the chasmogamous flowers have set seed, jewelweed (as well as violets and other species) produces tiny, self-fertilized (cleistogamous) flowers that never open (Figure 13.16). They have vestigial petals, no nectar, and few pollen grains and remain in a green, budlike stage. Thus, if outcrossing fails or never develops, the plants have ensured a next generation by self-fertilization.

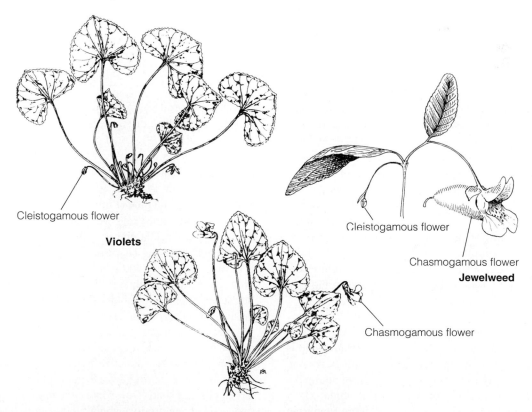

Cleistogamous flower

Violets

Cleistogamous flower

Chasmogamous flower

Jewelweed

Chasmogamous flower

FIGURE 13.16 The normal cross-pollinated flowers of jewelweed and violets are termed chasmogamous. They are conspicuous and produce obvious seeds. But hidden beneath the leaves of jewelweed and at ground level in violets are tiny self-fertilized cleistogamous flowers that never open but also produce seeds. The seeds of cleistogamous flowers are important for the maintenance of local adapted populations of these plants.

The above discussion applies to simultaneous hermaphroditism. More interesting is sequential hermaphroditism, in which individuals budget their reproductive costs by changing gender as they grow during their life cycle. Gender change, in which the individual switches between a costly female reproductive function and a less expensive male, maximizes individual reproductive success.

Gender change among animals, notably fish species, appears to be stimulated by a social change involving sex ratios in the population. Sex reversal among some species of marine fish can be initiated by the removal of one or more individuals of the other sex. Among some coral fish, removal of females from a social group stimulates an equal number of males to change sex and become females (Friche and Friche 1977). In other species, removal of males stimulates a one-to-one replacement of males by sex-reversing females (see Shapiro 1979, 1987). Among the mollusks, the Gastropoda (snails and slugs) and Bivalvia (clams and mussels) have sex-changing species, and almost all of them change from male to female (Wright 1988). For example, if a young slipper limpet (*Crepidula* spp.) of rocky shorelines settles on its adult habitat without others of its own kind, it grows to a large size and develops into a female. If it settles on an occupied rock, it will develop into a small male. The male, however, changes rapidly into a functional female if larger occupants are removed.

Plants also exhibit a gender change (Freeman et al. 1980). One such plant is jack-in-the-pulpit (*Arisaema triphyllum*) (Policansky 1981, Lovett et al. 1982, Bierzychudek 1982a, 1984), a clonal woodland herb whose genet is a perennial corm and whose ramet is a single annual shoot (Figure 13.17). Jack-in-the-pulpit produces staminate (male) flowers one year, an asexual vegetative shoot the next, and a carpellate (female) or monoecious shoot the next. Over its life span a jack-in-the-pulpit may produce both genders as well as an asexual vegetative shoot, but in no particular sequence. The sequence may be from asexual to male or female, from male to asexual to female, or from female to asexual to male. Usually an asexual stage follows a gender change. Gender change in jack-in-the-pulpit appears to be triggered by an excessive drain on the photosynthate by female flowers. If the plant is to survive, one carpellate flowering could not follow another. To avoid death, the plant reduces its reproductive effort the next year by changing its gender or becoming vegetative. Among jack-in-the-pulpits, the size of the plant is more important than its sexual state in determining its future sex. Large males are more likely to be female the next spring than are small females (Policansky 1981).

Age and Size For many species clutch size and fecundity are related to age and size of the parent. This relationship is especially strong in plants and poikilothermic animals. Among plants perennials delay flowering until they have attained a sufficiently large leaf area to support seed production. Many biennials living in poor environments also delay flowering beyond the usual two-year life span, until environmental conditions become more favorable. Annuals show no such relationship between leaf area and the percentage of energy

FIGURE 13.17 The jack-in-the-pulpit (*Arisaema triphyllum*) can change from asexual to male or female, from female to asexual to male, or from male to asexual to female, depending on energy allocations and reserves. The plant gets its name from the spadix enclosed in a hoodlike sheath. This fruiting plant is the female stage.

devoted to reproductive output, but size differences among annuals do result in differences in the number of seeds produced (Primack 1979). Small plants produce fewer seeds, even though the plants themselves may be contributing the same proportionate share to reproductive effort as larger plants.

Age at maturity and seed production vary among trees. Some, like Virginia pine (*Pinus virginianus*), start to produce seed at five to eight years when growing in open stands. In dense stands, Virginia pine may delay seed production for 50 years. Quaking aspen (*Populus tremuloides*) produces seeds at the age of 20 years and white oak not until 50 years. Acorn production varies with size. Trees 16 inches in diameter at breast height produce about 700 acorns; those of 24 to 26 dbh produce 2000 or more acorns (Downs and McQuilkin 1944).

Fecundity in fish increases with size, which in turn increases with age. Because early fecundity reduces both growth and later reproductive success, fish obtain a selective advantage by delaying sexual maturation until they grow larger. Gizzard shad (*Dorosoma cepedianum*) reproducing at two years of age produce about 59,000 eggs. Those delaying reproduction until the third year produce about 379,000 eggs. Among the gizzard shad only about 15 percent spawn at two years of age, about 80 percent at three years.

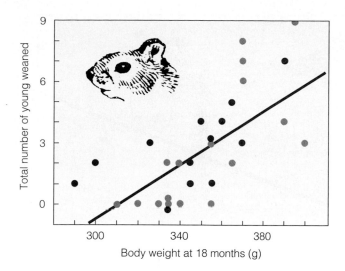

FIGURE 13.18 Lifetime reproductive success of the European red squirrel (*Sciurus vulgaris*) according to body weight in their first winter as an adult. Yellow, coniferous habitat; brown, deciduous habitat. (From Wauters and Dhondt 1989:645.)

The optimal clutch size of the loggerhead sea turtle (*Caretta caretta*) is constrained by the female's egg-carrying capacity, which is related to body size (Hays and Speakman 1991). Because of the high energy sea turtles expend on land, they may be under selective pressure to maximize clutch size to minimize the energy expended per egg.

An apparent relationship also exists between body size and fecundity among homeotherms. Heavier females are more successful in reproduction and experience increased survival (Sadlier 1969). For example, body weight of female European red squirrels (*Sciurus vulgaris*) in Belgian forests was strongly correlated with lifetime reproductive success (Wauters and Dhondt 1989) (Figure 13.18). Squirrels weighing less than 300 g did not come into estrus. Among red deer, young hinds and old hinds enjoyed the highest survival of their calves and middle-aged hinds the poorest (Clutton-Brock et al. 1982). This difference probably reflects the better body condition of young hinds and a greater energy investment in offspring by experienced old hinds. Such an investment by old hinds, however, is not reflected in another measure of fecundity, reproductive value, which usually declines with age.

GENDER ALLOCATION

Another significant aspect of reproduction is that perennial question: is it a girl or a boy? Because gender in many animals is randomly determined by the X and Y chromosomes (or some variation thereof), we expect on the average a 50:50 ratio. R. A. Fisher (1930) carried the explanation one step further to explain why sex ratios tend to be equal. Natural selection favors those parents who invest equally in their sons and daughters. When an average son and an average daughter are equally expensive to produce, the sex ratio of

the offspring at the end of parental care should be 1:1. The costs of rearing each sex, however, may not be equal. One sex may cost parents less to produce and rear than the other because it requires less food or other resources, has a higher mortality during the period of parental care (thus freeing additional resources for the remaining siblings), or imposes less stress on the mother. Eventually the sex ratio will become skewed toward the less expensive sex. If the sex ratio becomes skewed toward females, any factor that leads the bearer to produce more sons than daughters would be selected for. The reason is that sons would mate, on the average, with more than one female and thus more efficiently transmit genes to future generations. If the population sex ratio, however, is biased toward males, sons will mate, on the average, with less than one female, and so any gene that favors female offspring will be selected for. When the sex ratio is at equilibrium, selection for either sex ceases.

Much of the theory of sex ratio adjustment applies to polygamous mammals. It is based on the premise that variations in physical condition or social rank of the mother result in differential investment in sons and daughters. One model (Trivers and Willard 1973) predicts that a female in good condition would leave more surviving grandchildren if she produced more sons than daughters, whereas a female in poorer condition would maximize her fitness by producing more daughters. A second model, local resource competition (Clark 1978), is based on the predominance of male dispersal and female philopatry (remaining at home) in mammal populations. Its premise is that mothers of high social rank and associated access to resources should produce mostly daughters, because they would inherit the mother's rank and privileges. Low-ranking females should produce sons that could disperse to areas where they could maximize their reproductive success (Silk 1983).

How do these predictions hold up in natural populations? In the red deer (*Cervus elaphus*) social rank and body size affected the breeding success of both sexes. Dominant hinds were heavier, conceived earlier, and produced larger calves and significantly more sons than subordinate hinds (Figure 13.19). Stags born to females above median rank experienced higher levels of reproductive success than hinds, whereas hinds born to subordinate females accrued greater reproductive success than stags (Clutton-Brock et al. 1986). In the white-tailed deer (*Odocoileus virginianus*) the opposite was true. Sex ratios of offspring were related to the nutritional status and age of the doe, which reflect habitat quality. Fawns, yearling does, and undernourished does produced a preponderance of males, whereas heavier and older does produced significantly more daughters than sons (Ozoga and Verme 1986). The preponderance of females produced by older, well-nourished does fits the local resource competition model. Because daughters remain with their mothers until three years old, when they move to an area adjacent to the mother's home range and form a family group of their own (Ozoga et al. 1982), local competition for resources could develop, whereas male offspring disperse as yearlings. Does occupying poor range would benefit little by producing

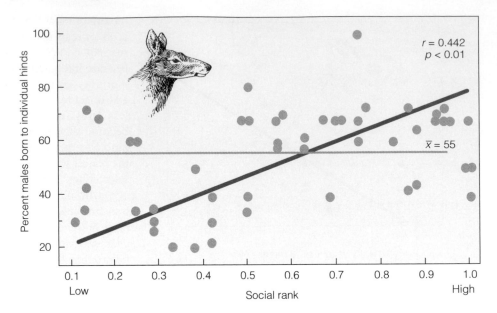

FIGURE 13.19 Percentage of male offspring born to individual red deer (*Cervus elaphus*) hinds differing in social rank. High-ranking females tend to have sons. (Clutton-Brock et al. 1986:464.)

potentially competing female offspring and would maximize their fitness by producing males.

Among some species of primates, high-ranking females have access to more resources than low-ranking females, and males disperse. Because daughters inherit their mother's rank and its associated access to resources, high-ranking females produce more daughters, whereas low-ranking females, with access to fewer resources, produce more sons.

If access to resources resulting in better-nourished, heavier, and larger females biases sex ratios, then perhaps sex ratios are influenced by environmental conditions. Such influences appear in some of the parasitoid wasps that lay a single egg per host. Host size correlates with the size of the wasp that emerges from it. Sex determination in wasps is through haplodiploidy. Fertilized eggs produce females; unfertilized eggs produce males. Thus females can control the sex of the offspring by deciding to deposit a fertilized or unfertilized egg. Because female size and fecundity are closely related, female wasps lay fertilized or female-producing eggs in large hosts and male eggs in small hosts. The relationship is somewhat more complex, but the point is made. A clearer relationship exists among many of those reptiles that bury their eggs in the ground. The sex of their offspring depends on temperatures during incubation. Among some species males develop at cool temperatures, among others at high temperatures (Charnov and Bull 1977). How and why this type of gender allocation occurs has no satisfactory explanation yet (Bull and Charnov 1989).

R-SELECTION AND *K*-SELECTION

There are obvious broad differences among reproductive patterns in organisms. Some species, such as weeds and insects, are small, have high reproductive rates, and live short lives. Others, like trees and deer, are large and have low reproduc-

tive rates and long lives. Ecologists call the former *r*-selected species and the latter *K*-selected species, after the two terms in the logistic equation. Populations of the former grow rapidly and do not seem to reach or remain at carrying capacity. The populations of the latter attain and more or less remain around carrying capacity.

The concept of *r* and *K* species originated with MacArthur and Wilson (1967). They used the terms *r-selection* and *K-selection* in their theory of island biogeography to distinguish between types of selection pressures on island colonists. Pianka (1970) elaborated on the distinction. Empty islands would be colonized by a variety of immigrants, the most successful of which would be species with the best mechanisms for dispersal and the best ability for rapid population growth in unfavorable habitats. Natural selection would favor individuals able to achieve rapid population growth in uncrowded resource-rich environments. However, as populations increased, conditions became crowded, and individuals competed for resources, natural selection would favor the most competitive individuals, those that were able to continue population growth at high sustained densities near carrying capacity. MacArthur and Wilson considered the former as *r*-selected, because environmental conditions keep growth of such populations on the rising part of the logistic curve. Mortality in these species is largely density-independent. They considered the latter as *K*-selected because they are able to maintain their densest populations at equilibrium (asymptote) or carrying capacity (*K*). *K* species can compete effectively for food and other resources in a crowded environment. Mortality in these species results mostly from density-related factors.

The theory of *r*- and *K*-selection predicts that species in these different environments will differ in life history traits such as size, fecundity, age at first reproduction, number of reproductive events during a lifetime, and total life span. Species popularly known as ***r*-strategists** are typically short-lived. Among *r* species, selection favors those genotypes that confer high reproductive rate at low population densities,

Chapter 13 Life History Patterns **235**

early and single-stage reproduction, rapid development, small body size, large number of offspring (but with low survival), and minimal parental care. They have the ability to make use of temporary habitats. Many inhabit unstable or unpredictable environments where catastrophic mortality is environmentally caused and relatively independent of population density. For them environmental resources are rarely limiting, and they are able to exploit relatively uncompetitive situations. Tough and adaptable, *r*-strategists, such as weedy species, have means of wide dispersal, are good colonizers, and respond rapidly to disturbance.

K-strategists are competitive species with stable populations of long-lived individuals. Among them selection favors genotypes that confer a slower growth rate at low populations, but the ability to maintain that growth rate at high densities. *K*-strategists have the ability to cope with physical and biotic pressures, possess both delayed and repeated reproduction, and have a larger body size and slower development. They produce few seeds, eggs, or young. Among animals, parents care for the young; among plants, seeds possess stored food that gives the seedlings a strong start. *K*-strategists exist in environments in which mortality relates more to density than to unpredictability of conditions. They are specialists, efficient users of a particular environment, but their populations are at or near carrying capacity and are resource-limited. These qualities, combined with their lack of means of wide dispersal, make *K*-strategists poor colonizers.

The original concept of *r*-selection and *K*-selection applied to the natural selection of those traits among individuals in populations. Later the concept was applied to species characteristics. Small species with rapid reproductive rates, such as insects and mice, were categorized as *r* species, and long-lived species like oak trees, deer, and elephants were considered *K* species. *r* and *K* species occupied endpoints on a continuum from *r*-selection to *K*-selection. Such a view tempts a classification of species either as *r*-selected or *K*-selected, but it is difficult to force species into such a classification. Under certain conditions individuals or populations will exhibit *r*-selected traits or *K*-selected traits. Meadow mice living in environments where dispersal can take place easily exhibit characteristics of *r*-selection, whereas those living under conditions in which there is no dispersal sink assume *K*-selected characteristics (Tamarin 1978). White-tailed deer exhibit *r*-selected traits when the species spreads into new habitat or is greatly reduced by hunting. However, at high densities, especially where females are not reduced, the population remains at *K* and the population exhibits *K*-selected traits with low overall recruitment (McCullough 1979).

The concept of *r*-selection and *K*-selection assumes deterministic environments: one that is unpredictable and another that is stable. Environments, however, randomly fluctuate over the life of an organism. Such conditions favor adult survival at the expense of present fecundity. Schaffer (1974), Stearns (1976), and others have called this adjustment bet-hedging. According to this hypothesis, in a variable environment in which adult mortality is high and juvenile mortality is low, selection should favor early maturity, larger reproductive effort, and more young to replace

the adults. Under the same conditions if adult mortality is low and juvenile survival is high, then natural selection should favor late maturity, for no need exists for either early maturity or high production of young. Thus a variable environment that affects juvenile survival more than adult survival should lead to lower reproduction each year and increased iteroparity (*K*-traits). Conversely, if the variable environment affects adult survival more than juvenile survival, then it should lead to higher reproduction each year and reduced iteroparity (*r*-traits). Thus one can arrive at the same strategies by considering the response of individuals to selection pressures only in terms of *r*.

In contrast to the use of *r*- and *K*-strategies in plants, J. Grime (1977, 1979) has proposed a three-endpoint system consisting of ruderal or *R*-strategists, competitive or *C*-strategists, and stress-tolerant or *S*-strategists (Figure 13.20). *R*-strategists, typically weedy species, occupy uncertain or disturbed habitats, have a short growth form, reproduce early in life, possess high fecundity, experience one lethal reproduction (semelparity), and have well-dispersed seeds. *C*-strategists and *S*-strategists occupy more stable environments and are long-lived, often drastically reducing the opportunity of seedling establishment, resulting in high juvenile mortality. Beyond these two characteristics, the two have evolved quite different life history strategies. *C*-strategists, such as grasses in an ungrazed grassland, live in competitive but productive and relatively undisturbed environments, attain maximum vegetative growth, reproduce early, and repeatedly utilize an annual expenditure of energy stored prior to seed production. *S*-strategists live in stressed environments, such

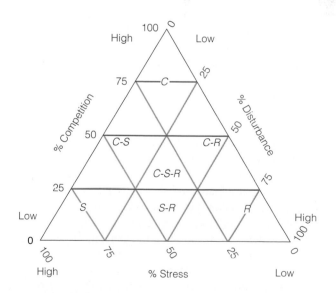

FIGURE 13.20 Equilibrium points among competition (*C*), stress (*S*), and disturbance (*R*) strategies. Strong competitors occupy the upper part of the triangle; the stress tolerators occupy the lower left-hand part of the triangle; and the ruderals, opportunistic species adapted to disturbance, occupy the lower right-hand part of the triangle. Species within the triangle are intermediate in their responses, depending on environmental conditions and life history traits. (From Grime 1977:1187.)

as highly disturbed sites or forest understory, have delayed maturity, intermittent reproductive activity, and long-term energy storage. Within these three endpoints are intermediate types: competitive ruderals (*C-R*), living in environments of low stress and moderate disturbance, such as fertilized moderately grazed grasslands; stress-tolerant ruderals (*S-R*), living in extreme environments such as rock crevices; *C-S-R* plants, found in habitats where competition is reduced by the combined effects of stress and disturbance, such as old fields; and *C-S* species living in areas of high stress and moderate disturbance. *R* species are equivalent to species on the *r* end of the *r-K* continuum, *C* species occupy the middle, and *S* species fill the *K* end of the continuum.

HABITAT SELECTION

Reproductive success depends heavily on choice of habitat. Settling on less than optimal habitat can result in reproductive failure. How are organisms able to assess the quality of the area in which they settle? What do they seek in a living place? Such questions have been intriguing ecologists for many years. Answering them is important in these times of diminishing habitats. To maintain, improve, and restore habitats, we need to know what features are important to the species occupying them (see Verner et al. 1986).

Habitat selection among vertebrates is partly genetic (Wecker 1963, Klopfer 1963) and partly psychological. Lack and Venables (1939) and Miller (1942) suggested that birds recognize their ancestral habitat by conspicuous though not necessarily essential features. The Nashville warbler (*Vermiforma ruficapilla*), a typical inhabitant of open heath edges of northern bogs, selects open stands of aspen and balsam fir and forest openings of blackberry and sweet fern (*Comptonia peregrina*) in the southern part of its range in New York. These habitats are visually suggestive of bog openings (Smith 1956). MacArthur and MacArthur (1961) and later many others (for example, MacArthur 1972, James 1971, Balda 1975) demonstrated a strong correlation between structural features of vegetation and the species of birds present.

Habitat selection probably involves a hierarchical approach (Hilden 1965, Wiens and Rotenberry 1981, Hutto 1985, Orians and Whittenberger 1991). Birds appear to assess initially the general features of the landscape—the type of terrain; presence of lakes, ponds, streams, and wetlands; gross vegetational features such as open grassland, shrubby areas, types and extent of forests; and homogeneous or patchy vegetational distribution. Once in a broad general area, the birds respond to more specific features of habitats, such as the structural configuration of vegetation, particularly the density of leaves at various elevations above the ground and degree of vegetational patchiness (Figure 13.21). James called this vegetational profile associated with the breeding territory of a particular species the "niche gestalt." Although once regarded as unimportant in habitat selection and utilization, floristics (the individual plant species present) also appears to be important.

(a)   Yellowthroat

(b)   Hooded warbler

(c)   Ovenbird

FIGURE 13.21 The vegetation structure that characterizes the habitat of three neotropical warblers. (a) Yellowthroat (*Geothlypis trichas*), a bird of shrubby margins of woodland and wetlands and brushy fields. (b) Hooded warbler (*Wilsonia citrina*), a bird of small forest openings. (c) Ovenbird (*Seiurus aurocapillus*), an inhabitant of deciduous or mixed conifer-deciduous forests with open forest floor. The labels 9–10 m, 18–30 m, and 15–30 m refer to height of vegetation. (After James 1971.)

The structural characteristics of the trees and shrubs may affect the foraging activities of birds. Various species of plants influence the type of herbivorous arthropods, and thus the levels of prey abundance (Holmes and Robinson 1981, Robinson and Holmes 1984, Wiens 1985).

Still other structural features determine a habitat's suitability (see Fish and Wildlife Service 1980, Verner et al. 1986). The lack of song perches may prevent some birds from colonizing an otherwise suitable habitat. Their introduction can stimulate the colonization of that area. For example, when telephone lines were strung across a treeless heath, tree pipits (*Anthus trivialus*), birds that require an elevated singing perch, moved into the area (Lack and Venables 1939). Woodcock (*Philohela minor*) will not utilize a singing ground, an opening in shrubby fields and forest from which the bird performs its courtship flights, unless the opening allows sufficient room for flight (Sheldon 1967). A small opening surrounded by tall trees is not suitable, but an opening of the same size surrounded by low shrubs is.

An adequate nesting site is another requirement. Animals require sufficient shelter to protect parents and young against enemies and adverse weather. Selection of small island sites, such as muskrat houses, by geese provides protection against predators. Cavity-nesting animals require suitable cavities, dead trees, or other substrate in which they can construct such cavities. In areas where such sites are absent, populations of birds and squirrels can be increased dramatically by providing nest boxes and den boxes.

There appears to be a relationship between food availability and habitat selection (Hutto 1985, Smith and Shugart 1987, Orians and Whittenberger 1991). This is particularly true in habitat selection in migratory birds as they arrive on their nesting grounds. For them the availability of food is more important than sharply defined structural features (Hutto 1985). Because food resources are not available at the time the birds have to make their decision, they do not know exactly what the habitat has to offer in the future. They need to rely on direct or indirect cues to assess quality of habitat and the use of space within a habitat.

For example, that question was explored by Smith and Shugart (1987) in a study of habitat selection by the ovenbird (*Sieurus aurocapillus*), a ground-foraging warbler of the eastern North American deciduous forest (Figure 13.21c). They hypothesized that structural habitat cues are the proximate factor determining territory settlement and size. To test the hypothesis, they examined the relationships among habitat structure, prey abundance, and intrapopulation variation in the ovenbird in a deciduous forest in Tennessee. They assessed vegetation structure, plotted territories, and determined prey abundance by sampling forest floor litter invertebrates from each of 115 sample vegetation plots contained within 23 territories of ovenbirds and from 100 plots outside of ovenbird territories during late April and early May (the time when arriving ovenbirds select their habitat within the woods). T. M. Smith and Shugart found a significantly higher prey abundance per unit area within the territories than out-

side the territories. Seventy-five percent of the variation in prey abundance was correlated with habitat structure, particularly the nature of the canopy. Forest litter invertebrates were most abundant in those areas of the forest with large trees, a closed canopy, and a sparse understory. In contrast, areas of more open canopy and dense understory had fewer invertebrates. The microclimate of the forest floor under a closed canopy, with its lower temperatures, less drastic diurnal fluctuations in temperature and moisture in forest litter, and increased relative humidity, provided superior habitats for forest litter invertebrates, major food of the ovenbird. The ovenbird apparently cued in on areas of closed canopy and sparse understory as microhabitats within the woods offering the greater abundance of food. This hypothesis was supported by a significant correlation between predicted prey abundance and territory size. However, variations in territory size were related to structural features of the habitat rather than to prey abundance or intraspecific competition.

Habitat selection is a common behavioral characteristic of vertebrates; fish, amphibians, reptiles, birds, and mammals furnish most of the examples. Garter snakes (*Thamnophis elegans*) living along the shores of Eagle Lake in sagebrush-ponderosa pine country in northeastern California selected as their retreat sites rocks of intermediate thickness (20 to 30 cm) over thinner and thicker rocks. Shelter under thin rocks became lethally hot; shelter under thicker rocks would not allow the snakes to warm to their preferred range of body temperature (T_b) (see Chapter 8). Under the rocks of preferred thickness, the snakes would never overheat, and they would achieve and maintain their preferred body temperature for a long period (Huey et al. 1989, Huey 1991). Insects, too, cue in habitat features. Whitham (1980) found that the gall-forming aphid *Pemphigus*, which parasitizes the narrowleaf cottonwood (*Populus angustifolia*), selects the largest leaves to colonize and discriminates against small leaves. Beyond that they select the best positions on the leaf (see Chapter 12). Occupancy of this particular habitat, which provides the best food source, produces individuals with the highest fitness.

Even though a given habitat may provide suitable cues, it still may not be selected. The presence of others of the same species may be necessary to attract more individuals. In social or colonial species like herring gulls (Dorst 1958), an animal will choose a site only if others of the same species are already there. On the other hand, the presence of predators and human activity may discourage a species from occupying otherwise suitable habitat. Human activities on northern lakes inhibit the nesting of loons (Reem 1976).

Most species exhibit some plasticity in habitat selection. Otherwise these animals would not settle in what appears to us as less suitable habitat or colonize new habitats. Often individuals are forced to make this choice. Available habitats range from optimal to submarginal; the optimal habitats, like good seats at a concert, fill up fast. The marginal habitats go next, and the latecomers or subdominant individuals are left with the poor habitats where they may have little chance of reproducing successfully (Fretwell and Lucas 1969) (see chapter 12).

The ability of some members of a particular species to select habitats that deviate from those of others must exist on both a phenotypic and genetic level. Ovenbirds in the northern part of their range select woodland habitats that are exactly the opposite of those in Tennessee. The Ontario birds prefer a more open canopy and denser understory (Stenger and Falls 1959). The differences in habitat may result from variations in plant species associated with the structural gradient and from their influence on the composition, microclimate, and chemistry of the forest floor, which affects prey abundance. Late successional stages in Ontario forests are dominated by mixed conifers and hardwoods, whereas those in Tennessee are deciduous. The contrast in ovenbird habitats suggests that the species may have evolved patterns of habitat selection to match patterns of productivity over its geographical range (Smith and Shugart 1987). Another example of strong contrast in choice of habitat types is found in the black-throated green warbler (*Dendroica virens*). It is associated with coniferous forests in the northern part of its range (Robiehaud and Villard 1999) and with drier oak forests in the middle and southern Appalachians (Collins 1983).

Do plants select habitats, and if so how? F. A. Bazzaz (1991) explored some answers. Plants can hardly get up and move about to find a suitable site. The only recourse plants have in habitat selection is to send out seeds in anticipation that they will arrive at some place suitable for seedling germination and survival.

Plants, like animals, fare better in certain habitat types, characterized by such environmental factors as light, moisture, nutrients, and presence of herbivores and symbionts. Habitat choice involves the ability of plants to disperse with the aid of wind, water, or animal agents to preferred patches of habitat, which more often than not involves an element of chance. The problem for plant dispersers is that many plants have the same set of requirements for light, nutrients, and the like, so that they may face competitors once the propagule arrives. A plant has a better chance of survival and growth if it is a habitat specialist. Once established, such plants can expand into surrounding areas by modular growth.

Consider the cattail, a habitat specialist inhabiting marshy places. Each year a cattail stand sends out billions of seeds that are carried miles by the wind. Most settle on highly unsuitable places. Just a few seeds drop into very wet roadside ditches where they quickly germinate, take root, and within a few years spread their rhizomes into surrounding wet ground.

The rapid loss and degradation of habitats that is causing the decline of many species of animals has stimulated intensive research since the 1970s into the relationship between organisms and their specific habitat requirements (see Verner et al. 1986). Such knowledge is imperative if we are to maintain and restore wildlife habitats. Habitat assessment and habitat modeling are two approaches to determining essential features of an organism's habitat. Unfortunately, a model that predicts habitat features for a species in one part of its range may not apply in other parts of its range. We must incorporate habitat assessment information into land use management, development, and zoning plans. Failure to understand the habitat requirements of animals and plants and to use such knowledge in land use planning leads to species extinctions.

Summary

1. Life history patterns encompass traits and behaviors that help individuals achieve fitness. These activities range from sexual reproduction and sexual selection to components of an individual's reproductive effort including parental care, investments, and expenditures.

2. Sexual reproduction requires the acquisition of mates, which involves both mating systems and sexual selection. Mating systems include two basic types, monogamy and polygamy. Two general kinds of polygamy are polygyny, in which the male acquires more than one female, and polyandry, in which the female acquires more than one male. The potential for competitive mating and sexual selection is higher in polygamy than in monogamy.

3. Sexual selection, a form of natural selection, involves intrasexual and intersexual selection. Intrasexual selection or male competition increases selective pressures for the evolution of horns, large size, or exaggerated morphological features such as elaborate plumages and bright colors that in some way influence female choice. Intersexual selection involves female choice among males offering resources or offering only genes. Sexual selection favors traits that enhance mating success even if they handicap the male by making him more vulnerable to predation.

4. Optimal fitness in sexual reproduction can be achieved only if parents balance energy allocated to present reproduction with that allocated to survival and future fecundity. One alternative is to invest a maximum amount of energy into a single reproductive effort in a lifetime, as exemplified by annual plants, many insects, and some species of salmon. The other alternative is to allocate less energy to each reproduction and repeat reproductive efforts through a lifetime. Organisms may invest reproductive effort into many small offspring and provide a minimal amount of parental care; or they may invest a similar amount of energy into fewer, larger individuals and extended parental care. A single reproductive effort, semelparity, or production of many young with minimal energy invested in each is characteristic where mortality of adults is high. Because little difference in potential for population increase exists between single and repeated reproduction, repeated reproduction, iteroparity, may be a response to an unpredictable survival of individuals from young to adult. By reproducing several times, an organism is more likely to assure reproductive success.

5. Reproduction has costs. Individuals with repeated reproduction face decreased future survival and decreased future fecundity with each present investment in reproduction. Heavy early investment in reproduction can reduce individual survival and thus future fecundity. Light investment can improve survival and improve future fecundity, but the individual still has potential lower life expectancy. Somewhere in between lies the best strategy for optimal lifetime reproductive success, the sum of present and future reproductive success.

6. Resource allocation may influence sex ratios in the offspring because of the differential costs of producing males and females. Natural selection often favors those parents who invest equally in sons and daughters. When both are equally expensive, the sex ratio will be 1:1. The costs of rearing, however, may not be equal, and the sex ratio will become skewed toward the less expensive sex.

7. Allocation of energy to reproduction is related in part to mortality of parents and offspring, its relation to the population, and predictability of the environment. Organisms living in an unpredictable environment or subject to heavy environmentally induced mortality tend to allocate a greater proportion of energy to reproduction, and to expand rapidly when conditions are favorable. Such organisms are said to be r-selected (from the r term in the logistic equation) because selection favors high productivity. Organisms that occupy a more predictable environment and are more subject to density-related mortality tend to allocate less energy to reproduction. They are said to be K-selected (from the K term in the logistic equation) because selection favors efficient use of the environment.

8. Because quality of habitat is essential to reproductive success, habitat selection is an important part of an organism's life history pattern. How animals cue in on habitat quality is not clearly understood, but features such as structure and diversity of vegetation cover, particularly as they relate to food, appear to be important. Quality may be modified by the availability of nesting sites, song perches, escape cover, and the like. Individuals across a species' range exhibit regional plasticity in selection of habitats.

Review Questions

1. What forms might sexual reproduction take? What are the stimuli for and the advantages of sexual reversals?

2. What are the advantages and disadvantages of monogamy and polygamy for males and females? What are the several types of polygamy?

3. What is sexual selection and how does it function? Which sex does the selecting? What are some cues?

4. What is the relationship between intrasexual selection and intersexual selection?

5. Distinguish between resource-based and genes-only sexual selection. Which one imposes the greater selection pressure on males?

6. What is unique about lek behavior? Discuss the three models: female choice, hotspot, and hotshot. Does one model really explain all lek behavior?

7. Distinguish between altricial and precocial young.

8. Define reproductive effort.

9. What are the costs of reproduction? What is the significance of how these costs are apportioned? How does brood reduction enter the picture?

10. What is the relationship between clutch size and fitness?

11. What is the relationship among the parent's age, size, and fecundity?

12. Report on the lek behavior of other species such as prairie chicken, black grouse, bird of paradise, Uganda kob, and hammer-headed bats. What model do they fit, if any?

13. In a randomly mating population, male and female offspring should be produced in equal numbers, and parents should invest equally in both sexes. Why are sex ratios in many natural populations skewed toward one sex or the other?

14. In the light of this chapter, discuss such human behavior as mate selection, family size, investment in young, and the relationship between age and fecundity. (See Buss 1994.)

Cross-References

Adaptation, 81; energy allocation, 485; modular populations, 161; reproductive values, 225; population genetics, 369; natural selection, 81; fitness, 81; intraspecific predation, 299; parasitism, 307.

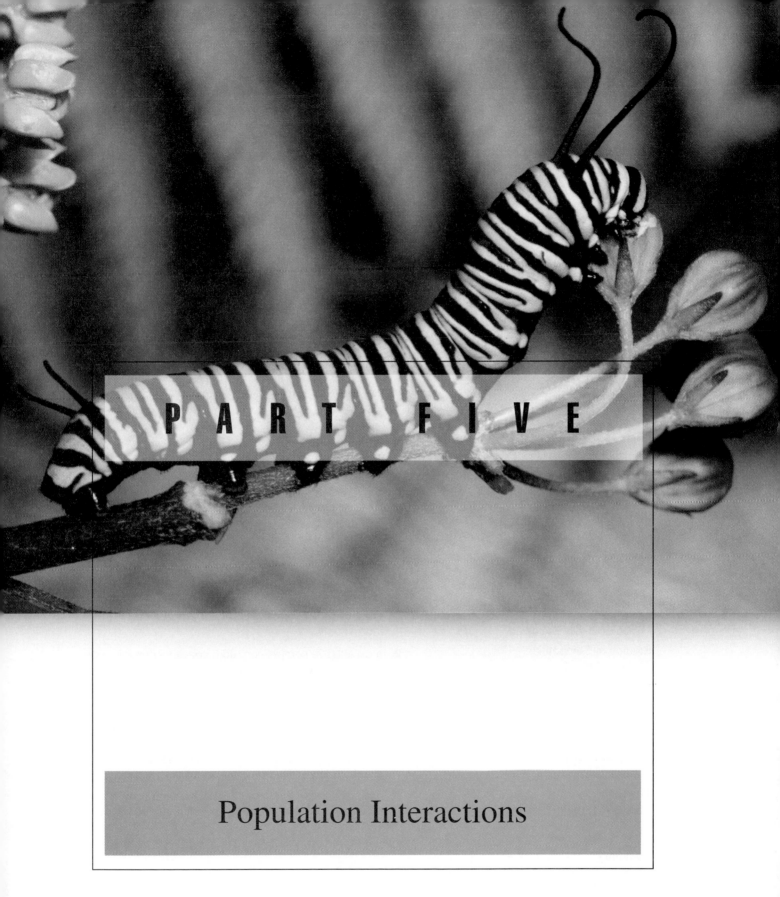

PART FIVE

Population Interactions

C H A P T E R 14

Interspecific Competition

CONCEPTS

1. Interspecific competition results when two or more different species seek the same resource that is in short supply.

2. In exploitative competition, both species consume the resource; in interference competition, one competitor denies access to a resource.

3. Interspecific competition may result in stable or unstable coexistence or in exclusion of a competitor.

4. Competition is reduced by the differential use of resources by potential competitors or resource partitioning and by environmental variability.

5. The niche is the functional role of an organism in a community. It may or may not be constrained by interspecific competition.

Up to this point the emphasis has been on single-species populations. Now we must broaden our scope. Individuals within a single-species population interact with individuals of different species as well with others of their own kind. The effects of these interactions on population growth can be positive, negative, or neutral (Table 14.1).

Neutral interactions (designated 0 0) have no effect on the growth of interacting populations. Positive interactions (+ +) benefit both populations. Such relationships are termed **mutualism** (Chapter 17). In some situations, a one-sided relationship develops in which one species benefits and the other is neither benefited nor harmed (+ 0). This relationship is known as **commensalism.** Examples of commensalism are epiphytes, plants that grow on the branches of trees. They depend on trees for support only; their roots draw nutrients from the humid air. In another one-sided relationship between two species, one population is negatively affected while the other remains unaffected (− 0). This relationship is called **amensalism.** A nebulous relationship, amensalism probably involves a chemical interaction such as the production of an antibiotic or allelochemical agent by one of the organisms. Other relations positively affect one population and are detrimental to the other (+ −). Such relationships are predation and parasitism. **Predation** (Chapter 15) is the killing and consumption of prey. **Parasitism** (Chapter 17) is an interaction in which one, usually small, organism (the parasite) lives in or on another (the host) from which it obtains food. Finally, relationships can have adverse effects on both populations (− −). Such relationships develop when a needed resource is in limited supply relative to the number seeking it. This interaction between species is **interspecific competition.**

Individuals within a population experience intraspecific competition with members of their own kind (Chapter 12). At the same time they experience to a greater or lesser degree interspecific competition with individuals of other species. To demonstrate that interspecific competition is important, we have to show that one species uses a resource to the extent that it limits the population size of another.

Interspecific competition may be exploitative or interference, two terms somewhat akin to the scramble and contest types of intraspecific competition. In **exploitative competition** the species use the same resource such as food. Use by one reduces availability for another. The outcome is determined by how efficiently each of the competitors uses the resource. It often results in reduced growth of all competitors. **Interference competition** is a direct interaction between competitors in which one interferes with or denies access to the resource by another. In animals, interference usually involves aggressive behavior.

Among individual plants and sessile animals, both fixed in space, interspecific competition is influenced by the degree of proximity. Each individual affects the environment of its neighbor by consuming resources in limited supply, by modifying environmental conditions (for example, shading and protecting plants from wind and predators), and by producing toxins. These changes can alter the rate of growth, biomass accumulation, and the growth form of individual plants.

CLASSIC COMPETITION THEORY

The Lotka-Volterra Model

In the early part of the twentieth century two mathematicians, the American Alfred Lotka and the Italian Vittora Volterra, independently arrived at mathematical expressions to describe the interaction between two species using the same resource. Both began with the logistic equation for population growth presented in Chapter 11. For two species designated as species 1 and species 2, these equations can be written as follows:

$$\frac{dN_1}{dt} = r_1 N_1 \left(\frac{K_1 - N_1}{K_1} \right)$$

$$\frac{dN_2}{dt} = r_2 N_2 \left(\frac{K_2 - N_2}{K_2} \right)$$

Here N_1 and N_2 are the population sizes for species 1 and species 2, r_1 and r_2 are the respective intrinsic rates of increase, and K_1 and K_2 are the population carrying capacities. They then added to the logistic equation for each species a coefficient to account for the competitive effect of one species on the population growth of another. For species 1, αN_2 is the coefficient that gives the competitive effect of species 2; N_2 is the number of species 2 individuals and α is competitive impact per individual of species 2 on species 1. This constant, in effect, converts the number of members of one species population (2) into an equivalent number of members of the other (1). Similarly for species 2, the competition coefficient is βN_1. The paired equations, which now consider both intraspecific and interspecific competition, are

Species 1: $\dfrac{dN_1}{dt} = r_1 N_1 \left(\dfrac{K_1 - N_1 - \alpha N_2}{K_1} \right)$

Species 2: $\dfrac{dN_2}{dt} = r_2 N_2 \left(\dfrac{K_2 - N_2 - \beta N_1}{K_2} \right)$

As you can see in the absence of any interspecific competition—either α or $N_2 = 0$ in the first equation, and β or $N_1 = 0$

TABLE 14.1 **Population Interactions, Two-Species System**

	RESPONSE	
Type of Interaction	A	B
Neutral	0	0
Mutualism	+	+
Commensalism	+	0
Amensalism	−	0
Parasitism	+	−
Predation	+	−
Competition	−	−

Note: 0 = no direct effect; + = positive effect on growth of population; − = negative effect on growth of population.

in the second—the population of each species grows logistically to equilibrium at its K or carrying capacity.

Remember that in the logistic equation, as the number of individuals in each population (N) increases toward its carrying capacity (K), the growth of the population (dN/dt) approaches 0. In two competing populations the inhibitory effect of the competing population must also be considered. For example, the carrying capacity for species 1 is K_1, and as N_1 approaches K_1 the population growth (dN_1/dt) approaches zero. However, since species 2 is also vying for the limited resource that determines K_1, we must consider the impact of species 2. Since α is the per capita effect of species 2 on species 1, the total effect of species 2 on species 1 is αN_2. The population density of species 2 that will exactly equal K_1 can be calculated as K_1/α. If the population of species 2 (N_2) is equal to K_1/α, the population of species 1 can never increase. Conversely, the population density of species 1 that exactly equals K_2 is K_2/β. If the population density of species 1 (N_1) is equal to K_2/β, the population of species 2 can never grow.

We consider both effects in calculating population growth. As the combined population effect ($N_1 + \alpha N_2$) approaches K_1, the growth rate of species 1 will approach zero. The greater the density of the competing species (N_2), the greater the influence on reducing the growth rate (dN_1/dt) of species 1. The lower the growth rate of species 1, the lower its density (N_1). The lower the density of species 1, the less its competitive effect on species 2 (βN_1). The less its competitive effect on species 2, the higher the growth rate of that species (dN_2/dt). The higher the growth rate, the greater the negative influence of species 2 on species 1 (αN_2). This cycle is opposed by that for the negative effect of species 1 on species 2.

The outcome of competition, then, depends on the relative values of K_1, K_2, α, and β. If $N_2 = K_1/\alpha$, N_1 can never increase; and if $N_1 = K_2/\beta$, N_2 can never increase. To state it differently, the presence of species 1 decreases the carrying capacity for species 2 at a certain rate. Similarly, species 2 decreases the carrying capacity for species 1 at a certain rate. The reason, of course, is that each species has to share limited resources with the other.

Depending on the combination of values for the Ks and for α and β, the Lotka-Volterra equations predict four different potential outcomes. In two situations one species wins out over the other. In one case species 1 inhibits further increase in species 2 while continuing to increase itself. In this case species 2 is driven to extinction. Conversely species 2 may inhibit further increase in species 1 while continuing to increase itself; then species 1 eventually disappears. In the third situation each species when abundant inhibits the growth of other species more than it inhibits its own growth. The outcome depends in part on which species is the most abundant. The two species may coexist for a while, but eventually one species wins. In real-life situations the outcome may depend on which of the two species has the competitive advantage in the face of environmental change over time. In the fourth situation the two species coexist. In that situation neither population can achieve a density capable of eliminating the other. Each species inhibits its own population growth more than it inhibits the population growth of the other species.

What the equations describe can be understood more easily with some graphic models provided in Figure 14.1. In each graph the x axis represents the population size of species 1, the y axis that of species 2. Two lines are plotted on each graph in Figures 14.1c–f, one representing each of the two species. The diagonal line for species 1 represents the combined population densities of species 1 and 2 that equal K_1; on this line, therefore, $dN_1/dt = 0$. The line for species 2 is the combination $N_1 + N_2$ that equals K_2. So for any point on the species 1 line, $N_1 + N_2 = K_1$. When $N_1 = K_1$, N_2 must then be zero. When $N_2 = K_1/\alpha$, N_1 must then be equal to zero. For the line describing species 2, the values where the line crosses the axes will be $N_2 = K_2$ and $N_1 = K_2/\beta$.

The diagonal lines are referred to as the zero growth isoclines. The zero growth isoclines for species 1 and 2 are shown in Figures 14.1a and b respectively. The space below the isocline for species 1 (Figure 14.1a) represents combinations of N_1 and N_2 below the carrying capacity; here, therefore, the population is increasing (dN_1/dt greater than zero). This is represented by the arrows parallel to the x axis and pointing in the direction of increasing values of N_1. The space above the isocline represents combinations of N_1 and N_2 that are greater than carrying capacity; here, therefore, the population is declining (dN_1/dt is negative). In this region the arrows point in the direction of decreasing population size.

Figures 14.1c–f represent the possible outcomes when the two isoclines are combined. In Figure 14.1c the isocline of species 2 is parallel to, and lies outside, the isocline of species 1. In this case, even when the population of species 1 is at its carrying capacity (K_1), the density cannot stop the population of species 2 from continuing to increase. As species 2 continues to increase, it eventually leads to the extinction of species 1. In Figure 14.1d the situation reverses and species 1 wins, leading to the exclusion of species 2.

FIGURE 14.1 (opposite) The Lotka-Volterra model of competition between two species. In (a) and (b) populations of species 1 and 2 in the absence of competition will increase in size and come to equilibrium at some point along the diagonal line or isocline of zero growth ($r = 0$). In the shaded area below the line, r is positive and the population increases (as indicated by arrows); above the line, the population decreases. In (c) species 1 and 2 are placed in competition. Because the isocline, the zero growth curve, of species 1 falls outside the isocline of species 2, species 1 always wins, leading to the exclusion of species 2. In (d) the situation is reversed, and species 2 wins, leading to the exclusion of species 1. In (e) and (f) the isoclines cross. Each species, depending on the circumstances, is able to inhibit the growth of the other. In (e) each species inhibits the growth of the other species more than it inhibits its own growth. Which species wins often depends on the initial proportion of the two species. In (f) neither species can exclude the other. Each inhibits the growth of its own population by intraspecific competition more than it inhibits the growth of the other population.

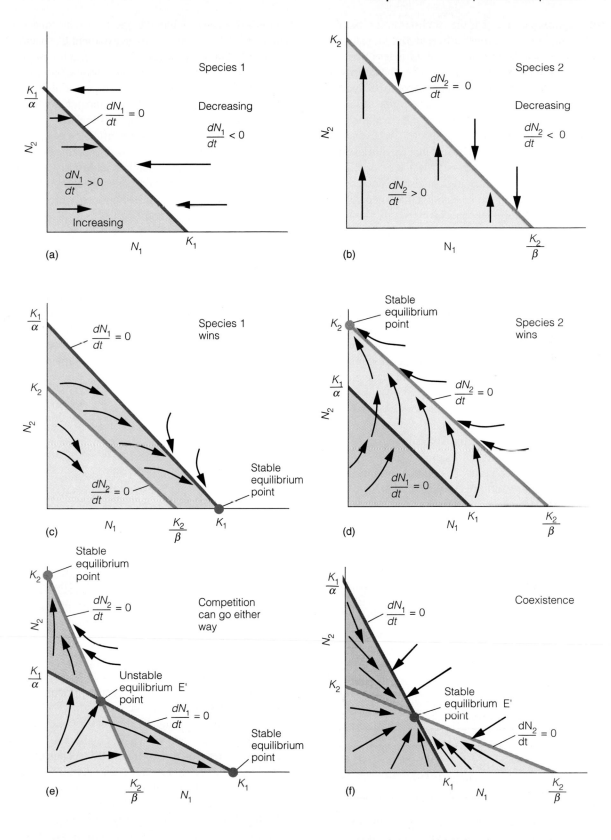

In both Figures 14.1e and f, the isoclines cross, but the outcomes of competition are quite different. In Figure 14.1e, note that along the x axis, the value of K_2/β is less than the value of K_1. Recall that the value $N_1 = K_2/\beta$ is the density of species 1 that exactly equals the carrying capacity of species 2 (K_2). The fact that K_2/β is less than K_1 means that species 1 can achieve population densities that would exceed the density required to drive the population of species 2 to extinction (produce negative growth rates). The fact that K_1/α is less that K_2 along the y axis shows that, likewise, species 2 can achieve population densities sufficiently high to drive species 1 to extinction. Which species "wins" the competition depends on the initial populations of the two species.

In Figure 14.1f the result of competition in quite different. In this case neither species can exclude the other, leading to coexistence. Note that in this case, K_1 is less than K_2/β. This means that the population density of species 1 can never reach a density sufficient to eliminate species 2. The reason is that for this to happen, the population of species 1 would have to reach $N_1 = K_2/\beta$. Likewise, K_2 is less than K_1/α, and so species 2 can never achieve a high enough population density to eliminate species 1. Intraspecific competition inhibits the growth of each population more than it inhibits the growth of the other species.

Some major assumptions lie behind the Lotka-Volterra model: (1) the environment is homogeneous and stable, without any fluctuations; (2) migration is unimportant; (3) the effect of competition is instantaneous; (4) coexistence requires a stable equilibrium point; (5) competition is the only important biological interaction (see Schoener 1982, Chesson and Case 1986, Roughgarden 1986).

Obviously these assumptions are somewhat unrealistic. Immigration can occur, competition is not instantaneous, and environmental changes can affect competitive outcomes. Consider, for example, two species (species 1 and species 2) that draw upon the same limiting food resources, seeds. Second, assume that the size distribution of seeds and their abundance varies as a function of environmental conditions. For example, in Figure 14.2 the average seed size increases from environment A to B and C. As the size distribution of seeds changes, so will the carrying capacity (K) for each species. Let us assume the following relationship:

Environment

	A	B	C
Species 1 (K_1)	225	150	75
Species 2 (K_2)	75	150	225

The interspecific competition coefficients will be a function of diet overlap and the relative consumption rates of the two species For this example let us assume

$$\alpha = 0.5$$
$$\beta = 0.5$$

The outcomes of competition in these three environments using the Lotka-Volterra equations are graphically depicted in

Figure 14.2. In environment A, species 1 wins; in environment B, species 1 and 2 coexist; in environment C, species 2 wins.

The Lotka-Volterra equation and graphic models based on them apply well to animal populations, but not to plants. The models are based on an animal's potential for increase in numbers. In many plant populations the potential for increase is in biomass. Accordingly, competitive relations among plants may be examined in terms of the influence of one species on the growth of another, where growth is defined in terms other than population density, such as yield of dry matter or production of tillers. Such experiments typically follow changes in the proportion of two competing species through planted in an initial mixture where both species are equally represented (in density, biomass, etc.). An alternative procedure is to start the experiment with mixtures of two species sown at a variety of proportions and to detect changes in each as time elapses. The proportions at the end of the period are then plotted against the initial proportions. This procedure can be helpful in revealing density-dependent interaction. The potential outcomes of such experiments can be described by four basic types of interactions (Figure 15.3), suggestive of those predicted by the Lotka-Volterra equations (Harper 1977):

1. The proportion of the two species remains unaltered through time. The balance of the two species is subject only to random variation. Such interactions are rarely if ever seen in nature.

2. One or the other species has the competitive advantage in the mixture at all proportions.

3. The minority species is always at an advantage. If a high proportion of species A is sown, species B gains, and if a high proportion of species B is sown, species A gains. Mixtures will always tend toward an equilibrium mixture.

4. The majority species is always at an advantage. If a high proportion of species A is sown, species B goes to extinction, and visa versa. The outcome of competition is dependent of the starting proportions and there is no equilibrium mixture.

Competitive Exclusion

The Lotka-Volterra equations predict, as graphs (c) and (d) in Figure 14.1 illustrate, that if one species in a competitive situation grows rapidly enough to prevent the population increase of another, it can reduce that population to extinction or exclude it from the area. This model led to **Gause's principle,** named for the Russian ecologist who demonstrated the concept experimentally. It states that two species with identical ecological requirements cannot occupy the same environment.

The idea was far from original with Gause (and he laid no claim to it). For example, the ornithologist Joseph Grinnell wrote in 1904: "Two species of approximately the same food habits are not likely to remain long evenly balanced in numbers in the same region. One will crowd the other out. The one longest exposed to local conditions, and hence best fitted, though ever so slightly, will survive to the exclusion of any less-favored would-be invader."

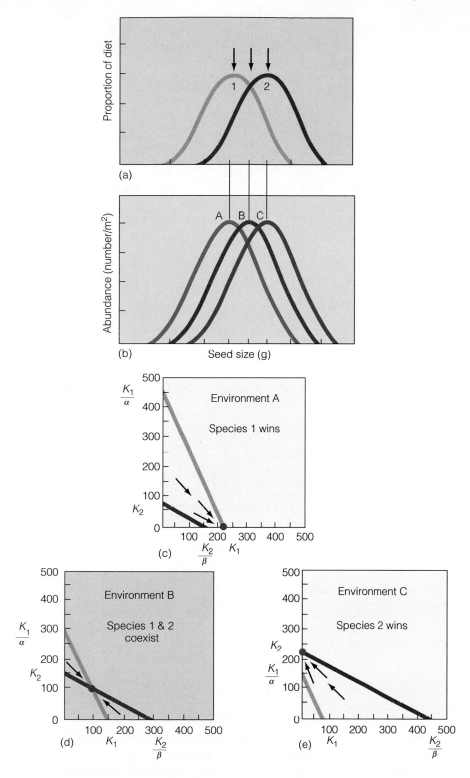

FIGURE 14.2 Hypothetical example of the interaction of two bird populations vying for a common food resource—seeds. The proportion of different seed sizes making up the diets of the two species is shown in (a); the relative abundance of the different seed sizes in three different environments (A, B and C) is shown in (b). If we assume the parameters for the Lotka-Volterra model given in the text, the outcomes for the competitive interactions of the two species in these three different environment are shown in (c)–(e).

The concept eventually gained the name **competitive exclusion principle.** Hardin (1960) wrote: "Complete competitors cannot coexist. Two competing species with identical ecological requirements cannot occupy the same area." However, this concept is not really a principle (Cole 1960). It is little more than an ecological definition of a species. A corollary of the statement is that if two species coexist, they must possess ecological differences. Obviously, two separate species cannot have identical requirements; being different

species, they must have somewhat different ecologies. However, two or more species can compete for some essential resource without being complete competitors.

Pielou (1974) provides a set of conditions in addition to the utilization of resources in short supply that should be met for competitive exclusion to take place: (1) competitors must remain genetically unchanged for a sufficiently long period of time for one species to exclude the other; (2) immigrants from areas with different conditions cannot move into the

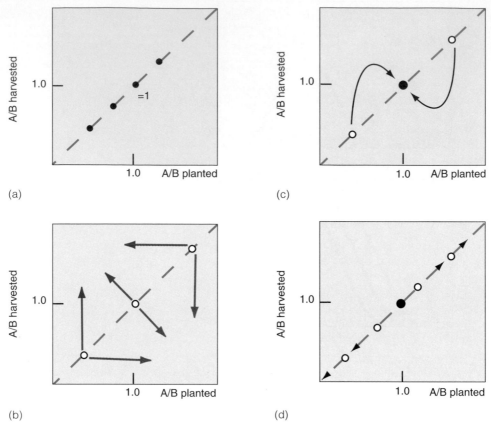

FIGURE 14.3 Ratio diagrams showing the possible outcomes of competition between two plant species (species A and B) planted at different initial proportions. The x-axis is the initial ratio of the two species at planting, while the y-axis is the ratio of the two species at harvest. Values below 1.0 on both axes indicate a higher abundance of species B relative to species A, and values greater than 1.0 a greater abundance of species A relative to species B. The dashed line represents the condition where the ratio at harvest is equal to that at planting. The points represent the ratio of the two species at the time of planting (x-axis), and the harvest ratio (y-axis) at the time of planting, which is assumed to be the same as the initial ratio. The arrows show the expected pattern of change (through time) in the ratio of the two species through time, from initial planting to final harvest. Four basic types of outcome are possible as a result of competitive interactions: a) the ratio of the two species remains unaltered, therefore there are no arrows suggesting a shift in the ratio through time; b) one species or the other wins independent of the initial proportions in which they are planted (blue arrows represent competition in favor of species B and red in favor of species A); c) the minority species is at a competitive advantage, with ratio converging on a stable equilibrium (ratio of 1.0); d) competitive advantage to the species with the highest initial proportion (ratio of 1.0 is an unstable equilibrium point).

population of the losing species; (3) environmental conditions must remain constant; and (4) competition must continue long enough for equilibrium to be reached. In the absence of any of these requirements, species usually coexist.

Competition, for simplicity, is usually considered on a one-dimensional gradient such as food size or water availability; but in natural situations competition is spread among species over a number of resources. In such situations, the combined effects of minimal competition on one species by a number of other species can be equivalent to strong competitive interaction for one resource from a single competing species. This relationship has been termed **diffuse competition** by MacArthur (1972). Theoretically, diffuse competition can exclude a species or greatly reduce its numbers through competitive interactions with a specific combination of other species, rather than with just one strong competitor.

STUDIES OF COMPETITION

Laboratory Studies

The best place to observe interspecific competition is in laboratory cultures of small invertebrates and microorganisms and in the greenhouse, isolated from environmental fluctuations and outside interference. Gause (1934) set out to test the Lotka-Volterra equations experimentally. He used two species of *Paramecium*, *P. aurelia* and *P. caudatum*. *Paramecium aurelia* has a higher rate of increase than *P. caudatum*. When both were introduced into one tube containing a fixed amount of bacterial food, *P. caudatum* died out. The population of *P. aurelia* interfered with the population growth of *P. caudatum* because of its higher rate of increase (Figure 14.4). (Compare Figures 14.1c and d.) In another experiment, Gause used *P. caudatum* and *P. bursaria*. Both species reached stability,

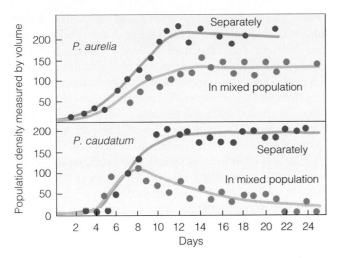

FIGURE 14.4 Competition experiments with two related ciliated protozoans, *Paramecium aurelia* and *P. caudatum,* grown separately and grown in a mixed culture. In a mixed culture *P. aurelia* outcompetes *P. caudatum,* and the result is competitive exclusion. (From Gause 1934.)

because *P. bursaria* confined its feeding to bacteria on the bottom of the tube, whereas *P. caudatum* fed on bacteria suspended in solution. Although the two used the same food supply, they occupied different parts of the culture. In effect, each utilized food unavailable to the other. (Compare Figure 14.1f.) Park (1948) and Crombie (1947) carried out competition experiments involving several species of flour beetle and obtained results similar to those of Gause.

Tilman and associates (Tilman et al. 1981) grew laboratory populations of two species of diatoms, *Asterionella formosa* and *Synedra ulna,* which require silica for the formation of cell walls. They monitored not only population growth and decline but also the level of silica. When grown alone in a liquid medium to which resource (silica) was continually added, both species kept silica at a low level (Figures 14.5a and b). When grown together, *S. ulna* took silica to a level below which *A. formosa* could not survive and reproduce (Figure 14.5c). In this experiment *S. ulna* competitively excluded *A. formosa* by reducing resource availability.

Such laboratory studies of two interacting species under controlled and manipulated conditions demonstrate the reality of interspecific competition. They provide the experimental data needed to test and modify mathematical models of competition and refine the concepts of competition. However, do the findings of laboratory experiments apply to the more complex competitive interactions in the real world of nature?

Field Studies

A major tenet of ecology holds that competitive interactions among taxa, genera, and trophic levels determine the nature and structure of natural communities (see Chapter 20). In a classic paper, Hairston, Smith, and Slobkin (1960) hypothesized that herbivores do not compete because they are predator-limited, whereas producers and carnivores experience strong competition. In 1976 Menge and Sutherland proposed

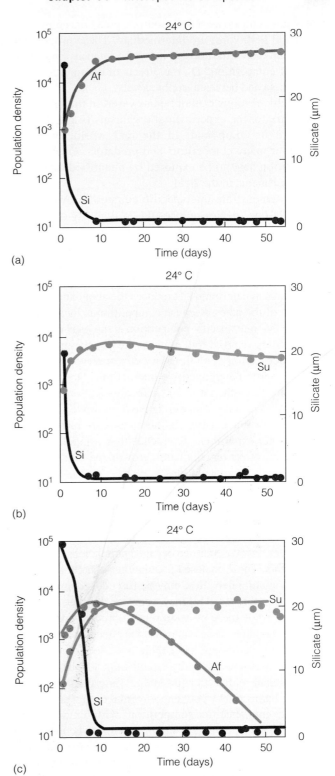

FIGURE 14.5 Competition between two species of diatoms, *Asterionella formosa* (Af) and *Synedra ulna* (Su), both of which require silica (Si) for the formation of cell walls. (a) When grown alone in a culture flask, *A. formosa* reaches a stable population level at which it keeps silica at a constant low level. (b) *Synedra ulna* does the same, except that it draws the silica down to an even lower level. (c) When the two species are grown together in a culture flask, *S. ulna* drives *A. formosa* to extinction, because *S. ulna* reduces the silica level to a point below which the other species cannot exist. (From Tilman et al. 1981:1025, 1027.)

that plants compete least for resources, predators compete the most, and herbivores are intermediate. These propositions opened up more questions. What is the intensity and overall effect of competition? Do the effects of competition differ among taxa and between trophic levels?

Such questions cannot be answered in the laboratory, where life for the experimental organisms is akin to being in a jail or compound. In the real world, organisms encounter many competitors and predators. Questions on competition have to be explored by quantified, manipulative experiments in the field.

To demonstrate interspecific competition we have to show that one species reduces the availability of a shared resource and that in doing so it affects the well-being of the other. Interspecific competition at the level of the individual may reduce breeding success or survival. To see whether such competition affects a species at the population level, we would have to determine if it reduces the abundance and distribution of the other. Because competition is localized, it is possible that interspecific competition at the level of the individual will have no overall effect at the population level.

Under natural conditions, we can observe fairly easily what appear to be competitive interactions. One species of bird replaces another at a winter feeding station; bluebirds, tree swallows, and starlings compete for a limited number of nest boxes or tree cavities. In these cases we have observations but not experimental evidence that competition has a pronounced effect on population growth and on the survival of individuals and populations.

Attempting to demonstrate interspecific competition under truly natural conditions is difficult. The experimental populations must live naturally, yet be held under some type of control amenable to experimental manipulations. Most field experiments involve enclosed or caged populations of assumed competitors. These enclosed populations, usually held above or below natural densities, may be forced to compete more actively than they would if they were free. Further, the organisms may enjoy more protection from predators that, under uncaged conditions, would alleviate the effects of competition.

Field experiments have their own sets of problems. Experimenters have difficulty establishing adequate controls and sufficient replicates and manipulating the populations. They are hampered by poor experimental design and by confounding interspecific competition with environmental effects. However, the task is not impossible. The hypotheses to be tested must be well defined, the experimental procedures must represent the natural situation, and the experiments must be replicated.

Well over 100 field studies involving several hundred experiments have attempted to determine the effects of interspecific competition on species assumed to be competitors and the relative importance of competition at different trophic levels. Most of these studies have involved organisms that have similar resource requirements, including fish, amphibians, marine mollusks, aquatic insects and snails, and marine and terrestrial plants. In his survey of the literature Schoener (1983b) found that 90 percent of the studies and 76 percent of the species involved demonstrated competitive interac-

tions. Connell (1983) in his survey found competition demonstrated in most studies, in 50 percent of the species, and in two-fifths of the experiments. Many studies, however, involved intraspecific competition as well. Goldberg and Barton (1992) found that among plants interspecific competition was stronger than intraspecific competition, and that competition did affect distribution patterns, relative abundance, and diversity. Most studies they reviewed, however, were short-term, lasting from several months to three years.

Surveys of individual studies considered independently tell us little about wider effects. Gurevitch and associates (1992) took a different approach to their analysis of studies between 1980 and 1989 covering 93 species in a variety of habitats. They employed a statistical test, called meta-analysis (Hunter et al. 1982), a statistical synthesis of results from a set of primary studies. It employs differences between means of two groups, control and experimental, but does not reanalyze original data. The studies analyzed showed that interspecific competition does have a large overall effect, but this effect varies widely among organisms and experimental approaches. They detected small effects of competition among marine mollusks and echinoderms, small to moderate effects of competition among carnivores and primary producers, large effects among frogs, toads, and arthropods of streams, and no effects among herbivorous terrestrial insects. Effects among small herbivores vary. Among herbivores interspecific competition is less than intraspecific competition. As we might expect, caged organisms experience greater competition than organisms living free. Among plants, competition does not differ between highly productive and poorly productive ecosystems.

Coexistence Most interspecific competitive relations are probably expressed as stable or unstable equilibrium, if indeed competition occurs at all. Classical competition theory assumes that the environment is stable and competition is continuous, but in reality interspecific competition is probably discontinuous (see Wiens 1977, Wiens et al. 1986, Chesson 1986), because environments are variable and populations are patchily distributed in space and time. Organisms using identical but limited resources coexist because of different responses to a fluctuating environment and differing life history traits.

In variable environments resource levels vary between superabundance and scarcity. Periods of scarcity, which occur irregularly, create ecological crises that can result in intense interspecific competition and act as a major selective force. On the small Isla Daphne in the Galapagos Islands, Grant (1986) and his associates followed the populations of two species of Darwin's finches—*Geospiza fortis,* the medium ground finch, and *G. scandens,* the cactus ground finch—over ten years (Figure 14.6a). In that time there was one long dry period, brought about by El Niño (see Chapter 2) from May 1976 to January 1978. During that period seed production declined precipitously (Figure 14.6c). Many seed-bearing plants died and were not replaced until 1983, the next wet year, keeping seed availability low for several years. This period of food scarcity was accompanied by a population crash of *G. fortis* and a less drastic decline in the population of *G. scandens.* Foraging diets and behavioral changes reduced overlaps in food. The diet of *G.*

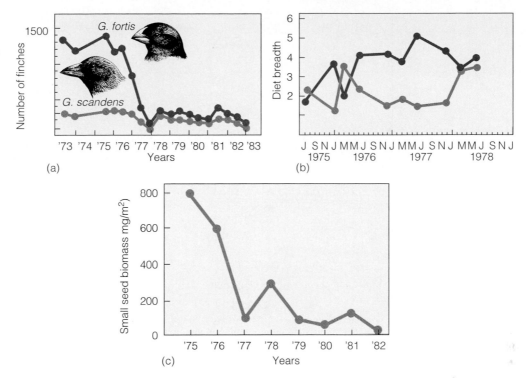

FIGURE 14.6 Change in (a) the population sizes and (b) breadth of diets of two of Darwin's finches, *Geospiza fortis* and *G. scandens,* in response to a drought on Isla Daphne. (c) The precipitous decline in the biomass of small seeds at the beginning of the dry season from 1976 to 1982. In response to this change in food availability, *G. scandens* decreased and *G. fortis* increased its diet breadth. After rainfall broke the drought in 1983, the changes were reversed. (From Grant 1986:180, 186.)

scandens, which specializes in the seeds of cactus, became narrower, and that of the generalist *G. fortis,* which feeds on a variety of seeds including those of cactus, became broader (Figure 14.6b). During the drought the diets of the two birds diverged and the overlap diminished, possibly because of the decline in jointly exploited foods.

Exclusion Because competitive relationships among individuals and species interact with the physical and biological environments, including physiological tolerances and predation, ecologists have a difficult time determining degrees of competitive exclusion. Connell (1961a, b) demonstrated some degree of competitive exclusion among two species of barnacles on the Scottish coast. Heller and Gates (1971) found some evidence of competitive exclusion among four species of chipmunks on an altitudinal gradient in the eastern slopes of the Sierra Nevada mountains of California.

In recent years the black duck (*Anas rubripes*) has been declining in northeastern North America and is being replaced by the mallard (*A. platyrhynchos*). To examine the hypothesis that mallards first invaded and then replaced black ducks on fertile wetlands in southern Ontario, Merendino et al. (1993) studied 131 wetlands. They surveyed occupancy of the wetlands for breeding populations of the two species over the previous 20 years; and they measured the characteristics of the wetlands, including size, shape of shoreline, percentage of emergent vegetation, distance from disturbance, and water chemistry.

The researchers divided the wetlands into seven categories: (1) wetlands invaded by mallards in the 1970s but never recorded as holding black ducks; (2) wetlands invaded by mallards in the 1980s but not used by black ducks; (3) wetlands where mallards replaced black ducks in the 1970s; (4) wetlands where mallards replaced black ducks in the 1980s; (5) wetlands inhabited by both species; (6) wetlands used only by black ducks; and (7) unused wetlands. Wetlands where the mallards first appeared were more fertile than the wetlands occupied later; and wetlands where mallards first replaced black ducks were more fertile than those occupied later. Black ducks currently are restricted to wetlands with very low fertility. The biologists concluded that by competitively excluding black ducks from productive wetlands (accompanied by some hybridization), mallards have contributed to the decline of the black duck in southern Ontario.

The decline and exclusion of black ducks from Ontario, western Quebec, and the mid-Atlantic and Great Lakes region, where mallards predominate, sometimes by a margin of 50 to 1, has been confirmed by an extensive survey by Ducks Unlimited research biologist Mark Petrie (1999). In contrast black duck populations are stable in eastern Quebec, the Atlantic provinces, and northern New England, where few mallard breed (Figure 14.7).

Competitive exclusion is most evident when exotic species successively invade new habitats and outcompete native species for space, nutrients, or other resources, eventually

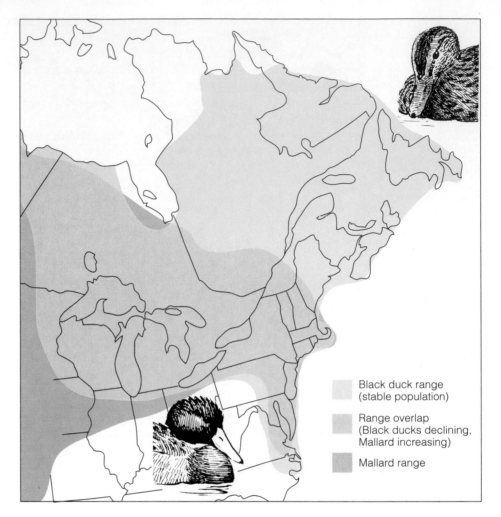

FIGURE 14.7 The mallard (*Anas platyrhynchos*), the Northern Hemisphere's most abundant duck, is a strong competitor of the black duck (*A. rubripes*). In recent years, the mallard populations have been increasing and expanding the periphery of their range into the northeastern North American range and coastal habitat of the black duck. Here, it is outcompeting the black duck, resulting in a strong decline in black duck populations. The black duck population remains stable in the northern part of its range, away from the mallard.

Black duck range (stable population)

Range overlap (Black ducks declining, Mallard increasing)

Mallard range

displacing them (Pimentel et al. 2000). The colorful purple loosestrife (*Lythrum salicaria*), a perennial from Europe, has invaded prime wetlands throughout the temperate regions of United States and Canada (Malecki et al. 1993) (Figure 14.8). It replaces native wetland species and eliminates natural food and cover essential to wetland wildlife. Its life history characteristics—including the production of 2.5 million long-lived easily dispersed seeds per plant and growth rates of established seedlings exceeding 1 cm/day—make it a formidable competitor to wetland plants. Similarly the Australian pine (*Casuarina*), introduced in southern Florida, is invading and displacing vegetation of the Everglades.

Among animal invaders, one of the most notorious is the zebra mussel (*Dreissena polymorpha*). A native of Russia, it invaded Lake Erie and Lake Ontario in 1985–86 via water ballasts in foreign ships. Lacking any ecological restraints, it spread dramatically to Lake Michigan, Lake Huron, and the Hudson, Illinois, Susquehanna, Mississippi, and Ohio rivers. It colonizes any hard surface, including pilings, boats, water intake pipes, and the shells of native mussels. The weight of zebra mussels prevents native mussels from opening their shells to feed. Highly efficient filter feeders, zebra mussels remove phytoplankton, small zooplankton, and suspended particles from the water, improving its clarity. But in doing so, zebra mussels remove a large percentage of primary produc-

FIGURE 14.8 The colorful growth of purple loosestrife (*Lythrum salicaria*) in this marsh belies its aggressive exclusion of native wetland plants.

tivity and thus reduce the population of consumers that depend on phytoplankton. Such interference in the food chain affects fish recruitment and growth (Ludyanski et al. 1993, Mills et al. 1994). The feeding activity of zebra mussels also reduces availability of edible particles for benthic consumers, mainly native bivalves, and increases the density of planktonic bacteria, not eaten by zebra mussels. In effect these bivalves divert resources from the pelagic and deep water zones to vegetated shallows and zebra mussel beds (Strayer et al. 1998)

Allelopathy A particular form of interference competition among plants is **allelopathy,** the production and release of chemical substances by one species that inhibit the growth of other species. These substances range from acids and bases to simple organic compounds that reduce competition for nutrients, light, and space. Produced in profusion in natural communities as secondary substances, most compounds remain innocuous, but a few may influence community structure. For example, broomsedge (*Andropogon virginicus*) produces chemicals that inhibit the invasion of old fields by shrubs, thereby maintaining its dominance (Rice 1972). Bracken fern (*Pteridium aquilinum*), the most widely distributed vascular plant in the world (Page 1982), produces plant poisons that accumulate in the upper surface of the soil. These phytotoxins kill the germinating seeds of many plants, especially conifers, and reduce growth of seedlings. These allelopathic effects along with the heavy, smothering overwinter accumulation of dead fronds allow bracken ferns to dominate competitively large areas of ground (Ferguson and Boyd 1988). Likewise, the black walnut (*Juglans nigra*) of the eastern North American deciduous forest is antagonistic to many plants (Brooks 1951).

In desert shrub communities a number of shrubs (*Larrea, Franseria,* and others) release toxic phenolic compounds to the soil through rainwater. Under laboratory conditions, at least, these substances inhibit germination and growth of seeds of annual herbs (McPherson and Muller 1969). Other desert shrubs (*Artemisia* and *Salvia*) that commonly invade desert grasslands release aromatic terpenes such as camphor to the air. These terpenes are adsorbed from the atmosphere onto soil particles. In certain clay soils these terpenes accumulate during the dry season in quantities sufficient to inhibit the germination and growth of herb seedlings. As a result, invading patches of shrubs are surrounded by belts devoid of herbs and by wider belts in which the growth of grassland plants is reduced (Muller et al. 1968). Allelopathy may not be the only reason for belts devoid of vegetation. Studies of plant-animal interactions suggest the impact of predation by hares and consumption of seeds by rodents, birds, and ants (Bartholomew 1970).

Diffuse Competition If interspecific competition between two species is difficult to establish, it is more difficult to demonstrate diffuse competition, the sum of weak competitive interactions among ecologically related organisms. Davidson (1985) carried out a five-year experiment with colonies of three species of harvester ants, the large *Pogonomyrmex rugosa,* the intermediate-sized *P. desertorum,* and the

small *Pheidole xerophila.* The intermediate *P. desertorum* increased on plots from which its interference competitor, the large species *P. rugosa,* was removed; and the small species, *P. xerophila,* its exploitative competitor, declined. Davidson attributed the decline in the small species to the absence of the large harvester ant. The large harvester ants apparently aided the small species indirectly by suppressing populations of the intermediate-sized ants.

RESOURCE PARTITIONING AND UTILIZATION

Observations of a number of species sharing the same habitat suggest that they coexist by utilizing different resources. Animals eat different sizes and kinds of food, or feed at different times or in different areas. Plants occupy a different position on a soil moisture gradient, require different proportions of nutrients, or have different tolerances for light and shade. Each species exploits a portion of the resources, which becomes unavailable or is unusable to others. As populations use resources they are also depleting the resource base, which may or may not be renewed. What becomes important in interspecific competition, then, is the rate of consumption versus the rate of renewal. Among plants, in particular, competition is not for just one resource, such as food among animals, but for several resources simultaneously, such as light, soil nutrients, and water. Thus interspecific competition involves both resource partitioning and differential resource utilization.

Resource Partitioning: Theoretical Considerations

Consider an animal species A, which in the absence of any competitor utilizes a range of different-sized food items (Figure 14.9a). We can picture that utilization as a bell-shaped curve on a graph, with food as the ordinate and fitness as the abscissa. Most individuals feed about the optimum. Individuals at either tail feed on larger or smaller food items, respectively. As population size increases, the range of food taken may increase, because intraspecific competition forces some individuals to seek food at two extremes. Such intraspecific competition fosters increased genetic variability in the population.

Now allow a second species, B, to enter the area. When its resource use curve is superimposed on the curve of species A, B shows considerable overlap (Figure 14.9b). Selective pressure from interspecific competition forces both species A and species B to narrow their range of resource use. Natural selection will favor those individuals living in areas of minimal or no overlap. Ultimately the two species will narrow their ranges of resource use. They will diverge, moving to the left and the right on the graph. Direct interspecific competition will be reduced, and the two species will coexist. Thus, while intraspecific competition favors expansion of the resource base, interspecific competition narrows the range. The populations involved have to arrive at some balance between the two.

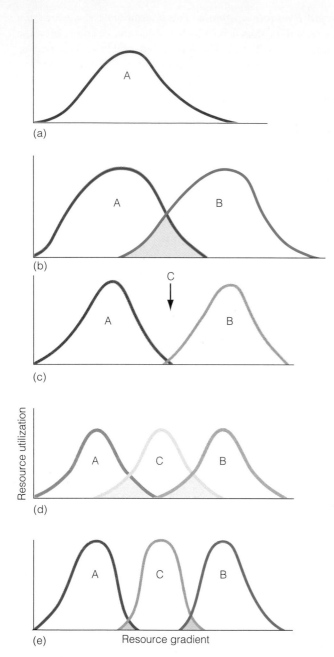

FIGURE 14.9   Theoretical resource gradient utilized by three competing species, A, B, and C. (a) A is the only species occupying the resource gradient. (b) Species B invades the resource gradient and partially competes with A. (c) In response to selective pressures, both A and B narrow their range of resource use to optimum, and C invades that portion used at less than an optimal level. (d) C competes with both A and B on parts of the resource gradient. (e) In response to selection pressure A, B, and C narrow their range of resource use to optimum.

Now allow a third species, C, to invade this resource gradient at a point between the utilization curves of A and B (Figure 14.9c). Species C can successfully invade if A and B are rare, if they are below carrying capacity, and if resources are abundant. Under these conditions competition will force each of the three to become more specialized in their resource utilization, to utilize optimal

resources, and to space themselves more narrowly on the resource gradient (Figures 14.9 d and e) (for theory, see MacArthur and Levins 1967).

Field Examples

Intensive field studies have turned up numerous examples of presumed resource partitioning. Lack (1971) noted that nine species of the genus *Parus* living in the broadleaf woods and coniferous forests of Europe and Great Britain feed in different parts of the tree canopy and consume different food throughout the year. Three of them, the blue tit, the great tit, and the marsh tit, inhabit the broadleaf woods. The blue tit, the most agile of the three, works high up in the trees gleaning insects, mostly 2 mm in size or smaller, from leaves, buds, and galls. The great tit, which is large and heavy, feeds mostly on the ground and seeks prey in the canopy only when taking caterpillars to feed its young. Its food consists of large insects 6 mm and over, supplemented with seeds and acorns. The marsh tit feeds largely on insects around 3 to 4 mm, which it gleans in the shrub layer and in twigs and limbs less than 6 m above the ground. It, too, feeds extensively on seeds and fruits. In the northern coniferous forests live the coal tit, the crested tit, and the willow tit. The more agile coal tit forages high up in the trees among the needles. There it seeks and feeds on aphids and spruce seeds. The willow tit consumes a high proportion of vegetable matter and feeds in the few available broadleaf trees. When in the conifers, it spends most of its time in the lower parts and on the branches rather than on the twigs. The crested tit is confined mostly to the upper and lower parts of the trees and the ground, but the bird does not feed in the herb layer. Thus by feeding in different areas and on different-sized insects, as well as different types of vegetable matter, these species divide the resources among themselves.

MacArthur (1958) observed a similar partitioning among five species of warblers inhabiting the spruce forests of the northeastern United States. Each fed in a different part of the canopy, and each was specialized behaviorally to forage in a somewhat different manner.

A similar partitioning of resources exists among plants. Plants experience strong abiotic and biotic selective pressures on gross morphology, both above and below ground. These pressures result in differing methods of exploiting light, water, and nutrients. Once plants are committed to a life form, they are committed to a particular mode of resource utilization. For this reason intraspecific competition can be more influential than interspecific competition.

This point is well illustrated in Cody's (1986) studies of life forms of desert plants. He found that below-ground root morphologies were much more important than above-ground structures to coexistence. Conspecifics with similar root structures tended to be widely spaced, and their nearest neighbors were species with different and complementary root systems (Figure 14.10). Species such as *Echinocereus,* the hedgehog cactus, have spreading roots within 15 cm of the surface. Others, such as *Ntymenoclea salsola,* the white burrobush, have deep taproots that extend 2 m or more

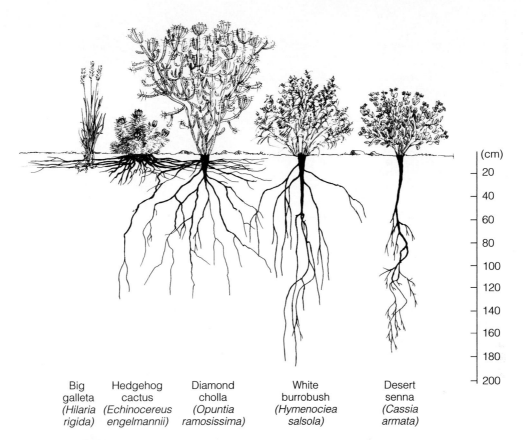

Big galleta (*Hilaria rigida*) Hedgehog cactus (*Echinocereus engelmannii*) Diamond cholla (*Opuntia ramosissima*) White burrobush (*Hymenociea salsola*) Desert senna (*Cassia armata*)

(cm)
20
40
60
80
100
120
140
160
180
200

FIGURE 14.10 Partitioning of the soil resource by a group of Mojave Desert plants. Root system morphology is species-specific. Species such as *Hilaria rigida* and *Echinocereus engelmannii* are shallow surface rooters, able to take up moisture quickly during occasional rains. *Opuntia* and *Hymenociea* employ more spreading roots at various intermediate depths. Plants such as *Cassia* have deep taproots. Plants with the same root morphology are not near neighbors. (After Cody 1986:386–387.)

below the surface to reach water and nutrients. Still others, such as the *Opuntia* cactus, have deep spreading roots rather than taproots. The deep-rooted species would conflict with conspecific near neighbors but not with shallow-rooted species. These shrub species coexist because they have different root systems that allow them to exploit water and nutrients in separated areas.

Differential Resource Utilization

Competition among plants takes on other dimensions, because they are competing for several resources at the same time, notably light, soil nutrients, and water along a resource gradients. Tilman (1980, 1982, 1986) has presented a model of plant competition that considers both species growth and resource levels. It is based on the theory that each plant species is a superior competitor along a resource gradient of light and nutrients or light and moisture. Changes in these resources should result in changes in competitive interaction between plant species.

The theory is best illustrated graphically, first for a single species then for two species. Consider a single species using two essential resources, soil nitrogen and light. The

resource availability can be plotted on two axes, with light on the *x* axis and soil nitrogen on the *y* axis (Figure 14.11). The population response of the species at zero growth on the two gradients is plotted as a solid line with a right-angle corner where the two gradients intersect. This line represents the resource-dependent zero net growth isocline. For all points along the isocline, population growth is zero. Above and to the right of the zero net growth isocline, populations increase. Below and to the left of the isocline, populations decline or fail to survive.

For any one point along the zero net growth isocline, there is only one point, the supply point, where resource supplies are constant. If a species uses a resource at a rate faster than it can be renewed, then the population will decline along the zero isocline (toward the bottom left of the diagram) until it establishes a new equilibrium point. Conversely, if renewal of the resource exceeds consumption, then the population will grow along the isocline until it establishes a new equilibrium point (Figure 14.12).

Now consider the interaction of two species, A and B, each of which has different requirements for light and soil nitrogen (Figure 14.13). The zero net growth isocline of A lies closer to the axes than does that of B. Thus A can survive

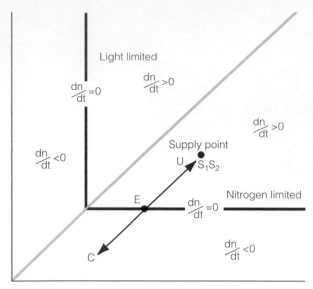

Light at soil surface

FIGURE 14.11 Zero net growth isocline for a plant species potentially limited by two resources, light at soil surface and available soil nitrogen. The vertical and horizontal lines intersecting at right angles are the resource-dependent growth isoclines for soil nitrogen and light at the soil surface, respectively. The zero net isocline is rectangular because both resources are essential. The diagonal line shows the proportions of the two resources, light and soil nitrogen for which the population is equally limited by the two. For zero net population growth, the plant population should use the two resources in the proportions shown. S_1, S_2 is the resource supply point for light and nitrogen. C is the consumption vector. It points away from the supply point because consumption draws down the resource supply. U is the resource supply vector, because it renews the resource supply. The point on the new growth supply isocline is the equilibrium point associated with the supply point shown. At this point, births equal deaths and resource supply balances consumption. (After Tilman 1986:362.)

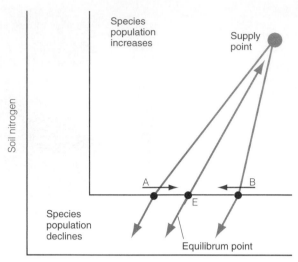

Light at soil surface

FIGURE 14.12 What happens if consumption vectors veer away from the equilibrium point? At A, the consumption rate is less than the renewal rate, and so the population will grow toward the equilibrium point E. If the consumption rate is greater than the renewal rate, as at B, then the population will decline to the equilibrium point.

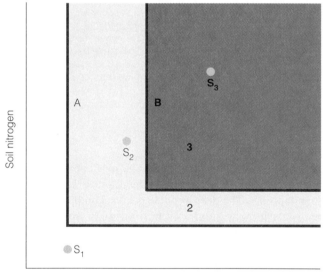

Light at soil surface

FIGURE 14.13 Competitive exclusion. The resource-dependent zero net growth isocline of species B lies farther away from the resource axes than that of species A. In area 1, neither species can exist because S_1, the supply point of light and nitrogen, lies outside the isoclines of both species. In area 2, only A can exist because the supply point S_2 falls within the isocline of species A but outside that of species B. In area 3, A also wins because the supply point S_3 falls within the region above the isoclines of both species, and so A reduces the resource concentrations along its own isocline to a point where B cannot exist, and B is competitively excluded. (After Tilman 1986.)

and grow at lower supply levels of both light and nitrogen than B. If the resource supply point (S) falls outside the isoclines of both A and B, neither species can exist. If the supply point falls in the region above and to the right of A's zero net growth isocline, but outside of B's isocline, only A can exist. If the supply point falls within the region above the isoclines of both species, then A reduces the resource concentrations along its own isocline to a point where B cannot exist, and B is competitively excluded.

In the usual situation, each species has a greater requirement for one resource than for another. In this case the isoclines will intersect (Figure 14.14). In our example A is the superior competitor for nitrogen and B is the superior competitor for light. The point at which their isoclines cross is the equilibrium point at which both species can coexist. Again, if the supply point falls outside either of the two isoclines, as at point S_1, neither species can survive. If the supply point falls at S_2 between A and B, then A wins and B loses. If it falls at S_3 between B and A, then B wins. If the supply point falls somewhere within the common region enclosed by both, then the outcome depends on the position of the supply point. If it falls at S_4, then B, with lower light requirements, should dis-

place A from habitats with low light availability. B does so because it reduces the light at the soil surface to a level below that required for species A to survive. At point S_6, which has a low supply of nitrogen, both species will be nitrogen-lim-

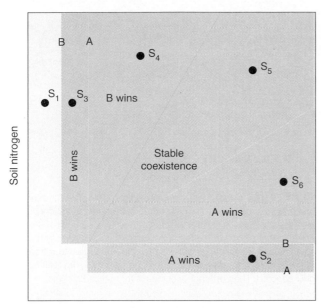

FIGURE 14.14 Competition for soil nitrogen versus light in two hypothetical species with a different need for each resource. (After Tilman 1986:363.)

ited. A, the superior competitor for nitrogen, should displace species B. In both situations, the reason for displacement is the greater use by each species of the resource that limits its own growth. In an intermediate habitat, point S₅, each species is limited by a different resource, B by nitrogen and A by light. Each species uses more of the resource that limits its growth and will be limited by the resource for which it is the poorer competitor. The equilibrium point of stable coexistence in habitat 5 will be that at which combined resource consumption by the two species equals resource supply. This concept relates to plant succession, discussed in Chapter 21.

This model of differential resource utilization appears to be more realistic than the classic competition model, because it takes into account not only population growth but also depletion and renewal of resources that affect growth. Like other ecological models, though, it will require extensive field testing under natural conditions.

THE NICHE

Closely associated with interspecific competition is the concept of the niche. **Niche** is one of those nebulous terms in ecology, its meaning colored by various interpretations that equate it with habitat, functional roles, food habits, and morphological traits. In everyday terms, a niche is a recess in a wall where you place something, usually an ornamental object; or it is a position in life suitable for a person. In ecology it means an organism's place and function in the environment—or does it?

One of the first to propose the idea of the niche was the ornithologist Joseph Grinnell (1917, 1924, 1928). In his study

of the California thrasher (*Toxostoma redivivum*) and other birds, he suggested that the niche be regarded as a subdivision of the environment occupied by a species, "the ultimate distributional unit within which each species is held by its structural and functional limitations." Essentially Grinnell was describing the habitat of the species.

Charles Elton (1927), in his classic *Animal Ecology,* considered the niche as the fundamental role of the organism in the community—what it does, its relation to its food and enemies. This idea stresses the occupational status of the species in the community.

Other definitions are variations on the same theme. Odum (1971) considers the habitat as the animal's address and the niche as its occupation. Whittaker et al. (1973) consider the niche only as a functional position, whereas the niche and habitat combined comprise the ecotope, "the ultimate evolutionary context of the species." Pianka (1978) regards the niche as embracing all the ways in which a given individual, population, or species conforms to its environment.

The definition that links the niche to competition was proposed by G. E. Hutchinson (1957). It is based on the competitive exclusion principle. According to this concept, an organism's niche consists of many physical and environmental variables, each of which can be considered a point in a multidimensional space. Hutchinson called that space the **hypervolume.**

We can visualize a multidimensional niche to a certain extent by creating a three-dimensional one. Consider three niche-related variables for a hypothetical organism: food size, foraging height, and humidity (Figure 14.15). Suppose the animal can handle only a certain range in food size. Food size, then, is one dimension of the niche (Figure 14.15a). Add the foraging height, the area to which it is limited seeking food. If we graph that on the second axis and enclose the space, we have a rectangle, representing a two-dimensional niche (Figure 14.15b). Suppose, too, that the animal can survive and reproduce only within a certain range of humidity. Humidity can be plotted on a third axis. Enclosing that space, we come up with a volume, a three-dimensional niche (Figure 14.15c). Of course, many more variables, both biotic and abiotic, influence a species' or an individual's fitness. A number of these dimensions, *n*—difficult to visualize and impossible to graph—make up the *n*-dimensional hypervolume that would be the species' niche. An individual or a species free from the interference of another could occupy the full hypervolume or range of variables to which it is adapted. That is the idealized **fundamental niche** of the species.

The fundamental niche of the species assumes the absence of competitors, but rarely is this the case. Competitive relationships may force the species to constrict a portion of the fundamental niche it could potentially occupy. In those parts its fitness might be reduced to zero. The conditions under which an organism actually exists are its **realized niche** (Figure 14.16). The niche may be further restricted by the absence of certain features of the niche at any given point in time and space. Like the fundamental niche, the realized niche is an abstraction. In their studies,

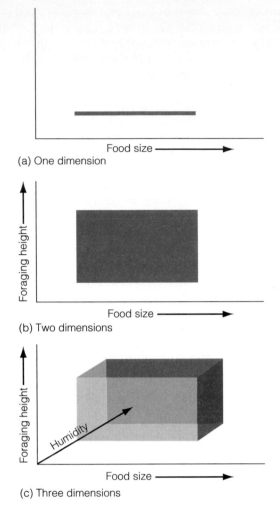

(a) One dimension

(b) Two dimensions

(c) Three dimensions

FIGURE 14.15 Models of niche dimension for a hypothetical animal. (a) A one-dimensional niche involving food size. The animal can live on food of intermediate size but not on food that is large or small. (b) A second dimension has been added, foraging height. Enclosing that space, we obtain a two-dimensional niche. (c) Suppose the organism can survive and reproduce only within a certain range of humidity, graphed as a third axis. By enclosing all those points, we arrive at a three-dimensional niche space or volume.

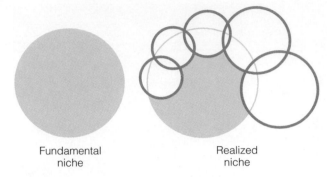

Fundamental niche

Realized niche

FIGURE 14.16 Fundamental and realized niches. The fundamental niche of a species is the full range of environmental conditions, biological and physical, under which it can exist. Under pressure of superior competitors whose niches overlap, the species may be displaced from part of its fundamental niche and forced to retreat to that portion of the fundamental niche hypervolume to which it most highly adapted. The portion it occupies is its realized niche.

ecologists usually confine themselves to one or two niche dimensions, such as a feeding niche, a space niche, or a tolerance niche.

Consider two examples. Root (1967) studied the exploitation of the feeding niche by the blue-gray gnatcatcher (*Polioptila caerulea*) in California oak woodlands. He characterized the niche of the bird by the size of its food and by the height above the ground at which it captured food (Figure 14.17). Simplified for the sake of example, the bird's fundamental niche could be described by a maximum range of size of prey between 1 and 14 mm in length and by a foraging area of ground level to 10 m. The gnatcatcher's niche center, indicated by frequency of capture and stomach content analysis, consists of insects 3 to 5 mm long taken 2.4 to 8.5 m above the ground. The further the height and food dimensions diverge from this center, the more the gnatcatcher's niche may overlap those of other species.

Putwain and Harper (1970) studied the population dynamics of two species of dock, *Rumex acetosa* and *R. acetosella,* each growing in hill grasslands in North Wales. *Rumex acetosa* grew in a grassland community dominated by velvet grass (*Holcus lanatus*) and red and sheep fescues (*Festuca rubra* and *F. ovina*); *R. acetosella* grew in a community dominated by sheep fescue and bedstraw (*Galium saxatile*). To determine interference and niches of the two dock species, Putwain and Harper treated the flora with specific herbicides to remove selectively in different plots (1) grasses, (2) forbs except *Rumex* species, and (3) the *Rumex* species. All species except *R. acetosella* spread rapidly after the grasses were removed; that species increased only after both grasses and nongrasses were removed.

The niches of these two plants are diagrammed in Figure 14.18. The fundamental niche of *R. acetosella* (R) overlaps the fundamental niches of both grasses (G) and other forbs (D). Only when these competitors are eliminated does this dock realize its fundamental niche. *Rumex acetosa,* however, overlaps only with the grasses, and only their removal is necessary to permit expansion of this dock throughout its fundamental niche. The seedlings, however, occupy niches different from the mature plant.

This fact emphasizes an important aspect of niches. The fundamental and realized niches of an organism can change with its growth and development. Insects with a complex life history may occupy one niche as larvae and an entirely different niche as an adult. As a larva, a butterfly feeds on green foliage; as an adult, it feeds on nectar. In other organisms, such as many fish, the niche space changes as the organisms matures in a size-dependent way. As body size of the a fish grows larger it may switch from feeding on plankton to eating small fish. With this change in diet comes a change in its interspecific relationships.

Niche Overlap

The example of *Rumex* brings up the question of niche overlap. What happens when two or more species use a portion of

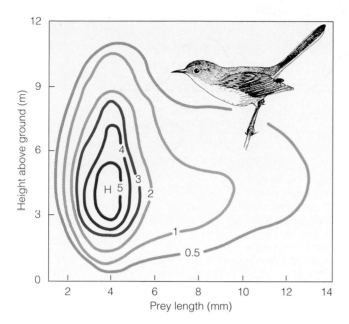

FIGURE 14.17 The feeding niche of the blue-gray gnatcatcher (*Polioptila caerulea*), based on of two variables, size of prey and feeding height. The contour lines map the feeding frequencies for adult gnatcatchers during the incubation period during July and August in California oak woodlands. The maximum response level is at H. Contour lines spreading out from this optimum represent decreasing response levels. The outer contour line probably represents the outer boundary of the realized niche for these two variables. (For discussion of such an analysis, see Maguire 1973, Hutchinson 1978.) (Diagram from Whittaker et al. 1974, based on data from Root 1967.)

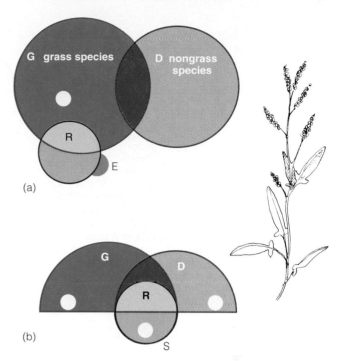

FIGURE 14.18 Niche relationships of (a) *Rumex acetosa* and (b) *R. acetosella* in mixed grassland swards. In each diagram the fundamental niches of grass species (G) and nongrass species (D) overlap. The fundamental niche of *Rumex* species (R) is shown as a continuous line, and the realized niche is shaded. E is that part of the fundamental niche of *R. acetosa* that is expressed in the presence of nongrass species and does not overlap the fundamental niches of G and D. The fundamental niches of seedlings (S), shown by the small colored circles, are contained within the fundamental niches of grasses, nongrasses, and mature *Rumex*. (From Putwain and Harper 1970.)

the same resource, such as food, simultaneously? The theoretical model of the niche assumes that competition is intense, that only one species can occupy a niche space, and that competitive exclusion takes place in areas of overlap. The amount of niche overlap is assumed to be proportional to the degree of competition for that resource. In a condition of minimal or no competition, niches may be adjacent to one another with no overlap, or they may be disjunct (Figure 14.19). At the other extreme, in a condition of intense competition, the fundamental niche of one species may be completely within or correspond exactly to another, as in the case of the seedling *Rumex*. In such instances there can be two outcomes. If the niche of species 1 contains the niche of species 2 and species 1 is competitively superior, species 2 will be eliminated entirely. If species 2 is competitively superior, it will eliminate species 1 from the part of the niche space species 2 occupies. The two species then coexist within the same fundamental niche.

When fundamental niches overlap, some niche space is shared and some is exclusive, enabling the two species to coexist (Figures 14.18 and 14.19). Considerable niche overlap does not necessarily mean high competitive interaction. In fact, the reverse may be true. Competition involves a resource in short supply. Extensive niche overlap may indicate that resources are abundant and little competition exists. Pianka (1972, 1975) has suggested that the maximum

tolerable overlap in niches should be lower in intensely competitive situations than in environments with low demand/supply ratios.

In fact, both high niche overlap and the absence of overlap may reflect other environmental and behavioral influences and not interspecific competition at all. To attribute niche overlap or the lack of it to interspecific competition may be to ignore real reasons. Consider a study of habitat partitioning by three species of grassland sparrows—the grasshopper sparrow (*Ammodramus savannarum*), Henslow's sparrow (*A. henslowii*), and the savannah sparrow (*Passerculus sandwichensis*)—on broad expanses of reclaimed surface mines in north-central Pennsylvania by Piehler and Whitmore (Piehler 1987). They mapped 65 sparrow territories (20 grasshopper, 23 savannah, and 22 Henslow's sparrows) and measured 14 structural variables of the vegetation, including basal and overhead cover, height, density, and litter depth. Territorial boundaries of all three species overlapped to a degree. Except for the semicolonial Henslow's sparrow, territorial boundaries within species were contiguous. Territorial boundaries among the three species showed varying degrees of overlap, the greatest being a 31 percent overlap between the grasshopper and savannah sparrows. The partitioning of the grassland habitat and thus habitat niche

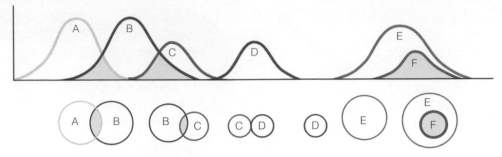

FIGURE 14.19 Niche relationships visualized as graphs on a resource gradient and as Venn diagrams. Species A and B have overlapping niches of equal breadth but are competitive at opposite ends for the resource gradient. B and C have overlapping niches of unequal breadth. Species C shares a greater proportion of its niche with B than B does with C. (However, B shares its niche also with A at the other end.) C and D occupy adjacent niches with little possibility of competition. D and E occupy disjunct niches and no competition exists. Species F has a niche contained within the niche of E. If F is superior to E competitively, it persists and E shares that part of its niche with F. Compare Figure 14.18. (Adapted from Pianka 1978.)

configurations could be explained by a gradient of vegetational structure (Figure 14.20) alone, based on increasing habitat richness (vegetation density, vegetation height, bare ground cover, and litter depth). Henslow's sparrows inhabited areas of tall, dense, thick grass. On the other end of the gradient, with some bare ground, thinner and shorter forb and grass cover, and less litter depth, was the grasshopper sparrow. Intermediate between the two was the savannah sparrow. In addition to the wide differences in the vegetational cover they inhabited, the two congeneric species, the grasshopper sparrow and the Henslow's sparrow, were very similar in morphological characters of bill width and bill length, whereas the savannah sparrow had significantly smaller ($p < .05$) bill width and length.

For simplicity, niche overlap is usually considered one-dimensional or two-dimensional. In reality, a niche involves many types of resources: food, a place to feed, cover, space, and so on. Rarely do two or more species possess exactly the same requirements. Species overlap on one gradient, but not on another; thus total competitive interactions may be less than the competition or niche overlap suggested by one gradient alone (Figure 14.21).

Niche Width

If we plotted the range of resources—for example, food size—used by an animal or the range of soil moisture conditions occupied by plants, the length of the axis intercepted by

FIGURE 14.20 Niche overlap and habitat partitioning among three species of grassland sparrows on a Pennsylvania study site. A two-dimensional plot of the first (habitat) and second (territory size) discriminant axes. Note the wide separation between the two congeneric species, the grasshopper and Henslow's sparrows. (From Piehler 1987.)

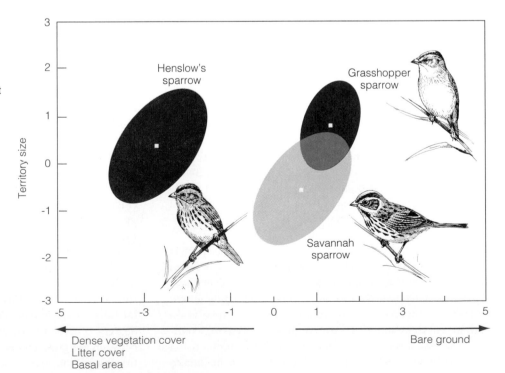

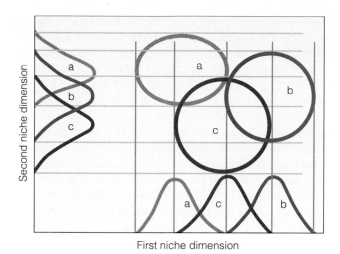

FIGURE 14.21 Niche relationships based on two gradients. Species may exhibit considerable overlap on one gradient and little or none on another. When niche dimensions are added, niche overlap may be reduced considerably. (Adapted from Pianka 1978.)

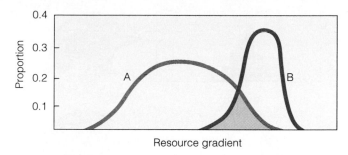

FIGURE 14.22 Hypothetical distribution of a species with a broad niche (A) and a species with a narrow niche (B) on a resource gradient. The niches overlap (shaded area). Species A overlaps a greater proportion of species B more than B overlaps A.

the curve would represent **niche width** (Figure 14.22). Theoretically, niche width (also called niche breadth and niche size) is the extent of the hypervolume occupied by the realized niche. A more practical definition is the sum total of the different resources exploited by an organism (Pianka 1975). Measurements of a niche usually involve the measure of some ecological variable such as food size or habitat space.

Niche widths are usually described as narrow or broad. The wider the niche, the more generalized the species is considered to be. The narrower the niche, the more specialized the species is. Most species have broad niches and sacrifice efficiency in the use of a narrow range of resources for the ability to use a wide range of resources. As competitors they are superior to specialists if resources are somewhat undependable. Specialists, equipped to exploit a specific set of resources, occupy narrow niches. As competitors they are superior to generalists if resources are dependable and renewable. A dependable resource is closely partitioned among specialists with low interspecific overlap (Roughgarden 1974). If resource availability is variable, generalist species are subject to invasion and close packing with other species during periods of resource abundance.

Niche Responses

Niche width provides some indication of resource utilization by a species. If a community made up of a number of species with broad niches is invaded by competitors, intense competition may force the original occupants to restrict or compress their utilization of space and to confine their feeding and other activities to those patches of habitat providing optimal resources. Competition that results in the contraction of habitat rather than a change in the type of food or resources utilized is called **niche compression** (MacArthur and Wilson 1967).

Conversely, if interspecific competition is reduced, a species may expand its niche, utilizing space previously

unavailable to it. Niche expansion in response to reduced interspecific competition is called **ecological release.** Ecological release may occur when a species invades an island that is free of competitors, moves into habitats it never occupied on the mainland, and increases in abundance (Cox and Ricklefs 1977). Such expansion may also follow when a competing species is removed from a community.

Interactions among three species of doves inhabiting New Guinea and associated islands serve as examples. Traveling inland from the coast on the island of New Guinea, you encounter the emerald ground dove (*Chalcophaps indica*) inhabiting the coastal shrub, then Stephan's dove (*C. stephani*) occupying second-growth forest, and finally the cinnamon ground dove (*Gallicolumba rufigula*) inhabiting the inland rain forest. On the island of Bagabag, where the cinnamon ground dove is absent, Stephan's doves expands into the rain forest, while the emerald dove remains restricted to the coastal scrub. On the islands of New Britain, Tolokiwa, and Karkar, where the cinnamon ground dove and emerald ground dove are absent, Stephan's dove occupies all three habitats. On the Espirito Santa, where both the cinnamon ground dove and Stephan's dove are absent, the emerald dove occupies all habitats from the coastal scrub through the rain forest (Diamond 1975).

Associated with compression and release is another response, **niche shift.** Niche shift is the adoption of changed behavioral and feeding patterns by two or more competing populations to reduce interspecific competition. The shift may be a short-term ecological response or a long-term evolutionary response involving some change in a basic behavioral or morphological trait. With the exception of evolutionary change, niche shift does not involve the establishment of a new niche, as the term might imply. Rather niche shift refers (or should refer) to a shifting of the realized niche within the range of the fundamental niche.

An example of short-term niche shift involves three cogeneric species of sunfish (Centrarchidae): the bluegill (*Lepomis maerochirus*), the pumpkinseed (*L. gibbosus*), and the green sunfish (*L. cynellus*). Werner and Hall (1976, 1977, 1979) stocked the three species in experimental ponds 30 m in diameter supporting natural stands of emergent and

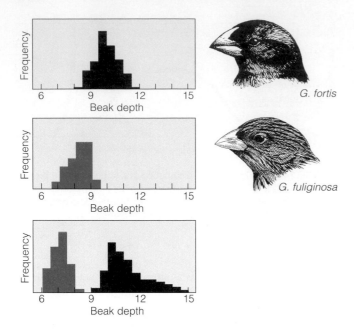

FIGURE 14.23 Apparent character displacement in beak size in populations of the Galapagos finches *Geospiza fortis,* the medium ground finch, and *Geospiza fuligenosa*, the small ground finch. (a) Beak size distribution for the medium ground finch on the island Daphne Major in the absence of the small ground finch. (b) Beak size distribution of the small ground finch on the island of Los Hermanos in absence of the medium ground finch. (c) Beak size distribution for the medium ground finch and small ground finch on Santa Cruz island, inhabited by both species. Beak size of the small ground finch is significantly smaller compared to the population on Los Hermanos and similarly the beak size of the medium ground finch is significantly larger compared to the population on Daphne Major. Such character displacement may be the outcome of natural selection induced by intraspecific competition for food, predominately seeds. (After Grant 1986).

submerged vegetation and their associated prey populations. When stocked in the ponds alone, each species inhabited the vegetation zones supporting the largest prey. When bluegill and green sunfish were confined together in equal densities in a pond, the bluegill shifted to open water; but in the absence of green sunfish, bluegills invaded the dense vegetation. Bluegills and pumpkinseeds have a dietary overlap of 50 to 55 percent. When the two were stocked together in a pond, the bluegill again moved to open water. Perhaps its generalized diet and its ability to move into different habitats as the situation requires accounts for the bluegill's position as the most common sunfish in ponds.

A long-term response to niche shift may involve **character displacement** that involves the gradual evolutionary separation of two species in morphology or physiology as an outcome of competition for a resource. Clear demonstrations of character displacement are rare because a definitive demonstration requires a great deal of evidence (see Arthur 1982, Taper and Case 1992, Thompson 1994). One of the better examples involves two Darwin's finches of the Galapagos Islands, the medium ground finch *Geospiza fortis* and the small ground finch *G. fuliginosa,* both of which feed on an

overlapping array of seed sizes. The medium ground finch is allopatric (lives separately) on Daphe Island and the small ground finch is allopatric on Los Hermanos. Populations of each on these two islands possess similar but overlapping beak size (Figure 14.23). On the island of Santa Cruz, where the two species are sympatric (live together), beak size (measured by depth) of the medium ground finch is much larger than on Daphe Major and the beak size of small ground finch is significantly smaller than on the island of Los Hermanos. In the face of competition for food, selection favored those individuals of medium ground finch with a large beak size that could effectively exploit larger seeds while favoring individuals of small ground finch that fed on smaller seeds. The outcome of this competition was a shift in feeding niches.

Summary

1. Relations between species may be positive (+) or beneficial; negative (−) or detrimental; or neutral (0). There are six possible interactions: (0 0), neutral; (+ +), in which both populations mutually benefit (mutualism); (− −), in which both populations are affected adversely (competition); (0 +), in which one population benefits and the other is unaffected (commensalism); (0 −), in which one population is harmed and the other is unaffected (amensalism); and (+ −), in which one population benefits and the other is harmed (predation, parasitism).

2. Interspecific competition—the seeking of a resource in short supply by individuals of two or more species, reducing the fitness of both—may be one of two kinds, interference and exploitative. Exploitative competition depletes resources to a level of little value to either population. Interference involves aggressive interactions (passive in plants, active in animals). A particular form of interference competition is allelopathy, the secretion of chemical substances that inhibit the growth of other organisms.

3. As described by the Lotka-Volterra equations, four outcomes of interspecific competition are possible. Species 1 may win over species 2, or species 2 may win over species 1. Both of these outcomes represent competitive exclusion. A third possibility is unstable equilibrium, in which the potential winner is the one most abundant at the outset or most able to respond to a changing environmental condition. A final possible outcome is stable equilibrium, in which two species coexist, but at lower population levels than if each existed in the absence of the other.

4. The competitive exclusion principle—two species with exactly the same ecological requirements cannot coexist—has conceptual difficulties. Competitive exclusion relates to competition between two species. In reality, competition may involve competitive interaction from a number of species on several gradients that may be equivalent to strong competitive interaction from one species. Such a relationship is known as diffuse competition.

5. Competitive interactions between two species have been demonstrated most clearly with laboratory populations. It is much more difficult to demonstrate strong competitive interactions between species in the field. Field experiments have their own set of problems. A number of them have demonstrated short-term competitive interactions among some animal and especially plant species at the experimental local scale, but relating the findings to larger, nonexperimental populations is difficult. Nevertheless, competition theory has stimulated crit-

ical examinations of competitive relationships, especially how species coexist and how resources are partitioned.

6. Coexistence occurs when organisms using identical but limited resources respond differently to a fluctuating environment because of differing life history traits. Because competitive relationships among individuals and species interact with physical and biological environments, it is difficult to determine the degree of competitive exclusion. Competitive exclusion is most evident when an exotic organism invades and outcompetes native organisms in their own environment.

7. Seemingly associated with interspecific competition is resource partitioning. Species sharing the same habitat coexist by using a portion of the resources, such as food or nutrients, that is unavailable or unusable to others. Resource partitioning among plants involves differential resource utilization. It is based on the premise that species compete simultaneously for several resources, such as light and nutrients, but differ in their requirements for the two limiting essential resources. Coexistence occurs when the combined resource competition of the two species equals the resource supply.

8. Closely associated with interspecific competition is the concept of the niche. A niche is the functional role of an organism in the community. It might be constrained by interspecific competition. In the absence of competition, an organism occupies its fundamental niche. In the presence of interspecific competition, the fundamental niche is reduced to a realized niche, the conditions under which an organism actually exists.

9. When two different organisms use a portion of the same resource, such as food, these niches are said to overlap. Overlap may or may not indicate competitive interaction.

10. The range of resources used by an organism suggests its niche width. Species with broad niches are generalists, whereas those with narrow niches are specialists.

11. In the absence of competition, the organism may expand its niche and experience ecological release. Organisms may also undergo niche shift by changing their behavioral or feeding patterns to reduce interspecific competition.

Review Questions

1. Distinguish among the following population interactions: neutral interactions, mutualism, commensalism, amensalism, parasitism, competition, predation.
2. According to the Lotka-Volterra theory of competition, what are the four possible outcomes of competition?
3. What conditions are necessary for competitive exclusion? Is allelopathy a mechanism for competitive exclusion among plants?
4. Why is it more difficult to demonstrate interspecific competition in the field than in the laboratory? Can you relate laboratory studies to field situations?
5. What is resource partitioning? How might it relate to interspecific competition?
6. Define the niche. What is a fundamental niche? A realized niche? Relate niche to interspecific competition and resource partitioning.
7. Read an experimental study of competition in the field critically. Analyze the experiment, considering experimental design, application of laboratory studies to the field situation, demonstration of interspecific competition, and applicability of the results to the natural community. Good choices are Kross 1980, Brown 1982, Wilbur and Alford 1985, Gurevitch 1985, and McGraw and Chapin 1989.

Cross References

Trophic levels, 501; food chain, 395–397, 493; population growth, 184; social dominance 201; territoriality, 202; chemical defense by plants, 286; parasitism, 394; competition and community structure, 391–392; predation and community structure, 393–394; plant succession, 391–392.

CHAPTER 15

Concepts of Predation

CONCEPTS

1. Predator-prey models predict oscillations in predator and prey populations.

2. Functional responses relate the amount of prey taken by predators to prey density.

3. Increase in number and fecundity of predators in relation to changes in prey density is numerical response.

4. Optimal foraging is a strategy employed by foragers to maximize food intake and minimize time and energy spent in acquiring it.

5. Risk-sensitive foraging relates to the forager's response to the mean and variance of its food distribution.

No type of population interaction is more misunderstood or stirs more emotional reaction than predation. Sympathy goes to the prey. When we look at it objectively, though, predation becomes a step in the transfer of energy in the ecosystem.

Predation is commonly associated with the strong attacking the weak, the lion pouncing on the deer, the hawk upon the sparrow. However, considered more broadly, predation also includes **parasitoidism,** a case of the weak attacking the strong. In this situation, one organism, the parasitoid, attacks the host (the prey) by laying its eggs in or on the body of the host. After the eggs hatch, the larvae feed on the tissues of the host until it dies. The effect is the same as that of predation. Another special form of predation is **cannibalism,** in which the predator and the prey are the same species. The concept of predation has been extended still further to include **herbivory,** in which grazing animals of all types feed on plants. Herbivores kill their prey when consuming seeds or the whole plant, or they function as parasites when they consume only part of the plant but do not destroy it. Thus predation in its broadest sense can be defined as one organism feeding on another living organism, or **biophagy.**

Ecologically, predation is more than just a transfer of energy and nutrients. It represents a direct and often complex interaction of two or more species, of the eaters and the eaten. The numbers of some predators may depend on the abundance of prey, and predation may be involved in the regulation of prey populations.

MODELS OF PREDATION

In 1925 A. J. Lotka, a mathematician and physical scientist, proposed the first model of predator-prey interactions in *Elements of Physical Biology.* In 1926 the Italian mathematician A. Volterra independently came up with a similar model. Neither of the two extended the logistic equation to the two-species system. Instead, Lotka adapted the chemical principle of mass action (Berryman 1992). Mass action assumes that individual predators and individual prey encounter each other randomly in the same way that molecules interact in a chemical solution. The responses of predator and prey populations are assumed to be proportional to the products of their population densities.

The Lotka-Volterra model involves paired equations, one for the prey population (P) and one for the predator population (C). The prey growth equation involves two components, the maximum rate of increase per individual and the predatory removal of prey from the population. The growth equation for the prey is $dP/dt = rP$. Predators, though, will remove prey at a rate that depends on the frequency of the predator-prey encounters. These encounters will increase as the number of predators (C) increases and as the number of prey (P) increases. The exact number of encounters (and successful capture) will depend on search and attack frequency (a'). Therefore

$$\frac{dP}{dt} = rP - a'CP$$

In the absence of food, individual predators starve. The assumption is the number of predators will decline exponentially through starvation in the absence of prey—$dC/dt = -qC$, where q is the mortality rate.

The birth rate of the predators depends on two factors: (1) the rate at which food is consumed ($a'CP$), and (2) the predators' efficiency at converting food consumed into offspring (f). Thus the predator birth rate is $fa'CP$. The predator growth rate then is

$$\frac{dC}{dt} = fa'CP - qC$$

These equations translate into isoclines for prey and the predator (Figure 15.1). For the prey:

$$rP = a'CP \rightarrow r = a'C \rightarrow C = \frac{r}{a}$$

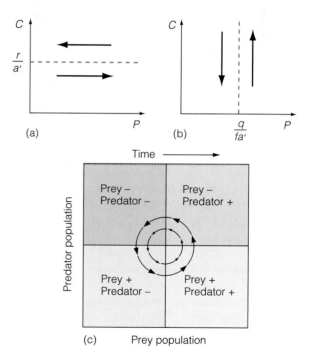

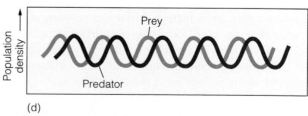

FIGURE 15.1 The zero isoclines for predator (a) and prey (b) populations. Note that the zero isoclines for both species are a fixed (constant) population size. The combined zero isoclines (c) result in a spiraling of the combined population sizes (P, C) through the four quadrants defined by the two isoclines. The resulting temporal dynamic of the two populations (d) is a cyclic oscillation of the two populations with the predator population lagging behind that of the prey.

Since r and a' are constants, the prey isocline is a line for which C is a constant (Figure 15.1a).

For the predator:

$$fa'CP = qC \rightarrow fa'P = q \rightarrow P = \frac{q}{fa'}$$

Since q, f, and a' are constants, the isocline is a vertical line along which P is constant (Figure 15.1b). The combined isoclines depict the Lotka-Volterra model graphically (Figure 15.1c). The ordinate C is the number of predators; the abscissa P is the number of prey. The horizontal straight line is the zero growth curve or isocline of the prey, and the vertical line is the zero isocline of the predator. In the area to the right of the vertical line, predators increase; to the left, they decrease. In the area below the horizontal line, the prey increase; above it they decrease. The circle of arrows represents the joint populations of predator and prey, and how the size of the population of each changes with it. If a point or arrow falls in the region to the left of the vertical line, the prey population is not large enough to support the predators and thus the predator population declines. If an arrow falls to the left of the vertical and above the horizontal, both populations are declining. The predator population decreases enough to permit the prey population to increase, moving the arrow left of the vertical below the horizontal. The increase in the prey population now permits predators to increase and thus the arrow moves to the right of the vertical below the horizontal. As the predator population increases, depressing the prey population, the arrow to the right of the vertical moves above the horizontal. This interaction between predator and prey results in reciprocal oscillations of both predator and prey with some time delay in the predator's response (Figure 15.1d). These regular oscillations or cycles will continue ad infinitum.

The Lotka-Volterra model is based on a number of underlying assumptions. These assumptions are (1) in the absence of predation, the prey experiences exponential growth; (2) the predator population declines exponentially in the absence of prey; (3) predators move at random among randomly distributed prey; (4) the proportion of encounters that result in the capture and consumption of prey are constant at all predator and prey densities; (5) the number of prey taken increases in direct proportion to the number of predators, a linear response; (6) all responses are instantaneous with no time lag for handling and ingesting prey; (7) energy input through predators is immediately converted to the birth of more predators. The model makes no allowance for age structure, for interaction of prey with their own food supply, and for density-dependent mortality of the predator.

A decade later an ecologist, A. J. Nicholson, and a mathematician, W. Bailey, recognized the deficiencies in the Lotka-Volterra model. They developed a model for a host-parasitoid relationship (Nicholson and Bailey 1935). Parasitoids differ from predators in that their attacks do not remove the host from the prey population. In fact, one host may be attacked many times. The number of parasites reared on the host depends on the number of attacks, not the number of hosts encountered.

Like the Lotka-Volterra model Nicholson and Bailey's model consists of two equations, but they are difference equations:

$$H_{t+1} = \lambda H_t \exp(-aP_t)$$
$$P_{t+1} = H_t[1 - \exp(-aP_t)]$$

Here H is the host, P the predator or parasitoid, t is time, λ is the finite rate of increase for the host population, and a is the rate of parasitism per parasitoid.

Nicholson and Bailey based their model on a set of assumptions somewhat different from those of the Lotka-Volterra model. Predators search randomly for static prey uniformly dispersed over a homogeneous landscape. The predators have a constant area of discovery and an insatiable appetite. The predators "sample" a certain proportion of the prey population. The time element is discrete, not continuous. The generations of both predator and prey have the same time span and are of the same length. Predator mortality is density-independent. The conversion of energy input by the predator into the birth of more predators is not immediate but delayed by one generation.

These features of the Nicholson-Bailey model allow an estimate of the prey in the next generation. If the number of hosts that the parasitoid removes within its sampling area is equal to the fraction of the prey that represents recruitment, then the base parental stock remains. If the parental stock remaining is sufficient to replace the individual prey taken and if it is sufficient to maintain the density of the parasitoids, then the two populations remain stable indefinitely. However, if the parasitoid removes part of the parental stock along with recruitment, the prey and ultimately the parasitoid populations decline. If much of the recruitment is left untouched, prey increase and the predators may not be up to the task of removing increased recruitment. In either of the last two cases, the two interacting populations will undergo oscillations with increased amplitude, according to the Lotka-Volterra equations.

The Nicholson-Bailey model does not respond in exactly the same manner. If parental stock of the prey is sufficient to replace the prey taken as well as to maintain the predator population, then the two populations will remain stable as in the Lotka-Volterra model. If, however, predators remove too many individuals or if the growth of the prey population outstrips the ability of the predators to reduce the prey population, the results are unstable populations with ever-increasing amplitudes and the possible extinction of one population or the other (Figure 15.2).

The Lotka-Volterra and Nicholson-Bailey models emphasize the influence of predators on prey populations. Genetic changes, stress, emigration, aggression, availability of cover and hiding places, difficulty of finding prey as they become more scarce, and other attributes also influence predator-prey relationships. To add more realism to predation models, Rosenzweig and MacArthur (1963) developed

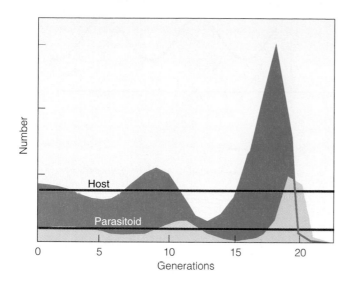

FIGURE 15.2 Population oscillations of host and parasitoid predicted by the Nicholson-Bailey equations. (After Nicholson 1933.)

a series of graphic models that consider a wider range of predator-prey interactions (Figure 15.3). Modifying the zero growth isocline of the prey to account for a low rate of growth at low and high densities, they plot the growth curve of the prey as convex rather than horizontal. The isocline of the predator levels off at high prey density, reflecting non-linearity, compared with the linear response of the Lotka-Volterra equations. The Rosenzweig-MacArthur model of a stable cycle of interaction is similar to the Lotka-Volterra model. Predator and prey populations increase if the joint abundance of predator and prey falls inside the convex curve to the right of the vertical. If the joint abundance falls outside the curve, the populations of both decline. Because the prey curve intersects the predator curve at right angles, the curve produces neutral, stable cycles in populations of predator and prey (Figure 15.3a). By moving the predator curve to the right or to the left so that the curves no longer intersect at right angles, we can produce damped and unstable cycles (Figures 15.3b and c).

These two models represent situations in which the prey has no refuge from the predator. Growth rates of both predator and prey populations are a function of the frequency with which the predator comes into contact with the prey. In nature a fraction of the prey escapes predation by hiding in refuges. Predators then decline from emigration, starvation, or failure to reproduce. As the prey population increases in the refuge area, the surplus repopulates the surrounding area (Figure 15.3d).

In these models the equilibrium density of the prey population does not increase, even if the prey population is increasing. All the gain in production goes to the predator (which you can demonstrate by moving the vertical predator isocline to the right and left). Thus the Rosenzweig-MacArthur models consider that prey equilibrium is independent of prey density and that prey stability depends on characteristics of the predator.

EXPERIMENTAL PREDATOR-PREY SYSTEMS

How well does the Lotka-Volterra model predict the behavior of predator-prey systems? The question stimulated experimental investigations into systems involving a single predator species and a single prey species.

The Russian biologist G. F. Gause (1934) reared together under constant environmental conditions a predatory ciliate, *Didinium nasutum,* and its prey, another ciliate, *Paramecium caudatum* (Figure 15.4). In one experiment, Gause introduced the predator into a clear nutritive medium in test tubes supporting a population of *Paramecium.* Unable to escape the predatory ciliates, the *Paramecium* succumbed to predation. With their food supply gone, the *Didinium* starved (Figure 15.4a). Gause discovered that regardless of the density of the two populations, the predator always exterminated the prey. Only by periodically introducing prey (immigrants) into the medium was Gause able to maintain the predator population (Figure 15.4c). In this manner he was able to maintain populations together and produce the regular fluctuations in both predicted by the Lotka-Volterra equations.

In another experiment, Gause introduced sediment into the bottom of the tubes. Here some of the *Paramecium* could find refuge from their predator. When the predators ate all the available prey, they died from the lack of food. Meanwhile the *Paramecium* that took refuge in the sediment multiplied, and eventually they took over the medium (Figure 15.4b). Gause could maintain both populations only when a portion of the prey population could escape predation by retreating to a refuge and when he periodically introduced a small number of predators into the medium. Gause's laboratory experiments did not support the predictions of the Lotka-Volterra predation model. Rather, his results suggested that the predator-prey relationship was one of overexploitation and annihilation of the prey followed by the collapse of the predator population. Only immigration from other prey populations could maintain the predator-prey system.

The Gause experiments took place in a simple environment. This fact prompted C. Huffaker, a University of California entomologist studying the biological control of crop pests, to investigate whether he could establish an adequately large and complex laboratory environment in which a predator-prey system would not be self-exterminating (Huffaker 1958). He chose as the prey the six-spotted mite, *Eotetranychus sexmaculatus,* which feeds on oranges and as predator another mite, *Typhlodromus occidentalis.* Both reproduce via parthenogenesis. Huffaker established experimental populations on trays holding 40 cells in four rows of ten each. He placed oranges in some of the positions and rubber balls of the same size in the remaining cells. By changing the number and placement of oranges on the tray, and by covering the oranges with paper and sealing wax to vary the amount of exposed surface, he was able to control both the total food available and the pattern of food dispersion. Thus he manipulated the size and distribution of the prey population. In most of his experiments he established

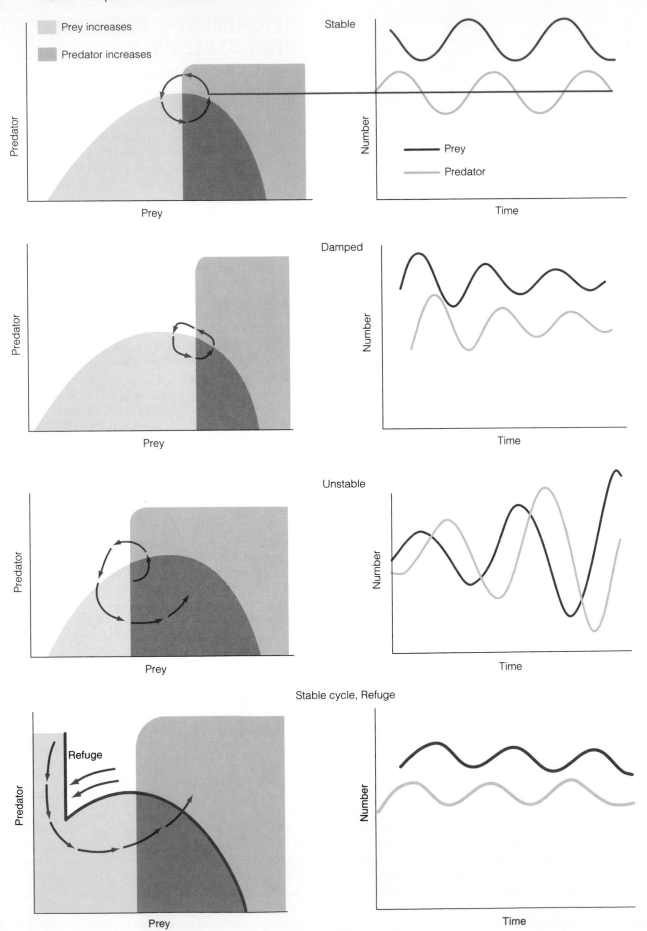

Prey increases

Predator increases

Stable

Predator

Prey

Number

Time

Prey

Predator

Damped

Predator

Prey

Number

Time

Unstable

Predator

Prey

Number

Time

Stable cycle, Refuge

Predator

Refuge

Prey

Number

Time

a prey population with 20 females and introduced a predatory population with 2 females 11 days later.

In one experiment Huffaker created a simple environment by concentrating the food. He placed four oranges in adjacent positions. The concentrated prey populations grew up to 650 individuals and survived 25 to 30 days before being eliminated by the predaceous mites. In another experiment Huffaker alternated 20 small areas of food with 20 foodless positions. The prey population grew to 2000 individuals (Figure 15.5). Because the scattered oranges provided a refuge that took the predators a longer time to discover, the population survived for 36 days.

In still another experiment Huffaker established a complex environment where food was widely dispersed. He used three 40-cell trays for a 120-cell universe in which he placed 120 oranges. Each orange had 1/20 of its surface exposed as a feeding area for the mites. Thus the available feeding area was equivalent to six oranges. Huffaker added small wooden pegs within the trays to speed the dispersal of the six-spotted mites. The wooden pegs provided dispersal launching points for six-spotted mites that float to new areas on strands of silk. He established a maze of Vaseline barriers across the trays to slow the dispersal of the predatory mites, which could travel only by foot. Huffaker then stocked the trays with 120 six-spotted mites, placing one mite on each of the 120 oranges, and added 27 female predatory mites five days later. He placed one predator on each of 27 oranges across the major sections of the trays.

In this experiment and the one using the 20 small areas of food, the prey and predator found plenty of available food for population growth. Density of the predators increased as the prey population increased. In the experiment in which the food was concentrated and dispersion of the prey population was minimal, predators readily found the prey, quickly responded to changes in prey density, and destroyed the prey rapidly. In fact, that situation was self-annihilative. In the experiment in which the primary food supply and the prey

were dispersed, predator and prey went through three oscillations before the system died out (Figure 15.6).

Several important conclusions resulted from the study. First, predators cannot survive when the prey population is low for a prolonged period relative to the longevity of the predators. Second, a self-sustaining predator-prey relationship cannot be maintained without immigration of prey. Third, the complexity of prey dispersal and predator-searching relationships, combined with a period of time for the prey to recover from the effects of predation and to repopulate the area, had more influence on the period of oscillation than did the intensity of predation. Thus Huffaker's experiments emphasized that the Lotka-Volterra model says nothing about the need for heterogeneity of the environment to maintain a predator-prey system. Only by introducing heterogeneity

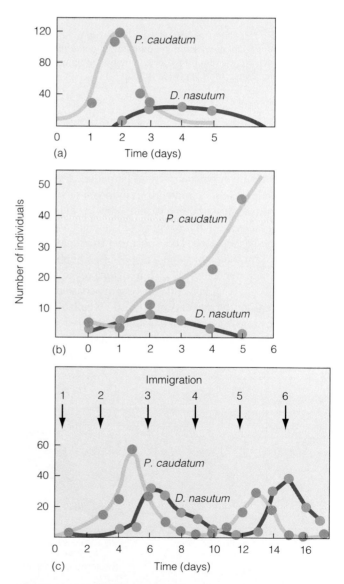

FIGURE 15.3 (opposite) Rosenzweig-MacArthur model of predator-prey interactions. The prey curve is convex rather than straight, involving density-dependent growth, unlike the predator curve. (a) Stable cycle. The prey population can increase if the joint abundances of predator and prey fall inside the area below the isocline of the prey. Prey population will decrease outside this area. The growth curves intersect at right angles as in the Lotka-Volterra model (Figure 15.1). (b) If the predator isocline is moved to the right or to the left, it will no longer intersect the prey isocline at right angles. When the predator's isocline is moved to the right, it intersects the descending part of the prey isocline, which dampens oscillations. (c) If the predator isocline intersects the ascending part of the prey isocline, it increases oscillations, producing an unstable system. Vectors spiral outward and the amplitude of the population increases steadily until a limit cycle is reached. Such situations can lead to the extinction of either the predator or the prey. (d) If prey have access to a refuge, a portion of the population escapes predation and supplies immigrants to the exploited area. This immigration limits oscillations and leads to a stable cycle. (See MacArthur and Connell 1966:154–156.)

FIGURE 15.4 Outcome of Gause's experiments of predator-prey interactions between the protozoans *Paramecium caudatum* and *Didinium nasutum* in three microcosms: (a) oat medium without sediment; (b) oat medium with sediment; (c) with immigration in oat medium without sediment. (After Gause 1934.)

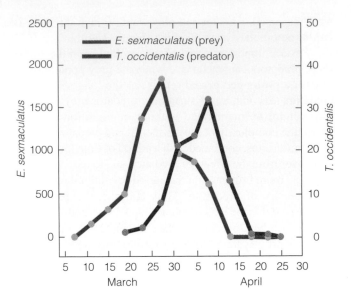

FIGURE 15.5   Densities per orange area of the prey, *Eotetranychus sexmaculatus,* and the predator, *Typhlodromus occidentalis,* with 20 small areas of food for the prey alternating with 20 foodless positions. (After Huffaker 1958:365.)

was Huffaker able to obtain empirical results that even approached the Lotka-Volterra model.

The degree of dispersion and the area employed were too restricted in Huffaker's experiment to perpetuate the system. Pimentel et al. (1963) attempted to provide an environment with a space-time structure that would allow the existence of a parasite-host system. They chose as subjects a parasitic wasp (*Niasonia vitripennis*) and a host fly (*Musca domestica*) and provided for the environment a special population cage, a group of interconnected cells. A predator-prey system living in 16 cells died out, but a 30-cell system persisted for over a year. Increasing the system from 16 to 30 cells decreased the average density of parasites and hosts per cell and increased the chances for survival of the system. The lower density was due to the breakup and sparseness of both para-

site and host populations. The greater number of individual colonies that remained following a severe decline of the host assured survival of the system, because these colonies provided a source of immigrants to repopulate the environment. Moreover, amplitude of the fluctuations of the host did not increase with time, as proposed by the Nicholson-Bailey model. Apparently the fluctuations were limited by intraspecific competition.

These laboratory experiments support studies made in the field. Sometime around 1839 prickly pear cactus (*Opuntia*) was introduced from America into Australia as an ornamental. As is often the case with introduced plants and animals, the cactus escaped from cultivation and rapidly spread to cover 60 million acres in Queensland and New South Wales. To combat the cacti, a South American cactus-feeding moth (*Cactoblastis cactorum*) was liberated. The moth multiplied, spread, and destroyed the cacti until plants existed only in small, sparse, widely distributed colonies.

The decline of the prickly pear also meant decline of the moth. Most of the caterpillars coming from moths that had bred on prickly pear in the previous generation died of starvation. In areas where only a few moths survived, not many plants were parasitized. As prickly pear increased, so did the moth, until the cactus colony was again destroyed. In areas where no moths survived the colony spread once more, but sooner or later it was found by moths from other areas and destroyed. However, seed scattered into new areas established new colonies that maintained the existence of the species and thereby maintained the predator-prey system.

The rate of establishment of prickly pear colonies is determined by the time available for the colonies to grow before they are found by moths. As a result an unsteady equilibrium exists between cactus and moth. Any increase in the distribution and abundance of the cactus leads to an increase in the number of moths and subsequent decline in the cactus. The maintenance of this predator-prey, or more accurately herbivore-plant, system depends on environmental discontinuity. The relative inaccessibility of host or prey in time and space limits the number of parasites and predators.

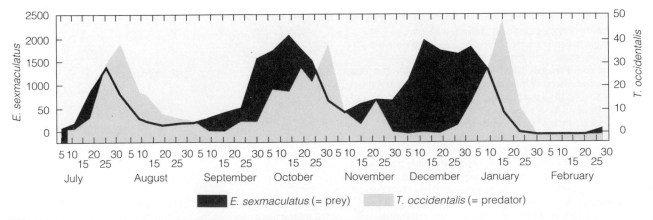

FIGURE 15.6   Three oscillations in predator and prey populations when the predatory mite *Typhlodromus occidentalis* preyed on the orange-feeding six-spotted mite, *Eotetranychus sexmaculatus,* in a complicated environment. (After Huffaker 1958:370.)

In further investigations of the moth-cactus relationship, J. Monro (1967) found that the moth may conserve food for succeeding generations of moths by limiting its own numbers. At high densities the moth clumps its egg sticks rather than laying them randomly on prickly pear. The clustering overloads certain plants of prickly pear with eggs, resulting in the destruction of the plants. In dense stands of prickly pear clustering initially has little influence on larval survival, for as an overloaded plant collapses, it falls on its neighbor and larvae can move to a new source of food. Later, as dense stands become broken up into isolated plants, the sedentary larvae are unable to cross the wide gaps and die of starvation. As mean density increases, the proportion of eggs wasted by clumping increases.

However, because the eggs are clustered rather than widely distributed, more plants escape infestation altogether or are subject to a lighter infestation than would be expected if eggs were laid at random. Monro found that this mechanism, which is employed most in the center of the range of the moths, acts to conserve the food supply for succeeding generations and to maintain a constant level of both the food resource and the moth population.

These examples to some extent illustrate predatory-prey interactions, both at the plant-herbivore and at the herbivore-carnivore level. Although for simplicity they are considered separately, predator-prey interactions at one trophic level influence predator-prey interactions at the next trophic level. Interactions at two or more trophic levels are often involved in predator-prey stability.

PREDATOR RESPONSES

The Lotka-Volterra equations of predation hint at two distinct responses of predators to changes in prey density. As prey density increases, each predator may take more prey or take them sooner, a **functional response;** or predators may become more numerous through increased reproduction or immigration, a **numerical response.**

Functional Responses

The idea of a functional response was introduced by Solomon (1949) and explored in detail by Holling (1959, 1961, 1966). Holling recognized three types of functional response (Figure 15.7). Type I, in which the number of prey eaten per predator increases linearly to a maximum as prey density increases (the Lotka-Volterra assumption); Type II, in which the number of prey eaten increases at a decreasing rate toward

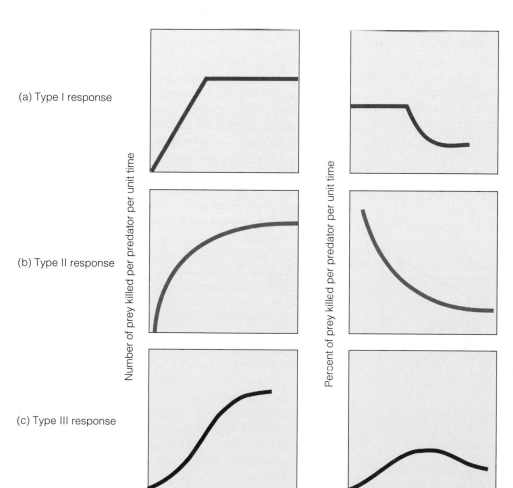

(a) Type I response

(b) Type II response

(c) Type III response

Number of prey killed per predator per unit time

Percent of prey killed per predator per unit time

Prey density

Prey density

FIGURE 15.7 Three types of functional response curves, which relate predation to prey density. (a) Type I. The number of prey taken per predator increases linearly to a maximum as prey density increases. Graphed as a percentage, predation declines relative to the growth of the prey population. (b) Type II. The number of prey taken rises at a decreasing rate to a maximum level. When considered as a percentage of prey taken, the rate of predation declines as the prey population grows. Type II predation cannot stabilize a prey population. (c) Type III. The number of prey taken is low at first and then increases in a sigmoid fashion, approaching an asymptote. When plotted as a percentage, the functional response still retains some of the sigmoid features, but declines slowly as the prey population increases. Type III functional response has the potential of stabilizing prey populations. (After Holling 1959.)

a maximum value; and Type III, in which the number of prey taken is low at first and then increases in a sigmoid fashion, approaching an asymptote.

Type I Response   Type I response is a specialized one, the sort assumed in the simple predation models and best demonstrated in the laboratory. Predators of any given abundance capture food at a rate proportional to their encounter with prey items up to the point of satiation. There is density-independent mortality of the prey up to that point. In northern Finland, where *Microtus* vole populations are cyclic, the European kestrel (*Falco tinnunculus*), short-eared owl (*Asio flammeus*), and long-eared owl (*A. otus*) exhibit a linear Type I functional response (Figure 15.8) (Korpimaki and Norrdahl 1991), except that the horizontal saturation level is never reached. The three raptors take voles in proportion to their availability. As the density of the voles increases, the kill rate increases but does not level off. This failure to level off suggest that the populations of the voles do not become so abundant that the predators become satiated.

Type II Response   The Type II response is described by the disk equation, named for an element in the experiment from which it was derived. In Holling's experiment, the predator was represented by a blindfolded person and the prey by sandpaper disks 4 cm in diameter thumbtacked in different densities to a 1 m square table. The "predator" tapped the table with a finger until a prey was found and then removed the disk. The "predator" continued to search and encounter (tapping, discovery, and removal) for one minute. Holling found that the number of disks the predator could pick up increased at a progressively decreasing rate as the density of disks increased. Predator efficiency rose rapidly as the density of the disks increased, up to the point where the predator spent so much time picking up and laying aside disks that the predator could handle only a maximum number at a time.

Holling repeated this experiment using insect predators such as the praying mantis and insect prey such as flies. He found the same relationship between prey density and handling time as he did with the disk-handling experiment. Handling time increased as the density of the prey increased. Although search time was short, the predator could handle only a limited number of prey in a given time. These experiments demonstrated several important components of predation: density of prey, attack rate of the predator, and handling time, including time spent pursuing, subduing, eating, and digesting prey.

Type II functional response is described by the disk equation

$$\frac{N_a}{P} = \frac{aNT}{1 + aT_hN}$$

where P_a is the number of prey or hosts killed or attacked, C is the number of predators or parasitoids, P is the number of prey, and P_a/C is number of prey eaten per predator; a is a constant representing the attack rate of the predator or the rate of successful search, T is total time predator and prey are exposed, T_h is handling time, and T_s is time spent by predator in search of prey. T is determined by the equation

$$T = T_s + T_hP_a$$

The Type II functional response curve derived by Holling is identical to the Michaelis-Menten equation of enzyme kinetics:

$$b(N) = m[N/w + N)]$$

Because handling time is the dominant component, rise in the number of prey taken per unit time decelerates to a plateau (Figure 15.9) while the number of prey is still increasing. For this reason Type II functional response cannot act as a stabilizing force on a prey population unless the prey occurs in patches. Thus Type II response is destabilizing (see Murdock and Oaten 1975).

The concept of functional response developed around the consumption of animal prey by predators and parasitoids. What about herbivorous predation on plants? Does this relationship involve a functional response? Recent studies indicate that herbivores exhibit a Type II function response relative to the plants they consume (Gross et al. 1993). However, there are some fundamental differences between the functional responses of carnivores and herbivores.

Consider the fox as the carnivore, the rabbit as the mammalian herbivore, and various grasses and herbs as the rabbit's food. The fox searches for, attacks, kills, eats, and digests the rabbit. Each process is discrete and on a one-to-one basis. The fox will not hunt for another rabbit until it has processed the one it has killed. The plant-eating rabbit, however, uses different processes. The rabbit's prey is a patch of plants, a large concentrated group of plants, or a part of a modular plant. The rabbit does not completely eat and rarely kills the plant, and so it is available for additional attacks. In this way the rabbit acts more like a parasite than a predator.

Having located a patch of food, the rabbit eats by cropping or taking bites of the plant. With each bite it takes a

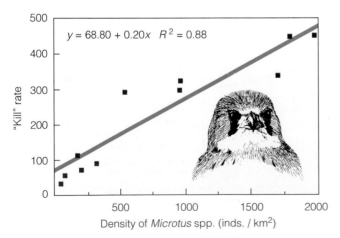

$y = 68.80 + 0.20x$ $R^2 = 0.88$

FIGURE 15.8   Type I linear functional response of a pair of European kestrels to *Microtus* vole densities during the breeding season. The curve did not reach a horizontal level, presumably because the vole population was not high enough for the predator to reach saturation level. (After Korpimaki and Norrdahl 1991.)

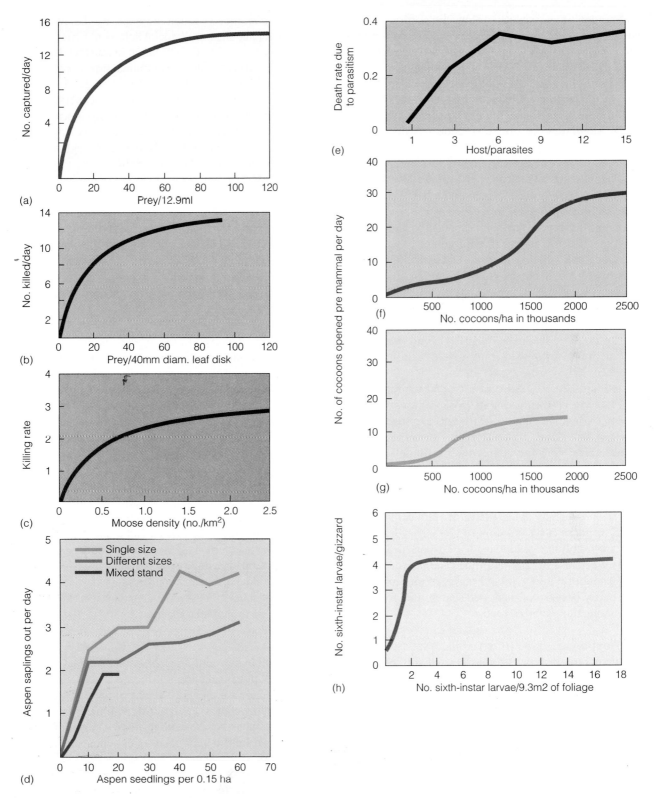

FIGURE 15.9 Examples of Type II and Type III functional response curves. (a) First instar *Linyphia triangulaus* (spider) feeding on *Drosophila*. (b) Numbers of amphipod crustaceans *Corophium* taken per minute by a shorebird, the redshank (*Tringa totanus*). The decline of prey eaten at high densities in this example probably results from interference among predators. Predators are discouraged from feeding in areas where a large number of individuals are already congregated and leave the area. (c) The response of wolves to changing moose density. (d) Mean daily rates of aspen cutting in relation to density in three experimental treatments. Type III functional response curves are graphed in e through h. (e) *Encarsia formosa* parasitizing *Trialeurodes vaporariorum*. (f) Shrew *Sorex* feeding on sawfly larvae (*Neodiprion*). (g) *Peromyscus* (deer mice) preying on sawfly larvae. (h) Bay-breasted warbler (*Dendroica castanea*) feeding on spruce budworm (*Choristoneura fumiferana*) larvae. (a, e from Hassell et al. 1976, b from Goss-Custard 1977, c from Messier 1994, d from Fryxell and Doucet 1993, f, g, h from Holling 1964.)

certain amount of plant tissue. How big a bite it takes depends on the size of its mouth and the nature of the plants (see Shipley et al. 1994). The rabbit can take a larger bite of dense growth of clover than it can of the single leaves of plantain. There is a point at which to eat a given amount of plant tissue, the rabbit will have to take two bites instead of one. As bite size declines, the intake rate declines. Bite size, then, regulates the intake of food (Spalinger and Hobbs 1992).

Food intake, in turn, depends on how fast a grazing herbivore, in our case the rabbit, can chew and swallow food. As the rabbit feeds, one process affects the other. The rate at which the rabbit crops food is influenced by the spacing and density of plants. Because the rabbit cannot crop and chew at the same time, in dense growth chewing competes with cropping and vice versa, which reduces the intake rate (Gros et al. 1993). However, as it moves between clumps, the rabbit chews what it has eaten, which can increase its intake rate. Because the rabbit may chew while searching for new plants, searching and handling time overlap. To account for the factors controlling food intake, ecologists have to modify the classic functional response equation (see Spalinger and Hobbs 1992).

Fryxell and Doucet (1993) investigated the functional responses of beavers (*Castor canadensis*) to various densities of trembling aspen (*Populus tremuloides*) saplings, a preferred food, in three kinds of stand: (1) pure stands of saplings of the same size; (2) pure stands with different sizes of saplings; and (3) mixed stands of aspen, alder (*Alnus rugosa*), and red maple (*Acer rubrum*) of various sizes. Beavers had the highest consumption or cutting rates in stands of aspen of the same size, especially at high densities, producing a classic Type II response curve (see Figure 15.9). They had intermediate consumption rates in stands of aspen of different sizes, partly because of handling time. The beaver had the lowest consumption rate in mixed stands of multiple-sized saplings, partly because of increased search time for aspen growing among nonpreferred foods.

Type III Response   Type III functional response is more complex than Type II. It has been associated with predators that can learn to concentrate on a prey when it becomes more abundant. In Type III response the number of prey taken per predator increases with increasing density of prey and then levels off to a plateau where the ratio of prey taken to prey available declines (see Figure 15.9). Because the amount of prey taken is density-dependent, Type III functional response is potentially stabilizing. It is difficult to apply the Type II Holling disk equation to Type III functional responses. The Michaelis-Menten equation, however, can be applied to Type III responses.

Type III responses invariably involve two or more prey species; the predator has a choice of prey. In the presence of several prey species, the predator may distribute its attacks among the prey in response to the relative density of the prey species. Predators may take most or all of the individuals of a prey species that are in excess of a certain minimum number, determined, perhaps, by the availability of prey cover and the prey's social behavior. Errington (1946), drawing on his studies of predation in muskrat and bobwhite quail populations, called this minimal prey population level, at which the predator no longer finds it profitable to hunt the prey species, its **threshold of security.** Type III responses have been called compensatory because as prey numbers increase above the threshold, the "doomed surplus" becomes vulnerable to predation through intraspecific competition. Below the threshold of security the prey species compensates for its losses through increased litter size and greater survival of young.

Although intuitively the concept of the threshold of security appears to be sound, it has never been rigorously tested in the field. Based on intrinsic density-dependent population regulation through intraspecific competition, the threshold of security concept minimizes the role of predators as an important extrinsic force in population regulation (Korpimäki and Krebs 1996).

Switching   Involved in the Type III response curve is the role of the facultative predator and alternative prey. Although the predator may have a strong preference for a certain prey, it can turn to an alternative, more abundant prey species that provides more profitable hunting. If rodents, for example, are more abundant than rabbits and quail, foxes and hawks will concentrate on the rodents. This idea was advanced early by Aldo Leopold in *Game Management* (1933), in which he described alternative prey species as buffer species because they stood between the predator on one hand and game species on the other. If the population of buffer prey is low, the predators turn to the game species: the foxes and hawks will concentrate on the rabbits and quail. This turning by a predator to an alternative, more abundant prey such as mice was later termed **switching** by Murdoch (1969). In switching the individual predator concentrates a disproportionate amount of attacks on the more abundant species and pays little attention to the rarer species (Figure 15.10). As the relative abundance of the two prey species changes, the predator changes its diet.

Switching, according to Murdoch and Oaten (1975), is caused by three behaviors of the predator: (1) changing its preference toward the more abundant prey as it eats it more frequently by choice; (2) ignoring rare prey; or (3) concentrating search in more rewarding areas. Any one of these three behaviors results in Type III response curves.

Search Image   The reason for the sigmoidal shape of Type III response is the subject of much study and debate (see Royama 1970, Croze 1970, Murdock and Oaten 1975, Curio 1976). One explanation was advanced by L. Tinbergen (1960), based on his studies of the relation between woodland birds and insect abundance. According to Tinbergen's hypothesis, when a new prey species appears in a given area, its risk of becoming prey is low. The birds have not yet acquired a **search image** for the species. A search image is a perceptual change in the ability of a predator to detect a

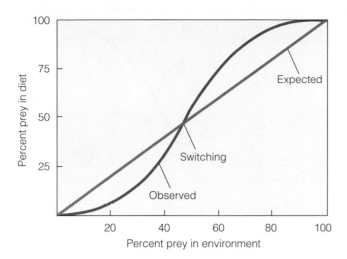

FIGURE 15.10 A model of switching. The straight line represents a situation in which the theoretical amount of prey is consumed at a rate proportional to its availability in the environment. It represents a constant preference with no switch. The curved line represents the proportion of prey species actually taken. At low densities the proportion taken is less than expected. Switching occurs at the point where the curved line crosses the straight.

familiar cryptic prey. Once the predator has secured a palatable item of prey, the predator finds it progressively easier to find others of the same kind. The more adept the predator becomes at securing a particular prey item, the longer and more intensely it concentrates on the item. In time the numbers of the prey species become so reduced or its population so dispersed that encounters between it and the predator lessen. The search image for that species begins to wane, and the predator begins to react to another species. The combination of increasing density of prey and establishment of a search image results in a sudden increase of the perceived prey species in the predator's diet, giving a sigmoid functional response curve.

Studies have shown that predator can acquire a search image from remarkably few experiences (Croze 1970, Dawkins 1971, Curio 1976). In losing an image the predator may simply not respond to the perceived stimulus or may in fact no longer perceive it—that is, no longer distinguish the properties of the prey from the background. The search image is maintained by rewards in the form of the acquisition of food. When rewards are no longer there, the bird turns to another image. In effect, the predator responds to changes in rewards. The extinction of an image tends to occur more slowly than its acquisition, and among some predators the search image may be retained for some time, even in the absence of rewards. Croze (1970) found that carrion crows (*Corvis corone*) retained their search image for eight days without reward. After that time the search image declined rapidly, but was still retained.

Ratio-Dependent Predation The functional response equation relates prey death rate (prey consumption) to prey density. Standing alone, the equations tell us little about the role of

functional response to population growth. We need to insert the functional response, $b(N)$, into the prey growth equation:

$$N/dt = d\ aN(1 - N/K) - b(N)P$$

The problem with this equation, however, is its time scale (Berryman 1992). The functional response equation is based on the hunting behavior of the predator and thus is on a fast time scale, whereas the prey growth equation reflects a slower time scale. To get around this problem, some ecologists have proposed a modified functional response equation dependent on the ratio of prey to predators rather than on prey density. They express this ratio-dependent functional response equation as

$$b(N/P) = m(N/P)(w + N/P) = mN(wP + N)$$

where m = maximum predator attack rate and w = prey density.

The insertion of this ratio-dependent functional response equation into the prey equation produces a humped prey zero-growth isocline and a slanted predator zero-growth isocline (Figure 15.11) in which predator density influences both prey density and prey equilibrium (Arditi and Berryman 1991). As prey abundance increases, predator abundance increases; and conversely as prey density decreases, predator abundance decreases. Thus both predator and prey equilibria rise and fall with prey productivity. The mechanism involved is based on resource sharing and interference that result from differences in time scales of feeding and reproduction, and from spatial heterogenity (Akcakaya et al. 1995).

The concept of ratio-dependence has its critics. They argue that this theory lacks any demonstration that it arises from biologically reasonable mechanisms (Murdoch 1994),

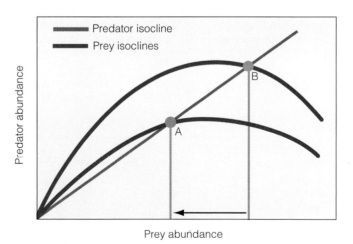

FIGURE 15.11 Predator-prey interaction when a ratio-dependent functional response curve is incorporated into the logistic model. The predator zero isocline is slanted rather than vertical. The prey zero isocline is convex. Prey increases below its zero isocline. The vertical lines indicate increased prey vulnerability to predation or a lower rate of increase. The points of intersection represent points of equilibrium relative to the predator populations. Prey equilibrium changes as its vulnerability to predation changes. (After Arditi and Berryman 1991.)

that it is not supported by data and lacks empirical evidence (Abrams 1994), and that the more standard density-dependent models of predator-prey interactions are preferable (Gleeson 1994). These criticisms have been strongly rejected by proponents of ratio-dependent theory (Berryman et al. 1995, Akcakaya et al. 1995).

Numerical Response

In addition to functional responses, predators may exhibit numerical responses. Numerical response refers to an increase in predators resulting from an increase in food supply. Numerical response may be short term as exhibited by **aggregative response,** in which predators tend to congregate in patches of high prey density. One or two members of the predator species discover and begin to feed on the prey item; other members of the species observe the feeding response and follow suit (see Curio 1976).

Intermediate levels of aggregation may increase the efficiency of predation, because a number of predators foraging together can locate areas of prey abundance more quickly than a few. At high levels of aggregation in areas of high prey density, interference among predators may be so great that the efficiency of predation declines. Encountering an individual of its own species, a predator may temporarily cease hunting or leave the area. Interference among predators reduces the proportion of total prey or hosts the predator or parasite encounters, because the predator's search time is reduced (Hassell et al. 1976).

Aggregative responses of predators to areas of high prey density may have a pronounced influence on stability of predator-prey interactions. Hassell and May (1974) presented an idealized general aggregative response curve for predator-to-prey distribution (Figure 15.12). The response curve exhibits a lower plateau of low prey density and an upper plateau of high prey density where predators do not distinguish between prey areas. The model predicts that predators

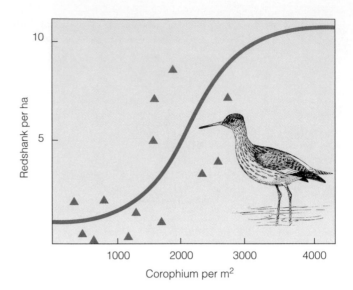

FIGURE 15.13 Aggregative response in the redshank (*Tringa totanus*). The curve plots the density of the predator (the redshank) against the average density of arthropod prey (*Corophium volutator*). (After Hassel and May 1974.)

FIGURE 15.12 Model of general aggregative response. At the lower plateau of prey density, predators do not distinguish among low (unprofitable) prey areas; at intermediate densities (shaded area), predators discriminate markedly; at the upper plateau, predators do not discriminate among high-density (profitable) areas. (Hassel and May 1974.)

discriminate markedly in areas of intermediate prey density and tend to congregate in areas of higher density. An example of this type of distribution is the response curve for the redshank (*Tringa totanus*), a shorebird that tends to concentrate in areas of its preferred food, the amphipod crustacean *Corophium volutator* (Figure 15.13). On the other hand, predators tend to avoid areas of low prey density, making those individuals much less vulnerable. Low-density areas provide the prey species with partial refuges against predation.

Long-term numerical response is of two types. One type is the immigration or emigration of predators in response to changes in prey density. Such movements, however, can be considered a true numerical response only when predators move into an area from some distance. For example, in their long-term study of the European kestrel, short-eared owl, and long-eared owl inhabiting a 47 km^2 farmland in western Finland, Korpimaki and Norrdahl (1991) found that the nesting pairs of these raptors fluctuated in close accordance with the spring density of *Microtus* voles (Figure 15.14a). The raptors were able to track the vole populations without a time lag because the birds were mobile and the vole cycles were asynchronous in the region. An increase in the vole population stimulated a rapid immigration of raptors to the area. There was just as rapid an emigration when the vole population declined.

The second type of long-term numerical response is an increase or decrease in the rate of predator natality and mortality in response to changing prey density. This type is influenced by the kind of predator involved: parasitoid or true predator. For true predators that require several prey to complete their development, numerical response depends on the rate at which predators can locate and consume suitable prey (Lawton et al. 1975).

Survival rates of arthropod instars and the young of nonarthropod predators depend directly on the density, size, and availability of prey. A scarcity of prey means a lack of food. A lack of food results in poor survival of young or instars. A poor survival rate has a direct bearing on numerical response of a predator population.

Some parasitoids feed only on one host. Among them adult fecundity is limited by the number of hosts the female can find. With them the relationship between prey density and fecundity is linear (Hassel and May 1973).

In all other situations nutrition, controlled by the amount of prey eaten during the adult stage, affects fecundity. Predators can use the energy remaining after meeting maintenance demands for reproduction. If predators face limited food and therefore have low energy, they experience low fecundity and thus a low numerical response. With increasing prey density, the predator's fecundity increases and with it, the numerical response. For example, the population of the European kestrel in western Finland increased from 2 during a period of low density of voles to 46 at a high density. The mean number of young produced ranged from 4.6 at high vole densities to 0 at low densities. Fecundity of the kestrel, however, made only a minimal contribution to its numerical response.

Numerical response, positive or negative, is not immediate in most cases, especially in situations where it depends on increased fecundity. There is necessarily a time lag between adequate nutritional intake, development and birth of young, and their maturation to reproducing individuals.

The "fugitive" Cape May (*Dendroica tigrina*) and bay-breasted (*D. castanea*) warblers of the northeastern North American forests provide examples of delayed numerical response. Outbreaks of spruce budworm apparently dictate the abundances of those birds (MacArthur 1958, Morris et al. 1958, Mook 1963) (Figure 15.14b). During outbreaks the two species increase more rapidly than other warblers because of extra-large clutches. During years between outbreaks they decline and even become extinct locally.

In general, then, numerical response takes three basic forms (Figure 15.15): (1) direct or positive response, in which the number of predators per unit area increases as the prey density increases; (2) no response, in which the predator population remains proportionately the same; (3) inverse or negative response, in which the predator population declines in relation to prey population (Hassell 1966).

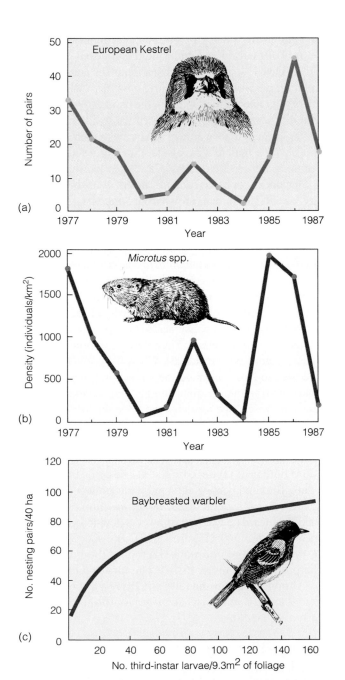

FIGURE 15.14 (a, b) Numerical response of nesting pairs of European kestrels to changes in density of *Microtus* voles at Alajaki, western Finland, during 1977–1987. (Adapted from Korpimaki and Norrdahl 1991.) (c) Numerical response of nesting bay-breasted warblers to density of spruce budworm. (After Mook 1963.)

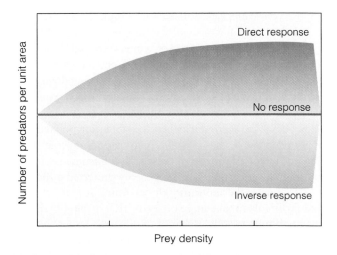

FIGURE 15.15 Basic forms of numerical response. No response means that the number of predators remains the same in the face of increasing prey density. A direct response implies that predators increase in response to an increasing prey density. Inverse response means that the number of predators per unit area declines as prey density increases. (From Hassell 1966.)

FORAGING THEORY

The Type III response curve may reflect the acquisition of a search image in some predators, but the same response curve could result from how predators must allocate energy to secure prey (Royama 1970). It is energetically unprofitable for predators to spend time where prey density is low. Predators must discover the most productive way to allocate their hunting time among different prey species of different abundances in different patches. Profitability is measured not by prey density, but by the amount of prey (preferably measured in terms of biomass) that a predator can harvest in a given time. This profitability of hunting is sufficient to produce Type III curves.

The profitability of hunting by a predator relates to the manner in which its prey is distributed. If prey were distributed in a fine-grained manner, the predator could pick and choose with a search image. Prey, however, are distributed in a coarse-grained manner in patches across the landscape. These patches vary in size and in the quality and quantity of resource, and so the predator must be able to locate profitable patches. This fact gave rise to the concept of the optimal use of patchy environments advanced by MacArthur and Pianka (1966), which later evolved into optimal foraging theory. This theory forms a basis against which actual foraging strategies can be compared.

A foraging animal wants to obtain the most energy from food intake relative to the energy expended in securing and eating the food. The difference is net energy gain. Its **optimal foraging strategy** provides a maximum net rate of energy gain, endowing the animal with the greatest fitness. It involves two separate but related components. One is optimal diet; the other is foraging efficiency. Theoretical ecologists have come up with certain rules (hypotheses) for optimal foraging.

Optimal Diet

Suppose you scatter black oilseed (a medium-sized sunflower seed) and white millet (a small seed) for winter birds. According to studies of preferred food of winter birds, black oilseeds are preferred by cardinals, house finches, and nuthatches, who extract the meat. The preferred food of mourning doves supposedly is millet, a close relative of its natural food. You soon discover that the doves choose oilseed over the millet. They will not crack the seeds as finches do—their thin pointed bills are not adequate to the task—but will eat them whole. Only after the day's allotment of oilseed is consumed will the mourning doves turn to their supposedly preferred millet.

Obviously, the mourning doves found it energetically more profitable to take large oilseeds, rich in carbohydrates, over small millet seeds. Because of oilseed's larger size, the doves could acquire more energy with less handling time, which meant remaining for a much shorter period in the food patch. When feeding on millet, the doves had to handle many more seeds providing much smaller packets of energy per unit effort. In effect, the doves made an optimal economic decision. They chose larger, more profitable seeds over smaller ones.

According to the "decision rules," a consumer should: (1) prefer the most profitable prey (items that yield the greatest net energy gain); (2) feed more selectively when profitable prey or food items are abundant; (3) include less profitable items in the diet when the most profitable foods are scarce; and (4) ignore unprofitable items, however common, when profitable prey are abundant. The mourning doves made all the "right" decisions, but they were operating under ideal conditions unwittingly provided: an abundance of food with a choice of only two items. Natural conditions in which they had to locate food patches that provided much smaller and more diverse prey items might have produced a different outcome.

Although the theory of optimal diet makes practical sense, it is difficult to test under field conditions. Not only would we have to know exactly what items the animals were consuming, we would also have to know the relative availability of all potential food items in the habitat, as well as the profitability index of each. Even then we would not necessarily know enough. Consumers might make their decisions on criteria other than relative availability (for a short discussion of this point see Taylor 1984:93–94). Some studies have been done under controlled conditions involving birds (great tit, Krebs et al. 1978), fish (brown trout *Salmo trutta*, Ringler 1979) and invertebrates (shore crabs *Carcinus maenas*, Elner and Hughes 1978).

Werner and Hall (1974) presented groups of ten bluegill sunfish with three sizes of *Daphnia* in a large aquarium. They allowed the fish to forage for a period of time, then killed them and examined their stomachs to determine the number and size of *Daphnia* taken. When the density of the prey presented was low, the fish consumed the three sizes according to the frequency encountered. They showed no preference for any size category (Figure 15.16). When the prey population was dense, the fish consumed the largest prey items. When presented with an intermediate number, the fish took the two largest size classes. The results of these feeding trials support the optimal foraging theory.

Davies (1977) studied the feeding behavior of the pied wagtail (*Montacilla alba*) and yellow wagtail (*M. flava*) in a pasture field near Oxford, England. The birds fed on various dung flies and beetles attracted to droppings. They had access to prey of several sizes: large, medium, and small flies and beetles. The wagtails showed a decided preference for medium-sized prey (Figure 15.17). The size of the prey corresponded to the optimum-sized prey the birds could handle profitably (Figure 15.18). The birds ignored small sizes. Although easy to handle, small prey did not return sufficient energy, and large sizes required too much time and effort to handle.

Foraging Efficiency

Most animals live in a heterogeneous or patchy environment. In feeding, they have to concentrate on the most productive food patches. This fact has given rise to another set of decision rules in optimal foraging theory. The consumer should: (1) concentrate foraging activity in the most productive patches; (2) stay with those patches until their profitability

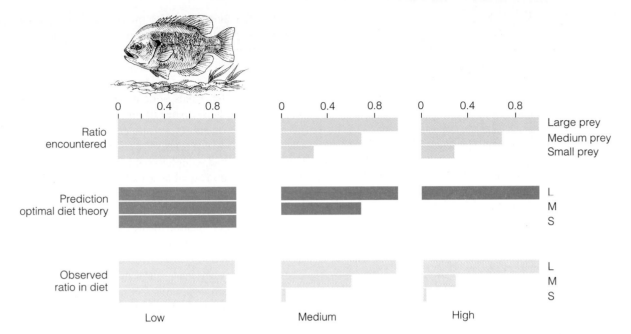

FIGURE 15.16   Optimal choice of diet in the bluegill sunfish preying on different sizes of *Daphnia*. The histograms show the ratio of encounter rates with each size class at three different densities, the prediction of optimal ratios in the diet, and observed ratios in the diet. Note the bluegill's preference for large prey. (After Werner and Hall 1974:1048.)

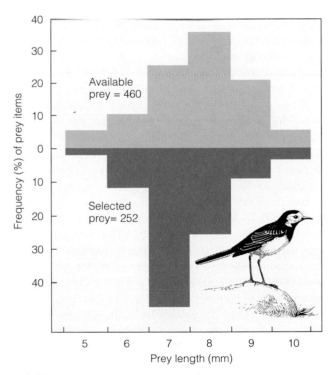

FIGURE 15.17   Pied wagtails show a definite preference for medium-sized prey, which are taken in amounts disproportionate to sizes of prey available in the environment. (Davies 1977:48.)

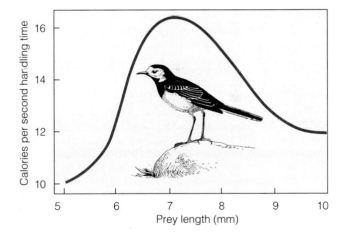

FIGURE 15.18   Prey size chosen by pied wagtails (*Montacilla alba*) is the optimal size for maximum energy per handling time. Small sizes provide too few calories. Large sizes require too much handling time. (Davies 1977:48.)

These rules are covered by the **marginal value theorem** (Charnov 1976, Parker and Stewart 1976), which gives the length of time a forager should profitably stay in a resource patch before it seeks another. The length of stay relates to the richness of the food patch, the time required to get there, and the time required to extract the resource. When a forager arrives on a patch, it initially has a high rate of extraction and energy gain (Figure 15.19); but as time progresses, the abundance of the resource and the rate of extraction decline, until on the average it is no longer profitable for the forager to remain. Too long a stay depletes the resource. Conversely, if

falls to a level equal to the average for the foraging area as a whole (Figure 15.19); (3) leave the patch once it has been reduced to a level of average productivity; and (4) ignore patches of low productivity.

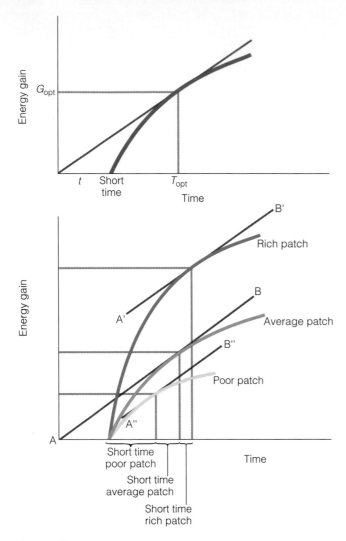

FIGURE 15.19   How long should a predator remain in a habitat patch seeking food? This graph provides a theoretical answer. Time spent in travel and time spent in a habitat patch are plotted against energy gain, which declines as food is depleted. The curve represents the cumulative amount of food harvested relative to time in the patch. The straight line represents average food intake per unit time for the habitat as a whole. Where the line touches the curve, the predator has reached average cumulative net food gain for the habitat as a whole. Beyond this point net food gain declines to below average. Thus the point represents the optimal time (T_{opt}) for the predator to seek a more profitable food patch. (After Krebs 1978:42.)

the forager leaves a patch too soon, it does not use the resource efficiently. Ideally the forager should leave for another patch at the point (indicated by the intersection of the straight-line tangent to the curve in Figure 15.19b) where energy gains start to diminish. The model predicts that foragers should remain in a rich food patch longer than in a poor one, and that as travel time between patches increases, it should remain in the patch longer to balance energy loss in travel. Overall the forager should leave all patches, regardless of their profitability, when they have been reduced to the same marginal value that is average for the environment as a whole.

Whether animals go by these rules has also been the object of experimentation both in the laboratory and in the field. Hubbard and Cook (1978) studied the foraging behavior of a parasitoid, the ichneumon wasp *Memeritis canescens,* in a laboratory arena containing patches of the host, larvae of *Ephestia cantella.* Hosts in densities of 64, 32, 16, 8, and 4 were placed in petri dishes filled to the brim with hardened plaster of Paris, the hosts' substrate. Space between the larvae was filled with wheat bran. One result of the experiment was that all patches were depleted of larvae to a common level of host abundance. The richest patches suffered the greatest depletion. As exploitation proceeded, the amount of time spent by the wasp in patches of highest density declined, and the proportion of time spent in the next-richest patches increased.

Zack and Falls (1976a–c) studied foraging strategy under more natural conditions using ovenbirds (*Seiurus aurocapillus*). The ovenbird is a ground-nesting, ground-foraging neotropical warbler of eastern deciduous and northern mixed deciduous-conifer forests. While foraging, it walks slowly and scans continuously for invertebrates in the leaf litter and low vegetation. In a series of four experiments, Zack and Falls exposed captive ovenbirds individually to a patchy food supply (mealworms) presented in natural outdoor pens in typical habitat.

In one of four experiments, Zack and Falls presented four fixed patch locations in which they interchanged prey densities. They found that the ovenbirds increased their search path exponentially with prey density. The birds rapidly shifted their search efforts as prey densities were interchanged. Because less search path was required per prey found in patches of dense prey, the birds concentrated their efforts in areas of high profitability and took a higher percentage of prey available on the sites. Ovenbirds did not always visit every patch location during the observation periods, but they always visited the high-density prey patches. This finding suggests that the birds' discovery of one or more profitable feeding areas discourages the sampling of other patches to assess their profitability. The ovenbird's tendency to quit searching after encountering one or more profitable patches and its ability to learn the location of and return to patches of high prey density may limit the number of patches the bird will exploit. The ovenbird may use other patches only if it discovers them by chance.

Zack and Falls (1979) then observed the foraging behavior of free-ranging male ovenbirds in a deciduous forest in Ontario with no control over the birds or their prey (Figure 15.20). They sampled prey density in the foraging areas of each bird and found the litter invertebrates to be patchily distributed. This study yielded three main results: (1) the areas used for foraging consistently had higher prey densities than areas not used for foraging; (2) the birds concentrated their search paths in areas of high prey density and returned to those areas; (3) search paths typically were directional; and (4) the birds foraged systematically and avoided inefficient random searching. However, the investigators could not determine whether the birds were foraging optimally because of the complexity of the situation and the lack of detailed information on prey types. Zack and Falls (1981) concluded

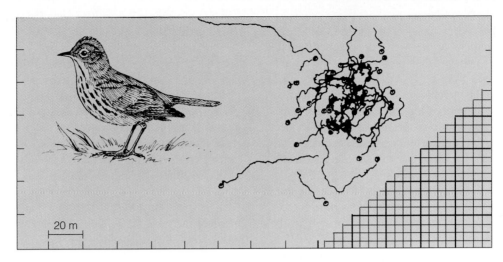

FIGURE 15.20 Search paths of a free-ranging foraging male ovenbird (*Seiurus aurocapillus*) in a deciduous forest in Ontario. (After Zack and Falls 1979.)

that the results of their series of controlled laboratory experiments were of slight value in evaluating wild birds' performance, because the laboratory situation was much simpler than the complex world of the forest.

These and other experimental studies, however, support the hypothesis of optimal foraging up to a point. It is not surprising that they should. Much optimal foraging theory concerns actions you would expect any mobile animal to take: forage in areas where food is abundant, leave when searching is no longer rewarding, select the larger and most palatable items of food, leave the poor items until last, and travel no farther than necessary to feed. Where the theory breaks down is in the expectation that animals will choose patches in the order of profits or take only optimal food items first and ignore the rest. Such choices may be characteristic of animals foraging in a stable laboratory environment. It is not necessarily the way animals behave in the wild.

This observation is recognized in the concept of **satisficing,** in which the decision maker is satisfied after meeting some minimal requirement (Simon 1956, Stephens and Krebs 1986). An analogy is your seeking a good restaurant in a new town. If you "foraged" optimally, you would wait to dine until you had checked out all the restaurants in town and found what you considered to be the best one. Rarely, however, would you "forage" in that manner. Instead, you would probably look at several restaurants, but not all, and settle on one that is acceptable. Animals appear to do the same.

Being opportunists, animals will take some less than optimal food items upon discovery, and they may quit before food items are reduced to some minimal level. Nor will they pass up certain profitable patches because they do not meet some theoretical expectation. Animals quickly learn where food is and where food is not, and they do not waste much time on a patch after it is depleted. Foragers, however, will stay with a patch as long as the rate of replenishment exceeds the rate of depletion.

Risk-Sensitive Foraging

The marginal value theorem model assumes that the animal knows the quality of food patches and expects a constant reward on each visit. In reality the quality of the patches varies randomly over space and time. In this situation the animal has to decide whether to go back to a patch that gives it a constant rate of return or visit a new patch where the return is unknown. The choice is important if the animal is to avoid a shortfall in energy needs. Animal behaviorists call such decision making **risk-sensitive foraging.**

The word *risk* here is borrowed from economics. Financial investors may have some knowledge about the probability of an event (such as a rise in interest rates or increased demand for a product). On the other hand, they may lack full knowledge and face a degree of uncertainty. This uncertainty or risk is what foragers face in their variable environment. The foragers must first learn the probability distribution of rewards derived from certain behaviors. Then they have to seek a strategy to exploit those distributions. That is where the risk comes in.

How animals make such decisions has been the subject of a number of behavioral experiments in the laboratory (Real and Caraco 1986). For example, Caraco and associates (1980) determined the daily energy requirements of a captive flock of yellow-eyed juncos (*Junco haeonotuis*). They provided food (millet) at two feeding stations separated by a partition in their aviary cage. The experimenters could manipulate the energy budgets of the birds by depriving the birds of food prior to any trials. In a given experiment one feeding station always offered a constant reward (risk-averse or low risk). The other feeding station offered an unpredictable reward—no seeds half the time; some seeds the other half of the time (risk-prone or constant risk). Thus the birds faced choices between a constant number of seeds and a random number of seeds; but always the mean of the variable reward equaled the mean of the constant reward.

When deprived of food for one hour in experimental tests and still in a positive energy balance, the juncos avoided risk by preferring the predictable site. When deprived of food for four hours, the birds switched their preference for the variable reward. They changed from being risk-averse to risk-prone. Under energy stress the variable site offered the possibility of providing 50 percent more food than the constant site, whereas the constant site would not provide sufficient food to meet energy needs. Of course, there was the 50 percent risk of finding no food. Nevertheless, in the face of high energy demand, risk-prone behavior maximized daily survival.

Animals living in natural conditions face such choices each day. They may start out risk-prone and as time goes on become risk-averse. This behavior has given rise to the **expected energy budget rule:** be risk-prone if the daily energy budget is negative; be risk-averse if it is positive (Stephens 1981).

Animal behaviorists may demonstrate risk-sensitive foraging in the laboratory, but do foragers in the wild practice it? Wild foragers do have a number of choices denied captive experimental animals. A few behaviorists have undertaken experimental studies of risk sensitivity under natural conditions (for example, Wunderle and Cotto-Navarro 1988, Barkan 1990, Cartar 1991). Cartar (1991) investigated the risk-sensitive foraging of three species of colonial bumblebees in coastal southwest British Columbia—*Bombus melanopygus, B. mixus,* and *B. sithensis.* These bees feed on the nectar and pollen of their two most common food plants, seablush (*Plectritis congesta*) and dwarf huckleberry (*Vaccinium caespitosum*). Both plants offer foraging bumblebees equivalent expected rates of return, but the huckleberry is more variable. The bees store the nectar they collect in open-topped honey pots. The nectars they accumulate during the day they use during the night.

Cartar manipulated the energy requirements of the bees by depleting the honey pots or enhancing them with 50 percent sucrose solution. He then censused the foragers from depleted and enhanced colonies visiting the two flower species. He hypothesized that if the bees were risk-sensitive, they should increase their relative use of the more variable huckleberry when the colonies were depleted of energy than when they were enhanced. The bees did so. Because the huckleberry had the higher probability of greater nectar returns, the bees accepted the gamble. They made their foraging decisions based on the energy requirements of the colony relative to expected intake of energy. Their behavior suggested that the bees were sensitive to the mean and variance of the energy rewards offered by the two plants.

Another, wholly unrelated type of risk-sensitive foraging relates to **predation risk.** Habitat cover and foraging areas both vary in their foraging profitability and predation risk. In deciding where it will feed, the forager must balance its energy gains against the risk of being eaten. If predators are about, then it may be to the forager's advantage not to visit a most profitable but predator-prone area and to remain in a less profitable but more secure part of the habitat. Ecologists have done many studies on how the presence of preda-

tors affects foraging, mostly in aquatic invertebrates and fish (for a review see Lima and Dill 1990).

In the coniferous forests of central Finland, flocks of willow tits (*Parus montanus*) and crested tits (*P. cristatus*) forage in spruce, pine, and birch trees in winter. Their major threat to survival is the Eurasian pigmy-owl (*Glaucidium passerinum*). The owl is a diurnal ambush or sit-and-wait hunter that pounces downward on its prey. Its major food is voles, and when vole populations are high, usually every three to five years, the predatory threat to these small birds declines. When vole populations are low, however, the small birds become the owl's major prey. During these lows the willow and crested tits forsake the outer branches and more open parts of the trees. They restrict their foraging to the dense inner parts of spruce and to the tops of the more open pine and the leafless birches (Suhonen 1993).

Foraging theory is an area of active interest among theoretical behavioral ecologists. They are attempting to develop models incorporating animal foraging decisions, resource quantity and quality, and environmental constraints to explain and to predict foraging behavior. Because of an animal's own individuality in its decision making, it is doubtful that foraging behavior can be reduced to sets of predictive mathematical equations. Foraging models, however, can provide valuable insights into ways in which animals utilize their environment. (For discussions see Krebs 1987, Kamil et al. 1986, Real and Caraco 1986.)

Summary

1. Predation is the consumption of one living organism by another, a relationship in which one organism benefits at the other's expense. In its broadest sense predation includes herbivory and parasitism.
2. Interactions between predator and prey have been described by the mathematical models of Lotka and Volterra and by subsequent modifications of their model by others. Essentially all these models predict oscillations of predator and prey populations. The oscillations may be stable, damped, or unstable.
3. Relationships between predator and prey populations result in two distinct responses. As density of prey increases, predators may take more of the prey, a functional response; or predators may become more numerous, a numerical response.
4. There are three types of functional responses. In Type I the number of prey taken per predator increases linearly to a maximum as prey density increases. In Type II the number of prey taken rises at a decreasing rate toward a maximum. It occurs in situations of varying densities of one-species prey. Type III involves two or more species of prey. In Type III the number of prey taken is low at first and then increases rapidly to an asymptote, resulting in a sigmoidal pattern. Type III responses involve a search image, in which the predator develops a facility for finding a particular prey item, and switching, in which the predator turns to an alternative, more abundant prey species for more profitable hunting. It takes that prey in a disproportionate amount relative to other prey species.
5. Numerical response refers to the increase of predators resulting from an increased food supply. Numerical response may involve an aggregative response, the influx of predators to a food-rich area, or more importantly, a change in the rate of growth.

6. The rate of growth of the predator population changes through modifications in developmental time, survival rates, and fecundity. Such changes produce a delayed numerical response, for a time lag necessarily exists between birth of young and maturation of reproducing individuals.

7. Because prey occurs in patches, the predator finds it more efficient to spend time in areas not necessarily where prey is most abundant, but where hunting is most profitable in terms of time allocated relative to net energy gained. Study of such behavior has given rise to the concept of optimal foraging, a strategy that obtains for the predator a maximum rate of net energy gain. There is a break-even point, above which foraging in a particular patch is profitable and below which it is not.

8. Optimal foraging involves an optimal diet, one that includes the most efficient size of prey for both handling and net energy return. Optimal foraging efficiency involves the concentration of activity in the most profitable patches of prey and the abandonment of those patches when they are reduced to the average profitability of the area as a whole.

9. Based on the mean and variance of foraging rewards, decisions about patch choice are risk-sensitive. Depending on their energy status, foragers may choose between a constant source of a reward or a variable one with the probability of a much greater reward. When confronted with an energy shortfall, foragers seem to gamble on the latter.

Review Questions

1. What do the Lotka-Volterra and Nicholson-Bailey models of predation predict? What are some weaknesses of these models?
2. How does the Rosenzweig-MacArthur model of predation differ from the Lotka-Volterra and Nicholson-Bailey models?
3. How does the ratio-dependent predation theory modify predation models?
4. What is functional response in predation? Distinguish among Type I, Type II, and Type III responses.
5. What is switching? How might search image relate to switching?
6. What is numerical response?
7. What is optimal foraging? How might risk-sensitive foraging relate to optimal foraging?
8. The models of predation and functional response are theoretical. How can we apply these models to the management of species populations, their food resources including vegetation, and pest management?

Cross-References

Population growth, 184; logistic equation, 185; carrying capacity, 186; intraspecific competition, 195; exploitative and interference competition, 195, 243.

C H A P T E R 16

Predators and Their Prey

CONCEPTS

1. Predation on plants involves their consumption by herbivores.

2. Herbivory can reduce plant fitness or it can stimulate biomass production.

3. Plants affect herbivore fitness through the quantity and quality of plant tissue and through defensive mechanisms.

4. Many herbivores have evolved ways of breaching plant defenses.

5. Animal prey and predators have evolved various tactics of defense and attack that vary according to their costs and benefits in time and energy.

6. Intraguild predation, the killing and eating of species using the same resources, results in an energy gain and a reduction of competition for the predator.

7. Cannibalism, the killing and eating of conspecifics, can have pronounced demographic effects within a population.

8. Predator-prey cycles seems to be driven by interactions of food shortages, malnutrition, and predation.

9. The ability of predators to regulate prey populations depends on the population dynamics of the prey.

The previous chapter introduced the theory of predation. It looked at the reciprocal relations between predator and prey as modeled by the Lotka-Volterra equations, the relationship between prey density and consumption by predators, and the role of optimal foraging in predator-prey relations. Little was said, however, about actual interactions between predator and prey that make up predator-prey systems. These systems involve three trophic levels: vegetation, herbivores, and carnivores. Predator-prey interactions involving plants and their herbivorous predators make up vegetation-herbivore systems. This system has a direct influence on the next system, herbivores and their predators, the carnivores. This chapter explores these interactions.

HERBIVORE PREDATION ON PLANTS

Predation on plants by herbivores includes both defoliation and the consumption of fruits and seeds. The results of the two forms of predation are different.

Defoliation is the destruction of plant tissue (leaf, bark, stem, sap, and roots). Some plant predators, such as aphids, do not eat tissue directly, but, acting as parasites, tap plant juices without killing the plant. Other herbivores consume tissue directly, destroying part or all of the plant. If grazers eat seedlings, they kill the plant. If they remove only part of the plant, its survival depends on the amount and continuation of grazing. Continued grazing may eventually kill the plant, but if grazing ceases, the plant may regenerate. Although grazed plants may persist and regenerate, defoliation still has an adverse effect. Grazing decreases plant biomass. Removal of leaves changes the plant's competitive position in the stand. Loss of foliage and subsequent death of some roots (root pruning) reduce the vigor of the plant, its competitive ability, and its fitness (Harper 1977).

The impact of seed predation is difficult to assess. If density-dependent processes let few individuals survive, seeds removed by predators represent that portion of the population that has no future. In such instances seed predation has no real impact. If predators remove seeds from an expanding population or from areas being colonized, predation reduces the rate of increase. On the other hand, if consumption of seeds is a mechanism for seed dispersal, as when seeds are contained in a palatable fruit and then carried in the gut of a fruit-eating herbivore, predation can be to the plant's advantage.

Effects on Plant Fitness

Removal of plant tissue—leaves, flowers, bark, stems, roots, sap—affects a plant's fitness and its ability to survive, even though it may not be killed outright (see Dirzo 1984). Loss of foliage and loss of roots decrease plant biomass, reduce the vigor of the plant, place it at a competitive disadvantage to surrounding vegetation, and lower its reproductive effort or fitness (Figure 16.1). These effects are especially strong in the juvenile stage, when the plant is most vulnerable.

FIGURE 16.1 Intense predation on oaks by gypsy moths. Such defoliation can kill weaker trees in the forest and reduce the growth of others. Increased light and nutrient input from droppings of caterpillars can increase understory growth.

Although the plant may be able to compensate for the loss of leaves by increasing photosynthetic assimilation in the remaining leaves, it may be adversely affected by the loss of nutrients, depending on the age of the tissues removed. Young leaves are dependent structures, because they import and consume nutrients drawn from reserves in roots and other plant tissues. As the leaf matures, it becomes a net exporter, reaching its peak before senescence sets in. Grazing herbivores such as sawfly and gypsy moth larvae, deer, and rabbits concentrate on more palatable, more nutritious leaves. They tend to reject older leaves because they are less palatable, being high in lignin and other secondary compounds (tannin, for example). If grazers concentrate on young leaves, they remove considerable quantities of nutrients.

Plants respond to defoliation with a flush of new growth that drains nutrients from reserves that otherwise would have gone into growth and reproduction. Defoliation also draws on the plants' chemical defenses, a costly response. Often the withdrawal of nutrients and phenols from roots exposes them to attack by root fungi while the plant marshals its defenses in the canopy (Parker 1981). If defoliation of trees is complete, as often happens during an outbreak of gypsy moths or fall cankerworms (*Alsophila pometaria*), replacement growth differs from the primary canopy removed. The leaves are smaller, and the total canopy area may be reduced by as much as 30 to 60 percent (Heichel and Turner 1976). Defoliation in a subsequent year may cause an even further reduction in leaf size and number. Some trees may end up with only 29 to 40 percent of the original leaf area to produce food in a shortened growing season.

Defoliation of coniferous trees results in their death. Conifers do not have the physiological traits that allow them to recover by forming new needles. That is why outbreaks of spruce budworm in spruces and adelgids in balsam fir devastate large areas dominated by these species.

Some plant predators, such as aphids, tap plant juices on new growth and young leaves rather than consume tissue directly. Sap suckers can decrease growth rates and biomass of woody plants by 25 percent.

Damage to the cambium and growing tips (apical meristem) is more destructive in some plants (Figure 16.2) Deer, mice, rabbits, and bark-burrowing insects feed on those parts, often killing the plant or more likely changing its growth form.

Moderate grazing, even in a forest canopy, can have a stimulating effect, increasing biomass production, but at some cost to vigor and at the expense of nutrients stored in the roots. The degree of stimulation depends on the nature of the plant, nutrient supply, and moisture. Although many biennial and perennial herbaceous species are adversely affected by herbivory, some species respond by producing more branch tissue than shoots.

For grasses, moderate grazing typically increases biomass, but severe grazing causes a decline in biomass production. Adverse effects are greatest when new growth is developing. (Andrzejewska and Gyllenberg 1980). Grasses, however, are well adapted to grazing and may benefit from it. Because the meristem is close to the ground, older rather than more expensive young tissue is consumed first. Grazing stimulates production by removing older tissue functioning at a lower rate of photosynthesis. It reduces the rate of leaf

FIGURE 16.2 Heavy browsing on low woody growth and herbaceous plants by white-tailed deer prevents any plant from escaping predation and achieving significant growth.

aging, prolonging active photosynthetic production, and it increases light intensity on underlying young leaves, among other things. Some grasses can maintain their fitness only under the pressure of grazing, even though defoliation reduces sexual reproduction (McNaughton 1979, Owen 1980, Owen and Wiegert 1981).

Some ecologists, however, challenge the idea that herbivory benefits grazed plants (Belsky 1986). Although abundant evidence exists that herbivory affects all aspects of plant growth, little evidence exists to show beneficial effects on fitness and population dynamics (Crawley 1989). Herbivory, especially grazing by large herbivores and outbreaks of foliage-consuming insects such as gypsy moth and spruce budworm, can sharply reduce or eliminate populations of certain trees and shrubs and stimulate the growth and spread of nonselected species. In the arctic, snow geese (*Chen caerulesescens*), grazers on tundra vegetation, have experienced an explosive increase in numbers from less than 1 million birds in 1970 to 3 million in 1998. Their grazing and grubbing for roots and rhizomes are destroying many plant species, which are being replaced by unpalatable salt-tolerant ones. These changes brought about by herbivory threaten irreversible changes in the tundra plant community and in the food base and breeding habitat for the snow geese themselves (Jefferies 1988, Ben-Atri 1998).

There are few experimental studies following the effects of herbivory on seed production and seedling survival. Paige and Whitman (1987) experimentally demonstrated the response of scarlet gilia (*Ipomopsis aggregata*), a tall showy biennial growing in extensive patches in Rocky Mountain meadows, to grazing by elk and mule deer. Fifty-six percent of the experimental population lost 95 percent of its above-ground biomass. The gilia overcompensated for this loss by producing multiple-flowering stems and up to three times as many flowers, fruits, and seeds as the nongrazed controls. The grazed and ungrazed populations did not differ in the number of seeds produced per fruit, seed weight, and germinating success and survival. The grazed plants, however, averaged a 2.4 fold increase in fitness over uneaten plants. These results suggest that for this plant there was a reproductive advantage to having parts removed by herbivores. A similar concept applies to the pinching back of horticultural plants such as chrysanthemums and certain annuals to stimulate flower production.

Plant Defenses Against Herbivores

Because plants are fixed in place, they are at a distinct disadvantage in any predator-prey relationship. To counteract predatory attack, plants have arrived at some modes of defense against their herbivorous predators throughout their evolutionary history. These defenses range from chemical methods, widespread among plants, to mimicry and structural features such as spines and the hardness of seeds.

Chemical Defenses Chemical defense is a first line of defense by plants against herbivores (see Levin 1976). The basis of chemical defense is an accumulation of secondary products (metabolic products not directly related to plant metabolism) ranging from alkaloids to terpenes, phenolics, resins, and steroidal, cyanogenic, and mustard oil glycosides and tannins. Phenolics, a by-product of amino acid metabolism, are ubiquitous to seed plants. Alkaloids, also amino acid derivatives, occur in several thousand species, and cyanogenic glycosides are present in a few hundred species. The secondary products may be stored within the cells and released only when cells are broken, or they may be stored and secreted by epidermal glands to function as a contact poison or as a volatile inhibitor.

Although production and storage of such metabolites are expensive to the plant, there is little evidence that plants evolved these metabolites for a defensive purpose. During the evolutionary history of plants, these secondary compounds may have resulted as metabolic by-products or have served some past (or perhaps even present) physiological function (see Futuyma 1983). In time these compounds became useful deterrents to herbivore predation, although they cannot defend against the full suite of enemies that affect herbivore fitness.

Chemical defense may be qualitative or quantitative. Qualitative defenses involve highly toxic secondary substances stored in discrete parts of the plant such as vacuoles, epidermal hairs, and latex-resin systems. The plants use these secondary substances, such as cardiac glycosides and alkaloids that interfere with metabolism or disrupt development of many insects. The plant can synthesize these substances quickly. They are effective at low concentrations, are readily transported to the site of attack, and work fast. They can be shuttled about the plants from growing tips to leaves, stems, roots, and seeds, and they can be transferred from seed to seedling. Such responses are **induced defenses.** They decrease the impact of the herbivory by reducing damage to the plant. Plants employing qualitative defense are usually short-lived, mostly annuals and perennials.

Quantitative defense depends on the plant tissue being eaten in large amount by the herbivore. The secondary compounds involved, including tannins and resins, are concentrated near the surface tissues of leaves, in bark, and in seeds. They form indigestible complexes of leaf proteins, reduce the rate of assimilation of dietary nitrogen, inhibit the ability of microorganisms to break down proteins in herbivore digestive systems, and lower palatability. Metabolites such as phenolic terpenes and saponins may be toxic, causing illness and occasionally death. Some, such as tannins, discourage consumption and depress growth rates (Robbins et al. 1987, Bernays et al. 1989).

Quantitative defenses are largely constitutive. They are found in varying quantities in the leaves, stems, and other tissues of plants. They can change over evolutionary time and even during the maturation of a plant, and they act independently of damage to the plant. The problem with quantitative defenses, however, is their slow response, because they are not easily mobilized. The plant produces these chemicals in large amounts only when stimulated to do so by a herbivore attack. Such a response is called induced (Hunter and Schultz 1995, Morris 1997) If the induced response affects herbivore

assimilation of the plant tissue, preference for the plant, reproductive output, or survival, the response is **induced resistance** (Karbon and Baldwin 1997). Such a response may persist for several years. For example a year after defoliation by gypsy moths, oaks increased tannin and phenolic content of their leaves and increased their toughness (Schultz and Baldwin 1982).

Secondary plant substances can affect the reproductive performance of some herbivorous mammals. Isoflavonoid in plants—usually concentrated in legumes, particularly alfalfa and ladino clover—mimic estrogenic hormones, especially progesterone. When consumed, these isoflavonoids exert an estrogenic effect and induce a hormonal imbalance in grazing herbivores that results in infertility, difficult labor, and reduced lactation.

Secondary compounds also serve as reproductive cues for some voles (Berger et al. 1981). A particular compound, 6-methoxybenzoxaxolinone (6-MBOA) found in grass, rapidly stimulates reproductive effort in montane voles (*Microtus montanus*). When voles feed on grass, they stimulate the injured plant tissue to release an enzyme that converts a precursor compound abundant in young growing tissue to 6-MBOA. The ingested chemical serves as a cue to the voles that the vegetative growing season has begun. Such a chemical cue allows the voles to produce offspring when food resources will be available to them. These chemical signals are important to the voles because they live in an environment where food resources are unpredictable and depend on the timing of snowmelt and other environmental conditions. Yearly differences in the appearance of new vegetative growth and of 6-MBOA may influence population fluctuations (Negus et al. 1977).

Secondary compounds may also function as warning odors, repellents, attractants, or in some cases, direct poisons. Volatile components advertise substances that insects and other herbivores would find repellent if they touched the plant. Bitter tastes imparted by tannins and cardiac glycosides can deter further consumption of both seeds (Janzen 1971) and foliage. Such repellents inhibit feeding on the plant possessing them and also add a measure of protection to associated plants. For example, grazing by cattle on bentgrasses (*Agrostris*) and fescue (*Festuca*) is reduced considerably in the presence of buttercup (*Ranunculus bulbosus*), which contains a powerful irritant of skin and mucous membranes (Phillips and Pfeiffer 1958). The presence of such plants can cause the herbivore to fail to locate the palatable plants or to reject them along with the repellent plant (Atsatt and O'Dowd 1976).

Herbivore Countermeasures to Chemical Defenses

Although plants may possess powerful chemical defenses that work well against generalist herbivores, they can be breached, especially by specialists. The main mechanism involved is detoxification of secondary compounds. The major detoxifying system is mixed function oxidase (MFO). Possessed by all animals, MFO metabolizes foreign, potentially toxic substances. In vertebrates the MFO activity is located in the liver; in insects it is in the gut, fat bodies, and

Malphigian tubules. By oxidation, reduction, and hydrolysis, MFO converts fat-concentrating (lipophilic) foreign chemicals into water-soluble molecules that can be eliminated by the excretory system.

The MFO system is a general detoxifying agent, nonspecific in character and induced into activity by a wide array of toxic compounds. It probably evolved in animals to degrade toxic by-products of animal metabolism and harmful compounds ingested. Thus animals, especially the insects, are preadapted to handle many toxic, chemically unrelated compounds. Because of the ubiquitous occurrence of MFO, adaptations to new specific toxic compounds require little genetic change, as witnessed by the rapidity with which insect pests become adapted to new insecticides. Thus insects discovering an abundant new source of food can adapt quickly to novel toxic compounds and become feeding specialists on certain families of plants. For example, some butterflies of the family Pieridae, notably the cabbage butterfly (*Pieris rapae*), and the large white, small white, and cabbage aphids (*Brevicoryne brassicae*) feed on members the Crucifer family. The allyl glucosinolate in these plants is toxic to all noncruciferous feeders. Larvae of the monarch butterfly feed on the highly toxic milkweed (*Asclepias syriaca,* Family Asclepiadaceae) and sequester its cardiac glycosides in their bodies as a chemical deterrent to predation. For such specialists, the volatile chemicals of the host plants may act as an attractant rather than a deterrent. Thus the females of specialists are programmed to seek out and lay eggs on plants on which the larvae can overcome chemical defense.

Other insects get around chemical defenses by stopping the flow of toxic sap to the leaf on which they will feed (Carroll and Hoffman 1980). Plants with such chemical defenses possess toxic latex and resins under pressure in secretory canals associated with leaf veins. In milkweeds and sumacs the secretory canals follow major leaf veins. On these plants, beetles, katydids, and certain caterpillars cut leaf veins to sever the secretory canals, blocking the flow of latex to the intended feeding sites (Dussourd and Eisner 1987, Dussourd and Denno 1991). Among plants in the aster and cucumber families, the canals are arranged in networks instead of only along the leaf veins. On these plants, caterpillars cut trenches across the leaf, severing all strands of the network (Figure 16.3). Among toxic plants in the morning glory family, the secretory canals are confined to the major leaf veins. Herbivores feeding on these plants dispense with any cutting and feed between the major veins.

Mimicry   Mimicry is usually considered an evolutionary response in animals (see below), but animals in search for food may have stimulated mimicry in the plant kingdom. L. E. Gilbert (1975) found evidence of plant mimicry in his study of the passionflower butterfly (*Heliconius*) (Figure 16.4) and its food plant, the passionflower (*Passiflora*). *Passiflora,* a vine of the New World tropics, comprising around 350 species, has a wide range of intraspecific and interspecific leaf shapes. The number of *Passiflora* species found in any one area is about 2 to 5 percent of the 350 species. Some 45 species of highly host-specific species of *Heliconius* butterflies use *Passiflora* species as an egg-laying site and as a source of larval food. Each species of *Heliconius* uses a limited group of plants. Visually sophisticated butterflies learn the position of the vines and return to them on repeated visits. Within a habitat, the leaf shapes of passionflowers vary among species. Under visual selection by butterflies, passionflowers apparently evolved leaf forms that make them more difficult to locate. Because the larval food niche is broader than that of the ovipositing females, there has been

FIGURE 16.3 Caterpillars avoid toxins by severing leaf veins and cutting trenches.

FIGURE 16.4 *Heliconius vicina,* one of the *Heliconius* butterflies that use *Passiflora* species as egg-laying sites and a source of larval food.

selective pressure for divergence among *Passiflora* species (Figure 16.5). Probably because of these selection pressures, *Passiflora* leaf shapes converge with those of associated tropical plants that *Heliconius* finds inedible. So close are the convergences that plant taxonomists have named some *Passiflora* species after the genus they resemble.

In addition two *Passiflora* species, *P. cyanea* and *P. auriculata,* have evolved glandular outgrowths on the stipules (structures near the leaf bases) that mimic the size, shape, and golden color of *Heliconius* eggs at the point of hatching. Because *Heliconius* females detect and then reject shoots that carry eggs and young of other females, *Passiflora* achieves a measure of protection by egg mimicry (Williams and Gilbert 1981).

Structural Defenses Some of the least costly defenses available to plants are structures that make penetration by predators difficult, if not impossible. These include tough leaves, spines, and epidermal hairs on leaves, which may trap, impale, or fence out insects and discourage browsing by vertebrate herbivores, as well as hard-coated seeds. These

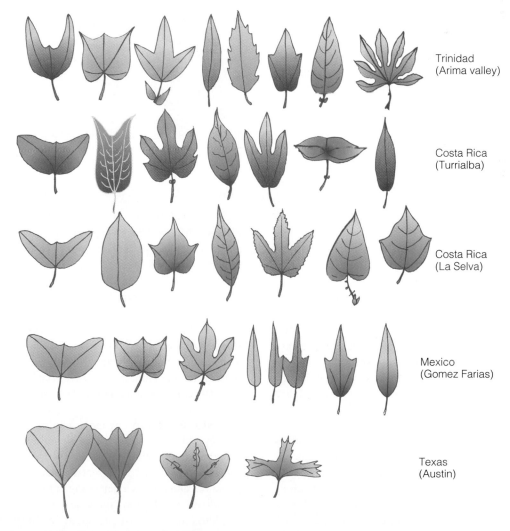

Trinidad
(Arima valley)

Costa Rica
(Turrialba)

Costa Rica
(La Selva)

Mexico
(Gomez Farias)

Texas
(Austin)

FIGURE 16.5 Variation in leaf shape among groups of sympatric species of *Passiflora*. The leaf shapes tend to converge with those of other common species of a number of genera, inedible to *Heliconius* butterflies. (After Gilbert 1975.)

structures may have evolved early in the history of the plants, when they might have been subject to even greater predatory pressure. Because they represent little investment, plants still retain them.

Many seeds have thick, hard seed coats that provide protection from seed-eating animals. The problem with such seed defense is that the seeds need to be scarified—the hard seed needs scratching or scoring of the seed coat to weaken it—so the seedling itself can escape. If the seed is not scarified, the seedling embryo is sealed in, never to germinate. Many plants, however, have turned seed predation in a mechanism for seed dispersal, such as the transport and caching of seeds by squirrels, jays, mice, and ants.

The role of structural defenses in plants is mostly presumed. Little experimental evidence exists to demonstrate the effectiveness of such apparent defensive structures against grazing herbivores. Cooper and Owen-Smith (1986) investigated experimentally the effects of plant spinescence on the feeding habits of three large browsing mammalian herbivores of Africa. These were the kudu (*Tragelapus strepsiceros*), a large African antelope attaining female weights of 180 kg; the impala (*Aepyceros melampus*), a medium-sized African antelope attaining a female body weight of 50 kg; and the Boer goat, a domestic ungulate weighing about 35 kg. The experimenters hand-reared the antelope from calves to allow close observation of feeding habits from very close range, 1 to 5 under natural conditions, to determine biting rates the animals employed. They converted the bites to dry biomass by collecting samples of leaves and shoots of a size similar to those eaten and drying them to a constant weight. They calculated eating rate as the product of bite size (dry mass) and biting rate.

One of the experiments involved a detailed study to examine the influence of spinescence. They selected ten plants each of five species of trees at a height accessible to impalas outside the enclosure. The woody plants exhibited three basic types of spinescence: (1) paired prickles or thorns situated in or close to the leaf axils; (2) short, sharp-tipped branchlets or spines, sometimes carrying small leaves; and (3) prickles of various kinds on leaves. Thorns were either straight and long, up to 70 mm, or short and sharply curved (hooked). On each tree, two branches were matched for size, shape, density of leaf cover, and ease of access to impalas. Cooper and Owen-Smith labeled the paired branches and removed the thorns from one of the pairs. Two months later they visually estimated the relative loss of foliage from browsing.

Results clearly showed that thorns and spines affected the feeding behavior of the three ungulates. These structures restricted bite sizes to mostly single leaves or leaf clusters, and hooked thorns retarded biting rates. Acceptability of leaves of those plant species offering small leaf size along with prickles was lower, at least for kudu, than those of other palatable plant species. The inhibitory effect of prickles was greater for impalas and goats than for kudu, which bit off the shoot ends despite the prickles (Figure 16.6). For certain straight-thorned species, kudu compensated partially for

FIGURE 16.6 Kudu (*Tragelapus strepsiceros*) browsing on *Acacia*.

their slow eating rates by spending more time gathering the leaves. Most spinescent species were similar to unarmed palatable species in their acceptability to the ungulates, even though the armed species had a higher crude protein to their foliage. Probably these spinescent species, especially species of *Acacia,* would be preferred over unarmed species but for the thorns.

The main effect of these armed structural defense features is to restrict bite size, thus increasing handling time. Thorns, spines, and prickles restrict foliage losses to large herbivores. In addition, the animals may incur scar tissue in the esophagus and scratches in the buccal and esophageal mucosa.

Predator Satiation A more subtle defense among plants and animals is the physiological timing of reproduction so that a maximum number of offspring are produced within one short period of time. The great abundance of prey satiates the predators and allows a percentage of the offspring to escape.

Predator satiation is a major strategy against predation in plants and is most prevalent in those species lacking strong chemical defenses. It involves four approaches (Janzen 1971). The first approach is to distribute seeds so that all of a seed crop is not equally available to seed predators. Seeds of most trees fall near the parents and the number of seeds declines rapidly as the distance from the tree increases. Seed predators concentrate about the parent plant. Many of the scattered seeds are missed by the searching predators, in part because of search image and unprofitability. These survivors must produce most of the recruitment. A second approach is to shorten the time of seed availability. If all seed matures and is available at one time, seed predators will not be able to use the entire crop before germination. A third approach is to produce a seed crop periodically rather than annually, as oaks do. The longer the time between seed crops, the less opportunity dependent seed predators have to maintain a large population between crops. Seed predators often experience local increases in density during good seed years, but decline rapidly when the food supply is depleted. This strategy reduces the number of predators available to exploit the next seed crop.

The production of a periodic seed crop depends on synchronization of seed production among individuals. Weather events, such as late frosts or protracted dry spells, and internal physiological constraints usually bring about this synchronization (Janzen 1971). As individuals of a tree species in a community become synchronized, strong selection pressures build against nonsynchronizing individuals, because they experience heavy seed predation between peak years. Such individuals either drop out of the community over evolutionary time or become synchronized.

Predator satiation may be further assured if during any fruiting season the timing of the seed crop of one species is the same as another, and both share seed predators. Seed predators may be attracted away from one species to another, reducing predatory pressure on both species.

Models of Plant-Herbivore Interaction

The interrelations of plants and herbivores have been examined theoretically by May (1973), Caughley (1976a, b), and Noy-Meir (1973), all of whom present mathematical approaches and analyses.

The growth of vegetation as a function of plant biomass can be described by an expression comparable to the logistic growth equation:

$$\frac{dV}{dt} = aV\left(1 - \frac{V}{K}\right)$$

Here V = biomass of vegetation, K = maximum sustained biomass (carrying capacity), and a = rate of increase. The rate of increase slows as interference and the competition for sunlight and moisture increase (Figure 16.7).

When a herbivore population feeds on previously ungrazed vegetation, the vegetation's rate of growth slowed by an amount proportional to the intensity of grazing (number of herbivores consuming plants multiplied by the rate at which the herbivores eat the vegetation). When vegetation is at maximum sustained biomass (K), herbivores can eat all they want, although the quantity is limited by the herbivore's

intake capacity. If the vegetation increases while the herbivore population remains the same, grazers increase the amount of vegetation they eat up to a point of saturation. If the vegetation declines, the amount herbivores consume also declines, because their intake of food is limited by the forage available. These conditions represent a Type II functional response curve (Figure 16.7a).

If herbivores increase, a numerical response, they reach a level where the overgraze the vegetation, as frequently happens with deer, snowshoe hares, and lemmings. If the vegetation is grazed to a point where little or no growth exists, the plant population may become extinct (Figure 16.7c, d). If the vegetation has an ungrazed reserve, the reserve may be used by the plants to attain a low-biomass steady state (Figure 16.7b). Depending on the population density of the grazing herbivores, the vegetation may attain and stabilize at a high biomass (V) or reach an unstable equilibrium point at which the vegetation may be able to restore itself if grazing pressure lets up, or it may slip to extinction. In some situations, especially where a Type III functional response is involved, the vegetation may exhibit two stable steady points, one at a high plant biomass and another at a low plant biomass (Figure 16.7e). Between the two is an unstable equilibrium point.

Interactions between various vegetation growth curves and various herbivore densities can result in a number of plant-herbivore relations (for some detailed examples and discussion, see Noy-Meir 1975). As herbivores increase, vegetation declines (Figure 16.8). The vegetation recovers, the herbivore population increases, and the two populations approach equilibrium, the vegetation with grazing pressure and the herbivore with its food supply (Caughley 1976b).

A notable vegetation-herbivore interaction involves the ten-year cycle of the snowshoe hare, as illustrated by a study of the food-hare relations in the Yukon (Smith et al. 1988) (Figure 16.9). As snowshoe hare populations increased toward peak densities, they depleted their winter food supply, dominated by the preferred woody browse species, bog birch (*Betula glandulosa*) and willow (*Salix glauca*). The hares had to subsist on large twigs and less preferred browse such as white spruce (*Picea glauca*). The situation became critical when essential browse fell below that needed to support the population overwinter, approximately 300 g per individual per day of stems 3 to 4 mm in diameter (Keith 1974, Pease et al. 1979, Wolf 1980). The decline in the peak densities of snowshoe hares was initiated by an overwinter shortage of winter food, which brought on malnutrition (Smith et al. 1988). Malnutrition at peak densities in a snowshoe hare population results in a high winter mortality of juvenile hares, a high overwinter loss of weight, low bone marrow fat, low levels of liver glycogen, and a late onset of reproduction the following summer. Malnourished hares become highly vulnerable to predation, and the population crashes.

In other regions the decline in hares is also related to chemical defenses of alder, birches, and some willows, which strongly influence the selection of winter forage among many subarctic browsing vertebrates (Bryant and Kuropat 1980). Hares avoid these more nutritious plants, especially juvenile

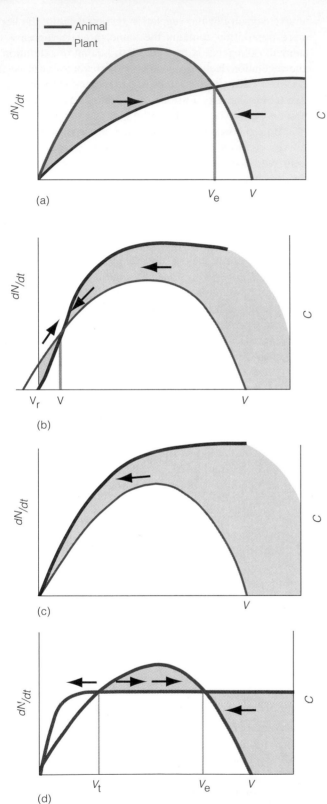

(a)

(b)

(c)

(d)

FIGURE 16.7 Logistic plant growth (dN/dt) as a function of plant biomass, over which is imposed consumption (C) per animal as a function of plant biomass. In green areas plant growth exceeds consumption; in purple areas consumption exceeds growth. (a) The consumption curve is below the growth curve at all biomass levels. Intersection of the two curves indicates the point of stable equilibrium (V_e) between plant growth and herbivore consumption. Deviation in either direction will cause net changes in V, tending to restore stable equilibrium. It is an undergrazed state as plant growth, animal consumption, and secondary production are below maximum. (b) Overgrazing to low-biomass steady state (V_e). Vegetation has some ungrazable reserve biomass (V_r) that prevents complete extinction. (c) Overgrazing to extinction. The consumption curve exceeds the plant growth curve at all levels of V. If no inaccessible plant reserves exist, the plant population becomes extinct. (d) Steady state and unstable turning point to extinction. This situation occurs if the consumption curve is steeper than in (a). The two curves intersect at two points, one at a steady state at high biomass (V_e) and the other at low biomass (V_t). Any deviation can lead to extinction if V becomes lower than V_t. (e) Two steady states, one at high biomass and the other at low biomass. This situation occurs where a plant is ungrazable. The two curves intersect three times, producing a stable steady state at a high plant biomass (V_e) and at a low plant biomass (V_l) and an unstable equilibrium or turning point between them (V_t). (After Noy-Meir 1976.)

growth and buds, because they contain more resin and phenolic glycosides (Reichardt et al. 1984, Palo 1984, Sinclair and Smith, 1984, Bryant et al. 1985). For example, hares avoid the juvenile internodes of Alaskan green alder (*Alnus crispus*). They contain three times the concentration of two deterrent secondary metabolites, pinosylvin and pinosylvin

methyl ether, found in mature twigs, which the hares did consume (Bryant et al. 1983, Clausen et al. 1986). Mountain (arctic) hares (*Lepus timidus*) selectively feed on mature over juvenile twigs of a number of species of willow and avoid low-growing species that have high levels of secondary metabolites (Takvanainen 1985). The decline in secondary

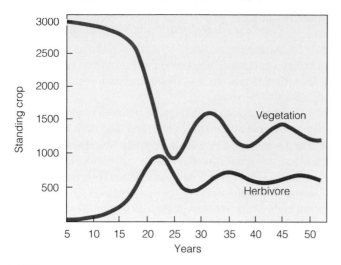

FIGURE 16.8 Trend of vegetation density and animal numbers after an herbivore eruption. Note that as the herbivore population increases, vegetation biomass decreases; and as herbivores decline, vegetation increases. Eventually vegetation growth and herbivore consumption reach a steady state. (After Caughley 1976.)

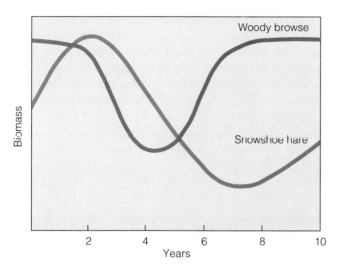

FIGURE 16.9 A vegetation-herbivore cycle involving woody vegetation, particularly aspen, and the snowshoe hare. Note the time lag between the cycle of vegetation recovery and the growth and decline of the snowshoe hare population. (Adapted from Keith 1974.)

metabolites in mature woody growth suggests that juvenile resistance can be an adaptation against mammal browsing at the ground level. This adaptation could further reduce available winter food and help trigger cycling in hares.

PREDATION ON ANIMALS

Herbivore predation on plants supports the carnivore, the third trophic level. Unlike herbivores, carnivores are not faced with a lack of quality in their food. It is quantity that is frequently lacking. That dictates a somewhat different relationship between the eater and the eaten. As pointed out in

Chapter 15, numbers of prey become important. Fitness of the predator depends on its ability to capture prey. Fitness of the prey depends on its ability to elude predation and, if a herbivore, at the same time to overcome plant defenses. That sort of combination puts a squeeze on herbivores.

In an evolutionary context, predator and prey play a sort of game (see Brodie and Brodie 1999). Prey evolve often elaborate means of defense. To capture the prey, the predator must come up with a way to breach the defense. The relationship between the two involves a flux of adaptive genetic change in each. All of these changes are variations within the context of several simple tactics with different time and energy costs and benefits (Malcolm 1990) (Figure 16.10). For the prey, defense involves hide, run, or fight. For the predator, hunting involves sit and wait, stalk, or active search and pursuit.

Over evolutionary time, predators generally seem to experience an adaptive gap between themselves and their prey (Bakker 1983, Brodie and Brodie 1999). Fossil records suggest that predators did not evolve rapidly enough to track the escape adaptations of their prey. Thus predators possess a suboptimal and hardly adequate efficiency in predation.

Animal Prey Defense

Prey have evolved an array of defensive tactics ranging from chemical defenses to simply overwhelming the predator. They affect the predator's breadth of diet and its ability to secure prey.

Chemical Defenses Chemical defenses are widespread among many groups of animals; and as with plants, they may have been borrowed from some other use. Venom, for example, protects snakes from enemies, but it is also the means by which the snakes capture prey.

There is an array of chemical defenses. Some species of fish release pheromones from the skin into the water that act as alarm substances and induce fright in other members of the same or related species (Pfeiffer 1962). The fish produce the pheromone in specialized cells in the skin that do not open to the surface, and so the pheromone is released only when the skin is broken. Fish in the vicinity receive the stimulus through the olfactory organs. Such alarm substances are most common among fish that are social, lack defensive structures, and are nonpredaceous.

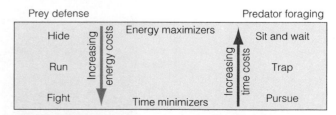

FIGURE 16.10 Basic categories of prey defense and predator offense and the costs and benefits in time and energy. (After Malcolm 1990:58.)

Arthropods, amphibians, and snakes employ secretions to repel predators. Many arthropods produce, often in copious amounts, strongly odorous, easily detected substances (Eisner and Meinwald 1966, Eisner 1970). They produce the secretions in glands with large saclike reservoirs that are essentially infoldings of the body wall and discharge it through small openings. The secretions may ooze on the animal's body surface, as in millipedes; be aired by the evagination of the gland, as in beetles; or be sprayed for distances of up to a meter, as in grasshoppers, earwigs, and stink bugs. These secretions effectively repel birds, mammals, and insects alike by their effect on the predator's face and mouth. Some mammals, such as shrews, skunks, and other mustelids, also possess secretions that discourage attacks by would-be predators.

Active components in the defensive secretions of many arthropods occur as toxic secondary substances, such as saponins, glossypol, and cyanogenic glycosides, used as chemical defenses by plants. Although these toxins inhibit herbivores from feeding on the plants, some arthropods can incorporate toxic substances ingested from the plants into their own tissues. In turn, the toxin protects the herbivore from its enemies. The monarch caterpillar, for example, feeds on milkweeds (page 288) containing a cardiac glycoside, a substance that causes illness in birds that eat the monarch (Figure 16.11) (Brower 1984, 1988; Brower and Fink 1985).

Warning Coloration and Mimicry   Animals that possess pronounced toxicity and other chemical defenses often possess warning coloration, bold colors with patterns that serve as warning to would-be predators. The black-and-white stripes of the skunk, the bright orange of the monarch butterfly, and the yellow-and-black coloration of many bees, wasps, and some snakes serve notice of danger to their predators (Figure 16.12). All their predators, however, must have had some unpleasant experience with the prey before they learn to associate the color pattern with unpalatability or pain.

The association of conspicuous coloration with unpalatability or other averse qualities has for a long time been attributed to the evolution of warning signals advertising that fact. However, unpalatability can evolve independently of warning coloration, as evidenced by shrews and other unpalatable animals. Were animals conspicuously colored for some other reason highly vulnerable to predation? Could the costs of production of toxins, which might have also evolved for different reasons, be outweighed by the protective advantage it confers together with conspicuous coloration?

Similarly, animals living in same habitats with inedible species sometimes evolve a similar mimetic or false warning coloration. That phenomenon was described some 100 years ago by the English naturalist H. W. Bates in his observations of tropical butterflies. The type of mimicry he described, now called Batesian, is the resemblance of an edible species, the mimic, to an inedible one, the model. Once the predator has learned to avoid the model, it avoids the mimic also. In Batesian mimicry the mimic benefits at the expense of both the model and the predator. The model may suffer because the number of its mimics may be greater than the number of models, in which case it will take longer the predator to learn to avoid the model. The model, as a result, will suffer greater losses in the learning process. The predator faces the cost of the loss of palatable food it otherwise could have eaten (Huheey 1988).

FIGURE 16.11   The unpalatable monarch butterfly (*Dannus plexippus*) resting on a toxic milkweed (*Asclepias syriaca*) that is eaten by the larval monarch caterpillar.

(a)

(b)

FIGURE 16.12 The warning coloration of the poisonous coral snake (*Micurus fulvius*) (a) is mimicked by the nonvenomous scarlet king snake (*Lampropeltis*) (b).

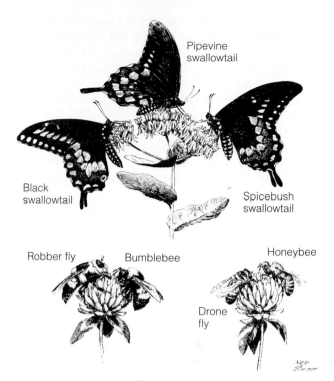

FIGURE 16.13 Mimicry in insects. The model, the distasteful pipevine swallowtail, has as its mimics the black swallowtail and the spicebush swallowtail. The black female tiger swallowtail is a third mimic. All these butterflies are found in the same habitat. The robber fly (Asilidae), a mimic of the bumblebee (Apinae), illustrates aggressive mimicry. The drone fly is a mimic of the bumblebee.

A second type of mimicry is Müllerian, described in 1879 by Fritz Müller. In this type of mimicry an unpalatable or venomous species mimics another. Such mimicry, which involves mutual warning, is advantageous to both. Because the predator associates distastefulness with the pattern without having to handle both species, both model and the mimic reduce the predation load on each. Müllerian mimicry differs from Batesian in that feedback from handling either species is negative, reinforcing the learning process on part of the predator.

Mimicry of various sorts is common among animals (Pasteur 1982). A familiar example among North American butterflies involves the viceroy butterfly (*Basilarchia archippus*), and the queen (*Danaus gilippus*) and monarch butterflies (*D. plexippus*), both distasteful to birds (Brower 1958) (Figure 16.13) . The models and mimic all have an orange ground color with white and black markings. They are remarkably alike, yet the viceroy's nonmimetic relatives are largely blue-black in color. The supposedly palatable viceroy

butterfly was long considered to a Batesian mimic of the monarch and related queen butterflies. However, the viceroy actually is as unpalatable as the monarch butterfly and more unpalatable than queen butterfly. Because both are unacceptable, the mimicry of the viceroy is more Müllerian than Batesian (Ritland and Brower 1991). In the relationship between the viceroy and the monarch and the viceroy and the queen, it appears that the highly unpalatable viceroy is the co-model for both the monarch and the queen and that the relationship is Müllerian. The chemical defense of the three butterflies depends on the food plants the butterfly larvae consume and the chemicals they store. In some instances the queens are less palatable than the viceroy and in other instances they are more palatable. Thus depending on the their diet, the queens shift between Batesian and Müllerian mimicry (Ritland 1994)

Mimicry is widespread among butterflies, tropical species in particular, but mimicry is not confined to mimicking models within a taxon. Some butterflies and butterfly larvae possess eyespot patterns that suggest the eyes of snakes or the eyes of large avian predators that attack small insectivorous passerine birds. Juvenile lizards and snakes mimic highly unpalatable large millipedes (Vitts 1992), insect larvae mimic snakes, and snakes even mimic snakes (see Pough 1988).

Cryptic Coloration   Another defense makes locating prey more difficult. Certain color patterns and behaviors evolved by prey enable them to hide from predators. Such cryptic colorations involve patterns, shapes, postures, movements, and behaviors that tend to make the prey less visible.

Some animals are protectively colored, blending into the background of their normal environment. Such protective coloration is common among fish, reptiles, and many ground-nesting birds. Countershading or obliterative coloration, in which the lower part of the body is light and the upper part is dark, reduces the contrast between the unshaded and shaded areas of the animal in bright sunlight.

Object resemblance is common among insects. For example, walking sticks (Phasmatidae) resemble twigs (Figure 16.14), and katydids (Pseudophyllinae) resemble leaves. Some animals possess eyespot markings, which intimidate potential predators, attract their attention away from the animal, or delude them into attacking a less vulnerable part of the body.

Associated with cryptic coloration is flashing coloration. Certain butterflies, grasshoppers, birds, and ungulates, such as the white-tailed deer, display extremely visible color patches when disturbed and put to flight. The flashing col-oration may distract and disorient predators; or as in the case of the white-tailed deer, it may serve as a signal to promote group cohesion when confronted by a predator (Smith 1991). When the animal comes to rest, the bright or white colors vanish, and the animal disappears into its surroundings (see Harvey and Greenwood 1978 for review).

Armor and Weapons   Some of the most effective means of defense involve protective armor (Figure 16.15). Clams, armadillos, turtles, and numerous beetles all withdraw into armor coats or shells when danger approaches. The associated problem is the animal's inability to assess the external environment. Is the predator large or small, still present or departed? How much foraging time should an animal sacrifice before daring to open up its defenses? Porcupines, hedgehogs, and echidnas have quills (modified hairs), which effectively discourage predators.

Behavioral Defenses   Some animals' defenses are behavioral. One is the alarm call, given at the moment of potential danger when a predator is sighted. High-pitched alarm calls are not species-specific. They are recognized by many different animals close by. But an unanswered question is to whom the calls are directed—the predator or the conspecific prey. If directed toward potential prey, the alarm call could be either altruistic or selfish (Chapter 19). If the alarm exposed the caller's position to the predator, the caller could draw the predator's attention away from conspecifics, including kin, or it could attract more conspecifics for cooperative defense and lower its risk of being taken. Alarm calls do function to warn close relatives, as in the case of Belding ground squirrels (Sherman 1977). Highly sedentary, closely related females live in close proximity to each other. Adult and one-year-old females living with relatives respond quickly to danger and give most of the alarm calls, which warn offspring

FIGURE 16.14 The walking stick (Phasmatidae), which feeds on the leaves of deciduous trees, strongly resembles a twig.

FIGURE 16.15 The armor-bearing armadillo (*Dasypus novemcinctus*) has strong bony plates overlain with horns that develop from the skin.

and other relatives. Beyond that there are few conclusive studies on the evolution and function of alarm calls.

Alarm calls often bring in numbers of potential prey that respond by mobbing or harassing the predator. An example is the harassment of an owl perched in a tree by many small birds attracted to the scene by general alarm calls. Mobbing may involve harassment at a safe distance or direct attack. The outcome for the prey is a reduction in the risk of predation to themselves and their offspring. But like alarm calls, the adaptive and evolutionary significances of mobbing are still obscure.

A distraction display diverts the attention of predators away from eggs or young. Distraction displays are most common among birds. Birds with precocious young, such as the killdeer (*Charadrius vociferus*), usually exhibit the most vehement distraction displays at the time the eggs hatch, and altricial birds, such as the vesper sparrow (*Puoecetes gramineus*), at the time the young fledge. Because the beneficiaries of distraction display are the immediate offspring, the behavior probably evolved through kin selection.

Living in groups may be the simplest defense for some prey species. Groups, especially in mobbing situations, would probably deter a predator that would not be so inhibited when facing only one or two prey individuals. Sudden explosive group flight can confuse a predator, unable to decide which individual to follow. A predator is less likely to find prey when individuals are grouped than if an equal number of individual prey were distributed as solitary prey. By keeping to a group, an individual reduces its chances of being taken. Collectively these two antipredator advantages of group living, called the **attack-abatement effect** by Turner and Pitcher (1986), reduce the risk to group prey.

Among some animals the size of the group may show a direct relationship to the density of specialized predators. For example, in winter, when the availability of forage was less abundant and more patchy, the group size of musk ox (*Bribos moschatus*) in the northwestern Yukon was about 1.7 times as large as in the summer (Heard 1992). During both seasons, however, variation in group size was directly related to wolf densities (Figure 16.16). The group size of musk ox appeared to be a trade-off between decreased predation risk for individuals in larger groups at high wolf densities in winter and a decreased benefit of group foraging in summer.

Predator Satiation A more subtle defense is the timing of reproduction so that most of the offspring are produced in a very short period of time. A good deal of this restricted seasonality of birth can be attributed to the selective advantage of producing young when food will be sufficiently abundant to support them. The other advantage is the synchronization of births to reduce predation on the newborn. This strategy is employed by such ungulates as the caribou (*Rongifer tarandus*) and wildebeest (*Connochaetes taurinus*) (Bergerud 1971, Schaller 1972, Rutberg 1987). Such reproductive synchrony can reduce predation in three ways: (1) the collective defense of the young by breeding adults; (2) interference with the predator's ability to pick out a prey individual; and (3) pro-

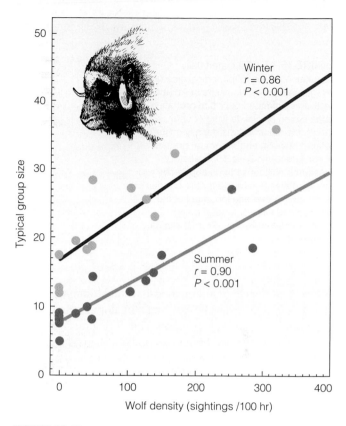

FIGURE 16.16 Increase in musk ox group size in winter ($y = 0.063x + 16.76$) and in summer ($y = 0.0546x + 7.76$) in the Yukon. (After Heard 1992.)

duction of such an abundance of prey that predators can take only a fraction of them. The remaining young quickly grow beyond a size easily handled by predators (Schaller 1972, Bergerud 1971). Such reproductive synchrony functions best against specialized predators with a Type II functional response. For prey faced with generalized predators, asynchronous reproduction may be the best strategy (Ims 1990).

The 13-year and 17-year appearances of the periodical cicadas (*Magicicada* spp.) function in much the same manner. By appearing suddenly in enormous numbers, they quickly satiate predators and do not need to evolve costly defensive mechanisms (Figure 16.17). Although huge numbers of adults succumb to predators, the losses hardly dent the total population (Williams et al. 1993). In other years, predators must seek alternative prey. Thus the cicadas' major defense is to prevent predators from ever evolving any dependence on them.

Predator Offense

As prey evolved ways of avoiding predators, predators by necessity had to evolve better ways of hunting and capturing prey (Bakker 1983).

Hunting Tactics Predators have three general methods of hunting: sit and wait or ambush, stalking, and search and pursuit (see Figure 16.10). Ambush hunting involves lying in wait

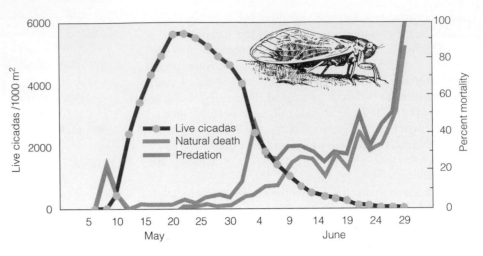

FIGURE 16.17 Estimated daily population density of periodical cicadas (Magicicada) on a study site in Arkansas (left) and estimated daily bird predation rates based on cicada wing counts (right). Maximum cicada density occurred around May 24, and maximum predation occurred around June 10. Predatory pressures built up as birds apparently acquired a search image for the cicadas. At the height of predation, most of the cicadas had already emerged and escaped bird predation. (From Williams et al. 1993:1148.)

for prey to appear. This method is typical among certain insects and some frogs, lizards, and alligators. Ambush hunting has a low frequency of success, but it requires a minimal expenditure of energy. Stalking, typical of herons and some cats, is a deliberate form of hunting with quick attack. The predator's search time may be great, but pursuit is minimal. Therefore it can afford to take smaller prey. Pursuit hunting, typical of many hawks, involves minimal search time. Because pursuit time is great, these predators must secure relatively large prey to compensate for the energy expended. Searchers spend more time and energy to encounter prey. Pursuers, theoretically, spend more time to capture and handle prey once they notice it.

Predators have energy requirements that can be met only by profitable foraging. Predators cannot afford to pursue prey too small to meet their energy requirements unless that prey is abundant and can be captured quickly. Predators also have a limit on upper size. The prey may be too large to consume or too difficult or dangerous to handle. In fact, some prey species become invulnerable to predation through body growth.

Some predators have evolved methods of killing prey much larger than themselves. Certain snakes, arthropods, and shrews use venom to kill large prey. Predators may hunt in groups to lessen the risks they face in attacking prey and to increase foraging efficiency. Wolves, African hunting dogs, jackals, African lions, and Harris's hawks (*Parabueto unicinctus*) (Bednarz 1988) are examples of predators that take large prey by hunting cooperatively. True cooperative hunting involves at least two members of a stable social unit pursuing, capturing, and sharing prey in a coordinated manner. Individual participation in group hunting, however, varies considerably (Packer and Ruttan 1988, Scheel and Packer 1991). Some get into the thick of the action; others refrain or hold back. More individuals tend to refrain from the final pursuit if the prey is small; most join in if the prey is large and dangerous.

Conversely, predators much larger than their prey have evolved ways of filtering organisms from their environments, particularly in aquatic communities. Examples are net-spinning caddisflies that feed on drift and baleen whales that feed on krill (Euphausiacea). Intermediate-sized preda-

tors are usually hunters, whereas very small predators are usually parasitoids.

Cryptic Coloration and Mimicry in Predators Animal prey use cryptic coloration, concealing coloration, and mimicry to their advantage; predators do the same. Cryptic and concealing coloration enable them to blend into the background or break up their outlines. Predators can deceive prey by resembling the host or prey, a deception called **aggressive mimicry** (Pasteur 1986). An example is the model bumblebee and the mimic robber fly (*Mallophora homboides*) (see Figure 16.13). Not only does the robber fly benefit from reduced predation, it also exploits the model for food. The robber fly preys on Hymenoptera by preference, and its resemblance to its prey allows it to escape notice until the bee finds it too late to flee or defend itself (Brower and Brower 1962). The females of certain species of fireflies imitate the flashing of other species, attracting to them males of those species, which they promptly kill and eat (Lloyd 1980). Among birds, the zone-tailed hawk (*Buteo albonotatus*) soars with groups of vultures (*Cathartes* spp.) that scavenge for food as it searches for prey on the ground below (Willis 1963). By mimicking nonpredatory vultures, the zone-tailed hawk can deceive its live prey to false security.

Adaptations for Hunting Predators have acquired various adaptations that improve their hunting ability in addition to such weapons as fangs and claws. Bats, for example, produce ultrasonic sounds through the nose and mouth that enable them to detect prey by echolocation (for a good summary see Vaughan 1986). Night-hunting owls can locate prey by hearing rather than by sight. Their feathered facial disks reflect the sounds of prey and direct them to the ears (Konishi 1973). The owl's large ear openings are positioned asymmetrically, enabling the bird to detect differences in the elevation of the prey. The owl's ability to fly noiselessly allows it to come upon the prey without alerting the victim. Day-hunting northern harriers (*Circus cyaneus*) flying over densely grown grass fields have similar facial disks and placement of ears that enable the hawks to locate by sound voles hidden in the grass (Rice 1982).

Cannibalism

A special form of predation is cannibalism, more euphemistically called intraspecific predation. Cannibalism, more widespread and important in the animal kingdom than many ecologists admit, is killing and eating an individual of the same species. Cannibalism is common to a wide range of animals, aquatic and terrestrial, from protozoans and rotifers through centipedes, mites, and insects to frogs, birds, and mammals, including humans (Elgar and Crespi 1992). About 50 percent of terrestrial cannibals, mostly insects, are normally herbivorous species, the ones most apt to encounter a shortage of protein. In freshwater habitats, most cannibalistic species are predaceous, as they are in all marine ecosystems (Fox 1975a–c).

Cannibalism has been found mainly in stressed populations, particularly those facing starvation. Some animals do not become cannibalistic until other food runs out; others do so when the availability of alternative foods declines and individuals in the population become nutritionally disadvantaged (Alm 1952). Cannibalism is probably initiated when hunger triggers search behavior, lowers the threshold of attack, increases the time spent foraging, and expands the foraging area. It is consummated when the individual encounters vulnerable prey of the same species (see Polis 1981). Other conditions that may promote cannibalism are (1) crowded conditions or dense populations, even when food is adequate; (2) stress, relegating some members to vulnerable low social rank; and (3) the presence of other vulnerable individuals, such as nestlings, eggs, or runts, even though food resources are adequate.

Whatever the cause, local conditions and the nature of local populations influence the intensity of cannibalism. In general, cannibalism fluctuates greatly over both long and short periods of time. Among some predaceous fish, such as walleye (*Stizostedium vitreum*) (Fortney 1974), and insects, such as freshwater backswimmers (*Notonecta hoffmanni*) (Fox 1975b, c), cannibalism is most prevalent in summer. This season coincides with a decrease in normal prey and a reduction of spatial refuges for the young.

Not all individuals in a population become cannibals (see Polis 1981, Pfenning 1997). Intraspecific predation is usually confined to older and larger individuals. Among some species, cannibalistic individuals identify kin through various methods and avoid eating those individuals, initially at least (Pfenning 1997). Individuals receiving the brunt of cannibalism are the small and the young, but not always. In some situations groups of smaller individuals will attack and devour larger individuals.

Demographic consequences of cannibalism depend on the age structure of the population and the feeding rates of various age classes. Even at very low rates, cannibalism can produce demographic effects. Three percent cannibalism in the diet of walleyes could account for 88 percent of mortality among young (Chevalier 1973). Cannibalism can account for 23 to 46 percent of the mortality among eggs and chicks of herring gulls (Parsons 1971), 8 percent of young Belding ground squirrels (Sherman 1981), and 25 percent of lion cubs (Bertram 1975). If a large proportion of either an entire population or a vulnerable age class is eaten, cannibalism can cause violent fluctuations in recruitment.

Cannibalism can become a mechanism of population control that rapidly decreases the number of intraspecific competitors as food becomes scarce. It is unlikely to bring about extinctions of local populations because of its short-term nature. It decreases as resources become more available to survivors and as vulnerable individuals become scarcer. By reducing intraspecific competition at times of resource shortages, cannibalism may actually reduce the probability of local extinction of a population. In the long term, however, cannibalism can be self-defeating because it runs counter to the second law of thermodynamics and trophic level dynamics. The exceptions are among those animals whose young feed at lower trophic levels than the adults. Then cannibalism would involve the harvesting of young grazers.

Cannibalism can provide a selective advantage to survivors. Survivors gain a meal and eliminate a potential competitor for food as well as a potential conspecific predator. With the population reduced, the survivor has more food, enhancing its chances of longer survival, rapid growth, and increased fecundity. Cannibalism may also be rewarding from a nutritional standpoint, leading to increased growth rates and reproduction of cannibalistic individuals over noncannibalistic ones. This may come about in part because the individuals consumed contain the proper proportion of nutrients necessary for growth, maintenance, and reproduction.

Cannibals can also increase their own fitness by reducing the fitness of competitors. By killing and eating other individuals of the same sex they reduce competition for mates. They can eat the offspring of a competitor, as adult Belding ground squirrels do. Among some animals, insects and spiders in particular, the females will kill and eat the male after mating, reducing the probability that other females will encounter a mate (Jackson 1992).

Cannibalism can be a selective disadvantage if individual survivors become too aggressive and destroy their own progeny or genotype completely, reduce their genotype faster than the genotypes of conspecific competitors, or reduce the chances of successful reproduction by eliminating suitable mates. One inherent danger in cannibalism is the transmission of disease from the victim to the cannibal. This is especially true because pathogens already adapted to a specific host and resistant to its immune system are more likely to infect a conspecific than a heterospecific individual (Freeland 1983).

Selection can balance advantages against disadvantages. In some situations the disadvantages of cannibalism are less severe than starvation and reproductive failure caused by inadequate nutrition. For example, parents cannibalizing some of their own offspring can increase the probability of survival and fitness of either parents or surviving offspring or both (Polis 1981, Rohwer 1978) and use rather than waste energy already invested in them. If starvation reduces a population, the survivors may be nutritionally stressed; but if cannibalistic individuals remove conspecifics early, they reduce population density early, and per capita food supply remains high. Survivors have improved their fitness because, being well fed as juveniles, they grow faster, survive better, and produce more young.

Cannibalism may be less costly to individuals, but is disadvantageous from an evolutionary viewpoint. For this reason it is highly improbable that strong selection exists for the trait. With a few exceptions cannibalistic individuals do not distinguish between conspecifics and other prey, but rather are opportunistic predators. Rarely is cannibalism a distinct behavioral trait.

Intraguild Predation

Typical food web construction shows predators occupying specific links. Species A is herbivorous; species B feeds on species A; species C and D both feed on species B. What if D also feeds on species C? That situation, fairly common to many communities, is **intraguild (IG) predation.** Intraguild predation is the killing and eating of a species that uses similar resources and thus is a potential competitor (Polis and Holt 1990) (Figure 16.18). What makes intraguild predation unique is that it combines elements of predation and competition. By feeding on a competitor, the predator acquires an energy gain. Simultaneously, the predator has reduced potential competition.

The simplest form of intraguild predation is a three-species system in which one of the competitors is also a predator and the other is its prey (Polis et al. 1989, Polis and Holt 1992). It occurs among species that eat the same foods. The potential prey species or age class has smaller body size than the predator, and it falls into its normal prey size. The young of many species are particularly vulnerable to species with whom the adults compete. An IG prey species may compete heavily for resources shared with the young of an IG predator; or the IG predator may prey heavily on the young of the IG competitor. Where the IG predator feeds on the young of competing IG prey, it can greatly reduce or even eliminate the local abundance of the IG prey. If such predation excludes or decreases the population of a more efficient IG competitor, the predator increases the supply of the resource for itself.

Intraguild predation can be severe enough to reduce local populations and to regulate or affect community structure. In the Caribbean islands lizards are the dominant vertebrate insectivores. They feed extensively on web spiders as well as on the prey taken by the spiders. Lizards are both predators on and competitors with the spiders. Populations of web spiders are about ten times as dense on islands without lizards as on islands with lizards (Schoener and Toft 1983).

To discover the process underlying this effect, Schoener and Spiller (1987) experimentally removed lizards from randomly selected plots on a very large island in the Bahamas. They found that spider densities in the removal plots were 2.5 times higher than on the control plots. The effects of removing the lizards showed down through the food web. Abundance of arthropod prey increased, as did the consumption of prey. This study, coupled with one by Palaca and Roughgarden (1984), demonstrated that lizards can significantly reduce spider populations on tropical areas. Contrary to most studies, in which predation increases species diversity, predator removal resulted in an increase in the number of spider species.

Intraguild predation has important implications in resource management, especially fisheries management. Fisheries management has a long history of introducing exotic species into lakes and streams either for sport fishery or for forage production. An example is the introduction of opossum shrimp (*Mysis relicta*), a voracious predator of zooplankton, especially cladocerans, as a forage species into the Flathead River-Lake ecosystem in Montana (Spencer et al. 1991). It was stocked to stimulate the production of kokanee salmon (*Oncorhynchus nerka*), a landlocked species of sockeye salmon. Kokanee salmon were also introduced to the system in 1916, when they replaced native populations of cutthroat trout (*O. clarki*). Shortly after the introduction of the shrimp populations, various species of cladocerans declined sharply or became extinct. Instead of increasing with an increased food supply, the salmon population declined, in part because the shrimp retreated to the benthic region during the day, escaping predation by the diurnal feeding salmon. Rather than feeding on shrimp, the salmon preferred cladocerans, which were also the major prey of young salmon. The shrimp, however, were superior competitors with the young salmon for the cladocerans. The result was a collapse of the kokanee salmon population in the lake. The demise of the salmon population reduced the autumnal congregation of bald eagles on the lake, as well as other wildlife attracted to the spawning run.

The moral of these studies is twofold. One, look beyond the current management concept that predator-prey relations, especially in fisheries, are on a one-to-one basis. Two, never introduce a new or exotic species into an ecosystem in which you do not understand feeding relationships. To do so can destroy its evolved predator-prey relationship, eliminate native species, and reduce biodiversity.

Predator-Prey Cycles

The concept of predator-prey cycles is an outgrowth of the neutral Lotka-Volterra predation equations, later modified as stable limit cycles, in which the prey cycle is driven by predation (see Korpimäki and Krebs 1996). The classic example of a predator-prey cycle is that of the snowshoe hare (*Lepus americanus*) and lynx (*Lynx lynx*). By using data obtained

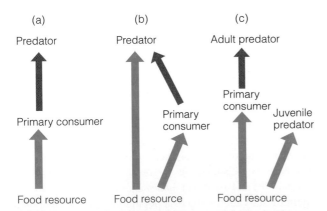

FIGURE 16.18 Food webs. (a) Web without intraguild predation. (b) A three-species food web with intraguild predation. (c) A three-species food web with an age-structured species whose juveniles compete with the consumer and whose adults eat the consumer. (After Polis and Holt 1992:152.)

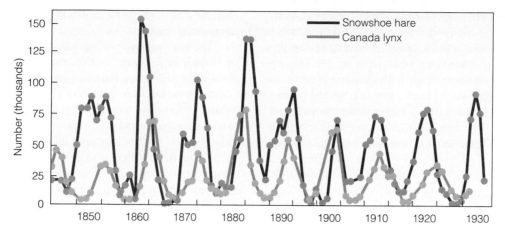

FIGURE 16.19 The classic nine- to ten-year snowshoe hare–lynx cycle in northern North America. Cycling is evident in the fur returns from the snowshoe hare and the lynx. Although the cycles are real, the interaction between predator and prey is an artifact. The lynx fur returns are from western Canada and the snowshoe hare returns are from the Hudson Bay region. The two cycles are not coupled. For this reason, in some years the lynx decline precedes the hare decline. If the predator were truly attacking the prey, the lynx decline would always follow the hare decline. (From MacLuilch 1937.)

from the fur returns of the Hudson Bay Company, MacLuich (1937) first plotted the snowshoe hare–lynx cycle. Elton and Nicholson (1942) further analyzed the cycle, and that analysis has become enshrined in the ecological literature. As first described, the snowshoe hare and lynx cycled in tandem with a time lag between the two. The lynx population peaked as the snowshoe population declined (Figure 16.19), and the snowshoe hare population recovered before that of the lynx. Although this cycling appeared to be an excellent example of a long-term predator-prey interaction, a more critical examination of the data suggested otherwise. The classic cycle was based on hare data from the Hudson Bay region of eastern Canada, whereas the lynx data were from western Canada (Finerty 1980). The two cycles were not exactly coupled.

As we saw, the nine- to ten-year snowshoe hare cycle appears to be related in part to a vegetation-hare interaction. As the number of snowshoe hares increases, they experience an increasing shortage of food over winter that leads to malnutrition. Malnutrition and low winter temperatures weaken the hares, making them extremely vulnerable to predation (Keith et al. 1984, Sinclair et al. 1988). Intense predation by many kinds of predators causes a rapid decline in the number of hares and for several years holds the population at a level much lower than the habitat could actually support. Facing a shortage of food, the predators fail to reproduce or to rear their young. With a decline in predatory pressure and a growing abundance of winter food, the hare population begins to rise sharply, starting another cycle (Figure 16.20).

Seeking causes of cycles, especially the ten-year cycle of hares and lynx, has long intrigued ecologists and wildlife biologists. The causes are especially puzzling because the ten-year cycle is synchronized across the boreal regions of North America. Suggested reasons have ranged from random variations in population oscillations (Cole 1951) to sunspots. Delury (1930) suggested that cycles are caused by sunspot

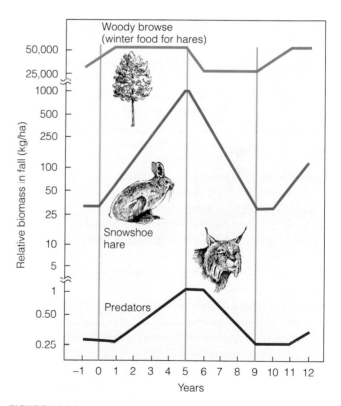

FIGURE 16.20 Model of the vegetation-herbivore-predator cycle involving the snowshoe hare and lynx. Note the time lag between the cycle of vegetation recovery, growth and decline of the snowshoe hare population, and the rise and fall of predator populations. (Adapted from Keith 1974.)

variations in solar radiation and weather. Now the idea that sunspots affect the snowshoe hare cycle has emerged again, not as a cause, but as a synchronizer.

Sinclair et al. (1993) reexamined the relationship of sunspot activity to snowshoe hare cycles. They first looked for clues in the growth rings of white spruce (*Picea glauca*) in the Yukon. At the highs of their cycle, snowshoe hares, deprived of birch and willow browse, turn to eating the nutrient-poor apical stems of white spruce, the loss of which affects the growth of spruce. These periods of stress were marked in cross sections of the trunk by dark marks in the growth rings of the trees. The experimenters found that the presence of tree ring marks among trees less than 50 years old in a 5 km transect correlated with the number of stems browsed. The frequency of the marks correlated with the density of hares in the same region over one well-documented hare cycle. With this information the biologists were able to determine the frequency of dark marks in tree rings within the first 50 years of life of trees germinating between 1751 and 1983. By cross-correlating this data with the hare fur records of the Hudson Bay Company, they found that the tree ring marks of low growth and highs of hare cycles were locked in phase. Next the biologists cross-correlated both tree ring marks and snowshoe hare cycles with sunspot numbers. The resulting correlogram showed a ten-year periodicity. Phase analysis of the data revealed that tree marks and sunspot numbers have periods of nearly constant phase difference during the years 1751–1787, 1838–1870, and 1948 to the present. These periods coincided with the time spans of sunspot maxima. The biologists then examined the data from a 102.5 m ice core from Mt. Logan in Kluane National Park, Yukon, to determine the net annual snow accumulation over the years. They found a nearly constant phase relationship between net annual snow accumulation (an indicator of climatic conditions), tree mark ratios, hare fur records before 1895, and sunspot numbers during high amplitudes in the hare cycle across the whole boreal region.

The synchrony of the ten-year cycle across the North American boreal region requires an external synchronizer that could be entrained by sunspot cycles. Thus solar activity can act indirectly as the synchronizer of hare cycles through climate and its effect on food availability (Sinclair and Gosline 1997). The role of weather and sunspot activity was challenged by Ranta et al. (1997), who failed to find such a synchrony in Finland. However, as pointed out by Sincalir and Gosline (1997), weather systems are in different phase relations with solar activity in different areas of the globe; thus solar-hare phase relations would also differ.

Climatic synchronization of cycles is given further credence by a study of the continent-wide synchronization of a crash of populations of ten taxa of herbivorous insects that occurred between 1987 and 1990 with a rebound in the early in the 1990s (Hawkins and Holyoak 1998). Ranging from gypsy moths to leaf miners and grasshoppers, they represented a wide range of life histories, habitats, and geographic regions of North America. Time-series data indicated relatively high populations of these insects in the early 1980s. In the late 1980s a drought across the continent severely affected the growth and survival of plants, which in turn affected the food supply of insects. Thus climatic

extremes can synchronize insect population crashes on a continental scale.

The ten-year cycle of the snowshoe hare is characteristic of the boreal region. South of the boreal region, some populations of snowshoe hares exhibit cyclic fluctuations and some do not. Dolbeer and Clark (1975), Tanner (1975), and Wolff (1980) have studied why. In coniferous forests and associated regions south of the boreal forests, snowshoe hares exhibit cyclic fluctuations only in a uniform environment of spruce and fir (which is also characteristic of the boreal forest). In regions where the environment is very patchy, where many kinds of vegetation patterns exist, cycles do not occur. There are several possible reasons. Fragmented habitats support a greater diversity of prey species, which adds stability to populations of facultative predators. These predators can maintain sustained predation on hares (Keith 1983). Hares occupying high-quality habitats are protected from predation, whereas hares living in areas of poor cover are subject to predation. These patches of high-quality habitat that provide excellent food and cover act as refuges from which surplus animals repopulate poorer habitat patches. Predators in turn eliminate these hares. Such dispersal and predation tend to hold the population of hares in better habitats at a level at which they do not overutilize their food supply, thus damping cyclic behavior.

In contrast to the ten-year cycle is the two- to five-year cycle of microtine voles. Like the snowshoe hare, some vole populations are cyclic; others are not. For example, voles cycle in northern Fennoscandia, but not in the south (Figure 16.21) (Hanski 1987). Paralleling the rise and fall of the voles are similar trends occurring in populations of shrews (*Sorex* spp.). Predation by least weasels (*Mustela nivalis*) appears to be the common factor synchronizing the crash of voles and shrews (Henttonen 1985). Following the crash of voles and shrews, the weasels also decline. During this time of food scarcity, the weasels turn to cavity-nesting passerine birds, ignored as prey when rodent densities are high (Järvinen 1985). Failure of the vole and shrew populations to cycle in southern Fennoscandia is probably related to the diversity of general predators exhibiting a Type III functional response. In the north the few specialized predators exhibit a numerical response.

Regulation

That predators have an adverse impact on the abundance of prey species is an ingrained idea hard to dislodge. Humans quickly blame predators for the decline in any species in which they have a vested interest, such as game species. However, long-term studies examining the effect of predation on vertebrate populations by Paul Errington (1943, 1945, 1946, 1963) suggest that predators feed on the "doomed surplus" and have little impact on the productivity of prey populations. The unstated premise in Errington's studies was that the predator would have to reduce *r*, the rate of increase of the prey population, to regulate it. To accom-

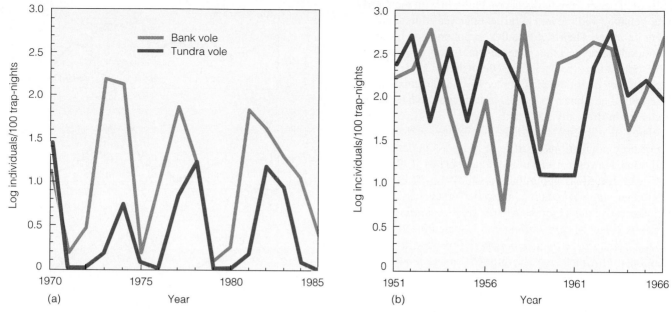

FIGURE 16.21   Two examples of fluctuations in co-occurring species of voles in northern and southern regions of Eurasia. (a) Synchronous cycles in bank vole (*Clethrionomys glareolus*) (yellow line) and tundra vole (*Microtus oeconomus*) (green line) in the north (Fennoscandia). (b) Nonsynchronous and noncyclic fluctuations in *C. glareolus* and *M. oeconomus* in the south (Tula region near Moscow). (After Hanski 1987:55.)

plish that, predators must remove a portion of the reproductive age classes (see Taylor 1984:125–139), which is not part of the doomed surplus.

There is some evidence that natural predation accomplishes such regulation, especially where specialized predators are involved. Larger predators, such as African lions and wolves, tend to take the most vulnerable individuals—the young, the old, and those in poor condition—and few

of the reproductive age classes (Figure 16.22). Such predation is compensatory; that is, the losses are made up by future reproduction.

Theory holds that predators interact with prey to produce either stable equilibria or cycles, but field evidence is scarce. Erlinge et al. (1984) studied the interactions of nine vertebrate predator species and their major prey, largely rodents and rabbits, in central Sweden. They found that the feeding

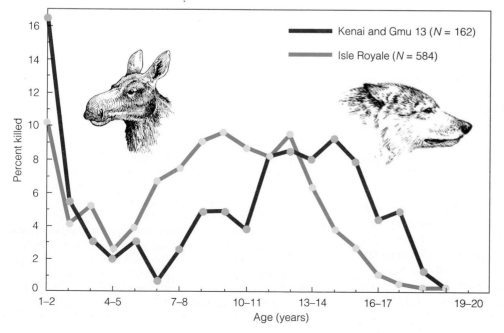

FIGURE 16.22   The ages of adult moose killed by wolves on the Kenai, Alaska, and on Isle Royale, Michigan. The yearlings and older individuals are most vulnerable, especially on the Kenai. A smaller moose population on Isle Royale apparently increased predatory pressure on younger moose, but the heaviest predation still fell on the old individuals. (Data from Peterson et al. 1984 and Ballard et al. 1987.)

habits of generalist predators in combination with their territorial behavior did prevent significant annual fluctuation in rodent numbers. Their field studies, supplemented by simulation models, suggest that generalist predators can maintain stable populations by switching their diets in response to changing prey densities (see Chapter 15). Such regulation can occur only if alternative prey is abundant and predator populations are intrinsically regulated.

The role of predators and alternative prey in the regulation of prey populations is difficult to assess under natural conditions. However, Lindstrom and associates (1994) were provided a fortuitous opportunity to study the effects of red fox (*Vulpes vulpes*) predation in Sweden on cyclic voles and their alternative prey, mountain hare (*Lepus timidus*) and forest grouse (black grouse *Tetrao tetrix*, capercaillie *T. urogallus*, and hazel grouse *Bonasa bonasia*). Fox prey heavily on voles, but during the low of the three- to four-year vole cycle, fox turn to their alternative prey and induce cyclic fluctuations in hare and grouse. If fox predation limited the populations of hare and grouse, then a decrease in the fox population should result in an increase in the prey and reduce or eliminate cycles in the alternative prey. In the late 1970s and 1980s an epizootic of sarcoptic mange (*Sarcoptes scabiei*) swept through fox populations in Sweden. The mange mite, discovered for the first time in Sweden in 1975, causes hair loss, skin deterioration, and the ultimate death of the host. The epizootic decimated fox populations and provided the opportunity for the Swedish biologists to monitor the populations of the fox and its prey over a 20-year period, from 1972 to 1992 (Figure 16.23). Even though the voles were the major prey of the fox, they were not affected by the loss of the fox and continued to exhibit cyclic behavior. Populations of hare and grouse, however, increased by 40 to 100 percent. The results suggest that predation by the red fox does limit the number of hares and grouse during the downside of the vole cycle. As the voles declined in their cycle, the fox turned to alternative prey and in effect transferred the cycle to the hare and grouse. Thus predatory pressure on alternative prey varied with the cyclic fluctuations of the primary prey species, the voles.

By contrast, in predator-ungulate systems the predators are loosely regulated by social interactions but lack any alternative prey. One prey species has to bear the brunt of a specialized predator. This situation can result in wide fluctuations in prey populations. Extensive studies of moose-wolf relationships point out that the role of predation depends on the ratio of prey to predator (Gasaway et al. 1983, Peterson and Page 1983, Ballard et al. 1987). At a ratio of more than 30 moose per wolf, predation can be significant. The moose population is likely to remain stable or increase if it is below carrying capacity, unless some other source of mortality such as a severe winter intervenes. At a ratio of 20 to 30 moose per wolf, predation can control a moose population. Whether the population remains stable or declines depends on the combined effects of hunting, food supply, and winter snows. At a ratio of 20 moose per wolf, predation can cause a decline in the moose population.

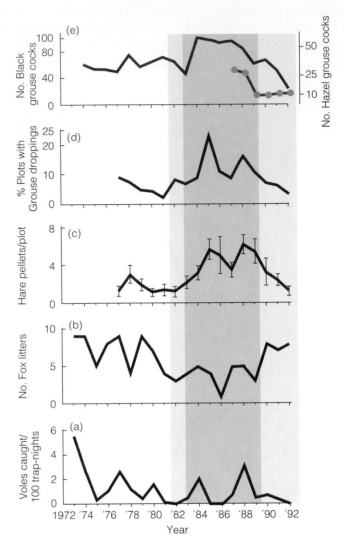

FIGURE 16.23 Population fluctuations in the Grimso Research Area in south-central Sweden: (a) voles; (b) red fox litters; (c) mountain hare (*Lepus timidus*); (d) all grouse; (e) displaying black grouse (*Tetrao tetrix*) cocks and hazel grouse (*Bonasa bonasia*) cocks. Shaded area indicates years of mange epizootic. Darker shading indicates a period of low fox densities as interpreted from the variation in the number of fox litters. (After Lindstrom et al. 1994:1046.)

Summary

1. Interaction of predators and their prey makes up a predator-prey system. Predator-prey interactions at one trophic level influence predator-prey interactions at the next trophic level. Involved are plant-herbivore systems and herbivore-carnivore systems.

2. Plant predation involves defoliation by grazers and consumption of seeds and fruits. Interactions between changes in plant biomass and herbivores result in oscillations of both plant and herbivore populations or, in an equilibrium situation, the vegetation with grazing pressure and herbivore with the vegetation.

3. Herbivory affects plant fitness by reducing the amount of photosynthate and the ability to produce more. In response to the selection pressures of herbivory, plants have evolved measures

of defense. Plants try to prevent losses by denying herbivores palatable or digestible food or by producing secondary compounds that interfere with growth and reproduction. These secondary compounds are involved in chemical defense, which involves distasteful or toxic substances that repel, warn, or inhibit would-be herbivores. These substances are secondary metabolic products such as alkaloids, phenolics, and cardiac glycosides. Chemical defense is most successful against generalist herbivores. Certain specialists breach the chemical defense and detoxify secretions or sequester the toxins in their own tissues as defense against predators.

4. Other forms of defense include mimicry among plants to hide from specialized herbivores or to attract seed dispersers and structural defense involving hairs, thorns, and spines. One form of defense is predator satiation. Reproduction is timed so that fruits and seeds are so abundant that seed predators can take only a fraction of them, leaving a number to escape and germinate.

5. Herbivore-carnivore systems involve interactions between the second and third trophic levels. These interactions influence the fitness of both predator and prey.

6. In response to the selective pressure of predation, animal prey species have also evolved measures of defense. Chemical defense involves a distasteful or toxic secretion that repels, warns, or inhibits a would-be attacker. Cryptic coloration and behavioral patterns enable a prey to escape detection or inhibit predators. Warning coloration signals to a predator that that the intended prey is distasteful or disagreeable in some manner. Usually the predator has to experience at least one encounter with such a prey to learn the significance of a color pattern. Some palatable species mimic unpalatable species, thus acquiring some protection from predators, and some mutually unpalatable species mimic each other, thereby increasing each other's protection from predation.

7. Another form of defense, similar to that of plants, is predator saturation. Individuals of some prey populations exhibit synchronized timing of reproduction, producing so many young in one short period of time that predators can take only a fraction of them. Those remaining escape by growing to a size too large for the predators to handle easily.

8. Although warning and cryptic coloration and mimicry are usually associated with prey species, such mechanisms are also employed by predators to increase hunting efficiency. Predators also use special hunting tactics and adaptations.

9. An unusual form of predation is cannibalism, in which predator and prey are the same species. Often associated with stressed populations, cannibalism can result in pronounced demographic effects within a population, including the loss of younger age classes and lowered reproduction. It can become an important form of population control.

10. Some predators kill and eat species that use similar resources. This is called intraguild predation. Although it combined elements of predation and competition, intraguild predation differs from competition because the predator gains energy and nutrients. Intraguild predation differs from classical predation because it reduces potential competition.

11. Predation may stabilize prey population, but more often the relationship results in unstable fluctuations. The ten-year snowshoe hare-lynx cycle involves a three-trophic-level interaction among vegetation, hares, and lynx. Recovery of hares follows a decline in lynx and a recovery of vegetation.

Review Questions

1. Differentiate between the forms of predation on plants.
2. What effect does herbivory have on plant fitness? How do plants respond to defoliation?
3. Why do grasses withstand herbivory much better than woody plants?
4. In what ways do plants affect herbivore fitness?
5. How do the following mechanisms allow plants to defend themselves against herbivores: mimicry, structural defenses, predator satiation?
6. What two types of chemical resistance are employed by plants? How do some herbivores breach them?
7. How do vegetation and herbivores interact as a predator-prey relationship? Discuss this in relation to a vegetation–snowshoe hare cycle. How does chemical defense by plants enter the picture?
8. Schultz (1988) makes the statement, "So little is known (especially by ecologists) of the regulation of plant development and biosynthesis of secondary compounds that is it reasonable to ask whether 'induced' responses represent anything other than the inevitable and incidental consequence of tissue loss." Discuss this statement relative to evolution or coevolution of defensive responses of plants to herbivory.
9. Discuss the effectiveness of the variety of defenses employed by prey species against predators. Do the same for the variety of countermeasures used by predators to overcome prey defenses. Relate your answer to specific examples.
10. What is mimicry? What is the model? The mimic? Contrast Batesian with Müllerian mimicry.
11. How can predator satiation reduce the impact of predation? What behavioral and physiological characteristics of a species can lead to this type of defense?
12. Can predators regulate animal prey species? If so, under what conditions?
13. What seems to trigger cannibalism in a species? What are some of the selective advantages? Disadvantages?
14. What is intraguild predation? How might this concept relate to the overexploitation of species by humans and to the introduction of an exotic species into the ecosystem?
15. What is the relationship between predator and prey cycles? If predation is such as integral component of microtine and hare cycles in northern regions, why are populations of similar or the same species not cyclic in southern regions?
16. Choose one of the three scenarios for the evolution of cryptic coloration and mimicry, and present arguments for its plausibility (refer to Guilford 1988, Turner 1988).
17. What ethical, economic, social, and political problems and controversies arise when deer or moose, preyed on by wolves, happen also to be a major game species, preyed on by humans?

Cross References

CHAPTER 17

Coevolutionary Interactions: Parasitism and Mutualism

CONCEPTS

1. Coevolution is reciprocal selection pressure on interacting populations.

2. Parasitism is an association between two organisms in which one, the parasite, benefits at the expense of the other, the host.

3. Parasites may be facultative or obligate.

4. Parasites can be grouped into two types, microparasites and macroparasites.

5. Transmission from host to host may be direct or indirect.

6. Host responses to parasites range from biochemical to behavioral changes.

7. The population dynamics of hosts and parasites are interactive.

8. Brood parasitism depends on the social structure of both host and parasite.

9. Mutualism is a positive relationship between individuals or populations of different species.

10. Some mutualisms are a form of symbiosis. Symbiosis involves a long-term physical contact between individuals of different species in which an individual of one species lives permanently on or within another.

11. Mutualism may be obligate symbiotic, obligate nonsymbiotic, nonobligate, direct, indirect, or defensive.

12. Mutualism may have evolved from predator-prey, parasite-host, or commensal relationships.

13. Because of positive relationships between populations, the population consequences of mutualism are difficult to define and to model.

In the previous chapters we explored the relationship of predation or herbivory on plants and predation or carnivory on animal prey. These relationships share one characteristic: selection pressures on one species in a broad sense influence selection pressures on the other. Selection pressures of grazing on plants and of predators on animal prey result in counteradaptations to the effects of one on the other. Predation on plants is met by structural and chemical defenses. Predation on animals is countered by increased speed, warning coloration, and chemical defense on part of the prey. In this chapter we meet two other types of more specialized reciprocal selection on interacting populations: **parasitism** and **mutualism.**

Parasites evolve ways of gaining access to and exiting from their hosts and means of counteracting immune responses, whereas hosts acquire improved immune responses. Such relationships are antagonistic compared with mutualism, in which the reciprocal selection pressures benefit each of the interacting species. Both types of relationships result in a reciprocal evolutionary change in interacting species, ending in some degree of specialization; this process is called **coevolution.** Certain traits of each species evolve in response to the traits of another (Ehrlich and Raven 1965, Janzen 1980, Thompson 1994). Any evolutionary change in one member that affects the relationship with another species may, in turn, change the selection forces acting on the other member. The interacting species play a game of adaptation and counteradaptation.

Such a restricted definition seems to imply that interacting species grew up together over evolutionary time and that they experienced complementary gene-to-gene coevolution. For example, a gene that confers a defensive capability to a host organism is countered by the distinct corresponding gene that confers the ability to the other organism to exploit its host, as is the case between winter wheat (*Triticum aestivum*) and its destructive pest, the Hessian fly (*Mayetiola destructor*) (Gallun 1977). Genes in the wheat that confer resistance are selected for by the attacks of the pest, and the genes in the fly that counter wheat defenses permit the fly to reproduce on selected plants. Such gene-to-gene coevolution appears to result between parasites and hosts.

Whether one-to-one or gene-to-gene coevolution between species implies that the two species interacted together over evolutionary times is one of the highly debated areas in ecology (Howe and Westley 1988, Futuyma and Slatkin 1983, Thompson 1994). Some argue that no proof exists that coevolution between species occurred on a one-to-one basis. The chances are that many supposedly coevolved pairs did not grow up together and that they presently inhabit environments different from those in which they evolved (Howe 1985, Herrara 1985). Instead the organisms probably evolved in different types of habitats through time, each with somewhat different selection pressures. When these plants or animals invaded new habitats, they were preadapted to the organisms at hand. If the traits they had already acquired fit the situation, then the two interacted in a manner suggesting long-term coevolution. If the relationship meshed, then further evolutionary changes were minor. What appears to be a highly coevolved system may not have involved an evolutionary change in either partner. The relationship would then continue to be selected for by current interaction.

Another approach to coevolution is to consider it less restrictively, as a more general response of one group of species to another. A particular trait may evolve in several species in one taxon in response to the selective pressures of a trait or a suite of traits in several species in another taxon. Plants might have evolved chemical and physical defenses against a diverse array of herbivorous insects (see Chapter 16). In turn, many insects evolved the ability to detoxify a wide range of plant chemicals. Similarly, animals might have evolved a generalized immune system in response to a wide range of parasites. Plants adapt to a suite of pollinators that visit their flowers. Such interactions are termed **diffuse coevolution** because the adaptive responses are spread over many interacting species.

The term *diffuse evolution* has been called vague and a catch-all for interacting systems involving more than a pair of species (Hougen-Eitzman and Rausher 1994, Iwio and Rausch 1997). Many relationships—for example, guilds of pollinators interacting with guilds of plants or herbivores interacting with plants—may appear to be diffuse because they are spread among many interacting species over a wide geographic range. What appears to be diffuse may be more pairwise coevolution on a local scale (Thompson 1994).

Thus coevolved relationships are often symbiotic. **Symbiosis** refers to any long-term, intimate relationship between two species. In symbiosis the relationship between the two species is obligate. Intestinal parasites of mammals and birds are examples. In **nonsymbiotic** relationships both members of the pair may gain some benefit, but they do not live together. Their relationship is facultative (opportunistic). Each member can live independently, but benefits accrue from the relationship.

PARASITISM

Parasitism is a condition in which two organisms live together but one derives its nourishment at the expense of another. Parasites, strictly speaking, draw nourishment from the live bodies of another species. Typically parasites do not kill their hosts as predators do, although the host may die from secondary infection, suffer from stunted growth, emaciation, or sterility, or in a weakened condition fall to predators. Many species of insects have a parasitic larval stage that draws nourishment from the tissues of their hosts. By the time the parasitic larvae have reached metamorphosis, they have completely consumed the soft tissues of the host. These parasites, known as **parasitoids,** essentially act as predators.

For decades, ecologists paid scant attention to the role of parasites in the population dynamics and the role that parasitism plays in community structure. A major reason was that in most natural populations the effects of parasitism are difficult to detect. Bodies of victims that die are quickly eaten by detritivores and scavengers, eliminating traces of disease.

About seventy years ago, Aldo Leopold, in his classic book *Game Management* (1933), remarked that the role of disease in wildlife populations was radically underestimated. He hypothesized that parasites control both predators and prey, that they may delimit the range of some species, limit population density, and become involved in population fluctuations. "Disease," he wrote, "does not yield to observational methods of study." Understanding begins only when field observations are combined with experimental study and laboratory techniques. In recent years ecologists have been undertaking the kind of studies that Leopold suggested. As a result we know more about disease in natural populations.

These studies are revealing that parasites do have a pervasive influence on plant and animal populations, just as Leopold surmised. Parasites can increase the death rates and decrease the birth rates of their hosts, and affect their nutritional status, energy demands, and growth rates. Parasites can alter host behavior, affect the outcome of intraspecific and interspecific competition, and increase the susceptibility of the host populations to predation. Parasites may influence mate choice, alter the sex ratios of host populations, and induce sterility (Minchella and Scott 1991). In effect, parasitism may have a greater influence on community structure and population dynamics than predation or interspecific competition.

Characteristics of Parasites

Parasites include viruses, many bacteria, fungi, and an array of invertebrate taxonomic groups, including the arthropods. The presence of a heavy load of parasites is considered an infection, and the outcome of an infection is a disease. A **disease** is any condition of a plant or animal that deviates from normal well-being. Not all parasites, however, are agents of disease.

Parasites exhibit a tremendous diversity in the ways they exploit their hosts (for an overview see Croll, 1966). They may parasitize plants or animals, or both. They may be **ectoparasites** that live on the outside of the host or **endoparasites** that live within the body of the host. Some are full-time parasites, others only part-time. Part-time parasites may be parasitic as adults and free-living as larvae, or the reverse.

Parasites have developed numerous ways to gain entrance to their hosts, even to the point of using several hosts as dispersal agents. They have evolved various means and degrees of mobility, ranging from free-swimming ciliated forms to ones totally dependent on other organisms for transport. They have developed diverse ways of securing themselves to the host to maintain position and means of surviving the biochemical hazards of living inside a host.

Parasites may be restricted to one or a limited number of species or genera of host. A number of parasites of birds, especially certain tapeworms, live only in one particular order or genera (see Baer 1951). Some parasites live their entire life cycle on one host, whereas others require several hosts.

The usual approach to typing parasites has been taxonomic: tapeworms, roundworms, and the like. May and Anderson (1979) have suggested that parasites be distinguished on the basis of size, as microparasites and macroparasites. **Microparasites** include the viruses, bacteria, and protozoans. They are characterized by very small size and a short generation time. They develop and multiply rapidly within the host and tend to induce immunity to reinfection in hosts that survive initial infection. The duration of the infection is short relative to the expected life span of the host. Transmission of these parasites from host to host is direct, although they may involve some other species as a vector.

Macroparasites are relatively large in size. In animals they include parasitic worms, the platyhelminths, acanthocephalans, roundworms, flukes, lice, fleas, ticks, mites, and fungi (Figure 17.1). Parasites of plants include fungi, rusts, smuts, and plants such as dodder, broomrape, and mistletoe. Macroparasites have a comparatively long generation time, and direct multiplication in the host is rare. The immune response macroparasites stimulate is of short duration and depends on the number of parasites in the host. Macroparasites persist in hosts by continual reinfection. They may spread by direct transmission from host to host or by indirect transmission, involving intermediate hosts and vectors.

Hosts as Habitat

Hosts are homes for parasites, and parasites have exploited every conceivable habitat on and within them. In fact, they represent the extreme in specialization for resource exploitation (Price 1980). Among animals some parasites live on the skin hidden within the protective covers of feathers and hair. Some burrow beneath the skin. Others live in the bloodstream, in the heart, brain, digestive tract, liver, spleen, mucosal lining of the stomach, spinal cord, or brain, in the nasal tract and lungs, in the gonads, bladder, muscle tissue, pancreas, eyes, or gills of fish, to mention some sites among many. Parasites of insects live on the legs, on the upper and lower body surfaces, and even on mouthparts.

Parasites within the host will colonize different sites within the same organ system. For example, different species of the coccidian protozoans of the genus *Eimeria* inhabit different regions in the guts of mammals: one in the duodenum, another in the lower duodenum and upper small intestine, a third in the lower small intestine, and a fourth in the caecum and rectum.

FIGURE 17.1 Lungworm in the air passages of a white-tailed deer.

Plant parasites, too, divide up the habitat. Some live on the surface of roots and stems; others penetrate the roots and bark to live in woody tissue beneath. Some live at the base of the stem (root collar) where the plant emerges from the soil. Others live within the leaves, on young leaves, on mature leaves, on flowers, pollen, or fruits.

A major problem for parasites, especially parasites of animals, is gaining access to and escape from the host. Parasites of the alimentary tracts of vertebrates enter the host orally and escape by the way of the rectum, a path used by other parasites. Parasites of the lungs enter orally or penetrate the skin and travel to the lungs by the pulmonary system. They escape mainly by being coughed up and swallowed into the alimentary tract. Liver parasites, exploiting one of the nutritionally richest habitats in the animal body, arrive there by way of the circulatory system, bile duct, and hepatic portal systems and escape by the same route. Parasites of the urogenital systems enter orally and travel through the gut to the site of residency and exit through the urinary system. Blood parasites enter and escape by way of the skin, but always with the aid of some obliging vector such as mosquitoes and ticks. Parasites then end up in muscle tissues, where they usually exit in capsules, reaching a blind end. For them the only way out is for their host to be killed and eaten by a predator.

Life Cycles

Parasites face unique problems moving from host to host. For the parasite, the host is both habitat and food. Many parasites have evolved complex life cycles geared to escape and relocation. For most parasites the hosts function as islands from which they must escape; then they must survive in a generally hostile external environment while locating other host islands, some of which are occupied and some of which are not. For animal parasites the host islands are movable.

Parasites cannot move from host to host at will. They can escape only during a transmission stage, no matter whether the means of dispersal is direct or indirect. Successful transmission depends on contact between the host and the infective stage. The infective stage is essential.

All parasites reach a stage in their life cycle within the host when they can develop no further. **The definitive host** is the one in which the parasite becomes an adult and reaches maturity. All others are **intermediate hosts,** which harbor some developmental phase. Parasites may require one, two, or even three intermediate hosts. Each infective stage can develop only when it is independent of the definite host, and it can continue its development only if it can find another intermediate or its definitive host. To this end many parasites exploit the hosts' feeding habits or behavioral patterns. Thus the population dynamics of a parasite population is closely tied to the population dynamics of the host.

Direct Transmission   Direct transmission is the transfer of the parasite from one host to another by direct contact or through a carrier or vector. The life cycle of the parasite does

not involve intermediate stages in a secondary host. Most microparasites are transmitted directly.

One such disease of wild mammals and humans is rabies, transmitted through the saliva by the bite of a rabid animal. The rabies virus follows the nerves from the infection point to the spinal column and brain before symptoms appear. The symptoms are both behavioral and physical. Infected animals are excitable and restless, exhibit convulsions, wander aimlessly, and show no fear of humans. Physically they are emaciated, exhausted, and partially paralyzed. Among wild animals, rabies occurs most commonly in coyotes, foxes, skunks, raccoons, and bats. Foxes, dogs, and raccoons are the primary transmitters of the disease.

Microparasites that infect plants are also transmitted directly when the plant comes into contact with the virus or resting spores in the soil. In the presence of suitable hosts the spores germinate and penetrate the roots. Or spores may be wind-carried and come to rest on and infect leaves of the plant. Others are carried by insect vectors. One example is elm phloem necrosis, caused by a mycoplasmal organism. It is transmitted by root grafts (a situation in which roots of one tree become grafted onto the roots of a neighbor) or by the widely distributed white-banded elm leafhopper (*Scaphoideus luteolus*). This insect ingests the mycoplasma while feeding on the sap of a diseased tree and transmits it while feeding on a new host tree.

Many important macroparasites of animals and plant move directly from infected to uninfected host by direct contact. Parasitic nematodes (*Ascaris*) live in the digestive tracts of mammals. There female roundworms lay thousands of eggs, which are then expelled with the feces. If they are swallowed by the host of the correct species, the eggs hatch in the intestines, bore their way into the blood vessels, and come to rest in the lungs. From here they ascend to the mouth, usually by causing the host to cough, and are swallowed again to reach the stomach, where they mature and enter the intestines.

Nematodes are important macroparasites of plants. The female of one species, the eelworm (*Heterodera schachti*), lives a parasitic existence in the roots of such plants as tomatoes and beets. Fertilized by the free-living male, the female sends thousands of larvae into the soil, ready to take up a parasitic existence in other plants.

Most important external and debilitating parasites of birds and mammals are spread by direct contact. They include lice, mange mites, ticks, fleas, and botfly larvae. Many of these parasites lay their eggs directly on the host, but fleas lay their eggs and their larvae hatch in the nests and bedding of the host (even in the shag rugs of homes with dogs and cats), and eventually leap onto nearby hosts.

Some fungal parasites of plants spread through root grafts, and others are carried by insect vectors. For example, an important fungal infection of white pine (*Pinus strobus*), *Fomes annosus,* spreads rapidly through pure stands of the tree by root grafts. The devastating Dutch elm disease, caused by the introduced fungus *Ceratocystis ulmi*, is spread from tree to tree by spore-carrying elm beetles, *Scolytus multistriatus* and *Hylurgopinus rufipes.*

Plant macroparasites even include a number of flowering plants themselves. One group are **holoparasites,** plants that lack chlorophyll and draw their water, nutrients, and carbon from the roots of host plants. Notable among them are members of the broomrape family, Orobanchaceae. Two are the familiar squawroot (*Conopholis americana*), which parasitizes the roots of oaks (Figure 17.2), and beechdrop (*Epifagus virginiana*), which parasitizes mostly beech trees.

Another group are the **hemiparasites.** They are photosynthetic, but they draw water and nutrients from their host plant. The most familiar hemiparasites are mistletoes, whose sticky seeds attach to limbs, send out roots that embrace the limb, and send roots into the sapwood. Mistletoe can reduce the growth of its host.

Indirect Transmission   Many parasites, both plant and animal, utilize indirect transmission. During different stages of the life cycle they require different hosts. The brainworm (*Parelaphostrongylus tenuis*) (in spite of its name the parasite is a lungworm) of the white-tailed deer has as its intermediate host during its larval stage a snail or slug that lives in the grass (Figure 17.3) (Anderson 1963, 1965). The deer ingests the infected

FIGURE 17.2   Squawroot (*Conopholis americana*), a member of the broomrape family, is a hemiparasite on the roots of oak.

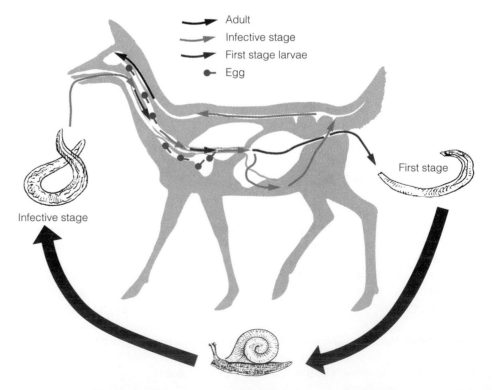

FIGURE 17.3   Life cycle of a macroparasite, the brainworm *Parelaphostrongylus tenuis,* with indirect transmission in white-tailed deer, moose, and elk. (After Strelive in Davidson 1981:141.)

Adult

Infective stage

First stage larvae

Egg

First stage

Infective stage

snail while grazing. In the deer's stomach, the larvae leave the snail, puncture the deer's stomach wall, enter the abdominal membranes, and follow the lumbar and other nerves to the vertebral canal carrying the spinal cord and on to the spaces surrounding the brain. Here, about 40 days after infection, the subadults mature and migrate forward to the cranium, where they deposit eggs in the venous circulatory system. Eggs pass through the bloodstream to the lungs, where they develop into first-stage larvae. The larvae move into the air sacs of the lungs, are coughed up by the deer, swallowed, and passed out through the feces. The larvae invade the foot of a snail as it moves across the feces. In snails the larvae grow, molt twice, and give rise to the infective stage that the deer acquire as they ingest the snails clinging to the vegetation. In white-tailed deer, infestations of this parasite produce minor or no symp-

toms. In moose and elk, infestations cause serious and often fatal neurological symptoms.

Other parasites need two intermediate hosts. An example is the black grub or black spot infection of minnows and sunfish (Figure 17.4), caused by a small white larvae of a fluke (*Trematoda*), *Uvulifer ambloplitis*. Once infected, the fish host lays down a black pigment around the thick-walled cysts, causing black spots on the body wall. The adult stage is attached by suckers to the mucosa of the intestine of the primary host, the kingfisher. The eggs pass through the intestine of the kingfisher into the water, where the miracidia hatch. The miracidia, moving through the water by cilia, seek an intermediate host, a snail, to which they gain entrance by secreting a tissue-dissolving substance. Within the snail, the miracidia transform into saclike sporocysts, which eventually

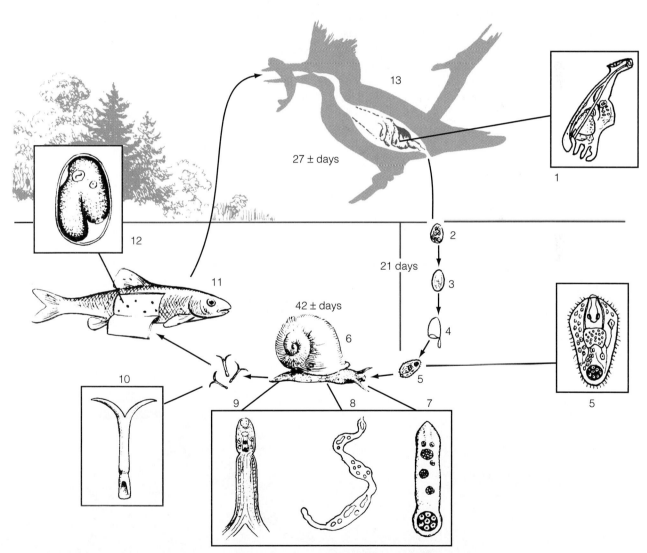

FIGURE 17.4 The life cycle of the black grub, *Uvulifer ambloplitis*. (1) Adult trematode in the intestine of the kingfisher; (2) immature egg; (3) mature egg; (4) empty shell; (5) miracidium, the ciliated larva of the fluke that parasitizes the snail; (6) the first intermediate host, the snail; (7) and (8) sporocysts; (9) cercaria, larval form of the fluke produced asexually in the snail; (10) free-swimming cercaria, which penetrates beneath the scales or into the fins of fish and encysts, producing black spots in about 22 days; (11) minnow with skin cut away to show black grubs; (12) sketch of parasite within inner cyst; (13) kingfisher. (Adapted from Hunter and Hunter 1934, New York State Conservation Department.)

give rise to tailed larvae called cercaria. These free-swimming larvae emerge from the snail and seek the next intermediate host, a fish, which they must find within two days or die. The cercaria penetrate the fish and encyst. The fish, in turn, is eaten by the kingfisher (*Megaceryle alcyon*), and the cercaria become free and mature in the bird's intestinal tract. The new adults start laying eggs to repeat the cycle.

Indirect transmission among plant macroparasites is uncommon except among the rusts, which employ the wind to carry the infective stages from primary and intermediate hosts. An example is white pine blister rust (*Cronartium ribicola*) (Figure 17.5). The rust produces spores in the diseased bark of pine in the spring. These spores, which cannot reinfect pines, are carried by the wind. Some may land on species of *Ribes*, or gooseberry shrubs. On the infected leaves of *Ribes* the rust produces two kinds of spores, early and late summer forms. The early summer form infects other *Ribes* only, intensifying infection on these plants during the summer. Late summer spores infect only white pine needles. Carried by the wind, the spores germinate on the needles and grow until they reach the bark. There the rust produces spindle-shaped diseased areas or cankers. Two to five years after infection, and annually thereafter, spores produced in the bark are carried by the wind to nearby *Ribes*. Like other parasites that require intermediate hosts, conditions for transmission are critical. These include a widespread infection of *Ribes*, an abundant production of pine-infecting spores, favorable temperature, moisture, and wind conditions, and hosts within 280 m of the infected *Ribes*, the maximum carrying distance of spores on the wind. The disease has been controlled by eliminating *Ribes* from within and near white pine stands.

Dynamics of Transmission

The impact of parasites on host populations relates in part to the nature of transmission and the density and dispersion of the host population. Microparasites, dependent for the most part on direct transmission, require a high host density to persist. For them ideal hosts live in groups or herds. To persist the parasites need a long-lived infective stage that does not induce long-term immunity in the host population. Immunity reduces populations of a parasite, if indeed it does not eliminate them. An example of a parasite in wild populations that does not confer long-term immunity is rabies; one that does confer immunity to animals that survive the disease is distemper. Outbreaks of microparasitic diseases appear to occur when the density of the host population is high, and they tend to sharply reduce host populations. Examples are viral distemper in raccoons, which can be significant in controlling their populations (Gorham 1966, Haberman et al. 1958).

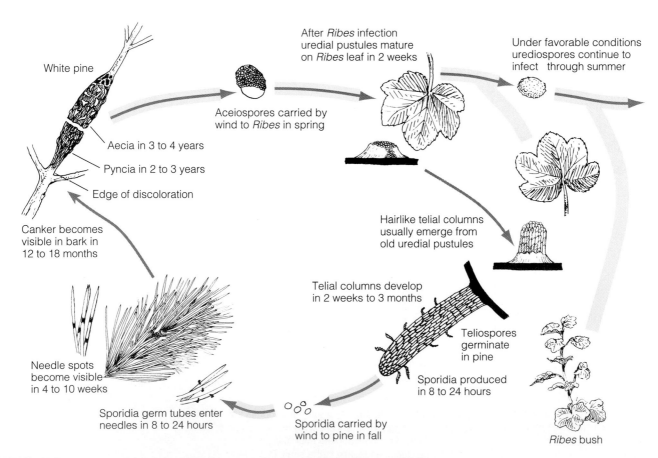

FIGURE 17.5   Life cycle of the white pine blister rust (*Cronartium ribicola*). The rust infection on the shrub gooseberry, *Ribes*, appears as yellowish to orange pustules of spores. An annual infection, it has little effect on *Ribes*. On the white pine the infection is perennial and appears as large orangish blisterlike cankers on the bark.

Transmission of parasites is most successful when the population of potential hosts is dense, particularly if the parasites depend on direct contacts among hosts. When host populations become dense and direct contact is frequent among individuals carrying the disease and susceptible ones, the parasite spreads rapidly through the population with high mortality. Such effects are most evident when parasites are introduced into a population with no evolved defenses. In such cases the disease may be density-independent. It can reduce populations, exterminate them locally, or restrict the distribution of the host. For example, the chestnut blight fungus (*Endothia parasitica*) spread rapidly through the American chestnut (*Castanea dentata*), eliminating it as a commercial timber species and changing the composition of the eastern deciduous forest. Rinderpest, a viral disease of cattle, introduced into Somalia by cattle importations from either Arabia or India between 1884 and 1889, spread swiftly through populations of native cattle, wildebeest (*Con-*

nochaetes taurinus), and African buffalo, decimating those species. After several periodic outbreaks, the disease was removed by vaccination of cattle (see Sinclair 1977, 1979). Such a rapid spread of viral and bacterial diseases in dense populations is called an **epizootic** in animal populations and an **epidemic** in humans. Such epizootics involve virulent forms that sweep through susceptible populations as an advancing front, exemplified by the spread of rabies through Europe (Figure 17.6).

Indirect transmission, typical of macroparasites, is more complex. To persist the parasite requires a highly effective transmission stage, which often involves a close association with food webs. Parasites with indirect transmission exist at low population levels, but because of efficient transmission they do well and persist for a long time in low-density populations of hosts. Transmission from host to host can occur only with the dispersal of an infective stage that lives independently of the definitive host. It can complete its life cycle

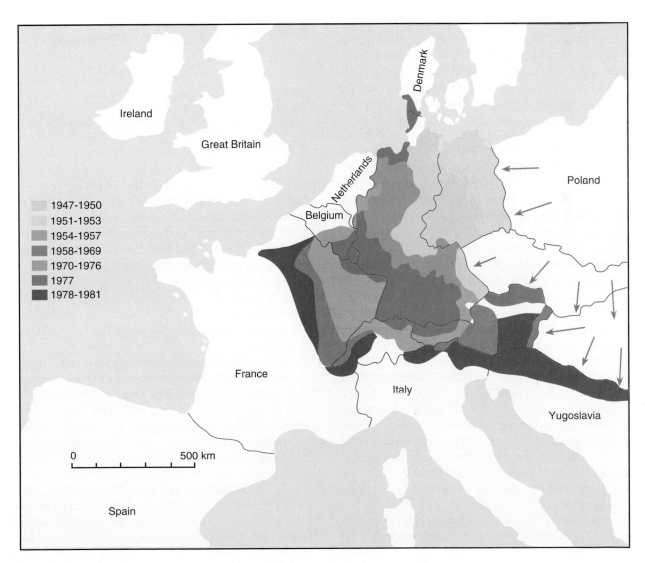

FIGURE 17.6 The spread of rabies in Europe. Originally diagnosed in Poland in 1939, it has spread westward over 1400 km of Europe at a rather constant rate of about 20–60 km per year. This epizootic developed independently of dogs and has spread across the continent with foxes as the principal vector. (From Steck 1982:62.)

only if it can infect another host, intermediate or definitive, depending on the stage of the parasite's life cycle. The sexual form of the malarial parasite cannot survive in the mosquito. It must transfer to a vertebrate host to develop into an adult stage. The brainworm of deer has to locate a snail as an intermediate host to continue its development to an infective stage that can be transmitted back to the deer. Even more complex is the trematode infecting the kingfisher and other fish-eating birds. It has to develop three infective stages. Each stage has to be uniquely adapted to its host. It has to be able to access the host, overcome the immune response of the host, and escape from the host. For many animal parasites this process means exploiting the feeding habits of the definitive host and adapting to the habits of the intermediate hosts. The brainworm of the deer exploits the snail's habit of crawling up grass stems, where it risks being eaten along with the grass by a grazing deer. Unless the snail is swallowed by the deer, the parasite will perish.

Some parasites depend on predator-prey relationships for transmission. The tapeworm *Echinococcus* (Cestoda) lives in the intestines of carnivores. It spends its larval stage as hydatid cysts in the muscles and tissues of any organ in its immediate hosts—herbivores and, under certain conditions, humans closely associated with the usual hosts. When a carnivore feeds on the flesh of infected herbivore prey, the hydatid cysts gain entrance to the digestive tract of the carnivore and develop into adult tapeworms. One species, *E. granulosus,* the dog tapeworm, is found in dogs and wild canids, especially the wolf. A second species, *E. multilocularis,* has a restricted range in the northern regions of Asia, Europe, and North America. Its definitive host is the arctic fox (*Alopex lagopus*), its intermediate host several species of voles. Both species of tapeworm appear to change the behavior of its intermediate host to ensure transmission back to its definitive host. Wolves preferably select moose infected with hydatid larvae, which concentrate in the lungs. The presence of larvae may be indicated by some odor on the breath of infected moose, by excretions, or by some aberrant behavior. Voles heavily infected with hydatid larvae of *E. multilocularis,* which rapidly grow in and destroy the liver, appear large and fat, but they have difficulty moving and are easy prey for the fox.

The distribution of macroparasites, especially those with indirect transmission, is highly clumped or overdispersed (Figure 17.7). Only a few members of the host population harbor major parasite loads and act as reservoirs of the infection. These individuals are the ones that are most likely to succumb to parasite-induced mortality, suffer reduced reproductive rates, or both (Figure 17.8). Uninfected hosts are widely scattered or intermixed among populations of other species, so that the probability of the parasite or its carrier coming into contact with the susceptible individuals is low. In other words transmission depends crucially on both the density of hosts and the distance between parasite and potential host.

Evidence is growing that parasites, once overlooked as a major force in community population dynamics, function like

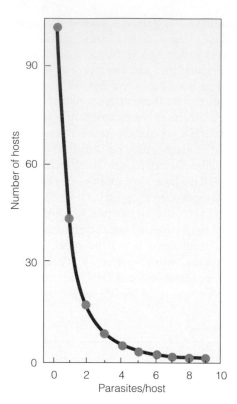

FIGURE 17.7 Overdispersion of the tick *Ixodes trianguliceps* (Birula) on a population of the European field mouse (*Apodemus sylvaticus*). Most of the individuals in the host population carry no ticks. A few individuals carry most of the parasite load. (From Randolph 1975:454.)

top predators (Grenfell 1992). This function is most evident in the dynamics of plant-herbivore systems involving directly transmitted parasites. In these systems the transmission of parasites is coupled with host food consumption. The levels of infection and pathogenicity increase as the density of the host population increases and the availability of food and level of nutrition decrease. There is evidence that overutilization of vegetation and accompanying increases in parasitism alter the stability of plant-herbivore interactions, causing cycles.

Examples are cyclic outbreaks of forest lepidopteran insects, especially the tent caterpillars (*Malacosoma* spp.), that appear to be induced by nuclear polyhedrosis viruses (*Baculovirus* spp.) (Myers 1993). The virus is so named because its DNA is surrounded by a polyhedral-shaped crystal during much of its life cycle. Caterpillars pick up this virus, which replicates in the nucleus of host cells, as they feed on virus-contaminated leaves. It takes four days to three weeks for the infected caterpillar to succumb to the virus. When it does, the caterpillar ruptures, releasing the polyhedra onto leaves and bark. Although the virus is killed by ultraviolet light, it can persist in protected areas, such as bark, for a number of years and spread to new localities by rain, wind, and insects. Gradual buildups of the virus accompanied by increasing densities of the caterpillars generate population cycles (Myers 1988, 1990, 1993).

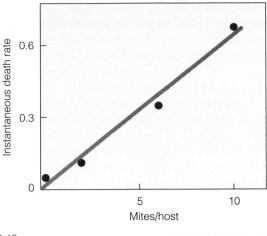

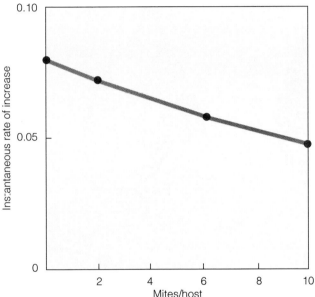

FIGURE 17.8 The influence of parasitism by the mite *Hydryphantes tenuabilis* on the water measurer (*Hydrometra myrae*), a common hemipteran on ponds. (a) The effects of parasite density in a host on the rate of parasite-induced mortality of the host. (b) The effect of parasitic infection on fecundity. (From Lanciana 1975:691.)

Host Response to Parasitism

Just as prey respond to predators, so hosts react to invasions of parasites. Reactions are physiological and behavioral. Hosts may successfully counter the invaders or succumb to the effects of the parasites.

Biochemical Responses Biochemical and cellular response are defensive; they attempt to prevent or reduce the effects of the parasitic invasion. One response is inflammation, provoked by the death or destruction of host cells. The infection stimulates the secretion of histamines and increased blood flow to the site, bringing in phagocytes, lymphocytes, and leukocytes to attack the infection. Scabs form on the skin by fibrosis, as in the case of heavy mange mite infestations on red fox and other canids (Figure 17.9). Internal reactions can produce calcareous cysts in muscle or skin that imprison the parasite. Examples are

FIGURE 17.9 Encrustations of mange, caused by a mite, on a fox.

the cysts of the infective stage of the trematode in its intermediate fish host (see Figure 17.4) and the cysts of the roundworm *Trichinella spiralis* (Nematoda), which causes trichinosis in humans and in the muscles of pigs and bears.

The ultimate biochemical response involves the immune system. Depending on the parasite, this response may produce short-term or life-long immunity to the infection. When a foreign protein or antigen enters the bloodstream, it is taken up by lymphocytes, which produce a molecular template corresponding to the antigen. This template is known as an antibody. The antibodies in the bloodstream neutralize invading antigens.

The immune response, however, can be breached. Antibodies specific to an infection normally are composed of proteins. If the animal suffers from poor nutrition and its protein deficiency is severe, its normal production of antibodies is inhibited. This depletion of energy reserves breaks down the immune system and allows the virus or other parasites to become pathogenic. The ultimate breakdown in the immune system occurs in humans infected with the human immunodeficiency virus (HIV), the causal agent of AIDS. The virus attacks the immune system itself, exposing the host to a range of deadly infections.

Abnormal Growths The host may also react to parasitic infections with abnormal growth. Plants respond to the invasion of their tissues by bacteria and fungi by cyst formation in the roots; scab formation in fruits and roots, which cuts off contact of the fungus with healthy tissue; root knot, a reaction to nematodes in tomatoes and other plants; black knot, a limb canker in black cherry (*Prunus serotina*) in response to the fungus *Apiosporina morbosa;* and root nodules in legumes. They react to attacks on leaf, stem, fruit, and seed by gall wasps, bees, and flies by forming abnormal growth structures unique to the particular gall insect (Figure 17.10). Production of galls does not benefit the plant. In fact, a heavy infestation of gall insects, particularly leaf galls, can weaken and kill the plant. However, gall formation can expose the larvae of some galls to predation. The conspicuous, swollen knobs of the goldenrod stem gall, for example, attract the downy woodpecker (*Picoides pubescens*), who excavates and eats the larva within the gall.

Malarial parasites in vertebrates can stimulate the spleen into producing many red blood cells and antibodies, resulting in its enlargement. Certain helminth parasites can produce cancerous tumors in their mammalian hosts. Larvae of warble flies (*Oestridae*) develop in boil-like swellings (warbles) just under the skin. These growths subside after the transformed larvae escape through a hole cut in the skin. Invertebrates, especially mollusks, respond to irritants or parasitic larvae by pearl formation. Pearl formation is usually associated with some inanimate foreign body, but often it is the response of an invertebrate host to a parasitic infection.

Sterility Parasitic infection of plants and certain invertebrates can destroy or alter a host's reproductive system. Consider the example of four fungus-eating *Drosophila* flies of

FIGURE 17.10 Galls are a morphogenic response by plants to an alien substance in their tissues, in this case the presence of a parasitic egg. The response involves a genetic transformation of the host's cells. (a) The pine cone gall, a bud gall caused by the gall midge, *Rhabdophaga strobiloides*. (b) The hickory leaf gall, induced by a gall aphid, *Phylloxera canyaeglobuli*. (c) Oak bullet gall, a woody globose gall caused by a gall wasp, *Disholcaspis globulus*. (d) The familiar goldenrod ball gall, a stem gall induced by a gallfly, *Eurosta solidagin*.

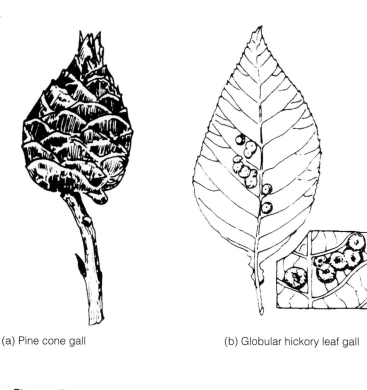

(a) Pine cone gall

(b) Globular hickory leaf gall

(c) Oak "fig root" gall

(d) Goldenrod ball gall

eastern North America: *D. putrida, D. falleni, D. recens, and D. testaca.* These flies are parasitized, often heavily, by the gut nematode *Howardula aoronymphium* (Allantonematidae). The female nematode, living in a mushroom, infects *Drosophila* larvae on the mushroom by penetrating the cuticle of the larvae. The female grows in the larvae, produces eggs, and releases larval nematodes in the hemocoel of adult flies. The larval nematodes exit through the anus or ovipositor when the fly visits a mushroom. Although the nematode does not interfere with male mating success, it does cause sterility in female flies. Of the infected females in the populations of several species studied by J. Jaenike (1992), one hundred percent of *D. putrida,* 90 percent of *D. testaca,* and about 50 percent of female *D. falleni* are completely sterile. This parasite-induced sterility can have a pronounced effect on the host populations. Nematode parasites can reduce the potential growth rate of a population of *D. testaca* by 24 percent, of *D. putrida* by 11 percent, and of *D. falleni* by 5 percent. Because all three of these flies feed on the same species of mushrooms (Jaenike and James 1991), parasitism can influence the outcome of interspecific interactions among them.

Infection of plants, notably grasses and sedges, by systemic endophytic fungi that grow within the leaves and stems and by nonsystemic smuts and rusts that grow externally can prevent the plants from reproducing sexually by "castration"—inhibiting floral development or aborting developing flowers (Clay 1991). Nonsystemic fungal parasites, the smuts and rusts, which make up most of the parasitic fungi of plants, colonize new hosts by contagious spread of spores, as in white pine blister rust. Most castrating fungi are systemic endophytes that spread by vegetative growth into tillers and tubers. Although inhibiting sexual reproduction by "castration," these endophytic fungi stimulate vegetative reproduction or cloning. By doing so, such fungi increase the availability of new hosts and prevent the evolution of resistant progeny through sexual reproduction. By interfering with sexual reproduction, such endophytic parasites run the risk of reducing the genetic diversity of their hosts, especially if those hosts are annual plants, making them vulnerable to adverse environmental changes.

Behavioral Changes Heavily parasitized animals often behave abnormally. Rabbits infected with the bacterial disease tularemia (*Pasteurella tularensis*), transmitted by the rabbit tick (*Haemaphysalis leporis-paulstris*), are sluggish and difficult to move from cover. Rabid foxes and raccoons may be overly aggressive or overly tame and unafraid of human contacts.

Behavior modification of a host induced by a parasite aids its transmission from the intermediate to the final hosts by making parasitized prey easier for predators to capture. The Pacific killifish (*Fundulus parvipinnisis*) is a second intermediate host for the trematode *Euhaplorchis californiensis* of southern California salt marshes. The parasite matures sexually in a number of predatory birds, the definitive hosts. Thus predation on a parasitized fish completes the parasite's life cycle. In the fish, the larval trematode seeks out the brain, causing behavioral changes. Infected fish contort,

shimmy, or jerk as they swim, making them conspicuous to their avian predators and thus increasing predation rates on intermediate host (Lafferty and Morris, 1996).

Birds and mammals respond defensively to ectoparasites by grooming. Among birds the major form of grooming is preening, which involves manipulation of plumage with the bill and scratching with the foot. Both activities remove adults and nymphs of lice from the plumage (Clayton 1991).

Mate Selection Parasites may affect mate selection in birds. The hypothesis has been advanced that parasitic infections or resistance to such infections can influence which mate a female chooses (Hamilton and Zuk 1982, Zuk 1991, 1992). Supposedly males possessing the brightest plumage or other ornaments are the most free of parasites. By choosing males with the brightest colors, females are selecting the healthiest males. This hypothesis is highly controversial (Real 1988). Evidence used to support it comes from correlation studies relating mating success with parasitic loads and physiological responses (Zuk 1991, Hudson and Dobson 1991, Hill, 1999).

Parasite Population Dynamics

There is a similarity between predator-prey relationships (Chapter 16) and parasite-host relations. The parasite acts out in its own way the role of a functional predator, but with a major difference. The host is at once food and habitat, so that the parasite is much more intimately related to its "prey" than the predator. Its fate is closely related to the density of the host population and the distribution of the individual parasites among the host. A host can support only so many adult parasites before the parasites experience intraspecific competition or parasite-induced mortality that takes the parasites along with their host.

The situation becomes magnified when the parasites possess two or more infective stages. Each stage behaves as an individual population subject to its own population stresses and the immune responses and longevity of the host. Premature death of a host results in the death of the parasites. Thus the parasite is faced with both the natural and parasite-induced mortality of the host.

Within the host the parasite faces its own intraspecific interactions, especially density-dependent interference competition. Growth of a parasitic population may be constrained by space within the host (Figure 17.11) as well as its own fecundity (Figure 17.12). Ackert and associates (1931) found a relationship between the amount of larvae and eggs taken in by chicks and growth rates of the parasite. An intake of a smaller amount of eggs gave a higher percentage of hatching and a greater growth rate of worms, whereas a high infestation resulted in a low percentage of hatching and slow growth rates.

Immune responses to high infestation can increase parasite mortality and reduce hyperinfestation. The nematode *Haemoinchus contortus* occurs as an adult in the stomach of sheep and is transmitted in the larval stage found on grass. Entering the stomach, the infecting larvae burrow into the mucosal lining of the host's gut. They remain there for a short

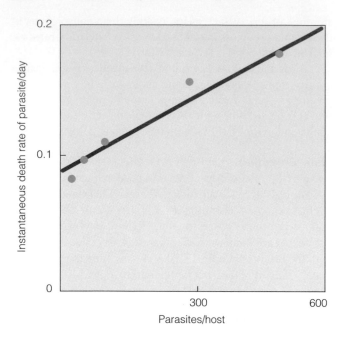

FIGURE 17.11 The relationship between parasite density and instantaneous parasite death rate within a host. The hosts are chickens and the parasite is the gut nematode *Ostertagia ostertagi*. The rate of mortality is linear with density. (From Ackert et al. 1931.)

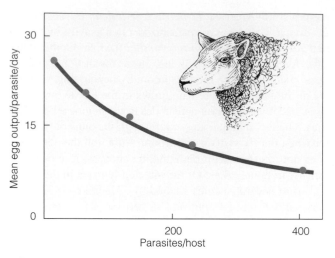

FIGURE 17.12 High density of parasites in a host, probably resulting in strong intraspecific competition, sharply reduces the production of transmission stages. The hosts are sheep infected with the liver fluke *Fasciola hepatica*. (From Boray 1969.)

time before returning to the lumen of the stomach as adults. Before their stay ends, the larvae induce a rise in the histamine concentration in the sheep's blood and evoke an antibody response. In the stomach the adult nematodes take up feeding on blood. Continuous reinfection of the sheep brings in more larvae, which attach themselves to the mucosa. Their presence induces further production of histamine and antibodies. They circulate in the blood and are taken up by the blood-feeding adults, killing them. With the expulsion and death of the adult nematodes, the newly invading larvae find a place to live. Such a reaction by the host prevents an overpopulation of parasites within it.

As in the study of the population dynamics of nonparasites, the most important parameter is R_o, the basic reproductive rate. The reproductive rate of parasites can be described by a model similar to that derived from the life table in Chapter 10. Among microparasites, R_o is the average number of newly infected hosts that arise from each infected host during its lifetime when introduced into a large susceptible population. The population cannot grow until it reaches some transmission threshold N_T at which $R_o = 1$. If R_o is less than 1, the parasite population will die out; if R_o is greater than 1, the parasite population increases (Anderson and May 1979). A large value of R_o means a high infection rate. Among macroparasites R_o is the average number of offspring produced by a mature parasite throughout its reproductive life span that successfully complete their life cycle and attain reproductive maturity.

R_o is given as $\beta N/(a + b + v)$, where N is host population size, β is transmission rate of the parasite, a is the instantaneous host death rate when mortality is due to the influence of the parasite, b is the natural instantaneous host

death rate, and v is the instantaneous birth rate of the parasitic transmission stage.

The transmission threshold N_T ($R_o = 1$) must be crossed if the disease or parasite is to spread. This threshold is given as $N_T = (a + b + v)/\beta$. Thus $R_o = N/N_T$.

This model is a starting point. The model becomes more complex as variables are introduced, depending on the parasites and hosts involved. Some of these variables are the expected life span of mature adult parasites, expected life span of the infective stage of the parasite, proportion of mature parasites within the host, and the proportion of the transmissive stage that become infective. The model is basic to studying and understanding the spread of disease. Determining the role of parasites and disease in the population dynamics of host populations is one of the most challenging areas of ecology (Price et al. 1986).

Evolutionary Responses

A premise in ecology is that parasitic species that do not harm their hosts have the best chance for long-term survival. A parasite that confers high virulence causes a high mortality of hosts. The parasite gains no advantage if it kills its host and thus itself. Parasites with low virulence ensure a long duration of infection. Thus natural selection would favor adaptive responses between parasite and host. As an outcome, host and parasite evolve a mutual tolerance or coadaptation.

Disease represents a lack of adaptation between parasite and host. Although the death of a host may appear to be detrimental to the parasite, there is evidence that natural selection may actually favor parasites with genes that confer virulence, especially among certain microparasites transmitted by vectors (Ewald 1983). Evolutionary pressure may favor increased severity in situations where competition exists between genetically different parasites of the same species. Highly productive, and therefore highly virulent, parasites would reach susceptible hosts much faster than those of lower productivity. Rapid transmission would occur, especially if arthropod vectors were involved. Once these virulent forms entered new hosts, the immune responses of the host might reduce the entry of less virulent forms. Thus highly virulent forms would outcompete the less virulent ones, but reduce the fitness of the host. As the host built up immune responses, it would reduce the fitness of the parasite. Then parasites with genes for lower virulence in the surviving resistant hosts would have the selective advantage. Reduced virulence would bring about increased survival of both parasites and host and ensure long-term propagation.

The parasite-host interaction between the European rabbit and the viral infection myxomatosis is an example of evolutionary forces at work, the adaptation and counteradaptation of parasites and hosts (see Holmes 1983, May and Anderson 1983). The myxomatosis virus is transmitted from host to host by an arthropod vector, mosquitoes.

To control the introduced rabbit, the Australian government introduced the rabbit's viral parasite into the population. Initially, the highly virulent strain had the competitive advantage and spread rapidly. The first epidemic of myxomatosis was fatal to 97 to 99 percent of the rabbits. The second resulted in a mortality of 85 to 95 percent; the third, 40 to 60 percent (Fenner and Ratcliffe 1965). The effect on the rabbit population was less severe with each succeeding epizootic, suggesting that the two populations were becoming more coadapted.

In this adjustment, attenuated genetic strains of virus, intermediate in virulence, tended to replace highly virulent strains. Too high a virulence killed off the host; too low a virulence allowed the rabbits to recover before the virus could be transmitted to another host. Also involved was a passive immunity to myxomatosis conferred on the young born to immune does. Finally a genetic strain arose in the rabbit population providing an intrinsic resistance to the disease.

The transmission of myxomatosis depends on *Aedes* and *Anopheles* mosquitoes, which feed only on living animals. Rabbits infected with the more virulent strain lived for a shorter period than those infected with a less virulent strain. Because the latter live for a longer period, the mosquitoes have access to that virus for a longer time. That gives the less virulent strain a competitive advantage over the more virulent.

In those regions where the less virulent strains have the competitive advantage, the rabbits are more abundant because fewer die. That means more total virus is present in those regions than in comparable areas where the more virulent strains exist. Thus, the virus with the greatest rate of increase and density within the rabbit population is not the one with the selective advantage. Instead the virus whose demands are balanced against supply has the greatest survival value.

Social Parasitism

Another form of parasitic relationship is **social parasitism,** in which one organism depends parasitically on the social organization of another. Two forms are brood parasitism and kleptoparasitism. **Brood parasitism** is foisting incubation of eggs or care of young onto surrogate parents. Brood parasitism may be temporary or permanent, facultative or obligatory (Wilson 1975). Facultative brood parasitism may occur within or among species. **Kleptoparasitism** is the forcible theft of prey by the parasite from the host.

Brood Parasitism Temporary facultative brood parasitism within a species is well developed among ants and wasps. For example, a newly mated queen of the wasp genus *Polistes* or *Vespa* will attack established colonies of her own species and displace the resident egg-carrying queen (Wilson 1975). Females of a number of species of birds—notably some of the Galliformes, ostriches, swallows, starlings, bluebirds, American robins (*Turdus migratorius*), moorhens (*Gallinula chloropus*), American coots (*Fulica americana*), and waterfowl—will lay some of their eggs in the nests of others of the same species (Petrie and Møller 1991). Twenty-one species of ducks are known to lay eggs in nests other than their own (Weller 1959). The redhead duck (*Aythya americana*) is an example. Approximately 5 to 10 percent of female redhead ducks are nonparasitic and nest early. Over 90 percent lay eggs parasitically at one time or another. More than half of these ducks are semiparasites and also build nests of their own. The remainder build no nests, and thus become temporary, obligatory parasites.

Although common in ants, the most outstanding examples of temporary, obligatory brood parasitism occur in birds. Brood parasitism has been carried to the ultimate by the cowbirds and Old World cuckoos, both of which have lost the acts of nest building, incubating the eggs, and caring for the young. They are obligatory parasites who pass off these duties to the host species by laying eggs in their nests. The brown-headed cowbird (*Molothrus ater*) of North America removes one egg from the nest of the intended host and the next day lays one of her own as a replacement (Figure 17.13). Some host birds counter by ejecting the eggs from the nest. Others hatch the egg and rear the young cowbird, usually to the detriment of their own offspring. The host's young may be pushed from the nest or die from lack of food because of the more aggressive nature and larger size of the young cowbird.

A third type of brood parasitism is permanent and obligatory between species. The parasitic form spends its entire life cycle in the nest of the host (Wilson 1975). This type of social parasitism is common among ants and wasps. In most cases the species are workers and queens have lost the ability to build nests and care for the young. The queen gains entrance to the nest of the host and either dominates the host queen or kills her outright and takes over the colony.

For temporary obligatory brood parasites, such as the cowbird and European cuckoo, the advantages of social parasitism are obvious. Foisting parental care of young on other species is the only way they can perpetuate their own. The hosts of such parasites are adversely affected, especially if they are experiencing their first contact with the parasite. Such a situation has developed with Kirtland's warbler (*Dendroica kirtlandi*), a relict species that inhabits extensive jack pine (*Pinus banksiana*) stands in a compact central homeland of about 250 km² in northern lower Michigan (Mayfield 1960). Before European settlers arrived, Kirtland's warbler apparently was isolated by a belt of unbroken forest 320 km wide from the parasitic brown-headed cowbird of the central

FIGURE 17.13 A newly hatched cowbird crowds the nestling of its barn swallow host. Note the two swallow eggs still unhatched.

plains, a bird closely associated with grazing animals. When settlers cleared the forest and brought grazing animals with them, cowbirds spread eastward and northward into jack pine country. Never associated with the cowbird, Kirtland's warbler did not evolve defenses against brood parasitism, such as ejecting an egg, building a new nest over the old, or rearing young successfully along with the cowbird. This parasitism has resulted in an alarming reduction in warblers fledged, a trend that is being reversed by a strong cowbird control program in this warbler's breeding range. However, with increasing fragmentation of the eastern deciduous forest, other warbler species of the forest interior are experiencing similar undefended parasitism by the cowbird.

Although obligatory brood parasitism has attracted the most attention (Payne 1998, Slagsvold 1998), nonobligatory brood parasitism, especially intraspecific parasitism, is widespread among birds (Rohwer and Freeman 1989). From an evolutionary viewpoint, it would seem that nonobligatory brood parasitism, in which individuals "decide" to be parasitic or not, would be maladaptive. The hosts are forced into misdirected parental care. Parasites capable of rearing their own broods risk lower reproductive success by entrusting the rearing of their offspring to others. Further, an evolutionary conflict of interest arises in such populations: the parasite can become a host and the host a parasite. Brood parasitism would seem to hold little advantage for either the host or the parasite.

What advantages do accrue for the nonobligatory brood parasite? There is no single or obvious answer, but a major reason among nonobligates simply is to make the best of a bad situation (Yom-Tov 1980, Petrie and Møller 1991, Sorensen 1993). Parasitic behavior may be strongest in those populations of a species occupying areas of uncertain environmental conditions, such as lack of nesting sites or food. Under those conditions the potential brood parasite has four options: (1) become a nonbreeder and join the floating population; (2) become a typical nester; (3) become a parasitic egg layer as a low-cost alternative, even though reproductive failure may be high; or (4) adopt a combined strategy of both parasitic egg laying and nesting. In the last option the individual runs the risk of becoming a host itself (Sorensen 1991). A classic example of mixed strategy is found in the European barn swallow (*Hirundo rustica*) (Møller 1987) and the cliff swallow (*H. pyrrhonota*) (Brown and Brown 1991) (Figure 17.14). Heavy ectoparasitic infections by nest mites, loss of poorly attached nests, and extreme weather conditions are uncertain environmental conditions that stimulate mixed strategy in these two species.

Brood parasitism could reduce the individual fitness of a host because part of the host's energy goes to rearing young of others. However, being parasitized may improve the host's reproductive success, because it spreads the risk of partial predation on the brood among unrelated young, especially in precocial species (Eadie and Lumsden 1985), whereas the parasite is adversely affected. Thus the improved survival of the host's own young may compensate for a reduced clutch and the cost of hatching and caring for the parasitic young. (Petrie and Møller 1991).

Many hosts have evolved some antiparasitic strategies. Among these is ejecting of strange eggs, especially if they are recognizable by different markings, guarding the nest, or even deserting it. The last is especially true with the wood duck (*Aix sponsa*), in which nest parasitism markedly increases the size of the clutch (Semel and Sherman 1986). In such situations the cost of rebuilding a nest and relaying eggs is less than that of rearing parasitic young. Because of the apparent ease with which hosts of some species become parasitized, the question arises why potential hosts have not evolved stronger defensive tactics. It could be that the cost of defending the nest is greater than that of rearing parasitic young.

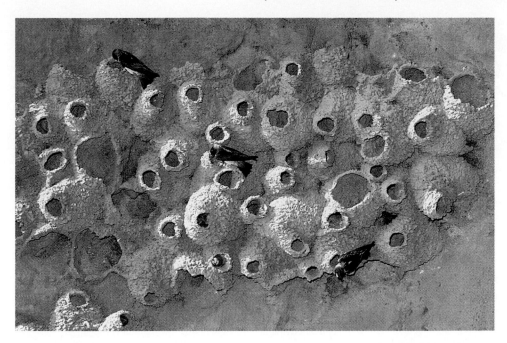

FIGURE 17.14 The colonial nesting habit of cliff swallows provides an ideal opportunity for intraspecific nest parasitism.

Kleptoparasitism Kleptoparasitism is a form of social parasitism in which the parasites obtain a substantial portion of their food by stealing it from the host. A bald eagle (*Haliaeetus leeuocephalus*), for example, forces an osprey (*Pandion haliaetus*) to drop the fish it caught, which the eagle retrieves and eats. Although kleptoparasitism is widespread among various taxa, it is most prevalent among two groups of birds, the Falciformes (eagles, falcons, and hawks) and the Charadriiformes (skuas, gulls, and waders). As with other forms of parasitism, a close relationship has evolved between the kleptoparasite and its host. Kleptoparasites depend on other species to locate and obtain prey that can then be stolen. Although kleptoparasitism involves interaction between two individuals, the kleptoparasites often invade the social structure of the hosts. Invasion of the social structure is most common in situations where potential hosts are aggregated into breeding colonies or feeding groups (Barnard 1984, Barnard and Thompson 1985), the availability of food is both temporally and spatially predictable, and the hosts offer little defense.

For example the black-headed gull (*Larus ridibundus*) parasitizes feeding flocks of golden plovers (*Pluvialis apricaria*) and lapwings (*Vanellus vanellus*) in the English countryside in winter (Barnard and Thompson 1985). Black-headed gulls exploit the social structure of plovers and lapwings adept at locating rich patches of forage to gain their food. The gulls select only unwary birds that have extracted a large worm from the soil. The gull then chases or harasses the bird until it drops the worm, which the gull catches or picks up from the ground.

Hosts do have some defensive tactics. They can engage in evasive flight. They can shift to smaller, less visible prey, and when foraging maintain a greater distance between themselves and the kleptoparasite. They can attempt to recover the prey. Or they can tolerate the situation and compensate by feeding at a higher rate, as the lapwings do.

MUTUALISM

In parasitic relationships one species obviously exploits the other. In the other highly coevolved relationship, mutualism, the relations are usually beneficial. Mutualism is defined as a reciprocal relationship at the individual or population level between two different species (Boucher et al. 1982). Out of this relationship, most obvious at the individual level, both species supposedly enhance their survival, growth, and fitness (Holmes 1983). However, evidence emerging from recent studies suggests that many mutualistic relationships are more reciprocal exploitation, or parasitism, than cooperative efforts between individuals, which the definition of mutualism implies (Barrette 1983).

Types of Mutualisms

Obligate Symbiotic Mutualism Some forms of relationship are so permanent and obligatory that the distinction between the two interacting populations becomes blurred. A good example is the fungi-algae symbiosis in the lichens. The basic structure of the lichen is a mass of fungal hyphae. Within this formation is a thin zone of algae that usually forms colonies of 2 to 32 cells (Figure 17.15). Some 27 different genera of algae have been associated with lichens, about 90 percent of them green algae and 10 percent cyanobacteria. The most common genus involved is *Trebouxia*. It is the only one not found in a free-living state (Holmes 1983). The fungi involved belong to diverse taxonomic groups, most of them being close relatives of free-living ascomycetes that are either parasites of plants and animals or decomposers. Many lichens produce and disperse spores, which form mats of mycelia that may live an independent, saprobic existence for a short time, until they capture algal cells. Depending on the species involved,

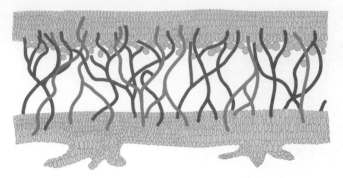

FIGURE 17.15 Section through a typical lichen body, showing the algae within the fungal mass. Chapter opening photo shows some examples of lichens.

FIGURE 17.16 Coral-like growth of ectomycorrhizae on the root of a woody plant.

the algae change the morphology of the affected fungi. So specific are these morphological changes to a particular algal-fungal relationship that it can be classified as a distinct species of lichen.

Algae and fungi supposedly live together for each other's benefit. The algae gain the protection of the fungal thallus; the lichens derive nutrition from the photosynthetic algae. Except for the genus *Trebouxia*, the relationship may be somewhat less than mutualistic. The lichen may be parasitizing the algae for nutritional gain, since cultured free-living forms of the lichen grow more slowly than those associated with algae. However, algal cells may leak metabolites to the surrounding soil rather than passing them on to the fungus. There is no evidence that the fungus provides anything for the algae other than protection from damaging solar radiation and desiccation (Ahmadjian 1970).

A more convincing example of mutualism is the intimate association of a plant's roots with fungal hyphae, called **mycorrhizae.** The plant supplies energy to the fungi, and the fungal hyphae take up mineral nutrients from the soil and transport them into the host's roots.

One form common to many trees of temperate and tropical forests is **ectomycorrhizae (EMC).** It consists of a well-developed fungal sheath or mantle around the root (Figure 17.16). This mantle is connected to the inside of the root by a network of hyphae called Hartig's net (Figure 17.17). The outside sheath develops into another network of hyphae that produces shortened and thickened roots that suggest coral. This outside network functions as extended root hairs (Harley and Smith 1983).

Another form is **endomycorrhizae,** notably the vesicular arbuscular mycorrhizae (VAM) (Figure 17.18). They are associated with a wide range of agricultural and native plants. The fungi involved belong to the genus *Endogone;* they have not been cultured independently of the host. The roots are infected by hyphae in the soil or from germ tubes that develop from spores. The hyphae penetrate the cells of the host to form a finely bunched network, called an **arbuscle,** the site of exchange between fungus and host, and oil-rich vesicles, fungal energy storage organs. The hyphae act as extended roots for the plant, drawing in phosphorus at distances beyond those reached by the roots and root hairs. The arbuscles have a definite life span. Their disappearance may be due to their

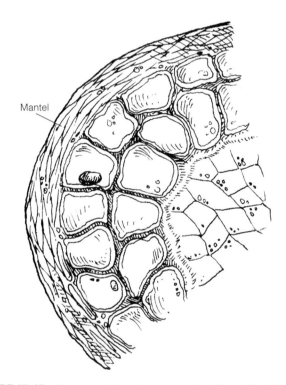

Mantel

FIGURE 17.17 Ectomycorrhizae form a mantle of fungi about the tips of rootlets. Hyphae invade the tissues of rootlets between the cells. This network is called Hartig's net.

digestion by the host plant. Unlike EMC, VAM do not change the shape or structure of the root.

Mycorrhizae, especially important in nutrient-poor soils, aid in the decomposition of litter and translocation of nutrients, especially nitrogen and phosphorus, from the soil into root tissue (Mosse et al. 1981, Newman 1988). Mycorrhizae increase the capacity of roots to absorb nutrients, provide

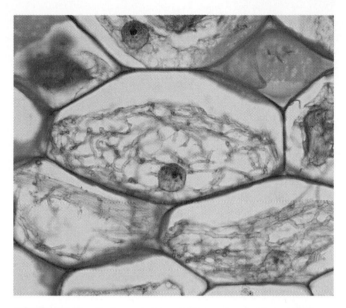

FIGURE 17.18 Endomycorrhizae in the root cells of an orchid.

selective ion accumulation and absorption, mobilize nutrients in infertile soil, make available certain nutrients bound up in silicate minerals (Voigt 1971), and improve the ability of the plant to extract water from the soil (Allen and Allen 1986). In addition, mycorrhizae reduce susceptibility of their hosts to invasion of pathogens by utilizing root carbohydrates and other chemicals attractive to pathogens (Marx 1971, Mosse et al. 1981). They provide a physical barrier to pathogens and stimulate the roots to elaborate chemical inhibitory substances. In return, the roots of the host provide support and a constant supply of carbohydrates. The association between the plant and the fungus can be tenuous. Any alteration in the availability of light or nutrients for the host creates a deficiency of carbohydrates and thiamin for the fungi. Interruption of photosynthesis causes a cessation of fruiting by mycorrhizae. So important are mycorrhizae to the growth of plants that foresters, restoration ecologists, and nursery operators inoculate nursery-grown shrubs and trees, especially pines to be planted in reforestation and restoration of highly disturbed areas. Lack of appropriate mycorrhizae in soil may inhibit the colonization of disturbed areas and early succession sites by forest trees.

Similar mutualistic relationships permitting organisms to exploit nutrient-poor environments are found in the sea. The oceans offer innumerable examples of all sorts of mutualistic interactions, especially among the coral reefs. Coral reefs, self-formed calcareous substrates occupied by anthozoans, are found largely in warm nutrient-poor tropical waters. Living within the cells of the endoderm layer of the oral cavity of the coralline anthozoans are photosynthetic dinoflagellate algae (zooxanthellae) (Figure 17.19). The heterotrophic anthozoans utilize the photosynthetic products of the algae. In turn the coral anthozoans remove, retain, and recycle essential nutrients from the water used by the zooxanthellae (Muscatine and Porter 1977). Although they are carnivorous suspension feeders capturing zooplankton from the surrounding water, anthozoans derive only about 10 percent of their daily energy requirement from zooplankton. They obtain 86 percent of their energy and caloric requirements from algal fixation of C and N and are able to survive and flourish in their nutrient-poor environment by recycling nutrients with their symbiont algae.

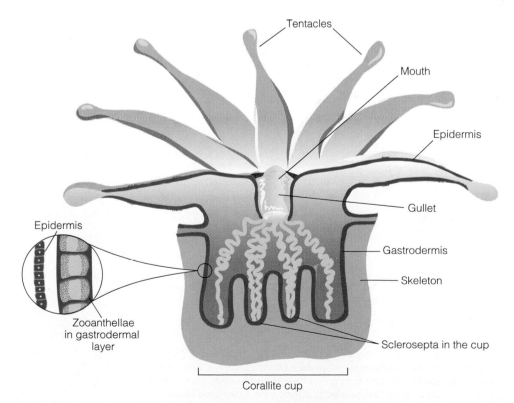

FIGURE 17.19 Anatomy of a coral polyp, showing location of the symbiotic zooxanthellae.

In addition, algal photosynthesis improves the ability of the anthozoans to lay down calcified coral structures, enabling them to build reefs fast enough to both counteract destruction and increase their benthic cover.

Obligate Nonsymbiotic Mutualism More common is nonsymbiotic obligate mutualism, in which the mutualists live physically separate lives yet cannot survive without each other. Pollination and seed dispersal systems offer many examples. Other obligatory relationships involve shelter, protection against predators, and reproduction.

Some of the most interesting cases exist between ants and fungi and between ants and plants. Fungus-growing attine ants depend on a slow-growing fungus that they "farm." Neither the ants nor the fungi can survive without each other (Martin 1970). A classic ant-plant relationship involves the ant-acacia mutualism (Janzen 1966, Hocking 1975). The Central American ants live in the swollen thorns of acacia (*Acacia* spp.), from which they derive shelter. They feed on protein-rich nodules growing at the tips of pinnate leaves, called Beltian bodies (named after the tropical naturalist Thomas Belt, who discovered their role). These nodules and sugar-secreting nectaries growing on the vegetative parts (extrafloral nectaries) provide an almost complete diet for all stages of the ants' development. In turn the ants protect the plants from herbivorous animals such as cattle. At the least disturbance, the ants swarm out of their shelters, emitting repulsive odors and attacking the intruder until it is driven away. Experimental data demonstrate the effectiveness of extrafloral nectaries and the ants they attract in *Acacia* and other plants possessing them in conferring protection against herbivores. Among other plants with extrafloral nectaries, the protective relationship is not so clear. (See Bentley 1977 for a review.)

Some mutualistic relationships involve a third member. Some ectomycorrhizae are epigeous—that is, they produce their sporocarps above ground and forcibly discharge their spores to the air. Hypogeous ectomycorrhizae produce their sporocarps below ground in structures popularly known as truffles. These hypogeous fungi are common to the coniferous forests of western North America and the Eucalyptus forest of Australia. In the coniferous forests these mycorrhizae depend on chipmunks and voles to disperse their spores (Maser et al. 1978), and in the Eucalyptus forests they depend especially on the marsupial rat-kangaroos (Johnson 1996). Attracted to the fruiting bodies by species-specific odors, these small mammals eat the sporocarps, which make up a significant part of their diet. When they defecate, the rodents spread the viable spores necessary for the survival and health of the forest trees (Figure 17.20).

Nonobligatory (Facultative) Mutualism Most mutualisms are nonobligatory and opportunistic (facultative), at least on one side of the mutualistic relationship. Facultative mutualisms are widely involved in seed dispersal and polli-

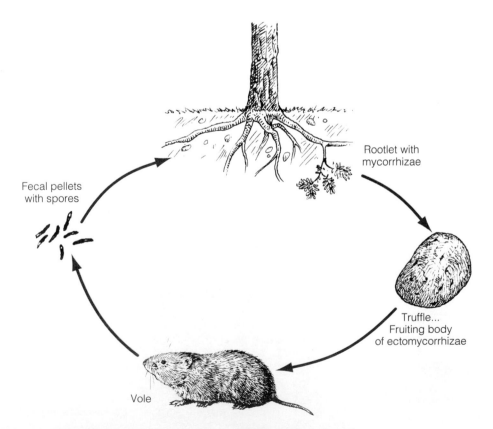

Fecal pellets with spores

Rootlet with mycorrhizae

Truffle... Fruiting body of ectomycorrhizae

Vole

FIGURE 17.20 Three-way relationship among hypogeous ectomycorrhizae, trees, and small mammals. The ectomycorrhizae need the tree for energy. The tree needs the ectomycorrhizae for uptake of nutrients from the soil. Small mammals, needed to disperse the spore, feed heavily on the truffles.

nation; the benefits are spread over many plants, pollinators, and seed dispersers, discussed later in this chapter. Because these mutualisms involve interactions among guilds of species, they are called diffuse.

Defensive Mutualism A major problem faced by many livestock producers is the toxic effects experienced by cattle grazing on certain grasses, particularly perennial rye grass (*Lolium perenne*) and tall fescue (*Festuca arundinacea*). These grasses are infected by certain fungal endophytes (Clavicipitaceae, Ascomycetes), notably *Acremonium ceonophialum,* that live inside plant tissue. The mycelium of this fungus grows as the leaves of the grass grow and extends through the intercellular space between plant cells (Figure 17.21), deriving nourishment from the intercellular fluids of the plant. At the same time the fungus produces physiologically active alkaloids in the plant's tissues that impart a bitter taste in the grass. The infected grass is poisonous to grazing mammals, particularly cattle and horses, and a number of insect herbivores. The toxic effects of this fungus in large herbivores include intolerance to heat, poor weight gain, and reproductive failure in mares. In more extreme cases in mammals, the alkaloids constrict small blood vessels to the brain, cause convulsions, tremors, stupor, gangrene of the extremities, and death. This symbiotic relationship between plant and fungus suggests a defensive mutualism in which the fungus provides a strong toxic defense against grazing (Clay 1988, Ball et al. 1993) There are costs to the plant. The fungal infection causes sterility in the host plant by inhibiting flowering or aborting seeds. Some plants have a few counteradaptations that restore fertility; but in most plants the loss of sexual reproduction is balanced by the greater vegetative growth of the infected plants and enhanced growth in the absence of herbivory.

Indirect Mutualism Mutualistic relationships described so far are direct: one species immediately benefits the other.

Other mutualistic-type relationships are indirect. For example, suppose that prey species A, eaten by predator X, strongly inhibits prey species B, eaten by predator Y. If the effect of predator X on prey A permits the expansion of prey B, X in effect provides more food for Y. The activities of X then indirectly benefit Y.

A similar mutualistic effect may occur in competitive situations. Suppose species A competes strongly with species B, but mildly with C; and that species C competes strongly with B, but mildly with A. The combined effects of A and C are to their mutual advantage against B. A number of such indirect mutualisms could influence community organization. (For more discussion of indirect mutualism see Vandermeer 1980, Wilson 1980, Waser and Real 1979, Lane 1985.)

There are few empirical examples demonstrating indirect mutualism. How indirect mutualism might function appeared in experimental aquatic food webs studied by H. Wilbur and J. Fauth (1990). This elaborate experiment involved the construction of 16 different food webs with four replicates in an array of 64 experimental ponds. The two predators were the larvae of the green darner dragonfly (*Anax junius*), voracious predators of anuran larvae, and a newt (*Notophthalmus viridescens*). The two prey species were tadpoles of the pickerel frog (*Rana palustris*) and the American toad (*Bufo americanus*), two competitors. The model for each one-predator, two-prey system predicted low and nearly equal survival of the two prey (4 percent for *Bufo* and 7 percent for *Rana*). At the end of the experiment, however, the observed survival in these systems for the toad was 50 and 33 percent and for the frog 75 and 35 percent. In the four-species system, 35 percent of the toads and 59 percent of the frogs survived. Even though the two prey species were competitors, the minor effects of competition were swamped by the beneficial effects of sharing the risks of predation. Under the pressure of predation the two anurans became mutualists and experienced higher survival than if each was the only prey.

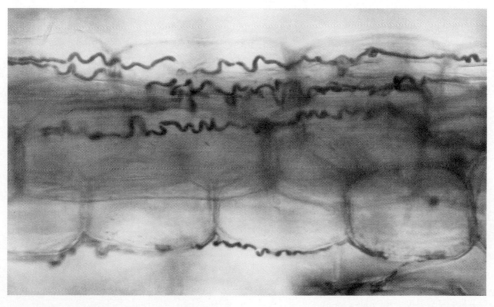

FIGURE 17.21 Endophytic fungi in a blade of fescue (*Festuca*).

The results of this experiment point to the subtle interrelations among species. If one member of an indirect relationship should go extinct, the other species would decline.

Pollination

The "goal" of pollination is specific and direct. The plant must transfer its pollen from the anthers of one plant to the stigma of a conspecific. Some plants simply disperse their pollen to the wind. This method of pollination works well and costs little when the plants grow in large homogeneous stands, as grasses and pine trees do. Wind dispersal is unreliable when conspecifics are scattered individually or in patches across a field or forest. These plants depend on animals for pollen transfer, mostly insects with some assistance from nectar-feeding birds and bats.

A number of plant-animal pollination systems involve obligates. Examples are the mutualistic relationship of several species of *Yucca* with associated species of yucca moths, *Tegeticula,* and of roughly 600 species of *Ficus* with their own specialized pollinators, minute fig wasps (Agaonidae) (Anstett et al 1997). Yuccas depend exclusively on yucca moths for pollination, so that the relationship is a nonsymbiotic obligate mutualism (Figure 17.22). The larvae of yucca moths live symbiotically in the yucca seed heads as obligate predators of their seeds. The adult moths gather pollen; yet they neither take nectar from the flowers nor use the pollen to feed their larvae. At the same time the females lay eggs in the ovaries of newly opened flowers and then deposit the pollen on the stigma. After the seeds have fully developed, each developing larva eats several seeds before dropping to the ground to form a cocoon.

In return for pollination, yuccas must pay out a certain amount of seeds as larval food, without which the yucca moth could not survive. Overall, yucca moth larvae decrease viable seed production by as much as 19 percent, and yuccas expend up to 30 percent of the benefits gained from pollination to support yucca moth larvae (Addicott 1986).

Some other pollination systems involve less complex nonsymbiotic obligate mutualism. Well-known examples are orchids in lowland neotropical forests. Scattered widely through the forests, these orchids depend entirely on male euglossine or golden bees (tribe Euglossini in the honeybee family), specialists in the pollination of orchids (Figure 17.23). Extremely fast fliers, these bees cover long distances between orchids in the tropical forests. However, the male bees obtain no food from the visit. The orchids could not afford to meet the bees' energy demands. Instead the male bees collect fragrances from secretory cells on the lips of the flowers. The male bees use these fragrances to develop their own pheromones to attract females. Female bees ignore the orchids entirely.

Most orchids offer no reward at all. To attract the euglossines, such orchids mimic other flowers that do; or they have flowers with the scent, shape, or color patterns that mimic female bees. When males attempt to copulate with the flower mimic, they pick up pollen.

The orchids contain their pollen in a single mass, the pollinium, with an attachment device. When the bee brushes against it, the pollinium becomes attached to the bee at a spe-

(a)

(b)

FIGURE 17.22   (a) Yuccas (*Yucca*) are conspicuous plants of the deserts of southwestern North America. (b) Yucca moth (*Tegeticula*), on which the yucca solely depends for pollination.

FIGURE 17.23 Tropical orchids depend on male euglossine bees for pollination.

cific location on its body. It remains there, for days or weeks if necessary, until the bee visits another orchid of the same species, whose stigma retrieves it. Bees may visit more than one species of orchid and pick up polliniums from each of them. To prevent wrong deliveries, each species of orchid has its pollinium so located within the flower that it adheres at its own specific location on the bee's body. It becomes detached only when the bee enters the correct orchid. This fact suggests that coevolution has occurred between bees and orchids and that the bees have been a strong selection force in the evolution of orchids (see Heinrich 1979, van der Pijl and Dodson 1966, Feinsinger 1983, Faegri and van der Pijl 1979).

Like the fragrance-collecting bees, nectar-feeding animals visit flowering plants to exploit them, not to pollinate them. Most nectivores are generalists. They find little advantage in specializing, except as temporary facultative specialists. Because of the short seasonal flowering of each species, often shorter than the availability of fruits, nectivores depend on a progression of flowering plants through the season. Nectivores cannot afford to commit themselves to one flower, but they do concentrate on one species while its flowers are available.

Rather, plants are the ones which have to specialize, to entice animals by color and odor, dust them with pollen, and then reward them with a rich source of food: sugar-rich nectar, protein-rich pollen, and fat-rich oil. Providing such rewards is expensive for plants. Nectar and oils are of no value to the plant except as attractants for potential pollinators. They represent an expenditure of energy that the plant otherwise might use in growth.

Many species of plants, such as blackberries, elderberries, cherries, and goldenrods, are generalists themselves. They

flower profusely and provide a glut of nectar that attracts a diversity of pollen-carrying insects, from bees and flies to beetles. Other plants are more selective, screening their visitors to ensure efficiency in pollen transfer. These plants may have long corollas allowing access only to insects and hummingbirds with long tongues and bills and keeping out small insects that eat nectar but do not outcross the plants. Some, such as the closed gentian, have petals that only large bees can pry open.

In addition to nectar, some plants provide oil as a floral reward (Vogel 1969, Buchmann 1987). Many genera in a number of families, including Iridaceae, Orchidaceae, Scrophulariaceae, Concurbitaceae, Solanaccac, and Primulaceae, mostly in neotropical savannas and forests, have specialized oil-secreting organs, called **elaiophores.** One type is epithelial elaiophores, which consist of small areas of secretory epidermal cells beneath a protective cuticle on the petals in which secreted lipids accumulate. A second type is trichome elaiophores, made up of hundreds to thousands of glandular trichomes that secrete lipids in a thin film of oil exposed to the air. In some plants, however, the lipids are protected within deep floral spurs. The flowers are visited by highly specialized bees in four families that use the energy-rich floral oils in place of or along with pollen as provisions for developing larvae. These bees possess modified cuticular and setal structures designed for mopping up, storing, and transporting oil to the nest (see Buchmann 1987).

The relationship between the common Central American plants *Heliconia* and hummingbirds illustrates the many factors that may be involved in such mutualisms. Growing in the openings of tropical forests or along the forest edge, *Heliconia* propagates vegetatively by rhizomes and usually forms large clumps. When two years of age or older, each individual *Heliconia* plant in the clump blooms. The bloom consists of several showy bracts, each of which encloses several flowers. The bracts open one after another over a period of days or weeks. Each flower within a bract lasts only a day. The flowers are tubular and vary in length and curvature depending on the species. Some have long curvaceous corollas, 32 mm or less (Stiles 1975). Some species bloom either in the wet season or the dry season, whereas others bloom throughout the year but have a wet or dry seasonal peak in flowering. All are pollinated by insects or birds and offer a supply of sugar-rich nectar as an inducement to their pollinators.

Stiles found that in his Costa Rican study area, nine species of hummingbirds visit nine species of *Heliconia*. Just as the flowers of *Heliconia* have straight or curved corollas, Stiles found that the hummingbirds, too, could be divided into two groups, hermit hummingbirds with long, curved bills and nonhermits with shorter, straight bills. Stiles observed that the five *Heliconia* species with the long, curved corollas are visited to a significantly greater extent by hermits than by nonhermits, whereas three of the species with short corollas are visited commonly by straight-billed nonhermits.

In return for nectar, hummingbirds pollinate the respective flowers. *Heliconia* depend on hummingbirds for pollen transfer. Hermit hummingbirds, probing long, curved corollas, carry pollen at the base of the bill or on the head. The nonhermits, or straight-billed hummingbirds, carry pollen on the chin or mandibles. If the short-corolla flowers are somewhat curved, but the path to the nectar is short and straight, allowing easy access, the hummingbird has pollen deposited on the bill.

Because of the number and types of plants and birds involved, some isolating mechanisms are essential. *Heliconia* select against hybridization by sequential and nonoverlapping peaks in flowering, by spatial isolation, and by promoting behavioral isolation in hummingbirds. The behavioral differences include responses of the hummingbird to visual cues of flowers and to caloric content of nectar. The mutualism thus depends not only on the morphological fit between bird bill and flower corolla, but also on flower phenology, energy content of nectar, and energy demands and behavioral responses of hummingbirds.

Although the relationship of plants and their pollinators has been investigated for years (Real 1983) and the literature is large, many questions remain. Do plants and pollinators coevolve? To get at that question, researchers need to incorporate systematics, morphology, and behavior into their studies. For example, we know little about the evolution of flower choice and flower visiting by animals. Can coevolution take place if plants benefit from constant visits by animals and animals benefit from exclusive access only over a short period of time? Does an exclusive, obligate coevolving relationship increase the fitness of the plant and animal? Do plants experience simultaneous selection from a suite of pollinators? What are the effects of different pollinators on seed set?

The answers are basic to conserving many species and maintaining the integrity of ecosystems, especially tropical ones. If essential relationships between plants and their pollinators are severed by deforestation, loss of habitat, or extinction of one member, then we may experience the loss of associated species and witness a sharp decline in biodiversity. Once any of the participants are gone, there is no way to restore the relationships.

Seed Dispersal

Plants with seeds too heavy to be dispersed by wind depend on animals to carry the seeds some distance from the parent plant and deposit them in sites favorable for seedling establishment. Some seed-dispersing animals are seed predators, consuming the seeds for their own nutrition. Plants depending on such animals must produce a tremendous number of seeds over their reproductive lifetimes and sacrifice most of them to ensure that a few will survive, come to rest on a suitable site, and germinate.

An example is the relationship between Clark's nutcracker (*Nucifraga columbiana*) and the whitebark pine (*Pinus albicaulis*) of western North America (Figure 17.24). Whitebark pine and a few other pines, such as piñon pine (*Pinus edulis*), produce large wingless seeds that can be dispersed away from the parent trees only by animals. The seeds of whitebark pine are eaten and hoarded by several species of rodents, including chipmunks, and by Steller's jays (*Cyanocitta stelleri*) and Clark nutcrackers. Only the nutcracker possesses the behavior appropriate to disperse the seed systematically and successfully away from the tree (Hutchins and Lanner 1982, Tomback 1982). Typical of jays, the bird carries seed in cheek pouches and caches the seeds deep enough in the soil of forests and open fields to reduce

FIGURE 17.24 This Clark nutcracker (*Nucifraga columbiana*), a jay of the high mountains of western North America, stores and eats the seeds of the whitebark pine (*Pinus albicaulis*) in which it is perched.

predation by rodents. The number of seeds cached per individual per year is enormous, about 98,800. The nutcrackers fail to retrieve enough of these seeds to allow a large number of them to establish seedlings. Although the cost is high, whitebark pine is virtually dependent on the bird for seed dispersal, and the ranges of the two species roughly coincide.

Some plants use a seed predator not only to disperse the seeds but also to protect them from other predators. In the deserts of the southwestern United States (O'Dowd and Hay 1980), in the sclerophyllous shrublands of Australia (Berg 1975), and in the deciduous forests of eastern North America (Beattie and Culver 1981, Handel 1978), a number of herbaceous plants, including many violets (*Viola* spp.), depend on ants to disperse their seeds. Such plants, called **myrmecochores,** have an ant-attracting food body on the seed coat, called an **elaiosome.** Appearing as shiny tissue, the elaiosome contains lipids and sterols essential to certain physiological functions in insects.

The ants carry seeds to their nests, where they sever the elaiosome and eat it or feed it to their larvae. The ants discard the intact seed within abandoned galleries of the nest. Ant nests, richer in nitrogen and phosphorus than surrounding soil (Culver and Beattie 1978), provide a suitable substrate for seedling emergence and establishment. Further, by taking seeds away from the parent plant, the ants significantly reduce rodent predation on them (O'Dowd and Hay 1980, Heithaus 1981). Heithaus (1981) found that rodents, particularly the white-footed mouse (*Peromyscus leucopus*), removed 84 percent of the seeds of bloodroot (*Sanguinaria canadensis*) when ants were excluded from the parent plants, but only 13 to 43 percent when ants were allowed access.

Plants have an alternative approach to seed dispersal: to enclose the seed in a nutritious fruit attractive to fruit-eating animals, the frugivores. Frugivores are not seed predators; they consume only the endocarp surrounding the seed. With some exceptions, they do not impair the vitality of the seed. Most frugivores do not depend exclusively on fruits, because fruits tend to be deficient in certain nutrients such as protein, and because they are only seasonally available.

To use frugivorous animals as agents of dispersal, plants must attract them but discourage the consumption of unripe fruit. Plants protect unripe fruit by cryptic coloration, such as green fruit among green leaves, and by unpalatable texture, repellent substances, and hard outer coats. When seeds mature, plants attract fruit-eating animals by presenting attractive odors (Howe 1980), altering the texture of fruits and seeds, improving succulence, acquiring a high content of sugar and oils, and "flagging" their fruits with colors—red, black, blue, yellow, white—to catch attention (Stiles 1982) (Figure 17.25).

Plants have two alternative approaches to seed dispersal by frugivores. One is to become opportunistic and evolve fruits

(a)

(b)

FIGURE 17.25 Color attracts fruit-eating birds and mammals to plants that depend on them for seed dispersal. (a) White berries of poison ivy (*Rhus radicans*). (b) Red fruits or hips of multiflora rose (*Rosa multiflora*), an exotic that has reached a pest status. (c) Blue fruits of wild grape (*Vitus* sp.). (d) Black fruit of Hercules club (*Aralia spinosa*).

(c)

(d)

that can be exploited by a large number of dispersal agents. Such plants opt for quantity dispersal, the scattering of a large number of seeds with the chance that a diversity of consumers will drop some seeds in a favorable site. Such a strategy is typical of but not exclusive to plants of the temperate regions. There most fruit-eating birds and mammals are opportunistic consumers, rarely specializing in any one kind of fruit and not depending exclusively on fruit for their basic sustenance. The fruits are usually succulent and rich in sugars and organic acids and contain small seeds that pass through the digestive tract unharmed (Stiles 1980). Large numbers of small seeds are so dispersed, but few are deposited on suitable sites. The length of time such seeds remain within the digestive tracts of some small birds is no more than 30 minutes, so the distance dispersed depends on how far the birds go right after eating (Stiles 1980). Such dispersal is a lottery in the truest sense.

In temperate regions fruits ripen in early and late summer, when the young of the year are no longer dependent on highly proteinaceous food and both adults and young can turn their attention to a growing abundance of fruits. Fruits of late summer and fall ripen when migrant birds come through. They congregate in flowering dogwoods, spicebush, and wild grape to feed on high-quality fruits with nutrient-rich flesh (Figure 17.26). Such fruits do not last long, and their seeds are scattered widely. Fruits of lower quality, with less fats and sugars, hang on well into winter (Stiles 1980). They are available to birds over a longer period of time and are consumed when more palatable, short-lived succulent fruits are gone. In such a manner those plants avoid competition for dispersers in early fall.

The second approach to seed dispersal is to depend on a small number of birds and mammals that are exclusively consumers of fruit. Such plants are mostly tropical forest species, 50 to 75 percent of which produce fleshy fruits whose seeds are dispersed by animals. Rarely are frugivores obligates of the fruits on which they feed. Exceptions include the oilbirds and a large number of tropical fruit-eating bats. Among these, the flying foxes of the Old World eat the fruits in place and drop the seeds beneath the tree (Figure 17.27). The smaller spear-nosed fruit bats of the New World pluck the fruits and fly to a safe perch some distance away to avoid predators waiting for them in the fruiting trees.

Most trees attract frugivores that consume many different fruits. Dispersers of the seeds of one plant are also dispersers for others, for several reasons. Fruits vary widely in their nutritional value; by eating a variety of them, frugivores tend to balance their diets. Plants have few means available to restrict consumption of their fruits to a certain few frugivores. However, because the fruits of plants are of various sizes, shapes, colors, aromas, nutrient contents, and palatability, some are consumed chiefly by mammals and others by birds (for review see Howe 1986).

Like other areas of mutualism, the study of the relationships among plants and their seed dispersers is still embryonic. It lacks strong empirical approaches in spite of the relatively large literature available (see Howe 1986). Much needs to be learned. What makes the task more difficult is the fact that seed dispersal is less highly linked to animals than pollination. One reason is that, aside from ants, most seed dispersers are long-lived vertebrates that cannot depend exclusively on one species of a fruit-producing plant. Another is the random fluctuations in fruit production and animal populations over time. Most fruit-producing plants evolved to exploit interchangeable sets of animals for seed dispersion. Nevertheless, many questions need to be investigated. Are all dispersers of the seeds of a particular plant or set of plants equally effective? What happens if the more effective dispersers disappear? Where do most of the seeds go? How is dispersion influenced by the digestive morphology and physiology of the animals involved? Even though they are less specialized than pollinators, the loss of animal dispersers could result in the loss of associated plant species.

Origins of Mutualism

How did mutualism arise? Population ecologists speculate that mutualism may have evolved from predator-prey, parasite-host, or commensal relationships. Initially one member of the relationship increased the stability of a resource level for the second. In time, energy benefits accrued to the second member, and perhaps its activities began to improve the fitness of the first. For example, suppose a host tolerant of a parasitic infection exploits the relationship for the host's own benefit. In time the two exploit each other, as in plant-mycorrhizal mutualism. Selection then favors mutual interaction

FIGURE 17.26 The frugivorous cedar waxwing (*Bombycilla cedrorum*) feeds on the red berries of mountain ash (*Sorbus*).

FIGURE 17.27 Fruit-eating bats, like the epulated fruit bat (*Epomorphus wahlbergi*) of Africa and the little tent-building bat (*Uroderma bilobatum*) of the neotropical rain forest, are important seed dispersers of tropical trees.

to the point that the two become totally dependent on the other, as in obligate symbiotic mutualism. At the extreme the two function as one individual, as the lichens do.

Among nonsymbiotic mutualists, obligate or facultative, the relationship may have begun with exploitation. Birds or insects came to the plants to feed on pollen or nectar and frugivores to feed on fruits. In the process they accidentally carried pollen to similar plants or dispersed seeds away from the parent plant. Such plants experienced improved fitness and ultimately adapted to exploit the visitors as a means of dispersing pollen or seeds.

Such mutualistic relationships evolved indirectly. Pollinators came to the plant to collect pollen or tap nectar supplies for food, not to aid the plant in completing its life cycle. Fruit-eating animals visited plants for fruits, not to disperse seed. Neither did the mutualisms arise as one-to-one relationships. Plants were visited by an array of hungry insects, birds, and mammals. Except in rare situations, these animals did not, and still do not, specialize on one plant; thus groups of plants were visited by groups of various taxa of animals, which differed locally. The result was **diffuse coevolution.** Guilds of related or unrelated taxa evolved the capacity to use a range of plant resources, resulting in novel ecological associations and diffuse rather than paired coevolution.

A number of constraints favor diffuse coevolution. One is multiple relationships. A flower may be served best by one or two species of pollinators, but it is visited by many. An insect, such as a bee, may visit only one or two species of flowers, but that same flower will be visited by other insects as well. A plant may have its seeds dispersed most efficiently by one type of frugivore, but the frugivore will eat a variety of different fruits seasonally. Therefore it is unlikely that close evolution occurs between just one plant and one pollinator or seed disperser. A second constraint is asymmetrical evolution. For instance, woody plants are much more ancient and have evolved more slowly than any of their pollinators or seed dispersers (Herrera 1985). As one animal group disappeared from the scene, its place was taken by new groups that exploited the plants in a similar manner. Plants, however, retained any traits that encouraged pollination or fruit consumption. Further restricting tight coevolution are variations in the distribution and abundance of plants and animals, seasonal differences in flowering and fruiting, and seasonal changes in animal abundance and distribution.

Population Effects

Mutualism is most easily appreciated at the individual level. It is fairly easy to comprehend the interaction between ectomycorrhizal fungi and their oak or pine host, to observe and quantify the dispersal of pine seeds by squirrels and jays, and to measure the cost of dispersal to plants in terms of seeds consumed. Mutualism may improve the fitness of the pine and the consumption of pine seeds may improve the fitness of the seed predators; but what about the consequences at the level of the population?

The population consequences of mutualism are considerably more difficult to define than those of predation and parasitism, because the relationship is more nebulous and harder to model. Mutualism exists at the population level only if the population growth rate of species A increases with the increasing density of species B, and vice versa.

The question is most relevant to obligate and facultative nonsymbiotic mutualists. For obligate symbiotic mutualists the relationship is straightforward. Remove species A and the population of species B no longer exists. If ectomycorrhizal spores fail to infect the rootlets of young pine, they will not develop. If the young pine invading a nutrient-poor old field fails to acquire its mycorrhizal symbiont, it will not grow well.

For nonsymbionts, obligate or facultative, the effect on populations may be limited to the extent that one species benefits another and to that part of each other's life history cycle involved in the mutualistic relationship. Consider the yucca and yucca moth mutualism in which the yucca depends on the moth for pollination. Aker (1982) studied the relationship of *Yucca whipplei* and its moth *Tegeticula maculata*. Throughout the flowering season adult yucca moths were distributed evenly among the available flowers. The number of pollinators on the flower heads was directly proportional to the number of flowers available. In turn the number of fruits set on the plant was directly proportional to the number of flowers produced. If too few moths were available, many flowers would go unpollinated. Most flowers were not fertilized and the fruit production of some plants was limited by pollination. In general, the relative abundance of moths to flowers was low enough that most plants were pollinator-limited.

Individual yuccas regulate mature fruit production by aborting excess fruit. Fruit abortion affects the survival of the seed-consuming moth larvae and their emergence from the capsules at the end of the summer. Thus the size of the adult moth population the following year is determined by the number of plants that flowered and matured fruit. The dropping of immature fruit (along with the larvae) matches the moth larval abundance to the number of plants reproducing in the current year.

These observations suggest that the yucca and the yucca moth have reciprocal influences on each other's population; but whether the yucca moth limits or increases population recruitment and growth of yucca is not clear. The yucca moth may limit seed production (Addicott 1985), but recruitment of yucca seedlings may be influenced more by such extrinsic conditions as insufficient rainfall for seed germination, by seed predation, and by animal browsing on seedlings. The number of moths visiting the yuccas may be influenced by weather conditions and by the spatial distribution of the plants. The vigor of the yucca as evidenced by the size of the basal rosettes influences the total number of flowers produced (Aker 1982). In other words, the population growth and density of yuccas and moths may be influenced by situations unrelated to their strong mutualistic relationship.

A more definitive example of the population consequences of mutualism is provided by a demographic analysis of an ant-seed mutualism by Hanzawa, Beattie, and Culver (1988). It involves a guild of ants and golden corydalis (*Corydalis aurea*), an annual or biennial widely distributed in open or disturbed sites in the northeastern and western continental United States, Canada, and Alaska. The three researchers compared the survivorship of both seeds and plants, fecundity, reproduction, and growth rates of two seed cohorts of

the plant. They relocated one cohort to ant nests undisturbed by ant foragers and hand-planted a control cohort of equal numbers near each nest. The ant-handled cohort had significantly higher survivorship than the control (Figure 17.28). The ant-handled cohort produced 90 percent more seeds than the control cohort; its net reproductive rate R_o was 8.0 and that of the control 4.2. The finite rate of increase of the ant-handled cohort was 2.83 per year, compared with 2.05 per year for the control. The ant-handled cohort experienced greater reproductive success, not because of any great difference in the fecundity of the plants but because of a significantly higher survival to reproductive age. Its higher survival was due largely to dispersal and to the protected microsites of the ant nests, not to the removal of seeds from the vicinity of parent plants nor to the distance moved.

Theoretical ecologists have been attempting to model the population dynamics of mutualism. Mutualism, like competition, involves interactions between two populations, but the interactions are positive rather than negative. The general approach to modeling mutualism has been a modification of the terms of the Lotka-Volterra equations for competition (Chapter 14) in which the negative alphas of competition become positive. An increase in one population directly influ-

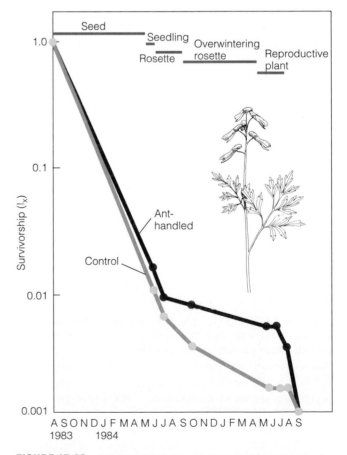

FIGURE 17.28   Survivorship curves for ant-handled (brown line) and control (yellow line) cohorts of golden corydalis (*Corydalis aurea*). The mutualism between ants and the plant results in a higher survivorship of overwintering rosettes and reproductive plants. (After Hanzawa et al. 1988:7.)

ences and is directly influenced by an increase in the other. Because each species benefits the other, the population growth equation of each must include a term for the rate of increase of the other. The maximum values of K do not enter into the relationship, and so K is changed to X (Pianka 1994):

$$\frac{dN_1}{dt} = r_1 N_1 \left(\frac{X_1 - N_1 + \alpha_{21} N_2}{X_1} \right)$$

$$\frac{dN_2}{dt} = r_2 N_2 \left(\frac{X_2 - N_2 + \alpha_{12} N_1}{X_2} \right)$$

where X_1 and X_2 are the equilibrium density of species 1 and species 2 in the absence of the other species, α_{12} is the beneficial effect of one individual of N_1 on N_2, and α_{21} is the beneficial effect of one individual of N_2 on N_1.

The pair of equations predicts that the population equilibrium density of each species is increased by the density of the other species. Such equations suggest that mutualism could produce a runaway positive feedback that would be destabilizing for both populations unless a strong negative feedback of density-dependent population regulation checked the positive feedback of mutualism.

Mutualistic relationships obviously are much more complex than the two basic equations describe. The equations fail to include any variable to stop unbounded growth. This weakness has led to the development of a number of more complex mathematical formulations, especially ones relative to stabilizing the population dynamics of the mutualistic species (see Vandermeer and Boucher 1979, Vandermeer 1980, Travis and Post 1979, Heithaus et al. 1980, Wolin and Lawlor 1984, Post et al. 1985).

To cap growth and to stabilize the system, a third species that is a competitor or a predator of one of the mutualists must be brought into the system, as often happens in real-life mutualisms involving seed dispersal. In a plant-pollinator system a certain population density of both interacting species is necessary before any equilibrium is possible and both populations reach maximum stable densities. If plant density is too low and pollinators have difficulty finding plants, the pollinators may decline below replacement level. Further, the environment may impose its own limits on population growth. The nature of many mutualistic systems—diffuse and involving arrays of species in very different taxa—makes it difficult to develop a realistic two-species model analogous to the predator-prey and interspecific competition models. Nevertheless, an understanding of the role of mutualism in population dynamics and community structure must rest on a stronger empirical foundation than we now have. The future of many species and communities depends on that understanding.

Summary

1. When predators and parasites seek prey and hosts, prey and hosts respond by escape or defense. It is possible, though, that when one party derives a benefit from its relations with a second party, this party, too, may benefit from the relationship. In all of these relationships, each individual of the interacting pair is imposing selection pressure on the other. These reciprocal selection pressures result in coevolution.

2. Many coevolved relationships involves symbiosis. Symbiosis is living together in a permanent, long-term association in which one organism lives within or on the other. The symbiosis is parasitism if one exploits its host, commensalism if the guest has no effect on its host, and mutualism if both guest and host benefit.

3. Parasitism is a situation in which two organisms live together, but one derives its nourishment at the expense of the other. A parasitic infection can result in disease, a state or condition of a plant or animal that deviates from normal well-being.

4. Parasites may be divided into microparasites and macroparasites. Microparasites include the viruses, bacteria, protozoa, and fungi. They are small in size, have a short generation time, multiply rapidly in the host, tend to produce immunity, and spread by direct transmission. They are usually associated with dense populations of the host. Macroparasites, relatively large in size, include parasitic worms, lice, ticks, fleas, fungi, and other forms. They may be ectoparasites, living outside the body of the host, or endoparasites, living inside the body of the host. Macroparasites have a comparatively long generation time, rarely multiply directly in the host, are persistent with continual reinfection, and spread by direct and indirect transmission.

5. Hosts are the habitat of parasites. The problem faced by parasites is to gain entrance into and escape from the host. The life cycles of parasites revolve about these two problems. The adult stages live in the definitive hosts, from which they escape by means of direct contact with other hosts or by means of vectors. Vectors are organisms that carry or transmit the parasite from one organism to another. Many vectors are hosts for some developmental or infective stage of the parasite. These vectors become intermediate hosts of the parasite, and transmission from definitive to intermediate and back to definitive hosts is considered indirect. Indirect transmission often involves the food chain.

6. Transmission of parasites, direct or indirect, is complicated by patchy or clumped distribution of the hosts. As a result parasites may become overdispersed, in which the greatest load of parasites is carried by relatively few individuals in the population, with most remaining free of infection.

7. Hosts respond to parasitic infection by biochemical responses, including an inflammatory process and immune reactions. Parasitic infections result in abnormal growth, sterility, and behavioral changes. There is evidence that parasitic infections may also influence mate selection in birds.

8. A high population of parasites within an individual can experience both intraspecific competition and immune responses from the host. These interact to increase mortality of parasites and reduce their fecundity. Because the death of a host does not benefit a parasite that depends on its host for both food and shelter, natural selection favors less virulent forms of the parasite that can live within or on the host without killing it. Both hosts and parasites develop a sort of mutual tolerance with a low-grade widespread infection.

9. Mathematical models of parasite and host relationships are important in studying the spread and effects of parasites and associated diseases. These models suggest the Lotka-Volterra equations describing predator-prey relationships, except that they are more complex. Depending on the parasite involved, these

models must incorporate such parameters as parasite density in the host, density of the host, proportion of parasites in the transmissive stage, rate of production of eggs and several life stages, and influence of parasite on host survival, among others.

10. Social parasitism involves brood parasitism and kleptoparasitism. In brood parasitism one organism parasitically depends on the social structure of another to rear its young. It may be temporary or permanent, facultative or obligatory, and interspecific or intraspecific. Brood parasitism occurs most commonly among ants, wasps, and birds.

11. Kleptoparasitism involves the forcible thievery of food by one individual from another. Although occurring among many taxa, kleptoparasitism is most prevalent among hawks and gulls, skuas, and wading birds. It is most common where potential hosts are aggregated in breeding colonies and feeding groups. Kleptoparasitism is energetically profitable for the parasite, but it adversely affects the foraging behavior and energy budgets of the host.

12. Mutualism is a positive reciprocal relationship between two species. Mutualism may be symbiotic or nonsymbiotic, obligatory or nonobligatory. Obligate symbiotic mutualists are physically dependent on each other, one usually living within the tissues of the other, such as occurs in lichens. Obligate nonsymbiotic mutualists depend on each other, but they lead independent lives, such as certain plants and their pollinators.

13. Nonobligate facultative mutualists are opportunistic and the relationship is mostly temporary. Nonobligate mutualisms usually involve interactions among guilds of species involved in seed dispersal and pollination. In exchange for dispersal of pollen and seeds, plants reward animals with food—fruit, nectar, and oil. To reduce wastage of pollen, some plants possess morphological structures that permit only certain animals to reach the nectar. To disperse their seeds some plants possess colorful, succulent fruits to attract frugivores that will disperse seeds through the gut. Other plants depend on seed predators that cache stores of seeds, some of which will go unretrieved and germinate.

14. Population effects of mutualism are difficult to model because so many mutualistic systems are diffuse, involving arrays of species in different taxa, rather than one-to-one relationships. Because an increase in the population of one species results in an increase in the population of the other, models must incorporate some variables to halt unbounded growth.

Review Questions

1. What is coevolution? How does the concept relate to mutualism? Predator-prey relationships? Parasite-host relationships?
2. What is parasitism? How does it relate to disease?

3. Characterize microparasites and macroparasites relative to the groups involved and types of immune responses each stimulates.
4. Distinguish between the definitive host and an intermediate host.
5. What is the relationship between (a) population density and (b) predation in the transmission of parasites?
6. From an evolutionary viewpoint, why should parasites and their hosts develop a mutual tolerance?
7. How would an approach to the control of a parasite that utilizes indirect transmission differ from that for one utilizing direct transmission?
8. What is kleptoparasitism? What is its effect on the host?
9. Robertson, Watson, and Cook (1992) commented that with the common eider (*Somateria mollissima*), intraspecific brood parasitism "is a well refined behavior and may be a viable means of obtaining reproductive output." Petrie and Møller (1991) suggest that a major consequence of intrabrood parasitism is depression of the average fitness of individuals in a population relative to a population without brood parasites. Discuss the apparent contradictions of these two ideas based on field observations.
10. What is mutualism? Symbiosis? Symbiosis is frequently equated with mutualism. What are the distinctions between the two?
11. Distinguish between nonobligate (facultative) and obligate mutualism; symbiotic and nonsymbiotic mutualism; defensive and indirect mutualism.
12. What are mycorrhizal mutualisms? What is the importance of these mutualisms? Why are they obligate?
13. In what ways are pollination and seed dispersal forms of facultative mutualism?
14. How would selective pressures differ between a plant visited by several species of bees and a plant visited by a single species of bee?
15. Why do fruit-eating animals fail to specialize on one species of fruit? What is an advantage of seed dispersal by the way of the animal gut?
16. Why is it difficult for coevolution to take place on a one-species-to-one-species basis? Are there any examples of one-to-one coevolution?

Cross References

Natural selection, 363; adaptation, 81; rumen metabolism, 121; function of temperate forests, 620; function of tropical forests 624; coral reefs, 683; interspecific competition models, 243; predator-prey interactions, 265; parasite-host interrelationships, 315; population growth, 184; sexual selection, 217; intraspecific competition, 195; predator-prey relations, 285–305; parasites and community structure, 394.

CHAPTER 18

Human Impacts on Populations

CONCEPTS

1. Successful exploitation of a natural population depends on sustained yield per unit time balanced with production per unit time.

2. Sustained yield is rarely achieved, because economic pressures override ecological considerations.

3. Habitat loss and fragmentation decimate flora and fauna.

4. Saving wildlife depends largely on habitat protection and restoration.

5. Alien species of plants and animals that invade an environment compete with and often displace native species.

6. Control of pests and weeds, largely alien species, involves reducing the populations below some economic threshold level.

The preceding chapters focused on the competitive and predatory relations among species and their influences on population growth and regulation. In these discussions, there was no indication of the effect of the dominant competitive and predatory species on Earth—*Homo sapiens*—on the populations of other organisms. Overexploitation of natural populations of plants and animals, especially when coupled with habitat loss and fragmentation, has resulted in a serious decline and local or global extinction of species. The accidental and intentional introduction of alien species has resulted in the competitive replacement of native species of plants and animals. The human homogenization and simplification of the environment has allowed the spread of pests and weeds; the economic and ecological damage they cause calls for the use of expensive and often environmentally damaging control measures. This chapter explores these interrelations between humans and other organisms and the impact of each upon the other.

POPULATION EXPLOITATION

A form of highly selective and intensive predation, often not related to the density of either predators or prey, is exploitation by humans. Humans have always exploited natural populations in their tenure on Earth, often intensively (Thomas 1956, Perlin 1991, Kay 1998). Some of this exploitation goes as far back as the Pleistocene, when it contributed to the extinction of large mammals of that era such as the mastodon. The deliberate overharvesting of buffalo, great auk, African ungulates, whales, and many pelagic fish are examples of short-sighted exploitation. On the other hand, populations of some species, such as white-tailed deer, are expanding. They are reaching pest status in many places, because of reduced human exploitation and the lack of natural predators, long since exterminated. Because of the high economic values of many natural populations, we have made attempts to exploit them wisely by maintaining an equilibrium between recruitment to the population and harvest, with various degrees of success and failure.

Sustained Yield

Although some of the terms used in defining exploitation of populations are similar to those used in productivity, the meanings are somewhat different. When fishery and wildlife biologists speak of yield, they refer to the individuals or biomass removed when the population is harvested. **Biomass yield** is the product of the number harvested times the average weight. (Yield may indicate weight without numbers, or vice versa.) The **standing crop** is the biomass present in a population at the time it is measured. Productivity is the difference between the biomass left in the population after harvesting at time t and the biomass present in the population just before harvesting at some subsequent time $t + 1$.

The objective of regulated exploitation of a population is sustained yield: the yield per unit time being equal to productivity per unit time. In its simplest form sustained yield is described by an equation first proposed by E. S. Russell for fishery exploitation. Although the equation was specifically developed for fisheries, it is applicable to any exploitable population. With some minor modifications this equation is

$$B_{t+1} = B_t + [A_{br} + G_{bi} - (C_{bf} + M_b)]$$

where B_{t+1} is total biomass of exploitable stock just before harvesting, at time $t + 1$; B_t is total biomass of exploitable stock just after the last harvest, at time t; A_{br} is biomass gained by the younger recruits just grown to exploitable stock; G_{bi} is biomass added by the growth of individuals in both B_t and A_{br}; C_{bf} is any biomass exploitatively removed during the harvest period; and M_b is biomass lost from exploitable stock by natural causes between times t and $t + 1$. The equation explicitly includes the productivity due to individuals that were born and individuals that died during the time interval from the end of one harvest period to the beginning of the next.

The equation is highly simplified. In an unexploited fish population A_{br}, C_{bf}, and M_b are interdependent. For example, in a stable environment largely undisturbed by humans, fish populations appear to be dominated by large species. In turn, each species population appears to be dominated by large old fish. When humans start to exploit such a population, significant changes take place. To compensate for exploitation directed first toward the largest members of the population (organisms that under natural conditions are normally secure from predation), the population exhibits an increased growth rate, a reduced age of sexual maturity, increased number of eggs per unit of body weight, and reduced mortality of small members of the population (Regier and Loftus 1972). Populations of other vertebrates react in a similar way.

Sustained yield clearly depends on the rate of increase. Sustained yield does not imply holding a population at ecological carrying capacity (K), for at that level the rate of increase equals zero. A population stable in the absence of harvesting can be harvested under sustained yield only after the plant-herbivore or the herbivore-carnivore system has been manipulated to raise r, the rate of increase. This change can be made in two ways: (1) improve food and cover to increase the carrying capacity by increasing available resources, fecundity, and survival; or (2) lower the density by removing a certain number and then stabilize the population at some lower density (Figure 18.1). Within limitations, the lower the population density falls beneath the carrying capacity, the higher its rate of increase. Thus a higher rate of harvest is needed to hold the reproductive population stable at some desired lower density.

The idea is to have the rate of harvest H equal to the rate of increase r. The rate of harvest should hold the rate of growth at zero. H would have to equal the rate at which the population would increase if the harvest were stopped (Caughley 1976).

Consider a deer population increasing at a rate of 20 percent a year. It has a finite rate of increase, $e^r = 1.20$ and $r = 0.182$. To hold the population stable, the herd must be harvested at the instantaneous rate of $H = 0.182$. If the population is harvested only during a certain season of the year, as is usual with deer, then the sustained yield is calculated from an isolated rate of harvest h, defined as $h = 1 - e - H$,

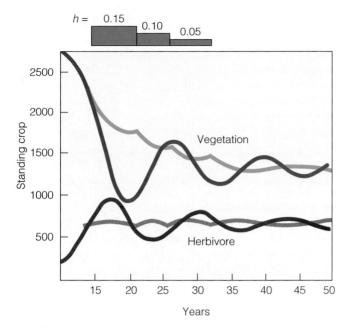

FIGURE 18.1 Effect of harvesting on a plant-herbivore system. Superimposed on the interaction diagram in Figure 16.8 is the path of standing crops of vegetation and herbivores after herbivores have been harvested over the intervals of time and at the rates per year designated in the rectangles above. As a result of management, fluctuations are reduced and the system moves to equilibrium. Such equilibrium results if harvest is initiated at a rate of about one-half of the population's intrinsic rate of increase when the animal population is well below the peak. This rate of harvest must be maintained until plant density levels off and begins to increase. (After Caughley 1976a.)

high wastage of production. To manage a population influenced by density-independent variables, such as climate or temperature, the objective is to reduce wastage by increasing the rate of exploitation. The role of harvesting is to take all individuals that otherwise would be lost to natural mortality. This type of exploitation is described by the expression

$$\text{maximum yield} = B_t - \min(R_t)$$

where B_t is biomass at time t and $\min(R_t)$ is the minimum number of reproducing individuals left at time t that will ensure replacement of maximum yield at time $t + 1$ (Watt 1968).

Such a population is often difficult to manage because the stock can be severely depleted unless there is repeated reproduction. An example is the Pacific sardine (Murphy 1966, 1967), a species in which there is little relationship between breeding stock and the subsequent number of progeny produced. Exploitation of the Pacific sardine population in the 1940s and 1950s shifted the age structure of the population to younger age classes. Prior to exploitation, 77 percent of the reproduction was distributed among the first five years. In the fished population, 77 percent of the reproduction occurred in the first two years of life. The population approached that of single-stage reproduction subject to pronounced oscillations (Figure 18.2). Two consecutive years of reproductive failures resulted in a collapse of the population, from which it has yet to recover.

in this case 0.167. This isolated rate of harvest would have taken into account natural mortality to the population occurring between the period of birth and the time of harvest. Assuming a deer population after the fawning season of 1000 animals and allowing a natural finite rate of mortality of 0.25 per year, we could remove 151 animals in the fifth month. This action would allow the number at the next fawning season to climb back to 1000. The addition of 375 young would compensate for hunting and natural mortality (for details on carrying out calculations, see Caughley 1976).

Although often considered as such, sustained yield is not a particular value for a given population. There may be a number of sustained yield values corresponding to different population levels and different management techniques. The level of sustained yield at which the population declines if exceeded is known as **maximum sustained yield.** Maximum sustained yield is not always the most efficient harvest level, because of other considerations such as species interactions, esthetics, land use problems, and the like. Harvesting may be aimed at the **optimum sustained yield,** the level of sustained yield determined by consideration of these other factors as well as maximum sustained yield.

The higher the r of a species, the higher the rate of harvest that yields the maximum biomass production. Species characterized by scramble competition (r-strategists) have a

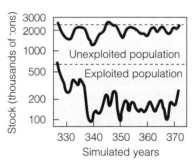

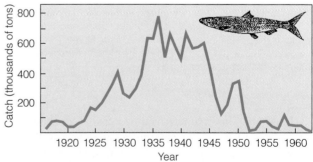

FIGURE 18.2 (a) Simulation of an exploited and an unexploited population of sardines, both subject to random environmental variation in reproductive success. The dashed lines indicate population size at *K*. Note how exploitation adds instability and how dangerously low the population can get. Compare this simulation with (b), the annual catch of the Pacific sardine (*Sardinops sagax*) along the Pacific coast of North America. Overfishing, environmental changes, and an increase in a competing fish, the anchovy (*Anchova* spp.), made the population collapse.

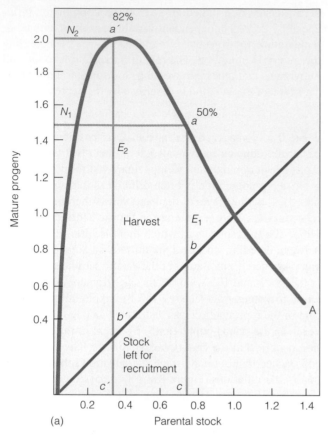

(a)

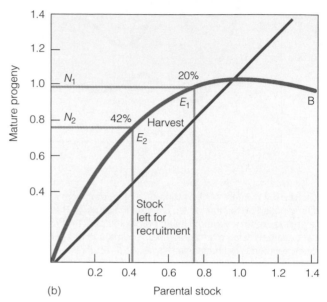

(b)

FIGURE 18.3 Reproduction curves illustrating rates of exploitation. Reproduction curves diagram the relationship between recruitment or net reproduction, a function of the number of mature progeny, and the density of parental stock. The 45° line represents the replacement level of the stock, along points where density dependence is absent. The dome-shaped curve is the plot of actual recruitment in relation to size of parental stock. The apex of the curve, lying above and to the left of the diagonal line, represents maximum replacement reproduction. The curve must cut the diagonal line at least once and usually only once. Where the curve and diagonal line intersect is the point at which parents are producing just enough progeny to replace current losses from reproductive units. (a) The reproduction curve typical of r-strategists suggests that low parental stock can be very productive. On curve A, a vertical line ac cuts the 45° line at b. Segment ab (E_1) is harvest; bc is stock left for recruitment. Line ac represents the point on the curve at which 50 percent of the mature population is harvested each period. The line $a'c'$ is the 82 percent point, the maximum surplus reproduction and the maximum rate of exploitation possible for this population. (b) Curve B is typical of K-strategists, among which a low density of parental stock is not very productive. The two points represent levels at which 20 and 42 percent, the maximum, can be harvested.

In curve A, $E_2 > E_1$ and $N_2 > N_1$. Under these conditions, the greater the standing crop, the greater is the sustained yield. In curve B, $E_2 > E_1$ yet $N_1 > N_2$; a high standing crop does not result in a greater sustained yield. A knowledge of parent-progeny relations is essential for exploitation of natural animal populations.

In populations characterized by density-dependent regulation (K-strategists), the maximum rate of harvest depends on age structure, frequency of harvest, number left behind after harvest, fluctuations in environment, and variations in fecundity. It also depends on density of the population to be harvested and the rate of harvest needed to stabilize the density at that level.

This type of harvest is described by the expression

$$P_b = \max B_{t+1} - B_t(X)$$

where P_b is biomass productivity from t to $t + 1$, B_t biomass at time t, B_{t+1} is biomass at time $t + 1$, and X represents the several variables that influence biomass production over time t to $t + 1$.

This relation is illustrated in Figure 18.3. For any position of the stock to the left of the 45° line, there is a rate of exploitation that will maintain the stock at that position. Maximum sustained yield does not necessarily require a large standing crop. Let a be any position on the curve and c a perpendicular line that cuts the 45° line at b. At equilibrium the portion bc of the recruitment must be used for the maintenance of the stock, for $bc = ac$; ab can be harvested. There is, however, a limit to exploitation, a limit that is influenced by the inflection point of the curve. For curve A, the maximum rate of harvest is about 82 percent; for curve B, 42 percent. In these curves the size of the reproductive stock that will give maximum sustained yield will not be greater than half of the replacement of the reproductive population. The greater the area of the reproduction curve above the 45° line, the greater the optimum rate of reproduction.

In summary, Caughley (1976) gives six points applicable to harvesting of populations:

1. A population stable in numbers must be reduced below a steady density to obtain a croppable surplus.

2. There is an appropriate sustained yield for each density to which a population is reduced.

3. For each level of sustained yield, there are two levels of density from which this sustained yield can be harvested (Figure 18.4).

4. Maximum sustained yield can be harvested at only one density (Figure 18.4).

5. If a constant number is harvested from a population each year, the population will decline from steady density and stabilize at the upper population size for which that number is the sustained yield. If this number exceeds the maximum sustained yield, the population declines to extinction.

6. If a constant percentage of the population is harvested each year (the percentage applying to the standing crop of that year), the population will decline and stabilize at a level at equilibrium with the rate of harvesting. This level may be above or below that generating maximum sustained yield.

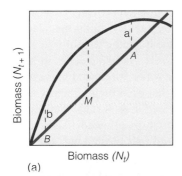
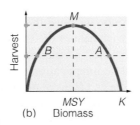

FIGURE 18.4 (a) A sustained yield model for *K*-strategists harvested under three regimes. The 45° line represents the replacement level of the population. Where it intercepts the curve, reproduction balances losses. In the first regime, the population is harvested down from a steady state to size $N_t = A$. The dashed line *a* represents the number that could be harvested each year to hold the population stable at *A*. In the second regime, the population is reduced to $N_t = B$. A number represented by the dashed line *b* could be harvested each year to hold the population stable at *B*. The yield in this case is a large proportion of a small population. Maximum sustained yield is at *M*, where the diagonal line and the curve have maximum separation. (b) A parabolic recruitment curve illustrates the concept in a different fashion. Maximum sustained yield is approximately *K*/2, represented at *M*. At *A* the equilibrium is stable at high density, much above MSY. *B* is an unstable equilibrium point, much below MSY, vulnerable to chance extinction.

Problems with Sustained Yield Management

Although sustained yield management (SYM), particularly multiple sustained yield (MSY), looks good on paper, it depends too heavily on the logistic equation. The problem is SYM views management as a numbers game. The number or biomass removed theoretically is replaced by an equal number of new recruits to the harvestable component. Under SYM the "surplus," based on density compensation, supposedly is reduced; future production is optimized by optimizing harvest. Thus SYM assumes, as the models presented indicate, that a relationship exists between stock and recruitment. No proof of this theory exists. The usual approach to SYM fails to consider adequately the following: size and age classes, differential rates of growth among them, sex ratio, survival, reproduction, and environmental uncertainties; all these data are difficult to obtain. To attempt MSY without such information is balancing the population on the edge of catastrophe. Yet fisheries proceed with inadequate and often erroneous data. The prediction of stocks and harvest is further compromised by species interaction. Fishery managers treat stocks as separate units rather than as part of a complete system. Adding to the problem is the common-property nature of the resource—because it belongs to no one, it belongs to everyone to use as each sees fit.

Several approaches to management of exploitable population are in current use. One is the **fixed quota,** in which a certain percentage is removed each harvest period based on MSY estimates. Harvesting supposedly matches recruitment. Often used in fisheries, such an approach is risky. Again lacking data on the size of the stock and recruitment, a fixed quota can drive a population to commercial, if not actual, extinction. Because populations fluctuate from harvest period to harvest period, the MSY will vary from year to year. If such fluctuations are not taken into account, there will be times of overharvest. Combined with environmental changes, overharvest has been responsible for the demise of some fisheries such as the Pacific sardine.

A second approach is the **dynamic pool model.** This is essentially the one described in the opening part of this section. It assumes a constant natural mortality rate that is independent of density and is same for all age classes. Growth rates are age-specific but unrelated to density. Animals removed replace those lost via density-related natural mortality, and no more. Many fishery biologists, for example, believe that density-related mortality is concentrated in the early life stages of the fish.

The flaws in such an approach should be obvious based on previous discussions of population dynamics. Fishing mortality can be additive to natural mortality; growth rates are affected by population density, as is recruitment to the population. In practice the dynamic pool model translates fishing mortality into fishing effort, based on type of equipment such as size-selective gill-nets that sort out age (size) classes, efficiency of equipment, and seasonal nature of exploitation. Few dynamic pool models have been developed, let alone put into practice. A general weakness of the model is the inability to estimate natural mortality accurately.

A third approach is **harvest effort,** often used in establishing seasons for sport hunting and fishing. The number of animals killed is manipulated by controlling hunting effort: the number of hunters in the field, the number of days of hunting (season length), and the size of the bag limit. To reduce the kill, hunting effort is decreased by reducing the allowable kill, shortening the season, or closing the season entirely. The reverse approach is used to increase the kill. In general, such a rule-of-thumb approach has been more successful in managing exploitable populations than the fixed-quota approach.

The permit system is a special variant of the harvest-effort approach used to control more tightly the number of animals killed. It is based on the maximum number of animals to be removed from any one defined area and the amount of hunting effort needed to achieve that goal. For example, it is well established that approximately 15 percent of deer hunters are successful in any one hunting season. To achieve a desired level of kill, the number of deer to be removed from any one area is used as the expansion ratio to set the number of permits issued. For example, if the desired harvest level is 100 animals, then approximately 700 permits should be issued. A modified form of the permit system is used in allotting the take of fish by commercial fishermen.

The three models above are biological, based on logistic growth, and have a major flaw. They fail to incorporate the most important component of population exploitation, economics (Figure 18.5) (see Hall et al. 1986, Walters 1986). Once exploitation of a natural resource becomes a commercial enterprise, the pressure builds to increase that exploitation to maintain the underlying economic infrastructure built on the exploitable population. Once in place, any attempts to reduce the rate of exploitation meets strong opposition, supported by arguments that reduction will result in unemployment and industrial bankruptcy. This, of course, is a short-term argument, because in the long run the resource will be depleted and the economic collapse will occur. That

fact is written across the country in abandoned fishery processing plants, rusting fishing fleets, and deserted logging towns. With conservative exploitation on lower economic and biological scales, the resource could still be exploited. Because of the failure of harvest regulations, nationally and internationally, and the common nature of the resource, exploitation efforts increase, even in the face of declining stocks. Instead of a reduced harvest effort, it is increased by technological improvements at finding and harvesting the remaining resource to a point that it collapses. That is the story of the whaling industry that led to the sharp decline and the near extinction of various species of whales (Figure 18.6).

Fisheries follow the same pattern, as exemplified by the collapse of the North Atlantic cod (*Gadus morhira*). Since the 1600s the North Atlantic cod fishery was an important source of food for western Europe in the form of salted cod. Settlers of eastern Canada and New England found a rich lode of cod off the coasts of New England, Nova Scotia, and Newfoundland. Early exploitation of the resource by fishermen using small ships gave way to larger schooners that enabled them to reach the fisheries of the Grand Banks. Improved equipment, such as the introduction of long trawl lines with baited hooks, increased the catches enormously. The belief even in the scientific community at that time was that fishing stocks could not be depleted and that natural checks would occur before overfishing took place. As the North Sea stocks became depleted, the fishing industry moved to the North Atlantic.

Two factors spelled the end of the North Atlantic cod fisheries. One was the development by Clarence Birdseye in 1924 of several methods of freezing cod fillets and sticks that could be shipped to the market as fresh. In face of a declining market for salt cod, the availability of frozen fresh cod stimulated a rapidly expanding demand for this fish (Kurlansky 1997). Large corporate fishing fleets, many government subsidized and dominated by factory ships that caught, processed, and froze fish without having to return to shore, displaced smaller fishing vessels. As fish catches in North Atlantic declined the answer was not reduced harvest and management of the resource, but even bigger factory ships. Efficiency of equipment increased. Use of sonar, helicopters, and communication between ships allowed fishermen to locate and focus in on an area of fish, clean it out, and move on to another. Large powerful ships allowed the use of miles-long drift nets and huge otter trawls specially equipped with chains to stir up the bottom and drive fish into nets. Few fish escape, and the bottom is virtually cleared of all benthic invertebrate life. As a result the fishing industry in New England and eastern Canada has virtually collapsed (Figure 18.7). Causes were threefold: (1) an overestimation of abundance and underestimation of fishery mortality; (2) ability to find and catch fish at low levels of abundance, an increased effort related to overcapacity in the fleet, and economic incentive to maintain a high catch; (3) increased discarding and nonreporting of small fish as population declined and fishing mortality increased (Myers et al. 1997).

As the major species declined, especially cod and haddock, the fishing industry turned mainly to smaller bony oily fish once

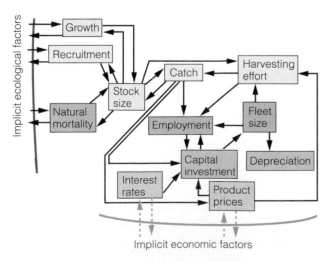

FIGURE 18.5 The relationship between sustained yield models and economic models of harvesting. Sustained yield models ignore powerful economic factors.

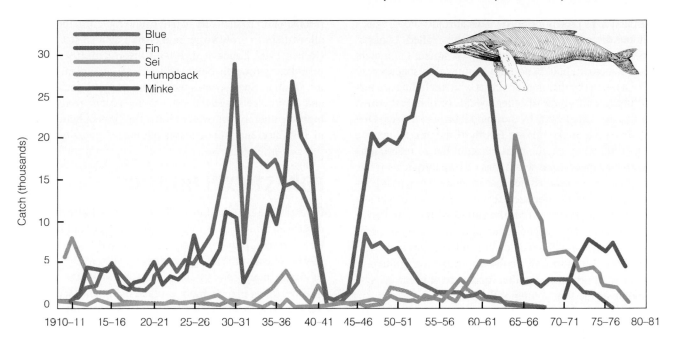

FIGURE 18.6   Catches of whales in the Southern Hemisphere, 1910–1977, the focal point of whaling after stocks in the Northern Hemisphere had been depleted. Note the virtual cessation of whaling during World War II years, 1941–1945. The precipitous decline in blue whales (*Balaenoptera musculus*) began before 1940. After World War II the fin whale (*Balaenoptera physalus*) bore the heaviest exploitation, but for a while blue whale take increased. The increase in harvesting intensity on the blue whale points out a truism in resource exploitation. A high harvesting effort on a declining stock can continue if some alternative resource is abundant enough to support that effort. The stock of fin whale supported the incidental harvesting of blue whales. Then the story was repeated with the fin whale. As that stock declined, whalers turned to the smaller sei and minke whales in an effort to maintain their investment in boats and equipment. Finally the whaling industry collapsed, but not until whales were close to extinction. (After Allen 1980.)

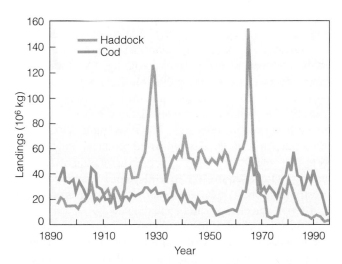

FIGURE 18.7   Landings of Atlantic cod (*Gadus morhua*) and haddock (*Melanogrammus aeglefinus*) from Georges Bank for the period 1893–1996. Note the increase in fishing intensity for both species in 1960 through 1980, followed by collapse. (Fogarty & Murawski.)

considered as trash and stock and species higher up on the food chain such as sharks and swordfish—with disastrous outcomes. Size of swordfish, taken on lines 50 km long baited with thousands of hooks, has dropped from 120 to 30 kg in the past 20

years. The breeding population of swordfish and sharks has been reduced by one-half off the southeastern coast of the United States. Modern vessels, tracking systems, and fishing nets strung 1 km below the surface have decimated orange roughy, a popular market fish found off New Zealand. Catches have plummeted 70 percent in the 6 years, with little hope of recovery, because the fish requires 25 years to reach breeding age (McGinn 1999).

Overexploited populations exhibit certain easily discernible symptoms (Watt 1968). Up to a certain rate of exploitation, the stock can replace itself. Beyond this critical point, certain changes point to impending disaster. Exploiters experience decreased catch per unit effort as well as a decreasing catch of one species relative to the catch of related species. There is a decreasing proportion of females pregnant, due both to sparse populations and to a high proportion of young nonreproducing animals. The species fails to increase its numbers rapidly after harvest. A change in productivity relative to age and age-specific survival shows that the ability of the population to replace harvested individuals has been impaired.

Another problem is that traditional population management, especially by fisheries, considers stocks of individual species as single biological units rather than components of a larger ecological system (Fujita et al. 1998). Each stock is managed to bring in a maximum economic return, overlooking the need to leave behind a certain portion to continue its ecological role as predator or prey. This attitude encourages a tremendous discard problem, euphemistically called "bypass." Employing large drift nets that encompass square kilometers of ocean, fishermen haul in not only commercial species they seek, but a range of other marine life as well, including sea turtles, dolphins, and scores of other species of fish. Fishermen bypass this unwanted catch by dumping it back to the sea. Discarded fish alone make up one-fourth of the annual marine catch. In 1995 in the Pacific Northwest, of the 27 metric tons of fish taken, 9 metric tons was bypass. The ecological effects of bypass can be enormous. Because much of the bypass consists of juvenile and undersized fish of commercial species, the practice can serious affect the future of those fisheries (Bricklayer et al. 1989). The removal or reduction of other species can interfere with predator-prey interactions in the sea, the dynamics of interspecific competition, and intraguild predation. Such disturbances can alter food webs of ocean ecosystems and upset the functioning of the pelagic ecosystem.

A similar economic attitude also prevails in the management of some game species. In too many instances, biologists have emphasized the increase of recreational opportunities for hunters over the welfare of the species involved. For example, rather than restrict human exploitation on moose populations, the approach is to kill wolves to reduce natural predatory loss (see Gasaway et al. 1983).

The reluctance to reduce seasons or to more tightly restrict bag limits relates in part to the economics of hunting. Most wildlife programs, directly or indirectly, depend on hunting license revenue. Loss of such revenue reduces income to wildlife agencies. This reduction has a two-edged effect. Loss of revenue from reduced hunting license sales may reduce pressure on some hunted species, but it also reduces money to support wildlife habitat restoration and acquisition, which benefits all species, including endangered ones, and other programs essential to wildlife welfare.

Management of exploitable populations depends too much on crisis management. No steps are taken to rescue a species until its population has fallen so low that the species becomes endangered. Then it is protected and expensive recovery programs initiated with the hope that populations will recover. The simple solution in the past was to manage the population more judiciously.

Reserves

The depleted fish stocks and the impact of dredging and trawling on the seas have stimulated international interest in the establishment of marine reserves. This concept goes back to the early days of game restoration in the United States (Leopold 1933, Trippensee 1948). The idea behind the early refuges, which were closed to hunting, was that the excess population would flow out and restock and enhance the productivity or abundance of game on the surrounding area. The success of such refuges depended in part on the enforcement of no hunting within the refuges.

The same concept lies behind the proposed marine reserves or refuges. Such reserves would maintain a sufficient reserve of reproductively active fish to replenish the stock and allow them to grow larger and ensure against stock collapse (Robert 1997, Lauck et al. 1998, Allison et al. 1998). Currents would carry progeny outside the reserves to replenish the fishing grounds. Such refuges would protect multispecies fisheries and *K*-type species such as sharks and orange roughy that are highly vulnerable to overexploitation, protect habitat, restore biodiversity, and reduce losses of genetic diversity.

FOREST EXPLOITATION

Forests have not fared much better than wildlife. They were destroyed rapidly over the centuries to clear land for agriculture and to supply building materials and firewood (Perlin 1991). Worsening shortages of wood stimulated some form of forest management in Great Britain in the 1600s. In Europe, forest management evolved in France, Germany, and Switzerland in the 1800s. Gifford Pinchot introduced the practices of forest management in 1892 on the Biltmore Estate in North Carolina. Because the forests seemed endless, the timber industry had little interest in forest management until exploitative logging, land clearing, and fires had decimated forests (Williams 1989). The same fate was befalling western forests until the federal government set much of the acreage aside, some in parks and the most in national forests. In spite of federal ownership, western national forests are being heavily clear-cut (Figure 18.8).

The goal of sustained yield in forestry is to achieve a balance between net growth and harvest. To accomplish this goal, the mature or unexploited forest must be cut in some fashion to stimulate regeneration, much in the same manner as an unexploited fish or wildlife population must be reduced to increase *r*, the rate of increase. Commercial foresters, who view forests as a commodity rather than a biological entity, use this argument to justify the clear-cutting of remaining old-growth forests (see Chapter 29). Management of forests, however, proceeds in a time frame different from the management of exploitable animal populations, because regenerating a forest requires decades. The time frame for the next harvest depends on the desired wood products. Pulp wood, posts, and poles require only a short-term rotation of 20 to 40 years; sawlogs require 65 to 100 years. Sustained yield forestry works best on large acreages where blocks of timber of different age classes can be maintained.

Sustained yield in forestry is much more sophisticated than in wildlife and fisheries, simply because it is easier to inventory, measure growth and potential yields of biomass, and manipulate populations or stands of trees. To achieve this end, foresters have an array of silvicultural and harvesting techniques from clear-cutting to selection cutting, with many variations in between. In recent years, research foresters and forest ecologists have been giving considerable attention to nutrient cycling, nutrient conservation, and biological diversity in managed forests. However, practicing foresters rarely incorporate these research findings in their management and harvesting plans.

FIGURE 18.8 A clear-cut in a Douglas fir stand in western Washington.

The concept of sustained yield is ingrained in forestry and is practiced to some degree by large timber companies and federal and state forestry agencies. But industrial forestry's approach to sustained yield is to grow trees as a crop rather than maintaining a forest ecosystem. Their management approach is a cornfield approach: clear-cut, spray herbicides, plant or seed the site to one species, and in the case of paper pulp, harvest in 20 to 40 years, clear-cut, and plant again. Clear-cutting practices in some national forests, especially in the Pacific Northwest and the Tongas National Forest in Alaska, hardly qualify as sustained yield management when below-cost timber sales are mandated by the government to meet politically determined harvesting quotas. Even more extensive clear-cutting of forests is taking place in the northern forests of Canada, especially British Columbia, and in large areas of Siberia. As the timber supply dwindles in the Pacific Northwest, the timber industry that moved west following the depletion of eastern hardwoods and Lake States pinery is moving east again to the exploit the regrown eastern hardwood forests, especially the rich and diverse central hardwoods forest. From Virginia and eastern Tennessee to Arkansas and Alabama, timber companies have built more than 140 highly automated chip mills that cut up trees of all sizes into chips for paper pulp and particle board (Louma 1997). Feeding the mills requires clear-cutting 500,000 hectares annually, producing ¾ million tons of chips. The growing demand for timber has boosted timber prices, stimulating more clear-cutting. The rate of harvest is wholly unsustainable. In face of growing timber demands, sustained yield management has hardly filtered down to smaller parcels of private land. Poor cutting practices with little concern for regeneration and species composition leave behind forests of poor quality and understocked stands.

The problem of sustained yield forestry, like that of sustained yield fisheries, is its economic focus on the resource with little concern for the forest as a biological community. Once an old-growth forest is cut, that particular ecosystem will not return. A carefully managed stand of trees, often reduced to one or two species, is not a forest in an ecological sense. Rarely will a naturally regenerated forest, and certainly not a planted one, ever reach an old-growth stage. By the time the trees reach economic or financial maturity—based on the type of rotation—they are cut again. Economic maturity is not the same as ecological maturity. For this reason, prospects of future old-growth forests are becoming as extinct as the passenger pigeon that once depended on them.

HABITAT LOSS AND FRAGMENTATION

Once I was talking to a group of grade-school children who were concerned how the removal of a small stand of trees and shrubs that stood in the way of school expansion would affect the chipmunks living there. I told them the truth. With their habitat gone, the chipmunks would disappear. They would leave the area seeking a new place to live and in the process they would probably die. The children had their first short lesson on the impact of human destruction of wildlife habitat on a very local and personal scale.

Few areas escape urban, suburban, and industrial expansion. Highways, housing construction, and industrial development continually eat away at forest, farmland, deserts, and grassland. As humans overtake the land, native inhabitants—plants, wildlife of all kinds, and even soil organisms—are forced into continuously shrinking parcels of habitat, separated from one another by a sea of development. As habitat shrinks, so do populations of wild things.

Habitat fragmentation affects species at both geographical and local scales (see Chapter 23). As human populations and the accompanying demand for land increase, they intrude into and compress and fragment the natural range of species. Species with large geographical ranges can withstand a greater degree of habitat loss and fragmentation than can those with small geographical ranges (Maurer and Nott 1998). Such species have the greatest proportion of local populations living within the center of their range. Species with small or restricted geographic ranges, especially endemic and ecologically specialized species, tend to have a higher proportion of local populations living near the periphery, where environmental conditions are more extreme. Once fragmented, these peripheral populations face a higher risk of going extinct and their range shrinks even further. This is the story of the prairie chicken (*Tympanuchus cupido*), the golden lion tamarin (*Leontopithecus rosalia*) of the eastern Atlantic Forest of Brazil, and numerous other species.

For migratory species, habitat loss and fragmentation present a major problem across their total range. Neotropical warblers experience increasing fragmentation of habitat by development and logging in their breeding range in North America, and they find their wintering habitat either destroyed or highly fragmented by logging, coffee plantations, fire, and expansive destruction of tropical forests for agriculture and settlement (Terborgh 1989). Between the two, migrant birds find coastal stopover areas taken over by development and industrialization; and on way the birds are faced with a maze of transmission towers that cause heavy mortality.

Habitat destruction and fragmentation are not confined to terrestrial systems. Aquatic habitats also are experiencing similar losses. Drainage of wetlands for agriculture and development eliminate habitat for organisms dependent on aquatic habitats. Logging, highway construction, development, and pollution destroy habitat for fish and aquatic invertebrates and plants. Dams isolate and fragment flowing water habitats (see Chapter 30), block movement of fish, impose lakelike conditions on free-flowing streams and rivers, alter water temperatures, and increase sediment loads (see Chapter 32).

Such habitat destruction is a major cause of the massive decline in fauna and flora. Two-thirds of the migratory birds in North America winter in the Caribbean and Latin America, and they have been experiencing a 4 percent decline annually.

Two out of every three bird species worldwide are declining and 11 percent are threatened with extinction (Tuxtill 1998). Twenty-five percent of all of the 4400 species of mammals are declining because of habitat loss; 11 percent are endangered. Of 26 species of threatened darters, small, colorful freshwater fishes, 60 percent are declining because of habitat alteration. One in eight plant species worldwide faces the same prospects.

Refuges: Protection and Management

Increasing threats to the future of wild species have prompted the establishment of refuges or reserves to maintain the viability of species populations. Early on, ecologists debated vigorously over the merits of large reserves versus several small ones. The debate, of course was largely academic because the size of any reserve established will be and has been influenced more by economics, politics, and geography. Nevertheless, differences exist in the utility of large and small refuges. Large reserves provide the greatest protection and support the largest populations, although even the largest reserves may be too small for certain large carnivores and herbivores. Small reserves, especially if associated with large reserves, have their advantages. They shelter portions of the population from environmental catastrophe that could threaten the populations confined to a large reserve and in total may support a higher biodiversity than one large reserve alone.

Although isolated and surrounded by other types of land use, small reserves do hold populations of species that represent different samples of the gene pool (see Chapter 19). Subject to somewhat different selection pressures and to some degree of genetic drift (see Chapter 19), these smaller populations help maintain genetic diversity in the species population. Reserves serve both to protect the species and to provide a surplus for colonizing surrounding areas, if they can move and habitat exists. Often the surplus has no available habitat to colonize. Elephants restricted to parks and reserves in Asia and Africa and the bison of Yellowstone National Park overpopulate the available habitat and move outside park boundaries, where they conflict with human interests. The reintroduction of the wolf to Yellowstone ecosystem faced opposition from those who fear the wolf populations may expand to areas outside the park (Smith et al. 1999). The survival of such species depends on management at varying levels of intensity to maintain their populations and ensure survival of their habitats.

Small populations, especially in reserves, are subject to many problems and dangers. Concentrated in small pockets, many species are more easily reduced by poaching or predation. Isolated from others of the same species, the populations may be too small to maintain the social cohesiveness necessary for successful mating and reproduction.

Small populations cannot survive unless both they and remaining available habitat are protected against such intrusions as development, agricultural land clearing, and logging. Some of our most endangered species occupy natural reserves, many of which are too small to support a viable population of the species. One approach is to increase the size of the reserve to provide habitat for an expanding population or to embrace the entire ecosystem. Most reserves do not. For example, the Yellowstone Park is only part of the Yellowstone ecosystem, which includes surrounding national forests subject to logging and grazing (Clark et al. 1999). The bison herd in the park has expanded beyond the capability of the park itself to support them, especially in winter. Part of the bison problem could be solved if public national forest lands open to grazing by domestic livestock were allotted to buffalo instead.

Human activities just outside park boundaries—logging, grazing, agriculture, housing, and recreational development—violate the integrity of many reserves. Many reserves are only a part of the total surrounding ecosystem. To protect the reserves, concentric buffer areas of land use should encircle the core area. Land use should be minimal about the immediate boundary, with land use activity increasing proportionally with distance.

Excessive human activity within their boundaries compromises many otherwise excellent parks and reserves. Intensive recreational development, trails, and human incursions into the back country that bring people in direct contact with large species of wildlife decrease the value of the reserve and actually further endanger the species. The case of the grizzly bear (*Ursus arctos*) in national parks is a good example. A certain segment of society would prefer to see the grizzly bear exterminated to make back country travel safe for humans.

Corridors of habitat interconnecting one habitat patch with another (see Chapter 23) can enhance the integrity and stability of fragmented populations (see Chapter 19). On a small scale, hedgerows act as corridors between woodlands. On a much larger scale, greenbelts in cities and suburbs, plantings along major highways, and vegetation along rivers and streams provide not only travel corridors for wild animals but also dispersal routes for woody plants. Of particular value are corridors of woody vegetation along rivers and streams because they provide water, food, and cover for a great diversity of species.

Because of isolation, impingement by development and other activities, lack of necessary natural disturbances, and other factors, reserves need some form of management to maintain their value as such. Management options range from habitat restoration to protection and restoration of populations. A current approach is to concentrate on restoration, maintenance, and preservation of ecosystems of which they are a part and not on threatened species as individuals. Such an approach ensures the maintenance of associated systems and their occupants. For example, protecting the old-growth forests of the Pacific Northwest also protects the habitat for the spotted owl (*Strix occidentalis*), marbled murrelet (*Brachyramphus marmoratus*), and red tree vole (*Arborimus longicaudis*).

Habitat restoration involves planting, protection, and maintenance through various management techniques. Eroded lands can be restored to grassland, cutover forest land replanted to trees, and drained wetlands reflooded and replanted with aquatic vegetation. Because some species require certain early, and thus ephemeral, stages of succession (see Chapter 21), those stages must be maintained by such techniques as cutting and burning. The endangered Kirtland warbler (*Dendroica kirtlandii*) (Figure 18.9), for example, requires early-stage stands of jack pine (*Pinus banksiana*). Such stands can be maintained only by periodic burning of blocks of overage pine to stimulate regeneration of young stands. In such cases, the management objective is to keep blocks of vegetation in various stages of succession to ensure continuance of acceptable habitat.

FIGURE 18.9 Kirkland warbler at its nest concealed on the ground in a jack pine stand.

WILDLIFE RESTORATION

Since North America (and African and Australia as well) was first settled by Europeans, its wildlife was decimated by market hunting, wanton killing, and habitat destruction (Matthiessen 1987). Species, such as the heath hen (*Tympanuchus cupido*), passenger pigeon (*Ectopistes migratorius*), and Carolina parakeet (*Conuropsis carolinesis*), became extinct. In the early 1900s, a number of individuals made strenuous efforts to halt the destruction of wildlife. One effort was the passage of the Lacey Act of 1900, which prohibited interstate transportation of wildlife, dead or alive. Another was the Migratory Bird Act of 1913, which gave international protection to migratory birds and ended market hunting for waterfowl. By the 1930s, when the United States was suffering from drought and severe soil erosion that resulted in the Dust Bowl, wildlife was still in serious trouble. Then wildlife restoration efforts began in earnest with the passage of the Pittman-Robertson or Federal Aid in Wildlife Restoration Act in 1937, financed by an excise tax on shooting equipment, which supported wildlife research and habitat restoration.

Comeback of such species as the white-tailed deer, pronghorn antelope (*Antilocapra americana*), and wild turkey (*Meleagris gallopavo*) was achieved by a series of actions that allowed low populations to expand. One was strict protection from hunting followed by highly regulated hunting seasons once the populations were near recovery. States and the federal government set aside refuges and reserves to protect both animals and habitat. They reintroduced wild individuals taken from pockets of abundance to areas of scarcity and empty habitats.

What made the restoration efforts most successful was the availability of large areas of empty habitat. During the Great Depression of the 1930s, abandoned farmland and rangeland reverted to natural vegetation. Devastated forest lands were growing back, providing outstanding food and cover for white-tailed deer and other species. So successful were restoration efforts that some of the species involved are approaching pest status. White-tailed deer are probably more abundant now than at the time of settlement. They are so numerous that they are killed on highways, invade suburban gardens, and destroy agricultural crops.

The wild turkey is a good example of restoration of a species from the brink of extinction (Figure 18.10). Originally the turkey's range included all or parts of the 39 states and extended into Ontario, Mexico, and Central America. By the mid-1800s the species had been eliminated from northeastern United States, and by 1900 from the midwestern states. In 1949 only small populations of eastern wild turkey survived on about 12 percent of their ancestral range. Most existed in more remote areas of the Appalachian Mountains. Intensive studies of the biology and ecology of the species, financed by Pittman-Robertson money, provided the informational base needed for restoration efforts. Basic to restoration was live-trapping social groups of turkeys from the wild and their release into empty habitat. Maturing forests that improved turkey habitat, continued intensive studies of the bird, and the wild turkey's unforeseen ability to adapt to habitats previously thought unsuitable aided large-scale restoration efforts. Today the wild turkey population is about 3 million birds. Like the

white-tailed deer, flocks in some regions have invaded and settled in wooded suburban areas.

Such restoration efforts involve transplantation and protection of wild individuals. Restoration of some populations relies on the introduction of individuals from captive-bred populations. Introduction of captive-bred individuals to the wild requires pre-release and post-release conditioning, including the acquisition of food, shelter finding, interaction with conspecifics, and fear and avoidance of humans. Numerous attempts have been made, with many failures. Some successful reintroductions among the birds include the whooping crane (*Grus americana*), masked bobwhite (*Colinus virginiana*), and Hawaiian goose or nene (*Branta sandwicensis*), peregrine falcon (*Falco peregrinus*), and California condor (*Gymnogyps californianus*) and among the mammals the wolf and European wisent (*Bison bonasus*). Problems with reintroductions of captive-bred individuals include high costs, logistical difficulties, shortage of habitat, and the questionable ability of the individuals to adapt to the wild.

Both wild and captive-bred individuals may be translocated to build up the numbers of individuals already present and to introduce new genetic material into populations. Such translocations must be done carefully. Not only is there danger of introducing disease, but the new individuals must be integrated into the social and breeding structure of the native population to achieve the desired results.

Just as important is the genetic background of the transplanted animals to ensure they are adapted to their new environment. If not, these individuals can weaken the resident stock. Attempts to increase depleted populations of the northern bobwhite (*Colinus virginianus*) in Pennsylvania and New England with birds of southern origin resulted in both the death of the introduced birds and their "hybrid" offspring during cold winters to which they were not adapted. The northern populations went with them because of reduced recruitment from the young.

It is an unfortunate commentary on human treatment of Earth that the only hope for many species of wild things is safety in captivity. Only a small number of a few endangered species can be reared in captivity. However, for some species it is only way they can be increased initially. Captive propagation, however, has its problems including small population size, potential inbreeding, incompatibility of captive individuals relative to mating, and lack of social interaction. Captive propagation programs cannot be carried out indefinitely. After a number of generations, depending on population size, the captive stock will begin to show signs of inbreeding depression (see Chapter 19) and domestication. For this reason it is important to consider opportunities for reintroduction when feasible.

Successful restoration, then, involves a number of facets. Scientific studies of the species' biology, ecology, and behavior provide the data needed to successfully manage the growing populations. Protection through strict law enforcement, adequate financing, and needed public concern and cooperation aid population expansion. Protected reserves, large areas of suitable habitat, and the adaptability of the species further ensure the species' recovery.

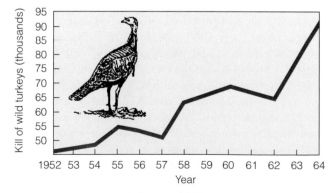

FIGURE 18.10 Growth of the wild turkey population in the United States following restoration (through 1964), as derived from hunting harvest data. Note the sharp increase in growth after 1962. Turkey populations are still increasing.

Management of Restored Populations

Burgeoning human populations, rapidly shrinking habitats, illegal hunting and commercial traffic in wildlife, and over-harvesting of some species create different problems for the protection and maintenance of dwindling populations from those faced by early restoration efforts. It is no longer possible for most species to exist as they did in presettlement days or even as in the recent past. Their survival depends on management at varying levels of intensity to maintain their populations and ensure survival of their habitats.

If restoration is successful, the species may reach a point where its population exceeds the ability of the habitat to support it or comes in conflict with human interests. For example, the restoration of white-tailed deer and Canada geese (*Branta canadensis*) have resulted in an overabundance of these two species to a point where they have become nuisances in many areas. Nonmigratory Canada geese in the eastern United States are common year-long residents of golf courses, parks, suburban housing developments—wherever grass and sufficient water are available to them (Conover and Chasco 1985, Chasko and Conover 1988) (Figure 18.11). They litter these sites with droppings and damage the grass. These geese descended from captive populations held as live decoys back in the 1930s and stopover migratory birds that stayed, attracted by the rich source of food. White-tailed deer

have experienced an explosive population growth during the past 20 years, brought on in part by failure to harvest does, reduced hunting pressure, and invasion into deer habitat by suburban developments that provided both protection and a rich source of highly palatable food. Deer abundance is resulting in the elimination of many herbaceous and woody plants from the forest, inhibiting forest regeneration (Waller and Alverson 1997), Deer in human-dominated landscapes (or vice versa) bring with them deer-vehicular accidents, destruction of ornamental plants, and Lyme disease. They easily become habituated to humans, whose ambivalent feelings toward the deer make control of the populations difficult (Stout et al. 1997, Green et al. 1997)

Overpopulation results in deterioration of habitat, a scarcity of food, and in an impairment of habitats for associated species. As the habitat is damaged beyond the ability to support a viable population, the species declines. Prevention of such damage may require the trapping and removal of or cropping the excess individuals. These surplus individuals can provide transplant stock for depleted habitats. For example, the highly productive white rhino (*Diceros simus*) population in the Umfolozi Game Reserve in South Africa has supplied a number of animals for reintroduction into other reserves.

Cropping requires some knowledge of the interrelations and functions of the species in the system. Are apparent overpopulation problems evident in some parks simply a part

FIGURE 18.11 Like white-tailed deer, Canada geese have adapted to a suburban way of life.

of natural plant-herbivore cycles, or does the situation require human intervention? For example, elephants, which feed on woody browse, are confined to parks and reserves where they are converting tree savannas to grassland that cannot provide sufficient food for them. The outcome is starvation of elephants. When the elephant population declines, the vegetation slowly returns. The question is whether to allow such a cycle to proceed or to reduce the elephant herd in order to try to achieve some stability between elephants and the trees. It is difficult to arrive at some answer because of the longevity of elephants and the time required for the trees to regenerate. The decision affects not only the elephants but the entire ecosystem.

Protection of reintroductions needs the long-term support of governments and, especially, the communities in the vicinity of the parks and reserves. In most places it is necessary to demonstrate to these communities in a tangible way economic benefits to be derived locally from parks and reserves and their wildlife so that they become collaborators with the program (Adams and McShane 1992). Protection must be accompanied by education programs regarding the species and its habitats from the national to local levels. Education is absolutely necessary at the local level to gain support of the communities about the region. Without such support, reintroduction programs will likely fail.

Ultimately the fate of wild creatures and the ecosystems of which they are a part rests with controlling the explosive growth of human populations. As the number of the dominant mammal on Earth increases, diversity of life decreases. Eventually we will be left with only those plants and animals that are tolerant of or share the human-dominated landscape. The outstanding ecosystems, the tropical forests, old-growth forest ecosystems, savannas, and grasslands, even marine environments and their unique forms of plants and animals, will exist only as memories preserved in books and films.

POLLUTION

Pesticides, sedimentation, excessive nutrients, toxic wastes, and other pollutants all affect populations. Chlorinated hydrocarbons from DDT and DDT-like pesticides to PCBs and other toxins pollute food webs worldwide (see Chapter 26), as well as affect reproduction and cause physical deformities in frogs and birds. Nutrient-rich runoff from agricultural enterprises in the Mississippi Basin enters the Gulf of Mexico. There it sparks algal blooms that deplete water of oxygen, creating a dead zone the size of New Jersey devoid of benthic fauna and fish. A somewhat similar situation exists in the Chesapeake Bay, which receives nutrient-rich water from the Potomac watershed involving 150 headwater streams that drain six states. Pollution in the Bay has resulted in the decline of water quality, and in the reduction of benthic and marine diatoms—a base of the food web. It has adversely affected the ability of the bay to function as a nursery for American shad (*Alosa sapidissima*), striped bass (*Marone saxatilis*), and blue crab (*Callinectes sapidus*).

Increasingly frequent oil spills from oil tankers and other ships kill numerous coastal birds, mammals, and intertidal biota from oil toxicity, hypothermia, smothering, or cleanup and result in alteration in food chains and community structure (Paine et al. 1996). Recovery from oil spills is slow. The *Exxon Valdez* spill in Prince William Sound, Alaska, occurred on March 24, 1989. Ten years later, only the bald eagle appeared to have recovered to pre-spill levels.

BIOINVADERS

Wreaking havoc on many native species through competition, predation, and introduction of diseases and parasites are alien species that in one manner or another find their way to other places (see Chapter 14). Once separated from distant places by geographical barriers such as oceans, inhospitable land masses, and mountain ranges, alien species find their way to new places with intentional or unintentional help from humans.

The Pacific Islands were among the first to feel the effects of alien invasions. Accompanied by dogs, pigs, and Polynesian rats, Polynesians from New Guinea were the first alien invaders to the Islands. Following these were European visitors and traders bringing with them black (*Rattus rattus*) and brown (*Rattus norvegicus*) rats that traveled along in sailing ships and escaped when ships docked. Unadapted to the competitive and predatory pressures the invaders brought with them, many native species quickly became extinct.

Island species suffer the most. In Hawaii, for example, during the past 200 years, 263 of island native creatures disappeared; 300 are listed as endangered or threatened; and 1400 life forms are in trouble or extinct. Among the island's 111 birds, 51 are extinct and 40 are endangered. Among the 1126 native flowering plants, 93 are extinct and 40 are endangered. Introduction of exotics is the cause behind 95 percent of Hawaii's species endangerments. On the Pacific island of Guam the brown tree snake (*Boiga irregularis*), a native of New Guinea, accidentally reached the island around 1950, probably aboard military equipment transported there for dismantling. The snake has eliminated 9 of 12 native bird species, 6 of 12 native lizards, and 2 of the 3 native fruit bats; it even invades houses on the island and bites sleeping infants (Rodda et al. 1997).

Causing even more problems has been the deliberate introduction of alien species without giving any consideration to the ecological effects such novel species could have on the native biota. In 1876 Professor Samuel Baird, who headed the newly formed U.S. Fish Commission, encouraged the introduction of carp (*Cyprinus carpio*) to bolster the overexploited native fishery in United States (Fritz 1988) for food production. He argued that the carp was harmless in relation to all other fishes and had the ability to populate waters to their greatest extent. He was right on the last statement and wrong on the former. The carp, a bottom feeder, destroys the habitat of and outcompetes native fish in many rivers, stream, and lakes, even though millions of pounds of carp are harvested for food each year. The furbearer nutria (*Myocaster coypus*) introduced from South America competes with the native muskrat (*Ondatra zibethicus*). In turn, the muskrat escaped from a captive population in Central Europe and spread across the continent. It is an important pest species in the Netherlands because it burrows into dikes. The North American gray squirrel (*Sciurus carolinensis*) introduced into Europe competes with and has caused a decline in the populations of the native European red squirrel (*Sciurus vulgaris*). The gray squirrel is a serious forest pest because of its habit of stripping bark and eating tree buds. A perennial herb, purple loosestrife (see Chapter 14), and the Australian tree *Maleleuca,* both introduced as ornamental plants, are rapidly displacing native wetland plants in North America to the detriment of wetland wildlife. *Myrica faya,* a nitrogen-fixing tree, was imported from the Canary Islands to the big island of Hawaii in 1900 to halt soil erosion. The tree is now spreading over 30,000 acres of the Hawaii National Park, displacing the native forest (Vitousic and Walker 1989).

Other introductions have wreaked even greater damage. One was stocking Nile perch (*Lates niloticus*) as food fish into Lake Victoria, home to approximately 60 endemic species of cichlid fish. This highly predatory fish has driven 60 percent of native cichlid species to extinction. Adding to Lake Victoria's problem is the invasion of water hyacinth (*Eichhornia crassipes*), a native of the Amazon Basin, imported into Africa in the nineteenth century as an ornamental. Drifting down rivers into Lake Victoria, the water-choking growth has covered 90 percent of the lake's shoreline, has ruined spawning grounds, and threatens a collapse of the lake's fisheries (Goldschmidt 1993, Bright 1998). In spite of the lessons learned from Lake Victoria, Nicaraguans in the 1980s stocked the African tilapia (*Oreochronis*) to increase fishery exports in Lake Nicaragua (McKaye et al. 1995). As the catch of tilapia increased, that of native food fish, the cichlids, decreased. Now the tilapia threatens the collapse of the lake ecosystem.

Transplanting fish for sport fishery has resulted in the demise of native species in lakes and stream. Introducing nonnative brook (*Salvelinus fontanialis*), rainbow (*Oncorhynchus mykiss*), and golden (*O. aguabonita*) trout in normally fishless lakes and ponds above 7000 feet in the High Sierra Mountains has resulted in the near disappearance of the mountain yellow-legged frog (*Rana muscosa*). Unlike other frogs, the tadpole stage of the mountain yellow-legged frog lasts between two and four years, making the species highly vulnerable to fish predation. Although the rainbow trout is a native western species, its introduction into other streams dominated by cutthroat trout (*O. clarki*) has resulted in either the extinction of the cutthroat trout in certain streams or hybridization with rainbow trout (Rhymer and Simberloff, 1996). Of the 11 native fish that made up the bulk of the Great Lakes fishery, 4 are extinct and the remaining are threatened by the influx of an array of 141 exotic species of various taxa, one-third of which arrived by the way of ships' ballasts. The introduction of exotics was behind the demise of 12 of the 24 extinct fish in the United States, and exotics are affecting 29 of the 54 threatened species.

Once bioinvaders have established themselves and become capable of overrunning native biota, the task of controlling them is difficult and expensive. The control and eventual elimination of the Mediterranean fruit fly (*Ceratitis capitata*) that was a threat to the citrus crops of California cost millions of dollars. Determination of the risks and benefits is a necessary precursor to the introduction of a nonindigenous species for any purpose. This determination involves studying both the characteristics of the species and unintended effects, as well as limiting or preventing future introductions (Ruesink et al. 1995).

PEST POPULATIONS AND CONTROL

Populations of most species are affected negatively by human population growth. Some species, however, thrive in highly disturbed, anthropogenic urban and suburban habitats. Extensive acreages of monocultural cropland and forest tree plantations provide huge areas of abundant food and cover, providing conditions for outbreaks of pest species. Many of these species cause serious economic damage. Others are more a nuisance value. What constitutes an animal pest or a plant weed depends on your viewpoints and values.

Pests are animals that humans consider undesirable, a classification that varies with time, place, circumstances, and individual attitudes. Some animals are obvious pests, especially the ones that are camp followers of humans such as rats, cockroaches, and fleas. Others became pests as agriculture invaded and replaced the habitats of native grazing herbivores and their predators. Herbivorous mammals, large and small, found an abundance of food in fields and gardens, and grain-feeding birds turned to large fields of grains, especially rice. Deprived of their natural prey, large predators found domestic animals easy prey.

Plants, too, became competitors with humans. Wild plants in fields and gardens compete with crops for space, light, and nutrients, reducing yields and becoming, in human terms, weeds. A **weed** is a plant growing anywhere not desired, a plant out of place. Violets growing in a thicket are wildflowers; violets growing in a lawn may be weeds, (depending on the lawn owner's value judgment). Electric companies regard trees growing in power line rights-of-way as weeds. To a naturalist, the same growth is wildlife habitat. Many foresters consider any trees growing in the forest that have no commercial value as weed trees; to a naturalist those trees may be beautiful or important for wildlife.

Most animal pests and weeds, native and nonnative, show a mix of r and K characteristics. Most familiar and obnoxious pests, such as house flies and fleas, possess certain r characteristics (see Chapter 11). They adapt well to conditions provided by humans, and spread rapidly where both food and shelter are abundant. Common weeds, such as dandelion (*Taraxacum* spp.) and ragweed (*Ambrosia* spp.), are much the same. They disperse easily, colonize highly disturbed sites, and are persistent.

Many native pest species, such as white-footed mice (*Peromyscus leucopus*) and ground moles (*Talpidae*), are more nuisance pests and are usually regulated by natural predators and parasites. This situation is not so true for exotic pests such as brown rats, house mice (*Mus musculus*), cockroaches, flies, and flour beetles. After centuries, it is evident that we cannot eradicate these pests. The best we can do is control their numbers. Holding their populations down involves some form of pest control, deliberating reducing their numbers by any means in contrast to the attempts to increase populations of declining and threatened species. Controls should reduce pests to a level below which they will not cause economic injury (Figure 18.12). At that point, costs of control are less or at best equal to the net increase in the value derived from such control. Even so, many people initiate control measures when no strong economic values are involved, such as spraying insecticides and herbicides on lawns. Control measures vary with life history and effect of the pest or weed.

Chemical Control

An ancient and popular means of pest control is chemical. The Sumerians used sulfur to combat crop pests, and the Chinese as long ago as 3000 B.C. used substances derived from plants as insecticides. By the early 1800s such chemicals as Paris green, Bordeaux mixture, and arsenic were used to combat insect and fungal pests. The major chemical weapons, however, appeared after World War II with the development of organic insecticides (containing carbon). The original impetus for their development was combating insect vectors of human disease, especially in tropical areas during World War II. The initial success of these insecticides encouraged their rapid use in agriculture, for which the chemical industry provides an arsenal of more than 500,000 biocides.

These organic chemicals, with varying degrees of toxicity and persistence, are either synthetic (of human manufacture) or botanic (derived from plants). Major groups of synthetic pesticides include the chlorinated hydrocarbons, organophosphates, and carbamates. All are broad-spectrum insecticides and in one manner or another affect the nervous system. Prominent among the botanicals are pyrethrums, toxicants extracted from the flower heads of chrysanthemum plants. They also affect the nervous system, but they are among the safest and least persistent insecticides. Organic herbicides are used to control weeds. One class is contact herbicides that kill foliage by interfering with photosynthesis. Systemic herbicides, absorbed by plants, overstimulate the growth hormones. The plants grow faster than they can absorb nutrients and die.

Chemical control has its dark side. The chemicals affect not just the target species but a wide array of nontarget species as well (see Chapter 26), often eliminating them and upsetting natural food webs, especially through the suppression of natural predatory species. The surviving pests then resurge in greater numbers than ever. Some herbicides contain dioxins, which have been linked with birth defects and cancers in humans.

Perhaps more insidious is that pesticides lose their effectiveness as the target species evolves resistance to them. As one pesticide replaces another, the pests acquire a resistance to them all. Some species, notably certain mosquitoes and cotton bollworms, have overcome the toxic effects of every pesticide to which they have been exposed. Insect pests need only about five years to evolve pesticide resistance; their predators do so much more slowly. Then pest outbreaks become even more disastrous. For example, insecticide applications in the rice fields of southeast Asia to control the rice stemburrower resulted in the emergence of a far worse pest, the rice brown planthopper (*Nilaparvata lugens*). Agriculturists soon discovered that it is an insecticide-induced resurgent pest whose damage is directly correlated to insecticide use. In response to the pest, agriculturists used new planthopper-resistant varieties together with insecticidal spraying of rice, but to little avail. Within several years these plants, too, were vulnerable to attack. Increased insecticide use only made the problem worse. Growers halted the use of insecticides. Research workers then initiated an experimental program to control the planthopper by encouraging its natural enemies—generalist predators that feed on planktivore feeders and detritivores in the rice fields early in the season. Increasing early season organic matter in tropical

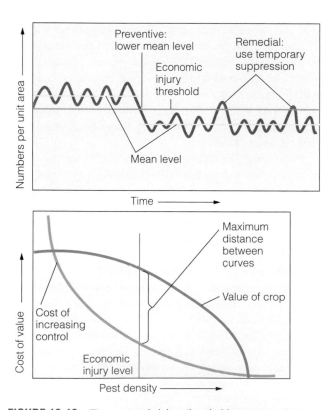

FIGURE 18.12 The economic injury threshold suggests when pest control measures should be taken. (a) When a pest exceeds the economic injury threshold, we should take strong measures to reduce the population. Below the threshold, temporary suppression measures suffice. (b) The economic injury level is that point at which the value added to the crop exceeds the cost of increasing control.

rice fields supported large populations of early season prey population for the generalist predator. This increased prey enabled populations of generalist predators to build, increasing predatory pressure on later season pests. Such a management program that promotes conservation and increase of native natural predators and reduced insecticide use is a form of biological control of pests.

Biological Control

Farmers long ago observed that natural enemies of pests act as controls. As early as A.D. 300, the Chinese were introducing predaceous ants into their citrus orchards to control leaf-eating caterpillars. In the late Renaissance, naturalists made important observations on insect parasitism that later proved used useful in understanding the role of biological control of pests and weeds.

Insect pests have their own array of enemies in their natural habitats. When an animal or plant is introduced, intentionally or unintentionally, into a new habitat outside of its natural range to which it can adapt, it leaves its enemies behind. Freed from predation and finding an abundance of resources, the species quickly becomes a pest or a weed. This fact has led to the search for natural enemies to introduce into populations of pests to reduce their populations.

Because the pest is usually an exotic, biological control involves the introduction of a nonindigenous predator or parasite to control the pest. There are some of classic examples of biological control (Debach 1974). One is the use of the pathogenic organism *Bacillus thuringiensis* (Bt). It is widely used in certain crops such as corn to control corn borers and in eastern oak forests to combat gypsy moths (*Porthetria dispar*). Because the organism is commercially produced, is not an active searcher, and is commercially sprayed on affected crops, it is frequently classified as an insecticide, not biological control.

Biological control has been effective against some weeds. The introduction of the cactus-eating moth, *Cactoblastis cactorum,* a native of Argentina, into Australia effectively reduced and controlled the rapidly spreading prickly pear (*Opuntia* spp.), which had been introduced into Australia in 1901 (Osmond and Monroe 1981). In 1900 the Klamath weed or St. John's wort (*Hypericum perforatum*), a native of Eurasia and northern Africa, was introduced along the Klamath River in California. An aggressive colonizer of overgrazed pastures and toxic to cattle and sheep, it soon occupied over 800,000 ha in the northwestern United States. Among the 600 species of insects that fed on the plant in its native environment was the leaf-eating beetle *Chrysolina quadrigema.* Introduced in 1945, this beetle has reduced Klamath weed to a rare roadside weed in California.

Biological control, like chemical control, can backfire (Simberloff and Stiling 1996). The success of the cactus-feeding moth *Cactoblastus* in controlling *Opuntia* in Australia encouraged its introduction to several West Indies islands to control *Opuntia* there. In time the moth made its way to Florida, where it now threatens the existence of several native *Opuntia* species (Pemberton 1995). The European lady beetle (*Coccinella septempunctata*) was introduced into the United States to control the Russian wheat aphid. The species was so successful that it has replaced many native American lady bird beetles; it has become the lady bug with which most of us are familiar (Elliot et al. 1996). The moral is that although using nonindigenous predators as biological controls can be effective, these species possess their own inherent dangers that must be assessed before they are released. They, too, can become alien invaders. Like the work in the Southeast Asian rice fields, it behooves biological control scientists to explore the employment of the array of native general predators.

Genetic Control

In recent years genetics has been employed in the struggle against many pests. Numerous wild plants and animals have evolved their own defenses against natural enemies (see Chapter 16). These defenses include allelopathic effects against plant competitors and toxins in tissues of both plants and animals that inhibit or suppress predation. One approach to pest control, then, is to make use of these defenses and breed further genetic resistance to pests into plants. This approach has been used successfully in some crop plants, such as corn, wheat, and rice. The process involves crossing of cultivars with wild relatives to capture the gene within the gene pool of the cultivated plant. Such an approach requires considerable effort in locating, recognizing, and studying wild relatives. Because most crop plants are tropical in origin, the destruction of tropical ecosystems results in the erosion of the genetic storehouse on which we can draw.

Genetic engineering, the use of recombinant DNA material to introduce the gene with the desired trait into the plant, is the latest technique to increase plant resistance to insect pests. This technique involves transferring into the plant a single gene that either confers resistance to viruses or to herbicides, or enables synthesis of endotoxins—poisons that inhibit the feeding activity of insect enemies.

Another technique is splicing genes of *Bacillus thuringiensis* (Bt) into crop plants, notably corn, cotton, and potatoes. This splicing creates transgenic varieties that produce stomach poisons fatal to crop-damaging pests. The use of these bioengineered seeds eliminates the need for spraying the fields with pesticides. So effective is this method of pest control that it kills all the particular pest involved such as corn borer or cotton boll weevil, but with a potentially major drawback. Because the insect pests are continually exposed to the toxin, the few that survive obviously can evolve a resistance to Bt. To slow the evolutionary process, the Environmental Protection Agency has mandated resistance management strategy to prevent or delay the evolution of Bt-resistant insects. This strategy involves the planting of refuge fields near Bt-protected crops in which the crops do not produce the Bt toxins. By escaping Bt annihilation in the refuges, large numbers of Bt-susceptible insects will mate with the few Bt-resistant survivors from the Bt-protected fields, thus slowing or preventing the evolution of Bt-resistant insects (Hargrove 1999).

Genetic engineering for pest control has other inherent dangers. Plants with engineered traits could evolve new genotypes with somewhat different life histories or physiological traits. Engineered crop plants might transfer genes over relatively long distances by hybridizing with related plants that differ in their life history characteristics. Many crops, such as celery, asparagus, and carrot, have weedy relatives with high reproductive output and efficient seed dispersal. If these weedy relatives acquire the engineered gene, then the gene could be spread through the range of the plants. This transfer would create weeds against which current herbicides would be ineffective, and insect-resistant plants could speed the evolution of even more resistant insect pests.

Another aspect of a genetic approach to insect suppression is sterile male release. Competitive sterile males reared in great numbers in the laboratory are introduced in sufficiently large numbers to ensure they will be involved in a high proportion of the matings in the field. If the number of sterile males is kept high as the population of the pest declines, the proportion of sterile to fertile matings increases. If the population of the pest is initially high, insecticides may be used to reduce the population before sterile males are released. Such a method works only if certain conditions are met: the pest population must be fairly isolated and not subject to immigration of wild males or emigration of sterile males; genetically different subpopulations are not involved; and no genetic changes occur in the reared population.

An example is the screwworm control program. The screwworm is the larvae of the blowfly *Cochliomyia hoiminivorax,* which lays its eggs in open wounds of warm-blooded animals. The larvae enter the wound and feed on the flesh of the animal. The screwworm became a major livestock pest in the southeastern United States and Texas, and a major eradication effort began in l950s involving the release of factory-reared sterile males. The program was highly successfully until 1972, when a major outbreak occurred. The cause was a major genetic difference between the sterile males released and the wild type. In 1977 the defective factory genetic strain was replaced, and control was regained. This experience points out the necessity of maintaining strict quality control to monitor genetic changes in the factory populations.

Mechanical and Cultural Control

Another approach to pest control is mechanical. For centuries agriculturists have used fencing and other barriers not only to keep livestock in, but to deter predators and to prevent herbivores such as deer and rabbits from feeding on crops. Horticulturalists and home owners resort to use of repellents applied on and about plants. Sticky traps wrapped around trees prevent caterpillars from climbing up to the crown. Traps baited with pheromones and sticky paper are effective in attracting and capturing males of specific insect pests such as gypsy moths. Light traps, although effective, are indiscriminant in the insects they attract and kill innumerable beneficial insects along with the insect pest. Cultivation, hoeing, and hand weeding are typical methods of eliminating competitive weeds from crops fields and gardens.

Homogeneous habitats provide the opportunity for many large outbreaks of pests. Agricultural crops and extensive stands of forest trees of the same and closely related species, such as southern pines and balsam fir, provide huge areas of abundant food and cover. Spruce budworm (*Choristoneura fumiferana*) and southern pine beetle (*Dendroctonus frontalis*), for example, have swept across hundreds of thousands of acres of forests. One deterrent to the rapid spread of pests is the creation of patchy environments in which homogeneous stands are broken up by other types of vegetation, such as interplanted row crops, interspersion of hardwoods with conifers, and hedgerows. These patches not only scatter the food supply, checking the spread of the pest and breaking its population into smaller units more vulnerable to predation, but they also harbor enemies of the pests.

Integrated Pest Management

Because chemical, biological, and other methods alone obviously are not the solution to pest control, entomologists have developed a holist approach to pest control called **integrated pest management (IPM).** Integrated pest management considers the biological, ecological, economic, social, and even aesthetic aspect of pest control and employs a variety of techniques. The objective of IPM is to meet the pest not at the point of a major outbreak but at the time when the size of the population is easiest to control. The approach is to rely first on natural mortality caused by weather and natural enemies, with as little disruption of the natural system as possible, and at the same time hold the pest below the economic injury level (Figure 18.13).

Successful integrated pest management requires the knowledge of the population ecology of each pest and its associated species and the dynamics of the host species. It involves considerable field work monitoring the pest species and its natural enemies by such techniques as egg counts, and trapping adults to acquire such information necessary to determine the necessity, timing, and intensity of control measures. These control measures may involve minimal chemical spraying at appropriate times and cultural techniques such as thinning and salvage operations in an affected forest. One goal of IPM is minimal use of chemicals to reduce the development of genetic resistance against pesticides.

The control methods must be adjusted to the situation, which may vary from one location to another. The intensity of control or no control is based on the degree of sustainable damage, the costs of control, and the benefits to be derived. It involves a series of management decisions (Figure 18.13) including assessment of size and activity of the pest at designated spots; potential effects of pest control on the resource; cost-effectiveness; the degree of and acceptability of envi-

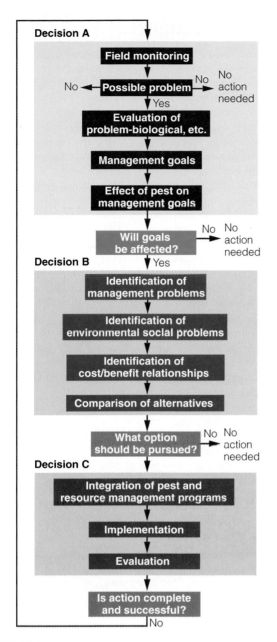

FIGURE 18.13 A decision-making model for integrated pest management. Note the inputs of biological, social, and environmental information.

Summary

1. If plant and animal populations exploited for human use are to be productive continuously, they must be managed for sustained yield. The yield per unit time should equal production per unit time. Such an approach to management necessarily differs between K-selected and r-selected species.

2. Based on the logistic equation, sustained yield models fail to take into consideration all aspects of population dynamics, including natural mortality and environmental uncertainty. Sustained yield management considers the resource as a single biological unit that is managed for maximum economic return, rather than as a component of a larger ecological system.

3. Economic considerations too often outweigh biological considerations, resulting in the eventual commercial and even biological extinction of a species. Economic interference in sustained yield management also encourages a wastage of marine life from juvenile fish to dolphins and turtles caught in large nets along with commercial species. Euphemistically called "bypass," this enormous tonnage of wasted marine life results in ecological disturbances to marine ecosystems and to the overall fishery resource.

4. Economic considerations often override sustainability in forestry, devastating forests. As in other resources, sustainability requires a balance between the volume of timber cut and the volume of timber growth.

5. Pollution involving pesticides, sediments, excessive nutrients, toxic wastes, and oil spills causes heavy mortality, interferes with reproduction, and destroys habitats.

6. Habitats of wild creatures and plants, both terrestrial and aquatic, are being fragmented and isolated, resulting in relatively small, isolated populations. Such habitat destruction is a major cause in the massive decline in fauna and flora.

7. Increasing threats to the future of wild species have prompted the establishment of refuges or reserves. There has been a vigorous debate over the relative merits of large versus many small reserves. Large reserves provide the greatest protection and support the largest populations. Small reserves support greater diversity of species and greater genetic diversity within a species, and they protect portions of a population from environmental catastrophe.

8. Populations in small reserves may be too small to remain viable. If they do reproduce successfully, their numbers may exceed the ability of the habitat fragment to support them. In such situations, we can use the surplus to recolonize empty habitats.

9. Species experiencing declining populations require strong efforts at restoration if they are not to become locally or globally extinct. Because the loss of habitat is the primary cause of the decline of threatened and endangered species, a major restoration effort involves the protection, restoration, and management of habitats.

10. The future of threatened populations depends on the establishment of protected parks and reserves. Surrounded by land developments of various sorts and extents, small reserves should be connected by corridors that allow movement and dispersal of individuals among habitat patches.

11. To reduce inbreeding in reserves, we may have to transfer individuals among reserves. In extreme cases of highly endangered species, individuals may have to be captured from the wild and held in captive breeding programs to save the species from extinction.

ronmental and social consequences; the use of cultural approaches; implementation of preventive or suppression tactics; and posttreatment evaluations. Because of its technical and scientific approach, integrated pest management requires expert knowledge and a field staff to acquire the information needed to arrive at management decisions. Integrated pest management has been employed in the control of spruce budworms in Canada and southern pine beetles in the United States, in apple orchards in the United States and Canada, in cotton fields in Texas, and in alfalfa-growing regions in the United States.

12. All attempts at the protection and restoration of threatened species must be accompanied by protection from poaching and by an educational program to gain local support.

13. Bioinvasion, the accidental or intentional introduction of exotic species, through intense competition and predatory pressures, results in the displacement and often extinction of many native species. Alien invaders compromise the integrity of ecosystems and as pest species cause extensive economic damage.

14. Pests are defined as undesirable animals and weeds as plants growing anywhere not desired. The objective in controlling pests is to hold the population below a level at which they cause economic injury.

15. There are a number of methods of controlling pests and weeds. One is the use of organic chemical pesticides and herbicides. Their use involves serious risks of toxic environmental pollution and of the evolution of resistant strains of pests and weeds. Because chemical pesticides are nonselective, they kill off natural predators and parasites, often allowing a pest population to increase dramatically.

16. Biological control is based on the introduction of natural predators and parasites into the pest population. Biological controls can be effective in keeping some pests below the economic injury level. Biological control, however, has its downside. Some of the natural enemies introduced to control a certain pest species may attack native species as well.

17. Breeding for genetic resistance to insect pests and weeds is another important and effective tool. Transplanted genes are used to produce plants resistant to viral diseases and insect attacks. Such genetic engineering, however, involves a risk that the gene may transfer to wild, weedy relatives that could develop into highly resistant strains of weeds.

18. In some situations, mechanical controls such as trapping pests and cultivating the ground to kill weeds are effective. A cultural approach is to increase habitat diversity or patchiness, reducing availability of food and making dispersal of pests more difficult.

19. The integrated pest management approach to pest control is more philosophy than a specific strategy. It combines appropriate chemical, biological, genetic, mechanical, and cultural control measures. Based on careful monitoring, it weighs environmental and social benefits and costs.

Review Questions

1. Explain sustained yield. What is the difference between maximum sustained yield and optimal sustained yield?

2. Report on the effects of modern commercial fishing on other resources and ocean life.

3. What indirect effect do you have on ocean life when you enjoy an "all you can eat" shrimp or crab legs dinner?

4. Why is there obvious unconcern about the destruction of the ocean when we show concern about destruction of tropical and Pacific Northwest forests?

5. How do economic interests block sustained yield practices? Apply your answers to commercial fisheries and cutting of tropical and old-growth forests.

6. The management and recovery of endangered species are often fraught with controversy. Report on the capture of all remaining California condors (*Gymnogyps californianus*) to undertake a captive breeding program and the degree of its success. Was the decision a wise one? What about the black-footed ferret (*Mustela nigripes*)?

7. Define pest and weed. Why are the two terms value-laden?

8. Under what conditions do plants and animals become weeds and pests?

9. Based on knowledge gained from previous chapters, describe the characteristics you think make up an effective biological control agent.

10. What are the ecological dangers of plants genetically engineered to resist pests?

11. How can a patchy environment reduce pest outbreaks?

12. What is integrated pest management?

13. How can integrated pest management channel public input into a local or regional pest control program?

Cross References

Population growth, 184; interspecific competition 243, extinction, 191; habitat selection, 236; inbreeding, 369; genetic drift, 373; patchy environment, 458; corridors, 455.

The Complete Recipe for Apple Pie

The apple pie, like baseball, Old Glory, and the Fourth of July, is an American icon. To make an apple pie you need flour, sugar, butter, and cinnamon, but the central ingredient is the apple. There is, however, one essential ingredient that appears in no cookbook or grandmother's recipe handed down through the generations. That ingredient is the bee. To make an apple you need bees.

Apples are pomes—simple fleshy fruits that grow around the seeds. All fruits arise from flowers and thus are found exclusively in the flowering plants. For the flower to produce seeds and fruit, it must be fertilized. To understand this process and the role of bees, we must first examine how flowering plants reproduce.

In flowering plants, the female reproductive cell, the egg, is housed in a structure known as the pistil. The male reproductive material needed to fertilize the egg is pollen. The pollen grains are produced in a part of the flower called the stamen and stored on exposed structures referred to as anthers (Figure V-A). Fertilization of the egg by pollen grains leads to the formation of seeds. In some cases, such as the apple, a fleshy body referred to as a fruit covers the seeds. For flowering plants to reproduce successfully pollen grains from one flower must reach the pistil of another flower—this process is called pollination.

One means by which pollen from one flower reaches the pistil of another is by way of the wind. Many types of plants, such as grasses, sedges, conifers, and a number of broad-leaved trees including walnuts, oaks, birches, and poplars, rely on wind for pollen transport. Plants that are wind-pollinated usually produce large amounts of small, buoyant pollen grains in small, inconspicuous flowers. In fact, some species' flowers, such as those of grasses, are so modest that they may not be recognized as flowers at all.

Wind can be a very effective transporter of pollen. Pollen grains have been collected in atmospheric surveys as high as 19,000 feet and observed as much as 3000 miles from their parent plant. However, wind-aided transport is unpredictable. The overwhelming majority of the millions of pollen grains produced by a single plant will never reach the flower of another individual and will be wasted. Because pollen is rich in proteins and oils, this represents a low return on a very significant energy investment by the plant.

To reduce this waste many plants have adopted a different means by which to transport their pollen grains. Some plants have evolved mutually beneficial relationships with animals that act as couriers, carrying the pollen from one flower to another.

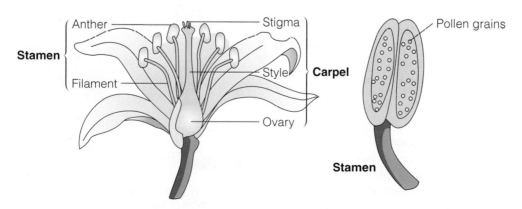

FIGURE V-A   Diagram of flower showing pistil, stamen, and anthers.

Although these animals, which include lizards, birds, bats, and insects, increase the likelihood of successful fertilization, they don't work for free. Pollen, because of its high concentration of proteins and fats, is very nutritious. Animals benefit from transporting pollen by having access to this high-quality food resource. Plants also pay their couriers with something that is much cheaper to synthesize—nectar. Nectar, little more than sweetened water, is produced from special glands called nectaries that hide deep within the structures of flowers. This deep position provides two advantages to a plant. It reduces nectar loss by evaporation or dilution by rain, and animals visiting the flowers are forced to brush past the plant's anthers. In doing so, they inadvertently collect some pollen for transport.

Of all the animal couriers of pollen, insects are the most numerous, widespread, and probably the first to serve in this capacity. The plants that employ insects to carry their pollen must first attract the insects to their flowers. To do so they tailor their advertisements to match insects' sensory abilities. Many insect-pollinated plant species produce large, colorful, fragrant flowers to advertise their stores of pollen and nectar. Insects use their senses of sight and smell to find them. As an insect moves about a flower searching for its payment of food, it occasionally brushes against the anthers and some of the sticky pollen grains adhere to its body and limbs. When the insect moves to another flower, pollen picked up at previously visited flowers may get brushed off and fall on the stigma—the upper portion of the pistil—fertilizing the flower.

Various types of insects including beetles, butterflies, moths, and even grasshoppers pollinate flowers. However, bees are the champion pollinators. Bees, along with wasps and ants, are members of a group of insects that are characterized by having wings fashioned of thin membranes (order Hymenoptera). There are about 20,000 species of bees in the world. Bees differ from the other Hymenopterans by having enlarged hind feet, which are typically equipped with baskets of long, stiff hairs for gathering pollen. Bees, both young and old, feed entirely on plant food, which for them consists of the pollen and nectar of flowers. They convert the latter into honey in the digestive tract. The long hairs on the legs of bees hold the sticky masses of pollen until the bees extract it. Bees also usually have a dense coat of hairlike bristles on the head and thorax that assists in picking up the pollen grains. In many species, the lips are fused to form a long tube used for sucking nectar.

Bees may be solitary or social. The social bees (which secrete wax) are the honeybees, the bumblebees, and stingless bees and number about 400 species. Solitary bees include the carpenter, plaster, leaf-cutting, burrowing, and mason bees, which are named from the method or materials they use to construct their nests.

The honeybee commonly raised to produce honey and wax in many parts of the world is *Apis mellifera* of Old World origin (Figure V-B). Honeybees secrete wax from glands in their abdomen to build nests, otherwise known as combs, for the larval bees. Worker bees store honey for future use in the hexagonal chambers of the combs. In the wild, honeybees construct nests in the hollows of trees or in caves. Beekeepers use nesting boxes, called hives, to house their bees. In addition to its value as a producer of honey and wax, the common honeybee is even more valuable as a pollinator of crops. This service by bees returns to agriculture between 10 and 20 times the total value of honey and wax they produce. To provide pollination services for agricultural crops, commercial beekeepers transport their colonies from place to place across the country. Until recently the domestic honey bee has been the only bee species whose colonies can be readily moved about.

Honeybees are of inestimable value as agents of cross-pollination. Many plants such as alfalfa and almonds are almost completely dependent upon honeybees for successful reproduction. Currently, the agricultural production of apples is totally dependent on honeybees as a pollinator. In the United States alone, honeybees pollinate approximately $10 billion worth of crops annually.

FIGURE V-B Honeybee (Apis mellifera) gathering nectar from flower. Note pollen on legs.

FIGURE V-C   A beekeeper inspecting honeybee hives.

Unfortunately, American honeybees these days are beset by a host of problems that seriously threaten their contribution to agriculture. Because of disease, virtually no wild bee colonies remain. In addition, the number of managed bee colonies in the United States has decreased from around six million in the 1940s to three million in 1996, with the number of American beekeepers dropping by about 20 percent since 1990. In part this decline is due to economics. The import of honey from countries with lower production costs has resulted in declining market prices. However, the roots of the decline in American honeybee populations go back to the mid to late 1980s when South American queen bees illegally shipped to a few Florida beekeepers carried with them two species of parasites, known as tracheal and varroa mites. Bees that have had a long period of contact with these mites have evolved various defenses.

The European honeybee, favored by nearly all North American beekeepers, has only recently been exposed to the parasitic mites and has not had enough time to develop a resistance. Varroa mites are the primary cause of the virtual elimination of wild honeybees in the northern two-thirds of North America. The damage to bee colonies, which is also observed in domesticated hives, is the work of the pinhead-sized adult females who use their piercing mouthparts to extract the circulatory fluid of the bees. Tracheal mites kill in a different manner. They enter the respiratory passages of bees and eventually become so numerous that they suffocate their hosts.

There is a good chance that American honeybees will ultimately acquire a resistance to the imported parasites as bees have in other countries. Resistance has spontaneously developed in some American mite-infested colonies. Bees become resistant to varroa mites by "learning" to simply clean the mites and mite larvae off themselves and out of their hives. Bees are also able to acquire resistance to tracheal mites but the mechanism by which this occurs is at this time unknown. The pressing question concerning the acquisition of resistance is whether it will occur in time to avert the potentially enormous losses that loom ahead for those crops, such as apples, that are dependent upon bee-assisted pollination.

With American honeybees' numbers falling as a result of natural enemies and economic woes, researchers are exploring other pollination options such as using non-honey-producing solitary bees that are immune to the effects of the mites. Of the 20,000 or so species of bees (3500 of these inhabit the U.S.), 400 are social, and of those, only six are honeybees. The rest are solitary bees. Solitary bees are much better pollinators than honeybees. The hornfaced bee, *Osmia cornifrons*, used commercially for several decades in Japan, can visit 15 flowers a minute, setting 2450 apples a day—compared to the 50 flowers a honeybee can set in a day. At these rates, apple growers would need only 500 to 600 hornfaced bees per hectare (2.47 acres) compared to the thousands of honeybees needed to pollinate the same area. In addition to being more numerous and better pollinators, solitary bees, because they do not need to protect a high-quality energy store like honey, are much more docile than honeybees. This attribute makes raising and transporting solitary bees a much easier and safer endeavor.

The stage is now set for a transition in American agriculture from the use of honeybees to solitary bees as the major pollinators of many of our important crops, including the apple. We will need to wait to see if this transition occurs in time to avoid major financial loss to farmers. This substitution of an essential ingredient to the all-American apple pie will go largely unnoticed by most of us. But the next time you cut into a slice of apple pie, recall the ingredient that appears in no recipe book—the bee.

PART SIX

Ecological Genetics

C H A P T E R   19

Population Genetics

CONCEPTS

1. Genetic information is coded in the DNA molecule.

2. Major sources of genetic variation are mutations and the segregation and recombination of genes.

3. Natural selection can change gene frequency.

4. Natural selection may be stabilizing, directional, or disruptive.

5. Maintenance of altruistic traits may involve group or kin selection.

6. Small populations are vulnerable to inbreeding and genetic drift, which reduce genetic variability.

7. The maintenance of genetic variability depends on effective population size, which may be lower than the actual population.

8. A minimum viable population size is necessary to ensure the persistence of a population over time.

As a rapidly expanding human population steadily encroaches on natural areas, large, expansive populations of many plants and animals species are being reduced to small isolated populations. These populations are forced to exist in fragments of their former habitat. Not only are these small populations now vulnerable to chance demographic changes in birth rates and death rates and chance variations in the environment; they are also subject to genetic deterioration. How much genetic variation have these small populations lost or retained from the original population? What is the probability of the loss of genetic diversity? How would such a loss affect the fate of a small population? What must be done to maintain genetic diversity in small populations? How can genetic variation be maintained in small captive populations of endangered species that will provide the source of stock for reintroduction? These and related questions are explored in this chapter.

GENETIC VARIATION

Wherever you go along the seashore, whether it be on long stretches of beach or harbors and docks, you see and hear the gulls, especially the ubiquitous herring gull. Even a moderately alert observer will detect differences among herring gulls. Most conspicuous are the adults with their bluish-gray back, their white head, neck, underparts, and tail, their black-tipped primary wing feathers, and their yellow bill with a bright red spot near the tip of the lower mandible. Among the adult gulls are younger birds with a different pattern. Some are darkish brown-gray, mottled and barred on the back with white and grayish-buff. Others are lighter in tone with some gray on the back. Still others are similar to adults but with some dusky spotting on the tail and wings. Their bills may have only a suggestion of the red spot.

If you examine the gulls more closely, you will detect other subtle differences. Size, shades and patterns of gray on the back, length of the bill, shape of the red spot, length of the wing, and other characteristics vary among the birds. In fact, so widespread are these smaller differences that if you look carefully at the birds and become acquainted with a colony, you can distinguish one bird from another, just as you can tell one person from another.

Types of Variation

The most obvious variations among the members of a population are **discontinuous,** that is, variations in a specific character or sets of characters that separate individuals into discrete categories, such as male or female or age classes. Thus differences in patterns of plumage enable ornithologists to classify individuals in the gull population as first-year birds, second-year birds, third-year birds, and mature adults. Another type of discontinuous variation is morphological, such as male and female or the red and gray phase of the screech owl. Other discontinuous differences are biochemical, such as blood groups in humans, or even behavioral, such as song dialects in birds.

A second type of variation is **continuous,** a variation in a character that can be placed along a range of values. Characters subject to continuous variation can be measured—for example, tail length of a species of mouse, number of scales on the belly of a snake, rows of kernels on an ear of corn, and shapes and sizes of sepals and petals. The measurements of such a character or set of characters for several individuals in a population can be tabulated as a frequency distribution and arranged graphically as a histogram (Figure 19.1).

Genotypes and Phenotypes

Inherited characteristics of a species and variations in individuals are transmitted from parent to offspring. The sum of the hereditary information carried by the individual is the **genotype.** The genotype directs the development of the individual and underlies the morphological, physiological, and behavioral characteristics of the individual. The external or observable expression of the genotype is the **phenotype.** Some phenotypic expressions result from an interaction of the genotype and the environment. For example, a seedling with a gene for the formation of chlorophyll will develop the normal green color if germinated in the light, but it will be white if germinated in the dark. The gene directs the character of the green color, but its expression is affected by environmental conditions, because chlorophyll develops only in the light.

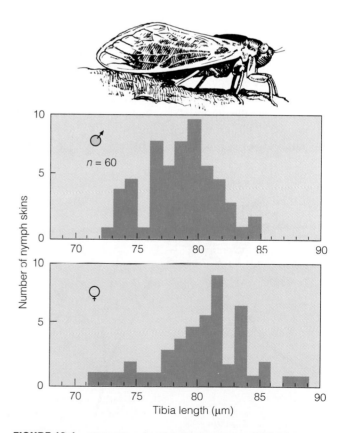

FIGURE 19.1 Histogram showing the frequency distribution of the hind tibia lengths of nymphal exuvia (shed skin) of the periodical cicada *Magicicada septendecim.* (After Dybas and Lloyd 1962.)

The ability of a genotype to give rise to a range of phenotypic expressions under different environmental conditions is known as **phenotypic plasticity.** Some genotypes have a narrow range of reaction to environmental conditions and therefore give rise to a fairly constant phenotypic expression. However, many plants and animals that can survive under a wide range of environmental conditions possess variable and diverse phenotypic responses. Some of the best examples of such phenotypic plasticity are found among plants. The size of plants, the amount of reproductive tissue, and even the shape of the leaf may vary widely at different levels of nutrition, light, and moisture (Figure 19.2). Lacking the mobility of animals, plants must possess more flexibility in their response to environmental conditions to survive. Phenotypic plasticity represents nongenetic variation. An environmentally induced modification of a character is not inherited. However, the ability to modify such a character under certain environmental conditions is inherited.

Sources of Variation

The primary genetic control mechanism, found within the nucleus of every cell in the organism, is deoxyribonucleic acid—DNA. DNA, the information template from which all cells in the organism are copied, is a complex molecule in the shape of a double helix, resembling a twisted ladder. The long strands, comparable to the uprights of a ladder, are formed by an alternating sequence of deoxyribose sugar and phosphate groups. The connections between the strands, or the rungs, consist of pairs of the nitrogen bases adenine, guanine, cytosine, and thymine. In the formation of the rungs, adenine is always paired with thymine and cytosine is always paired with guanine. The DNA molecule is divided into smaller units, called nucleotides, consisting of three elements: phosphate, deoxyribose, and one of the nitrogen bases bonded to the strand at the deoxyribose. The information of heredity is coded in the sequential pattern in which the base pairs occur. Each species

is unique in that its base pairs are arranged in a different order and probably in different proportions from every other species.

In eukaryotic cells DNA is present in larger units called chromosomes, which are found in most living organisms. Each species has a characteristic number of chromosomes in every cell, and the chromosomes occur in pairs. When cells reproduce—a process of division called **mitosis**—each resulting cell nucleus receives the full complement of chromosomes, or the **diploid** number (for example, 46 in humans). In organisms that reproduce sexually, the germ cells or gametes (sperm and egg) result from a different process of cell division, **meiosis,** in which the pairs of chromosomes are split, so that each resulting cell nucleus receives only one-half the full complement, or the **haploid** number (23 in humans). When egg and sperm unite to form a new individual, the diploid number is restored. The chromosomes recombine in a great array of combinations. This segregation and recombination of chromosomes and the hereditary information they carry are the primary sources of variation.

Each chromosome carries units of heredity called **genes,** the informational units of the DNA molecule. Because chromosomes are paired, genes are also paired in the body cells. The position a gene occupies on a chromosome is known as a **locus.** Genes occupying the same locus on a pair of chromosomes are termed **alleles.** If each member of the pair of alleles affects a given trait in the same manner, the two alleles are called **homozygous.** If each affects a given trait in a different manner, the pair is called **heterozygous.** During meiosis the alleles are separated as the chromosomes separate. At the time of fertilization the alleles, one from the sperm and one from the egg, recombine as the chromosomes recombine.

Major interest lies in sources of new genetic variations. One source is the reassortment and recombination of existing genes, both at the level of the gene and at the level of the chromosomes. The other source is mutation, or change in the gene or chromosome.

Recombination of Genetic Material   The major source of genetic (and thus phenotypic) variation among individuals in a population is the recombination of genetic material during sexual reproduction. Each egg or sperm, being haploid, carries one of the two members of a pair of alleles. When the two gametes unite to form a zygote, the alleles are recombined. The number of possible recombinations is extremely large. Recombination does not result in any change in genetic information, as mutation does, but it does provide different combinations of genes upon which selection can act. Some combinations of genes are more adaptive than others. Selection determines the variations or new types that survive in the population. The poorer combinations are eliminated.

The amount of recombinations in a population is limited by a number of characteristics of the species. One limitation is the number of chromosomes and the number of genes. Another is the frequency of crossing over, the exchange of corresponding segments of homologous chromosomes during meiosis. Others include gene flow between populations, the length of generation time, and the type of breeding—for example, single versus multiple broods in a season among animals and self-pollination versus cross-pollination in plants.

FIGURE 19.2   Environmental plasticity in the growth of leaves of the yellow water-buttercup (*Ranuculus flabellaris*). The submerged leaves are divided into threadlike segments. The floating leaves are much broader and less divided.

Mutation A **mutation** is an inheritable change of genetic material in the gene or chromosome. Organisms that possess such changes are called mutants.

Mutations come in two broad types: macromutations and micromutations. **Macromutations** are chromosomal mutations that result from a change in the number of chromosomes or a change in the structure of the chromosome. Of greatest evolutionary interest is a mutation resulting from a change in the number of chromosomes. Such a change can arise in two ways: the complete or partial duplication of the diploid number rather than the transmission of the haploid number, or a deletion of some of the chromosomes.

Polyploidy is the duplication of entire sets of chromosomes. It can arise from an irregularity in meiosis or from the failure of the whole cell to divide at the end of the meiotic division of the nucleus. Forms of polyploidy are triploid (3*n* or three haploid sets), tetraploid (4*n*), and so on.

Polyploidy exists mostly in plants. The condition is rare in animals, because an increase in sex chromosomes would interfere with the mechanism of sex determination and the animal would be sterile. Polyploid plants, on the other hand, can reproduce and spread asexually. Polyploid plants differ from the normal diploid individuals of the same species in appearance and are usually larger, more vigorous, and occasionally more productive.

Another form of macromutation involving sets of chromosomes is duplication or deletion of part of a normal complement of chromosomes. Such deletions or duplications result in abnormal phenotypic conditions. (One such condition is Down syndrome in humans, in which there are three copies of chromosome 21.)

Macromutations also arise with a change in the physical structure of a chromosome involving the deletion, duplication, inversion, or translocation of segments of the chromosome (Figure 19.3). **Deletion** is the loss of a part of a

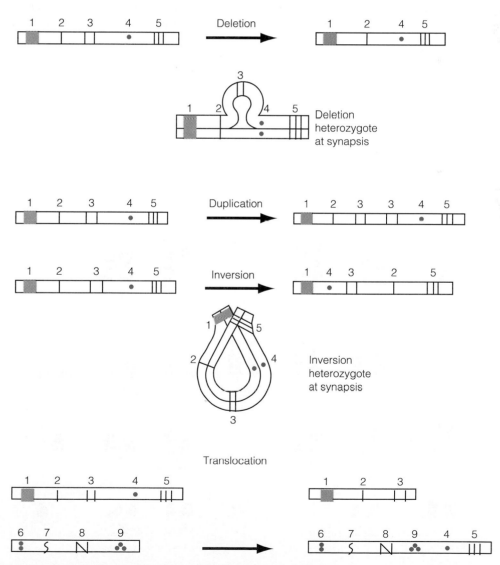

FIGURE 19.3 Types of chromosomal aberrations. When these altered chromosomes join with normal homologues during the first meiotic division, they assume characteristic configurations that allow locus-by-locus matching. Synaptic configurations of deletion and inversion are shown. (Diagrams after Wilson and Bossert 1971.)

chromosome; a definite segment and the genes thereon are missing in the offspring cell. **Duplication** is an addition to the chromosome due to internal doubling. **Inversion** is an alteration of the sequence of genes in the chromosomes. It may occur when a chromosome breaks in two places and the segment between the breaks becomes turned around. Such a break reverses the order of genes with respect to an unbroken chromosome, and it interferes with pairing in a heterozygous individual. **Translocation** is the exchange of segments between two nonpaired (nonhomologous) chromosomes in a heterozygous individual. The genes in the translocated segment become linked to those of the recipient chromosome. Most mutations arising from physical changes in chromosomes are lethal, produce abnormal, nonviable gametes, or have a marked effect on the development of an individual.

Micromutations are gene mutations. They are alterations in the DNA sequence of one or a few nucleotides. During meiosis the gene at a given locus usually is copied exactly and eventually becomes part of the egg or sperm. On occasion the precision of this duplication process breaks down, and the offspring DNA is not an exact replication of the parent DNA. The alteration may be a change in the order of nucleotide pairs, the substitution of one nucleotide pair for another, the deletion of a pair, or various kinds of transpositions.

The rate of mutation in general is low. Most common mutations involve the change of one allele into another. Consider a pair of alleles: one **dominant,** *A;* and the other **recessive,** *a.* In a population homozygous for gene *A,* for example, *A* eventually will mutate to *a* in some of the gametes; and in a population having both genes, mutations may be forward to *a* or backward to *A.* If *A* mutates to *a* faster than *a* to *A,* the frequency of *A* decreases. Rarely is one of the alleles lost to the population, for reversibility prevents a long-term or permanent loss. Eventually such mutations arrive at an equilibrium. Even if one allele is lost from the population, it may reappear by mutation.

Determining Genetic Variation

Phenotypes reflect some of the genetic variation that exists among individuals. Only by exposing the genetic variation buried within and among individuals and populations can we get some measure of the variation that exists.

One method of obtaining a broad picture of genetic variation is the determination of chromosomal variation among individuals. Researchers accomplish this by arresting cell division in white blood cells at metaphase, when chromosomes are dividing, and then staining them to reveal band patterns. The metaphase chromosomes are photographed, and the photographs are enlarged. Individual chromosomes are cut out and arranged by size and shape. This display produces an individual's **karyotype,** a picture of the number and morphology of the chromosomes (Figure 19.4). Comparing karyotypes provides an initial assessment of genetic variation within and among individual in a population.

The bands in the chromosomes represent loci of genes. Major genetic differences relate to the number of loci that exhibit variation. The loci are coded for proteins or enzymes

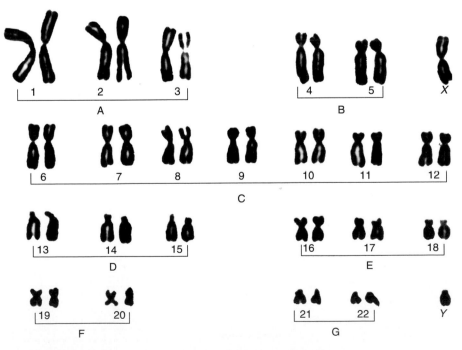

FIGURE 19.4 Human male metaphase chromosomes arranged as a karyotype. Note that groups of chromosomes with similar morphologies are arranged under letter designations (A through G). This arrangement is based on the size of the condensed chromosomes. In the male all chromosomes except the X and Y sex chromosomes are present in pairs.

possessed by the different alleles at that locus. The genetic variants of these proteins are called allozymes. Two individuals with the same allozymes at a locus are considered identical. Distinguishing the different genetic forms of these allozymes involves a technique called gel electrophoresis (for an accessible description of this and other procedures, see Futuyma 1998, Campbell, 1996, Russell 1996). This technique separates macromolecules, either nucleic acids or proteins, on the basis of size by subjecting a homogenized sample of tissue to an electrical charge that moves through a gel substrate. The procedure ends with the staining of the gel to reveal bands of separate DNA molecules across the gel. These bands, each representing an allele at a particular locus, allow the identification of particular DNA molecules (Figure 19.5). The position of the stained proteins provides a comparison of the genetic variability in a population. If the protein bands of one individual do not line up with the protein bands of another, the individuals are considered genetically different.

The DNA itself can be used to explore genetic variation. Some of the DNA consists of encoding sets of alleles. The rest of DNA consists of noncoding regions. DNA is extracted from the cells of an individual and cut in fragments at specific sites with a restricting enzyme. Restricting enzymes, isolated mainly from bacteria, recognize specific short base pair sequences, usually four or five, and cut the DNA within these sequences. Samples of DNA fragments from each individual are placed in an electrophoretic gel to be separated by size and electrical charge. Each sample of fragments of different sizes moves through the gel to form a characteristic pattern of bands. The DNA on the gel is denatured by a chemical treatment, and the DNA strands of particular interest are transferred onto a special paper. Next a radioactive probe — using a labeled nucleic acid single-strand DNA molecule that complements the DNA sequence of the genetic marker of interest—is added to the DNA bands. The probe attaches to the DNA by base pairing to the restriction fragments. Next an X-ray film is placed over the membrane and exposed. Each labeled hybrid shows on the exposed film as band—each individual's unique DNA fingerprint.

Aiding in DNA fingerprinting is the use of the polymerase chain reaction (PCR) technique to make millions of copies of any piece of DNA in a test tube. DNA is mixed with a DNA replication enzyme—DNA polymerase—and nucleotide monomers. In this mixture, DNA replicates two daughter molecules that in turn replace two and so on, as long as nucleotides remain, producing millions of similar molecules in a few hours. It enables researchers to use material obtained from very small samples in the genetic fingerprinting discussed above. Such techniques enable researchers to study genetic variation of mummified bodies, woolly mammoths frozen in glacial ice, plant fossils, and other fossil material and to obtain the DNA sequence of a wide variety of organisms quickly.

Hardy-Weinberg Equilibrium

If a gene occurs in two forms, A and a, then any individual carrying it can fall into three possible diploid classes: AA, aa, and Aa. Individuals in which the alleles are the same, AA or aa, are called **homozygous;** and those in which the alleles are different, Aa, are called **heterozygous.** The haploid gametes produced by homozygous individuals are either all A or all a;

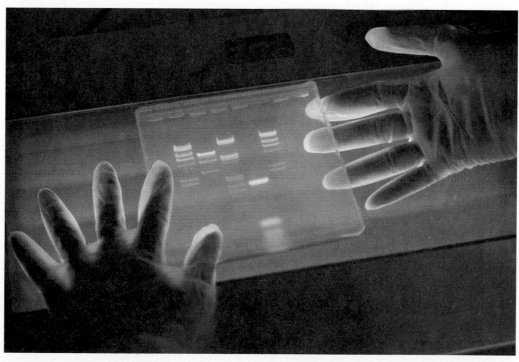

FIGURE 19.5 DNA molecules, separated by electrophoreses and revealed in electrophoretic gels, are being studied under ultraviolet light.

and those by the heterozygous, half *A* and half *a*. They can recombine in sexual reproduction in three possible ways: *AA*, *aa*, and *Aa*. The proportion of the gametes carrying *A* and *a* is determined by the individual genotypes, the genes received from the parents. Eggs and sperm unite at random, and so the proportion of offspring of different genotypes can be predicted based on parental genotypes.

Assume that a population homozygous for the dominant *AA* is mixed with an equal population homozygous for the recessive *aa*. Their offspring, the F_1 generation, will consist of .25 *AA*, .25 *aa*, and .50 *Aa* (Figure 19.6). These proportions are called **genotypic frequencies.** The **allele frequencies,** of course, are .50 of *A* and .50 of *a*.

Will this proportion be maintained through successive generations of a bisexual population? We can phrase this question in the form of a null hypothesis that states the proportions will be maintained through successive generations of a bisexual population under certain conditions: (1) mating is random; (2) mutations do not occur, or if they do, the rate of mutation from *A* to *a* is the same as from *a* to *A*; (3) the population is closed (gene flow from one population to another does not exist); (4) population size is infinite; and (5) no natural selection occurs.

Let us go back to that hypothetical population of homozygous males and females that produced the F_1 generation. The frequency of allele *A* can be designated as *p* (whose value is between 0 and 1) and the frequency of *a* as *q*, equal to $1 - p$ (whose value is between 0 and 1). If mating is random, then the probability that an offspring will receive an *A* allele from the father is *p* and an *A* allele from the mother is also *p*, so the probability that the offspring will be homozygous for *AA* is p^2. The same argument shows that the probability that the offspring will be homozygous for *aa* is q^2. But the probability also exists that the offspring will receive an *A* allele from its father and an *a* allele from its mother: $p \times q$. Or it could receive an *a* allele from the father and an *A* allele from its mother: $q \times p$. The combined probability is $2pq$. Thus the predicted genotypic frequency in the offspring in the population can be expressed as

$$p^2 + 2pq + q^2 = 1$$

This expression is known as the Hardy-Weinberg law.

Simply stated, the Hardy-Weinberg law says the following: if *p* is the frequency of allele *A*, the dominant, and *q* is the frequency of allele *a*, the recessive, so that $p + q = 1$, then the genotypic frequency will be $p^2 + 2pq + q^2 = 1$, where p^2 is the frequency of homozygous individuals *AA*, q^2 is the frequency of homozygous individuals *aa*, and $2pq$ is the frequency of heterozygous individuals *Aa*. In the hypothetical population above, the proportion of the genotypes in the F_1 generation will be $(.5)^2 + 2(.5 \times .5) + (.5)^2$. The same genotypic frequency will be maintained into the F_2 generation if the conditions of the Hardy-Weinberg law hold (Figure 19.7).

The same tendency toward stable ratios of alleles can be demonstrated even if the ratio is not the classical Mendelian ratio. Normally both *p* and *q* = .5. Imagine a population in which the ratio of *A* alleles (*p*) to *a* alleles (*q*) is .6 to .4 (Figure 19.8). The frequency of the genotypes in the F_1 generation will be .36 *AA*, .48 *Aa*, and .16 *aa;* and the allele frequency will be $(.6)^2 + 2(.6 \times .4) + (.4)^2$. We can conclude that all succeeding generations will carry the same proportions of the three genotypes, provided the assumptions mentioned earlier are met.

The stated assumptions, of course, are never met perfectly in any real population. If not, then what is the value of the Hardy-Weinberg law? First, it serves as a null hypothesis against which to test the departure of frequencies away from the Hardy-Weinberg equilibrium. A departure indicates that one or more of the assumptions is invalid, a starting point for further investigation.

In certain cases the Hardy-Weinberg equation can be used to calculate genotypic frequencies, especially when the character is one for which mating is usually random, such as certain blood types. For example, about 1 in every 10,000 newborn Caucasians suffers the effects of phenylketonuria, the inability to metabolize phenylalanine, which can lead to severe mental retardation. The gene causing the

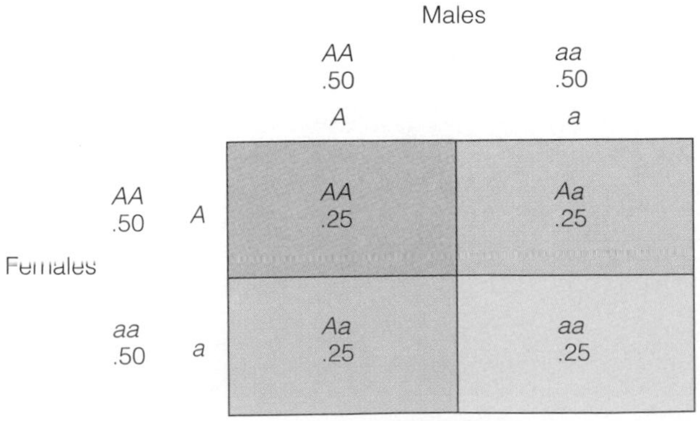

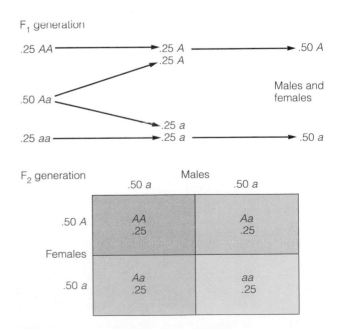

FIGURE 19.6   Mixing two homozygous populations.

FIGURE 19.7   Proportions in the F_2 generation.

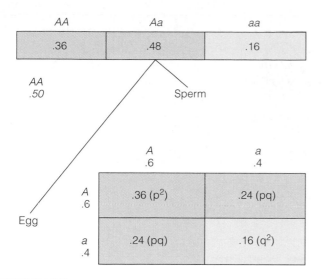

FIGURE 19.8 The Hardy-Weinberg law applied to a hypothetical case.

condition obviously is the recessive q, so the genotypic frequency $q^2 = 1/10,000$; therefore $q = (1/10,000)^{1/2} = .01$, and $p = 1.00 - .01 = .99$. Carriers for the disease would be the heterozygotes $2pq = 2(.01 \times .99) = .0198$. Thus about 1 person in 50 $(1/.0198 \approx 50)$ would carry the allele. These heterozygotes are phenotypically indistinguishable from the noncarriers.

NATURAL SELECTION

Nonrandom Reproduction

Variation in a population seldom is constant from generation to generation. One reason is gene mutation, the ultimate source of genetic variation. Of more immediate consequence is nonrandomness of reproduction within a population. Not all members contribute their genetic characteristics to the same degree to the next generation. It is this selectivity that is natural selection.

Before a given individual in a population can contribute to the succeeding generation, it must first survive to reproduce. Survival begins from the time of fertilization through the periods of development, growth, and sexual maturation. Fertilized eggs may fail to develop fully and die from physiological or environmental causes. Disease, predation, and accidents eliminate those young not quite as swift or as strong as their siblings. In such survival genetic variation plays a key role, for natural selection influences the frequency of alleles in a population. If a mutation arises that places its carrier at a disadvantage when expressed, selective pressures eliminate the individual; on the other hand, and advantageous mutation is retained.

An example of such selection can be found among the flies. When DDT was first used as an insecticide against houseflies, the chemical was highly effective, destroying the bulk of local populations. Among the flies were a few that did not die, that carried an allele or a combination of genes that made them resistant to the spray. Resistance in one strain of flies was due to a recessive gene. Flies homozygous for this gene tolerated a high concentration of DDT, while homozygous dominants and heterozygotes were killed. These flies survived to multiply. Many of their offspring were as resistant to the sprays as the parents; some were even more resistant. The least resistant were selected against; the most highly resistant were retained in the reproductive population. Later applications of DDT continually selected for a combination of genes most resistant to the insecticide. As a result DDT became ineffective in fly control, and newer, stronger sprays were required. Eventually even these sprays select resistant strains of flies, which will become adapted to the new environmental conditions. However, to acquire this resistance the flies pay a price. In the absence of DDT the resistant flies are inferior competitors to the nonresistant flies, which have a shorter development time (Pimentel et al. 1951). If the spraying is stopped, evolution will be reversed and the resistance will largely disappear from the fly population.

Similar selective processes operate on all organisms. Once surviving young reach reproductive age, more individuals are eliminated from the parental population. The maintenance of genetic equilibrium is based on random mating, but mating is not random. Many species of animals, particularly among birds, fish, and some insects, have elaborate courtship and mating rituals (Chapter 13). Any courtship pattern that deviates from the commonly accepted pattern is selected against, and the deviating individual and its genes are eliminated from the reproductive population. On the other hand, animals possessing a color pattern or movement that accents the typical pattern are selected for. Any new mutations that improve on courtship, mating signals, and ritual would possess a favored position in subsequent generations. Among polygamous species, in particular, the majority of males go mateless, for the females mate with dominant males that tolerate no interference from younger or less aggressive males. States of psychological and physiological readiness also are involved in mate selection. Unless both male and female are at the same state of sexual readiness, mating will not occur.

Neither is fecundity random. Some families or lines increase in number through time; others fade away. Obviously those who produce more offspring increase the frequency of their genes in a population and affect natural selection. For example, if individuals with allele A produce ten offspring to every one produced by those with allele a, the proportion of A in the population will increase. There is a limit, however, for natural selection does not always favor fecundity. If an increased number of young per female results in reduced maternal care, survival of offspring may be reduced, particularly among those animals whose chances of individual survival are high. Those organisms whose chances of individual survival are low—for example, ground-nesting game birds, oceanic fish, parasites, and marine invertebrates—have become very fecund.

Fitness and Modes of Selection

If an organism can tolerate a given set of conditions so that it can leave fertile progeny, thus contributing its genetic traits to the population gene pool, it can be said to be adapted to its environment. If an organism survives only as an individual and leaves few or no mature, reproducing progeny, thus contributing little or nothing to the gene pool of the population, it is poorly adapted. Those individuals that contribute the most to the gene pool are said to be the most fit, and those that contribute little or nothing are the least fit. The **fitness** of the individual is measured by its reproducing offspring. Natural selection is not a measure of individual survival, but of differential reproduction, the ability to leave the most reproducing offspring.

In simplest terms fitness is measured by comparing the number of reproducing offspring produced by one genotype with the number produced by another. Suppose genotype *AA* produces 250 reproducing offspring and genotype *BB* produces 200. The reproductive success of genotype *BB* compared with *AA* is reduced by 50; or expressed in fractional terms, 50/250 = .20. Obviously *AA* is the more successful genotype.

In measuring selection or the adaptive value of a genotype, fitness is frequently designated as *W*, the value of which ranges from 1 for the most productive genotype to 0 for no reproduction (lethal genes). In our simple example the value of *W* for *AA* would be designated as 1; the value of *W* for *BB* would be 1.00 − .20, or .80.

The selective pressure acting on a genotype is designated as a **selection coefficient,** *s*. It can be stated as the difference between 1 and the fitness value. In the example, the selection coefficient for *AA* is 0; for *BB* it is .20. Thus, *W* (fitness) = 1 − *s*; similarly, *s* = 1 − *W*. For *BB*, *W* = 1 − .20 = .80; *s* = 1 − .80 = .20. This example is simplified. The calculations used to determine fitness values and selection coefficients are more complex, but not difficult. Good discussions are given in Haldane (1954), Wallace (1968) Harle (1988), and Ricklefs (1990).

Within a population selection may act in three ways. Given an optimum intermediate genotype, **stabilizing selection** favors the average expression of the phenotype at the expense of both extremes. This type of selection takes place in all populations. **Directional selection** favors one extreme phenotype at the expense of all others. The mean phenotype is shifted toward the extreme, provided that heritable variations of an effective kind are present. **Disruptive selection** favors both extremes over a range of phenotypes, although not necessarily to the same extent, at the expense of the average type (Figure 19.9).

Disruptive selection is most apt to occur in a population living in a heterogeneous environment in which there is a strong selection for adaptability to a developing environmental condition (see Chapter 13). As a result the population may subdivide into two phenotypes. This division would give rise either to polymorphism or to separation into populations with different characteristics. The latter is most likely to take place in areas where selection is intense and where optimum habitat adjoins or is penetrated by less than optimum habitat. Organisms settling in these habitats will adapt to the local environment. If disruptive selection is strong enough, it will lead to preferential mating and eventually to genetic divergence of two or more groups.

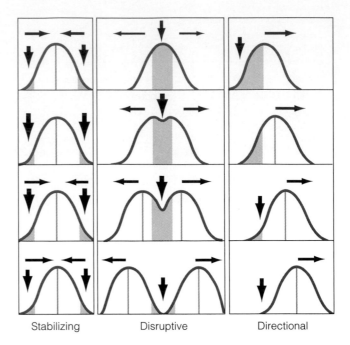

Stabilizing   Disruptive   Directional

FIGURE 19.9   Three modes of selection. Stabilizing selection favors organisms with values close to the population mean. Consequently, little or no change is produced in the population. Directional selection accounts for most of the change observed in evolution. Disruptive selection increases frequencies in the extremes. The curves represent the frequency of organisms with a certain range of values. Upward arrows indicate favorable selection, downward arrows adverse selection. Left and right arrows indicate evolutionary change. Shaded areas represent the phenotypes being eliminated by selection.

An example of the development of a polymorphic species through disruptive selection is the peppered moth (*Biston betularia*) in England. Before the middle of the nineteenth century the moth, as far as is known, was always white with black speckling in the wings and body (Figure 19.10). In 1850 near the manufacturing center of Manchester, a melanistic form of the species was caught for the first time. The black form, *carbonaria,* increased steadily through the years until it became extremely common, often reaching a frequency of 95 percent or more in Manchester and other industrial areas. From these places *carbonaria* spread mostly westward into rural areas far from the industrial cities. The black form came about by the spread of dominant and semidominant mutant genes, none of which are recessive. This increased frequency and spread was brought about by natural selection. The normal form of the peppered moth has a color pattern that renders it inconspicuous when it rests on lichen-covered tree trunks; but the grime and soot of industrial areas carried great distances over the English countryside by prevailing westerly winds killed or reduced the lichen on trees and turned the bark of the trees a nearly uniform black. The dark form is conspicuous against the lichen-covered trunk, but inconspicuous against the black.

A British biologist, H. R. D. Kettlewell (1961; see also Kettlewell 1965) experimentally demonstrated the role of nat-

FIGURE 19.10 Normal and melanistic forms of the polymorphic peppered moth *Biston betularia* at rest on a lichen-covered tree. The spread of the melanistic form *carbonaria* in the industrial areas is associated in part with improved concealment of black individuals on soot-darkened, lichen-free tree trunks. Away from industrial areas the normal color is most frequent because black individuals resting on lichen-covered trunks are subject to heavy predation by birds.

ural selection in the spread of the dark form. He reared, marked, and released melanistic and light forms in polluted woods. The melanistic form had a better survival rate than the light form and therefore left more offspring. To confirm the role of natural selection, Kettlewell released light and dark forms in unpolluted woods. There the light form survived better. The reason was selective predation. In woods with lichen-covered trees, the melanistic form was more easily seen by several species of insect-feeding birds and was therefore subject to heavier predation. In polluted woods, the light form bore the brunt of predation. For this reason the normal form has virtually disappeared from polluted country, but it is still common in the unpolluted areas in western and northern Great Britain.

With the passage of the Clean Air Act in Britain in 1965, however, the mean concentration of sulfur dioxide pollution fell from 300 $\mu g\ m^{-3}$ in 1960 to 50 $\mu g\ m^{-3}$ in 1975 and has remained fairly constant since then. Selection has continued against the black form to the point that that the melanistic form has declined from 90 percent in 1975 to 26 percent in 1989 (Clarke et al. 1990).

Although selective predation is regarded as the major influence in maintaining polymorphism in the peppered moth, the polymorphism may also be maintained by another factor independent of predation (Lees and Creed 1975). The dark-colored individuals apparently have a physiological advantage over the nonmelanistic form in withstanding the effects of air pollution. The dark form will increase in those areas where it has the physiological advantage. In some cases visual advantage is less than the physiological advantage.

Group and Kin Selection

According to evolutionary theory, selection operates on the level of the individual and acts on phenotypes. Certain traits, however, such as warning calls, warning coloration, and behaviors that benefit a group are often at the expense of the individual. Such acts and traits are called **altruistic.** Altruism,

strictly defined, is the sacrifice of one's own well-being in the service of another. In genetic terms, the altruist contributes to the genetic fitness of another individual, the recipient of the altruistic act, while decreasing its own fitness (Michod 1982).

For many years, evolutionary ecologists have debated this puzzling question. How can such traits be maintained in a population when the altruist and thus the altruistic gene are selected against, whereas the nonaltruistic or selfish gene is at a selective advantage? Is it possible to pass on a trait that benefits the species but harms the organism that carries it?

Group Selection   To explain such traits, some population and behavioral ecologists have suggested that **group selection** operates on the differential productivity of local populations (Wilson 1980). The characteristic selected improves the fitness of the group, though it may decrease the fitness of any individual in the group. Any genetic differences among local populations that decrease the likelihood of local extinction, or increase the likelihood that one local population will produce emigrants or colonists who will affect the genetic composition of another, will favor that group.

The idea of group selection was suggested by Darwin in *The Origin of Species* (1859) and by Sewall Wright (1931a, b, 1935), who called it **interdemic selection** (Figure 19.11). The concept began to receive serious and controversial attention when a Scottish ecologist, V. C. Wynne-Edwards, published in 1962 *Animal Dispersion in Relation to Social Behavior.* In that book Wynne-Edwards advanced the idea that animals tend to avoid overexploitation of their habitat, especially the food supply, by altruistic restraint in population growth, either by reducing or refraining from reproduction. Restraint was achieved through the mechanism of social behavior in which displays provided information about the local population. Local populations that restrained reproduction were more likely to survive than populations that grew beyond the ability of the resource to support them. Such populations would decline or go extinct, leaving empty habitats. These habitats would then be colonized by dispersers from altruistic populations.

The idea of group selection was vigorously challenged by many evolutionary ecologists (see Williams 1966, Wines 1966, Lack 1966, Ghiselin 1974, Maynard Smith 1976), who argued that natural selection could act only on individuals. It was accepted in modified forms by others (Lewontin 1965, Emlen 1973, Gilpin 1975, Alexander and Borgia 1978). D. S. Wilson (1975, 1977, 1979, 1980, 1983) framed the idea of group selection in a somewhat different context. He views group selection as a component of natural selection that operates on the differential productivity of local populations within a global system. Local populations or **demes** are groups of individuals that interact with one another sometime in their life cycle, giving any two of them an equal opportunity of becoming neighbors (Mayr 1963, Wilson 1977). Within demes are genetically distinct individuals that form trait groups (Wilson 1979, 1980) for a part of a generation or for many generations. These trait groups interact in a number of ways, such as mating, competition, defense against predation, and the like. We can

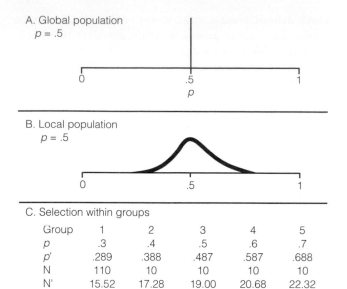

A. Global population
p = .5

0   .5   1
             p

B. Local population
p = .5

0   .5   1

C. Selection within groups

Group	1	2	3	4	5
p	.3	.4	.5	.6	.7
p'	.289	.388	.487	.587	.688
N	110	10	10	10	10
N'	15.52	17.28	19.00	20.68	22.32

D. Global population
p = .5056

0   .5   1

FIGURE 19.11   A numerical example of interdemic selection. The
x axis of the graph shows the given frequencies (p) of the A allele
and the y axis the relative abundance of s groups within a given
frequency p. (a) Global population represented by a single
frequency of A_1. (b) Local groups that vary in p values. (c) Table
showing selection within five representative groups, using Wright's
model (see below) with $b = 2$ and $s = .05$. (d) Global population
after selection. Wright's model for the socially advantageous A_1
allele (frequency = p) that benefits the group (b term) at the
expense of itself (s term) is

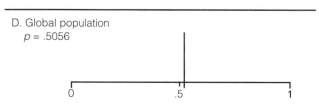

Genotype	Frequency	Selective value
AA	p^2	$(1 + bp)(1 - 2s)$
Aa	$2p(1 - p)$	$(1 + bp)(1 - s)$
aa	$(1 - p)^2$	$(1 + bp)$

(Wilson, 1983.)

consider a deme not only as a population of individuals but
also as a population of groups isolated relative to certain
traits. Thus the population of a given organism has levels.
Total population over a large given area, the global population, consists of numerous local populations, or demes, isolated in varying degrees from one another; and each deme
consists of a number of trait groups. Within these trait
groups, intrademic group selection as well as individual
selection takes place.

Individual selection is that component of natural selection
that operates on the differential fitness of individuals within
local populations. Traits promoted by individual selection are
considered selfish, and those promoted by group selection,
altruistic. Altruistic or group-benefit traits increase the relative
productivity of local populations within a global system. Altruistic traits may be weak—that is, not strongly sacrificial—or
strong. Selection for an altruistic trait might occur in species
with small freely interbreeding local populations sufficiently

isolated to allow some differentiation in gene frequency, and
with little gene flow between other local groups.

A model of intrademic group selection illustrates how an
altruistic trait, although selected against in an individual, may
be selected for in a group (Wilson 1983). Assume that an
altruistic allele A benefiting the group at the expense of itself
exists in a global population with a frequency of $p = 0.5$.
Individuals, however, are distributed in local populations in
which p varies from 0.3 to 0.7, and selection operates within
each group as outlined in Figure 19.11. Note that in each local
population, the A allele declines in frequency because of individual selection, but at the same time populations N' increase
relative to p. When we weigh the new global frequency of A
by group size, the frequency of p actually increases, suggesting that an altruistic allele can evolve, even though it is
selected against within each group. As with individual selection, the model suggests that group fitness involves some heritable phenotypic variation that influences group productivity
or persistence. Natural selection will favor that characteristic, just as genetic variation influences individual selection.
Only within groups can an allele can have low relative individual fitness and still be selected. Thus an altruistic allele in
a population selected against in mixed groups can persist in
a population by intrademic group selection.

Consider a local population or deme possessing a high
frequency of an allele that decreases mortality or increases
reproduction that is surrounded by demes possessing the allele
at lower frequencies. The deme possessing a high frequency
of the adaptive allele will develop some selective advantage
over neighboring demes. It will produce a greater surplus population that will emigrate to surrounding demes or into empty
habitat patches (created perhaps by local extinctions of populations possessing the adaptive allele in low frequencies). In
fact, group selection may be particularly important for traits
affecting dispersal (see Slatkin 1987). Group selection can
then be defined as changes in allele frequencies resulting from
differential extinction or productivity of groups. Because the
groups themselves are short-lived, extinction of groups is not
as relevant as differential productivity of groups or the contribution of groups to the mating pool (Michod 1982).

Kin Selection   Evolution of an altruistic trait may take place
more easily when groups involved are made up of closely
related individuals sharing genes by common descent. In such
a setting an individual may increase its own fitness and the fitness of close relatives in the long run by decreasing its fitness
in the short run. This idea, advanced by W. D. Hamilton in
1964, was called **kin selection** by Maynard Smith (1964).

The cost and benefit of an altruistic trait depends on the
closeness of the relationship. Genes for altruism can be selected
only when the benefit to the recipient is greater than the reciprocal of the coefficient of relationship between the altruist and
the recipient. This requirement was defined by Hamilton:

$$k > 1/r$$

Here r is the proportion of the genes of two individuals identical because of common descent and k is the relationship of

recipient's benefit (*b*) to the altruist's cost (*c*): $k = b/c$. The value of *r* varies from 1 for identical twins to 0 for no relationship. For parent-offspring the value is .5, because the offspring receives one-half of its genes from each parent; and conversely, a parent contributes one-half of its genes to an individual offspring. The value for a full sibling would be .5; for a half sib, .25; for an uncle, aunt, niece, or nephew, .25; and for a first cousin, .125. The closer the relationship, the higher the value of *r*. If the value of *r* is large, that is, the relationship is close, the value of $1/r$ is small. If *k*, the ratio of fitness to the loss of fitness, exceeds $1/r$, then the altruistic gene would be selected for. If the value of $1/r$ is small, then the ratio of benefits *b* to cost *c*, or *k*, can be large and still favor the altruistic gene. In other words, an altruist can take greater risk to help a close relative than a distant one. Even in that situation the benefit/cost ratio *k* would have to exceed a value of 2 (1/.5).

Consider an individual A who, facing 100 percent chance of death, gives a warning call to a full sibling. If this alarm warns only a single sibling, then the benefit (survival of 1) relative to the cost (death of 1), would be $1/1 < 1/.5$, and altruism would be selected against. If the warning saves four full sibs, then $4/1 > 1/.5$, and the trait would be selected for. If the act incurs only 50 percent chance of death, then the benefit, 1/.5, would be the same as $1/r$, 1/.5, and there would be no benefit from the act. By the same token, the altruist in the initial situation above would have to save more than eight first cousins before the trait would be selected for.

Two aspects of fitness are involved in kin selection. One is individual fitness, which involves the genes in the individual's own offspring (classic selection). The other is the additional fitness acquired by improving the fitness of very close relatives, especially parents and sibs (kin selection). An individual's own direct reproductive success added to that of its close relatives possessing replication of its own genes is **inclusive fitness** (Hamilton 1964), a concept used to explain many aspects of social behavior.

Kin selection, then, is the evolution of a genetic trait expressed by one individual that affects the genotypic fitness of one or more directly related individuals (Michod 1982). It is favored when an increase in fitness of closely related individuals is great enough to compensate for the loss in fitness of the altruistic individual. If the dispersers and helpers are very close kin, as they usually are, those individuals serve to increase the fitness of the genetic traits they hold in common.

An altruistic act need not mean sacrifice of life. Any act counts that improves reproductive opportunities for the remaining sibs or parents. An example is dispersal from a kin group, as in black bear (*Ursus americanus*) (see Rogers 1987) and ground squirrel (*Spermophilus* spp.) (Holekamp and Sherman 1989), or helping parents raise full or half sibs.

The most widespread examples of altruism are eusociality and cooperative breeding. **Eusociality** is the extreme in social living (Andersson 1984). It is characterized by (1) cooperative caring for the young, (2) division of labor with more or less sterile workers caring for individuals engaged in reproduction, and (3) an overlap of at least two generations of life stages able to contribute to colony labor. Such euso-

ciality is found among bees, wasps, and termites, and one species of mammal, the naked mole rat.

Among the Hymenoptera there are various levels of eusociality. It is most highly developed among bees and termites, where it gives real meaning to the term kin selection. In the bees, the workers are nonreproducing or sterile females waiting on a sister, the reproductive queen. In these insects, males arise from unfertilized eggs and possess one set of chromosomes from the mother (haploidy). The females have one set from each parent (diploidy). Such sex determination is called haplodiploidy. The females have all the genes from their father ($r = 1$) and one-half their mother's genes ($r = .5$), so that the female offspring have three-quarters of their genes in common. Thus the females are more closely related to full sisters than to their mother and their own offspring. Nonreproductive females can propagate their own genes most effectively and increase their own fitness best by helping create more sisters.

The most remarkable example of eusociality among vertebrates is the naked mole rat (*Heterocephalus glaber*), a member of the family Bathyergidae, the mole rats of Africa. The eusocial naked mole rat, restricted to the hot, arid Horn of Africa, lives in colonies ruled by a queen waited on by nonreproducing individuals, both male and female. The queen, with two or three large male consorts, breeds year-long and produces litters of up to 12. The queen and the breeding males are concerned with the handling, care, and grooming of the young. Larger nonbreeding males and females defend the colony and remove dirt from the tunnels; smaller nonbreeders transport food, gather nest material, and clear tunnels (Honeycutt 1992, Sherman et al. 1991).

A much less complex form of reproductive altruism is cooperative breeding behavior, known to be expressed in 220 species of birds (Brown 1987) and 120 species of mammals (Riedman 1982). In cooperative breeding, adult individuals regularly help the genetic parents in the rearing of the young. The cooperative breeding groups are typically family groups formed through the retention of grown offspring. Such breeding systems can develop complex extended families involving more than one breeding pair (Emlen 1991, Koenig and Mumme 1987). Two of the simpler ones are those of the Florida scrub jay (*Aphelocoma coerulescens*) and the black-backed jackal (*Canis mesomelas*) of Africa.

The basic social unit of the Florida scrub jay (Figure 19.12) is a monogamous breeding pair together with some of their young from the previous one or two years (Woolfenden 1975, Woolfenden and Fitzpatrick 1984). These mature young help in group defense of the territory (McGowan and Woolfenden 1989). During the breeding season they bring food to the nestlings, tend and guard the nest, and continue to provide food and protection to the fledgling out of the nest. Female helpers rarely remain for more than two years because they join the breeding population. Unmated males may be around longer.

The black-backed jackal (Figure 19.13) has a similar social structure (Moehlman 1979, 1983, 1986). Some of the young of the year remain with their parents through the following breeding season, forming groups of three to five adults. The mature offspring help the parents rear the next

FIGURE 19.12   The Florida scrub jay (*Aphelocoma coerulescens*) inhabits the thickets of sand pine, scrub oaks, and palmettos along the east and west coasts of Florida. Young scrub jays help their parents feed and care for young for several years.

FIGURE 19.13   The black-backed jackal (*Canis mesomelas*) inhabits savannas and open woodlands of southern Africa. Some young remain with the monogamous parents, do not breed, and assist with the next litter.

year's litter. They bring food to the nursing female, regurgitate food for the young, play with and groom them, and guard the pups when the parents are away from the den.

In these cooperative breeding systems, the helpers do enhance the fitness of the breeders and indirectly their own fitness (Emlen 1991). Pairs with helpers have higher reproductive success than those without them (Figure 19.14). Helpers increase the success of reproductive attempts by the mated pair by helping to feed and care for the young. Helpers also reduce the workload of the parents, enabling them to produce a second litter. Because they defend against predators and reduce reproductive stress, helpers improve the breeders' survivorship, increasing their inclusive fitness.

If these helpers are so successful at aiding the parental pair, why don't they go out on their own? Several hypotheses have been advanced to explain why young adults stay at home. One hypothesis is that the habitat is saturated and suitable territories are limited. The mature offspring have to wait for a suitable territory to open, and waiting is best done at home in familiar surroundings. Another hypothesis is that although vacant territories are available, they are at best marginal, so why not wait for a high-quality opening? Because of the helpers' assistance, the family size of the pair increases. As the family group expands, its territory expands. Eventually a dominant male helper may claim a part of the enlarged territory as his own, or inherit it and become a breeder, or hold a competitive edge for nearby openings. By helping, the male has come out ahead. He has

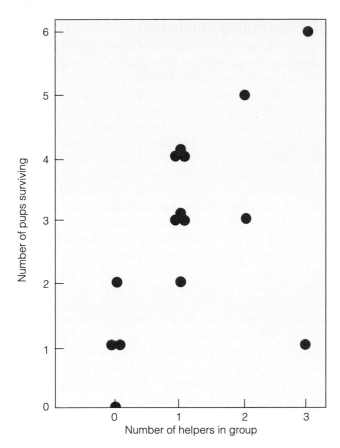

FIGURE 19.14 Correlation between the number of parental helpers and reproductive success in the black-backed jackal. (From Moehlman 1979.)

increased his inclusive fitness and improved the chances of his own direct fitness by increasing his opportunity to become a breeder. A third hypothesis is that dispersal into unknown situations is risky, especially if the only habitat available is marginal. Early dispersers face an increased risk of mortality, a reduced probability of securing a mate, and the high costs of independent reproduction, especially in a fluctuating environment.

In the long run, then, what does a helper gain by staying at home for a while? First, it increases the probability of its own survival. By belonging to a larger group it has increased vigilance against predators and access to social and physical resources of the natal territory. Second, the helper enhances its own likelihood of becoming a successful breeder by gaining access to a superior territory and thus increased probability of obtaining a mate. Third, the helper can increase its reproductive success because of its experience at parenting, and it will gain a similar set of helpers. Finally, helping increases the survival of related breeders whose production of young will increase the helper's own inclusive fitness through their reproductive efforts. Thus in a way, a helper may not be altruistic at all, but practicing a form of selfish individual selection under the guise of cooperation (Lignon 1981).

INBREEDING

Although a local population of a species consists of all individuals occupying a given area, it really is made up of a number of semi-isolated subpopulations or metapopulations, often with minimal interchange (Ralls et al. 1986). Isolation becomes most pronounced when subpopulations occupy patches or islands of habitats, as do house mice restricted to individual barns or white-footed mice confined to islands of forest growth scattered through agricultural or suburban landscapes. Such small populations may be subject to inbreeding.

Inbreeding, simply defined, is breeding between relatives. More precisely inbreeding is the mating of relatives that carry alleles identical by descent from a common ancestor or ancestors. Identical by descent means that two alleles are copies of the same DNA sequence found in a common ancestor (Figure 19.15). With inbreeding, mates on the average are more closely related than they would be if they had been chosen at random from the population. A broader definition of inbreeding is the mating of relatives that share a greater common ancestry than if they had been drawn at random from a large deme with N greater than 1000 (Shields 1993). Thus inbreeding forms a continuum from extreme inbreeding between sibs or between parents and sibs to extreme outbreeding (discussed later) (Table 19.1). As the population size decreases, the probability exists that two parents are related, even if mating at random. Some reasons for inbreeding are small isolated populations, close proximity of potential mates, ecological preferences, morphological resemblance among individuals, and the like. The principal effect of inbreeding is an increased frequency of homozygous genotypes (Figure 19.16).

The extreme in inbreeding is self-fertilization, which may occur in plants. Such inbreeding provides a measure of comparison for degrees of inbreeding. Consider the population described by the Hardy-Weinberg equilibrium earlier, in

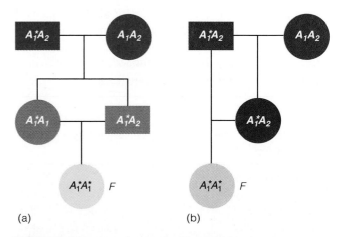

FIGURE 19.15 Development of autozygosity, possession of alleles identical by descent. The pedigrees chart inbreeding due to (a) sib mating and (b) parent-offspring mating. The A_1^* allele is obtained by direct descent. The inbreeding coefficient, F, is 1/4 in both cases.

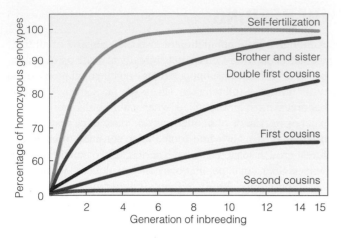

FIGURE 19.16 The percentage of homozygous offspring from systematic matings with different degrees of inbreeding. Note how rapidly homozygosity declines as the relationship of offspring becomes further removed from the original parent stock.

which p (A) = q (a) and both are equal to .5. Thus p^2 = .25, $2pq$ = .50, and q^2 = .25. In our hypothetical population, all breeding involves selfing from the start. Within the population, all homozygotes AA and aa breed true. Offspring from the heterozygotes will be one-half heterozygotes and one-half homozygotes each generation. The homozygotes produced will be added to the pool of homozygotes in the population; the remaining heterozygotes in the next generation again will produce offspring one-half homozygotes and one-half heterozygotes. Eventually the self-fertilizing population will become exclusively homozygous.

Within one generation the gene frequencies will be [(1/4)(1) + (1/2)(1/4)], (1/2)(1/2), [(1/4)(1) + (1/2)(1/4)], or 3/8 AA, 2/8 Aa, 3/8 aa (Table 19.2). Already the degree of heterozygosity has declined from 1/2 to 1/4. The expected frequency of heterozygosity in a self-fertilizing population is $2pq \times (1/2)^n$, where $2pq$ is the initial frequency of heterozygotes and n is the number of consecutive generations. Note that with inbreeding, the frequency of alleles remains the same, as predicted by the Hardy-Weinberg equilibrium, but homozygosity increases at the expense of heterozygosity. In effect, variation originally partitioned within individuals becomes partitioned among individuals.

The Inbreeding Coefficient

Exclusively self-fertilizing populations, the extremes of inbreeding, are uncommon. Even populations of self-fertilizing plants experience some periods of cross-fertilization or outbreeding. Close inbreeding usually involves mating between brother and sister, parent and offspring, or more frequently cousins. Sometimes a homozygote's alleles are identical because of independent mutations. Such individuals are **allozygous** for the alleles involved. Inbred individuals possess alleles that are identical by descent; they can be traced back to a common ancestor (see Figure 19.15). Such homozygotes are **autozygous.** Thus one outcome of inbreeding is an increase in the frequency of autozygous individuals.

The degree of inbreeding depends on relationship. Brothers and sisters and parents and offspring share one-half of their genes, so the relationship is very close. First cousins share one-quarter of their genetic heritage. As relationships diverge, the frequency of shared genes diverges (see Figure 19.16). The amount of inbreeding between relatives is measured by the coefficient of inbreeding (F). F is the probability that an individual receives at a given locus two genes that are identical by descent; stated differently, it is the amount of heterozygosity that has been lost. For a self-compatible population, the inbreeding coefficient is

$$F = \frac{H_0 - H}{H_0}$$

where H_0 = $2pq$ and H is the actual frequency of heterozygous genotypes in the population. F is a measure of the fractional reduction of heterozygosity in an inbreeding population relative to reduction in a randomly mating population with the same frequency of alleles.

Consider a heterozygous population possessing two kinds of gametes in which no two alleles are alike. The total number of kinds of gametes produced in the population is $2N_0$. When the gametes unite to form the zygotes of the next generation, a probability of $1/2N_0$ exists that two identical gametes will unite to form a homozygote. Within an inbreed-

TABLE 19.1 The Inbreeding/Outbreeding Continuum

Extreme inbreeding	$N_e \leq 4$
Moderate inbreeding	$5 \leq N_e \leq 100$
Mild inbreeding	$101 \leq N_e \leq 1,000$
Mild outbreeding	$1,001 \leq N_e \leq 10,000$
Moderate outbreeding	$10,001 \leq N_e$
Extreme outbreeding	Interpopulation or interspecies

N_e = effective breeding population size—see p.374.
Source: Shields 1993.

TABLE 19.2 Decrease in Heterozygosity Under Systematic Self-Fertilization Starting with an Equilibrium Population ($p = q = 1/2$)

	GENOTYPIC FREQUENCIES				
Generations	A/A	A/a	a/a	F	q
0	$1/4$	$1/2$	$1/4$	0	$1/2$
1	$3/8$	$1/4$	$3/8$	$1/2$	$1/2$
2	$7/16$	$1/8$	$7/16$	$3/4$	$1/2$
3	$15/32$	$1/16$	$15/32$	$7/8$	$1/2$
4	$31/64$	$1/32$	$31/64$	$15/16$	$1/2$
n	$\dfrac{1 - (1/2)^n}{2}$	$(1/2)^n$	$\dfrac{1 - (1/2)^n}{2}$	$1 - (1/2)^n$	$1/2$
∞	$1/2$	0	$1/2$	1	$1/2$

ing population, the probability that an individual in the first filial or F_1 generation will have two alleles identical by descent is the same as the probability of having two identical alleles, $1/2N_0$. Thus the inbreeding coefficient of the first generation is $F_1 = 1/2N_0$.

The second inbreeding generation will not consist only of homozygotes produced by union of two gametes with alleles identical by descent from the first generation. The probability also exists that two gametes from different homozygous individuals in the F_1 generation but descending from the same ancestors in generation 0 will also be present. Thus the population consists not only of homozygotes produced by new breeding but also homozygotes attributed to previous inbreeding (Table 19.3):

$$F_1 = \frac{1}{2N}$$

$$F_2 = \frac{1}{2N} + \left(1 - \frac{1}{2N}\right)F_1$$

$$F_3 = \frac{1}{2N_2} + \left(1 - \frac{1}{2N_2}\right)F_2$$

$$F_n = \frac{1}{2N_{n-1}} + \left(1 - \frac{1}{2N_{n-1}}\right)F_{n-1}$$

The inbreeding coefficient consists of two parts. The first part, $1/2N_{n-1}$ is derived from new inbreeding. The second part, $(1 - 1/2N_{n-1})F_{n-1}$, is attributed to previous inbreeding.

The above coefficients relate to an idealized population. This constraint can be removed if the term $1/2N_{n-1}$ is replaced by F, the rate at which heterozygosity is lost or alternatively the rate at which genes become fixed:

$$F = \frac{F_n - F_{n-1}}{1 - F_{n-1}}$$

When $F = 0$ no inbreeding occurs, and when $F = 1$ complete inbreeding occurs. In summary, upon inbreeding the Hardy-Weinberg frequencies change to

$$[p^2(1 - F) + pF] + 2pq(1 - F) + [q^2(1 - F) + qF]$$

Models of inbreeding have been the domain of plant and animal breeders. They apply these concepts to develop new varieties of horticultural plants and crops and to improve livestock breeds from broiler chickens to racehorses. Human geneticists also use such models and concepts in the study of genetically carried defects such as susceptibility to certain diseases. Conservation geneticists are employing inbreeding models in studies of endangered species.

The use of inbreeding models requires the analysis of pedigrees. From pedigrees, geneticists can determine lineage and thus the degree of inbreeding among related individuals. Conservation geneticists have been developing pedigrees of individuals in the zoo populations of many endangered species. They use this information to plan breeding programs designed to increase genetic diversity in the captive populations that will provide animals for reintroduction to the wild.

Obtaining pedigrees of individuals in natural populations is almost impossible without following marked breeding individuals and their offspring. In his 14-year study of black-tailed prairie dogs (*Cynomys ludovicianus*), John Hoogland (1992) developed pedigrees of the Wind Cave population in South Dakota. He found that individuals avoid breeding with close kin—parents, offspring, and full and half siblings. However, they inbreed regularly with more distant kin such as full and half first cousins, full and half first cousins once removed, full and half second cousins, and so on. Keller and Arcese (1988) constructed pedigrees for 16 generations in the population of nonmigratory song sparrows (*Melospiza melodia*) on the small (6 ha) Mandarte Island, British Columbia. Pedigrees revealed that 59 percent of all mating takes place between relatives, about 4 percent between full sibs. Lacking such information for most species of interest, conservation geneticists make use of DNA fingerprinting (discussed earlier). It enables them to study the degree of inbreeding within local populations of a particular species and to assess the degree of genetic diversity among the various separate populations of that species.

Consequences of Inbreeding

In normally outcrossing populations, close inbreeding is detrimental. For one, it increases autosomal homozygosity with all its attendant problems. Rare recessive deleterious genes become expressed in a homozygous state. This condition can cause any one of many effects: premature death, decreased mating success, decreased fertility, decreased fecundity, small body size, loss of vigor, reduction in growth, reduced pollen and seed fertility in plants, and various meiotic abnormalities such as poor chromosomal pairing. These consequences are termed **inbreeding depression.**

TABLE 19.3 Genotypic Frequencies with Inbreeding

Genotype	With Inbreeding Coefficient F			With F = 0 (Random Mating)	With F = 1 (Complete Inbreeding)
	Allozygous genes		*Autozygous genes*		
AA	$p^2(1 - F)$	$+$	pF	p^2	p
Aa	$2pq(1 - F)$			$2pq$	0
aa	$q^2(1 - F)$	$+$	qF	q^2	q

Two hypotheses exist relative to the expression of inbreeding depression. One is the **partial dominance hypothesis** that defines inbreeding as the expression of deleterious genes masked in the heterozygous condition. Inbreeding results in an increase in the frequency and expression of deleterious or lethal recessive genes. The hypothesis predicts that in some inbreeding lines the deleterious alleles will be purged, leaving them fitter than the original. Other lines will eventually go extinct. The **overdominance hypothesis** predicts that heterozygotes possess superior fitness. Inbreeding results in a steady decline in fitness because of the loss of heterozygosity (Roff 1997). Population geneticists find it difficult to definitively attribute the mechanisms underlying inbreeding depression to either hypothesis (Mitton 1993, Waser 1993, Roff 1997).

Close inbreeding in nature is rare, less than 2 percent in natural populations of vertebrates for which there are data (Ralls et al. 1986); inbreeding, however, to some degree is common to many populations. Inbreeding depression is most evident in some plants, domestic animals, zoo animals, and some natural populations of fish (Waldman and McKinnon 1993). Little data or strong evidence exists on inbreeding depression in natural populations, but there are several examples. One is the lions of Ngorongoro Crater, East Africa (O'Brian and Evermann 1988, Packer et al. 1988). The population consists of about 100 lions, 30 of which are adults, in six prides. All of them descended from 15 survivors of an outbreak of a fly plague. They have lost 10 percent of their genetic diversity and possess only one-half of the genetic diversity of the neighboring Serengeti lions. No new male lions have entered Ngorongoro Crater breeding population since 1969, resulting in inbreeding for five generations. The lion population exhibits abnormal sperm in males and a decline in reproductive performance. Similarly, offspring from eggs produced by full sib matings in the Mandarte Island song sparrow population experience 49 percent reduction in survival to breeding age (Keller and Arcese 1998). Inbreeding depression is also characteristic of tropical cooperative spiders (Reichert and Roeloffs 1993). These spiders possess a high level of population subdivision with little to no dispersal and experience habitual close inbreeding with high full sib relatedness. They experience low viability, smaller egg masses, and a high rate of colony extinction compared with outbred cogenitors.

Inbreeding is not necessarily detrimental for a population. Occasional inbreeding will maintain both certain rare alleles and coadapted gene complexes that otherwise might be lost with continual outbreeding. Inbreeding can expose recessive genes that can be purged from the population through selection pressure. Close inbreeding under certain artificial situations is used in the breeding of domestic plants and animals to fix certain desirable genes that will breed true, in spite of the expression of deleterious ones. These inbred lines, however, are then outcrossed to produce "hybrid" vigor.

Paradoxically, the outcrossing of mildly inbred populations in natural populations (as happens when new individuals introduced into local populations to augment small populations) can result in **outbreeding** and accompanying **outbreeding depression.**

Outbreeding depression occurs when individuals of a species population adapted to its local environmental conditions mate with individuals from a different local population adapted to its own local environment. Hybridization between the two populations can destroy the coadapted gene complexes, making the offspring maladapted to local conditions, resulting in a decrease in the fitness of the progeny. As with inbreeding depression, few data exist on outbreeding depression in natural populations (Shields 1993, Waldman and McKinnan 1993). One of the best recorded examples involves the ibex in Czechoslovakia. When the Tatra Mountain ibex (*Capra capra ibex*) became extinct because of overhunting, ibex from Austria, adapted to a similar environment, were transplanted into the region to restore the population. Later bezoars (*C. ibex aegagrus*) from Turkey and Niberian ibex from (*C. ibex nubiana*) from Sinai were introduced to the Tatra herd to augment the local population. The resulting fertile hybrids rutted in the early fall rather than in winter as the native ibex did. The resulting offspring were born in February, the coldest month of the year, and died from exposure. The population again became extinct (Greig 1979). Similar but undocumented outcomes resulted when cottontail rabbits (*Sylvilagus floridanus*) from Missouri were introduced into local populations of rabbits in Pennsylvania and southern bobwhite quail were introduced into local populations of northern bobwhite quail in New England.

Much has been written about the natural mechanisms that may exist to reduce inbreeding. However, few studies exist to document the levels of inbreeding in populations or the means to prevent it. One hypothesis with some supporting evidence is sex-biased dispersal of individuals from the natal group or site (Greenwood 1980, Holmes and Sherman 1983). One sex stays behind, the other leaves. Among birds, juvenile females most frequently leave the home area; the males tend to return to the vicinity of their birth. Separation is further enhanced in many species by monogamous mating habits and the frequent loss of a mate during the nesting season or between years. In the complex cooperatively breeding social structure of acorn woodpeckers (*Melanerpes formicivorus*), reproductive vacancies, breeder male or female, are filled by immigrants from other groups and not by helpers within the group (Koenig et al. 1998). Among mammals young females stay close to the home place (philopatry), while young males, often driven away by the females, seek new places to live. Adult male lions defend groups of females from competitors, but in a few years they are forced to relinquish the pride to vigorous younger males from the outside who challenge and defeat resident males (Pusey and Packer 1986). Sex-biased dispersal among black-tailed prairie dogs (*Cynomys ludovicianus*) reduces breeding between sibs and half sibs relatives by 90 percent (Hoogland 1992). Dispersal involves virtually all males. Young males leave the family group or coterie before breeding, and young females remain behind. Adult male prairie dogs are more likely to move to new breeding groups if adult daughters are in the home colony. Male Townsend voles (*Microtus townsendii*) are less likely to join the breeding population and disperse further if their mothers and sisters are still in their natal home range (Lambin 1994). Although dispersal does reduce inbreeding, sex-biased dispersal may have evolved for other reasons, such as enhanced reproductive success, increased access to food resources and space, and other amenities (Greenwood 1980).

A second mechanism to reduce inbreeding is the deliberate avoidance of kin. Such a choice would necessarily involve some form of kin recognition (Holmes and Sherman 1982, 1983). Because of their close association during early life, siblings of some species of mammals recognize one another over time. Females of both the Belding's ground squirrel (*Spermophyllus beldingi*) and arctic ground squirrel (*S. phyllus*) can distinguish between full and half sisters, although this discrimination fades over a relatively long period of absence. Females mate with unrelated males, leave the group if a related male returns, or fail to come into estrus, especially if the father is in the group (Hoogland 1982). In some species, young experience delayed maturation or suppressed reproduction in the presence of parents. For example, white-footed mice mature more quickly if the opposite sex is removed (Wolff 1992).

Some species of birds avoid inbreeding by extra-pair copulations. Philandering males intrude territories and solicit mating, often successfully, with paired, socially monogamous females (Westneat et al. 1990, Parker and Burley 1998). Electrophoresis studies showed that among the splendid fairy wrens (*Malurus splendens*) of Australia 25 percent of breeding pairs were close relatives, though 60 percent of progeny in nests were not sired by resident males, but from extra-pair copulations. Inbreeding amounted to only 5 percent (Rowley and Russell 1990, Rowley and Brooker 1993).

Few data exist on kin recognition in birds, with even less evidence that birds exhibit any strong type of inbreeding avoidance (Keller and Arcese 1998, Rowley et al. 1993). Inbreeding avoidance in most species appears to be an outcome of high survival rates and relatively large populations in which the probability of close inbreeding is low. Small populations experiencing low survival rates exhibit greater degrees of inbreeding.

GENETIC DRIFT

In sexual reproduction only a few of the gametes produced actually form a new generation. In general, all an individual's genes will be represented somewhere among its gametes, but not in any two of them. Under conditions of stable population size, two gametes are about all that an average individual can leave behind. For a heterozygote *Aa*, there is a 50:50 chance that the two gametes will either be both *A* or both *a*, assuming no natural selection. Thus a 50:50 chance exists that a heterozygote will fail to pass on one of

its genes. In a whole population these losses tend to balance each other, so that the gene frequencies of the filial generation are a replica, but never an exact one, of the parents' gene frequencies. This is simply the familiar law of averages at work. The larger the population, the more closely the gene frequencies of each generation will resemble those of the previous generation.

However, large populations consist of numerous metapopulations, often more or less isolated from one another, like white-footed mice in a small woodlot. Because of habitat fragmentation, more and more populations of species are being isolated to varying degrees. In these cases, the subpopulations represent random sampling among gametes not representative of the gene pool of the larger population. Each subpopulation has its own distinct sample. Chance fluctuation in allele frequencies in these small populations as a result of random sampling is called **genetic drift.** Over time some genes will continue to segregate, while others will become fixed. After some time, the population will become fixed or homozygous for some alleles, and other alleles will be lost. Over time, allele frequencies spread out progressively as the proportion of fixed genes steadily increases, ultimately resulting in homozygous populations (Figure 19.17).

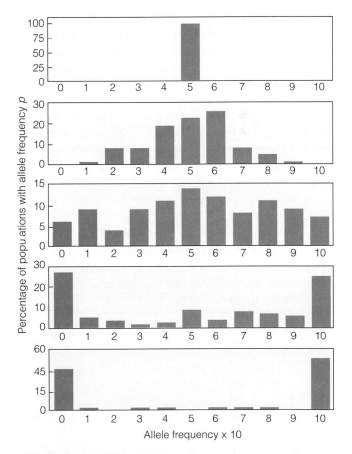

FIGURE 19.17 Changes over time in allele frequency due to drift (from top to bottom, generation number = 0, 1, 5, 10, 30). There are initially 100 populations of size 10 with an allele frequency of .5. At each generation, 10 offspring are produced by random selection of parents from the population. (After Roff 1997.)

How rapidly this outcome occurs depends on the size of the population. Consider the data in Table 19.4, which compares the variances over generations for small and large populations when $p = q = .5$, the number of individuals is constant, and the generations are discrete. The variance at any given time depends on the size of the population, the number of generations elapsed, and initial gene frequencies. Note that a population of 500 shows little variance over 50 generations and minimal variance after 100 generations. But allele frequencies in natural populations behave so erratically that one cannot predict the probability of ultimate fixation.

Genetic drift mimics inbreeding. Inbreeding comes from nonrandom mating, whereas genetic drift comes from random mating. However, both processes increase homozygosity.

The measure of genetic drift is the **fixation index,** the reduction of heterozygosity of a subpopulation due to random genetic drift. It measures the amount of inbreeding due solely to population subdivision. It is similar to the inbreeding coefficient presented earlier:

$$F_{ST} = \frac{H_T - H_S}{H_T}$$

Here H_S represents the heterozygosity of a randomly mating subpopulation and H_T represents the heterozygosity in an equivalent randomly mating total population.

Thus the fixation index F_{ST} is the probability that two alleles chosen at random in the same subpopulation are identical by descent (for a full discussion of the derivation of the fixation index, see Hartyl 1988). There exists, however, a hierarchy of inbreeding coefficients at the individual, subpopulation, and total population levels (Table 19.5). Subpopulations consist of social subdivisions made up of mated pairs and polygamous groups. These social groups diverge genetically, especially at the local matriarchal level, thus increasing genetic variation of the subpopulation. If the subpopulation

TABLE 19.5 Hierarchy of Inbreeding Fixation Indices

F_{IS}	Proportion of variation within individuals relative to that expected in the subpopulation	$F_{IS} = \dfrac{(H_s - H_I)}{H_s}$
F_{ST}	Proportion of variation in a subpopulation relative to the variance in total population	$F_{SI} = \dfrac{(H_T - H_S)}{H_T}$
F_{IT}	Deviation of actual heterozygosity in the total population from that expected if alleles combined randomly	$F_{IT} = \dfrac{(H_T - H_I)}{H_T}$

H_I = heterozygosity within individuals
H_S = expected heterozygosity within subpopulations
H_T = expected heterozygosity of the total population

is sampled without taking the social structure into consideration, the sample will consist of an admixture of genes possessed by the social subdivisions (Suggs et al. 1996). The result will be a higher than expected level of homozygosity.

As in inbreeding, the value of F will change as genetic drift continues generation after generation. The average value of the fixation index in subpopulations in generation t, or F_t, is

$$F_t = 1 - \left(1 - \frac{1}{2N}\right)$$

The value of F_t will range from 0 for no homozygosity to 1 for complete homozygosity.

Effective Population Size

The values of F_t are based on an ideal population characterized by a constant population size, equal sex ratios, equal probability of mating among all individuals, and a constant dispersal rate. Most populations are not "ideal." There are age-related differences in reproduction. Particularly in polygamous populations, the ratio of breeding males to females is unequal. In such populations the number of males is more important than the number of females in determining the amount of random drift. For these reasons the actual size of a small subpopulation is of little meaning. Of greatest importance is the genetically **effective population size, N_e.**

N_e is not the same as the actual number of breeding individuals. The effective population size equals the number of adults contributing gametes to the next generation. If the sexes are equal in number and all have an equal probability of producing offspring, N_e equals the number of breeding adults in the population. If the number of breeding males and females in the population is not equal, then N_e is less than N. N_e is defined as the size of an ideal population subject to the same degree of genetic drift as a particular real population. The ideal population is a randomly breeding one with a 1:1

TABLE 19.4 Variances over Generations for Populations of Different Sizes with Discrete Generations and a Constant Number of Individuals, All with $p = q = .5$

Number of Generations	SIZE OF POPULATION					
	6	10	50	100	500	1000
1	.02	.01				
2	.04	.02	.01			
3	.06	.04	.01	.01		
4	.07	.05	.01	.01		
5	.09	.06	.01	.01		
10	.15	.10	.02	.01		
50	.25	.23	.10	.06	.01	
100	.25	.25	.16	.10	.05	.01

Source: Mettler and Gregg 1969:51.

sex ratio and with the number of progeny per family randomly distributed (Poisson).

Unequal Sex Ratios   In a monogamous population in which one male mates with one female, all offspring are less closely related than in a polygamous population with, say, a ratio of one breeding male to five females. In the latter situation the offspring would be half or full sibs. The chance of an allele becoming lost or fixed (and thus the amount of genetic drift) is much greater in such a population.

The effective population size is given by

$$N_e = \frac{4N_m N_f}{N_m + N_f}$$

where N_m and N_f are the numbers of breeding males and females respectively. As the disparity in the ratio of males to females widens, the effective population diminishes (Figure 19.18).

Consider a population of white-tailed deer consisting of 100 adult does and 50 adult bucks. The actual size of the population is 150. Because of the unequal sex ratio we might assume that the effective population size is $4(50 \times 100)/150 = 133$. However, the white-tailed deer is a polygamous species with a dominance hierarchy among the males. Only the dominant males breed, and subdominant, potentially breeding males cannot contribute their genes to the next generation during any particular breeding season. Thus the effective population size is even smaller. Assume that 25 dominant males mate with 100 does, an average of four does per breeding male, not an unrealistic breeding ratio. Under these conditions the actual breeding population consists of 100 does and 25 bucks. The effective breeding population, N_e, now is $4(25 \times 100)/150 = 66$. What N_e tells us is that the sampling error or genetic drift in our deer population of 150 animals with a sex ratio of 1:4 is equal to that of a population of 66 with an equal number of males and females. In other words, in our deer population the 25 males are not genetically equivalent to the 100 females. The males contribute disproportionately to the gene pool of the next generation. Each male contributes $1/2 \times 1/25 = .02$ of the genes to the next generation, whereas each female contributes $1/2 \times 1/100 = .005$ to the next generation.

Effective population size becomes more significant in exploited populations, especially those in which trophy animals, usually dominant males, are selected by hunters and removed from the population. For example, antler size in the white-tailed deer is genetic (Harmel 1983, Smith et al. 1983), controlled by a single set of dominant-recessive alleles (Templeton et al. 1983). A single dominant allele has a major effect on the phenotypic expression of five to ten points and a single recessive determines the phenotypic expression of two to four points. If these males are selected against by sport hunting (as they are in North America), then subdominant males become the dominants and genes for smaller antlers could become fixed in the local population.

Isolation by Distance   The concept of effective population size also applies to a population whose individuals are more or less continuously distributed over a large area, larger than the greatest dispersal distance of the species. This is known as the neighborhood area for the species. An estimate of the neighborhood area is the root-mean-square of dispersal distance of individuals about their natal origin (Dobzhansky and Wright 1947). The effective breeding size of such populations, often called **neighborhood size,** depends on the number of breeding individuals per unit area and the amount of dispersion between an individual's birthplace and the birthplace of its offspring. The latter is denoted by σ^2, the one-way variance (half the standard deviation squared) of the distance between birth and the site of first breeding, or the root-mean-square dispersal distance. The variance is estimated as

$$s^2 = \frac{1}{N} \sum_{i=1}^{N} (d - d_i)^2$$

where d_i is the location of individual i's breeding site, d is the individual's birthplace, and N is the number of individuals in the sample. Since d, the birthplace, is at distance 0, the mean-square distance is

$$s^2 = \frac{1}{N} \sum_{i=1}^{N} d_i^2$$

In terms of dispersal distance, σ, and the number of breeding individuals per unit area, δ, the effective or neighborhood population size is given by $N_e = 4\pi\sigma^2\delta$ (where π is 3.14).

The equation can be applied to data on the bannertail kangaroo rat (*Dipodomys spectabilis*) obtained by W. T. Jones (1987), who studied the dispersal of this rodent on a 36 ha study area in southeastern Arizona. The kangaroo rat lives in mounds that persist for and are occupied over many generations. Dispersing juveniles settle in vacant mounds. Jones determined that the adult breeding population consisted of 70 males and 72 females, for a density of 2.27/ha.

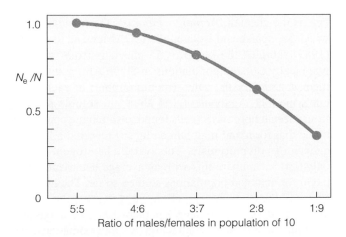

FIGURE 19.18   The depression of effective population size N_e due to disparity in the sex ratio (m/f) of reproducing animals. Note that as the ratio widens with fewer males relative to the number of females, as it may in polygamous species, especially where males are hunted, the effective population size drops dramatically. (From Foose and Foose 1984.)

Two hundred and eighteen juveniles (107 males, 111 females) moved a mean distance of 50 m from natal site to breeding site. The estimated mean dispersal distance d for the juvenile population,

$$\sum_{t=1}^{n} d f(d_i)$$

where d is dispersal distance and $f(d_i)$ is the fraction of individuals dispersing distance d_i (see Moore and Dolbeer 1989), is 140.25 m, and the mean-square dispersal distance σ^2,

$$\sum_{t=1}^{n} d^2 f(d_i)$$

is 10,943 m^2/2 = 5472 m^2 or 0.5472 ha. The estimated effective population size is $N_e = 4\pi(0.5476)(2.27) = 15.75$. The neighborhood size is small because of the very limited dispersal of the kangaroo rat.

If dispersion follows a normal curve, then 39 percent of all individuals will have their offspring within a radius of σ centered at their own birthplace; 87 percent will have their offspring within a radius of 2σ; and 99 percent will have their offspring within a radius of 3σ (Wright 1969, 1978). Therefore even within large areas of habitat, the overall population is a mosaic of subpopulations restricted by dispersal distance, which may promote a degree of inbreeding. Even northern white-tailed deer subpopulations occupying winter deer yards inhabit largely exclusive summer ranges that rarely overlap with other subpopulations (Nelson and Mech 1987). Such social groups created by restricted dispersal and mating patterns result in the genetically divergent breeding groups, discussed earlier, that could influence effective population sizes.

Population Fluctuations   Actual populations are dynamic; they fluctuate through time. In fact, numbers may change dramatically. Under adverse environmental conditions or sudden loss of habitat, the population may decline sharply or "crash" (Figure 19.19). The survivors of the crash, the progenitors of future populations, possess only a sample of the original gene pool. Suppose that the effective size of a particular population for generation 1 is N_1, for generation 2, N_2, and so on. The overall increase in F_{ST} is given as

$$1 - F_{ST} = \left(1 - \frac{1}{2N_e}\right)^t$$

where

$$\frac{1}{N_e} = \left(\frac{1}{t}\right)\left(\frac{1}{N_1} + \frac{1}{N_2} + \cdots \frac{1}{N_t}\right)$$

The effective size of a population fluctuating over time is calculated as the reciprocal of the average of the sum of reciprocals of the effective number of each generation. This special sort of average is called the harmonic mean. This harmonic mean is strongly influenced by the smaller values (Crow and Kimura 1970). Thus a population crash would

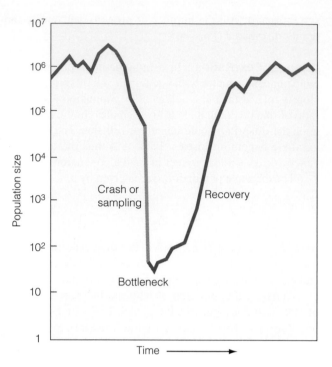

FIGURE 19.19   A population faced with an environmental catastrophe or overexploitation enters a bottleneck in which the surviving population consists of only a sample of the total gene pool. If the population makes a complete recovery, it may lack the genetic diversity found in the original population. If the population remains small, as in captive or isolated situations, it is subject to random genetic drift. (From Frankel and Soulé 1981:32.)

tend to reduce the average effective size over time. The sharp reduction in numbers creates a severe population **bottleneck.** During a bottleneck some genes may be lost from the gene pool as a result of chance. This loss, which can severely reduce genetic diversity in the remaining population, will be carried through into future generations.

There are several examples from natural populations. One is the cheetah (*Acinonyx jubatus*), the populations of which are sparse and isolated. S. T. O'Brien and associates (1983) sampled the blood of 55 cheetahs from two geographically isolated populations in South Africa and found them genetically the same (monomorphic) at each of 44 allozyme loci. Analysis of 155 abundant soluble proteins from cheetah fibroblasts (cells found in vertebrate connective tissue that form and maintain collagen) revealed a low frequency of polymorphism. The average heterogeneity was only 0.013. Compare this with an average heterogeneity of 0.036 for mammal populations studied so far. The homozygosity of the cheetah populations suggests that the species experienced a severe bottleneck perhaps 100 generations ago, resulting a severe range contraction. A similar situation exists with the northern elephant seal (*Mirounga angustirostris*) (Bonnell and Selander 1974).

A bottleneck also occurs when a small group of emigrants from one subpopulation founds a new subpopulation and when small populations of animals, such as wild turkey or otter, are introduced into new or empty habitats. The emigrant or introduced population carries only a sample of genes from the parent population, and so it, too, is subject to random genetic drift. This drift is known as **founder's effect** (Figure 19.20). The gene pool of the future population is derived from the genes in the original founding population.

Gene Flow, Mating Strategies, and Genetic Drift

Unless they are separated by a wide expanse of inhospitable habitat, some interchange takes place among subpopulations. A few white-footed mice move from one woodlot to another, and home ranges of subpopulations of deer overlap at the peripheries. If immigrants or dispersers become part of the breeding population, they introduce a different genetic sample that tends to reduce or slow random genetic drift and helps maintain genetic diversity.

The degree to which dispersal influences genetic drift depends on effective population size, amount of dispersal or immigration, and the degree to which individuals in the population are monogamous or polygamous. In a small population of constant size, monogamous species should experience inbreeding depression more slowly than polygamous species, because of their maximum effective population size. The greater the sex disparity in polygamous species and the larger the harem, the greater the rate of inbreeding. Where immigration or dispersal brings genetic interchange, loss in homozygosity declines (Figure 19.21). Little genetic interchange is needed to prevent significant random genetic drift among subpopulations and to maintain genetic diversity. Change in homozygosity in subpopulations experiencing some immigration is given by

$$F = \frac{1}{4N_m + 1}$$

where N_m is the actual number of migrants per generation.

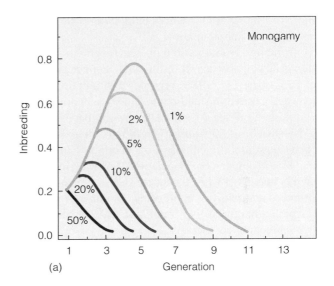
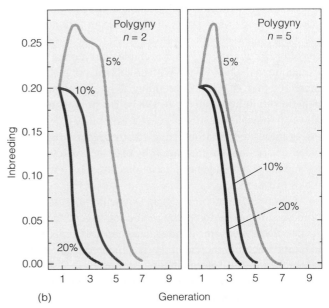

(a)

(b)

FIGURE 19.21 Alleviation of inbreeding over time. (a) Monogamous populations with either sex dispersing. (b) Polygamous populations with males dispersing. Immigrant males were always considered successful in polygamous matings. Numbers on the curves represent dispersal rates. Note that when dispersal rates in monogamous populations are low relative to population size, inbreeding continues to increase rapidly because of slow diffusion of new alleles into the population. Contrast the rapid alleviation of inbreeding depression in polygamous matings with male dispersal and a harem of five. Even in polygamous populations that reach large inbreeding coefficients, only a few dispersing males are necessary to reduce genetic drift. (From Chesser 1983:74, 75.)

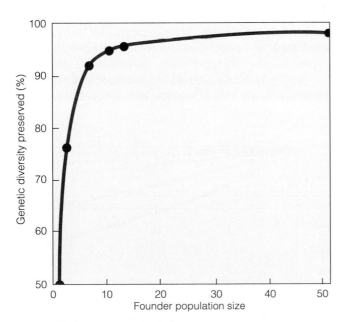

FIGURE 19.20 Genetic diversity in founder populations of various sizes. Founder populations of 10 may hold 90 percent of the genetic diversity found in the parent population. Populations of 50 may contain a nearly complete sample. That genetic diversity can be maintained only if the population expands. These facts are important in the introduction of a species into vacant or new habitat. (From Foose 1983:388.)

F_{ST} decreases as the number of immigrants increases. For example, one immigrant per generation would give a value of $F = (1/4)(1) + 1 = 1/5 = .2$. Five immigrants per generation would greatly reduce F: $F = (1/4)(5) + 1 = 1/21 = .05$. Conservation geneticists believe that one reproductively successful immigrant per generation is the minimal number needed to slow genetic drift and five immigrants the maximum (Frankel and Soulé 1981:129), although the percentage of immigrants needed may depend more on whether the population is monogamous or polygamous. Too many immigrants could swamp the genetic character of the subpopulation.

Information from the field on genetic effects of immigration is lacking, because it is difficult to collect data. However, R. K. Chesser (1983) provides a simulation model, which suggests that in small monogamous subpopulations a slow immigration rate does not alleviate inbreeding depression, because of the slow diffusion of unrelated alleles into the gene pool and the equally slow decline of inbreeding among kin. In polygamous populations, however, inbreeding is quickly alleviated when the harem size is large and immigrant males are involved in breeding. Dominant immigrant males with a harem of five (see Figure 19.21) can rapidly reduce inbreeding coefficients.

Remember that certain advantages and disadvantages derive from monogamy and polygamy. Although inbreeding may be more severe in small monogamous populations, the effective size is maximized (assuming equal numbers of both sexes). Monogamous populations do retain rare alleles and great qualitative genetic variation. The largely monogamous sea otter (*Enhydra lutris*), for example, was hunted to virtual extinction in the late 1800s. With protection, populations have increased to over 100,000, retaining about 80 percent of their genetic variability (Ralls et al. 1983). The polygamous seals, however, have low levels of genetic variability (Testa 1986). Polygamous populations may retain greater quantitative genetic variation, but they are subject to the loss of rare alleles and the potential genetic variation carried by nonbreeding males (Testa 1986).

Given sufficient numbers and some immigration, cycles of inbreeding and outbreeding should retain genetic variation. As populations become divided into subpopulations and their habitats become increasingly fragmented, the need for research into the role of mating strategies and gene flow between populations in the maintenance of genetic variation in small populations increases.

MINIMUM VIABLE POPULATIONS

The effective population statistic tells us something about the nature of the breeding population relative to the total population—but we need to know more. Breeding populations experience mortality and dispersal, and these lost individuals must be replaced by younger animals. For this reason, the total population must have an adequate age structure and be above a critical size to maintain itself. Below that size and structure, inbreeding and the loss of selectable genetic variation become a problem for continued survival.

The threshold number of individuals that will ensure the persistence of a subpopulation in a viable state for a given interval of time, say a hundred years, is the **minimum viable population (MVP)** (Schaffer 1981, Gilpin and Soulé 1986). The minimum viable population has to be large enough to cope with chance variations in individual births and deaths, random series of environmental changes, and random changes in allele frequencies or genetic drift.

How large must a viable population be to balance the rate of loss of genetic variability by the rate of gain via new mutations? Based on studies of the mutation–genetic drift equilibrium in fruitflies (*Drosophila*), conservation geneticists estimate that the rate of added genetic variance by mutation that balances the genetic variance reduced by genetic drift for a given trait is on the order of 10^{-3} (0.001) (Lande and Barrowclough 1987, Stacey and Taper 1992). This rate translates into a minimum effective population size of 500 individuals that is needed to retard the effects of genetic drift. An absolute minimum N_e of 50 is needed to avoid inbreeding depression (Franklin 1980). Any smaller population is subject to serious genetic drift, to loss of the genetic variability necessary to track environmental changes, and to high probability of stochastic extinction. These estimates (see Table 19.4 and Figure 19.22) have given rise to the 50/500 rule (Franklin 1980).

This 50/500 rule has created a number of problems in the management of endangered species and has built-in dangers and inefficiencies. Relying on variations in mutation rates is risky for two reasons: we do not know the reliability of those rates; and many mutations are deleterious. Moreover, we rarely know the effective population size for the population concerned; and genetic formulas estimating the MVP are appropriate only for populations with discrete generations.

Applying the 50/500 rule is a dangerous numbers game for species survival. If we consider a population of 500

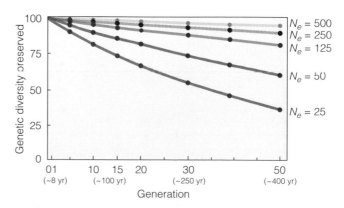

FIGURE 19.22 Loss of genetic diversity as measured by heterozygosity due to random genetic drift for various effective population sizes, based on a rate of decline in heterozygosity of $1/2N_e \times 100\%$ per generation. Note how rapidly genetic diversity declines when N_e is small. Even with an N_e of 500, some loss of genetic diversity takes place. (From Foose 1983:376.)

individuals as sufficient to maintain all species, we may be condemning certain species to extinction. If we consider a population of 50 as minimal for sustaining subpopulations, we may also be condemning them to the same fate. In assessing the fate of 120 bighorn sheep populations over a 70-year span, Joel Berger (1990) found that 100 percent of the populations with fewer than 50 individuals went extinct within 50 years, and populations with more than 100 individuals went extinct within 70 years. He concluded that 50 individuals is not a viable population size for bighorn sheep. On the other hand, if we had considered a population of 50 as minimal for a species' recovery, we would have given up on the whooping crane (*Grus americana*), peregrine falcon (*Falco peregrinus*), black-footed ferret (*Mustela nigripes*), European bison (*Bison bonatus*), Pere David deer (*Elaphurus davidianus*), and other comeback species whose remnant populations were considerably below 50. The point to remember is that MVP is only a general guideline for the genetic management of endangered species (Hedrick and Miller 1992).

Immediate efforts to save endangered species depend more on demography than on genetics. What is important initially is the size of the remnant population. Sizes of MVP vary among species (Pimm 1991). Those species in which reproduction is highly density-dependent and that live in a more or less constant environment may persist for a long time in spite of a decline in genetic diversity. Species with highly variable population sizes, such as lagomorphs, may need large populations to counteract environmental stochastic effects on population growth. Thus MVP should be a guide to the size of a population that should be maintained, but it should never be used as a precise rule in deciding a species' fate.

Summary

1. The raw material of evolution and adaptation to local environments is the genetic variability of individuals in local populations. Variations may be discontinuous, reflected by differences in color or blood groups, or continuous, possessing a range of values.
2. Observed variations are phenotypes, the external or observable expressions of the genotypes, which are the sum of the hereditary information carried by the individual. The ability of a genotype to give rise to a range of phenotypic expressions under different environmental conditions is known as phenotypic plasticity.
3. Sources of variation are imbedded in DNA molecules. In eukaryotic cells, DNA exists in large units, the chromosomes. Each chromosome carries units of heredity called genes, the informational units of the DNA molecule. Genes occupying the same position or locus on a pair of chromosomes are termed alleles. If each member of the pair of alleles affects a given trait in the same manner, the two alleles are called homozygous. If each member of the pair affects a given trait in a different manner, the pair is called heterozygous.
4. Two major sources of sources of genetic variation are (1) the reproductive recombination of genes provided by parents in bisexual populations and (2) mutations, inheritable changes of genetic material in the gene or chromosome. Chromosomal mutations that result from a change in the structure or number of chromosomes are macromutations. Gene mutations that result from alterations in the DNA sequence of one or a few nucleotides are micromutations.
5. Theoretically, variations in biparental populations, as reflected in gene frequencies and genotypic ratios, remain in Hardy-Weinberg equilibrium given as $p^2 + 2pq + q^2 = 1$. This equilibrium gives the expected genotypic frequencies of *AA, Aa,* and *aa* respectively, if the following conditions exist: random mating; equilibrium in mutation; lack of selection; and a relatively large, closed population. In nature, such conditions rarely occur and there is a departure from genetic equilibrium.
6. Natural selection acts on genetic variability, reducing the influence of less fit and favoring the more fit. Fitness is measured by the number of reproducing offspring contributed to the next generation. Natural selection works in three ways: nonrandom mating, nonrandom fecundity, and nonrandom survival. Certain genetic combinations are more fit than others, and these transmit more genes to future generations than less fit combinations.
7. The direction evolution takes depends on the genetic characteristics of those individuals in the population that survive and leave behind reproducing progeny. Natural selection may be stabilizing, maintaining the current genetic equilibrium and favoring intermediate phenotypes; directional, favoring phenotypes at one extreme of the range; or disruptive, favoring genotypes at both extremes of the range at the same time.
8. Although natural selection seems to impinge only on individuals, the evolution of certain altruistic traits seems to require some form of group and/or kin selection. Group selection, which operates on the differential production of local populations, favors characteristics that improve the fitness of the group; these characteristics may decrease the fitness of any individual within the group.
9. Kin selection involves the evolution of altruistic traits in a group of closely related individuals possessing genes by common descent. It involves the evolution of a genetic trait expressed by one individual that affects the genotypic fitness of one or more directly related individuals. Individual fitness plus any additional fitness acquired by improving the fitness of very close relatives is inclusive fitness.
10. Because of habitat fragmentation and human exploitation of the landscape, populations of many species of plants and animals are being reduced to isolated or semi-isolated small populations. These small populations carry only a sample of the genetic variability of the total population. Such situations can lead to inbreeding and genetic drift.
11. Inbreeding, mating between relatives, brings out hidden genetic variation, increases homozygous genotypes at the expense of heterozygous ones, and reveals the effects of homozygous rare alleles that often result in reduced fecundity, decreased viability, and even death.
12. The degree of inbreeding is given by the inbreeding coefficient, *F,* which measures the probability that two alleles at a locus in an individual are identical by descent—that is, derived by replication of a single allele in an ancestral population.
13. Small populations are subject to random genetic drift, chance fluctuations in allele frequency as a result of random sampling among gametes. Drift results in the fixation of alleles and populations lacking in genetic variability. Genetic drift mimics inbreeding by increasing homozygosity, but it is the result of random mating and the alleles involved are not necessarily identical by descent.

14. The effects of random drift are measured in terms of heterozygosity of individuals, subpopulations, and the total population. The value $F_{ST} = (H_T \times H_S)/H_T$ measures the degree of genetic structuring among subpopulations.

15. Influencing the degree of random genetic drift is the effective population size, N_e, the size of an ideal population having the same rate of increase in F_{ST} as the population in question. In monogamous populations with an equal number of breeding males and females, the actual population size and the effective population size are the same. In polygamous populations a wide disparity in sex ratios can strongly reduce the effective population size and increase genetic drift. The effective population size is given by $4N_m N_f/(N_m + N_f)$, where N_m and N_f are the number of males and females respectively.

16. If the population is spread out more or less uniformly across the landscape, the effective size becomes $4\pi\sigma^2 d$, where d is the density of breeding individuals per unit area and σ^2 is the root-mean-square of dispersal distance. If the population fluctuates over time, the effective population is given by the harmonic mean of the various population values.

17. Exchange of individuals among populations (emigration and immigration) can reduce genetic drift and inhibit genetic divergence among subpopulations. The equilibrium value of F_{ST} for immigration or dispersal is given by $1/(4N_m + 1)$, where N_m is the number of migrants per generation. Only a few immigrants per generation are needed to reduce random genetic drift and keep F_{ST} below .10.

18. If a subpopulation is to persist over time, it has a threshold number below which the population must not fall to maintain genetic diversity. This is called the minimum viable population, the size at which the rate of loss of genetic variance is balanced by the rate of mutation. Although a figure of 500 is generally given, this may be much too low because such a value fails to consider age structure and sex ratios of the population and chance variations in the environment. Because of the adverse effects of population fragmentation, population genetics is becoming very important part in the preservation and management of wild species in the face of human development of planet Earth.

Review Questions

1. Distinguish between genotype and phenotype. Upon which does natural selection work?

2. Define allele, locus, homozygous, heterozygous, diploid, haploid, and gene pool.
3. What are the major sources of variation in the gene pool?
4. What is the Hardy-Weinberg law? What is its significance?
5. Contrast stabilizing, directional, and disruptive selection.
6. What is meant by fitness, direct fitness, indirect fitness, and inclusive fitness?
7. What is altruism? Relate this trait to the theories of group selection and kin selection.
8. What is inbreeding? Why does inbreeding increase homozygosity? Contrast allozygous with autozygous genes.
9. Define the coefficient of inbreeding. What does it measure?
10. What are the end values of F, the inbreeding coefficient, and what do the values tell us?
11. What is inbreeding depression? Outbreeding depression? How do they affect small populations?
12. To what degree and under what conditions does inbreeding occur in natural populations?
13. What is genetic drift? What is the difference between inbreeding and genetic drift?
14. What is effective population size? How does it relate to inbreeding and genetic drift? To the sex ratios of reproducing animals in a population?
15. What is the relationship between dispersal distance and effective population size in a continuously distributed population?
16. What is a population bottleneck? What is its significance genetically?
17. What is the founder's effect?
18. What is meant by minimum viable population?
19. Show that a breeding population with a ratio of 2 males to 10 females is genetically less variable than a population with a ratio of 6 males to 10 females.
20. Argue why it is unwise to base the minimum size of a population necessary for species survival on genetics alone.
21. Suggest ways in which gene flow between isolated populations of a species could be maintained. What factors must be considered?

Cross References

PART SEVEN

The Community

CHAPTER 20

Community Structure

CONCEPTS

1. The community is an assemblage of interacting species occupying a given area.

2. Communities are characterized by both a physical and biological structure.

3. Communities possess a vertical structure and form a horizontal pattern across the landscape.

4. The mix of species that makes up a community defines its biological structure.

5. The population interactions of competition, predation, parasitism, and mutualism influence community organization.

6. The structure of the community can be defined in terms of feeding relationships or food webs.

7. The classification of communities is scale-dependent.

A saying goes, "You can't see the forest for the trees." When we walk through a forest, an oak, maple, or pine tree catches the eye; we may notice a squirrel, deer, or bird. As we walk on, we realize that collections of such individuals make up populations of species in the area. It is harder to see that these populations in the forest are all in some way influencing each other. In doing so, they form another recognizable ecological unit, the community. Forest, grassland, hedgerow, lawn, stream, marsh—these and other units in the landscape contain unique groupings of organisms known as biotic communities.

THE COMMUNITY DEFINED

Plant and animal populations do not live alone as separate entities. They share the same environments and habitats, and interact with one another in various ways. This collection of plant and animal populations interacting directly or indirectly is referred to as a **community.** Such a definition embraces the idea of community in its broadest sense. Within the community some species may interact more strongly among themselves than with others, utilizing habitat or food resources in a similar manner. These groups are called **guilds.** For example, birds feeding mostly on insects would make up an insect-feeding guild.

Other commonly used definitions of the community are more restrictive in their meaning. Zoologists may apply the term *community* to a certain group of species, such as a bird community or a mammal community of a particular forest or grassland. Botanists use the term **association** for a plant community possessing a definitive species composition (mix of species). Ecologists may recognize and contrast communities as **heterotrophic** and **autotrophic.** Forests and grasslands are examples of autotrophic communities, which require only the energy provided by the sun to drive the process of photosynthesis. Groups of organisms that inhabit such habitats as a fallen log, a tiny pool of water in the hollow of a tree, or a cave are examples of heterotrophic communities, which depend on the autotrophic community for their energy source. Although ecologists may define different categories or classifications of communities, all communities have certain characteristics that define their biological and physical structure. These characteristics vary in both space and time.

PHYSICAL STRUCTURE

Communities are characterized not only by the mix of species (the biological structure), but also by physical features of the biotic and abiotic environment. The physical structure of the community reflects abiotic factors, such as the depth and flow of water in aquatic environments. It also reflects biotic factors, such as the spatial configuration of organisms. In a forest, for example, the size and height of the trees and the density and dispersion of their populations define the physical attributes of the community.

Life Forms

The form and structure of terrestrial communities are largely defined by the vegetation. The plants may be tall or short, evergreen or deciduous, herbaceous or woody. Such characteristics can describe growth forms. Thus we might speak of shrubs, trees, and herbs and further subdivide the categories into needle-leaf evergreens, broadleaf evergreens, broadleaf deciduous, thorn trees and shrubs, dwarf shrubs, ferns, grasses, forbs, mosses, and lichens.

A more useful system is the one designed in 1903 by the Danish botanist Christen Raunkiaer. Instead of considering a plant's growth form, he classified plant life by the relation of the embryonic or meristemic tissues that remain inactive over the winter or prolonged dry period—the **perennating tissue**—to their height above ground. Such perennating tissue includes buds, bulbs, tubers, roots, and seeds. Raunkiaer recognized six principal life forms, which are summarized in Table 20.1 and Figure 20.1. All of the species within a community can be grouped into these six classes: therophytes, cryptophytes (geophytes), hemicryptophytes, chamaephytes, phanerophytes, and epiphytes. For example, trees and shrubs greater than 25 cm in height are classified as **phanerophytes** because their leaf-producing buds are elevated above the ground on stems. In contrast, grasses are classified as **cryptophytes**, because the above-ground tissues die back in winter or during prolonged dry periods. A community with a high percentage of perennating tissue well above ground (such as phanerophytes) would be characteristic of warmer, wetter climates. A community where most of the plants are classified as cryptophytes and hemicryptophytes would be characteristic of colder or drier environments. When the species within a community are classified into life forms and each life form is expressed as a percentage, the result is a life form spectrum that reflects the plants' adaptations to the environment, particularly climate (Figure 20.2). Such a system of classification provides a standard means of describing the structure of a community.

TABLE 20.1 Raunkiaer's Life Forms

Name	Description
Therophytes	Annuals survive unfavorable periods as seeds. Complete life cycle from seed to seed in one season.
Geophytes (Cryptophytes)	Buds buried in the ground on a bulb or rhizome.
Hemicryptophytes	Perennial shoots or buds close to the surface of the ground; often covered with litter.
Chamaephytes	Perennial shoots or buds on the surface of the ground to about 25 cm above the surface.
Phanerophytes	Perennial buds carried well up in the air, over 25 cm. Trees, shrubs, and vines.
Epiphytes	Plants growing on other plants; roots up in the air.

FIGURE 20.1 Raunkiaer's life forms: (a) phanerophytes; (b) chamaephytes; (c) hemicryptophytes; (d) geophytes (cryptophytes); (e) therophytes; (f) epiphytes. The parts of the plant that die back are unshaded; the persistent parts with buds (or seeds in the case of therophytes) are dark.

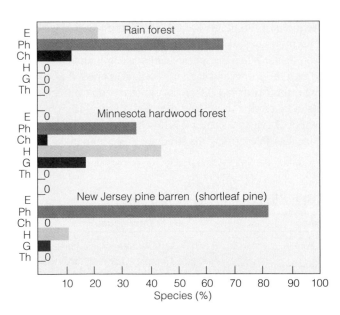

FIGURE 20.2 Life form spectra of a tropical rain forest (adapted from Richards 1952), a Minnesota hardwood forest (data from Buell and Wilbur 1948), and a New Jersey pine barren (data from Stern and Buell 1951). Note the absence of hemicryptophytes, geophytes, and therophytes from the tropical rain forest and the prominence of epiphytes. The pine barrens are dominated by phanerophytes.

Vertical Stratification

A distinctive feature of a community is **vertical structure** (Figure 20.3). On land, vertical structure is determined largely by the life form of the plants—their size, branching, and leaves—which in turn influences and is influenced by the vertical gradient of light. The vertical structure of the plant community provides the physical structure in which many forms of animal life are adapted to live. A well-developed forest ecosystem, for example, has several layers of vegetation. From top to bottom, they are the **canopy,** the **understory,** the **shrub layer,** the **herb** or **ground layer,** and the **forest floor.** We could even continue down in to the root layer and soil strata. The tropical rain forest has one additional stratum, **emergents,** trees that rise above the general canopy of the forest (see Chapter 29).

The **canopy,** which is the primary site of energy fixation through photosynthesis, has a major influence on the rest of the forest. If it is fairly open, considerable sunlight will reach the lower layers and the understory and shrub strata will be well developed if water and nutrients are ample. If the canopy is dense and closed, light levels are low and the understory and shrub layers will be poorly developed.

In the forests of the eastern United States, the **understory** consists of tall shrubs such as witch hobble (*Viburnum alnifolium*), understory trees such as dogwood (*Cornus* spp.) and hornbeam (*Carpinus caroliniana*), and younger trees, some of which are the same species as those in the canopy. Species that are unable to tolerate shade will die; others will eventually grow to reach the canopy after some of the older trees die or are harvested.

The nature of the **herb layer** will depend on the soil moisture and nutrient conditions, the slope position, the density of the canopy and understory, and the aspect of the slope, all of which vary from place to place throughout the forest.

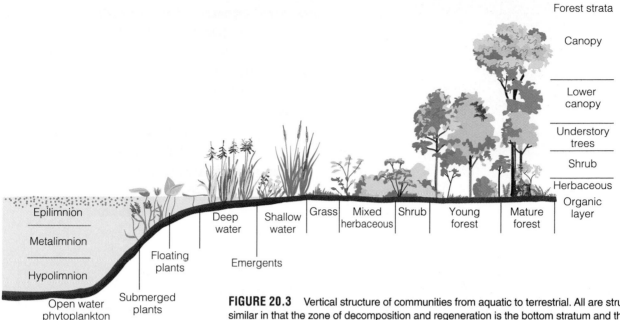

FIGURE 20.3   Vertical structure of communities from aquatic to terrestrial. All are structurally similar in that the zone of decomposition and regeneration is the bottom stratum and the zone of energy fixation is the upper stratum. In the sequence from the aquatic to the terrestrial, stratification and complexity of the community become greater. Stratification in aquatic communities depends on physical factors, the gradients of oxygen, temperature, and light. Stratification in terrestrial communities is largely biological. Dominant vegetation affects the physical structure of the community and microclimatic conditions of temperature, moisture, and light. Because the forest often has five or six strata, it can support a greater diversity of life than a grassland with three strata. Floating and emergent aquatic plant communities can support a greater diversity of life than open water.

The final layer, the **forest floor,** is the site where the important process of decomposition (Chapter 9) takes place and where nutrients are released from decaying organic matter for reuse by the forest plants.

Aquatic ecosystems such as lakes and oceans have strata determined by light penetration, and profiles of temperature and oxygen (see Chapter 30). In the summer, well-stratified lakes have a layer of freely circulating water, the **epilimnion;** a second layer, the **metalimnion,** which is characterized by a **thermocline** (a very steep and rapid decline in temperature); the **hypolimnion,** a deep, cold layer of dense water about 4°C, often low in oxygen; and a layer of bottom mud (see Figure 20.3). In addition, two other structural layers are recognized, based on light penetration: an upper **photic zone** roughly corresponding to the epilimnion, which is dominated by autotrophic phytoplankton and is the site of photosynthesis; and a lower layer, the **benthic zone,** in which decomposition is most active. The lower layer roughly corresponds to the hypolimnion and the bottom mud.

Communities, whether terrestrial or aquatic, have similar biological structures. They possess an autotrophic layer, which fixes the energy of the sun and manufactures food from inorganic substances. It consists of the area where the light is most available: the canopy of the forest, the herbaceous layer of grassland, and the upper layer of water of lakes and seas. Communities also possess a heterotrophic layer that utilizes the food stored by autotrophs, transfers energy, and circulates nutrients by predation in the broadest sense and by decomposition.

Each vertical layer in the community is inhabited by characteristic organisms (Figure 20.4). Although considerable interchange takes place among the vertical strata, many highly mobile animals confine themselves to only a few layers, particularly during the breeding season. Occupants of a vertical stratum may change during the day or season. Such changes reflect daily and seasonal variations in the physical environment, such as humidity, temperature, light, and oxygen content of water; shifts in the abundance of essential resources such as food; or different requirements of organisms for the completion of their life cycles. For example, D. L. Pearson (1971) found that birds occupying the upper strata of a tropical dry forest in Peru moved to the lower strata during the middle of the day for several reasons: to secure food (insects move to lower levels), to escape high levels of solar radiation and associated heat stress, and to conserve moisture.

In general, the greater the vertical stratification of a community, the more diverse its animal life. The variety of life in a terrestrial community is heavily influenced by the number and development of layers of vegetation. If a certain stratum is absent, the animal life it normally shelters and supports is also missing. Therefore grassland, with few strata, is poorer in species than a highly stratified forest ecosystem (see Karr and Roth 1971). Within forest ecosystems, a strong relationship exists between the degree of vertical stratification (called foliage height diversity) and the number of bird species present (Figure 20.5). Likewise in aquatic communities, the greater the variation along vertical gradients of light, temperature, and oxygen, the greater is the diversity of life.

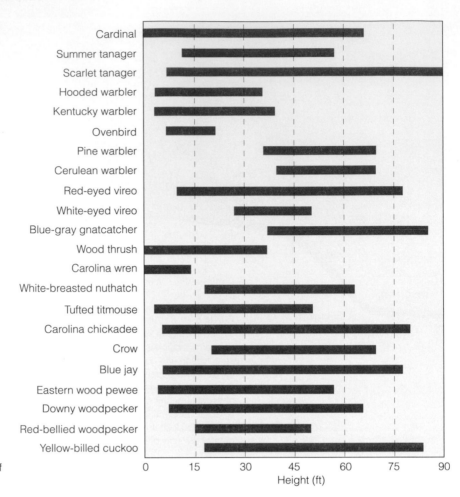

FIGURE 20.4 Vertical distribution of major breeding bird species on Walker Branch watershed based on the total observations of birds during the breeding season regardless of activity. (After Anderson and Shugart 1974.)

Horizontal Structure

Walking across a typical old field in eastern North America, we move through patches of open grass, clumps of goldenrods, tangles of blackberry, and small thickets of sumac and other tall shrubs. Continuing into an adjacent woodland, we cross through an open understory and patches of shade-tolerant undergrowth of laurel and viburnum. Then we come upon open **gaps,** openings in the canopy caused by the death of a canopy tree, where dense thickets of new growth have claimed the sunlit openings. We continue our walk to the top of an open hill and view the pattern of vegetation across a larger landscape. We note the quiltlike patches of forests, fields, croplands, roads, and human settlements. These large patches, spatially separated from one another, produce a horizontal pattern that increases the physical and ecological complexity of the environment.

The patchiness of vegetation across the landscape exists on different scales: we discovered within-patch heterogeneity in the old field, between-patch heterogeneity when we passed from field to forest, and finally the patterned landscape. At all levels, the sizes, shapes, and dispersion of patches affect their colonization by individuals, the persistence of these individu-

als on the patch, the population dynamics, and the number of species in an area. Horizontal patterns influence dispersal and foraging by animals. Within the old field the size and number of patches of goldenrods (*Solidago* spp.) are important to the goldenrod beetles (*Diabrotica*). On a larger scale the presence of patches of low woody vegetation is essential to the rabbit. For the deer heterogeneity on an even larger scale, involving large patches of woods, old fields, and grassland, is essential. Each species and individual requires its own scale of patchiness, or horizontal structure, within the matrix of the landscape.

For plants opportunities for recruitment, growth, reproduction, and survival change with spatial variations in soils, nutrients, moisture, light, and competition from other plants. The type of plant reproduction and the timing and patterns of seed production over time influence vegetation patchiness. Plants with wind-dispersed and animal-dispersed seeds can travel far and have a wider distribution across the landscape. The distribution of these species may be more continuous across the landscape, exhibiting less patchiness than plants with poor dispersal mechanisms. Vegetative or clonal reproduction produces distinctive clumps of certain plants in an otherwise homogeneous environment. Allelopathic effects (see Chapter 14) and

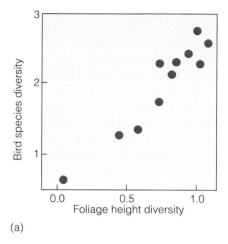

(a)

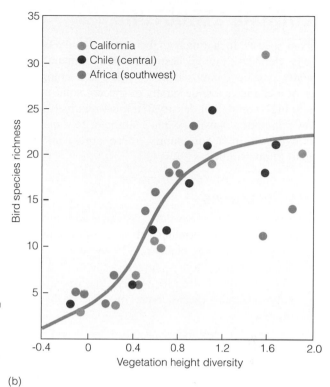

(b)

FIGURE 20.5 Bird species become more diverse as vertical stratification increases in forest ecosystems. (a) The vertical distribution of some bird species in a spruce-fir forest in Wyoming. (b) The relationship between bird species diversity and foliage height diversity for areas of deciduous forest in eastern North America. Foliage height diversity is a measure of the vertical structure of the forest. The greater the number of vertical layers of vegetation, the greater the diversity. ((a) MacArthur and MacArthur 1961. (b) Cody, 1975.)

shading lead to the suppression of some plant species and to the development and growth of others.

The patterns we see at all scales in the landscape are the product of an array of physical and biological influences (Figure 20.6). Soil structure, soil fertility, moisture conditions, degree of slope, and direction of slope (aspect) influence the microdistribution of plants. Patterns of light and shade shape the development of understory vegetation. Runoff and small variations in topography and microclimate produce well-defined patterns of plant growth. Grazing animals have subtle but important effects on the spatial patterning of vegetation, as do windthrow, fire, and human disturbances (see Chapter 23). A patchy environment in turn influences the distributional pattern of animal life across the landscape (Wiens 1976).

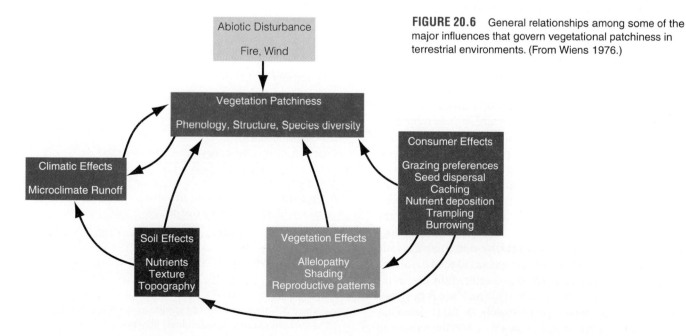

FIGURE 20.6 General relationships among some of the major influences that govern vegetational patchiness in terrestrial environments. (From Wiens 1976.)

BIOLOGICAL STRUCTURE

The mix of species, including both their number and relative abundance, define the biological structure of a community. A community can be composed of very few but common species; or it can have a wide variety of species, some common with high population density, but most relatively rare with low population density. When a single or few species predominate within a community, these organisms are referred to as **dominants.**

Species Dominance

It is not easy to determine what constitutes a dominant species or, in fact, to determine which are the dominant species. Dominants in a community may be the most numerous, possess the highest biomass, preempt the most space, make the largest contribution to energy flow or mineral cycling, or by some other means control or influence the rest of the community.

You can assign dominance to numerically superior organisms, but numerical abundance alone is not sufficient. A species of plant, for example, may be widely represented yet exert little influence on the community as a whole. In the forest the small or understory trees may be numerically superior; yet the nature of the community is controlled by the fewer large trees that overshadow the smaller ones. In such a situation the dominant organisms are not those with the largest numbers but those that have the greatest biomass or that preempt most of the canopy space and thus control the distribution of light.

Local differences in the environment, such as moisture availability, nutrient levels, and topographic positions, can create patches of different dominants within the community, further complicating the dominance relationships within it.

In other cases a scarce organism can dominate by its activity within the community. The predatory starfish *Pisaster,* for example, preys on a number of associated species that share a common habitat and resources. By reducing population numbers of these species through predation, the starfish reduces competitive interactions among them, and so they are able to coexist (Paine 1966). If the starfish is removed, a number of prey species disappear and one becomes dominant. In effect, the predator controls the structure of the community and so must be regarded as the dominant or **keystone species.** Keystone species are those whose presence is critical to the integrity of the community.

To determine dominance plant ecologists have used several approaches (Table 20.2 and Appendix B). They can measure **relative abundance** of species, comparing the numerical abundance of one species with the total abundance of all species. They can measure **relative dominance,** a ratio of the basal area or biomass of one species to total basal area (cross-sectional area of a tree at a fixed height above the ground) or biomass for all species. This measure is best used when species involved are approximately the same size. They can use **relative frequency,** an index based on the number of sample points or plots in which a species is found to occur

relative to the total number of samples taken. This index is best used when the species involved are very different in size. Often all three measurements are combined to arrive at an **importance value** for each species. Most species do not achieve a high level of importance in the community, but those that do serve as **index species.** Once species have been assigned importance values, the stands can be grouped or classified based on the dominant species. Such techniques are useful in the comparison of communities or relating the species composition of communities to underlying features of the physical environment, such as soil or elevation.

Species Diversity

Among the array of species that make up a community, few are abundant. Indeed, individuals of most species make up only a small proportion of the total population in the community. Consider the structure of the tree component of a mature woodland in West Virginia consisting of 24 species over 10 cm dbh (diameter at breast height), as presented in Table 20.3. Individuals of two tree species, yellow-poplar and white oak, make up nearly 44 percent of the stand. The next most abundant trees—black oak, sugar maple, red maple, and American beech—each make up from under 7 percent to just over 5 percent of the stand. Nine species range from 1.2 to 4.7 percent,

---

TABLE 20.2 Measures of Dominance

1. Dominance =
$$\frac{\text{basal area or aerial coverage, species A}}{\text{area sampled}}$$

2. Relative dominance =
$$\frac{\text{basal area or coverage, species A}}{\text{total basal area or coverage, all species}}$$

3. Relative density =
$$\frac{\text{total individuals, species A}}{\text{total individuals, all species}}$$

4. Frequency =
$$\frac{\text{intervals or points where species A occurs}}{\text{total number of sample plots or points}}$$

5. Relative frequency =
$$\frac{\text{frequency value, species A}}{\text{total frequency value, all species}}$$

6. Importance value = relative frequency + relative dominance + relative density

   All the above results may be multiplied by 100.

7. Simpsons index of dominance:
$$\text{dominance} = \frac{\Sigma n_i(n_i - 1)}{N(N - 1)}$$

   where N is total number of individuals of all species and n_i is the total number of individuals of species A

TABLE 20.3 Structure of Vegetation of a Mature Deciduous Forest in West Virginia

Species	Number	Percentage of Stand	Species	Number	Percentage of Stand
Yellow-poplar (*Liriodendron tulipifera*)	76	29.7	Bitternut hickory (*Carya cordiformis*)	5	2.0
White oak (*Quercus alba*)	36	14.1	Pignut hickory (*Carya glabra*)	3	1.2
Black oak (*Quercus velutina*)	17	6.6	Flowering dogwood (*Cornus florida*)	3	1.2
Sugar maple (*Acer saccharum*)	14	5.4	White ash (*Fraxinus americana*)	2	.8
Red maple (*Acer rubrum*)	14	5.4	Hornbeam (*Carpinus caroliniana*)	2	.8
American beech (*Fagus grandifolia*)	13	5.1	Cucumber magnolia (*Magnolia grandiflora*)	2	.8
Sassafras (*Sassafras albidum*)	12	4.7	American elm (*Ulmus americana*)	1	.39
Red oak (*Quercus rubra*)	12	4.7	Black walnut (*Juglans nigra*)	1	.39
Mockernut hickory (*Carya tomentosa*)	11	4.3	Black maple (*Acer nigrum*)	1	.39
Black cherry (*Prunus serotina*)	11	4.3	Black locust (*Robinia pseudoacacia*)	1	.39
Slippery elm (*Ulmus rubra*)	10	3.9	Sourwood (*Oxydendrum arboreum*)	1	.39
Shagbark hickory (*Carya ovata*)	7	2.7	Tree of heaven (*Ailanthus altissima*)	1	.39
				256	100.00

and the nine remaining species as a group represent less than 5 percent of the stand. A second West Virginia woodland, described in Table 20.4, has a different composition. That community consists of ten species of which two, yellow-poplar and sassafras, make up nearly 84 percent of the stand.

These two forest communities illustrate a typical pattern in temperate forests—a few common species associated with less abundant ones. They also illustrate two other characteristics of the distribution of species within a community—**species richness,** the number of species, and **evenness,** the relative abundance of individuals among the species. The more equitable the distribution, the greater is evenness. Species diversity, which considers both the species richness and evenness, increases as the numbers of individuals in the total population are more equitably distributed among the species. The stand described in Table 20.3 is richer in species (24 vs. 10) and has a greater evenness than that described in Table 20.4.

In order to quantify species diversity for the purposes of comparison, a number of indices have been proposed (Pielou 1975). One of the most widely used is the Shannon Index (also called Shannon-Weiner), which considers both richness and evenness. The Shannon Index measures diversity by the formula

$$H = -\sum_{i=1}^{s} (p_i)(\log_2 p_i)$$

where H is the diversity index, s is the number of species, $\log_2$ is the log base 2, and p_i is the proportion of individuals of the total sample belonging to the ith species.

This index, based on information theory, is a measure of uncertainty. The higher the value of H, the greater is the uncertainty, or the probability that the next individual chosen at random from a collection of species containing N individuals will not belong to the same species as the previous one (in our example, the species of tree). The lower the value of H, the greater the probability that the next individual encoun-

tered will be the same species as the previous one. In the forest described in Table 20.4, whose diversity index is 1.87, the probability is high that in sampling the trees, the next tree picked at random will be a yellow-poplar or a sassafras. In the forest described in Table 20.3, the diversity index, 3.59, is higher and the chance that the next tree encountered at random will be a yellow-poplar is considerably less.

We can gain a better appreciation of evenness or equability by comparing the relative species abundances in the community with the maximum possible evenness. The maximum possible species diversity (H_{max}) for a community of s species would be the condition where the individuals composing the community were evenly distributed among all s species. This

TABLE 20.4 Structure of Vegetation of a Second Deciduous Forest in West Virginia

Species	Number	Percentage of Stand
Yellow-poplar (*Liriodendron tulipifera*)	122	44.5
Sassafras (*Sassafras albidum*)	107	39.0
Black cherry (*Prunus serotina*)	12	4.4
Cucumber magnolia (*Magnolia grandiflora*)	11	4.0
Red maple (*Acer rubrum*)	10	3.6
Red oak (*Quercus rubra*)	8	2.9
Butternut (*Juglans cinerea*)	1	.4
Shagbark hickory (*Carya ovata*)	1	.4
American beech (*Fagus grandifolia*)	1	.4
Sugar maple (*Acer saccharum*)	1	.4
	274	100.0

is the condition of maximum evenness. By dividing the actual species diversity of the community (H) by the maximum possible diversity for the community (H_{max}), an index of species evenness (J) can be derived. The maximum possible value of J will be 1, when the community exhibits the maximum possible evenness ($H = H_{max}$). The value of J will approach zero as the community becomes increasingly dominated by a single species. For the Shannon Index, this measure of evenness is given by

$$J = \frac{H}{H_{max}} = \frac{-\sum p_i \ln p_i}{\ln s}$$

The stand described in Table 20.3 has an evenness index of 0.78 ($H_{max} = 3.17$). The stand in Table 20.4 has an evenness index of 0.56 ($H_{max} = 2.26$).

In communities composed of organisms with a wide range of sizes, an index may lead us to underestimate the importance of fewer but larger individuals and overestimate that of the more common species. One of the distinctive failures of the indices is the inability to distinguish between the abundant and the rarer species, which contribute little to the index. Nevertheless, diversity indices do provide one measure of the biological structure of community that is convenient for making comparisons. (For a discussion of indices, see Hurlbert 1971, Peet 1974.) Diversity indices may be used to compare species diversity within a community (**alpha diversity**, α), between communities or habitats (beta diversity, β), and among communities over a geographical area (**gamma diversity**, γ) (Whittaker 1972).

Species Abundance

Describing a community using a single index of diversity is convenient for purposes of comparison; however, it is not a unique description of community composition. Two communities with the same diversity index value do not necessarily have the same exact species richness and evenness. A complete picture of the distribution of species abundances in a community must involve an examination of the relative abundance of all species in some systematic fashion. This picture is obtained by plotting the relative abundance of each species against rank, where rank is defined by relative abundance. Thus, the most abundant species is plotted first along the x axis, with the corresponding value on the y axis being the value of relative abundance. This process is continued until all species are plotted. The resulting graph is called a **rank-abundance diagram.** The rank-abundance curves for the two forest stands presented in Tables 20.3 and 20.4 are shown in Figure 20.7.

Theoretically, the relative abundances of the species exhibited in the rank-abundance diagram represent the manner in which species divide the resources or environmental space. In a community where the organisms are of similar size, abundant species preempt a larger proportion of space and resources than do less abundant species. In this way, the rank-abundance curves tell us something about how resources are partitioned within a community. Using the model of niche breadth and overlap among competing species presented in Chapter 14, we

can view the rank-abundance curves as a model of niche partitioning. In an attempt to understand the mechanisms underlying the community organization, three statistical models have been developed to describe the patterns of relative abundance.

The **random niche** or **broken stick model** (MacArthur 1960) views abundance as a random partitioning of resources distributed along a continuum. The analogy is that of a stick on which are randomly marked a number of points that represent niche boundaries. Then the stick is broken at each point into segments. The length of each segment represents the abundance of species. If the segments, representing the relative abundances of the species, are plotted in a rank-abundance diagram with abundance (y axis) expressed on a log scale, then a curve like A in Figure 20.8 will result. The model assumes that species in the community use the critical resource with no overlap between species. This model is rarely realistic (Hairston 1969). It produces the highest evenness of the three models. This type of rank-abundance curve is observed only in small samples of taxonomically related animals with stable populations and long life cycles occupying a small homogeneous community, such as the sampling of nesting birds in a forest during the breeding season.

The **niche preemption** or **geometric distribution hypothesis** supposes that the most successful or dominant species preempts the most space. The next most successful claims the next largest share of space, and so on, with the least successful occupying what little space is left. The resulting rank-abundance curve is a straight line (curve B in Figure 20.8), and the distribution of the species forms a geometric series. This model produces the highest dominance and lowest evenness of the three models. Such a distribution is achieved only by plant communities containing few species and occupying severe environments such as a desert. In most plant and animal communities, species overlap in the use of space and resources (see discussion of niche in Chapter 14).

The third model, the **log-normal hypothesis** (Preston 1962), supposes that the relative abundance of each species

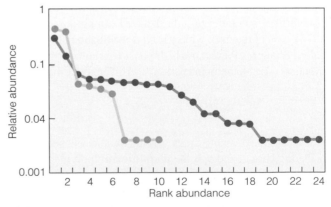

FIGURE 20.7   Rank-abundance curves for the two forest communities described in Tables 20.3 and 20.4. Rank abundance is the species ranking based on relative abundance. Relative abundance (y axis) is expressed on a $\log_{10}$ axis. Note that the forest community of Table 20.3 has a higher species richness and evenness than the forest stand of Table 20.4.

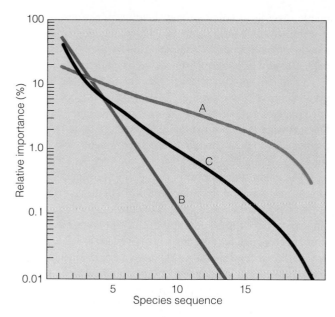

FIGURE 20.8 Graphical representation of species abundance hypotheses. In these graphs, the importance values (the percentage of all species that a particular species represents), expressed as total density, total biomass, total frequency, total productivity, or some other measurement, are plotted against a ranked sequence of species. According to the random niche hypothesis (curve A), the boundaries are located at random positions along the line. By the niche preemption or geometric distribution hypothesis (curve B), the relative importance of each species plotted in sequence on a log scale yields a straight line. The log-normal hypothesis (curve C) supposes that species distribution is determined by a large number of variables that affect the competitive abilities of the species involved. (After Whittaker 1965.)

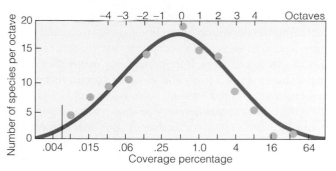

FIGURE 20.9 A bell-shaped curve of importance values resulting from the log-normal distribution of plant species. The Importance value in this example is determined by the percent ground surface covered by the species. It is represented on the horizontal scale (logarithmic) by octaves in which the species are grouped by doubling. (After Whittaker 1965.)

is determined by a variety of conditions, such as food, space, microclimate, and other environmental variables that directly or indirectly affect species success. If a variety of environmental factors are influencing the abundance of species in different and often independent ways, then the resulting pattern may appear rather random (see May 1981). The rank abundance curve produced by this model will fall somewhere between the random niche and geometric distribution curves (curve C in Figure 20.8).

As in the broken stick model, the line representing the niche space can be divided into segments. In this case each segment represents a range of relative abundance values grouped into frequency distribution classes called octaves (for methodology see Ludwig and Reynolds 1988, Krebs 1989, Poole 1974). The octaves are defined so that each successive octave represents a doubling of the number of individuals in the previous octave. For example, 1–2, 2–4, 4–8, and so on. This doubling translates into taking logarithms of abundance to the base 2. The octave with the greatest number of individuals (the mode) is given the value of 0. The rest of the octaves are numbered in plus and minus directions away from the mode. A plot of the relative abundance values results in a bell-shaped or normal curve (Figure 20.9). The log-normal distribution most closely approximates the distribution of relative abundances obtained from communities rich in species.

As with the rank abundance diagrams, the log-normal distribution plots are most useful in summarizing observed abundance relationships within and among communities.

INFLUENCES OF POPULATION INTERACTIONS ON COMMUNITY STRUCTURE

The three models presented above describe patterns of species abundances, but they are of little value in determining the underlying causes for the observed abundance relationships. The biological structure of the community is a result of a rich array of factors relating to both the physical and biological environment. These include the variety of species interactions presented in the preceding chapters. Any conceptual model of how communities are structured—the patterns of species distribution and abundance—must explicitly address the influence of species interactions.

Competition

Since Darwin (1859), ecologists have considered interspecific competition, especially competitive exclusion, as the cornerstone of community structure. Lack (1954, 1971), Hutchinson (1959), and MacArthur (1960, 1972), among others, have stressed the role of competition in shaping community organization, including species distribution, resource allocation, niche segregation, and the like (see Chapter 14).

Salisbury back in 1929 pointed out the role of competition in shaping plant communities. Since that time, numerous studies have been undertaken to demonstrate the role of interspecific competition in determining community structure, with emphasis on the animal component (see Connell 1983, Schoener 1983). Many of these studies suggest, if they do not prove with certainty, that competition exists between certain species pairs and within guilds of species (for specific examples see Chapter 14). But do these studies demonstrate that interspecific competition exerts a strong influence on community organization?

The answer to this question depends on many things, including the definition of community. As we stated in the introduction to this chapter, ecologists use the term *community* in a variety of ways. If the community in question is limited to members of a particular group of species that exploit resources in a similar fashion, such as the guild of ground-foraging birds, competition may be shown to be important in controlling their relative abundances. But does competition within a guild of ground-foraging birds influence the structure of the entire bird community, or the community in its fullest sense—the entire group of plants and animals inhabiting the forest? Determining the effects of interspecific competition on shaping the whole community is difficult at best.

Detecting the role of competition in structuring communities is still more complex, because competition may be important only at certain times. The availability of essential resources may vary in time as a result of variations in the physical environment. For example, the production of seeds in a grassland may be related to rainfall. In most years rainfall may be sufficient to provide seeds in excess of those required by seed-eating animals such as birds, insects, and small mammals. A study of competition in these years may reveal no limitation of the common resource, and therefore no competitive interactions. However, in a drought year, seed production may be extremely low. During this period, competition for the limited seed resource may be quite severe. Although competition may not be active most of the time, these episodes of low resource availability may significantly influence community structure (Wiens 1977).

Despite the difficulties associated with determining the importance of competition in structuring communities, interspecific competition unquestionably does play a role in the organization of at least some communities. Anyone who has a garden experiences the effects of competition between garden plants and weeds. A great deal of competition for light, moisture, nutrients, and space exists among plants from forests to deserts. The outcome of competitive interactions among plant species has a pronounced influence on the physical structure of the terrestrial communities, and in turn on distribution and abundance of animal life.

Among animal populations in the community, competition is usually most evident among species of the same guild. For example, in autumn many animals, especially jays, wild turkeys, deer mice, squirrels, black bears, and white-tailed deer, feed on acorns. If the acorn crop is poor, these animals compete for this limited resource. Because squirrels depend more heavily on acorns than most other animal species, they may feel the effects of competition more keenly than the more generalist consumers (Allen 1943).

The nature of competitive interactions may be influenced by other types of species interactions within the community. For example, when one species is placed at a disadvantage by the interaction with another species, competitors can take advantage of resources made available by the decline in competition. Mortality of oak trees defoliated by gypsy moths on nutrient-rich mesic sites is higher than that observed on nutrient-poor or more xeric sites (Statler and Serro 1983, Hicks and Fosbroke 1987). One hypothesis for oak mortality on mesic sites is that yellow-poplar and sugar maple, usually ignored by oak-feeding gypsy moth larvae, maintain their canopy and use the moisture and nutrient resources that normally would be used by the now defoliated oaks, thereby gaining a competitive advantage.

Competition for space can be a major factor influencing the structure of communities, bestowing on the occupant the advantage of sole or increased access to essential resources. Fast-growing trees, for example, intercept light and shade out plants beneath their canopy. Mussels completely cover rocky substrates along the shore.

The clearest examples of competition influencing community structure are the effects of introductions or invasions of exotic species. Exotic species of birds introduced into the Hawaiian islands have not only excluded native species with similar ecological requirements but have also excluded or prevented the successful establishment of other exotics (Moulton and Pimm 1986). When two introduced species of plankton-feeding fish, the alewife (*Alosa pseudoharengus*) and rainbow smelt (*Osmerus mordax*), proliferated in Lake Michigan, seven native species of fish with similar food habits declined drastically (Crowder et al. 1981). escapee, and multiflora rose (*Rosa multiflora*), widely planted in the past for soil conservation purposes, have invaded old fields and forest edges in North America, crowding out native plants and affecting the structure and composition of animal life. Since arriving in the ballast tanks of freighters from Europe in 1985, zebra mussels have spread from the Great Lakes to the Mississippi River and its tributaries (see Chapter 14). Growing on the tops of native freshwater mussels and killing them, zebra mussels threaten the extinction of native species.

In summary, the importance of competition in community structure no doubt varies from community to community. Within any given community, competition is most pronounced among sessile organisms such as plants, or among members of the same guilds. Even when competition is found among species, determining the degree to which competitive interactions influence the relative abundance of species within a community is at best difficult. This difficulty arises because patterns that appear to be the consequence of competition may have alternative explanations, like variations in environmental factors that have a direct impact on population dynamics, such as climate, or other types of species interactions that can directly influence population size, such as predation (Andrewartha and Birch 1984).

Predation

Whereas the influence of competition on community structure is somewhat obscure, the influence of predation is more demonstrable. Because all heterotrophs derive their energy and nutrients from the consumption of other organisms, the influence of predation can more readily be noticed throughout a community. The direct influence of predation on the population density of prey species can have the additional impact of influencing the interactions among prey species, particularly on competitive relationships.

Hairston, Smith, and Slobodkin (1960) proposed that within a community, carnivores compete strongly among themselves and severely exploit their resources. Herbivores are regulated by predators and have little impact on vegetation. Plant populations, not greatly reduced by herbivory, compete for light, water, and nutrients. In this model, the relative abundance of heterotrophs is largely regulated by predator-prey interactions (see Chapters 15 and 16).

If predation has a major influence on community structure, then the removal of a major predator should have an impact on the relative abundances of species within the community. R. T. Paine (1966, 1969) experimentally removed the top carnivore, the starfish *Pisaster ochraceus,* from a rocky intertidal community on the Pacific coast of Washington. This community, in addition to the starfish, consisted of four species of algae, a sponge, filter-feeding barnacles and mussels, browsing limpets and chitons, and a predatory whelk. Both barnacles and mussels, when given the opportunity, aggressively and tenaciously take over space and exclude other sessile organisms. The starfish fed on sessile barnacles and mussels as well as on limpets and chitons. For two years Paine excluded the starfish from a swath 8 m long and 2 m deep on a rocky intertidal area. In the undisturbed area, the control, the community remained intact. On the swath where the starfish were removed, barnacles settled successfully but were soon crowded out by the mussels. All but one species of algae disappeared for lack of space, browsers moved away for lack of food, and the number of species dropped from 15 to 8. Apparently the predaceous starfish, by feeding on barnacles and mussels, made space available for the less competitive species, thus helping to maintain species diversity.

In this example, *Pisaster* functioned as a keystone species in the community; it had a major effect on the structure of the community. Without it, certain species disappeared in the face of strong competitors and the diversity of the community was reduced. Thus certain species have a pervasive influence on the rest of the community, and the effects of their removal cascade down through the community.

Keystone species are not necessarily top predators. Herbivory affects the structure of plant populations and influences competition and productivity. A classic example is the influence of rabbits on the diversity of species in English pastures (Tansley and Adamson 1925, Zeevalking and Fresco 1977). A sharp reduction in the rabbit population in southern England from the myxomatosis virus resulted in an aggressive growth of meadow grass in fields inhabited by the spectacular large blue butterfly, *Maculinea arion.* The heavy growth of grass resulted in the extinction of open-ground ant colonies, the nests of which were utilized by large blue caterpillars. As a result the large blue is nearly extinct. The loss of one keystone grazing herbivore, the rabbit, resulted in the local extinctions of two other species as well.

Grazing herbivores, from prairie dogs (Whicker and Detling 1988) and pocket gophers (Huntley and Inouye 1988) to large ungulates (McNaughton et al. 1988), influence the structure of grassland communities from the North American plains to East Africa. Grazing ungulates, in particular, reduce canopy height, stimulate tiller growth, and favor low-growing grasses and herbaceous plants at the expense of taller grasses. Intensive grazing favors plants that are prostrate, have small

leaves, or grow in rosettes close to the ground and selects against taller forms. When grazing pressure is relaxed, tall-growing grasses assume dominance at the expense of lower-growing forms, reducing plant diversity (McNaughton 1984, 1985).

Woody vegetation is not exempt from the effects of herbivores, notably browsing ungulates. Browsing by white-tailed deer in many forests of eastern North America has eliminated understory herbaceous and woody plants, including seedlings and sprouts of trees. It has converted the understory to ferns and created browse lines on the trees (Marquis 1981, Marquis and Grisez 1978). Heavy browsing by moose (*Alces alces*) on Isle Royale has favored spruce over deciduous trees. The change in quantity and nature of litterfall has resulted in a decrease in microbial biomass and nitrogen availability (McInnes et al. 1988).

The rocky intertidal regions of the New England coast are inhabited by a number of marine algae, including the ephemeral species *Enteromorpha, Ulva,* and *Porphyra* and the perennial fucoid algae *Fucus vesticulosus* and *F. distichus,* and by a grazing herbivore, the snail *Littorina littorea.* Lubenchenco (1983, 1986) excluded snails from rocks sheltered against wave action. She discovered that the ephemeral species of algae colonized and become established on the rocks. They monopolized the space and inhibited colonization by the perennial fucoid algae. Where snails were present, grazing pressure reduced competition for space among the three ephemerals. The snails prevented any one of them from becoming dominant, although each has the potential of excluding the others. In the absence of grazers in the herbivore exclusion cages, the perennial *Fucus* was able to colonize the rocks and assume dominance. Although *Littorina* snails eat small *Fucus,* they avoid larger individuals if ephemerals are present. *Fucus* has chemical and structural defenses against snails, but the ephemerals have no known defenses. They persist where grazers are absent, exist at very low levels, or are inactive (during winter months). Because snails are rare at sites exposed to heavy wave action, exposed rocky intertidal sites become havens for ephemerals. These experiments suggest that by controlling competitive interactions, herbivory increases species richness and influences the community structure of rocky intertidal shores.

Such studies, and there are many of them, suggest some basic principles about the influence of grazing on plants. If the herbivore selectively grazes the dominant species, then plant diversity increases. Overgrazing in winter and spring followed by undergrazing in summer produces maximum floral richness. (This is usually the case in natural situations because deer, elk, and others can be pressed for food in winter, but find plenty in summer.) The most species-rich communities are developed by continuous grazing with a maintained population of generalist (nonselective) herbivores.

The impacts of herbivory on community structure are related to the intensity of grazing and the selectivity of the grazers. If the dominant plant is highly palatable, overgrazing will reduce it and allow other, less palatable species to occupy the area; but if the dominant species are unpalatable, then grazing only serves to increase their dominance. If the herbivore is regulated by food supply and not by predators, then it can significantly reduce the population of a plant species on which it feeds, allowing for invasion by other species. As the intensity of grazing increases and the populations of palatable grasses are further reduced, more and more unpalatable species are grazed, until eventually only unpalatable species remain. This condition is often seen in farm pastures in eastern North America, where overgrazing by domestic livestock results in conspicuous growths of unpalatable plant species, such as ironweed (*Veronia* spp.) and thistle (*Cirsium altissimum*), late in the growing season.

By selectively eating seeds of certain plants over others, seed-eating predators influence plant community composition. Using exclosures and plots sown with known quantities of seeds of annual plants, Borchert and Jain (1978) found that meadow mice and house mice consumed 75 percent of wild oats (*Avena fatua*) seed, 44 percent of wild barley (*Hordeum leporinum*) seed, and 37 percent of ripgut brome (*Bromus diandrus*) seed. Showing a strong preference for *Avena* and *Hordeum,* the mice reduced the numbers of these plants by 62 percent and 30 percent respectively, stimulating competitive release of *Bromus*. Mice also influence the populations of plants by cutting mature plants and eating seedlings (Batzli and Pitelka 1970).

Parasites and Disease

Parasites are an integral part of natural communities. Their overall effect on community structure becomes most apparent when an outbreak of disease decimates or reduces an affected population. With the clearing of forests, the brown-headed cowbird (*Molothrus ater*), a nest parasite (see Chapter 17), has expanded its range through North America. It has invaded forest edge and even the forest interior where it parasitizes birds, such as hooded warblers (*Wilsonia citrina*), that were never exposed to nest parasitism in their evolutionary history. Lacking antiparasitic behavior, these birds are experiencing parasite-induced declines in their populations (Goldwasser et al. 1980, Walkinshaw 1983).

The introduction of rinderpest, a viral disease, to East African ungulates (see Chapter 17) decimated herds of African buffalo (*Syncerus caffer*) and wildebeest (*Connochaetes taurinus*). The introduction of avian malaria carried by introduced mosquitoes eliminated most native Hawaiian birds below 1000 m, the altitude above which mosquito cannot persist. Introduced bird species that are resistant to malaria have assumed dominance in the absence of native species (Warner 1968). Outbreaks of mange periodically reduce red fox (*Vulpes vulpes*) populations in New York and other areas.

Virulent tree diseases have markedly changed the composition of North American forests. The chestnut blight (*Endothia parasitica*), introduced into North America from Europe, nearly exterminated the American chestnut (*Castanea dentata*) and removed it as a major component of the forests of eastern North America. With its demise oaks and birch increased. Dutch elm disease, caused by a fungus (*Graphium ulmi*) spread by beetles, has nearly removed both the American elm (*Ulmus americana*) from North America and the English elm (*U. glabra*) from Great Britain.

How parasitism can influence interactions within a community beyond a simple host-parasite relationship is illustrated in the moorland ecosystem of Scotland. Two plant species, heather (*Calluna vulgaris*) and bracken (*Pteridium aquilinum*), dominate the Scottish moorlands. The moorlands are the major habitat of the red grouse (*Lagopus scoticus*) as well as grazing land for domestic sheep. The red grouse, which feeds on young shoots of heather, is the host for several parasites, internal and external, including the tick *Ixodes ricinus,* which carries a virus, louping ill. The virus normally is associated with sheep, but ticks also transmit the disease to grouse chicks (Duncan et al. 1978).

The problem of disease transmission is intensified by management of the moorlands for livestock. Sheep grazing on the moorlands provide the reservoir of the virus. In addition, heather is being replaced by bracken (*Pteridium aquilinum*), a species once managed and harvested for livestock bedding. Mature heather resists invasion; but burning of moorlands to provide young nutritious shoots for the grouse reduces mature heather and its ability to resist invasion of bracken with its fast-growing rhizome system. Bracken produces a mat of damp litter that provides an ideal habitat for ticks. The shift in dominance from heather to bracken has increased exposure of the grouse to the tick and consequently to the louping ill virus. The virus has been associated with the long-term decline in grouse densities (Dobson and Hudson 1986). Because the virus is so pathogenic in grouse, the birds cannot maintain the virus without the presence of sheep, the vaccination of which should theoretically eliminate the disease.

Mutualism

Mutualism is too often overlooked as a mechanism in community structure. Although it is difficult to demonstrate, it may be more significant than either competition or predation. The importance of mutualistic relationships among conifers, mycorrhizae, and voles in the forests of the Pacific Northwest is one example. The conifers depend on mycorrhizal fungi associated with the root system for nutrient acquisition from the soil. In return, the mycorrhizae depend on the conifers for energy in the form of carbon (see Chapter 17 for more detailed discussion). The mycorrhizae also have a mutualistic relationship with voles, which feed on the fungi and disperse the spores, infecting the root systems of other conifer trees.

Mutualistic relationships are common in plant reproduction, where plant species often depend on animal species for pollination, seed dispersal, or germination (see examples in Chapter 17). Temple (1977) suggests that the tree *Calvaria major* on the island of Mauritius has failed to produce seedlings in spite of adequate seed production because its seeds had to pass through the gut of the dodo (*Raphus cucullatus*), a now-extinct species of bird related to the ostrich. Relationships between pollinators and certain flowers are so close that loss of one could result in the extinction of the other (Pimm and Pimm 1982).

The extent to which mutualistic relationships structure communities can be masked because mutualistic interactions can arise in which the affected species never come into contact, influencing each other's fitness or population growth rate indirectly through a third species (Boucher et al. 1982). For example, the dietary patterns of three coexisting species of seed-eating birds are shown in Figure 20.10. Species A and species C select relatively small and large seeds respectively, and therefore exhibit only a minor overlap in resource selection. As a result, A and C are only minor competitors. Species B selects seeds of intermediate size and overlaps extensively with both species A and C. The joint competitive pressure of A and C would act to reduce the population of species B, their shared competitor. A and C benefit each other indirectly by reducing the fitness of a strong competitor of the other species. In another situation species A and C may be mutualistic with B, in which case A and C benefit each other indirectly through their positive effect on species B.

Boucher et al. (1982) provide a number of examples of indirect mutualism. One example comes from a study of subalpine ponds in Colorado (Dodson 1970). It involves the relationships among two species of herbivorous species of *Daphnia* and their predators, a midge larva (*Chaoborus*) and a larval salamander (*Ambystoma*). *Ambystoma* prey on the larger of the two *Daphnia* species, while the midge larva prey on the small species. Where salamander larvae were present, the number of large *Daphnia* was low and the number of small *Daphnia* high. But in ponds in which the salamander larvae were absent, small *Daphnia* were absent and midges could not survive. The two species of *Daphnia* apparently compete for the same resources. When the salamander larvae are not present, the larger of the two *Daphnia* species can outcompete the smaller. With the salamander larvae present, predation reduces the population growth rate of the larger *Daphnia,* allowing for the coexistence of the two species. In this example, two indirect mutualistic relationships arise. The salamander larvae indirectly benefits the smaller species of *Daphnia* by reducing the population size of its competitor. Subsequently, the midge

apparently depends on the presence of salamander larvae for its survival in the pond. This possible indirect mutualism can be demonstrated only under controlled experiments involving population manipulations of the four species.

The idea of indirect mutualism is highly speculative, requiring strong experimental demonstration in natural communities. It does suggest that mutualism, as well as competition and predation, can be an integrating force in the structuring of natural communities. Mutualism, long overshadowed by interspecific competition, is now receiving increased attention (see Howe and Westley 1988).

FOOD WEBS AND COMMUNITY STRUCTURE

Perhaps the most fundamental process in nature is the acquisition of food—the energy and nutrients required for assimilation. As we have seen in Chapters 15 through 17, the variety of species interactions—predation, parasitism, competition, and mutualism—are all involved in the acquisition of food resources. For this reason, ecologists studying the structure of communities often focus on the feeding relationships of the component species—the manner in which species interact in the process of acquiring food resources.

The most abstract representation of feeding relationships within a community is the food chain. A **food chain** is a descriptive diagram—a series of arrows, each pointing from one species to another for which it is a source of food. For example, grasshoppers eat grass, clay-colored sparrows eat grasshoppers, and marsh hawks prey on the sparrows. We write this relationship as follows:

$$\text{grass} \longrightarrow \text{grasshopper} \longrightarrow \text{sparrow} \longrightarrow \text{hawk}$$

Feeding relationships in nature, however, are not simple, straight-line food chains. Rather they involve numerous food chains meshed into a complex **food web** with links leading from primary producers through an array of consumers: herbivores, carnivores, and omnivores (Figure 20.11). Such food webs are highly interwoven, with linkages representing a wide variety of species interactions. As discussed previously, the structure of communities cannot be understood solely in terms of the direct interactions between species. Two species drawing on a common food resource may benefit each other by reducing the population of a common competitor. A predator may reduce competition between two prey species by controlling their population sizes below carrying capacity. An analysis of the mechanisms controlling community structure must include these indirect effects represented by the structure of the food web.

Scores of papers and several books (for example Pimm 1982, Polis and Winemiller 1995) have been written on food webs and community structure, and a review of this literature reveals a variety of interesting and applicable questions. For example, what determines the size and complexity of food webs? How are food webs organized and structured? How are food webs affected by the removal of species or the successful invasion of a new species? Are complex food webs more

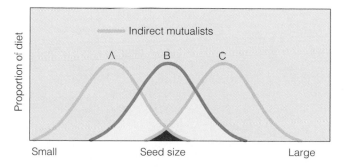

FIGURE 20.10 How indirect mutualism might work. In this hypothetical situation, species A, B, and C utilize a shared food resource—seeds. Species A and species C compete moderately, and both compete strongly with B. When all three species occur together, A and C together exert strong competition on B, reducing the intensity of competition by B. Thus in the presence of B, A and C have an indirect beneficial effect on each other.

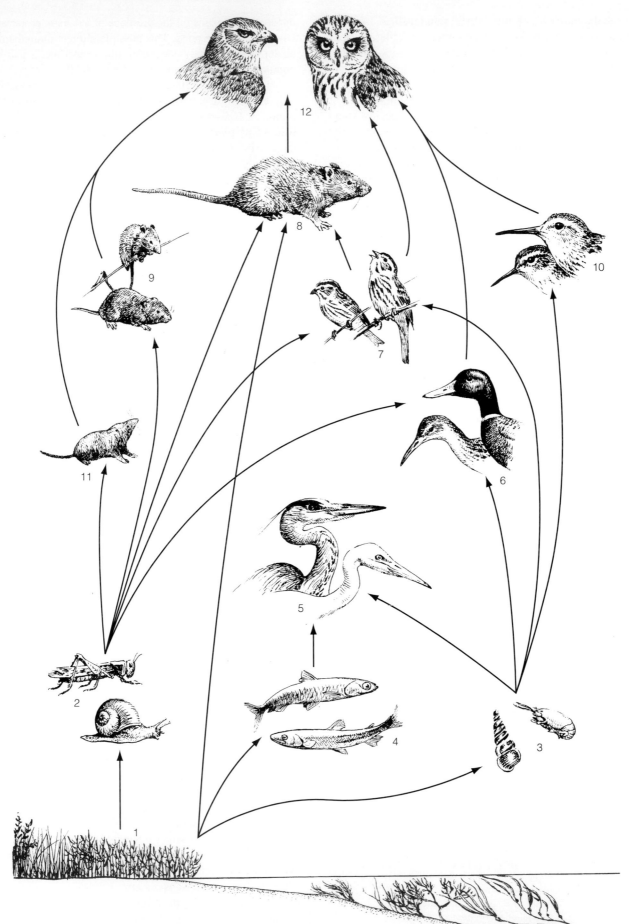

stable than simple ones? What processes, including populations dynamics, determine the patterns of food webs?

Experiments involving the removal or addition of species from a community have been extremely useful in understanding the structure of food webs and the role of species interactions in structuring communities. The removal of the predatory starfish *Pisaster ochraceus* from a rocky intertidal community (see page 392) decreased species diversity by changing the relationships among a suite of competing species. When the peacock bass (*Cichla ocellaris*), native to the Amazon River, was introduced as a game fish into a tributary of Panama's Lake Gatun around 1967, this voracious predator eliminated 8 of the 11 principal native fish species and reduced the others by 75 to 90 percent. With these fish gone, another species, the algae-feeding *Cichlasoma maculicauda,* whose juveniles formed the prey of some of the species eliminated, increased. With plankton-feeding fish gone, zooplankton that had been their preferred prey increased. With their prey species largely eliminated, herons, terns, kingfishers, and tarpon decreased (Zaret and Paine 1973). Thus the structure of a food web can be affected, often adversely, when a new or exotic species, such as a carnivore or a grazing herbivore, is introduced into an ecosystem or when a key species is removed from the food web.

The study of food web structure in a variety of communities, terrestrial, marine, and freshwater, has resulted in the emergence of a number of general patterns. Analyses suggest that food webs in fluctuating environments—ones characterized by variations in temperature, salinity, pH, moisture, and other conditions—tend to have shorter food chains with fewer links than those in more constant environments. Food chains in constant environments, such as pelagic regions of the ocean, are characterized by a greater species richness and more links to the food chain.

Environmental variability alone, however, does not appear to constrain the average or maximum length of a food chain. Highly stratified environments such as a forest and pelagic water column have longer food chains than those in poorly stratified habitats such as grassland, tundra, and stream bottoms. The widest food webs, those with the greatest number of herbivores, were the shortest. In contrast, narrow food webs had the greatest fraction of top carnivores.

The analysis of food webs has led to the development of two views of how community structure is regulated. One is **bottom-up regulation.** This view emphasizes the limitations imposed by the availability of food resources (species popu-

lations) at the next lower level and the role of competition among species that draw on those food resources. The level above is influence by the resources provided by the level below (autotrophs limit herbivores, herbivores limit carnivores, and so on). In the second view, referred to as **top-down regulation,** the abundance at each level is controlled by consumers (predators) at the top of the food chain. When carnivores suppress the number of herbivores, plants experience a release from grazing and flourish. If carnivores decrease, herbivores increase and plants decrease (Hairston et al. 1960, Hairston and Hairston, 1993). These two views of processes controlling community structure are fundamentally different and still the topic of much research and debate.

ASSEMBLY RULES

In the previous discussion we have explored what might happen when a species is removed or added to a food web. These experiments are critical to understanding how the interrelationships among species influence the current structure of communities. They have shown us that in many cases, the removal or addition of a species can have profound consequences on the structure and function of the community. These results emphasize the importance of the interactions, both direct and indirect, among the species making up the community. Given this interconnectedness and, in many cases, interdependence among species within a community, one is faced with the inevitable question of how communities come into being. Or to state this question more simply: how do the species become assembled to form a community? What factors might be involved in the assemblage of species and how those species fit together to form community structure is one of the most intriguing problems in ecology.

At one level, the answer to this question seems obvious. For a herbivore to survive, plants must be present, and likewise, for a population of predators to persist, there must be prey. However, in the above discussion of food webs, we have seen examples where the factors involved are far from obvious, such as when the presence of a predator allows for the coexistence of two competing species on which it preys.

Ecologists have used two approaches to investigate community development and structure. One, already alluded to, is to remove a keystone species and study how the community restructures itself—what species become the new dominants

FIGURE 20.11 (Opposite) A midwinter food web in a *Salicornia* salt marsh (San Francisco Bay area). Producer organisms are marsh and aquatic plants. Marsh plants (1) are consumed by terrestrial herbivorous invertebrates, represented by the grasshopper and snail (2), and by herbivorous marine and intertidal invertebrates (3). Fish, represented by the smelt (*Osmerus mordax*) and anchovy (*Anchova* spp.) (4), feed on vegetable matter from both ecosystems. The fish in turn are eaten by first-level carnivores, represented by the great blue heron (*Ardea herodias*) and common egret (*Casmerodius albus*) (5). Continuing through the web are the following omnivores: clapper rail (*Rallus longirostrus*) and mallard duck (6), savanna (*Passerculus sandwichensis*) and song (*Melospiza melodia*) sparrows (7), Norway rat (8), California vole (*Microtus californicus*) and salt marsh harvest mouse (*Reithrodontomys raviventris*) (9), and the least (*Calidris minutilla*) and western (*C. mauri*) sandpipers (10). The vagrant shrew (*Sorex vagrans*) (11) is a first-level carnivore. Top carnivores are the harrier (*Circus cyaneus*) and short-eared owl (*Asio flammeus*) (12). (Adapted from Johnson 1956:99.)

and what species increases or decrease. The second approach is to attempt to reconstruct the sequence in which species were added when the community was formed, and to figure out what colonization sequences may or may not be possible (Diamond 1975, Drake 1990). In effect, this second approach tries to establish the rules that govern the assembly of species to form communities. It is much like fitting together a jigsaw puzzle, with one significant difference. The assembly rule for putting together a jigsaw puzzle is fixed by the relative shapes of the pieces. Assembly rules for a community involve alternative pieces that also fit but change the configuration of the puzzle (Figure 20.12).

Consider species A that initially colonizes a site, increases in population size, and dominates space and food resources. Then two other species, B and C, invade the area. Both of these species can effectively compete with and share the site with A, but B and C are strong competitors. Each can potentially outcompete and displace the other. In this case, the advantage goes to the species that colonizes the site first, allowing its population to increase in the absence of competition from the other (see case 3 of the Lotka-Volterra model, Figure 14.2). Thus depending on their order of arrival, the community now will consist of AB or AC (Figure 20.12). Now a fourth species D arrives on the site. It occupies a somewhat different niche, but still competes with C. If the community species D invades is AB, it can take its place in the community, which will now consist of ABD. If the combination is AC, then D cannot invade the community.

This example is very simplistic. Many more species and types of species interactions are involved in the colonization

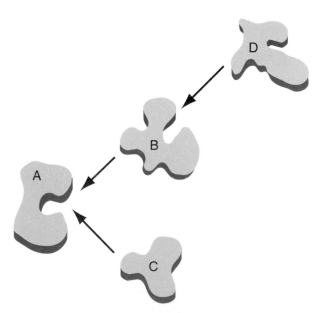

FIGURE 20.12   Pieces of a jigsaw puzzle that can be fitted together in two combinations represent an analogy of community assembly. If A colonizes first, then either B or C, competing species, can colonize. If B is the successful colonizer, then D can also fit in. If C is the successful colonizer, D will not be able to colonize, because the community will resist invasion; piece D will not fit. (After Drake 1990.)

and establishment of communities. However, theoretical studies examining the assembly of communities do provide some insights into how communities develop. Perhaps the most important insight gained is the importance of understanding the historical context in which a community arises (Ricklefs and Schluter 1993). Although the same species may be found in similar communities, each may have colonized the area at different times and in different sequences, which would change species relationships. Each community develops within a historical context that will influence species interactions. Thus different assembly routes will produce differences in community organization.

Of what value is this highly theoretical concept? Assembly rules become quite important in conservation, and particularly in the restoration of ecosystems. If you are attempting to reconstruct or restore a prairie from agricultural lands, for example, you will want to give some thought to the species of plants to be planted and what animals might be introduced or colonize the area naturally. Should you introduce prairie dogs or buffalo? What impact might the introduction of wolves into a wolfless wilderness have on the current community? How will certain species of fish fit into a pond or a restored stream ecosystem? Too often, assembly rules are given little consideration in such restoration work.

CLASSIFICATION OF COMMUNITIES

If a community is defined by its physical and biological structure, how different must two adjacent areas be to be referred to as different and separate communities? This is not a simple question. Moving across the landscape, we notice that the nature of the physical and biological structure of the community changes. Often these changes are small, subtle ones in the species composition or height of the vegetation. However, as we travel further and further, these changes become more pronounced. For example, the region of central Virginia just east of the Blue Ridge Mountains is a landscape of rolling hills. The area is a mosaic of forest and field. As we walk through a forest, the appearance remains much the same. That is, the physical structure of the community is very similar; the canopy, understory, shrub layer, and forest floor appear much the same. However, the biological structure, the mix of species that compose the community, may change quite dramatically. As we move from a hilltop to the bottomland area along the stream, the mix of species changes from oaks (*Quercus* spp.) and hickory (*Carya* spp.) to tree species that are associated with much wetter environments such as sycamore (*Platanus occidentalis*), hornbeam (*Carprinus caroliniana*), and sweetgum (*Liqudambar styraciflua*) (Figure 20.13). These changes in the physical and biological structure of communities as we move across spatial gradients (the landscape) are referred to as **zonation.**

Given the differences in the species composition of the forests found on the hilltop and bottomland, most ecologists would define these two forest stands as different vegetation

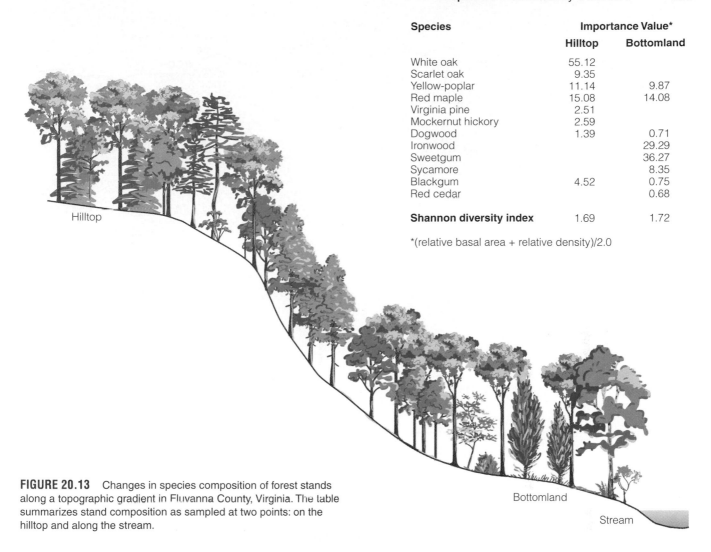

Species	Importance Value*	
	Hilltop	**Bottomland**
White oak	55.12	
Scarlet oak	9.35	
Yellow-poplar	11.14	9.87
Red maple	15.08	14.08
Virginia pine	2.51	
Mockernut hickory	2.59	
Dogwood	1.39	0.71
Ironwood		29.29
Sweetgum		36.27
Sycamore		8.35
Blackgum	4.52	0.75
Red cedar		0.68
Shannon diversity index	1.69	1.72

*(relative basal area + relative density)/2.0

FIGURE 20.13 Changes in species composition of forest stands along a topographic gradient in Fluvanna County, Virginia. The table summarizes stand composition as sampled at two points: on the hilltop and along the stream.

communities. But as you walk between these two areas, the distinction may not seem so straightforward. If the transition between the two communities is abrupt and distinct, there may be no problem in defining the community boundaries. However, the differences in the species composition and patterns of dominance observed in the two communities may occur gradually over the distance from hilltop to stream. In this case, the boundary is not so clear.

Classification

Although it is easy to describe the similarities and differences between two areas in terms of species composition and structure (see Appendix B for examples of similarity indices), the actual classification of areas into distinct communities or groups of communities involves a degree of subjectivity, and it often depends on the objectives of the individual study. There are a number of approaches to community classification, each somewhat arbitrary and each suited to a particular need or viewpoint (Mueller-Dombois and Ellenberg 1974). The most commonly used classification systems are based on physiognomy, species composition, dominance, and habitat.

Physiognomy (the general appearance, vertical structure, and growth form of vegetation) is a highly useful method of naming and delineating communities, particularly in surveying large areas and in subdividing major types into their component communities. Because animal distribution appears to correlate with vegetational communities, classification by physiognomy will relate to both plant and animal life. Communities so classified are often named after the dominant form of life, usually plants, such as coniferous or deciduous forest, sagebrush, or shortgrass prairie. A few are named after animals, such as the barnacle–blue mussel (*Balanus-Mytilus*) community of the rocky tidal zone. One community, of course, may grade into another. In these cases the classification may be based on arbitrary, although specific, criteria. Where habitat boundaries are well defined, communities may be classified by physical features such as tidal flats, sand dunes, cliffs, ponds, and streams.

Finer subdivisions may be based on species composition, a system that works much better with plants alone than with animals. Such a classification requires a detailed study of the individual community (see Mueller-Dombois and Ellenberg 1974), covering frequency, dominance, constancy, presence, and fidelity. In this system, areas in which similar

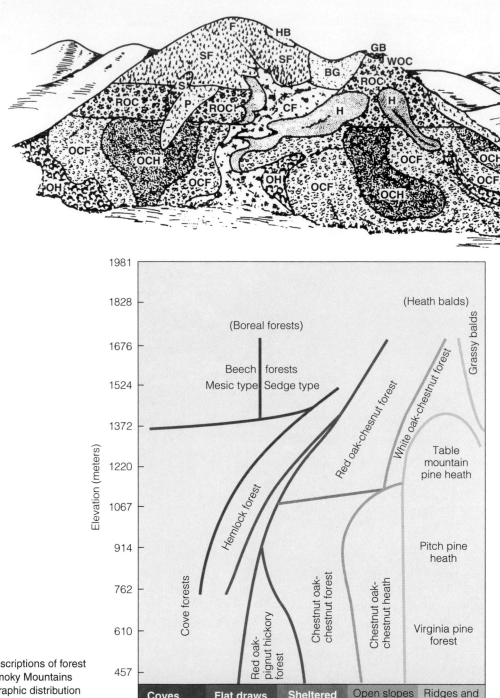

FIGURE 20.14 Two descriptions of forest communities in Great Smoky Mountains National Park. (a) Topographic distribution of vegetation types on an idealized west-facing mountain and valley. (b) Idealized arrangement of community types according to elevation and aspect. (Whittaker 1956.)

combinations of species occur can be classified as the same community type. The type is named after the dominant organisms, or the ones with the highest frequency. Examples are the *Quercus-Carya* association, or oak-hickory forest; the *Stipa-Bouteloua* association, or mixed prairie; and the animal-dominated *Balanus-Mytilus*, or barnacle–blue mussel, community.

European ecologists have developed a floristic classification with emphasis on dominance, constancy, and diag-

nostic species (Braun-Blanquet 1965). They group communities into classes, orders, alliances, and associations (see Poore 1962, Whittaker 1962, Mueller-Dombois and Ellenberg 1974). Such a classification involves fidelity, the faithfulness of a species to a community type. Species with low fidelity occur in a number of different communities and those with high fidelity in only a few. Seldom, if ever, are the latter found away from certain closely associated plants and animals. The greater is the constancy, the ratio of species always associated with the community type to the total number of species, the more homogeneous is the community and the more sharply it can be delineated. Often, however, this close association merely reflects a group of species unable to grow successfully under a wide range of ecological conditions or with other species. Species can be grouped as **exclusive,** those completely or nearly confined to one type of community; **characteristic,** those most closely identified with a certain community; and **ubiquitous,** those with no particular affinity for any community. The species grouped as characteristic, high in both constancy and dominance, are the ones that define the community type.

Ordination

A different approach to comparing the structure of communities involves ordinational techniques. **Ordination** is the arrangement of communities along a linear axis according to their similarity in species composition. The relative position of the sample units on the axis and to each other provides maximum information about their ecological similarities (Austin et al. 1984, Ludwig and Reynolds 1988). Ordination is an exploratory data analysis technique designed to seek patterns or trends.

The earliest of these ordination methods in plant ecology was the polar ordination of Bray and Curtis (1957), a technique specifically developed to analyze plant community data. It involves the use of an index of community similarity to ordinate stands. Using the importance values describing the relative contribution of each species to the forest stand, an index of similarity between all possible combinations of stands can be calculated. The two most dissimilar stands then define the endpoints of the axis, and all other stands are positioned between these two based on their relative similarities to the two endpoints (see Ludwig and Reynolds 1988, Beals 1984). A detailed example of the Bray-Curtis ordination technique is presented in Appendix B. Although subjective, polar ordination remains a useful method.

Since the early development of community ordination and classification, more mathematically sophisticated and complex multivariate methods have been developed to classify and ordinate communities. These methods include principal component analysis, cluster analysis, discriminate analysis, correspondence analysis, and nonlinear ordinations. The value of these methods is their ability to summarize a large, complex set of variables. They enable ecologists to evaluate the relationships between communities based on patterns of species composition, and develop new hypotheses about environmental effects on species distribution and community patterns.

Scale and Community Classification

The example of distinguishing between forest communities in central Virginia presented in Figure 20.13 involved a small spatial area, with changes in species composition from hilltop to bottomland over a distance on the order of hundreds of meters to a kilometer. By expanding the spatial dimension to include

a larger and larger area, the observed differences in the community structure, both physical and biological, will increase. An example is the pattern of forest zonation in the Great Smoky Mountains National Park presented in Figure 20.14 (Whittaker 1956). The zonation of forest communities is a complex pattern related to elevation, and slope position and aspect. Note that the description of the forest communities in the park contains few species names (e.g., hemlock forest, beech mesic type). The names given to these communities are not meant to suggest a lack of species diversity; rather, they present a shorthand method of naming communities based on the dominant species. Each community could be described by providing a complete list of species and their relative population sizes and contribution to the total biomass as was done in Figure 20.13. However, such descriptions would prove to be impractical and unnecessary to communicate the major changes that are occurring in the structure of these communities across the landscape. In fact, as the area of interest (spatial dimension) is expanded even further to include the entire eastern United States, the nomenclature of classifying (describing the variation in structure among) forest communities becomes even broader (Figure 20.15). Note that in the broad-scale description of forest zonation in the eastern United States shown in Figure 20.15 (Braun 1950), all of

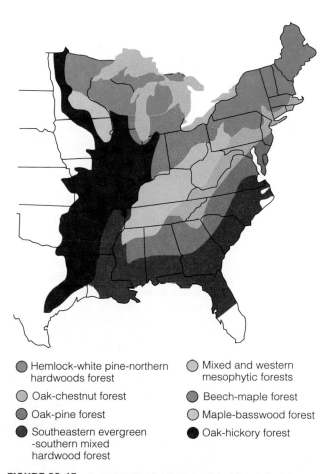

- ● Hemlock-white pine-northern hardwoods forest
- ● Oak-chestnut forest
- ● Oak-pine forest
- ● Southeastern evergreen -southern mixed hardwood forest
- ● Mixed and western mesophytic forests
- ● Beech-maple forest
- ● Maple-basswood forest
- ● Oak-hickory forest

FIGURE 20.15 Large-scale distribution of deciduous forest communities in the eastern United States is defined by eight regions. (Braun 1950.)

the Great Smoky Mountains National Park shown in Figure 20.14 (located in southeastern Tennessee and northwestern North Carolina) is described as a single forest community type or region: oak-chestnut, a type that extends from New York to Georgia.

The examples of zonation presented above make an important point to which we shall return to when we examine the processes that are responsible for these spatial changes in community structure: our very definition of community is a spatial concept. Just as with the biological definition of population (Chapter 10), the community is defined as a spatial unit that occupies a given area. As shall be seen later, the difficulty in delineating communities as discrete spatial units has led to problems in understanding the processes responsible for the patterns of zonation seen in the world around us.

Summary

1. A biotic community is a naturally occurring assemblage of plants and animals living in the same environment, mutually sustaining and interacting directly or indirectly with one another.
2. Communities exhibit some form of layering or vertical stratification. In terrestrial communities, stratification is determined by the life forms of plants and influences the nature and distribution of animal life. In addition to vertical stratification, terrestrial and shallow freshwater communities may also exhibit horizontal structure produced by variation in environmental conditions, resulting in a spatial heterogeneity or patchiness within the community. The communities that are highly stratified vertically and are patchy hold the richest variety of animal life, for they contain a great assortment of microhabitats and environmental conditions.
3. Dominant species may control the character of communities. The dominant species may be the most numerous, possess the highest biomass, preempt the most space, or make the largest contribution to energy flow. But the dominant species may not necessarily be the most important species in the community. But the dominant species in a community may be difficult to define or determine.
4. The biological structure of a community is defined by its species composition. The species diversity of a community is defined by both the species richness, the number of species, and the evenness of the distribution of individuals among the species (that is, the number of individuals of each species). One can compute a species diversity index, which is useful only on a comparative basis, either within a single community over time or among communities.
5. The structure or organization of a community is influenced by interactions among its member species. Interspecific competition may exclude or reduce the abundance of species or the number of individuals within a species. Predators may reduce the population sizes of competing prey species, permitting their coexistence. Parasites and disease may reduce or eliminate cer-

tain species from the community, opening it for invasion by others. Finally, mutualistic relationships, both direct and indirect, may integrate community structure in ways that are now only beginning to be appreciated.
6. Community structure can be analyzed in terms of feeding relationships. A food chain is a descriptive diagram—a series of arrows, each pointing from one species to another for which it is a source of food. Communities are composed of numerous food chains meshed into a complex food web with links leading from primary producers through an array of consumers: herbivores, carnivores, and omnivores. There are two views of community regulation based on the study of food webs: bottom-up regulation emphasizes resource availability and competition; top-down regulation emphasizes the effects of predation on regulating prey populations.
7. Delineating and classifying communities can be difficult unless the transition between adjacent communities is abrupt. The classification of communities often can be arbitrary and involves a variety of statistical approaches. The simplest classification scheme is based on physiognomy and dominant species. More detailed schemes may involve methodical studies of floristics. The classification of communities is made more difficult because, like populations, the community is a spatial concept, and defining the boundary or spatial extent of a community is often arbitrary and dependent on the purposes of the particular study.

Review Questions

1. What is a community?
2. What is the major factor(s) that defines the physical structure of terrestrial communities?
3. What determines the vertical structure of aquatic communities?
4. What defines the biological structure of a community?
5. Distinguish among species dominance, species richness, and species diversity. How are these concepts interrelated?
6. Relate the concepts of food chain and food web.
7. How might two species have a positive influence on each other (mutualism) indirectly through their shared interaction with a third species?
8. How might the presence of a predator influence the interactions between two competing prey species on which it feeds?
9. Why is it often difficult to delineate and classify communities?

Cross References

Adaptation, 81; competition, 243; mutualism, 321; predator-prey interactions, 293; vegetation-herbivore interactions, 286; parasitism, 307; food chains and webs, 395, 493; patterns of species richness, 388; forest structure, 613; vertical structure of open water ecosystems, 633; grassland structure, 568; mycorrhizae, 322; succession, 404; community similarity, Appendix B; importance value, Appendix B; ordination, 401.

CHAPTER 21

Community Dynamics

CONCEPTS

1. Spatial changes in species composition and community structure and function are referred to as zonation.

2. Changes in species composition and community structure and function over time at a given location are called succession.

3. Succession may be primary, starting on sites previously unoccupied by plants, or secondary, starting on sites where plants have previously grown.

4. Succession within a community can appear to be unidirectional with a defined endpoint—the climax—or as a mosaic of patches, each undergoing a continuous sequence of cyclic replacement.

5. Animals depend on certain successional stages, and therefore animal life changes as plant communities change.

6. Community patterns observed today are products of evolution and plant succession over geological time.

The physical and biological structure described in the preceding chapter are not static characteristics of the community. Both change temporally and spatially. The vertical structure of the community changes with time as plants become established, grow, and die. The birth and death rates of species change in response to environmental conditions, resulting in a shifting pattern of species dominance and diversity.

As environmental conditions change in time and space, the structure of the community, both physical and biological, likewise changes. The result is a dynamic mosaic of communities on the landscape. It is this changing pattern of community structure that is the focus of community ecology, and the topic of this chapter.

ZONATION: SPATIAL VARIATION IN COMMUNITY STRUCTURE

As we move across a landscape, the biological and physical structure of the community changes. At first these changes may be subtle, the presence of a new species not encountered in the adjacent area, or a change in the vertical structure, such as when one encounters an opening in a forest. As we walk further, the differences in community structure may become more obvious, with noticeable changes in the dominant species. For example, Robert Whittaker (1956) examined changes in the dominant tree species along an elevational gradient on the dry, south-facing slopes of the Great Smoky Mountains of Tennessee (Figure 21.1). At the lower elevations, Virginia pine (*Pinus virginiana*) is the dominant species, with pitch pine (*P. rigida*) being less common. Between 2000 and 2500 feet in elevation, there is a marked shift in species composition, with pitch pine replacing Virginia pine as the dominant species. Continuing up in elevation, Virginia pine disappears from the community and table-mountain pine (*P. pungens*)

becomes the codominant with pitch pine. The abundance of table-mountain pine continues to increase with altitude while pitch pine declines. Although an ecologist is likely to distinguish three community types (corresponding to the three pine species) along this gradient, no distinct boundaries exist.

Patterns of spatial variation in community structure or zonation are common to all environments, aquatic and terrestrial. Figure 21.2 provides an example of zonation in a salt marsh. Note the variations in both the physical and biological structure of the communities as we move from the tidal zone (shoreline) through the marsh to the upland terrestrial environments. The dominant plant growth forms in the marsh are grasses and sedges. These growth forms give way to shrubs and trees as we move to dry land, and the depth to the water table increases. Within the zone dominated by grasses and sedges, the dominant species change as we move back from the tidal areas. A variety of environmental factors change as we move along this spatial gradient, including microtopography, water depth, and salinity (see Chapter 30). Distinct communities are recognizable in these differing environments, in terms of both the species that dominate and the structural features of the vegetation such as height, density, and dispersion that define the structure of the community.

SUCCESSION: TEMPORAL VARIATION IN COMMUNITY STRUCTURE

Community structure varies not only in space but also in time. Suppose that rather than moving both horizontally and vertically in space, as with the examples of zonation, we stand in one position and observe the area as time passes. For example, abandoned cropland and pastureland are common sights in agricultural regions in once-forested areas in eastern North America. No longer tended, the land quickly grows up in grasses, goldenrod, and weedy herbaceous plants. In a few years these same weedy fields are invaded by shrubby growth—blackberries, sumac, and hawthorn. These shrubs are followed by fire cherry, pine, and aspen. Many years later this abandoned land supports a forest of maple, oak, cherry, or pine (Figure 21.3). The process we would have observed, the gradual and seemingly directional change in the structure of the community through time from field to forest, is called **succession.** Succession in its most general definition is the temporal change in community structure through time. In contrast to zonation, succession refers to a given point in space—a single location.

The sequence of communities from grass to shrub to forest historically has been called a **sere** (from the word *series*), and each of the changes is a seral stage. Although each **seral stage** is a point in a continuum of vegetation through time, it is recognizable as a distinct community. Each has its characteristic structure and species composition. A seral stage may last only one or two years or several decades. Some stages may be missed completely, or they may appear only in abbre-

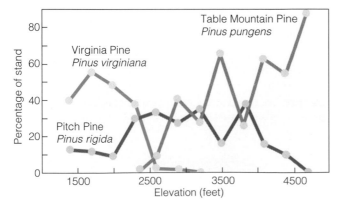

FIGURE 21.1   Distribution of tree species along an elevation transect (dry, south-facing slope) in the Smoky Mountains of Tennessee. No distinct boundaries separate what ecologists would likely describe as three communities along this gradient: Virginia pine forest at low elevations, pitch pine heath at middle elevations, and table-mountain pine heathland at high elevations. (Data from Whittaker 1956.)

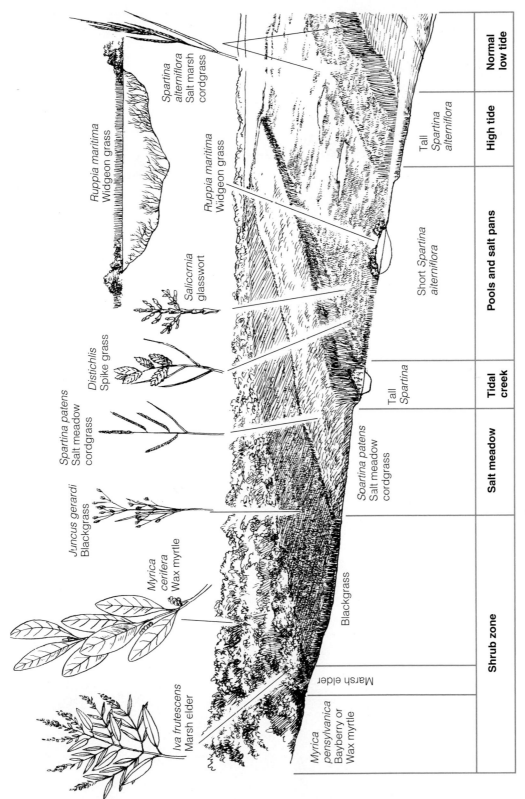

FIGURE 21.2 Patterns of zonation in an idealized New England salt marsh, showing the relationship of plant distribution to microtopography and tidal submergence.

viated or altered form. For example, when an abandoned field grows up immediately in forest trees, as in Figure 21.3, the shrub stage appears to have been bypassed; but structurally its place is taken by the incoming young trees.

Like zonation, the process of succession is generally common to all environments, both terrestrial and aquatic. The ecologist William Sousa carried out a very interesting experiment to examine the process of succession in a rocky intertidal algal community in southern California. A major form of natural disturbance in these communities is the overturning of rocks by the action of waves. Algae populations then recolonize these cleared surfaces. To examine this process,

FIGURE 21.3 Successional changes in an old field in western Pennsylvania over 50 years.
(a) The field as it appeared in 1942, when it was moderately grazed. (b) The same area in 1963.
(c) A close view of the rail fence in the left background of (a). (d) The same area 20 years later.
The rail fence has rotted, and white pine and aspen grow in the area. (e) The field in 1972, when
aspen has claimed much of the ground. (f) In 1992, the field is covered with a young forest
dominated by quaking aspen and red maple.

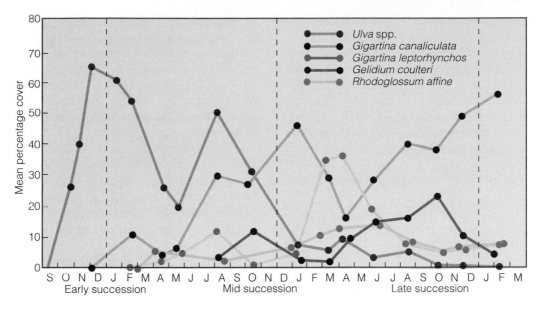

FIGURE 21.4 Changes in the cover of five algal species that colonized blocks introduced to the intertidal zone in September 1974. Note the shifting pattern of species dominance as time progresses. (Data from Sousa 1979).

Sousa placed concrete blocks in the water to provide new surfaces for colonization. The results of the study show a pattern of colonization and extinction, with populations that initially colonize the concrete blocks being displaced by other species as time progresses (Figure 21.4). This is the process of succession. The initial, or **early successional species,** often referred to as **pioneer species,** are usually characterized by a high growth rates, smaller size, high degree of dispersal, and high rates of population growth (*r*-selected species, see Chapter 13). In contrast, the **late successional species** generally have lower rates of dispersal and colonization, slower growth rates, and are larger and longer-lived (*K*-selected species, see Chapter 13). As the names early and later succession imply, the patterns of species replacement with time are not random. In fact, if the experiment were to be repeated tomorrow, one would expect the resulting patterns of colonization and extinction, the **successional sequence,** to be very similar to that presented in Figure 21.4.

A similar pattern of succession occurs in terrestrial plant communities. The patterns of species replacement following forest clearing (clear-cutting) at the Hubbard Brook Experimental Forest in New Hampshire are shown in Figure 21.5. Prior to forest clearing, the understory was dominated by seedlings and saplings of beech (*Fagus grandifolia*) and sugar maple (*Acer saccharum*). Large individuals of these two tree species dominated the canopy, and the seedlings represent successful reproduction of the parent trees. Following the removal of the larger trees (timber harvest) in 1970, the numbers of beech and maple seedlings declined and were soon replaced by raspberry thickets and seedlings of sun-adapted (shade intolerant), fast-growing, early successional tree species such as pin cherry (*Prunus pensylvanicum*) and yellow birch (*Betula alleghaniensis*). After many years, these species will eventually be replaced by the

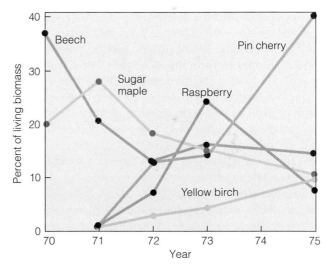

FIGURE 21.5 Changes in the relative abundance (percentage of living biomass) of woody species in the Hubbard Brook Experimental Forest after a clear-cut in 1970. (From F. Bormann and G. E. Likens, *Pattern and Process in a Forested Ecosystem*, p. 116, Fig. 4.4b. New York: Springer-Verlag, 1979. Copyright © 1979 Springer-Verlag. Used by permission.)

late successional species of beech and sugar maple that previously dominated the site.

The two studies presented above point out the similar nature of successional dynamics in two very different environments. They also present examples of two different types of succession: primary and secondary. **Primary succession** occurs on a site previously unoccupied by a community—a newly exposed surface such as the cement blocks in rocky intertidal environment. In contrast, the study at Hubbard Brook following forest clearing is an example of **secondary**

succession. In contrast to primary succession, secondary succession occurs on previously occupied (vegetated) sites following disturbance. In this case, disturbance is defined as any process that results in the removal (either partial or complete) of the existing vegetation (community). As seen in the Hubbard Brook example, the disturbance does not always result in the removal of all individuals. In these cases, the amount (density and biomass) and composition of the surviving community will have a major influence on the subsequent successional dynamics. A more detailed discussion of disturbance and its role in structuring communities is presented in Chapter 23.

PRIMARY SUCCESSION

Primary succession begins on sites that never have supported life. Such places are rock outcrops and cliffs (Burbanck and Platt 1964, Shure and Ragsdale 1977), sand dunes (Cowles 1899, Olson 1958), and newly exposed glacial till (Crocker and Major 1955, Lawrence 1958).

For example, consider primary succession on a very inhospitable site, a sand dune. A product of pulverized rock, sand is deposited by wind and water. Where deposits are extensive, as along the shores of lakes and oceans and on inland sand barrens, sand particles may be piled in long windward slopes to form dunes (Figure 21.6). Under the forces of wind and water, the dunes can shift, often covering existing vegetation or buildings. The establishment and growth of plant cover acts to stabilize the dunes. Colonization of sand dunes and the progressive development of vegetation was originally described by H. C. Cowles (1899) in a pioneering classic study of plant succession on the dunes about Lake Michigan.

Grasses, especially beach grass (*Ammophila breviligulata*), are the most successful pioneering plants, and they function to stabilize the dunes. When they and such associated plants as beach pea (*Lathyrus japonicus*) stabilize the dunes at least partly, mat-forming shrubs invade the area.

From this point the vegetation may pass to pine and then oak, or to oak directly without an intervening pine stage. Because of the low moisture reserves in the sand, oak is rarely replaced by more moisture-demanding (mesophytic) trees (Olson 1958). Only on the more favorable leeward slopes and in depressions, where microclimate is more moderate and where moisture can accumulate, does succession proceed to more mesophytic trees such as sugar maple (*Acer saccharum*), basswood (*Tilia americana*), and red oak (*Quercus rubra*). Because these trees shade the soil and accumulate litter on the soil surface, they act to improve nutrients and moisture conditions. On such sites a mesophytic forest may become established without going through the oak and pine stages. This example emphasizes one aspect of primary succession: the colonizing species ameliorate the environment, paving the way for invasion of other species.

Newly deposited alluvial soil on a floodplain represents another example of primary succession. Over the past 200 years, the glacier that once covered the entire region of Glacier Bay National Park, Alaska, has been retreating (melting) (Figure 21.7). As the glacier retreats, the newly exposed landscape is initially colonized by a variety of species, such as alder (*Alnus* spp.) and cottonwood (*Populus* spp.). Eventually these early successional species are replaced by the later successional tree species of spruce (*Picea* spp.) and hemlock (*Tsuga canadensis*) (Figure 21.8), and the resulting forest resembles the forest communities in the surrounding landscape.

The island of Hawaii in the Hawaiian islands is known for its active volcanoes. The lava flows that result represent newly formed landscapes on which primary succession occurs (Figure 21.9). The previous plant communities in these areas have been destroyed by the fire and intense heat, and the existing soils have been covered by a new layer of igneous rock. In addition to covering the existing landscape, the lava flows create new land mass, expanding the extent of the island. These new landscapes formed by lava flows also extend to the underwater environment, providing new substrates for colonization of aquatic life, especially algae. Doty (1957) followed algal recolonization after lava flow in

FIGURE 21.6 Primary succession on a coastal sand dune colonized by beach grass (*Ammophila breviligulata*).

(a)

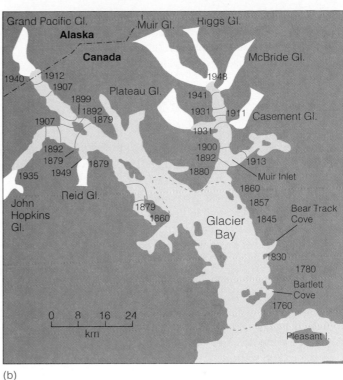

(b)

FIGURE 21.7   (a) Primary succession along riverine environments of Glacier Bay National Park, Alaska. (b) The Glacier Bay fjord complex in southeastern Alaska, showing the rate of ice recession since 1760. As the ice retreats, it leaves moraines along the edge of the bay, in which primary succession occurs.

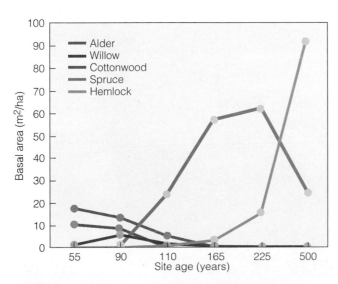

FIGURE 21.8   Changes in community composition with stand age for sites in Glacier Bay, Alaska, ranging in age from 55 to 500 years. Species abundance expressed as basal area per hectare, where basal area is the cross-sectional area of tree at a defined height above ground. (Source: Eric Alan Hobbie "Nitrogen Cycling During Succession," Masters thesis, University of Virginia 1994.)

FIGURE 21.9   Trees and ferns colonize a lava flow on the island of Hawaii.

Hawaii. A series of successional stages of microalgae and macroalgae occurred on the sterile, newly formed rock. Within 10 years, a community similar to that found on 100-year-old lava flows had become established.

SECONDARY SUCCESSION

One of the classic examples of secondary succession in terrestrial environments is the study of old-field succession in the Piedmont of North Carolina (Billings 1938, Keever 1950). The year a crop field is abandoned, the ground is claimed by annual crabgrass (*Digitaria sanguinalis*), whose seeds, lying dormant in the soil, respond to light and moisture and germinate. But the crabgrass's claim to the ground is short-lived. In late summer the seeds of horseweed (*Lactuca canadensis*), a winter annual, ripen. Carried by the wind, they settle on the old field, germinate, and by early winter have produced rosettes. The following spring, horseweed, off to a head start over crabgrass, quickly claims the field. During the summer the field is invaded by other plants—white aster (*Aster ericoides*) and ragweed (*Ambrosia artemissifolia*).

By the third summer broomsedge (*Andropogon virginicus*), a perennial bunchgrass, invades the field. Abundant organic matter and ability to exploit soil moisture efficiently permit broomsedge to dominate the field. About this time pine seedlings, finding room to grow in open places among the clumps of broomsedge, invade the field. Within five to ten years the pines are tall enough to shade the broomsedge. Eventually hardwood species, such as oaks and ash, grow up through the pines and, as the pines die, take over the field (Figure 21.10). Development of the hardwood forest continues as shade tolerant trees and shrubs—dogwood (*Cornus florida*), redbud (*Cercis canadensis*), sourwood (*Oxydendrum arboreum*), hydrangea (*Hydrangea arborescens*), and others—fill the understory.

Studies of physical disturbance in marine environments have demonstrated secondary succession in seaweed, salt marsh, mangrove, seagrass, and coral reef communities. Dug-

gins (1980) examined the process of succession following disturbance in the subtidal kelp forests of Torch Bay, Alaska. One year after the removal of the kelp forest, a mixed canopy of kelps (*Nereocystis luetkeana* and *Alaria fistulosa*) formed, and an understory of *Costaria costata* and *Laminaria dentigera* developed. During the second and third years, continuous stands of *L. setchelli* and *L. groenlandica* developed and the community had returned to its original composition. A similar pattern has been observed in the subtidal kelp forests off the California coast (Figure 21.11) by Reed and Foster (1984).

Secondary succession in seagrass communities has been described for a variety of locations, including the shallow tropical waters of Australia (Shepherd and Womersley 1981, Birch and Birch 1984) and the Caribbean (Williams 1990). In the U.S. Virgin Islands the green algae (*Halymenia, Penicillus, Udotea, Caulerpa*) are early colonizers following disturbance, stabilizing and adding organic matter to the sediments. The seagrasses *Syringodium filiforme* and *Thalassia testudinum* then grow into the area, forming a community similar to that observed prior to disturbance.

Secondary succession in the seagrass communities of Florida Bay have been described in detail by Zieman and Zieman 1989 (Figure 21.12). The removal of grass cover from an area by the wave action associated with storms or by heavy grazing by sea turtles and urchins creates openings in the grass cover, exposing the underlying sediments. Erosion on the down-

FIGURE 21.11   Giant kelp (*Macrocystis pyrifera*) off the California coast. Individuals can grow to 30 m in length.

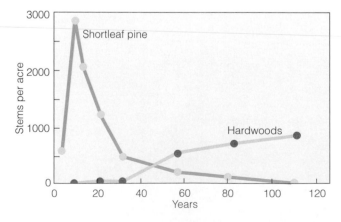

FIGURE 21.10   Decline in the abundance of shortleaf pine and increase in the density of hardwood tree seedlings during succession on abandoned farmland in the Piedmont area of North Carolina. (After Billings 1938.)

FIGURE 21.12 Disturbed areas (light-colored areas) within seagrass communities in the Florida Bay. These areas, called blowouts, undergo a process of recovery that involves a shift in species dominance from macroalgae to seagrasses (see Figure 21.13). Inset shows a Thalassia testudinum dominated community in the Florida Bay. (Photo courtesy of J. Zieman.)

current side of these openings results in localized disturbances called blowouts (Figure 21.13). The initial colonizers on these disturbed sites are typically rhizophytic macroalgea, in which species of *Halimeda* and *Penicillus* are the most common. These algae have some sediment-binding capability, but their ability to stabilize the sediments is minimal and their major function in the early successional stage seems to be the contribution of sedimentary particles as they die and decompose.

Halodule wrightii, the local pioneer species of seagrass, colonizes readily from either seed or rapid vegetative branching. The carpet of cover laid by *Halodule* further stabilizes the sediment surface. Eventually *Thalassia testudinum* will begin to colonize the region. Its larger leaves and extensive rhizome and root system effectively trap and retain particles, increasing the organic matter of the sediment. The once-disturbed area now again resembles the surrounding seagrass community.

FIGURE 21.13 Idealized pattern of secondary succession in seagrass communities of Florida Bay following disturbance (see Figure 21.12). The area is initially colonized by rhizophytic macroalgae. These algae species are soon displaced by the early successional seagrass species, *Halodule wrightii.* This species is eventually displaced by the larger species, *Thalassia testudinum.*

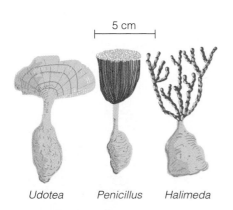

5 cm

Udotea   *Penicillus*   *Halimeda*

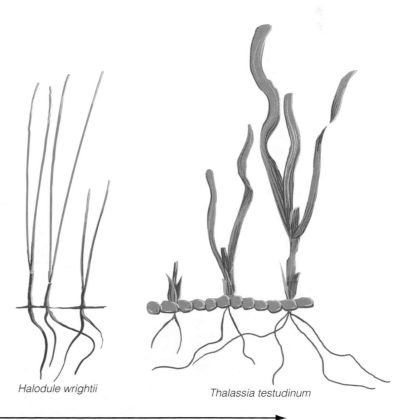

Halodule wrightii   *Thalassia testudinum*

Time following disturbance

TIME AND DIRECTION IN SUCCESSION

Time is the integral component of succession. As time passes the physical and biological structure of the community changes, going through a series of successional stages or seres. These changes are orderly, not haphazard, and barring disturbance by humans or natural events the reappearance of a community resembling that which previously occupied the site is predicted.

Climax

The process of succession has the appearance of direction in time, eventually arriving at some equilibrium or steady state with the physical and biotic environment. This apparent end-point of succession was termed the **climax** by Clements (1916). In his theory, Clements recognized only one climax for a region, whose characteristics are determined solely by climate. Successional processes and modifications of the environment overcome the effects of differences in topography, parent material of the soil, and other factors. Given sufficient time, all seral communities in a region will converge to and stabilize at a single climax. The whole landscape will be clothed with a uniform plant community. Communities other than the climax are related to it by successional development. This climax view of succession was later modified by Tansley (1935) and Whittaker (1953), but has since fallen out of favor among ecologists, although the term *climax community* is still used to describe communities such as old-growth forests.

Cyclic Replacement

As one observes a landscape undergoing succession, there is often a general appearance of uniformity. Although spatial heterogeneity occurs, there is an apparent sequence of stages, each different from the previous one. Eventually the stages lead to a community that persists on the landscape until a disturbance once again resets the clock, repeating the process of succession. It is this large-scale view of succession in the prairies of the midwestern United States that led Clements to his theory of the climax. However, when observed from the perspective of a much smaller spatial scale—say a single location within the grassland or forest—the view can be quite different. As canopy individuals die within the forest, openings are created. These openings are characterized by higher light levels, higher temperatures, and reduced competition for essential resources in the absence of the canopy tree(s). In a way, the opening caused by the death of a single or few canopy trees resembles the type of disturbance that initiates secondary succession in the examples provided above, only on a much smaller spatial scale. Such changes are constantly occurring in patches across the community.

This perspective of small-scale dynamics within the community led A. S. Watt (1947) in a classic paper, "Pattern and Process in the Plant Community," to describe succession as a cyclic rather than linear process leading to some defined endpoint. Watt suggested that successional stages within a community that appear to be directional are often phases in a cycle of vegetation replacement (Figure 21.14). Death of vegetation or periodic disturbance starts regeneration again at some particular stage. Such changes usually occur continuously on a small scale within a community, and they are repeated over the course of time throughout the entire community.

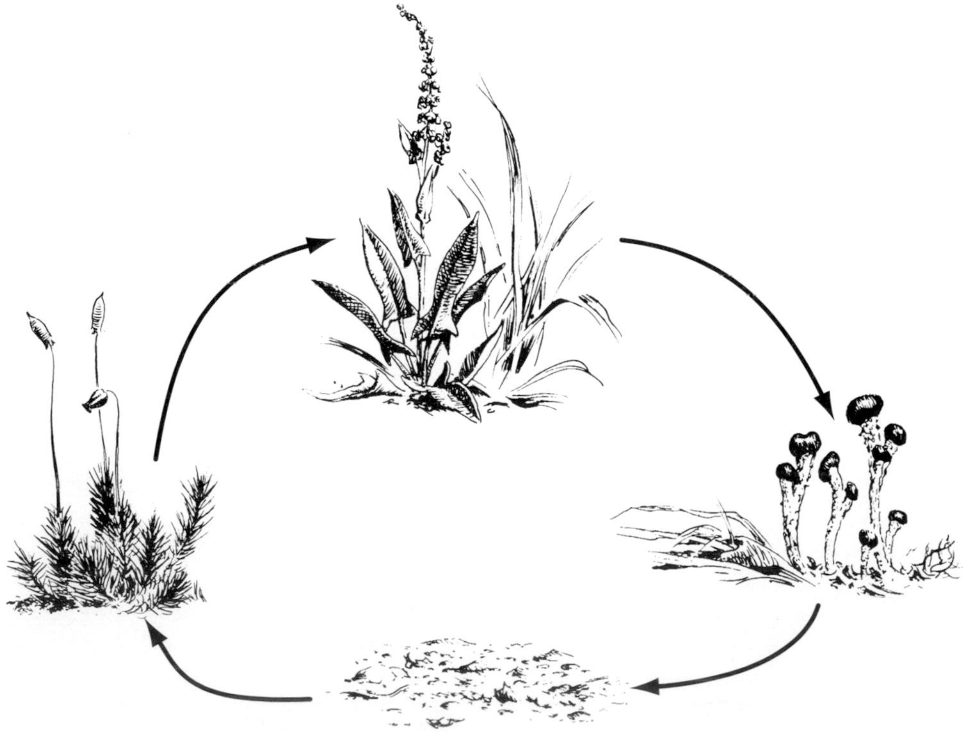

FIGURE 21.14 Cyclic replacement in an old-field community in Michigan. The bare areas at the bottom of the downgrade are invaded by moss to start the upgrade series. Mosses are invaded by Canada bluegrass (*Poa compressa*) and dock (*Rumex acetosella*). The accumulated dead leaves of these plants are covered by lichens that crowd out the grass. Rain, frost, and wind destroy the lichens to start the cycle with bare ground again.

The process of cyclic replacement has been used to describe succession in a wide variety of plant communities, such as old fields (Evans and Cain 1952), desert scrub (Yeaton 1978), coastal tundra in Alaska (Webber et al. 1980), and dwarf heather in Scotland (Watt 1955). The process of cyclic replacement has been used to reinterpret the apparent climax forest community that inhabits much of the northeastern United States. Shade tolerant sugar maple and beech and the early successional yellow birch (*Betula alleghaniensis*) are three dominants in the northern hardwood forests of New Hampshire. Rarely do seedlings and saplings of each of these species grow beneath overstory parents. Beech seedlings and saplings are most commonly associated with a sugar maple overstory, and sugar maple with a yellow birch overstory. Yellow birch seedlings are rather widely distributed throughout the forest. Forcier (1975) has hypothesized that a process of cyclic replacement occurs within the forest. When a beech or sugar maple dies, its place in the canopy will not be taken immediately by another beech or sugar maple. Instead, yellow birch will fill the opening in the canopy. When the short-lived yellow birch dies, sugar maple will fill the gap. Eventually it will be replaced by beech. Such small-scale succession within a forest has also been called **reciprocal replacement.** It gives rise to a cycle of replacement: a successional sequence of yellow birch–sugar maple–beech that occurs continuously on smaller scale of forest patches, driven by the mortality of canopy trees.

Shifting Mosaic

The idea of cyclic replacement changes the concept of succession from a unidirectional series of stages finally reaching some endpoint, to that of the larger community being composed of a variety of patches undergoing different stages in a process of cyclic replacement. In effect, the process of succession is neverending, brought about by the continuous processes of birth, growth, and death of individuals within the community. This view of succession suggests a **shifting-mosaic steady state,** a concept advanced by Bormann and Likens (1979). The term *shifting-mosaic* refers to the community being composed of a mosaic of patches, each in a phase of successional development (Figure 21.15). The term *steady state* is a statistical description of the collection of patches, the average state of the forest. In other words, the mosaic of patches shown in Figure 21.15 is not static. Each patch is continuously changing, going through some successional sequence such as in the model of cyclic replacement presented above. Patches in the mosaic that are currently classified as late successional revert back to early successional as a result of the death of a canopy tree(s). Patches currently classified as early successional undergo shifts in species composition and dominance by later successional species. Although the mosaic is continuously changing, the average composition of the forest (average over all patches) may remain fairly constant—in a steady state. This example of a continuously changing population of patches that remains fairly constant when viewed collectively rather than individually is very similar to the concept

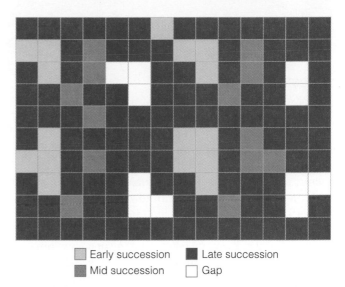

FIGURE 21.15 Representation of a forested landscape as a mosaic of patches in various stages of successional development. Although each patch is continuously changing, the average characteristics of the forest may remain relatively constant—in a steady state.

Early succession Late succession Mid succession Gap

of a stable age distribution presented in Chapter 11. In a population with a stable age distribution, the proportion of individuals in each age class remains constant even though individuals are continuously entering and leaving the population through births and deaths.

Although at first consideration the concepts of shifting mosaic steady state and climax appear to be two opposing views of community dynamics, the concept of the community as a mosaic steady state is not inconsistent with the view of succession moving unidirectionally to some endpoint—the **climax** (Smith and Urban 1988). As often happens in both science and everyday life, two persons observing the same phenomenon can have quite opposing interpretations of what occurred merely as a result of observing the action from different perspectives. For example, if you were to examine the dynamics of the community shown in Figure 21.15 from the perspective of a single patch—a single sample plot within the forest—you would view the dynamics as a cyclic process. However, if you were to describe the dynamics of the forest in terms of the changes in species composition over the entire landscape, the composition would appear quite stable through time.

If the entire forested landscape represented by the mosaic of patches in Figure 21.15 were to undergo a disturbance that removed most of the trees (such as a clear-cut or a severe fire), all of the patches would begin again in an early successional stage. With time the patches would move together through the successional sequence, eventually arriving at a stage where the forest is dominated by later successional tree species. At this point the landscape would appear to have arrived at some endpoint in the successional process, the climax. However, as individual canopy trees die, the process of cyclic replacement begins, and the landscape becomes a mosaic of patches in various stages of the succession. From an overall perspective, the composition of the forest appears quite stable, but succession continues at the scale of the patch.

Autosuccession

The observed similarity in pattern of species colonization and extinction through time for sites across a wide range of aquatic and terrestrial environments suggests that succession is a common feature of most communities. However, some communities, particularly those found in extreme environments, are often characterized by an absence of temporal shifts in species composition following disturbance (Noy Meir 1973, Zedler 1981, Peet and Loucks 1977). For example, Hanes (1971) described the patterns of vegetation dynamics in arid chaparral (shrubland) plant communities of California as **autosuccession,** referring to the self-replacing nature of the vegetation. The sequence of vegetation change in chaparral after fire is unusual, in that shrubs composing the mature community (the mature shrubs removed by fire) are present in the vegetation during the first year of recovery. Also, invasion of the chaparral community once the initial first-year population is established is uncommon. Thus, succession in chaparral communities differs from the usual sequence of changes common to secondary succession in most plant communities. Succession in chaparral is more a gradual elimination of individuals present from the onset rather than a replacement of initial shrubs by new species.

Fluctuation: Nonsuccessional Dynamics

Fluctuations are nonsuccessional or short-term reversible changes (Rabotnov 1974). Fluctuations differ from succession in that although the relative abundance of the species making up the community may change over time, the species composing the community remain the same. No new species invade the site, and changes in dominants may be reversible. These changes in species abundance result from seasonal or annual variations in environmental conditions such as soil moisture or temperature, or preferential selection of one species over another by grazers.

Dye and Spear (1982) examined annual variations in the structure of a savanna community in southwest Zimbabwe. Over the period from 1971 to 1981 the dominant grass species shifted from *Urochloa mosambicensis* to *Heteropogon contortus* (Figure 21.16a). This shift in species dominance is similar to that seen in the examples presented in Figures 21.5, 21.8, and 21.9; however, the change in dominance is not a result of succession. The shift in dominance is a result of variations in annual rainfall (Figure 21.16b). Rainfall during the 1971–2 and 1972–3 rainy seasons was much lower than average. *Urochloa mosambicensis* can tolerate dry conditions much better than can *Heteropogon contortus*. With the return to higher rainfall during the remainder of the decade, *Heteropogon contortus* became the dominant grass species. Annual rainfall in this semiarid region of southern Africa is highly variable, and fluctuations in species composition such as those shown in Figure 21.16 are a common feature of the community.

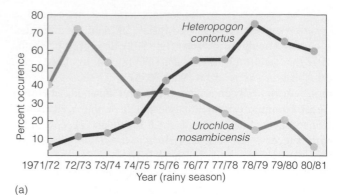

(a)

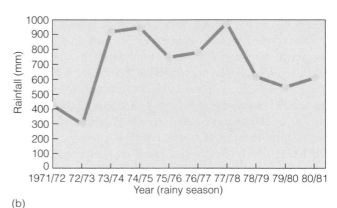

(b)

FIGURE 21.16   (a) Shift in the dominant grass species in a savanna community in southwest Zimbabwe over the period 1971–1981. The shift is in response to changing patterns of precipitation over the same period (b). *Urochloa mosambicensis* was able to tolerate the drier conditions during the 1971–2 and 1972–3 rainy seasons. With the increase in rainfall beginning in the 1973–4 season, *Heteropogon contortus* came to dominate the site. (Dye and Spear 1982.)

Fluctuations may also involve replacement of one age class by another within a species. Sprugle (1976) describes a wave regeneration pattern in balsam fir (*Abies balsamae*) forests in the northeastern United States. Trees die off continually at the edge of a "wave" and are replaced by vigorous stands of young balsam fir (Figure 21.17). Openings in the forest, exposing trees to the wind on the leeward side of the opening, initiate the cycle. Desiccation of the canopy foliage by winter winds and the loss of branches and needles in winter from frost damage are among the major causes of tree mortality. Their death exposes the trees behind them to the same lethal conditions, and they likewise die. This process continues so that a wave of dying trees through the forest is followed by a wave of vigorous reproduction.

These regeneration waves follow each other at intervals of about 60 years. The process is so regular that all stages of degeneration and regeneration are found in the forest at all times, provided the stand is not cut. The cycle results in a steady state similar to that discussed for the shifting mosaic, because the degenerative changes in one part of the forest are balanced by regenerative stages in another.

FIGURE 21.17 Cross section through a regeneration wave in a balsam fir (*Abies balsamae*) forest. The wave is initiated at the location of standing dead trees, with mature trees beyond and vigorous reproduction below. In the area where dead trees have fallen, a crop of young fir seedlings is developing. Beyond them is a dense stand of fir saplings, followed by a mature fir forest and then by a second wave of dying trees. (After Sprugle 1976.)

DEGRADATIVE SUCCESSION

The previous discussion of succession has focused on communities that are largely defined by the structure and dynamics of photosynthetic plants. These communities are referred to as **autotrophic communities.** Autotrophic communities derive their energy directly from solar radiation and utilize atmospheric CO_2 to form organic carbon compounds through the process of photosynthesis. In contrast, other communities are not dominated by autotrophs, instead deriving their energy through the breakdown of organic carbon compounds contained in living and dead plant and animal tissues. These **heterotrophic communities** are typically involved in the process of decomposition. Dead trees, animal carcasses and droppings, plant galls, tree holes—all furnish a substrate on which groups of plants and animals live, succeed each other, and eventually disappear. In these instances succession is characterized by early dominance of fungi and invertebrates that feed on dead organic matter. Available energy and nutrients are most abundant in the early stages of succession, and decline steadily as succession proceeds. One such example of this type of **degradative** or **heterotrophic succession,** as it is often referred to, can be found within a single acorn in the forest (Figure 21.18).

An acorn supports a tiny parade of life from the time it drops from the tree until it becomes a part of the humus (Winston 1956). Succession often begins while the acorn still hangs on the tree. The acorn may be invaded by insects, which carry to the interior pathogenic fungi fatal to the embryo. Most often the insect that invades the acorn is the acorn weevil (*Curculio rectus*). The adult female burrows through the pericarp into the embryo and deposits its eggs. Upon hatching, the larvae tunnel through to the embryo and consume about half of it. If fungi (*Penicillium* and *Fusarium*) invade the acorn simultaneously with the weevil or alone, they utilize the material. The embryo then turns brown and leathery, and the weevil larvae become stunted and fail to develop. These organisms represent the pioneer stage.

When the embryo is destroyed, partially or completely, by the pioneering organisms, other animals and fungi enter the acorn. Weevil larvae cut through the outer shell and leave the acorn. Through this exit hole fungi-feeders and scav-

FIGURE 21.18 An acorn in the early stages of degradative succession. Its outer shell has been breached, exposing its interior to invasion by a parade of detritvores.

engers enter. Most important is the moth *Valentinia glandenella,* which lays its eggs on or in the exit hole, mostly during the fall. Upon hatching, the larvae enter the acorn, spin a tough web over the opening, and proceed to feed on the remainder of the embryo and the feces of the previous occupants. At the same time several species of fungi enter and grow inside the acorn, only to be eaten by another occupant, the cheese mites (*Tryophagus* and *Rhyzoglyphus*). By the time the remaining embryo tissues are reduced to feces, the acorn is invaded by cellulose-consuming fungi. The fruiting bodies of these fungi, as well as the surface of the acorn, are eaten by other mites and collembolans and, if moist, by cheese mites as well. At this time predaceous mites enter the acorn, particularly *Gamasellus,* which is extremely flattened and is capable of following smaller mites and collembolans into crevices within the acorn. Outside, on the acorn, cellulose and lignin-consuming fungi soften the outer shell and bind the acorn to twigs and leaves on the forest floor.

As the acorn shell becomes more fragile, holes other than the weevils' exits appear. One of the earliest appears at the base of the acorn, where the hilum (the scar marking the attachment point of the seed) falls out. Through this hole, larger animals such as centipedes, millipedes, ants, and collembolans enter, although they contribute nothing to the decay of the acorn. The amount of soil in the cavity increases, and the greatly softened shell eventually collapses into a mound and gradually becomes incorporated into the humus.

The pattern of succession exhibited in the degradation of the acorn is typical of that found in any type of fresh organic matter, whether it is a log, an animal carcass, or dung. The organisms that first colonize the site are ones that can feed on fresh organic matter. Their feeding activities bring about physical and chemical changes in the substrate. After they have exploited the energy and nutrients accessible to them, they disappear. Replacing them is a group able to extract nutrients and energy left in a less accessible form. One group of organisms follows another. Each group changes the substrate to a point that it can no longer survive there. Eventually the organic matter is degraded. Thus heterotrophic succession follows a pattern similar to that of autotrophic succession, in which changes in the substrate are brought about by the organisms themselves. It differs from autotrophic succession, however, in that energy is degraded and not accumulated in organic biomass.

SUCCESSION AND ANIMAL LIFE

As the biological and physical structure of vegetation changes during the process of succession, animal life also changes (Figure 21.19). Animal life is influenced more by structural characteristics of vegetation than by species composition. For this reason successional stages of animal life might not correspond to the successional stages identified by plant ecologists. For example, a young stand of yellow-poplar or balsam fir under 6 m tall functions as a shrub stage community for many species of animals, whereas a plant ecologist would classify the yellow-poplar stand as an intolerant tree stage and fir as a tolerant tree stage.

Early terrestrial successional stages in eastern North America support animals of grasslands and old fields, such as meadowlarks (*Sturnella magna*), meadow voles, and grasshoppers.

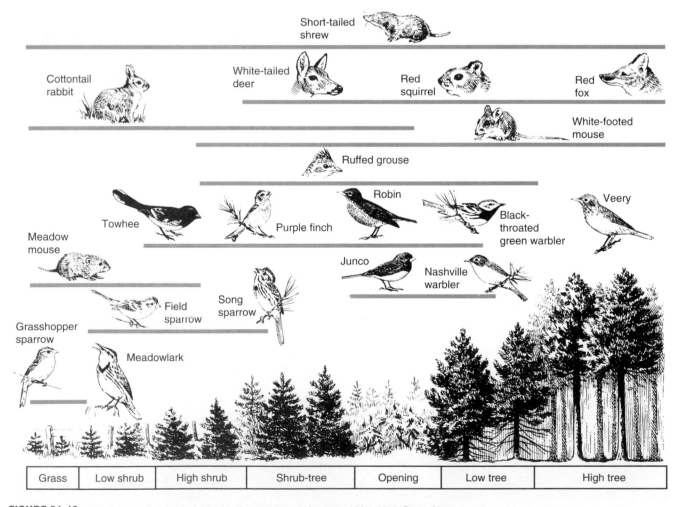

| Grass | Low shrub | High shrub | Shrub-tree | Opening | Low tree | High tree |

FIGURE 21.19   Wildlife succession in large conifer plantations in central New York State. Note how some species appear and disappear as vegetation density and height change. Other species are common to all stages. (After Smith 1960.)

Invasion of scattered woody plants adds a new structural element. Meadowlarks decline, and field sparrows (*Spizella pusilla*) and song sparrows (*Melospiza melodia*) appear. When tall woody vegetation, whether shrubs or young trees, eventually claims the area, shrubland animals move in. Field sparrows decline, and the thickets are claimed by towhees (*Pipilo erthyrophthalmus*), catbirds (*Dumetella carolinensis*), brown thrashers (*Toxostoma rufum*), and goldfinches (*Carduelis tristis*). Meadow mice give way to white-footed mice. When woody growth exceeds 6 m in height and the canopy closes, shrub-inhabiting species decline, to be replaced by birds and insects of the forest canopy. As the community matures and more structural elements are added, new species appear, such as tree squirrels, woodpeckers, and birds of the forest understory like hooded warblers (*Wilsonia citrina*) and wood thrushes (*Catharus mustelinos*).

A number of animal species are highly dependent on specific successional stages. Grasshopper sparrows (*Ammodramus savannurum*) need large areas of open grassland. The future of this bird depends on maintenance of grassland. The golden-winged warbler (*Vermivora chrysoptera*) depends on early successional shrubland that ultimately converts to forest. As the shrubland disappears, so does the warbler (Confer and Knapp 1981). The American woodcock (*Philohela minor*) requires three early successional types: abandoned fields and forest openings; alder, aspen, and young hardwood thickets; and young, open second-growth hardwoods (Sheldon 1967). Decline and loss of these successional stages is reducing the woodcock population. Other species, such as the spotted owl (*Strix occidentalis*), inhabit only mature or old-growth forests. When they are cut, the animal life they support goes, replaced by species associated with early successional stages. Other animals, particularly mammals such as short-tailed shrews and deer mice (*Peromyscus* spp.), are ubiquitous, inhabiting a range of successional stages.

Diversity of animal life across a range of seral stages varies with the nature of each individual community. The key to diversity of wildlife in a given area is the maintenance of a heterogeneous landscape with habitat patches of various successional stages.

PALEOSUCCESSION

Succession as experienced by humans takes place over a very short period in terms of Earth's history. Changes also take place on a grander time scale. Over the past 4 billion years, Earth has changed profoundly. Land masses drifted northward and broke into continents. Mountains emerged and eroded; seas rose and fell; ice sheets advanced to cover large expanses of the Northern and Southern Hemispheres and retreated. All these changes affected climate and other environmental conditions from one region of Earth to another. Many species of plants and animals evolved, disappeared, and were replaced by others. There were major shifts in vegetation patterns. What life was like in the past and under what conditions it existed, we infer from present-day conditions. Conversely, as the geologist Charles Lyell remarked, the past

is the key to the present. The key to distributions of animals and plants today may be found in the study of the past.

Records of plants and animals composing past communities lie buried as fossils: bones, insect exoskeletons, plant parts, and pollen grains. These fossils enable us to determine plant and animal associations of the past and, in a broad way, the climatic changes that brought about the gradual destruction of one type of community and the emergence of another. Such interpretation is based on the assumption that organisms of the past possessed ecological requirements similar to those of related species living today. For example, if modern palms and broadleaf evergreens are tropical plants, we assume that their ancient prototypes also lived in a tropical climate. The study of the relationships of ancient flora and fauna to their environment is **paleoecology.**

The Pleistocene Epoch

Of particular interest to the paleoecologist are the climatic and vegetation changes that followed the advance and retreat of glaciers during the Pleistocene. Changes in postglacial vegetation and climate are recorded in the bottoms of lakes and bogs. As glaciers retreated, they scooped out holes and dammed up rivers and streams, which filled with water to form lakes. Organic debris accumulated on the bottom to form peat, marl, and mud. Pollen, spores, and small invertebrates that blew in from adjacent vegetation settled on the water and sank. Microscopic examination of samples of organic deposits obtained at regular intervals reveals the fossil remains of these organisms. Various genera of fossil pollen can be identified by comparison with pollen growing today. Radiocarbon dating reveals the age of the sediments back as far as 35,000 years in which the pollen occurs. The relative abundance of pollen of several genera indicates the predominant vegetation at the specified depth of deposition (Figure 21.20).

Pollen investigation can indicate only trends in vegetation and climate through the past. At present it is impossible to determine the exact structure and composition of the prevailing vegetation during any one time period. Tree species that produced more pollen than others will appear more abundant than they really were. Some pollen might have been carried some distance by the wind or perhaps buried deeper by soil invertebrates. Insect-pollinated plants might not be represented at all. Many pollen grains can be identified only to genera and not to species. For example, the pollen of different types of oak cannot be distinguished to give a clue to the particular type of oak forest existing at a particular time. However, modern paleoecological techniques that take these problems into account and improved identification procedures are enabling paleoecologists to provide a rather accurate picture of postglacial vegetation.

The Pleistocene, which began some 2 million years ago, marked the end of the Tertiary period of the Holocene and the beginning of the Quaternary by ushering in the Ice Age. However, recent studies of deep sea sediments, the geophysics of ocean bottoms, and the Antarctic ice suggest that ice caps have been part of Earth's geological and ecological history at least since the Miocene and even earlier. Thus the Pleistocene

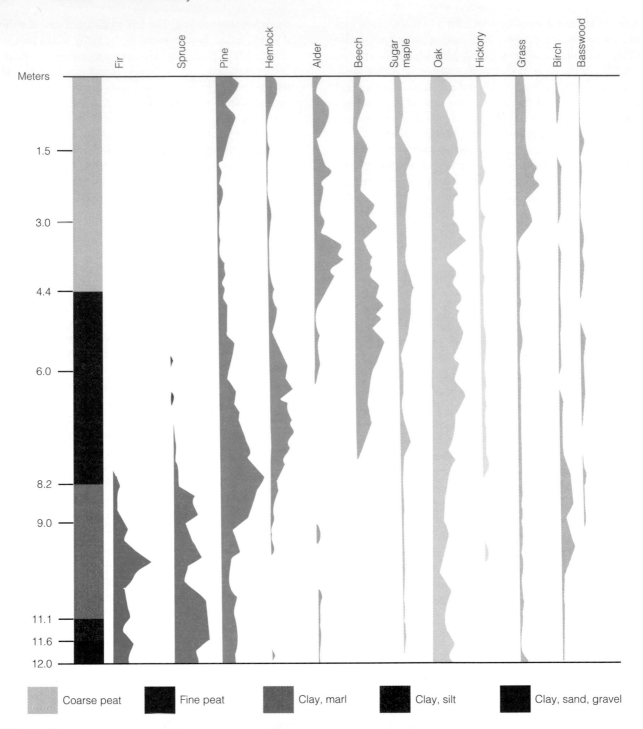

FIGURE 21.20   Pollen diagram for Crystal Lake, Hartstown Bog, Crawford County, Pennsylvania, at the edge of the maximum advance of the Wisconsin glaciation. The graphs indicate the various genera based on counts of 20 pollen grains for each spectrum level. Grass pollen counts are expressed as percentages of total free pollen. Note the five major forest successions from bottom to top. (1) Initial spruce-fir forest together with some pine and oak invaded as the glacier retreated. Alder, which today precedes spruce, followed it then. (2) As the climate warmed, spruce and fir gave way to jack pine forest with some oak and birch. (3) Later forest surrounding the pond was dominated by oak with hickory, beech, and hemlock. (4) Oaks next dominated the forest, with some hickory, sugar maple, and beech. Grass became an important component in the forest openings. (5) Forests were dominated by oaks and pine (possibly white pine) with sugar maple and beech. Hemlock again became important. Compare this diagram with Figures 21.26 and 21.27. (Adapted from Walker and Hartman 1960:463.)

is simply a stage in a continuum of ice and vegetation through the Cenozoic. The present distribution of plants and animals can be appreciated only in the context of longer successional development in the past.

Pre-Pleistocene Development

At the beginning of the Cenozoic era, some 70 million years ago, most of present-day continental North America and Europe was joined by land. By the beginning of the Miocene epoch (Table 21.1), forests closely related to the present-day deciduous forest existed with little variation across the northern continents. Known as the Arcto-Tertiary forest, it was a mixture of broadleaf and coniferous species, roughly divided into boreal and temperate elements. The boreal elements consisted of pines, spruces, cypress, birches, and willows. Because

tropical and subtropical climates existed far north of the present positions, neotropical and Paleotropical-Tertiary forests, ancestors of today's tropical forests, covered most of central North America. Probably in the Eocene, a mixed woodland, the Madro-Tertiary flora, developed on the Mexican plateau.

From the Miocene on, the climate began to cool. The western mountain system in North America rose; climatic zones and their biota were pushed southward, and tropical forests were driven into Central America. The American portion of the Arcto-Tertiary forest was separated from that of Europe and Asia by continental drift, and certain species, such as *Metasequoia, Ailanthus,* and *Ginkgo,* became extinct in North America. As the mountains rose, the broad rain shadow on their lee side wiped out the Arcto-Tertiary forest and stimulated the development of grassland in the central part of North America. A relict Arcto-Tertiary forest, poor in

TABLE 21.1 Geological Time Scale

Era	Period	Epoch	Age (Millions of Years)	DOMINANT LIFE Plants	DOMINANT LIFE Animals
Cenozoic: the age of mammals	Quaternary	Recent	0.01	Agricultural plants	Domesticated animals
		Pleistocene	2		Ice Age—first true humans; mixture and then thinning out of mammalian fauna
	Tertiary	Pliocene	10	Herbaceous plants rise; forests spread	Culmination of mammals; radiation of apes
		Miocene	25	First extensive grasslands	
		Oligocene	35		Modernization of mammals; mammals become dominant
		Eocene	55		Mammals become conspicuous
		Paleocene	70		Expansion of mammals; extinction of dinosaurs
Mesozoic: the age of reptiles	Cretaceous		135	Angiosperms rise; gymnosperms decline	Dinosaurs reach peak; first snakes appear
	Jurassic		180	Cycads prevalent	First birds and mammals appear
	Triassic		230	Gymnosperms rise; seed ferns die out	First dinosaurs; reptiles prominent
Paleozoic	Permian		280	Conifers become forest trees; cycads important	Great expansion of primitive reptiles
	Carboniferous				
	Pennsylvanian		310	Lepidodendron, sigillaria, and calamites dominant; the swamp forest	Age of cockroaches; first reptiles
	Mississippian		345	Lycopods and seed ferns abundant	Peak of crinoids and bryozoans
	Devonian		405	First spread of forests	First amphibians; insects and spiders
	Silurian		425	First known land plants	First land animals (scorpions)
	Ordovician		500	Algae, fungi, bacteria	Earliest known fishes; peak of trilobites
	Cambrian		600	Algae, fungi, bacteria; lichens on land	Trilobites and brachiopods; marine invertebrates
Precambrian	Late			Algae, fungi, bacteria	First known fossils
	Early		4500	Bacteria	No fossils found

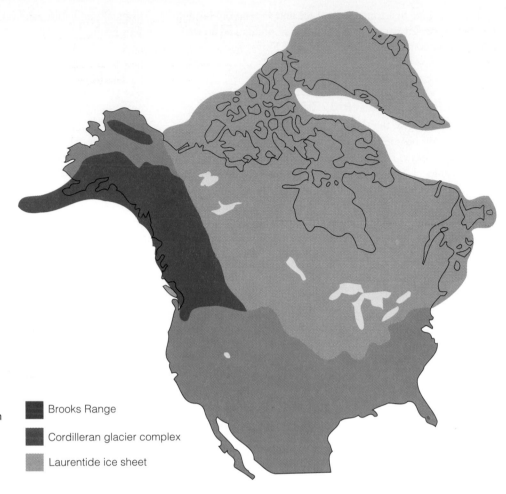

FIGURE 21.21 Glaciation in North America. In the Pleistocene, northern North America was covered by at least four ice sheets. The last and most significant was the Wisconsin ice sheet.

■ Brooks Range

■ Cordilleran glacier complex

■ Laurentide ice sheet

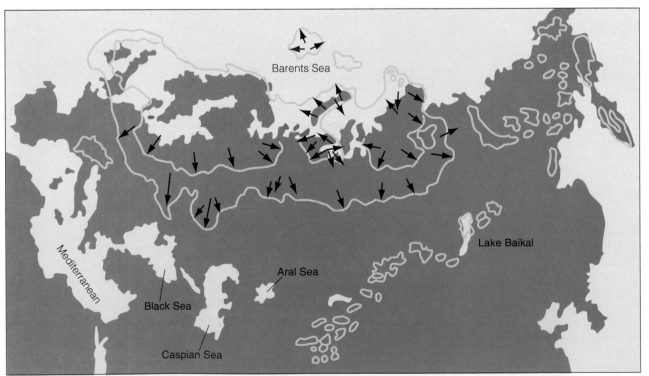

FIGURE 21.22 Glaciation in Europe. In the Pleistocene, northern Eurasia was covered by ice sheets similar to those covering North America. The most important was the last, called the Weichselian. Note the disjunct glacier in the region of the Alps. (After Flint 1971:545.)

species but including the sequoias and redwoods, was left in the Pacific Northwest. Elements of the Madro-Tertiary forest moved northward to occupy dry lands vacated by the Arcto-Tertiary forest and eastward to form sclerophyllous-pine woodlands ancestral to the Southern oak-pine woodlands of today. In the late Pliocene a continuing climatic cooling accompanied by mountain building brought on continental glaciation.

Glacial Periods

The Pleistocene was an epoch of great climatic fluctuations throughout the world. At least four times during the Pleistocene ice sheets advanced and retreated in North America (Figure 21.21), and at least three times in Europe (Figure 21.22). Four times in North America and three times in Europe the biota retreated and advanced, each advance having a somewhat different mix of species.

Each glacial period was followed by an interglacial period (Table 21.2 and Figure 21.23). The climate in each

TABLE 21.2 Glacial and Interglacial Stages

Britain	Northern Europe	North America	Climate
Flandrian (postglacial)	Weichselian		temperate
Devensian (last glaciation)	Weichselian	Wisconsin	cold glacial
Ipswichian (last interglacial)		Sangamon	temperate
Wolstonian	Saalian	Illinoian	cold glacial
Hoxnaian	Holstein	Yarmouth	temperate
Anglian	Elsterian	Kansas	cold glacial
Comerian		Aftonian	temperate
Beestonian		Nebraskan	cold
Pastonian			temperate
Baventian			cold

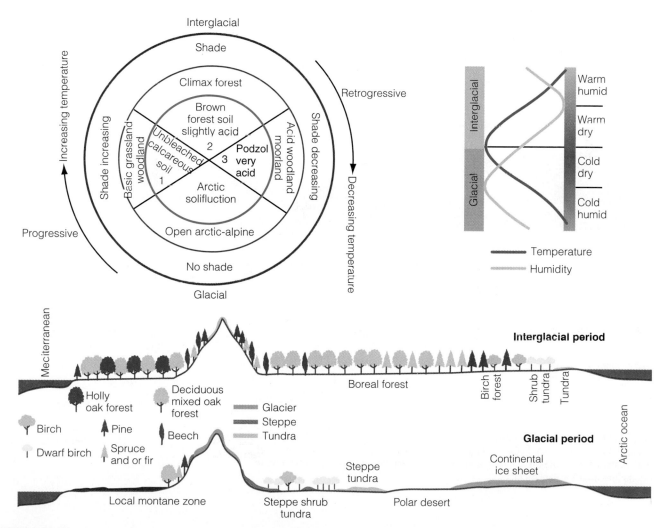

FIGURE 21.23 The glacial-interglacial cycle in northwestern Europe. (After van der Hammer et al. 1974.)

stage oscillated between cold and temperate. During the cold stage tundralike vegetation and boreal species of fir and spruce dominated the landscape. As glaciers retreated and the climate ameliorated, easily dispersed, light-demanding forest trees such as pine and birch advanced northward. As the soil improved and the climate continued to warm, these trees were replaced by the slower dispersing, shade tolerant trees such as oak and ash. The development of the next glacial period brought in spruce, fir, and tundra vegetation. This vegetation changed the soil from mull to acid mor (see Chapter 4). As both climate and soil began to deteriorate, heaths dominated the vegetation and forest trees disappeared.

Major differences in vegetation bordering the glacier existed between Europe and North America. In Britain and Europe a wide belt of tundra edged the glacier. The Alps, with their large ice cap, diverted the westerly flow of warm air southward. As a result, the Arcto-Tertiary flora was decimated by the early cold stages of the Pleistocene. Temperate genera were forced southward, but the retreat was blocked by mountains, deserts, and seas (Figure 21.24). Only the hardy boreal genera could survive. In North America no such barrier existed. The glacier extended farther south into a warmer zone, across which the flow of air was unimpeded (Figure 21.25). As a result, spruce forests grew virtually to the edge of the ice

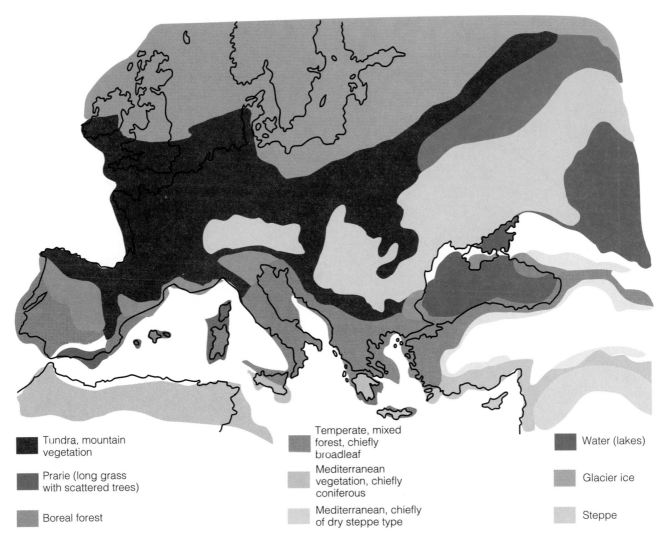

Tundra, mountain vegetation

Prarie (long grass with scattered trees)

Boreal forest

Temperate, mixed forest, chiefly broadleaf

Mediterranean vegetation, chiefly coniferous

Mediterranean, chiefly of dry steppe type

Water (lakes)

Glacier ice

Steppe

FIGURE 21.24 Assumed distribution of vegetation in Europe at the Weichsel/Würm maximum. The Black Sea and the Caspian Sea are interconnected lakes. Note the predominance of tundra vegetation, the patches of boreal forest, and the highly restricted distribution of temperate deciduous forest. (After Flint 1971.)

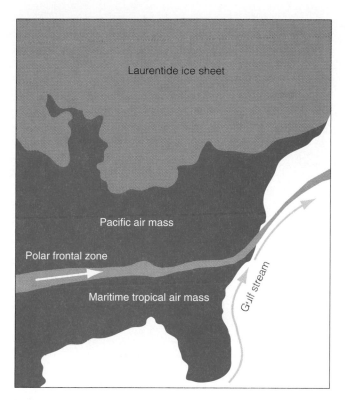

FIGURE 21.25 At the full glacial interval 18,000 BP in North America, the boreal forest was separated from the temperate deciduous forest further south by a narrow belt of cool, temperate coniferous and deciduous tree species (see Figure 21.26, top middle). This narrow belt of vegetation probably represents the mean annual position of a strong zonal circulation of air, the polar frontal zone, which extended eastward across the western North Atlantic Ocean. The narrow stable climatic boundary separated the Pacific air mass immediately to the north of the polar frontal zone from the maritime air mass to the south. (Adapted from Delcourt and Delcourt 1984:277.)

and Tertiary precursors of present-day deciduous forest survived to spread north.

The last great ice sheet, the Laurentide reached its maximum advance about 20,000 to 18,000 years ago during the Wisconsin glaciation stage in North America. Canada was under ice. A narrow belt of tundra about 60 to 100 km wide bordered the edge of the ice sheet. It probably extended southward into the high Appalachians, where a few relict examples exist today, and into the high peaks of the Adirondacks (Wright 1970, Delcourt and Delcourt 1981, 1985). Boreal forest, dominated by spruce and jack pine, covered most of the eastern and central United States as far as western Kansas. Its southern limit was about 1200 km south of the modern southern border of boreal forest in Canada. West of Kansas lay uninterrupted sand dunes, a treeless landscape shaped by the intense winds generated by the nearby ice

sheet. South of the boreal forest was a transition belt of conifers and mixed hardwoods that separated the boreal forest from the oak-pine forests to the south. Mesic, temperate hardwood species—beech, sugar maple, basswood, walnut, buckeye, yellow-poplar, chestnut, hickory, and oak—found a refuge in the loess-capped uplands of the Mississippi Valley, in dissected valley slopes along major southern river systems, ravines, and irregular karst terrain, and perhaps along the southern Atlantic and Gulf coastal regions exposed by seas that were 300 m lower than today (Figure 21.26) (Delcourt and Delcourt 1985).

The climatic changes that accelerated the retreat of the Laurentide ice sheet in the late Wisconsin also brought a sudden end to the boreal forest in unglaciated North America. In the western part of the glacial region, spruce was replaced by prairie grass. Farther east, closer to the edge of the present prairie region, spruce gave way to pine, birch, and alder (Amundson and Wright 1979). In the southern Appalachians oak and pine replaced spruce, except for relict stands at high elevations. Pines moved northward rapidly from their Appalachian refuge in the Carolinas and dominated much of the region about the newly formed Great Lakes (Figure 21.27) (Davis 1981, 1983). Hemlock and other species appeared in the southern Appalachians, from which they invaded the deglaciated areas. Other species, such as chestnut, moved much more slowly, taking 3000 years to reach New England from the central Appalachians (Davis 1981). In the Western Cordillera (parallel chains of mountains) during the Wisconsin glaciation, the tree line and tundra vegetation moved downslope 800 to 1000 m lower than today. Many of the modern desert basins were shrub steppes dominated in part by sagebrush. (For an excellent description of the continent after the Ice Age in North America, see Pielou 1992.)

The end of the Pleistocene witnessed a major transition in vegetation and associated animal life. As the climate slowly warmed, southern species advanced slowly northward, overtaking northern species growing along the glacial edge. How far one southern species advanced, and how far south more northern species remained, depended on the physiological adaptations of the species and local climatic conditions.

The long-term perspective provided by paleoecological studies is critical to understanding the current nature of ecological communities. It is all too easy to assume that the current distribution of communities over the landscape and the observed patterns of successional dynamics represent some fixed relationship with the surrounding physical environment, and that the current mix of species that define the biological structure of communities have always coexisted. Through the study of the past we come to appreciate the dynamic nature of the distribution and abundance of plants and animal species and, consequently, the communities they form.

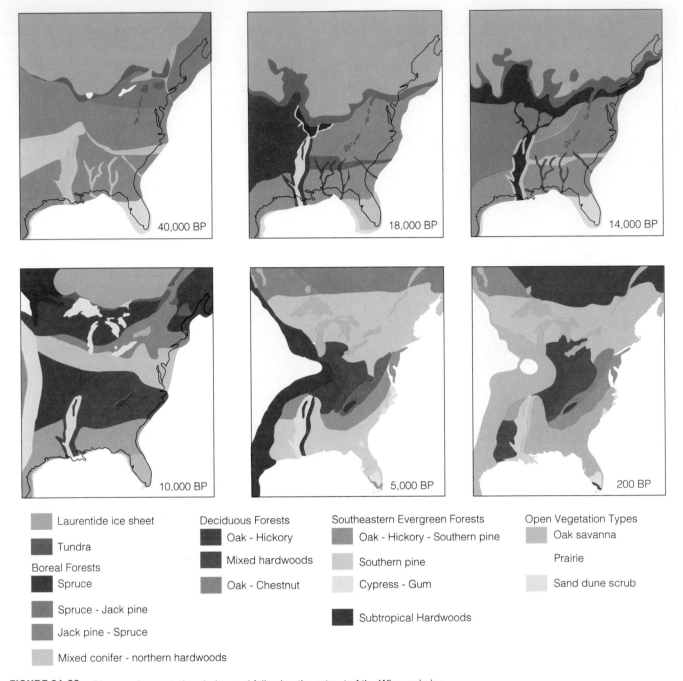

Laurentide ice sheet

Tundra

Boreal Forests
- Spruce
- Spruce - Jack pine
- Jack pine - Spruce
- Mixed conifer - northern hardwoods

Deciduous Forests
- Oak - Hickory
- Mixed hardwoods
- Oak - Chestnut

Southeastern Evergreen Forests
- Oak - Hickory - Southern pine
- Southern pine
- Cypress - Gum
- Subtropical Hardwoods

Open Vegetation Types
- Oak savanna
- Prairie
- Sand dune scrub

FIGURE 21.26   Changes in vegetation during and following the retreat of the Wisconsin ice sheet, reconstructed from pollen analysis at sites throughout eastern North America. (Adapted from Delcourt and Delcourt 1981.)

Summary

1. The structure of communities, physical and biological, changes in both space and time. Spatial change in the structure of communities is called zonation and reflects shifts in the dominant species. The gradual sequential change in the relative abundances of dominant species in a community through time is called succession.

2. Succession that begins on sites devoid of or unchanged by organisms is termed primary; and succession that proceeds from a state in which other organisms were already present is called secondary.

3. The process of succession is often represented as unidirectional, with the community progressing to some defined endpoint—or climax—that reflects the local environment.

4. Successional stages within a community that appear to be directional, moving toward a defined endpoint, are often phases in a cycle of vegetation replacement. Death of vegetation or periodic disturbance starts regeneration again at some particular stage. Such changes usually occur continuously on a small

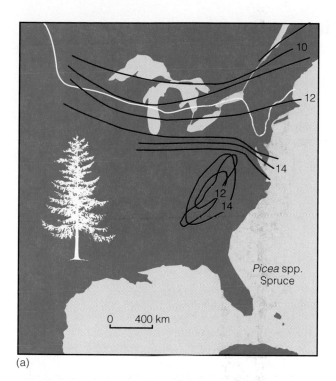

(a)

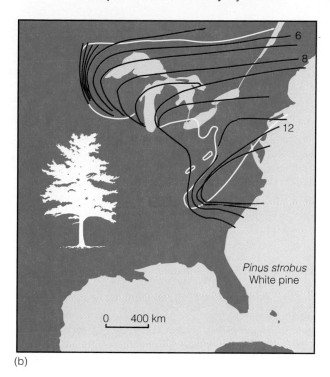

(b)

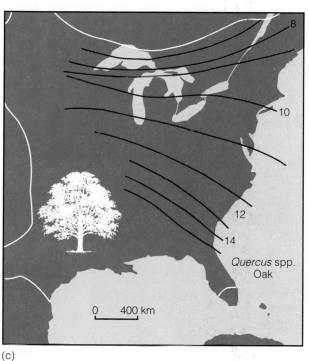

(c)

FIGURE 21.27 Postglacial migration of three tree species. The lines represent the leading edge of the expanding population. (a) An aggressive pioneer, spruce moved quickly into the tundra and took over the landscape newly exposed by the melting ice sheet. Its speed of northward movement varied. It moved into the Great Lakes region shortly after the ice receded, but 2000 years intervened until it arrived in New England. Spruce preceded alder, the opposite of successional sequences on glacial material today. (b) White pine found a refuge during the height of the ice sheet along the East Coast and in the foothills of the Appalachians, where it was mixed with stands of hardwoods. The species was absent from the full glacial sites in the Mississippi Valley and Florida. Paleoecological evidence of white pine appeared first in Virginia, then expanded rapidly north westward to Minnesota 7000 years ago. Its westward expansion was blocked by dry climates. White pine extended northward, reaching sites north of its present range. There it occurred briefly in large populations before competitive reduction by hardwoods. (c) Oaks, widespread in the southern United States during full glaciation, spread rapidly northward between 10,000 and 9000 BP and reached the full limits of their range by 7000 BP. The northern limit of oak coincides with the northern limit of white pine. (Adapted from Davis 1981:138, 144, 145.)

scale within a community, and they are repeated over the course of time throughout the entire community. This processes is called cyclic replacement.

5. The concept of a shifting-mosaic steady state offers an alternative to the concept of climax as a description of succession. The term *shifting-mosaic* refers to the community being composed of a mosaic of patches, each in a phase of successional development. The term *steady state* is a statistical description of the collection of patches, the average state of the forest.

6. Dead trees, animal carcasses and droppings, plant galls, tree holes—all furnish a substrate on which groups of plants and animals live, succeed each other, and eventually disappear. In these instances succession is characterized by early dominance of fungi and invertebrates that feed on dead organic matter. Available energy and nutrients are most abundant in the early stages of succession, and they decline steadily as succession proceeds.

7. As vegetational communities change, animal life dependent on each seral stage also changes. Many forms of animal life are

specific to each stage. As succession progresses, animals characteristic of earlier stages or communities disappear.

8. Although we experience succession on a short-term temporal
scale, succession and changes in Earth's vegetation and life
have taken place on a grander scale over geological time. The
distribution of life we observe today has been influenced heavily by long-term dynamic changes in Earth. Over the course
of its existence, Earth has experienced the breakup of large
land masses into continents that drifted apart, the buildup of
mountains, the rise and fall of sea levels, and advance and
retreat of ice sheets. With these physical changes came climatic
changes—warming and cooling, aridity and heavy precipitation. Patterns of vegetation changed as climates changed.

9. Some of the most pronounced changes, as evidenced by pollen
profiles from bogs and pond and lake bottoms, occurred during the Pleistocene, the Ice Age. At that time several advances
and retreats of ice sheets eliminated vegetation over much of
the northern part of the Northern Hemisphere and pushed
plants and animals southward. With each northward retreat of
the ice sheet, vegetation moved north. Trees experienced different survival patterns from one glaciation to the next, producing different forest communities during each glacial
interval. In Europe, tree species found their southward retreats
blocked by the glacial ice of the Alps. Many species became
extinct, reducing present-day species diversity of the European
deciduous forest. In North America no such barriers existed.

10. During the last great ice sheet of the Wisconsin glacial period,
18,000–20,000 years ago, pockets of tundra existed adjacent
to the ice sheet, boreal forests extended south to mid-continent,
oak-pine forests covered the south, and the species-rich mesic
forest was confined to the Mississippi Valley, the Gulf Plain,
and other scattered refuges. As the ice sheet retreated, the climate warmed rapidly and trees moved northward. Boreal forest shifted into Canada, replaced to the south by oak-pine
forest and mixed deciduous forest. Mesic species moved north-
and northeastward out of their confined refuges at different
rates and arrived at the northern limits of their ranges at different times. The nature of vegetational communities today
reflects the evolutionary impact of changing conditions during
the Pleistocene.

Review Questions

1. Define succession. Distinguish between primary and secondary
succession. Provide some local examples.

2. How do zonation and succession differ?

3. What is cyclic succession?

4. What is the shifting mosaic concept? How does it relate to the
climax concept?

5. What major events in the Pleistocene shaped vegetation patterns evident today?

6. Locate an area with which you have been familiar over time. What
vegetation changes have taken place? What might have brought
them about: logging, land abandonment, suburban development?
Old aerial photographs or a series of them will provide insights.

7. Select an assortment of wildlife species—birds or small mammals—and relate their habitat requirements to a successional or
seral stage. Then determine what happens to these species as
succession advances. How might their habitat be retained? What
conflicts arise when species have differing habitat requirements?

8. Some ecologists argue that the concept of the climax is dead.
What arguments might they use to defend their position?

9. How does the concept of the climax as generally perceived
color our perception of what the vegetation might have been
like before European settlement?

10. What distinguishes degradative succession from succession
involving changes in plant communities?

11. What significance does the study of plant migration in the postglacial period have for the study of current vegetation patterns
and potential changes that might result from global climate
change? (For help see Clark et al. 1998.)

Cross References

Shade tolerance, 92–93; process of decomposition, 448; rocky intertidal zones, 676; vertical structure of communities, 384; patch
dynamics, 458.

CHAPTER 22

Processes Controlling Community Dynamics

CONCEPTS

1. The concept of the community has evolved through time from that of an association of species to that of individual species responding independently to environmental gradients.

2. A fundamental niche of a species functions as a primary constraint on its distribution and abundance.

3. As environmental conditions change in time and space, patterns of species distribution and abundance will likewise change.

4. Succession can be viewed as temporal shifts in species composition in response to autogenic changes in environmental conditions.

5. A number of theoretical models explain patterns of succession by examining trade-offs involved in species adaptations to varying environmental conditions.

6. Allogenic changes in environmental conditions across the landscape give rise to zonation.

7. Both species' growth rates and disturbance frequency influence patterns of diversity during succession.

8. Herbivores can have both direct and indirect effects on community dynamics.

9. The community is a spatial unit, whereas the continuum concept is based on species response to underlying environmental gradients that vary spatially across the landscape.

As we have learned in the previous chapter, community structure does not vary randomly across the landscape (zonation) or through time (succession); rather, it exhibits repeatable, often predictable patterns. In fact, the observed similarity in pattern of species colonization and extinction through time for sites across a wide range of environmental conditions suggests to ecologists a common mechanism or mechanisms influencing the process of succession. The search for these underlying mechanisms controlling the dynamics of communities has been and continues to be a major focus of ecology. Our current understanding of processes structuring communities is built on a rich history of ecological research that spans the past two centuries.

A HISTORICAL OVERVIEW

Early descriptions of plant succession can be found in the writings of nineteenth-century naturalists, such as those published by Henry David Thoreau (1860) describing plant succession following logging in New England. It was H. C. Cowles (1899), though, in his study of vegetation development on the sand dunes of Lake Michigan, who first described explicitly the process of plant succession. Later, F. E. Clements (1916) developed a descriptive theory of succession and advanced it as a general ecological concept. His theory of succession had a powerful influence (or inhibition, as some critics say) on ecological thought.

Clements viewed succession as a process involving several phases, to each of which he gave his own terminology. Succession began with the development of a bare site, called nudation (disturbance). Nudation was followed by migration, the arrival of propagules onto the area. Migration was followed by the establishment and initial growth of vegetation. Clements called this phase ecesis (growth). As vegetation became well established, grew, and spread, various species began to compete for space, light, and nutrients. This phase Clements called competition. Competition was followed by reaction, a period in which the plants themselves alter the environmental conditions. The outcome of this process was the replacement of one plant community by another, eventually reaching a phase called stabilization, where one community (mix of species) persists.

Clements's theory of succession dominated ecology for years, almost becoming dogma, because it provided an orderly, logical explanation for the development of plant communities. However, it did have critics. Their problem with Clements's view of succession was not in the description of the processes outlined above, but rather in his more fundamental concept of community. To understand Clements's view of the community, we must go back to the very definition of plant communities established at the International Botanical Congress of 1910.

When you walk through most forests, you see a variety of plant and animal species—a community. If you walk far enough, the dominant plant and animal species that you see will change (for an example, see Figure 20.13). As you move from hilltop to valley, the structure of the community will dif-

fer. But what if you continue your walk over the next hilltop and into the adjacent valley? You will most likely notice that although the communities on the hilltop and valley are quite distinct, the communities on the two hilltops or valleys are quite similar. As a botanist might put it, they exhibit relatively consistent floristic composition. At the International Botanical Congress of 1910, botanists adopted the term *association* to describe this phenomenon. An association is a type of community with (1) relatively consistent species composition, (2) a uniform physical (physiognomic) structure, and (3) a distribution that is characteristic of a particular habitat, such as the example of a hilltop or valley. Whenever a particular habitat (or set of environmental conditions) repeats itself in a given region, the same group of species occurs.

It was this view of the plant community as an association that so influenced Clements. The logic went that if clusters or groups of species repeatedly associated together, that is evidence for either positive or neutral interactions among them. Such evidence favors a view of communities as integrated units. Based on this logic, Clements developed what has become known as the organismal concept of communities.

Clements viewed the species in an association as having similar environmental requirements and therefore similar distributional limits along important environmental gradients (Figure 22.1a). The boundaries between adjacent associations are narrow, with very few species in common. This view of the community suggests a common evolutionary history and similar fundamental responses and tolerances (see Chapter 3) for the component species. Mutualism and coevolution (see Chapter 17) play an important role in the evolution of species making up the association. The community has evolved as an integrated whole.

Because Clements considered the community as an integrated whole, he looked on the community as a superorganism, the ultimate expression of which was the climax (see Chapter 21). The climax was an assemblage of vegetation that belonged to the highest type of vegetation community possible under the prevailing climate. According to Clements, it can reproduce itself, "repeating with essential fidelity the stages of its development." To Clements each stage of succession represented a step in the development of a superorganism, the climax. Each seral stage so modifies the environment that plants of that stage eventually can no longer exist there. Instead they prepare the site for the replacement plants of the next stage. The process continues until the vegetation arrives at the self-reproducing climax. That marked the end of succession.

Contrasting with Clements's organismal view of communities was that of the botanist H. A. Gleason (1917, 1926). Gleason's view of the community was almost exactly the opposite of that put forward by Clements. Gleason regarded the community as consisting of individual species that respond independently to environmental conditions. He argued, "the vegetation of an area is merely the resultant of two factors, the fluctuation and fortuitous immigration of plants and an equally fluctuating and variable environment."

The individualistic concept put forward by Gleason emphasizes the species rather than the community as the

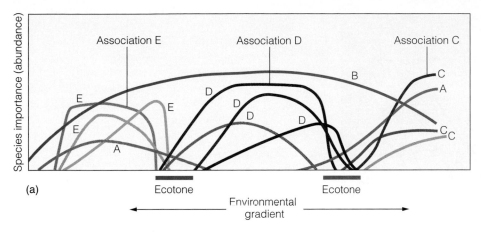

(a)

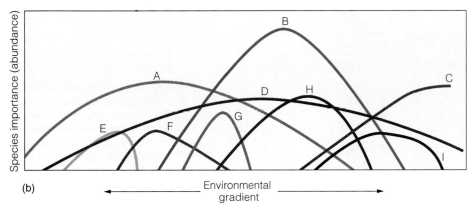

(b)

FIGURE 22.1 Two models of community. (a) The organismal or discrete view of communities proposed by Clements. Clusters of species (C's, D's, and E's) show similar distribution limits and peaks in abundance. Each cluster defines an association. A few species (for example, A) have sufficiently broad ranges of tolerance that they occur in adjacent associations, but in low numbers. A few other species (for example, B) are ubiquitous. (b) The individualistic or continuum view of communities proposed by Gleason. Clusters of species do not exist. Peaks of abundance of dominant species, such as A, B, and C, are merely arbitrary segments along a continuum. This view of community is similar to the groups of species associated with environments e_1 and e_2 in Figure 22.5.

essential unit. Succession results from the individual responses of different species to the prevailing environmental conditions. Species respond independently to the environment according to their own genetic characteristics. Plants involved in succession are those that arrive first on the site and are able to establish themselves under prevailing environmental conditions. As time passes, plants modify the environment and competition and other interactions among species determine the final outcome. His view became known as the individualistic continuum concept.

The continuum concept states that the relationship between coexisting species (species within a community) is a result of similarities in their requirements and tolerances, not a result of strong interactions or common evolutionary history. In fact, Gleason concluded that changes in species abundance along environmental gradients occur so gradually that it is not practical to divide the vegetation (species) into associations. In contrast to Clements's view, species distributions along environmental gradients do not form clusters, but represent the independent response of species (Figure 22.1b). Boundaries between communities are gradual and difficult to identify. What is referred to as the community is merely the group of species found to coexist under any particular set of environmental conditions. The major difference between these two views is the importance of interactions, both evolutionary and current, in the structuring of communities.

The continuum concept put forward by Gleason was a major influence on the development of plant ecology. By the

1950s plant ecologists had abandoned many of the central principles of the Clementsian view of communities. Instead of the "complex organism" analogy, ecologists were emphasizing the individualistic nature of species' responses to the environment. Plant ecologists such as Curtis (1959) and Whittaker (1956, 1960, 1967) were viewing communities in the context of species responses to underlying environmental gradients—an approach called gradient analysis. The classification of vegetation in the Smoky Mountain National Park presented in Figure 20.14 is an example of describing vegetation in terms of communities or associations of plant species that are linked with specific habitats. An alternative view of the distribution of vegetation in the Park can be seen in Figure 22.2. Rather than examining the distribution of plant associations, the abundance of each species can be plotted relative to soil moisture (dry to wet), as it varies with aspect and slope position (Figure 22.2a). The distributions of the dominant tree species on which the classification of communities shown in Figure 22.2a is based form a continuum along the moisture gradient (Figure 22.2b). The species distributions overlap extensively, and no distinct grouping of species into communities or associations are visible. The observed pattern resembles that presented in Figure 22.1b describing the individualistic continuum concept proposed by Gleason.

In 1977 Connell and Slatyer proposed a theoretical framework for understanding succession that included three different models: facilitation, inhibition, and tolerance. This

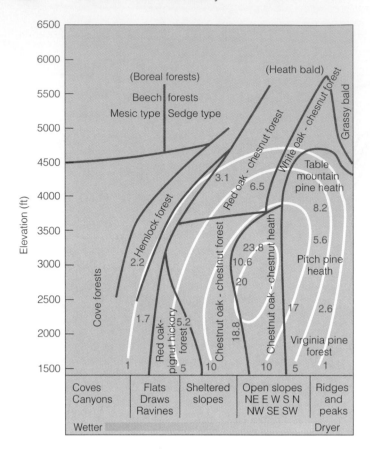

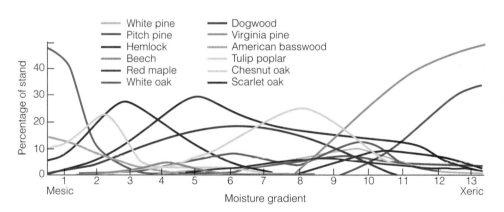

FIGURE 22.2 (a) Distribution of plant communities (associations) in the Smoky Mountain National Park in relation to elevation (*y* axis), and slope position and aspect (*x* axis). Communities are classified and named based on the dominant plant (predominantly tree) species. Abundance (expressed as percent of total stems over 1 cm dbh) of chestnut oak (*Quercus prinus*) overlaid on the distribution of plant communities. Note that the Chestnut Oak Forest and Chestnut Oak Heath communities correspond to the maximum distribution; however, the species distribution extends over a wide variety of plant communities. (b) Abundance of major tree species that make up vegetation communities plotted along the gradient of moisture availability, which varies as a function of slope position and aspect. (Whittaker 1956.)

framework focused on species interactions, and in many ways it was a more formal description of the processes first presented by Clements.

In the facilitation model, the organisms themselves bring about changes within the community. Early successional species modify the environment in such a way that it prepares the site for later successional species, thus facilitating their success. For example, early successional species that colonize the newly exposed sediments in Glacier Bay, Alaska (see Figures 21.7 and 21.8), have *Rhizobium* bacteria (see Chapter 17) associated with their root systems. These bacteria fix atmospheric nitrogen, increasing the nitrogen available in the soil. In this way, early successional species, such as alder (*Alnus*), improve the environmental conditions for later successional species.

The species interactions in the inhibition model are purely competitive. No species is competitively superior to another. The site belongs to those species that become established first and are able to hold their position against all invaders. They make the site less suitable for both early and late successional species through consumption of resources and modification of the environment. As long as these species persist, they maintain their position. Ultimately, those species that are long-lived come to dominate even though early successional species may suppress their growth for a long time. Such succession is not orderly and is less predictable than that observed under the facilitation model.

The tolerance model involves the interaction of competition and life history traits. It suggests that later successional species are neither inhibited nor aided by species of earlier stages. Later-stage species can invade a site, become established, and grow to maturity in the presence of those preceding them. They can do so because these later species have a greater tolerance for the lower level of resources created by earlier species. As time progresses, the early successional species decline in abundance and the community is dominated by the tolerant species.

A difficulty with the framework proposed by Connell and Slatyer is that most successional sequences cannot be categorized into any one of their models. Rather, elements of all three models may be involved in any one successional sequence.

In 1981 Noble and Slatyer proposed a framework for understanding succession that differed from preceding models. Their approach placed the focus on species life history characteristics that determine the place of a species in a succession rather than species interactions. They referred to these species characteristics as vital attributes. Vital attributes fall into three categories: (1) ability and method by which a species recovers following disturbance; (2) the ability of a species to grow and reproduce under competition; and (3) species longevity. Species within an area are classified based on their vital attributes, and predictions about successional sequences are possible.

The vital attributes model represents a departure from previous models in that it sought to explain succession in terms of species characteristics, and as such it is solidly grounded in the Gleasonian view of communities. This approach was in keeping with the work by Bazzaz and colleagues (see Bazzaz 1979 for review) that sought to understand succession in terms of differences in the physiological characteristics of early and late successional species (Table 22.1). The work of Bazzaz and other ecophysiologists during the 1970s represents a critical step toward understanding the

structure of communities by examining the basic physiological response of species to key environmental factors. As we shall see, this work laid the foundation for current theories of community structure and dynamics.

A SIMPLE MODEL OF COMMUNITY DYNAMICS

A framework for examining community dynamics (patterns of zonation and succession) can be constructed based on the following four premises:

1. The fundamental niche of a species (see Chapter 14) acts as a primary constraint on its distribution and abundance.

2. Species vary in their fundamental niches (environmental tolerances).

3. Environmental conditions change in time and space.

4. The fundamental niche is modified by species interactions (realized niche).

A simple model based on these premises provides a framework for linking patterns of community dynamics in space and time. The examples provided will focus on communities of relatively simple structure. The objective is to provide a foundation for examining current theories of community structure, presented later in the chapter, that attempt to understand more complex patterns of community dynamics.

In Chapter 5 we defined the response of an organism to the physical environment in terms of a bell-shaped curve

TABLE 22.1 Physiological Characteristics of Early and Late Successional Plants

Attribute	Early Successional Plants	Late Successional Plants
Seed dispersal in time	Well dispersed	Poorly dispersed
Seed germination:		
Enhanced by		
light	Yes	No
fluctuating temperatures	Yes	No
high NO_3	Yes	No
Inhibited by		
far red light	Yes	No
high CO_2 concentration	Yes	No
Light saturation intensity	High	Low
Light compensation point	High	Low
Efficiency at low light	Low	High
Photosynthetic rates	High	Low
Respiration rates	High	Low
Transpiration rates	High	Low
Stomatal and mesophyll resistances	Low	High
Resistance to water transport	Low	High
Recovery from resource limitation	Fast	Slow
Resource acquisition rates	Fast	Slow

Source: After Bazzaz 1979.

relating some measure of performance to variations in environmental conditions (see Figure 5.1). The range of environmental conditions under which a species can survive, grow, and reproduce—its fundamental niche (see Chapters 5 and 14)—represents a constraint on its distribution and abundance. For example, Grace and Wetzel (1981) designed an experiment to examine the influence of water depth on the abundance of the cattail species *Typha latifolia,* which grows along the marshy edges of ponds and lakes. Individuals were planted in experimental ponds at the Kellogg Biological Station, Michigan, along a gradient of water depth ranging from dry land to a depth of 1.2 m. The response of *T. latifolia* along this gradient is shown in Figure 22.3a. This species grows best at water depths between 20 and 60 cm. It can grow on dry land where the water table is 20 cm below ground, but it does not grow at water depths greater than 80 cm. The response of *T. latifolia* to water depth described in Figure 22.3a limits the distribution and abundance of the species to the edges of ponds and lakes (Figure 22.3b). The species' fundamental niche acts as a primary constraint on its distribution and abundance.

The second premise states that species differ in their environmental responses—their fundamental niches. The reason for this diversity of environmental tolerances is the focus of Part 3 (Chapters 5–8) and can summarized by the simple premise that *the characteristics that allow an organism to prosper under one set of environmental conditions often limit its ability to do equally well under differing environmental conditions.* For an example, we can return to the experiments of Grace and Wetzel. The experiments examining the response of *T. latifolia* to water depth actually included a sec-

ond species of cattail found in the region, *Typha angustifolia.* The response of the two species of cattail to the experimental gradient of water depth is shown in Figure 22.4a. Although the curves defining the responses of the two species overlap, there are differences in the range of conditions that can be tolerated. Note that both species can survive in the shallow waters; however, only the narrow-leaved cattail, *T. angustifolia,* can grow in water deeper than 80 cm depth. This brings us to our third premise—environmental conditions change in time and space.

The third premise is obvious; yet it is a critical element to understanding the dynamics of communities. As the environmental conditions change, in this case the water depth, the possible distribution and abundance of these two species will change. It is *possible* for both species of cattails to survive in the shallow water near shore, and therefore form a component of the community. However, the broad-leaved cattail, *T. latifolia,* will never be a part of the community in the deeper waters because of its inability to tolerate water deeper than 80 cm.

In this example, the environmental gradient—water depth—changes largely as a function of distance from the shoreline of the pond or lake. However, the shoreline and subsequent profile of water depth may change because of a variety of factors such as rainfall, rate of water input to the lake from streams, or increased evaporation resulting from high air temperatures. In these examples, the change in environmental conditions (water depth) is not a direct function of the organisms, but a feature of the physical environment. This type of environmental change is called allogenic. In contrast, other types of environmental change are a direct result of the organism within the community. For example, the vertical profile of

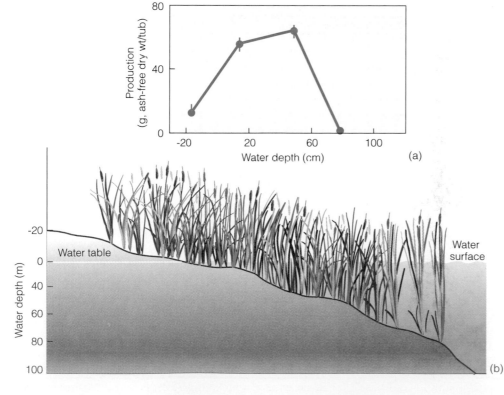

FIGURE 22.3   (a) Distribution of the cattail species *Typha latifolia* along an experimental gradient of water depth. (b) Hypothetical distribution of *T. latifolia* along the edge of a pond resulting from the response to water depth presented in (a). (Adapted from Gene and Wetzel 1981.)

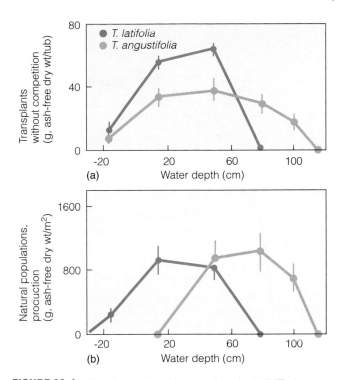

(a)

(b)

FIGURE 22.4 The distribution of two species of cattail (*Typha latifolia* and *T. angustifolia*) along a gradient of water depth: (a) grown separately in an experiment; (b) grown together in natural populations. The response of the two species reflects their fundamental niche (physiological tolerances) in the absence of competition. The response of each species is altered by the presence of the other. They are forced to occupy only their realized niches. (Adapted from Gene and Wetzel 1981.)

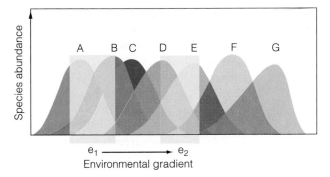

FIGURE 22.5 Fundamental niches of seven hypothetical species along an environmental gradient (e.g., moisture, temperature, or elevation) in the absence of competition from other species. The species all have bell-shaped responses to the gradient, but each has different tolerance limits defined by a minimum and maximum value along the gradient. As conditions change (in either time or space), for example from e_1 to e_2, the set of species that can potentially occur in the community changes. These changes can occur in either time or space.

light within a forest is a direct result of the interception and reflection of solar radiation by the trees (see Figure 3.3). This type of environmental change is called autogenic. As environmental conditions change, whether allogenically or autogenically, the structure of the community will change.

Thus far, we have developed a model of changes in community structure based on differences in the environmental response of species (their fundamental niche) and changes in environmental conditions. This relationship is summarized by the graphic presented in Figure 22.5. We can represent the fundamental niches of a variety of species by defining a number of bell-shaped curves along some environmental gradient, such as the availability of water or light for plants, or habitat type for animal species. The response of each species is defined in terms of abundance. Although the curves defining the fundamental niches of the species overlap, there are differences in the range of conditions that can be tolerated (i.e., beyond which the species is not found to survive). The distribution of fundamental niches along the environmental gradient represents a primary constraint on the structure of communities. For any given range of environmental conditions, only a subset of the species can survive, grow, and reproduce. As the environmental conditions change in either space or time, the possible distribution and abundance of species will change.

The model thus far, however, considers only the responses of species to environmental conditions in isola-

tion—without considering possible interactions among the species that can co-occur under any set of environmental conditions. As we have seen in Part 3, there are a variety of ways in which species within the community interact and modify patterns of distribution and abundance. This brings us to the fourth premise; the fundamental niche is modified by species interactions (realized niche; see Chapter 14).

Organisms can interact in either of two ways, through direct contact (interactions such as competition and predation) or indirectly, through the modification of the physical environment. Direct interactions, such as physical conflict or predation, are more easily observed and therefore fairly obvious. Indirect interaction through the modification of the physical environment, however, is by far the more common means by which populations and individuals interact and influence the structure and dynamics of communities. Numerous examples have been presented in the previous chapters. In plant communities, taller plants intercept solar radiation, consequently reducing the availability of photosynthetically active radiation (light) to smaller plants, reducing their rates of photosynthesis and growth. The uptake of water and nutrients by the root systems of one plant decreases the availability of those essential resources to neighboring plants. The consumption of seeds by one species of seed-eating bird reduces the availability of food to other species that share the same seed resource.

These are all examples of competition resulting from the sharing of a limiting resource. Individuals of one species modify the environment, in these cases the availability of some essential resource, for other species inhabiting the same area. These competitive interactions may limit the distribution and abundance of the species by reducing rates of population growth or increasing rates of mortality. As an example of competition modifying the distribution and abundance of a species, we can return to the case of the two species of cattails presented in Figure 22.4a. The distribution of the two species along the gradient of water depth shown in Figure 22.4a is in the absence of competition. That is, in the experiment each species was grown along the water depth gradient in the absence of the other. These responses therefore represent the potential response of the species—the fundamental niche. When the two species are grown together along the same gradient of water depth, their distributions, or realized niches, are changed (Figure 22.4b). Note that even though *Typha angustifolia* can grow in the shallow waters (0 to 20 cm depth) and above the shoreline (−20 to 0 cm depth), in the presence of *Typha latifolia*, it is limited to water depths of 20 cm or deeper. Individuals of *T. latifolia* outcompete individuals of *T. angustifolia* for the limited resources of nutrients, light, and space in shallow water, thus limiting *T. angustifolia* to the deeper waters where *T. latifolia* cannot survive.

In other situations, one species may modify the environment so as to make it more favorable for another species. An example is the mutualistic association between green plants and mycorrhizal fungi discussed in Chapter 17. Recall that the plant provides the fungi with carbon (food), while the fungi provide nitrogen and phosphorus that may not be otherwise available to the plant. Therefore, the fungi enable the plants to grow in environments where otherwise the low availability of these nutrients would limit growth and survival.

An example of how the fundamental niche of a tree species, the Brazilian sour orange (*Citrus aurantium*), is modified by its association with mycorrhizal fungi appears in Figure 22.6. In this experiment, the availability of phosphorus in the soil was controlled by the addition of fertilizer. When the amount of phosphorus added to the soil is high, both individuals with and without the fungi grow equally well. Because of the high availability of phosphorus to the plants, the uptake of phosphorus does not limit plant growth and the fungi provide no added advantage to the plants. However, when no phosphorus fertilizer is added to the soil, the plants with fungi grow ten times as large and fast as the plants without the fungi. The fungi enable the trees to grow in areas of low nutrient availability, where they would otherwise be unable to grow and survive.

In both of these examples, competition and mutualism, interactions between two populations modify the response of species to environmental conditions, influencing their distribution and abundance. The fundamental niche constrains the distribution and abundance of species as environmental conditions vary; however, the response of a species can be modified by its interactions with other species. Our simple model just discussed provides the framework for interpreting community dynamics.

AUTOGENIC ENVIRONMENTAL CHANGE

In Chapter 21, succession was defined as changes in community structure though time, more specifically, as temporal changes in species dominance. One species initially colonizes an area, but as time progresses its population declines only to

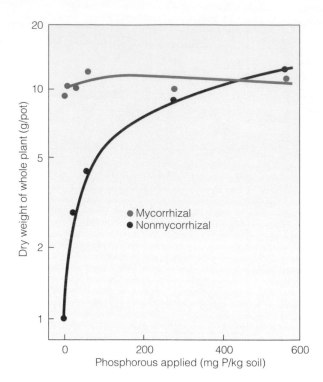

FIGURE 22.6 The effects of mycorrhizal fungi on the growth of Brazilian sour orange trees under different levels of phosphorus fertilization. Plants with mycorrhizae growing in soil with low phosphorus levels have a high growth rate, because mycorrhizae increase phosphorus available to the plant. The mycorrhizal plants do well at all phosphorus levels in the soil, whereas nonmycorrhizal plants respond dramatically to phosphorus fertilization. (Abott and Robsen 1985)

be replaced by another species (see examples in Figures 21.4, 21.5, 21.8, 21.10, and 21.13). This general pattern of changing species dominance as time progresses is seen in most environments, which suggests a common underlying mechanism or mechanisms.

One feature common to all plant successions is autogenic environmental change. In both primary and secondary succession, the colonization of an area by plants alters the environmental conditions. One clear example is the alteration of the light environment. As discussed in Chapter 3, the vertical profile of light within a plant community is a direct result of the leaves reflecting and intercepting solar radiation. As you move from the top of the plant canopy to ground level, the light available to drive the processes of photosynthesis declines (see Figure 3.3). In the initial period of colonization during succession, few if any plants are present. In the case of primary succession, the newly exposed site has not been occupied previously. In the case of secondary succession, plants have been either killed or removed by some disturbance. Under these circumstances, the availability of light at ground level is high, and seedlings can germinate and grow. As plants grow, their leaves intercept sunlight, reducing the availability of light to shorter stature plants (Figure 22.7). This reduction in available light will most likely reduce rates of photosynthesis, slowing the growth of

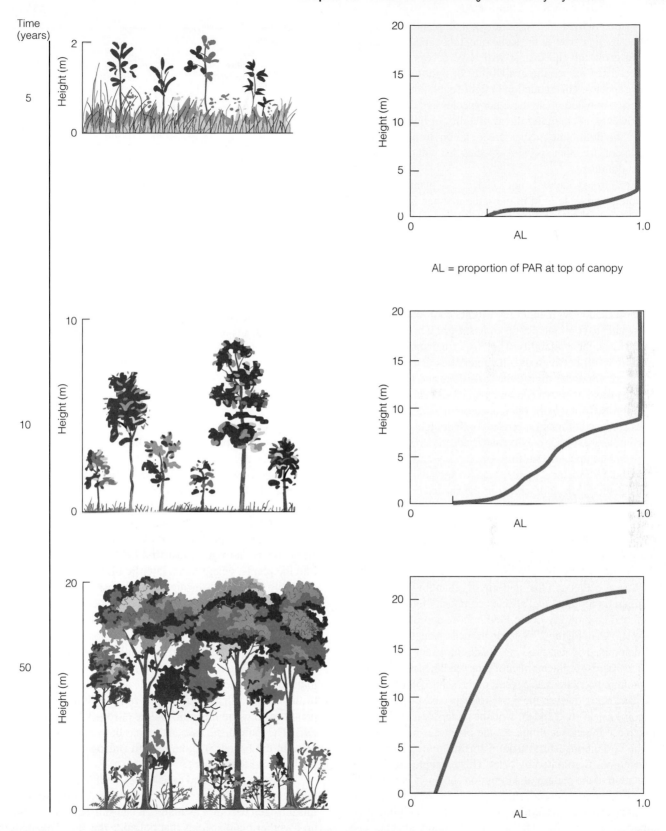

AL = proportion of PAR at top of canopy

FIGURE 22.7 Changing vertical light profile through succession. Following disturbance, the site is dominated by low-stature herbaceous vegetation. As time progresses, the stature of vegetation increases as woody plants come to dominate the site. As the height of the canopy increases, the vertical profile of light changes to reflect the increased leaf area and the range of height over which the leaf area is distributed.

these shaded individuals. Assuming not all plant species photosynthesize and grow at the same rate, those species of plants that can grow tall the fastest will have access to the light resource. They reduce the availability of light to the slower-growing species. This reduction in light enables the fast-growing species to outcompete the other species and dominate the site. However, in changing the availability of light below the canopy, the dominant species create an environment that is more suitable for other species—species that will later displace them as dominants.

Recall from Chapter 6, not all species of plants respond to variation in available light in the same manner. Sun-adapted, shade-intolerant plants exhibit a very different response to light than do shade-adapted, shade-tolerant species. Shade-intolerant species exhibit high rates of photosynthesis and growth in high-light environments. Under low light levels, they cannot continue photosynthesis, growth, and survival. In contrast, shade-tolerant plant species exhibit much lower rates of photosynthesis and growth under high-light conditions. However, they are able to continue photosynthesis, growth, and survival under lower light availability. There is a fundamental physiological trade-off between the characteristics that enable high rates of growth under high-light conditions and the ability to continue growth and survival under shaded conditions.

In the early stages of plant succession, shade-intolerant species come to dominate as a result of their high growth rates. Shade-intolerant species overtop and shade the slower-growing, shade-tolerant species. As time progresses and light levels decline below the canopy, seedlings of the shade-intolerant species cannot grow and survive in the shaded conditions. At this time, although the shade-intolerant species dominate the canopy, no new individuals are being recruited into the populations. In contrast, shade-tolerant species are able to germinate and grow under the canopy. As the shade-intolerant plants that make up the canopy die, shade-tolerant species in the understory replace them. This pattern of changing population recruitment, mortality, and species composition through time for a forest community in the Piedmont region of North Carolina is shown in Figure 22.8. Early in succession (time) the forest is dominated by fast-growing, shade-intolerant pine species. As time progresses, the number of pine seedlings recruited into the community declines as a result of the shaded conditions on the forest floor. However, the shade-tolerant oak species are able to become established and grow in the understory, indicated by the increase in oak seedlings. As the pine trees in the canopy die, the community shifts from a forest dominated by pine species to one dominated by oaks. This example provides an explanation of the pattern of succession shown in Figure 20.8.

In the above example, succession is the result of changing competitive ability under autogenically changing environmental conditions. The shade-intolerant species can successfully compete during the early stages of succession because of their ability to grow quickly in the high-light environment. However, as the autogenic changes in the light environment occur, the ability to tolerate and grow under shaded conditions enables the shade-tolerant species to successfully compete during the later stages of succession.

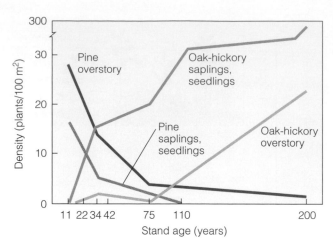

FIGURE 22.8   Dominance shift of overstory and understory (seedlings and saplings) pines, oaks, and hickories during secondary succession in the Piedmont region of North Carolina. Early succession pine species initially dominate the site. Pine seedling regeneration declines as the light decreases in the understory. Shade-tolerant oak and hickory seedlings establish themselves under the reduced light conditions. As pine trees in the overstory die, oak and hickory replace them as the dominant species in the canopy. (Billings 1938.)

Light is not the only environmental factor that changes over the course of succession. Other autogenic changes in environmental conditions affect the relative competitive abilities of plant species and influence patterns of succession. In Chapter 21, the colonization of newly deposited glacial sediments and lava flows provided an example of primary succession. Because of the absence of a well-developed soil, very little nitrogen is present in these newly exposed surfaces. Atmospheric nitrogen must first be fixed and incorporated into the plants, where it can later be broken down and made available in the soil through the process of decomposition (see Chapters 9 and 25). The virtual absence of mineral nitrogen restricts the establishment, growth, and survival of most plant species on these sites. However, those terrestrial plant species having a mutualistic association with nitrogen-fixing bacteria *Rhizobium,* such as alder (*Alnus*), can grow and dominate the site. These plants provide a source of carbon (food) to the bacteria that inhabit their root system. In return the plant has access to the atmospheric nitrogen fixed by the bacteria. The mutualistic association has both a cost (carbon supplied to the bacteria) and a benefit (nitrogen acquired from the bacteria) to the plant.

As these plants shed their leaves or die, the nitrogen that they contain is released to the soil through decomposition. As this nitrogen becomes available, it can subsequently be taken up by other plant species that colonize the site. This pattern of increasing accumulation of litter on the soil surface, organic carbon in the soil, and soil nitrogen during primary succession in Glacier Bay, Alaska (see Figure 21.7), is shown in Figure 22.9. As nitrogen becomes available in the soil, species that do not have the added cost of this mutualistic association, and exhibit faster rates of growth and recruit-

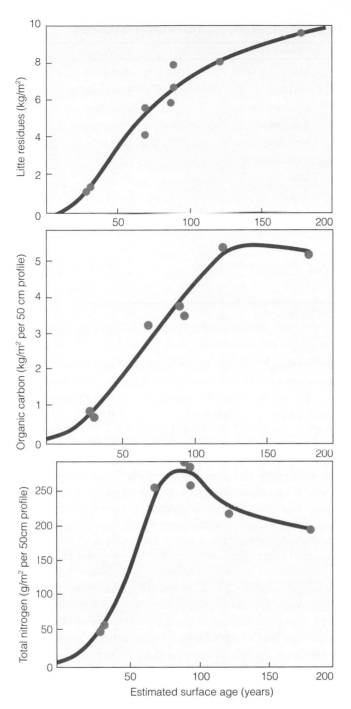

FIGURE 22.9 Changes in soil properties during succession on glacial debris at Glacier Bay, Alaska (see Figure 21.7). (a) Increase in dead organic matter at the soil surface (plant litter residue); (b) buildup of soil organic matter; (c) changes in total soil nitrogen. (Crocker and Major 1955.)

ment, come to dominate the site (see Figure 21.8). As in the last example, the changing pattern of species dominance through time—succession—is a result of autogenic changes in the environment and the associated changes in the relative competitive ability of the species colonizing the site.

Although competition is viewed as a population process (see Chapters 12 and 14), it actually takes place among individuals. The ability of an individual to compete is constrained

by individual traits based on a suite of life history and physiological attributes as outlined by Bazzaz (1979) in Table 22.1. There are two components of plant response to the environment that are critical for understanding community dynamics. One is the response of the individual to the prevailing environment, such as light, nutrients, and moisture. The other is how individuals modify the environment—autogenic environmental change. It is the combination of these two plant responses to the environment that give rise to the dynamics of communities across the landscape.

CURRENT MODELS OF COMMUNITY DYNAMICS

The simple model presented above explains the shifting patterns of species dominance through succession in terms of differences in population recruitment, growth, and competitive ability. These population responses reflect the aggregated properties of individual plants interacting with changing environmental conditions in time and space. This simple model forms the basis for a variety of current theories of community dynamics. We will examine three such models that focus on how the physiology, morphology, and life history traits of individual species influence species interactions and ultimately species distribution and abundance under changing environmental conditions (Grime 1979, Tilman 1987, 1988, Huston and Smith 1987, 1989).

Plant Strategies and Vegetation Processes

Grime (1977, 1979) expanded the r and K concept of life history classification discussed in Chapter 13 to include three primary plant strategies (R, C, and S) relating plant adaptations to different habitats (Table 22.2; for detailed discussion of Grime's classification of plant strategies see Chapter 13, page 235). Species exhibiting the R, or ruderal, strategy rapidly colonize disturbed sites but are small in stature and short-lived. Allocation of resources is primary to reproduction, with characteristics allowing for a wide dispersal of propagules to newly disturbed sites. Predictable habitats with abundant resources favor species that allocate resources to growth, favoring resource acquisition and competitive ability (C species). Habitats where resources are limited favor stress-tolerant species (S species) that allocate resources to maintenance.

Grime's theory views succession as a shift in the dominance of these three plant strategies in response to changing environmental conditions (habitats). Following the disturbance that initiates secondary succession, essential resources (light, nutrients, and water) are abundant, selecting for ruderal (R) species that can quickly colonize the site. As time progresses and plant biomass increases, competition for resources will occur, selecting for competitive (C) species. As resources become depleted as a result of high demand by growing plant populations, the C species are eventually replaced by stress-tolerant species (S) that are able to persist under the low-resource conditions. This pattern of changing dominance in

TABLE 22.2 Some Traits Correlated with Typical Competitive, Stress-Tolerant, and Ruderal Plant Species

Trait Type	Competitive	Stress-Tolerant	Ruderal
Life form	Variable	Variable	Herbs
Shoot morphology	Dense canopy	Variable	Small stature
Leaf form	Variable	Leathery, needles	Variable
Leaves	Deciduous	Evergreen	Deciduous
Longevity	Long or short	Very long	Very short
Flowering	Annual	Intermittent	Annual
Reproductive maturity	Late	Late	Early
Reproductive effort	Small	Small	Large
Perennation	Buds, seeds	Leaves and roots	Seeds
Growth rate	Rapid	Slow	Rapid
Stress response	Rapid	Slow	Reproduces
Litter	Persistent, copious	Persistent, sparse	Not persistent, sparse
Palatability to herbivores	Variable	Low	Often high

Source: Modified from J. P. Grime, 1979. *Plant Strategies and Vegetation Processes,* Wiley, New York.

plant strategies in response to changing environmental conditions is shown in Figure 22.10a. In environments (habitats) where resource availability is inherently low, the role of competitive species that demand high resource abundance may be diminished. In this case the succession make proceed directly from ruderal to stress-tolerant species (curve R_2).

In Grime's framework, changes in species (plant strategy) dominance result from autogenic changes in resource availability. The changes in resource availability are a direct result of resource consumption by the plants, with resource abundance decreasing as succession progresses (Figure 22.10b). This decline in resource abundance with succession includes both above- (light) and below-ground (nutrient and water) resources.

Resource-Ratio Model

The resource-ratio model advanced by Tilman (1985, 1988) is based on the trade-off in characteristics that enable plants to compete for the essential resources of nitrogen and light (Figure 22.11a). The ability to effectively compete for light is associated with allocation of carbon to the production of above-ground tissues—leaves and stem. Conversely, the ability to effectively compete for nitrogen is associated with the production of root tissues. This pattern of changing allocation of carbon under varying nutrient and light resources is discussed in detail in Chapters 6 and 7.

Succession comes about as the relative availability of nitrogen and light change through time. In Tilman's model, the availability of these two essential plant resources (light and nitrogen) is inversely related. Environmental conditions range from habitats with soils poor in nutrients but with a high availability of light at the soil surface to habitats with nutrient-rich soils and low availability of light. Community composition changes along this gradient as the ratio of nitrogen and light change. Species reach an equilibrium with the supply rates of

the limiting resources. In doing so, they lower the available resources to a point at which other species cannot invade.

In early primary succession, the colonizing species are those adapted to a regime of low soil nitrogen and high light. As biogeochemical processes (see Figure 22.9, for example) make more soil nutrients available, plant growth increases, reducing the availability of light at the soil surface. The changing ratio of soil nitrogen to light leads to the replacement of one plant species by another (Figure 22.11b). Over time these changes favor plants adapted to high nutrient and low light availability at the soil surface. Differences in competitive abilities are the result of trade-offs in characteristics required to exploit above-ground or below-ground resources.

Secondary succession flows in much the same pattern, but at a more rapid rate. The disturbed site is characterized by high light availability at the ground, but a low availability of nutrients. Plant species that allocate carbon to the production of roots can best compete under the high-light, low-nitrogen conditions. As time progresses plant biomass accumulates, decreasing available light. The fixation of nitrogen, the accumulation of litterfall and soil organic matter, and the release of mineralized nitrogen in the process of decomposition increases nitrogen supply rate with time. As succession progresses, species dominance shifts to species that allocate carbon to stem and leaf production and are able to compete under the high-nitrogen, low-light conditions.

Species composition and rapidity of change during secondary succession depend upon at which point on the gradient the species colonize the area. If soil nitrogen is initially high following disturbance, the magnitude of species change will be less, with succession proceeding directly to species that are superior competitors for the rapidly declining light resource.

Both Grime's and Tilman's models of succession predict changes in the resource supply and resulting shifts in species adapted to the changing environmental conditions. In both

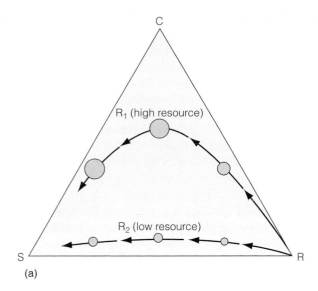

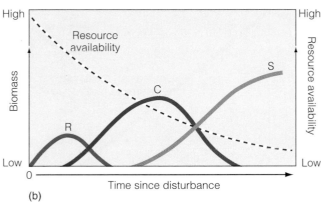

FIGURE 22.10 (a) Grime's triangular model of plant strategies showing the pathway of secondary succession under conditions of high and low initial resource availability (e.g., high- vs. low-nutrient site). The circles indicate the total plant biomass at each stage of succession. The letters at each corner of the triangle represent the three primary plant strategies (*R*, *C*, and *S*). Points at any location within the triangle represent intermediate species (see Chapter 13, page 235 for detailed description of model). The trajectory for R_1 (high resource) shows the greater importance of competitive species when resource availability is high. The role of *C* species under low resources (curve R_2) would be diminished or absent. (b) Shifts in the dominance of the three primary plant strategies and changing patterns of resource availability as succession proceeds. (Adapted from Grime 1979.)

models these changes are autogenic. However, these two models differ in the nature of environmental change and in their views of competition. In Grime's model the availability of all resources, both above and below ground, decrease as succession progresses. This decline is a result of increasing resource demand by the plants as the stand develops. As a result, the plant characteristics that allow a species to compete effectively do not change depending on the resource. This is the basis for defining a single competitive strategy (*C* species). In contrast, Tilman's model is based on the inverse relationship (shifting ratio) between the availability of nitrogen and light. It is the trade-off in characteristics necessary to effectively compete for these two resources that defines the range of plant strategies

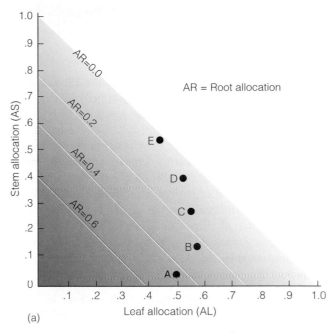

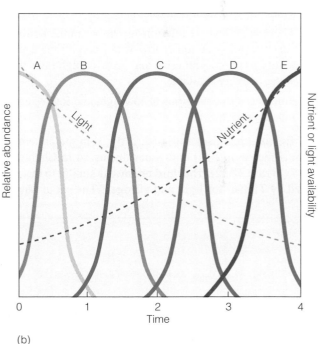

FIGURE 22.11 (a) Allocation triangle showing possible patterns of photosynthate allocation to leaves (AL), stems (AS), and roots (AR) by plants. Note that AL+AS+AR=1. Each diagonal line is a line of equal allocation to roots. Species with allocation patterns that fall near the origin have a high proportion of their biomass in roots. Those near the apex of the triangle have a high allocation to stem. The allocation patterns of five species (A–E) are shown. These allocation patterns reflect a shift in plant strategies from adaptation to low nutrients and high light (species A), to adaptation to high nutrients and low light (E). (Adapted from Tilman 1994.) (b) Tilman's resource-ratio hypothesis. During the initial stages of succession, light availability is high and nitrogen availability is low. As time progresses, light availability declines while nitrogen availability increases. The shift in species dominance over this period reflects the changing relative competitive abilities of the five hypothetical plant species defined in (a). (Adapted from Tilman 1988.)

based on patterns of carbon allocation to the production of leaves (and stem) versus roots.

Individual-Based Model

Huston and Smith (1987; Smith and Huston 1989, Huston 1994) proposed a third model of community dynamics based on plant adaptations to environmental gradients. Their model is based on the cost-benefit concept that plant adaptations for the simultaneous use of two or more resources are limited by physiological and life history constraints. Their model focuses on the resources of light and water. The plants themselves largely influence variations in available light within the community (autogenic), while the availability of water is largely a function of climate and soils (allogenic).

The consequences of constraints on the simultaneous use of light and water by individual plants are summarized by the following three premises:

1. There is an inverse relationship between the ability to survive and grow under low-light conditions and the ability to photosynthesize and grow at high rates when the availability of light is high.

2. There is an inverse relationship between the ability to survive and grow under low-water conditions and the ability to photosynthesize and grow at high rates when water is freely available.

3. Tolerances to conditions of low light and low water are interdependent.

The first two premises follow directly from the patterns of plant response to light and water discussed in Chapters 6 and 7 (Figure 22.12). The third premise is similar to that proposed by Tilman for light and nitrogen. The set of physio-

logical and morphological characteristics that enable a species to survive and grow under shaded conditions (e.g., allocation of carbon to the production of leaves and stem) is in direct conflict with its ability to tolerate low water availability (e.g., high allocation to the production of roots).

Huston and Smith (1987) investigated the consequences of these premises using an individual-based model for plant succession. Their model is derived from two widely used individual-based forest models (Botkin et al. 1972, Shugart and West 1977). The individual-based model simulates annual population dynamics of plants on a defined area by considering birth, growth, and death of individual plants. Each species is defined by given species-specific life history traits. The authors defined a set of hypothetical plant species that spanned the range of possible responses to light and moisture (Figure 22.13). They then examined the resulting patterns of succession under varying water availability (xeric to mesic sites).

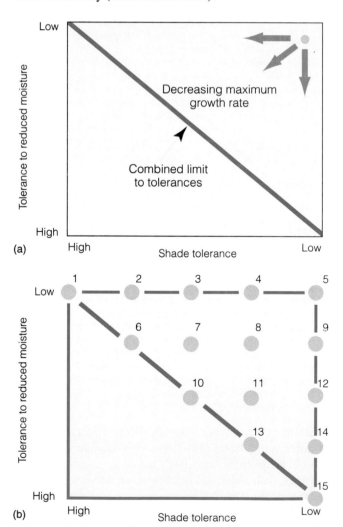

(a)

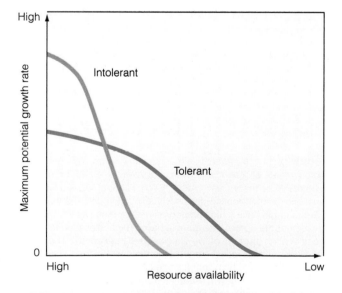

FIGURE 22.12 Growth rate in relation to resource availability for plants of two degrees of tolerance. Note the inverse relationship between the resource level where the growth rate is zero (*x* intercept of the curves) and the maximum rate of growth achieved under high-resource conditions. (From Smith and Huston 1989; based on Larcher 1975, Bazzaz 1979, Orians and Solbrig 1977, Chapin et al. 1986.)

FIGURE 22.13 (a) Possible plant strategies for light and water use, illustrating some of the consequences of the trade-offs described in the three premises. The highest rate of growth (carbon gain) is in the upper right corner of the figure, and growth decreases with increasing tolerance to low levels of light and/or water. (b) Division of the continuum plant strategies into the 15 discrete functional types used in the computer simulations. Each functional type is defined by maximum growth rate, shade tolerance, and tolerance to low moisture levels. (Smith and Houston 1989.)

Succession is interpreted as a temporal shift in species dominance, primarily in response to autogenic changes in light availability (Figures 22.14a–c). The pattern of species dominance shifts from fast-growing, shade-intolerant species early in succession to slower-growing, shade-tolerant species later in succession. As site conditions change from wet (mesic) to dry (xeric), the species involved in the succession shift to more xerophytic species (plant types); however, the mechanisms underlying the process of succession remain the same. Zonation is interpreted as a spatial shift in species dominance, primarily in response to the effect of allogenic changes in water availability on the dynamics of competition for light (Figure 22.14d).

The Huston and Smith model, like that of Tilman, assumes that physiological and energetic constraints prevent any one species from being competitive under all environmental conditions. The result is that the relative competitive abilities of species vary across a range of environmental conditions. This view is in contrast to that of Grime. However, the Huston and Smith model does differ from that of Tilman in that it does not assume any fixed (inverse) relationship between above- and below-ground resources. The absence of this constraint allows for a much richer array of plant responses and resulting community dynamics.

ALLOGENIC ENVIRONMENTAL CHANGE

The focus on succession thus far has been on shifting patterns of community structure in response to autogenic changes in environmental conditions. However, purely abiotic environmental (allogenic) change can produce patterns of succession over time scales ranging from days to millen-

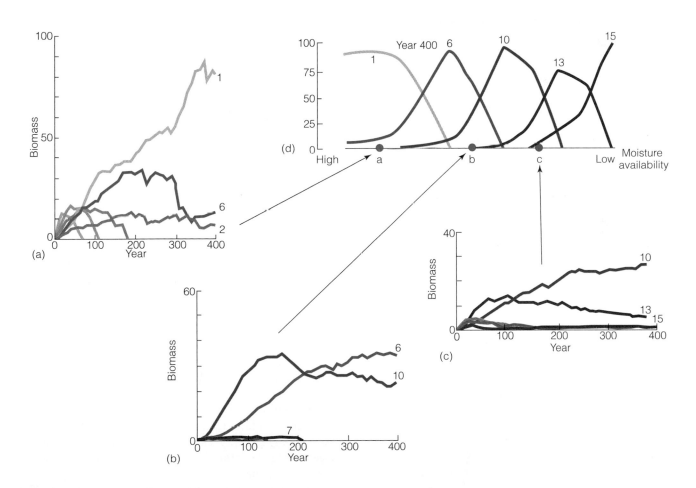

FIGURE 22.14 (a–c) Successional sequences resulting from the same functional plant types (hypothetical species) under three different moisture conditions. The plant types used in the model simulations are those presented in Figure 22.13. All 15 plant types were included in each simulation. Changes in the set of species involved in succession at each site reflect differences in species adaptations to changing moisture conditions. Note that the role of a plant species in succession changes under different moisture conditions. For example, species 6 is a subdominant in the succession under wet conditions, but it is the late successional dominant under drier conditions. (d) Species zonation along a moisture gradient. The curves represent the relative biomass of the species at the end of each successional simulation. For example, the relative abundance of species at the end of the three simulations shown in (a–c) are identified by the points on the moisture gradient (x axis) that correspond to the soil moisture conditions used in the simulation of succession. (Adapted from Smith and Houston 1989.)

nia or longer. Fluctuations in the environment that occur repeatedly over the lifetime of an organism are unlikely to influence patterns of succession among species with that general life span. For example, annual fluctuations in temperature and precipitation will influence the relative growth responses of different species in a forest community, but they will have little influence on the general patterns of secondary succession outlined in Figures 21.10 and 22.9. In contrast, shifts in environmental conditions that occur at periods as long or longer than the organisms' life span are likely to result in shifts in species dominance—succession (Huston 1994). For example, seasonal changes in temperature, photoperiod, and light intensity produce a well-known succession of dominant phytoplankton in freshwater lakes, which is repeated with very little variation each year (Hutchinson 1967, Lewis 1978, Wetzel 1983). Seasonal succession of phytoplankton in Lawrence Lake, a small temperate lake in Michigan, is presented in Figure 22.15 (Crumpton and Wetzel 1982). Periods of dominance are correlated with species' optimal temperature, nutrient, and light requirements. Competition and seasonal patterns of predation by herbivorous zooplankton also interact to influence the temporal patterns of species composition.

Over a much longer time scale of decades to centuries, patterns of sediment deposition can have a major influence on the successional dynamics of coastal estuarine communities. The marshlands of the River Fal in Cornwall, England, have expanded seaward some 800 m over the past century (Ranwell 1974). The seaward expansion is a result of silt deposition lowering water depths. On the landward side, woodland plant species invade the marshlands, leading to a successional sequence from marsh to woodland over time.

A similar pattern of long-term transition in community structure resulting from sediment deposition occurs in freshwater environments. Ponds and small lakes act as a settling basin for inputs of sediment from the surrounding watershed. These sediments form an oozy layer that provides a substrate for rooted aquatics such as the branching green algae (*Chara*) and pondweeds (*Potamogeton*). These plants bind the loose matrix of bottom sediments and add materially to the accumulation of organic matter. Rapid addition of organic matter and sediments reduces water depth and increases the colonization of the basin by submerged and emergent vegetation. That, in turn, enriches the water with nutrients and organic matter. This enrichment further stimulates plant growth and sedimentation, and it expands the surface area available for colonization by macrophytes (see Carpenter 1981). Eventually the substrate, supporting emergent vegetation such as sedges and cattails, develops into a marsh. As drainage improves and the land builds higher, emergent plants disappear. Meadow grasses invade to form a marsh meadow in forested regions and wet prairie in grass country. Depending on the region, the area may pass into grassland, swamp woodland of hardwoods or conifers, or a peat bog.

Over an even longer time scale, changes in regional climate have a direct influence on the temporal dynamics of communities. The shifting distribution of tree species (see Figure 21.7) and forest communities (see Figure 21.6) over the past 16,000 years following the last glacial maximum in eastern North America is an example of how long-term allogenic changes in the environment can directly influence patterns of both succession and zonation on a local, regional, and even global scale.

Allogenic environmental change is the dominant influence on spatial variations in community structure—zonation. Patterns of temperature and moisture resulting from regional variations in climate (see Chapter 2) are the major determinant of regional and global patterns of vegetation distribution, and they form the basis of most vegetation classification systems (see Chapter 27). On a more local scale, climate interacts with soils and topography to influence patterns of temperature and soil moisture (see Chapter 4). The patterns of zonation in the Smoky Mountains of Tennessee shown in Figures 21.1 and 22.2 result from variations in temperature and soil moisture related to altitude, aspect, and slope.

The underlying geology of an area interacts with climate to influence soil characteristics such as texture (see Chapter 4). In turn, texture has a direct effect on soil moisture-holding capacity, cation exchange, and base saturation (see Chapter 4), influencing the moisture and nutrient environment of plants.

In aquatic environments, water depth (see Figure 22.4) and salinity (see Figure 21.2) are examples of allogenic environmental gradients that directly influence the distribution and dynamics of communities.

The distinction between autogenic and allogenic environmental change is often blurred when abiotic (allogenic) environmental factors influence the manner in which organisms modify the environment. Temperature, soil pH, water depth, salinity, and a wide variety of other environmental factors that are largely allogenic have a direct influence on plant

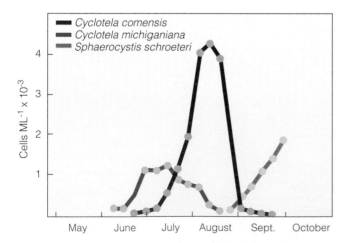

FIGURE 22.15 Temporal changes in the abundance of dominant phytoplankton species over the period of May through October in Lawrence Lake, Michigan (1979). The mean generation time of species is in the range of 1–10 days. (Adapted from Crumpton and Wetzel 1982.)

processes that determine survival, growth, and reproduction. By affecting rates of survival, growth, and reproduction, allogenic variations in environmental conditions influence competition for resources. Fast growth rate is a competitive advantage in gaining access to light and nutrients, resources that are both under strong autogenic control.

Pastor and colleagues (1984) examined patterns of nitrogen availability in different forest stands on Blackhawk Island, Wisconsin. The investigators found that the differences in nitrogen availability among the forest stands on the island are directly related to differences in the rate of nitrogen mineralization and release during decomposition (see Chapter 9). The observed variation in mineralization rates was a direct result of differences in the chemical and structural properties of the leaf litter produced by the species that dominate the different sites (a topic discussed in detail in Chapter 9). The authors go on to explain that differences in leaf characteristics among tree species influencing decomposition and nitrogen mineralization reflect different adaptations to soil water availability, which varies across the island as a function of soil texture. Species adapted to the low water availability of the coarse-textured sandy soil on the island produce leaf litter that decomposes slowly and is low in nitrogen content. In contrast, species on the heavier textured soils produce leaves that decompose readily and contain a higher level of nitrogen. In this example, allogenic variations in soil texture and water availability across the island influence not only patterns of zonation, but also the characteristics of the litter produced by the species that inhabit the different plant communities. In turn, the differences in litter quality directly influence the patterns of soil nitrogen availability on the island, an example of autogenic variation in environmental conditions. This example shows the complex interactions among environmental factors that directly influence plant processes and subsequently community dynamics.

SUCCESSION AND SPECIES DIVERSITY

Patterns of diversity through succession have been investigated by comparing sites within an area that are at different stages of succession; such groups of sites are known as chronoseries or chronosequences. In a study of secondary succession in old-field communities, Bazzaz (1975) found that species diversity, both species richness and equitability, increased during the first 40 years following abandonment. Tilman (1988) reports a similar pattern for 22 old-field sites at Cedar Creek, Minnesota, ranging from fields abandoned in the 1920s to those abandoned in the 1980s. Plant diversity increases with site age (time since abandonment).

Nicholson and Monk (1974) examined abandoned agricultural sites in the Piedmont of Georgia covering a sequence of 200 years that includes the shift in dominance from herbaceous to woody plants—field to forest. Diversity increased

for the first 80–100 years, remaining fairly constant thereafter. Whittaker (1975; Whittaker and Woodwell 1968) observed a different temporal pattern of species diversity for sites in upstate New York (Figure 22.16). Species richness increases into the late herb stages and then decreases into shrub stages. Species richness then increases again in young forest, only to decrease as the forest ages.

The processes of species colonization and replacement drive succession. To understand patterns of species richness and diversity during succession, we need to understand how these two processes vary in time. Colonization increases species richness. Species replacement typically results from competition, or an inability of the species to tolerate the changing environmental conditions. Species replacement acts to decrease species richness. We can examine how these two processes interact to influence changing patterns of species diversity by returning to our earlier example of terrestrial succession resulting from autogenic changes in the light environment.

Let us define a hypothetical successional sequence involving five species, A through E (Figure 22.17a). Following disturbance the site is initially colonized by species A, an annual herbaceous plant. Soon thereafter, the site is colonized by perennial herbaceous species (B), increasing species richness. Eventually, shrubs colonize the site (C), further increasing species richness. With the colonization of the site by shade-intolerant trees (D), the density of woody vegetation increases to the point where herbaceous species found in the earlier stages of succession cannot successfully reproduce in the shaded conditions created by the closed canopy of young trees. With time, shade-tolerant tree species (E) become established, and with the mortality of shade-intolerant species in the canopy, the site becomes dominated by shade-tolerant trees. The resulting pattern of species diversity corresponding to this succession is presented in Figure 22.17b.

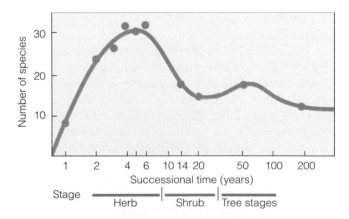

FIGURE 22.16 Changes in plant diversity during secondary succession of an oak-pine forest in Brookhaven, New York. Diversity is reported as species richness in 0.3 ha samples. Species richness increases into the late herbaceous stages, declines into the shrub stage, increases once again into the early forest stages, and declines thereafter. (Adapted from Whittaker 1975, Whittaker and Woodman 1968.)

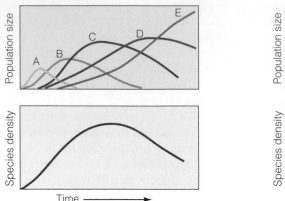
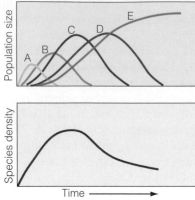

FIGURE 22.17 (a) Hypothetical succession involving five plant species (A–E), and (b) the associated temporal pattern of species diversity. Note that species diversity increases initially as new species colonize the site. However, as autogenically changing environmental conditions and competition result in the replacement of early successional species, diversity declines. (c) When the growth rates of the five species are doubled, the succession progresses more quickly, and (d) the pattern of species diversity reflects the earlier onset of competition and the more rapid replacement of early successional species. As a result, the period over which species diversity is at its maximum is reduced. (Adapted from Hoston 1994.)

Note that during the early phases of succession, diversity increases as new species colonize the site. However, as time progress, species become displaced—replaced as dominants by slower-growing, more shade-tolerant species. The peak in diversity during the middle stages of succession corresponds to the transition period, after the arrival of later successional species but prior to the decline (replacement) of early successional species. This pattern is seen in the level of diversity during succession in upstate New York (see Figure 22.16), with the exception that there are two periods of transition and two peaks in species diversity during the successional sequence. The two peaks correspond to the transitions between the herbaceous- and shrub-dominated phases, when both groups of plants were present, and the transition between early and later stages of woody plant succession. Species diversity declines as shade-intolerant tree species displace the earlier successional trees and shrubs.

The rate at which displacement occurs will be influenced by the growth rates of species involved in the succession. If growth rates are slow, the process will move slowly; if growth rates are fast, the process of displacement will occur more quickly (Figure 22.17c). This observation led Huston (1979, 1994) to conclude that patterns of diversity through succession will vary with environmental conditions (particularly resource availability) that directly influence the rates of plant growth. By slowing the population growth rate of competitors that will eventually displace earlier successional species, the period of coexistence is extended and species diversity will remain high. This hypothesis has the interesting consequence of predicting the highest diversity at low to intermediate levels of resource availability by extending the period over which species coexist (Figure 22.17d).

Disturbance can have a similar effect to that of reduced growth rates by extending the period over which species

coexist (Connell 1978, Huston 1979, Huston 1994) (Figure 22.18). In the simplest sense, disturbance acts to reset the clock in succession. By reducing or eliminating plant populations, the site is once again colonized by early successional species, and the process of colonization and species replacement begins again. If the frequency of disturbance (defined by the time interval between disturbances; see Chapter 23) is high, then later successional species will never have the opportunity to colonize the site. Under this scenario diversity will remain low. In the absence of disturbance, later succession species will eventually displace earlier successional species and species diversity will decline (see Figure 22.16). At an intermediate frequency of disturbance, colonization can occur, but competitive displacement is held to a minimum. The pattern of high diversity at intermediate frequencies of disturbance is referred to as the intermediate disturbance hypothesis (Connell 1978).

INFLUENCE OF HERBIVORES ON COMMUNITY DYNAMICS

Thus far, the framework that we have developed for understanding community dynamics, more specifically vegetation dynamics, has focused on adaptive characteristics of plants, and interactions between plant populations and the physical environment. However, heterotrophic organisms play a crucial role in a variety of processes related to vegetation dynamics, both zonation and succession. Animal life is critical in the pollination, dispersal, and germination of many plant species (see Chapter 17). In addition, decomposers break down organic matter, making nutrients available for uptake by plants (see Chapters 9 and 25). Herbivores modify community dynamics, directly and indirectly.

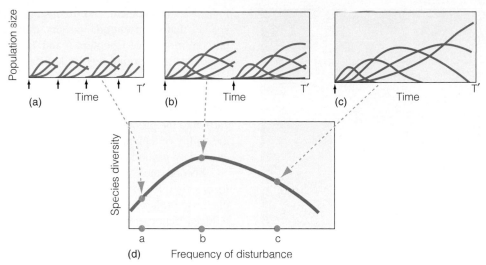

FIGURE 22.18 Patterns of succession for five plant species shown in Figure 22.17a under three levels of disturbance frequency: frequent, intermediate, and no disturbance. Time of disturbance is shown as an arrow on the *x* axis (time). Note the differences in species diversity after a specified period of time (shown as *T'* on the *x* axis). (a) Under a high frequency of disturbance, the absence of later successional species reduces overall diversity. (b) At an intermediate frequency of disturbance, all species coexist and diversity is at its maximum. (c) When disturbance is absent, later successional species eventually displace the earlier successional species and once again diversity is low. (d) The general form of the relationship between species diversity and the frequency of disturbance is a hump-shaped curve with maximum species diversity at intermediate frequency and magnitude of disturbance. The three examples shown in a–c are labeled on the *x* axis. (Adapted from Hoston 1994.)

By selecting certain plant species, herbivores directly influence mortality and recruitment, favoring the population growth of one species over another. Moose on Isle Royale in Lake Superior selectively feed on the seedlings and saplings of deciduous hardwood tree species of aspen, birch, ash, and maple, ignoring the conifer species of spruce and fir (Pastor et al. 1988). Long-term experiments using exclosures to exclude moose from certain areas have demonstrated that selective patterns of herbivory have changed the community structure of the island. In the exclosures, the abundance of the deciduous hardwood species is much higher.

By selectively feeding on seedlings and saplings of oak, yellow-poplar, and maple, and avoiding other species such as black cherry (*Prunus serotina*), the expanding populations of white-tailed deer are influencing the future composition of the forest in eastern North America. In other places deer eliminate woody reproduction and convert the site to ferns, grass, and goldenrod.

The mountain pine beetle (*Dendroctonus ponderosae*) has a major impact on forest communities in the Rocky Mountains of the western United States. Populations periodically build up, with beetles preferentially attacking large individuals of lodgepole pine (*Pinus contorta*). Female beetles lay eggs within the bark at the base of the tree. After hatching, the larvae feed in the inner bark (phloem), destroying the vascular system. Individuals eventually emerge as adults through exit holes chewed in the outer bark, moving on to attack healthy trees.

Following fire, lodgepole pine grows at a rapid rate and occupies the dominant position in the stand. Lodgepole pine is eventually succeeded by the more shade-tolerant species, consisting primarily of Douglas fir (*Pseudotsuga menziesii*) at the lower elevations and subalpine fir (*Abies lasiocarpa*) and Englemen spruce (*Picea engelmannii*) at the higher elevations throughout most of the Rocky Mountains. The role played by the mountain pine beetle is to periodically remove the large, dominant pines (Figure 22.19). This removal leaves growing space for subalpine and Douglas fir, thus hastening succession by these species (Amman 1977).

The southern pine bark beetle (*Dendroctonus frontalis*) has also affected community dynamics in the eastern United States. Increased mortality of pine species has hastened the process of secondary succession outlined in Figures 21.10 and 22.8. The death of pine trees in the canopy increases light availability to hardwood tree species in the understory and shrub layers. Increased light, together with reduced competition for other essential resources, promotes higher growth rates, hastening the dominance of hardwood species.

Herbivory also has indirect effects on community dynamics. Outbreaks of insect herbivores that feed on the forest canopy have a major influence on the forest environment. Defoliation of the canopy increases temperatures and available light at the forest floor, favoring the growth of understory trees and shrubs, as well as seedlings and herbaceous plants.

Outbreaks of the gypsy moth (*Porthetria dispar*) larvae in the northeastern United States have almost completely

FIGURE 22.19 Subalpine fir and Douglas fir seedlings growing in openings created when mountain pine beetles killed some of the larger dominant lodgepole pines (trees on the ground).

defoliated the canopy in oak forests. Not all of the leaf material is consumed. The larvae are not efficient consumers, and many of the green leaves fall to the forest floor. These green leaves are high in nutrients and decompose rapidly, resulting in an increase in the availability of nutrients in the soil. The increase in nutrients, combined with a reduced competition from oak species, favors the growth of plant species that are not selected by the larvae. However, the reduction in nutrient demand resulting from the defoliation of the dominant tree species results in an overall decrease in the uptake of nutrients. Nutrients are leached from the soil into the streams and transported out of the forest. This loss of nutrients can have long-term effects on the growth and composition of the forest. Further examples of the influence of herbivores on the nutrient dynamics in plant communities are presented in Chapter 25.

TOWARD A GENERAL MODEL OF COMMUNITY DYNAMICS

We began this chapter with a simple model of community dynamics based on the concept of the species' niche and changing environmental conditions. Although the three models discussed above (Grime 1979, Tilman 1987, 1988, Huston and Smith 1987, 1989) are all variations of this simple framework, they differ significantly in their interpretations of plant adaptations to the environment, competition, and the nature of autogenic environmental change through succession. At first these differences may seem troubling in that there is no general consensus, no single accepted model of community dynamics in the ecological literature. However, the debate of ideas that these differences represent is an essential feature of scientific development.

Despite these differences, certain similarities emerge that begin to form a general framework for understanding community dynamics. First is the view of community dynamics based on variations in species attributes: physio-

logical, morphological, and life history characteristics of the species that represent adaptations to different environmental conditions (see Chapters 6 and 7). A consensus among ecologists is emerging that trade-offs exist in the characteristics that determine the response of organisms to a variety of environmental factors. Numerous examples are presented in Chapters 6 and 7. For example, the trade-off in allocating carbon to the production of either roots or leaves influences the ability of a plant to acquire critical above-ground (light) and below-ground (water and nutrients) resources.

Second, there is a correlation among characteristics that form a complex of adaptation to the environment. For example, shade tolerance, the ability to survive and grow under reduced light conditions, involves a suite of characteristics related to leaf-level physiology (low light compensation point, reduced leaf respiration), morphology (allocation to the production of leaves), growth rate, and longevity (also see Table 22.1). In addition, within a given plant growth form (trees, shrubs, grasses, etc.), shade-tolerant species tend on average to have larger seeds than shade-intolerant species, a characteristic that not only aids in their initial survival and growth under low-light conditions (Grime and Jeffrey 1965), but also influences the ability of the species to disperse on the landscape (Goldberg and Werner 1983).

Although the dynamics of resource availability during succession and the nature of competition for those changing resources remain as areas of debate, the importance of autogenic environmental change in the process of succession has emerged as a general principle. The trade-offs in species characteristics limit the ability of any one species or group of species to dominate under all environmental conditions, and the autogenic changes in environmental conditions result in a shifting pattern of dominance through time. As environmental conditions change in time and space, the set of species capable of exploiting those conditions will change.

Since the development of the climax concept by Clements in the early part of the twentieth century, the view of succession as a unidirectional process of vegetation change toward some endpoint or equilibrium has been supplanted. A new view of community dynamics as a nonequilibrium process has emerged to replace it. An equilibrium among competing species is rarely achieved; however, a steady-state dynamic equilibrium (see discussion of steady-state shifting mosaic, Chapter 21) that represents a balance between local disturbances and successional dynamics can occur (Pickett 1976, White 1979, Doyle 1981, Shugart 1984).

Theoretical models in all areas of science represent a search for generality, an attempt to identify common mechanisms to explain a certain class of observations. This is true of the models we have examined in this chapter. Their objective is to identify and explain features common to many or all communities. This task is quite different from attempting to describe and explain the dynamics of any one, specific community. In most cases, a model that can predict the general behavior of many different communities is not adequate for describing the variety of factors that influence the dynamics at any given site. Factors such as prior land use or state of the site prior to disturbance, soils, topography, microclimate, and herbivore populations will interact to influence the suite of species that colonize and succeed at any location. For example, the patterns of algal succession in the rocky intertidal zone presented in Figure 21.4 vary as a function of the season during which the site becomes available for colonization (Sousa 1979). Results suggest that the process of colonization is influenced by temperatures at the time it occurs.

The context of the site within the surrounding landscape will also play a critical role on community dynamics (see Chapter 23). The structure of the surrounding community will influence the physical environment as well as determine the local seed pool for recruitment of new individuals and colonization of the site following disturbance. In this case, the species of organisms (particularly plants) that are available to colonize the site will be directly influenced by the current composition and successional state of the surrounding landscape.

CONCEPT OF THE COMMUNITY REVISITED

Our initial discussion of the processes influencing community structure and dynamics contrasted two views of the community. The organismal view of Clements stressed the community as an entity made up of interdependent species.

In contrast, in the individualistic or continuum view of Gleason, the community is an arbitrary concept. Each species responds independently to the underlying features of the environment. Research reveals that, like most polarized debates, the reality lies somewhere in the middle, and our viewpoints are often colored by our perspective. The organismal community is a spatial concept. As you stand in the forest you see a variety of plant and animal species, interacting and influencing the overall structure of the forest. The continuum view is a population concept, focusing on the responses of the component species to the underlying features of the environment. A simple example is presented in Figure 22.20 representing a transect up a mountain in an area with four plant species present. The distribution of the four plant species is presented in two ways. In one view, the species distribution is plotted as a function of altitude or elevation. Note that the four plants exhibit a continuum of species regularly replacing each other in a sequence of A, B, C, and D with increasing altitude. This view is very similar to that presented for the individualistic view of communities. The second view of species distribution is a function of distance along the altitudinal gradient (mountainside). As you move up the mountainside, the distributions of the four species are not continuous. As a result, a number of species associations might be recognized as you walk along the transect. These associations are identified by different symbols representing the combination of species.

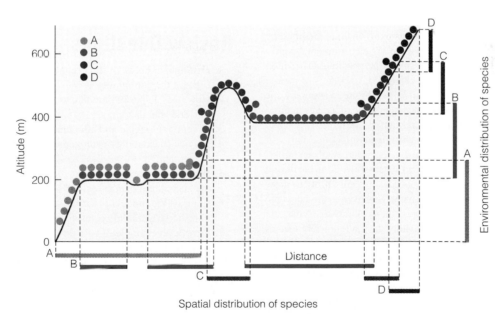

FIGURE 22.20 Patterns of co-occurrence for four species on a landscape along a gradient of altitude. The environmental distribution of the four species is presented in two ways: (1) the spatial distribution of the species along a transect of the mountainside, and (2) as a function of their response to altitude. Note that the species' responses to the altitudinal gradient are continuous, but their spatial distributions along the transect are discontinuous. The patterns of species composition (and community composition) along the mountain gradient are a result of the spatial pattern of environmental conditions (altitude) and the individual responses of the species. The responses of the species to the environmental gradient are consistent with the individualistic or continuum view of communities proposed by Gleason (see Figure 22.1b). However, consistent patterns of species co-occurrence across the landscape are a function of the spatial distribution of environmental conditions. Repeatable patterns of species co-occurrence in similar habitats is consistent with the idea of plant associations as supported by Clements (see Figure 21.1a).

These communities composed of coexisting species are a consequence of the spatial pattern of the landscape.

The two views are quite different yet consistent. Each species has a continuous response along an environmental gradient, elevation. Yet it is the spatial distribution of that environmental variable across the landscape that determines the overlapping patterns of distribution—the composition of the community.

The simple example presented in Figure 22.20 examines only one feature of the environment—elevation. Yet the structure of communities is the product of a complex interaction of pattern and process. Species respond to a wide array of environmental factors that vary spatially and temporally across the landscape, and the interactions among organisms influence the nature of those responses.

Summary

1. Historically, there have been two contrasting concepts of the community. The organismal concept views the community as a unit, the association, in which each species is a component of the integrated whole. The individualistic concept views the relationship among coexisting species as a result of similarities in environmental requirements and tolerances.

2. A simple framework for examining community dynamics (patterns of zonation and succession) can be constructed based on the following four premises: (a) The fundamental niche of a species acts as a primary constraint on its distribution and abundance; (b) species vary in their fundamental niches (environmental tolerances); (c) environmental conditions change in time and space; and (d) the fundamental niche is modified by species interactions (realized niche). As environmental conditions change in both time and space, the possible distribution and abundance of species will change.

3. Changes in environmental conditions brought about by the organisms themselves are termed autogenic. Changes in environmental conditions that are purely abiotic are termed allogenic.

4. Succession can result from the changing competitive ability of species under autogenically changing environmental conditions. Shade-intolerant plant species can successfully compete during the early stages of succession because of their ability to grow quickly in the high-light environment. However, as the autogenic changes in the light environment occur, the ability to tolerate and grow under shaded conditions enables the shade-tolerant species to successfully compete during the later stages of succession.

5. A number of theoretical models have been proposed to explain community dynamics. These models focus on how the physiological, morphological, and life history traits of individual species influence species interactions and, ultimately, species distribution and abundance under changing environmental conditions. Although all three models examined in the chapter are based on patterns of plant adaptation to the environment as a common mechanism of community dynamics, the models differ significantly in their interpretations of specific plant adaptations to the environment, competition, and the nature of autogenic environmental change through succession.

6. Fluctuations in the environment that occur repeatedly over the lifetime of an organism are unlikely to influence patterns of succession among species with that general life span. Purely abiotic environmental changes (allogenic environmental change) that occur over time scales greater than the longevity of the dominant organisms can produce patterns of succession over time scales ranging from days to millennia or longer. Allogenic variations in environmental conditions across the landscape are a major influence on patterns of zonation.

7. Species diversity typically increases during the early stages of succession. The pattern of diversity as succession (time) progresses is a function of the rates of species colonization and replacement. The rate at which these processes proceed, and subsequently the patterns of diversity, will be influenced by both the growth rates of the species involved and the frequency of disturbance.

8. Herbivores can have both a direct and an indirect effect on community dynamics. By selecting certain species over others, herbivores can directly influence patterns of species composition in both time (succession) and space (zonation). Herbivores can also influence vegetation dynamics indirectly, by changing environmental conditions, such as the vertical light profile in a forest through defoliation of the canopy.

9. A variety of theoretical models have been proposed to explain community dynamics. Although these models disagree on many points, they generally agree on the importance of trade-offs in the characteristics that determine the response of organisms to a variety of environmental factors, and the importance of autogenic environmental change in the process of succession.

10. The community is a spatial concept; the individualistic continuum is a population concept. Each species has a continuous response along an environmental gradient. The pattern of spatial variation in the physical environment across the landscape interacts with species responses to determine distribution and abundance.

Review Questions

1. Distinguish between the organismal and individualistic concepts of the community.
2. How does a species' fundamental niche constrain its distribution and abundance?
3. Contrast autogenic and allogenic environmental change.
4. Provide an example of autogenic environmental change during succession.
5. Why do shade-tolerant plant species tend to dominate during the later stages of succession?
6. Contrast the models of Grime and Tilman regarding changes in resource availability during succession.
7. Provide an example of both direct and indirect impacts of herbivores on patterns of succession.
8. Reconcile the organismal concept with the individualistic concept of community.

Cross References

Interspecific competition 243; mutualism 321; realized niche 257; shade-tolerant/shade-intolerant 92–93; plant adaptations to nutrient availability 112–117; plant adaptation to water availability 104–107; succession 405; zonation 404; vertical structure of community 384; light attenuation 44–48; nitrogen fixation 517; *Rhizobium* bacteria 517; mycorrhizal fungi 322; herbivore-plant interactions 291; species diversity 388; Grime model of plant strategies 235; adaptation 81–83.

C H A P T E R 2 3

Landscape Ecology

CONCEPTS

1. The study of the causes and consequences of spatial pattern in the landscape is landscape ecology.

2. Patterned landscape consists of a mosaic of patches of different origins embedded in a matrix.

3. Patches and corridors influence physical flows and organismal movements across a landscape.

4. Patchy environment relates peripherally to island biogeography theory that states species equilibrium, influenced by island size, is attained when immigration rates balance extinction.

5. Fragmentation of the environment results in habitat patches of various sizes.

6. A major source of landscape heterogeneity is disturbance.

7. Disturbances vary in intensity, frequency, and spatial scale that in turn influence landscape heterogeneity.

Consider the view of the Virginia countryside in Figure 23.1. It is a quilt-work of forest, fields, golf course, hedgerows, pine plantation, pond, and human habitations. This quilt-work of different types of land cover is called a **landscape mosaic**—clusters of local communities repeated in a similar manner over a wide area (Turner 1999). In a variety of ways, each element or patch in this mosaic relates to and interacts with the others. The study of the causes and consequences of spatial patterns on the landscape, as shown in Figure 23.1, is the realm of **landscape ecology** (Turner 1989, 1998).

LANDSCAPE: PATTERNS AND PATCHES

The landscape can be viewed as a heterogeneous area composed of a variety of different communities. The distinct communities making up this mosaic are called landscape elements or **patches.** Patches are relatively homogeneous areas that differ from their surroundings: the pond, fields, or pine plantations in Figure 23.1. They vary in size, shape, and type and are embedded in a **matrix**—surrounding areas that differ in species structure or composition (Forman and Gordon 1986, Turner 1998).

The basic landscape patterns you view in Figure 23.1 are products of a wide variety of environmental factors operating at many levels. The broad-scale distribution of natural communities on the landscape is dictated by the regional patterns of geology, topography, soils, and climate. This particular landscape lies on the flat but ascending land of the Piedmont, where the coastal plain meets the Blue Ridge Mountains to the west. The vegetation growing there is strongly influenced by changing climatic conditions as influenced by topography. The Piedmont, wetter than the east-facing slopes of the Blue Ridge, supports mesic forests of red and white oak, tulip poplar, sweetgum, and red maple. As you advance up the slopes of the Blue Ridge, the geology, soils, and microclimate change. The drier slopes of the Blue Ridge support a more xeric type of vegetation: chestnut oak (*Quercus prinus*), sassafras (*Sassafras albidum*), and table mountain pine (*Pinus pungens*). Variations in the geomorphology and microclimates of the Blue Ridge create a rich mosaic within the expanse of the forest seen in the distance of Figure 23.1.

Upon this stage of natural communities on the landscape, human activities and disturbances at various scales have and continue to make their mark on the land, in one way diversifying the landscape pattern and in another way diminishing it. These activities transform one community into another—for example, the clearing of forest to create a golf course or home site. Many of the landscape patterns we observe today reflect the early land survey methods developed in the United

FIGURE 23.1   A view of a Virginia landscape showing a mosaic of vegetation patches.

States, which were based on straight lines and rectangles. Historically, American land surveys were set on straight lines east to west (consider the Mason-Dixon line or many state and county boundaries), paying no attention to topography or natural regions. Early surveys divided the land into sections, half sections, and finally quarter sections of 160 acres (Johnson 1976). The straight-line survey is reflected in the square corners where woods and fields, croplands and developments, and other landscape elements meet (note the square corners of the fields and the pine plantation in Figure 23.1). This straight-line checkerboard pattern, often overlooked by the casual observer, has a lasting impact on the landscape.

Let us turn our attention again to the landscape in Figure 23.1. Within this landscape mosaic, the various patches have their own unique origins. The small separate woodlots may represent areas or the original forest cover that were not cleared. Such areas, referred to as **remnant patches,** are surrounded by areas that have been modified to provide for different types of land use. These altered areas, which make up most of the foreground valley, are referred to as **introduced patches.** These consist of the maintenance-requiring golf course in the lower right-hand corner, the pine plantation and pasture field in the middle foreground, and human habitation. Introduced patches involve the elimination of natural ecosystems, the introduction of exotic species from plants to pets, and maintenance activities, such as mowing.

For the most part, human activities dictate the size and shape of landscape patches, as evident in Figure 23.1. However, natural variations in the landscape features such as soil conditions and topography, and disturbance events such as fire and grazing by herbivores, constrain and interact with human activities to determine the size and shape of many patches. Patches may be square, elongate, round, or convoluted, large—covering many hectares—or small. The size, area, shape, and orientation of landscape patches have an important influence on many physical and ecological processes that directly influence their structure, such as the flow of wind, the dispersal of seeds, and the movement of animals, and on their suitability as habitats for plants and animal. All of these are topics that we will examine later.

EDGES

Edges are one of the most conspicuous features of the landscape. They mark the boundaries between a housing development and the forest from which it was carved, between a roadside and the adjacent woods, or between a pond and a field. The place where the edge of one patch meets the edge of another is called a **border** (Figure 23.2). The two edges and the border combined make up the **boundary** (Forman 1995). Some boundaries between landscape patches are abrupt with a sharp contrast between the adjoining patches, such as between a forest and an adjacent agricultural field. Others are less distinct, with low contrast between adjoining edges, such as between two forest communities. In this case, the vegetation of one patch blends with the other to form a

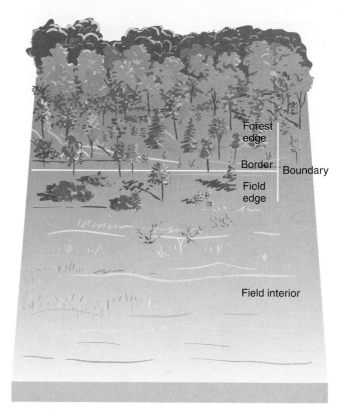

FIGURE 23.2 Spatial relationships of border, boundary, edge, and interior in a landscape. (Based on Forman 1995.)

sort of transition zone called an **ecotone** (Figure 23.3). Plant and animal species adapted to environmental conditions existing in the edges advance as far into either community as their abilities will allow. Thus in the ecotone, species common to each community mingle with species common to the edge, often resulting in a highly diverse and unique community within these boundary environments.

Edge Development

Some edges result from abrupt changes in soil type, topography, geomorphic features (such as rock outcrops), and microclimate. Under such conditions, long-term natural features of the physical environment determine adjoining vegetation types. Such edges, referred to as **inherent,** are usually stable and permanent (Thomas et al. 1979) (Figure 23.4). Other edges result from such natural disturbances as fire, storms, and floods or from such human-induced disturbances as livestock grazing, timber harvesting, agriculture, and suburban development. Such edges, maintained only by periodic disturbances, are called **induced** (Figure 23.4). Unless maintained, these disturbed areas will tend to revert to their original state, such as with the process of secondary succession on abandoned agricultural lands (see Chapter 21 for examples), and the edge will disappear with time.

Edges, like all communities, are dynamic and change in space and time (see Chapters 21 and 22). Figure 23.5 shows typical changes in the boundary between two communities (landscape patches) as time progresses. In the initial creation of an induced edge, such as by the clearing of a forested area

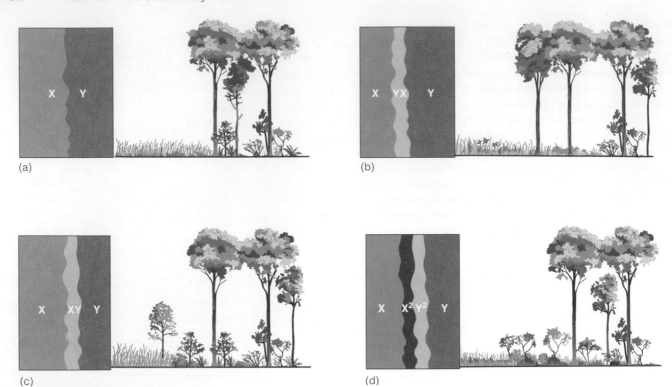

FIGURE 23.3 Types of ecotones. (a) Abrupt, narrow edge with no development of an ecotone. (b) Narrow ecotone developed by advancement of community Y into community X to produce ecotone YX. (c) Community X advances into community Y to produce ecotone XY. (d) Ideal ecotone development in which plants from both communities invade each other to create a wide ecotone X^2Y^2. This type of ecotone will support the most edge species.

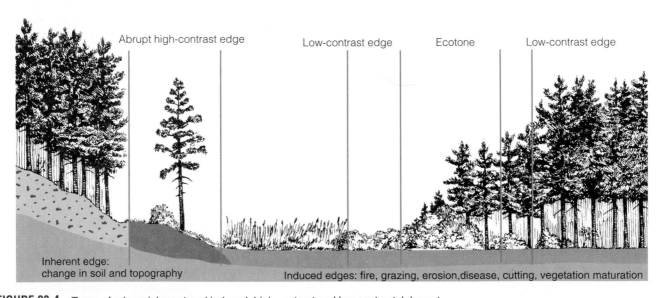

FIGURE 23.4 Types of edges: inherent and induced, high contrast and low contrast. Inherent edges are mostly abrupt. Edges of high contrast exist between widely different adjacent communities, such as shrub and mature forest. Edges of low contrast involve two closely related successional stages, such as shrubs and sapling growth. (After Thomas et al. 1979.)

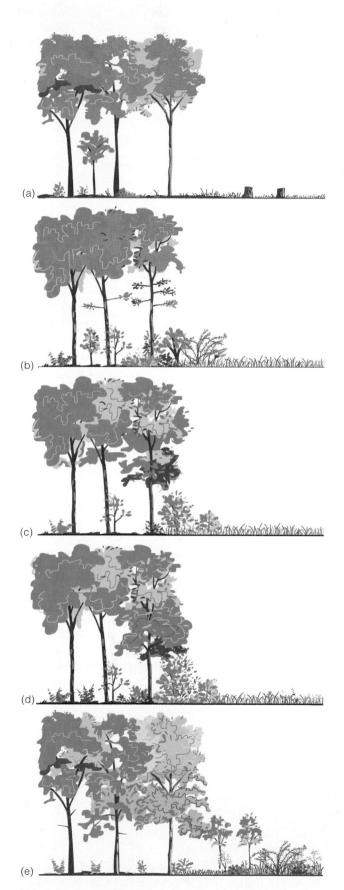

FIGURE 23.5 Stages in edge development. See text for details. (After Ranney et al. 1981:83.)

for agriculture, there is an increase in light penetration into the adjacent woodland (Figure 23.5a). This increased penetration of solar radiation into the adjacent forest will raise air temperatures, increase evaporation, and result in the development of xeric conditions. In time, vegetation responds to changed conditions (Figure 23.5b). Some trees along the edge exposed to increased light expand their crowns; others species develop new branches, called epicormic, along the trunk. These new epicormic branches extend the canopy downward toward the ground. Some shade-tolerant understory trees, unable to withstand the shock of sudden increased light and heat, may die. High light intensity and xeric conditions, however, favor highly competitive, shade-intolerant plant species, such as blackberries and sumac. Eventually, a dense understory of shade-intolerant woody seedlings and herbaceous plants develops. Because of open conditions, edge vegetation invades the woodland.

In time, increased expansion of tree crowns, increased growth of epicormic branches, and growth of edge understory plants close the gap between crown and edge vegetation and reduce the penetration of solar radiation into the adjacent forest (Figure 23.5c). Some species disappear, and edge growth into the woods declines. At this point, edge vegetation experiences increased competition for light (Figure 23.5d). Some species survive while others disappear, depending on their competitive ability. The number of dominant and codominant species in the edge declines, and such edge-oriented species as hawthorn, hickory, aspen, and oak replace shrubs such as blackberry (Figure 23.5e). At this stage, a few large trees with branching close to the ground dominate the edge. Only minimal edge understory remains (Figure 23.5f). When conditions permit, woody species may invade the adjacent field, developing an ecotone (Figure 23.5g). Eventually, the edge becomes a mixture of shade-tolerant and shade-intolerant species.

This successional process that occurs in edge communities arises because environmental conditions in the newly formed edge are different from those of the adjacent vegetation communities, especially in the case of forests (Ranney 1977). Environmentally, such edges reflect steep gradients of wind flow, moisture, temperature, and solar radiation between extremes of open land and forest interior. Wind velocity is greater at the forest's edge than within the forest, creating higher rates of evaporation and xeric conditions in and around the edge. With increased temperatures transpiration increases, placing greater demands on soil moisture by plants. Because changes in the penetration of solar radiation are influenced by aspect (position relative to the direction of incoming solar radiation), north-facing and south-facing edges will differ in environmental conditions (Figure 23.6). In the Northern Hemisphere, a south-facing edge may receive three to ten times more hours of sunshine a month during midsummer than a north-facing edge, making it much warmer and drier. Although the depth to which sunlight penetrates the vertical edge of the forest depends on a variety of factors, including solar angle, edge aspect, density and height of vegetation, latitude, season, and time of day, in general the

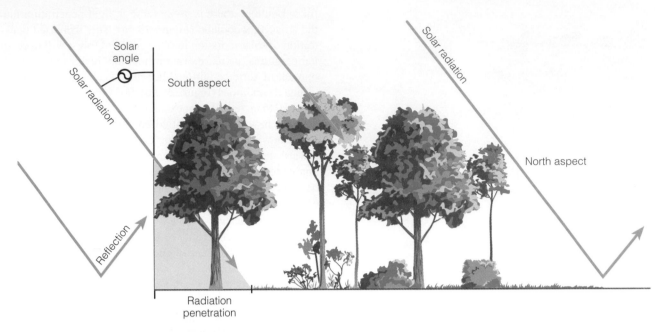

FIGURE 23.6   Influence of solar radiation on the edge between a field and a forest. Solar radiation does not affect all edge aspects equally. North-facing edges receive almost no direct sunlight and limited reflection from nearby fields, whereas south-facing edges receive both. The depth to which solar radiation penetrates forest edge on level ground depends on the height of the canopy and solar angle. Within the zone of radiation penetration, light intensifies and summer daytime temperatures are higher than in forest interiors sheltered by tree canopies. (After Ranney 1977.)

edge effect extends about 50 m into the forest (Wales 1972, Ambuel and Temple 1983, Blake and Karr 1984, Freemark and Merriam 1986).

Edge Habitat

Because the edge environment is often different in both structure and composition from the adjacent communities, it represents a unique environment, as well as a location with access to a variety of different communities. Animals characteristic of the edge are often those that require two or more vegetation communities. For example, one edge species in eastern North America, the ruffed grouse (*Bonasa umbellus*), requires forest openings with an abundance of herbaceous plants and low shrubs, dense sapling stands to shelter drumming or display logs, small timber for nesting cover, and mature forests for winter food and cover. Because the ruffed grouse spends its entire life in an area of 5 to 10 ha, all these different types of vegetation structure must be found within a small area. Some species, such as the indigo bunting (*Passerina cyanea*), are restricted exclusively to edge environments (Figure 23.7). Because of diversity of environmental conditions within the boundary between communities, the variety and density of life are often greatest in and about edges and ecotones. This phenomenon has been called the **edge effect** (Leopold 1933).

The edge effect is influenced by the amount of edge available—its length, width, and degree of contrast between adjoining vegetation communities (Patton 1975, Sisk et al.

1997). Width can be measured in number of ways, but it is best defined as distance between the border and the point where physical conditions (such as microclimate) and vegetation do not differ significantly from those in the interior of the patch. The greater the contrast between adjoining plant communities, the greater species richness should be. An edge between a forest and grassland in general supports more species than an edge between a young and a mature forest (McElveen 1977, Harris and McElveen 1981, Harris 1989). The larger the adjoining communities, the more opportunity exists for plants and animals of each of those communities to occupy the area along with the species of the edge. If patches of vegetation are too small to support species requiring larger areas of habitat, then the area will be dominated by edge species.

Edge Interactions

Edge effect may increase species diversity, but it can also create ecological problems (Temple 1986, Reese and Ratti 1988, Robinson 1988, Yahner 1988). Edges, especially abrupt ones, attract mammalian and avian predators. Raccoons, opossums, and foxes use edges as corridors or travel lanes, and avian predators, especially crows and jays, can easily locate birds nesting within the habitat block.

Edges further alter species interactions by differentially restricting or facilitating the movements of animals and the dispersal of plants across a landscape (Forman 1995, Sisk et al. 1997, Fahrig 1994, Farina 1998). Thorny plants and dense

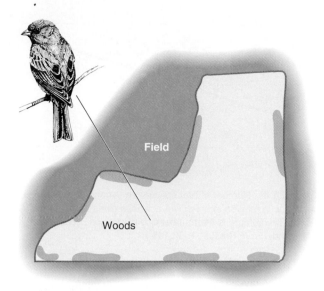

FIGURE 23.7 Map of territories of a true edge species, the indigo bunting (*Passerina cyanea*), which inhabits woodland edges, larger gaps in the forest creating edge conditions, hedgerows, and roadside thickets. The male requires tall, open song perches and the female a dense thicket in which to build a nest. (After Whitcomb et al. 1981:143.)

low growth at the boundary of forest and field can inhibit the movement of animals into or out of the forest patches. Human alteration of the landscape for urban development and agriculture causes fragmentation and presents barriers to dispersal for plants and animals; furthermore, these barriers are terrestrial habitats with their own sets of species, including domestic animals such as cats, dogs, sheep, and cattle and species of wildlife highly adaptable to human habitations. Some species, like cats, weasels, raccoons, and crows, invade the edges and move into the interior, increasing predatory pressure on animal species inhabiting the forest interior (Gates and Gysel 1978, Chusko and Gates 1982, Wilcove 1985).

Aldo Leopold in his classic text *Game Management* (1933) formalized the concept of the edge effect, as the *law of the edge.* He stated that the potential abundance of wildlife species with small home ranges that require two or more vegetation types is roughly proportional to the sum of the edges. Thus the density of an edge species varies as a constant proportion of the edge. The amount of edge depends on the heterogencity of the landscape.

Leopold applied the edge principle only to resident, edge-obligate species with restricted home ranges, not to mobile species. However, wildlife biologists and ecologists recognized that animal and plant life, in general, were more abundant about the edges than in the interior of forests and grassland. Soon they were extending the idea beyond Leopold's original concept. The way to increase wildlife diversity was to increase edge habitat became a tenant of ecology and wildlife and forest management. Application of this concept involved creating openings in the forest and breaking up large open fields with plantings of shrubs and trees, all designed to increase the edge effect. Carried to an extreme,

this management for diversity results in the breaking up of large homogeneous areas of vegetation into smaller and smaller patches or fragments that eliminate the presence of wildlife requiring larger areas or habitats provided by the forest interior. At this point further fragmentation and creation of edge reduces heterogeneity, with the landscape itself becoming a homogeneous community dominated by edge and early and mid-successional species (Guthery and Bingham 1992).

CORRIDORS

Corridors are strips of vegetation linking one patch with another on the landscape. The vegetation of the corridor is similar to the patches it connects but different from the surrounding landscape in which they are set. (For examples, note the various types of corridors in Figure 23.1.) Usually, corridors originate from human disturbance or development and are remnants of largely undisturbed land between agricultural fields or developments. Some may be narrow-line corridors, such as lines of trees planted as windbreaks, hedgerows, roads and roadside strips, and drainage ditches (Forman and Godron 1981, 1986, Forman 1995). Wider bands of vegetation are called strip corridors, consisting of both interior and edge environments. Such corridors may be wide strips of woodlands or power line rights-of-way and belts of vegetation along streams and rivers.

Corridors have two functional roles in the landscape; they provide an often unique habitat for a variety of plant and animal species, and they function as biological corridors—travel ways between habitat patches (Rosenberg et al. 1997). Many corridors along streams and rivers provide important riparian habitat for animal life. In suburban and urban settings, corridors provide habitat for edge species and act as stopover habitat for migrating birds. In Europe the long history of hedgerows in the landscape has allowed the development of typical hedgerow animal and plant communities (Pollard et al. 1974). Corridors also provide protective cover for prey or scouting positions for predators, allowing them to remain concealed while hunting in adjacent vegetation patches.

Corridors also act as conduits, providing dispersal routes and travel lanes for species between habitat patches, especially when corridors interconnect to form networks. They probably function best as travel lanes for individuals moving within the bounds of their home range. By facilitating the movement of individuals among different patches, corridors can facilitate gene flow between subpopulations occupying those patches, or the reestablishment of species in habitats that have experienced local extinctions. Corridors can also act as filters, providing dispersal routes for some species but not others. Various sized gaps in corridors allow certain organisms to cross and restrict others—the **filter effect.** Corridors can also have a negative impact on some populations by creating avenues for the spread of disease between patches or allowing the movement of predators.

Roads, corridors designed as dispersal routes for humans, dissect the landscape and have a variety of impacts on the patches they adjoin (Forman and Alexander 1998).

Two- to four-lane high-speed roads are a major source of mortality for wildlife ranging in size from large mammals to insects, and they effectively divide populations of many species. All types of roads alter or in some way affect roadside vegetation. Salt spread on highways to remove ice and snow kills adjacent salt-intolerant plants. Particulate matter from tires and diesel exhaust, and chemical pollutants, along with salt, alter roadside vegetation. Dust from dirt and gravel roads and litter drift over roadsides. Water runoff during storms and snowmelt carries pollutants and debris into adjacent patches. Noise from passing traffic discourages wildlife from occupying otherwise suitable habitat. Most important perhaps, road corridors allow people to access remote areas with often disastrous ecological effects, as exemplified by logging roads cut in tropical forest. Where roads invade, people and development follow.

The effectiveness of biological corridors as a means of stimulating immigration between habitat patches has never been explicitly demonstrated. Beyond general observations of corridor use, there is little experimental evidence of the role of corridors in species dispersal. The use of corridors relates to the nature of the landscape mosaic in which it is set and to the probability of the disperser finding the corridor, using it, and successfully traversing it. Many dispersing individuals may ignore the corridor and move directly across the landscape (Rosenberg et al. 1977). Nevertheless, the use or nonuse of corridors as avenues of dispersal in no way negates their value as habitat.

ISLAND BIOGEOGRAPHY THEORY

The various patches, large and small, that form the vegetation patterns across the landscape suggest islands of different sizes. Some are near to each other; others are remote and isolated. A patch of forest, for example, may sit within a sea of cropland or housing developments, isolated from other forest patches on the landscape. The size of these patches and their distances from each other on the landscape have a pronounced influence on the nature and diversity of life they hold. The influence of area on species richness did not escape the notice of early naturalist-explorers and biogeographers, who noted that large islands hold more species than small islands. The zoogeographer P. Darlington (1957) later suggested a rule of thumb: a tenfold increase in area leads to a doubling of the number of species. In 1962 F. W. Preston formalized the relationship between the area of an island and the number of species present. When the two values are plotted as logarithms, the number of species varies linearly with island size (Figure 23.8):

$$\log S = \log c + z \log A$$

or

$$S = cA^z$$

Here S = the number of species, A = area of the island, c = a constant measuring the number of species per unit area, and z = a constant measuring the slope of the line relat-

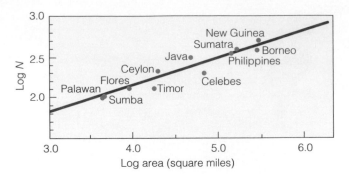

FIGURE 23.8   Number of bird species on various islands of the East Indies in relation to area. The abscissa gives areas of the islands. The ordinate is the number of bird species breeding on each island. (From Preston 1962:195.)

ing S and A. The slope, z, is a measure of the change in species richness per unit area. The steeper the slope of the line, the larger the increases in species richness per unit increase in island size.

MacArthur and Wilson (1963) presented this relationship between area and species richness for island environments as a theory of island biogeography. Briefly, it states that the number of species of a given taxon established on an island represents a dynamic equilibrium between the rate of immigration of new colonizing species and the rate of extinction of previously established ones. The process by which this dynamic equilibrium is achieved is outlined in Figure 23.9.

Consider an island empty of life some distance from a mainland. The mainland is a source of potential colonists for the island. Not all mainland species are equal in their ability to disperse to the island. The better dispersers rapidly occupy the island first, followed eventually by poorer dispersers. As the number of resident species increases, the

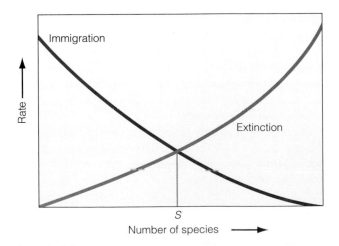

FIGURE 23.9   Equilibrium model of species on a single island. The point at which the curve for rate of immigration intersects the curve for rate of extinction determines the equilibrium number of species in a given taxon on the island. S represents the equilibrium number of species.

number of new immigrant species arriving on the island decreases because fewer new potential immigrants are available from the source pool.

At the same time, some island inhabitants go extinct. As the number of species residing on the island increases, the extinction rate increases. This increased extinction rate results from three factors: (1) the increase in species richness means more possible species to go extinct whatever the cause may be; (2) as species richness increases, the potential for competitive interactions increases; and (3) given a finite resource base, as the number of species on the island increases, the population size of each must decline. The equilibrium species richness is achieved when the rates of immigration and extinction are the same—at the value of species richness in Figure 23.9 where the lines representing the rates of immigration and extinction cross. This value of species richness is an equilibrium point. If species richness increases above this value, the rate of extinction exceeds the rate of immigration and species richness declines. If the number of species on the island falls below this equilibrium, immigration exceeds extinction and species richness increases. At equilibrium between immigration and extinction the number of species remains stable, although the composition of species may change. In this case, the rate at which one species is lost and a replacement gained is called the **turnover rate.**

Both the size and proximity of an island to the mainland strongly influence the rates of immigration, extinction, and turnover and consequently the predicted equilibrium species richness (Figure 23.10). Extinction rates are higher on smaller islands than on larger ones, because large islands can hold bigger populations of any particular species. Immigration rates decrease with increasing distance from the mainland; thus, islands closer to the mainland have a higher immigration rates than distant islands. Considering both immigration and extinction rates, we can hypothesize that near islands

reach equilibrium with more species than distant islands, and small islands reach equilibrium with fewer species than large islands. Interactions of size and distance from the mainland and equilibrium species richness are shown in Figure 23.10. Turnover rates are greater for near islands than for distant islands of the same size, because a source of replacement immigrants is closer. However, turnover rates for near, small islands are greater than for near, large islands because extinction rates are greatest on small islands.

In an attempt to test the predictions of the island biogeography model, Simberloff and Wilson undertook a number of experimental studies on small mangrove islands in the Florida Keys (Simberloff and Wilson 1969, 1970, Wilson and Simberloff 1969). After surveying the arthropod populations on selected islands, several of the islands were defaunated by enclosing them in tents and fumigating with methyl bromide, killing all arthropods. Patterns of recolonization of the islands by arthropod populations tended to verify the MacArthur-Wilson dynamic equilibrium model. The islands richest and poorest in species prior to defaunation were also the richest and poorest after defaunation and subsequent colonization. Islands with the greatest number of species were those closest to the main stands of mangroves. However, species found frequently were not the same as those present before the experiment, and species turnover was high.

Similar experiments were undertaken in small islets of salt marsh at Oyster Bay, Florida (Ray 1981, Strong and Ray 1982). In these experiments, recolonization and extinction rates of arthropod populations corresponded to the general predictions of the island biogeography model, with immigration rates negatively correlated and extinction rates positively correlated with species richness. However, high variability in the observed rates of both immigration and extinction suggests that these demographic processes are influenced by a variety of life history and trophic characteristics of the species, as well as specific interactions among resident species.

Equilibrium species richness and turnover rates on already occupied islands can be assessed only by long-term annual censuses on each island. An example is the seasonal census of confirmed nesting birds over 26 years in the 16 ha Eastern Wood of Bookham Common, an isolated woodland near Surrey, England. Over the 26-year period, 44 species of birds appeared in the woods. Of these, 6 apparently did not nest, leaving 38 breeding species. Of the species that nested on the island, 4 had territories extending beyond the woods, and 9 species never had more than two pairs nesting in the woods. Eleven more species nested in fewer than 5 years during the 26-year census period. They had to be considered casual species, counted as immigrants and extinctions in calculating annual turnover. Only 14 species were continuous residents. During the 26 years, Eastern Wood experienced a considerable turnover of species with an average of three immigrations and three extinctions annually. Equilibrium species richness, as determined by graphing immigration and extinction rates (Figure 23.11), is predicted as 32 species, somewhat higher than the 27 species that Eastern Woods supported on average over the census years (Williamson 1981).

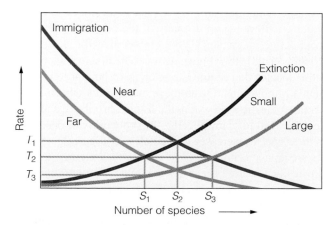

FIGURE 23.10 Graphical representation of the island biogeography theory, involving both distance and area. Equilibrium species densities are labeled by corresponding value of S. Immigration rates decrease with increasing distance from a source area. Thus distant islands attain species equilibrium with fewer species than near islands, all else being equal ($S_3 > S_2$ for large islands; $S_2 > S_1$ for small islands). Extinction rates increase as the size of the island becomes smaller.

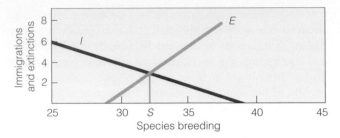

FIGURE 23.11 Immigration and extinction curves for Eastern Wood, Bookham Common, Surrey, England, a 16 ha oak woodland. The immigration curve is a regression line (the points have been omitted) that cuts the abscissa at 39 species. The maximum number of species that bred at one time or another in Eastern Wood was actually 44. The extinction line is at 45°, indicating one extinction for every species present over 29.

The major limitation of the equilibrium theory of island biogeography is that it examines species richness only. The model makes no assumptions about species composition; it treats all species in a taxon as equivalents. It does not address limitations related to the life history or habitat requirements of the species involved and assumes that the probabilities of extinctions and immigrations are the same for all species. Degree of isolation is relative, depending on the taxon involved. What is a short distance for a bird may be an insurmountable distance for a mammal or a lizard. In addition, the model's assumption that extinctions relate to an island's area and distance does not consider the fact that immigrations and extinction may not be independent. Extinction of a dwindling population may be slowed or even halted by an influx of immigrants, the **rescue effect** (Brown and Kodrich-Brown 1977).

An alternative approach to island biography is the habitat diversity theory. This theory suggests that it is the diversity of habitats that supports species richness, not area per se. Larger islands may have lower extinction rates and support more species than smaller islands because they have more diverse habitats. There is considerable evidence that habitat heterogeneity can override the influence of island size, with smaller islands having high habitat heterogeneity supporting a greater species richness than larger, more homogeneous islands (Rigby and Lawton 1981, Blair and Karr 1984, Freemark and Merriam 1986).

Although island theory was applied initially to oceanic islands, there are many other types of islands. Mountaintops (Brown 1971, 1978), bogs, ponds, dunes, areas fragmented by human land use, host plants and their insects (Janzen 1968, Strong et al. 1984), and animals and their parasites—all are essentially island habitats. As Simberloff (1974:162) put it: "Any patch of habitat isolated from similar habitat by different, relatively inhospitable terrain traversed only with difficulty by organisms of the habitat patch may be considered an island." Exactly how can island theory relate to landscape patches?

Landscape patches or islands differ considerably from the oceanic islands on which the theory was developed (Wilcox 1980). Oceanic islands are terrestrial environments surrounded by an aquatic barrier to dispersal. They are inhabited by organisms of various taxa that arrived there by chance dispersal over a long period of time or represent remnant populations that existed on the area long before isolation. By contrast, the organisms that inhabit landscape patches are samples of populations that extend over a much larger area. These samples (patch communities) contain fewer species than are found over the larger area and more species represented by only a few individuals. Unlike oceanic islands, terrestrial landscape patches are in one way or another interconnected by other terrestrial environments, which often present less of a barrier to movement and dispersal among patches.

PATCH DYNAMICS

The mosaic of patches that define the landscape is constantly changing. Natural processes such as secondary succession can expand or contract vegetation patches, or further fragmentation can result from human disturbances.

Habitat Fragmentation

Fragmentation in large tracts of forest and grassland may begin intrusively with human creation of openings or clearings within the otherwise contiguous vegetation cover, such as by clear-cuts, wildlife food plots, and housing development (Harris and Silva-Lopez 1992). Although considerable forest or grassland remains, the structural integrity of the habitat declines. If continued, localized clearing eventually results in a fragmented forest, grassland, or shrubland, creating a mosaic of unconnected patches (Figure 23.12). These patches or fragments are separated by other types of habitats, particularly urban and suburban developments and agricultural lands (Figure 23.13) and now function as habitat islands.

A scale-dependent process, fragmentation up to a point results in no loss of species. As the process continues, however, the remaining area is reduced to a critical size below which it will not support many of the original species, and local extinction occurs (Whitcomb et al. 1981, Karr, 1982, Robbins et al. 1989) (Figure 23.14). The first to disappear are those species that require large areas of habitat to maintain viable populations. Often these are species whose habitat begins some distance within the forest, shrubland, and grassland. Termed **interior species,** the probability of occurrence of these species increases with patch size (Figure 23.15). Others are area-sensitive species because they require large territories or foraging areas. As fragmentation continues and these species disappear, edge and **area-insensitive species,** ones at home in small to large units of habitat, increase in abundance. Although the number of species in a fragmented habitat may increase initially with the creation of edge environments, the number of species it contains eventually will decline as fragmentation continues.

What size of habitat patches are needed to maintain regional populations and satisfy habitat requirements of the species concerned? What size of patch maintains greatest species diversity? At what size of habitat patch do interior and

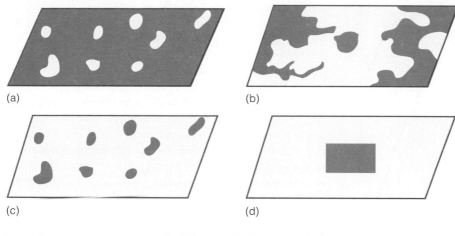

(a)

(b)

(c)

(d)

FIGURE 23.12 (a–c) How a forest becomes fragmented. (d) A larger forest tract becomes isolated when the surrounding land has been cleared, a situation common to parks and reserves. (After Harris and Silva-Lopez 1992.)

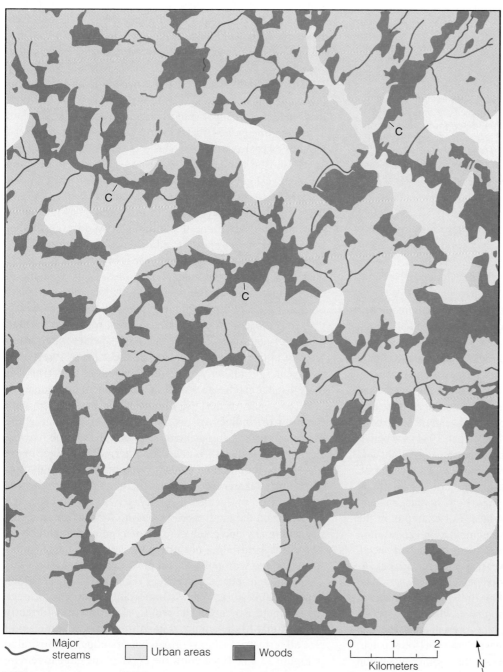

	Major streams		Urban areas		Woods
	Open water		Open fields + shrub-hedgerow		C = Corridor example

0   1   2
Kilometers

N

FIGURE 23.13 At the time of European settlement, this section of eastern Maryland was mostly forested. Over the course of several hundred years, much of the land was cleared for agriculture, fragmenting the remaining forest land into isolated patches. In the past quarter century, urban and suburban developments have encroached on the area, further fragmenting and isolating the forest, and doing the same to agricultural lands. Note the corridors that still remain, connecting various forest islands. (Adapted from Whitcomb et al. 1981:127.)

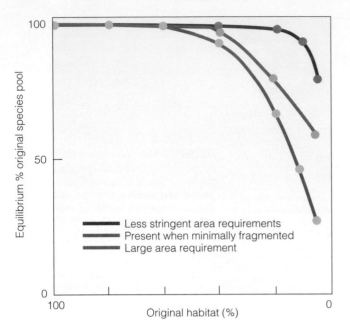

FIGURE 23.14   A model of the number of species remaining in each species pool as fragmentation proceeds. Blue dots and line show the pool of species with larger area requirements and low freedom or movement; brown dots and line show the species pool with less stringent area requirements. The green dots and line depict the proportion of the first species that would be present when the habitat is minimally fragmented. (From McLellen et al. 1986.)

achieved a maximum balance between edge and interior species. Beyond that size species richness declined, leveling off at 9.3 ha. The decline was associated with the loss of edge species. Similar patterns of species diversity have been observed in patches of prairie vegetation in North America (Simberloff and Gotelii 1984), and in patches of Yorkshire limestone pavements, chalk quarry reserves, and lowland heaths in Great Britain (Higgs and Usher 1980). Thus there appears to be a general pattern of maximum species diversity with patches of intermediate size. This pattern results from the negative correlation between edge species and the size of habitat patches combined with the positive correlation between interior species and increased area.

Several studies that have examined bird species diversity in forest patches reveal a pattern of increasing species richness with patch size, but only up to a point (Moore and Hooper 1975, Galli et al. 1976, Ambuel and Temple 1983, Blake and Karr 1984, Freemark and Merram 1986). Whitcomb and colleagues studied patterns of species diversity in forest patches in western New Jersey (Whitcomb et al. 1981, Lynch and Whitcomb 1977). Their findings suggested that maximum bird diversity is achieved with woodlands 24 ha in size. However, those woodlands held no true forest interior species, such as the ovenbird (*Seiurus aurocapillus*), that is highly sensitive to patch size and requires extensive areas of woods (Figure 23.16).

Both Black and Karr (1984) and Freemark and Merriam (1986) investigated the species composition of bird communities in large and small forest habitat islands (from 3 to 7620 ha) in agricultural regions in Illinois and Ontario, Canada, respectively. They found that although two or more smaller forest habitats supported more species, long-distance migrants and interior species, typical of larger tracts, were missing or poorly represented. Large forest tracts were important for increasing the number of forest interior neotropical migrants and certain resident species such as hairy woodpecker (*Picoides villosus*). Large forest tracts with a high degree of heterogeneity held the most species, supporting birds of both edge and interior. The presence of forest interior species in smaller woodlands depends on the nearness of those habitat patches to a pool of replacement individuals.

The results of these studies point out the problem with focusing on indices of diversity alone in evaluating the impacts of habitat fragmentation. Although maximum diversity is achieved with patches of intermediate size, many species that require larger areas are excluded. Although fragmentation of larger forest patches may not result in a significant decline in species diversity, it will eliminate many species from the landscape. Managing for maximum diversity within any one patch may not maximize the total diversity of the landscape, nor provide adequate habitat for all species found within the region prior to fragmentation.

Although species diversity is related to area, area alone is not the full story. What is important is the ratio of edge (or perimeter) to area. The length of perimeter is directly proportional to the square root of the area. At some small size all terrestrial islands or patches are edge, as illustrated in Figure 23.16a. If we allow the depth of the edge to remain

area-sensitive species disappear? Such questions have stimulated studies of the response of both plants and animals to habitat fragmentation and patch size (see Wilcove 1986, Lovejoy et al., 1986, McLellan 1986, Higgs 1981, Cutler 1991, Heikert 1994, Knick and Rotenberry 1995, Redpath 1995).

The minimum size of habitat needed to maintain interior species differs between plants and animals. For plants, patch size, per se, is not as important in species persistence and extinction as environmental conditions (Weaver and Kellman 1981). For many shade-tolerant plant species found in the forest interior, the minimum area depends on the patch size required to allow for moisture and light conditions typical of the interior. That area depends in part on the nature of the edge about the stand (Levenson 1981, Raney et al. 1981)—whether it is open or closed (reducing the penetration of light and wind)—on canopy closure, and on the ratio of edge to interior. If the stand is too small or too open, the interior environment becomes so xeric that mesic species, both herbaceous and woody, cannot survive and reproduce. For example, in the northeastern United States, forest fragmentation can result in the decline of mesic species such as sugar maple and beech, encouraging the growth of more xeric species such as oak.

In a study of forest patches (woodlots) in southeastern Wisconsin, Levenson (1981) found that the species richness of plants was the greatest in edge environments where edge species coexist with species characteristic of the forest interior. The total number of woody species increased with woodlot (forest patch) size up to approximately 2.3 ha. At that point, vegetation

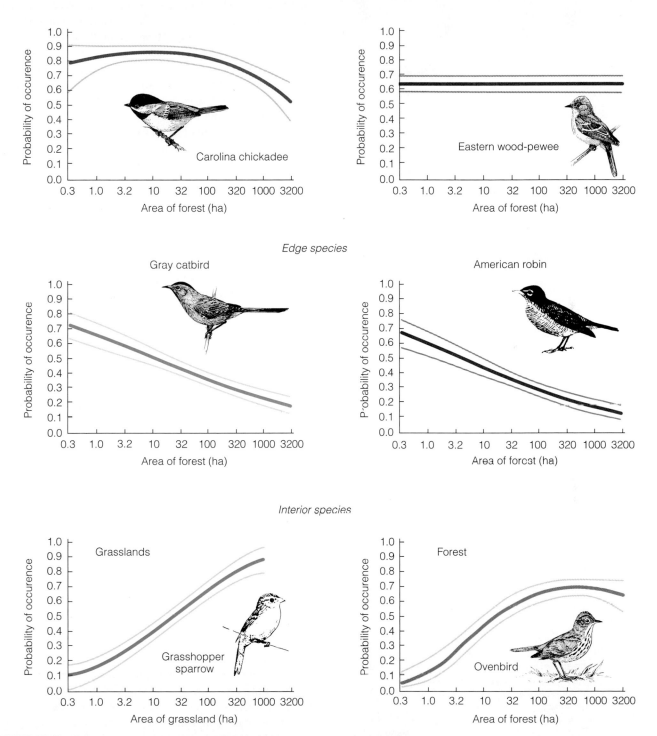

FIGURE 23.15 Differences in habitat responses among edge, area-insensitive, and area-sensitive for interior bird species. The graphs indicate the probability of detecting these species from a random point in forests and grasslands of various sizes. The dotted lines indicate 95 percent confidence intervals for the predicted probabilities. (Adapted from Robbins et al. 1989:18; grasshopper sparrow: Herkert 1994.)

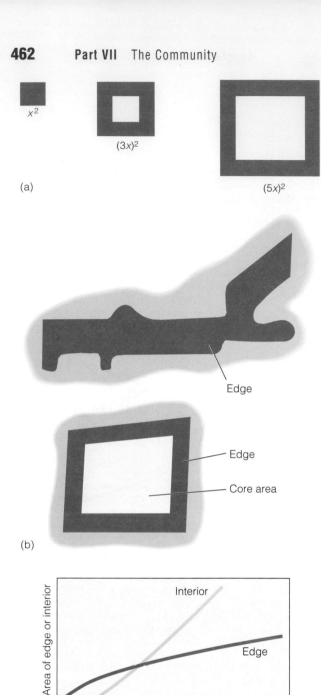

FIGURE 23.16 Relationship of island or fragment size to edge and interior conditions. (a) Functionally, all small islands of habitat are edge. Allowing the depth of edges to remain constant, the ratio of edge to interior decreases as island size increases. When the island size is large enough to maintain mesic and interior conditions, an interior begins to develop. (b) Size alone is not a determinant of interior-edge conditions. Shape of the island is also critical. A long or rectangular habitat patch is all edge. (c) A graph shows the relationship between edge and interior as island size increases. Below point *A*, the woodland is all edge. As its size increases, interior area increases and the ratio of edge to interior decreases. This relationship of size to edge holds for circular or square islands. Long narrow woodland islands as in the top part of (b), whose width does not exceed the depth of the edge, would be edge communities, even though their area might be the same as that of square or circular ones. (a and c after Levenson 1981:32; b after Temple 1986:304.)

constant while increasing area, the ratio of edge to interior decreases as the habitat island size increases. When the island size becomes large enough to maintain interior conditions, an interior begins to develop. However, size alone is not a primary determinant of edge-interior conditions. Configuration or shape of the island is also critical. In Figure 23.16b the rectangular wooded island contains 39 ha, yet because of its shape it is entirely edge; its width does not exceed the depth of its edge, or to put it differently it has a high ratio of edge perimeter to area. Out of its 16 species, none are interior. The square woodland of 47 ha has a core interior area of 20 ha, and 6 of its 16 species are interior species sensitive to fragmentation. This relationship between patch size and the relative abundance of edge and interior habitats is further illustrated in Figure 23.16c. For small patches (size *A* or less on the *x* axis), only edge habitats are found and the patch cannot support interior species. As patch size increases, the interior area increases and the ratio of edge to interior decreases. This relationship of area to edge holds for circular or square patches, but not for irregularly shaped or rectangular islands. Long narrow woodland islands as that in Figure 23.16c, whose width does not exceed the depth of the edge, would be edge communities, even though their area might be the same as that of square or circular ones.

The above examples show that the perennial argument over size of habitat patches and the maintenance of species diversity is not as simple as it appears (Schwartz 1999). Given that a number of relatively small patches will hold a higher species diversity in toto than large patches, an overemphasis on managing for species diversity can lead to a decline and ultimate extinction of interior and area-sensitive species that exist in large or more homogeneous habitats. We must consider both species diversity and species composition in the conservation and management of lands.

Patch Contraction and Expansion

Landscape patches may be further reduced by fragmentation pressures encroaching from one, several, or all sides. Increasing contraction has severe effects on the fragment by reducing size and increasing isolation and edge effects. Associated with such contraction are highways, roads, power line rights-of-way, and ski slopes that bisect a habitat and create a barrier that effectively divides populations of many organisms into fractions on either side. The impact of such divisive fragmentation varies with the species. For some large or highly mobile animals, a highway may not be so much a physical barrier as a dangerous one, resulting in high mortality as evidenced by road kills. For most small mammals, reptiles, amphibians, and invertebrates of low mobility, roads effectively fragment populations by imposing a barrier individuals cannot cross (Mader 1984, Forman 1995). Even in large tracts of seemingly unfragmented forests, dirt roads and hiking trails can have a divisive effect on forest floor inhabitants. The endangered Cheat Mountain salamander (*Plethodon nettingi*), endemic to West Virginia, exists in isolated populations across its restricted range in forested areas above 960 m. These populations are further subdivided in their high-elevation forest habitat by hiking trails that individuals will not cross, restricting gene flow within the population (Pauley 1992, personal communication).

Patches often expand if undisturbed. For example, declining agriculture in the eastern United States over the early to middle twentieth century resulted in the abandonment of many agricultural fields. Adjacent forest has encroached on these areas, a process discussed in Chapter 21 (secondary succession). The advancing forest patch, which in effect is an advancing edge, rarely occurs in a straight line. Vegetation moves into the surrounding landscape as protruding fingers with concave boundaries. Plant species dispersed by wind and animals invade areas well beyond the advancing boundary, giving rise to new patches of vegetation, which eventually merge as they grow and expand.

Patch Metapopulations

Habitats scattered as landscape patches, large and small, are inhabited by spatially separated subpopulations or metapopulations (see Chapter 14). To varying degrees, these subpopulations are interconnected by individuals dispersing among patches. As predicted by island biogeography theory, large patches or extensive areas of grassland and forest hold higher populations and a greater diversity of species than small patches. They provide a source of individuals to repopulate small patches. Populations in small patches are vulnerable to local extinction. If the habitat quality is suitable, individuals from source (larger) patches can recolonize the patch. Recolonization can be slow if the patches are separated by poor or unsuitable habitat and corridors are lacking. How rapidly colonization occurs depends on connectivity—the presence of corridors, the size of the patch providing colonizers, the distances between patches (Grubb and Doherty 1999), and the ability of species to disperse.

LANDSCAPE DISTURBANCE

The patterns we see on the landscape are heavily influenced by disturbance, past and present. **Disturbance** is any relatively discrete event—such as fire, wind, floods, extremely cold temperatures, drought, and epidemics—that damages or disrupts community structure and function, resulting in the mortality of organisms and loss of biomass (Huston 1994, Pickett and White 1985).

Characteristics of Landscape Disturbance

Disturbances have both spatial and temporal characteristics. These include intensity, frequency, and spatial extent or scale.

Intensity Intensity of a disturbance is measured by the proportion of the total biomass or of the population of a particular species that is killed or removed. Intensity is influenced by at least three factors. One is the magnitude of the physical force involved, such as the strength of the wind or the amount of energy released during a fire. A second is morphological and physiological characteristics of the organisms that influence their response to the disturbance. A third is the nature of the substrate, especially as it relates to sessile organisms on rocky shores.

Frequency Frequency is the mean number of disturbances that occur within a particular time interval (Figure 23.17). The return interval is the inverse of frequency, or the mean time between disturbances for a given area. It is often difficult to separate frequency of disturbance from the scale and intensity of a disturbance. In temperate and tropical forests, in which natural disturbances occur on a small scale, such as the death of individual trees, the frequency of disturbance within a given stand is high. The frequency of more intense disturbances, such as fire, is low, between 0.5 and 2.0 percent per year. The return interval is thus between 50 to 200 years (Brow 1985, Runnel 1985). The long interval between disturbances (low frequency) allows biomass to accumulate between disturbances, and the resulting intensity of disturbance can be high. Runkle (1985) estimates that over the period of a decade, 4 to 14 percent of the total land area of the Great Smoky Mountain National Park is subject to disturbance (see Figure 20.14). This slow rate of disturbance and replacement maintains diversity in a mature forest over a period of many years and allows the coexistence of species with different life history characteristics (Chapter 22). Such infrequent disturbances give the time-locked observer the impression that the forest is unchanging.

In many regions fire is a major determinant of landscape patterns. How frequently a fire burns over a given area—its return rate—is influenced by the occurrence of droughts, accumulation and flammability of the fuel (biomass), the resulting intensity of the burn, and human interference. Prior to European settlement of North America, fires occurred about every three years in the grasslands of the Midwest. This amount of time was needed for sufficient mulch, dead stems, and leaves to accumulate. In forest ecosystems, the frequency of fires varies greatly, depending on the type of forest (Heinselman 1981a). Frequent light fires that burn only the surface layer may have a return interval of 1 to 25 years, whereas stand-destroying fires may have a return interval of 25, 100, or even 300 years.

Various forest ecosystems appear to burn and develop under certain fire frequencies. Low-intensity surface fires every 5 to 20 years were typical in presettlement forest of ponderosa pine (*Pinus ponderosa*) in western North America. Such fires prevented the buildup of a heavy fuel load, thinned the stand, eliminated the later successional shade-tolerant conifer species that could carry fire up into the crown, and encouraged an open, grassy understory. Red pine (*Pinus resinosa*) and white pine (*P. strobus*) forests of the Great

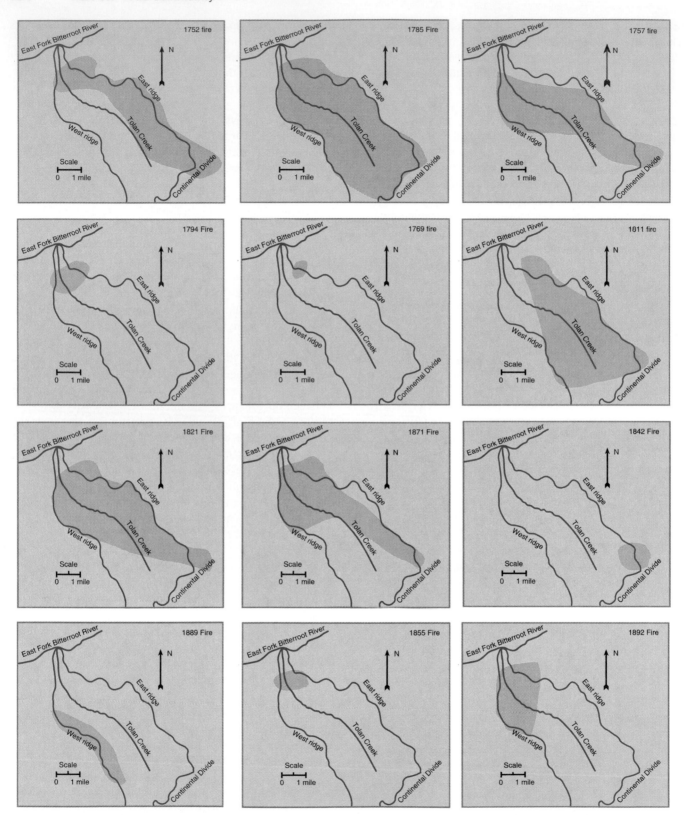

FIGURE 23.17   Frequency of disturbance in a community as illustrated by the fire history of the Tolan watershed in the Bitterroot National Forest between 1734 and 1900. Twelve of the 23 fires are illustrated in this series. No fires burned after 1900 because of fire suppression. The watershed supports ponderosa pine (*Pinus ponderosa*), Douglas-fir (*Pseudotsuga menziesii*), western larch, lodgepole pine (*Pinus contorta*), and white-barked pine (*Pinus albicaulis*). The general pattern was one of frequent fires leaving substantial remnants of older trees. Most fires burned lightly on the drier slopes at lower elevations, perpetuating ponderosa pine as the dominant species in open strands. The fires killed much of the ponderosa pine regeneration and most of the invading Douglas fir. Fires burned with greater intensity on north-facing slopes, where dense young growth of Douglas fir provided more opportunity for crown fires. Even here, however, much of the old growth survived. In the lower subalpine forest, dominated by lodgepole pine, fires were often of low or medium intensity, spreading mostly on the forest floor. (From Arno 1976:23–29.)

Lakes region experienced infrequent surface fires of low intensity with a return interval of about 28 to 35 years. They were punctuated at longer intervals of 150 to 300 years by severe surface and crown fires that destroyed the stand.

Communities dependent on infrequent large-scale disturbances for regeneration respond negatively to disturbances that come at too frequent intervals. Under such conditions, successional dynamics often results in the replacement of these species and the shift to some other community. For example, short fire intervals in chaparral vegetation in California result in their changing to a relatively permanent, degraded shrubland and annual grassland (Zedler et al. 1983).

For species or communities that depend on fire for regeneration or suppression of later successional species, fire cannot be considered a disturbance; rather, suppression of fire becomes the disturbance. In lodgepole pine or mixed coniferous forests of western North America, a natural fire regime involves reoccurring ground fires at intervals of 7 to 25 or more years. These fires have a significant role in maintaining conditions for tree regeneration and the suppression of later successional shade-tolerant tree species. They also create and maintain openings in the forest canopy, in effect creating natural firebreaks. Fire suppression allows the accumulation of fuel, crown closure, development of an understory that can carry fire to the crown, and senescence and death of individual trees. Such protection against fires adds to the fuel load and greatly lengthens the fire cycle. When fires do occur, they are intense, causing extensive tree damage and mortality over a very large area (Holling 1980, Kilgore 1973). The long years of fire suppression in Yellowstone National Park until 1975 (Romme and Knight 1982) accompanied by an unexpected period of drought set the stage for extensive fires there in 1988 (Figure 23.18).

A similar effect occurs when natural insect predators of trees are suppressed to protect stands of susceptible timber. Spruce budworm (*Choristoneura fumiferana*) infestations, like short-interval ground fires, create a patchy environment. Extensive spraying of spruce and fir forest to eliminate the budworm protects a timber stand from short-term timber losses, but it sets the stage for a widespread outbreak of the insects covering hundreds or thousands of hectares.

A hurricane is another powerful recurring event of high intensity but low frequency. The frequency (average time interval between occurrences) of hurricanes with the intensity of Hurricane Hugo in 1989—winds over 166 km/hr—is once every 50 to 60 years in the Caribbean. The return interval (affecting the same area) of all hurricanes is about 21 years (Scatena and Larsen 1991).

Scale The impact of a disturbance on community structure and landscape is also a matter of scale in the size of area affected. Disturbances range from very small, frequent ones such as the death of a single tree in a forest to large-scale, less frequent disturbances that embrace extensive areas swept by fire, buried under volcanic ash, torn by landslides, or denuded by human land-clearing schemes. What constitutes a small disturbance depends on the scale of the landscape in which it occurs. For example, small-scale disturbances that cause the death of indi-

FIGURE 23.18 In 1988 massive crown fires swept through Yellowstone National Park. The fire was intense in part because of the buildup of fuel brought about by fire suppression in the past.

viduals or groups of trees, thereby opening the canopy, would have more impact on a small woodland than on a large forest.

Abrasive action of waves tears away mussels and algae from tidal rocks. In grasslands, digging by badgers and groundhogs exposes small patches of mineral soil that are colonized by herbaceous plants. A dominant tree dies or falls in a forest, opening up the canopy. The outcome of such disturbances is the creation of a **gap**, a term originally applied by Watt (1947) to openings that become localized sites of vegetative regeneration and growth.

Consider a forest. A single tree struck by lightning or killed by a fungal infection that remains standing as a snag creates a simple gap. A dominant tree uprooted by a windstorm creates a larger, more complex one (Figure 23.19). When it falls, it opens a space in the canopy. Its upturned roots tear up the soil and form a pit and a mound of exposed mineral soil open for colonization. Its falling trunk slices through its neighbors, snapping the limbs of nearby canopy and understory trees. The fallen crown crushes the understory beneath it. Once created, some gaps in the forest continue to grow as trees about the gap succumb to winds and insect damage.

Within the gaps, the microclimate differs from that of the rest of the forest. Light and soil temperature increase, while soil moisture and relative humidity decrease. In small gaps, the response is typically vegetation reorganization. The crowns of trees about the edge of the gap expand to fill the opening. Their closure of the canopy inhibits any initial growth response understory plants may have made (Trimble and Tryon 1966).

Large gaps encourage the growth of intolerant species. Such species require openings of at least 0.1 ha or one of size at least twice the height of the surrounding trees. Even then, the most responsive growth is in the center of the gap (Tryon and Trimble 1969). In eastern deciduous forests of North America, intolerant species such as yellow-poplar (*Liriodendron tulipifera*) and black cherry (*Prunus serotina*) respond to larger gaps. Their ability to do so accounts for their conspicuous and continued presence in mature forests dominated by tolerant species.

A vigorous response of herbaceous plants and ferns to disturbance can adversely affect the growth of seedlings of forest trees. The density of tree seedlings can be much greater in places on the forest floor with little herbaceous cover than in patches of herbaceous growth (Maguire and Forman 1983). The competitive and allelopathic effects of dense patches of herbs and ferns inhibit forest tree regeneration (Horsley 1977, Ferguson and Boyd 1988, George and Bazzaz 1999).

Large-scale disturbances induced by fire, logging, land clearing, and other such events to be discussed result in responses that go beyond vegetation reorganization and involve colonization by opportunistic species. Some of these species may already be present on the site as seeds of woody and herbaceous plants, root stocks, stump sprouts, and surviving seedlings and saplings. Other colonizing species are wind and animal dispersed. Long-term recovery involves a series of successional stages in which long-lived species characteristic of the original community eventually replace the short-lived, opportunistic, early colonizing species (see Chapter 21).

Sources of Landscape Disturbance

Many of the most powerful sources of disturbance are natural, such as lightning-set fires, hurricanes, and floods. Some of the most lasting disturbances are human-induced, such as land clearing for crops and development.

Fire   Fire is a major natural phenomenon that has an important role in the evolution and maintenance of ecosystems worldwide. Globally, large regions are characterized by vegetation that evolved under fire: the grasslands of North America, the shrublands of the southwestern United States and the Mediterranean region, the African grasslands and savannas, the eucalyptus forests of Australia, the southern pinelands of the United States, and even-aged stands of coniferous forests of western North America. Up to 95 percent of the virgin forests of Wisconsin were burned during the five centuries prior to the land being settled by white Europeans (Curtis 1959). These fires not only enabled such species as yellow birch (*Betula alleghaniensis*), hemlock (*Tsuga canadensis*), pines, and oaks to persist, but also were normal and necessary to perpetuate those forests (Maisurow 1941, Curtis 1959). In Alaska, fires have converted white spruce stands

FIGURE 23.19   Trees uprooted by ice and windstorms create new microreliefs on the forest floor to be colonized by herbaceous woodland plants.

into treeless herbaceous and shrub communities of fireweed and grass or dwarf birch and willow (Lutz 1956). Their growth is so thick that forest trees cannot become established.

Fires fall into three types: surface, ground, or crown. Each leaves its own distinctive imprint on the landscape. Type and behavior depend on the kind and amount of fuel, moisture, wind, and other meteorological conditions, season of the year, and the nature of the vegetation. **Surface fire,** the most common type, feeds on the litter layer. In grasslands it consumes dead grass and mulch, converting organic matter to ash. Usually fire does not harm the basal portions of roots, stalks, tubers, and underground buds, but it does kill most of the invading woody vegetation. In the forest, surface fires consume leaves, needles, debris, and humus. They kill herbaceous plants and seedlings and scorch the bases and occasionally the crowns of trees. Surface fires may kill thin-barked trees like maples by scorching the cambium layer. Thick-barked trees, like oaks and pines, are better protected.

If the fuel load is high and the wind strong, surface fires may leap into the forest canopy to cause a **crown fire,** one that sweeps through the canopy of the forest. Crown fires are most prevalent in coniferous forests because of the flammability of the foliage. If the canopy is unbroken, the fire may sweep across it and tops and branches fall to the ground to further feed the fire. A crown fire kills most above-ground vegetation through which it burns, skips, and hops, leaving patches unburned, creating a mosaic of unburned and regenerating vegetation (Figure 23.20). Thus fire or the lack of it creates or changes the structure and pattern of vegetation and develops a mosaic of different stand ages.

Ground fire that consumes organic matter down to the mineral substrate or bare rock is the most destructive (Figure 23.21). It is most prevalent in areas of deep, dried-out peat and of extremely dry, light organic matter such as an accumulation of conifer needles. Such a fire is flameless with extremely high temperatures, and it persists until all available fuel is consumed. In spruce and pine forests, with their heavy accumulation of fine litter, a ground fire can burn down to

expose rocks and mineral soil. Such fires eliminate any opportunity for that vegetation type to return, irreversibly changing the nature of the landscape.

As an agent of disturbance, fire sets into motion the process of regeneration for some fire-adapted species by stimulating germination of seeds. The seed may be stored on the plant or in the soil, awaiting the fire to release seeds or stimulate germination. Jack pine and lodgepole pine are two coniferous species that retain unripened cones for many years on trees. Seeds remain viable within the cones until a crown fire destroys the stand. Then the heat opens the cones and releases the seeds—an event called **serotiny**—to a newly prepared seedbed. Other species such as the chaparral species, ceanothus (*Ceanothus*) and manzanita (*Arctostaphylos*), rely on increased temperature of fire-heated soil to crack the hard seed coat, which is impervious to water and other softening agents.

Other fire-adapted species respond to fire by resprouting. Although fire kills the tops and foliage, new growth appears as bud sprouts and root sprouts. Certain trees, particularly a number of *Eucalyptus* species in Australia, possess buds protected beneath the thick bark of larger branches. The buds survive crown fires and break out to develop new foliage. Other plants sprout from buds on roots, rhizomes, root collars, and specialized structures called lignotubers. Shrubs, such as blackberries (*Rubus*) and blueberries (*Vaccinium*), and trees, such as aspen (*Populus*), sprout vigorously from roots. Such sprouting is a very effective means of survival and increasing a plant population.

Fire prepares the seedbed for some species of trees by exposing mineral soil, eliminating competition for soil moisture and nutrients from fire-sensitive and shade-tolerant species. Periodic surface fires thin some coniferous stands, such as ponderosa (*Pinus ponderosa*) and longleaf (*P. palustris*) pine. Importantly, fire acts as a sanitizer, terminating outbreaks of insects and such parasites as mistletoe by destroying senescent stands and deadwood and providing conditions for regeneration of vigorous young trees. Fire in grassland consumes the standing dead material and accumulation of litter. It

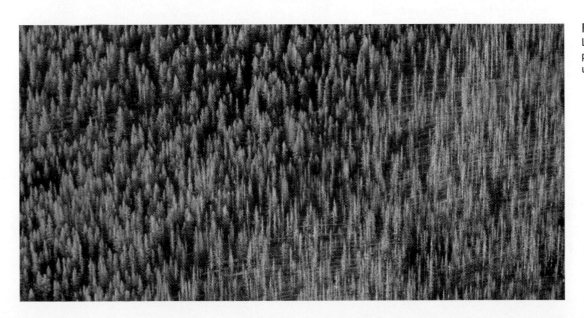

FIGURE 23.20
Landscape mosaic of patches of burned and unburned forest.

FIGURE 23.21 Fires of great intensity can have a profound influence on ecosystems. After the spruce forest located in the Allegheny Plateau area in West Virginia known as Dolly Sods was cut in the 1860s, intense ground fires burned, fed by piles of logging debris. Fire consumed the peatlike ground layer to bedrock and mineral soil. The forest never recovered, and the plateau is now a boulder-strewn landscape with intermittent patches of blueberry, dwarfed birches (*Betula*), mountain ash (*Sorbus americana*), and bracken fern (*Pteridium aquilinum*).

exposes the soil to the heat of the sun and speeds the recycling of nutrients, both of which stimulate the growth of new grasses.

Wind and Ice   Wind can inflict a major disturbance to vegetation. It shapes the canopies of trees exposed to prevailing winds, affects their growth of wood, and uproots them from the ground (windthrow). Mature trees, whose trunks lack the suppleness of youth, are especially vulnerable to windthrow. So, too, are trees growing on shallow and poorly drained soils whose roots, spreading along the ground, are not well anchored in the soil. Trees weakened by fungal disease, insect damage, and lightning strikes and tropical forest trees carrying a heavy load of epiphytes in their crowns are also candidates for windthrow. The impact of wind is accentuated when strong winds accompany heavy snowfall that weights down trees or heavy rains that soften the soil about the roots of trees. The position trees occupy in the forest also affects their vulnerability to wind damage. Trees bordering ragged forest gaps and those growing along forest edge, roads, and power lines are more likely to blow down than trees in the forest interior.

Widespread ice storms can devastate large areas of forest. Such an ice storm covered much of the northeastern United States in winter of 1998 and remained on the trees for several weeks. This storm flattened a vast acreage of spruce and fir in Maine and jack pine forest in extreme northeastern New York State.

Hurricanes have a devastating impact on ecosystems, especially those with wind speeds in excess of 166 km/hr and associated rainfall in excess of 200 mm. The force of a hurricane extends about 40 km about the center of the storm. (Anthes 1982, Scatena and Larsen 1991). Hurricanes cause landslides in hilly and mountainous country and massive defoliation and blowdowns of timber. They greatly increase the total nutrient input to the forest floor and alter nutrient cycling, especially of nitrogen and phosphorus (Lodge et al. 1991). In addition hurricanes are a major force role in maintaining diversity in mon-

tane rain forests of the Caribbean (Doyle 1981, Weaver 1989). Hurricane Hugo, which swept through southern United States and Puerto Rico in 1989, and Hurricane Andrew in 1992 devastated much of the vegetation in their paths and adversely affected certain animal species (Figure 23.22). For example, Hurricane Hugo destroyed much of the remaining old-growth stands of longleaf pine, habitat of the endangered red-cockaded woodpecker (*Picoides borealis*). This bird depends completely

FIGURE 23.22 Hurricane Hugo destroyed many pine forests on the coastal plain in South Carolina.

on old-growth pines affected with a fungus-caused soft rot of the heartwood for nesting sites.

Moving Water   Moving water is a powerful agent of disturbance. Storm floods scour stream bottoms, cut away banks, change the courses of streams and rivers, move and deposit sediments, and bury or carry away aquatic organisms. Strong waves on rocky intertidal and subtidal shores overturn boulders and dislodge sessile organisms. This action clears patches of hard substrate available for recolonization and maintains local diversity (Sousa 1979, 1985, Connell and Keough 1985). High storm tides break down barrier dunes, allowing seawater to invade behind the dunes, and change the geomorphology of barrier islands.

Drought   Prolonged drought can have a pronounced effect on vegetation composition and structure. On the grasslands of western Kansas during the drought years of the 1930s, blue grama with its physiological ability to resist dry conditions became two times as dense as the drought-sensitive buffalo grass. When the rains came, buffalo grass quickly responded and in two years reversed its position, becoming five times as dense as blue grama. After ten years without drought conditions, the two species were codominants (Coupland 1958). In temperate forests, prolonged drought can result in heavy mortality of shallow-rooted tree species, such as hemlock and yellow birch, as well as understory trees and shrubs (Hough and Forbes 1943, Bjorkbom and Larson 1977). Drought dries up wetlands, causing crisis conditions among waterfowl and other wetland birds, muskrats, and amphibians and greatly reducing their populations.

Animals   A walk into a forest inhabited by a high population of deer or across an overgrazed grassland provides visual evidence of the impact that herbivorous animals can have on the landscape. Overgrazing native rangelands of the southwestern United States, for example, has reduced the organic mat and thus the incidence of fire. By dispersing seeds of mesquite (*Prosipis* spp.) and other shrubs through their droppings, cattle have encouraged the invasion of these woody plants onto overgrazed rangelands (Phillips 1965, Box et al. 1967). In many parts of eastern North America, large populations of white-tailed deer have eliminated certain trees, such as white cedar (*Thuja occidentalis*) and American yew (*Taxus canadensis*), from the forest. They have destroyed forest reproduction, and developed a browse line—the upper limits on a tree at which deer can reach foliage (Figure 23.23). In cutover areas of hardwood forests on the Allegheny Plateau in Pennsylvania, deer greatly reduced pin cherry and blackberry. They selectively reduced sugar maple and favored the expansion of ferns and grasses, which inhibit the regeneration of oaks and other trees (Marquis 1974, 1981, Marquis and Grisez 1978, George and Bazzaz 1999).

The African elephant (*Loxodonta africana*) has long been considered a major influence on the development of savanna vegetation. When their numbers are in balance with the vegetation and their movements are not restricted, elephants have an important role in creating and maintaining the woodlands. When elephants exceed the capacity of their habitat to support

FIGURE 23.23   An overpopulation of deer results in the elimination of forest understory plants and the creation of a browse line in the forest trees. The browse line is the highest point that hungry deer can reach for food.

them, their feeding habits combined with fires devastate flora, fauna, and soils. Elephant destruction of trees (Figure 23.24) acts as a catalyst for fires, which are the primary cause of converting woodland to grassland (Wing and Buss 1970).

However, interactions between key herbivores and community structure are not quite that simple. Interaction and the long-term equilibrium result from relationships among vegetation, climate, fire, and large herbivores. Consider the *Acacia* woodlands of the central Savuti channel of Botswana's Chobe National Park in southern Africa. There the *Acacia* woodlands are declining as mature, even-aged trees die from old age and elephant damage. The *Acacia* woodlands became established in the late 1800s, when the viral disease rinderpest killed off many of the large ungulate species, elephants were decimated by ivory hunters, and the Savuti channel dried from drought. The channel refilled in the 1950s, only to dry again in the 1980s. A return of the large woods will require another such combination of events that would reduce browsing pressure long enough to allow the seedlings to reach a safe size, perhaps 10 to 15 years (Walker 1989). Such a combination of events might have occurred again in Africa during the recent prolonged droughts, with their great losses in ungulates.

Beaver (*Castor canadensis*) modify many forested areas in North America and Europe. By damming streams, they alter the structure and dynamics of flowing water ecosystems

FIGURE 23.24 Consuming great quantities of woody vegetation and uprooting trees, elephants of the African savanna have an important influence on ecosystem succession and stability. The life cycles of certain trees and the maintenance of the savanna ecosystem depend in part on the disturbance regime of elephants. Too many elephants destroy the ecosystem; too few allow bush encroachment.

(Naiman et al. 1986). Pools behind dams become catchments for sediments and sites for organic decomposition. By flooding lowland areas, beavers convert forested stands into wetlands. By feeding on aspen, willow, and birch, beavers maintain stands of these trees, which otherwise would be replaced by later successional species. Thus, the action of beavers creates a diversity of patches—pools, open meadows, and thickets of willow and aspen interspersed with later successional species.

Birds may appear to be an unlikely cause of major vegetation changes. In the lowlands along the west coast of Hudson Bay, however, large numbers of the lesser snow goose (*Chen caerulescens caerulescens*) have affected the brackish and freshwater marshes. Snow geese grub for roots and rhizomes of graminoid plants in early spring and graze intensively on leaves of grasses and sedges in summer. The dramatic increase in the number of geese have stripped large areas of their vegetation, resulting in the erosion of peat and the exposure of underlying glacial gravels. There is little likelihood that the vegetation that newly establishes will closely resemble the original composition (Kerbes et al. 1990).

Outbreaks of insects such as gypsy moths and spruce budworms defoliate large areas of forest and result in the death or reduced growth of affected trees. The degree of mortality may range from 10 to 50 percent in hardwood forests infested by gypsy moths to 100 percent in spruce and fir stands infested by spruce budworms. Outbreaks of bark beetles have much the same effect in pine forests. The impact of spruce budworm, bark beetles, and other major forest insects

is most intense in large expanses of homogeneous forested landscapes where natural fires have been suppressed for long periods of times, allowing late-stage stagnated stands to develop; such stands are highly susceptible to spruce budworm outbreaks (Wolf and Cates 1987). In their own way these insects act to regenerate senescent or stagnated forests.

Timber Harvesting One of the major large-scale disturbances to the world's forests is timber harvesting. Disturbance by logging depends on the methods of timber harvesting employed: selection cutting or uneven-aged management, some form of even-aged management or clear-cutting, or wholesale exploitation with little regard for the future (Figure 23.25). In selection cutting, mature single trees or groups of trees scattered through the stand are removed. The cuts per hectare are relatively light, and the regeneration of new trees to replace those cut are grown as understory to the older remaining trees. Selection cutting produces only gaps in the forest canopy and favors reproduction of shade-tolerant over shade-intolerant trees. Forest composition essentially remains unchanged (Trimble 1973, Johnson 1984, Lorimer 1989).

Even-aged management involves removal of the forest and reversion to an early stage of succession (Figure 23.26a). Unless followed by fire or badly disturbed by erosion and logging activities (Figure 23.26b), the cutover area fills in rapidly with herbs, shrubs, sprout growth, and seedlings of trees present as advanced regeneration in the understory. The area passes quickly through the shrub stage to an even-aged young forest. Because many of the most valuable timber trees are shade-intolerant to mid-tolerant species, they can be regenerated only by removal of mature trees, exposing the ground to sunlight.

FIGURE 23.25 Clearcutting on steep slopes of western mountains results in forest destruction and the potential for massive erosion.

(a)

(b)

FIGURE 23.26 (a) Block clear-cutting in a western coniferous forest. Such cutting fragments the forest. (b) Unless carefully managed, clear-cutting can cause severe disturbance to a forest ecosystem.

There are three approaches to even-aged management (Smith 1986). One is clear-cutting 11- to 44-ha blocks of timber within large forest tracts. A second method is strip-cutting, which involves the removal of all merchantable timber and remaining trees in strips 15 to 30 m wide. Every third strip is removed, followed by removal of the remaining strips in two cuttings two to four years apart. A third method is shelterwood cutting, which leaves 10 to 70 percent of the stand remaining after cutting. When new growth is well under way, the remaining trees are removed. The first two methods of even-aged management favor regeneration of intolerant tree species. Shelterwood cutting retains some of the characteristics of the original forest yet permits intolerant species to regenerate.

Foresters often modify the regenerating forest to meet their requirements, introducing a form of disturbance during the development of the stand. Early in the life of a new forest, foresters may remove tree species not desired for timber or individuals of poor form. This improves, by economic but not necessarily ecological standards, the composition of the stand and the quality of the trees. Thinning can encourage the maximum growth of crop trees. Increased space between the trees stimulates crown expansion and increases growth.

Response of vegetation to the sudden removal of a forest canopy in timber harvesting is often rapid, as seeds and seedlings of intolerant woody plants take advantage of changed environmental conditions (Bormann and Likens 1979). One example in the northern hardwood forests is pin cherry (*Prunus pensylvanica*). Its seeds are carried to a forest by birds and small mammals or are deposited on the forest floor by an earlier stand of cherry. Pin cherry seeds can remain dormant for up to 50 years. When the forest canopy is removed and moisture, temperature, and light conditions

become favorable, pin cherry seeds germinate. Young trees quickly dominate the site, crowding out the associated blackberry, which also colonizes the area mostly from resident seeds (Marks 1974). If the seedling growth is dense, a pin cherry canopy can close in four years. Canopy closure eliminates other species except highly shade-tolerant seedlings of sugar maple or beech (Figure 23.27). If seedling growth is moderately dense, species with wind-disseminated seeds, such as yellow birch and paper birch (*Betula papyrifera*), will also occupy the site. Within 30 to 40 years, pin cherry dies out, allowing birch, sugar maple, and beech to dominate the gap. However, during its period of tenure, pin cherry has contributed numerous seeds to the forest floor, ready to reclaim the site when another disturbance provides the opportunity.

Sites may be colonized by species seeding-in from outside the area. Yellow-poplar and black birch (*Betula lenta*) are two such species in the eastern deciduous forest. The nature and success of their colonization depend on a number of conditions including distance from the site, size of the seed crop, timing of seed arrival, a sufficiently large number of seeds dispersed to the area to ensure germination and seedling survival, and exposed mineral soil as a seedbed.

Although the response of woody vegetation to large-scale removal of timber is rather well understood, little attention has been given to the response of the shade-tolerant understory herbaceous plants to disturbance. Ash and Barkham (1974) studied the response of the herbaceous understory layer of an English coppice forest after cutting. (A coppice forest is one in which the stump sprouts or root suckers are maintained as a main source of regeneration, with cutting rotations between 20 and 40 years.) Cutting coppice involves complete canopy removal, resulting in increased surface temperature on the forest floor and full exposure to light. Typically, a number of

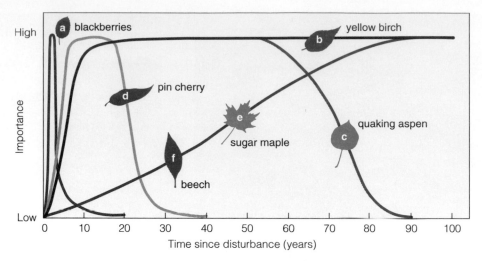

FIGURE 23.27   The importance of different species along a gradient of time following logging of a typical northern hardwoods forest. Immediately after disturbance blackberries (a) dominate the site, but they quickly give way to yellow birch (b), quaking aspen (c), and pin cherry (d). Intolerant pin cherry assumes dominance early, but within 30 years it fades from the forest. Yellow birch, an intermediate species, assumes early dominance, which it retains into the mature or climax stand. Quaking aspen, an intolerant species, begins to drop out after 50 years. Meanwhile sugar maple (e) and beech (f), highly tolerant species, slowly gain dominance through time. In about 100 years the mature forest is dominated by beech, sugar maple, and birch. (After Marks 1974:75.)

open-habitat or opportunistic species germinate and become established (Figure 23.28) but are soon excluded by the developing canopy cover. In spite of the disturbance, characteristic woodland species persist throughout the cycle. Adapted to a high-light regime in spring before the leaves are out, these plants have the ability to tolerate high light intensity. At the same time, they can coexist with annuals and open-habitat perennials that cast a shade on the ground like a tree cover. As opportunistic species disappear, woodland herbs again assume dominance, often developing into monospecific stands. There is evidence that logging practices in the broadleaf deciduous forests of the eastern United States, involving clear-cutting, frequency of cutting, soil compaction, and microclimatic changes on the forest floor, adversely affect survival and recolonization of forest understory species (Duffy and Meier 1992, Bratton 1994, Matlack 1994).

Cultivation   Human activity has a more profound impact on ecosystems than natural disturbances because we have the ability to radically change the natural environment. One of the more permanent and radical changes in vegetation communities comes about when we remove natural systems and replace then with cultivated cropland (Figure 23.29).

Many of us are burdened with the idea that humans converted natural ecosystems to cultivation only in relatively recent times. In fact prehistoric human populations converted land to pasture and cultivation as long as 5000 years BP (before present). Their activities changed the pattern of the landscape, extended or reduced the ranges of woody and herbaceous plants, allowed the invasion and spread of opportunistic weedy species, and changed the dominance structure in woodlands (Delcourt 1987). While the greatest changes

took place in Europe, the same effects occurred in North American and South America. Prehistoric native Americans in the Tennessee River Valley, for example, intensively cultivated the bottomlands (Smith 1978).

Cultivated plant communities are simple, highly artificial, and consist mainly of introduced genetically altered species well adapted to grow on disturbed sites. Because of the very simple and homogeneous ecosystem involved, tillage brings with it new pests destructive to both cultivated and natural vegetation. Tillage disturbs the structure of the soil and exposes it to water and wind erosion. In temperate forest regions, abandoned and degraded agricultural lands eventually return to some form of forest. In tropical regions, the degraded cultivated lands may or may not do so. Intermediate-sized disturbances, typified by slash-and-burn or tropical swidden agriculture, in which intensive cultivation rarely lasts more than three years, return to forest, provided the plots are relatively close to a seed source and the soil and not been heavily eroded. Where slash-and-burn is replaced by more intensive cultivation, abandoned lands rarely return to any semblance of the original forest (see Kowal 1966, Jordan 1986). On high-intensity disturbances involving large areas of land, as exemplified by the land clearing and conversion to grassland in the Amazon, the areas are so degraded that the abandoned sites may be permanently altered to heath-like communities.

Surface Mining   Surface mining accounts for a high percentage of coal production and for most other mineral extraction such as of iron, copper, gold ores, gravel, and limestone. The impact and magnitude of damage vary with the region and the degree and success of reclamation efforts. The effects

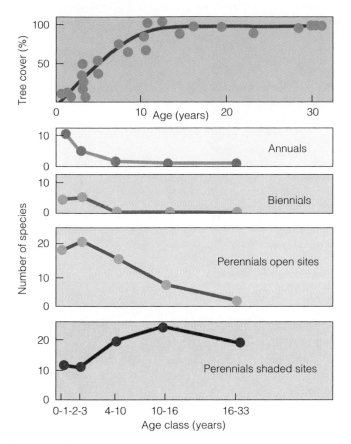

FIGURE 23.28 Response of understory herbaceous vegetation to disturbance in a coppice stand in England. The graphs show changes in the percentage of cover produced by growth of the canopy and the number of herb and shrub species in the field or ground layer at different times after coppicing. Not all species were present during the 30-year period of growth. Total numbers for each type are indicated. Note the rapid decline of annuals, biennials, and open-site perennials as the canopy closes. Open-site perennials dominate the field layers shortly after coppicing. Perennials of shaded sites show a sigmoidal growth response as the canopy closes. (From Ash and Barkham 1976:706.)

FIGURE 23.29 Most of the tallgrass prairie and oak woodland in the United States has been converted to fields of corn and soybeans.

are most pronounced in mountainous regions, as in the Appalachians. There, surface mining follows the contour of the mountain slopes or involves mountaintop removal and valley fill, destroying miles of streams (Figure 23.30). Whatever the method, surface mining does violence to the land. Deep, unweathered rock strata are broken and brought to the surface, where the material is subject to rapid weathering, releasing heavy metals such iron and zinc in toxic amounts. Carried away in high concentration by water coming off mined sites, these elements reduce water quality downstream for both aquatic life and humans. Unless expensive precautions are taken, sediment deposition in stream can be extremely high.

Surface mining can have serious impacts on groundwater. Water tables, once deep in the underlying rock strata, are exposed and flow freely to the newly created surface. Large quantities of water that would have been taken up by trees and lost to the atmosphere are added to the surface runoff. During heavy storms this runoff intensifies the height and damage of flash floods.

Some strip mine reclamation aims at restoration of the original community. Although the reclamation of preexisting communities may be of limited success in some western shrubland and midwestern grasslands communities, in the forested regions of the Appalachians, reclamation efforts concentrate on the establishment of grasslands. These expanding grasslands replace forests and some farmland, reducing landscape diversity. The redeeming feature is that some of these grasslands are providing new habitat for rapidly declining populations of grassland plant and animal species. For example, meadowlarks (*Sturnella magna*) and Henslow (*Ammodramus henslowii*) and grasshopper sparrows are among grassland bird species that face shrinking habitat availability within the eastern United States as abandoned crop fields succeed to forest.

ANIMAL RESPONSE TO LANDSCAPE DISTURBANCE

How does animal life respond to disturbances of various magnitudes? Over a short term the effect may be negative; over a long term it may be positive, depending on the species involved. With disturbance it is clearly a case of "One person's meat is another's poison." Small-scale disturbances, particularly treefall gaps, result in a positive response. The new vegetative growth filling in the gap provides low ground cover attractive to such gap species as hooded warblers (*Wilsonia citrina*) and Kentucky warblers (*Oporornis formosus*). The canopy gaps provide open areas needed by flycatchers (Tryannidae) for hawking insects, yet the openings have no effect on canopy-dwelling species. Clear-cut areas in a forest create the early successional habitats needed by the opportunistic or ephemeral species such as prairie warblers (*Dendroica discolor*), chestnut-sided warblers (*D. pensylvanica*), and woodcock (*Scolopax minor*). Such species require recurring disturbances to maintain their habitat. Large clear-cut areas, on the other hand, eliminate habitat of canopy-dwelling vertebrates and invertebrates, who will not return for several decades. Generally, ground-dwelling small

FIGURE 23.30 Contour multiple-seam mining in the steep-sided Appalachian Mountains forever changes the mountains, destroys the forests, eliminates many forms of wildlife, and disrupts the natural water regimes.

mammals are little affected by timber harvesting. Cutting of old-growth forest eliminates old-growth dependent species such as spotted owls (*Strix occidentalis*) and red tree voles (*Phenacomys longicaudus*) of the Pacific Northwest.

A view of its immediate aftermath may give the impression that fire is a major agent of destruction to wildlife. On a short-term basis fire destroys or partially destroys habitat and reduces available forage for ungulates (Singer and Mack 1999). It may cause some injury and death, either directly by fire or indirectly by predators who take advantage of prey suddenly driven from or deprived of cover. On the African savanna, kites and other birds follow grass fires, hunting insects driven to flight by the advancing fires. Many flying insects, such as grasshoppers and moths, fly in front of the flames and are often engulfed by wind-driven gas clouds, but a surprising number go through the fire unscathed. Unless nesting, birds are rarely directly affected by fire other than short-term loss of habitat. Some species of birds may decrease following fires, but ground-foraging birds seem to increase (Wright and Bailey 1982).

Many mammals, especially those that live in burrows, survive fires (Bradley et al. 1992). Large mammals are quite adept at keeping ahead of flames and working their way back through gaps and unburned patches to burned-over areas behind the flames (Main 1981, Boyce and Merrill 1991, Singer and Mack 1999). The major problem faced by these mammals is the short-term lack of food and cover. High populations of post-fire grazing herbivores feeding on newly regenerating plants may overgraze or eliminate palatable species.

For fire-intolerant species, such as tree-dwelling squirrels, severe fires destroy habitat and eliminate for some time those species dependent on it. At the same time, fire improves habitat for a fire-impervious species, especially those that favor open and shrubby land. Many animals favor both pre- and post-fire conditions and in fact depend on such fluctuations in habitat. Fire produces a mosaic of shrubs, timber, and open land. Such patchy environment is essential for such species as snowshoe hare, black bear, white-tailed deer, and ruffed grouse.

FIGURE 23.31 The endangered Kirtland's warbler is a fire-dependent species inhabiting the jack pine forests of Michigan.

A few species are fire-dependent. They require the periodic disturbance of fire to maintain their habitat. One is the endangered Kirtland's warbler (*Dendroica kirtlandii*) (Figure 23.31). Restricted to the jack pine forests of the lower peninsula of Michigan, the warbler requires large blocks (40+ ha) of even-aged stands of pine 1.5 to 4.5 m tall with branches close to the ground. Smaller or larger trees are unacceptable. Intervals of fire are needed to maintain blocks of young jack pine habitat.

Another group of fire-dependent birds is the *Sylvia* warblers that occupy the Mediterranean shrublands of Sardinia (Walter 1977). Of the five species, two are fire-dependent. One species, the Dartford warbler (*S. undata*), can occupy a habitat patch that has been burned within 6 years. Another species, Marmora's warbler (*S. sarda*), occupies only 18- to 20-year-old patches with tall, shrubby growth. Two of the remaining species are fire-adapted, and one is fire-tolerant.

Summary

1. Landscapes consist of mosaics of patches related to and interacting with each other. The study of the causes and consequences of these spatial patterns is landscape ecology.
2. The various patches in the landscape have their own unique origin, from remnant patches of original vegetation to introduced patches requiring human maintenance. Natural disturbance and human activity determine their size and shape.
3. The places where the edges of two or more different patches meet are the boundaries. The two edges combine to make up the borders, which may be abrupt or diffuse. An edge may be inherent, produced by a sharp environmental change such as a topographical feature; or it may be induced, created by some form of disturbance that is limited in extent and changes through time.
4. Often the vegetation of one patch blends with another to form an ecotone. Typically, ecotones have a high species richness because they support selected species of adjoining communities and also a group of opportunistic species adapted to edges. At the same time, edges attract predators, facilitate movements of animals, and act as a barrier between patches.
5. Linking one patch to another are corridors, strips of vegetation similar to the patch but different from the surrounding matrix. Corridors act as conduits or travel lanes, function as filters or barriers, and provide dispersal routes among patches.
6. A positive relationship exists between species diversity and area. Generally, large areas support more species than small areas. This species-area relationship underlies the theory of island biogeography, which proposes that the number of species an island holds represents a balance between immigration and extinction. Immigration rates to an island are influenced by its distance from a mainland or source of potential immigrants. Thus, islands most distant from a mainland would receive fewer immigrants than islands closer to the mainland. Extinction rates are influenced by the area of the island. Because small islands hold smaller numbers of individuals of a species and have less variation in habitat, they experience higher extinction rates than large islands.
7. Some aspects of island biogeography theory relate to the fragmentation of natural habitats such as forests that become isolated by surrounding agricultural and urban lands.
8. Although smaller fragments hold fewer species than larger fragments, size is not the sole criterion for the value of such fragments for wildlife. Important is the ratio of amount of interior to amount of edge. Many species are area-sensitive—that is, they require large unbroken blocks of habitat. Unless the habitat parcel is large enough to hold interior species, the area is inhabited only by edge and area-insensitive species.
9. Disturbance is any physical force, such as fire or wind, that damages natural systems and results in the mortality of organisms. It influences community structure and contributes to biodiversity.
10. Disturbances vary in intensity, frequency, and spatial scale. Intensity is measured by the proportion of the biomass or populations killed or removed. Frequency is the mean number of disturbances that take place within a given time interval.

Inversely related to frequency is the return interval, the mean time between disturbances on the same piece of ground.
11. Important to the nature of disturbances is spatial scale. Small-scale disturbances are typical of rocky intertidal shores and temperate and tropical forests. Wave action and moving water create gaps among sessile organisms on rocky substrates. Treefall and removal of individual trees by logging create small gaps in forests. Response to gap formation involves canopy closure, invasion by opportunistic species, and growth response by tolerant species, depending on biotic conditions.
12. Large-scale disturbances induced by major events such as fire, hurricanes, logging, insect outbreaks, and land clearing for cultivation result in responses that go beyond vegetational organization and involve colonization by opportunistic species. Such disturbances can shape and modify the nature of the system by favoring certain species and eliminating others, or they can ensure regeneration of the system itself.
13. Of great ecological importance are the frequency and return interval of disturbances. Small-scale disturbances have a high frequency within the system, but the rate of disturbance is low and the return interval is between 50 to 200 years. Natural large-scale disturbances have low frequency and a return interval of 25 to several hundred years. Frequency of large-scale disturbances relates to the life span of the longest living species. Too frequent disturbances relative to the life span of the species involved can eliminate certain species by destroying plants before they have had time to mature and seed. Too long a time between disturbances can result in the elimination of mid-tolerant species, reduce system diversity, and set the stage for highly destructive disturbances.
14. Fire is the major natural large-scale disturbance of terrestrial ecosystems. It has both beneficial and adverse effects. It results in loss of soil nutrients but also makes nutrients available. It sets into motion regeneration of fire-adapted systems by stimulating root sprouting and germination of seeds. It can favor fire-resistant species and eliminate fire-sensitive ones, thereby influencing composition and structure of forest systems.
15. Other major disturbance regimes include wind—especially hurricanes—flooding, and drought. Major human-induced disturbances involve logging, surface mining, cultivation, and development; these produce profound changes, often permanent, in the landscape.
16. Response of the animal component to disturbance depends on the species involved. Short-term impacts involve the loss of food and cover. Long-term effects may be the loss of habitat for some species and the gain of habitat for others. Some species depend on disturbance for the maintenance of their habitat, especially those associated with the more ephemeral stages of early succession. Other species depend on periodic fires to maintain their habitat and to provide a mosaic of vegetation types required in their life cycle. A few fire-dependent species would go extinct without periodic fires to maintain their habitat.

Review Questions

1. What is landscape ecology?
2. What major factors are behind the basic landscape patterns?
3. Distinguish between an edge and an ecotone.

4. Explain the edge effect.

5. What is island biogeography theory?

6. What is habitat fragmentation? What forces are behind it?

7. How do habitat patches in the landscape mosaic differ structurally and functionally from oceanic islands?

8. What aspects of island biogeography relate to patches in a landscape mosaic?

9. Show the relationship among patch size, interior species, and species diversity.

10. Differentiate between a landscape and an ecosystem.

11. Survey the size and shape of patches in your local landscape.

12. What types of corridors can you find in Figure 23.1? What is their origin?

13. Contrast interior species with edge species. Are all species occupying a small patch edge species?

14. How could an abrupt edge be modified to attract more edge species?

15. Discuss the use of a patchy environment by an edge species such as the cottontail rabbit (*Sylvilagus floridanus*) or tiger swallowtail butterfly (*Papilio glaucus*). What vegetation patches do they need? Why are they found only in an edge situation? What is the role of disturbance in maintenance of edge habitat?

16. Define disturbance.

17. How do intensity and frequency of a type of disturbance influence its effects?

18. How do small-scale disturbances relate to diversity and long-term stability of a community?

19. What constitutes large-scale disturbances? What are some major natural forces behind such disturbances?

20. How does fire differ in its ecological effects from other disturbances?

21. What major microclimatic changes relative to surrounding forest would you expect to discover in a small gap in the forest? Consider light, air temperature, soil temperature, and humidity. Discuss these changes relative to opportunistic species. (Refer to Chapters 4, 6, 7, and 8.)

22. Controlled burning of the understory of longleaf pine at regular intervals is a common management practice. What is the purpose of such burns? What regulates their frequency? What effect do they have on understory vegetation? What would be the outcome if such disturbances were eliminated?

23. Refer the pattern of forest fires in Figure 23.17. Assume that the 12 fires diagrammed rather then the actual 23 represent the total occurrence of fire from 1734 to 1900. Select several points at random in the watershed, and determine the frequency and return interval of fire for those points.

Cross References

Soil types, 69; sun and shade plants, 91; microclimates, 38; metapopulations, 163; population sources and sinks, 166, 200; genetics of small populations, 365; succession, 404; forests, 602; grassland, 563; shrubland, 577; rocky intertidal shores, 676.

Asteroids, Bulldozers, and Biodiversity

In the early morning of June 30, 1908, a fireball visible for hundreds of miles exploded in the atmosphere some six miles above the headwaters of the Tunguska River in the Siberian region of Russia. Thirty-five miles from the site, people were knocked to the ground, unconscious. The detonation was heard at least 600 miles away. Light was reflected from the dust scattered high into the atmosphere from the blast, brightening the night skies throughout western Europe. Russian scientists who visited the remote region 20 years later found trees blown down in a parallel line more than 20 miles from the site. The cause of this destruction is unclear—scientists are still divided as to whether it was a meteorite or fragment of a comet nucleus that exploded before striking Earth's surface.

Approximately 65 million years ago, at the end of the Cretaceous period, in the region of the Yucatan Peninsula, scientists now believe that a similar, but much larger, event occurred. A large meteorite struck Earth's surface, leaving a crater 180 kilometers in diameter under the waters of the Caribbean. Evidence from deep-sea sediment cores reveals a remarkable record of the meteorite's impact and the resulting debris. The debris, blasted high into the atmosphere, may have triggered a major decline in Earth's temperature. Scientists now believe that the assault by this massive asteroid or comet was in large part responsible for the extinction of 70 percent of all species that inhabited Earth at that time, including the dinosaurs. In the period that followed, the various species that eventually would come to dominate the oceans and land surface changed dramatically from the previous inhabitants. For example, on land, mammals diversified and increased in size, a process that eventually gave rise to a particular species of primate—*Homo sapiens*, otherwise known as humans.

Paleontologists refer to the loss of species at the end of the Cretaceous as a "mass extinction event." It was not the only such event; Earth has undergone a number of mass extinctions, such as the one during the Permian (250 million years ago) when over 50 percent of Earth's species disappeared from the fossil record, including 96 percent of all marine species. The fossil record teaches us that these mass extinction events change the course of evolution, inducing a dramatic shift in the types of organisms that inhabit the planet.

Earth is currently undergoing a mass extinction on par with previous events, with an estimated annual loss of species in the tens of thousands. The current mass extinction event, however, is different from those of the past. The cause is not extraterrestrial, such as from a meteor or comet, or a change in sea level or climate. The destruction is due to bulldozers, agriculture, and urbanization— the destruction of habitat.

Although in North America's recent past, hunting for food and other goods has led to the extinction of many mammal and bird species, when compared to current losses, the number of species extinguished by overkill is relatively small. Hunting has led to the extermination of marine mammals such as Stellar's sea cow (extinct about 1767), the New England sea mink (about 1880), and the Caribbean monk seal (about 1952). Overkill has been found to be the main cause in virtually all 46 modern extinctions of large terrestrial mammals. Among birds, about 15 percent of the 88 modern species extinctions and 83 subspecies extinctions have been attributed to overkill. Affected species include the great auk (1844) and the passenger pigeon (1914). In some instances overkill has resulted from the often mistaken belief that a wild species was a threat to gardens or to domestic animals. Victims of this belief include the Carolina parakeet (Figure VII-A), the only native U.S. parakeet (1914).

Other causes of recent extinction include the introduction of predators and competitors to systems with isolated endemic species unable to cope with the pressures imposed by the new arrivals. Predators introduced by humans have wiped out many species, particularly in island habitats that

FIGURE VII-A   Carolina parakeet.

Although accounting for only 7 percent of the total land area, tropical rain forests are home to more than 50 percent of all terrestrial plant and animal species. Over the last several decades, extensive land clearing in the tropical rain forests of Asia, Africa, and South America has greatly decreased their global distribution and resulted in the extinction of an inestimable number of plant and animal species. In the Amazon region of Brazil, the rate of forest clearing during the 1980s exceeded 1 percent per year, decreasing the extent of forest from 1 million km^2 to current estimates of just over 50,000 km^2. Forest clearing in Madagascar has resulted in the removal of over 90 percent of the original forest cover (Figure VI-B). This loss of habitat has resulted in the large-scale extinction of endemic species, including six genera of lemurs. Since 1960, 95 percent of the rain forest cover in Ecuador has been destroyed, resulting in an estimated loss of over 200,000 species. These are but a few examples.

The changes in land use that bring about this destruction and large-scale extinction are not limited to the tropics or to terrestrial environments. Pollution of our inland waterways, dredging and filling of coastal wetlands, and the destruction of coral reefs by pollution and siltation are having a similar effect to forest clearing on Earth's freshwater and coastal environments.

To truly understand the impact that humans are having on the diversity of species inhabiting Earth, we must place the current rates of species extinction into the context of the total biological diversity of the planet—its biodiversity. What is the present state of biodiversity on Earth? According to E. O. Wilson, Frank B. Baird Professor of Science at Harvard University, about 1.4 million living species of all kinds of organisms have been described by science. Approximately 750,000 are insects, 41,000 are vertebrates, and 250,000 are plants (vascular plants and bryophytes). The remainder consist of a complex array of invertebrates, fungi, algae, and microorganisms. Scientists who catalogue plant and animal species, called systematists, say that this picture of species diversity is still very incomplete, except for a few well-studied groups such as the vertebrates and flowering plants. Recent investigations in the canopy of Peruvian rain forests suggest that the total number of species inhabiting Earth could be much higher. Such studies are identifying thousands of new species each year. (Other major habitats that remain poorly explored include the coral reefs, the floor of the deep sea, and the soil of tropical forests and savannas.) According to Wilson, we do not know the true number of species on Earth, not even to the nearest order of magnitude. His own guess is that the absolute number falls somewhere between 5 and 30 million.

Why should we care about how many other organisms there are or how humans affect the biodiversity of the planet? The term *biodiversity* (*bio* = biological, *diversity* = variety) did not exist before the mid-1980s. In large part the concern is a direct product of technology. It is no coincidence that the first Earth Day celebration in the spring of 1970 came only

previously had been relatively free of predators. The largest number of such extinctions have been caused by rats, carried across the Pacific by Polynesians and Micronesians over the past few thousand years, and spread throughout the rest of the world by Europeans over the past 500 years. Examples include the five species of birds killed off on Lord Howe Island shortly after rats arrived with a ship in 1918 and the extinction of four species of birds, a bat, and numerous invertebrates after rats arrived on New Zealand's Big South Cape Island in 1964. Although competition can also cause extinction, interspecific competition most commonly acts to reduce the geographic range and abundance of native species, thereby making them more vulnerable to other causes of extinction. In North America, the introduction of the European house sparrow and the starling has constricted the ranges of such native birds as the purple finch and the eastern bluebird, but has not yet led to their extinction.

By far the most important cause of extinction by humans is the destruction of natural habitats in order to create farms and ranches for raising crops and livestock. Such habitat alteration can begin the slide of a species to extinction long before the last vestige of its habitat is eliminated. Fragmentation of the original home ranges, reduction of breeding sites and feeding grounds, and other factors can cause population size to decline to critically low numbers well before a species habitat has been completely destroyed. The largest number of current species' extinctions is occurring in the tropics, particularly the tropical rain forests.

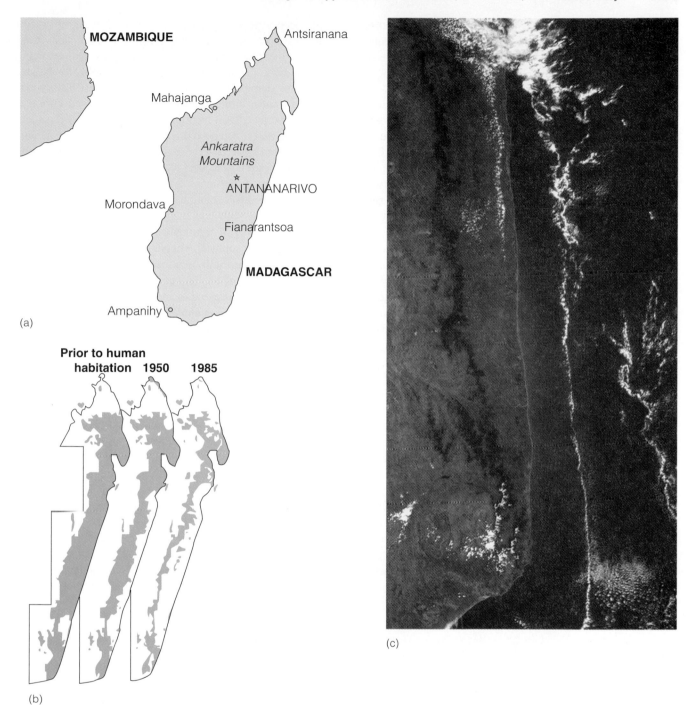

FIGURE VI-B Decline of tropical rain forest in Madagascar. (a) Geographic location of Madagascar off the eastern coast of Southern Africa. (b) Decline in forest cover in eastern, coastal region of the island (forest shown in green). (c) Photo of eastern coastal Madagascar taken from space shuttle. Forested areas appear as dark green band running from north to south.

nine months following *Apollo* 11's historic first landing on the moon. Our first view of Earthrise from the lunar surface put the finite nature of our shared planet into perspective.

The growing use of satellite remote sensing (see Figure VII-B) in the late 1970s and early 1980s allowed us for the first time to quantify the decline of natural ecosystems, particularly in remote regions of the globe. Once the true extent of the problem was realized, the concern moved from academic circles to society at large. By the 1990s this awareness and concern extended into the international community with the convening of the United Nations Earth Summit in Rio de Janeiro in 1992. At the Earth Summit, over 150 nations signed the Convention on Biodiversity, making the preservation of the world's biodiversity an international priority. Although there are many reasons voiced, the arguments about the importance of maintaining biodiversity can be grouped into three categories: economic, evolutionary, and moral.

The economic argument is based largely on self-interest. Many of the products we use come from organisms with which we share this planet. Obviously, the foods we eat are all derived from other organisms. Every time we buy a drug or other pharmaceutical, there is almost a 50-50 chance that we can attribute some of the essential constituents to a wild species. The value of medicinal products derived from such sources is now over $40 billion every year. (Of course, there is no way to assign value to the lives that are saved by the use of medicines obtained from natural products.) We derive rubber, solvents, and paper from trees. Cotton, flax, leather, and a host of other natural materials are used to clothe us. Modern industrial society owes a lot to Earth's genetic resources that in one way or another contribute to a number of products that better our standard of living. Today's benefits from nature's cornucopia are astonishing enough, but they represent only the tip of the iceberg. At this point, scientists have taken only a preliminary look at some 10 percent of the 250,000 plant species, a considerable number of which have already proved to be of enormous economic importance. Additionally, we have scarcely scratched the surface of the potential of the products derived from the animal kingdom. As these species are lost through extinction, so is their potential for human exploitation.

The second argument for the preservation of biodiversity is based on genetics (see Chapter 19). The current patterns of biodiversity are a product of ecological and evolutionary processes that have acted on species that existed in the past. The processes of mutation, mixing of genetic information through sexual reproduction, and natural selection, together with the essential ingredient of time, give rise to new species.

All species eventually go extinct, many leaving no trace of their past presence other than fossilized impressions buried deep in the earth. Others, however, fade into extinction after having given rise to new species. For example, it is believed that all modern birds can trace their evolutionary history to the earliest known bird, *Archaeopteryx*, which lived during the Jurassic period (fossil record dates to 145 million years ago). If *Archaeopteryx* had been driven to extinction before acting as the evolutionary seed of more modern birds, the variety of life at our backyard bird feeders would be quite different than it is today. Likewise, the mass extinction of modern-day species limits the potential evolution of species diversity in the future.

The third category of arguments in support of conserving biodiversity is philosophical. Humans are but one of millions of species inhabiting Earth, relative newcomers to the long evolutionary history of life on our planet. It is the nature of all organisms to both respond to and alter their surrounding environment. However, it is unlikely that any other species in the history of Earth has had such a dramatic impact on its environment in such a short period of time. The fundamental question facing humanity is a moral one. To what degree will we allow human activities to continue to result in the extinction of tens of thousands of species with whom we share this planet? It is on this question that the debate on the value of biodiversity will center. Arguments based on economics will fall to the wayside as technology allows us to synthesize medicines and other products currently made from plant and animal products, and concerns for the evolutionary future of our planet appear all too abstract when balanced against the needs of our growing human population. Science can work to identify and quantify the problem, but its solution lies outside the realm of science. It involves social, economic, and ethical issues that influence all our lives. Unlike so many problems facing society that science is called upon to solve, this is one that science can identify but that members of society—including you—must help decide how to solve.

PART EIGHT

The Ecosystem

C H A P T E R 2 4

Ecosystem Productivity

CONCEPTS

1. The ecosystem is the biotic community and the physical environment functioning together as a system.

2. Energy, potential and kinetic, is the ability to do work.

3. The laws of thermodynamics govern energy flow in ecosystems.

4. Light energy is fixed by autotrophs as primary production and transferred to and stored by heterotrophs as secondary production.

5. Net primary production is influenced by the variety of environmental factors that affect rates of photosynthesis.

6. Secondary productivity is limited by net primary productivity.

7. Efficiencies of energy transfer vary widely among autotrophs and poikilothermic and homeothermic heteotrophs.

8. The major pathways of energy through the ecosystem are the detrital and grazing food chains and interactions between the two.

9. Models of energy flow involve trophic or feeding levels.

The distribution and abundance of species—the biological structure of the community—vary in response to changes in environmental conditions. However, it is equally true that the organisms themselves, in part, define the physical environment. An example is the role of autogenic changes in light and nutrient availability in driving the process of plant succession discussed in Chapter 22. It is this inseparable link between the biological environment (the community) and the physical environment that led A. G. Tansley to coin the term **ecosystem** in an article written in the journal *Ecology* in 1935. In the article, Tansley writes:

> The more fundamental conception is ... the whole system (in the sense of physics) including not only the organism-complex, but also the whole complex of physical factors forming what we call the environment.... We cannot separate (the organisms) from their special environment with which they form one physical system.... It is the system so formed which (provides) the basic units of nature on the face of the earth.... These *ecosystems*, as we may call them, are of the most various kinds and sizes.

The concept of the ecosystem arose from the debate at the time concerning the nature of biological communities (see discussion in Chapter 22). Tansley believed that the concept of the community was incomplete, in that it did not include the physical environment. It was his premise that the biological and physical components interact to such a degree that it is not meaningful to separate the two when studying natural environments. The organisms that make up the community both respond to and influence the physical environment. If we are to understand pattern and process in communities, we must include the physical environment.

COMPONENTS OF ECOLOGICAL SYSTEMS

Although every ecosystem, forest, grassland, or lake, is different, all ecosystems are composed of two parts, the biotic and the abiotic (Figure 24.1). The biotic part consists of all interacting organisms living in the area, the **community.** The abiotic part embraces the physical environment with which the organisms of the community interact. The biotic and abiotic exchange energy and materials.

In simplest terms all ecosystems, aquatic and terrestrial, consist of three basic structural and functional components: autotrophs, heterotrophs, and inorganic and dead organic matter. The autotrophs, the energy-capturing base of the system, are largely green plants and algae (see Chapter 6). They fix the energy of the sun and manufacture food from simple inorganic and organic substances (see Chapters 6 and 7). Autotrophic metabolism is greatest in the upper layers of the ecosystem—the canopy of the forest and the surface water of lakes and oceans.

The heterotrophs utilize the food stored by the autotrophs, rearrange it into other organic compounds, and finally decompose the complex materials into simple, inorganic substances (see Chapter 8 and 9). In this role they influence the rate of energy and nutrient flow through the ecosystem.

The heterotrophic component is often subdivided into two subsystems: consumers and decomposers (see Chapters 8 and 9). The consumers feed largely on living tissue, whereas the decomposers break down dead matter or **detritus** into inorganic substances. No matter how they are classified, all heterotrophic organisms are consumers, and all in some way act directly or indirectly as decomposers.

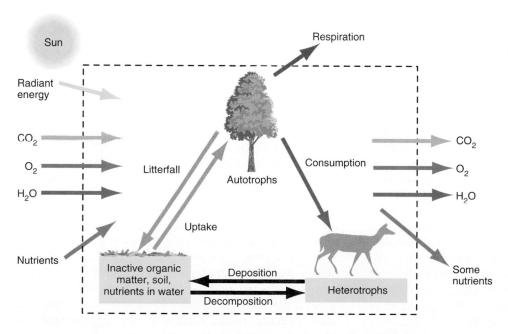

FIGURE 24.1 Schematic diagram of an ecosystem. The dashed line represents the boundary of the system. (Adapted from O'Neill 1976.)

Heterotrophic activity in the ecosystem is most intense where organic matter accumulates—in the upper layer of the soil, in the litter of terrestrial ecosystems, and in the sediments of aquatic ecosystems.

The third component consists of detritus, dissolved organic and inorganic substances in aquatic systems, and the soil matrix. Such matter is the basis of the internal cycling of nutrients in the ecosystem.

Inputs into the system are both biotic and abiotic. The **abiotic inputs** are energy, inorganic substances (CO_2, N, O_2, and mineral nutrients derived from the weathering of rocks), organic compounds (proteins, carbohydrates, humic acids, and organic matter), and precipitation. Radiant energy, both heat and light, imposes constraints on the system by influencing temperature and moisture regimes, seasonality, and photosynthetic activity. Temperature and moisture determine, in part, what organisms can live in a system and affect its productive capability. The **biotic inputs** include other organisms that move into the ecosystem as well as influences imposed by other ecosystems in the landscape.

The driving force of the system is the energy of the sun, which causes all other inputs to circulate through the system. Outflows from one subsystem become inflows to another. Whereas energy is used and dissipated as heat of respiration, the chemical elements from the environment are being recycled. Consumers regulate the speed at which nutrients recycle through the system.

An ecosystem is also a spatial entity; it occupies space. For some ecosystems, such as lakes, ponds, pastureland, golf courses, patches of forest isolated in the landscape, and small watersheds, the boundaries are relatively distinct. They can be isolated as objects (called ecotopes), manipulated relative to inputs and flows among components, and studied (Bormann and Likens 1979, Swank and Crossley 1988). Most ecosystems, however, possess no distinct boundaries. Boundaries described in most cases are arbitrary subdivisions of a continuous gradation of communities.

Ecosystems are dynamic; they change with time. The environment that influences the inputs of energy and matter is continuously variable, and the biotic components live, die, and adapt to the environmental changes. Ecosystems are subject to disturbances (see Chapter 23) such as fire, floods, hurricanes, volcanic activity, and human disruptions that change their structure and function. Thus a given ecosystem on any particular site is not a permanent entity, but part of a shifting pattern on the landscape. Biotic and abiotic components making up the ecosystem structure may change, biomass may accumulate or decline, but functional processes still operate, although perhaps at a different level.

The ecosystem concept is not without its critics. Much of the criticism of the concept is leveled at idea advanced by some ecologists that the ecosystem is a well-defined physical, living entity—a superorganism—that is deterministic and goal-oriented, seeking an equilibrium with the abiotic environment. Out of this concept arises such vague ideas as ecosystem health and ecosystem integrity.

In spite of critics, ecosystems, however ill-defined on the landscape, possess certain genuine properties that are a mix of biological and physical processes. Ecosystems involve the import and export of nutrients, and physical and chemical processes that retain and release nutrients from soil and water. Ecosystems possess trophic structure—the sum of the feeding activities of individual organisms in the ecosystem—that provides the pathway for the movement of materials between the abiotic and biotic environment. Thus, the primary focus of ecosystem ecology is the exchange of energy and matter. This forms the basis of the discussion in the chapters that follow.

THE NATURE OF ENERGY

The sunlight that floods Earth is the ultimate source of energy that keeps the planet functioning. Solar energy arrives as light in packets of energy known as photons (see Chapter 3). When these photons reach the atmosphere, land, and water, they are transformed into another form of energy, heat, which warms Earth, warms the atmosphere (see Chapter 2), drives the water cycle (Chapter 3), and causes currents of air and water (Chapter 2). Those photons that reach plants are transformed into photochemical energy used in photosynthesis (see Chapter 6). That photochemical energy, fixed in carbohydrates and other organic compounds, becomes the source of energy for other living organisms.

Energy Defined

Energy is defined as the ability to do work. Work is what happens when a force acts through a distance, expressed as force × distance. Energy is either potential or kinetic. Potential energy is energy at rest. It is capable of and available for work. Kinetic energy is energy in motion, doing work. Work that results from the expenditure of energy can either store or concentrate energy (as potential energy) or arrange or order matter without storing energy.

Energy is measured in several units. The SI (Système Internationale) measure of a unit of energy is the *joule;* a joule equals 4.168 one-gram calories. A *calorie* is the amount of heat necessary to raise 1 gram (= 1 ml) of water 1°C at 15°C. When large quantities of energy are involved, a *kilogram calorie* (kcal) or Calorie is more appropriate. A kilogram calorie is the amount of heat required to raise 1 kilogram (or 1 liter) of water 1°C at 15°C. Another often-used unit of measure is the *British thermal unit* (Btu), the amount of heat necessary to raise 1 pound of water 1°F. One Btu equals 252 g cal.

The transfer of energy involves the movement or flow of energy from one point to another, known as flux or power, measured in watts per square meter per second (1 $W \cdot m^{-2} = 4.1868 \times 10^3$ or 4186.8 kcal/sec). This flux of energy requires an energy source, the point from which energy flows, and an energy sink or receiver, the point to which it flows. Without a sink for heat energy, the sun could not be an energy source. Earth receives energy from the sun, absorbs a part of it, and gives up energy as heat to another sink, outer space.

Energy content and flow depend on the actions of individual molecules. Characteristically, thermal energy is distributed rapidly among all molecules in a system without necessarily causing a chemical reaction. The effect of thermal energy is to set the molecules into a state of random motion and vibration. The hotter the object, the more the molecules are moving, vibrating, and rotating. These motions spread from a hot body to a cooler one, transferring energy in molecular terms from one to the other.

The energy of light waves, on the other hand, causes electronic transitions within atoms and molecules, called excitations. These excitations can lead to photochemical reactions as in the light reactions of photosynthesis (see Chapter 6).

Laws of Thermodynamics

The laws or rules that describe the expenditure and storage of energy are called the laws of thermodynamics. The **first law of thermodynamics** concerns the conservation of energy. It states that energy is neither created nor destroyed. Energy may change forms, pass from one place to another, or act on matter, transforming it to energy in various ways. However, regardless of what transfers and transformations take place, no gain or loss in total energy of a system occurs. All of it can be accounted for. Energy is simply transformed from one form or place to another.

For example, when wood is burned to ash, the potential energy present in the wood equals the kinetic energy released. Heat is evolved to the surroundings, and so we call this type of reaction **exothermic.** On the other hand, energy from the surroundings may be drawn into a reaction. Here, too, the first law holds true. In photosynthesis, for example, the molecules of the products (glucose) store more energy than the reactants (carbon dioxide and water). The extra energy comes from sunlight. Again there is no gain or loss in total energy. When energy from the outside flows into a system to raise it to a higher energy state, the reaction is **endothermic.**

Although the total amount of energy in any reaction, such as burning wood, does not increase or decrease, much of the potential energy is degraded in quality and becomes unable to perform further work. This energy ends up as heat, serving to disorganize or randomly disperse the molecules. The measure of this relative disorder is called **entropy.**

This degradation of energy is the subject of the **second law of thermodynamics,** which makes an important generalization about energy transfer. It states that when energy is transferred or transformed, part of the energy is lost as waste; it assumes a form that cannot be passed on any further. When coal is burned in a boiler to produce steam, some of the energy creates steam that performs work, but part of the energy is dispersed as heat to the surrounding environment. The same thing happens to energy in an ecosystem. As energy is transferred from one organism to another in the form of food, a large part of the energy is degraded as heat through metabolic activity, with a net increase in entropy. The remainder is stored as living tissue. However, biological systems do not seem to conform to the second law, for the tendency of life is to produce order out of disorder, to decrease rather than increase entropy.

The second law, theoretically, applies to the isolated, closed system, in which there is no exchange of energy or matter between the system and its surroundings. Closed systems are largely physical laboratory creations. Closed systems approach thermodynamic equilibrium, a point at which all energy has assumed a form that cannot do work. A closed system tends toward a state of minimum free energy (energy available to do work) and maximum entropy, whereas an open system maintains a state of higher free energy and lower entropy. In other words, the closed system tends to run down; the open one does not. As long as there is a constant input of free energy and matter to the system and a constant outflow of entropy (in the form of heat and waste), the system maintains a steady state. Life is an open system maintained in a steady state.

Energy entering the biosphere as visible light is stored during photosynthesis in energetic covalent bonds found mainly in glucose. From that point, energy and materials move through the ecosystem. As they do, biochemical changes reduce matter into compounds of less potential chemical energy. The chemical changes are accompanied by the production of metabolic heat, which eventually goes into the energy sink. This loss of heat is accompanied by a loss of carbon dioxide, water, and nitrogenous compounds that are cycled through the biosphere. Although some energy is irrevocably lost from the biosphere into outer space, some of it is stored in the system.

Cycling of Energy in Ecosystems

The ultimate source of energy in natural ecosystems is solar radiation. The range of wavelengths between 400 and 700 nm, known as photosynthetically active radiation (see Chapter 3), is used by green plants in photosynthesis (see Chapter 6). Autotrophs use the energy harvested from the sun to convert inorganic carbon compounds, primarily CO_2, into simple sugars—organic carbon compounds. These carbon compounds are the source of energy for plant respiration allowing for maintenance and the synthesis of new tissues, as well as the source of food energy for heterotrophs. Energy stored in the chemical bonds of organic carbon-based compounds forms the basis of energy flow through ecosystems. Thus, any discussion of energy flow through ecosystems is fundamentally a discussion of solar energy and carbon and must begin with an understanding of the initial fixation of energy by autotrophs.

PRIMARY PRODUCTION

The flow of energy through the ecosystem starts with the process of photosynthesis (see Chapter 6). During photosynthesis, pigments within the chloroplasts intercept solar energy. It is this energy (stored as the chemical products of the light reactions; see Chapter 6) that is utilized to fuel the chemical reactions that produce simple organic carbon compounds from carbon dioxide and water. The bonds of these organic molecules are energy-rich, in effect storing energy derived from the sun. Thus, all organic compounds of plants—cellulose, starch, nucleic acids, proteins, and the various other compounds—contain high amounts of chemical bond energy. Energy accumulated by plants is called **production,** or more specifically **primary production,** because it is the first and most basic form of energy storage in an ecosystem.

All of the energy that is assimilated in photosynthesis represents **gross primary production (GPP).** Like all living organisms, plants require energy for the maintenance of basic metabolic processes. The energy for these needs is provided by through the oxidation of organic compounds in the process of **respiration (R)** (see Chapter 6). Energy remaining after respiration and stored as organic matter is **net primary production (NPP).** This flow of energy can be expressed by the formula

$$\text{net primary production (NPP)} = \text{gross primary production}$$
$$\text{(GPP)} - \text{respiration (R)}$$

Production, both gross and net, is usually measured as the rate at which energy or biomass is produced per unit area per unit time, or **productivity.** This rate is expressed in such terms as kilocalories per square meter per year ($kcal/m^2/yr$), a measure of energy, or grams per square meter per year (g dry wt/m^2/yr), a measure of biomass. If you know the caloric value of the plant material, you can convert dry weight per unit area to kilocalories.

The accumulated organic matter found on a given area at a given time is the **standing crop biomass.** Biomass is usually expressed as grams dry weight of organic matter per unit area (g dry wt/m^2). Biomass differs from productivity, which is the rate at which organic matter (biomass) is produced. Nor is the biomass present at any given time the same as total production. Biomass does not include any plant tissue (production) eaten by herbivores or lost through death prior to the period of measurement.

High standing crop biomass need not imply high productivity, because the size of a standing crop represents accumulated biomass, not a rate of production. A low-productivity ecosystem may have a high biomass because it has accumulated living material over a long period of time. In contrast, a highly productive ecosystem may have a low standing biomass if it has undergone an intense disturbance, such as fire, in the recent past. The size and longevity of the organisms within the ecosystem will also influence the standing crop biomass. Smaller organisms may have short life expectancies, show high turnover rates, and accumulate less biomass than longer-lived, larger organisms. For example, phytoplankton populations in a lake might have a turnover rate (generation time) from days to a week. In addition, predation by zooplankton reduces their standing biomass. Although they are highly productive, the accumulated biomass is low. The size of their standing crop has little relationship to their productivity.

Although the size of the standing crop is not synonymous with high productivity, the size of the standing crop does influence the capacity to produce. A pond with too few fish or a for-

est with too few trees does not have the capacity to utilize the energy available. On the other hand, too many fish or too many trees results in less energy available to each individual. Crowding lowers the efficiency of resource use and influences the storage and transfer of energy through the ecosystem.

Environmental Controls on Primary Productivity

Productivity of terrestrial ecosystems is influenced by climate. Measured estimates of net primary productivity for a variety of different terrestrial ecosystems are plotted in Figure 24.2 as a function of annual precipitation and mean daily temperature for each site. As can be seen from the graphs, net primary productivity increases with increasing temperature and rainfall. These relationships between primary productivity and the environmental factors of temperature and precipitation are a result of their influence on the rates of photosynthesis, the amount of leaf area that can be supported, and the duration of the growing season (see Chapter 7).

Primary productivity is a function of the rate of photosynthesis (energy capture) and the total surface area of leaves that are photosynthesizing. Extremely cold and hot tempera-

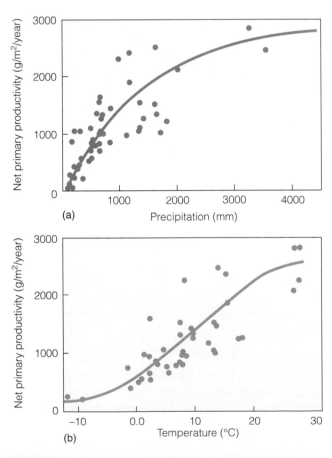

FIGURE 24.2 Net primary productivity for a variety of terrestrial ecosystems (a) as a function of mean annual precipitation, and (b) as a function of mean annual temperature. (Adapted from Reichle 1970.)

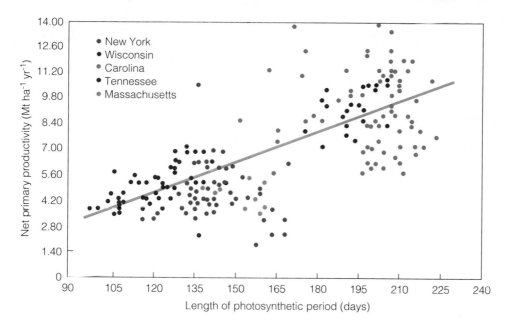

FIGURE 24.3 Relationship between net primary productivity of deciduous forest and the length of the growing season. (Leith 1975.)

tures limit the rate of photosynthesis. Within the range of temperatures that are tolerated, rates of photosynthesis rise with temperature. In addition, with the exception of the tropical zone, the length of the growing season is defined as the period when temperatures are sufficiently warm to support photosynthesis and net primary productivity. As a result, warmer temperatures typically support both higher rates of photosynthesis and a longer growing season (Figure 24.3), resulting in a higher net primary productivity.

As discussed in Chapter 6, for photosynthesis and productivity to occur in terrestrial ecosystems, the plant must open its stomata to take in CO_2. When the stomata are open, water is lost from the leaf to the surrounding air. For the plant to keep its stomata open, water must be available for uptake by the roots to replace that lost through transpiration. The higher the rainfall, the more water is available to plants for transpiration. The amount of water available to the plant will therefore limit both the rates of photosynthesis and the amount of leaves (surface area that is transpiring) that can be supported.

Although the two graphs in Figures 24.2a and b show the independent effects of temperature and precipitation on primary productivity, in reality the influence of these two factors is very much interrelated. Warm temperatures result in high water demand (see Chapter 7). If temperatures are warm but water availability (precipitation) is low, productivity will also be low. Likewise, if water availability is high but temperatures are low, productivity will be low. It is the combination of warm temperatures and adequate water supply to meet the demands of transpiration that results in the highest values of primary productivity. This pattern is reflected in Figure 24.4, which relates the rate of net primary productivity of various ecosystems with the measured value of actual evapotranspiration. The combined value of surface evaporation and transpiration, evapotranspiration reflects both the

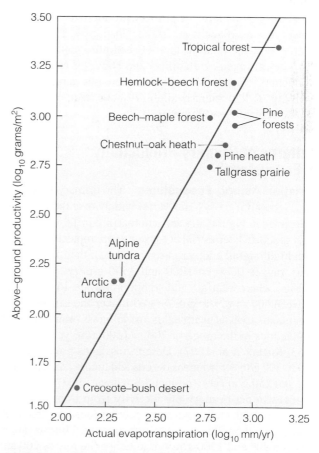

FIGURE 24.4 Range of above-ground net productivities of world ecosystems based on evapotranspiration, which depends on both precipitation and temperature. (From R. H. MacArthur and J. H. Connell, *The Biology of Populations*, p. 175, Fig. 7.3. New York: John Wiley, 1966. Used by permission.)

demand and supply of water to the ecosystem. The demand
is a function of incoming radiation and temperature, while
supply is a function of precipitation.

The pattern of net primary productivity as a function of
actual evapotranspiration presented in Figure 24.4 reflects the
direct link between ecosystem water balance and primary
productivity. The relationship between ecosystem water balance and primary productivity is clearly demonstrated in the
work of Grier and Running (1978) and Gholz (1982) in the
Pacific Northwest. Vegetation was sampled along a transect
from the Pacific Coast to the east slopes of the Cascade
Mountains (Figure 24.5a). The vegetation along the transect
represents 12 major ecosystem zones found in Oregon and
Washington states. Observed patterns of maximum leaf area
index, standing biomass, and net primary productivity are all
strongly related to a simple index of growing season water
balance (Figures 24.5b and c). The water balance index is a
measure of the difference between precipitation and potential evapotranspiration (water demand) during the growing
season (May–October).

In addition to climate, the availability of essential nutrients required for plant growth (see Table 3.1 and Chapter 7)
has a direct effect on ecosystem productivity. John Pastor and
colleagues (Pastor et al. 1984) examined the role of nitrogen
availability on patterns of primary productivity in different
forest types on Blackhawk Island, Wisconsin. Their results
clearly show the relationship between nitrogen mineralization rate (nitrogen availability) and above-ground primary
productivity (Figure 24.6). Similar increases in primary productivity with nutrient availability have been observed in
aquatic ecosystems (Figure 24.7).

Patterns of Primary Productivity in Ecosystems

Variation Among Ecosystems   The primary productivity of terrestrial ecosystems varies widely over the globe and
is depicted in Figure 24.8 and summarized in Table 24.1. The
most productive terrestrial ecosystems are tropical rain forests
with high rainfall and warm temperatures; their net productivity ranges between 1000 and 3500 g/m²/yr. Temperate
forests, where rainfall and temperature are lower, range
between 600 and 2500 g/m²/yr (Whittaker and Likens 1975).
Shrublands such as heath balds and tallgrass prairie have net
productions in the range of 700 to 1500 g/m²/yr (Whittaker
1963, Kucera et al. 1967). Desert grasslands produce about
200 to 300 g/m²/yr, whereas deserts and tundra range between
100 and 250 g/m²/yr (Rodin and Bazilevic 1967). These differences in net primary productivity from tropic to arctic
regions are reflected in litter production (dead organic matter reaching the soil surface), which in tropical forests ranges
between 900 and 1500 g/m²/yr, in temperate forests 200 and
600 g/m²/yr, and in arctic and alpine regions 0 and 200
g/m²/yr (Bray and Gorham 1964).

Net productivity of the open sea is generally quite low
(Figure 24.9). Productivity in the open waters of the cool temperate oceans tends to be higher than those of the tropical
waters as a function of nutrient supply (see discussion in

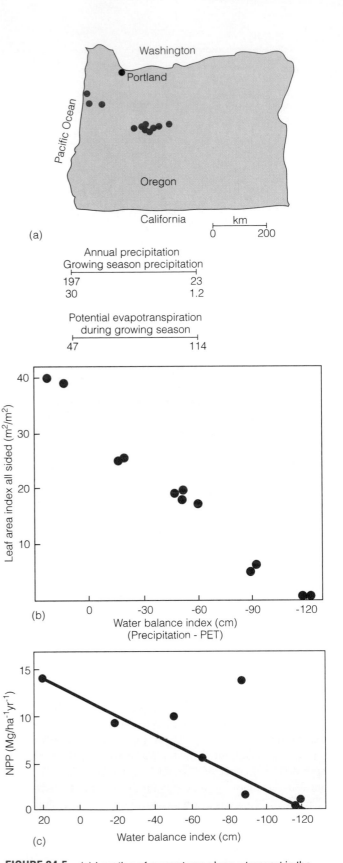

FIGURE 24.5   (a) Location of ecosystems along a transect in the
Pacific Northwest. Patterns of annual growing season (May–October)
precipitation and potential evapotranspiration (PET) for the transect
are shown. (b) Maximum observed leaf area index of ecosystems
along transect expressed as a function of water balance index.
(c) Annual net primary productivity for 7 of the 12 ecosystems along
the transect expressed as a function of water balance index.
(Adapted from Grier and Running 1978 and Gholz 1982.)

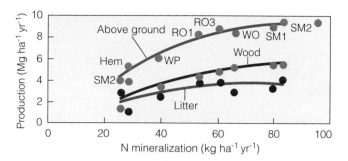

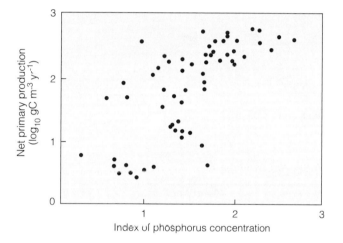

FIGURE 24.6 Relationship between above-ground net primary production (total, wood and litter) and nitrogen mineralization rate (annual) for a variety of forest ecosystems on Blackhawk Island, Wisconsin. The abbreviations refer to the dominant trees in each stand: Hem=hemlock; RO=red oak; WO=white oak; SM=sugar maple; WP=white pine.

FIGURE 24.7 Relationship between net primary production of phytoplankton in lakes throughout the world and estimated steady-state concentration of phosphorus. (After Schindler 1978.)

Chapter 25). However, in some areas of tropical upwelling (see Chapter 2), such as the Peru Currents (seen as a band of high productivity in the tropical ocean waters off the west coast of South America), net productivity can reach 1000 g/m²/yr. Coastal ecosystems and the continental shelves generally have higher productivity than the open waters. Shallow waters and the input of nutrients from terrestrial ecosystems via rivers function to increase plant productivity. Swamps and marshes, ecosystems at the interface of land and water, have net productivity ranging from 900 to 3300 g/m²/yr. Estuaries, because of input of nutrients from rivers and tides, and coral reefs, because of input from changing tides, have a net productivity between 1000 and 2500 g/m²/yr.

Consistently high productivity usually results from an energy subsidy to the system. This subsidy may be a warmer temperature, greater rainfall, circulating or moving water that carries in food or additional nutrients, or in the case of agricultural crops, the use of fossil fuel for cultivation and irrigation, the application of fertilizer, and the control of pests. Among agricultural ecosystems, sugar cane has a net productivity of 1700 to 1800 g/m²/yr, hybrid corn 1000 g/m²/yr, and some tropical crops 3000 g/m²/yr.

Temporal Changes Within Ecosystems In any ecosystem, annual net production (accumulation of biomass) changes with age. For example, a Scots pine (*Pinus sylvestris*) plantation studied by Ovington (1961) achieved maximum production of 22×10^3 kg/ha at the age of 20; it then declined to 12×10^3 kg/ha at 30 years of age. In general, primary productivity in terrestrial ecosystems increases initially during succession or stand development, followed by a decline as time progresses (Figure 24.10). Growth is initially slow, increases as leaf area develops, peaks as leaf area reaches its maximum, and declines thereafter.

The decline in productivity of forest ecosystems has commonly been attributed to a changing balance between photosynthesis and respiration (Yoda et al. 1965, Whittaker and Woodwell 1967). This hypothesis states that leaf area will reach some maximum limited by the resources of light, water, and nutrients at the site. From that point forward, stand photosynthesis will remain constant as a result of the constant leaf area. However, respiration will continue to increase as biomass accumulates with the growth of nonphotosynthetic tissues (stems and roots). Since the net primary productivity of the stand is the difference between gross photosynthesis and respiration, productivity will decline. Although accepted almost universally until recent times, recent evidence does not support this hypothesis as the explanation of declining productivity with ecosystem age (Ryan et al. 1997). Estimates of woody-tissue respiration suggest that maintenance respiration is too small to account for the decreased productivity (Ryan and Waring 1992). Ryan and colleagues (1977) suggest an alternative explanation that involves a variety of potential causes including reduced photosynthesis with plant age, decreased nutrient supply with stand age, and increased reproductive effort and increased mortality of older individuals. The decline in primary productivity is a combination of reductions in both photosynthesis and respiration with stand age (Figure 24.11).

Productivity within an ecosystem varies not only systematically through time, but also from year to year as a result of variations in the physical environment. As previously discussed, productivity is influenced by such factors as nutrient availability, precipitation, temperature, length of the growing season, herbivory, and disturbances such as fire. For example, the net primary productivity of a grassland ecosystem may vary by a factor of eight between wet and dry years (Weaver and Albertson 1956). Overgrazing of grasslands by cattle and sheep or defoliation of forests by such insects as the gypsy moth can seriously reduce net production. Fire in grasslands may increase productivity if rainfall is average or above, but reduce it if precipitation is low (Kucera et al. 1967). An insufficient supply of nutrients, especially nitrogen and phosphorus, can limit net productivity, as can atmospheric pollution.

Energy Allocation

Net primary production represents the storage of organic matter in plant tissue in excess of respiration. Plants budget this fixed energy or net income for different uses. A portion is allocated to growth, the buildup of components such as stems

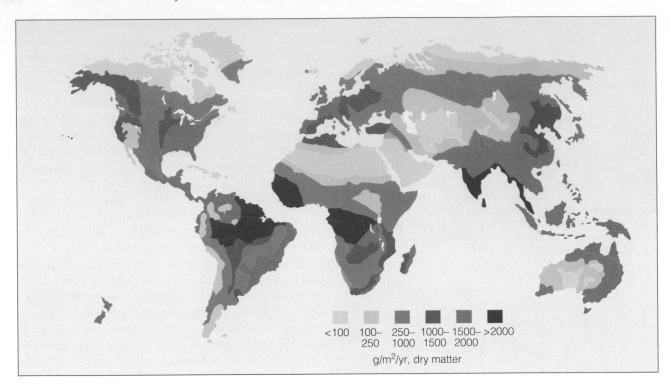

FIGURE 24.8   A map of world terrestrial primary production. Note the high productivity of tropical regions. (Based on Golley and Leigh 1972.)

TABLE 24.1   Net Primary Productivity and Plant Biomass of World Ecosystems

Ecosystems (in Order of Productivity)	Area (106 km2)	Mean Net Primary Productivity per Unit Area (g/m²/yr)	World Net Primary Productivity (10⁹ mtr/yr)	Mean Biomass per Unit Area (kg/m²)
CONTINENTAL				
Tropical rain forest	17.0	2000.0	34.00	44.00
Tropical seasonal forest	7.5	1500.0	11.30	36.00
Temperate evergreen forest	5.0	1300.0	6.40	36.00
Temperate deciduous forest	7.0	1200.0	8.40	30.00
Boreal forest	12.0	800.0	9.50	20.00
Savanna	15.0	700.0	10.40	4.00
Cultivated land	14.0	644.0	9.10	1.10
Woodland and shrubland	8.0	600.0	4.90	6.80
Temperate grassland	9.0	500.0	4.40	1.60
Tundra and alpine meadow	8.0	144.0	1.10	0.67
Desert shrub	18.0	71.0	1.30	0.67
Rock, ice, sand	24.0	3.3	0.09	0.02
Swamp and marsh	2.0	2500.0	4.90	15.00
Lake and stream	2.5	500.0	1.30	0.02
Total continental	149.0	720.0	107.09	12.30
MARINE				
Algal beds and reefs	0.6	2000.0	1.10	2.00
Estuaries	1.4	1800.0	2.40	1.00
Upwelling zones	0.4	500.0	0.22	0.02
Continental shelf	26.6	360.0	9.60	0.01
Open ocean	332.0	127.0	42.00	0.003
Total marine	361.0	153.0	55.32	0.01
World total	510.0	320.0	162.41	3.62

Source: Adapted from Whittaker and Likens 1973.

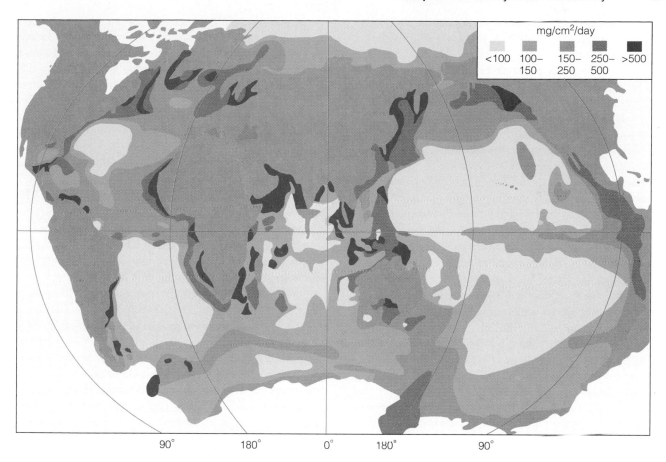

FIGURE 24.9 Geographical variations in the primary productivity of the ocean. Note that the highest productivity is near the coastal zones; the lowest productivity is in the open sea.

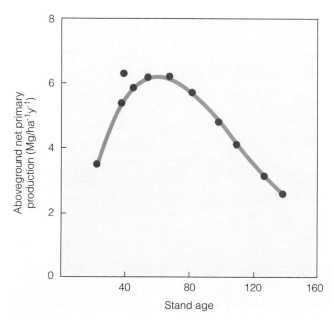

FIGURE 24.10 Changes in above-ground net primary productivity with age for a stand of white spruce (*Picea abies*) in the region of Karelia, Russia. (From Ryan et al. 1997, adapted from data in DeAngelis et al. 1980.)

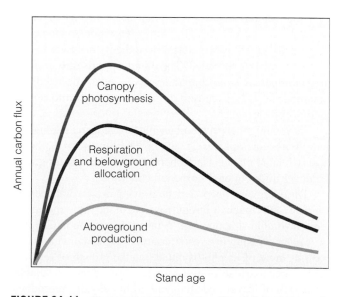

FIGURE 24.11 Proposed model of carbon flux and above-ground net primary productivity during forest stand development. Canopy photosynthesis increases as leaf area develops in young forest stands. As the stand ages, rates of photosynthesis and respiration decline, resulting in a reduction in productivity. (From Ryan et al. 1997.)

and leaves that promote the further acquisition of energy and nutrients (Chapin et al. 1990). A portion goes to storage, which is photosynthate built up in the plant for future growth and other functions (Chapin et al. 1990).

This storage involves accumulation, reserve formation, and recycling. Accumulation is the increase of compounds that do not directly support growth. These include carbon compounds such as such as starch and fructose, nitrogen as specialized storage proteins, and mineral ions. Accumulation occurs when the resource supply exceeds demands of growth and maintenance. It is most prevalent in species with inherently slow growth rates, such as trees. Reserve formation involves the synthesis of storage compounds from resources that otherwise would be allocated directly to promote growth. By recycling material from aging tissue to new growth, the plant can retain compounds that otherwise would be lost to litter (see discussion of nutrient retranslocation in Chapter 7). How plants budget their energy resources tells much about their life history and the way they respond to environmental conditions and stresses.

With a limited energy budget and changing needs through its life history, a plant has to divert resources from one use to another during the year. Failure to maintain a balanced energy budget kills the plant. Early in its life history, a plant spends net production on growth—new leaves and stems—that will add to its earning power of increased photosynthesis (see Chapter 6). Once a plant has built up capital in vegetative growth, it diverts energy from and often at the expense of growth to a reserve budget from which it can draw energy when needed, such as the replacement of leaves lost to herbivores. Later in the life cycle, the plant spends a considerable amount of its photosynthate on flowers and then fruits. The plant withdraws resources from its reserve budget and vegetative capital to spend on reproduction, for its fitness depends on such allocations.

This withdrawal is evident in maturing annual and perennial plants. Because the plant has not only diverted resources away from the large lower leaves on its stem, but has actually withdrawn material from them, the lower leaves become brown or shriveled. The plant supports only a sufficient number of small upper leaves to meet the needs of maintenance. All of the remaining energy goes into reproduction and storage.

As you might expect, budget allocations differ among the many types and species of plants. They even vary among closely related plants or members of the same species growing under different environmental conditions.

The energy budget of a plant growing in a favorable environment looks much different from one of the same species in a less favorable situation (see detailed discussion in Chapter 6). A plant growing in a resource-impoverished environment expends more of its photosynthate on the growth of roots to reach scarce nutrients or moisture. It has less energy available for growth of leaves and stem and, ultimately, for reproduction. The same species living on a nutrient-rich environment expends less on roots and more on vegetative growth. Some plants use more nutrients than they need, a strategy called luxury consumption. In the presence of an abundance of nitrogen,

many plants increase their uptake of that element, decrease their allocation to roots, increase their investments in leaves and stems, accumulate above-ground biomass, and even fail to allocate sufficient energy to reproduction (see Fitter 1986).

Resource allocations differ considerably among annuals, perennial herbs, and woody plants. An annual plant (Figure 24.12) gets its initial start from the minimal amount of energy contained in the seed. Once the first seedling leaves begin photosynthesis, the plant invests up to 60 percent of its production to growth of above-ground vegetative biomass. When the plant has sufficient growth in leaf, stem, and root, the annual plant diverts production to flowers; only about 10 to 20 percent now goes to leaves. As the fertilized flowers develop into seeds, the plant withdraws energy from leaves and roots and sends 90 percent of its current photosynthetic production into growth and ripening of seeds, its only means of ensuring survival to the next year (Larcher 1995).

Perennials (Figure 24.13) have a different lifestyle, and their energy budget reflects it. Many, such as asters, goldenrods, and daisies, begin growth in the spring from energy reserves stored in roots and, in some species such as saxifrage (*Saxifraga*) and roundleaf ragwort (*Senecio obovatus*), the rosettes of basal leaves. Once growth begins, the plant puts all of its production into vegetative structure, laying down these reserves as capital to draw on for flowering. Once the plant is ready to flower, it diverts its production away from storage to production of flowers and seeds, even to the point of sacrificing some of the lower leaves on the stalk. Near the end of the season, perennial plants translocate energy from their aging leaves to the roots, where the energy will be available for initiating next season's growth (Larcher 1995).

Other herbaceous plants, such as white trillium (*Trillium grandiflorum*) (Figure 24.14) and spring beauty (*Claytonia* spp.), take a different approach. They draw on energy reserves

FIGURE 24.12   Annual plants, such as foxtail (*Setaria*), allocate most of their production to growth and ripening of seeds. Seeds are their only means of overwinter survival.

FIGURE 24.13 Perennial plants, like this ox-eye daisy (*Chrysanthemum leucanthemum*), start growth in spring from energy stored in roots, to which they return energy in the fall. During the flowering period, the plant diverts most of its energy to reproduction.

FIGURE 24.14 White trillium (*Trillium grandiflorum*), an early spring flower, draws on energy reserves in the roots to support early blooms and early vegetative growth. Then they send their production back into the roots to build up reserves for the following spring.

stored in the root to support early flowers or other reproductive structures and early leaves in the spring. In late spring and early summer, they produce more leaves that generate photosynthate, which is sent back into the roots to build up capital for next year's blooms. When the plant has accomplished that, it ceases production and above-ground biomass dies.

Woody plants must invest their energy into more items, with proportionately more going to woody tissue and roots. In spring the tree draws on and uses up at least one-third of the reserves to get its new leaves started. Once photosynthesis begins, the deciduous trees budget their energy pro-

duction for growth of leaves, then flowers, fruit, new cambium, new buds, and final deposits of starch in roots and bark, roughly in that order. A tree's most expensive drain on its annual energy budget is the production of fruit. A pine tree expends 5 to 15 percent of its annual photosynthetic production on cones; beech and oaks may spend more than 20 percent of their net energy income on nuts and acorns; and apple trees can use up 35 percent on apples (Larcher 1995). In fact, fruit production is so expensive that most trees can afford the luxury of a good crop of fruit only once every three to seven years. When they do, trees draw down their carbohydrate reserves and limit the growth of other components. At the end of the growing season, deciduous trees withdraw carbohydrates and nutrients from the leaves and send them back to the roots.

The patterns of energy allocation for trees change with the age and size of the individual (Figure 24.15). Initially, individuals allocate a greater percentage of energy fixed in photosynthesis to the production of current twigs and leaves, which function to further increase the capture of energy in photosynthesis and promote growth. As the tree increases in size, a greater proportion of the energy is allocated to the production of stems and then branches. As a result the relative proportion of total biomass contained in leaves and twigs declines with tree age and size. On an ecosystem-wide basis, individuals of a variety of species with different life histories and annual variations in reproductive activity coexist and the overall patterns of energy allocation represent a weighted average of the component species.

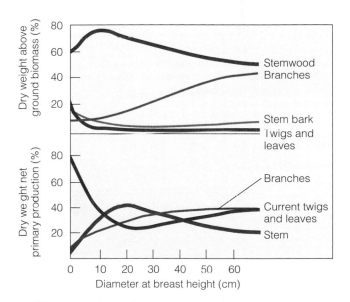

FIGURE 24.15 Relation of above-ground biomass and production to size of tree for 63 sample trees of three major species common to the deciduous forest of eastern North America—sugar maple, yellow birch, and beech. Note that as trees increase in size, the proportion of biomass in stemwood and branches increases. In smaller trees, current leaves and twigs account for the greater percentage of primary production. This percentage declines rather rapidly as trees approach pole stage (20 to 30 cm dbh), then increases as trees mature. (From Whittaker et al. 1974:239.)

Root-to-Shoot Ratios

Overall, the proportionate allocation of net primary production to below-ground and above-ground biomass tells much. A high root-to-shoot ratio (R/S) indicates that most of the production goes into the supportive function of plants and that most of their active biomass is below ground. A large root biomass enables plants to reach water and nutrients. Thus, plants with a large root biomass are more effective competitors for water and nutrients and can survive more successfully in infertile and harsh environments. Plants with a low R/S ratio have most of their biomass above ground and assimilate more light energy, resulting in higher productivity.

Differences in root-to-shoot ratios characterize various ecosystems. Sedge and grass meadows of the tundra, plants characteristic of an environment with a long, cold winter and a short growing season, have R/S ratios ranging from 5 to 11. Tundra shrubs may range from 4 to 10. Prairie grasses have an R/S ratio of about 3, indicative of cold winters and low moisture supply. Forest ecosystems, with their high above-

ground biomass, have a low R/S ratio. For the Hubbard Brook Forest in New Hampshire, the R/S ratio (based on data of Gosz et al. 1976) for trees is 0.213, for shrubs 0.5, and for herbs 1.0. As we could predict, the R/S ratio increased from the trees down to the herbaceous layer.

Changes in R/S ratios reflect the response of plants to stress. Grasses may respond to grazing stress by concentrating more of their net production in roots (Andrews et al. 1975). In heavily grazed plots on shortgrass prairie, grasses allocated 69 percent of net primary production to roots, 19 percent to shoots, and 12 percent to basal crowns. In contrast, lightly grazed prairie allocated 60 percent of net primary production to roots, 22 percent to shoots, and 18 percent to crowns.

Biomass Distribution

Above-ground biomass is distributed vertically in the ecosystem (Figure 24.16). The vertical distribution of leaf biomass in terrestrial systems, and of floating-leaf vegetation and the concentration of phytoplankton in the surface waters of aquatic systems, influences the penetration of light (see

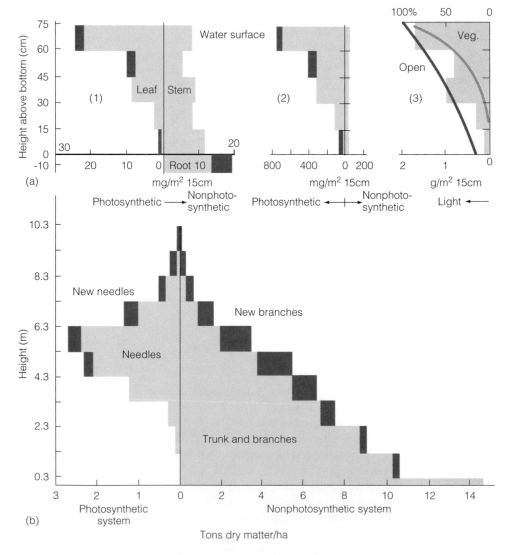

FIGURE 24.16 Vertical distribution of production and biomass in aquatic and terrestrial communities. (a) Three graphs for the pondweed (*Potamogeton*) community: (1) division of biomass into leaf, stem, and root (solid areas represent winter buds); (2) concentration of chlorophyll in the plant community; (3) leaf area and light profile. The orange line is light in the community; the blue line represents light in open water. Note that less light reaches greater depths in the community because of vegetation. (After Ikusima 1965.) (b) Structure and productive system (foliage) of a pine-spruce-fir forest in Japan. (After Monsi 1968.)

Chapter 3). The degree of light penetration, in turn, influences the distribution of production in the ecosystem.

The region of maximum productivity in aquatic ecosystems is not the upper strongly sunlit surface (strong sunlight inhibits photosynthesis) but some depth below, depending on the clarity of the water and the density of plankton growth. As depth increases, light intensity decreases until it reaches a point at which the light received by the phytoplankton is just sufficient to meet respiratory needs and production equals respiration. This point is known as **compensation intensity.**

In the forest ecosystem a similar situation exists. The greatest amount of photosynthetic biomass as well as the highest net photosynthesis is not at the top of the canopy, but at some point below maximum light intensity. In spite of wide differences in plant species and types of ecosystems, the vertical profiles of biomass are similar.

SECONDARY PRODUCTION

Net primary production is the energy available to the heterotrophic components of the ecosystem. Theoretically, at least, all of it is available to the herbivores or even to the decomposers, but rarely is it all utilized in this manner. The net primary production of any given ecosystem may be dispersed outside of the ecosystem by humans, wind, or water. Much of the living material is physically unavailable to the grazers—they cannot reach many plants or plant parts. The organic matter of live organisms is unavailable to decomposers and detritus-feeders, and grazers may not feed on dead plants. The amount of net production available to herbivores may vary from year to year and from place to place. The quantity consumed varies with the type of herbivore and the density of the population. Once consumed, a considerable portion of the plant material, again depending on the kind of plant and the digestive efficiency of the herbivore, may pass through the animal's body undigested. A grasshopper assimilates only about 30 percent of the grass it consumes, leaving 70 percent available for the decomposers (Smalley 1960). Mice, on the other hand, assimilate about 85 to 90 percent of what they consume (Golley 1960, Smith 1962).

Energy, once consumed, either is diverted to maintenance, growth, and reproduction or is passed from the body as waste products—feces, urine, and fermentation gases (Figure 24.17). The energy content of feces is transferred to the detritivores. Of the energy left after these losses, part is utilized as heat increment, which is heat required for metabolism above basal or resting metabolism. The remainder of the energy is **net energy,** available for maintenance, production, and reproduction. It includes energy involved in capturing or harvesting food, muscular work expended in the animal's daily routine, and energy needed to keep up with the wear and tear on the animal's body. The energy used for maintenance is lost as heat.

Maintenance costs, highest in small, active, warm-blooded animals (see Chapter 8), are fixed or irreducible. In small invertebrates, energy costs vary with temperature, and

a positive energy balance exists only within a fairly narrow range of temperatures. Below 5°C, spiders become sluggish, cease feeding, and have to utilize stored energy to meet metabolic needs. At approximately 5°C, energy absorbed into living cells (assimilated) approaches energy lost through respiration. From 5° to 20.5°C, spiders assimilate more energy than they respire. Above 25°C, their ability to maintain a positive energy balance declines rapidly (Van Hooke 1971).

Energy remaining from maintenance and respiration—net energy—goes into **secondary** or consumer production—fat, growth, and the birth of new individuals. (Within secondary production there is no portion known as gross production. What is analogous to gross production is actually assimilation.) The utilization and fate of available energy in the form of exploitable biomass by consumers is outlined in Figure 24.18

A consumer's energy budget can be represented as

$$C = A + (F + U)$$

where C is the energy ingested or consumed, A is the energy assimilated, and F and U are the energy lost through feces and nitrogenous wastes.

The term A can be refined further as

$$A = P + R$$

where P is secondary production and R is energy lost through respiration. U, representing nitrogenous wastes, should be included as part of A ($A = P + R + U$), because the production of such wastes is involved in the homeostasis of organisms.

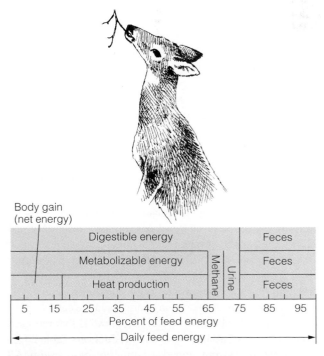

FIGURE 24.17 Relative values of the end products of energy metabolism in white-tailed deer. Note the small amount of net energy gained (body weight) in relation to that lost as heat, gas, urine, and feces. The deer is a herbivore, a first-level consumer. (After Cowan 1962:5.)

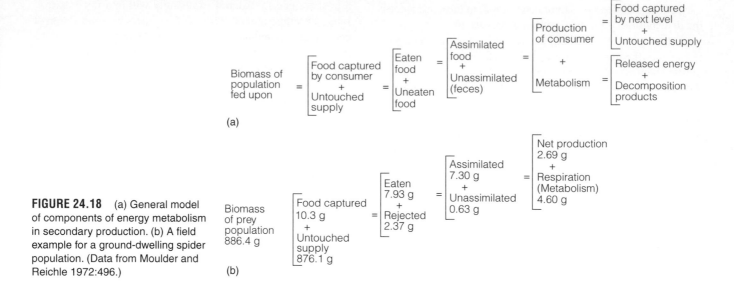

FIGURE 24.18   (a) General model of components of energy metabolism in secondary production. (b) A field example for a ground-dwelling spider population. (Data from Moulder and Reichle 1972:496.)

Because of the difficulty of separating nitrogenous from fecal wastes, U typically is not included in A. Thus,

$$C = P + R + (F + U)$$

or secondary production is

$$P = C - R - (F + U)$$

Secondary production depends on the on quantity, quality (including nutrient status and digestibility), and availability of net production primary productivity as a source of energy. Therefore, any of the environmental constraints on primary productivity, such as climate, will also act to constrain secondary productivity within the ecosystem. For example, Figure 24.19a shows the observed relationship between mean annual rainfall and the productivity of large herbivores in African ecosystems. The increase in large herbivore production with increasing rainfall is a direct result of the corresponding increase in net primary productivity shown in Figure 24.2a, which provides the food source for the herbivore populations. A similar relationship between phytoplankton production (primary productivity) and zooplankton production (secondary production) in lake ecosystems is shown in Figure 24.19b.

Although the model of consumption and fate of energy in Figure 24.18 pertains to all consumers (secondary producers), the variety of heterotrophs inhabiting an ecosystem can differ quite markedly in their rates of consumption, assimilation, and production. These differences are critical to understanding the flow of energy through ecosystems. Differences among heterotrophs in energy balance can be examined from the viewpoint of two different ratios that represent the efficiencies of assimilation and production (Table 24.2).

The ratio of assimilation to consumption or ingestion, A/I, is the **assimilation efficiency.** This index is a measure of the efficiency of the consumer at extracting energy from the food it consumes. It relates to food quality and effectiveness of digestion. The ratio of production to assimilation, P/A, is

called the **production efficiency.** This is a measure of the efficiency of a consumer in incorporating assimilated energy into new tissue (growth), or secondary production. The production efficiency reflects the relative balance of the energy allocation to production and respiration. These two ratios, the assimilation and production efficiencies, underlie the magnitude of a third ratio: production to consumption, P/I. This index indicates how much energy consumed by the animal is converted into production. It is a measure of the efficiency with which energy is made available to the next group of consumers.

The ability of the consumer population to use the energy it ingests varies with the type of consumer and the species (Table 24.2). Homeotherms use about 98 percent of their assimilated energy in metabolism and only about 2 percent in secondary production. Poikilotherms, on the average, convert about 44 percent of their assimilated energy to secondary production. They turn a greater proportion of their assimilated energy (A) into biomass (P). However, there is a major difference in assimilation efficiency between poikilotherms and homeotherms. Poikilotherms have an efficiency of about 42 percent, whereas homeotherms have an average efficiency of over 70 percent. Therefore, the poikilotherm has to consume more calories than a homeotherm to obtain sufficient energy for maintenance, growth, and reproduction.

A wide range of conversion efficiencies exists among various feeding groups (Table 24.3). Production efficiency in plants (net production/light absorbed) is low, ranging from 0.34 percent in some phytoplankton to 0.8 to 0.9 percent in grassland vegetation. Herbivores use plant production with varying degrees of efficiency, depending on whether they are poikilotherms or homeotherms. Because they eat foods already converted to animal tissue, carnivores, both poikilothermic and homeothermic, have high assimilation efficiencies. On the North American midwestern grasslands, average herbivore production efficiency, involving mostly poikilotherms, ranges from 5 to 16 percent. Carnivores have production efficiencies ranging from 13 to 24 percent.

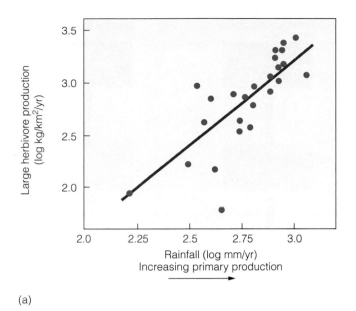

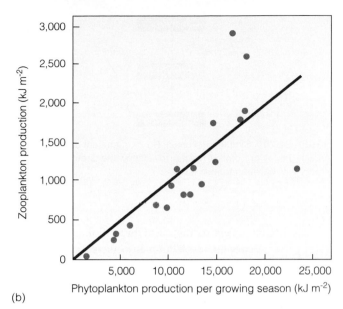

(a)

(b)

FIGURE 24.19 The relationship between primary and secondary productivity: (a) rainfall (which affects primary productivity; see Figure 24.2a) and secondary productivity of large mammalian herbivores in Africa (adapted from Coe et al 1976); (b) phytoplankton and zooplankton production in lake ecosystems. (Adapted from Brylinsky and Mann 1973.)

TABLE 24.2 Assimilation Efficiency and Production Efficiency for Homeotherms and Poikilotherms

Efficiency	All Homeotherms	Grazing Arthropods	Sap-Feeding Herbivores	Lepidoptera	All Poikilotherms
Assimilation					
A/I	77.5 ± 6.4	37.7 ± 3.5	48.9 ± 4.5	46.2 ± 4.0	41.9 ± 2.3
Production					
P/A	2.46 ± 0.46	45.0 ± 1.9	29.2 ± 4.8	50.0 ± 3.9	44.6 ± 2.1
P/I	2.0 ± 0.46	16.6 ± 1.2	13.5 ± 1.8	22.8 ± 1.4	17.7 ± 1

Source: Based on data from Andrzejewska and Gyllenberg 1980.

TABLE 24.3 Consumer Efficiency (Secondary Production/Secondary Consumption)

Habitat	Growing Season (days)	PRODUCERS Production	PRODUCERS Efficiency (%)	HERBIVORES Production	HERBIVORES Efficiency (%)	CARNIVORES Production	CARNIVORES Efficiency (%)
Shortgrass plains	206	3.767	0.8	53	11.9	6	13.2
Midgrass prairie	200	3.591	0.9	127	16.5	37	23.7
Tallgrass prairie	275	5.022	0.9	162	5.3	15	13.9

Source: Based on data from Andrzejewska and Gyllenberg 1980.

FOOD CHAINS

Energy fixed by plants is the base on which the rest of life on Earth depends. This energy stored by plants is passed along through the ecosystem in a series of steps of eating and being eaten known as a **food chain.** Feeding relationships within a food chain are defined in terms of trophic or consumer levels. From a functional rather than a species point of view, all organisms that obtain their energy in the same number of steps from the autotrophs or primary producers belong to the same **trophic level** in the ecosystem. The first trophic level belongs to the primary producers, the second level to the herbivores,

FIGURE 24.20 Simple food chain. Boxes represent different feeding groups or trophic levels. Arrows show the direction of energy flow between trophic levels.

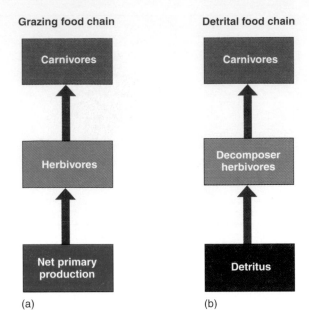

FIGURE 24.21 Two parts of any ecosystem: (a) a grazing food chain and (b) a detrital food chain.

and the higher levels to the carnivores. Some consumers occupy a single trophic level, but many others, such as omnivores, occupy more than one trophic level.

Food chains are descriptive. They represent a more abstract expression of the food webs presented in Chapter 20. Major feeding group are defined based on a common source of energy, such as autotrophs, herbivores, and carnivores. Each feeding group is then linked to others in a manner that represents the flow of energy. A simple food chain is presented in Figure 24.20. Three feeding groups—autotrophs, herbivores, and carnivores—are represented by boxes. The arrows linking the boxes represent the direction in which energy flows.

Major Food Chains

Within any ecosystem there are two major food chains, the grazing food chain and the detrital food chain (Figure 24.21). The two food chains are distinguished by their source of energy or food for the initial consumers. In grazing ecosystems autotrophs, or living plant tissues, are the primary source of energy for the initial consumers, the herbivores. In the detrital food chain, the initial consumers, primarily bacteria and fungi, use dead organic matter, detritus, as their source of energy.

Grazing Food Chains The grazing food chain is the one more obvious to us. Cattle grazing on pastureland, deer browsing in the forest, rabbits feeding in old fields, and insect pests feeding on garden crops represent the basic consumer groups of the grazing food chain. Although highly conspicuous, the grazing food chain is not the major food chain in terrestrial and many aquatic ecosystems. Only in some aquatic ecosystems do the grazing herbivores play a dominant role in energy flow.

Voluminous data exist on phytoplankton productivity, filtration rates by grazing zooplankton, and production efficiencies of zooplankton. Few data, however, are available on the flow of energy, rate of grazing, biomass turnover rates for phytoplankton, and turnover of zooplankton biomass within the same aquatic system. Hillbricht-Ilkowska (1974) studied the relationship between primary production of phytoplankton and consumer production in freshwater lakes. He found that the direct utilization of primary production by filter-feeding zooplankton was intense. Its efficiency of energy transfer was high

when the size and structure of phytoplankton were favorable for filter feeding. In lakes where net-forming phytoplankton was dominant, direct utilization was low and energy transfer inefficient. As an example, in lakes in which the productivity of phytoplankton ranged from 200 to 1200 kcal/m^2/yr, productivity of filter-feeding zooplankton amounted to 10 to 250 kcal/m^2/yr with a transfer efficiency of 2 to 30 percent. In lakes in which filter-feeding zooplankton productivity ranged from 50 to 150 kcal/m^2/yr, the productivity of predaceous zooplankton ranged from 2 to 50 kcal/m^2/yr with an energy transfer efficiency of 5 to 40 percent.

In terrestrial systems, a small proportion of primary production goes by way of the grazing food chain. Over a three-year period, only 2.6 percent of net primary production of a yellow-poplar forest was used by grazing herbivores, although the holes made in the growing leaves resulted in a loss of 7.2 percent of photosynthetic surface (Reichle et al. 1973). In a study of energy flow through a shortgrass prairie ecosystem, involving ungrazed, lightly grazed, and heavily grazed plots, Andrews et al. (1974) found that on the heavily grazed prairie cattle consumed 30 to 50 percent of aboveground net primary production. In both lightly and heavily grazed plots, about 40 to 50 percent of energy consumed by the cattle is returned to the ecosystem and the detrital food chain as feces (Dean et al. 1975). In the heavily grazed Serengeti plains of East Africa, herbivores consume 33 to 66 percent of annual above-ground production. In some instances consumption may reach 99 percent (McNaughton 1985, McNaughton and Georgiadis 1988).

Although the above-ground herbivores are the conspicuous grazers, below-ground herbivores can have a pronounced impact on primary production and the grazing food chain. Andrews et al. (1975) found that below-ground herbivores consisting mainly of nematodes (Nematoda), scarab beetles (Scarabaeidae), and adult ground beetles (Carabidae) accounted for 81.7 percent of total herbivore assimilation on the ungrazed shortgrass prairie, 49.5 percent on the lightly grazed prairie, and 29.1 percent on the heavily grazed prairie. Ninety percent of the invertebrate herbivore consumption took place below ground. Nematodes processed 50 percent of the total energy. On the lightly grazed plots, prairie cattle consumed 46 kcal/m^2 during the grazing season and the below-ground invertebrates consumed 43 kcal/m^2. When a nematicide was added to a midgrass prairie, above-ground net production increased 30 to 60 percent. Thus, below-ground herbivores can impose a greater stress on the ecosystem than above-ground herbivores.

Size of both food items and consumers has considerable influence on the direction a food chain takes, because there are upper and lower limits to the size of food an animal can capture. Some animals are large enough to defend themselves successfully. Some foods are too small to be collected economically; it takes too long to secure enough to meet the animal's metabolic needs. Thus the upper limit to the size of an animal's food is determined by its ability to handle and process, and the lower limit by its ability to secure enough.

There are exceptions, of course. By injecting poisons, spiders and snakes kill prey much larger than themselves; wolves hunting in packs can kill an elk or a caribou. The idea that food chains involve animals of progressively larger sizes is true only in a general way. Where parasites are involved, the opposite situation exists. The larger animals are at the base, and as the number of links increases, the size of the parasites becomes smaller.

Golley (1960) carefully worked out a grazing food chain for old-field vegetation, meadow mice, and weasels (Figure 24.22). The mice are almost exclusively herbivorous, and the weasels live mainly on mice. The vegetation converts about 1 percent of the solar energy into net production, or plant tissue. The mice consume about 2 percent of the plant food available to them, and the weasels about 31 percent of the mice. Of the energy assimilated, the plants lose about 15 percent through respiration, the mice 68 percent, and the weasels 93 percent. The weasels use so much of their assimilated energy in maintenance that a carnivore preying on weasels could not exist.

In a very general way, energy transformed through the ecosystem by way of the grazing chain is reduced by a magnitude of 10 from one level to another. Thus if an average of 1000 kcal of plant energy is consumed by herbivores, about 100 kcal is converted to herbivore tissue, 10 kcal to first-level carnivore production, and 1 kcal to second-level

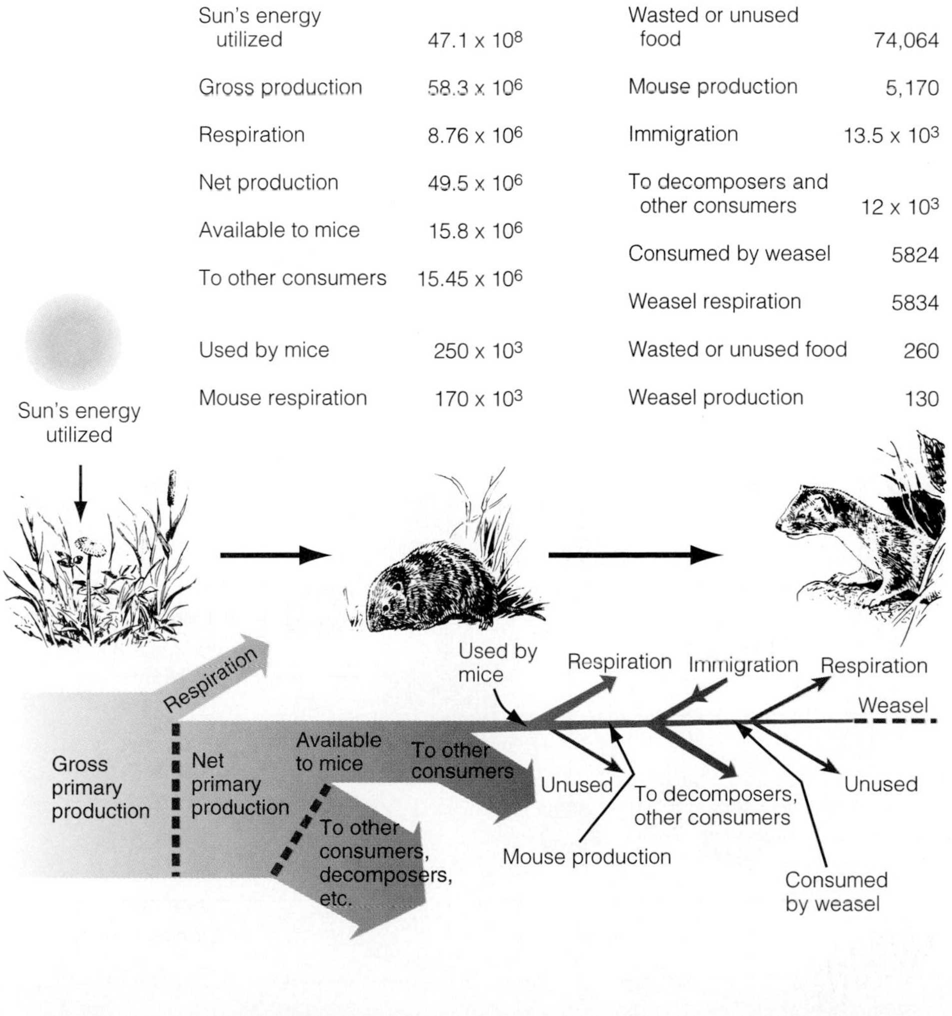
Sun's energy utilized 47.1 x 10^8
Gross production 58.3 x 10^6
Respiration 8.76 x 10^6
Net production 49.5 x 10^6
Available to mice 15.8 x 10^6
To other consumers 15.45 x 10^6
Used by mice 250 x 10^3
Mouse respiration 170 x 10^3
Wasted or unused food 74,064
Mouse production 5,170
Immigration 13.5 x 10^3
To decomposers and other consumers 12 x 10^3
Consumed by weasel 5824
Weasel respiration 5834
Wasted or unused food 260
Weasel production 130
Sun's energy utilized
Respiration
Gross primary production
Net primary production
Available to mice
To other consumers
To other consumers, decomposers, etc.
Used by mice
Respiration
Immigration
Respiration
Weasel
Unused
To decomposers, other consumers
Unused
Mouse production
Consumed by weasel

FIGURE 24.22 Energy flow through a food chain in an old-field community in southern Michigan. The relative size of the blocks suggests the quantity of energy flowing through each channel. Values are in cal/ha/yr. (Based on data from Golley 1960.)

carnivores. The amount of energy available to second- and third-level carnivores is so small that few organisms could be supported if they depended on that source alone. For all practical purposes, each food chain has from three to four links, rarely five (Pimm 1982). The fifth link is distinctly a luxury item in the ecosystem.

Detrital Food Chains   The detrital food chain is common to all ecosystems, but in terrestrial and littoral ecosystems it is the major pathway of energy flow. In yellow-poplar (*Liriodendron tulipifera*) forests, 50 percent of gross primary production goes into maintenance and respiration, 13 percent is accumulated as new tissue, 2 percent is consumed by herbivores, and 35 percent goes to the detrital food chain (Edwards, unpublished, cited by O'Neill 1976). Two-thirds to three-fourths of the energy stored in a grassland ecosystem that is ungrazed by cattle is returned to the soil as dead plant material, and less than one-fourth is consumed by herbivores (Hyder 1969). Of the quantity consumed by herbivores, about one-half is returned to the soil as feces. In the salt marsh ecosystem, the dominant grazing herbivore, the grasshopper, consumes just 2 percent of the net production available to it (Smalley 1960).

Energy flow through a detrital food chain is difficult to measure, although the use of radioactive tracers gives some idea of energy flow through selected food webs. Andrews et al. (1975), basing their estimates on a number of studies, determined that microbial activity accounted for 99 percent of total saprophytic assimilation in a shortgrass prairie ecosystem. Saprophagic grazers such as nematodes accounted for the remaining 1 percent.

Gist and Crossley's (1975) study of a selected invertebrate population living in forest litter provides an example of a detrital food chain (Figure 24.23). Although they did not collect data for energy flow, the flux of radioactive calcium and phosphorus permitted the construction of a food web. The quantity of elements provided some idea of the energy flow. The litter is consumed by five "herbivorous" groups: millipedes (Diploda), orbatid mites (Cryptostigmata), springtails (Collembola), cave crickets (Orthoptera), and pulmonate snails (Pulmonata). The mites and springtails are the most important of these litter feeders. Small spiders (Araneidae) and predatory mites (Mesostigmata) prey. The predatory mites feed on annelids, mollusks, insects, and other arthropods. Spiders feed on predatory mites, springtails, pulmonate snails, and small spiders; Carabid beetles and medium-sized spiders prey on cave crickets and other insects. The medium-sized spiders, in turn, become additional items in the diet of the beetles. Beetles, spiders, and snails are eaten by birds and small mammals, members of the grazing food chain not shown on the original web. Through predation, detrital food webs feed into grazing food chains at higher consumer levels. Detrital and grazing food webs are separate compartments in an ecosystem only at the detritus and primary consumer levels.

Supplementary Food Chains   Other feeding groups, such as the parasites and scavengers, form supplementary food chains in the community. Parasitic food chains are highly complicated because of the life cycle of the parasites. Some parasites are passed from one host to another by predators in the food chain. External parasites (ectoparasites) may move from one host to another. Other parasites are transmitted by insects from one host to another through the bloodstream or plant fluids.

FIGURE 24.23   A detrital food web involving litter-dwelling invertebrates of a hardwood forest in the southeastern United States. The dashed line represents the boundaries of the system. Note that the detritus food chain involves a herbivorous component—millipedes and mites to the left and snails, crickets, and springtails in the center. The herbivores support a carnivorous component. The detrital food web, like the grazing food web, can become complex. (After Gist and Crossley 1975:86.)

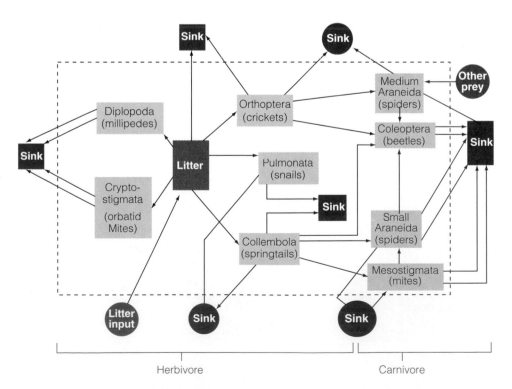

Food chains also exist among parasites themselves. Fleas that parasitize mammals and birds are in turn parasitized by a protozoan, *Leptomonas*. Chalcid wasps (Chalcedectidac) lay eggs in the ichneumon or tachinid fly grub (Ichneumonidae), which in turn is parasitic on other insect larvae.

Interactions Between Major Food Chains

The two major food chains, grazer and detrital, are combined in Figure 24.24 to produce a generalized model of trophic structure and energy flow through an ecosystem. The two food chains are linked, with the initial source of energy for the decomposer food chain being the input of waste materials and dead organic matter from the grazing food chain, indicated as a series of arrows from each of the trophic levels in the grazing food chain going toward the box designated as detritus (dead organic matter). There is one notable difference in the flow of energy between trophic levels in the grazing and decomposer food chains. In the grazing food chain the flow is unidirectional, with net primary production providing the energy source for herbivores, herbivores providing the energy for carnivores, and so forth. In the decomposer food chain, the flow of energy is not unidirectional. The waste materials and detritus (organisms) in each of the consumer trophic levels are recycled, returning as an input to the detritus box at the base of the food chain.

In addition to the flow of detritus and waste materials from the grazer to the decomposer food chains, these two food chains are also interconnected via the process of predation. In addition to breaking down dead organic matter, decomposer organisms are also food to numerous other animals. Slugs eat the larvae of certain flies and beetles, which live in the heads of fungi and feed on the soft material. Mammals, particularly the red squirrel and chipmunks, eat woodland fungi. Dead plant remains are food sources for springtails and mites, which in turn are eaten by carnivorous insects and spiders. They in turn are energy sources for insectivorous birds and small mammals. Blowflies (Calliphoridae) lay their eggs in dead animals, and within 24 hours the maggot larvae hatch. Unable to eat solid tissue, they reduce the flesh to a fetid mass by enzymatic action in which they feed on the proteinaceous material. These insects are food for other organisms.

MODELS OF ENERGY FLOW

The concept of energy flow in ecological systems is one of the cornerstones of ecology. The model was first developed by Raymond Lindeman in 1942 in a study of the trophic dynamic structure of Lake Mendota in Wisconsin. The trophic dynamic concept is based on the assumptions that the laws of thermodynamics hold for plants and animals, that plants and animals can be arranged into feeding groups or trophic levels, that at least three trophic levels—primary producer, herbivore, and carnivore—exist, that the net energy content of one trophic level is passed on to the next trophic level above, and that the system is in equilibrium.

Whether at the producer or consumer level, energy flow through the ecosystem is mediated by the individual (Figure 24.25a). A quantity of energy in food is consumed. Part of it is assimilated and part is lost as feces, urine, and other waste products. Part of the assimilated energy is used for respiration, and part is stored as new tissue that can be used to some extent as an energy reserve. Some is used for growth of the individual and some for reproduction.

A model of energy flow through a population (Figure 24.25b) shows some additions. Growth in individual biomass becomes changes in population size or standing crop. Part of the biomass goes to predators and parasites, and there are gains and losses of energy and biomass from and to other ecosystems.

The population boxes can be fitted into several trophic levels and linked to form a model of energy flow through the ecosystem (Figure 24.25c). In this particular model there is no attempt to separate out the decomposers and detritus-feeders, because they must fall into one of the several trophic levels (herbivore, carnivore, etc.) when the food web is collapsed into this most simple form.

The linked model of grazing and detrital food chains introduced in Figure 24.24 is an expanded form of the Lindeman's trophic dynamic model presented in Figure 24.25. Together the two major food chains, grazer and decomposer, combine to produce a generalized model of trophic structure and energy flow through an ecosystem. To quantify the flux of energy through the ecosystem requires that consumption, ingestion, assimilation, respiration, and production at each trophic level be evaluated.

The procedure can be demonstrated using the diagram of a single trophic compartment as in Figure 24.26a. The energy available to a given trophic level (designated as n) is the production of the next lower level ($n - 1$); for example, net primary production (trophic level $n - 1$) is the available energy

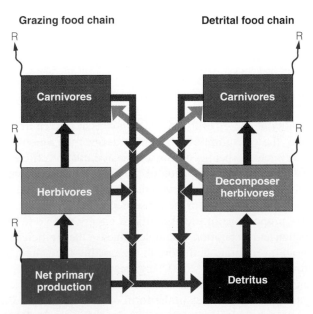

FIGURE 24.24 Grazing and detrital food chains from Figure 24.21 combined, showing their connections. R = respiration.

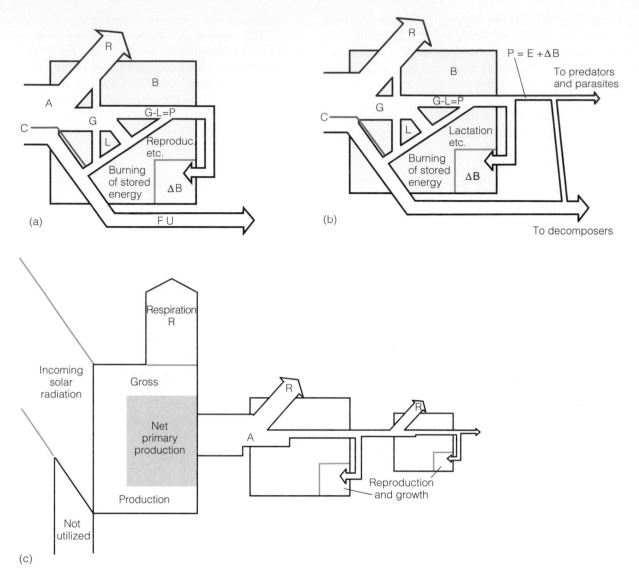

FIGURE 24.25   Models of energy flow: (a) through the individual organism, (b) through the population, and (c) through the ecosystem. Note the losses and the portion of energy accumulated as growth in the organism. The ecosystem model considers the decomposers as occupying one of several trophic levels rather than as on a separate trophic pathway. *A*=assimilation, *B*=biomass, ΔB = changes in biomass, *C*=consumption, *E*=material removed from population to ecosystem, *FU*=rejects, *G*=total biomass growth, *L*=decrement of biomass through weight loss and death, *R*=respiration, and *P*=production.

for grazing herbivores (trophic level n). Some proportion of that production is consumed or ingested (I). The remainder eventually makes its way to the dead organic matter of the detritus food chain. Of the energy consumed, some portion is assimilated by the organisms (A) and the remainder is lost as waste materials (W) that make their way to the detritus food chain. Of the energy assimilated, some is lost to respiration, shown as the arrow labeled R, which is leaving the upper left corner of the box, and the remainder goes to herbivore production (P_n).

This flow can be quantified using the efficiency values presented earlier, together with an additional efficiency value, the **consumption efficiency,** the ratio of ingestion to production of the trophic level below (I/P_{n-1}). Given a

quantity of energy available to organisms at a given trophic level, the consumption efficiency defines the amount of energy actually being consumed. The assimilation efficiency (A/I) defines the proportion of the energy ingested that is assimilated and the proportion that is lost as waste material. Production efficiency of the consumer (P_n/A) defines the relative proportions of the assimilated energy that go to production and respiration. Example values of these efficiencies for an invertebrate herbivore in the grazing food chain are provided in Figure 24.26b. Using these efficiency values, we can track the fate of a known quantity of energy (1000 kcal) available to herbivores in the form of net primary productivity through the herbivore trophic level (Figure 24.26b).

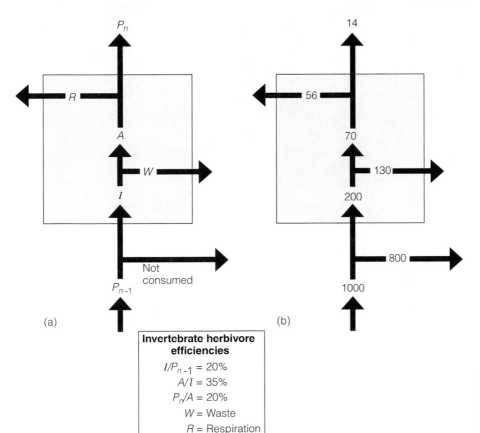

Invertebrate herbivore
efficiencies

$I/P_{n-1} = 20\%$

$A/I = 35\%$

$P_n/A = 20\%$

W = Waste

R = Respiration

FIGURE 24.26 (a) Energy flow within a single trophic compartment. (b) A quantified example of energy flow through that compartment for an invertebrate herbivore. Values are in kcal.

By defining the efficiency values shown in Figure 24.26 for each trophic level in the grazing and herbivore food chains, the flow of energy from net primary productivity through the ecosystem can be calculated in the same manner as shown for the single trophic level. The production from each trophic level provides the input to the next higher level, while the unconsumed production, waste products, and dead individuals from each trophic level provide the input of energy into the dead organic matter compartment. Ultimately, the entire flow of energy through the ecosystem is a function of the initial transformation of solar energy into net primary productivity. Eventually, all energy entering the ecosystem as net primary production is lost through respiration.

Energy Flow in Different Ecosystems

Although the general model of energy flow presented in Figure 24.24 pertains to all ecosystems, the relative importance of the two major food chains and the rate at which energy flows through the various trophic levels can vary widely among different types of ecosystems. In most terrestrial and shallow water ecosystems, with their high standing crop and relatively low harvest of primary production, the detrital food chain is dominant. In deep water aquatic ecosystems, with their low biomass, rapid turnover of organisms, and high rate of harvest, the grazing food chain may be dominant.

The amount of energy moving through these two routes varies among ecosystems. In an intertidal salt marsh, less than 10 percent of living plant material is consumed by herbivores and 90 percent goes the way of the detritus-feeders and decomposers (Teal 1962). In fact, most of the organisms of the intertidal salt marsh obtain the bulk of their energy from dead plant material. In a Scots pine plantation, almost 50 percent of the energy fixed annually is utilized by decomposers (Figure 24.27). The remainder is removed as yield or is stored in tree trunks (Ovington 1961). In some communities, particularly those that are undergrazed, unconsumed organic matter may accumulate and remain out of circulation for some time, especially when conditions are not favorable for microbial activity. The decomposer or detritus food chain receives additional materials from the waste products and dead bodies of both the herbivores and carnivores.

A comparison of the general patterns of energy flow through four distinct ecosystem types is presented in Figure 24.28. The relative size of each box represents the amount of energy in each trophic level of the food chain, and the arrows represent the relative flow of energy between trophic levels. The open water ecosystem and associated phytoplankton community have the highest consumption efficiency (proportion of NPP consumed), with the grazing food chain playing a greater role in energy flow than in the other three ecosystem types. In terrestrial ecosystems, the grazing food chain is much more important in grassland than in forest

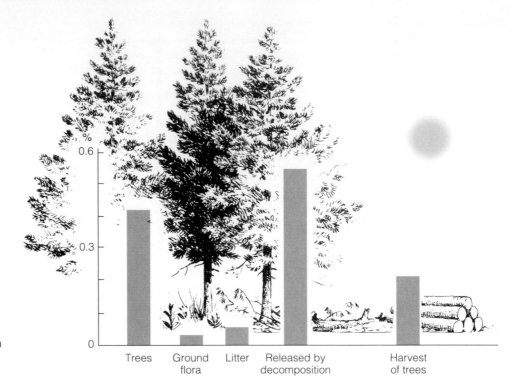

FIGURE 24.27   The fate of the 1.3 percent of solar energy assimilated as net production by a 23-year-old Scots pine plantation. (After Ovington 1961.)

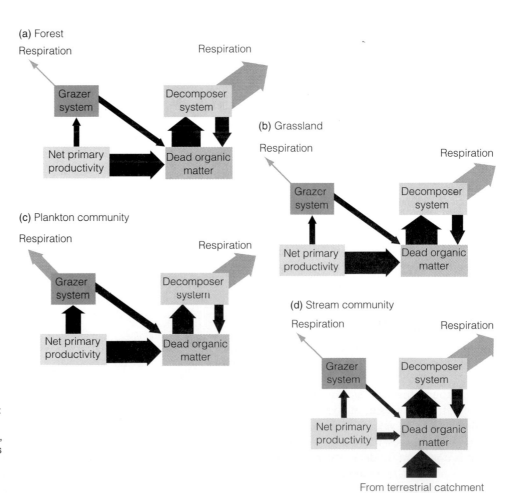

FIGURE 24.28   General patterns of energy flow through four ecosystems: (a) forest, (b) terrestrial grassland, (c) ocean (phytoplankton community), and (d) stream. Relative size of boxes and arrows are proportional to the relative magnitude of compartments and flow. (From Begon et al. 1986.)

ecosystems (see Chapter 28). In forest ecosystems, the vast majority of net primary productivity is not consumed as living tissues by herbivores; rather, it is stored as woody standing crop biomass that eventually makes its way to the detrital food chain as dead organic matter (see Chapter 29). Stream ecosystems have extremely low net primary productivity, and the grazing food chain is minor (see Chapter 30). The detrital food chain dominates and depends on inputs of dead organic matter from adjacent terrestrial ecosystems.

Ecological Pyramids

Charles Elton advanced the concept of examining an ecosystem in terms of trophic structure, or consumer layers, in his classic book *Animal Ecology* (1927). He pointed out the great difference in the numbers of organisms involved in each step of the food chain. Consumers at the lower end of the chain are the most abundant. In successive links of consumption, carnivores decrease rapidly in number and increase in size until there are very few carnivores at the top. This concept is known as the **pyramid of numbers** (Figure 24.29a). However, consumer levels based on numbers is misleading. Although consumers at the base of the pyramid may be most abundant, there is great deal of variation in their size and numbers. Plants are eaten by herbivores ranging in size from very small arthropods to elephants.

More realistic is a **pyramid of biomass** (Figure 24.29b). If the numbers of consumers at each feeding level are multiplied by their weight, we arrive at a pyramid of biomass. This pyramid indicates by weight or other measurement of living material the total bulk of organisms of fixed energy present at any one time—the standing crop. Because some energy or material is lost through respiration at each link, the total mass supported at each level is limited by the energy stored or the rate of energy storage in the level below. In general, the biomass of producers is greater than the biomass of the herbivores they support, and the biomass of the herbivores is greater than that of the carnivores. This relationship results in a gradually sloping pyramid, particularly for terrestrial and shallow water communities where the producers are large, accumulation of organic matter is great, life cycles are long, and the rate of harvesting is low.

However, for some ecosystems, the pyramid of biomass is inverted. Primary production in aquatic ecosystems, such as lakes and open seas, is concentrated in phytoplankton. Planktonic algae have a short life cycle, multiply rapidly, accumulate little organic matter, and are heavily exploited by herbivorous zooplankton. At any point the standing crop is low. As a result, the base of the pyramid is much smaller than the structure it supports.

Pyramids of numbers and biomass tell us something about structure, but little about function. When production, amount of biomass accumulated, is considered in terms of energy, the pyramid suggests the amount of energy flow from one level to another. The basis of the pyramid of energy (Figure 24.29c) is the quantity of energy fixed, stored, and passed on to the next trophic level. Such an approach eliminates the

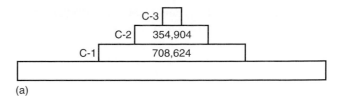

(a)

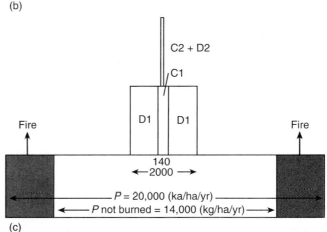

(b)

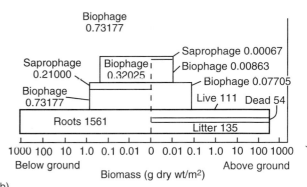

(c)

FIGURE 24.29 Three kinds of ecological pyramids. (a) A composite pyramid of numbers based on vegetation in an old field in Michigan (Evans and Cain 1952) and an animal census of two pastures and a meadow in New York (Wolcott 1937). (After Odum 1983:152.) (b) A pyramid of biomass for a northern shortgrass prairie for July. The base of the pyramid represents biomass (g dry weight/m²) of producers; the second (middle) level, primary consumers; and the top, secondary consumers. Above-ground biomass (right) and below-ground biomass (left) are separated by a dashed vertical line. The trophic level magnitudes are plotted on a horizontal logarithmic scale. The compartments are divided on a vertical linear scale according to live, standing dead, and litter biomass or biophagic and saprophagic consumer biomass. Unlike the conventional pyramids, this one recognizes the detrital as well as the grazing food chain components on the same trophic levels. (After French et al. 1979.) (c) A pyramid of energy for the Lamto Savanna, Ivory Coast. *P* is primary production; C1, primary consumers; C2, secondary consumers; D1, decomposers of vegetable matter; D2, decomposers of animal matter. Again, the detrital and grazing food chains have been collapsed into the same trophic levels. (After La Motte 1975:216.)

biases of numbers and size and emphasizes the role organisms play in the transfer of energy. For example, some organisms have a small biomass, but they assimilate and pass on considerably more energy than some other organisms with larger biomass. On a pyramid of biomass these smaller organisms would appear much less important in the ecosystem than they really are.

The concept of trophic levels has weaknesses. The typical trophic structure is based on plants or primary production and the transfer of energy from primary production to higher levels. It discounts detrital material, decomposers, and saprophages. In other words, it basically is a description of the grazing food chain, although there have been attempts to incorporate biophages and saprophages into pyramids of biomass and of energy (Figure 24.29b). Consumers, especially those above the herbivore level, often occupy more than one trophic level, and their contribution to the biomass and energy flow at each level must be apportioned. Moreover, the trophic level concept does not take into account the availability of energy. All of the energy at any level is not available to consumers. Among green plants, for example, there is a difference in energy available to consumers from various plant parts (Odum and Biever 1984). Certain structures, because of chemical or morphological characteristics, are not eaten. The concept also gives the false impression that energy does not cycle through ecosystems. Actually, a unit of energy reaching the decomposer bacteria and fungi is not necessarily dissipated, as the concept implies; some of it becomes part of the decomposer biomass to be recycled up through the food web again.

ECOSYSTEM PRODUCTIVITY

Net primary productivity is the measure of net energy gain by autotrophs. As presented in this chapter, NPP is the source of energy for the heterotrophic component of the ecosystem (consumers and decomposers). Net primary productivity limits the flow of energy through the grazer and detrital food chains, placing an upper limit on secondary productivity. A measure of total ecosystem productivity must incorporate both primary and secondary productivity. Net ecosystem productivity (NEP) is the sum of energy stored (net gain) in live plant biomass (NPP), live animal biomass, and dead organic matter. As with net primary productivity, net ecosystem productivity is usually measured as the rate at which energy or mass (both living and dead organic matter) is produced per unit area per unit time. This rate is expressed in such terms as kilocalories per square meter per year (kcal/m^2/yr), a measure of energy, or grams per square meter per year (g dry wt/m^2/yr), a measure of biomass.

Net ecosystem productivity can be either positive or negative. Negative values reflect a net loss of energy from the ecosystem. This loss occurs when net energy gain in primary productivity is lower than energy loss through respiration by the decomposers, resulting in a net loss of energy stored in dead organic matter within the ecosystem. This set of conditions occurs in older ecosystems, such as old-growth forests, where the majority of gross primary production is expended on respiration, resulting in low net primary productivity and litterfall. Under this set of conditions, the rates of decomposition can exceed the annual input of dead organic matter.

Summary

1. The biotic community together with the physical environment functioning together as a system is called an ecosystem.
2. The ecosystem is an energy-processing system, receiving abiotic and biotic inputs. The driving force is the energy of the sun. Abiotic inputs include oxygen, carbon dioxide, and nutrients from the weathering of Earth's crust and from precipitation. Biotic inputs include organic material from surrounding ecosystems.
3. The ecosystem itself consists of three components: (a) the autotrophs, producers that fix energy of the sun; (b) the heterotrophs, consumers and decomposers that use the energy and nutrients fixed by the producers and return nutrients to the system; (c) dead organic material and inorganic substrate that act as short-term nutrient pools and maintain the cycling of nutrients within the system.
4. Energy is the ability to do work. Energy is either potential energy capable of and available for work, or kinetic energy, energy in motion or doing work. The laws of thermodynamics describe the expenditure and storage of energy. The first law, conservation of energy, states that energy is neither created nor destroyed. The second law states that at each transfer or transformation of energy, part of that energy assumes a form that cannot be passed on any further because it is randomly dispersed, often as heat. Randomly dispersed energy is called entropy.
5. Ultimately, closed systems run down. Open systems, such as ecosystems, however, with their constant input and outflow maintain a steady state, in spite of the second law of thermodynamics.
6. The flow of energy through an ecosystem begins with the fixation of solar energy by autotrophs in the process of photosynthesis. Total energy fixed as organic molecules by autotrophs is gross primary production. Energy remaining after respiratory costs accumulates as fixed energy or net primary production. The rate at which energy is fixed is productivity.
7. Net primary productivity is a function of the rate of photosynthesis and the total photosynthetic surface. Environmental factors that influence the rate of photosynthesis and leaf area will directly affect the rate of primary productivity. Climate has a direct influence on primary productivity. In general, productivity increases with increasing precipitation and temperature, resulting in a positive correlation between actual evapotranspiration and net primary productivity across ecosystems. Nutrient availability also has a direct control on rates of primary productivity.
8. Plants allocate the energy they fix/capture to above-ground and below-ground vegetative growth—roots leaves, stems, flowers, fruits, and seeds. The amount allocated varies with the type of plant and season. Annuals eventually send 90 percent of their photosynthetic production to growth and ripening of seeds. At growing season's end, perennials translocate energy remaining after seed and flower production to roots. Bulbous plants use stored energy for flower and seed production, then send current photo-

synthate to roots to build up capital for next year's blooms. Woody plants invest energy into more items including growth of leaves, flowers, fruit, new cambium, new buds, and final deposits of starch in roots and bark. The allocation of net primary production to below-ground and above-ground biomass is expressed as root-to-shoot ratio. A high ratio indicates that most of the production goes to supportive tissue and large root biomass below ground.

9. Primary production flows through the heterotrophic component of the ecosystem to be fixed as secondary production. The level of primary productivity limits the maximum secondary productivity of the ecosystem. Energy consumed and assimilated by heterotrophs goes to maintenance, growth, and reproduction. Among homeotherms, most of the energy assimilated goes to maintenance. Among poikilotherms, maintenance costs are much less, and so most of the assimilated energy goes to growth. However, homeotherms assimilate much more of the energy they consume than do poikilotherms.

10. Energy flow in ecosystems may take two routes. One goes through the grazing food chain; the other goes through the detritus food chain, in which the bulk of production is used as dead organic matter by decomposers and saprovores. Much of the energy flowing into the detrital food chain is recycled through the upper levels of both the detrital and grazing chains before its final dissipation.

11. The loss of energy at each transfer limits the number of trophic levels in the food chain to three, four, or rarely, five. Biomass declines at each level. The plot of the weight of individuals at each successive trophic level yields a sloping pyramid. In certain aquatic situations, where there is a rapid turnover of small aquatic consumers, the pyramid of biomass may be inverted. Energy flow, however, always decreases from one trophic level to the next.

12. The ratio of energy flow from one trophic level of an ecosystem to the next or between individual organisms is ecological efficiency. Because efficiencies are dimensionless, several different ratios can be determined. Some of the most useful are assimilation efficiencies, growth efficiencies, and utilization efficiencies.

13. Net ecosystem productivity is the net uptake or loss of energy (biomass or carbon) by the ecosystem, which includes both primary and secondary productivity.

14. Net ecosystem productivity is the net energy or biomass gain by the ecosystem. It differs from net primary productivity in that it includes the production and respiration of the heterotrophic component of the ecosystem.

Review Questions

1. What is meant by an ecosystem? By whom and when was the term coined?
2. What are the three major components of an ecosystem? How are they related?
3. What is the driving force of an ecosystem?
4. Distinguish between potential and kinetic energy.
5. What conditions are necessary for energy flow?
6. What are primary production, primary productivity, gross primary production, standing crop, and respiration?
7. What is the significance of root-to-shoot ratios for production in terrestrial ecosystems?
8. In what manner do plants allocate energy during the year?
9. What is secondary production?
10. What is the difference in energy allocation and energy transfer efficiencies between homeotherms and poikilotherms?
11. What are the two major food chains? How do they interrelate?
12. What is a trophic level? Relate trophic levels to three kinds of ecological pyramids.
13. What problems are associated with the concept of trophic level?
14. Consider your diet for one week and the work out a food chain of which you were a part. On how many trophic levels did you function? What was the original source of the food?

Cross References

Biotic influences on soil, 67–68; the living soil, 64–65; process of photosynthesis, 86–99; process of decomposition, 148–149; plant-herbivore interactions, 291; plant defenses, 286–287; predator-prey relations, 302–304; intraguild predation, 300.

CHAPTER 25

Biogeochemistry I: Nutrient Cycling

OUTLINE

Types of Nutrient Cycles

Model of Nutrient Cycles
    Inputs
    Internal Cycling
    Outputs

Contrasting Nutrient Cycling in Terrestrial and Aquatic Ecosystems

Major Biogeochemical Cycles
    The Carbon Cycle
    The Nitrogen Cycle
    The Sulfur Cycle
    The Phosphorus Cycle

Linkages Among Biogeochemical Cycles

Summary
Review Questions
Cross References

CONCEPTS

1. Nutrients move through the ecosystem in biogeochemical cycles that involve biological, geological, and chemical components.

2. A general model of nutrient cycling in ecosystems includes inputs, outputs, and internal cycling.

3. The internal cycling of nutrients within the ecosystem is closely tied to the rates of two key processes: net primary productivity and decomposition.

4. Environmental factors that influence the rates of net primary productivity and decomposition will directly influence the cycling of nutrients through the ecosystem.

5. The cycling of nutrients in open water aquatic ecosystems is influenced by the seasonal patterns of the vertical temperature profile.

6. The carbon cycle is closely tied to the cycling of energy through the ecosystem.

7. The nitrogen cycle is driven by microbial processes, including the initial fixation of nitrogen from the atmosphere.

8. The phosphorus cycle is wholly sedimentary, with reserves coming largely from phosphate rocks.

9. The sulfur cycle has both a gaseous and sedimentary component.

10. The major biogeochemical cycles are linked, because they all involve constituents of organic matter.

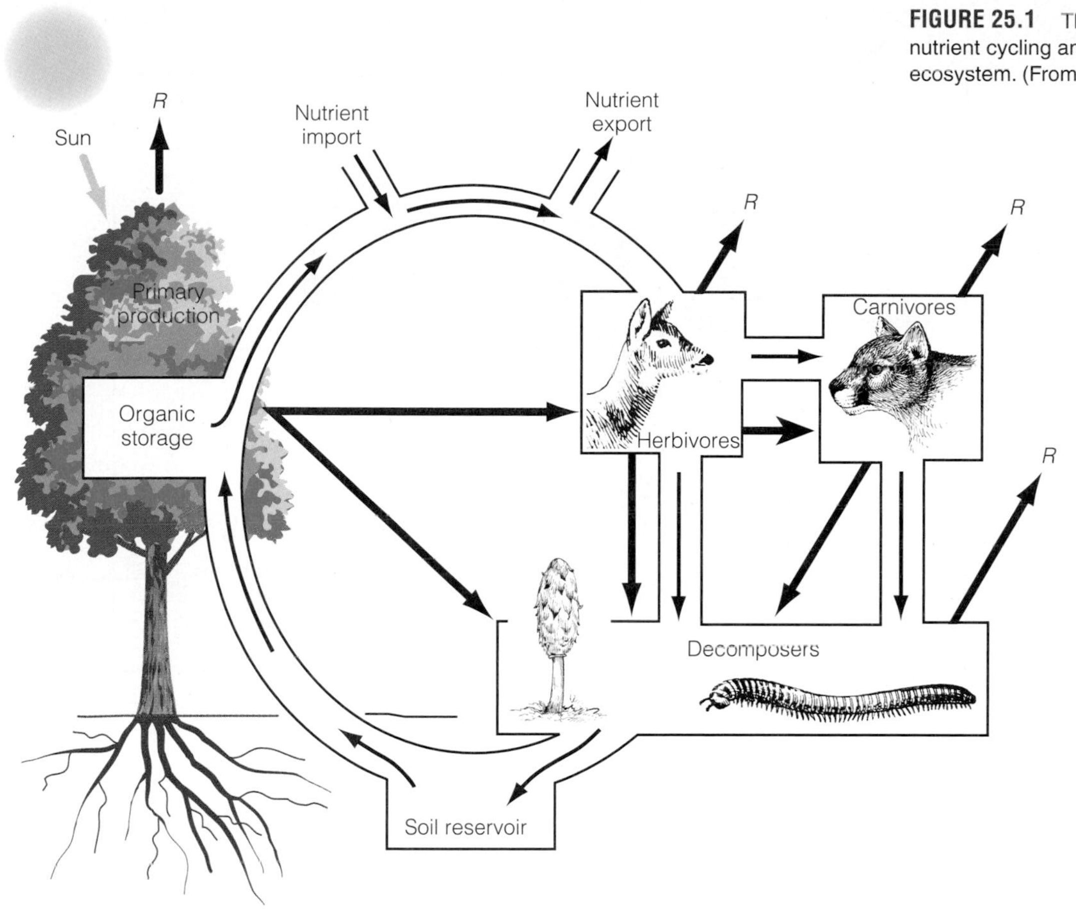

FIGURE 25.1 The interrelationship between nutrient cycling and energy flow in the ecosystem. (From Smith 1976:16.)

The living world depends on the flow of energy and the circulation of matter through ecosystems. Both influence the abundance of organisms, the rate of their metabolism, and the complexity and structure of the ecosystem. Energy and matter flow through the ecosystem together as organic matter; one cannot be separated from the other (Figure 25.1). The link between energy and matter begins in the process of photosynthesis. Solar energy is utilized in the fixation of CO_2 into organic carbon compounds. Organic matter, the tissues of plants and animals, is composed not only of carbon, but a variety of essential nutrients outlined in Table 3.1. Because of this link between energy and matter, the general model of energy flow through an ecosystem presented in Chapter 24 provides a basic framework for examining the flow of matter though ecosystems.

TYPES OF NUTRIENT CYCLES

Nutrients move through the ecosystem in **biogeochemical cycles.** These cycles involve chemical exchanges of elements among atmosphere, rocks of Earth's crust, water, and living things. "Bio" in the term refers to living organisms; "geo" refers to the rocks, air, and water; and "chemical" refers to the chemical interactions involved. Biogeochemical cycles are typically classified as belonging to one of two general types: gaseous or sedimentary. This classification is based on the primary source of nutrient input to the ecosystem. Both types of cycles involve biological and nonbiological agents, both are driven by the flow of energy, and both are tied to the water cycle.

The main sources of nutrients possessing a **gaseous cycle** are the atmosphere and the ocean, and for that reason such nutrients have pronouncedly global circulation patterns. The main reservoirs of nutrients with **sedimentary cycles** are the soil and the rocks of Earth's crust. Sedimentary cycles vary from one element to another, but essentially each has two abiotic phases: the salt solution phase and the rock phase. Mineral elements found in rocks come directly from Earth's crust and are released slowly by weathering. Then they enter the water cycle as soluble salts. Unless these mineral elements are absorbed by plants, they move through the soil to streams and lakes and eventually reach the seas, where they remain indefinitely. Other salts return to Earth's crust through sedimentation. They become incorporated into salt beds, silts, and limestone; after weathering, they again enter the cycle.

MODEL OF NUTRIENT CYCLES

Although the biogeochemical cycles of the various essential nutrients required by autotrophs and heterotrophs differ in detail, from the perspective of the ecosystem, all biogeochemical cycles have a common structure, sharing three basic components: inputs, internal cycling, and outputs

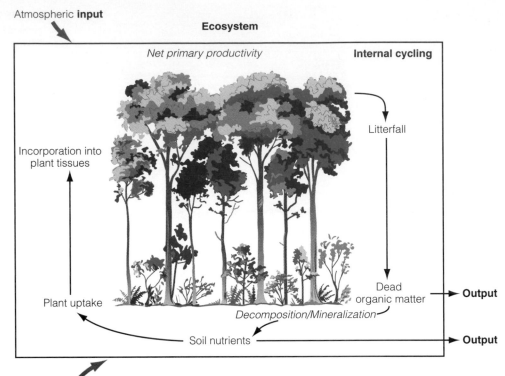

FIGURE 25.2 A generalized model of nutrient cycling in a terrestrial ecosystem. The three common components of inputs, internal cycling, and outputs are shown in bold. The key ecosystem processes of net primary productivity and decomposition are italicized.

(Figure 25.2). We shall look at this general model of biogeochemical cycling in ecosystems first, and then examine specific biogeochemical cycles in more detail.

Inputs

The input of nutrients to the ecosystem depends on the type of biogeochemical cycle. Nutrients with a gaseous cycle, such as carbon and nitrogen, enter the ecosystem via the atmosphere. In contrast, nutrients such as calcium and phosphorus have sedimentary cycles, with inputs dependent on the weathering of rocks and minerals (see Chapter 4). The process of soil formation and the resulting soil characteristics have a major influence on processes involved in nutrient release and retention (see Chapter 4). Many soil materials are deficient in nutrients on which plants depend, affecting both plants and herbivores.

Supplementing nutrients in the soil are nutrients carried by rain, snow, air currents, and animals. Precipitation brings appreciable quantities of nutrients, called **wetfall** (Eaton et al. 1973, Patterson 1975, Likens et al. 1985). Some of these nutrients, such as tiny dust particles of calcium and sea salt, form the nuclei of raindrops; others are solutions of trace gases. Additional nutrients, such as sulfates and nitrates, wash out of the atmosphere as the rain falls. For some nutrients the amount brought in by airborne particles and aerosols, collectively called **dryfall**, may exceed that carried in by precipitation (Table 25.1). Dust and vapors were found to supply well over half of the input of calcium, potassium, nitrates, and sulfate to the Walker Branch experimental forest at Oak Ridge, Tennessee (Lindberg et al. 1986).

Between 70 and 90 percent of rainfall striking the forest canopy reaches the forest floor. As it drips through the canopy (throughfall) and runs down the stems (stemflow), rainwater picks up and carries with it nutrients deposited as dust on leaves and stems together with nutrients leached from them. Therefore, rainfall reaching the forest floor is richer in calcium, sodium, potassium, and other nutrients than rain falling in the open at the same time (Tamm 1951, Madgwick and Ovington 1959, Eaton et al. 1973, Patterson 1975). Throughfall in an English oak woodland accounted for 17 percent of the nitrogen, 37 percent of the phosphorus, 72 percent of the potassium, and 97 percent of the sodium added by the canopy to the soil; the remainder was added by fallen leaves (Carlisle et al. 1966).

Although stemflow contributes only 5 percent of the total rainfall reaching the forest floor, it is so concentrated about the trunk that for some species the moisture it supplies is five to ten times as great as the rainfall nearby. The amount of stemflow reaching the ground varies with the species. Smooth-barked beeches have considerably more stemflow than oaks, whose bark absorbs water (Patterson 1975). In North American forests of balsam fir (*Abies balsamea*), epiphytic lichens remove nitrogen and add calcium and magnesium to stemflow (Lang et al. 1976). Although throughfall provides more nutrients because of its large volume of flow, stemflow provides a more concentrated nutrient solution. Stemflow concentration, like volume, is species-specific. Beech and hickories return considerably more calcium and potassium than oaks; and pines return smaller amounts of calcium, magnesium, potassium, and manganese than hardwoods (Patterson 1975).

TABLE 25.1 Annual Atmospheric Deposition of Major Ions into an Oak *(Quercus)* Forest, Walker Branch, Oak Ridge, Tennessee

Process	ATMOSPHERIC DEPOSITION (mEq/m²/yr)					
	SO_4^{2-}	NO_3^-	H^+	NH_4^+	Ca_2^+	K^+
Precipitation	70 ± 5	20 ± 2	69 ± 5	12 ± 1	12 ± 2	0.9 ± 0.1
Dry deposition						
Fine particles	7 ± 2	0.1 ± 0.02	2.0 ± 0.9	3.6 ± 1.3	1.0 ± 0.2	0.1 ± 0.05
Coarse particles	19 ± 2	8.3 ± 0.8	0.5 ± 0.2	0.8 ± 0.3	30 ± 3	1.2 ± 0.2
Gas	62 ± 7	26 ± 4	85 ± 8	1.3	0	0
Total deposition	160 ± 9	54 ± 4	160 ± 9	18 ± 2	43 ± 4	2.2 ± 0.3

Source: From Lindberg et al. 1986.

The major sources of nutrients for aquatic life are inputs from the surrounding land in the form of drainage water, detritus, and sediment, and from precipitation. Flowing water aquatic systems are highly dependent on a steady input of detrital material from the watersheds through which they flow (Chapter 30).

Internal Cycling

Primary productivity in ecosystems depends on the uptake of essential mineral (inorganic) nutrients by plants and their incorporation into living tissues. Nutrients in organic form, stored in living tissues, represent a significant proportion of the total nutrient pool in most ecosystems. As these living tissues senesce, the nutrients are returned to the soil or sediments in the form of dead organic matter. Various microbial decomposers transform the organic nutrients into a mineral form, a process called mineralization (see Chapter 9), and the nutrients are once again available to the plants for uptake and incorporation into new tissues. This process is called

internal cycling and is an essential feature of all ecosystems. It represents a recycling of nutrients within the ecosystem.

Witherspoon et al. (1962, 1964) demonstrated how such internal cycling works in a pioneering study using radioisotopes of elements to quantify the cycling of nutrients through an ecosystem. The object was to follow the pathway of a radiolabeled trace element through a forest ecosystem. Cesium behaves like potassium. It is highly mobile, cycles rapidly in an ionic form, and is easily leached from plant surfaces by rainfall. Moreover, because a known quantity of the element could be traced, the amounts of the element apportioned to wood, twigs, and leaves could be determined.

Witherspoon inoculated the trunks (boles) of 12 white oak *(Quercus alba)* trees with 20 microcuries (μC) of ^{134}Cs. He followed gains, losses, and transfers of this isotope in the trees and soil. About 40 percent of the ^{134}Cs inoculated into the oaks in April moved into the leaves in early June (Figure 25.3). Leaching of radiocesium from the leaves began when the first

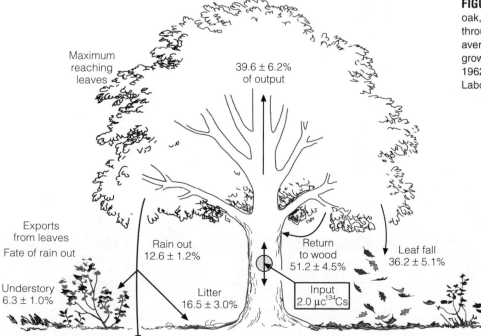

FIGURE 25.3 The cycle of ^{134}Cs in white oak, an example of the pathways of nutrients through plants. The figures are for an average of 12 trees at the end of the 1960 growing season. (After Witherspoon et al. 1962; courtesy Oak Ridge National Laboratory.)

rains fell after inoculation. By September this loss amounted to 13 percent of the maximum concentration in the leaves. Seventy percent of this rainwater loss reached the mineral soil; the remaining 30 percent found its way into the litter and understory. When the leaves fell in autumn, they carried with them twice as much radiocesium as had leached from the crown. Over the winter, half was leached to the mineral soil. Of the radiocesium in the soil, 92 percent still remained in the upper 10 cm nearly two years after inoculation. Eight percent of the cesium was confined to an area within the crown perimeter, and 19 percent was located in a small area about the trunk. In spring, cesium retained over winter in the wood and minimal transfers from the soil and litter moved back into the leaves. This quantified study provided early insights and a general model of internal cycling and retention of elements in forest trees.

Only a small fraction of the nutrient pool is involved in short-term annual cycling of nutrients in the forest ecosystem. Nutrients taken up by trees are returned to the forest floor by litterfall, throughfall, and stemflow (Figure 25.4). A significant portion of the nutrient uptake is stored in tree limbs, trunk, bark, and roots as accumulated biomass. This portion is effectively removed from short-term cycling. Some of the nutrients accu-

mulate in the litter and in the living biomass of consumer organisms, including decomposers of the forest floor, from which they are recycled at various rates. Nutrients accumulated in soil organic matter have a key role in recycling because they prevent rapid losses from the system. Large quantities of nutrients are bound tightly in this organic matter structure; they are not readily available until released by activities of decomposers.

Open water ecosystems, such as lakes and ponds, lack the long-term biological retention of nutrients typical of forested systems. Nutrient availability depends heavily on the turnover of nutrients in phytoplankton and zooplankton. Major long-term storage takes place in deep bottom sediments, where nutrients may be unavailable for a long time. Retention of nutrients in flowing water ecosystems is difficult, but it is aided by logs and rocks, which hold detritus in place, and by algal uptake of nutrients.

Ecosystem Processes Influencing the Rate of Nutrient Cycling You can see from Figure 25.2 that cycling of nutrients through the ecosystem depends on the processes of primary productivity and decomposition. Primary productivity determines the rate of nutrient transfer from inorganic to organic form

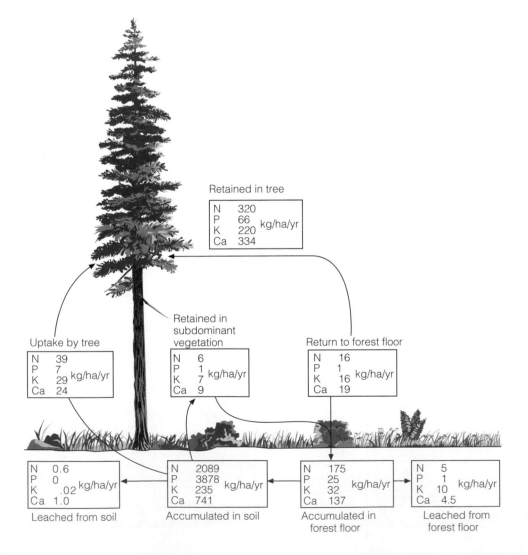

FIGURE 25.4 Nutrient cycling of nitrogen, phosphorus, potassium, and calcium in a second-growth Douglas-fir (*Pseudotsuga menziesii*) forest. (Based on data from Cole et al. 1967.)

Retained in tree

N	320	
P	66	kg/ha/yr
K	220	
Ca	334	

Uptake by tree

N	39	
P	7	kg/ha/yr
K	29	
Ca	24	

Retained in subdominant vegetation

N	6	
P	1	kg/ha/yr
K	7	
Ca	9	

Return to forest floor

N	16	
P	1	kg/ha/yr
K	16	
Ca	19	

Leached from soil

N	0.6	
P	0	kg/ha/yr
K	.02	
Ca	1.0	

Accumulated in soil

N	2089	
P	3878	kg/ha/yr
K	235	
Ca	741	

Accumulated in forest floor

N	175	
P	25	kg/ha/yr
K	32	
Ca	137	

Leached from forest floor

N	5	
P	1	kg/ha/yr
K	10	
Ca	4.5	

(nutrient uptake), and decomposition determines the rate of transformation of organic nutrients into inorganic form (nutrient release). Therefore, the rates at which nutrients cycle through the ecosystem will be directly related to the rates at which these two processes occur. But how do these two processes interact to limit the rate of the internal cycling of nutrients through the ecosystem? The answer lies in the interdependence of these two key processes. As an example, we can examine the cycling of nitrogen, an essential nutrient for plant growth.

The direct link between soil nitrogen availability, rate of nitrogen uptake by plant roots, and the resulting leaf nitrogen concentrations was presented in Chapter 7 (see Figure 7.25). The maximum rate of photosynthesis is strongly correlated with leaf nitrogen concentrations, because a large portion of leaf nitrogen is contained within compounds directly involved in photosynthesis (e.g., rubisco and chlorophyll). The availability of nitrogen in the soil will therefore directly affect rates of ecosystem primary productivity via its influence on photosynthesis and carbon uptake.

A low availability of soil nitrogen reduces not only net primary production (NPP, the total production of plant tissues), but also the nitrogen concentration of the plant tissues that are produced (see Chapter 7). Thus, the reduced availability of soil nitrogen influences the input of dead organic matter to the decomposer food chain (see Chapter 9) in two important ways by reducing (1) the total quantity of dead organic matter produced (as a direct result of lower NPP), and (2) its nutrient concentration. The net effect is a lower input of nitrogen in the form of dead organic matter.

The rate of decomposition and nitrogen mineralization (nutrient release) are directly related to both the quantity and quality of organic matter as a food source for decomposers (Chapter 9). Lower nutrient concentrations in the dead organic matter promote immobilization of nutrients from the soil and water to meet the nutrient demands of the decomposer populations. This immobilization effectively reduces nutrient availability to the plants, adversely affecting primary productivity.

You can now begin to see the feedback system that exists in the internal cycling of nutrients within an ecosystem (Figure 25.5). Reduced nutrient availability can have the combined effect of reducing both the nutrient concentration of plant tissues (primarily leaf tissues) and net primary productivity. This reduction lowers the total amount of nutrients returned to the soil in dead organic matter. The reduced quantity and quality (nutrient concentration) of organic matter entering the decomposer food chain increases immobilization and reduces the availability of nutrients for uptake by plants. In effect, low nutrient availability begets low nutrient availability. Conversely, high nutrient availability encourages high plant tissue concentrations and high net primary productivity. In turn, the associated high quantity and quality of dead organic matter encourages high rates of net mineralization and nutrient supply in the soil.

Climate In addition to its role in weathering of rocks and minerals and soil formation (see Chapter 4), climate directly affects the rate of nutrient cycling in ecosystems

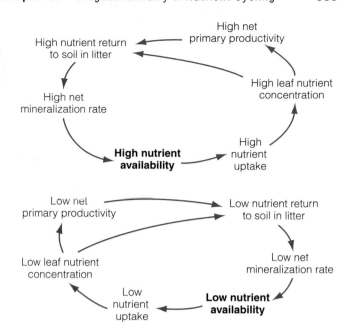

FIGURE 25.5 Feedback that occurs between nutrient availability, net primary productivity, and nutrient release in decomposition for initial conditions of low and high nutrient availability. (Chapin 1980.)

through its influence on rates of primary productivity (see Figure 24.2) and decomposition. Both increase as conditions become warmer and wetter. This relationship is reflected in the positive correlation between actual evapotranspiration and both net primary productivity and rate of decomposition as conditions become warmer and wetter (see Figure 24.4 and 9.11). The net effect is a faster rate of nutrient cycling in warm, wet environments, such as a tropical rain forest, than in cooler (temperate or boreal forest) or drier (grassland) ecosystems.

Although both net primary productivity and decomposition rate increase with increasing actual evapotranspiration, an examination of rate of input and standing mass of dead organic matter on the soil surface in a variety of forest ecosystems shows that these two processes exhibit slightly different responses to climate. This difference has a major influence on the cycling of nutrients in these ecosystems. The annual input of dead organic matter (plant litter) and the standing mass of dead organic matter on the soil surface for a number of forest ecosystems are plotted in Figure 25.6. Note that there is an inverse relationship between the production of dead organic matter (litter) and the amount of dead organic matter on the forest floor. The warmer, wetter conditions of the tropical rain forest result in high rates of net primary productivity and subsequently higher annual rates of litter input to the forest floor. However, these ecosystems are characterized by a very low mass of litter on the forest floor. This difference between rate of input and standing mass is a direct result of high rates of decomposition, or to put it another way, the organic matter is being consumed by decomposers at about the same rate at which it is falling to the forest floor. In contrast, boreal forests, which are found in much cooler,

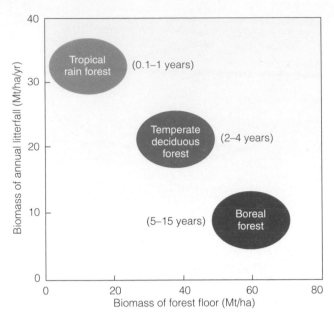

FIGURE 25.6 Relationship between the standing mass of dead organic matter (biomass) on the forest floor and the annual rate of litterfall (dead organic matter input) for three types of forest ecosystems. Tropical rain forests, found in wet, warm environments, have a high rate of litterfall but low biomass on the forest floor, the result of high rates of decomposition. In contrast, the boreal forests found in cool, wet environments have a low annual input of litterfall but a high forest floor biomass, because of the very low rates of decomposition and the accumulation of dead organic matter on the forest floor. Values in parentheses are the average turnover rates of litter.

northern latitudes (see Chapter 29), have very low rates of net primary productivity, and subsequently, low rates of litter input. However, the extremely low rates of decomposition result in a buildup of dead organic matter on the forest floor.

At first, the inverse relationship between rate of input and standing mass of dead organic material shown in Figure 25.6 might seem odd in that both rate of net primary productivity (and associated litter input to the soil surface) and decomposition decrease moving in latitude from the tropical rain forest to the temperate and boreal forest regions. However, this inverse relationship results from a greater slowdown of decomposition from declining temperatures relative to the slowing of net primary productivity (and litter input). As a consequence of this difference between the rate of litter input and the rate of decomposition, more nutrients are tied up in dead organic matter in the cooler, more northern forest ecosystems, resulting in a slower rate of internal nutrient cycling.

Species Characteristics Nutrient cycling in an ecosystem is also influenced by the nature of its organisms themselves. Organisms such as phytoplankton and zooplankton in aquatic systems are short-lived, grow rapidly, absorb nutrients quickly, and just as quickly return them to the available nutrient pool. Other organisms, such as forest trees, are large, grow slowly, and store large quantities of nutrients in their biomass for much longer periods. This portion is effectively removed from short-term cycling.

Each species contributes differently to the overall nutrient cycling within an ecosystem. Trees and shrubs, for exam-

ple, sequester varying amounts of elements in short- and long-term nutrient pools in their structural components— wood, bark, twigs, roots, and leaves. Short- and long-term are defined by the rate at which the different structural components decompose. Nutrients sequestered in slower decomposing tissues such as bark and wood are released more slowly than those contained in leaves, twigs, and fine roots.

Species also differ in their concentrations of nutrients and ability to recycle them. These differences were noted by Day and McGinty (1975) in a study of nutrient cycling by several species of trees on the Coweeta Watershed at Franklin, North Carolina (Table 25.2). One was a canopy tree, chestnut oak (*Quercus prinus*); the second, an understory species, flowering dogwood (*Cornus florida*); and the third, an understory evergreen shrub, rhododendron (*Rhododendron maximum*).

Chestnut oak had the largest standing crop of nutrients, as we would expect of a large tree. Within the tree, leaves held the most potassium and magnesium, and bark had the largest concentration of calcium. Among the three species, the leaves of chestnut oak had the highest standing crop of nitrogen.

The evergreen rhododendron had the largest standing crop of leaf biomass of the three species. Its thick, long-lived leaves held more calcium and magnesium than its other parts. Because its leaves are evergreen, the nutrients are recycled over a period of seven years (average leaf longevity) instead of the one year typical of the deciduous chestnut oak and flowering dogwood.

Dogwood, a small tree, had the smallest biomass, but it had a high leaf-to-wood ratio and the distinction of possessing the highest concentration of calcium in its leaves. Dogwood concentrated over three times as much calcium per unit leaf biomass as did chestnut oak, and one and a half times as much as rhododendron. The small dogwood recycled 66 percent as much calcium as the chestnut oak and 150 percent more than rhododendron. Thus, differences in the nutrient content of various species in an ecosystem, the size of the short- and long-term nutrient pools, and the rates at which the plants recycle them have a pronounced influence on nutrient cycling.

Outputs

The export of nutrients from the ecosystem represents a loss that must be offset by inputs if a net decline is not to occur. Export can occur in a variety of ways, depending on the nature of the specific biogeochemical cycle. Carbon is exported to the atmosphere in the form of CO_2 via the process of respiration by all living organisms. Likewise, a variety of microbial and plant processes result in the transformation of organic and inorganic nutrients to a gaseous phase that can subsequently be transported from the ecosystem in the atmosphere. Examples of these processes will be provided in the following sections, which examine specific biogeochemical cycles.

Transport of nutrients from the ecosystem can also occur in the form of organic matter. Organic matter from a forested watershed can be carried from the ecosystem through surface flow of water in streams and rivers. The input of organic carbon from terrestrial ecosystems constitutes the majority of

TABLE 25.2 Mean Nutrient Concentrations in Three Forest Trees, Coweeta Hydrologic Laboratory, Franklin, NC

Species	Bark	Wood	Twigs	Leaves
	POTASSIUM (% DRY WEIGHT)			
Chestnut oak	0.13 ± 0.04	0.18 ± 0.04	0.39 ± 0.03	1.26 ± 0.06
Flowering dogwood	0.34 ± 0.06	0.18 ± 0.02	0.72 ± 0.04	1.44 ± 0.11
Rhododendron	0.13 ± 0.06	0.24 ± 0.04	1.55 ± 0.77	0.52 ± 0.26
	CALCIUM (% DRY WEIGHT)			
Chestnut oak	1.25 ± 0.17	0.09 ± 0.01	0.68 ± 0.06	0.58 ± 0.07
Flowering dogwood	2.36 + 0.26	0.11 ± 0.01	0.80 ± 0.06	1.85 ± 0.11
Rhododendron	0.30 ± 0.10	0.07 ± 0.31	0.99 ± 0.24	1.20 ± 0.29
	MAGNESIUM (% DRY WEIGHT)			
Chestnut oak	0.07 ± 0.01	0.01 ± 0.001	0.08 ± 0.02	0.14 ± 0.01
Flowering dogwood	0.18 ± 0.03	0.04 ± 0.01	0.13 ± 0.01	0.42 ± 0.03
Rhododendron	0.02 ± 0.02	0.01 ± 0.01	0.15 ± 0.05	0.19 ± 0.06

Species	Twigs	Leaves	Twigs	Leaves
	NITROGEN (% DRY WEIGHT)		*PHOSPHORUS (% DRY WEIGHT)*	
Chestnut oak	0.43 ± 0.04	2.30 ± 0.06	0.15 ± 0.01	0.16 ± 0.004
Flowering dogwood	0.22 ± 0.02	2.11 ± 0.06	0.10 ± 0.002	0.16 ± 0.01
Rhododendron	1.01 ± 0.14	1.25 ± 1.47	0.21 ± 0.04	0.13 ± 0.02

*Error terms are ± 1 standard error.
Source: From Day and McGinty 1975:741.

energy input into stream ecosystems (see Figure 30.24). Organic matter can also be transferred between ecosystems by herbivores. Moose feeding on aquatic plants can transport and deposit nutrients to adjacent terrestrial ecosystems in the form of feces. Conversely, the hippopotamus (*Hippopotamus amphibius*) feeds at night on herbaceous vegetation adjacent to the body of water in which it resides. Large quantities of nutrients are then transported in the form of feces and other wastes to the water.

Organic matter has a key role in recycling nutrients because it prevents rapid losses from the system. Large quantities of nutrients are bound tightly in organic matter structure; they are not readily available until released by activities of

decomposers. However, some nutrients are leached from the soil and carried out of the ecosystem by underground water flow to streams. These losses may be balanced by inputs to the ecosystem, such as the withdrawal of nutrients from deep soil reserves and by the weathering of parent rock material.

Considerable quantities of nutrients are withdrawn permanently from ecosystems by harvesting (Table 25.3), especially in farming and logging, as biomass is directly removed from the ecosystem. In such ecosystems these losses must be replaced by the application of fertilizer; otherwise, the ecosystem becomes impoverished. In addition to the nutrients removed directly through biomass removal, logging can also result in the transport of nutrients from the ecosystem by altering processes

TABLE 25.3 Comparison of the Average Annual Yield and Nutrient Removal by a 16-Year-Old Loblolly Pine Plantation with That of Agricultural Crops

Crop	Yield (Tons/ha)	REMOVAL (kg/ha)			
		N	P	K	Ca
Loblolly pine, whole tree	14.5	17.5	2.4	12.6	12.8
Loblolly pine, pulpwood	8.75	6.5	0.9	5.1	6.4
Corn (grain)	11.75	130.5	29.7	37.3	—
Soybeans (beans)	3.0	145.0	14.85	46.7	—
Alfalfa (forage)	10.0	212.5	23.25	185.6	75.4

Source: Data from Jorgensen and Wells 1986.

involved in internal cycling. The removal of trees in clear-cutting and other silvicultural (forest management) practices increases the amount of radiation (including direct sunlight) reaching the soil surface. The resulting increase in soil temperatures promotes decomposition (see Chapter 9) and results in an increase in net mineralization rates (Figure 25.7). This increase in nutrient availability in the soil occurs at the same time that demand for nutrients is low because plants have been removed and net primary productivity is low. As a result, there is a dramatic increase in the leaching of nutrients from the ecosystem in surface waters (Figure 25.8). This export of nutrients from the ecosystem results from decoupling the two processes of nutrient release in decomposition and nutrient uptake in net primary productivity.

Depending on its intensity, fire kills vegetation and converts varying proportions of the biomass and soil organic matter to ash (see Chapter 23 for discussion of fire). In addition to the loss of nutrients through volatilization and airborne particulate, the addition of ash changes the chemical and biological properties of the soil (Raison 1979). Many nutrients become readily available, and nitrogen in ash is subject to rapid mineralization. If not taken up by vegetation during recovery, nutrients may be lost from the ecosystem through leaching and erosion (Christensen 1977, Uhl and Jordan 1984). Streamwater runoff is often greatest after fire because of reduced water demand for transpiration. High nutrient availability in the soil coupled with high runoff can lead to large nutrient losses from the ecosystem.

CONTRASTING NUTRIENT CYCLING IN TERRESTRIAL AND AQUATIC ECOSYSTEMS

The process of nutrient cycling is an essential feature of all ecosystems and represents a direct (cyclic) link between net primary productivity and decomposition. However, the nature of this link varies among ecosystems, particularly between terrestrial and aquatic ecosystems.

In virtually all ecosystems there is a vertical separation between the zones of production (photosynthesis) and decomposition (Figure 25.9). In terrestrial ecosystems, the plants themselves bridge this physical separation between the zone of decomposition at the soil surface and the zone of productivity in the plant canopy—the plants physically exist in both zones. The root systems provide access to the nutrients made available in the soil through decomposition, and the vascular system within the plant transports these nutrients to the sites of production. In aquatic ecosystems, this is not always the case.

In shallow water environments of the shoreline, emergent vegetation such as cattails, cordgrasses (*Spartina*), and sedges are rooted in the sediments. Here, as in terrestrial ecosystems, the zone of decomposition and production are linked directly by the plants. Likewise, submerged vegetation, such as seagrasses and kelps, is rooted in the sediments with the plants extending up the water column into the photic zone (see Chapter 31), the shallower waters where light levels support higher productivity. However, as water depths increase, primary production is dominated by free-floating phytoplankton, which migrates vertically within the upper waters (photic zone). Here exists a physical separation between the zones of decomposition in the bottom, or benthic zone, and the surface waters where temperatures and light availability support primary productivity. This physical separation between the zones where nutrients become available through decomposition and the zone of productivity where nutrients are needed to support photosynthesis and plant growth is a major factor controlling the productivity of open water ecosystems.

To understand how nutrients are transported vertically from the deeper waters to the surface, where temperature and light conditions can support primary productivity, we must first examine the vertical structure of the physical environment in open water ecosystems.

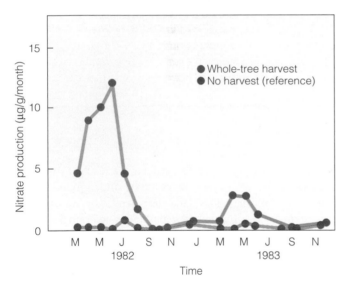

FIGURE 25.7 Comparison of nitrate (NO_3^-) production following logging for a loblolly pine (*Pinus taeda*) plantation in the southeastern United States. Data for the reference stand (no harvest) are compared with those of a whole-tree harvest clear-cut. (Adapted from Vitousek 1992.)

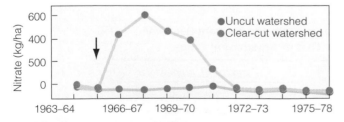

FIGURE 25.8 Temporal changes in the nitrate concentration of streamwater for two forested watersheds in Hubbard Brook, New Hampshire. The forest on one watershed was clear-cut, while the other forest remained undisturbed. Note the large increase in concentrations of nitrate in the stream on the clear-cut watershed. This increase is due to increased decomposition and nitrogen mineralization following the removal of trees. The nitrogen was then leached into the surface and groundwater. (Adapted from Likens and Borman 1995.)

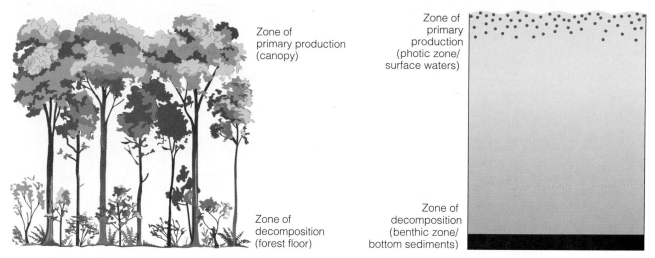

Zone of
primary production
(canopy)

Zone of
decomposition
(forest floor)

Zone of
primary
production
(photic zone/
surface waters)

Zone of
decomposition
(benthic zone/
bottom sediments)

FIGURE 25.9 Comparison of the vertical zones of production and decomposition in (a) a terrestrial (forest) and (b) an open water (lake) ecosystem. Note that in the terrestrial ecosystem the two zones are linked by the vegetation (trees). However, this is not the case in the lake ecosystem.

As presented briefly in Chapter 20 and in detail in Chapter 30, the vertical structure of open water ecosystems, such as lakes or oceans, can be divided into three rather distinct zones: the epilimnion, the metalimnion, and the hypolimnion (Figure 25.10). The **epilimnion,** or surface water, is relatively warm as a result of the interception of solar radiation. In addition, the oxygen content is relatively high due to the diffusion of oxygen from the atmosphere into the surface waters. In

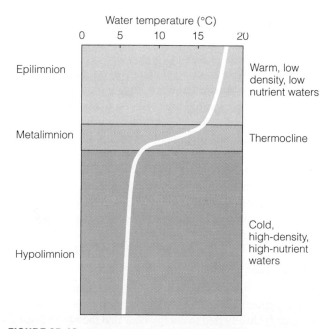

Water temperature (°C)

| 0 | 5 | 10 | 15 | 20 |

Epilimnion — Warm, low density, low nutrient waters

Metalimnion — Thermocline

Hypolimnion — Cold, high-density, high-nutrient waters

FIGURE 25.10 Vertical structure of an open water ecosystem such as a lake or an ocean. The vertical profile can be divided into three distinct layers. The epilimnion or surface water is a layer of warm, oxygen-rich water. This hypolimnion is the deep, cold oxygen-poor layer. The transition zone between these two layers is called the metalimnion and is characterized by a dramatic shift in temperature, called the thermocline.

contrast, the **hypolimnion,** or deep water, is cold and relatively low in oxygen. The **metalimnion** is a transition zone between the surface and deep waters and is characterized by a steep temperature gradient called the **thermocline.** The thermocline is the transition zone between the warmer surface waters and the deeper cold waters. In effect, the vertical structure can be represented as a warm, low-density, surface layer of water on top of a denser, cold layer of deep water, separated by the rather thin zone of the thermocline. This vertical structure and physical separation of the epilimnion and hypolimnion have an important influence on the distribution of nutrients and subsequent patterns of primary productivity in aquatic ecosystems. The colder, deep waters are relatively nutrient-rich, but temperature and light conditions cannot support high productivity. In contrast, the surface waters are relatively nutrient-poor; however, this is the zone where temperatures and light can support high productivity.

Although winds blowing over the water surface cause turbulence that mixes the waters of the epilimnion, this mixing does not extend into the colder, deeper waters because of the thermocline. As autumn and winter approach in the temperate and polar zones, the amount of solar radiation reaching the water surface decreases and the temperature of the surface water declines. As the water temperature of the epilimnion approaches that of the hypolimnion, the thermocline breaks down and mixing throughout the profile can take place (Figure 25.11). If surface waters become cooler than the deeper waters, they will begin to sink, displacing deep waters to the surface. This process is called **turnover.** With the breakdown of the thermocline and mixing of the water column, nutrients are brought up from the bottom to the surface waters. With the onset of spring, increasing temperatures and light in the epilimnion give rise to a peak in productivity with the increased availability of nutrients in the surface waters. As the spring and summer progress, the nutrients in the surface water are used, reducing the nutrient content of the

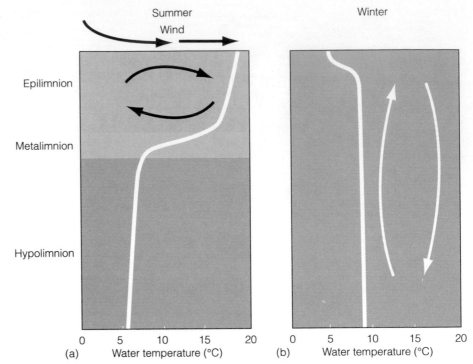

Summer

Wind

Winter

Epilimnion

Metalimnion

Hypolimnion

0   5   10   15   20
(a)   Water temperature (°C)

0   5   10   15   20
(b)   Water temperature (°C)

FIGURE 25.11   Seasonal dynamics in the vertical structure of an open water aquatic ecosystem in the Temperate Zone. Winds mix the waters within the epilimnion during the summer (a), but the thermocline isolates this mixing to the surface waters. With the breakdown of the thermocline during the winter months (b), turnover occurs, allowing the entire water column to become mixed. This mixing allows nutrients in the epilimnion to be brought up to the surface waters.

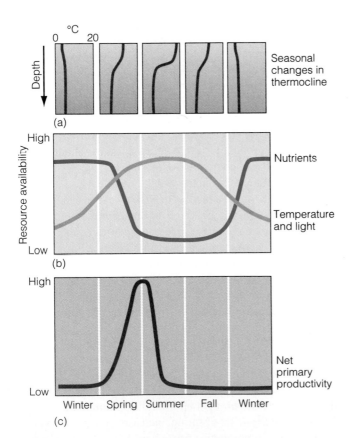

°C
0   20

Depth

Seasonal changes in thermocline

(a)

High

Resource availability

Nutrients

Temperature and light

Low
(b)

High

Low
Winter   Spring   Summer   Fall   Winter

Net primary productivity

(c)

FIGURE 25.12   Seasonal dynamics of (a) the thermocline and associated changes in (b) the availability of light and nutrients, and (c) net primary productivity of the surface waters.

water, and a subsequent decline in productivity occurs. The resulting annual cycle of productivity in these ecosystems (Figure 25.12) is a direct function of the dynamics of the thermocline and the resulting behavior of the vertical distribution of nutrients.

MAJOR BIOGEOCHEMICAL CYCLES

Nutrient cycling in all ecosystems involves the three basic components of inputs, internal cycling, and outputs outlined earlier. In addition, the feature common to all biogeochemical cycles of essential macro- and micronutrients (see Chapter 3) is that they are components of living tissues. As such, they share a common journey from soil and sediment, the incorporation into plant and animal tissues, and the release once again through the process of mineralization by decomposers. The specifics of this journey, however, vary from element to element as a function of their origins and specific chemical properties.

The Carbon Cycle

Because it is a basic constituent of all organic compounds and a major element in the fixation of energy by photosynthesis, carbon is so closely tied to energy flow that the two are inseparable. In fact, the measurement of productivity is commonly expressed in terms of grams of carbon fixed per square meter per year. The source of all fixed carbon, both in living organisms and fossil deposits, is carbon dioxide (CO_2), found in the atmosphere and dissolved in the waters of Earth. To trace its cycling through the ecosystem is to redescribe photosynthesis and energy flow (Figure 25.13).

The cycling of carbon dioxide involves assimilation by plants and conversion to glucose. From glucose, plants synthesize polysaccharides, proteins, and fat and store them in the form of plant tissue. When digesting plant tissue, herbivores synthesize these compounds into other carbon compounds. Meat-eating animals feed on herbivores, and the carbon compounds are redigested and resynthesized into other forms. Some of the carbon is returned directly by both plants and animals in the form of CO_2 as a by-product of respiration. The remainder for a time becomes incorporated in the living biomass.

Assorted decomposer organisms release carbon contained in animal wastes and in the cells of all plants and animals following death. The rate of release depends on environmental conditions such as soil moisture, temperature, and precipitation (see Chapter 9). In tropical forests, most of the carbon in plant detritus is quickly recycled, for there is little accumulation in the soil. In drier regions, such

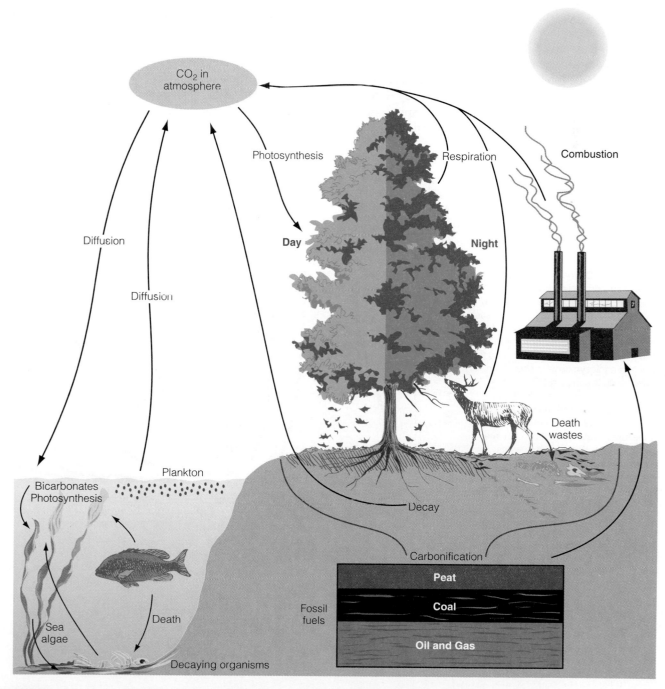

FIGURE 25.13   The carbon cycle. Although the main reservoir is the gas CO_2, considerable quantities are tied up in organic and inorganic compounds of carbon in the biosphere.

as grasslands, considerable quantities of carbon are stored as humus. In swamps and marshes, where dead material falls into the water, organic carbon is not completely mineralized; it is stored as raw humus or peat and is only slowly broken down.

Similar cycling takes place in the freshwater and marine environments. Carbon dioxide diffuses into the upper layers of water according to equilibrium reactions that depend on pH:

$$CO_2 + H_2O \rightleftharpoons H^+ + HCO_3^- \rightleftharpoons 2H^+ + CO_3^{2-}$$

At pH < 4.3, most carbon dioxide is found as a dissolved gas, between 4.3 and 8.3 as bicarbonate (HCO_3^-), and > 8.3 as carbonate (CO_3^{2-}). Together these forms are known as dissolved inorganic carbon (DIC). Phytoplankton utilizes the dissolved inorganic carbon and converts it into carbohydrates. The carbohydrates so produced pass through the aquatic food chains. Plankton are consumed by zooplankton, and eventually all are consumed by decomposers. The carbon dioxide produced by respiration is reutilized by the phytoplankton in the production of more carbohydrates.

Significant portions of carbon, bound as carbonates in the shells of mollusks and foraminifers, become buried in the bottom mud at varying depths when the organisms die. Isolated from biotic activity, that carbon is removed from the short-term cycling and becomes incorporated into bottom sediments, which through geological time may appear on the surface as limestone rocks or as coral reefs.

Local Patterns The concentration of carbon dioxide in the atmosphere around plant life fluctuates throughout the day (Figure 25.14a). At daylight, when photosynthesis begins, plants start to withdraw carbon dioxide from the air and the concentration declines sharply, by as much as 25 percent. By afternoon, when the temperature is increasing and the humidity is decreasing, the respiration rate of plants increases, and the assimilation rate of carbon dioxide increases. At sunset the light phase of photosynthesis ceases, and the light-independent phase (Calvin cycle) ceases shortly thereafter. Carbon dioxide is no longer being withdrawn from the atmosphere, and its concentration in the atmosphere increases sharply. A similar diurnal fluctuation takes place in aquatic ecosystems.

There is also an annual course in the production and utilization of carbon dioxide (Figure 25.14b). This seasonal change relates both to temperature and to growing seasons. In the spring, when land is greening and phytoplankton is actively growing, the daily production of carbon dioxide is high. Measured by nocturnal accumulation in spring and summer, the rate of carbon dioxide production by respiration

FIGURE 25.14 Diurnal and seasonal patterns of carbon cycling in a forest. (a) Daily flux of CO_2. Note the consistently high level of CO_2 on the forest floor, the site of microbial respiration. Atmospheric CO_2 in the forest is lowest from midmorning to late afternoon. CO_2 levels are highest during the night, when photosynthesis is shut down and respiration is pumping CO_2 into the atmosphere. (From Baumgartner 1968, after Miller and Rusch 1960.) (b) Changes in CO_2 production over the course of a year in a pine forest, based on respiration rates during inversions. Respiration rates as measured by CO_2 production are highest during the summer, when photosynthesis and decomposition are the greatest. (From Woodwell and Dykeman 1966.)

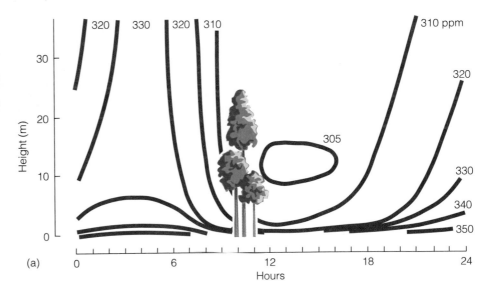

(a)

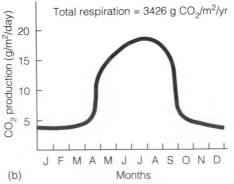

(b)

may be two to three times as high as winter rates at the same temperature. The rate increases dramatically about the time of the opening of buds and falls off just as rapidly about the time the leaves of deciduous trees start to drop in the fall. This increased production of CO_2 does not accumulate in the atmosphere, however, because of the high rate of CO_2 fixation by photosynthesis.

The Nitrogen Cycle

Nitrogen is an essential constituent of protein, a building block of all living material. It is also a major constituent, about 79 percent, of the atmosphere. The paradox is that in its gaseous state, nitrogen is unavailable to most life. It must first be converted to some chemically usable form. Getting it into that form makes up a major part of the nitrogen cycle.

The nitrogen cycle consists of four processes (Figure 25.15). These are fixation, the conversion of nitrogen in its gaseous state to a usable form; mineralization or ammonification, the conversion of amino acids in organic matter to ammonia; nitrification, the oxidation of ammonia to nitrites and nitrates; and denitrification, the reduction of nitrates to gaseous nitrogen. Most of the nitrogen cycle is driven by microbes.

Fixation converts gaseous nitrogen (N_2) to ammonia and nitrates. Ammonia (NH_3) is the product of biological fixation. Nitrates are the product of high-energy fixation by lightning; occasionally cosmic radiation and oxygen in the atmosphere combine into nitrates, which come to Earth in rainwater as nitric acid (H_2NO_3). Estimates suggest that less than 35 mg N/ha/yr arrives on Earth by high-energy fixation.

Biological fixation, the more important method, makes available 1.4–7.0 kg N/ha/yr in natural ecosystems, with values reaching as high as 200 kg/ha/yr in some agricultural systems. Biological fixation accounts for roughly 90 percent of the fixed nitrogen contributed to Earth each year. In biological fixation, molecular (or gaseous) nitrogen is split into two atoms:

$$N_2 \longrightarrow 2N$$

This step is energy-expensive, because the two nitrogen atoms are connected by a triple bond. It requires an input of 160 kcal for each mole (28 g) of nitrogen. The free N atoms can combine with hydrogen to form ammonia, with the release of about 13 kcal of energy:

$$2N + 3H_2 \longrightarrow 2NH_3$$

This fixation is accomplished by mutualistic bacteria living in association with leguminous and root-noduled nonleguminous plants, by free-living bacteria, and by cyanobacteria (blue-green algae). In agricultural ecosystems, approximately 200 species of nodulated legumes are the preeminent nitrogen fixers. In nonagricultural systems, some 12,000 species of organisms, from bacteria and cyanobacteria to nodule-bearing plants, are responsible for nitrogen fixation.

Legumes, the most conspicuous of the nitrogen-fixing plants, have a mutualistic relationship with members of the bacterial genus *Rhizobium*. Rhizobia are aerobic, non-spore-forming, rod-shaped bacteria (Figure 25.16). They live in the immediate surroundings of plant roots, called the **rhizosphere.** Stimulated by secretions and enzymes from the legumes, swarming rhizobia enter the root hairs, where they multiply and increase in size. This invasion and growth results in swollen, infected root hair cells that make up the central tissues of the root nodules. Inside the nodules, the bacteria change from rod-shaped to a nonmobile form that carries on nitrogen fixation. Although the process is well known, the mechanisms are still the subject of intensive study.

A large number of nonleguminous nodule-bearing plants, most of them early pioneering species, grow on sites where soil is low in nitrogen. These plants make significant contributions of nitrogen to wildlands. Among such plants are alder (*Alnus*), New Jersey tea (*Ceanothus*), and Russian olive (*Elaeagnus*).

Also contributing to the fixation of nitrogen are free-living soil bacteria. The most prominent of the 15 known genera are the aerobic *Azotobacter* and the anaerobic *Clostridium*. *Azotobacter* prefers soils with a pH of 6 to 7 that are rich in mineral salts and low in nitrogen. *Clostridium* is ubiquitous, found in nearly all soils. Both genera produce ammonia as the first stable end product. Free-living and symbiotic bacteria both require molybdenum as activators and are inhibited by an accumulation of nitrates and ammonia in the soil.

Cyanobacteria are another important group of largely nonsymbiotic nitrogen-fixers. Of some 40 known species the most common are in the genera *Nostoc* and *Calothrix*, found in both soil and aquatic habitats. Cyanobacteria are often pioneers on bare mineral soil. Especially successful in waterlogged soils, they are nitrogen-fixers in the rice paddies of Asia. Cyanobacteria are perhaps the only fixers of nitrogen over a wide range of temperatures in aquatic habitats from Arctic to Antarctic seas to freshwater ponds and hot springs. Like bacteria, cyanobacteria require molybdenum for nitrogen fixation.

Other plants may be involved in nitrogen fixation. In humid tropical forests, epiphytes growing on tree branches and bacteria and algae growing on leaves may fix appreciable amounts of nitrogen. Certain lichens have been implicated in

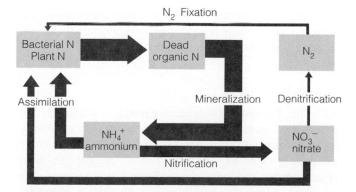

FIGURE 25.15 The bacterial processes involved in nitrogen cycling. The width of each arrow is an approximation of the process rate. (From Blackburn 1983:64.)

(a)

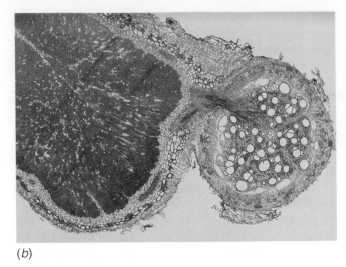

(b)

FIGURE 25.16   (a) Nodules of *Rhizobium* bacteria on the roots of a legume. (b) Cross-section of a root nodule showing *Rhizobium* bacteria.

nitrogen fixation (Henriksson and Simu 1971). Lichens with nitrogen-fixing ability possess nitrogen-fixing cyanobacteria species as their algal component.

Once made available by either high-energy or biological fixation, nitrogen can be utilized by plants and incorporated into living tissues (Figure 25.17). As plants and animals die, the nitrogen contained in organic compounds is returned to the soil as dead organic matter. Nitrogen in organic compounds is unavailable to plants and must first be transformed by decomposers into an inorganic or mineral form.

Mineralization or **ammonification** of organic nitrogen is the major step in the nitrogen cycle. In this process, proteins in dead plant and animal material are broken down by bacteria and fungi to amino acids. The amino acids are oxidized to carbon dioxide, water, and ammonia, with a yield of energy; for example:

$$CH_2NH_2COOH + 1\tfrac{1}{2}O_2 \longrightarrow 2CO_2 + H_2O + NH_3 + 178 \text{ kcal}$$

Ammonium, or the ammonia ion, is absorbed directly by plant roots, incorporated into amino acids, and passed through the food chain. Some of the ammonia is dissolved in water and part is bound in the soil and sediments.

Nitrification is a biological process in which ammonia is oxidized to nitrate and nitrite, yielding energy. Two groups of organisms are involved. *Nitrosomonas* bacteria use the ammonia in the soil as their sole source of energy. They can promote its oxidation first to nitrous acid and water and then to nitrite:

$$NH_3 + 1\tfrac{1}{2}O_2 \longrightarrow HNO_2 + H_2 + 165 \text{ kcal}$$
$$HNO_2 \longrightarrow H^+ + NO_2^-$$

Energy left in the nitrite ion is exploited by another group of bacteria, the *Nitrobacter*, which oxidizes the nitrite ion to nitrate with a release of a small amount of energy:

$$NO_2^- + 1\tfrac{1}{2}O_2 \longrightarrow NO_3^-$$

In nitrification, *Nitrosomonas* oxidizes 35 moles of nitrogen for each mole of CO_2 assimilated; *Nitrobacter* oxidizes 100 moles.

Like ammonium, nitrates are absorbed by plants and incorporated into organic matter. With senescence, nitrogen tied up in organic compounds is returned to the soil in organic matter, where it once again undergoes the process of mineralization during decomposition. Nitrification is generally beneficial, with many plant species preferentially utilizing nitrates over ammonium. However, nitrates are more readily leached from the soil than ammonium. Most soils have a greater capacity to exchange cations than anions (see Chapter 4). If quantities of nitrates are large enough and sufficient water percolates through the soil, nitrates can be removed faster than plant roots can take them up. This situation results in a significant loss to aquatic ecosystems (see Figure 25.8). An abundance of nitrates also leads to increased losses of gaseous nitrogen.

Nitrates are a necessary substrate for **denitrification,** in which the nitrates are reduced to gaseous nitrogen by certain organisms to obtain oxygen. The denitrifiers, represented by fungi and the bacteria *Pseudomonas,* are facultative anaerobes. They prefer an oxygenated environment, but if oxygen is limited, they can use NO_3^- instead of O_2 as the hydrogen acceptor. In doing so, they release N_2 in the gaseous state as a by-product:

$$C_6H_{12}O_6 + 4NO_3^- \longrightarrow 6CO_2 + 6H_2O + 2N_2$$

Once in gaseous form, nitrogen is lost from the ecosystem to the atmosphere, where it once again must undergo fixation to be made available for utilization in primary productivity.

The Sulfur Cycle

Sulfur has a long-term sedimentary phase, in which it is tied up in organic material (coal, oil, and peat) and inorganic form (pyritic rocks and sulfur deposits). Sulfur is released by the weathering of rocks, erosional runoff, decomposition of organic matter, and industrial production, and it is carried to terrestrial and aquatic ecosystems in a salt solution. However, the bulk of sulfur first appears in the gaseous phase as a volatile gas, hydrogen sulfide (H_2S), in the atmosphere.

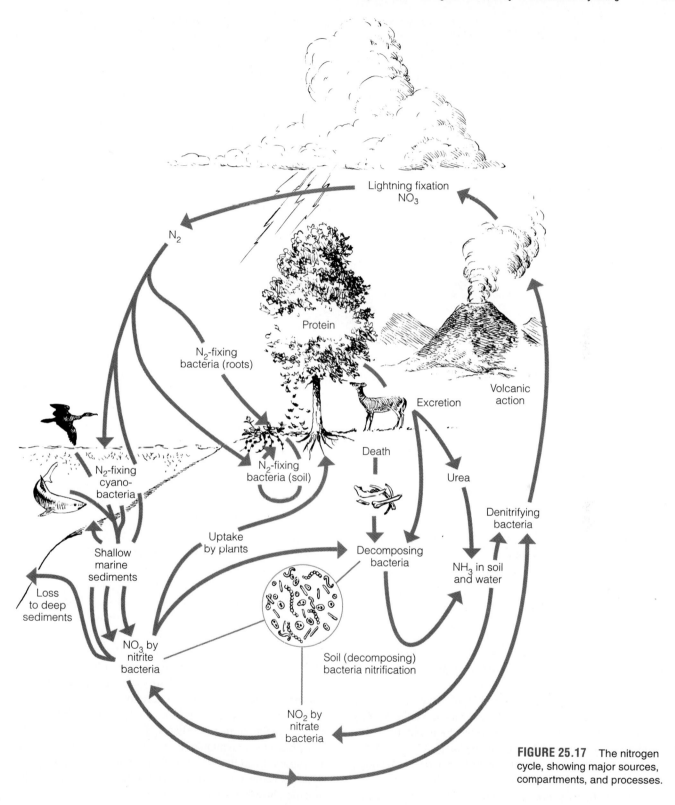

FIGURE 25.17 The nitrogen cycle, showing major sources, compartments, and processes.

Hydrogen sulfide comes from several sources: the combustion of fossil fuels, volcanic eruptions, and gases released in terrestrial and aquatic decomposition. It quickly oxidizes into another volatile form, sulfur dioxide (SO_2). Atmospheric sulfur dioxide, soluble in water, is carried back to Earth in rainwater as weak sulfuric acid (H_2SO_4).

The oceans are a major source of another gaseous form of sulfur, dimethylsulfide [$(CH_3)_2S$]. Dimethylsulfide (DMS) is produced during the decomposition of phytoplankton (Andreae 1990, Kiene 1990). The majority of DMS is degraded by microbes in the surface waters. Although only a small percentage of the total production in

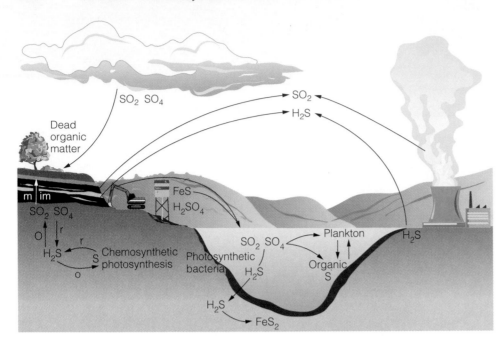

FIGURE 25.18   The sulfur cycle. Note the two components, sedimentary and gaseous. Major sources from human activity are the burning of fossil fuel and acidic drainage from coal mines. Major natural sources are volcanic eruptions and decomposition of organic matter. See text for description of cycle.

of DMS is lost to the atmosphere, recent estimates suggest it constitutes the largest atmospheric emission of sulfur gas (Spiro et al. 1992). In the atmosphere, DMS is rapidly oxidized by OH radicals, forming sulfate aerosols that are deposited in precipitation.

Whatever the source, sulfur in soluble form is taken up by plants. Starting with photosynthesis, it is incorporated through a series of metabolic processes into such sulfur-bearing amino acids as cysteine. From the producers, the sulfur in amino acids is transferred to consumers (Figure 25.18).

Excretions and death carry sulfur in living material back to the soil and to the bottoms of ponds, lakes, and seas, where sulfate-reducing bacteria release it as hydrogen sulfide or as a sulfate. Other microorganisms in the forest soil convert inorganic sulfate to organic forms and incorporate it into forest ecosystems (Swank et al. 1983). Colorless sulfur bacteria both reduce hydrogen sulfide to elemental sulfur and oxidize it to sulfuric acid. Photosynthetic green and purple bacteria utilize hydrogen sulfide in the presence of light as an oxygen acceptor in the reduction of carbon dioxide. Best known are the purple bacteria found in salt marshes and in mud flats of estuaries. These organisms can carry the oxidation of hydrogen sulfide as far as sulfate, which may be recirculated and taken up by the producers or may be used by sulfate-reducing bacteria. The green sulfur bacteria can carry the reduction of hydrogen sulfide to elemental sulfur.

Sulfur, in the presence of iron and under anaerobic conditions, will precipitate as ferrous sulfide (FeS_2). This compound is highly insoluble under neutral and alkaline conditions and is firmly held in mud and wet soil.

The Phosphorus Cycle

Phosphorus, unlike sulfur, occurs in only very minute amounts in the atmosphere, and none of its known com-

pounds have appreciable vapor pressure. Therefore, the phosphorus cycle can follow the hydrological cycle only partway, from land to sea (Figure 25.19).

Under undisturbed natural conditions the source of phosphorus in the soil is the mineral apatite, a phosphate of calcium [$Ca_5(PO_4)_3(F, Cl, OH)$]. Phosphorus also occurs in secondary forms as compounds of calcium and iron, and in organic combination. Under the best of conditions, the amount of available phosphorus is small. Phosphates, such as calcium triphosphate [$Ca_3(PO_4)_2$], are freely soluble only in acid solutions and under reducing conditions. Its natural limitation in aquatic ecosystems is emphasized by the explosive growth of algae in water with heavy discharges of phosphorus-rich wastes (Stevenson and Stoermer 1982).

The main reservoirs of phosphorus in the biosphere are rock and natural phosphate deposits, from which the element is released by weathering, by leaching, by erosion, and by mining for agricultural use. Some of it passes through terrestrial and aquatic ecosystems as organic phosphorus by the way of plants, grazers, predators, and parasites; and it is returned to the ecosystem by excretion, death, and decay. In terrestrial ecosystems, organic phosphates are reduced by bacteria to inorganic phosphates. Some are recycled to plants, some become unavailable in chemical compounds, and some are immobilized in microorganisms. Some of the phosphorus of terrestrial ecosystems escapes to lakes and seas, both as organic phosphates and as particulate organic matter.

In marine and freshwater systems, the phosphorus cycle involves three major fractions: (1) particulate organic phosphorus (POP), which includes phosphorus contained in both dead particulate matter and phytoplankton; (2) dissolved inorganic phosphates (DIP), mostly soluble orthophosphate (PO_4^{3-}) that comes from various sources, both aquatic and terrestrial; and (3) dissolved organic phosphorus (DOP), which is excreted by organisms, especially zooplankton. In

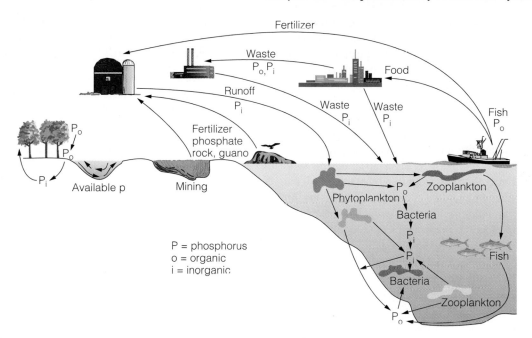

FIGURE 25.19   The phosphorus cycle in terrestrial and aquatic ecosystems. See text for description of cycle

most aquatic environments, particulate phosphorus is in greatest abundance, followed by DIP and then DOP.

Uptake by primary producers and bacteria is responsible for the low phosphate concentrations typical of surface waters. Phosphorus in phytoplankton may be ingested by zooplankton or by detritus-feeding organisms as dead organic matter. The principal means of regenerating DIP (phosphates) is by the decay of POP and by animals. Zooplankton may excrete as much phosphorus daily as is stored in its biomass (Pomeroy et al. 1963). By excreting phosphorus, zooplankton is instrumental in keeping the aquatic cycle going. More than half of the phosphorus zooplankton excretes is DIP, which is taken up by phytoplankton. The zooplankton in the Central Pacific Ocean, for example, may release 55–183 percent of the daily requirements of phosphorus for the phytoplankton (Perry and Eppley 1981). The remainder of the phosphorus in aquatic ecosystems is in organic forms (DOP). These are utilized by bacteria, which fail to regenerate much dissolved inorganic phosphate (DIP) themselves. Instead the bacteria are eaten by microbial grazers, who then excrete the phosphate they ingest (Johannes 1968).

Part of the phosphorus in aquatic ecosystems is deposited in deep and shallow sediments. Precipitated largely as calcium compounds, much of it becomes stored for long periods of time in bottom sediments. In marine ecosystems, seasonal overturns and upwellings return some of the phosphorus to surface waters, where it is available to phytoplankton. During the growing season, much of the phosphorus is tied up in organic matter, and only by a rapid turnover in its populations can phytoplankton meet its phosphorus requirements.

The role of organisms in the cycling of phosphorus in aquatic ecosystems is further illustrated by cycling in the tidal marshes. In the intertidal marshes of the southern United States, the marsh grass *Spartina alternifolia* withdraws phosphorus

from subsurface sediments. Half of the phosphorus withdrawn is fixed in plant tissue. The other half is leached from leaves by rain and tides and carried out to sea by the tide (Reimold 1972). When the grass dies, abundant animal life in the marsh and adjacent waters and tidal creeks use the detritus as food.

LINKAGES AMONG BIOGEOCHEMICAL CYCLES

Although in introducing each of the major biogeochemical cycles we have presented them independently, they are all linked in various ways. In specific cases they are linked through common membership in compounds that form an important component of their cycles, such as the link between calcium and phosphorus in the mineral apatite, a phosphate of calcium. In general, the cycled nutrients are all components of living organisms, constituents of organic matter. As a result, they travel together through the process of internal cycling.

Table 3.1 (page 55) lists the macro- and micronutrients required by autotrophs and heterotrophs. These nutrients are required in different proportions for different processes. For example, the equation in Chapter 6 that describes the process of photosynthesis dictates the utilization of six moles of water (H_2O) and the production of six moles of oxygen (O_2) for every six moles of CO_2 that is transformed into one mole of sugar $(CH_2O)_6$. The proportions of hydrogen, oxygen, and carbon involved in the process of photosynthesis are fixed. Likewise, a fixed quantity of nitrogen is required to produce a mole of rubisco, the enzyme that catalyzes the fixation of CO_2 in photosynthesis (see Chapter 6). Therefore, the nitrogen content of a rubisco molecule is the same in every plant, independent of species or environment. The same is true for the variety of amino acids, proteins, and other nitrogenous compounds that are essential for the synthesis of plant cells and tissues. The branch of chemistry dealing with the quantitative relationships of elements in combination is called **stoichiometry.** The stoichiometric relationships among various elements involved in processes related to carbon uptake and plant growth have an important influence on the cycling of nutrients in ecosystems.

Because of the stoichiometric relationship among the variety of macro- and micronutrients required by plants, the limitation of one nutrient can affect the cycling of all the others. As an example, we can examine the link between carbon and nitrogen presented earlier in this chapter. Although the nitrogen content of a rubisco molecule is the same in every plant, independent of species or environment, plants can differ in the concentration of rubisco found in their leaves and, therefore, in their concentration of nitrogen (g N/g dry weight). Plants growing under low nitrogen availability will have a lower rate of nitrogen uptake and less nitrogen for the production of rubisco and other essential nitrogen-based compounds. Differences in the concentration of rubisco in the leaves of different plant species is in part responsible for the differences in leaf nitrogen concentration shown in Figure 7.25 and the differences in the ratio of carbon to nitrogen for the leaf litter of different tree species shown in Table 9.1. In turn, the lower concentrations of rubisco result in lower rates of photosynthesis and carbon gain; and the lower concentrations of nitrogen in the leaf litter influence the relative rates of immobilization and mineralization and subsequent nitrogen release to the soil in the process of decomposition. In this manner, nitrogen availability and uptake by plants influence the rate at which carbon and other essential plant nutrients cycle through the ecosystem.

Conversely, the variety of other essential nutrients and environmental factors that directly influence the rate of primary productivity, and thus the demand for nitrogen, will influence the rate of nitrogen cycling through the ecosystem. In fact, the cycles of all essential nutrients for plant and animal growth are linked due to the stoichiometric relationships that define the mixture of chemicals that make up all living matter.

Summary

1. Nutrients flow from the living to the nonliving components of the ecosystem and back in a perpetual cycle. By means of these cycles, plants and animals obtain nutrients necessary for their survival and growth.
2. There are two basic types of biogeochemical cycles: the gaseous, represented by oxygen, carbon, and nitrogen, whose major pools are in the atmosphere; and the sedimentary, represented by the sulfur and phosphorus cycles, whose major pools are in Earth's crust. The sedimentary cycles involve two phases, salt solution and rock. Minerals become available through the weathering of Earth's crust, enter the water cycle as a salt solution, take diverse pathways through the ecosystem, and ultimately return to Earth's crust through sedimentation.
3. All nutrient cycles have a common structure, sharing three basic components: inputs, internal cycling, and outputs.
4. The input of nutrients to the ecosystem depends on the type of biogeochemical cycle, gaseous or sedimentary. The availability of essential nutrients in terrestrial ecosystems depends heavily on the nature of the soil. The major source of nutrients for aquatic life are inputs from the surrounding land in the form of drainage water, detritus, and sediments.
5. As plants take up nutrients from the soil or water, they become incorporated into living tissues, organic matter. As the tissues senesce, the dead organic matter is returned to the soil or sediment surface. Various decomposers transform the organic nutrients into mineral form, and they are once again available for uptake by plants. This process is called internal cycling. The rate at which nutrients cycle through the ecosystem is directly related to the rates of primary productivity (nutrient uptake) and decomposition (nutrient release). Environmental factors that influence these two processes will affect the rate at which nutrients cycle through the ecosystem.
6. The export of nutrients from the ecosystem represents a loss that must be offset by inputs if a net decline is not to occur. Export can occur in a variety of ways, depending on the nature of the specific biogeochemical cycle. A major means of transportation is in the form of organic matter carried by surface flow of water in streams and rivers. Leaching of dissolved nutrients from soils into surface and groundwater also represents a significant export in some ecosystems. Harvesting of biomass in forestry and agriculture represents a permanent withdrawal from the ecosystem. Fire is also a major source of nutrient export in some terrestrial ecosystems.
7. The carbon cycle is inseparably tied to energy flow. Carbon in the form of carbon dioxide (and carbonates in aquatic ecosystems) is fixed into organic compounds in the process of photosynthesis, passes through the food chain, and is eventually returned to the atmosphere through the process of respiration. The carbon dioxide cycle exhibits both annual and diurnal fluctuations.
8. The nitrogen cycle is characterized by the fixation of atmospheric nitrogen by mutualistic nitrogen-fixing bacteria associated with roots of many plants, largely legumes, and cyanobacteria. Other processes are ammonification, the breakdown of amino acids by decomposer organisms to produce ammonia; nitrification, the bacterial oxidation of ammonia to nitrate and nitrates; and denitrification, the reduction of nitrates to gaseous nitrogen.
9. The phosphorus cycle is wholly sedimentary, with reserves coming largely from phosphate rock. Once in the soil, inorganic phosphorus in taken up by plants and incorporated into organic compounds. With the senescence of living tissues the organic phosphorus is returned to the soil, where it is once again made available through mineralization during the process of decomposition. In aquatic ecosystems, phosphorus exists in three major states: particulate organic phosphorus (POP), dissolved organic phosphorus (DOP), and dissolved inorganic phosphorus (DIP). Transfer among these states is controlled by the processes of uptake by primary producers, decomposition, and excretion of DIP by zooplankton.
10. The sulfur cycle is a combination of gaseous and sedimentary cycles, because sulfur has reservoirs in Earth's crust and in the atmosphere. The sulfur cycle involves a long-term sedimentary phase in which sulfur is tied up in organic and inorganic deposits, is released by weathering and decomposition, and is

carried to terrestrial and aquatic ecosystems in salt solution. The bulk of sulfur first appears in the gaseous phase as a volatile gas, hydrogen sulfide (H_2S), in the atmosphere, which quickly oxidizes to form sulfur dioxide. Once in soluble form, sulfur is taken up by plants and incorporated into organic compounds. Excretions and death carry sulfur in living material back to the soil and to the bottoms of ponds, lakes, and seas, where sulfate-reducing bacteria release it as hydrogen sulfide or as a sulfate.

11. All of the major biogeochemical cycles are linked, as the nutrients that cycle are all components of living organisms, constituents of organic matter. The stoichiometric relationships among various elements involved in plant processes related to carbon uptake and plant growth have an important influence on the cycling of nutrients in ecosystems.

Review Questions

1. What are the two types of biogeochemical cycles? What are their distinguishing characteristics?
2. Why is the salt solution phase so important in the sedimentary cycle?
3. What are some of the inputs and outputs in a nutrient cycle?
4. What is the relationship among net primary production, decomposition, and nutrient cycling?
5. Contrast nutrient cycling in terrestrial and aquatic ecosystems. What is the outstanding difference?
6. What is the significance of the carbon cycle?
7. What is the relationship among photosynthesis, decomposition, and the carbon cycle?
8. Characterize the following processes in the nitrogen cycle: Fixation, ammonification, nitrification, and denitrification.
9. What biological and nonbiological mechanisms are responsible for nitrogen fixation?
10. What are sources of input into the nitrogen cycle? Sources of output?
11. What is the source of sulfur in the sulfur cycle? Why does the sulfur cycle have characteristics of both sedimentary and gaseous cycles?
12. What is the source of phosphorus in the phosphorus cycle? The sink?
13. Through what three compartments does the phosphorus cycle move in aquatic ecosystems?

Cross References

Solar radiation, 21–22; atmospheric circulation, 25–28; ocean gyres, 28; water molecule, 56–57; nutrients, 54–55; soil chemistry, 62–63; soil organic matter, 67–68; decomposition 148–149; mineralization, 152–154; C:N ratio, 152; seasonal overturn in lakes, 630; nutrient spiraling in streams, 649; peatlands 658; nitrogen cycling: grassland, 572; savanna, 597; desert, 587; tundra, 598; forest, 621; mutualism, 321.

CHAPTER 26

Biogeochemistry II:
Global Cycles and Human Impacts

CONCEPTS

1. The current atmospheric pool of oxygen is maintained in a dynamic equilibrium between the production of oxygen in photosynthesis and its consumption in respiration.

2. The major components of the global carbon budget involve the exchange of carbon dioxide between the atmosphere and the land surface and oceans.

3. One outcome of nitrogenous atmospheric pollution is an increased deposition of nitrogen, which can have both positive and negative effects on ecosystems.

4. By far the largest source of sulfur gases to the atmosphere is that of direct emissions from human industrial activities.

5. The human inputs of nitrogen and sulfur oxides to the atmosphere through industrial emissions has the effect of lowering the pH of rainfall—commonly referred to as acid rain.

In the previous chapter, we presented a general model of nutrient cycling in ecosystems and its application to specific biogeochemical cycles. The cycling of nutrients and energy occurs within all ecosystems, and it is most often studied as a local process—that is, the internal cycling of nutrients within the ecosystem and the identification of exchanges, both to (inputs) and from (outputs) the ecosystem. Through these processes of exchange, the biogeochemical cycles of differing ecosystems are linked. Often the output from one ecosystem represents an input to another, as with the case of the export of nutrients from terrestrial to aquatic ecosystems. These processes of exchange of nutrients among ecosystems require that we view the biogeochemical cycles from a much broader spatial framework than that of a single ecosystem. This is particularly true of those nutrients that possess a gaseous cycle. Because the main pools of these nutrients are the atmosphere and the ocean, they have pronouncedly global circulation patterns. In this chapter, we will expand our model of biogeochemical cycling to provide a global framework for understanding the cycling of oxygen, carbon, nitrogen, phosphorus, and sulfur. In addition, we will examine how human activities have altered certain biogeochemical cycles, affecting both ecosystems and human populations.

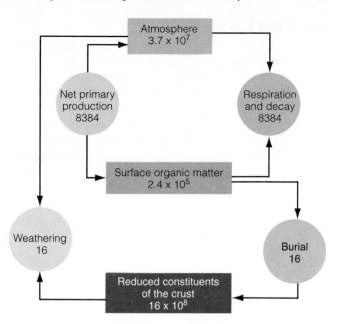

FIGURE 26.1 A simple model for the global biogeochemical cycle of O_2. Data are expressed in units of 10^{12} moles of O_2 per year or the equivalent amount of reduced compounds. Note that a small misbalance in the ratio of photosynthesis to respiration can result in a net storage of reduced organic materials in the crust and an accumulation of O_2 in the atmosphere. (From Schlesinger 1997.)

THE GLOBAL OXYGEN CYCLE

Main Features

Oxygen, the byproduct of photosynthesis, is very active chemically. It combines with a wide range of chemicals in the Earth's crust, and it reacts spontaneously with organic compounds and reduced substances. It is involved in the oxidation of carbohydrates in the process of respiration, releasing energy, carbon dioxide, and water (see Chapters 6 and 9). Its primary role in biological oxidation is that of an electron acceptor. The breakdown and decomposition of organic molecules proceed primarily by dehydrogenation. Hydrogen is removed by enzymatic action from organic molecules in a series of reactions and is ultimately accepted by oxygen, forming water.

The major supply of free oxygen supporting life is in the atmosphere (Figure 26.1). It comes from two significant sources. One is the photodissociation of water vapor, in which most of the hydrogen released escapes into outer space. The other source is photosynthesis, active only since life began on Earth. The current atmospheric pool of oxygen is maintained in a dynamic equilibrium between the production of oxygen in photosynthesis and its consumption in respiration. The residence time of atmospheric oxygen before its removal by respiration and decay and its replacement by photosynthesis is on the order of 4500 years (Walker 1984).

Because photosynthesis and respiration are complementary, involving the release and capture of oxygen, one would seem to balance the other, so that no significant quantity of oxygen would accumulate in the atmosphere. At some time in Earth's history, the amount of oxygen introduced into the atmosphere had to exceed the amount involved in the decay of

organic matter (respiration of microbial decomposers) and the oxidation of sedimentary rocks. Part of the atmospheric oxygen pool is due to the burial and storage of partially decomposed organic matter in the form of coal, oil, natural gas (the fossil fuels; see The Global Carbon Cycle), and organic carbon in sedimentary rocks (Kasting 1993, Knoll 1986).

Because oxygen is so reactive, its cycle is complex. It reacts rapidly with reduced organic matter in the soil and reduced mineral constituents of Earth's crust. The uplift and weathering of organic carbon in sedimentary rocks represents a small flux of oxygen to the atmosphere. Buried organic matter in the ocean that escapes oxidation conserves atmospheric oxygen by contributing its portion to replace the oxygen consumed by weathering of rocks, thus stabilizing the atmospheric reservoir (Figure 26.1).

The oxygen cycle is directly linked to other biogeochemical cycles. As a constituent of carbon dioxide, oxygen circulates freely throughout the biosphere. Some carbon dioxide combines with calcium to form carbonates. Oxygen combines with nitrogen to form nitrates, with iron to form ferric oxides, and with many other minerals to form various other oxides. Assuming that half the annual circulation of nitrogen on land and 15 percent of the nitrogen cycle in oceans involves the uptake of nitrogen by plants in the form of nitrate (NO_3^-), approximately 3 percent of the annual production of oxygen by photosynthesis is used to oxidize ammonium (NH_4^+) in nitrification reactions (see Chapter 25).

Part of the atmospheric oxygen that reaches the higher levels of the troposphere reduces to ozone (O_3) in the presence of high-energy ultraviolet radiation.

The Ozone Layer

Most of the ozone resides in the stratosphere, 10 to 50 km above Earth (Figure 26.2). The stratosphere is characterized by increasing temperature in contrast to the colder temperatures below, because the upper relatively thin layer of ozone absorbs most of the ultraviolet radiation and radiates infrared wavelengths to outer space. A downward intrusion of stratospheric air introduces sufficient ozone to the troposphere to initiate photochemical processes in the lower atmosphere. Most importantly, it shields Earth against biologically harmful solar ultraviolet radiation.

Ozone in the outer atmosphere is maintained by a cyclic photolytic reaction. Production of ozone requires the breakage of the O—O bond in O_2. Once freed, the O atoms rapidly combine with O_2 to form O_3:

$$O + O_2 \longrightarrow O_3$$

In the stratosphere, the O_2 bond is broken (dissociated) by solar radiation:

$$O_2 + hv \longrightarrow O + O$$

Here h is Planck's constant (cal/sec) and v is frequency of the light. This stratospheric reaction consumes a large amount of solar energy.

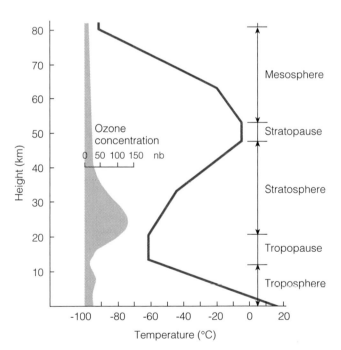

FIGURE 26.2 The ozone layer in the stratosphere. Absorption of longer ultraviolet wavelengths in the atmosphere acts on oxygen molecules (O_2) to produce ozone (O_3), most of which is concentrated in a layer in the stratosphere at approximately 20 to 35 km above Earth. The layer shows both seasonal changes in height (highest in summer and lowest in winter) and latitudinal changes in height (higher in low latitudes). The tropopause and stratopause are areas in Earth's atmosphere where temperature is constant with height. (After Gates 1993:190.)

At the same time, a reverse reaction consumes O_3:

$$O_3 + hv \longrightarrow O + O_2$$
$$O + O_3 \longrightarrow 2O_2$$
$$\text{Net reaction: } 2O_3 + hv \longrightarrow 3O_2$$

A number of other catalytic reactions tend to destroy O_3 and maintain a balance between O_2 and O_3, especially reactions involving the gaseous oxides of nitrogen and hydrogen. For example, NO consumes O_3 in the following catalytic cycle:

$$NO + O_3 \longrightarrow NO_2 + O_2$$
$$O_3 + hv \longrightarrow O + O_2$$
$$NO_2 + O \longrightarrow NO + O_2$$
$$\text{Net reaction: } 2O_3 + hv \longrightarrow 3O_2$$

The NO consumed in the first reaction is replaced by another molecule available for further destruction of ozone in the final reaction, and O_3 is consumed.

Under natural conditions in the atmosphere, a balance exists between the rate of ozone formation and its rate of destruction. This balance exists largely in the stratosphere. However, in recent times a number of anthropogenic and even biological catalysts injected into the stratosphere are reactive enough to cause a decrease in stratospheric ozone (Lal and Holt 1991, McElroy and Salawitch 1989, Cicerone 1987). Among them are chlorofluorocarbons (CCl_2F_2 and CCl_3F), methane (CH_4), both natural and anthropogenic, and nitrous oxide (NO_2) from denitrification and synthetic nitrogen fertilizer. Of particular concern is chlorine monoxide (ClO) derived from chlorofluorocarbons (CFCs) in aerosol spray propellants (banned in the United States and becoming banned worldwide), refrigerants, solvents, and other sources. Its chlorine can be incorporated in harmless chlorine nitrates ($ClNO_3$) or it can be involved in the catalytic destruction of ozone (Figure 26.3):

$$Cl + O_3 \longrightarrow ClO + O_2$$
$$O_3 + hv \longrightarrow O + O_2$$
$$ClO + O \longrightarrow Cl + O_2$$
$$\text{Net reaction: } 2O_3 + hv \longrightarrow 3O_2$$

In 1975 atmospheric scientists first discovered a pronounced springtime thinning in the ozone over the South Pole. Since then the thinning has developed into an ozone hole, where the ozone has been depleted 50 to 95 percent. The detection of chlorine monoxide in the hole suggested that ClO was the problem. Tiny ice particles in the cold lower stratosphere catalyzed chemical reactions that converted the chlorine in CFCs into ozone-destroying ClO (Jones 1989). In 1990 other atmospheric scientists noted a similar thinning in the ozone layers over the Arctic (Hofmann and Deshler 1991). This ozone loss was attributed not only to CFCs but also to natural aerosols of sulfuric acid that immobilize nitrogen oxides.

Nitrogen oxides protect the ozone by tying up chlorine into harmless forms. With nitrogen oxides immobilized, chlorine forms ClO. More surprising was the discovery in 1991 of sheets of ClO as far south as Cuba (Kerr 1992). The presence of ClO in the north temperate latitudes could be attributed to its escape

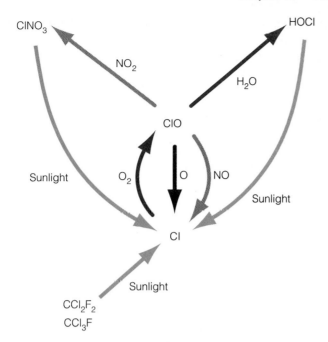

FIGURE 26.3   The cycle of ozone destruction induced by chlorine monoxide (ClO), derived from human use of chlorofluorocarbons. The chlorine can either be incorporated in harmless chlorine nitrates or be involved in the photolytic dissociation of ozone. (After Kerr 1987:1182.)

from the Arctic air mass. Some atmospheric scientists, however, believe that small ice particles associated with the strong updrafts of tropical air masses catalyze the production ClO and account for summertime as well as wintertime ozone depletion. Whatever the causes, the ozone layer in temperate latitudes is eroding at the rate of 3 percent per decade over the latitude of New York, and 5 percent over Buenos Aires and Sydney.

Scientists are concerned over the reduced amount of ozone, because it allows disproportionate amounts of ultraviolet radiation to penetrate Earth's atmosphere. For example, under the conditions of an overhead sun and average amounts of O_3, a 1 percent decrease in ozone would result in a 20 percent increase in ultraviolet penetration at wavelengths of 250 nm, a 250 percent increase at 290 nm, and a 500 percent increase at 287 nm (Walker 1977). Although solar ultraviolet wavelengths of less than 240 nm are absorbed by atmospheric O_2 and O_3, only O_3 is effective for wavelengths between 240 and 320 nm (Cicerone 1987). Such reductions in the ozone layer can have adverse ecological effects such as altering DNA, causing severe sunburn, and increasing skin cancer in humans, and inhibiting photosynthesis in some terrestrial plants and phytoplankton.

THE GLOBAL CARBON CYCLE
Main Features

The carbon budget of the Earth is closely linked to the atmosphere, land, and oceans and to the mass movements of air around the planet. The Earth contains about 10^{23} g of carbon. All but a small fraction is buried in sedimentary rocks and is not actively involved in the global carbon cycle. The carbon pool involved in the global carbon cycle (Figure 26.4) amounts to an estimated 55×10^{18} g or 55,000 Gt (Gt is a gigaton, equal to 1 billion [10^9] metric tons or 10^{15} g). Fossil fuels, created by the burial of partially decomposed organic matter, account for an estimated 10,000 Gt. The oceans contain the vast majority of the active carbon pool, about 38,000 Gt, mostly as bicarbonate ions (HCO_3^-) and carbonate ions (CO_3^-). Dead organic matter in the oceans accounts for 1650 Gt of carbon, and living matter, mostly phytoplankton, 3 Gt.

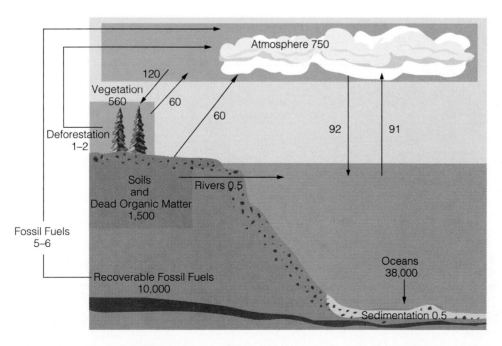

FIGURE 26.4   The global carbon cycle. Boxes show the sizes of the major pools of carbon, and arrows indicate the major changes among them. All values are in Gt of carbon and exchanges are on an annual time scale. The largest pool of carbon, geologic, is not included because of the slow rates (geologic time scale) of transfer with other active pools.

The terrestrial biosphere contains an estimated 1500 Gt of carbon as dead organic matter and 500 Gt as living matter (biomass). The atmosphere, the major coupling mechanism in the cycling of CO_2, holds about 750 Gt of carbon (Baes et al. 1977, Schimel et al. 1995, Houghton et al. 1996).

In the ocean, the surface water acts as the site of main exchange of carbon between atmosphere and ocean. The ability of the surface waters to take up CO_2 is governed largely by the reaction of CO_2 with the carbonate ion to form bicarbonates (see Carbon Cycle, Chapter 25). In the surface water, carbon circulates physically by means of currents and biologically through assimilation by phytoplankton and movement through the food chain. The net exchange of CO_2 between the oceans and atmosphere due to both physical and biological processes results in a net uptake of 2 Gt per year by the oceans, while burial in sediments accounts for a net loss of 0.1 Gt of carbon per year.

The uptake of CO_2 from the atmosphere by terrestrial ecosystems is governed by gross production (photosynthesis). Losses are a function of autotrophic and heterotrophic respiration, the latter being dominated by microbial decomposers. Until recently, exchanges of CO_2 between the land mass and the atmosphere (uptake in photosynthesis and release by respiration/decomposition) were believed to be nearly in equilibrium (Figure 26.4). However, more recent research suggests that the terrestrial surface is acting as a carbon sink, with a net uptake of CO_2 from the atmosphere (Houghton 1995).

Of considerable importance in the terrestrial carbon cycle is the relative proportions of carbon stored in soils and in living vegetation (biomass). Carbon stored in soils includes dead organic matter on the soil surface and in the underlying mineral soil. Estimates place the amount of soil carbon at 1500 Gt, compared with 560 Gt for the total world terrestrial biomass. Much of this carbon stored as dead organic matter is in the lower soil profile, where it exceeds carbon in the surface soil by a factor of 25. The lower soil layers of tropical forests and both tropical and temperate grasslands contain a large percentage of soil carbon, whereas soil carbon in temperate forests and Arctic regions is confined largely to the surface soil layers.

The average amount of carbon per unit of soil profile increases from the tropical regions poleward to the boreal forest and tundra (Table 26.1). Low values for the tropical forest reflect high rates of decomposition, which compensate for high productivity and litterfall (see Figure 25.5). Frozen tundra soil and waterlogged soils of swamps and marshes have the greatest accumulation of dead organic matter, because moisture, low temperature, or both inhibit decay. Soil carbon greatly exceeds biomass in boreal forests and tundra. Soil carbon likewise exceeds biomass in temperate grasslands, but in tropical savannas the reverse situation exists, probably because of recurring fires. World output of carbon from respiration resulting from the decomposition of soil carbon in terrestrial ecosystems amounts to approximately 60 Gt per year.

The global cycle of atmospheric carbon dioxide, like local cycles, exhibits a marked annual variation, particularly in the land-dominated Northern Hemisphere (Figure 26.5). North of the 30°N latitude, carbon dioxide content of the atmosphere up to the lower stratosphere begins to decline markedly during early spring, when photosynthetic activity and carbon storage in plant growth withdraw more carbon dioxide from the atmosphere than is replaced by respiration. The concentration of carbon dioxide reaches a low in August. By October the level increases again as photosynthesis declines and respiration dominates the exchange between the atmosphere and terrestrial ecosystems. The difference in the magnitude of the seasonal oscillation of CO_2 between the Northern and Southern Hemispheres seen in Figure 26.5 reflects the difference in the land mass of the two hemispheres.

TABLE 26.1 **Distribution of Detritus and Biomass by Ecosystem Types**

Ecosystem Type	Mean Total Profile Detritus (kg C/m^2)	CV* (%)	World Area (10^9 ha)	Total World Detritus (10^9 mtn C)	Total World Biomass (10^9 mtn C)
Woodland and shrubland	6.9	59	8.5	59	22.0
Tropical savanna	3.7	87	15.0	56	27.0
Tropical forest	10.4	44	24.5	255	460.0
Temperate forest	11.8	35	12.0	142	175.0
Boreal forest	14.9	53	12.0	179	108.0
Temperate grassland	19.2	25	9.0	173	6.3
Tundra and alpine	21.6	68	8.0	173	2.4
Desert scrub	5.6	38	18.0	101	5.4
Extreme desert	0.1	—	24.0	3	0.2
Cultivated land	12.7	—	14.0	178	7.0
Swamp and marsh	68.6	63	2.0	137	13.6
Total			147.0	1456	826.9

*CV = coefficient of variation = standard deviation/mean × 100.
Source: Adapted from Schlesinger 1977.

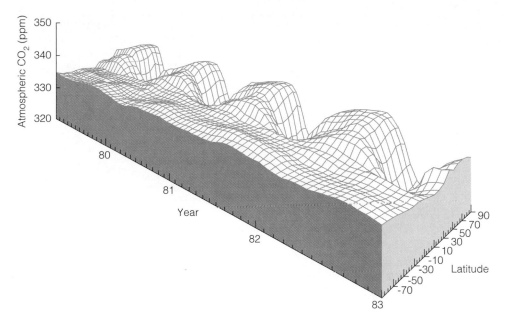

FIGURE 26.5 Seasonal fluctuations in the concentration of atmospheric CO_2 (1981–1984) shown as a function of latitude (10° belts). Note the seasonal shifts in concentration at any given latitude as well as the smaller amplitude of fluctuations in the Southern Hemisphere. (Conway et al. 1988.)

Human Impact on the Carbon Cycle

The CO_2 flux among land, sea, and atmosphere has been disturbed by a rapid injection of carbon dioxide in the atmosphere from the burning of fossil fuels and from the clearing of forests. Clearing increases the input of CO_2 from burning trees and decomposing organic matter.

Adding to the problem of increasing carbon dioxide are increases in other atmospheric gases, especially methane (CH_4). Its major sources are ruminant animals, microbial decomposition in swamps, marshes, and tundra, and industrial gases released to the atmosphere. Over time (resident time of 3.2 years) methane oxidizes to H_2O, especially in the stratosphere. Atmospheric methane has approximately doubled over the past 200 years (Figure 26.6). This increase is linked to human population growth and to increased cattle ranching and rice paddy production.

The rising atmospheric concentrations of CO_2 and methane have the potential to alter the global energy balance and subsequently the global climate system. The implication of rising atmospheric concentrations of CO_2 and possible climate change are the focus of Chapter 32.

THE GLOBAL NITROGEN CYCLE
Main Features

The global nitrogen cycle follows the pathway of the local nitrogen cycle, only on a grander scale. The global cycle is diagrammed in Figure 26.7. The atmosphere is the largest pool, containing 3.9×10^{21} g. Comparatively small amounts of nitrogen are found in the biomass ($3.5-10^{15}$ g) and soils ($95–140 \times 10^{15}$ g) of terrestrial ecosystems (Post et al. 1985, Batjes 1996).

The major sources of fixed nitrogen are biological (140×10^{12} g yr^{-1}) and high-energy fixation of atmospheric nitrogen by lightning $> 3 \times 10^{12}$ g yr^{-1}) (Burns and Hardy 1975, Ridley et al. 1996). Additional sources of nitrogen include inorganic nitrogen in rain, from such sources as fixed juvenile nitrogen from volcanic activity; absorption of ammonia from the atmosphere by plants and soil; and nitrogen accretion from windblown aerosols, which contain both organic and inorganic forms of nitrogen. In terrestrial ecosystems, ammonia and nitrates are taken up by plants and converted to organic nitrogen (amino acids) and moved through the food chain. Dead organic matter is broken down into ammonia, then into nitrates and nitrites (see Figure 25.14). Nitrates may be taken up directly by plants, immobilized by microbes, stored in decomposing humus, or leached into streams, lakes, and eventually seas. The uptake of inorganic nitrogen by plants is so rapid that at any given time, the pool of inorganic nitrogen

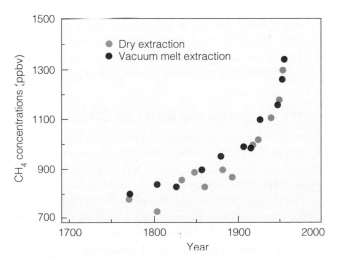

FIGURE 26.6 Concentration of atmospheric methane over the past 200 years as indicated by air samples taken from ice cores at Siple Station, Antarctica. (From Stauffer et al. 1991:221.)

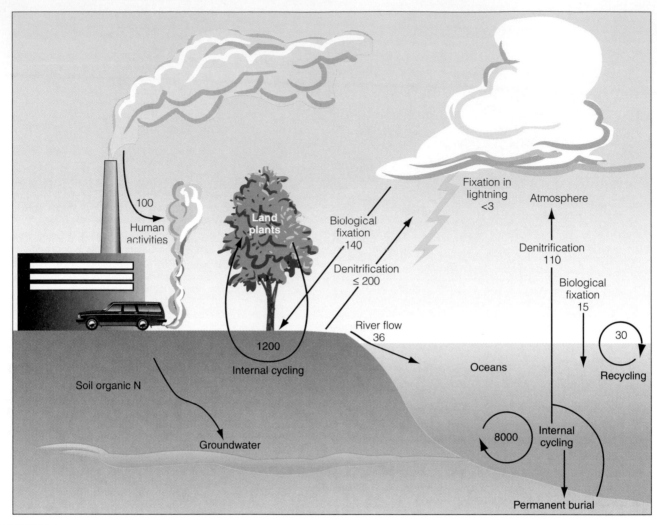

FIGURE 26.7   The global nitrogen cycle. Each flux is shown in units of 10^{12} g N/yr. (From Schlesinger 1997.)

(NH_4^+ and NO_3^-) in soils is very small. Most of the nitrogen stored in terrestrial ecosystems is tied up in organic matter, both living and dead.

Global estimates of denitrification in terrestrial ecosystems vary widely but are on the order of $\leq 200 \times 10^{12}$ g yr^{-1}, of which over half occurs in wetland ecosystems (Bowden 1986).

The major sources of nitrogen to the world's oceans are dissolved forms in the freshwater drainage from rivers (36×10^{12} g yr^{-1}) and inputs in precipitation (30×10^{12} g yr^{-1}) (Meybeck 1982, 1993, Duce et al. 1991). Biological fixation accounts for another 15×10^{12} g yr^{-1} (Carpenter and Romans 1991). In aquatic ecosystems, nitrogen is cycled internally in a similar manner to that discussed for terrestrial ecosystems, except that the large reserves, such as those contained in soil humus, are lacking. Life in the water contributes organic matter and dead organisms that undergo decomposition and subsequent release of ammonia and nitrates.

Tracer studies with ^{15}N, a stable, nonradioactive isotope, show that in marine ecosystems, ammonia is recycled rapidly by phytoplankton (Dugdale and Goering 1967). As a result little ammonia exists in natural waters, and nitrate is used

only in the virtual absence of ammonia. There are small but steady losses from the biosphere to the deep sediments of the ocean and to sedimentary rocks. In return there is a small addition of new nitrogen from the weathering of igneous rocks and juvenile nitrogen from volcanic activity.

Denitrification accounts for an estimated flux of 110×10^{12} g N yr^{-1} from the world's oceans to the atmosphere.

Human Impact on the Nitrogen Cycle

Under different circumstances, human intrusions into the nitrogen cycle result in either a reduction or an increase in nitrogen availability. Conversion of forests and grasslands to cropland results in a steady decline in nitrogen content of the soil. Mixing and breaking up the soil exposes more organic matter to rapid decomposition and leaching from the soil (see Figure 25.6). Removal of nitrogen through harvested crops and grazing causes additional losses (see Table 25.3).

On the other hand, excessive amounts of nitrogen may be added to the system, creating a variety of problems. Heavy applications of commercial fertilizer disturb the natural bal-

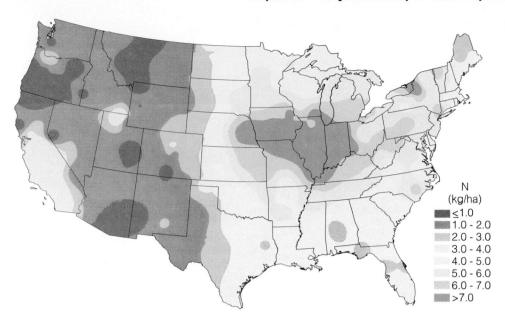

FIGURE 26.8 Estimated inorganic nitrogen deposition from nitrate and ammonium, 1998. (National Atmosphere Deposition Program.)

N
(kg/ha)
≤1.0
1.0 - 2.0
2.0 - 3.0
3.0 - 4.0
4.0 - 5.0
5.0 - 6.0
6.0 - 7.0
>7.0

ance between fixation and denitrification by reducing the former and increasing the latter. A considerable portion of the added nitrogen may be leached to the groundwater as nitrates. Animal wastes, especially from livestock concentrated in large feedlots, are another source of nitrates in groundwater. Excessive quantities of nitrates in groundwater become health-endangering pollutants.

A third source of nitrate pollution is human waste, particularly sewage, treated or otherwise, released into rivers, lakes, and estuaries. In spite of the magnitude of water pollution from agricultural sources, human effluents contribute even heavier loads, especially from municipal sewage treatment.

Automobiles and power plants are the major sources of nitrogenous pollutants in the atmosphere, particularly nitrogen oxides. Current estimates place nitrogen emissions due to automobiles, factories, power plants, and other combustion processes at 20×10^{12} g N yr^{-1}. The primary nitrogenous air pollutant is nitrogen dioxide (NO_2), a pungent gas that produces a brownish haze. In the atmosphere, nitrogen dioxide is reduced by ultraviolet light to nitrogen monoxide and atomic oxygen:

$$NO_2 \longrightarrow NO + O$$

Atomic oxygen reacts with molecular oxygen to form ozone (see page 526):

$$O_2 + O \longrightarrow O_3$$

Ozone, in a neverending cycle, reacts with nitrogen monoxide to form nitrogen dioxide and oxygen:

$$NO + O_3 \longrightarrow NO_2 + O_2$$

This cycle illustrates only a few of the reactions that nitrogen oxides undergo or trigger. In the presence of sunlight, atomic oxygen from nitrogen dioxide also reacts with a number of reactive hydrocarbons to form radicals. These radicals then take part in a series of reactions to form still more radicals that combine with oxygen, hydrocarbons, and nitrogen dioxide. As a result, nitrogen dioxide is regenerated, nitrogen monoxide disappears, ozone accumulates, and a number of secondary pollutants form, including formaldehydes, aldehydes, and peroxyacylnitrates, known as PAN. All of these pollutants, especially PAN and ozone, are very toxic and injure many forms of plant life exposed to them.

One outcome of nitrogenous atmospheric pollution is an increased deposition of nitrogen (Figure 26.8), which benefits ecosystems that are traditionally nitrogen-limited, especially northern and high-altitude forests. Such ecosystems, however, can suffer from too much of a good thing. Because under natural conditions nitrogen is limiting, these forests are efficient at retaining and recycling nitrogen from precipitation and organic matter. Significant export of nitrates to streams is seldom seen. However, many of these forests are receiving more nitrogen in the form of ammonium and nitrates than the trees and their associated microbial populations can handle and accumulate, a condition referred to as **nitrogen saturation.** Under these conditions, excess nitrogen, particularly in the form of nitrate, is leached into surface waters.

The first response to increased availability of nitrogen in a nitrogen-limited ecosystem is increased growth of both canopy and stems. Evidence suggests, however, that increased levels of nitrogen can lead to the decline and dieback of coniferous forests at high elevations (Figure 26.9) (Aber et al. 1989). If the increased growth in foliage continues into late summer, the late new growth may not have time to become frost-hardened and is killed during the winter. Overstimulated by nitrogen, tree growth exceeds the availability of other necessary nutrients in the soil, particularly phosphorus, and the tree begins to experience nutrient deficiencies. Experimental evidence suggests that the production of fine roots and ectomycorrhizae, which take up nutrients from the soil, is lower on sites rich in nutrients, especially nitrogen, than in nutrient-poor soils, and root turnover is higher. Trees on nutrient-poor soils have a longer-lived and

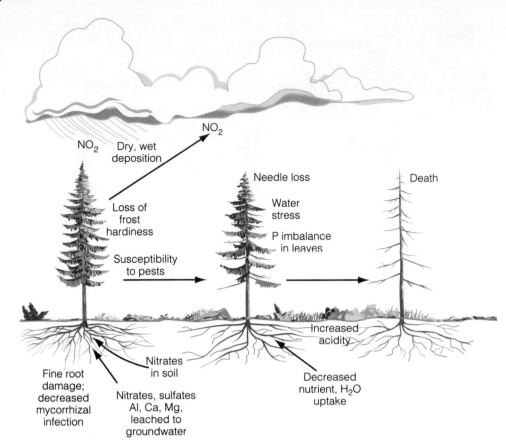

FIGURE 26.9 Excess inputs of nitrogen in forest ecosystems, according to the nutrient limitation hypothesis, affect many processes, resulting in decline and death of forest trees and converting the ecosystem from a nitrogen sink to a nitrogen source.

higher density root system and a lower turnover of root biomass. These conditions help the trees to scavenge nutrients from poor soils. As nitrogen levels increase, root biomass decreases because sufficient nitrogen is nearby. A smaller root biomass then inhibits the uptake of nutrients other than nitrogen and impairs the ability of trees to pull water from the soil during periods of drought.

As nitrogen deposition increases, ammonium levels in the soil increase. Excess ammonium in the soil stimulates nitrification, denitrification, and mobility of nitrates in the soil, even at a low pH. As excess nitrates are leached from the soil, they increase anion movement through the soil, releasing aluminum (Al^{3+}), as has been observed in the spruce stands of the Great Smoky Mountains. This aluminum finds its way to aquatic ecosystems, increasing their acidification.

THE GLOBAL PHOSPHORUS CYCLE

Main Features

The global phosphorus cycle (Figure 26.10) is unique among the major biogeochemical cycles in having no significant atmospheric component, although airborne transport of P in soil dust and sea spray is on the order of 1×10^{12} g P yr^{-1} (Graham and Duce 1979). Nearly all of the phosphorus in terrestrial ecosystems is derived from the weathering of calcium phosphate minerals. In most soils, only a small fraction of the total phosphorus is available to plants and the internal cycling of phosphorus from organic to inorganic forms is the major process regulating phosphorus availability for net primary productivity.

Rivers transport approximately 21×10^{12} g P yr^{-1} to the oceans (Meybeck 1982), but only about 10 percent of this amount is available for net primary productivity. The remainder is deposited in sediments. The concentration of phosphorus in the ocean waters is low, but the large volume of water results in a significant global pool of phosphorus. The turnover of organic phosphorus in the surface waters is on the order of days, and the vast majority of phosphorus taken up in primary productivity is decomposed and mineralized (internally cycled) in the surface waters. However, approximately 2×10^{12} g yr^{-1} is deposited in the ocean sediments. On a geological time scale, uplifting and subsequent weathering return this phosphorus to the active cycle.

Human Impact on the Phosphorus Cycle

Human activities have altered the phosphorus cycle. To maintain phosphorus levels in the soil, agriculturists must apply phosphate fertilizer. The source of that fertilizer is phosphate rock. Because of the abundance of calcium, iron, and ammonium in the soil, most of the phosphate applied as fertilizer combines with these elements to form insoluble salts, and little escapes in runoff.

Part of the phosphorus used as fertilizer is removed in crops when harvested. Transported far from the point of

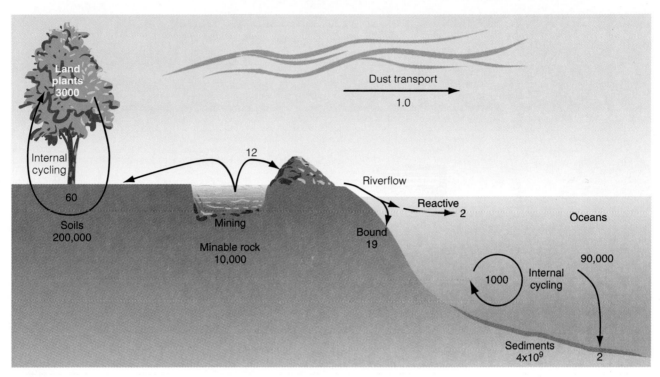

FIGURE 26.10   The global phosphorus cycle. Each flux is shown in units of 10^{12} g P/yr. (From Schlesinger 1997.)

application, this phosphorus eventually is released as waste in the processing and consumption of food. Concentration of phosphorus in wastes of food-processing plants and of feed-lots adds a quantity of phosphates to natural waters. Even greater amounts come from urban areas, where phosphates are concentrated in sewage effluents.

Most of the phosphorus enrichment of aquatic ecosystems comes from sewage disposal plants. Primary sewage treatment removes only 10 percent of the total phosphorus. Secondary treatment removes only 30 percent at best. Feedlots contribute runoff, but phosphorus has such a strong affinity for binding to soil particles that the problem of leaching is not as severe. Sewage contributes nearly all of the phosphorus reaching rivers and lakes.

Phosphorus is more of a problem with the overenrichment or **eutrophication** of freshwater ecosystems than nitrogen. Of the three nutrients required for aquatic plant growth, potassium is usually present in excess, nitrogen is supplemented by fixation, and phosphorus tends to be precipitated in the sediments and cannot be supplemented biologically. Therefore phosphorus is usually limiting, and in the presence of a fuller supply, algae respond with luxuriant growth.

THE GLOBAL SULFUR CYCLE

Main Features

The global sulfur cycle is presented in Figure 26.11. Although a great deal of research now focuses on the sulfur cycle, par-

ticularly the role of anthropogenic inputs, our understanding of the global sulfur cycle is primitive.

The gaseous phase of the sulfur cycle permits circulation on a global scale. The annual flux of sulfur compounds through the atmosphere is on the order of 300×10^{12} g. The atmosphere contains not only sulfur dioxide and hydrogen sulfide but sulfate particles as well. The sulfate particles become part of dry deposition; the gaseous forms combine with moisture to be recirculated by precipitation (see Figure 25.18, 26.13).

The oceans are a large source of aerosols that contain sulfate (SO_4); however, most are redeposited in the oceans as precipitation and dryfall (Figure 26.11). Dimethylsulfide is the major biogenic gas emitted from the oceans. Estimates of 16×10^{12} g S yr^{-1} make it the largest natural source of sulfur gases released to the atmosphere.

A variety of terrestrial biogenic sources of sulfur exist, but collectively these represent a minor flux to the atmosphere. The dominant sulfur gas emitted from freshwater wetlands and anoxic soils is hydrogen sulfide (H_2S). Emissions from plants are poorly understood, but forest fires emit on the order of 3×10^{12} g S annually (Andreae 1991). It is almost impossible to estimate the biological turnover of sulfur dioxide, because of the complicated cycling within the biosphere. Erikisson (1963) estimates that the net annual assimilation of sulfur by marine plants is on the order of 130×10^{12} g. Adding the anaerobic oxidation of organic matter brings the total to an estimated 200×10^{12} g.

Volcanic activity also contributes to the global biogeochemical cycle of sulfur. Major events such as Mt. Pinatubo

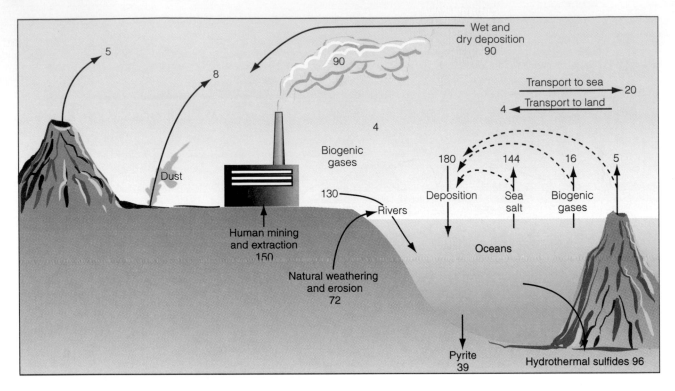

FIGURE 26.11   The global sulfur cycle. Each flux is shown in units of 10^{12} g S/yr. (From Schlesinger 1997.)

(1991) release on the order of $5–10 \times 10^{12}$ g S. When volcanic activity is averaged over a long record, the annual global flux is on the order of 10×10^{12} g S.

Human activity plays a dominant role in the biogeochemical cycle of sulfur. Thus, to complete the picture of the global sulfur cycle, we must examine the inputs due to industrial activity.

Human Impact on the Sulfur Cycle

By far the largest source of sulfur gases to the atmosphere is that of direct emissions from human industrial activities. Annually we pour into the atmosphere some $50–100 \times 10^{12}$ g of sulfur dioxide. Of this, 70 percent comes from the burning of coal. Due to the reactivity of sulfur gases in the atmosphere, most of the anthropogenic emission of sulfur dioxide (SO_2) is deposited locally as precipitation and dryfall. Deposition of sulfur on land is estimated at $90–10^{12}$ g yr^{-1} (Schlesinger 1997). The majority of emissions from land are redeposited on the land surface; the remainder undergoes long-distance transport in the atmosphere and is deposited in the oceans.

Human activities are also the major source sulfur transported in rivers. Rivers represent a significant input of sulfur to the oceans (130×10^{12} g S yr^{-1}). In addition to the weathering of sulfur-bearing rocks and minerals, at least 25 percent of the sulfate content of rivers is derived from air pollution, mining, erosion, and other human activities.

Once in the atmosphere, gaseous sulfur dioxide reacts with moisture to form sulfuric acid. Sulfuric acid in the atmosphere has a number of effects. It is irritating to the respiratory tract in concentrations of a few parts per million. In

a fine mist or absorbed in small particles, it can be carried deep into the lungs to attack sensitive tissue. High concentrations of sulfur dioxide (over 1000 micromilligrams/m^3) have been implicated as a prime cause in many air pollution disasters characterized by higher than expected death rates and increased incidence of bronchial asthma.

Plants exposed to atmospheric sulfur are injured or killed outright. Injury to plants is caused largely by acidic aerosols during periods of foggy weather, during light rains, or during periods of high relative humidity and moderate temperatures. Pines, more susceptible than broadleaf trees, react by partial defoliation and reduced growth. Exposure of plants to sulfur dioxide at concentrations as low as 0.3 ppm for 8 hours can produce both acute and chronic injury.

Sedimentary rocks containing ferrous sulfide (FeS_2), called pyritic rocks, may overlie coal deposits. Exposed to the air in deep and surface mining, the ferrous sulfide oxidizes and in the presence of water produces ferrous sulfate ($FeSO_4$) and sulfuric acid (H_2SO_4):

$$2FeS_2 + 7O_2 + 2H_2O \longrightarrow 2FeSO_4 + 2H_2SO_4$$

Other reactions produce ferric sulfate (Fe_2SO_4) and ferrous hydroxide ($FeOH_3$):

$$12FeSO_4 + 3O_2 + 6H_2O \longrightarrow 4Fe_2(SO_4)_3 + 4Fe(OH_3)$$

In this manner sulfur in pyritic rocks, suddenly exposed to weathering by human activities, discharges heavy slugs of sulfuric acid, ferric sulfate, and ferrous hydroxide into aquatic ecosystems. These compounds destroy aquatic life. Mining has converted hundreds of miles of streams and rivers in the eastern United States to highly acidic water (Figure 26.12).

FIGURE 26.12 Streams polluted by acid mine drainage are easily identified by the "yellow boy" or ferric sulfates. Such acidification has been the fate of most streams in coal-mining regions.

ACID DEPOSITION

Ever since the Industrial Revolution the inputs of anthropogenic sulfur dioxide and nitrogen oxides have been increasing, until such inputs equal the amount injected naturally, about 75 to 100 million tons annually. Most of the inputs are over the Northern Hemisphere. The major sources are power plants, industrial complexes, motor vehicles, and internal combustion engines. The latter are the major sources of nitrogen oxides (NO_x).

Once in the atmosphere, sulfur dioxide and nitrogen oxides become mixed. Some of the pollutants soon return to Earth as particulate matter and airborne gases, known as **dry deposition;** a major portion is transported far away. The pathway it takes is strongly influenced by the general atmospheric circulation (National Research Council 1983). During their atmospheric transport, SO_2 and NO_2 and their oxidative products become involved in complex chemical reactions involving hydrogen chloride and other compounds, oxygen, and water vapor. These reactions produce dilute solutions of strong acids, notably nitric and sulfuric acids (Figure 26.13). Ultimately they come to Earth in acid rain, snow, and fog, known as **wet deposition.**

The acidity of precipitation downwind from major industrial centers in eastern North America and northern and

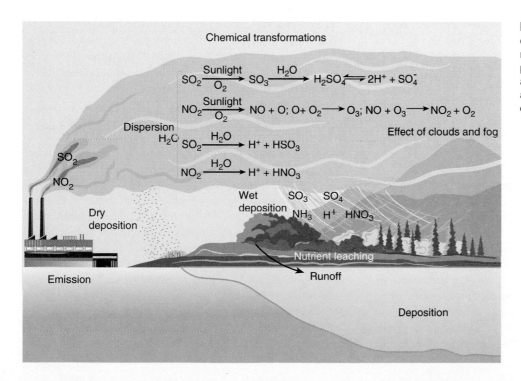

FIGURE 26.13 Formation of acid depositions. Excessive sulfur and nitrogen in several forms are being poured into the atmosphere. They are converted to sulfates, sulfides, and sulfuric and nitric acids and carried to Earth.

FIGURE 26.14 Hydrogen ion concentration as pH from measurements made at the field laboratories, 1998. (National Atmosphere Data Program.)

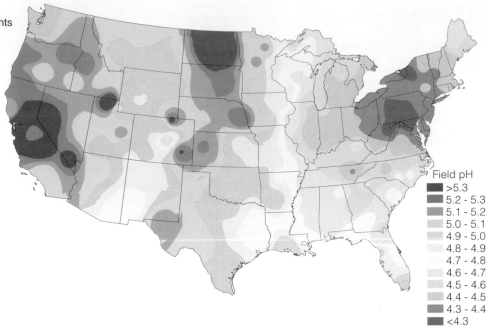

Field pH
>5.3
5.2 - 5.3
5.1 - 5.2
5.0 - 5.1
4.9 - 5.0
4.8 - 4.9
4.7 - 4.8
4.6 - 4.7
4.5 - 4.6
4.4 - 4.5
4.3 - 4.4
<4.3

central Europe has increased at least 2 to 16 times over that of precipitation in geographical areas remote from industrial pollution (Galloway et al. 1984). The annual pH of rain and snow in those continental regions averages between 4 and 4.5 (Figure 26.14), and it may range from 2.3 to 4.6, compared with a more natural pH of 5.6 (Likens and Bormann 1974). Added to precipitation is acidic fog whose pH may range from 2.2 to 4.0 (Waldman et al. 1982). In spite of the downward trends in sulfur emissions, the acidity of precipitation does not seem to be decreasing (Scherbatskoy et al. 1999).

Aquatic Ecosystems

Acid deposition, wet or dry, can have a pronounced effect on aquatic and terrestrial ecosystems. Those that are not well buffered are highly sensitive to acid inputs. Acid precipitation has been implicated in the acidification of mountain streams and lakes in the northern United States, especially the Adirondack Mountains, eastern Canada, and Scandinavia (National Research Council of Canada 1981, Swedish Ministry of Agriculture 1982, Likens 1992).

Acidic inputs into aquatic ecosystems come from rainfall, snowmelt, and groundwater leaching from adjacent watersheds. Of particular importance may be snowfall that accumulates over winter and melts quickly in the spring, discharging a huge dose of acidic water into streams and lakes (Hornbeck et al. 1976). This sudden, heavy input can drop the pH levels of aquatic ecosystems quickly. More important, it releases aluminum ions, which at the levels of 0.1 to 0.3 mg/l retard growth, gonadal development, and egg production of fish and increase fish mortality. The aluminum, rather than acidity, which some species of fish can tolerate, may be the most significant pollutant (Schofield and Trojnar 1980), although later studies suggest that acidity alone is sufficient

to cause mortality in developing brook trout (*Salvelinus fontinalis*) (Hunn et al. 1987).

Acidification of lakes reduces bacterial activity in sediments (Rao et al. 1984), inhibiting decomposition and nutrient regeneration. Reduced nutrient regeneration affects aquatic food webs by reducing phytoplankton and invertebrate populations on which fish depend for food. Loss of fish reduces survival and reproductive success of such fish-eating birds as herons (Ardeidae) and loons (*Gavia* spp.).

Acidic runoff from snowmelt that fills small basins in the soil creates temporary breeding pools for amphibians. Even in regions not experiencing acidification of lakes and streams, it inhibits the reproduction of frogs and salamanders (Pierce 1985). Most vulnerable to acidity is the fertilization stage, because sperm apparently disintegrate at low pHs (Schlichter 1981). This effect may be a cause of the rapid decline of amphibians. Tolerance to acidity among amphibians tends to increase through the larval to adult stages.

Terrestrial Ecosystems

The effects of acid precipitation on terrestrial ecosystems are more difficult to document or demonstrate, except in very localized vicinities of smelters or acid mine drainage. Soils throughout much of eastern North America appear to be sufficiently buffered to neutralize acidic precipitation. However, the naturally acidic soils of northeastern North America, exposed to acidic precipitation of a period of years, are experiencing adverse effects (Likens et al. 1996, 1998). Substantial acid-deposition-induced leaching of nutrients on these soils has shrunk the pool of calcium in the soil complex by as much as 50 percent during the past 45 years. It has increased the solubility of aluminum ions, and reduced nutrient availability, storage of calcium in the soil, and its availability for root uptake

(Lawrence et al. 1995). Free aluminum affects the structure and function of fine roots, reducing their ability to take up nutrients and moisture. Further, acid rainfall can inhibit the activity of fungi and bacteria in the soil, reducing the rate of humus production, mineralization, and fixation of nutrients. All of these interactions result in nutrient-deficient soils.

Little evidence exists to show that acid rain has any direct effect on most terrestrial plants, but it has substantial indirect effects. Acid rain, intercepted by vegetation, leaches nutrients, particularly cations like calcium, magnesium, and potassium, from the leaves and needles. This leaching, a normal process in nutrient cycling, has little effect on the trees' health provided they can replace the nutrients lost by uptake from the soil. However, forests at high elevations, notably spruce, are frequently enveloped by mists and fog that are more acidic than rainfall (DeHays et al. 1999).

Immersed in acidic clouds and mist, needle-leaf conifers, particular spruce and fir, effectively comb moisture out of the atmosphere. Studies of high-elevation red spruce (*Picea rubra*) forests show that this acidic mist induces the leaching of calcium ions from the needles, particularly the current year's foliage (DeHays et al. 1999). This loss alters the calcium physiology of the foliage. Membrane-associated calcium, mCa, is associated with response of leaf cells to changing environmental conditions. It stabilizes membrane functions and permeability and sends low-temperature signals across the plasma membranes of mesophyll cells. Membrane calcium interacts with biomolecules that help regulate physiological responses to the stress of freezing. Direct acidic deposition on red spruce foliage preferentially displaces calcium ions specifically associated with plasma membranes. A reduction in mCa that destabilizes plasma membranes and depletes the messenger calcium is the link between acid deposition and freezing injury to the needles of the current year. Growth reductions associated with red spruce decline in montane forests appear to be a direct result of freezing, rather than exposure to acidic conditions per se. Controlled environmental studies showed that red spruce exposed to acid mist exhibited improved growth *in the absence of stresses such as freezing* (Kohut 1990). The effects of acid-induced calcium leaching of foliage may extend to other north temperate forest tree species such as sugar maple (*Acer saccharum*), yellow birch (*Betula alleghaniensis*), and white pine (*Pinus strobus*).

Forests, especially coniferous ones, are declining in North America and Europe. Over central Europe, Norway spruce, pine, beech, and oak suffer from chlorosis and defoliation, which now may be attributed in part to acid-induced alteration of membrane-associated calcium. Fourteen percent of all Swiss forests, 22 percent of Austria's forests, and 29 percent of Holland's forests show symptoms of decline. In North America, similar symptoms have developed in the spruce and fir forests of the high Appalachians and in the pine forests of southern California.

Forest decline is not a new phenomenon. During the past 100 to 200 years, our forests have experienced several declines with different species affected. What sets the current decline apart from all others is differences among the symptoms in the past and the similarity of symptoms among species today. Past declines could be attributed to natural stresses, such as drought and disease. What causes forest decline and dieback today is not established, but the widespread similarity of symptoms suggests a common cause brought on by air pollution. All of the affected forests are in the path of pollutants from industrial and urban sources (Smith 1990).

Besides the economic and aesthetic loss of trees, air pollution and acidic rain alter succession by changing the species composition of the affected forests. Just as the chestnut blight shifted dominance in the central hardwoods forest from chestnut to oaks, so air pollution is shifting dominance from pines and other conifers to deciduous trees more tolerant of air pollution.

OTHER HUMAN IMPACTS ON BIOGEOCHEMICAL CYCLES

The accumulation of toxic materials in the biosphere has brought biogeochemical cycles forcibly to the attention of the public. A variety of elements and compounds that are found naturally in the environment at relatively low and harmless concentrations are now released into the environment at toxic concentrations as a direct result of human activities.

Heavy Metals: Lead

Heavy metals such as lead, mercury, and cadmium have always cycled through natural ecosystems in trace amounts. Human activities, however, have markedly increased their concentration in the environment. These metals then pass through the food chain in such a way that they become increasingly concentrated in organisms occupying the upper trophic levels.

Lead is an excellent example of a heavy metal injected into the ecosystem by human activity. Like sulfur and nitrogen, it moves great distances through the atmosphere from its point of origin. Automobiles burning leaded gasoline poured most of the lead into the air, until the use of leaded gas was banned in the United States, Japan, Brazil, and European Common Market countries. Mining, smelting, and refining of lead, lead-consuming industries, coal combustion, burning of refuse and sewage sludge, and the burning and decay of lead-painted surfaces add additional quantities to the atmosphere.

Because lead is emitted into the air as very small particles, less than 0.5 μm in diameter, it is widely distributed to all parts of Earth. Concentrations are highest near point sources of pollution. When automobiles used leaded gasoline, roadsides received heavier depositions than remote areas. Urban areas, close to industrial sources, may have a flux rate greater than 3000 g/ha/yr, whereas remote areas may experience a flux rate of less than 20 g/ha/yr.

Lead particles settle on the surface of soil and on vegetation. Forest canopies are particularly efficient at collecting lead from the atmosphere as dry deposition (Smith 1990). Lead accumulates in the canopies during the summer and is carried to the ground by rain as throughfall and stemflow and by leaf-fall in autumn. Between 1975 and 1984, the Hubbard Brook Experimental Forest in New Hampshire, remote from any major sources of lead contamination, experienced an input of 190 g/ha/yr. It had an output of only 6 g/ha/yr, leaving the rest behind in the soil. A closed-canopy spruce forest in Germany experienced a lead flux of 756 g/ha/yr in precipitation collected beneath the canopy, compared with 405/g/ha/yr in adjacent open fields.

Small amounts of lead occur naturally in some soil. In uncontaminated areas, this background lead amounts to 10 to 20 μg per gram of dry soil. In the forest soil, lead becomes bound to organic matter in the litter layer and reacts with sulfate, phosphate, and carbonate anions in the soil. In such an insoluble form, lead moves slowly, if at all, into the lower horizons. Its residence time in the upper soil layer is around 5000 years.

Once in the soil and on plants, lead enters the food chain. Plant roots take up lead from the soil. Leaves pick it up from contaminated air or from particulate matter resting on the leaf. Lead is then taken up by herbivorous insects and grazing mammals, who pass it on to higher consumers. This uptake through the food chain is most pronounced along highway roadsides where residual quantities persist. Microbial systems also pick up lead and immobilize substantial quantities of it.

The long-term increase in concentrations of atmospheric lead in the industrialized areas of Earth has resulted in significant increases of lead in humans. The average body burden of lead among adults and children in the United States is 100 times greater than the natural burden, and existing rates of lead absorption are 30 times the level in preindustrial society. An intake of lead can cause mental retardation, partial paralysis, loss of hearing, and death.

Chlorinated Hydrocarbons

Of all human intrusions into biogeochemical cycles, none has done more to call attention to materials cycling than the widespread application of DDT (dichlorodiphenyltrichloroethane). Its use is now banned in the United States. During World War II this newly developed pesticide was used in huge quantities to control disease-carrying and crop-destroying insects, particularly mosquitoes and boll weevils. As early as 1946 Clarence Cottam of the U.S. Fish and Wildlife Service called attention to DDT's damaging effects on ecosystems and nontarget species. The impact of pesticides on ecosystems, however, remained obscure until Rachel Carson wrote *Silent Spring* (1962), a book that exposed the dangers of hydrocarbons. The detection of DDT in tissues of animals in the Antarctic (Risebrough et al. 1976), far removed from any applied source of the insecticide, emphasized the fact that DDT does indeed enter the global biogeochemical cycle and become dispersed around Earth.

In the late 1960s scientists discovered that another chlorinated hydrocarbon, PCB, was also accumulating in foodstuffs, many species of fish and birds, and humans. PCB, a

major toxic waste, is a generic name for a number of synthetic organic compounds characterized by biphenyl molecules containing chlorine atoms. They are widely used in the electronics industry and in plastics, solvents, and printing inks. PCBs are discharged into the environment through sewage outfalls and industrial discharges.

DDT, PCBs, and other chlorinated hydrocarbons have certain characteristics that enable them to enter global circulation. Because they are highly soluble in lipids or fats and not very soluble in water, they tend to accumulate in the lipids of plants and animals. They are persistent and stable and have a half-life of approximately 20 years. They have a vapor pressure high enough to ensure direct losses from plants. They can become adsorbed by particles or remain as a vapor. In either state they can be transported by atmospheric circulation and then return to land and sea with rainwater.

DDT, like most insecticides, is applied on a large scale by aerial spraying (Figure 26.15). Half or more of a toxicant applied in this manner is dispersed to the atmosphere and never reaches the ground (Tarrant 1971). On the ground or on the water's surface, the pesticide is subject to further dispersion. Pesticides reaching the soil are lost through volatilization, chemical degradation, bacterial decomposition, runoff, and the harvest of crops.

In flowing water, chlorinated hydrocarbons are subject to further distribution and dilution as they move downstream. Insecticides released in oil solution penetrate to the bottom and cause mortality of fish and aquatic invertebrates (see reviews in Pimentel 1971a, Cope 1971). Trapped in the bottom rubble and mud, the insecticide may continue to circulate locally and kill for some days.

In lakes and ponds emulsifiable forms of DDT tend to disperse through the water, but not necessarily in a uniform way. DDT in oil solutions tends to float on the surface and move about in response to the wind. PCBs, however, accumulate in bottom sediments, adsorbed in silt and fine particles, which hold them in the aquatic environment.

Eventually these chlorinated hydrocarbons reach the ocean and associate with surface slicks, where their concentration may be 10,000 times as great as in lower waters. These slicks, which attract plankton, are carried by ocean currents. In the oceans, part of the residues may circulate in the mixed layer, some may be transferred below the thermocline to the abyssal waters, and more may be lost through sedimentation of organic matter.

Although considerable amounts of DDT and other chlorinated hydrocarbons are transported by water, these amounts are insignificant from the viewpoint of global circulation. The major transport of chlorinated hydrocarbon residues takes place in the atmosphere.

Not only does the atmosphere receive the bulk of pesticide sprays (well over 50 percent of that applied), it also picks up that fraction volatilized from soils, vegetation, and water. The adsorption of DDT residues to airborne particulate matter increases the atmosphere's capacity to hold DDT. Thus, the atmosphere becomes a large circulating reservoir of DDT and other chlorinated hydrocarbons (SCEP 1970). This fact is especially important because DDT is still widely used in other parts of the world.

Although the quantity of residues of some chlorinated hydrocarbons, especially the banned DDT, may be small, the

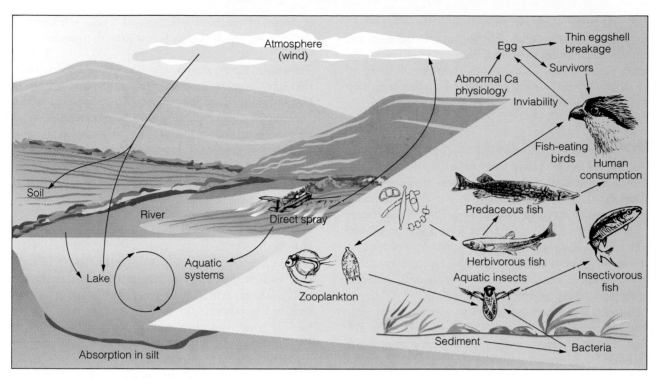

FIGURE 26.15 The movement of chlorinated hydrocarbons in terrestrial and aquatic ecosystems. The initial input comes from spraying on vegetation. A large portion fails to reach the ground and is carried on water droplets and particulate matter through the atmosphere.

residual concentrations are still sufficient to have a deleterious effect on marine, terrestrial, and freshwater ecosystems (Beyer and Gish 1980, Matthiessen 1985). DDT, its degradation product DDE, and PCBs tend to concentrate in the fatty tissues of living organisms, where they undergo little degradation (see Menzie 1969, Bitman 1970, Peakall 1970).

In the animal body, these chlorinated hydrocarbons interfere with calcium metabolism. They block ion transport by inhibiting the enzyme ATPase, which makes energy available to the cell. The reduced transport of ionic calcium across membranes can cause death. Chlorinated hydrocarbons also inhibit the enzyme carbonic anhydrase, which in birds is essential for the deposition of calcium carbonate in the eggshell and for the maintenance of a pH gradient across the membranes of the shell gland.

The high solubility of chlorinated hydrocarbons in lipids magnifies their concentration. Most of the chlorinated hydrocarbons contained in food is retained in the fatty tissue of the consumers. Because they break down slowly, chlorinated hydrocarbons accumulate to high and even toxic levels. They are passed on to consumers in the next trophic level, where again they are retained and accumulated. The carnivores in the top level of the food chain receive massive amounts of chlorinated hydrocarbons.

As long as the rate of accumulation of DDT is not too great, birds tend to eliminate some of it, which reduces the body burden. Fish, however, continuously exposed to PCBs in bottom sediments, concentrate them in their tissues (Stein et al. 1987), where the chemical residue is often higher than that of DDT. In some polluted waters, such as the Hudson River in New York

State and parts of the Great Lakes, fish have such high levels of PCBs that they are not fit for human consumption.

Since the prohibition of DDT in the United States in 1972, birds of prey are recovering from low reproductive rates induced by DDT. The recovery of the bald eagle (*Haliaeetus leucocephalus*), osprey (*Pandion haliaetus*), brown pelican (*Pelicanus occidentalis*), and other birds occupying high trophic levels attest to the highly detrimental effects chlorinated hydrocarbons have had on the reproduction of birds.

Despite strict laws regulating the use of DDT, the ecological threat of pesticides has not lessened. Chlorinated hydrocarbons, including DDT, are used extensively in other parts of the world, notably Central and South America and Asia. There, the problems associated with these insecticides still persist. They affect not only the native fauna, but also migratory birds from the Northern Hemisphere that come in contact with the insecticides on their wintering grounds. Quantities of pesticides are sent back north to the United States, Canada, and Europe on fruits and vegetables grown for the winter market.

More than 500 million kilograms of a wide range of chemical pest control materials (collectively called pesticides) are used annually in the United States. Herbicides make up 60 percent of this total, insecticides 24 percent, and fungicides 16 percent. Of these pesticides 341 million kilograms are used on agricultural crops and pastures, 55 million kg by government and industry, 4 million kg on forests, and a surprising 55 million kg in and around urban and suburban homes. The most concentrated use is about the home grounds, not in agricultural fields. The annual dosage of pesticides about homes is 14 kg/ha compared with 3 kg/ha on agricultural crops. As much as one-third of these household pesticides are never used and are thrown into the trash and ultimately into the environment. These excess pesticides, together with losses from croplands and roadsides, expose pests to widespread selective pressures, increasing their resistance to pesticides.

Radionuclides

Ever since the atomic bomb ushered in the atomic age, the impact of nuclear radiation on life on Earth has been a major concern. Involved are high-energy, short-wavelength radiations, known as **ionizing radiations.** They are so called because they can remove electrons from some atoms and attract them to other atoms, producing positive and negative ion pairs. Of greatest interest is ionizing electromagnetic or gamma radiation, which has a short wavelength, travels a great distance, and penetrates matter easily.

Sources of gamma radiation are atomic blasts from weapons testing, nuclear reactors, and radioactive wastes. By-products of both weapons testing and nuclear reactors, such as zinc-65 (^{65}Zn), strontium-90 (^{90}Sr), cesium-137 (^{137}Cs), iodine-131 (^{131}I), and phosphorus-32 (^{32}P), are radioactive. When the uranium atom is split or fissioned into smaller parts, it produces, in addition to tremendous quantities of energy, a number of new elements or fission products, including strontium, cesium, barium, and iodine. Some of these fission products last only a few seconds; others can remain active for several thousand years. These radioactive elements enter the food chain and become incorporated in living organisms, in which they can cause cancer and genetic defects.

In the same atomic reaction, some particles with no electrical charges, called neutrons, get in the way of high-energy particles. Nonfission products are the result. They include the radioisotopes of such biologically important elements as carbon, zinc, iron, and phosphorus, which are useful in tracer studies.

Both fission and nonfission products, together called **radionuclides,** are released to the atmosphere by nuclear testing and by wastes from nuclear reactors, hospitals, and research laboratories unless carefully handled. Later they return to Earth along with rain, dust, and other material as radioactive fallout. Once the isotopes reach the Earth, they enter the food chain and become concentrated in organisms in amounts that exceed by many times the quantities in the surrounding environment. In effect, local radiation fields develop in the tissues of plants and animals.

Of particular concern is the radioactive output of nuclear power plants, especially since the Three Mile Island and Chernobyl accidents. Pressurized water reactors, commonly used in nuclear power plants, release low levels of radioactivity to the air and condenser water. When water passes through the intense neutron flux of the reactor, it is contaminated by radioactivity. Trace elements in the water are activated, producing radioisotopes. Added to them are radioactive corrosive products from the surface of metal cooling tubes. Except for tritium, most of the radioisotopes are removed in a radioactive waste removal process, which creates additional problems of nuclear waste disposal. Those left have a short half-life and rapidly decay below detection levels.

Terrestrial Ecosystems   Radionuclides disperse in terrestrial ecosystems by gaseous, particulate, and aerosol deposition and in liquid and solid wastes. Plants intercept particulate radionuclide contaminants, absorb them through their foliage, and take them up from soil and litter. From the plants, radionuclides move through the ecosystem along the food chain.

Strontium-90 and cesium-137 are two of the most destructive radioactive materials released into the biogeochemical cycle (Figure 26.16). Ecologically, they behave like calcium and follow it in the cycling of nutrients. Both easily enter the grazing food chain, especially in regions with high rainfall or abundant soil moisture and with low levels of calcium and other mineral nutrients in the soil. One such region, the Arctic tundra, has been subject to heavy nuclear fallout from weapons testing in the past. It received another input as fallout from the Chernobyl nuclear power plant explosion in 1986.

Lichens, the dominant plants of the tundra, absorb virtually 100 percent of the radioactive particles and gases drifting onto them. From lichens the contaminants ^{90}Sr and ^{137}Cs travel up the food chain, from caribou and reindeer to wild carnivores and humans.

Early studies of the effects of nuclear fallout on human food chains involved Eskimos and northern Alaskan Indians (Palmer et al. 1963, Hanson 1971). Caribou are a major food source for northern Alaskan natives, who kill the animals during the northward migration in spring and stockpile the meat for late spring and early summer food. Because the caribou feed all winter on lichens, their flesh in spring during the study contained three to six times as much ^{137}Cs as it did in the fall. In spring the Indians showed a corresponding rise in ^{137}Cs level, often 50 percent (Figure 26.17). The level decreased when the people changed to a diet of fish in the summer.

The accident at Chernobyl produced a nuclear cloud that drifted northward over the Arctic, depositing much of its contamination over arctic Eurasia (Davidson et al. 1987) and causing a repeat of the story of the 1960s. Radioactive-contaminated lichens passed strontium and cesium on to the reindeer, making both meat and milk unsuitable for human consumption, severely affecting the Laplanders' economy and their future.

In vertebrate food chains, strontium and cesium usually accumulate throughout the body at higher trophic levels, but cobalt, ruthenium, iodine, and some other radionuclides do not. Some, such as iodine, concentrate in certain tissues. In arthropod food chains, potassium, sodium, and phosphorus accumulate, whereas strontium and cesium do not.

Aquatic Ecosystems   Radionuclides contaminate aquatic ecosystems largely through waste from nuclear power plants and from the nuclear processing industry. Radioactive materials that enter the water become incorporated in bottom sediments and circulate between mud and water. Bottom-dwelling insects and fish downstream from the source absorb some. In fact, they may be exposed to chronic low-level radiation. Under such conditions, a sort of equilibrium is estab-

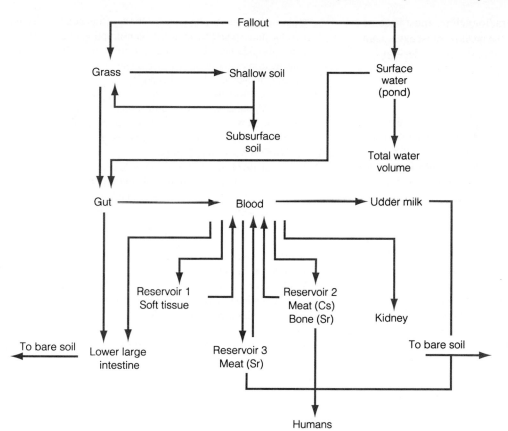

FIGURE 26.16 Radionuclide cycling through the food chain. Strontium and cesium are transferred through fallout to the grazing food chain and on to humans through meat and milk from reindeer, cattle, and sheep. This cycling occurred during the early days of nuclear weapon testing and more recently following the nuclear plant explosion at Chernobyl. (Courtesy Oak Ridge National Laboratory.)

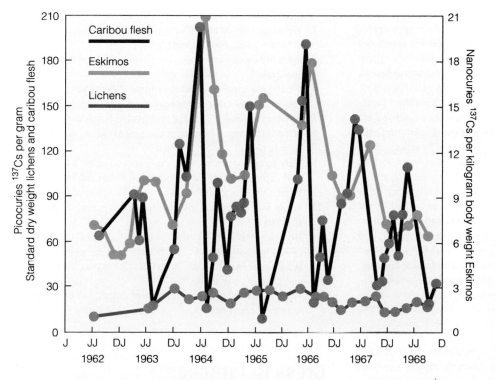

FIGURE 26.17 Cesium-137 concentrations in lichens, caribou flesh, and Eskimos at Anaktuvuk Pass, Alaska, during the period 1962–1968. Note the relationship between the concentration of cesium-137 in caribou flesh and the amount in humans. As the concentration in caribou declined seasonally, so did the concentration in humans. (From Hanson 1971.)

lished among retention in the organisms, the bottom sediments, daily input, and decay.

The concentration of radionuclides does not necessarily increase consistently through the food chain. In many situations the concentration decreases at higher trophic levels, because radionuclides tend to concentrate in the bones of fish and the shells of mollusks and do not move up into higher trophic levels.

In spite of considerable study, we still know little about the uptake, assimilation, distribution in tissues, turnover rates, and equilibrium levels of radionuclides in various ecosystems. As nuclear power plants and radioactive medical wastes increase, our knowledge of the behavior of radionuclides must become more sophisticated. Knowledge of the hazards must extend not only to humans, but also to the biota on which we depend.

Summary

1. Oxygen, the byproduct of photosynthesis, is very active chemically. It combines with a wide range of chemicals in Earth's crust, and it reacts spontaneously with organic compounds and reduced substances. It is involved in the oxidation of carbohydrates in the process of respiration, releasing energy, carbon dioxide, and water. The current atmospheric pool of oxygen is maintained in a dynamic equilibrium between the production of oxygen in photosynthesis and its consumption in respiration. An important constituent of the atmospheric reservoir of oxygen is ozone (O_3). The ozone layer blocks out much of the ultraviolet portion of solar radiation. Injection of chlorofluorocarbons, nitrogen oxides, and other oxides into the atmosphere is causing destruction of stratospheric ozone, with serious ecological implications.

2. The carbon budget of the Earth is closely linked to the atmosphere, land, and oceans and to the mass movements of air around the planet. In the ocean, the surface water acts as the main site of exchange of carbon between atmosphere and ocean. The ability of the surface waters to take up CO_2 is governed largely by the reaction of CO_2 with the carbonate ion to form bicarbonates. The uptake of CO_2 from the atmosphere by terrestrial ecosystems is governed by gross production (photosynthesis). Losses are a function of autotrophic and heterotrophic respiration, the latter being dominated by microbial decomposers.

3. Humans pour nitrogen dioxide into the atmosphere and nitrates into aquatic ecosystems. The major sources of nitrogen oxides are automobiles and power plants. Nitrogen dioxide is reduced by ultraviolet light to nitrogen monoxide and atomic oxygen. These substances react with hydrocarbons in the atmosphere to produce a number of pollutants, including ozone and PAN. Both PAN and ozone make up photochemical smog, a pollutant harmful to plants and animals. Excessive quantities of nitrates are added to aquatic ecosystems by improper use of nitrogen fertilizer on agricultural crops, by animal wastes, and by sewage effluents. Nitrogen oxides have also been implicated in forest decline.

4. The global phosphorus cycle is unique among the major biogeochemical cycles in having no significant atmospheric component, although airborne transport of P occurs in the form of soil dust and sea spray. Nearly all of the phosphorus in terrestrial ecosystems is derived from the weathering of calcium phosphate minerals. The transfer of phosphorus from terrestrial to aquatic ecosystems is low under natural conditions; however, the large-scale application of phosphate fertilizers and the disposal of sewage and wastewater to aquatic ecosystems results in a large input of phosphorus to aquatic ecosystems. This input of phosphorus from terrestrial sources is the major cause of eutrophication in freshwater ecosystems.

5. A considerable portion of sulfur is cycled in the gaseous state, circulating on a global scale. Sulfur enters the atmosphere from volcanic eruptions, the surface of the ocean, gases released by decomposition, and the combustion of fossil fuels. Entering the gaseous cycle initially as hydrogen sulfide, sulfur quickly oxidizes to sulfur dioxide. Sulfur dioxide, soluble in water, is carried to Earth as weak sulfuric acid. Whatever the source, sulfur is taken up by plants and incorporated into sulfur-bearing amino acids, later to be released by decomposition. Injected into the atmosphere by industrial consumption of fossil fuels, sulfur dioxide has become a major pollutant, killing plants, causing respiratory afflictions in humans and animals, and contributing to wet and dry acid deposition.

6. Acid deposition, by cloud droplets, precipitation, and dry particles, is implicated in the acidification of lakes and streams in northeastern North America and Scandinavia, where the soils are poorly buffered, and in the decline of spruce forests in eastern North America and Europe.

7. Industrial use of such heavy metals as lead and mercury, always present at low levels in the biosphere, has significantly increased their occurrence. These and other heavy metals pose health problems as they enter the food chain.

8. Of still more consequence globally are the chlorinated hydrocarbons. Used in insect control and for industrial purposes, these hydrocarbons have contaminated global ecosystems and entered food chains. Because they become concentrated at higher trophic levels, chlorinated hydrocarbons harm predaceous animals the most, often interfering with their reproductive capabilities.

9. Radionuclides from nuclear weapons testing, nuclear power plants, and hospital wastes can enter and become concentrated in food chains, particularly grazing ones, and transfer to higher trophic levels.

Review Questions

1. What are the two major biological processes involved in the production of oxygen to, and its uptake from, the atmosphere?
2. What is ozone? What influence does it have on the flux of solar radiation to the Earth's surface?
3. What are the major pools of carbon involved in the global carbon cycle?
4. What biological processes control the exchange of carbon between terrestrial ecosystems and the atmosphere?
5. What are the major inputs of phosphorus to aquatic ecosystems?
6. What is the major form of sulfur gas(es) in the atmosphere?
7. How is acid deposition formed?
8. Why do ecologists have difficulty accepting the hypothesis that acid deposition is the direct cause of forest decline?
9. What is the impact of lead in terrestrial and aquatic ecosystems?
10. What is the impact of chlorinated hydrocarbons on the ecosystem?
11. How are DDT, PCB, and other chlorinated hydrocarbons cycled in ecosystems?
12. What is the major source of radionuclides that enter terrestrial and aquatic ecosystems?
13. What is the most common route of dispersion of nuclides in terrestrial ecosystems?

Cross References

Nutrient cycling, 505; photosynthesis, 86; respiration, 88; solar radiation, 21; eutrophication, 638; carbon cycle, 514; nitrogen cycle, 517; sulfur cycle, 517; phosphorus cycle, 520; forest clearing, 701; global climate change, 707; food chain, 395, 493; trophic levels, 501; ion exchange, 62.

C H A P T E R   27

Biogeography of Ecosystems

CONCEPTS

1. Climate changes along latitudinal and altitudinal gradients result in zonation of vegetation, a basis of vegetation classification.

2. Major vegetation units reflecting climatic patterns are biomes.

3. Life zones arranged according to latitudinal regions, altitudinal belts, and humidity are the basis of the Holdridge system of classification.

4. Ecoregions, large portions of Earth's surface sharing common ecosystems, are the basis of a hierarchical classification.

5. Biogeographical regions provide the foundation for classifying the global distribution of animal life.

6. Species richness increases from the poles to the equator, and it decreases with elevation.

7. Patterns of species richness are correlated with features of the climate that enhance primary productivity.

A view from the window of a transcontinental flight from Boston to San Francisco is revealing to an ecology-minded passenger. Below, the pattern of vegetation changes from the mixed coniferous-hardwood forests of the Northeast to the oak forests of the central Appalachians with patches of high-elevation spruce forest. The forest cover, however fragmented, merges with midwestern cropland of corn, soybean, and wheat, land that was once the domain of tallgrass prairie. Wheat fields yield to high-elevation shortgrass plains, and the plains give way to the coniferous forests of the Rocky Mountains, capped by tundra and snowfields. Beyond the mountains to the Southwest lie tan-colored deserts.

On a trip of less than eight hours over any continent, the airborne ecologist can observe a wide range of vegetation that took months and years for early plant geographers and explorers, such as Friedrich Humboldt, Charles Darwin, Alfred Wallace, and the botanist Joseph Hooker, to discover (see Chapter 1). These naturalists noted that although the flora of various regions were quite distinct taxonomically, their physiognomic or physical structure was remarkably similar. Consider the photos of three desert regions in Australia, Africa, and south-central Asia shown in Figure 27.1. These three regions have a similar climate with low, seasonal precipitation. Although the species that dominate in these three shrub-desert ecosystems are quite distinct, their structure and function (ecosystem processes) are similar.

It is this similarity in vegetation structure that led plant geographers, notably J. F. Schouw (1823) and A. F. W. Schimper (1898, 1903), to correlate the distribution of vegetation to climate. They noted that the world could be divided into great blocks of physiognomically similar vegetation—deserts, grassland, and coniferous, temperate, and tropical forests—that correspond to similar blocks of climate. They called the divisions **formations.** These observations led to the development of a general understanding of the factors controlling the distribution of vegetation at a global scale.

These early studies of the association of the world distribution of plant life and later the association of animal life with climate evolved into the field of biogeography. **Biogeography** is the study of distributions of organisms, both past and present. Its goal is to describe and understand the many patterns in the distribution of species and larger taxonomic groups (Brown and Lomolino 1998). Biogeography involves a number of facets. Historical biogeography is concerned with the reconstruction of the origin, dispersal, and extinction of various taxa. Ecological biogeography is concerned with the study of the present distribution of life and the interaction between the organism and the environment.

VEGETATION ZONATION

The climate-induced patterns of vegetation are influenced by (1) latitude, (2) the location of regions within a continental land mass, which affects the amount of moisture they receive, and (3) altitude, in which mountains modify the climate pattern. As you move latitudinally from the equator to the Arctic or the

(a)

(b)

(c)

FIGURE 27.1   Shrub desert ecosystems: (a) Karoo Desert in Southern Africa, (b) Kara-Kum Desert in Turkmenistan, Central Asia, and (c) western New South Wales, Australia. All three sites are characterized by low, seasonal precipitation.

Antarctic, you cross tropical forests, deserts, grasslands, temperate forests, coniferous forests, tundra, and ice fields. A similar change in vegetation takes place altitudinally in mountain country. As mountains extend upward they reach into colder air, and the vegetation reflects the upward change in climatic conditions. Approximately every 300 meters (1000 ft) rise in elevation is equivalent to 160 km (100 miles) in latitude. Thus altitudinal changes in patterns of vegetation mirror analogous latitudinal changes.

Continental zonation of vegetation is pronounced in North America (Figure 27.2). In northern North America, two broad belts of vegetation stretching east and west, the tundra and boreal forest, are influenced by temperature.

Below these two belts the vegetation patterns are controlled mostly by precipitation and evaporation, the latter influenced considerably by temperature (Thornthwaite 1927). Because available moisture becomes less from east to west, vegetation follows a similar pattern with belts running north and south. Humid regions along the coast support forest vegetation. This zone is broadest in the east. West of the eastern forest region is a subhumid zone where precipitation is less than evaporation. Here the ratio of precipitation to evaporation is about 0.6 to 0.8 and the land supports a tallgrass prairie. Further west is semiarid country, where the precipitation to evaporation ratio is 0.2 to 0.4, which supports a shortgrass steppe. Beyond the shortgrass steppe

FIGURE 27.2 Vegetational zones of North America. (Courtesy of U.S. Department of Interior, Fish and Wildlife Service.)

are desert grasslands and the deserts of the Southwest. Western North America is dominated by mountains and montane vegetation.

In Eurasia, the zonation of natural vegetation, highly modified in places by centuries of human disturbance, more nearly follows that of temperature, resulting in broad belts of vegetation running east and west. In Africa, zonation also consists of broad east-west belts of mediterranean shrubland, desert, tropical grassland and savanna, tropical forest (western Africa), and tropical shrub forest.

In mountainous country on all continents, vegetation zones reflect climatic changes on an altitudinal gradient (Figure 27.3). The vegetation in the lowlands is characteristic of the region. Above the base region is the montane level, where the vegetation changes from shrub, woodland, chaparral, or deciduous forest (depending on the geographical location of the mountain range), to montane coniferous forests, alpine tundra, and at the summit of the very highest mountains, perpetual ice and snow. There are, of course, many variations among the continents, but this basic pattern persists.

FIGURE 27.3 Altitudinal zonation of vegetation in mountains. (a) Mount Marcy in the Adirondack Mountains of New York State. (b) The Rocky Mountains (generalized). (c) Complex cross section of the Andes at the latitude of Lima, Peru, from the coast to the high mountains. (After Walter 1979.)

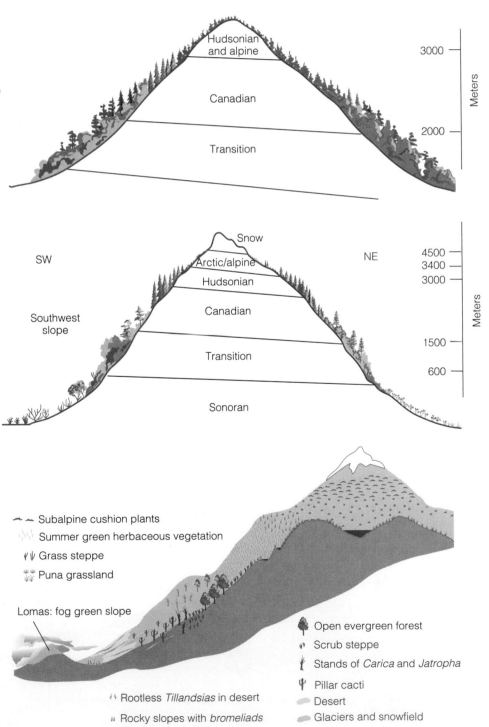

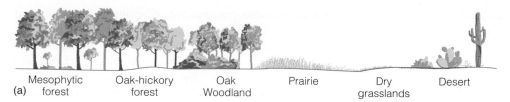

(a)
Mesophytic forest   Oak-hickory forest   Oak Woodland   Prairie   Dry grasslands   Desert

FIGURE 27.4 Gradients of vegetation in North America from east to west and north to south. (a) The east-west gradient reflects precipitation. The transect does not cut across the Rocky Mountains. (b) The south-north gradient reflects temperatures.

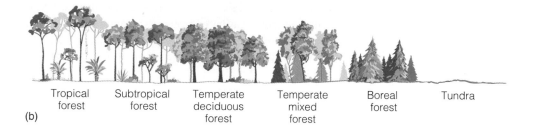

(b)
Tropical forest   Subtropical forest   Temperate deciduous forest   Temperate mixed forest   Boreal forest   Tundra

BIOMES

For many years ecologists have related world plant formations to climatic patterns. European ecologists, like the early plant geographers, consider the broad vegetation patterns of the world as plant formations. The ecologists F. E. Clements and V. E. Shelford (1939) introduced the concept of the biome as a classification of world vegetation patterns. They considered the broad plant formations and their associated animal life as biotic units. Clements and Shelford called these biotic units **biomes,** each of which is characterized by a uniform life form of vegetation, such as grassland or coniferous forest. Boundaries between biomes, of course, are broad and indistinct as one vegetation type blends into another.

If we were to sample vegetation along a transect cutting across a number of biome-types, such as up a high mountain slope or north to south across a continent, we would discover that plant communities change gradually. If we were to run this transect across mid-continental North America beginning in the moist, species-rich forests of the Appalachians and extending into the desert, the vegetation types would follow a gradient of climatic moisture. It would begin in the mesophytic forests of the Appalachians through oak-hickory forests, oak woodlands with a grassy understory, tallgrass prairies (now corn land), mixed prairies (wheat land), shortgrass plains, desert grasslands, and finally, desert shrublands (Figure 27.4a). Likewise, a transect from southern Florida to the Arctic would move along a climatic temperature gradient from a subtropical forest through temperature deciduous forest, temperate mixed forest, to boreal coniferous forest and tundra (Figure 27.4b).

In addition to gradual changes in vegetation, there are gradual changes in other ecosystem characteristics. From highly mesic situations and warm temperatures to xeric situations and cold temperatures, productivity, species diversity, and the amount of organic matter decrease. There is a corresponding decline in the complexity and organization of ecosystems, in the size of plants, and in the number of strata to vegetation. Growth form changes. The tropical rain forest is dominated by phanerophytes and epiphytes, the Arctic tundra by hemicryptophytes and geophytes. Wherever similar environments are present on Earth, the same growth forms exist, even though species differences may be great. Thus, different continents tend to have communities of similar physiognomy.

Depending on how finely you want to classify them, there are at least nine major terrestrial biomes or biome-types: tundra, taiga, temperate forest, temperate rain forest, tropical rain forest, savanna, temperate grasslands, chaparral, and desert. When Whittaker (1975) plotted these various biome-types on gradients of mean annual temperature and mean annual precipitation, they formed a distinctive climatic pattern (Figure 27.5). The graph is a reasonable approximation of global vegetation patterns. As the colored lines indicate, many types intergrade with one another. Soil, exposure to fire, and nature of regional climate can influence which one of the several biome-types occupies an area.

LIFE ZONES
Merriam Life Zones

Before the biome concept became generally accepted by ecologists, G. Hart Merriam (1884) became the first ecologist to define precisely the relationship between climate and vegetation. He developed his life zone system after observing the sharp zonation of vegetation on San Francisco Mountain in Arizona. Impressed with the importance of temperature as an influence on plant and animal distribution, he formulated his two laws regarding temperature control. The first law states that the northward distribution of terrestrial plants and animals is governed by the sum of the positive temperatures as an influence for the entire season of growth and reproduction. In determining the sum of positive or effective temperatures, Merriam assumed that a minimum of 6°C marked the beginning of the period of physiological activity in plants and reproductive activity in animals. The second law states that southward distribution is governed by the mean temperature of a brief period during the hottest part of the year. Using these laws, Merriam divided the North American continent into three primary transcontinental regions: the Boreal, the Austral, and the Tropical. He then subdivided these regions into zones.

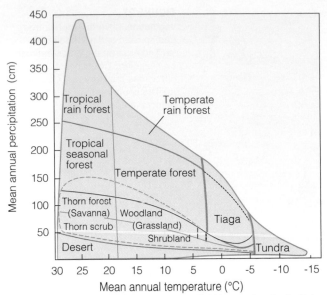

FIGURE 27.5 Pattern of world plant formations or biomes in relation to temperature and moisture. Where climate varies, soil can shift the balance between such types as woodland, shrubland, and grass. The colored line encloses a wide range of environments in which either grassland or one of the types dominated by woody plants prevail in different areas. (After Whittaker 1970a:167. See also map inside back cover.)

Although not used today, his life zones are reflected in the temperature hardiness zones mapped and widely used in horticultural and planting guides.

Holdridge Life Zone System

A more sophisticated approach to the relationship of vegetation patterns to climate is the Holdridge life zone system (Holdridge 1964, 1967, Holdridge et al. 1971). The Holdridge system involves three levels of classification: (1) climatically defined life zones; (2) associations, which are subdivisions of life zones based on local environmental conditions; and (3) further local subdivisions based on actual cover or land use. The term *association,* as used by Holdridge, means a unique ecosystem, a distinctive habitat or physical environment and its naturally evolved community of plants and animals. Holdridge based this classification on three assumptions: (1) interaction between temperature and rainfall determine vegetation patterns; (2) geographic boundaries of vegetation correspond closely to boundaries between climatic zones; and (3) mature, stable vegetational types represent discrete vegetation units recognizable throughout the world (see inside front cover).

The Holdridge life zone classification is a bioclimatic model that divides the world into life zones arranged according to latitudinal regions, altitudinal belts, and humidity provinces (Figure 27.6). The subdivisions relate the distribution of natural vegetation associations or zones to indices of biotemperature, mean annual precipitation, and potential evapotranspiration ratio (PET ratio). The life zones are depicted by a series of hexagons in a triangular coordinate system. Two variables, average annual biotemperature and precipitation, determine the classification. Average annual biotemperature is calculated by summing the average daily temperature over the year for those days where average daily temperature is above 0°C. The sum is then divided by 365 to provide a mean daily value for the year. Biotemperature provides an index similar to growing degree-days discussed in Chapter 7. It assumes that basic plant processes relating to growth and development do not occur at temperatures below 0°C, and that the sum (integration) of the temperatures above 0°C provides an index of energy for basic plant and ecosystem processes.

Identical axes of average annual precipitation form two sides (right and base) of the equilateral triangle. The potential evapotranspiration ratio forms the third axis of the triangle (left), and an axis for mean annual biotemperature is oriented perpendicular to its base. Striking equal intervals on these logarithmic axes forms hexagons that designate the Holdridge life zones.

The potential evapotranspiration ratio (PET ratio) is the quotient of PET and average annual precipitation. It is an index of aridity. PET is the potential evapotranspiration based on the input of radiation at a particular location. As such, it represents the demand for water by plants created through evaporation and transpiration. If the value of the PET ratio is 1.0, then demand (PET) is equal to supply (precipitation). For values less than 1.0, supply is greater than demand. For increasing values greater than 1.0, demand is progressively exceeding supply and conditions become more arid. Holdridge (1959) assumes, on the basis of data from a number of ecosystems, that PET is proportional to biotemperature (constant of proportionality = 58.93). The PET ratio therefore depends on the two primary variables, annual precipitation and biotemperature.

The Holdridge life zone classification differs from many other ecosystem or biome classifications in that it explicitly defines a relationship between climate and vegetation/ecosystem distribution. The global map of Holdridge life zones presented in Apendix C of the book represents an application of the general model presented in Figure 27.6. The map was generated using a global climate data set at a spatial resolution of 0.5° latitude and longitude (Leemans and Cramer 1990). For each location, long-term vales of mean monthly temperature and precipitation were used to calculate the variables required for the classification. The map, therefore, represents potential life zone (ecosystem) distribution based on the Holdridge classification and the global climate data set. The advantage of this and other classifications that explicitly define a relationship between climate and vegetation distribution is their potential use in examining the response of ecosystem distribution under changing climate conditions, either past or present (see application of Holdridge model in Figure 32.16).

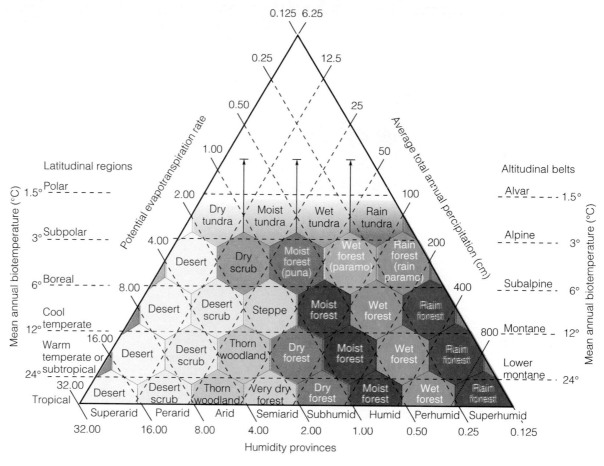

FIGURE 27.6   The Holdridge life zone system for classifying plant formations. The life zones are determined by a gradient of mean annual biotemperatures with latitude and altitude, the ratio of potential evapotranspiration to annual precipitation, and total annual precipitation. (After Holdridge 1947. See also map inside front cover.)

ECOREGIONS

A recent approach to the classification and mapping of the biotic world is the concept of the ecoregion, developed largely by R. W. Bailey (1976, 1995, 1998) of the U.S. Forest Service. He based it in part on the ecoregion concept of a Canadian geographer, J. M. Crowley (1967). **Ecoregions** are major ecosystems that result from predictable patterns of climate as influenced by latitude, global position, and altitude (Bailey 1998). Each ecoregion is a continuous geographical area across which the interactions of climate, soil, and topography are sufficiently uniform to permit the development of similar types or forms of vegetation. Thus, the ecoregion approach provides a means of mapping any large portion of Earth's surface over which the ecosystems have certain characteristics in common. On this basis, ecoregions occur in predictable locations in different parts of the world (see inside back cover and Figure 27.7).

The classification scheme involves a synthesis of climate types, vegetation associations, and soil types into a single, geographic, hierarchical classification that reflects both ecological properties and spatial patterns (Tables 27.1

and 27.2). The largest category is the **domain,** a subcontinental area of broad climatic similarity (Figure 27.8). There are four domains: polar, humid temperate, humid tropical, and dry. The first three are thermally defined; dry is defined by moisture. Domains are broken down into 14 **divisions,** based on the seasonality of precipitation or degree of coldness and dryness (Table 27.2, p.552). Divisions are subdivided into **provinces** that correspond to broad vegetation regions having a uniform regional climate and the same distinct vegetation and soils (Figure 27.9, p.552). Provinces are further subdivided into smaller and smaller categories useful in local areas (see Table 27.1). Boundaries between ecoregions, however, are imprecise. Some are narrow transition zones; others may be fairly abrupt.

Whereas the biome and Holdridge life zone system concentrate terrestrial ecosystems, the ecoregion concept embraces both the terrestrial and the oceanic. Ocean ecoregions result from the interaction of macroclimates and large-scale ocean currents. These interactions determine major hydrologic units that are the base of the three domains polar, temperate, and tropical (Table 27.3 and Figure 27.10a, p.553).

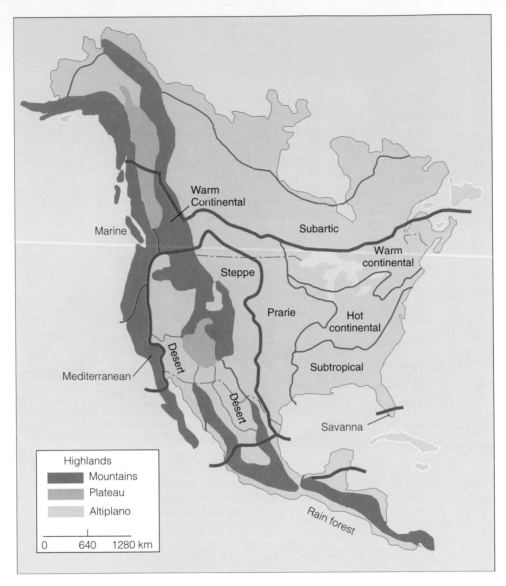

FIGURE 27.7 Ecoregions of North America. The heavy lines outline the domains: polar, dry, humid temperate, and humid tropical. The thin lines demarcate divisions (see Table 4.2). (From Bailey 1978.)

TABLE 27.1 A Hierarchy for Ecosystems

Name	Character
1. Domain	Subcontinental area of related climates
2. Division	Single regional climate at the level of Köppen's types (Trewartha 1943)
3. Province	Broad vegetation region with the same type or types of zonal soils
4. Section	Climatic climax at the level of Küchler's potential
5. District	Part of a section having uniform geomorphology at the level of Hammond's land-surface form regions (1964)
6. Landtype association	Group of neighboring landtypes with recurring pattern of landforms, lithology, soils, and vegetation associations
7. Landtype	Group of neighboring phases with similar soil series or families with similar plant communities at the level of Daubenmire's habitat types (1968)
8. Landtype phase	Group of neighboring sites belonging to the same soil series with closely related habitat types
9. Site	Single soil type or phase and single habitat type or phase

Source: From Bailey 1878, accompanying map; adapted from Crowley 1967 and Wertz and Arnold 1972.

TABLE 27.2 Characteristics of Some First-Order and Second-Order Ecoregions

Domain	Division	Temperature	Rainfall	Vegetation	Soil
Polar	Tundra	Mean temperature of warmest month <10°C	Water deficient during the cold season	Moss, grasses, and small shrubs	Tundra soils (entisols, inceptisols, and associated histosols)
	Subarctic	Mean temperature of summer is 10°C; of winter, −3°C	Rain even throughout the year	Forest, parklands	Podzols (spodosols and associated histosols)
Humid temperate	Warm continental	Coldest month below 0°C, warmest month	Adequate throughout the year	Seasonal forests, mixed coniferous-deciduous forests	Gray-brown podzolic (spodosols, alfisols)
	Hot continental	Coldest month below 0°C, warmest month >22°C	Summer maximum	Deciduous forests	Gray-brown podzolic (alfisols)
	Subtropical	Coldest month between 18°C and −3°C, warmest month >22°C	Adequate throughout the year	Coniferous and mixed coniferous-deciduous forests	Red and yellow podzolic (ultisols)
	Marine	Coldest month between 18°C and −3°C, warmest month <22°C	Maximum in winter	Coniferous forests	Brown forest and gray-brown podzolic (alfisols)
	Prairie	Variable	Adequate all year, excepting dry years; maximum in summer	Tallgrass, parklands	Prairie soils, chernozems (mollisols)
	Mediterranean	Coldest month between 18°C and −3°C, warmest month >22°C	Dry summers, rainy winters	Evergreen woodlands and shrubs	Mostly immature soils
Dry	Steppe	Variable, winters cold	Rain <50 cm/yr	Shortgrass, shrubs	Chestnut, brown soils, and sierozems (mollisols, aridisols)
	Desert	High summer temperature, mild winters	Very dry in all seasons	Shrubs or sparse grasses	Desert soils (aridisols)
Humid tropical	Savanna	Coldest month >18°C, annual variation <12°C	Dry season with <6 cm/yr	Open grassland, scattered trees	Latosols (oxisols)
	Rain forest	Coldest month >18°C, annual variation <3°C	Heavy rain, minimum 6 cm/month	Dense forest, heavy undergrowth	Latosols (oxisols)

Note: Names in parentheses in the soil column are soil taxonomy orders (USDA Soil Survey Staff 1975); see Table 4.2.
Source: Bailey 1978, map.

The polar domain is characterized by water that is rich in plankton and thus greenish in color, low in temperature, and low in salt content. The temperate domain embraces the middle latitudes between the poleward limits of the tropics and the equatorward limits of pack ice in winter. It is characterized by mixed waters whose currents correspond to wind direction around the subtropical atmospheric high-pressure cells (see Figure 2.8). The tropical domain is characterized by ocean water that is blue, high in temperature, high in salt content, and low in organic forms (Figure 27.11). Distribution of life relates to latitude, major wind systems, and precipitation and evaporation (Bailey 1998). These three domains consist of 14 divisions differentiated by combinations of circulation, tem-

perature, salinity, and the presence of upwellings. Like terrestrial divisions, each oceanic division occurs in several parts of the world that are broadly similar in physical and biological characteristics (Figures 27.10b and 27.11).

The ecoregion approach to classification is ecologically more useful than the other classifications because it relates to management strategy in its mapping levels. Because all systems operate within the context of larger systems, knowledge of larger systems allows us to better understand the smaller systems. A better understanding of smaller systems allows us to predict the outcome of land management and natural resource development. Its provides a foundation for ecological management of resources. Because each region has its

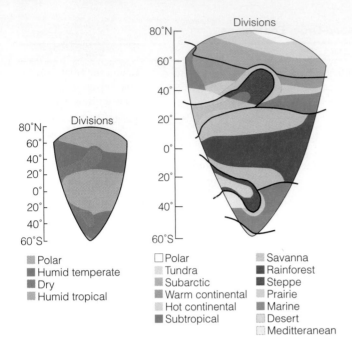

Polar
Humid temperate
Dry
Humid tropical

Polar
Tundra
Subarctic
Warm continental
Hot continental
Subtropical

Savanna
Rainforest
Steppe
Prairie
Marine
Desert
Meditteranean

FIGURE 27.8   Patterns of ecoregions that might occur on a hypothetical continent of low uniform elevation. (From Bailey 1999.)

Long Term Ecological Research (LTER) program takes place within an ecoregion framework.

BIOGEOGRAPHICAL REGIONS

Zoogeographers lagged behind plant geographers in their study of animal distribution. Complicating their studies were the great number of animal species and the lack of any clear-cut relationship between animal distribution and climate. Ultimately, zoogeographers did accumulate the basic information on the global distribution of animals. What they needed to do next—and this was a great enough task—was to arrange the facts and draw some general conclusions. Philip Sclater did this for birds in 1878. Alfred Wallace, also known for having developed the same general theory of evolution as Darwin, drew on Sclater's work, expanded it, and provided the major synthesis of animal distribution. He divided Earth into six biogeographical regions or realms based on the distributions of terrestrial animals. These realms are the Palearctic, Nearctic, Neotropical, Ethiopian, Oriental, and Australian (Figure 27.12). Because some zoogeographers consider the Neotropical and Australian regions to be so different from the rest of the world, they group the four other regions together to make a more general classification: Neogea (Neotropical), Notogea (Australia), and Megagea (the rest of the world). Each realm more or less embraces a major continental land mass and each is separated by oceans, mountain ranges, or deserts. Each region possesses certain distinctions and uniformity in the taxonomic units it contains, and each to a greater or

own distinctive flora, fauna, climate, soil, and landform, each requires its own approach to management. The ecoregion approach provides a means of studying the problems of resource management on a regional basis, and it provides a framework for the organization and retrieval of data gathered in resource inventory data. The National Science Foundation

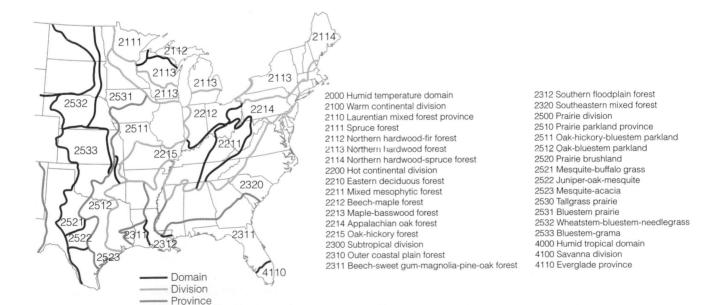

2000 Humid temperature domain
2100 Warm continental division
2110 Laurentian mixed forest province
2111 Spruce forest
2112 Northern hardwood-fir forest
2113 Northern hardwood forest
2114 Northern hardwood-spruce forest
2200 Hot continental division
2210 Eastern deciduous forest
2211 Mixed mesophytic forest
2212 Beech-maple forest
2213 Maple-basswood forest
2214 Appalachian oak forest
2215 Oak-hickory forest
2300 Subtropical division
2310 Outer coastal plain forest
2311 Beech-sweet gum-magnolia-pine-oak forest

2312 Southern floodplain forest
2320 Southeastern mixed forest
2500 Prairie division
2510 Prairie parkland province
2511 Oak-hickory-bluestem parkland
2512 Oak-bluestem parkland
2520 Prairie brushland
2521 Mesquite-buffalo grass
2522 Juniper-oak-mesquite
2523 Mesquite-acacia
2530 Tallgrass prairie
2531 Bluestem prairie
2532 Wheatstem-bluestem-needlegrass
2533 Bluestem-grama
4000 Humid tropical domain
4100 Savanna division
4110 Everglade province

—— Domain
—— Division
—— Province
—— Section

FIGURE 27.9   The subdivision of ecoregions in the eastern and central United States to the edge of the Humid Temperate Domain. (From Bailey 1978.)

TABLE 27.3 Oceanic Regions

500 Polar domain
510 Inner polar division
520 Outer polar division
600 Temperate domain
610 poleward westerlies division
620 Equatorward westerlies division
630 Subtropical division
640 High salinity subtropical division
650 Jet stream division
660 Poleward monsoon division
700 Tropical domain
710 Tropical monsoon division
720 High salinity tropical monsoon division
730 Poleward trades division
740 Trade winds division
750 Equatorward trades division
760 Equatorial countercurrent division
Shelf

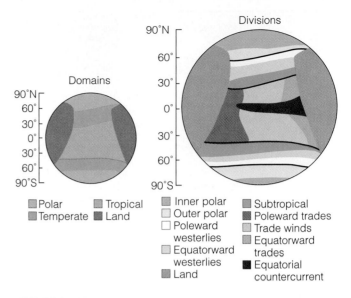

FIGURE 27.10 Ecoregion domains and divisions in a hypothetical ocean basin. (From Bailey 1999.)

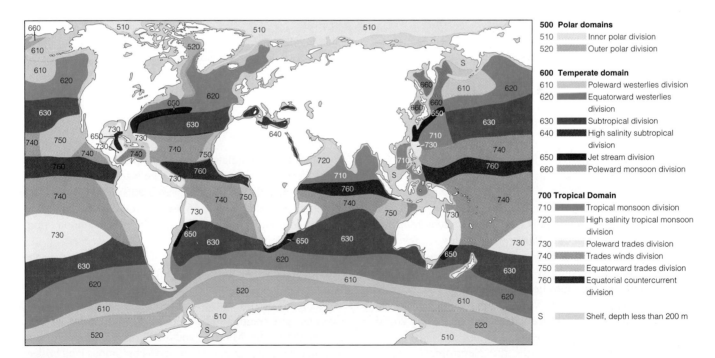

FIGURE 27.11 Divisions of the oceanic domains. (From Bailey 1999.)

lesser degree shares some of the families of animals with other regions. Each at one time or another in the history of Earth has had some land connection with another across which animals and plants could pass, Australia more briefly than the others.

Two regions, the Palearctic and the Nearctic, are closely related; in fact, the two are often considered as one, the Holarctic. The Nearctic contains the North American continent south to the Tropic of Cancer. The Palearctic region contains the whole of Europe, all of Asia north of the Himalayas, northern Arabia, and a narrow strip of coastal North Africa. The regions are similar in climate and vegetation. They are alike in their faunal composition, and they share, particularly in the north, similar animals, such as the

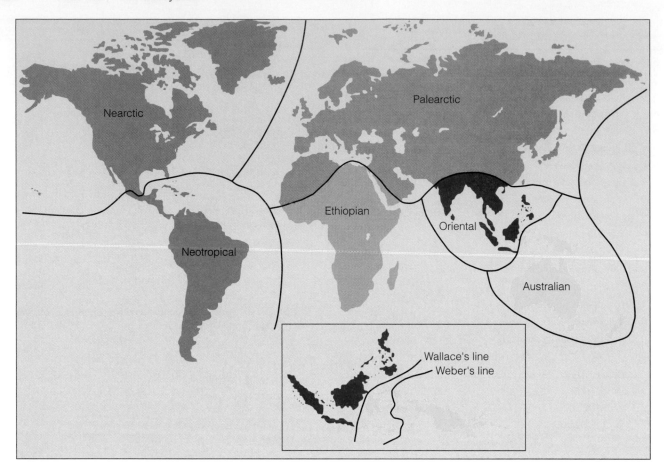

FIGURE 27.12   Biogeographical realms.

wolf, the hare, the moose (called elk in Europe), the stag (called elk in North America), the caribou, the wolverine (*Gulo gulo*), and the bison.

Below the coniferous forest belt, the two regions become more distinct. The Palearctic is not rich in vertebrate fauna, of which few are endemic. Palearctic reptiles are few and usually are related to those of the African and Oriental tropics. The Nearctic, in contrast, is the home of many reptiles and has more endemic families of vertebrates. The Nearctic fauna is a complex of New World tropical and Old World temperate families; the Palearctic is a complex of Old World tropical and New World temperate families.

South of the Nearctic lies the Neotropical, which includes all of South America, part of Mexico, and the West Indies. It is joined to the Nearctic by the Central American isthmus and is surrounded by the sea. Isolated until 15 million years ago, the fauna of the Neotropical is most distinctive and varied. In fact, about half of the South American mammals, such as the tapir and llama, are descendants of North American invaders, whereas the only South American mammals to survive in North America are the armadillo, opossum, and porcupine. Lacking in the Neotropical is a

well-developed ungulate fauna of the plains, so characteristic of North America and Africa. The Neotropical, however, is rich in endemic families of vertebrates. Of 32 families of mammals, excluding bats, 16 are restricted to the Neotropical. In addition, 5 families of bats, including the famous vampire, are endemic.

The Old World counterpart of the Neotropical is the Ethiopian, which includes the continent of Africa south of the Atlas Mountains and Sahara Desert and the southern corner of Arabia. It embraces tropical forests in central Africa and savanna, grasslands, and desert in the mountains of East Africa. During the Miocene and the Pliocene, Africa, Arabia, and India shared a moist climate and a continuous land bridge, which allowed the animals to move freely among them. This connection accounts for some similarity in the fauna between the Ethiopian and the Oriental regions. Of all the regions the Ethiopian contains the most varied vertebrate fauna; it is second only to the Neotropical in endemic families.

Lush forests cover much of the Oriental region, which includes India, Indochina, South China, Malaysia, and the western islands of the Malaysian Archipelago. It is bounded by the Himalayas, the Indian Ocean, and the Pacific Ocean.

On the southeast corner, where the islands of the Malaysian Archipelago stretch out toward Australia, there is no definite boundary. However, Wallace noted a sharp faunal gap between the Philippines and the Moluccas in the north and between the islands of Bali and Lombok on the south, where many Southeast Asian species reach their distributional limits and are replaced by forms from Australiasia. Here he drew a line to separate the Oriental from the Australian region. Later, a second line, Weber's, was drawn east of Wallace's line; it separates the islands with a majority of Oriental animals from those with a majority of Australian ones. Because the islands between these two lines represent a transition between the Oriental and Australian realms, some zoogeographers call the area Wallacea.

Of the tropical regions, the Oriental possesses the fewest endemic species and lacks a variety of widespread families. It is rich in primate species, including two families confined to the region, the tree shrews (Tupaiidae) and the tarsiers (Tarsiidae).

The most interesting and the strangest region, and certainly the most impoverished in vertebrate species, is the Australian. It includes Australia, Tasmania, New Guinea, and a few smaller islands of the Malaysian Archipelago. New Zealand and the Pacific Islands are excluded, for they are regarded as oceanic islands, separate from the major faunal regions. Partly tropical and party south temperate, the Australian is noted for its lack of a land connection with other regions, the poverty of freshwater fish, amphibians, and reptiles, the absence of placental mammals, and the dominance of marsupials. Included are the monotremes with two egg-laying species, the duck-billed platypus (*Ornithorhynchus anatinus*) and the short-beaked echidna (*Tachyglossus aculeatus*). The marsupials have become diverse and have evolved ways of life similar to those of the placental mammals of other regions.

PATTERNS OF SPECIES RICHNESS

The biogeographical realms of Earth support an amazing diversity of species. In fact scientists have identified and named approximately 1.4 million species (Figure 27.13); see Ecological Application). Because new species are continuously being discovered, quantifying the actual number of species inhabiting the Earth is an ongoing exercise. Some scientists, such as Harvard biologist E. O. Wilson, believe the actual number of species may be closer in number to 10 million (Wilson 1999).

The 1.4 million species that have been identified are not distributed equally across Earth's surface. There are distinct geographic patterns of species richness. The most notable is the decrease in species richness as you move from the equator toward the poles and from low to high altitudes in mountainous country. Mountainous regions generally support more species than flatlands because of topographic diversity, and peninsulas have fewer species than adjoining continental areas. From east to west in North America, the number of species of trees (Figure 27.14d) (Monk 1967), breeding land birds (Figure 27.14e) (MacArthur and Wilson 1967), and mammals (Simpson 1964) increases (Figure 27.14a–c). Amphibians are more abundant and diverse in eastern North America (Figure 27.14c) than in the western part of the continent, whereas reptiles are more diverse in the hot arid regions of western North America (Figure 27.14b) (Kiester 1971). This increased diversity on an east-west gradient relates to an increased diversity of the environment both horizontally and altitudinally. At the same time, species richness for all these groups decreases from south to north, for reasons discussed below. The species richness of marine life increases from the continental shelf, where food is abundant but the environment is changeable, to deep water, where food is less abundant but the environment is more stable.

Although the mechanisms underlying this geographic pattern in species diversity are not known, a variety of hypotheses

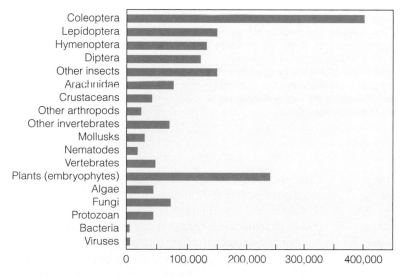

FIGURE 27.13 Number of living species of all kinds of organisms currently known. Species are classified by major taxonomic groups. Insects and plants dominate the diversity of living organisms. (Wilson 1999.)

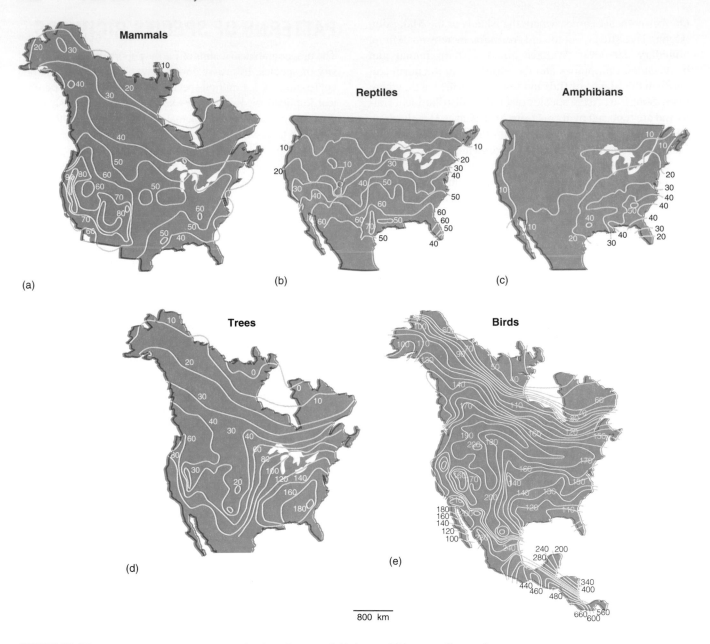

FIGURE 27.14   Latitudinal variation in the distribution of mammal, birds, amphibians, reptiles, and trees. Contour lines connect points with about the same number of species. The diversity of mammals (a) and the diversity of birds (e) reflect both latitudinal and altitudinal variations across the continent (after Currie 1991). The most pronounced latitudinal variations occur among amphibians (c) and reptiles (b) (Kiester 1971). Poikilothermic and endothermic, reptiles have their greatest diversity in the hot desert regions and lower latitudes of North America. Being not only poikilothermic but also highly sensitive to moisture conditions, amphibians reach their greatest diversity in the central Appalachians and decrease northward, southward, and westward. Distribution of trees reflect actual evapotranspiration. (From Currie and Paquin 1987.)

have been put forward. These hypotheses relate to a wide range of factors including the age (in an evolutionary sense) of the community (Fisher 1960, Simpson 1964), spatial heterogeneity of the environment (Simpson 1964), temporal stability of the climate (Fischer 1960, Connell and Orias 1964), and ecosystem productivity (Connell and Orias 1964), to name only a few. Of the variety of hypotheses proposed to account for global patterns of species diversity, those explicitly relating to features

of the environment known to directly influence basic plant and animal processes are the most easily interpreted.

Currie and Paquin (1987) examined the relationship between patterns of tree species richness in North America and a number of variables describing regional differences in climate. They found that although variation in tree species richness correlates to such climatic factors such as integrated measures of annual temperature, solar radiation, and precipitation, it most strongly correlates with estimates of actual evapotranspiration (Figure 27.15). Recall from earlier discussions in Chapters 7 and 23 that actual evapotranspiration (AET) is the flux of water to the atmosphere from the terrestrial surface through the combined processes of evaporation and transpiration. As such, it is a function of both the demand for water by plants brought about by the input of solar energy to the surface and the supply of water furnished through precipitation.

The pattern of increasing species richness as a function of AET parallels the pattern of increasing net primary productivity with AET discussed in Chapter 23. This pattern suggests a relationship between plant diversity and primary productivity. Increased resources and environmental conditions favorable for photosynthesis and plant growth may well give rise to increased plant diversity over evolutionary time; however, the relationship between productivity and plant diversity at a local scale is not so clear. Recall from Chapter 21 that high resource availability encourages high species growth rates, decreasing species diversity during succession (see Figure). The relationship between primary productivity and species richness, an ongoing area of investigation, emphasizes the need to understand the mechanisms that influence both the evolution and maintenance of diversity within any region.

Currie (1991) demonstrated a correlation of regional species richness of vertebrates across North America with potential evapotranspiration (PET), an index of integrated energy availability (Figure 27.16). The lower PET conditions are one reason that amphibians are more abundant and diverse in eastern North America (Figure 27.14c) than in the western part of the continent, while reptiles are more diverse

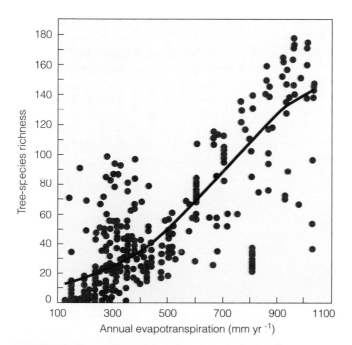

FIGURE 27.15 Relationship between annual measure of actual evapotranspiration (AET) and tree species richness for North America. (From Currie and Paquin 1987.)

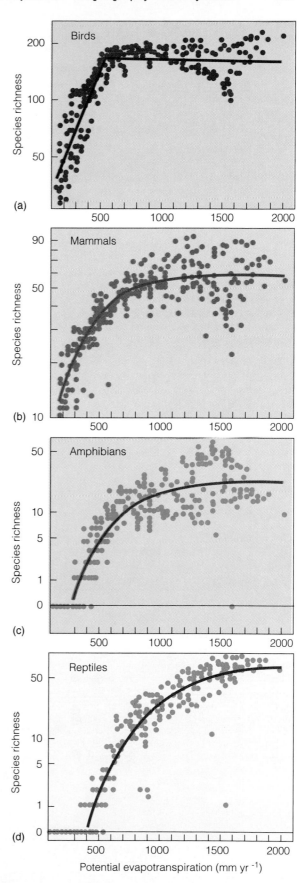

FIGURE 27.16 Relationship between annual estimate of potential evapotranspiration and species richness of (a) birds, (b) mammals, (c) amphibians, and (d) reptiles in North America. (From Currie 1991.)

FIGURE 27.17 The number of plant species endemic to each of the United States. For example, 1517 plant species are found in California and nowhere else. (From Gentry 1986.)

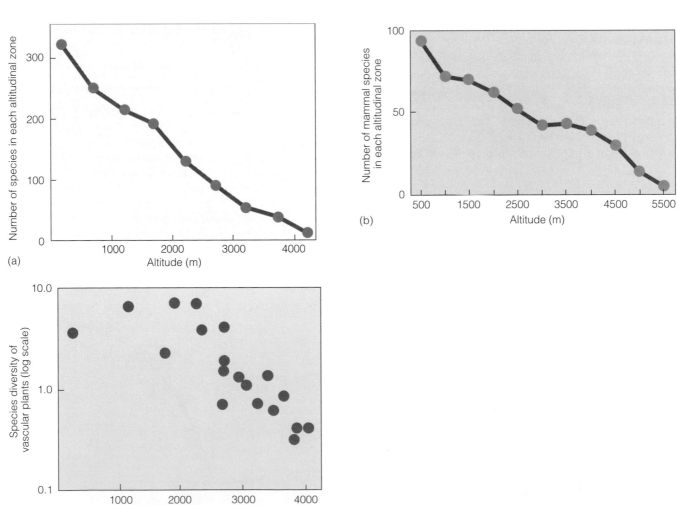

FIGURE 27.18 Relationship between species richness and altitude for (a) bird species in New Guinea (from Kikkawa and Williams 1971), (b) mammal species in the Himalayas (from Hunter and Yonzon 1992), and (c) vascular plants in the Himalayas. (From Whittaker 1977.)

in the hot arid regions of western North America (Kiester 1971) (Figure 28.14b), where PET conditions are high. The higher air temperatures associated with increasing PET would have a direct benefit to ectothermic species in maintaining body temperature, activity, and acquisition of food resources. In the case of endothermic species, the increased temperatures would result in a decreased need for energy from food to maintain body temperature (see Chapter 8).

In addition to the correlation of diversity with climate, Currie (1991) found an additional relationship between vertebrate diversity and plant species diversity. Given the correlation between the latter and net primary productivity, this observation may well be a reflection of the relationship between primary and secondary productivity presented in Chapter 23. This relationship between diversity and productivity for secondary producers may be similar to that observed for autotrophs.

Animal diversity is also linked to plant diversity through the relationship between vegetation and suitable habitat for animal species. Increased structural diversity, as measured by foliage height diversity, presented in Chapter 20 (Figure 20.5), provides a wider range of microhabitats and associated resources. Consequently, it is correlated with increased animal species diversity.

The relationship between habitat heterogeneity and species diversity is not limited to animals. Environmental heterogeneity also gives rise to increased plant species diversity. The number of **endemic** plant species—those restricted to a specific geographic region—in each of the United States is shown in Figure 27.17. The high diversity of endemic plant species in areas like California and Texas is related to the diverse landscapes and environmental conditions in these areas.

A similar pattern to that of decreasing species richness moving north from the equatorial zone is observed with increasing altitude. Figure 27.18 shows patterns of species richness with increasing elevation for birds species in New Guinea, and mammals and vascular plants in the Himalayan Mountains. The mechanisms underlying the declining species richness with elevation may be similar to those for latitude. Variations in temperature, PET, AET, and vegetation structure with increasing elevation parallel those observed with increasing latitude. However, the pattern of declining species with elevation may also be confounded by the fact that high-altitude communities generally occupy a smaller spatial area than corresponding lowland areas at equivalent latitudes. Likewise, these high-elevation communities are more likely to be isolated from similar communities, suggesting the importance of immigration rates in maintaining the long-term persistence of populations (see discussion of Island Biogeography in Chapter 23).

Summary

1. Climate has a pronounced influence on the latitudinal and altitudinal distribution of vegetation. Latitudinal changes result in the continental zonation of vegetation, most pronounced in North America and Africa. Altitudinal changes in patterns of vegetation mimic latitudinal changes.

2. Ecologists have developed different schemes for classifying these broad patterns of vegetation. One is the biome system, which groups plants and animals of the world into integral units characterized by distinctive life forms in a climax community. Boundaries of biomes, or major life zones as they are known in Europe, coincide with the boundaries of major plant formations of the world. By including both plants and animals as a total unit that evolved together, the biome permits recognition of the close relationship among all living things.

3. More refined is the Holdridge life zone system. It divides the vegetation of he world into zones defined by mean annual precipitation and mean annual biotemperature.

4. The more recent ecoregion scheme refines the biome types into hierarchies. It classifies terrestrial and ocean ecosystems on the basis of large geographical areas or domains, which are further subdivided into 14 divisions. Subdivisions of terrestrial ecosystems are based on the interaction of climate, soil, continental position, and topography. Oceanic ecoregions are based on the interaction of macroclimate, large-scale ocean currents, and salinity.

5. Zoogeographers divided the world into biogeographical or faunal realms. The three realms are subdivided into six regions. Each region is separated by a barrier of oceans, mountain ranges, or deserts that prevent free dispersal of animals, and each possesses its own distinctive forms of life. Each region is further subdivided by secondary barriers such as vegetation types and topography. This classification provides an ecological foundation for the management of regional resources.

6. The approximately 1.4 million species that scientists have identified are not distributed evenly over Earth's surface. In general, species richness decreases from the equator toward the poles, and with increasing elevation. A variety of hypotheses have been put forward to explain these patterns, including the role of climate. Regionally, plant species richness is correlated with actual evapotranspiration, suggesting a positive relationship between species richness and net primary productivity. Species richness of terrestrial vertebrates is correlated with potential evapotranspiration, an integrated measure of energy input to the ecosystem.

Review Questions

1. How does the pattern of temperature and moisture influence global vegetation? Mountain vegetation? How is this reflected in the vegetation of North America? Europe? Africa?
2. Contrast the biome, ecoregion, and Holdridge life zone concepts of the classification of major vegetation types.
3. Discuss situations in which the application of one of the vegetation classification systems would be preferable to the other.
4. Name some animals that characterize each of the six biogeographical regions.
5. How does species richness vary with latitude? With elevation?
6. How does plant species richness vary with climate?
7. How is climate correlated with regional patterns of vertebrate diversity?
8. How does environmental heterogeneity influence patterns of species diversity?

Time to Rethink the Lawn

It's any weekend day from April to October. America heads for the garage or garden shed to start the lawn mower. The tasks are well defined: mow, weed, fertilize, lime, aerate, and irrigate. The enemy is identified: crabgrass, dandelions, grubs, moles, ants, and Japanese beetles are among their ranks. The attack begins with herbicides and insecticides. It is an ongoing crusade, and the Holy Grail is the perfect lawn—the envy of the neighborhood.

The reason for America's obsession with the lawn is not clear, but theories abound. Some say it's rooted in early nineteenth-century British ideas of natural beauty. Evolutionary psychologists suggest that the love of lawns is genetically encoded—a molecular memory of the time when progenitors of the human species left their arboreal dwellings for Africa's savannas (Figure VIII-A). Whatever the reason, the lawns of 58 million American households combine to cover 20 million acres of the United States—an area roughly the size of Pennsylvania. Nor is this obsession inexpensive. Americans spend an estimated $30 billion per year on lawn care (enough to provide 1.5 million families with an income above poverty level) (Figure VIII-B). Despite the enormous economic expenditures, the real cost of creating and maintaining our lawns is perhaps best measured not in dollars but by the cumulative effects on environmental and human health.

Lawns are human-made ecosystems. To create a lawn, a bare piece of land is required. In the eastern regions of the United States, this means clearing forests. Most of the nitrogen and other essential nutrients in a forest are in the living vegetation and dead organic matter on the soil surface. When these are removed (by bulldozing or burning), chemical fertilizers must be applied in order for the topsoil to support a lawn. The nutrients in synthetic fertilizers are in an inorganic form that is readily available for uptake by grass plants. However, because these nutrients are not easily stored in the soil, they leach into ground or surface water. In contrast, nutrients in the organic matter that was previously on the surface of forest soil became available slowly through decomposition and nutrient mineralization (the transformation of nutrients from organic to inorganic form). Very little was lost from the ecosystem through leaching.

The problem of nutrients leaching from denuded forest soil is compounded when homeowners remove mowed grass clippings for esthetic reasons. Removal of clippings may result in a loss of up to 100 pounds of nitrogen per acre of lawn per year. Without the natural recycling of nutrients via decomposition of grass clippings, additional fertilizer needs to be added to maintain the lawn. Furthermore, to ensure maximum productivity, lawn owners often add more fertilizer than plants are capable of assimilating. This practice often leads to beautiful lawns, but it can create serious environmental problems.

Excess nutrients have a variety of negative effects on a lawn. Too much nitrogen can increase a grass plant's vulnerability to disease, reduce its ability to withstand extreme temperatures and drought, and discourage the activity of beneficial soil microorganisms. In addition, some synthetic fertilizers acidify the soil, reducing the uptake of magnesium, calcium, and potassium (elements important to biological and chemical processes) in plant and soil organisms.

The environmental impact of overfertilization is not limited to the lawns; it also affects ecosystems that communicate with lawns through the exchange of energy and matter. Nitrous oxide, a product of the breakdown of ammonia (a form of nitrogen commonly used as lawn fertilizer), is implicated in contributing to global warming. When leached from the soil into waterways, the nitrogen and phosphorus in fertilizers can lead to the eutrophication of aquatic ecosystems. This process begins with excessive growth of water plants and ends in a smelly body of water

FIGURE VIII-A   A well-manicured lawn and the African savanna, home of our early ancestors.

deprived of oxygen and the loss of many life forms (see the section on Eutrophic Systems, p. 638). Chemical contamination of wells is a notable problem for humans, with nitrate from lawn fertilizers being the most common pollutant. Recent EPA surveys indicate that 1.2 percent of community water systems and 2.4 percent of rural domestic wells nationwide contain concentrations of nitrate that exceed public health standards. High concentrations of nitrate in drinking water may cause birth defects, cancer, nervous system impairments, and "blue baby syndrome," a condition in which the oxygen content in the infant's blood falls to dangerously low levels.

Further environmental damage occurs from lawn owners' adversarial relationships with nature (Figure VIII-C). In a lawn owner's eyes, three groups of organisms threaten the lawn: animals (such as moles and insects), weeds, and fungi. They may bring disease or simply change the appearance of the lawn, disrupting the smooth, even carpet of green. Since the 1950s many homeowners have waged war on these enemies with pesticides, specifically with rodenticides, insecticides, herbicides, and fungicides. In contrast, naturally occurring ecosystems protect themselves against disease and insect outbreaks in many different ways. Some plants, such as milkweed, produce chemicals in their leaves that

FIGURE VIII-B   A sample of the array of chemicals used to attack lawn pests.

make them unpalatable. Many insect populations are held in check by predators or diseases that are absent—due to pesticide use—in lawn ecosystems. Although pesticides may hold down populations of unwanted lawn pests for a time, they also result in the death of beneficial organisms—such as birds, earthworms, some insects, bacteria, and fungi—important to the health of the lawn. In addition, pesticides may persist in the environment for a long time with their lethal capabilities intact, even as they travel by wind, surface runoff, or seepage through the soil to wells and reservoirs used for public water supplies, wetlands, streams, rivers, and lakes, and even to marine environments.

Pesticide contamination of groundwater is less well documented than fertilizer pollution but is of growing concern. Detectable levels of pesticides or their chemical breakdown products have been found in 10 percent of the wells in community water systems. Of the various types of pesticides, the most is known about insecticides. Many insecticides work by block-ing communication between cells of the nervous system, but the newest ones, such as Bt (for *Bacillus thuringiensis*), prevent insects from maturing. The latter are considered to be nontoxic to vertebrates because they are naturally occurring insecticides that function by affecting insect's growth hormones. However, we do not know their long-term effects when used liberally, and ought to keep in mind that many invertebrate hormones are similar or identical in molecular structure to vertebrate hormones. Furthermore, history has shown that when we use nature's products for our own purposes, rarely do we understand the full implications of our interference with the environment.

Left unattended, our lawns would quickly be invaded by a variety of native plant species, eventually giving way to the ecosystem that formerly occupied the site. To stave off the inexorable processes of succession, herbicides are applied to lawns. Only recently has it become apparent that herbicides, like fertilizers and insecticides, are also cycling throughout the environment. Because the data are both scarce and new, the public health implications of these findings are unclear. Nevertheless, herbicides are among the suspects for the suddenly high incidence of deformities observed in amphibians.

Only 5 percent of total yearly herbicide use in the United States is attributable to homeowners, but this 5 percent amounts to an astounding 33 million pounds. Herbicides come in a multitude of formulations. Some are selective, killing either broad-leaved plants or grasses, but not both; others are nonselective, killing all vegetation. The most toxic herbicides cause death if ingested by damaging cellular components; others prevent the germination of seeds. Herbicides work by a variety of means including inhibiting cell division, electron transport systems, or the synthesis of lipids, chlorophyll, or key enzymes. Because traces of herbicide residues are being detected in unexpected and

unwanted places, including our drinking water, we might be wise to exercise caution in their application until enough time has passed for us to understand their cycling in the environment and their potential toxicity.

Another problem with lawns is that they need watering. The United States is a nation where population increases have combined with increased per capita consumption of water to generate a water crisis. Since 1950, the rate of public water use (which excludes agricultural and self-supplied industrial use) grew at more than twice the rate of our population increase. Water tables are falling, and stream flow is decreasing in many river basins. For instance, water tables in the Dallas–Fort Worth area have fallen more than 400 feet in the last 30 years. Some states in the West are now battling over water rights among themselves and with Mexico, and water shortages are now also frequent on the typically much wetter East Coast. With lawn watering accounting for up to 60 percent of urban water use in the West and 30 percent in the East, it's not surprising that an increasing number of communities are restricting the use of water for lawns.

Grass grows from below ground in what is probably an evolutionary adaptation to grazing animals. We respond by using mechanized grazers—lawnmowers—to keep the grass at a uniformly short height. This, among other reasons, is why lawns are solidly linked to environmental issues that surround fossil fuel consumption, including smog, acid rain, oil spills, destruction of the ozone layer, and global warming. The most obvious use of fossil fuels for lawn management is in running the fleet of mechanized equipment, including mowers, aerators, leaf blowers, weed whackers, and edgers. A staggering 580 million gallons of fuel is consumed annually just by gasoline-powered mowers alone.*

Fossil fuels are also employed to manufacture and transport inorganic fertilizers and pesticides. The principal nutrients in lawn fertilizers are nitrogen, phosphorus, and potassium. Natural gas is a reagent in the production of ammonia, the most common source of nitrogen in inorganic fertilizers. Furthermore, industrial nitrogen fixation requires a great deal of heat and pressure, which are supplied by energy from fossil fuels. Fossil fuels are also used to mine and refine potassium and phosphorus. The bag of fertilizer you might buy in Virginia may have originated, in part, in Peru, Utah, and Saudi Arabia. Still more fossil fuel is required to power the ships, trains, trucks, and cars to move these products from their source to the lawn. The same is true for pesticides. Thus, calculation of the environmental costs of these products involves not only their direct chemical effects but also their hidden costs, such as consumption of fossil fuels.

When we think of managed ecosystems, we typically envision agricultural fields or forest plantations, yet the American lawn is the most expensive and management intensive of all ecosystems. The endless nature of our lawn maintenance cycle is clearly evident: water and nutrients promote growth; herbicides lessen competition from weeds; rapid growth requires frequent mowing; pesticides inhibit not only grass predators but also organisms that decompose clippings; to achieve the desired appearance, clippings are removed; the removal of clippings requires fertilizers to replace nutrients, which promotes growth, etc. Changes are under way, with landscapers attempting to incorporate more native species of grasses and other herbaceous vegetation adapted to the local environmental conditions and requiring less maintenance. It is now becoming fashionable to create meadows and mixed-plant gardens that will attract wildlife in place of unbroken expanses of lawn (Figure VIII-C). Research is ongoing to find nontoxic pesticides that do not easily cycle through the environment. Growing concerns about environmental pollution, energy conservation, and scarcity of water are leading to restrictions on water use, engine emissions, and pesticide applications. Although the weekend ritual of firing up the lawn mower to once again do battle with the forces of nature is likely to continue for some generations to come, the pressure is on to reduce, if not altogether abandon, the crusade for the perfect lawn.

* Interestingly, Public Service of New Hampshire, a utility company, is currently experimenting with renting flocks of sheep to do the mowing around their power lines. Some highway maintenance departments have ceased mowing median strips, to the delight of many motorists, who find the resulting wildflower meadows scenic.

FIGURE VIII-C   An alternative to the lawn.

PART NINE

Comparative Ecosystem Ecology

CHAPTER 28

Grassland to Tundra

CONCEPTS

1. Native grasslands occupy regions subject to periodic drought, with rainfall between 25 and 75 cm/yr, and characterized by an accumulation of soil organic matter.

2. Primary productivity of grasslands is directly related to precipitation and is heavily influenced by grazing.

3. Savannas, characterized by a continuous well-developed layer of grass and an open discontinuous layer of shrubs or trees, occur in regions with wet and dry seasons and rainfall of 25 to 200 cm/yr.

4. Shrublands, diverse in nature, range from mediterranean types in arid and semiarid regions to heathlands of cool-to-temperate climates and successional shrublands.

5. Deserts occur where annual rainfall ranges between 7 and 40 cm, potential evaporation exceeds rainfall, and productivity is low.

6. Tundras are cold-dominated ecosystems of high latitudes and high altitudes. Low temperatures, short growing seasons, and low availability of nutrients dictate the nature of tundra life.

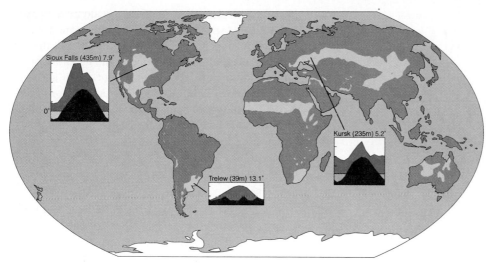

FIGURE 28.1 Location and climate of grasslands of the world. (Climate diagrams from Walter et al 1975.)

GRASSLANDS

When the European explorers looked out across the prairies for the first time, they witnessed a scene they had never before experienced. Nowhere in all western Europe had they seen anything similar. Lacking any other name, the explorers named these grasslands "prairie," from the French for "meadows."

This land was the North American prairie and plains, the climax grassland that occupied the mid-continent. It was one of several great grassland regions in the world, including the steppes of Russia, the pusztas of Hungary, the South African veldt, and the South American pampas (see inside cover). In fact, at one time grasslands covered about 42 percent of the land surface of the world, but today much of that area is under cultivation.

All grasslands have in common a climate characterized by high rates of evaporation, periodic severe droughts, a rolling-to-flat terrain, and animal life that is dominated by grazing and burrowing species. They occur largely where rainfall is between 25 and 75 cm/yr, too light to support a heavy forest growth and too heavy to encourage a desert (Figure 28.1). Grasslands, however, are not exclusively a climatic formation. Most of them require periodic fires for maintenance, renewal, and elimination of incoming woody growth.

Grassland Vegetation

Grasslands are dominated by grasses, members of the family Poaceae (Gramineae). Grasses are distinctive because their stems, called culms, produce narrow leaves that grow from their bases. This growth form allows grasses to be grazed or mowed with minimal mortality.

Grasses are either sod-formers or bunchgrasses. As the names imply, the former develop a solid mat of grass over the ground and the latter grow in bunches (Figure 28.2). The space between bunchgrasses is often occupied by other plants, usually herbs. Orchard grass (*Dactylis glomerata*), broomsedge (*Andropogon virginicus*), crested wheatgrass (*Agropyron desertorum*), and little bluestem (*Andropogon scoparius*) are typical bunchgrasses. They form clumps by the erect growth of all the culms and spread at the base by shoots that develop from underground stems (tillering). Sod-forming grasses, which include such species as introduced Kentucky bluegrass (*Poa pratensis*) and western wheatgrass (*Agropyron smithii*), spread by underground stems. Some grasses may be either sod or bunch depending on the local environment. Big bluestem (*Andropogon gerardii*) will develop a sod on a rich, moist soil and form bunches on a dry soil. Grasses are also categorized as cool season or warm season grasses. Cool season grasses, like Kentucky bluegrass, put on growth in the spring and become senescent in summer. Warm season grasses, like switch grass (*Panicum virgatum*), put on most growth and flower in the summer.

Associated with grasses are a variety of forbs, especially legumes. Cultivated haylands and pastures usually are planted to a mixture of grasses and such legumes as alfalfa (*Medicago sativa*) and red clover (*Trifolium pratense*). With them may grow unwanted plants such as mustard (*Brassica* spp.), dandelion (*Taraxacum officinale*), and daisy (*Chrysanthemum leucanthemum*). Successional grasslands often consist of native grasses and introduced species such as timothy (*Phleum pratense*) and bluegrass, and an assortment of herbaceous plants, including cinquefoil (*Potentilla* spp.), dewberry (*Rubus* spp.), and goldenrod (*Solidago* spp.). On the prairie legumes and asters (*Aster* spp.) are important components of native grasslands (Weaver 1954).

Grassland Types

Grassland is the largest of the four major vegetational formations, accounting for about 24 percent of Earth's vegetation. Much of the world's natural grasslands have been converted to cropland, supporting such cultivated grains (grasses) as corn, wheat, oats, and barley, or to cultivated grasslands—hay and pasture. In North America grasslands dominate much of the mid-continent with a gradation from mesic grasslands beginning at the eastern forest grassland ecotone to the semi-arid grasslands of the west (Figure 28.3). For convenience,

FIGURE 28.2 Growth forms and root penetration (maximum depth of about 2.5 m) of a bunchgrass (left) and a sod grass (right).

ecologists have grouped the North American grasslands associations into three major types: tallgrass, mixed-grass, and shortgrass prairies. Considerable diversity exists within each group, on both an east-west and a north-south gradient.

Cultivated and Successional Grasslands   Grasslands in normally forested regions are either cultivated (Figure 28.4a) or successional (Figure 28.4b). In highly developed agricultural areas, such as eastern and central North America, Great Britain, and Europe, cultivated grasslands are the major representatives of their class, although a few natural types do exist. By clearing the forests, humans developed grasslands principally as a source of food for livestock and secondarily for landscaping. In some agricultural regions, notably New England and the Great Lakes states in North America, grasslands abandoned by agriculture have reverted to forest. In other regions, especially Britain, some grasslands have persisted for centuries, becoming a sort of climax community supporting its own distinctive vegetation (see Duffey 1974).

Cultivated grasslands can be classified as permanent, in grass for over seven years and managed for hay, pasture, lawns, and golf courses; temporary or rotational, plowed every three to five years for crop production; and rough, marginal, unimproved, semiwild lands used principally for grazing.

Ecologically, permanent hayfields and grazing lands differ from rotational hayfields. Permanent haylands, more common in Britain than in North America, and permanent grazing lands consist of species that are adapted to periodic defoliation by cutting and grazing and produce their maximum growth in spring.

Rotational or temporary hayfields are dominated by two or three cultivated species, usually two grasses and a legume. Such hayfields support denser and heavier growth than permanent and successional grasslands. Management of such grasslands involves fertilization, mowing for hay, and at regular intervals, plowing to grow other crops, such as corn or small grains in the crop rotation. Such hayfields can provide an excellent habitat for grassland wildlife, but early mowing destroys nesting cover at the beginning or height of the nesting season (Figure 28.5a).

Tallgrass Prairie   The tallgrass prairie extends from the forest-grassland ecotone in Wisconsin, Indiana, and Illinois where fire controlled the encroachment of oak forests into the grassland to Minnesota, eastern South Dakota, Nebraska, Kansas, and Oklahoma. Together with the mixed-grass prairie, it occupies the Prairie Division of the Temperate Domain. Except for scattered remnants, such as the Sand Hills of Nebraska and the Osage and Flint Hills of Oklahoma and Kansas, the tallgrass prairie has been converted to cornfields and other cropland. Big bluestem is the dominant grass of the tallgrass prairie, particularly in moist lowlands. Associates include switch grass and Indian grass (*Sorghastrum nutans*) and a diversity of forbs, especially legumes and composites (Compositae). Drier uplands in the tallgrass prairie are dominated by bunch-forming needlegrass (*Stipa* spp.), side oats grama (*Bouteloua curtipendula*), and the prairie dropseed (*Sporobolus heterolepis*) (Kuchera 1992). Much of this grassland is interspersed with trees and shrubs, especially along streams and lower slopes of rolling hills.

Mixed-Grass Prairie   West of the tallgrass prairie region extending from the southern parts of the central Canadian Provinces through the Dakotas, western Kansas, and Texas is the mixed-grass prairie (Figure 28.5b) (Sims 1988). It is an ecotone between the tallgrass prairie and the shortgrass prairie or plains. Mid-height grasses occupy lowlands and shortgrass the higher elevations. The mixed prairie embraces largely the needlegrass-grama grass associations with needlegrass-wheatgrass associations dominating lower topographic positions on medium-textured soils (Coupland 1950, 1992). Because the mixed prairie experiences great variability in precipitation, its aspect varies widely from year to year. In moist years midheight grasses are prevalent, whereas in dry years shortgrasses and forbs are dominant. The grasses are largely cool season bunch species. Like the tallgrass prairie, most of the mixed prairie has been converted to grassland and cropland.

Shortgrass Steppe   South and west of the mixed prairie and grading into the desert is the shortgrass plains or steppe, a country too dry for most midgrasses. It extends from the Nebraska panhandle, southeastern Wyoming, and eastern Colorado southward to New Mexico and Texas (Figure 28.6). It makes up the Temperate Steppe Division of the Dry Domain. The shortgrass plain reflects a climate where the

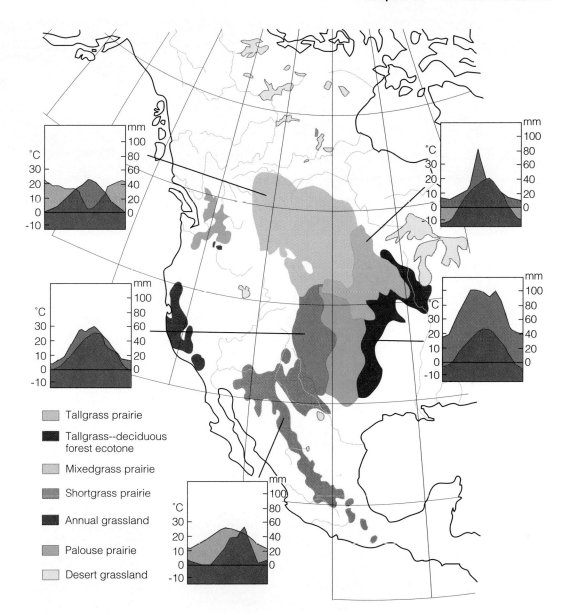

FIGURE 28.3 Distribution of the major grasslands of North America. (Adapted from Sims 1988.)

Tallgrass prairie

Tallgrass--deciduous forest ecotone

Mixedgrass prairie

Shortgrass prairie

Annual grassland

Palouse prairie

Desert grassland

(a)

(b)

FIGURE 28.4 (a) Pastureland dominated by perennial ryegrass (*Lolium perenne*) in England.
(b) Successional grassland in New York state with invading herbaceous plants and scattered shrubs.

(a)

(b)

FIGURE 28.5 (a) A remnant tallgrass prairie in Iowa. (b) The mixed-grass prairie has been called "daisy land" because of the diversity of its wildflowers.

FIGURE 28.6 Shortgrass steppe in western Wyoming.

rainfall is slight and infrequent (25 to 43 cm in the west, 51 cm in the east), the humidity low, the winds high, and evaporation exceeds precipitation. Shallow-rooted, the shortgrasses use moisture in the upper soil layers, beneath which is a permanent dry zone into which the roots do not penetrate. Sod-forming blue grama (*Bouteloua gracilis*) and buffalo grass (*Buchloe dactyloides*) dominate the shortgrass plains. On wet bottomlands, switch grass, Canada wild rye (*Elymus canadensis*), and western wheatgrass (*Agropyron smithii*) replace blue grama and buffalo grass. Because of the dense sod, fewer forbs grow on the plains (Lavenroth and Milehunas 1992).

Just as the tallgrass prairie was destroyed by the plow, so has much of the shortgrass plains been ruined by overgrazing and by plowing for wheat. Because of the low available moisture, the region could not support wheat farming. Drought, lack of a tight sod cover, and winds turned much of the short-

grass plains into the Dust Bowl of the 1930s, recovery from which has taken decades.

California Prairie California prairie or, more appropriately, California annual grassland is confined largely to the Central Valley of California. Part of the Mediterranean Division of the Humid Temperate Domain, it is associated with a mediterranean-type climate characterized by winter precipitation and hot dry summers. The original vegetation was perennial bunchgrasses, dominated by purple needlegrass (*Stipa pulchra*), and an array of other perennial and annual grasses and forbs. Overgrazing by livestock dating back to 1769, early introduction of exotic annual species well adapted to the mediterranean-type climate, and recurrent drought rapidly changed the nature of the grassland. The California prairie is now dominated by annual grasses, notably brome (*Bromus* spp.) and in places wild oats (*Avena fatua*) (Heady et al. 1992). Growth occurs during early spring and most plants are dormant in summer, turning the hills to a dry tan color accented by the deep foliage of scattered California oaks (*Quercus* spp.).

Palouse Prairie The intermountain region between the Rocky Mountains and the Cascade Range, extending from Southwest Canada, eastern Washington, and Oregon, supports another type of grassland, the Palouse prairie. It occupies the Temperate Steppe Division of the Dry Domain, characterized by winter wetness and summer drought. The Palouse prairie, named after the Palouse Indians, grows on Pleistocene loess, enriched in places with volcanic ash. Because the region is dominated by cool season short bunchgrasses and sagebrush (*Artemisia* spp.), it is often called shrub steppe. Originally, the dominant grasses were bluebunch wheatgrass (*Agropyron spicatum*) and bluebunch fescue (*Festuca idahoensis*). Highly sensitive to grazing, these grasses were largely replaced by cheatgrass (*Bromus tectorum*), Kentucky bluegrass (*Poa pratensis*), and such forbs as sunflower (*Helianthus* spp.) and prickly lettuce (*Lactuca* spp.). Thus, like other grasslands, the Palouse prairie has been greatly altered by cultivation, grazing, and plant introductions.

Desert Grassland Desert, or arid or desert grassland, extends from southeastern Texas to southern Arizona and far into Mexico. In many respects it is similar to the shortgrass steppe except that grama grasses (*Bouteloua* spp.) replace buffalo grass. Dominated by bunchgrasses, arid grasslands are widely interspersed with other vegetation, notably mesquite (*Prosopis* spp.). The climate is hot and dry. Rain falls only during the summer (July and August) and winter (December to February), in amounts that vary from 300 to 410 mm in the western parts to 510 mm in the east. Evaporation is high, up to 203 mm a year. Some annual grasses germinate and grow only during the summer rainy season, whereas annual forbs grow mostly in the cool winter and spring months. Desert

grasslands have been highly disturbed, beginning as far back as the eighteenth century when cattle ranching accompanied the Spanish missions and settlements (Schmutz et al. 1992). Overgrazing by domestic livestock, drought, and suppression of grassland fires have resulted in severe range deterioration. The result has been invasion by such aggressive species as mesquite and larrea (*Larrea tridentata*), and exotic grasses as African lovegrass (*Eragrostis* spp.) and buffelgrass (*Pennisetum ciliare*) (Bock and Bock 1998).

Other Grasslands At one time the great grasslands of the Eurasian continent extended from eastern Europe to western Siberia and south to Kazakhstan. The Eurasian steppes, treeless except for ribbons and patches of forest, are divided into four latitudinal belts from mesic meadow steppes in the north to semiarid grasslands in the south. The meadow steppes, extending south from the boreal forest, occupy a region of chernozem soil where the rainfall is 500 to 600 mm. Once outstandingly beautiful in spring and summer, little remains of the meadow steppes, turned under the plow for cereal grains. Further south where rainfall is 40 to 50 cm, grasslands are dominated by bunchgrasses, mostly *Stipa* species. In the central Asian steppes with their cold, dry spring, grasses are replaced by woody and herbaceous species of *Artemisia*. About the Black Sea and in Kazakhstan, where the humidity is higher, steppe vegetation is dominated by feather grasses (*Stipa* spp.) and fescues (*Festuca* spp.).

In the Southern Hemisphere, the major grasslands are in southern Africa and in South America. All belong to the Temperate Steppe Division of the Dry Domain. The Orinoco River basin in Venezuela is a vast grassy plain called the **llano.** Dominated by a variety of tropical grasses, it supports an extensive cattle industry (Figure 28.7). Further south the **pampas** extend westward in a large semicircle from Buenos Aires to cover about 15 percent of Argentina. In the eastern part of the pampas, dominated by tallgrasses, rainfall exceeds 90 cm well distributed

(a)

(b)

FIGURE 28.7 (a) Humid tropical grassland in Brazil with gallery forest in the background. (b) Dry tropical grassland in Ethiopia.

throughout the year. South and west, where rainfall is about 45 cm, semidesert vegetation becomes prominent. South into Patagonia, where precipitation averages about 25 cm, the pampas change to open steppe grasses dominated by *Stipa* and *Festuca*. These pampas have been modified by the introduction of European forage grasses and alfalfa. The eastern tallgrass pampas have been converted to wheat and corn.

The grasslands of southern Africa (not to be confused with savannas) occupy the eastern part of a plateau 1500 to 2000 m high in the Transvaal and the Orange Free State. Most of the rainfall comes in the summer, brought in by moist air masses from the Indian Ocean. The heaviest rainfall is in the east; the lowest is in the west, where grasslands grade into semiarid shrubland known as the Karoo. Once inhabited by great herds of antelopes, these grasslands are now agricultural lands.

Australia has four types of steppe. Arid bunchgrass occupies the northern part of the continent, where the rainfall averages between 20 and 50 cm, mostly in the summer. Areas with less than 20 cm rainfall support arid hummock grasslands. Coastal grasslands, dominated by *Sporobolus,* occupy the tropical summer rainfall region. Coastal areas where the precipitation is between 50 and 100 cm support subhumid grasslands. Most of these grasslands have been changed by fertilization, introduced grasses and legumes, and sheep grazing.

Structure

Vegetation   Grasslands possess three layers—roots, ground layer, and herbaceous layer (Figure 28.8). The root layer is more pronounced in grasslands than in any other major ecosystem. Half or more of the plants' biomass is hidden in the soil. Most of the root biomass rather uniformly occupies the upper 16 cm of the soil profile and decreases with depth. The depth to which roots of grasses extend is considerable. Little bluestem reaches 1.3 to 1.7 m and forms a dense mat to 0.8 m (Weaver 1954). Roots of blue grama and buffalo grass penetrate vertically to 1 m. In addition, many grasses possess underground stems, or rhizomes, that serve both to propagate the plants and to store food. On the end of the rhizome, which has both nodes and scalelike leaves, is a terminal bud that develops into aerial stems or new rhizomes. Rhizomes of most species grow at shallow depths, not over 10 to 14 cm deep. Forbs such as goldenrods and asters possess large woody rhizomes and fibrous roots that add to the root mat in the soil. Some, such as snakeroot (*Eupatorium*), have extensive taproots 5 m long. Among hayland plants, alfalfa possesses a taproot that grows to considerable depth.

The roots of all grassland plants are not confined to the same general area of the soil but develop in three or more zones. Shallow-rooted species seldom extend much below 60 cm. Others grow well below the shallow-rooted species, but seldom deeper than 1.5 m. Deep-rooted plants extend even further into the soil and absorb relatively little moisture from the surface soils (Weaver 1954, Wieland and Bazzaz 1975).

The ground layer is dominated by mosses and plants with a rosette-type growth form, such as dandelions. It is characterized by low light intensity during the growing season, temperatures cooler than ambient on the ground, and reduced wind flow. Even though the grass tops may move like waves of water, air close to the ground is calm. Conditions on grazed lands are different. Because the grass cover is closely cropped, the ground layer is subject to higher temperatures and greater wind velocity.

FIGURE 28.8   (a) Profile of a grassland showing physical stratification during the summer. (b) Leaf area indices, as indicated by bar graphs, at different levels in a grassland for both green and dead plant structures.

(a)

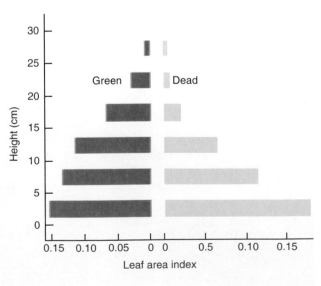

(b)

TABLE 28.1 Mean Standing Crop of Selected Grasslands, Growing Season (g/m², oven dry)

Standing Crop	Desert Ungrazed	Desert Grazed	Shortgrass Plains Ungrazed	Shortgrass Plains Grazed	Mixed Prairie Ungrazed	Mixed Prairie Grazed	True Prairie Ungrazed	True Prairie Grazed
Above ground	81.2	49.5	69.0	42.9	154.2	94.6	256.1	220.8
Cool season grass	0	0	6.6	2.6	101.6	9.5	4.6	24.2
Warm season grass	51.3	9.4	42.6	27.1	45.8	82.1	227.1	152.6
Old dead	0.9	1.1	25.9	18.3	80.2	58.5	204.0	44.0
Mulch	68.7	52.3	97.5	89.3	448.3	239.2	111.2	206.5
Below ground	197.0	185.0	1600.0	1800.0	1213.0	2086.0	893.0	781.0

Source: Adapted from Lewis 1971.

The seasonally variable herbaceous layer consists of three or more strata, more or less variable in height, according to grassland type. Ground layer plants, such as wild strawberry, dandelion, cinquefoil, and violets, grow upward to make the first stratum. As the growing season progresses, these plants become hidden beneath the middle and upper layers. The middle layer consists of shorter grasses and such herbs as wild mustard, coneflower (*Ratibida columnaris*), and asters. The upper layer consists of the flowering heads of tall grasses and forbs, such as goldenrod, most conspicuous in late summer.

Mulch Grasslands accumulate a layer of mulch on the ground. The oldest layer consists of decayed and humic remains of fresh mulch. Leafy fresh mulch consists of dead, often standing, residual herbage (thatch to lawn owners). Three or four years must pass before natural grassland herbage will decompose completely. The amount of accumulated mulch often is enormous. On a tallgrass prairie it may be two to three times the amount of annual production (Knapp and Seastedt 1986).

Grazing reduces mulch, as do fire and mowing (see Table 28.1). Light grazing tends to increase the weight of humic mulch at the expense of fresh (Dix 1960); moderate grazing results in increased compaction of mulch by grazing animals, which favors increase in microbial activity and a subsequent reduction in both fresh and humic mulch. Burning reduces both fresh and humic mulch, but the mulch structure returns two or three years after a fire on lightly grazed and ungrazed lands (Tester and Marshall 1961, Hadley and Kieckhefer 1963). Mowing greatly reduces fresh mulch and, in time, humic mulch as well. An unmowed prairie accumulated 10 metric tons of humic per hectare; a similar prairie, mowed, had less than 4 metric tons (Dhysterhuis and Schmutz 1947).

An accumulation of mulch has both positive and negative effects. It conserves soil moisture by reducing evaporation, but it also retards infiltration of precipitation. It alters the microclimate and physiology of emerging shoots, reducing CO_2 uptake. By insulating the soil surface from solar radiation, mulch lowers root zone temperature, thus reducing root productivity and inhibiting the activity of soil microbes and invertebrates. It provides nesting cover for grassland birds and offers cover for small mammals.

Animal Life All types of grasslands support similar forms of animal life. Many forms of this animal life exist within the strata of vegetation: roots, ground layer, and herb cover. Invertebrates, particularly insects, occupy all strata at some time during the year. In winter, insect life is confined largely to the soil, litter, and grass crowns as pupae and eggs. In spring, soil occupants are mostly earthworms and ants. Ground and litter layers harbor scavenger carabid beetles and predaceous spiders. Because this layer provides only limited supports for webs, there are more hunters than web builders among spiders.

Life in the herbaceous layer varies as the strata become more pronounced from summer to fall. Here invertebrate life is most diverse and abundant. Homoptera, Coleoptera, Orthoptera, Diptera, Hymenoptera, and Hemiptera are all represented. Insect life reaches two highs during the year, a major peak in summer and a less defined one in fall.

Mammals are the most conspicuous vertebrates of the grasslands, and the majority of them are herbivores. A large and rich ungulate fauna evolved on the grasslands. The bison (*Bison bison*) and pronghorn antelope (*Antilocarpa americana*) of North America were surpassed only by the richer and more diverse ungulate fauna of the East African plains. In Australia, marsupial kangaroos (Figure 28.9) are the grazing equivalent of the placental ungulates. South America lacks an ungulate fauna. Their closest equivalents are the large grazing rodents that live in herds, the capybara (*Hydrochaeris hydrochaeris*). Today herds of cattle and flocks of sheep have replaced most of the native grazing herbivores.

Grassland animals share some outstanding traits. Hopping or leaping is a common method of locomotion. Strong hind legs enable a variety of animals, such as grasshoppers, jumping mice, jackrabbits, and gazelles, to rise above the grass where visibility is unimpeded. Speed, too, is well developed. Some of the world's fastest mammals, such as antelopes and the cheetah (*Acinonyx jubatus*), live in grasslands. Many of the rodents are adapted to digging and burrowing. Because of dense grass and lack of trees for song perches, some grassland birds have conspicuous flight songs that advertise territory and attract mates.

Animal life in cultivated and successional grasslands depends on human management for the maintenance of habitat. Mowing for hay, a major management tool, destroys habi-

FIGURE 28.9 The eastern gray kangaroo (*Macropus giganteus*) is one of the dominant grazing marsupials of Australian grasslands.

tat at a critical time of year. Nests of rabbits, mice, and birds are exposed at the height of the nesting season. Losses from both mechanical injury and predation are often heavy, although most species remain to complete or reattempt nesting activity. Pasturelands more often than not are so badly overgrazed that they support little vertebrate life.

Successional grasslands, usually lacking a heavy cover of grass, do not usually support as wide a variety of life as hayfields. Such grasslands are inhabited by grasshopper sparrows (*Ammodramus savannarum*), vesper sparrows (*Pooecetes gramineus*), and meadow mice (*Microtus* spp.); but deep-grass species such as meadowlarks (*Sturnella* spp.) and bobolinks (*Dolichonyx oryzivorus*) are rare, if not absent.

Function

Grassland types, species composition, productivity, and other attributes are directly related to rainfall and temperature, with an inverse relationship between the two. In forested ecosystems (see Chapter 29), a positive relationship exists between temperature and net primary productivity. This relationship is evident in increasing productivity along a latitudinal gradient in eastern North America from boreal forest to the subtropical forest (Figure 28.10a). In the North American grassland, the relationship reverses (Figure 28.10b). The reason resides in the interaction between temperature and precipitation in the grassland ecosystem. Grasslands occupy regions of relatively low precipitation compared with forest ecosystems. As with forest ecosystems, net primary production increases with increasing mean annual precipitation (Figure 28.11). Whereas in forested ecosystems high temperatures stimulate high productivity, high temperatures have a negative on productivity of grasslands (Epstein et al. 1966, 1997). High temperatures increase evaporative losses, increasing the demand for water by plants. Highest grassland productivity takes place in cooler environments dominated by C_3 grasses. There, potential evaporation is lower relative to the water supply and net N mineralization is higher than in warm drier environments dominated by C_4 grasses (Epstein et al. 1998). The relationship reflects in part the reduction of water stress and the enhancement of nutrient uptake, and in part differences in water use efficiency (see Chapter 7).

In semiarid regions where water is limiting, plants have evolved adaptations to make the maximum use of water. In humid, tallgrass country, the vegetative canopy is dense and intercepts most of the solar radiation, and so adaptations of plants are for efficient capture of light at the expense of efficient use of water.

Production is mirrored in biomass accumulation (Table 28.1). Above-ground green biomass among an array of grass-

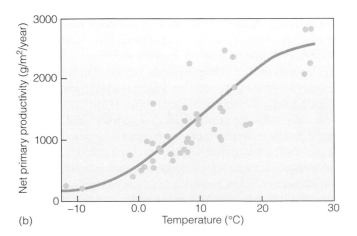

FIGURE 28.10

(a) Above-ground annual net primary production (ANPP) along a gradient of mean annual temperatures in the Great Plains of the United States. (Epstein et al. 1996.) (b) The relationship between net primary production (NPP) versus temperature for a variety of forest ecosystems. (Reichle 1970.) Note that net primary production in forests increases with increasing temperature, whereas net primary production in grasslands declines with increasing temperature. See text for details.

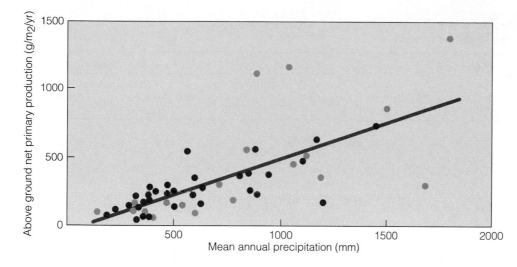

FIGURE 28.11 Relationship between above-ground primary production and mean annual precipitation for 52 grassland sites around the world. North American grasslands are indicated by dark green dots. (From Lauenroth 1979:10.)

lands studied ranges from 50 g/m² in arid grasslands to 827 g/m² in subhumid grasslands. Mean underground biomass ranges from 45 to 4707 g/m². In most tropical grasslands, that biomass is usually less than 1000 g/m² (median 200 g/m²). Except for a short seasonal period of maximum above-ground biomass, below-ground biomass is two to three times that above ground.

The relationship of below-ground to above-ground biomass is best expressed in terms of root-to-shoot ratios. Among grasslands the ratios range from 0.2 to 10.3. The ratio is smallest in tropical grasslands, a mean of 0.8. The ratio for most temperate grasslands is between 4 and 5; for the tallgrass and mixed-grass prairie, the root-to-shoot ratio may be 6 or 7.

Primary productivity ranges from 82 g/m²/yr in semiarid regions to 3396 g/m²/yr in subhumid tropical grasslands. In natural and seminatural temperate grasslands, production ranges from 98 to 2430 g/m²/yr. Much of the net production of grassland does not appear as above-ground biomass. In ungrazed grasslands, 75 to 85 percent of the photosynthate is translocated to the roots for storage below ground. In grazed grasslands, reduced allocation of carbon below ground and increased allocation to above-ground production results in decreased root biomass.

Grasslands have the highest index of productivity per unit of standing crop (20 to 55 percent) among all terrestrial ecosystems. The tundra during its short growing season has an index of 10 to 20 percent. The index for the boreal forest is only 2 to 5 percent and for the temperate deciduous forest, 8 to 10 percent. Of course, a great deal of energy in woody ecosystems goes into the maintenance of high standing crops, which grasslands do not support. Efficiency of energy capture by grasslands is 0.75 percent for net primary production. Light to moderate grazing increases that efficiency to 0.87 percent.

Grasslands evolved under grazing pressures of ungulates since the Cenozoic. Grasses have adapted to grazing by compensatory growth for tissues damaged or removed by grazing (McNaughton 1979, 1984, Risser 1985, Frank et al. 1998). Such compensation involves the removal of phenologically older, less productive tissue and increased photosynthetic

rates in residual tissue; increased light to young tissue stimulating shoot growth at the ground surface; reduction in leaf senescence, thus prolonging the photosynthetic period of remaining tissue; conservation of soil moisture by reducing transpiration surface; and recycling nutrients through dung and urine (McNaughton 1979).

Numerous experimental studies of herbivore-grass interactions by plant ecologists and range managers have demonstrated that moderate grazing stimulates primary production of grasslands (Dyer et al. 1993). McNaughton (1985) points out that moderate grazing on the Serengeti grasslands stimulated production up to twice the levels in ungrazed control plots. Other grassland studies examined grazed and ungrazed systems from a functional point of view. In general, total net primary production was greater on grazed than on ungrazed sites. However, grazed grasslands channeled more carbon below ground than did ungrazed grasslands (Table 28.2). Grazed grasslands allocated about 63 percent of total net primary production to below-ground net production and about 37 percent to above-ground net production. Ungrazed systems sent about 52 percent of total net primary production below ground and about 48 percent above. In contrast, ungrazed treatments had an above-ground production efficiency of 0.26 percent and the grazed 0.33 percent (Sims and Singh 1971).

Most of these studies involved native ungulates that seasonally moved from patch to patch, thus avoiding serious overgrazing. Domestic grazers—cattle and sheep—however, are sedentary, held at higher densities than native herbivores, and restrained in their movements by fencing and stable water sources. The resulting concentrated grazing can negate any beneficial effects of grazing on plants (McNaughton 1993, Briske 1995).

Another response of grasslands to grazing is a change in species composition (Sims 1988). Some grasses and forbs are sensitive to intensive grazing pressure and tend to disappear, while other species increase (Table 28.3) (Briske 1996, Bullock 1996). On desert grasslands of North America, black grama (*Bouteloua eriopoda*) is replaced by desert shrubs; on

TABLE 28.2 Net Primary Production and Efficiency of Energy Capture by Ungrazed and Grazed Grasslands Within the Growing Season

	Above Ground		Below Ground		Total	
Type	Net Production (kcal/m²)	Efficiency (%)	Net Production (kcal/m²)	Efficiency (%)	Net Production (kcal/m²)	Efficiency (%)
Grazed						
Desert	456	.06	625	.09	1081	.13
Shortgrass plains	476	.10	3285	.70	3761	.80
Mixed prairie	444	.10	1810	.41	2254	.51
True prairie	2048	.42	1701	.35	3749	.77
Ungrazed						
Desert	688	.09	489	.07	1177	.16
Shortgrass plain	568	.12	2153	.45	2721	.57
Mixed prairie	788	.18	1264	.29	2052	.47
True prairie	1348	.27	872	.17	2220	.44

Source: Adapted from Sims and Singh 1971.

TABLE 28.3 Response of Grassland Species to Grazing Distrubance, Grassland Biome, 1970

		% Community Standing Crop	
Grassland	Species	Ungrazed	Grazed
Desert	Black grama (*Bouteloua eriopoda*)	53	8
	Broom snakeweed (*Xanthocephalum sarothrae*)	15	26
	Russian thistle (*Salsola kali*)	3	25
Shortgrass plains	Blue grama (*Bouteloua gracilis*)	63	72
	Prickly pear (*Opuntia polyacantha*)	5	12
Mixed prairie	Wheatgrass (*Agropyron smithii*)	43	9
	Blue grama (*Bouteloua gracilis*)	9	37
	Buffalo grass (*Buchloe dactyloides*)	28	44
True prairie	Little bluestem (*Andropogon scoparius*)	69	23
	Indian grass (*Sorghastrum nutans*)	7	t
	Japanese chess (*Bromus japonicus*)	0	22
	Tall dropseed (*Sporobolus asper*)	2	37

Source: Lewis 1971.

shortgrass plains, which are the most stable under grazing pressure, blue grama (Opuntia) and prickly pear increase; on mixed-grass prairies, midgrasses decrease and shortgrasses and sedges increase. On tallgrass sites, tallgrasses disappear and little bluestem and tall dropseed increase; if grazing pressure is heavy, the site may be invaded by the weedy Japanese brome or chess (*Bromus japonicus*) (Lewis 1971).

Although large herbivores are the most conspicuous grazers, invertebrates are the most important ones. The above-ground biomass of invertebrates—including plant consumers, saprovores, and predators—ranges from 1 to 50 g/m², while grazing mammals amount to about 2 to 5 g/m². Below-ground invertebrates exceed 135 g/m². The major above-ground con-

sumers are grasshoppers, and the major below-ground consumers are nematodes. Nematodes account for 90, 95, and 93 percent of all below-ground herbivory, carnivory, and saprophagous activity, respectively. They account for 46 to 67 percent of root and crown consumption, 23 to 85 percent of fungal consumption, and 43 to 88 percent of below-ground predation. Above-ground herbivores consume 2 to 7 percent of primary production, while below-ground herbivores consume 13 to 46 percent.

Energy flow through small mammal populations in grasslands may be high relative to their biomass but low relative to the standing crop biomass (Risser et al. 1981). Small mammals consume only about 4 percent of the herbage available. Energy flow through breeding bird populations may be even smaller. Birds probably exert little influence on energy flow and storage of energy and nutrients in grassland ecosystems (Wiens 1973).

Not only is a large proportion of primary production consumed below ground, but a greater proportion is used at each trophic level (Figure 28.12). Invertebrates convert about 9 to 25 percent of ingested energy to animal tissue, whereas vertebrate grazers, such as sheep and cattle, convert 3 to 15 percent to animal tissue. However, most of the primary production, above ground especially, goes to the decomposers, dominated by fungi, whose biomass is two to seven times that of bacteria. Overall, the decomposer biomass exceeds that of invertebrates.

Green plant consumers are important in the cycling of nutrients in grassland ecosystems. Invertebrate consumers are highly inefficient in assimilating ingested material. Much of their intake is deposited as feces or frass. Because the nutrients in the frass are in a highly soluble form, they are returned rapidly to the system. Large grazing herbivores return a por-

tion of their intake as dung, which becomes a major pathway of nutrient cycling (Figure 28.13). They harvest nutrients over a large area and concentrate them into a small area (McNaughton 1985). A well-developed coprophagous fauna speeds the decay of manure and accelerates the activity of bacteria in feces.

Nitrogen provides an example of nutrient cycling in grasslands (Figure 28.14). About 90 percent of nitrogen in grasslands is tied up in soil organic matter, 2 percent in litter, 5 percent in live and dead plant cover, 1 percent in dead shoots, about 0.8 percent in soil microflora, and a very small amount in above- and below-ground invertebrates. When nitrogen is limiting, most of it is shunted to green herbage, where it remains during the growing season. Some of this is consumed by herbivores. Part is translocated below ground to roots. Much of this nitrogen is moved above ground to new growth the following season. Some is retained in standing dead leaves and returned to the litter and soil surface, where it is acted on by fungi and bacteria. Part becomes immobilized in microbial biomass but is quickly mineralized upon the death of microbes and reenters the nitrogen cycle. Still another fraction is tied up in humus and resists decomposition. Turnover, however, is fairly rapid, and most of the nitrogen that enters a green plant one growing season will reenter another the following year.

Central to nutrient cycling in grasslands is the accumulation of mulch or detritus. A large standing crop of mulch can have a detrimental effect on nitrogen cycling, particularly in tallgrass ecosystems (Knapp and Seastedt 1986). Mulch intercepts rainfall, from which microbes can assimilate inorganic nitrogen directly, before it reaches plant roots. The mulch inhibits nitrogen fixation by cyanobacteria and free-living nitrogen-fixing microbes. Periodic grassland fires clear away the mulch layer and release other nutrients in detritus to the soil, but nitrogen, equal to about two years of nitrogen inputs to the system through rainfall, is lost to the atmosphere. Periodic fires, however, stimulate the growth of nitrogen-fixing leguminous forbs and improve conditions for earthworms.

Because of their great economic importance as grazing land, grasslands have been the subject of intensive research. This research investigates grassland ecosystem functioning, the evolutionary adaptations of grasses to grazing, and the application of ecological principles to range and grassland management. In spite of years of study, our knowledge of grassland ecology is still limited. We know little about the relationship between above-ground and below-ground consumers and grassland productivity. We lack good techniques to measure below-ground production. Still debated is the relationship of overgrazing to grassland productivity. There are other questions. What is the effect of grazing, especially in semiarid and arid grasslands, on fossorial animals and on key species of plants? Does their loss impair ecosystem functioning? What is the relationship between periodic droughts and overgrazing in converting arid grasslands to desert? These and other questions await answers in grasslands.

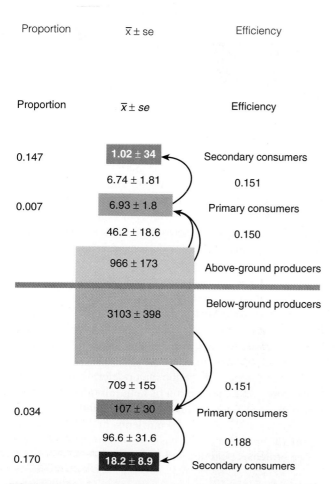

Proportion	$\bar{x} \pm se$	Efficiency
0.147	1.02 ± 34	Secondary consumers
	6.74 ± 1.81	0.151
0.007	6.93 ± 1.8	Primary consumers
	46.2 ± 18.6	0.150
	966 ± 173	Above-ground producers
	3103 ± 398	Below-ground producers
	709 ± 155	0.151
0.034	107 ± 30	Primary consumers
	96.6 ± 31.6	0.188
0.170	18.2 ± 8.9	Secondary consumers

FIGURE 28.12 Grassland production (in boxes) and consumption (between boxes), representing means and standard errors for desert, tallgrass, mixed, and shortgrass sites in North American grasslands. Production on one trophic level is represented as a proportion of production at the next trophic level and efficiency as the ratio of production to consumption. (From French et al. 1979:185.)

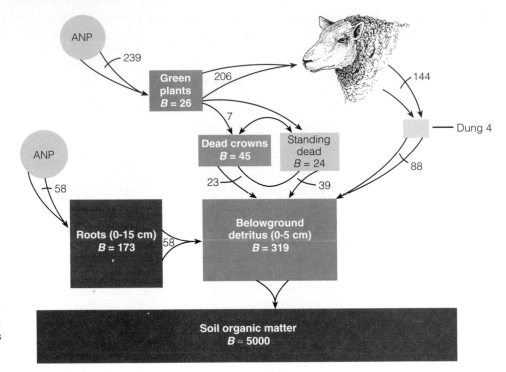

FIGURE 28.13 Carbon (energy) flow in a grazed pasture, indicating the role of grazing herbivores in the functioning of grassland ecosystems. Boxes represent the retention of carbon (g/m^2). Arrows represent the flow of carbon (g/m^2/yr). ANP=annual net productivity; B=biomass. Note the high proportion of energy that passes through the grazing herbivore. (From Breymeyer 1980:803.)

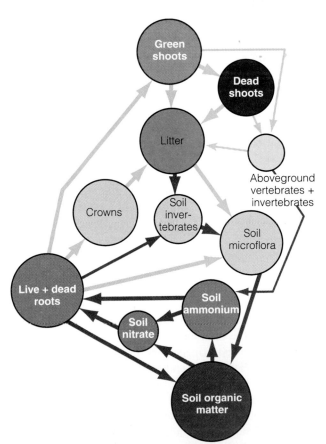

FIGURE 28.14 The nitrogen cycle in a shortgrass prairie. Circle size is proportional to the logarithm of the average standing crop in each compartment in grams of nitrogen per square meter. The width of the arrows is proportional to the average annual flows between compartments. The black arrows are rough estimates; we lack knowledge of below-ground processes. (Adapted from *Frontiers,* Academy of Natural Sciences of Philadelphia, 1978.)

TROPICAL SAVANNAS

The one ecosystem that defies general description is the tropical **savanna,** which makes up the Tropical Savanna Division of the Hot Tropical Domain. The problem is an old one, involving even its name (Hills 1965, Sarmiento and Monasterrio 1975, Bourliere and Hadley 1983). The word in its several origins, largely Spanish, referred to grasslands or plains. Over time the word was applied to an array of vegetation types representing a continuum of increasing cover of woody vegetation, from open grassland to widely spaced shrubs or trees to closed woodland with a grass understory (Figure 28.15).

Tropical savannas cover much of central and southern Africa, western India, northern Australia, large areas of northern and east-central South America, and some of Malaysia (Figure 28.16). Some savannas are natural. Others are seminatural or anthropogenic, brought about and still maintained by centuries of human interference.

Moisture appears to be the major determinant of savannas (Figure 28.17), a function of both rainfall (amount and distribution) and soil texture (Sarmiento and Monasterrio 1975, Tinley 1982). Fire, soil nutrients, and herbivores are important modifiers of the basic vegetation that the water regime promotes (Huntley and Walker 1982).

Characteristics

Savannas, in spite of their vegetational differences, exhibit some common characteristics. Savannas occur on land surfaces of little relief, often old alluvial plains. Because the soils are low in nutrients and moisture, they cannot support a forest. Precipitation exhibits extreme seasonal fluctuations; in South

FIGURE 28.15 Savanna ecosystems in southern Africa: (a) grass savanna; (b) shrub savanna; (c) tree savanna. (d) A tree savanna in Venezuela.

American savannas the soil water regime may fluctuate from excessively wet to extremely dry, often below the permanent wilting point. Savannas are subject to recurrent fires, and the dominant vegetation is fire-adapted. Grass cover with or without woody vegetation is always present. When present, the woody component is short-lived, with individuals seldom surviving for more than several decades (except for the African baobab trees, *Adansonia digitata*). Detrital-processing termites are a conspicuous component of savanna animal life.

Structure

The major and most essential stratum of the savanna ecosystem is grass, mostly bunch, with no vertical structure. The woody component adds one or two more vertical layers, ranging from about 50 to 80 cm when small woody shrubs are pre-

sent to about 8 m in the tree savannas. Highly developed root systems make up the larger part of the living herbaceous biomass. The root system is concentrated in the upper 10 cm but extends down to about 30 cm. Savanna trees have extensive horizontal roots that go below the layer of grass roots. They have a high root-to-shoot ratio (R/S) in sharp contrast to the low R/S ratio of typical forest trees. Competition may exist between grass and woody vegetation for soil moisture, but more intense competition takes place between trees, accounting for the spacing patterns of woody vegetation (Gutierrez and Fuentes 1979, Smith and Walker 1983, Smith and Grant 1986).

In contrast to the poorly developed vertical structure is a well-developed, although often unapparent, horizontal structure. Bunchgrasses form an array of clumps set in a matrix of open ground, creating patches of low vegetation with frequent

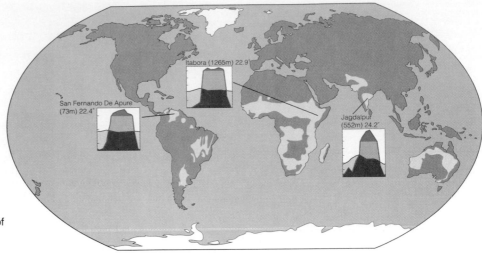

FIGURE 28.16   Location and climate of tropical savanna ecosystems. (Climate diagrams from Walter et al 1975.)

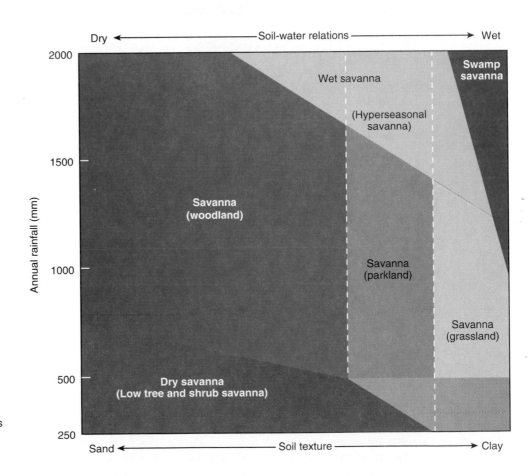

FIGURE 28.17   Classification of savannas of the world, based on annual rainfall and soil moisture as influenced by soil texture. (After Johnson and Tothill 1985.)

changes in microclimatic conditions. Widely spaced trees lend horizontal structure to the soil, brought about by changing physical conditions under the trees, especially *Acacia* (Belsky and Canham 1994). Shading from the trees reduces light, evapotranspiration, and soil temperature, providing environmental conditions favorable for colonization by shrubs adapted to lower light conditions. Seeds of such shrubs, introduced by animals, develop into shrubby thickets beneath the trees. Litterfall from trees and dung and urine from large grazing ungulates seeking shade enrich the soil beneath with organic matter and nutrients, especially total available nitrogen and phosphorus. All of these characteristics decline with distance from the trees (Figure 28.18).

Savannas support or are capable of supporting a large and often varied assemblage of herbivores, both grazing and browsing, particularly in Africa. This element of large graz-

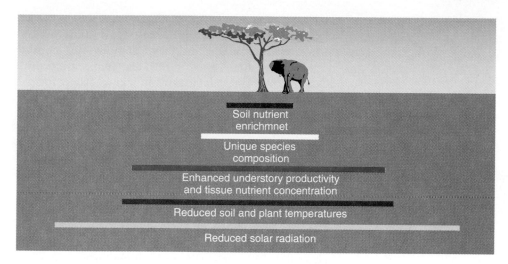

FIGURE 28.18 Zone of altered environmental characteristics surrounding an isolated *Acacia tortilis* in the African Savanna (Tsavo National Park, Kenya). (Adapted from Belsky and Canham 1994.)

ing herbivores is missing from the South American savannas, where it is replaced in part by a number of deer, tapirs (*Tapirus*), and capybara, the largest living rodent (Ojasti 1983). In spite of their visual dominance, large ungulates consume only about 10 percent of primary production. Dominant herbivores are the invertebrates, including acrid mites, acridid grasshoppers, seed-eating ants, and detrital-feeding dung beetles and termites (Lamotte 1975, Gillon 1983).

Function

Primary production in savannas is not well documented. Because of the wide diversity of savanna types, it is difficult to make any strong generalizations. Primary production is initiated at the beginning of the wet season when moisture releases nutrients from materials accumulated in the dry season and stimulates nutrient translocation from the roots to new growth (Lamotte and Bourliere 1983). A quick flush of growth into grass and woody plants follows. Large root systems efficiently transfer nutrients from the soil to above-ground biomass.

The nutrient pool, especially nitrogen, is low. Its circulation in a tree savanna (Bernhard-Reversat 1982, Sarmiento 1984) provides some insight into the function of savanna ecosystems. Nitrogen and organic matter content of the soil are low. Both are concentrated largely in the upper 10 cm of soil, with the highest concentration beneath the scattered trees. The first rains of the wet season trigger mineralization. Its rate, 5 to 8 percent of the total N per year, is fairly high. It is the function of the number of days the soil was wet for a given temperature and N content. Bernard-Reversat (1981) found that the nitrogen flux between soil and vegetation was higher under the trees than in the open, because of reduced evaporation (which kept the soil moist), more organic matter accumulation, and more herbaceous growth. Savanna trees, especially the African acacias, exhibited tight internal cycling. Nitrogen concentration in the leaves decreased as the dry season approached, with maximum withdrawal before leaf-fall. The trees transferred some of the nitrogen into new woody growth, but much of it went to the root reserve, where N accumulated for the flush of new season growth.

A similar tight circulation exists in neotropical savannas (Sarmiento 1984). An important portion of nitrogen, 66 percent of it, is translocated to the perennial roots, while another portion remains in the dry standing biomass (Figure 28.19). Most of the nitrogen in the dry above-ground biomass is lost to the atmosphere by volatilization if fire sweeps the savanna; otherwise, a fraction will be transferred to the soil through leaching effects of rainwater.

SHRUBLANDS

Covering large portions of the arid and semiarid world is climax shrubby vegetation (see inside cover). In addition climax shrubland exists in parts of temperate regions, because historical disturbances of landscapes have seriously affected their potential to support forest vegetation (Eyre 1963). Among such shrub-dominated plagioclimaxes—brought about by ecosystems and maintained by human disturbances—are the moors of Scotland and the macchia of South America. Outside these regions, shrublands represent a stage in the successional sequence leading back to forest. There they exist as second-class citizens of the plant world (McGinnes 1972), given little attention by botanists, who tend to emphasize dominant plants. Too little work has been done on successional shrub communities.

Characteristics of Shrubs

Shrubs are difficult to characterize. They have, as McGinnes (1972) points out, a "problem in establishing their identity." They constitute neither a taxonomic nor an evolutionary category (Stebbins 1972). One definition says that a **shrub** is a plant with woody persistent stems, no central trunk, and a height of 4.5 to 6 m. Size does not set shrubs apart, though, because under severe environmental conditions many trees will not exceed that size. Some trees, particularly coppice stands, are multiple-stemmed, and some shrubs have large single stems. Shrubs may have evolved either from trees or herbs (for detailed discussion on evolution of shrubs, see Stebbins 1972).

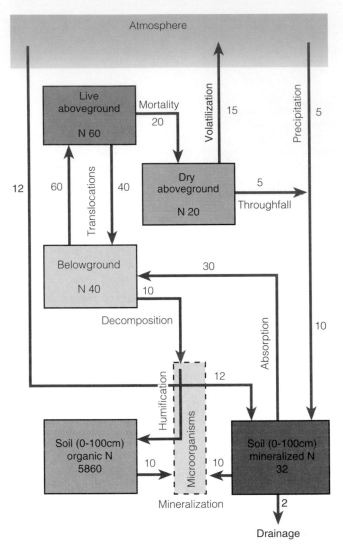

FIGURE 28.19 The annual nitrogen cycle in the seasonal savanna at Barinas, Venezuela. The above-ground biomass at maximum density accumulates 60 kg N/ha. As the above-ground growth dries, about 66 percent of the nitrogen translocates to the roots; the remainder remains in dry standing crop. Of this N, an estimated 25 percent is incorporated into the mineral nitrogen pool of the soil by throughfall. Some of the nitrogen in the dry above-ground material is lost through volatilization when burned. In the roots 25 percent of the biomass is recycled through decomposition. In this situation about 10 kg/ha/yr passes through soil microorganisms to the organic nitrogen pool. The value of nitrogen accumulated in the pools are in tons/ha, and the values of the flows are in tons/ha/yr. (After Sarmiento 1984:191.)

The success of shrubs depends largely on their abilities to compete for nutrients, energy, and space (West and Tueller 1972). They dominate in habitats stressed by aridity, nutrient-impoverished soils, cold winters, short growing seasons, and wind. In such environments shrubs have many advantages. They have less energetic and nutrient investment in above-ground parts than trees. Their stemmy growth affects light interception, heat dissipation, and evaporative losses. The multistemmed forms influence interception and stem-flow of

moisture, increasing or decreasing infiltration into the soil (Mooney and Dunn 1970b). Because most shrubs quickly form extensive root systems, they can utilize soil moisture deep in the profile. This feature gives them a competitive advantage over trees and grasses in regions where the soil moisture recharge comes during the nongrowing season. Because they have a high root-to-shoot ratio, shrubs draw less nutrient input into above-ground biomass and more into roots. Their perennial nature allows immobilization of limiting nutrients and slows nutrient recycling, favoring further shrub invasion of grasslands.

Subject to strong competition from herbs, some climax shrubs, such as chamise (*Adenostoma fasciculatum*), may inhibit the growth of herbs by allelopathy, the secretion of substances toxic to other plants (see Chapter 16) (McPherson and Muller 1969). Only when fire destroys mature shrubs and degrades the toxins do herbs appear in great numbers. As the shrubs recover, herbs decline. Seeds of herb species affected apparently have evolved the ability to lie dormant in the soil until released from suppression by fire.

Types of Shrublands

Mediterranean-Type Shrublands In five regions of the world lying for the most part between 32° and 40° north and south of the equator are areas characterized by a mediterranean climate: the semiarid regions of western North America, the regions bordering the Mediterranean, central Chile, the Cape region of South Africa, and southwestern and southern Australia (Figure 28.20). They make up the Mediterranean Division of the Humid Temperature Domain map. The mediterranean climate is characterized by hot, dry summers with at least one month of protracted drought and cool, moist winters. At least 65 percent of the annual precipitation falls during the winter months and for at least one month the temperature remains below 15°C (Aschmann 1973).

All five areas (see Appendix C) support physiognomically similar communities of xeric hardleaf evergreen shrubs and dwarf trees, known as broad sclerophyll vegetation, with herbaceous understory (for detailed descriptions see McKell et al. 1972, di Castri and Mooney 1973, di Castri et al. 1981). Although vegetation in all mediterranean-type ecosystems exhibits strong convergence (see Mooney 1977), each has evolved its own distinctive flora and fauna. Mediterranean vegetation evolved from tropical floras and developed in dry summer climates that did not exist until the Pleistocene (Raven 1973).

In addition to similar forms, vegetation in each of the mediterranean systems also shows similar adaptations to fire (see Mooney and Conrad 1977) and to low nutrient levels in the soil. In the mediterranean systems of the Northern Hemisphere, annuals make up 50 percent of the species and 10 percent of the plant genera; 40 percent of the species are endemics.

There are variations of the basic mediterranean-type ecosystems (see di Castri 1981). In the Mediterranean region, shrub vegetation often results from forest degradation and falls into three major types. The **garrigue,** resulting from

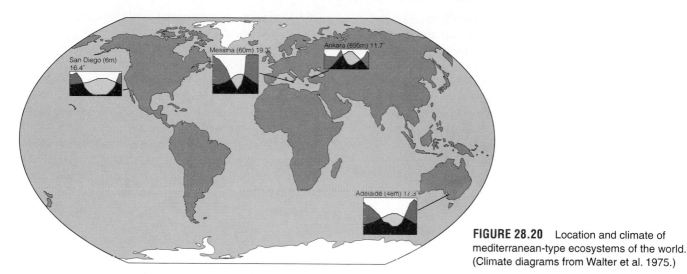

FIGURE 28.20   Location and climate of mediterranean-type ecosystems of the world. (Climate diagrams from Walter et al. 1975.)

degradation of pine forests, is low, open shrubland on well-drained to dry calcareous soil. The **maquis,** replacing cork oak forests, is higher thick shrubland in areas of more rainfall. The **mattoral,** further subdivided into high, middle, low, and scattered (Tamales 1981), appears to be equivalent to the North American chaparral (Soriano 1972, Tomaselle 1981a). The Chilean mediterranean system, also called mattoral, varies across topographic positions from the coast to slopes of coastal ranges and the foothills of the Andean cordillera (Rundel 1981). Much of the South African mediterranean shrubland is typically heathlands known as **fynbos,** discussed

later. The more typical mediterranean-type shrubland, dominated by broad-sclerophyll woody shrubs, goes by the names of **strandveld** and **renosterveld.**

In southwest Australia the mediterranean shrub country, known as **mallee,** is dominated by low-growing *Eucalyptus,* 5 to 8 m high, with broad sclerophyllous leaves. There are six types of mallee ecosystems, which intergrade. Of these, three fall into mediterranean-type ecosystems, with a typically grassy and herbaceous understory (Figure 28.21). The other three types occur on oligotrophic soils and fall under the category of heathland shrubs. Razed by fire at irregular inter-

FIGURE 28.21   Tall shrub mallee in Victoria, Australia, is an example of a mediterranean-type shrubland dominated by *Eucalyptus.* Note the canopy structure and the open understory of grass at the beginning of the spring rains. This type of vegetation supports a rich diversity of bird life.

vals, the mallee ecosystem differs markedly from typical mediterranean-type ecosystems in its summer growth rhythm. Mediterranean-type shrublands typically initiate new growth during the spring; but the mallee retains its summer growth rhythm from the subtropical Tertiary, out of phase with the mediterranean climate of the area. Growth takes place during the summer, the driest part of the year, drawing on water conserved in the soil during wet winter and spring showers (Specht 1981).

In North America the sclerophyllic shrub community is known as **chaparral,** a word of Spanish origin meaning a thicket of shrubby evergreen oaks. California chaparral (Figure 28.22) is dominated by scrub oak (*Quercus dumosa*) and chamise (*Adenostoma fasciculatum*). Another shrub type, also designated as chaparral, is associated with the Rocky Mountain foothills in Arizona, New Mexico, and Nevada, and elsewhere. It differs from California chaparral in two ways. It is dominated by Gambel oak (*Quercus gambelii*) and other species and lacks chamise; and it is summer-active and winter-deciduous, whereas California chaparral is evergreen, winter-active, and summer-dormant (see Hanes 1981).

For the most part, mediterranean-type shrublands lack understory and ground litter, are highly inflammable, and are heavy seeders. Many species require the heat and scarring action of fire to induce germination. Others sprout vigorously from root collars after a fire.

For centuries periodic fires have roared through mediterranean-type vegetation, clearing away the old growth, making way for the new, and recycling nutrients through the ecosystem. When humans intruded into this type of vegetation, they changed the fire regime, either by attempting to exclude fire completely or by overburning (see Trabaud 1981). In the absence of fire, chaparral grows tall and dense and yearly adds more leaves and twigs to those already on the

ground. During the dry season the shrubs, even though alive, nearly explode when ignited.

After fire the land returns either to lush green sprouts coming up from buried root crowns or to grass, if a seed source is nearby. Grass and vigorous young sprouts are excellent food for deer, sheep, and cattle. As the sprout growth matures, chaparral becomes dense, the canopy closes, the litter accumulates, and the stage is set for another fire.

Because of the rough terrain characteristic of the mediterranean-type ecosystem, some areas have remained relatively undisturbed, especially in California and South Africa. But in Australia and the Mediterranean basin, human activity, especially livestock grazing and fruit and vegetable farming, have degraded the broadleaf sclerophyllous vegetation (see Trabaud 1981).

Northern Desert Shrub   In the Great Basin of North America, the northern, cool, arid region lying west of the Rocky Mountains, is the **northern desert shrub.** The climate is continental, with warm summers and prolonged cold winters. Although this region is perhaps more appropriately considered a desert and is classified as a Temperate Desert Division of the Dry Domain, it is one of the most important shrublands in North America (Figure 28.23). Its physiognomy differs greatly from the southern hot desert, and the dominant vegetation is shrub. The vegetation falls into two main associations: one is sagebrush, dominated by big sagebrush (*Artemisia tridentata*), which often forms pure stands; the other is shadscale (*Atriplex confertifolia*), a C_4 species, and other chenopods (see Caldwell 1985), halophytes tolerant of saline soils. Inhabiting this shrubland are pocket and kangaroo mice (Heteromyidae), lizards, sage grouse (*Centrocerus urophasianus*), sage thrasher (*Oreoscoptes montanus*), sage sparrow (*Amphispiza belli*), and Brewer's

FIGURE 28.22   Southern California mixed chaparral is highly prone to fire.

FIGURE 28.23 Northern desert shrubland in Wyoming is dominated by sagebrush. Although classified as cool desert (see page 585), sagebrush forms one of the most important shrub types in North America.

sparrow (*Spizella breweri*), four birds that depend on sagebrush for their existence.

A similar type of shrubland exists in the semiarid interior of southwestern Australia. Numerous chenopod species, particularly the saltbushes of the genera *Atriplex* and *Maireana,* form extensive low shrublands on low riverine plains (Figure 28.24).

Heathlands Typically **heathlands** have been associated with cool to cold temperate climatic regions of northwestern Europe. The word **heath** comes from the German *Heide,* meaning "an uncultivated stretch of land," regardless of the vegetation. It just so happened that in parts of Germany uncultivated or waste land was dominated by Ericaceae. In reality heathlands are found in all parts of the world, from the tropics to polar regions and from lowland to subalpine and alpine altitudes.

Heathland vegetation is an assemblage of dense to mid-dense growth of ancient or primitive genera of angiosperms: evergreen sclerophyllous shrubs and subshrubs such as

FIGURE 28.24 Saltbush shrubland in Victoria, Australia, is dominated by *Atriplex.* It is an ecological equivalent to the shrublands of the Great Basin in North America.

Daphne heath (*Brachyloma daphnoides*), all adapted to fire (Figure 28.25). Heathland shrubs have leaves with thick cuticles, sunken stomata, thick-walled cells, and hard and waxy upper surfaces. Many species have leaves with small surface area—less than 25 mm and termed **leptophyll**—and others roll their edges in toward the midrib. Although mostly associated with heathlands, many heathland shrubs such as laurel and blueberries (*Vaccinium*) are present as shrubby understory in other ecosystems such as the deciduous forest.

Heathlands invariably occur on oligotrophic soils especially deficient in phosphorus and nitrogen. Heathlands are most extensive in the Arctic regions. Other extensive areas occur in the mediterranean-type regions of South Africa, where they are known as fynbos (Kruger 1979), and in southeastern and western Australia (see George et al. 1979). In subtropical to tropical climates, true heathlands are confined to alpine areas and to lowland, oligotrophic soils subject to seasonal waterlogging. Some heathlands, such as the heather-dominated moors of Scotland, are human-induced and can be maintained only by periodic fires (Gill and Groves 1981, Gimingham 1981).

There are two distinct heathland ecosystems: dry heathlands and wet heathlands. Dry heathlands are on well-drained soils and subject to seasonal drought; wet heathlands are subject to seasonal waterlogging. In wet heathlands, graminoids (grasses and sedges) may become codominant with heathland shrubs; and in extreme wet heathlands, the grass component is suppressed by *Sphagnum* moss. Both height and foliage cover of heathlands vary considerably with the habitat. They are divided according to height of the uppermost stratum:

shrubs > 2 m, scrub; shrubs 1–2 m, tall heathland; shrubs 25–100 cm, heathland; and shrubs < 25 cm, dwarf heathland.

Successional Shrublands On drier uplands, shrubs rarely exert complete dominance over herbs and grass. Instead, the plants are scattered or clumped in grassy fields. The open areas between are filled with the seedlings of forest trees, which in the sapling stage occupy the same ecological position as tall shrubs (Figure 28.26). Typical are the hazelnut (*Corylus* spp.), forming thickets in places, sumacs (*Rhus* spp.), chokecherry (*Prunus virginiana*), and shrub dogwoods (*Cornus* spp.).

On wet ground, the plant community often is dominated by tall shrubs and contains an understory intermediate between that of a meadow and a forest (Curtis 1959). In northern regions the common tall shrub community found along streams and around lakes is the alder thicket, composed of alder (*Alnus* spp.) possibly with a mixture of other species such as willow (*Salix* spp.) and red osier dogwood (*Cornus stolonifera*). Alder thickets persist for some time before being replaced by the forest. Out of the alder country, the shrub or carr community (*carr* is an English name for wet-ground shrub communities) occupies the low places. Dogwoods are some of the most important species in the carr. Growing with them are a number of willows, which as a group usually dominate the community.

Shrub thickets are valued as food and cover for game; and many shrubs, such as hawthorn (*Crategus* spp.), blackberry (*Rubus*), sweetbrier (*Rosa eglanteria*), and dogwoods, rank high as game food. However, the overall value of dif-

FIGURE 28.25 Heathlands, dominated by ericaceous shrubs, have a similar physiognomy around the world. Typical is this heathland along the southern coast of Australia.

FIGURE 28.26 In eastern North America and in northern and western Europe, shrublands are usually successional communities. This old field in southern New York state is in the early stages of shrub invasion.

ferent types of shrub cover, its composition, quality, and the minimum amounts needed, have never been assessed. There is some evidence that shrubs can form a stable community that will persist for many years (Egler 1953, Niering and Egler 1955). If incoming tree growth is removed either by selective spraying or by cutting, shrubs eventually form a closed community resistant to further invasion by trees. This fact could have wide application to the management of power line rights-of-way.

Structure

Shrub ecosystems, successional or mature, are characterized by woody structure, increased stratification over grasslands, dense branching on a fine scale, and low height, under 8 m. Typically there are three layers: a broken upper canopy, an irregular low shrub canopy, and a grass/herbaceous layer. This structure, of course, may vary. Dense shrubland may have only a canopy layer; and stratification often decreases as the shrubs reach maximum height, particularly in seral shrublands. Patchy distribution of shrubs varies with the vegetation type. Heathlands may exhibit minimal patchiness.

Mediterranean-type shrublands, notably the mattoral and the mallee, may have pronounced horizontal stratification. The greatest patchiness probably occurs in successional shrublands with scattered clumps of invading shrubs and trees.

Shrub communities have their own distinctive animal life. Successional shrub communities support not only species common to shrubby edges of forests and shrubby borders of fields but a number of species dependent on them, such as northern bobwhite (*Colinus virginianus*), cottontail rabbit (*Sylvilagus floridanus*), prairie warbler (*Dendroica discolor*), and yellow-breasted chat (*Ictera virens*). In Great Britain, some shrub communities, especially hedgerows, have been stable for centuries, and many forms of animal life, invertebrate and vertebrate, have become adapted to or dependent on them. Among these species are the whitethroat (*Sylvia communis*), linnet (*Acanthis cannabina*), blackbird (*Turdus merula*), and hedgehog (*Erinaceus europaeus*).

Stable shrub communities have a complex of animal life that varies with the region (Bigalke 1979, Dwyer et al. 1979, Blondel 1981, Cody 1973, Schoddee 1981). Within the mediterranean-type shrublands and heathlands, similarity in habitat structure and in the nature and number of niches has resulted in pronounced parallel and convergent evolution among bird species (Blondel 1981, Cody 1973).

Function

The functional aspects of shrubland ecosystems are poorly studied, but data from a few mediterranean-type systems provide some interesting insights into nutrient cycling.

In the California chaparral, precipitation falls mostly in the cool winter months. Most of the plant growth and flowering is concentrated in spring; 75 percent of the flowering plants, half of them annuals, bloom in May at the end of the rainy season. However, the evergreen dominants can fix CO_2 throughout the year. The greatest daily carbon gain is in the spring; the lowest during summer drought; and in winter with its short photoperiod, carbon fixation is minimal (Mooney 1981, Mooney and Miller 1985). The amount fixed in summer depends on the amount of rainfall received during the previous winter. Yearly accumulation of above-ground biomass for a sclerophyllous plant of intermediate age is about 1000 kg/ha (Mooney and Parsons 1973).

The soils of mediterranean-type ecosystems are low in nutrients and are especially deficient in nitrogen and phosphorus. The cycling of these nutrients appears to be tight, conservative, and seasonal (Mooney and Miller 1985), reflecting rainfall (or the lack of it) and associated microbial activity, high during the wet season. A flush of microbial activity, which involves decomposition of humus and mineralization of nitrogen and carbon, follows wetting of dry soil. Nitrate accumulation depends on a progressive drying period after rains, when the topsoil gradually dries out to an increasing depth. Improved soil aeration and increasing soil temperature favor rapid bacterial nitrification and the retention of nitrate ions in the soil. As the dry cycle continues, nitrates accumulate and remain fixed in the topsoil along with other

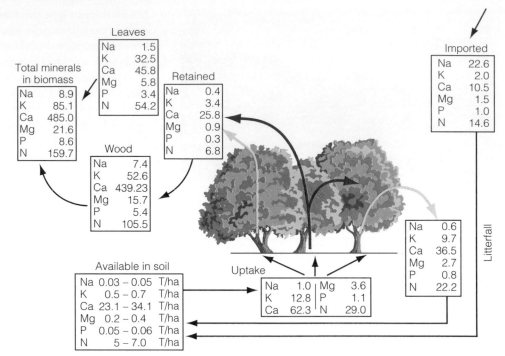

FIGURE 28.27 Annual turnover of macronutrients in the *Quercus coccifera* garrigue at Le Puech de Juge near Montpellier in Southern France. (After Lossaint 1973.)

nutrients (Schaefer 1973). When the rains arrive and wet the soil, the concentration of nutrients stimulates a flush of growth. If heavy rains suddenly enter dry topsoil, quantities of nutrients may be lost by leaching. In the California chaparral much of the nitrogen returned to the soil is lost through erosion (Mooney and Parsons 1973).

Some plants of the mediterranean systems exhibit nutrient conservation mechanisms. *Ceanothus,* an early successional species in the California chaparral, is a nonleguminous nitrogen fixer (Mooney and Parsons 1973). In the Australian mallee, *Atriplex vesticana,* a dominant plant, lowers the nitrogen content of the surrounding soil and concentrates nitrogen through litterfall in the soil directly beneath the plant. More nitrogen, however, is withdrawn from the soil than is returned by litterfall, which represents about 10 percent of the total plant nitrogen. Litter has lower nitrogen and phosphorus content than fresh leaves, which suggests that the plant withdraws nitrogen and phosphorus from the leaf into the stem before the leaf falls.

Australian sclerophyllous shrubs also exhibit phosphorus conservation, involving three mechanisms. First, in aging leaves phosphorus, like nitrogen, is recirculated through the plant with minimal losses to litter. Second, a fine mat of mycorrhizal roots penetrates the decomposing litter and takes up phosphorus. Third, polyphosphate forming and hydrolyzing enzymes are present in the roots of sclerophyllous plants. As orthophosphate is released from decomposing litter in spring, it is stored in roots as long-chain polyphosphate. When growth begins, the polyphosphate seems to be hydrolyzed back to orthophosphate and transported to the growing shoots (Specht 1973).

The best available data for mineral cycling in a mediterranean-type ecosystem are for a 17-year-old *Quercus coc-*

cifera garrigue in southern France (Figure 28.27). The total above-ground mineral mass is 773 kg/ha; 629 kg is in wood, 144 kg in leaves. Calcium is the most important element, amounting to 485 kg/ha, followed by nitrogen, 160 kg/ha, and potassium, 85 kg/ha. The mean annual incorporation of nutrients into perennial organs amounts to 38 kg/ha, of which two-thirds is calcium. Annual return of elements by the way of litter amounts to 75 kg/ha. Net primary productivity is on the order of 3.4 tons/ha/yr with a mean annual production of litter of 2.3 tons/ha (Lossaint 1973, Rapp and Lossaint 1981).

The ecology of mediterranean shrublands has been studied intensively over the past quarter century. Nevertheless, a number of questions remain, especially relating to management. What is the effect on native plants and animals of seeding grass on chaparral hills after fire? What is the relationship of these unnatural grasslands to the fire ecology of the region? Does the present unnatural fire frequency of once every two or three decades alter nutrient pools and eliminate certain species? What is the effect of absence of fire on long-term community structure, productivity, and species diversity? Answers to such questions are important in the maintenance of mediterranean ecosystems and the welfare of the human populations occupying them.

DESERTS

Geographers define **deserts** as land where evaporation exceeds rainfall. Deserts, which occupy about 26 percent of the continental area, occur in two distinct belts between 15° and 35° latitude in both the Northern and Southern Hemispheres—around the Tropic of Cancer and the Tropic of

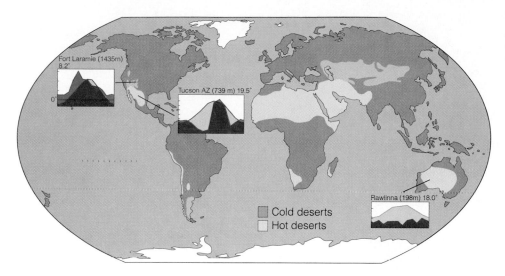

FIGURE 28.28 Hot deserts and cool deserts and arid shrublands of the world. (Climate diagrams from Walter et al. 1975.)

Capricorn (Figure 28.28). Cooler semiarid deserts, making up the Temperate Division of the Dry Domain, occur in the interior of continents. They include the North America Great Basin, and the Gobi, Takla Makan, and Turkestan deserts of Asia. They are characterized by low rainfall somewhere between 200 and 400 mm, long cold winters, and strong daily temperature fluctuations. Cool deserts are dominated by *Artemisia* and chenopod shrubs. (These semiarid deserts are considered under Shrublands.) Hot deserts, which make up the Tropical/Subtropical Desert Division of the Dry Domain, fall into two groups: true or arid deserts, in which precipitation is less than 250 mm; and extremely arid deserts, areas with rainfall less than 100 mm per year. Arid deserts include the North African deserts from the Sahara to the Thar, the Great Australian Desert, and the deserts of southwestern North America and Mexico. These deserts experience extreme daily fluctuations in day and night temperatures because of heating during the day and outgoing radiation at night. They range from no or scattered vegetation to some combination of chenopods, dwarf shrubs, and succulents. Extremely arid deserts include the Atacama Desert of Chile and Peru and the Namib Desert of coastal southwest Africa. Also very arid are polar deserts. Although they experience very low precipitation, they also experience very low total annual evaporation, allowing snow to accumulate over time. For this reason polar deserts are not considered true deserts. (For a detailed review of world deserts, see Evanari et al. 1985, 1986.)

Deserts are the result of several forces. One force that leads to the formation of deserts and the broad climatic regions of Earth is the movement of air masses over the planet. High-pressure areas alter the course of rain. The high-pressure cell off the coast of California and Mexico deflects rainstorms moving south from Alaska to the east and prevents moisture from reaching the southwest. In winter high-pressure areas move southward, allowing winter rains to reach southern California and parts of the North American desert. Winds blowing over cold waters become cold also; they carry little moisture and produce little rain. Thus the west coast of California and Baja California, the Namib Desert on coastal southwest Africa, and the coastal edge of the Atacama in Chile may be shrouded in mist, yet remain extremely dry.

Mountain ranges also play a role in desert formation by causing a rain shadow on their lee side. The High Sierras and the Cascade Mountains intercept rain from the Pacific and help maintain the arid conditions of the North American desert. The low eastern highlands of Australia block the southeast tradewinds from the interior. Other deserts, such as the Gobi and the interior of the Sahara, are so remote from the source of oceanic moisture that all the water has been wrung from the winds by the time they reach those regions.

Characteristics

The topography of the desert, unprotected by vegetation, is stark and, paradoxically, partially shaped by water. Unprotected, the soil erodes easily during violent storms. Sediment-bearing waters rush down slopes into water-cut canyons and river channels called **wadis** (**arroyos** in the southwestern United States). Receiving these waters are low basins called **playas.** These basins hold temporary lakes after rains, but the water soon evaporates, leaving behind a dry bed of glistening salt.

Structure

Woody-stemmed and soft brittle-stemmed shrubs are characteristic desert plants (Figure 28.29). In a matrix of shrubs grows a wide assortment of other plants—the yucca, cacti, small trees, and ephemerals. In the Sonoran, Peru-Chilean, South African Karoo, and southern Namib deserts, large succulents rise above the shrub level and change the aspect of the desert far out of proportion to their numbers. The giant saguaro, the most massive of all cacti, grows on the bajadas of the Sonoran Desert (Figure 28.30). Ironwood (*Olneya tesota*), smoketree (*Psorothamnus spinosa*), and paloverde (*Cercidium* spp.) grow best along the banks of intermittent streams.

(a)

(b)

FIGURE 28.29 Two hot deserts dominated by woody, brittle-stemmed shrubs. (a) Mojave Desert in Nevada. (b) Desert in the Northern Territory of Australia. The sparseness and spacing of the shrubs reflect the extreme aridity of these areas.

Both plants and animals are adapted to a scarcity of water either by drought evasion or drought resistance (Noy-Meir 1974, 1975). Drought-evading plants may be annuals or perennial. Annuals, known as ephemerals, persist as seeds during the drought periods, ready to sprout, flower, and produce seeds when moisture and temperature are favorable. Drought-evading perennials are largely geophytes and hemicryptophytes (see Chapter 20). Bulbs send up active growth during the rainy periods, but they may bloom independently of rainfall. There are two periods of flowering in the North American deserts: after winter rains come in from the Pacific Northwest, and after summer rains move up to the southwest out of the Gulf of Mexico. Some species flower only after summer rains, and a few bloom during both seasons. If rains fail, these ephemeral species do not bloom. Drought-evading animals, like their plant counterparts, adopt an annual lifestyle or go into estivation or some other stage of dormancy during the dry season. If extreme drought develops during the breeding season, birds fail to nest and lizards do not reproduce.

Drought-resisting plants or persistents are all perennials. Some shed leaves and rootlets in the upper dry area during the

FIGURE 28.30 Saguaro (*Carnegiea gigantea*) dominates the aspect of this portion of the Sonoran Desert in the southwestern United States.

dry season to reduce water loss. Others maintain evergreen leaves throughout the year. These are mainly deep-rooted shrubs, such as mesquite (*Prosopis* spp.) and tamarisk (*Tamarix*), whose taproots reach the water table, rendering them independent of water supplied by rainfall. Other shrubs, such as *Larrea* and *Atriplex,* are deep-rooted perennials with laterals that extend as far as 15 to 30 m from the stems. Other perennials such as cacti are succulents. They have a very shallow root system that can pick up moisture quickly from the rain and accumulate a large internal water reserve to carry them through droughts.

The desert floor is stark, a raw mineral substrate devoid of a continuous litter layer. Dead leaves, bud scales, and dead twigs accumulate in wind-protected areas beneath the plants and in depressions in the soil. Covering soil surface are crusts consisting of lichens, bryophytes, microfungi, algae, and cyanobacteria. This crust binds soil particles and thereby reduces soil erosion by protecting the soil from wind and the impact of raindrops. It roughens the soil, producing small catchments to collect water. Cyanobacteria and lichens supporting cyanobacteria fix atmospheric nitrogen, enhancing nitrogen cycling in desert plants. Breaking of this crust by livestock, vehicles, and other such disturbances affects the desert nitrogen cycle and accelerates erosion.

Function

The desert ecosystem differs from other ecosystems in at least one important way: the input of precipitation is highly periodic. It comes in pulses as clusters of rainy days 3 to 15 times a year, of which only 1 to 6 may be large enough to stimulate biologic activity. Thus, the desert ecosystem experiences periods of inactive steady states broken by periods of production and reproduction. Both processes are stimulated by rain and continue through short periods of adequate moisture; when water is scarce again, both production and biomass return to some low steady state.

Primary production in the desert depends on the proportion of available water used and the efficiency of its use. Data from various deserts in the world (for a summary, see Noy-Meir 1973, 1974) suggest that annual net primary production of above-ground vegetation varies from 30 to 200 g/m^2.

The amount of biomass that accumulates and rate of turnover (ratio of production to biomass) depend on the dominant type of vegetation. In deserts such as the Sonoran, where trees, shrubs, and cacti dominate, annual productivity is about 10 to 20 percent of the above-ground standing crop biomass of 300 to 1000 g/m^2. In deserts dominated by perennials, annual production is 20 to 40 percent of biomass of 150 to 600 g/m^2. Annual or ephemeral communities have a 100 percent turnover of both roots and above-ground foliage; annual production approximates peak biomass. These turnover rates are higher than those of forests and tundra. The ratio of below-ground biomass to above-ground for perennial grasses and forbs is between 1 and 20, and for shrubs between 1 and 3. In general, warm desert plants do not have a high root-to-shoot ratio. The root biomass is relatively small. Cool desert plants have a higher root-to-shoot ratio.

Nutrient cycling in arid ecosystems is tight, and two major nutrients, phosphorus and nitrogen, are in short supply. Much of the nutrient supply is tied up in plant biomass, living and dead. The tissues of desert plants have higher concentrations of nutrients than plants of mesic environments, and they tend to translocate certain elements before shedding any parts. For example, the nitrogen and phosphorus of fallen phyllodes (the enlarged and commonly flattened leaf stalks that function as leaves) of *Acacia aneura* in the northern Australian desert decrease markedly at the time of major phyllode fall, but the contents of potassium, calcium, and magnesium show little change. The desert plant *Artemisia* translocates phosphorus and potassium back to the twig before shedding its leaves. The nutrients retained in the shed parts collect beneath the plants, where microclimate conditions created by the shrubs favor biological activity. The soil is further enriched by animals attracted to the shade (Binet 1981, West 1979, 1981, West and Skuiins 1978). The plants, in effect, create islands of fertility beneath themselves.

This fact is illustrated in Figure 28.31, which shows the organic carbon and nutrient levels beneath and away from the base of an Egyptian desert shrub, *Anabasis aretioides*. The percentages of carbon and nitrogen are over 3 times those of the open soil, and the levels of phosphorus and potassium 5 and 21 times as great, respectively. Adding to

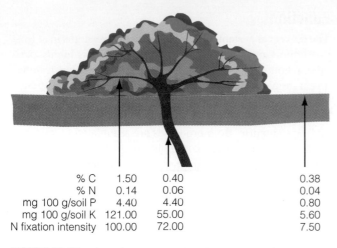

% C	1.50	0.40	0.38
% N	0.14	0.06	0.04
mg 100 g/soil P	4.40	4.40	0.80
mg 100 g/soil K	121.00	55.00	5.60
N fixation intensity	100.00	72.00	7.50

FIGURE 28.31   Organic content and carbon content of desert soil beneath and away from the base of an Egyptian desert shrub, *Anabasis aretioides.* A value of 100 is attributed to the soil surface, under the plant canopy. (From Binet 1981:328.)

primary production in the desert are lichens, green algae, and cyanobacteria (blue-green algae), abundant in soil crusts. Their biomass ranges up to 240 kg/ha, and they have an unusually high rate of nitrogen fixation. As pointed out in Figure 28.31, the intensity of fixation is greatest beneath the desert shrub, over 10 times greater than in the open.

In spite of their aridity, desert ecosystems support a surprising diversity of animal life, notably herbivorous species (Figure 28.32). Grazing herbivores of the desert tend to be generalists and opportunists in their mode of feeding. They consume a wide range of species, plant types, and parts. Desert sheep feed on succulents and ephemerals when available and then switch to woody browse during the dry period.

As a last resort herbivores consume dead litter and lichens. Small herbivores—the desert rodents, particularly the family Heteromyidae and ants—tend to be **granivores,** feeding largely on seeds.

These granivores are important in the dynamics of desert ecosystems. One of the notable small herbivores of the Sonoran Desert is the harvester ant (*Pogonomyrmex occidentalis*), which lives on seeds stored in underground granaries. During periods of drought, these ants gather mainly the seeds of several species of perennials. During winter rains, when annual plants flower and seed, ants gather the seeds of these plants. By using different plants, the ant obtains a constant food source throughout the desert year.

Herbivores can have a pronounced effect on primary producers, especially if they are more abundant than the land's capacity to support them. Once grazers have utilized annual production, they consume plant reserves, especially in long dry periods. Overgrazing and overbrowsing can so weaken the plant that the vegetation is destroyed or irreparably damaged. Areas protected from grazing, especially by cattle and goats, have a higher biomass and a greater percentage of palatable species than grazed areas.

Herbivores in a shrubby desert, under most conditions, consume only a small part of the above-ground primary production, but seed-eating herbivores can consume most of the seed production. Chew and Chew (1970) found that small grazing herbivores (the jackrabbit, *Lepus alleni,* and kangaroo rats, *Dipodomys* spp.) used only about 2 percent of the above-ground net primary production, but ate 87 percent of the seed production, a rate of consumption that could have a pronounced effect on the composition of plant populations. How pronounced has been demonstrated by J. R. Brown and Heske (1990). For 12 years they removed three species of

FIGURE 28.32   Large herbivores inhabit the less environmentally extreme regions of the Kalahari Desert of southern Africa.

FIGURE 28.33   The wide expanse of the Arctic tundra.

kangaroo rats from plots of a Chihuahuan Desert shrub habitat near a zone of a natural transition from desert to grassland. These rodents feed heavily on larger seeds and burrow in the soil. Their removal resulted in a threefold increase in density of tall perennials and annual grasses, a sharp decrease in desert shrubs, and invasion of rodent species typical of arid grasslands.

Desert carnivores, like the herbivores, are opportunistic feeders, with few specialists. Most desert carnivores, such as foxes and coyotes, have mixed diets that include leaves and fruits, and even insectivorous birds and rodents eat plant foods. Omnivory rather than carnivory and complex food webs seem to be the rule in desert ecosystems.

The detrital food chain seems to be less conspicuous in the desert than in other ecosystems, but nevertheless it is important. Most functional and taxonomic groups of soil microorganisms exist in the desert, from bacteria, fungi, and actinomycetes to nematodes, mites, and termites. In some deserts considerable amounts of nutrients may be tied up in termite structures, to be released when structures are destroyed. In the North American warm deserts, rates of decomposition of both surface and buried litter seem to be independent of rainfall, in part because soil fauna in the desert are active in the absence of moisture (Whitford et al. 1981, Santos et al. 1984). Bacterial and fungal decomposition is important. Its rate appears to be influenced by free-living soil nematodes. Their populations in turn are influenced by predatory mites (Santos et al. 1984). Although short-term litter decomposition does not seem to be related to yearly fluctuations in rainfall, in North American warm deserts it is highly correlated with long-term rainfall patterns, because it shapes the structure of the desert soil community. We know too little about the structure and the functioning of desert ecosystems, and the relationship of desert life to physical factors of the environment in the various desert types.

TUNDRAS

Encircling the northern pole of Earth is a region of cold, often desertlike conditions, the Arctic **tundra.** The word comes from the Finnish *tunturia,* meaning "a treeless plain" (Figure 28.33). At lower latitudes similar landscapes, the alpine tundra, occur at the peaks of tall mountains. In the Antarctic a well-developed tundra is mostly lacking because of a small land area and deep ice sheets. Arctic or alpine, the tundra is characterized by low temperatures, low precipitation (cold air can carry very little water vapor), and a short growing season ranging from 50 to 60 days in the high Arctic tundra to 180 days in low-altitude alpine tundra.

The Tundra Division of the Polar Domain is a land dotted with lakes and transected by streams. Its vegetation ranges from tall shrubs (2–5 m high), to dwarf shrub heath (5–20 cm high), graminoids, and mosses (Bliss 1988). Where the ground is low and moist, extensive bogs can exist. On higher, drier areas and places exposed to the wind, vegetation is scant and scattered, and the ground is bare and rock-covered. These regions are the **fellfields,** an anglicization of the Danish *fjoeldmark,* or rock deserts. Bliss (1981) divides the Arctic tundra into the polar desert, with less than 5 percent cover, polar semidesert, and tundra with 100 percent cover and wet to moist soil (Figure 28.34).

Characteristics

Frost molds the tundra landscape. Alternate freezing and thawing and the presence of a permanent frozen layer in the ground, the **permafrost,** create conditions unique to the Arctic tundra. The sublayer of soil is subject to annual thawing in spring and summer and freezing in fall and winter. The depth of thaw may vary from a few centimeters in some

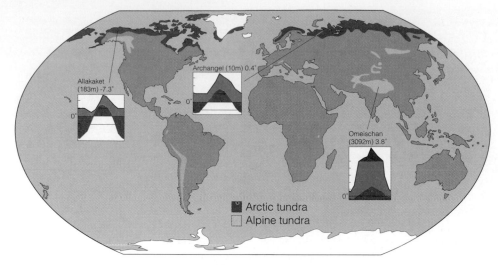

FIGURE 28.34 Arctic and alpine tundras of the world. (Climate diagrams from Walter et al. 1975.)

places to half a meter in others. Below the thaw depth, the ground is always frozen solid and is impenetrable to both water and roots. Because the water cannot drain away, flatlands of the Arctic are wet and covered with shallow lakes and graminoid bogs. This reservoir of water lying on top of the permafrost enables plants to exist in the driest parts of the Arctic.

The symmetrically patterned landforms so typical of the tundra result from frost. The fine soil materials and clays, which hold more moisture than the coarser materials, expand while freezing and then contract upon thawing. This action tends to push the larger material upward and outward from the mass to form the patterned surface.

Typical nonsorted patterns associated with seasonally high water tables are frost hummocks, frost boils, and earth stripes (Figure 28.35). Frost hummocks are small earthen mounds up to 1.5 m in diameter and 1.3 m high, which may or may not contain peat. Frost boils are formed when the surface freezes across the top, trapping the still unfrozen muck beneath. As this bulge chills and expands, the mud is forced up through the crust. Raised earth stripes, found on moderate slopes, appear as lines or small ridges flowing downhill. They

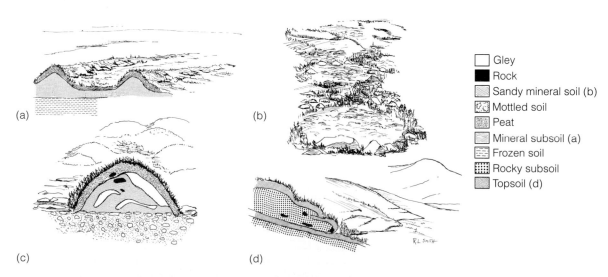

Legend:
- ☐ Gley
- ■ Rock
- Sandy mineral soil (b)
- Mottled soil
- Peat
- Mineral subsoil (a)
- Frozen soil
- Rocky subsoil
- Topsoil (d)

FIGURE 28.35 Patterned land forms of the tundra region: (a) unsorted earth stripes; (b) frost hummocks; (c) sorted stone nets and polygons; (d) solifluction terrace. (Diagrams adapted from Johnson and Billings 1963.)

apparently are produced by a downward creep or flow of wet soil across the surface of the permafrost.

Sorted patterns are characteristic of better-drained sites. The best known are the stone polygons, the size of which is related to frost intensity and the size of the material (Johnson and Billings 1962). The larger stones are forced out to a peripheral position, and the smaller and finer material, either small stones or soil, occupies the center. The polygon shape may result from an accumulation of rocks in desiccation cracks formed during drier periods. These cracks appear as the surface of the soil dries out, in much the same way as cracks appear in bare, dry, compacted clay surfaces in temperate regions. On the slopes, creep, frost thrusting, and downward flow of soil change polygons into sorted stripes running downhill. Mass movement of supersaturated soil over the permafrost forms solifluction terraces, or "flowing soil." This gradual downward creep of soils and rocks eventually rounds off ridges and other irregularities in topography. This molding of the landscape by frost action, called **cryoplanation,** is far more important than erosion in wearing down the Arctic landscape.

In the alpine tundra, permafrost exists only at very high elevations and in the far north, but frost-induced patterns—small solifluction terraces and stone polygons—are still present. The lack of permafrost results in drier soils; only in alpine wet meadows and bogs do soil moisture conditions compare to the Arctic. Precipitation, especially snowfall and humidity, is higher in the alpine than in the Arctic tundra, but the steep topography results in a rapid runoff of water.

The Arctic and alpine regions share many features that characterize the tundra biome. The vegetation of the tundra is structurally simple (Figure 28.36). The number of species tends to be small, growth is low, and most of the biomass and functional activity are confined to a few groups of plants. Most of the vegetation is perennial and reproduces vegetatively rather than by seed. Although it appears homogeneous, the pattern of vegetation is diverse. Small variations in microtopography result in a steep gradient of moisture. The combination of microrelief, snowmelt, frost heaving, and aspect, among other conditions, produces an almost endless change in plant associations from place to place (Polunin 1955).

Although the environment of the Arctic and alpine regions is somewhat similar, the vegetation of the two differs in species composition and adaptation to light. Mooney and Billings (1961) conducted a series of studies to compare physiological and growth responses in Arctic and alpine ecotypes of alpine sorrel (*Oxyria digyna*) (Figure 28.37). They found that the plant exhibits increased production of flowers and decreased production of rhizomes in the southern portion of its range. Northern ecotypes of the plant have a higher photosynthetic rate at lower temperatures and attain a maximum rate at lower temperatures. Alpine plants reach the light saturation point at higher intensities than Arctic plants, which are adapted to lower light intensities. Arctic plants require longer periods of daylight for photosynthetic activity and growth. The further north the geographic origin of the plant, the more slowly the plant grows under short photoperiod.

South of the Arctic tundra and below the alpine tundra is a transition zone between forest and tundra, the edge of which is marked by the tree line. The tree line is a zone of gnarled trees, stunted by the desiccating force of the wind in alpine areas, an area known as the **krummholz** or "crooked wood" (Figure 28.38). The krummholz in the North American alpine region is best developed in the Appalachian and Adirondack mountains. On the high ridges, trees begin to show signs of stunting far below the timber line. The stunting increases with altitude until spruces and birches, deformed and semiprostrate,

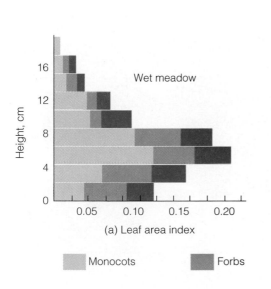

(a) Leaf area index

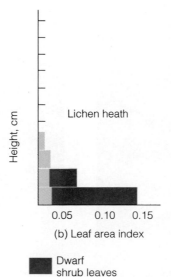

(b) Leaf area index

Monocots | Forbs | Dwarf shrub leaves

FIGURE 28.36 Structure of vegetation in Arctic tundra. (a) Maximum leaf area indices (m2/m2) at different plant heights (single surface) for a wet meadow. (b) Leaf area indices for a lichen heath. Note the structural simplicity. (After Berg et al. 1975.)

FIGURE 28.37   Alpine sorrel (*Oxyria digyna*) growing on the Arctic tundra.

(a)

(b)

FIGURE 28.38   The krummholz. (a) In the Rocky Mountains the tree line is sharply defined. (b) Dwarf spruce and fir grow in narrow pockets that hold snow in winter.

(a)

(b)

FIGURE 28.39   (a) The tree line in the Australian Alps in the Brindabella Range. (b) Low twisted growth of snow gum (*Eucalyptus pauciflora*) growing in protected pockets gives way to low-growing heaths.

form carpets 0.6 to 1 m high, impossible to walk through but often dense enough to walk upon. Where strong winds come in from a constant direction, the trees are sheared until the tops resemble close-cropped heads. However, the trees on the lee side of the clumps grow taller than those on the windward side.

In the Rocky Mountains, the krummholz is much less marked, for there the timber line ends almost abruptly with little lessening of height (Figure 28.38a). Most of the trees are flagged; that is, the branches remain only on the lee side. In the alpine regions of Europe, the krummholz is characterized by dwarf mountain pine (*Pinus mugo*), which forms dense thickets 1 to 2 m high on calcareous soils. On acid soil dwarf mountain pine is replaced by dwarf juniper (*Juniperus communis nana*). In the Australian Brindabella Range, the krummholz is marked by pockets of low, twisted snow gum (*Eucalyptus pauciflora*) (Figure 28.39).

Wind, cold, and winter desiccation are regarded as the cause of the dwarfed and misshapen condition of the trees. The ability of some tree species, such as mountain (mugo) pine, to show a krummholz effect may be genetically determined. Eventually conditions become too severe even for the prostrate forms, and trees drop out completely except for those that have taken root behind the protection of high rocks. These causes are only speculative. Much needs to be learned about the interaction of growth form, tolerance to cold, and spatial distribution of limited nutrients and water near the tree line (Stevens and Fox 1991). All of these factors could be involved.

Structure

Arctic Tundra   In spite of its distinctive climate and many endemic species, the tundra does not possess a vegetation type unique to itself. Structurally the tundra is a grassland and mixed shrubland.

In the Arctic only those species able to withstand constant disturbance of the soil, buffeting by the wind, and abrasion from wind-carried particles of soil and ice can survive. Although such an environment produces a vegetation that appears homogeneous, it is not. A combination of microrelief, snowmelt, frost heaving, and aspect, among other conditions, produces an endless change in plant associations from spot to spot. In the Arctic tundra, low ground is covered with cottongrass (*Eriophorum virginicum*) and other sedges, dwarf heath, and a sphagnum moss complex. Well-drained sites support heath shrubs, dwarf willows, and birches, dryland sedges and rushes, herbs, mosses, and lichens. The driest and most exposed sites—the flat-topped domes, rolling hills, and low-lying terraces, all usually covered with coarse, rocky material and subject to extreme action by frost—support only sparse vegetation. Plant cover consists of scattered heaths and mats of mountain avens (*Dryas*), saxifrages (Saxifragaceae), and other cushion plants, as well as crustose and foliose lichens.

Topographic location and snow cover delimit a number of Arctic plant communities. Steep, south-facing slopes and river bottoms support the most luxuriant and tallest shrubs, grasses, and legumes. Cottongrass (*Eriophorum angustifolium*) dominates the gentle north-facing and south-facing slopes, reflecting higher air and soil temperatures and greater snow depth. Pockets of heavy snow cover create two types of plant habitats, the snow patch and the snowbed. Snow patch communities occur where wind-driven snow collects in shallow depressions and protects the plants beneath. Snowbeds, typical of both Arctic and alpine situations, are found where larger masses of snow accumulate because of topographic peculiarities.

The conditions unique to the Arctic result in part from three interacting forces: permafrost, vegetation, and the transfer of heat. Permafrost is sensitive to temperature changes. Any natural or human disturbance, however slight, can cause the permafrost to melt. Because the permafrost itself is impervious to water, it forces all the water to move above it. Thus, surface water becomes conspicuous even though precipitation is low (see Brown and Johnson 1964, Brown 1970). Vegetation protects the permafrost by shading, which reduces the heating of the soil. It retards warming and thawing of the soil in summer and increases the average temperature in winter. If the vegetation is removed, the depth of thaw is 1.5 to 3 times that of the area still retaining the vegetation. Accumulated organic matter and dead vegetation further retard the warming of the soil in summer, even more than a vegetative cover. Thus vegetation and organic matter, acting as positive feedback, impede the thawing of permafrost and act to conserve it (Pruitt 1970, Bliss et al. 1973).

In turn permafrost, acting as negative feedback, chills the soil, retarding the general growth of both above-ground and below-ground parts of plants and the activity of soil microorganisms. It also impoverishes the aeration and nutrient content of the soil (Tyrtikov 1959). The effect is more pronounced the closer the permafrost comes to the formation of shallow root systems.

The tundra holds some fascinating animals, even though the diversity of species is low. Animals of the Arctic are mostly circumpolar in distribution. They include the caribou or reindeer (*Rangifera tarandus*), musk ox (*Ovibos moschatus*), arctic hare (*Lepus arcticus*), and arctic ground squirrel (*Alopex lagopus*). Some 75 percent of the species of birds of the North American tundra are common to the European tundra (Udvardy 1958).

Musk ox and caribou are the dominant large herbivores. Musk ox (Figure 28.40) are intensive grazers with low herd movements. In summer, they feed on grasses, sedges, and dwarf willow in the valleys and plains; in winter, they move up to the windswept ridges where snow cover is scant. Caribou are extensive grazers, spreading over the tundra in summer to feed on sedges and grasses, and migrating southward to the taiga in winter to feed on lichens.

Intermediate-sized herbivores include arctic hares, which feed on willows. The smallest and dominant herbivore over much of the Arctic tundra are the cyclic lemmings (*Lemmus* spp.), which feed on fresh green sedges and grasses. Species of herbivorous birds are few, dominated by ptarmigan (*Lagopus*) and migratory geese.

The major Arctic carnivore is the wolf (*Canis lupis*). (The polar bear is a marine predator.) The wolf preys on musk ox, caribou, and lemmings when they are abundant. Medium-sized predators include the arctic fox, which feeds on the arctic hare. The smallest mammalian predators, the least weasel (*Mustela rixosa*) and the short-tailed weasel (*M. erminea*), feed principally on lemmings and the nestlings and eggs of birds. Major avian predators, the snowy owl (*Nyctea scandiaca*) and the hawklike jaeger (*Stercorarius*), prey heavily on lemmings. Except for the wolf, whose population remains relatively stable when free from hunting by humans, the fortunes of most Arctic carnivores rise and fall with the flood and ebb of lemming life.

The Arctic tundra, with its expanse of ponds and boggy ground, is the haunt of myriad waterfowl, sandpipers (Scolopacidae), and plovers (Charadriidae), which arrive when the ice is out and return south before winter sets in.

Invertebrate life is scarce, as are amphibians and reptiles, for whom the short summers and cold are barriers. Some snails

FIGURE 28.40 The gregarious musk ox grazes on sedges and grasses in summer and browses on willows and other shrubs in winter.

are found in the Arctic tundra about Hudson Bay. Insects, reduced to a few genera, are nevertheless abundant, especially in mid-July. The insect horde consists of blackflies (Simuliidae), deer flies (Tabinidae), and mosquitoes (Culicidae).

Alpine Tundra The alpine tundra is a land of rock-strewn slopes, bogs, alpine meadows dominated by hair grass (*Deschampsia*), and shrubby thickets (Figure 28.41a). Cushion and mat-forming plants are important in the alpine tundra. Low and ground-hugging, they are able to withstand the buffeting of the wind. The cushionlike blanket traps heat, and the interior of the cushion may be 20°C warmer than the surrounding air, a microclimate that is used by insects.

In spite of similar conditions, however, only about 20 percent of the plant species of the Arctic and alpine tundra are the same, and they are of different ecotypes (see page 117). Lacking in the Rocky Mountain alpine tundra are heaths and a heavy growth of lichens and mosses between other plants; lichens are confined mostly to rocks, and the ground is bare between the plants. The alpine tundras of the Appalachian Mountains of North America and the Alps of Europe, however, are dominated by heaths and sedge meadows, and mosses are common. The Australian alpine region supports a growth of heaths on rocky sites, and wet areas are covered with sphagnum bogs, cushion heaths, and sod tussock grasslands (Figure 28.41b).

Alpine tundras also occur above the tree line of the high mountains in tropical regions: Central America, South America, Africa, Borneo, New Guinea, Java, Sumatra, and Hawaii. Tropical alpine regions undergo great seasonal variation in rainfall and cloud cover but experience little variation in mean daily temperature. Seasonal evenness in temperature is balanced by a strong daily fluctuation in temperature during the dry season from below-freezing conditions at night to hot summerlike temperatures during the day. Such a diurnal freeze-thaw cycle is unique to tropical alpine regions.

Tropical alpine tundras support tussock (bunch) grasses, small-leafed shrubs, and heaths; but the one feature that sets tropical alpine vegetation apart from the temperate alpine is the presence of unbranched to little-branched, giant, treelike rosette plants (Figure 28.42) (Smith and Young 1987). Although the genera and species differ among tropical alpine areas (for example, *Senecio* species in Africa and *Espeletia* species in the Andes), their growth forms and physiology are strikingly similar, suggesting convergent evolution. These species are the antithesis of the usual low-growing tundra species. The higher the elevation, the taller these plants grow, reaching up to 6 m. Many of these giant rosettes have well-developed water-storing pith in the center of the stem. They retain dead rosette leaves about the stem, which apparently act as insulation against the cold, and secrete a mucilaginous fluid about the bases of leaves that seems to function as a heat storage device. The leaves also possess dense pubescent hairs that reduce convective loss of heat (Meinzer and Goldstein 1984).

The alpine tundra, extending upward like islands in the mountain ranges of the world, is small in area and holds few characteristic animal species. The alpine regions of western North America are inhabited by pikas (*Ochotona* spp.), marmots (*Marmota* spp.), mountain goats (*Oreamnos americanus*) (not a goat at all but related to the South American chamois *Rupicapra rupicapra*), mountain sheep (*Ovis* spp.), and elk (*Cervus* spp.). Sheep and elk spend their summers in the high alpine meadows and winter on the lower slopes. The marmot, a mountain woodchuck, hibernates over winter, whereas the pika cuts and stores grass in piles to dry for winter. Some rodents, such as voles and pocket gophers (*Thomomys*), remain under the snow and ground during winter.

(a)

(b)

FIGURE 28.41 Alpine tundra (a) in the Rocky Mountains and (b) in the Australian Alps. Although the species of vegetation are different, the growth forms are convergent. The physiognomy is similar.

FIGURE 28.42 Tropical alpine tundra on Mt. Kenya, East Africa. Such alpine plants are able to resist freezing temperatures. Plant height tends to increase with elevation.

The alpine regions contain a fair representation of insect life. Flies and mosquitoes are scarce, but springtails, beetles, grasshoppers, and butterflies are common. Because of ever-present winds, butterflies fly close to the ground; other insects have short wings or no wings at all. Insect development is slow; some butterflies take more than two years to mature, and grasshoppers three.

Function

Primary Production Net annual primary production varies markedly across the tundra, depending on the plant community considered (Bliss 1975, Wielgolaski 1975a). Because of the microenvironmental conditions influenced by soil, slope, aspect, exposure to wind, snow depth, and drainage conditions, plant communities change rapidly over short distances.

Different components of the plant community make different contributions to net productivity, depending on the site. In the hummocky sedge-moss meadow of the Devon Island site, Canada, above-ground net primary productivity of vascular plants amounted to 44.7 $g/m^2/yr$, whereas mosses contributed 33 $g/m^2/yr$ and lichens contributed nothing. In the raised beach community dominated by lichens and cushion plants, vascular plants contributed 17.8 $g/m^2/yr$, mosses 2 $g/m^2/yr$, and lichens 25 $g/m^2/yr$. In an alpine wet meadow, vascular plants contributed 254 $g/m^2/yr$ and mosses 173 $g/m^2/yr$.

Much of the primary production of vascular plants is below ground rather than above. The data emphasize this important functional aspect of the tundra. The net annual above-ground primary production of vascular plants in a wet sedge-moss meadow at Devon Island was 45 g/m^2 above ground and 129 g/m^2 below; a wet meadow on an alpine tundra at the Hardangervidda site in Norway had a biomass of 254 g/m^2 above and 1316 g/m^2 below ground. (For summaries see Bliss et al. 1973, Wielgolaski et al. 1981.)

Although total production of the tundra is low because of a short growing season, daily primary production rates of 0.9 to 1.9 g/m^2 in the Arctic tundra and 2.2 g/m^2 in some alpine tundras are comparable to those of some temperate grasslands. The efficiency of primary production of the Devon Island tundra (Table 28.4) ranges from 0.03 percent for the polar desert to 1 percent for the hummocky sedge-moss meadow (Bliss 1975). Efficiency of a Norwegian tundra ranges from 0.6 percent for a dwarf heath community to 2.4 percent for a willow thicket (Wielgolaski and Kjelvik 1975). By comparison, efficiency of temperate grassland ecosystems ranges from 0.33 to 3.8 percent.

Biomass Biomass reflects the pattern of net production. At Devon Island, for example, biomass for the polar desert amounts to 270 g/m^2 and for the hummocky sedge-moss meadow 3208 g/m^2. More live biomass exists below ground than above in most tundra communities. On Devon Island the hummocky sedge-moss community has a root-to-shoot ratio of 1:12, the cushion plant–lichen community 1:0.06; and the wet sedge communities of both the Arctic and alpine tundra, 1:8.9.

TABLE 28.4 Net Annual Production and Efficiency for Various Tundra Plant Communities and Components (Growing Season of 50 to 60 Days)

Component	Net Production (kJ/m²)	Total Radiation (kJ/m²)	Efficiency (%)	
			Total Radiation	PAR*
Polar desert (plateau)	138	11.72×10^5	0.01	0.03
Polar semidesert (raised beach)	711	11.13×10^5	0.06	0.16
Sedge-moss meadows (all meadows)	2845	9.04×10^5	0.31	0.79
Hummocky sedge-moss meadow	3506	9.04×10^5	0.39	1.03
Total lowland	1916	9.62×10^5	0.20	0.50

*PAR (photosynthetically active radiation) = 40 percent of total radiation; efficiency was calculated on the basis of radiation received during the growing season for each component part and the contribution (%) of that component to the total.
Source: Bliss 1975.

Dead biomass on the tundra also accumulates below rather than above ground. The entire monocot portion of the vegetation grows and dies each season. Most of the herbaceous and woody plants accumulate little above ground and rapidly turn over energy in the living portion of the system. The below-ground portion is more persistent, with roots lasting for two to ten years, reflecting low mortality and slow decomposition. Because decomposition is slowest on wet sites, the difference between above-ground and below-ground biomass is greatest there. Decomposition is faster on well-drained and on nutrient-rich sites. The greatest accumulation of below-ground biomass is found, paradoxically, in those sites where net productivity is the lowest.

Decomposers Most of the production of the tundra enters the detrital food chain (Figure 28.43). At Devon Island, for example, grazing by lemmings accounts for only about 3 to 4 percent of the above-ground standing crop, except during cyclic highs. On the hummocky sedge-moss meadow, 2 percent of the plant production is used by herbivores; 98 percent of the primary production is channeled to microbivores and saprovores. The latter are a wide variety of soil organisms, dominated by protozoans. Among important soil invertebrates are potworms (Enchytraeidae), nematodes, fly larvae, and various crustaceans (McLean 1980). Annual production and consumption among these organisms are greater than in the herbivore system, even in a lemming high (Bunnell et al. 1975).

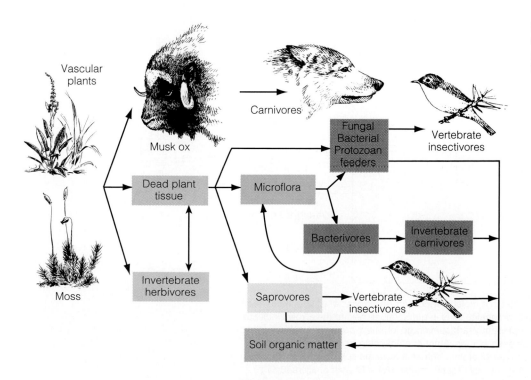

FIGURE 28.43 Energy flow diagram for a sedge-moss meadow system. (From Bliss 1975.)

However, the major detrital consumers are bacteria and fungi (Bunnell et al. 1980, Flanagan and Bunnell 1980). Tundra soils contain a diversity of soil bacteria. They occur in about the same abundance as in temperate soils. Fungi mycelia in tundra soils are as abundant or more so than in temperate mull and mor soils. As in other terrestrial systems, fungi appear to be more important than bacteria. Like primary producers, decomposers are restricted in their activities by the cold (Bliss et al. 1973, Bliss 1975, Bunnell et al. 1975, Rosswall 1975).

Consumers   Major herbivores of the tundra include waterfowl, ptarmigan, lemmings, hares, musk ox, and caribou (reindeer). In some parts of the Arctic tundra, lemmings are the dominant herbivores. During cyclic highs, these rodents may eat over 25 percent of the above-ground primary production. Of the food they eat lemmings return about 70 percent as feces, which accumulate in winter and release quantities of nutrients during the spring thaw (Bunnell et al. 1975). During cyclic highs, lemmings can hold primary production to 3 to 48 g/m². By reducing the litter layer, these animals can also reduce the insulation of the soil and increase the depth of thawing. In doing so, lemmings can influence the composition and nature of the tundra plant community.

In parts of the Arctic tundra, lemmings are not significant grazing herbivores. Their place is taken by musk ox and caribou. These two ungulates have an average standing crop of 0.17 kg/km² in the Canadian and Alaskan tundra, low compared with the ungulate biomass of 140 kg/km² of the African savanna. Musk ox remove less than 1 percent of potential primary production. However, musk ox are selective grazers. On restricted areas grazed on Devon Island, the animals removed 80 to 85 percent of the herbage available. Decomposition of their dung is slow, requiring five to ten years (Bliss 1975).

Preying on the herbivores are a number of carnivores. Arctic foxes consume 23 to 44 kcal/kg body weight, mostly lemmings, in winter. An efficient assimilator, the fox rarely passes as feces more than 5 percent of the total energy ingested. Extremely effective as predators, weasels may eat up to 20 percent of the lemming population. The snowy owl, another lemming predator, has a winter food requirement of four to seven lemmings a day. Not only can the predators act as a force to drive the lemming population to a low, but the predators themselves are affected by a scarcity of lemmings. When lemming populations are low, arctic foxes (*Alopex lagopus*) experience reproductive failures and snowy owls cannot exist on the tundra. When forced southward, the owls face an uncertain fate.

Nutrients   Arctic and alpine tundras are low in nutrients because of a short growing season, cold temperatures, low

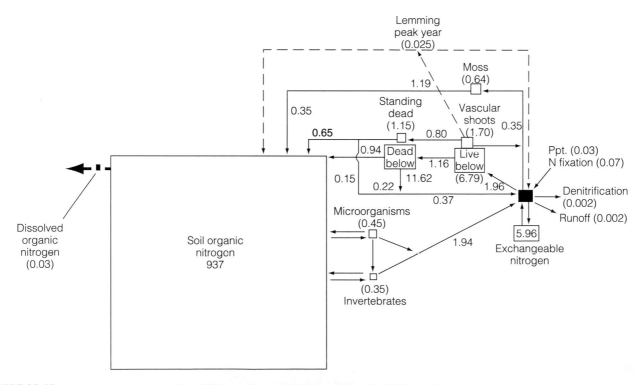

FIGURE 28.44   Annual nitrogen budget in a moist meadow on Arctic coastal tundra to the depth of 20 cm. The area of each box is proportional to its N in g/m². Values next to the arrows indicate annual fluxes in g/m²/yr. Solid lines represent years of average lemming populations. Years of high lemming populations are indicated by dashed lines. Amount of leaching from vascular plants is unknown; and mosses are assumed to get all of their nutrients from the soil (which they probably do not). Note the high proportion of N in the soil organic matter. (After Chapin et al. 1980:471.)

precipitation, and restricted decomposition. Of all terrestrial ecosystems, the Arctic tundra has the smallest proportion of its nutrient capital in live biomass. Dead organic material functions as a nutrient pool. Most of this nutrient capital is not directly available to plants. Pools of soluble soil nutrients, especially nitrogen and phosphorus, are small relative to exchangeable pools, which in turn are small relative to nonexchangeable pools (Figure 28.44) (Bunnell et al. 1975, Chapin et al. 1980). As a result, nutrient cycling in the tundra is conservative. Vascular plants retain and reincorporate nutrients, especially nitrogen, phosphorus, potassium, and calcium, in their tissues.

Leaching of nutrients is minimal, occurring mostly in the spring at the beginning of the growing season. As the snow melts, runoff increases, carrying with it animal debris that accumulates over winter. Spring rains leach senescent plant material produced the previous growing season. Rising temperatures stimulate decomposition. Because 60 percent of the active roots are in the upper 5 cm of the soil, the root mass once thawed takes up most of the nutrients. However, early in the season, plants have to compete with microbes, whose uptake of nutrients can exceed that of plant roots (Chapin et al. 1980). In summer vascular plants leak nutrients, especially phosphorus and potassium, from the cuticle of the leaves to the surface, where they are washed off by summer rains. These nutrients are often taken up by bryophytes (mosses), which absorb or adsorb nutrients before they reach the ground. Nutrients incorporated into mosses are slowly released. Thus, bryophytes function as a temporary nutrient sink.

In some tundra ecosystems, lemmings become involved in nutrient cycling. Foraging in meadows and ridges, but building their nests and defecating in troughs, lemmings transport nutrients from one place to another. Higher levels of soil nutrients, especially phosphorus, in polygon troughs, and higher rates of decomposition in these microsites may reflect lemming activity (Bunnell et al. 1975).

Two nutrients, nitrogen and phosphorus, are most limiting. Two major sources are precipitation and biological fixation. Nitrogen fixation is accomplished by both anaerobic and aerobic free-living bacteria; by cyanobacteria in soil, in water, and especially in mosses where they live epiphytically; and by lichens, particularly in the Fennoscandian tundra (Granhall and Lid-Torsvik 1975, Kallio and Kallio 1975). The input from precipitation is usually lower than the input from biological fixation, but at some places the two are equal.

Nitrogen accumulates in tissues in greater quantities than other nutrients, making much of it unavailable for recycling. At Barrow Research Station in Alaska about 65 percent of the gross input, 59 mg/m^2, is stored in the system as living and dead organic matter (Bunnell et al. 1975). Circulation in the system is restricted by decomposers, whereas flux rates are more or less controlled by nitrogen-fixing microorganisms. Small quantities of nitrogen are lost through leaching.

Phosphorus can be very limiting in the tundra ecosystem. Phosphorus apparently controls the rate of production

of new leaves by controlling the rate at which nutrients are removed from older leaves and translocated to new ones (Bunnell et al. 1975). Although the accumulations of nitrogen, potassium, and calcium are similar from site to site across the tundra, the accumulation of phosphorus is highest at the most productive sites.

Because available pools of both nitrogen and phosphorus are small, a constant turnover between exchangeable and soluble pools is necessary to replenish the quantity absorbed by plants. Bunnell et al. (1975) estimate that at the Barrow site soluble and exchangeable nitrogen must turn over 11 times during a growing season to meet plant needs, and phosphorus must be replenished 200 times during the growing season or 3 times during the day. Plants depend on nutrients released by decomposition, the uptake of which is aided by mycorrhizae.

Summary

1. Natural grasslands occupy regions where rainfall is between 25 and 70 cm a year. Many persist through the intervention of fire and human activity. Once covering extensive areas of the globe, native grasslands have shrunk to a fraction of their original size because of conversion to cropland and grazing lands. Disappearing along with the native grasslands have been the native grazing herbivores, replaced in part by domestic livestock.

2. Conversion of forests into agricultural lands, planting of hay and pasture fields, and development of successional grasslands on disturbed sites have extended the range of some grassland animals into once-forested regions.

3. Successional and climax grasslands are dominated by sod-formers, bunchgrasses, or both. Depending on their fire history and degree of grazing, grasslands accumulate a layer of mulch that retains moisture, influences the character and composition of plant life, and provides shelter and nesting sites for some animals.

4. Productivity varies considerably, influenced by precipitation. It ranges from 82 g/m^2/yr in semiarid grasslands to 40 times that much in subhumid, tame, and cultivated grasslands. The bulk of primary production goes underground to the roots. To a point, grazing stimulates primary production.

5. Although the most conspicuous grazers are the large herbivores, the major consumers are invertebrates. The heaviest consumption takes place below ground, where the dominant herbivores are nematodes. Most of the primary production goes to decomposers. Nutrients are recycled rapidly. A significant quantity goes to the roots, to be moved above ground to next year's growth.

6. Savannas are grasslands with woody vegetation. They are characteristic of regions with alternating wet and dry seasons. Difficult to characterize precisely, savannas range from grass with an occasional tree to shrub and tree savannas. The latter grade into woodland with an understory of grass. Much of the nutrient pool is tied up in plant and animal biomass, but nutrient turnover is high with little accumulation of organic matter.

7. Shrublands, which go by different names in various parts of the world, dominate regions with a mediterranean-type climate in which winters are mild and wet and summers are long, hot, and dry. Successional shrublands occupy land in transition from grassland to forest. Such shrubland may remain stable for years.

8. Shrublands characteristically have a densely branched woody structure and low height. The success of shrubs depends on their ability to compete for nutrients, energy, and space. In semiarid situations, shrubs have numerous competitive advantages, including structural modifications that affect light interception, heat losses, and evaporative losses.

9. Growth in mediterranean-type shrublands is concentrated at the end of the wet season, when nutrients in solution and a relative abundance of moisture produce a flush of vegetation. Nutrient cycling, especially of nitrogen and phosphorus, is tight. Many plants translocate nutrients from leaves to stem and roots before leaf-fall; others concentrate nitrogen in litterfall, which the plants take up again quickly in the wet season.

10. Deserts occupy about one-quarter of Earth's land surface and are largely confined to two worldwide belts, around the Tropic of Cancer and the Tropic of Capricorn. Deserts result largely from the movement of air masses, rain-blocking mountain ranges, and remoteness from sources of oceanic moisture. Two basic types of deserts exist: cool deserts, exemplified by the Great Basin of North America, and hot deserts.

11. The desert is a harsh environment in which plants and animals have evolved to become drought evaders or drought resisters. Functionally, deserts are characterized by low net production, opportunistic feeding patterns for herbivores and carnivores, and a detrital food chain that is less important than in other ecosystems. Crustlike growths of blue-green algae on the desert floor fix quantities of nitrogen, but most of it is lost to the atmosphere.

12. The alpine tundra of the high mountain ranges in lower latitudes and the Arctic tundra that extends beyond the tree line of the far north are at once similar and dissimilar. Both have low temperatures, low precipitation, and a short growing season. Both possess a frost-molded landscape and plant species whose growth rates are slow. The Arctic tundra has a permafrost layer; rarely does the alpine tundra. Arctic plants require longer periods of daylight than alpine plants and reproduce vegetatively, whereas alpine plants propagate themselves by seed.

13. Over much of the Arctic, the dominant vegetation is cottongrass, sedge, and dwarf heaths. In the alpine tundra, cushion and mat-forming plants, able to withstand buffeting by the wind, dominate exposed sites, while cottongrass and other tundra plants are confined to protected sites.

14. Net primary production is low because of a short growing season, although daily primary production rates are comparable to those of temperate grasslands. Biomass accumulates below rather than above ground. In spite of an assemblage of grazing ungulates and rodents, most production goes to decomposers. Major detrital consumers are bacteria and fungi.

15. Nutrient levels are low, and nutrients tend to accumulate and become stored in living and dead plant material unavailable for recycling. Circulation is restricted by limited activity of decomposers, while the flux rates are controlled by nitrogen-fixing organisms. Because pools of both nitrogen and phosphorus are small, constant turnover between exchangeable and soluble pools is necessary to replenish the quantity absorbed and retained by plants.

Review Questions

1. What characteristics do all grasslands have in common?
2. Distinguish between a bunchgrass and a sod grass.
3. What are the distinguishing characteristics of cultivated, annual, and successional grasslands?
4. Why is the root system so important in the grassland ecosystem?
5. What is the role of mulch in a grassland ecosystem?
6. What forms of animal life are characteristic of grasslands? What are some of their behavioral adaptations?
7. Discuss net production in grasslands. Where is net production the greatest within a grassland?
8. What is the impact of grazing on grasslands? How does above-ground herbivory compare with below-ground herbivory?
9. What distinguishes savannas from grasslands in structure and in function? Under what climatic conditions have they evolved?
10. Describe the role of fire and moisture in savanna ecosystems.
11. What is the role of trees, shrubs, and grass in the horizontal structure of savannas?
12. What is the relationship among trees, grass, and nitrogen cycling in savanna ecosystems?
13. What features are unique to shrublands?
14. What are mediterranean-type shrublands? What is their relationship to fire?
15. What are heathlands? Where do they occur?
16. What is the relationship between rainfall patterns and the nitrogen cycle in mediterranean-type ecosystems?
17. What climatic forces lead to deserts?
18. What are the major features of deserts?
19. What is a cool desert? A hot desert?
20. How are plants adapted to survive in the desert? What are drought resisters? Drought evaders?
21. Upon what does primary production in the desert depend?
22. Why is nitrogen limiting in desert ecosystems?
23. What physical and biological features characterize the tundra?
24. Define frost hummock, stone polygon, solifluction terrace, and permafrost.
25. Contrast adaptations of plants to light in Arctic and alpine tundras. Why the difference?
26. What is the krummholz?
27. Contrast the structural features of the Arctic tundra with those of the alpine tundra.
28. What is the relationship between tundra vegetation and permafrost?
29. Where is the bulk of primary production in the Arctic tundra? Why?
30. What is the role of grazing herbivores in the Arctic tundra?
31. What is unique about the nitrogen cycle in the Arctic tundra?

Cross References

Climate, 21–28; microclimate, 38–40; climate and vegetation, 27; water use efficiency, 108; plant adaptations to aridity, 108–109; animal adaptations to aridity, 139–140; adaptations to cold, 131–132, 134–137; nutrient cycling, 505–51; C_4 plants, 88–89; decomposition, 148–157; herbivores, 121–122; nitrogen cycle, 517–518; population cycles, 188–191, 300–302; plant defenses, 286–291; grazing disturbance, 467–470; succession, 286–291.

Forests

CONCEPTS

1. World forests fall into three general but highly diverse types—coniferous, temperate deciduous, and tropical—determined by broad environmental conditions.

2. Structurally, forests in general consist of four vertical layers of woody vegetation.

3. Vertical stratification in forests influences environmental conditions and diversity of life.

4. Dead and decaying wood are important components of the forest.

5. Old-growth forests possess distinctive structural features that set them apart from younger forests.

6. Nutrient cycling in forests is marked by the internal cycling of nutrients among leaves, stems, and roots, the release of nutrients from the litter to the soil by microbial decomposition, and uptake by roots. Although the processes in all forests are similar, differences exist among coniferous, deciduous, and tropical forests.

Of all the vegetation types of the world, none is more widespread or more diverse than the forest (see inside cover). Forests grow in distinct bands around the Northern Hemisphere. Progressing southward from the tundra are consecutive belts of coniferous, temperate deciduous, and about the equator, tropical forests. The latter are most extensive in the Southern Hemisphere. Within each band of global forest is a diversity of forest types.

NATURAL TYPES OF FORESTS

Coniferous Forests

Conifers are cone-bearing gymnosperms—resinous trees with dark green, needlelike or scalelike leaves. Except for the deciduous larches, bald cypress, and the unlikely gingko tree, conifers are evergreen, retaining their foliage year-round. This trait enables them to carry on photosynthesis throughout the year, whenever conditions are favorable. Dating back to the Carboniferous, conifers are much older than hardwood trees. Conifers rank as one of our most important natural resources, a major source of lumber and paper pulp.

Taiga and Boreal Forest The taiga (Russian literally for "land of little sticks") and associated boreal forest form a circumpolar belt across the Northern Hemisphere (Figure 29.1). It embraces the Subarctic Division of the Polar Domain between 50° and 70° North latitude. Winter, the dominant season, is severe and the small amount of precipitation the region receives comes in the three warm months. However, because of the cold that inhibits evaporation, the region is moist all year. The needle-leaf evergreens are adapted to cold winters, and they respond rapidly to the short summer.

The Subarctic Division has four major vegetation zones: (1) the forest-tundra ecotone characterized by open stands of stunted spruce, lichens, moss, cold lakes, bogs, rivers, and alder thickets; (2) open boreal woodland, or taiga, characterized by lichen-black spruce woodland (Figure 29.2); (3) the

FIGURE 29.1 Location of major coniferous forests of the world and associated climate diagrams. (Climate diagrams from Walter et al. 1975.)

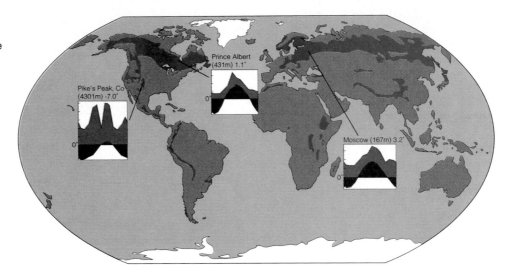

Pike's Peak, Co (4301m) -7.0°

Prince Albert (431m) 1.1°

Moscow (167m) 3.2°

(a)

(b)

FIGURE 29.2 The boreal forest extends from the taiga (a) dominated by black spruce at its northern limits to the spruce-birch forest (b) at its southern limits.

main boreal forest, with continuous stands of coniferous trees and moss and a low shrub understory broken up by poplar and birch on disturbed areas; and (4) the boreal-mixed forest ecotone that grades into the mixed hardwood-conifer forest of southern Canada and the northern United States (Oechel and Lawrence 1985).

Although the taiga and boreal forest throughout the world may have the same general appearance, the vegetation exhibits important regional differences. In Eurasia, the taiga and boreal forest begin in Scandinavia and extend across the continent to northern Japan. In Europe, the forest is dominated by Norway spruce (*Picea abies*); in Siberia, by Siberian spruce (*P. obovata*), Siberian stone pine (*Pinus sibirica*), and larch (*Larix sibirica*); and in the Far East, by Yeddo spruce (*Picea jaezoensis*). In North America, the boreal forest extends from Labrador across Canada and through Alaska to the Brooks Range. It is dominated by four genera of conifers, *Picea, Abies, Pinus,* and *Larix,* and two genera of deciduous trees, *Populus* and *Betula.* Dominant trees include black spruce (*Picea mariana*) and jack pine (*Pinus banksiana*).

Throughout much of its extent, the boreal forest experiences great seasonal fluctuations of temperature. These freeze-thaw cycles affect the growth and stability of shallow-rooted trees and, depending on the region, the presence or absence of permafrost. Occupying for the most part glaciated land, the taiga is also a region of cold lakes, bogs, rivers, and alder thickets.

Temperate Needle-Leaf Rain Forest Occupying the continental west coast of North America between 40° and 60°N from southern Alaska to California is the Marine Division of the Humid Temperate Domain, dominated by the maritime polar air masses delivering abundant rainfall. Moisture-laden prevailing westerly winds move in from the Pacific, meet the barrier of the Coast Range, and rise abruptly. Suddenly cooled by this upward thrust into the atmosphere, moisture in the air is released as rain and snow in amounts that can exceed 600 mm/yr. During the summer, when winds shift to the northwest, the air is cooled over chilly northern seas. Although rainfall is low, cool air brings in heavy fog, which collects on forest foliage and drips to the ground, adding 120 to 130 mm or more of moisture. This land of superabundant moisture, high humidity, and warm temperatures supports the temperate coniferous rain forest. Its luxuriant vegetation, including mosses, epiphytes, and ferns that cover branches and fallen logs, is dominated by a variety of conifers well adapted to wet mild winters, dry warm summers, and nutrient-poor soils (see Franklin and Dyrness 1973, Waring and Franklin 1979, Franklin and Waring 1979, Lassoi et al. 1985). These forests (Figure 29.3) are dominated by western hemlock (*Tsuga heterophylla*), mountain hemlock (*T. mertensiaa*), Pacific silver fir (*Abies amabilis*), and Douglas-fir, forming the densest coniferous forests containing some of the world's largest and tallest trees. Further south, where precipitation remains high, is the coast redwood (*Sequoia sempervirens*) forest, occupying a strip of land about 724 km long.

Woodlands In parts of western North America, where the climate is too dry for montane coniferous forests, we find piñon-juniper woodlands with which *Pinus* and *Juniperus* are

FIGURE 29.3 The temperate rain forest embraces the redwood and sequoia forests of California and the Douglas-fir–western hemlock (*Tsuga heterophylla*)–Sitka spruce (*Picea sitchensis*) forests of the northwest Pacific Coast. Typical is this redwood stand in northern California with characteristic undergrowth of sorrel (*Oxalis* spp.).

always associated (Figure 29.4). These ecosystems, a part of the Temperate Desert Division of the Dry Domain, are characterized by open growth small trees with a well-developed understory of grass and shrubs. The tree crowns rarely touch, and the root systems extend two to three times the diameter of the crowns. Piñon-juniper woodlands occur from the Front Range of the Rocky Mountains to the eastern slopes of the Sierra Nevada foothills. These woodlands have been greatly disturbed by livestock grazing, fuel harvesting, and exclusion of fire. Such disturbances the increase the density of trees and reduce the cover of grass (West 1988).

Southern Pine Forests The pine forests of southeastern United States fall into the Subtropical Division of the Humid Temperate Domain, marked by hot, humid summers and mild winters. They fall into two groups: the outer coastal plain mixed forest or temperate broadleaf evergreen forest (see page 609), and the southeastern mixed forest of southern pine and oak. The extensive pine forests of the outer coastal plain (Figure 29.5) are mostly secondary forests resulting from fire, deforestation, or deliberate forest management. Large areas of pine forests in the southeastern mixed forest likewise are successional, maintained by management practices for the

FIGURE 29.4 A piñon-juniper woodland in Utah.

FIGURE 29.5 Homogeneous stands of pine cover much of the lowland coastal plain and Piedmont of the southern United States. Many of these stands are dominated by loblolly pine.

production of southern pine lumber and forest products. At the northern end of the coastal pine forest in New Jersey, pitch pine (*Pinus rigida*) is the dominant species. There, a scrub ecotype of pitch pine and a dwarf form of blackjack oak (*Quercus marilandica*) rarely exceeding 3 m in height form extensive stands known as the pine barrens. Further south, loblolly (*Pinus taeda*), long-leaf (*P. australis*), slash (*P. elliotii*), and sand pine (*P. clausa*) are most abundant. In the southeastern mixed forest, loblolly and short-leaf pine (*P. echinata*) dominate. Where fires are allowed to burn or controlled burning is practiced, the understory is open and dominated by wiregrass (*Aristida*). In the deep south, saw palmetto (*Sereona repens*) and scrub palmetto (*Sabal etonia*) are common in the understory.

Montane Coniferous Forests The air masses that drop their moisture on the western slopes of the Coast Range then descend the eastern slopes, gain heat, and absorb moisture, creating the conditions that produce the Great Basin Desert (see Chapter 28). The same air rises up the western slopes of the Rocky Mountains, cools, and drops moisture again, although far less than on the Coast Range. Here, in the Cascade, Wasatch, Rocky, Sierra Nevada, and Cascade mountains, several coniferous forest associations develop (see Franklin and Dyrness 1973, Pfister et al. 1977). In the southwestern United States these coniferous forests occur between 2500 and 4200 m elevation, and in the northern United States and Canada between 1700 and 3500 m elevation. At high elevations in the Rocky Mountains, where winters are long and snowfall is heavy, grows the subalpine forest, dominated by Engelmann spruce (*Picea engelmannii*) and subalpine fir (*Abies lasiocarpa*) (Figure 29.6a). Middle elevations have stands of Douglas-fir, and lower elevations are dominated by open stands of ponderosa pine (*Pinus ponderosa*) (Figure 29.6b) and are open

FIGURE 29.6 (a) Subalpine mountain forest in the Rocky Mountains is dominated by subalpine fir (*Abies lasiocarpa*). (b) The drier lower slopes support ponderosa pine.

(a)

(b)

to thick stands of the early successional pioneering conifer, lodgepole pine (*P. contorta*).

Similar forests grow in the Sierra Nevada and Cascade mountains. There, high-elevation forests consist largely of mountain hemlock, red fir (*Abies magnifica*), lodgepole pine, sugar pine (*Pinus lambertiana*), incense cedar (*Calocedrus decurrens*), and the largest, if not the tallest, tree of all, the giant sequoia (*Sequoiadendron giganteum*), which grows only in scattered groves on the west slopes of the California Sierras.

A deciduous successional, occasionally permanent, species common both to the montane and the boreal forest is quaking aspen (*Populus tremuloides*). It is the most widespread tree of North America (Figure 29.7).

Montane coniferous forests dominated by pine also occur in Mexico and Guatemala. In Europe, forests of Scots pine (*Pinus sylvestris*) (Figure 29.8a) and Norway spruce (*Picea abies*) (Figure 29.8b) extend across the mountain ranges to the Spanish Sierra Nevadas. Both species are grown as plantation trees and as ornamentals in the United States.

Temperate Broadleaf Forests

Temperate forests, in spite of their name, do not exist in a temperate environment. They face extreme fluctuations in daily and seasonal temperatures that stress the physiological

FIGURE 29.7 Quaking aspen (*Populus tremuloides*) is the dominant deciduous tree in the western montane forest.

FIGURE 29.8 (a) A Scots pine forest in Scandinavia. (b) A Norway spruce forest in the Carpathian Mountains.

activity of plants and animals. Deciduous forests are leafless during the winter and in northern regions remain leafless for most of the year. They are exposed to droughts and, in places, flooding. In spite of their intemperate environment, temperate forest ecosystems are able to maintain high productivity.

Temperate broadleaf forests can be classified into temperate deciduous forests, temperate woodlands, and temperate evergreen forests. Like grasslands, the temperate broadleaf forests have been highly disturbed and modified by humans. Nearly all of the original forests have been cut for lumber or cleared for agriculture, mining, and urban development. Many of the forests we see today are second- and third-growth stands, or they are new stands that have reclaimed abandoned agricultural lands through planting and successional processes. Some of these forests are unmanaged, with natural processes predominating. Others are modified to some degree by managed processes, typically thinning, and by the removal of undesirable species and poorly formed trees.

Temperate Deciduous Forests Temperate deciduous forests of the Humid Temperate Domain once covered large areas of Europe, China, parts of South America, the Middle American highlands, and eastern North America (Figure 29.9). The deciduous forests of Europe and Asia have largely disappeared, cleared for agriculture and settlement. The dominant trees include European beech (*Fagus sylvatica*) pedunculate oak (*Quercus robur*), ashes (*Fraxinus* spp.), birches (*Betula*

spp.), and elms (*Ulmus* spp.) (Figure 29.10). Because of glacial history, the diversity of tree species in European deciduous forests is lower than that of North America and China.

In eastern North America, the temperate deciduous forest consists of a number of forest types that intergrade. The northern segment of the deciduous forest complex occupies the Warm Continental Division, subject to strong seasonal contrasts in temperature. It consists of a mixed forest of hemlock-white pine-northern hardwoods that occupies southeastern Canada and extends southward through the northern United States and along the high Appalachians into North Carolina and Tennessee. Beech (*Fagus grandiflora*), sugar maple (*Acer saccharum*), black cherry (*Prunus serotina*), red oak (*Quercus rubra*), and white pine (*Pinus strobus*) are the chief components.

On relatively flat glaciated country, with its deep, rich soil, grow two somewhat similar forests: the beech-sugar maple forest found mostly from southern Indiana north to Central Minnesota and east to central New York; and the sugar maple-basswood (*Acer saccharum–Tilia americana*) forest found in Wisconsin to Minnesota.

Occupying much of eastern central North America, the Hot Continental Division, is the extensive central hardwoods forest. The central hardwood can be divided into three major types: (1) The cove, or mixed mesophytic, forest consists of a large number of species dominated by yellow-poplar (*Liriodendron tulipifera*) (Figure 29.11a). This forest, which reaches its best development on the northern slopes and deep coves of

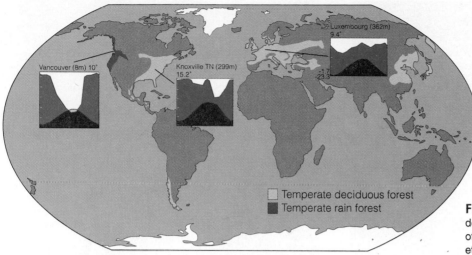

FIGURE 29.9 Location of major temperate deciduous forests and temperate rainforesrs of the world. (Climate diagrams from Walter et al. 1975.)

FIGURE 29.10 A European deciduous forest dominated by oak and ash.

the southern Appalachians, is one of the most magnificent in the world. Much of its original grandeur has been destroyed by fire and high-grading—removing the largest, commercially useful trees and leaving the smallest ones and unwanted species. However, even in second- and third-growth stands, its richness is apparent. (2) On dry sites and south-facing slopes grows the oak-chestnut forest. The chestnut (*Castanea dentata*) was eliminated as a timber tree by the exotic chestnut blight. Oaks have taken its place, and so this distinctive forest type is now largely oak forest. (3) Oak-hickory forests (Figure 29.11b) dominate the western edge of the central hardwoods in the Ozarks and the forests along the prairie river systems.

Temperate Woodlands In southwestern United States, Mexico, and California, open oak and oak-pine woodlands are transitional from coniferous forests to prairies, steppes, and desert. Belonging to the Dry Domain, woodlands in southern Arizona, New Mexico, and northern Mexico are dominated by oaks, especially Emory oak (*Quercus emoryi*), junipers, and pine, or they are pure oak woodlands. Forming a ring about the Central Valley of California is still another type—evergreen oak woodlands, dominated by blue oak (*Quercus douglasii*) and coast live oak (*Quercus aquifolia*), a common component of California valley foothill woodland, with a grassy undergrowth (Figure 29.12).

(a)

(b)

FIGURE 29.11 (a) A cove hardwood forest in Virginia. Redbud (*Cercis canadensis*) and dogwoods (*Cornus florida*) blooming in the understory add color to this forest dominated by yellow-poplar. (b) An oak-hickory forest in the Ozarks.

FIGURE 29.12 A blue oak woodland in California.

(a)

(b)

FIGURE 29.13 (a) An Australian mixed eucalyptus forest in the Grampian Range in western Victoria, Australia. (b) False beech (*Nothofagus*) forest in New Zealand.

Temperate Broadleaf Evergreen Rain Forest In several subtropical areas of the world are extensive forests of broadleaf evergreens. Such forests include many species of eucalyptus in Australia that may reach heights of 100 m (Figure 29.13a), anacardia gallery forests of South America and New Caledonia, and southern beech (*Nothofagus* spp.) forests in Patagonia and New Zealand (Figure 29.13b). Temperate broadleaf evergreen forests also occur in the Caribbean and on the North American continent along the Atlantic and Gulf coastal plains. Depending on location, these forests are characterized by live oak (*Quercus virginiana*), magnolias (*Magnolia* spp.), and redbay (*Persea borbonia*) with a low stratum of tree ferns and royal palm (*Roystonea elala*) and cabbage palm. The oaks and other trees support an abundance of epiphytes and the lichen, Spanish moss (*Tillandsia usneoides*).

Tropical Forests

Tropical Forests belong to the Rain Forest Division of the Humid Tropical Domain. Tropical forests experience a steady year-round temperature of about 23°C and a wide variation in rainfall, climatic conditions that are reflected in a diversity of vegetation patterns. In a very broad way, tropical forests can be classified tropical rain forest, montane rain forest, tropical seasonal forest, and tropical dry forest. Within these broad groups, however, are a diversity of vegetation types

unequaled anywhere else in the world. The tropical region contains as twice as many Holdridge life zones as the temperate zone and seven times the number of the boreal zone (Lugo and Brown 1995). Whereas a boreal forest type may be representative of a region, the same is not true of tropical forests. They range from the humid, wet lowland tropical forests to seasonal dry forests. Tropical forests once covered 20 percent of land surface; two-thirds of this has been degraded or eliminated by lumbering, clearing for plantations, settlements, and agriculture.

Tropical Rain Forests The tropical rain forest (Figure 29.14), so named in 1898 by the German botanist A. F. W. Schimper (see Chapter 1), comes in at least 30 to 40 types. These types include the monsoon forest, the evergreen savanna forest, the evergreen mountain forest, the tropical evergreen alluvial forest, as well as the true equatorial lowland tropical rain forest (see Odum 1970, Richards 1972, Whitmore 1984, 1990, Mabberley 1983, Terborgh 1992). They once formed a worldwide belt about the equator. The largest continuous rain forest is found in the Amazon basin of South America. West and central Africa and the Indo-Malaysian regions are other major locations of tropical rain forest. Smaller types of rain forest, now virtually gone, occur on the eastern coast of Australia, the windward side of the Hawaiian Islands, and the east coast of Madagascar.

FIGURE 29.14   (a) Amazonian tropical rain forest. Note the canopy emergents. (b) Lowland tropical rain forest in peninsular Malaysia. The prominent canopy trees are *Shorea,* representative of the dipterocarp family. (c) Eucalyptus rain forest in Australia. This rain forest is largely confined to scattered stands in Queensland. (d) Rain forest in Hawaii.

In all of these areas, rain forests grade into temperate and subtemperate rain forest.

Tropical rain forests grow where temperatures are constantly high and every month is wet (Figure 29.15). The mean annual temperature is about 26°C, the mean minimum rarely goes below 25°C, and the difference in temperature through the year is less than 4°C. Heavy rainfall occurs throughout the year, not less than 100 mm in any month for two out of every three years. What constitutes heavy rainfall varies with the region. The tropical rain forest regions in Latin America receive about 4000 mm rain

annually and those in Africa about 1500 mm. Under such conditions plant activity continues uninterrupted, resulting in luxuriant growth.

Tree species number in the thousands. A 10 square kilometer area of tropical rain forest may contain 1500 species of flowering plants and up to 750 species of trees. The richest is the lowland tropical forest of peninsular Malaysia, which contains some 7900 species. There, one of the major families, Dipterocarpaceae, contains 9 genera and 155 species, of which 27 are endemic. (The Asian dipterocarps have 12 genera and 470 species.) The few rain forest communities with

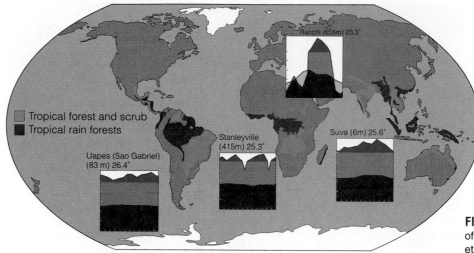

FIGURE 29.15 Location of tropical forests of the world. (Climate diagrams from Walter et al. 1975.)

single dominants are limited to areas of particular combinations of soils and topography.

The interior of a tropical rain forest is impressive. The tree trunks are straight, smooth, and slender, often buttressed with planklike extensions of roots (Figure 29.16), and reach 25 to 30 m before expanding into crowns with large, leathery, simple leaves (for the complex architecture of tropical trees, see Hallee et al. 1978). Climbing plants called lianas— long, thick, and woody—hang from trees like cables, and epiphytes grow on trunks and limbs. Undergrowth of the dark interior is often sparse, consisting of shrubs, herbs, and ferns. Litter decays so rapidly that the clay soil, more often than not, is bare. The tangled vegetation popularly known as jungle is secondary (second-growth) forest that develops where primary forest has been disturbed. Tropical rain forests account for millions of species of flora and fauna, and one-half of all known plant and animal species. Rain forests support 5 to 10 million species of arthropods, with more to be discovered.

Tropical Montane Rain Forests   In mountainous regions, the lowland tropical rain forest gives way altitudinally first to the lower montane and then to the upper montane forest (Figure 29.17). The lower montane forest, found only on taller mountains, is an ecotone between the mesophyll (broad leafed)-dominated lowland forest with an uneven billowing canopy and the microphyll (small leafed)-dominated upper montane forests. On small mountains, the transition between the lowland forest and the upper montane forest is abrupt. The montane forest features a lower, more even canopy, a dense lower canopy, and smaller trees with gnarled limbs heavily swathed in bryophytes and ferns.

FIGURE 29.16 Plantlike buttresses help support tall rain forest trees.

FIGURE 29.17 A mist-shrouded cloud forest in Costa Rica.

The summits of tropical mountains from southern Mexico to Peru, southeastern Brazil, west-central Africa, southeast Asia, and East and West Indies support cloud forests. These forests, usually 10 m tall or less, often called dwarf forests or elfin woods, are damp, dense, and dripping with moisture. They occupy an environment of frequent rainfall, persistent winds. and cool temperatures. The trees are festooned with mosses, leafy liverworts, epiphytic vascular plants, ferns, and gelatinous colonies of algae and are the habitat of many colorful birds.

Tropical Seasonal Forests   Tropical rain forests grade into semi-evergreen and semideciduous seasonal forests (Figure 29.18). They are characterized by less rainfall, more variable temperatures, and a dry season during which about 30 percent of the upper canopy tree species lose their leaves. In the Indo-Malaysian forests, new leaves emerge about a month after the coming of the monsoon rains. In other regions, leaves emerge about a month before the rainy season. Such forests are most common in southeastern Asia, India, South America, and Africa, and along the Pacific side of Mexico and Central America.

Tropical Dry Forests   Dry forests make up approximately 42 percent of all tropical forests (Figure 29.19). We overlook them and their importance in our concern about tropical rain forests (Murphy and Lugo 1986). Africa and tropical islands hold the largest proportion of dry tropical forest. Such forests make up about 22 percent of South American and 55 percent of Central American forested areas. Most of the original dry forests are gone, especially in Central America and India. Many of them have been converted to agricultural and grazing land, or they have regressed through disturbance to thorn woodland, savanna, and grassland.

Tropical dry forests experience a dry period, the length of which varies with latitude. The more distant the forest is from

FIGURE 29.18 A tropical seasonal forest on a mountain slope on Kauai, Hawaii.

FIGURE 29.19 A tropical dry forest in Africa. The plants in the middle distance are aloes, succulent herbaceous plants of the lily family.

the equator, the longer is the dry season, up to eight months. During the dry period, trees and shrubs drop their leaves. Before the start of the rainy season, which may be wetter than the wettest time in the rain forest, the trees begin to leaf. During the rainy season, the landscape becomes uniformly green.

STRUCTURE

The structural features of a forest are built upon vertical stratification created by the amount of space occupied by trunks, branches, twigs, and leaves at different levels. The amount of light spread uniformly at various levels controls the growth of woody plants adapted to grow at those heights (Terbrough 1991). In general there are four recognizable strata (presented in Chapter 20): the canopy that intercepts the bulk of solar radiation; the lower or understory tree canopy; the understory shrub layer; and the herbaceous or forest floor layer. All layers are not necessarily found in all forests. These several layers of vegetation influence microclimatic conditions within the forest, including light, moisture, temperature, and wind.

The highest temperatures are in the upper canopy, because this stratum intercepts solar radiation. Temperatures tend to decrease through the lower strata and change through the 24-hour period. At night the temperatures are more or less uniform from the canopy to the floor. In temperate deciduous forests, temperature stratification varies seasonally. In fall, when the leaves drop and the canopy thins, temperatures fluctuate more widely at the various levels. Maximum temperatures decrease from the canopy downward, but rise again at the litter surface. The soil, no longer shaded by an overhead canopy, absorbs and radiates more heat than in summer. This late winter and early spring warming of the forest soil, in part, stimulates the blooming of early spring woodland flowers.

The lowest humidity in the forest is a few feet above the canopy, where air circulation is best. The highest humidity is near the forest floor, the result of evaporation of moisture from the ground and settling of cold air from the strata above. Humidity in the temperate deciduous forest interior is highest in summer because of plant transpiration and poor air circulation. In a tropical rain forest, humidity increases and temperature and evaporation decrease. From the ground to 1 m, humidity stands at 90 percent; temperature on the average is 6°C cooler than outside forest cover, and it experiences a strong nocturnal inversion (Bourgeron 1983).

Bathed in full sunlight, the uppermost layer of the canopy is the brightest part of the forest. Down through the forest strata, light intensity dims to only a fraction of full sunlight. In an oak forest, only about 6 percent of the total midday sunlight reaches the forest floor; brightness of light at the forest floor is about 0.4 percent of that of the upper canopy. The amount of light that reaches the floor of a Malaysian rain forest is about 2 to 3 percent of incident radiation; half of that comes from sun flecks, about 6 percent from breaks in the canopy, and 44 percent from reflected and transmitted light (Mabberley 1983).

Dead wood in the form of large standing dead trees or snags (Figure 29.20), or downed trunks and limbs and dead branches on living trees, makes up an important component of the deciduous and coniferous forest ecosystems (Maser and Trappe 1984, Maser et al. 1979, Harmon et al. 1986). The mass of dead wood waxes and wanes with tree mortality and disturbances (Lang 1985) and never achieves equilibrium within a stand (Lang 1985, Long 1982). Woody litterfall usually increases over time (Long 1982) and becomes most conspicuous in old temperate forests, both deciduous and coniferous. Temperate broadleaf deciduous forests have a lower input, a faster decay rate, and thus a lower accumula-

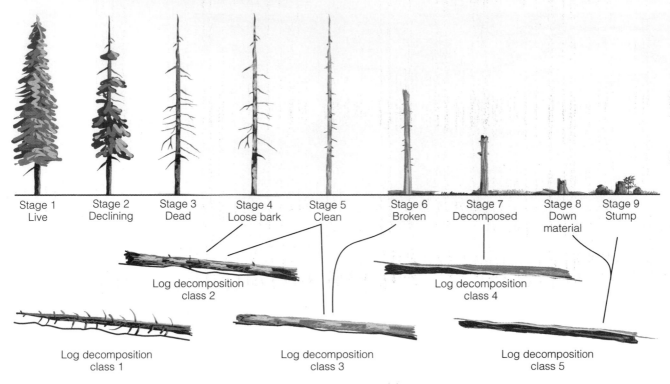

FIGURE 29.20 An important component of all forest ecosystems is dead wood, both standing and down. Snags or dead trees go through a process of decay, each stage of which supports its own group of animal life. When dead trees fall, they enter one of the last four decomposition classes.

tion of dead wood and nutrients on the forest floor than does coniferous forest, especially in the Pacific Northwest. There the cooler, wetter climate slows woody decomposition. Fallen trees and large limbs made up over 71 percent of the forest floor mass in a 250-year-old oak forest in New Jersey (Lang and Forman 1975), and over 60 percent in old-growth Douglas-fir stands (Grier and Logan 1977, Sollins et al. 1980).

Standing dead trees provide essential nesting and den sites for cavity-nesting birds and mammals. Eighty-five species of North American birds are cavity nesters. Some birds, notably the woodpeckers, are primary cavity nesters; they excavate nesting and roosting cavities. Other cavity nesters such as screech owls (*Otus asio*), wrens (*Thryothorus* spp.), and bluebirds (*Sialia* spp.) are secondary cavity nesters. Unable to excavate their own nests, they depend on those abandoned by woodpeckers and natural cavities in trees. All the cavity-dwelling mammals, such as squirrels, depend on natural cavities or holes excavated by woodpeckers for year-round places to sleep and rear young. Some birds, especially the creepers (*Certhia americana*), and some bats use cavities behind loose bark on dead trees as nesting sites or roosting sites.

Decaying logs, which may make up 10 to 20 percent of the ground surface in forests, are an important rooting medium for seedling trees, especially conifers. Decaying wood adsorbs both water and nutrients. Water retention capacity of decaying logs increases as decomposition increases. Thus, decaying wood provides a favorable medium for the establishment of coniferous seedlings (Ponge et al.

1988, Hornberg et al. 1998). Dead wood is inhabited by a host of insects living beneath the bark and in decaying wood—bark beetles (Scolitidae), wood-boring beetles (Bupestridae, Cerambycidae), carpenter ants (*Camponatus*), and termites (Isopoda)—that are preyed on by insectivorous vertebrate and invertebrate predators. Rodents, reptiles, and amphibians, especially salamanders, use well-decayed logs as burrows and runways. Thus, the elimination of standing and fallen dead trees, as is often done, can impoverish animal life in the forest.

Coniferous Forests

Coniferous forests fall into three broad classes according to growth form that influences their structure (Figure 29.21). These groups are (1) pines with straight, cylindrical trunks, whorled spreading branches, and a crown density that varies from the dense crowns of red and white pine to the open thin crowns of Virginia, jack, Scots, and lodgepole pine; (2) spire-shaped evergreens, including spruce, fir, Douglas-fir, and (with some exceptions) the cedars, with more or less tall pyramidal open crowns, gradually tapering trunks, and whorled, horizontal branches; and (3) deciduous conifers, such as larch and bald cypress, that have pyramidal, open crowns and shed their needles annually.

Vertical stratification in many coniferous forests is not well developed. Because of a high crown density and deep shade, the lower strata are often poorly developed in spruce and fir forests. The ground layer consists largely of ferns and mosses with few herbs. The maximum canopy development

Canopy

Understory
Shrubs
Ground cover

Spire-shaped conifer   Dense-crowned pines   Open-crowned pines   Tamarack

FIGURE 29.21 Vertical structure and stratification in coniferoous forests.

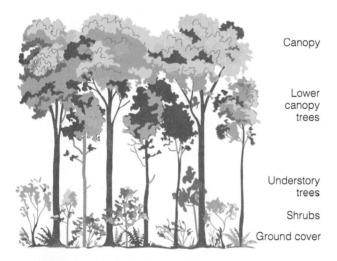

Canopy

Lower canopy trees

Understory trees

Shrubs

Ground cover

FIGURE 29.22 Stratification in an eastern deciduous forest.

in spire-shaped conifers is below the open upper crown. This feature gives such forests a profile different from that of pines. Pine forests with a well-developed high canopy may lack lower strata. However, stands of older growth and forests of open-crowned pines may have three strata: an upper canopy, a shrub layer, and a thin herbaceous layer. The litter layer in coniferous forests is usually deep, poorly decomposed, and lies on top of instead of being mixed with the mineral soil (see Chapters 4 and 9). It can inhibit the germination and development of woody understory and herbaceous plants.

Temperate Deciduous Forests

Highly developed, uneven-aged and older even-aged deciduous forests usually consist of four strata (Figures 29.22 and 29.23). The upper canopy (A) consists of dominant and codominant trees, below which is the lower tree canopy of

FIGURE 29.23 Leafless, this deciduous forest in late fall reveals its vertical structure.

saplings and understory trees (B), and the shrub layer (C). The ground layer (D) consists of herbs, ferns, and mosses. The litter layer is variable, depending on the lignin-to-nitrogen ratio. It ranges from a thin, rapidly decomposing layer typical of sugar maple and yellow-poplar stands to more slowly decomposing oak leaves.

Even-aged stands, the results of fire, clear-cut logging, and other large-scale disturbances, often have poorly developed strata early in their development because of dense shade. The low tree and shrub strata are thin, and the ground layer may be poorly developed, except in gaps, until the stand approaches maturity.

Old-Growth Coniferous and Deciduous Forests

Most forests with which we are familiar are relatively young second-, third-, or even fourth-growth (secondary) forests. They differ from those 150 to 450+ years old, known as old-growth (formerly called virgin). Old-growth forests, nearly gone and the subject of a bitter controversy in the Pacific Northwest, have certain characteristics in general (Franklin and Spies 1991, Peterken 1996). A late stage in successional development (Figure 29.24), they are often compositionally and always structurally different from earlier successional stages. Old growth consists of a wide range of tree sizes and spacing that results in a high degree of patchiness and heterogeneity at the scale of an individual stand. This patchiness results from some degree of disturbance over time within the stand. Old-growth stands are dominated by long-lived individuals and contain significant amounts of woody debris and downed logs (Figures 29.24 and 29.25). They possess more diversity in structure and function than secondary forests, and they are characterized by highly evolved complex relationships between plants and animals. Old-growth coniferous forests, especially boreal and western coniferous forests, support microcommunities of algae and lichens in the canopy (Lang et al. 1976, Johnson et al. 1982, Carroll 1979). These rather complex microcommunities include primary producers, consumers, and decomposers (Figure 29.26). Cyanophycophyllous lichens fix atmospheric nitrogen. Organic nitrogen lost through leaching from lichens combines with canopy moisture to form a dilute organic solution that in turn is taken up by microorganisms and other canopy epiphytes. Part of this microbial production is consumed by canopy arthropods.

Old-growth forests support certain species closely associated with them. In the Pacific Northwest these include the spotted owl, the marbled murrelet, and the red tree vole (*Phenacomys longicaudus*). Old-growth long-leaf pine is essential for the red cockaded woodpecker (*Picoides borealis*). The cutting of old-growth swamp forest in Louisiana in the 1950s brought about the extinction of the last remaining population of the ivory-billed woodpecker (*Campephilus principalis*). All of these relationships are destroyed when old-growth forests are cut.

No single definition applies all types of old-growth forest. Each has a different form, species composition, and stature at maturity. Some forests, particularly eastern deciduous forests, may achieve old-growth stage at 150–200 years. Other species and stands may not arrive at old-growth stage until much later. Because of the demise of old-growth forests, there are attempts to simulate the development of such forests

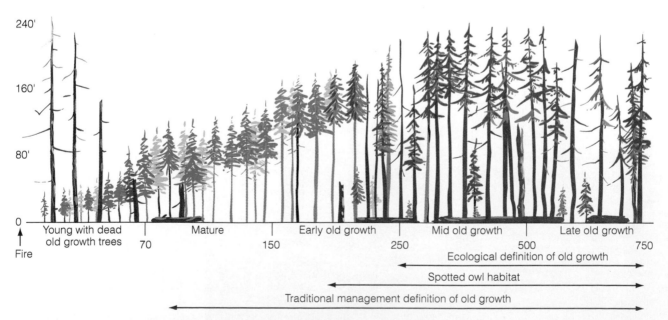

FIGURE 29.24 Successional development of an old-growth Douglas-fir forest after total destruction by fire. Note the changing structure during the various successional stages through old growth. Ecologists have a more restrictive concept of old growth than foresters, who consider the earlier stage of a mature forest as old growth.

FIGURE 29.25 Large amounts of dead wood and fallen logs characterize old growth forests. Old logs become a substrate for seedling trees.

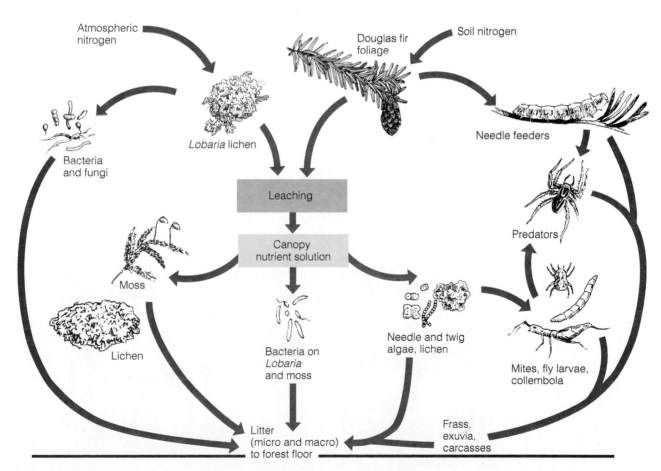

FIGURE 29.26 Nitrogen cycle in the canopy of an old-growth Douglas-fir. Such microecosystems probably exist in other forests, including tropical rain forests. Communities of primary producers (lichens) and biophage and saprophage consumers conserve and recycle nutrients such as nitrogen and increase return to the forest floor by leaching and throughfall. (After Johnson et al. 1982:193.)

through harvesting techniques in older even-aged stands (Franklin and Spies 1991, Peterken 1996).

Tropical Rain Forest

The tropical rain forest conventionally has been divided into five general layers (Richard 1965). The uppermost or emergent layer (E) consists of trees 50 to 60 m or more high. Their crowns rise above the rest of the forest to form a discontinuous canopy (Figure 29.27). The second layer consists of trees 24 to 36 m high (A). More or less continuous, it makes up the main canopy. A third layer (B), the lower tree stratum 15 to 24 m high is also continuous; the fourth layer (C), usually poorly developed in deep shade, consists of shrubs and young trees; and the ground layer (D) is composed of tall herbs and ferns. Unlike the temperate deciduous forest, stratification is often poorly defined; no clear demarcation exists between one layer and the other, the emergent tree stratum excepted (Hallee et al. 1978, Tomlinson 1983, Brunig 1983, Whitmore 1990, Terbrough 1991). The reasons are several. Many tree species have the same growth plan but differ in size. Different tree species mature at different heights. Other individuals become dwarfed because of unfavorable growth. This great variability in growth patterns and growth cycles can mask any strong stratification in some forests.

A conspicuous part of the rain forest is plant life dependent on trees for support. Such plants include epiphytes, climbers, and stranglers. Epiphytes, such as orchids, bromeliads, and aroids (members of the arum family), attach themselves to a host tree and obtain their nutrients from air, rainwater, and organic debris trapped by their aerial roots or leafless bases. Some epiphytes are important in recycling minerals leached from the canopy. Climbers are vinelike plants that reach the tops of trees and expand into the form and size of a tree crown. Climbers grow prolifically in openings, giving rise to the image of the impenetrable jungle. Stranglers start life as epiphytes. They send their roots to the ground and increase their stems in number and girth until the stranglers encompass the host tree and claim the crown limbs as support for their own leafy growth.

The mature tropical forest, like the mature temperate forest, is a mosaic of continually changing vegetation. Death of tall trees, brought about by senescence, lightning, wind storms, hurricanes, defoliation by caterpillars, and other causes, creates gaps that shade-intolerant pioneer species quickly fill (Poore 1968, Hartshorn 1978, Doyle 1981, Lugo and Scatena 1995). These trees are replaced eventually by shade-tolerant late successional species; but continuous, random disturbances across the forest ensures persistence of shade-intolerant species in the mature forest. A high frequency of treefall—Poore (1968) estimated that in his 12 ha Malaysian study area, one-half had experienced gap formation—may account for the low density of large trees (1 m+ dbh) in mature rain forests.

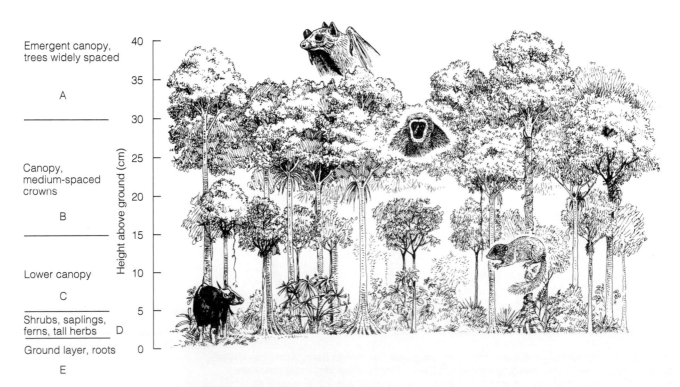

FIGURE 29.27 Stratification of vegetation and animal life in a Malaysian rain forest.

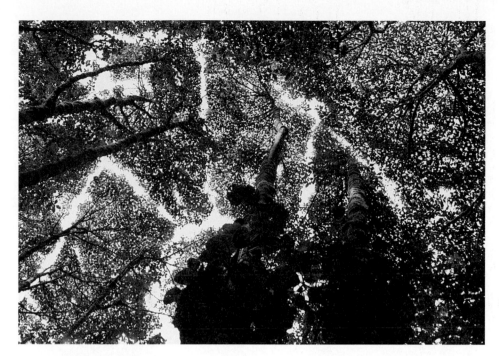

FIGURE 29.28 A view up through the canopy of a tropical rain forest reveals crown shyness. The upper canopy has sympodial crown structure.

Most tropical rain forest trees complete height growth when they have achieved only about one-third to one-half of their final trunk diameter. Thus, stratification or layering results when a group of species of similar height at maturity dominate a stand (Whitmore 1984, Terbrough 1991). Layering is also influenced by crown shape, which in turn is correlated with tree growth. Young trees still growing in height have a single stem and a tall narrow crown; they are monopodial. Mature trees have a number of large limbs diverging from the upper stem or trunk; they are sympodial. This change occurs when the bud of the main stem axis ceases to grow and the lateral buds take over their role. This process repeats itself, adding to crown growth and producing a pattern that suggests the spokes of an umbrella. Looking up into the canopy, the observer gains the impression that the crowns of the tree fit together like a jigsaw puzzle with the pieces about a meter apart (Figure 29.28). This growth pattern is called crown shyness.

Stratification and Animal life

The distribution of animal life in the forest is strongly influenced by stratification and microclimate. In the coniferous forest animal life forest varies widely, depending on the nature of the stand. Mites dominate the soil invertebrate litter fauna. Earthworm species are few and their numbers low. Insect populations, although not diverse, are high in numbers and, encouraged by the homogeneity of the stands, are often destructive. Spruce budworm (*Choristonema fumiferana*), found throughout the boreal forest, attacks balsam fir. Related species attack jack, red, and lodgepole pine. Sawflies (*Neodiprion*) attack a wide variety of pines, including pitch, Virginia, short-leaf, and loblolly.

A number of bird species are closely associated with coniferous forests. In North America, these include chickadees (Paridae), kinglets (Sylviidae), pine siskins (*Carduelis pinus*), crossbills (*Loxia* spp.), purple finches (*Carpodacus purpureus*), and hermit thrushes (*Hylocichla guttata*) (Figure 29.29). Related species, the tits and grosbeaks, are common to European coniferous forests.

Species diversity varies. In general, coniferous forests of northeastern and southeastern North America and the Sierra Nevada Mountains support the richest avifaunas (Wiens 1975). Bird densities are highest in the Pacific Northwest and lowest in immature northeastern coniferous forests. In the latter, more than 50 percent of the individuals are migratory neotropical warblers, whereas in western coniferous forests less than 10 percent are warblers. Foliage-gleaning insectivorous birds are dominant in all types of coniferous forests.

Except for strictly boreal species, such as the pine marten (*Martes americana*) and lynx (*Felis lynx*), mammals have much less affinity for coniferous forests. Most are associated with both coniferous and deciduous forests. The white-tailed deer, moose, black bear, and mountain lion (*Felis concolor*) are examples. Their north-south distribution seems to be limited more by climate, especially temperature, than by vegetation. The red squirrel (*Tamiasciurus hudsonicus*), associated with coniferous forests, is quite common in deciduous woodlands in the southern part of its range.

In the temperate deciduous forest, the greatest concentration and diversity of life in the forest occurs on and just below the ground layer. Many animals, particularly the soil and litter invertebrates, remain in the upper levels of the soil. Others, such as mice, shrews, ground squirrels, and forest

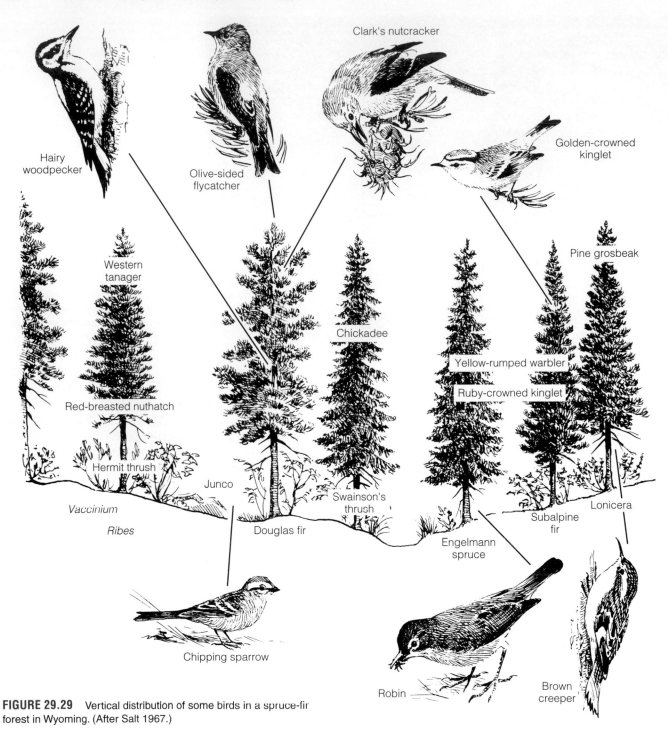

FIGURE 29.29 Vertical distribution of some birds in a spruce-fir forest in Wyoming. (After Salt 1967.)

salamanders, burrow into the soil or litter for shelter and food. Larger mammals, such as deer, live on the ground layer and feed on herbs, shrubs, and low trees. Birds move rather freely among several strata, but they favor one layer over another. Ruffed grouse (*Bonasa umbellus*) and ovenbirds (*Seiurus aurocapillus*) occupy the ground layer but move into the upper strata to feed, roost, or advertise territory.

Other species occupy the upper strata—shrub, low tree, and canopy layers. In the eastern deciduous forest, for exam-

ple, the red-eyed vireo (*Vireo olivaceus*) inhabits the lower tree stratum and the wood pewee (*Contopus virens*) the lower canopy. The black-throated green warbler (*Dendroica virens*) and scarlet tanager (*Piranga olivacea*) live in the upper canopy. Squirrels are mammalian inhabitants of the canopy, and woodpeckers, nuthatches, and creepers range up and down tree trunks in a space bounded by shrubs below and the canopy above. The absence of any one of the major strata will influence the diversity of forest life.

Stratification of animal life in the tropical rain forest, however, is pronounced (see Figure 29.27). Harrison (1962) recognized six distinct feeding strata. (1) A group feeding above the canopy consists largely of insectivorous and some carnivorous birds and bats. (2) A top of the canopy group—a large variety of birds, fruit bats, and other species of mammals—feeds on leaves, fruit, and nectar. A few are insectivorous and mixed feeders. (3) Below the canopy, a zone of tree trunks, is a world of flying animals—birds and insectivorous bats. (4) Also in the middle canopy are scansorial mammals that range up and down the trunks, entering the canopy and the ground zone to feed on the fruits of epiphytes growing on tree trunks, on insects, and on other animals. (5) Large ground animals make up the fifth feeding group. This includes large mammals and a few birds, living on the ground and lacking climbing ability, that can reach up into the canopy or cover a large area of forest. They include the large herbivores and their attendant carnivores. (6) The final feeding stratum includes the small ground and undergrowth animals, birds and small mammals capable of some climbing, that search the ground litter and lower parts of tree trunks for food. This stratum includes insectivorous, herbivorous, carnivorous, and mixed feeders.

Animal life in the tropical rain forest is largely hidden, either by dense foliage of the upper strata or by the cover of night. Birds are largely arboreal and, although brightly colored, remain hidden in the dense foliage. Ground birds are small and dark-colored, difficult to observe. Mammals appear scarcer than they really are, for they are largely nocturnal or arboreal. Ground-dwelling mammals are small and secretive. Tree frogs and insects are most conspicuous at evening, when their tremendous choruses are at full volume. Insects are most diverse at forest openings, along streams, and at forest margins, where light is more intense, temperatures fluctuate, and air circulates freely. Strongly colored butterflies, beetles, and bees are common. Among the unseen invertebrates, hidden in loose bark and in axils of leaves, are snails, worms, millipedes, centipedes, scorpions, spiders, and land planarians. Termites are abundant in the rain forest and play a vital role in the decomposition of woody plant material. Together with ants, they are the dominant insect life. Ants are found everywhere in the rain forest, from the upper canopy to the forest floor, although in common with other rain forest life, the majority tend to be arboreal.

Many specialized interactions among plants and animals exist in tropical rain forests. Plants, often widely dispersed in the forest, depend for pollination on birds, bats, and insects, especially beetles, bees, moths, and butterflies (see Procter and Yeo 1973, Baker et al. 1983, Howe and Westley 1988), and for seed dispersal on fruit-eating birds, bats, rodents, and primates. Heavy predation on seeds by insects may result in wide dispersal of tree species (Janzen 1971). Other interactions involve repellent toxins to discourage predation by herbivores and even insect-plant mutualisms in which insects such as ants live in hollow stems and prevent other insects from gaining entrance or feeding on the plant (Janzen 1967) (see Chapter 17).

FUNCTION

The structure of a forest, built from the accumulation of carbon (net primary production) at the various levels, provides the framework for the functional processes critical to net primary production—nutrient cycling.

Although nutrient cycling was discussed in some detail in Chapter 25, a brief retelling here as it relates to the forest is in order. The various process are outlined in Figure 29.30. They include (1) atmospheric inputs of dust, precipitation, and nitrogen fixation and outputs in drainage water; (2) uptake of nutrients from the soil and return of nutrients to the soil by plants through leaching, litterfall, and death of individuals; and (3) redistribution of nutrients within the plant, or internal cycling. The uptake of nutrients includes those nutrients returned to soil made available through leaching and decomposition, and nutrients from soil reserves.

Nutrient cycling varies through three general stages of forest development (Atwill 1986). The early stage involves the growth of biomass in which most of net primary production goes into the increase of photosynthetic tissues (the leaves) and the metabolic transport system (phloem and sapwood). In this stage of growth, up to 20 years, uptake of nutrients of soil is the greatest. From this point up to 50 or more years, depending on the species, most of the net primary production goes into the development of support tissue (the heartwood) and large amounts of nutrients are recycled through the tree, reducing dependence on uptake from the soil. As the tree matures, most of the net primary production

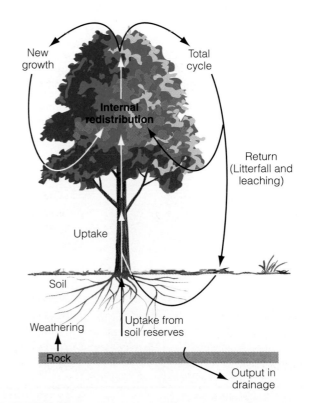

FIGURE 29.30 A schematic diagram of nutrient cycling in a forest.

goes into maintenance and litterfall. Older forests accumulate more nutrients in biomass than younger stands, and the rate of increase in nutrient storage declines (Ryan et al. 1997). Return of nutrients from plant to soil is maximum.

Loblolly pine (*Pinus taeda*) of the coastal plains of the southeastern United States provides an example of the uptake and accumulation of nutrients through the various stages of growth. Loblolly pine grows on a wide range of soils, from poorly drained lowland sites, where it does best, to dry, poor upland sites. This pine grows rapidly, achieving a diameter of 20 cm and a height of 18 m by age 20. For example, a 16-year-old stand of loblolly pine accumulated 232,490 kg/ha of biomass (Jorgensen and Wells 1986). Of this, 57 percent was in stemwood, 12 percent in branches, and 4 percent in foliage (Table 29.1). In this same stand, loblolly pine accumulated 26 percent of its nitrogen in needle biomass, 24 percent in stemwood, and 20 percent in roots (Table 29.2). Loblolly pine returned to the forest floor through litterfall and throughfall 37.5 kg/ha/yr of nitrogen and took up 54.4 kg/ha/yr (Table 29.3). The return amounted to 69 percent of the pines total requirement of 66 kg/ha/yr. Total uptake met 82 percent of the tree's requirement, whereas internal distribution amounted to 18 percent. A young growing stand, the pine obtained most of its nutrient requirement of nitrogen through uptake, and concentrated most of its nutrient uptake in biomass. Thus, at an early age loblolly pine exhibits a rapid accumulation of nutrients. From 20 to 40 years, nutrient accumulation slows as internal cycling increases. After 40 years, accumulation of nutrients declines as tree growth slows, becomes less dependent on uptake, and returns a greater portion to the soil through litterfall (Figure 29.31).

Data, summarized in Tables 29.1 and 29.2, for a 42-year-old and a 450-year old stand of Douglas-fir in the Cascade Mountains of Oregon and Washington provide an age-related contrast between nutrient accumulation and nutrient cycling (Cole et al. 1969, Cole and Rapp 1981, Johnson et al. 1982). The old stand has a considerably larger total biomass, both living and dead, than the younger stand. Although the two have similar foliar biomass, the young stand has 4 percent of its living biomass in foliage, higher than the old-growth stand with 2 percent. Most of the living biomass in both age classes is in branches and trunk. This biomass amounts to about 80 percent in the young stand and 86 percent in the 450-year-old stand. Sixteen percent of the living biomass of the young stand is in roots, compared with 12 percent of the old stand. Litter in the young stand accounts for 18 percent of the detrital organic matter and the soil 82 percent. In sharp contrast, litter in the old stand accounts for 55 percent of detrital organic matter and the soil 45 percent. Litter organic matter in the old stand exceeds soil organic matter because of the accumulation and slow decomposition of needles and the long-term decomposition of large fallen trunks and limbs.

The nitrogen budget of internal cycling and storage differs between young and old forest stands. The young stand has 32 percent of its N in foliage, whereas the old stand has only 14 percent in foliage. The old-growth stand has 66 percent of its nitrogen tied up in stemwood and bark compared with 39 percent in the young stand (see Table 29.2). The 450-year-old stand accumulates considerably more nitrogen, but it has a lower percentage of N in the soil. The young stand has 98 percent of its detrital nitrogen accumulated in the soil. The old stand returns somewhat more nitrogen to the forest floor

TABLE 29.1 Organic Matter Distribution, kg/ha (% in parentheses) for Young and Old-Growth Douglas-Fir and Young Loblolly Pine Ecosystems

Component	Douglas-Fir 42-Yr-Old	Douglas-Fir 450-Yr-Old	Loblolly Pine 16-Yr-Old
Overstory			
Foliage	9,097 (4)	8,906 (2)	9,700 (4)
Branches	22,031 (11)	48,543 (8)	27,900 (12)
Stemwood	121,687 (60)	472,593 (78)	132,180 (57)
Bark	18,728 (9)	*	18,460 (8)
Roots	32,986 (16)	74,328 (12)	44,550 (19)
Total	204,529	604,370	232,490
Subordinate vegetation	1,010	9,864	
Forest floor			
Wood	6,345 (1)	55,200 (1)	
Litter and humus	16,427 (17)	43,350 (54)	
Total	22,772	98,550	
Soil	111,552 (82)	79,250 (45)	
Total ecosystem	339,863	792,034	

*Value for bark included in stemwood.

Source: Data for Douglas-fir from Johnson et al. 1982; for loblolly pine from Jorgensen and Wells 1986.

TABLE 29.2 Nitrogen Distribution (% in parentheses), kg/ha, for Young and Old-Growth Douglas-Fir and Young Loblolly Pine Ecosystems

Component	Douglas-Fir 42-Yr-Old	Douglas-Fir 450-Yr-Old	Loblolly Pine 16-Yr-Old
Overstory			
Foliage	102 (32)	75 (14)	81 (26)
Branches	61 (19)	49 (9)	60 (19)
Stemwood	77 (24)	189 (36)	78 (24)
Bark	48 (15)	162 (30)	36 (11)
Roots	32 (10)	58 (11)	64 (20)
Total	320	533	319
Subordinate vegetation	6		
Forest floor			
Wood	14	132	
Litter and humus	161	434	
Total	175	566	307
Soil	2809	4300	1750
Total ecosystem	3310	5399	2376

Source: Data for Douglas-fir from Johnson et al. 1982; for loblolly pine from Jorgensen and Wells 1986.

TABLE 29.3 Nitrogen Transfers, (kg/ha/yr) for Young and Old-Growth Douglas-Fir and Young Loblolly Pine Ecosystems

Component	Douglas-Fir 42-Yr-Old	Douglas-Fir 450-Yr-Old	Loblolly Pine 16-Yr-Old
Input	1.67	2.0	5.4
Return to forest floor			
Throughfall	0.53	3.4	4.1
Litterfall	25.4	25.6	33.4
Total	25.93	29.0	37.5
Within vegetation			
Requirement	45.8	33.3	66.0
Redistribution	20.7	18.5	11.6
Uptake	25.1	14.8	54.4

Source: Data for Douglas-fir from Johnson et al. 1982; for loblolly pine from Jorgensen and Wells 1986.

than the young stand, mostly because of greater stemflow and throughfall (Table 29.3). The young stand has a considerably greater N requirement and a larger uptake than the old stand, 25.1 versus 14.8 kg/ha/yr. This uptake amounts to about 55 percent of the young stand's requirement and 44 percent of that for the old stand. In both stands, the deficiency of uptake is compensated for by recycling nitrogen within the biomass.

Ground moss is an important but overlooked component of some coniferous forests. Binkley and Graham (1981) found that in an old-growth Douglas-fir stand, ground-layer mosses account for only 0.13 percent of the above-ground biomass. In spite of this very low biomass, mosses add 5 percent to the estimated above-ground primary production, 5 percent to the uptake of calcium and potassium, and 10 percent to the uptake of nitrogen and phosphorus.

Nutrient cycling in the temperate deciduous forest differs from that in coniferous forest because trees annually return foliage as litter to the forest floor. There is no accumulation of nutrients in foliage over time. Nutrient cycling has been studied extensively in a number of temperate deciduous forest ecosystems in North America and Europe (Whittaker and Woodwell 1969, Likens et al. 1971, Likens 1976, Bormann and Likens 1979, Duvigneaud and Denaeyer-DeSmet 1970, Cole and Rapp 1981). The Walker Branch Mesic Hardwoods Forest at Oak Ridge, Tennessee, will serve as an example (Cole and Rapp 1981). The budget suggests

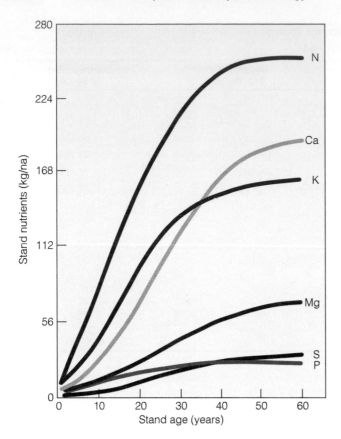

FIGURE 29.31 Accumulation of macronutrients in whole trees in loblolly pine stands on fertile sites. Note the rapid accumulation of nitrogen and calcium. (Adapted from Morgensen and Wells 1986:9.)

TABLE 29.4 Annual Element Balance of a 30- to 80-Year-Old Yellow-Poplar–Oak Forest, Oak Ridge, Tennessee (kg/ha/yr)

	N	P	K	Ca	Mg
Requirement	87.9	6.3	47.5	82.6	21.7
Uptake	58.1	3.4	40.0	87.8	12.4
Internal recycling	29.8	2.9	7.5	−5.2	9.3

Source: Data from Cole and Rapp 1981:359.

ticularly the foliage. The foliage, in turn, translocates a considerable portion of its nutrients back to stem before leaf-fall. However, mineral cycling can be maintained only if nutrients are pumped from soil reserves or are released through the weathering of parent materials.

The efficiency of nutrient cycling is reflected in the balance between inputs to the biological system and outputs or losses from the system through streamflow from the watershed. The difference represents the amount of recharge to the system from the soil pool. Table 29.5 summarizes the input from precipitation and output through leaching to the watershed for nitrogen and the four cations phosphorus, potassium, calcium, and magnesium. Nitrogen and phosphorus were rather tightly retained in the forest, whereas outputs exceed inputs for potassium, calcium, and magnesium. Input and output, of course, vary widely among ecosystems, influenced by climate and geology. The high output of calcium and magnesium reflects the high calcium content of the soil. In contrast, the annual output of calcium and magnesium from the low-calcium soils of Hubbard Brook Forest in New Hampshire amounted to 8.0 and 2.6 kg/ha respectively (Table 29.6) (Likens et al. 1967).

Coniferous and deciduous forests are relatively homogeneous compared with the highly diverse tropical forests that cover more than half the world's life zones. This wide variation precludes any sweeping generalizations about tropical forest function or comparison between tropical and temperate forests (Lugo and Brown 1991). Although both temperate and tropical forests appear to achieve similar biomass at maturity, the tropical forests approach steady state much earlier than the temperate forest (Lugo and Brown 1991).

Mineral concentrations vary widely among tropical forests, influenced by site, climate, and a wide range of soil types (Table 29.7) (Richter and Babbar 1991, Vitousek and Sanford 1986). Concentrations of all major nutrients are higher on more fertile soils, exemplified by the Panamanian forests, than on the infertile Oxisol/Ultisol soils that are especially low in phosphorus, potassium, and calcium. It follows that tropical forests on more fertile soils return more litter with higher concentrations of nutrients to the forest floor than do forests on other soils (Table 29.8). Phosphorus, always of low availability because of long-term weathering, rapid adsorption of the element by sesquioxide clays, and immobilization by decomposers, is most limiting on infertile soils.

several characteristics of mineral cycling in temperate deciduous forests. One is that the uptake of nutrients (annual increment of elements associated with trunk and branch wood minus annual loss through litterfall, leaf wash, and stemflow) does not meet requirements (annual increment of elements associated with trunk and branch wood plus current foliage production) (Table 29.4). The uptake of nitrogen amounts to 58.1 kg/ha/yr, 66 percent of requirement. Deficiencies have to be met by cycling nutrients within the tree biomass. Nutrient accumulations in tree biomass (Table 29.5) form a considerable pool, most of which is unavailable for short-term recycling. Forty-nine percent of N, 35 percent of P, 42 percent of K, 62 percent of Ca, and 49 percent of Mg are incorporated into woody biomass; 31, 51, 43, 28, and 24 percent respectively in root biomass; and 20, 14, 15, 10, and 27 percent respectively in foliage.

Considering the whole Walker Branch yellow-poplar–oak forest ecosystem, 7 percent of N, 3 percent of P, 12 percent of Ca, and 1 percent each of K and Mg are stored in forest floor vegetation. The litter layer is the most important nutrient pool, because it is quickly decomposed (average turnover time of four years) and its nutrients recycled, although the bulk of the nutrient pool is in mineral soil. Nutrients stored in living biomass, especially in roots, are translocated and recycled through the living biomass, par-

TABLE 29.5 Nutrient Balance Sheet for 30- to 80-Year-Old Yellow-Poplar–Oak Forest, Oak Ridge, Tennessee (kg/ha)

	N	P	K	Ca	Mg
Input/output					
Atmosphere	8.7	0.54	1.0	9.1	1.1
Leaching to watershed	1.8	0.02	6.8	147.5	77.1
Loss/gain	+6.9	+0.52	−5.8	−138.4	−76.0
Internal cycling					
Litterfall	36.2	2.7	19.1	58.3	8.3
Throughfall	12.0	0.4	18.4	21.9	3.4
Total	48.2	3.1	37.5	80.2	11.7
Live accumulation					
Woody biomass	189	15	127	462	38
Roots	122	22	132	211	19
Foliage	78	6	45	75	21
Total	389	43	304	748	78
Detrital accumulation					
Forest litter	187	11	14	294	22
Soil rooting zone	7,300	1,400	36,000	6,300	8,700
Total	7,487	1,411	36,014	6,594	8,722

Source: Data from Cole and Rapp 1981:394.

TABLE 29.6 Inputs and Losses of Nutrients of the Hubbard Brook Forest (kg/ha) New Hampshire

	Ca	Mg	Na	K
Input	3.0 ± 0	0.7 ± 0	1.0 ± 0	2.5 ± 0
Output	8.0 ± 0.5	2.6 ± 0.06	5.9 ± 0.3	1.8 ± 0.1
Loss/Gain	−5.0 ± 0.5	−1.9 ± 0.06	−4.9 ± 0.3	+0.7 ± 0.1

Source: Data from Likens et al. 1967.

TABLE 29.7 Nutrient Content of Above-Ground Biomass in Moist Tropical Forest Ecosystems

Site	Biomass (tons/ha)	NUTRIENTS (kg/ha)				
		N	P	K	Ca	Mg
Panama	316		158	3020	3900	403
Brazil	406	2430	59	435	432	201
Venezuela	335	1084	40	302	260	69

Source: Data from Vitousek and Sanford 1986.

Forests on infertile soils appear to cycle phosphorus more efficiently than forests on moderately fertile sites, by retranslocating phosphorus from leaves prior to litterfall. Montane tropical forests fit no single pattern because of the variability of their mountainous sites. Throughfall is the major pathway for cycling of potassium in all forests.

Except for nitrogen, nutrient budgets, like those of the temperate forest, can be assessed by comparing inputs from rainfall with outputs by stream discharge or by using soil lysimeters. The difference between the two represents inputs from the weathering of the soil and soil materials. Data for a tropical moist forest in Costa Rica indicate that

TABLE 29.8 Dry Mass and Nutrient Content of Litterfall

Site	Litterfall (tons/ha)	NUTRIENTS (kg/ha)				
		N	P	K	Ca	Mg
Moderately fertile soil						
Panama	11.1	195	15	47	216	26
Costa Rica	8.1	135	6	20	59	16
Sarawak	11.5	110	4	26	240	20
Infertile oxisols/ultisols						
Ivory Coast	11.9	170	8	28	61	51
Brazil	7.3	106	2	13	18	14
Malaysia	8.9	100	2	32	70	18
Montane forest						
Papua New Guinea	7.6	90	5	28	95	19
Sarawak	11	86	3	13	21	16
Hawaii	5.2	37	2	12	84	10

Source: Adapted from Vitousek and Sanford 1986:153.

inputs of phosphorus and potassium by precipitation exceed outputs, but considerably more calcium and magnesium are lost from the system than are gained by rainfall (Table 29.9). This finding and others suggest considerable input of these two nutrients from the soil reservoir and conservation of phosphorus and potassium by rapid internal cycling. Internal cycling is aided by the rapid return of nutrients leached by throughfall.

As in temperate forests, bacteria and fungi, the main agents of decay, release nutrients directly into the mineral soil, where they are subject to leaching. The fine roots of trees are woven into the matrix of mineral soil. On very poor sites, however, roots are concentrated in the well-aerated upper 2 to 15 cm of humus, and only a few roots penetrate the upper layer of mineral soil (Cornforth 1970b, Jordan 1979, 1982a, b). The roots of tropical forests support an abundance of mycorrhizal fungi that attach fine roots to dead organic matter by hyphae and rhizomorph tissue, but the full extent of the benefits of mycorrhizal fungi to forest trees and the manner in which they work is uncertain (Alexander 1989).

TABLE 29.9 Input-Output Balances for Moderately Fertile Tropical Moist Forest, Costa Rica (kg/ha/yr)

Element	Precipitation Input	Hydrologic Output
P	0.17	0
K	5.4	3.6
Ca	3.1	5.7
Mg	2.6	8.5

FATE OF FORESTS

For 5000 years humans have exploited forests for fuel and building materials, giving little thought to environmental consequences or future forest growth. The Bronze Age Minoans and Mycenaeans, the Greeks, and the Romans exploited the rich forests about them for fuel, ship building, construction, metal smelting and glass, and cleared large areas for cropland and pastures (see Perlin 1991). This loss of forest cover resulted in massive soil erosion and, ultimately, the conversion of Mediterranean forest lands to bare, rocky soil and degraded Mediterranean shrubland. The spread of Roman civilization into heart of Europe brought about clearing of land for agriculture and the heavy exploitation of forest for charcoal, smelting of ores, and expansive, elaborate building. Cutting wood for fuel and harvesting large areas of old trees to build large navies consumed massive amounts of timber. Eventually, this overexploitation caused timber shortages, especially in the 1600s and 1700s. At that time Ireland, once abundantly wooded, was deforested for casks and iron smelting (see Darby 1956, Perlin 1991). Through the millennia, like today, persons concerned with the environment decried forest destruction. Aristotle in Greece and Pliny and Cicero in Rome pointed out the dire environmental consequences already evident in the classical landscape and urged forest protection. In 1662 John Evelyn in England strongly condemned forest destruction in a report to the Royal Society that rallied public interest for forest conservation and reforestation. For the most part, however, such warnings were ignored by the political powers that controlled the fate of the forests.

Lessons from past forest destruction throughout Europe meant little to the settlers of North America. The rich, seemingly endless forest appeared inexhaustible, and loggers cut as if it were. In the 1800s North America, especially the United States, experienced the most rapid and massive deforestation of a continent by humans in Earth's history (Williams 1989). Well over 80 percent of the rich hardwood forests were destroyed through lumbering, and clearing for agriculture and settlements. The magnificent pine forests of the Great Lakes region experienced a devastation from which the land has never recovered. The extent of this deforestation in much of eastern United States is masked today by the return of the forest on previously cutover and abandoned agricultural land, although the replacement forests lack the diversity and grandeur of the originals.

Currently, the tremendous demand for lumber is accelerating the harvesting of timber. Vast areas of boreal forest are being clear-cut to meet the demand for construction timber and pulp wood, with little regard for future production. Cutover boreal forests require more than a hundred years to recover. Hardwood stands in eastern North America, cut in the 1930s, have reached saw timber size and are much in demand for flooring, furniture, and fine construction. Because large sawlogs are too valuable to use in heavy construction, timber industry now clear-cuts and chips younger stands for chipboard, used in building construction and in prefabricating structural joists.

The current heavy cutting of today's forests raises concern about the growth and nature of future forests. Growing large timber for sawlogs requires 65 to 100 years, depending on the species. Many landowners and foresters may view such long-term management as giving poor economic returns on land. Because of the economics involved, forest owners and managers may choose short-term cutting cycles of 20 to 40 years to produce fiber for the manufacture of chipboard and for pulp wood. Employment of such short-term and even longer harvesting cycles will result in management of forests as tree farms of native species, with the erosion of species diversity and loss of specific habitat for vulnerable species. In many regions, particularly in the Pacific Northwest and in the southern United States, the forest industry converts cutover native forest forests to simplified monocultural plantations of fast-growing pines and Douglas-fir (Figure 29.32). Such conversion involves clear-cutting of all trees and snags, prescribed burning or use of herbicides to control competing vegetation, and replanting the area to a single species.

On many cutover forests, public or private, regenerating new stands of native forests face problems. Just as second-growth stands differ in species composition from the primary stand, so third-growth stands differ from second-growth. Because of the widespread failure of oak reproduction, and the impact of selective browsing by white-tailed deer, many regenerating eastern hardwood forests will be dominated by red maple rather than a diversity of oaks.

Of greater significance to future of forests is the conversion of cutover areas to suburban building lots and industrial expansion, and the subdivision of older stands into wooded

FIGURE 29.32   This monocultural plantation of fast-growing loblolly pine lacks the complexity of natural pine stands. Such plantations make up industrial forests. Usually managed on short rotations, plantations provide the pulpwood, poles, and treated lumber used in construction.

home sites. In both cases forest land is permanently removed from future forest production, possibly exacerbating the future undersupply of timber and reducing wildlife habitat.

The tropical forest regions of the planet are experiencing a rapid rate of logging and often subsequent deforestation (Lanly 1995) (Figure 29.33). This includes most of the rain forests of Southeast Asia, as well as patches in Australia, West Africa, and Madagascar. At the current rate of destruction, little will remain of the Amazonian rain forest within 25 years.

Lost in the degradation or destruction of the tropical rain forest is its high diversity of plant and animal life. Although occupying only 7 percent of Earth's surface, tropical rain forests hold 50 to 80 percent of the world's plant species. Most of the animal species are endemic, resident, and nonmigratory. Once the habitat is destroyed, the species go extinct. Not only is the fauna endangered; for the first time in human history, hundreds of plant species also face extinction. They are at risk because of habitat destruction and the loss of fauna on which they depend for pollination and seed dispersal (Terbrough 1991, 1995) (see Chapter 17). The diverse tropical regions are a source of food, medicinal plants little studied and utilized, genetic diversity, and economic opportunity poorly managed for the future (see Myers 1983, Wilson 1988, Collins 1990).

FIGURE 29.33 The tropical rain forests of the world face destruction by land-cleaning schemes, the fate of this piece of Amazonian forest.

Summary

1. Coniferous, deciduous, and tropical forests are the three major types of forest ecosystems. Coniferous forests occupy a range of environments. The most extensive is the boreal forest that forms a vast belt encircling the northern part of the Northern Hemisphere. It is typical of regions where the summers are short and winters are long and cold. Pacific coastal coniferous forests grow in an area of superabundant moisture, high humidity, and warm temperatures. Pines, spruce, and fir dominate the montane coniferous forests in North America and Eurasia. Successional pine forests occupy coastal plains of the southern United States.

2. The deciduous forest, richly developed in North America, Europe, and eastern Asia, arises in a region of moderate precipitation and mild temperatures during the growing season. There are a number of variations influenced by soils, local climates, and topography. Temperate woodlands have widely spaced trees with a grassy understory. Subtropical climates in North America, South America, and Australia support temperate evergreen forests.

3. Diverse tropical forests occur in equatorial regions under environmental conditions that range from high humidity, heavy rainfall, minimal seasonal changes, and an annual mean temperature of about 28°C to seasonal dryness and an annual mean temperatures of about 17°C. Although widely variable, tropical forests can be grouped into four broad types: tropical rain forest, montane tropical rain forest, tropical seasonal forests, and tropical dry forests.

4. All forests are more or less stratified into layers of vegetation, including a canopy, subcanopy, and ground layer. Accompanying this vegetative stratification is stratification of light, temperature, and moisture. The canopy intercepts light and precipitation; the forest floor is shaded through the year in most coniferous and tropical rain forests, and in late spring and summer in deciduous forests.

5. Standing dead trees and decaying logs are an important component of many forests. Both are essential habitats for many species of animals, provide a substrate for seedling growth, and function as long-term storage of slowly available nutrients.

6. Old-growth forests, structurally distinct from younger stands, are characterized by large trees, a relatively large amount of standing and fallen dead wood, and a high degree of patchiness. Many support species dependent on old-growth forests. Few old-growth forests remain.

7. Coniferous and deciduous forests hold different species of animal life, but seasonal adaptations are similar. The greatest concentration and diversity of life are on and just below the ground layer. Other animals live in the various strata from low shrubs to the canopy. The tropical rain forest has pronounced feeding strata from above the canopy to the forest floor, and many of its animals are strictly arboreal. Whatever the forest, the trees that compose it create different environments that ultimately dictate the kinds of plants and animals that can live within it.

8. Mineral cycling is tight. Nutrients accumulate in woody biomass to form a pool unavailable for short-term cycling. Although the bulk of the nutrients is in mineral soil, important pools in mineral cycling are litterfall and foliar leaching. Internal cycling of some nutrients, especially nitrogen and phosphorus, is important in nutrient conservation. In most forest systems, only a small fraction of nutrients is lost from the system through streamflow.

Review Questions

1. Compare and contrast coniferous, deciduous, and tropical forests in structure and nutrient cycling.

2. What is the importance of dead wood in a forest stand? What are the ecological effects of removing dead wood for firewood and other uses, as is usual in European forests?

3. Name and characterize the several broad types of tropical forests.

4. If, as common belief holds, tropical forests soils are so nutrient-poor, how can they support such a high plant biomass and diversity?

5. Why has the tropical dry forest been most heavily affected by humans? (Review Murphy and Lugo 1986.)

6. Lugo and Brown (1991) point out that comparisons of ecosystem functions between tropical and temperate forests are often poor and inaccurate. Why should this be so?

Cross References

Relative humidity, 22; Holdridge life zones, 548; light, 43; internal nutrient cycling, 507; soil chemistry, 62; mor humus, 67; tropical soils, 70; soil groups, 69; decomposition, 148; primary production, 481; nitrogen cycle, 517; phosphorus cycle, 520; mycorrhizae, 322; mutualism, 321.

CHAPTER 30

Freshwater Ecosystems

CONCEPTS

1. Freshwater ecosystems fall into two broad groups: lentic ecosystems contained in a basin, and lotic or flowing water.

2. Nutrient cycling and energy flow in lentic ecosystems are largely contained within the basin, but they are heavily influenced by inputs from the surrounding landscape.

3. The structure of lotic ecosystems is shaped by flowing water.

4. Nutrient cycles in lotic systems involve spiraling, the retention and uptake of nutrients moving downstream.

5. Wetlands are best defined by considering vegetation, hydrological conditions, and soil.

6. The nature of a wetland is influenced by its hydroperiod.

Global aquatic ecosystems fall into two broad classes defined by salinity—freshwater ecosystems and saltwater ecosystems. The latter include inland brackish and salt water, as well as marine and estuarine habitats. Freshwater ecosystems, the study of which is known as **limnology,** are conveniently divided into two groups: **lentic** or standing water habitats, and **lotic** or running water habitats.

LENTIC ECOSYSTEMS

Lakes and ponds are inland depressions containing standing water. They vary in size from small ponds of less than one hectare to large seas covering thousands of square kilometers. They range in depth from 1 meter to over 2000 meters.

What is the origin of the lakes and ponds in the landscape? In the glaciated Northern Hemisphere, retreating glaciers left behind scoured-out basins. Glacial abrasion of slopes in high mountain valleys carved basins, which filled with water from rain and melting snow to produce tarns (Figure 30.1a, b). Retreating valley glaciers left behind crescent-shaped ridges of rock that dammed up water behind them. Numerous shallow kettle lakes and potholes formed in glaciated areas in northeastern North America and northwestern Europe (Figure 30.1c). Deposition of silt, driftwood, and other debris in the beds of slow-flowing streams formed other lakes. Loops of streams that meander over flat valleys and flood plains often become cut off, forming crescent-shaped oxbow lakes (Figure 30.1d). Shifts in Earth's crust, either by uplifting of mountains or the breaking and displacement of rock strata, causing part of a valley to sink, and craters of extinct volcanoes form depressions that fill with water (Figure 30.1e). Landslides can block streams and valleys to form new lakes and ponds. In a given area, all natural lakes and ponds have the same geological origin and the same general characteristics; but because of varying depths at the time of origin, they may represent several stages of development.

Many lakes and ponds are formed by the activities of two mammalian species—beavers and humans. Beavers dam up streams to make shallow but often extensive ponds. Humans create artificial lakes by damming rivers and streams for power, irrigation, and water storage, by constructing small ponds and marshes for water, fishing, and wildlife, and by allowing quarries and strip mine pits to fill with water.

Physical Characteristics

Temperature   Because of the high specific heat of water, aquatic environments do not experience the sharp daily fluctuations in temperature so characteristic of most terrestrial environments, and seasonal changes come on gradually. Slow warming and cooling of the surface water result in seasonal stratification of temperature and mixing of water in many lakes and ponds, and to a certain degree in seas (Figure 30.2).

In summer the intensity of solar radiation heats the surface water. The higher the temperature of the surface water, the lighter it becomes, so that it floats on the colder, denser water beneath. The warm upper layer is not easily mixed with the denser water below, creating a barrier between them. The freely circulating warm surface water is the **epilimnion** (Greek for "upper lake") (Figure 30.2). Below the thermal barrier is a middle mass of water, the **metalimnion** ("middle lake"). This middle mass is characterized by a steep and rapid decline in temperature of about 1°C for each meter of depth. This temperature gradient is known as the **thermocline.** Unfortunately, the two terms are used interchangeably to define the middle layer. Below these two layers is the **hypolimnion** ("lower lake"), a deep cold layer cut off from the air above.

With the coming of autumn, the air temperature falls. The surface water loses heat to the atmosphere through convection, conduction, and evaporation. The temperature of the surface water drops, and the metalimnion sinks. The epilimnion increases until it includes the entire lake. The temperature is now uniform from top to bottom, the lake waters circulate, and oxygen and nutrients are recharged throughout the lake. This seasonal mixing is called the **overturn.** The fall overturn, stirred by the wind, may last until ice forms.

As the surface water cools below 4°C, it becomes lighter and remains on the surface. If the climate is cold enough, the surface water freezes; otherwise, it remains close to 0°C. A slight inverse stratification may develop, in which the water becomes warmer up to 4°C with depth. The water immediately beneath the ice may be warmed by solar radiation through the ice. Because this warming increases its density, this water subsequently drops to the bottom, where it mixes with water warmed by heat conducted from bottom mud. The result is a higher temperature at the bottom, although the overall stability of the water is undisturbed. As the ice melts in spring, the surface water again reaches 4°C and streams downward, aiding the wind in mixing the water. After the entire water mass is 4°C, even the slightest winds can cause a complete circulation of water between the surface and the bottom. During this spring overturn, nutrients sequestered on the bottom, the oxygen in the surface waters, and the plankton within are mixed. Later in the season the surface water warms, and summer stratification develops.

This general picture of seasonal changes in temperature stratification does not apply to every lake and pond. In shallow lakes and ponds, temporary stratification of short duration may occur; in others, stratification may exist without a thermocline. In some very deep lakes, the metalimnion simply descends during periods of overturn, but does not disappear. In such lakes the bottom water never becomes mixed with the top layers.

Oxygen   Rarely is oxygen limiting in terrestrial environments; in aquatic environments the supply of oxygen, even at saturation levels, is meager and problematic. Oxygen enters the water by absorption from the atmosphere and by photosynthesis. The amount of oxygen and other gases water can hold depends on temperature, pressure, and salinity. Cold water holds more oxygen than warm water because the solubility of a gas in water decreases as the temperature rises. However, solubility increases as atmospheric pressure

(a)

(b)

(c)

(d)

(e)

FIGURE 30.1   Some types of lakes. (a) A cirque, a steep-sided, semicircular, glacial lake. (b) A rock basin glacial lake or tarn in the Rocky Mountains. (c) Potholes dot the glacial landscape of the north central United States. (d) An oxbow lake formed when a bend in a river was cut off from the main channel. (e) Lake formed by filling of a volcanic crater.

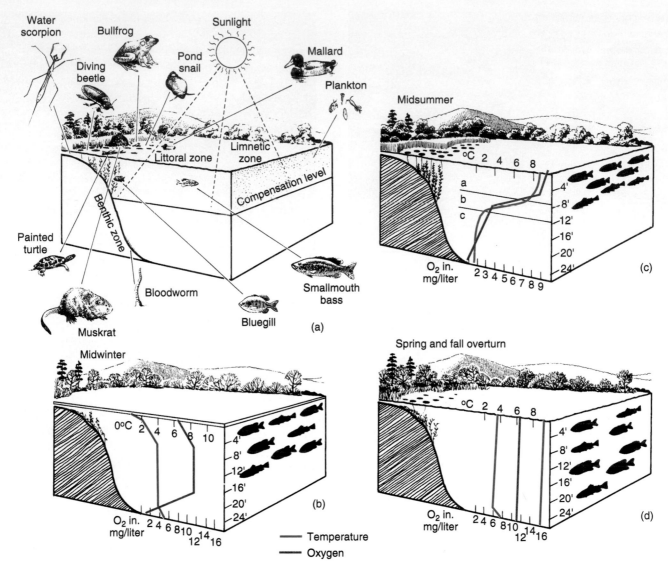

FIGURE 30.2 Seasonal variations in the stratification of oxygen and temperature and the distribution of aquatic life in lake ecosystems. (a) This generalized picture of a lake in midsummer shows the major zones—littoral, limnetic, profundal, and benthic. The compensation level is the depth at which light is too low for photosynthesis. Surrounding the lake is a variety of organisms typical of a lake community. (b, c) The distribution of oxygen and temperature during the different seasons affects the distribution of fish life. The narrow fish silhouettes represent trout, or cold-water species. The wider silhouettes are bass, or warm-water species. Note the strong horizontal stratification in midsummer, with a pronounced epilimnion (a), metalimnion (b), and hypolimnion (C). (d) Oxygen and temperature curves are nearly vertical during the spring and fall overturns.

increases and it decreases as salinity increases, which is of little consequence in fresh water.

Oxygen absorbed by surface water is mixed with deeper water by turbulence and internal currents. In shallow, rapidly flowing water and in wind-driven sprays, oxygen may reach and maintain saturation and even supersaturated levels, because of the increase of absorptive surfaces at the air-water interface. Water loses its oxygen through increased temperatures, increased respiration of aquatic life, and aerobic decomposition.

During the summer, oxygen, like temperature, may become stratified in lakes and ponds. The amount of oxygen usually is greatest near the surface, where an interchange between water and atmosphere, further stimulated by the stirring action of the wind, takes place. The quantity of oxygen decreases with depth because of decomposition in the bottom sediments. In some lakes, oxygen varies little from top to bottom; every layer is saturated for its temperature and pressure. Water in some lakes is so clear that light penetrates below the depth of the thermocline and encourages the growth of phytoplankton. Because of photosynthesis, the oxygen content may be greater in deep water than on the surface.

During spring and fall overturn, when water recirculates through the lake, oxygen becomes replenished in deep water.

In winter the reduction of oxygen in unfrozen water is slight, because bacterial decomposition is reduced by the cold and water at low temperatures holds the maximum amount of oxygen. Under ice, however, oxygen depletion may be serious, causing a heavy winterkill of fish.

Carbon Dioxide, Alkalinity, and pH Carbon dioxide, another atmospheric gas in water, behaves much differently from oxygen. Unlike oxygen, carbon dioxide combines chemically with water, and so it occurs both in free and bound states. Like oxygen, free carbon dioxide in fast-flowing water is in equilibrium with the atmosphere, according to the laws of the behavior of gases. Because the concentration of CO_2 in the atmosphere is low, the concentration of CO_2 is also low in water. Considerably more CO_2 is held as carbonate and bicarbonate ions.

CO_2 dissolved in water combines with water to form carbonic acid:

$$CO_2 + H_2O \rightleftharpoons H_2CO_3$$
$$H_2CO_3 \rightleftharpoons HCO_3^- + H^+$$
$$HCO_3 \rightleftharpoons H^+ + CO_3^{2-}$$

Carbon dioxide in simple solution and as H_2CO_3 is free carbon dioxide. Carbon dioxide in bicarbonate ions (HCO_3^-) and in carbonate ions (CO_3^{2-}) is called combined or bound. In the presence of an acid, combined carbon dioxide is converted to free CO_2.

The amount of acid needed to free CO_2 is the measure of the water's **alkalinity,** a gauge of buffering capacity relative to acids. The major contributor to the buffering capacity of the system is the carbonate-bicarbonate-CO_2 equilibrium. If a proton (H^+) is added to the water, it reacts with CO_3^{2-} to form bicarbonate (HCO_3^-), and the solution remains stable.

If H_2CO_3 dissociates, it releases the hydrogen ion, thus affecting pH. At neutrality (pH 7), most of the CO_2 is present as HCO_3^- (Figure 30.3). At a high pH, more CO_2 is present as CO_3^{2-} than at a low pH, where more CO_2 occurs in the free condition. Addition or removal of CO_2 affects pH, and a change in pH affects CO_2.

Seawater and hard fresh water (having much calcium and a high pH) are highly buffered. The pH of fresh water varies from 3 to 10. Rainwater has a pH of 5 to 6. The usual range

in streams and lakes is 6.5 to 8.5, but many lakes and streams have a pH as low as 3.5 seasonally, depending on the watershed. Waters draining from watersheds dominated geologically by limestone will have a much higher pH and be well buffered compared with waters from watersheds dominated by acid sandstone and granite. The pH of soft waters (without much calcium) can fluctuate widely, especially when receiving acid rain.

In soft waters, carbon dioxide may be reduced by photosynthesis of aquatic plants. If photosynthesis is high, corresponding respiration of organic matter, absorption from the atmosphere, and absorption from groundwater may not be enough to balance the losses. This fact can affect the physiology of aquatic invertebrates, especially in the oxygen affinity and alkalinity of the blood and the development of the exoskeleton. The limited supplies of Ca^{2+} and CO_3^{2-} in soft water restrict the distribution of some aquatic invertebrates.

A close relationship exists between the formation and dissolution of calcium carbonate and the photosynthetic and respiratory activity of aquatic plants:

$$Ca(HCO_3)_2 \rightleftharpoons \underset{\text{precipitated}}{CaCO_3} + H_2O + \overset{\text{assimilated}}{CO_2}$$

In hard water (high pH), plant activity removes carbon dioxide and causes the precipitation of calcium carbonate. In soft water, Ca^{2+} and CO_3^{2-} tend to remain in solution.

Structure

Unlike most terrestrial ecosystems, lentic ecosystems have well-defined boundaries—the shoreline, the sides of the basin, the surface of the water, and the bottom sediment. Within these boundaries, gradations of light, oxygen, and temperature profoundly influence life in the lake, its distribution, and its adaptations.

Lentic ecosystems can be subdivided into vertical strata and horizontal zones based on photosynthetic activity. The **littoral zone,** or shallow water zone, is the one in which light penetrates to the bottom (Figure 30.4). The area is occupied by rooted plants, such as wild rice, burreed, cattails, and sedges. Plants and animals found here vary with water depth, and a distinct zonation of life exists from deeper water to shore. Floating aquatics such as pond lilies or spatterdeck and water milfoil colonize the deeper water. In shallow water beyond the zone of floating plants grow the **emergents,** plants whose roots and lower stems are immersed in water and whose upper stems and leaves stand above water. Among these emergents are plants with narrow tubular or linear leaves such as bulrushes and cattails (Figure 30.5a). Associated with them are such broadleaf emergents as pickerelweed and arrow arum (Figure 30.5b). The distribution and diversity of plants vary with water depth and its fluctuations. Within the sheltering beds of emergent plants, animal life is abundant.

The **limnetic zone,** or open water zone, extends to the compensation level. It is inhabited by suspended organisms, the **plankton,** and free-swimming organisms, or **nekton,** such as fish.

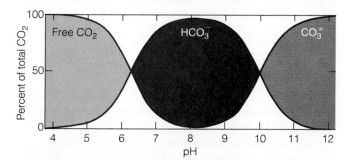

FIGURE 30.3 Theoretical percentages of CO_2 in each of its three forms in water in relation to pH.

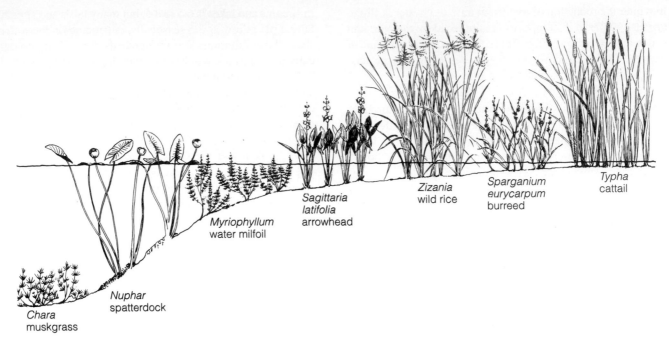

FIGURE 30.4 Zonation of emergent, floating, and submerged vegetation at the edge of a lake or pond. Such zonation does not necessarily reflect successional stages but rather response to water depth.

(a)

(b)

FIGURE 30.5 (a) Cattails dominate the littoral zone of many ponds to the exclusion of other emergent species. (b) A stand of arrow arum (*Peltandra virginica*) marks the edge of the littoral zone and the beginning of deep water.

Dominating the limnetic zone is phytoplankton (Figure 30.6a) including diatoms, desmids, and filamentous algae. Because these tiny plants carry on photosynthesis in open water, they are the base on which limnetic life depends. Suspended with the phytoplankton are small animals, or zooplankton (Figure 30.6b), which graze on the minute

plants. These animals form an important link in energy flow in the limnetic zone.

Phytoplankton, responding to temperature, oxygen, and light, are distributed vertically in the water column. Light, of course, sets the lower depth at which phytoplankton can exist. By their own growth, phytoplankton limit light penetration

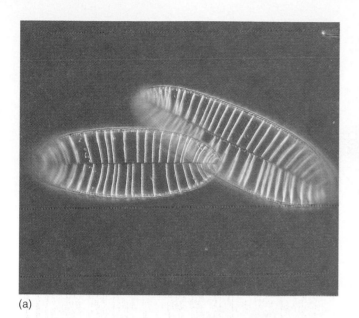

(a)

(b)

FIGURE 30.6 (a) Phytoplankton. (b) Zooplankton.

and thus reduce the depth at which they can live. Within these limits, the depth at which various species of phytoplankton live is influenced by optimum conditions for their development. Some phytoplankton species live just beneath the water's surface; others are more abundant a few feet beneath, whereas those requiring colder temperatures live deeper still.

Because animal plankton feeds on phytoplankton, most of it, too, is concentrated in the epilimnion. Like phytoplankton, animal plankton are often stratified; but their stratification changes seasonally, because most zooplankton are capable of independent movement. In winter, some planktonic forms are distributed evenly to considerable depths; in summer, they concentrate in those layers most favorable to them and their stage of development. At that season zooplankton undertake a vertical migration during some part of the 24-hour period. Depending on the species, these minute animals spend the day or night in the deep water or on the bottom and move up to the surface during the alternate period to feed on phytoplankton.

During the spring and fall, the overturn carries plankton downward. At the same time, nutrients, released by decomposition in the hypolimnion, are carried upward into the impoverished upper layers. In spring, when surface waters warm and stratification again develops, phytoplankton have access to both nutrients and light. A spring bloom develops, followed by a rapid depletion of nutrients and a reduction in planktonic populations, especially in shallow water.

Feeding on both phytoplankton and zooplankton are nekton organisms. In the limnetic zone, fish make up the bulk of the biomass. Their distribution is influenced mostly by temperature, oxygen, and food supply. For example, during the summer, bass and pike inhabit the warmer eplimnion water where food is abundant. In winter they retreat to deeper and warmer water. Lake trout (*Salvelinus namaycush*), on the other hand, move to cooler depths as summer advances. Dur-

ing the spring and fall overturns, when oxygen and temperature are fairly uniform throughout, both cold-water and warm-water species occupy all levels.

The sediments that rest on the bottom of lakes and ponds make up the **benthic zone.** The organisms that inhabit the benthic zone collectively are known as the **benthos.** Influencing the diversity and distribution of benthic organism are the differential deposition and mixing of types of sediments, the growth and death of roots, activities of benthic organisms themselves, and chemical gradients of oxygen and hydrogen sulfide (Covich et al. 1999). The bottom ooze is a region of great biological activity—so great that oxygen curves for lakes and ponds show a sharp drop in the water just above the bottom. When amounts of organic matter reaching the bottom are greater than can be used by bottom fauna, odoriferous muck rich in hydrogen sulfide and methane results. Because such organic muck lacks oxygen, the dominant organisms there are anaerobic bacteria. Under anaerobic conditions, decomposition cannot proceed to inorganic end products. Therefore, lakes and ponds with highly productive limnetic and littoral zones often have an impoverished fauna on and near the bottom. Life in the bottom ooze is most abundant in lakes with a deep hypolimnion, in which oxygen is still available.

Function

Although a lake might be considered a self-contained ecosystem, it is strongly influenced by inputs of nutrients from sources outside the basin (Figure 30.7). Nutrients and other substances move across the boundaries of the lentic system along biological, geological, and meteorological pathways (Likens and Bormann 1975).

Wind-borne particulate matter, dissolved substances in rain and snow, and atmospheric gases represent meteorological

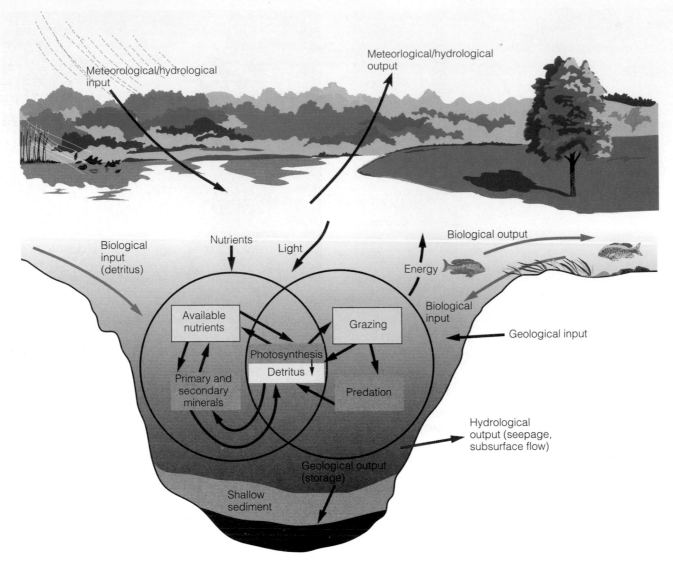

FIGURE 30.7 Model for nutrient cycling and energy flow in a lake ecosystem. Meteorological, geological, and biological inputs enter from the watershed. Nutrients and energy move through a number of pathways. Part accumulates in bottom sediments. (Based on Likens and Bormann 1974 and Rich and Wetzel 1978.)

inputs to the system. Meteorological outputs are small, mainly spray aerosols and gases such as carbon dioxide and methane. Geological inputs include nutrients dissolved in groundwater and inflowing streams and particulate matter washed into the basin from the surrounding terrestrial watershed. Geological outputs include dissolved and particulate matter carried out of the basin by outflowing waters and nutrients incorporated in deep sediments, which may be removed from circulation for a long period of time. Biological inputs and outputs, relatively small, include animals such as fish that move into and out of the basin. Energy input is largely sunlight, and energy output is heat of respiration (R). The lentic ecosystem receives its hydrological input from precipitation and the drainage of surface waters. Outputs involve seepage through walls of the lake basin, subsurface flows, evaporation, and evapotranspiration. Within the lentic ecosystem, nutrients move among three compartments:

dissolved organic matter, particulate organic matter, and primary and secondary minerals. Nutrients and energy move through the system by the grazing and detrital food chains.

Although past studies of lake metabolism have emphasized the phytoplankton-zooplankton grazing food chain, in reality lakes, like terrestrial communities, are dominated by the detrital food chain. The reason for an emphasis on the grazing food chain in lakes was a preoccupation with the open water zone and a disregard of the littoral zone, which adds significantly to lake productivity and supplies substantial quantities of detritus to the system. Detritus is all dead organic carbon. It includes particulate and dissolved organic carbon (POC and DOC) from external sources that enter and cycle within the system, and organic matter lost to a particular trophic level by such nonpredatory losses as egestion, excretion, and secretion (Rich and Wetzel 1978, Wetzel 1975).

Lake ecosystems function mostly within a framework of organic carbon transfer. The central pool, which comes from both internal and external sources, represents the major flow through the system. Particulate organic carbon comes from three sources: (1) imports, such as leaf fragments, into the system from the outside; (2) the littoral zone; (3) the limnetic zone. Most of the detrital metabolism takes place in the benthic zone, where particulate matter is decomposed, and in the limnetic zone during sedimentation.

Primary production is carried out in the limnetic zone by phytoplankton and in the littoral zone by macrophytes. The ratio of the contributions from these two sources varies among lentic systems. Phytoplankton production is influenced by nutrient availability in the water column. If nutrients are not limiting and the only losses are respiratory, the rate of net photosynthesis and biomass accumulation is high. In fact, a linear relationship exists between phytoplankton production and phytoplankton biomass (Brylinsky 1980). However, as phytoplankton biomass increases, shading and respiration per unit surface increases, and net photosynthesis and thus production declines. When nutrients are low, respiration and mortality increase, reducing net photosynthesis and thus biomass. However, if zooplankton grazing and bacterial decomposition are high, nutrients are recycled rapidly, resulting in a high rate of net photosynthesis even though the concentration of nutrients and biomass accumulation are low.

Rooted aquatic plants, the macrophytes, also contribute heavily to lake production. In highly fertile lakes, a heavy growth of phytoplankton shades out macrophytes and reduces their contribution. In less fertile lakes where phytoplankton production is low, light penetrates the water and rooted aquatics thrive. Macrophytes are little affected by nutrient exchange in the water column, because rooted aquatics draw on nutrients from the sediments rather than from open water.

Nutrient transfers within lentic ecosystems take place largely between the water column and sediments, and involve uptake by phytoplankton, zooplankton, bacteria, and other consumers as well as sedimentation in both the water column and benthic muds (Figure 30.8). In spring when phytoplankton bloom is at its height, nitrogen and phosphorus become depleted in the trophogenic zone, in part because of the high rate of photosynthesis, the high rate of sinking of dead phytoplankton, and the high rate of sedimentation.

In summer, conditions change. Because of a decline in phytoplankton in the eplimnion and a slower sinking rate, as much N and P enters solution as is taken up by phytoplankton in photosynthesis. In the bottom zone, N and P increase in the dissolved and particulate pools and in sediments. Phosphorus, in particular, becomes trapped in the hypolimnion and remains there, unavailable to phytoplankton until the fall overturn.

Macrophytes, however, can influence this transfer of phosphorus from sediments to the water column to phytoplankton, and at the same time increase the accretion of bottom sediments. Rooted macrophytes draw phosphorus from the sediment. S. Carpenter (1981), in a study in Lake Wingra, Wisconsin, found that macrophytes doubled the amount of sedimentary phosphorus made available to phytoplankton that they otherwise would have had to obtain via direct release from the sediment. Macrophytes obtained 73 percent of their shoot phosphorus from sediments, eventually making much of it available to phytoplankton. This uptake stimulates the production of both macrophytic and phytoplankton biomass, adding to sediment accumulation on the bottom. Sediment accumulation provides new areas for colonization

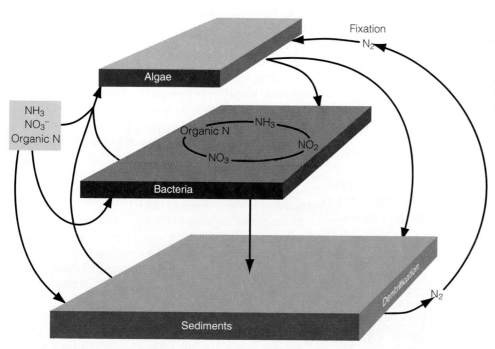

FIGURE 30.8 A model of nitrogen cycling in a lentic ecosystem showing the relationship between water column and sediments. Sediments are both a storehouse and a source of nutrients in the lentic system. (Based on Golterman and Kouwe 1980:138.)

and additional phosphorus for phytoplankton. Macrophytes enhance the recycling of phosphorus by mobilizing it from the sediments. Such mobilization and recycling accelerate the enrichment of lakes.

Although the productivity of a lake is related to the nutrient richness of its waters, other internal forces play a role. Two lakes may possess the same nutrient richness, yet differ in productivity. The differences relate to size-dependent phytoplankton metabolic activity and nutrient recycling (Carpenter and Kitchell 1984). Small phytoplankton, for example, have higher maximum growth rates overall, higher maximum growth at low levels of nutrients, and a lower sinking rate than large phytoplankton species.

Grazing on the phytoplankton are zooplankton, essential to the recycling of nutrients, particularly N and P. Depending on the size relationship of the dominant zooplankton to phytoplankton, these herbivorous zooplankton can influence the species composition and size structure of the phytoplankton community. These zooplankton, in turn, are preyed on by insects, crustaceans, and small fish; and these in turn become prey for larger fish. Interactions among these feeding groups flow down through the food web, influencing productivity at each trophic level.

Nutrient Status

A close relationship exists between land and water ecosystems. Primarily through the hydrological cycle, one feeds upon the other. The water that falls on land runs from the surface or moves through the soil to enter streams, springs, and eventually lakes. The water carries with it silt, clay, organic matter, and nutrients in solution, all of which enrich aquatic ecosystems. Human activities, including road building, logging, mining, construction, and agriculture, add an additional heavy load of silt, organic matter, and nutrients, especially nitrogen and phosphorus. The outcome is nutrient enrichment of aquatic systems, or **eutrophication.**

The term *eutrophy* (from the Greek, "well nourished") means a condition of being nutrient-rich. The opposite of eutrophy is **oligotrophy,** the condition of being nutrient-poor. The terms were introduced by the German limnologist C. A. Weber in 1907, for the development of peat bogs. E. Naumann later associated the terms with phytoplankton production in lakes: Eutrophic lakes found in fertile lowland regions hold high populations of phytoplankton; oligotrophic lakes, typical of rocky upland regions, contain little plankton. However, this concept of oligotrophy and eutrophy ignores the input of highly productive littoral zones.

Oligotrophic Systems Oligotrophic lakes (Figure 30.9a) are characterized by a low surface-to-volume ratio, water that is clear and appears blue to blue-green in the sunlight, bottom sediments that are largely inorganic, and a high oxygen concentration through the hypolimnion (Figure 30.10). The nutrient content of the water is low; although nitrogen may be abundant, phosphorus is highly limiting. Low nutrient availability is due to a low input of nutrients from external sources. This factor in turn causes a low production of organic matter, particularly phytoplankton. Low organic production results in low rate of decomposition and high oxygen concentration in the hypolimnion. These oxidizing conditions are responsible for low nutrient release from the sediments. The lack of decomposable organic substances results in low bacterial populations and slow rates of microbial metabolism. Although the number of organisms in oligotrophic lakes may be low, the diversity of species is often high. Fish life is dominated by members of the salmon family (Salmonidae).

When nutrients in moderate amounts are added to oligotrophic lakes, they are rapidly taken up and circulated. If nitrogen and carbon are in excess and phosphorus is limiting, the addition of phosphorus would stimulate growth; if nitrogen is limiting, the addition of that element would do the same. In most oligotrophic lakes, phosphorus rather than nitrogen is limiting (Vallentyne 1974, Wetzel 1975, Lampert and Sommer 1997). As increasing quantities of nutrients are added to the lake, it begins to change from oligotrophic to **mesotrophic** (having a moderate amount of nutrients) to eutrophic. This change has been happening at an increasing rate to clear oligotrophic lakes around the world.

Eutrophic Systems A typical eutrophic lake (Figure 30.10) has a high surface-to-volume ratio; that is, the surface area is large relative to depth. It has an abundance of nutrients, especially nitrogen and phosphorus, that stimulate a heavy growth of algae and other aquatic plants (Figure 30.9b). Increased photosynthetic production leads to increased regeneration of nutrients and organic compounds, stimulating further growth. Phytoplankton become concentrated in the upper layer of the water, giving it a murky green cast. The turbidity reduces light penetration and restricts biological productivity to a narrow zone of surface water. Algae, inflowing organic debris and sediment, and the remains of rooted plants drift to the bottom, adding to the highly organic sediments. On the bottom, bacteria partially convert dead matter into inorganic substances. The activities of these decomposers deplete the oxygen supply of the bottom sediments and deep water to a point where the deeper parts of the lake cannot support aerobic forms of life, and aerobic decomposition ceases. The amount of oxygen needed for oxidative decomposition is greater than the amount of oxygen available. The needed amount is called **biochemical oxygen demand (BOD).**

As the basin continues to fill, the volume decreases, and shallowness speeds the cycling of available nutrients and further increases plant production. Positive feedback carries the system to eventual extinction—the filling in of the basin and the development of a marsh, swamp, and ultimately a terrestrial community.

In fact, "galloping eutrophication" has been changing naturally eutrophic lakes into **hypertrophic** ones. An excessive nutrient content results from a heavy influx of wastes, raw sewage, drainage from agricultural lands, river basin development, runoff from urban areas, and burning of fossil

(a)

(b)

FIGURE 30.9 (a) An oligotrophic lake in Maine. (b) A eutrophic lake. Note the floating algal mats on the water.

fuels. This accelerated enrichment, which results in chemical and environmental changes in the system and causes major shifts in plant and animal life, has been called **cultural eutrophication** (Hasler 1969).

Dystrophic Systems Lakes that receive large amounts of organic matter from surrounding watersheds, particularly in the form of humic materials that stain the water brown, are called **dystrophic.** Although the productivity of dystrophic

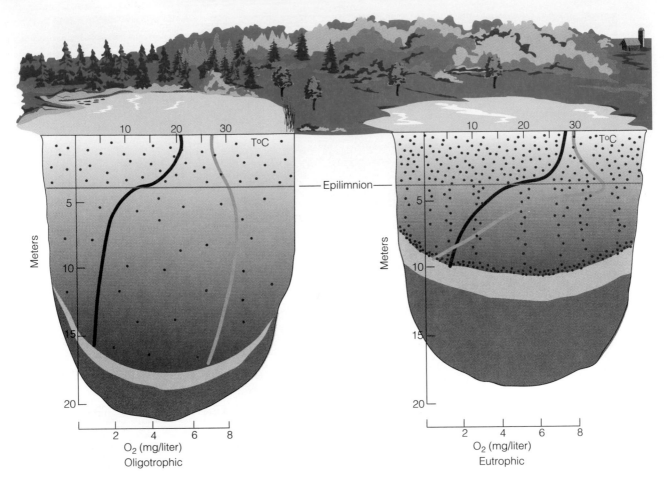

FIGURE 30.10 Comparison of oligotrophic and eutrophic lakes.

lakes is considered low, this refers only to planktonic production. Dystrophic lakes generally have highly productive littoral zones, particularly those that develop bog flora. This littoral vegetation dominates the metabolism of the lake ecosystem, providing a source of both dissolved and particulate organic matter (see Wetzel and Allen 1970, Wetzel 1975).

Marl Systems A fourth type of system, the **marl lake,** contains extremely hard water due to inputs of calcium over a long period of time. A hard-water lake is relatively unproductive and remains so because of the reduced availability of nutrients, even though carbonates remain high. Under certain conditions in these lakes, phosphorus, iron, magnesium, and other nutrients form insoluble compounds and are lost to the system. Sodium and potassium are low, but nitrogenous compounds are high. Because of low photosynthetic productivity, nitrogen remains unutilized. Calcium is often supersaturated, and carbonates, especially calcium carbonate, and humic dissolved organic matter precipitate to form marl deposits on the bottom. If carbonate inputs from drainage become reduced or depleted or if sediments build up high enough to support littoral vegetation and the growth of *Sphagnum,* the marl lake gradually develops into a bog (for details see Wetzel 1972, 1975.)

LOTIC ECOSYSTEMS

Flowing water is the dominant feature of lotic ecosystems from brooks to rivers. Velocity of the current molds the character of the stream, cuts its channels, and influences the lives of organisms inhabiting flowing waters. Velocity varies from stream to stream and within the stream itself. It is influenced by the size, shape, and gradient of the stream channel, the roughness of the bottom, the depth, and precipitation.

The velocity of flow affects the degree of silt deposition and the nature of the bottom. High water increases the velocity; it moves bottom stones, scours the streambed, and cuts new banks and channels. In very steep streambeds, the current may remove all but very large rocks and leave a boulder-strewn stream. Where current slows, silt particles drop out and settle on the bottom.

Flowing water transports nutrients to and carries waste products away from many aquatic organisms and may even sweep them away. Balancing this depletion of bottom fauna, the current continuously reintroduces bottom fauna from areas upstream. Similarly, as nutrients are washed downstream, more are carried in from above. For this reason, the productivity of primary producers in streams is 6 to 30 times that of those in standing water (Nelson and Scott 1962). The transport and removal action of flowing water benefits such continuous processes as photosynthesis.

Physical Characteristics

Streams may begin as outlets of ponds or lakes, or they may arise from springs and seepage areas (Figure 30.11a). As water drains away from its source, it flows in a direction dictated by the lay of the land and the underlying rock formations. Its course may be determined by the original slope and its regularities; or the water, seeking the least resistant route to lower land, may follow the joints and fissures in bedrock near the surface and shallow depressions in the ground.

(a)

(b)

FIGURE 30.11 (a) This small lake marks the beginning of a stream. (b) A satellite view of the delta of the Mississippi River, which marks the end of a river system that had its beginning source as a small stream.

Whatever its direction, water is concentrated into rills that erode small furrows, which soon grow into gullies.

Water, moving downstream, especially where the gradient is steep, carries with it a load of debris that cuts the channel wider and deeper and that sooner or later is deposited within or along the stream. At the same time, erosion continues at the head of the gully, cutting backward into the slope and increasing its drainage area.

Just below its source, the stream may be small, straight, and swift, with waterfalls and rapids. Further downstream, where the gradient is less and the velocity decreases, meanders become common. They are formed when the current, deflected by some obstacle on the floor of the channel, by projecting rocks and debris, or by the entrance of swifter currents, strikes the opposite bank. As the water moves downstream, it is thrown back to the other side again. These abrasive forces create a curve in the stream, on the inside of which the velocity is slowed and the water drops its load. Such cutting and deposition often cause valley streams to change course and to cut off the meanders to form oxbow lakes. When the water reaches level land, its velocity is greatly reduced, and the load it carries is deposited as silt, sand, or mud.

At flood time, the material carried by the stream is dropped on the level lands over which the water spreads to form floodplain deposits. These floodplains, which humans have settled so extensively, are a part of the channel that the river takes back at flood time, a fact that few people recognize. The current at flood time is swiftest in the normal channel of the stream and slowest on the floodplain. Along the margin of the channel and the floodplain, where the rapid water meets the slow, the current is checked and all but the fine sediments are dropped on the edges of the channel. Thus the deposits on the floodplain, which become alluvial soils, are higher on the immediate border and slope off gradually toward the valley side.

When a stream or river flows into a lake or sea, the velocity of the water is suddenly checked and the load of sediment is deposited in a fan-shaped area at the inlet point to form a delta (Figure 30.11b). Here the course of the water is broken into a number of channels, which are blocked or opened with subsequent deposits. As a result, the delta is characterized by small lakes and swampy or marshy islands. Material not deposited at the mouth is carried further out to open water, where it settles on the bottom. Eventually, the sediments build up above the water to form new land surface.

Structure

Fast Water   Fast or swiftly flowing streams are, roughly, all those whose velocity of flow is 50 cm/sec or higher (Nielsen 1950). At this velocity, the current will remove all particles less than 5 mm in diameter and will leave behind a stony bottom. The fast stream is often a series of three different but related habitats, the turbulent riffle, the run, a stretch of relatively smooth water leading to the quiet pool (Figure 30.12). The waters of the pool are influenced by processes occurring in the rapids above, and the waters of the rapids are influenced by events in the pool.

FIGURE 30.12   Different but related habitats in a stream: riffles (middle), run (foreground), and a pool (background).

Riffles are the sites of primary production in the stream (see Nelson and Scott 1962). Here the aufwuchs or periphyton assume dominance and occupy a position of the same importance as the phytoplankton of lakes and ponds. The **aufwuchs** consist chiefly of diatoms, cyanobacteria and green algae, and water moss (*Fontinalis*). Extensive stands of algae grow over rocks and rubble on the streambed and form a slippery covering. Growth during favorable periods may be so rapid that the stream bottom is covered in ten days or less (Blum 1960). Many small algal species are epiphytes growing on top of or among other algae.

The outstanding feature of much of this algal growth is its ephemeral nature. Scouring action of water and the debris it carries tears away larger growth, epiphytes and all, and sends the algae downstream. As a result, there is a constant contribution from upstream to the downstream sequence.

Below the riffles is the run where some sediment drops out, and below that are the pools. Here the environment differs in chemistry, intensity of current, and depth. Just as the riffles are the sites of organic production, so the pools are the sites of decomposition. They are the catch basins of organic materials, for here the velocity of the current is reduced enough to allow a part of the load to settle out. Pools are the major sites for free carbon dioxide production during the summer and fall.

Overall production in a stream is influenced in part by the nature of the bottom. Pools with sandy bottoms are the least productive, because they offer little substrate for either aufwuchs or animals. Bedrock, although a solid substrate, is so exposed to currents that only the most tenacious organisms can maintain themselves. Gravel and rubble bottoms associated with fast streams support the most abundant life because they have the greatest surface area for the aufwuchs, provide many crannies and protected places for insect larvae, and are the most stable (Figure 30.13a). Food production decreases as the particles become larger or smaller than rubble. Insect larvae, on the other hand, differ in abundance on the several substrates. Mayfly (Ephemeroptera) nymphs are most abundant on rubble, caddisfly (Trichoptera) larvae on

bedrock, and Diptera larvae on bedrock and gravel (Pennak and Van Gerpen 1947).

The width of the stream also influences overall production. Bottom production in streams 6 m wide decreases by one-half from the sides to the center; in streams 30 m wide, it decreases by one-third (Pate 1933). Streams 2 m or less in width are four times as rich in bottom organisms as those 6 to 7 m wide. This is one reason that headwater streams make such excellent trout nurseries.

Because precipitation, the source of all runoff and subsurface water, varies seasonally and annually, the rate and volume of streamflow may fluctuate widely over months, weeks, or even days from high-discharge floods to dry channels. These fluctuations have a pronounced effect on the structure the stream. During periods of dry weather, water in the channel of streams shrinks both sideways and downstream. Pools decrease in their length, width, and depth. Taking on the aspects of a pond, the pool becomes the refuge of

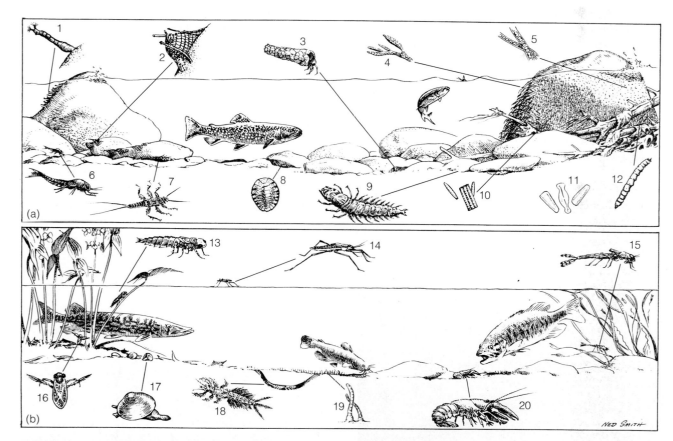

FIGURE 30.13 Comparison of life in a fast streams (a) and a slow stream (b). (1) Blackfly larvae (Simulildae); (2) net-spinning caddisfly (*Hydropsyche* spp.); (3) stone case of caddisfly; (4) water moss (*Fontinalis*); (5) algae (*Ulothrix*); (6) mayfly nymph (*Isonychia*); (7) stonefly nymph (*Perla* spp.); (8) water penny (*Psephenus*); (9) hellgrammite (dobsonfly larva, *Corydalis cornuta*); (10) diatoms (*Diatoma*); (11) diatoms (*Gomphonema*); (12) cranefly larva (Tipulidae); (13) dragonfly nymph (Odonata, Anisoptera); (14) water strider (*Gerris*); (15) damselfly larva (Odonata, Zygoptera); (16) water boatman (Corixidae); (17) fingernail clam (*Sphaerium*); (18) burrowing mayfly nymph (*Hexegenia*); (19) bloodworm (Oligochaeta, *Tubifex* spp.); (20) crayfish (*Cambarus* spp.). The fish in the fast stream is a brook trout (*Salvelinus fontinalis*). The fish in the slow stream are, from left to right, northern pike (*Esox lucius*), bullhead (*Ameiurus melas*), and smallmouth bass (*Micropterus dolommieu*).

larger stream fish. As the stream dries, submerged rocks emerge from the riffles, and remaining water becomes trapped in depressions throughout the riffle. The stream has now become a fragmented habitat, with each riffle fragment holding its own complement of refugees of small fish and stream invertebrates (Stanley et al. 1997, Grimm 1987) until normal of volume of water returns. As the runs dry, some stream bottom inhabitants, such as crustaceans, flatworms, and stream insects, seek refuge in the wet subsurface sediments, known as the **hyporheic zone** (Boulton et al. 1998). This zone is a sort of ecotone between groundwater and streamwater. In times of low water, decreasing stream depth forces surface water down into the sediments (downwelling) and pushes water in the pore spaces toward the surface (upwelling). This downward and upwater movement of water brings about an exchange of nutrients and organic matter that can influence the productivity of the stream.

Slow Water   As the current slows, a noticeable change takes place in streams (Figure 30.14). Silt and decaying organic matter accumulate on the bottom, and fine detritus from upstream is the main source of energy. Faunal organisms are able to move about to obtain their food, and a plankton population develops. The composition and configuration of the stream community approaches that of standing water.

With increasing temperatures, decreasing current, and accumulating bottom silt, organisms of the fast water are replaced by organisms adapted to these conditions (Figure 30.13b). Brook trout (*Salvelinus fontinalis*) and sculpin (Cottidae) give way to smallmouth bass (*Micropterus dolomieu*) and rock bass (*Ambloplites rupestris*), the dace (*Rhinichthep*) to shiners (*Phenacobius*) and darters (*Percina*). With current at a minimum, many resident fish lack the strong lateral mus-

cles typical of the trout and have compressed bodies that permit them to move with ease through masses of aquatic plants. Mollusks, particularly *Sphaerium* and *Pisidium,* and pulmonate snails, crustaceans, and burrowing mayflies replace the rubble-dwelling insect larvae. Only in occasional stretches of fast water in the center of the stream do remnants of headwater-stream organisms still live.

As the volume of water increases, as the current becomes even slower, and as the silt deposits become heavier, detritus-feeders, particularly mollusks, increase. Rooted aquatics appear. Emergent vegetation grows along the riverbanks, and duckweeds (Lemnacea) float on the surface. Indeed, the whole aspect approaches that of lakes and ponds, even to zonation along the river margin.

Abundant decaying matter promotes the growth of protozoan and other plankton populations. Scarce in fast water, plankton increase in numbers and species in slow water. Rivers have no typical plankton of their own. Those found there originate mainly from backwaters and lakes. In general, plankton populations in rivers are not nearly as dense as those in lakes. Time is too short for much multiplication of plankton, because little time is needed for a given quantity of water to flow from its source to the sea. Also occasional river rapids, often long, kill many plankton organisms by violent impact against suspended particles and the bottom. Aquatic vegetation filters out this minute life as the current sweeps it along.

Stream Order   Because streams become larger on their course to a river and are joined along the way by many others, streams at any one point in the landscape can be classified according to order (Figure 30.15). A small headwater stream without any tributaries is a first-order stream. When two streams of the same order join, the stream becomes one of

FIGURE 30.14   A slow stream meanders through a growth of willows.

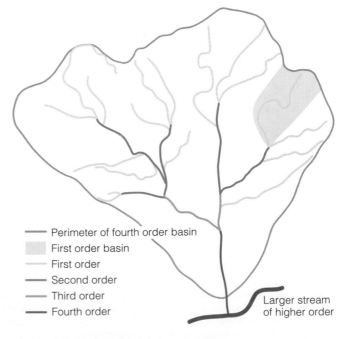

— Perimeter of fourth order basin
First order basin
- - - First order
— Second order
— Third order
— Fourth order
Larger stream of higher order

FIGURE 30.15   Stream orders and basins.

higher order. If two first-order streams unite, the resulting stream becomes a second-order one; and when two second-order streams unite, the stream becomes a third-order one. A stream can increase in order only when a stream of the same order joins it. Order cannot be increased with the entry of a lower order stream. In general, headwater streams are orders 1 to 3; medium-sized streams, 4 to 6; and rivers, greater than 6.

Streams receive runoff from the surrounding land as overland flow. The land area contributing to the flow for any one stream is its **basin** (Figure 30.15). A first-order stream collects overland flow from the slopes of a first-order basin. A second-order stream receives flow from first-order streams, and thus first-order basins, as well as its own direct overland

flow. The same pattern repeats for third and higher order streams. These basins are also called watersheds. A **watershed** is a body of land bounded above by a ridge or water divide and below by the level at which the water drains from the basin. The size of a watershed tends to increase geometrically with the mean basin areas of successive stream orders.

Function

The lotic or flowing water system is open and largely heterotrophic, especially in headwater streams. A major source of energy and nutrients is detrital material from the outside (Figure 30.16). Much of this organic matter comes in the form

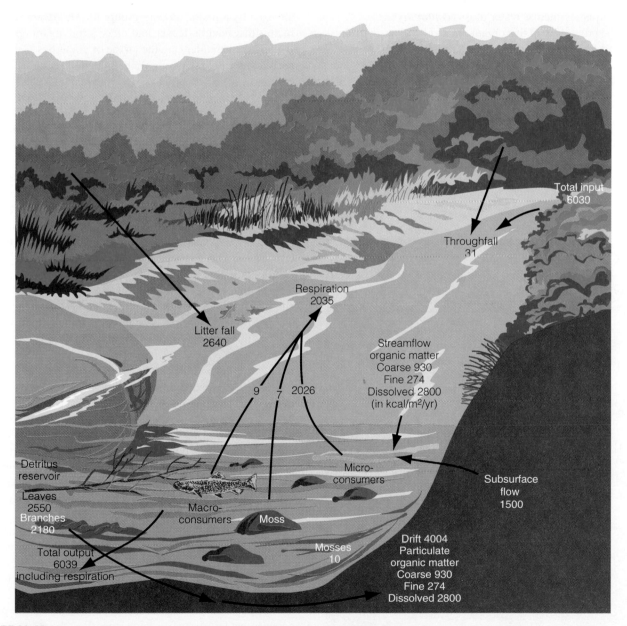

FIGURE 30.16 Energy flow in a stream ecosystem in kcal/m²/yr. Note the dependence on materials from terrestrial sources and inflow from upstream and the roles of coarse particulate organic matter (CPOM), fine particulate organic matter (FPOM), and dissolved organic matter (DOM). Primary production contributes little to energy flow. Energy values are based on Bear Brook, Hubbard Forest, New Hampshire. (Data from Fisher and Likens 1973.)

of leaves and woody debris dropped from streamside vegetation, collectively called **coarse particulate organic matter (CPOM)** (particles larger than 1 mm) (Figure 30.17). Another type of organic input is **fine particulate organic matter (FPOM),** material less than 1 mm, including leaf fragments, invertebrate feces, and precipitated dissolved organic matter. A third input is **dissolved organic matter (DOM),** material less than 0.5 μm in solution. One source of DOM is rainwater dropping through overhanging leaves, dissolving the nutrient-rich exudates on them. Another is subsurface seepage, which brings nutrients leached from adjoining forest, agricultural, and residential lands. Many streams receive inputs from dumping of industrial and residential effluents. Supplementing this detrital input is autotrophic production in streams by diatomaceous algae growing on rocks and by rooted aquatics such as water moss (*Fontinalis*).

The processing of this organic matter involves both physical and biological mechanisms (Cummins 1992). In fall, leaves drift down from overhanging trees, settle on the water, float downstream, and lodge against banks, debris, and stones. Soaked with water, the leaves sink to the bottom, where they quickly lose 5 to 50 percent of their dry matter as water leaches soluble organic matter from their tissues. Much of this DOM is either incorporated onto detrital particles or precipitated to become part of FPOM. Another part is incorporated into microbial biomass. Once softened, leaves and

other debris are processed by a number of species of invertebrates, which can be placed into several functional groups (Figure 30.18).

Functional Groups Within a week or two, depending on the temperature and the plant species involved, the surfaces of the leaves are colonized by bacteria and fungi, largely aquatic hyphomycetes. Fungi are more important on CPOM, because large particles offer more surface for mycelial development (Cummins and Klug 1970). Bacteria are associated more with FPOM. Microorganisms degrade cellulose and metabolize lignin. Their populations form a layer on the surfaces of leaves and detrital particles that is much richer nutritionally than the detrital particles themselves (Anderson and Cummins 1979). Leaves and other detrital particles are attacked by a major feeding group, the ***shredders*, invertebrates** that feed on leaves and other large organic particles. Among the shredders are the larvae of craneflies (Tipulidae) and caddisflies (Trichoptera), the nymphs of stoneflies (Plecoptera), and crayfish. They break down the CPOM, feeding on the material not so much for the energy it contains but for the bacteria and fungi growing on it (Cummins 1974, Cummins and Klug 1979, Cummins et al. 1989). In so doing, they skeletonize the leaf by feeding on the softer portions. Shredders assimilate about 40 percent of the material they ingest and pass off 60 percent as feces.

Broken up by shredders and partially decomposed by microbes, the leaf material along with fecal material becomes part of the FPOM, which also includes some precipitated DOM. Drifting downstream and settling on the stream bottom, FPOM is picked up by another feeding group of stream invertebrates, the *filtering* and *gathering* **collectors**. The filtering collectors include, among others, the larvae of blackflies (Simuliidae) with filtering fans, and net-spinning caddisflies, including *Hydropsyche*. Gathering collectors, such as the larvae of midges, pick up particles from stream bottom sediments. Collectors obtain much of their nutrition from bacteria associated with fine detrital particles.

Whereas shredders and collectors feed on detrital material, another group feeds on algal coating of stones and rubble. They are the ***scrapers*** (see Figure 30.19), which include the beetle larvae, popularly known as the water penny (*Psephenus* spp.), and mobile caddisfly larvae. Much of the material they scrape loose enters the drift as FPOM. Behaviorally and morphologically, scrapers are adapted to maintain their position in the current either by flattening to avoid the main force of flow or by weighting down with heavy mineral cases.

Feeding on mosses and filamentous algae are the ***piercers*** (see Figure 30.19), consisting largely of microcaddisflies. Another group, associated with woody debris, are the *gougers* (see Figure 30.19), invertebrates that burrow into waterlogged limbs and trunks of fallen trees.

Feeding on the detrital feeders and scrapers are predaceous insect larvae such as large stoneflies, the powerful dobsonfly larvae (*Corydalus cornuta*), stream salamanders, and fish such as sculpin (*Cottus* spp.) and trout. Invertebrate predators employ either ambush or searching strategies. Some

FIGURE 30.17 Much of the detrital input in this first-order stream comes with leaf-fall in autumn. Leaves cover the surface of the pool.

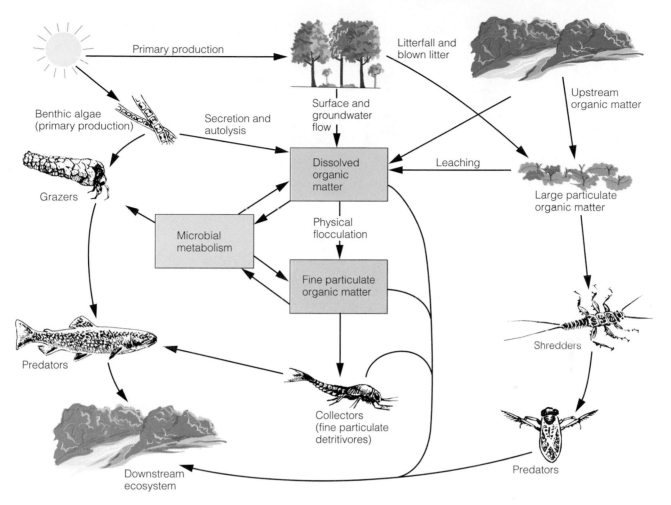

FIGURE 30.18 Model of structure and function in a stream, showing the processing of leaves, other particulate matter, and dissolved organic matter. (Adapted from Cummins 1974:663.)

engulf the prey whole or in pieces; others pierce the body of prey and suck out all or part of the contents. Predaceous aquatic insects detect prey largely by mechanical cues (Peckarsky 1982), and they are size-selective opportunists. Prey species in turn have evolved morphological and behavioral defenses, including flattened body shapes, protective cases (caddisflies), and cryptic coloration. Predators, especially those in headwater streams, do not depend solely on aquatic insects; they also feed heavily on terrestrial invertebrates that fall or are washed into the stream.

Because of current, quantities of CPOM and FPOM and invertebrates tend to move downstream to form a traveling benthos called **drift.** This process is normal in streams, even in the absence of high water and abnormal currents (for reviews see Hynes 1970, Waters 1972). Drift is so characteristic of streams that a mean rate of drift can serve as an index of the production rate (Pearson and Kramer 1972). This drift is essential to the production processes of downstream systems (Wallace et al. 1982).

Energy Flow and Nutrient Cycling Energy flow and nutrient cycling in a lotic ecosystem have been documented

for the well-studied, small, forested Bear Brook in Hubbard Forest of northern New Hampshire (Fisher and Likens 1973). That budget is summarized in Figure 30.16. Over 90 percent of the energy input comes from the surrounding forested watershed or from upstream. Primary production by mosses accounts for less than 1 percent of the total energy supply. Algae are absent from the brook. Inputs from litter and throughfall account for 44 percent of the energy supply, and geological inputs account for 56 percent. Energy is introduced in three forms: CPOM, represented by leaves and other debris; FPOM, represented by drift and small particles; and DOM. In Bear Brook 83 percent of the geologic input from sediments and particulate mineral matter and 47 percent of the total energy input are in the form of DOM. Sixty-six percent of the organic input is exported downstream, leaving 34 percent to be utilized locally.

Nutrient cycling is more difficult to assess, because of the very open nature of the lotic system. Triska et al. (1984) estimated the nitrogen budget of a small stream draining a 10.1 ha watershed in the old-growth Douglas-fir H. J. Andrews Experimental Forest in Oregon (Table 30.1). The major annual input of dissolved nitrogen, amounting to 15.3 g/m², came from two

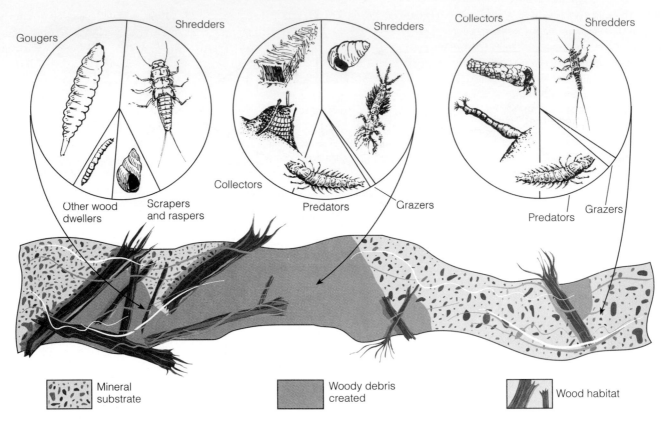

FIGURE 30.19 Large woody debris modifies the stream habitat, slowing the flow, retaining organic debris for processing by invertebrates, and adding to the diversity of biota. Note the addition of another functional group, the gougers, where woody debris is prominent. Woody debris habitat favors collectors and shredders, especially on small-order streams.

major sources: hydrological and biological. Hydrological inputs, largely subsurface flow in the form of seeps, accounted for most of the input, 11.1 g/m². Biological inputs accounted for 4.2 g/m². Particulate organic nitrogen contributed 11.9 g, of which CPOM made up 7.2 g and FPOM 4.8 g. Total output of

nitrogen was 11.4 g/m². Of this amount 2.5 g was particulate organic matter dominated by CPOM (1.0 g), and 8.8 g was dissolved N, less than the input because of biological uptake. The stream intercepted and transported nearly all of the nitrogen lost from the 10 ha watershed, yet effectively retained and processed

TABLE 30.1 Nitrogen Budget (g/m²) for a Small Coniferous Forest Stream (Watershed 10, H. J. Andrews Experimental Forest, Oregon)

Inputs		*Outputs*	
Dissolved organic N pool	15.25	Total nitrogen output	11.36
Hydrological	11.06	Dissolved N	8.81
DON	10.56	DON	8.38
NO^3-N	0.50	NO^3-N	0.43
Biological	4.19	Particulate organic N	2.53
N fixation	0.76	FPOM	1.66
Throughfall	0.30	CPOM	0.87
Litterfall	1.35	Insect emergence	0.02
Blown leaves	1.78		
Particulate organic N pool	11.93		
FPOM	4.77		
CPOM	7.16		

Source: Data from Triska et al. 1984.

a considerable portion of it. Nitrogen transported downstream would be utilized there for microbial consumption.

A major problem for flowing water ecosystems is retention of nutrients upstream. Although nutrient cycling is downhill in all ecosystems, the problem in lotic systems is the constant movement of the substrate—water—away from the system. Nutrients in terrestrial and lentic systems are recycled more or less in place. An ion of a nutrient passes from soil or water column to plants and consumers back to soil or water in the form of detrital material or exudates. It is then recycled within the same segment of the system, although losses do occur. Cycling is essentially temporal.

Lotic systems have an added spatial cycle. Nutrients in the form of DOM and POM move continually downstream. How rapidly they are carried depends on water flow and physical and biological retention in place. The greater the flow, the more rapid the loss. Physical retention involves storage in wood detritus such as logs and snags in the stream, accumulation of debris in pools formed behind boulders, leaf sediments, and beds of macrophytes. Recycling is biological, controlled by uptake and storage in animal and plant tissue.

The process of nutrient uptake, transformation, and release with downstream transport may be pictured as a longitudinal spiral (Figure 30.20). The combined processes of nutrient cycling and downstream transport are termed **spiraling** (Wallace 1977, Newbold et al. 1982). One cycle in the spiral is the distance required for the uptake of an ion of a nutrient by consumer organisms, its transformation into particulate form, and its return to water, where it is available for reutilization. The distance an ion travels in dissolved form

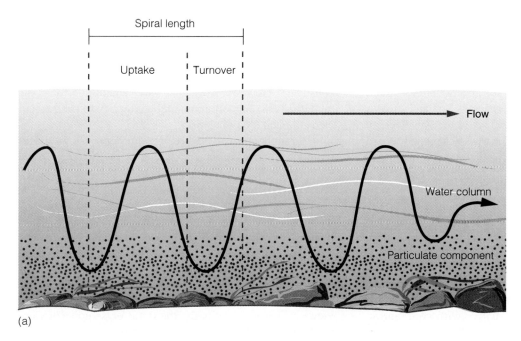

(a)

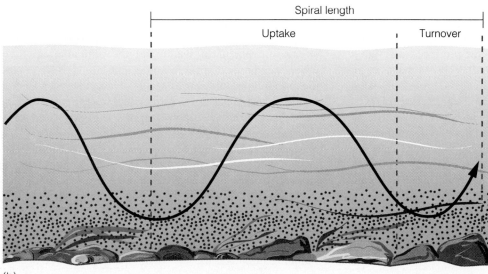

(b)

FIGURE 30.20 Nutrient spiraling in a lotic ecosystem between particulate organic matter and the water column. Uptake and turnover take place as nutrients flow downstream. (a) represents tight spiraling, and (b) more open spiraling. The tighter the spiraling, the more efficient. (Adapted from Newbold et al. 1982:630.)

before being taken up is its *uptake length*. The distance it travels in particulate form before being released to DOM is its *turnover length*. The uptake length plus the turnover length is the *spiral length*. The longer the spiral length, the less efficient the stream is in retaining nutrients; the shorter the length, the more efficient.

Tight spiraling of nutrients in flowing water ecosystems depends on the retention of leafy detritus in place long enough to allow the biological component of the stream, especially the shredders and microbes, to process organic matter. This delay is especially important in rapidly flowing headwater streams (Bilby 1989). Logs, snags, and other woody debris in the channel and along the banks, plus large rocks, act as dams, intercepting leafy detritus and forming pools that collect sediments. These debris dams create a diversity of physical habitats within the stream and influence the nature of the invertebrate community (Figure 30.19). In small streams (orders 0 to 2), wood and debris and sediments stored behind them may make up 50 percent of the stream area, and in larger streams (orders 3 to 4), 25 percent of the stream area (Triska et al. 1982).

What are the measured lengths of spirals? Webster et al. (1991) measured the short-term spiral lengths of marked phosphorus in three small heavily shaded woodland streams at the Cowecta Hydrologic Laboratory in the Appalachian Mountains of North Carolina. Uptake lengths ranged from 3 to 81 m. They were influenced by velocity and by temperature of the water. Nutrient uptake lengths are much shorter in the warmer water of summer than in the colder water of winter (D'Angelo et al. 1991).

The River Continuum

Lotic ecosystems involve a continuum of physical and biological conditions from the headwaters to the mouth (Vannote et al. 1980, Minshall et al. 1983). The upper reaches of the headwaters (orders 1 to 3) are usually swift, cold, and in forested regions, shaded. Riparian vegetation reduces light, inhibiting autotrophic production (Hawkins et al. 1982), and contributes more than 90 percent of organic input into streams as terrestrial detritus. Even when headwater streams are exposed to sunlight and autotrophic production exceeds heterotrophic inputs, organic matter produced invariably enters detrital food chains (Minshall 1978, Minshall et al. 1983). Dominant organisms are shredders, processing large litter and feeding on CPOM, and collectors, processors of FPOM. Populations of grazers are minimal, reflecting the small amount of autotrophic production, and predators are mostly small fish—sculpins, darters, and trout. Headwater streams are accumulators, processors, and transporters of particulate organic matter from terrestrial systems. As a result, the ratio of gross primary production to community respiration is less than 1. Consumer organisms utilize long-chain and short-chain organic compounds for transport downstream. Organisms of headwater streams are adapted to a narrow temperature range, to a reduced nutrient regime, and to maintenance of their position in the current.

As streams increase in width to medium-sized creeks and rivers (orders 4 to 6), the importance of riparian vegetation and its detrital input decreases. The lack of shading results in higher temperature. As the gradient declines, the current slows and becomes more variable. The diversity of microenvironments supports a greater range of organisms. An increase in light and temperature and a decrease in terrestrial input encourage a shift from heterotrophy to autotrophy, relying on algal and rooted plant production. Gross primary production now exceeds community respiration. Because of the lack of CPOM, shredders disappear, and collectors, feeding on FPOM transported downstream, and grazers, feeding on autotrophic production, become the dominant consumers. Predators show little increase in biomass but shift from cold-water to warm-water species.

As the stream order increases from 6 through 10 and higher, riverine conditions develop. The channel is wider and deeper. The volume of flow increases, and the current becomes slower. Sediments accumulate on the bottom. Both riparian and autotrophic production decrease, with a gradual shift back to heterotrophy. A basic energy source is FPOM, utilized by bottom-dwelling collectors, now the dominant consumers. However, slow, deep water and dissolved organic matter (DOM) support a minimal phytoplankton and associated zooplankton population.

Throughout the downstream continuum, the lotic community capitalizes on upstream feeding inefficiency. Downstream adjustments in production and the physical environment are reflected in changes in consumer groups (Figure 30.21). Through the continuum the lotic ecosystem achieves some balance between the forces of stability, such as natural obstructions in flow that aid in retention of nutrients upstream, and the forces of instability, such as drought, flooding, and temperature fluctuations.

Flood Pulse Concept Whereas the river continuum views a river longitudinally, the flood pulse concept views the larger river both laterally and longitudinally, including both the river channel and its associated floodplain (Johnson et al. 1995). The **flood pulse concept** holds that periodic flooding is a natural event to which the biota is adapted. The river's most important hydrological feature is not the downstream movement of water, but the annual predictable advance and retreat of flood waters that extend the river onto the floodplain (Junk et al. 1989). Thus, the river system includes the main channel, off-channel water bodies, and the floodplain. The floodplain supports a highly productive riparian forest, a variety of aquatic habitats, and a gradient of plant species adapted to seasonal degrees of flooding and drying. During flood time, flood waters bring nutrients and suspended sediments as well as dissolving nutrients previously mineralized on the dry floodplain (Figure 30.22). Coming with the flood waters are young fish and aquatic invertebrates for which the flooded areas provide nursery grounds. Flood waters drown existing terrestrial vegetation. Flooding stimulates the growth of aquatic plants and decomposition of terrestrial vegetation; receding water speeds the

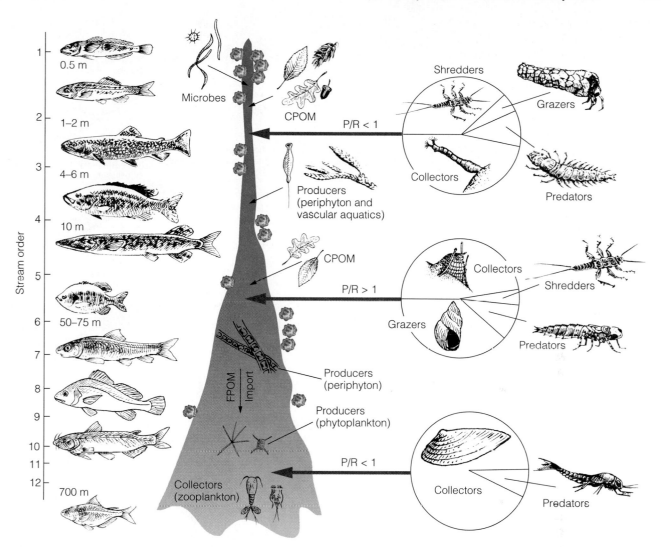

FIGURE 30.21 Changes in consumer groups along the river continuum. The headwater stream is strongly heterotrophic, dependent on terrestrial input of detritus, and the dominant consumers are shredders and collectors. As stream size increases, the input of organic matter shifts from particulate organic matter to primary production from algae and rooted vascular plants. The major consumer groups are now collectors and grazers. The zone at which the shift occurs depends on the degree of shading. As the stream increases to river, the lotic system shifts back to heterotrophy, supported by inputs of fine particulate organic matter and dissolved organic matter. A phytoplankton population may develop. The bottom consumers are collectors, mostly bottom-dwelling organisms. (After Vannote et al. 1980.)

decomposition of aquatic vegetation and regrowth of terrestrial grasses and shrubs.

The river continuum and the flood pulse concepts both apply to pristine conditions. Dams and pollution that change the nature of the downstream water compromise the river continuum. The channelization of rivers and the construction of levees designed to hold back water from the floodplain compromise the flood pulse concept. Both concepts, however, provide a basic understanding the nature of lotic ecosystems. Such an understanding is essential for restoring natural processes on selected sections of river and emulating natural hydrological regimes (Gore and Shields 1995).

Regulated Rivers and Streams

Few of the world's rivers are free-flowing. Most have been dammed, many repeatedly, channelized, or otherwise altered, modifying the river continuum (Cummins 1988). Such rivers, whose water levels and water flows are controlled, whose channels have been straightened, and whose waters have been isolated from their floodplains by levees, are termed **regulated rivers** (Figure 30.23).

Regulated rivers differ significantly from free-flowing ones (Liqon et al. 1995). Under normal conditions, free-flowing streams and rivers experience seasonal fluctuations in flow. Snowmelt and early spring rains bring scouring high

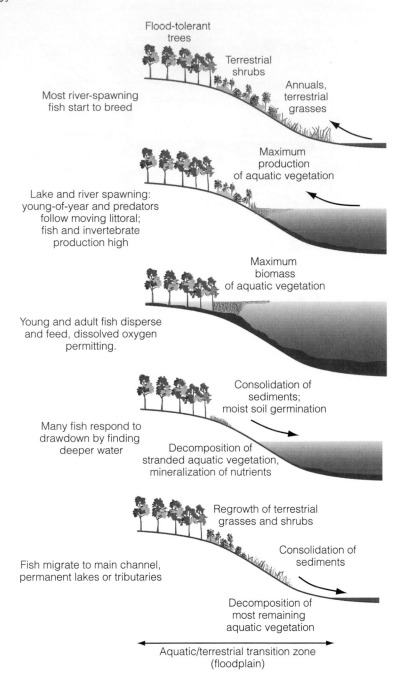

Flood-tolerant
trees

Terrestrial
shrubs

Annuals,
terrestrial
grasses

Most river-spawning
fish start to breed

Maximum
production
of aquatic vegetation

Lake and river spawning:
young-of-year and predators
follow moving littoral;
fish and invertebrate
production high

Maximum
biomass
of aquatic vegetation

Young and adult fish disperse
and feed, dissolved oxygen
permitting.

Consolidation of
sediments;
moist soil germination

Many fish respond to
drawdown by finding
deeper water

Decomposition of
stranded aquatic vegetation,
mineralization of nutrients

Regrowth of terrestrial
grasses and shrubs

Consolidation of
sediments

Fish migrate to main channel,
permanent lakes or tributaries

Decomposition of
most remaining
aquatic vegetation

Aquatic/terrestrial transition zone
(floodplain)

FIGURE 30.22   A schematic representation of the flood pulse concept. The five panels, showing a vertically exaggerated section of a floodplain, depict points in the annual hydrological cycle. Shown are some physical and biological responses to the advance and retreat of flood water in a relatively undisturbed floodplain. Dams, levees, and drainage disrupt this natural cycle on the floodplain. (Adapted from Bayley 1995, after Junk et al. 1989.)

water; summer brings low water levels that expose some of the streambed and speed decomposition of organic matter along the edges. Life of the lotic environment has adapted its life cycle to these seasonal changes. Damming a river or stream interrupts both nutrient spiraling from upstream and the lotic continuum. Downstream flow is greatly reduced as a pool of water fills behind the dam. This pool develops characteristics similar to those of a natural lake, yet retains some features of the lotic system, such as a constant inflow of water. Heavily fertilized by decaying detrital material on the newly flooded land, the lake develops a heavy bloom of phytoplankton and, in tropical regions, dense growths of floating

plants. Species of fish, often exotics adapted to lakelike conditions and introduced to provide a commercial or sport fishery, replace fish of flowing water.

The type of pool allowed to develop behind the dam depends on the purpose of the dam and has a strong effect on downstream conditions. Some are single-purpose dams for flood control or water storage; others are multiple-purpose dams, providing hydroelectric power, irrigation water, and recreation, among other uses. Flood control dams have a minimum pool; the dam fills only during a flood, at which time inflow exceeds outflow. Engineers release the water slowly from the dam so that outflow exceeds inflow, but downstream

FIGURE 30.23 Dammed river.

flooding is minimized. In time, water in the dam slowly recedes to original pool depth. During flood and post-flood periods, the river below carries a strong flow for some time, scouring the riverbed. During normal times, flow below the dam is stabilized. If the dam is for water storage, the reservoir holds its maximum pool; but during periods of water shortage and drought, drawdown of the pool can be considerable, exposing large expanses of shoreline for a long time and stressing or killing littoral life. Only a minimal quantity of water is released downstream, usually an amount required by law, if such exists. Hydroelectric and multiple-purpose dams hold a variable amount of water, determined by demand for power. Pools move up and down except during periods of power production, which involves pulsed releases strong enough to wipe out or dislodge benthic life downstream.

Dams with a large pool of water become stratified with a well-developed epilimnion, metalimnion, and hypolimnion. If water is discharged from the upper stratified layer of the reservoir, the effect of the flow downstream is similar to that of a natural lake. Warm, nutrient-rich, well-oxygenated water creates highly favorable conditions for some species of fish below the spillway and on downstream. If the discharge is from the cold hypolimnion, downstream receives cold, oxygen-poor water carrying an accumulation of iron and other minerals and a concentration of soluble organic materials. Such conditions, inimical to stream life, may persist for hundreds of kilometers downstream before the river reaches anything near normal conditions. The use of gated selective withdrawal structures or induced artificial circulation to increase oxygen concentration reduces such problems at some dams.

Impacts of dams on lotic systems are compounded when a number of multipurpose dams are built on a river. The amount of water released and moving downstream becomes less with each dam, until eventually all available water is consumed and the river simply dries up. That is the situation on the Colorado River, the most regulated river in the world. The river is nearly dry by the time it reaches Mexico.

Effects of dams go beyond simply changing the nature of the lotic system. Large dams on such rivers as the Columbia in North America interfere with the migratory patterns of anadromous fish, such as salmon, which come in from the sea and ascend rivers to spawn. Dams obstruct the upstream movement of fish. Although fish ladders provide some assistance, many local populations have been excluded from their traditional spawning streams. Even if spawning is successful, unnatural timing of high and low flows in the rivers may induce premature seaward migration of the young or lengthen their downstream passage, exposing them to high temperatures. Nearly 90 percent of juveniles perish on the journey from their home stream to the estuary.

WETLANDS

What is a wetland? The answer to this question is critical, for upon it rests the fate of many wetlands, among the most endangered ecosystems (for a worldwide review, see Williams 1990). Viewed as wastelands to be put to more productive uses, wetlands have been drained for agriculture, urban expansion, industrial sites, home sites, and dumps. Since settlement the United States has lost over 53 percent of its wetlands. Although wetlands make up only 5 percent of the land surface of the lower 48 states and 12 percent if Alaska and Hawaii are included, strong pressure to drain much of the remaining wetlands persists. A number of states and the federal government have instituted regulations to protect remaining wetlands for the values they provide. This protective legislation centers about the definition of a wetland.

Defining a **wetland** would seem to be simple enough. Wetlands do have certain characteristics: (1) the presence of water at or near the surface; (2) soils that differ from those of adjacent uplands; and (3) vegetation adapted to wet conditions (Mitsch and Gosselink 1993). Such an answer, however, is not enough. The problem is that wetlands are a halfway world between aquatic and terrestrial ecosystems and exhibit some characteristics of each. Combining the various characteristics and using them to draw a line between wetlands and uplands to obtain a precise definition is difficult. More often than not, attempts to define wetlands involve less science and more politics and economics. A piece of land classified as a wetland becomes off-limits to development.

Some wetlands are easy to distinguish. A water area supporting submerged plants such as pondweed (*Potamogeton*), floating plants such as pond lily, and emergents such as cattails and sedges is unquestionably a wetland. But what about a piece of ground where the soil is more or less permanently wet and supports some ferns and such trees as red maple (*Acer rubrum*), which also grow on the uplands? Where do you draw the line between wetlands and uplands on this gradient of soil wetness? On the aquatic side, at what point docs a wetland end and a true aquatic habitat with open water begin?

A number of wetland definitions exist (for a review see Mitsch and Gosselink 1993). A widely accepted scientific definition is presented in *Classification of Wetlands and Deepwater Habitats of the United States* (Cowardin et al. 1979):

> Wetlands are transitional between terrestrial and aquatic systems where the water table is usually at or near the surface or the land is covered by shallow water…. Wetlands must have one or more of the following three attributes: (1) at least periodically, the land supports predominately hydrophytes; (2) the substrate is predominately undrained hydric soil; and (3) the substrate is nonsoil and is saturated with water or covered by shallow water at some time during the growing season of each year.

A regulatory definition is used by the U.S. Army Corps of Engineers (1984):

> The term "wetlands" means those areas that are inundated or saturated by surface or ground water at a frequency and duration sufficient to support, and under normal circumstances do support, a prevalence of vegetation typically adapted for life in saturated soil conditions. Wetlands generally include swamps, marshes, bogs, and similar areas.

A Canadian regulatory definition states:

> Wetland is defined as land having the water table at, near, or above the land surface or which is saturated for a long enough period to promote wetland or aquatic processes as

indicated by hydric soils, hydrophytic vegetation, and various kinds of biological activity which are adapted to the wet environment.

The U.S. Army Corps of Engineers emphasizes vegetation cover. The Canadian definition emphasizes wet soil conditions, especially during the growing season. It is obvious that vegetation alone should not define a wetland. Vegetation serves as an indicator, within limitations, but also involved are soils and the hydrological conditions that created the wetland in the first place. Wetlands range along a gradient of permanent to periodically flooded conditions to permanently to periodically saturated soil at some time during the growing season and support hydrophytic (water-loving) vegetation (Figure 30.24) (Tiner 1991, Lyon 1993). Wetland soils are hydric; that is, they are saturated, flooded, or ponded long enough during the growing season to develop anaerobic conditions in the upper part (Soil Conservation Service 1987). These soils are either mineral or organic. Mineral soils have less than 25 to 35 percent organic matter on a dry weight basis and have a developed soil profile with a gray to blue-gray color brought about by gleization (see Chapter 4). Organic soils have an organic content greater than 25 to 35 percent, are composed primarily of plant materials in various stages of decomposition, are dark in color, and have higher water-holding capacities than mineral soils.

Hydrophytic vegetation consists of those plants adapted to grow in water or on soil or a substrate that is periodically deficient in oxygen because of excess water. Hydrophytic

FIGURE 30.24 Location of wetlands along a soil moisture gradient. (After Tiner 1991.)

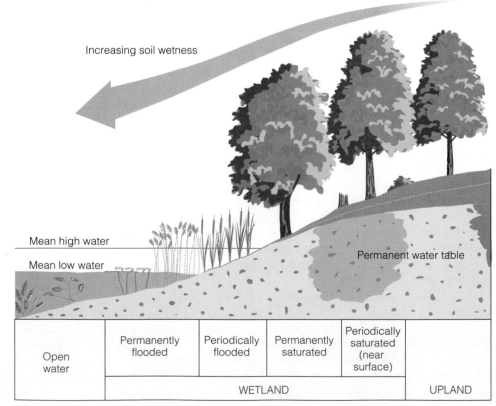

plants include several groups: (1) obligate wetland plants, such as the submerged pondweeds, floating pond lily, and emergent cattails and bulrushes, and trees such as baldcypress (*Taxodium distichum*); (2) facultative wetland plants, or amphibious plants, such as certain sedges and alder that can grow in standing water or saturated soil and rarely grow in other situations; (3) facultative species, such as red maple, that have about a 50:50 probability of growing in either wetland or nonwetland situations; and (4) facultative upland species, such as beech and Kentucky bluegrass, that have a 1 to 30 percent probability of growing in a wetland. It is these last plants that are critical in determining the upper limit of a wetland on the soil moisture gradient—a situation where species designation alone is insufficient.

For example, some species of trees usually associated with uplands adapt and grow quite well in wetland environments. One from eastern North America is red maple (*Acer rubrum*). It thrives in the drier uplands, but is also a conspicuous species in forested wetlands. In the uplands, red maple has a deep taproot; in wetland situations, the tree has a shallow root system that enables it to avoid anaerobic stress. The red maple is a facultative species that evolved ecotypes adapted to different soil moisture conditions. Black gum (*Nyssa sylvatica*), too, grows in both upland and wetland situations. Pitch pine (*Pinus rigida*), associated with the drier ridgetops of the southern Appalachians, has ecotypes that grow in poorly drained soils and muck of swamps. In fact, pitch pine is a dominant species in the wetlands of the extensive New Jersey pine barrens. Hemlock (*Tsuga canadensis*), a shallow-rooted species, and white pine (*Pinus strobus*) are also at home in wetland situations.

The point is that individual species of vegetation alone do not define a wetland. Although important indicators, especially in the wettest situations, ecotypes of upland species confuse the situation and make it difficult to draw a line on the gradient between wetland and upland. It is essential to consider the hydrologic conditions and soil properties along with the vegetation to identify and delineate a wetland.

Types of Wetlands

A wide variety of wetlands from fresh water to salt exists, and classifying them for management and conservation has presented problems. An older, shorter, but still generally useful classification appears in Table 30.2. A much more comprehensive hierarchical classification is presented by Cowardin et al. (1979).

Wetlands most commonly occur in three topographic and hydrologic situations. Many developed in shallow basins ranging from upland topographic depressions to filled-in lakes and ponds. These are **basin wetlands.** Other wetlands developed along shallow and periodically flooded banks of rivers and streams. These are **riverine wetlands.** A third type occurs along the coastal areas of larger lakes and oceans. They are known as **fringe wetlands.** Some of the best developed fringe wetlands are mangrove communities associated with tropical marine environments.

What separates the three types is the direction of water flow. Water flow in the basin wetlands is vertical, involving precipitation and capillary flow. In riverine wetlands, water flow is unidirectional. In fringe wetlands, flow goes in both directions because it involves rising lake levels or tidal action. The flows may bring in and carry away nutrients, and they may physically stress systems by exporting or importing excessive amounts of nutrients and sediments.

Wetlands may have deep or shallow water, or occupy soils that are water-saturated to various degrees. Wetlands dominated by emergent herbaceous vegetation are **marshes** (Figure 30.25). Supporting reeds, sedges, grasses, and cattails, marshes are essentially wet grasslands. Wetlands dominated by woody vegetation, or forested wetlands, are commonly called **swamps.** Deep water swamps are dominated by baldcypress, tupelo (*Nyssa sylvatica*), and swamp oak (*Quercus bicolor*) (Figure 30.26a). Shrub swamps dominated by alder and willows are known as carrs or scrub-shrub swamps (Figure 30.26b). Along large river systems may occur extensive tracts of **riparian woodlands** (Figure 30.27) that are occasionally or seasonally flooded by river waters but are dry for most of the growing season (see Gregory et al. 1991).

Seasonal shallow wetlands ranging from 50 m^2 to 0.5 ha, flooded during the winter and spring and dry in summer and fall, are called **vernal pools.** Associated with a mediterranean-type climate with wet winters and dry summers, vernal pools are most common in the United States west of the Sierra Nevada Mountains and in the Central Valley of California. Vegetation is dominated by annuals and seasonal invertebrate life. Virtually ignored as wetlands in other parts of the continent, especially the northeast, are similar vernal pools, usually called temporary ponds. In winter these depressions may hold ice-covered water collected from melting snow, but most fill in early spring with snowmelt and rainwater and dry up in summer. Invertebrate life is dominated by fairy shrimp (*Eubranchipus vernalis*) and other small crustaceans that survive the dry summer and winter as eggs or cysts (Kenk 1949). These pools are important breeding sites for frogs, toads, and salamanders.

Wetlands in which considerable amounts of water are retained by an accumulation of partially decayed organic matter are **peatlands** or **mires.** Mires fed by water moving through mineral soil, from which they obtain most of their nutrients, and dominated by sedges are known as **fens** (Figure 30.28a). Such peatlands are called **mincrotrophic.** Mires dependent largely on precipitation for their water supply and nutrients and dominated by *Sphagnum* moss are **bogs** (Figure 30.28b). Mires that develop on upland situations where decomposed, compressed peat forms a barrier to the downward movement of water, resulting in a perched water table above mineral soil, are **blanket mires** (Figure 30.29, p. 659). Blanket mires are more popularly known as **moors.** Because bogs depend on precipitation for nutrient inputs, they are highly deficient in mineral salts and low in pH. For this reason, such bogs are termed **ombrotrophic** ("food from the sky").

Bogs also develop when a lake basin fills with sediments and organic matter carried by inflowing water. These sediments divert water around the lake basin and raise the surface

TABLE 30.2 Types of Wetlands

Type	Site Characteristics	Plant and Animal Populations
Inland Fresh Areas		
Seasonally flooded basins or flats	Soil covered with water or waterlogged during variable periods, but well drained during much of the growing season; in upland depressions and bottomlands	Bottomland hardwoods to herbaceous growth
Fresh meadows	Without standing water during growing season; waterlogged to within a few inches of surface	Grasses, sedges, rushes, broadleaf plants
Shallow fresh marshes	Soil waterlogged during growing season; often covered with 15 cm or more of water	Grasses, bulrushes, spike rushes, cattails, arrowhead, smartweed, pickerelweed; a major waterfowl production area
Deep fresh marshes	Soil covered with 15 cm to 1 m of water	Cattails, reeds, bulrushes, spike rushes, wild rice; principal duck-breeding area
Open fresh water	Water less than 3 m deep	Bordered by emergent vegetation such as pondweed, naiads, wild celery, water lily; brooding, feeding, nesting area for ducks
Shrub swamps	Soil waterlogged; often covered with 15 cm or more of water	Alder, willow, buttonbush, dogwoods; nesting and feeding area for ducks to limited extent
Wooded swamps	Soil waterlogged; often covered with 0.3 m of water; along sluggish streams, flat uplands, shallow lake basins	North: tamarack, arborvitae, spruce, red maple, silver maple; South: water oak, overcup oak, tupelo, swamp black gum, cypress
Bogs	Soil waterlogged; spongy covering of mosses	Heath shrubs, *Sphagnum,* sedges
Coastal Fresh Areas		
Shallow fresh marsh	Soil waterlogged during growing season; at high tide as much as 15 cm of water; on landward side, deep marshes along tidal rivers, sounds, deltas	Grasses and sedges; important waterfowl areas
Deep fresh marshes	At high tide covered with 15 cm to 1 m of water; along tidal rivers and bays	Cattails, wild rice, giant cutgrass
Open fresh water	Shallow portions of open water along fresh tidal rivers and sounds	Vegetation scarce or absent; important waterfowl areas
Inland Saline Areas		
Saline flats	Flooded after periods of heavy precipitation; waterlogged within few inches of surface during the growing season	Seablite, salt grass, saltbush; fall waterfowl-feeding areas
Saline marshes	Soil waterlogged during growing season; often covered with 0.61 to 1 m of water; shallow lake basins	Alkali hard-stemmed bulrush, wigeon grass, sago pondweed; valuable waterfowl areas
Open saline water	Permanent areas of shallow saline water; depth variable	Sago pondweed, muskgrasses; important waterfowl-feeding areas
Coastal Saline Areas		
Salt flats	Soil waterlogged during growing season; sites occasionally to fairly regularly covered by high tide; landward sides or islands within salt meadows and marshes	Salt grass, seablite, saltwort
Salt meadows	Soil waterlogged during growing season; rarely covered with tide water; landward side of salt marshes	Cord grass, salt grass, black rush, waterfowl-feeding areas
Irregularly flooded salt marshes	Covered by wind tides at irregular intervals during the growing season; along shores of nearly enclosed bays, sounds, etc.	Needlerush, waterfowl cover area
Regularly flooded salt marshes	Covered at average high tide with 15 cm or more of water; along open ocean and along sounds	Atlantic: salt marsh cord grass; Pacific: alkali bulrush, glassworts; feeding area for ducks and geese,
Sounds and bays	Portions of saltwater sounds and bays shallow enough to be diked and filled; all water landward from average low-tide line	Wintering areas for waterfowl
Mangrove swamps	Soil covered at average high tide with 15 cm to 1 m of water; along coast of southern Florida	Red and black mangroves

Source: Adapted from Shaw and Fredine 1956.

FIGURE 30.27 A riparian woodland of bottomland hardwoods along a small stream in western United States.

FIGURE 30.25 A glacial prairie marsh supporting a heavy stand of emergent vegetation.

(a)

(b)

FIGURE 30.26 (a) A cypress deep water swamp in the southern United States. (b) A scrub-shrub swamp dominated by alder (*Alnus*) with a herbaceous understory of skunk cabbage (*Symplocarpus foetidus*).

(a)

(b)

FIGURE 30.28 (a) Morning fog rises over this fen in Michigan. (b) An upland black spruce (*Picea mariana*)-tamarack (*Larix laricina*) bog in the Adirondack Mountains of New York.

of the mire above the influence of groundwater. These are **raised bogs** (Figure 30.29). Other bogs form when a lake basin fills in from above rather than from below (Figure 30.30), producing a floating mat of peat over open water. Such bogs are often termed **quaking bogs.**

Structure

The structure of a wetland is influenced by the phenomenon that creates it—its hydrology. Hydrology has two components. One is the physical aspects of water and its movement: precipitation, surface and subsurface flow, direction and kinetic energy of water, and the chemistry of the water. The other is the **hydroperiod,** the duration, frequency, depth, and season of flooding. Length of the hydroperiod varies among types of wetlands. Basin wetlands have a longer hydroperiod and usually experience flooding during periods of high rainfall, and drawdown during dry periods. Both phenomena appear to be essential to the long-term existence of wetlands. Riverine wetlands have a short period of flooding associated with peak stream flow. The hydroperiod of fringe wetlands, influenced by wind and lake waves, may be short and regular, and it does not undergo the type of seasonal fluctuations characteristic of many basin marshes.

Hydroperiod influences plant composition, for it affects seedling germination, survival, and mortality at various stages of the plant's life cycle. The effect of hydroperiod is most pronounced in basin wetlands, especially those of the prairie regions of North America. In basins (called **potholes** in the prairie region) deep enough to have standing water through-

out periods of drought, the dominant plants will be submergents. If the wetland goes dry annually or during a period of drought, tall or mid-height emergent species such as cattails will dominate the marsh. If the pothole is shallow and flooded only briefly in the spring, then grasses, sedges, and forbs will make up a wet meadow community.

If the basin is sufficiently deep toward its center and large enough, then zones of vegetation may develop, ranging from submerged plants to deep water emergents such as cattails and bulrushes, shallow water emergents, and wet ground species such as spikerush. Zonation reflects the response of plants to hydroperiod. Those areas of wetland subjected to a long hydroperiod will support submerged and deep water emergents; those with a short hydroperiod and shallow water are occupied by shallow water emergents and wet ground plants (van der Valk 1981).

Periods of drought and wetness can induce vegetation cycles associated with changes in water levels. Periods of above-normal precipitation can raise the water level and drown the emergents to create a lake marsh dominated by submergents. During a drought the marsh bottom is exposed by receding water, stimulating the germination of seeds of emergents and mudflat annuals. When water levels rise again, the mudflat species drown, and the emergents survive and spread vegetatively (van der Valk and Davis 1978).

Peatlands differ from other freshwater wetlands in that their rate of organic production exceeds the rate of decomposition; much of the production accumulates as peat. Peatlands are more characteristic of northern regions where inflow and precipitation balance outflow and retention, or

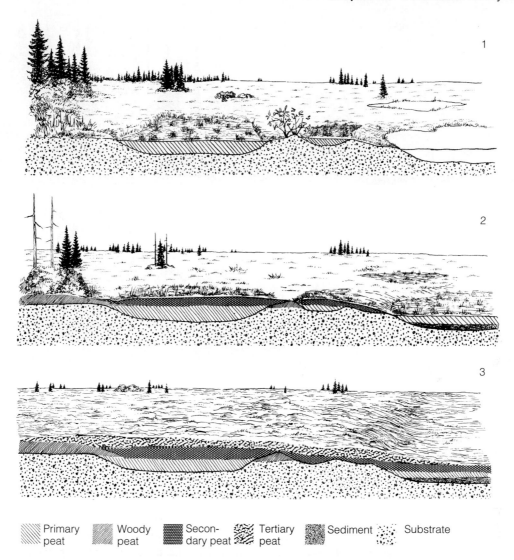

| Primary peat | Woody peat | Secondary peat | Tertiary peat | Sediment | Substrate |

FIGURE 30.29 Development of a raised bog and blanket mire. In the first stage, primary peat forms in basins and depressions. It reduces the surface retention of the reservoir. After peat fills the basin, it may continue to develop beyond the confines of the basin to form a raised bog. The raised bog consists of secondary peat, which acts as a reservoir, increasing surface retention of water on the area. Peat that develops above the physical limits of the groundwater and blankets even terrestrial situations is tertiary peat. It acts as a reservoir, holding a volume of water by capillary action above the level of the main groundwater drainage. The landscape acts as a perched water table, fed by precipitation falling on it. As the area becomes more wet and acidic, tree growth dies out and the landscape is covered by sphagnum and sedges.

where permafrost impedes the drainage of water and the bulk of summer precipitation is retained.

Upland bogs begin when *Sphagnum* moss invades higher ground surrounding a lake basin, when beaver back up water over streamside forests, or when forests are cut on wet sites. Sphagnum moss has the ability to absorb ions out of nutrient-poor water by exchanging hydrogen ions in its tissues for cations in solution. This exchange, which results in an increase of hydrogen and sulfate ions in the water, increases the acidity of the system, further decreasing the rate of decomposition and increasing the net accumulation of organic matter. Sphagnum moss adds new growth on top of the accumulating remains of past moss generations; and its spongelike ability to hold water increases water retention on the site. As the peat blanket thickens, the water-saturated mat of moss and associated vegetation is raised above and insulated from mineral soil. The peat mat then becomes its own reservoir of water, creating a perched water table. This process is called **paludification.**

Flowthrough bogs develop along streams or on gentle slopes and benches. Inflowing ground and surface water carry in sediments that accumulate near the water source, supporting marshy vegetation. Over time, enough peat develops to impede water flow and divert it around the peat into new

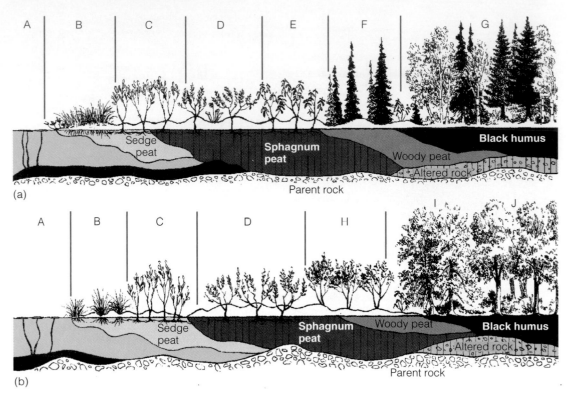

FIGURE 30.30   (a) Transect through a quaking bog, showing zones of vegetation, sphagnum mounds, peat deposits, and floating mats. A, pond lily in open water; B, buckbean (*Menyanthes trifoliata*) and sedge; C, sweetgale (*Myrica gale*); D, leatherleaf (*Chamaedaphne calyculata*); E, Labrador tea (*Ledum groenlandicum*); F, black spruce; G, birch-balsam fir-black spruce forest. (b) Alternative vegetational sequence: H, alder; I, aspen, red maple; J, mixed deciduous forest. (Adapted from Dansereau and Segadas-Vianna 1952.)

areas. Development of a raised bogs follows a somewhat similar pattern. Inflowing water into a pond or small lake basin carries in sediments and debris, the buildup of which supports the growth of sphagnum and the accumulation of peat. This accumulation diverts the flow of water away from the basin. The major source of water is now precipitation, surface runoff, and seepage from the surrounding area. In time, the water-saturated mat of peat and sphagnum expands and rises above the general water table to form a raised bog. Between the mat of peat and surrounding higher land is a moatlike area of shallow water dominated by sedges, called a **lagg.**

Although usually associated with and most abundant in boreal regions of the Northern Hemisphere, peatlands also exist in tropical and subtropical regions. They develop in mountainous regions or in lowland or estuarine regions where hydrological situations encourage an accumulation of partly decayed organic matter. Examples are the Everglades in Florida, the tropical coal swamps of the Carboniferous period, the pocosins of the southern United States coastal plains, and small bogs in the northern Appalachians.

Biologically, wetlands are among the richest and most interesting ecosystems. They support a diverse community of benthic, limnetic, and littoral invertebrates, especially crustaceans and insects. These invertebrates, along with small fishes, provide a basic food base for waterfowl, herons, gulls, and other

birds, and supply the fat-rich nutrients needed by ducks for egg production and the growth of young. Wetlands support a diversity of amphibians and reptiles, notably frogs, toads, and turtles, and such mammalian inhabitants as beaver, muskrats (*Ondatra zibethicus*), mink (*Mustela vison*), and otter (*Lutra canadensis*).

Herbivores make up a conspicuous component of animal life. Microcrustaceans filter algae from the water column. Snails eat algae growing on the leaves and litter; geese graze on new emergent growth; coots and mallards and other surface-feeding ducks feed on algal mats. The dominant herbivore in the prairie marshes is the muskrat. During population highs, muskrats can eliminate emergent vegetation, creating "eat-outs," and transform an emergent-dominated marsh into an open water one. Introduced into Eurasia, the muskrat has become the major herbivore in many marshes on that continent. Conversely, the nutria (*Myocaster coypus*), introduced into Louisiana in the 1930s as a furbearer, has spread throughout the southeast into the coastal marshes of Maryland and New Jersey. Its numbers, size, and herbivorous feeding habits make it a major threat to coastal marshes by its destruction of marsh vegetation. Muskrats are the major prey for mink, the dominant carnivore on the marshes. Other predators include raccoon, fox, weasel, and skunk, which can seriously reduce the reproductive success of waterfowl on small marshes surrounded by agricultural land.

Function

In recent years ecologists have given serious attention to the function of freshwater wetlands (Good et al. 1978, Greeson et al. 1979, Weller 1981, Gore 1983, Odum et al. 1984, Mitsch and Gosselink 1993). Much of this work has concerned glacial prairie wetlands, freshwater tidal wetlands, and southern deep water swamps. Two wetland types will serve as examples, freshwater marshes and peatlands.

Freshwater Marshes Productivity of freshwater marshes is influenced by hydrological regimes: groundwater, surface runoff, precipitation, drought cycles, and the like. It is also affected by the nature of the watershed in which the wetland lies, soils, nutrient availability, types of vegetation, and the life history of the plant species.

Annual and perennial vegetation differ in their contributions to biomass through the growing season (Figure 30.31). Annual emergents exhibit a linear increase in biomass, reaching a maximum in late summer. Perennials increase their biomass the first part of the growing season, then decline or level off as leaves become senescent (Whigham et al. 1978). However, the average maximum standing crop of biomass usually matches annual above-ground productivity. For example, a sedge wetland had a maximum standing crop biomass greater than 1000 g/m² and a net primary productivity greater than 1000 g/m²/yr. However, if the life history of the plants—including mortality, regeneration, and winter activity—was considered, net productivity was 1600 g/m²/yr (Bernard and Gorham 1978).

Below-ground production, much more difficult to estimate, appears for some species to be highest in summer, at the same time the peak above-ground biomass is achieved. Others, such as *Typha* and *Scirpus,* reach peak production in the fall, when nutrients are stored in the roots. Such species may have minimal root biomass in the summer because of nutrient transfer to above-ground biomass.

The cycling of phosphorus in a cattail marsh on Lake Mendota, Wisconsin (Prentki et al. 1978) provides an example (Figure 30.32). In spring, the growing new shoots of cattail draw on the phosphorus reserves in the rootstocks for their initial growth before mobilizing P from the soil. By June, cattails are accumulating P at a rate higher than they are accumulating biomass. By midsummer, *Typha* has accumulated over 4 g/P/m² in the shoots, 78 percent of total P in biomass. At the same time, the below-ground pool is minimal. In late summer and fall, P is remobilized in the rhizomes before the shoots die, but at a rate slower than accumulation of below-ground biomass. In fall, a large amount of P is lost through leaching and death of the shoots. Only about 28 percent of the summer accumulation is returned to the rhizomes, to be rapidly depleted again by spring growth.

To balance losses, cattails and other emergent plants must draw on a P supply from the soil. By doing so, the plants act as a nutrient pump, drawing nutrients from the soil, translocating them into the shoots, and then during the growing and post-growing season releasing them to the surface soil by leaching and death of the shoots. In this way, marsh plants make nutrients sequestered in the soil available for growth.

Nutrients accumulated in plant biomass become available for use through decomposition. Initial decomposition in wetlands is high, as leaching removes soluble compounds

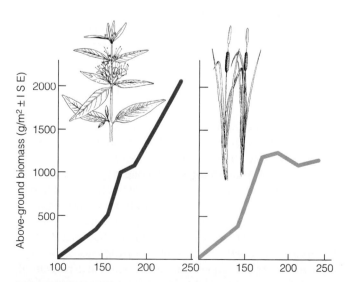

FIGURE 30.31 Pattern of above-ground biomass accumulation through the growing season for a freshwater annual, loosestrife (*Lythrum*), and a perennial, cattail (*Typha*). Note the linear increase in biomass in the annual and the sigmoid growth of the perennial. (From Whigham et al. 1978:12.)

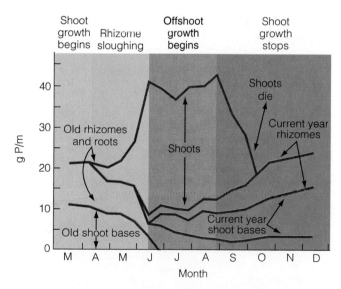

FIGURE 30.32 Seasonal stocks of phosphorus in cattail (*Typha latifolia*) plant parts and below-ground deficit in a marsh at Lake Mendota, Wisconsin. Note the sharp rise in P in the shoots during the summer months and its equally rapid decline in the fall when the shoots die. Much P is lost through leaching, but note the accumulation of some P in the current-year rhizomes. Below-ground deficit is highest during the summer, when most of the P is in the shoots, but the deficit is erased by winter. (After Prentki et al. 1978:176.)

that enter DOM (Figure 30.33). After initial leaching, decomposition proceeds more slowly. Permanently submerged leaves decompose more rapidly than those on the marsh surface, because they are more accessible to detritivores (Smock and Harlowe 1983) and because the stable physical environment is more favorable for microbes.

Peatlands   In a bog, nutrient input comes largely from precipitation because vegetation is not in contact with mineral soil and because inflow of groundwater is blocked. Nutrient availability, especially nitrogen, phosphorus, and potassium, is low, and most of the nutrients fixed in plant tissue are removed from circulation in the accumulation of peat. The amount of nutrients received by a bog depends on the location of peatland. Bogs near the sea, such as English blanket mires, receive considerably more nutrients, especially magnesium and potassium, than inland bogs. Few data exist on nutrient cycles in ombrotrophic bogs, but evidence suggests that some bog plants possess mechanisms to conserve nutrients.

In general, nitrogen is available from three sources: (1) precipitation, the major source; (2) nitrogen fixation by cyanobacteria living in close association with bog mosses and by bog myrtle, if the pH of the substrate is at least 3.5, below which nitrogen-fixing root nodule bacteria are inhibited; and (3) carnivory, by which certain plants such as sundews (*Drosera* spp.) and pitcher plants (*Sarracenia purpurea*) extract nitrogen from captured and digested insects. Cyanobacteria may fix 0.23 to 0.9 g/N/m^2/yr. Rosswall has calculated a nitrogen budget and nitrogen cycle for an ombrotrophic bog dominated by the trailing ground plant *Rubus chamaemorus* (Figure 30.34). Yearly demand is about 1 to 2 g/m^2.

The phosphorus cycle in a bog holds losses to a minimum and is more closed than the nitrogen cycle. *Rubus chamaemorus* increases its uptake of phosphorus prior to bud break. After bud break, the plant increases phosphorus in stem, leaf, and root. This increase correlates with an increase in peat temperature. After the plant completes shoot growth in summer, its phosphorus level declines in the shoots as it mobilizes phosphorus in developing fruits and in roots and rhizomes. In advanced senescence, the plant rapidly loses phosphorus from the shoots and accumulates it in roots and winter buds.

Energy flow in the mire system differs from that of other systems, because decomposition is impaired. In most ecosystems material that enters the detrital food web eventually is recycled, and energy is liberated or stored in living material. In mire systems about 10 percent of primary production accumulates in an undecomposed state, and energy is locked up in peat until environmental conditions change, favoring decomposition, or the material is burned.

Because of low temperatures, acidity, and nutrient immobilization, primary production in peatlands is low. In the Stordalen mire in Sweden, primary production amounted to 70 g/m^2/yr by bryophytes and 83 gm/m^2/yr by dwarf shrubs, forbs, and monocots collectively (Rosswall 1975). In an English blanket mire, average production was 635 g/m^2/yr, with sphagnum on wet sites contributing 300 g/m^2/yr (Heal et al. 1975).

Herbivore utilization of bog vegetation is low, in part because of unpalatability. Herbivores may include red grouse, willow grouse, hares, bog lemmings, and sheep. With these herbivores consumption is selective, usually confined to current shoots, opening buds, and fruits. In most bogs the dominant herbivores are insects (Heal et al. 1975, Moore et al. 1975). Predators include rodent-consuming weasels, harriers (*Circus* spp.), short-eared owls (*Asio flammeus*), insect-consuming frogs, shrews, pipits (*Anthus* spp.) (Europe), Nashville warblers (*Vermivora ruficapilla*) (North America), and invertebrates such as spiders and ground beetles.

Decomposer organisms are low in number, and the dominant species vary from year to year. In a Swedish mire, total fungal biomass was 58 g/m^2 and the bacterial biomass, largely anaerobic, was 22 g/m^2 (Rosswall et al. 1975). Invertebrate detritivores include rotifers, tardigrades, mites, and nematodes. In an English mire a single species of enchytraeid, *Cognettia sphagnetorum,* accounted for 70 to 75 percent of the total energy assimilated by the detritivores (Heal et al. 1975).

Rates of decomposition vary among bog vegetation. Litter of *Rubus chamaemorus* lost 2 percent of its weight in one year and 50 percent in three years. Shrub litter decomposes more slowly, while sphagnum decomposes hardly at all.

In an English mire (Moor House National Nature Reserve), the vegetational standing crop has an annual turnover of about 0.3 percent, and only about 1 percent of production is consumed by herbivores. The remainder of the vegetation enters the decomposer food web, where about 5 percent is assimilated by decomposers and 10 percent passes below the water table. About 85 percent of the production is

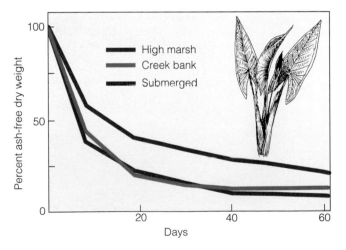

FIGURE 30.33 Decomposition of leaves of arrow arum in a tidal freshwater marsh, as measured by the percentage of original ash-free dry weight remaining in litter bags under three conditions: irregularly flooded high marsh exposed to alternate wetting, creek bed flooded two times daily, and permanently submerged. Note that the detrital material consistently wet showed the highest rate of decomposition, although the overall pattern of decomposition is similar. (From Odum and Haywood 1978:92.)

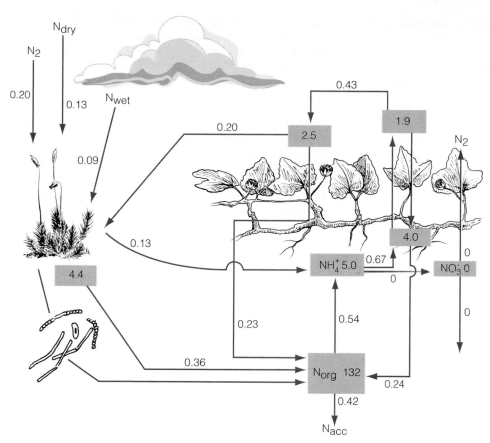

FIGURE 30.34 Nitrogen budget for an ombrotrophic mire in Stordalen, Sweden. Quantities of nitrogen are expressed as g N/m^2, flows as g $N/m^2/yr$. The vegetation has been divided into above-ground parts exemplified by *Rubus chamaemorus* and lichens (1.6 g N/m^2), below-ground parts of vascular plants (3.0 g N/m^2), above-ground litter (3.2 g N/m^2), and mosses (4.4 g N/m^2). N_{org} = organic nitrogen; N_{acc} = nitrogen accumulated. (After Rosswall 1975.)

decomposed by microflora; but the rate of decomposition is slow, with 95 percent turnover time of about 3000 years. In the top 20 cm, 95 percent turnover time is about 70 years (Heal et al. 1975).

Summary

1. Freshwater ecosystems are lentic (still water) or lotic (flowing water). Lentic ecosystems are lakes and ponds, standing bodies of water that fill a depression in the landscape. Geologically speaking, lakes and ponds are ephemeral features of the landscape. In time they fill in, draw smaller, and finally are replaced by a terrestrial community.

2. A lake exhibits gradients in light, temperature, and dissolved gases, resulting in seasonal stratification. In summer a top layer of warm, light water, the epilimnion, floats on top of a middle mass of water, the metalimnion, characterized by a steep decline in temperature. Below the metalimnion is a bottom layer of cold water, the hypolimnion. These gradients influence biological stratification. The area where light penetrates to the bottom of the lake, a zone called the littoral, is occupied by rooted plants. Beyond is the open water or limnetic zone, inhabited by plant and animal plankton and fish. Below the depth of effective light penetration, the diversity of life varies with temperature and oxygen supply. The bottom or benthic zone is a place of intense biological activity, for here decomposition of organic matter takes place. Anaerobic bacteria are dominant on the bottom beneath the profundal water, whereas the benthic zone of the littoral is rich in decomposer organisms and detritus-feeders. Although lake ecosystems are often considered as autotrophic systems dominated by phytoplankton and the grazing food web, lakes are strongly dependent on the detrital food web. Much of that detrital input comes from the littoral zone.

3. Lakes may be classified as eutrophic or nutrient-rich, oligotrophic, or dystrophic, acidic and rich in humic material. Most lakes are subject to cultural eutrophication, which is the rapid addition of nutrients, especially nitrogen and phosphorus, from sewage and industrial wastes. Cultural eutrophication has produced significant biological changes, mostly detrimental.

4. Lotic ecosystems are characterized by inputs of detrital material from terrestrial sources and currents of varying velocities that carry nutrients and other materials downstream. Lotic ecosystems exhibit a continuum of physical and ecological variables from source to mouth. There is a longitudinal gradient in temperature, depth, and width of the channel, velocity of the current, and nature of the bottom.

5. Changes in physical conditions are reflected in biotic structure. Headwater streams in forested regions are shaded and strongly heterotrophic, dependent on inputs of detritus. This detrital material is processed by a number of invertebrates, classified functionally as shredders, collectors, grazers, and gougers. These organisms, algae, and detritus slow the downstream movement of nutrients, known as spiraling. Larger streams,

open to sunlight, shift from heterotrophic to autotrophic. Primary production from algae and rooted aquatics becomes an important energy source. Large rivers return to a heterotrophic condition. They depend on fine particulate organic matter and dissolved organic matter as sources of nutrients and energy. Downstream systems depend on the inefficiencies of energy and nutrient processing upstream.

6. Closely associated with lakes and streams are wetlands, areas where water is at, near, or above the level of the ground and occupied by hydrophytic vegetation. Wetlands dominated by grasses are marshes; those dominated by woody vegetation are swamps. Wetlands characterized by an accumulation of peat are mires. Mires fed by water moving through the mineral soil (minerotrophic) and dominated by sedges are fens; those dominated by sphagnum moss and dependent on precipitation for moisture and nutrients (ombrotrophic) are bogs. Bogs are characterized by blocked drainage, accumulation of peat, and low productivity. A significant portion of nutrients fixed in plants is removed from circulation and stored in accumulated peat. The stored energy remains locked in peat until environmental conditions change to favor decomposition or until the material is burned. In contrast, marsh ecosystems are heavily influenced by nutrients stored in the soil organic material. Marsh vegetation acts as a nutrient pump, drawing nutrients from the substrate, translocating them to the shoots, and depositing them on the surface or losing them through leaching to the water.

Review Questions

1. What are the epilimnion, metalimnion, and hypolimnion? What is their significance?
2. What is the spring and fall overturn? What is its ecological importance to life in a lake?
3. Contrast the limnetic region with the littoral region of a lake. What is the contribution of the littoral zone to a lake ecosystem?
4. Compare and contrast the roles of phytoplankton and macrophytes in the function of lentic ecosystems.
5. Compare and contrast nutrient cycling in a lake, a stream, a freshwater marsh, and a bog.
6. Contrast between oligotrophy, eutrophy, and dystrophy.
7. What is the relationship of current to the nature of a stream?
8. Contrast the function of riffles and pools in a stream ecosystem.
9. What is the major source of nutrients in streams?
10. What are the meanings of DOM, FPOM, and CPOM?
11. Characterize the major functional groups of stream invertebrates. Describe their role in the flowing water ecosystem.
12. What is spiraling? How does it affect nutrient cycling and retention in streams?
13. How do downstream lotic systems relate to upstream systems?
14. Contrast the river continuum concept with the flood pulse concept. How do the two interrelate?
15. What are regulated rivers? How do they affect the river continuum?
16. What is the weakness of defining a wetland by vegetation alone?
17. Distinguish between obligate and facultative wetland plants.
18. What is hydroperiod? How does it relate to the structure of wetlands?
19. Distinguish between an ombrotrophic and a minerotrophic peatland.
20. Contrast phosphorous cycling in a cattail marsh and in a tundra sphagnum bog.
21. Consider the concept of a wetland (refer to Tiner 1992, Lyon 1993, Mitsch and Gosselink 1993). Then relate how the following would interpret a wetland from their own perspective and why each has a vested interest in how wetlands are defined: a wildlife and conservation biologist, a developer, a Midwest agriculturist, the Army Corps of Engineers, the Soil Conservation Service.
22. Study a topographic map of your area. Locate and mark drainage patterns and watershed or basin boundaries, starting with first-order streams. Then visit them in the field to study their gradient, size, rate of flow, and condition. Look for pollution, siltation, streambank vegetation, and proximity to human development. What major changes do you observe?

Cross References

Structure of water, 50; specific heat of water, 51; water cycle, 53; light, 43; nutrients, 54; hydric soil, 61; decomposition in aquatic environment, 156; energy flow, 481; food webs, 395, 493; oxygen cycle, 525; nitrogen cycle, 517; phosphorus cycle, 520.

CHAPTER 31

Saltwater Ecosystems

CONCEPTS

1. Temperature, salinity, tides, waves, and currents dictate the structure and function of marine ecosystems.

2. At the base of food webs in relatively unproductive open seas are phytoplankton, nanoflagellates, and cyanobacteria.

3. The most productive areas of the sea are coastal waters and coral reefs.

4. Rocky and sandy intertidal ecosystems are shaped by tidal cycles and the action of waves.

5. Coral reef ecosystems are created by coral organisms that make massive deposits of calcium carbonate in shallow tropical seas.

6. Estuarine ecosystems are strongly influenced by the inflow of fresh water from rivers.

7. The high production of salt marshes and mangrove swamps results from tidal subsidies of nutrients and energy.

Freshwater rivers eventually empty into the oceans, and terrestrial ecosystems end at the edge of the sea. For some distance beyond, there is a region of transition at the land and water interface. Rivers enter the saline waters of the ocean, creating a gradient of salinity. Coastal regions exposed to the open sea are dominated by tides. Beyond them lies the open ocean—shallow seas overlying continental shelves and the deep oceans.

PHYSICAL FEATURES

Stratification

Just as lakes exhibit vertical stratification, so do the seas. The ocean is divided into two main layers, the **pelagic region** or whole body of water, and the **benthic region** or bottom (Figure 31.1). Within the pelagic region there are three vertical layers. From the surface to about 200 meters is the **photic zone,** in which there are sharp gradients of light, temperature, and salinity. Ranging from 200 to 1000 m, where little light penetrates and where the temperature gradient is more gradual and without much seasonal variation, is the **mesopelagic** layer. It contains an oxygen-minimum layer and often a maximum concentration of nitrate and phosphate. Below the mesopelagic is the **bathypelagic,** where darkness reigns except for bioluminescence, the temperature is low, and pressure is great. Depending on its global location, it lies between 100 to 700 m and 2000 to 4000 m.

Horizontally, the pelagic region is subdivided into the **neritic province,** water that overlies the continental shelf, and the **oceanic province** (Figure 31.1). Lying over the major plains of the ocean, down to about 6000 m, is the **abyssal pelagic** or **benthipelagic province.** Water of the deep oceanic trenches is called the **hadalpelagic.**

The sea bottom is called the **benthic region.** It, too, can be divided into three zones, which differ in the nature of their bottom life. The **bathyal zone** covers the continental shelves and down to 4000 m. The **abyssal zone** embraces the broad ocean plains down to 6000 m, and the **hadal zone** is the benthic zone of oceanic trenches between 6000 and 10,000 m. The benthic zone underlying the neritic pelagic zone on the continental shelf is the **sublittoral** or **shelf zone.** The intertidal zone, like the margins of lakes, is called the **littoral.**

Temperature

The photic layer exhibits a stratification of temperature (Figure 31.2). In polar regions conditions remain fairly constant throughout the year. In middle latitudes temperatures vary with the season, as the climate changes. In summer the surface waters become warmer and lighter, forming a strong thermocline that decays with coming of winter. Tropical oceans, whose surface waters are constantly heated, have a strong permanent thermocline between 100 and 200 m.

Salinity

Two elements, sodium and chlorine, make up about 86 percent of sea salts. Along with other such major elements as sulfur, magnesium, potassium, and calcium, they make up 99 percent of sea salts. Seawater, however, differs from a simple sodium chloride solution in that the amounts of cations and anions are not balanced against each other (Table 31.1). The cations exceed the anions by 2.38 milliequivalents/kg. As a result, seawater is weakly alkaline (pH 8.0 to 8.3) and strongly buffered, a condition that is biologically important.

The amount of dissolved salt in seawater was usually expressed as parts per thousand ($\%_0$). The new term now used by marine biologists is **practical salinity units (psu).** A 1000 gm (1 kilogram) sample of seawater contains about 35 gm of dissolved compounds dominated by sodium and chlorine, and so its salinity is 35 psu.

The salinity of parts of the ocean is variable because of physical processes affecting it. Salinity is affected by evaporation, precipitation, movement and mixing of water masses of different salinities, formation of insoluble precipitates that sink to the ocean floor, and diffusion of one water mass to another. Salinities are most variable near the interface of sea and air.

FIGURE 31.1 Regions of the ocean.

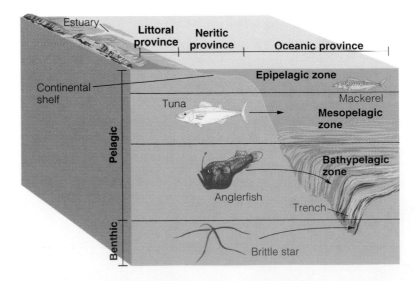

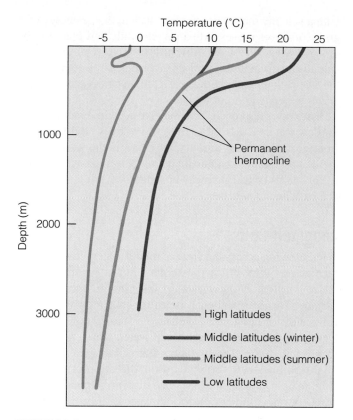

FIGURE 31.2 Temperature stratification in the oceans at high, middle, and low latitudes. (After Talt 1968.)

iodides. A major source of chlorine is hydrochloric acid (HCl) from submarine volcanoes in the midocean ridges. The hydrochloric acid reacts with bicarbonates to form water and carbon dioxide, freeing the chloride ion. The chloride ion then replaces the bicarbonate ion as the anion, balancing the sodium cation. (For the chemistry of seawater see Macintyre 1970.)

Salinity, like temperature, exhibits a vertical gradient in the sea (Figure 31.3). This gradient or **halocline** is especially pronounced at higher latitudes. There, abundant precipitation reduces surface salinity, whereas in tropical regions, evaporation increases salinity. In the middle latitudes, salinity together with temperature produces a marked gradient in density. Because the density of seawater increases with depth but does not reach its greatest density at 4°C, as with fresh water, there is no seasonal overturn. However, the rapid change in temperature, which produces a thermocline, is paralleled with a rapid change in the density of seawater. This zone of rapid change in density is the **pycnocline.**

Salinity imposes restrictions on life that inhabits the oceans (see Kinne 1971). Fish and marine invertebrates in marine, estuarine, and tidal environments have to maintain osmotic pressure, often under conditions of changing salinities. Most marine species are osmoconformers, lacking the ability to move into less saline water. Others, such as striped bass (*Morone saxatilis*) and eels (Ophichthidae), are osmoregulators and have the ability, at least during some part of their life cycle, to control the concentration of salts or water in their bodies. They can move into waters of lower than normal oceanic salinity.

Pressure

Another aspect of the marine environment is pressure. Pressure in the ocean varies from 1 atmosphere at the surface to

The salinity of the ocean has been ascribed to the accumulation of salts derived from the weathering of Earth's crust. This source can account for much of the sodium, but it cannot account for chlorine, boron, and salts such as bromides and

TABLE 31.1 Composition of Seawater of 35 psu

Elements	g/kg	Millimole/kg	Milliequivalent/kg
CATIONS			
Sodium	10.752	467.56	467.56
Potassium	0.395	10.10	10.10
Magnesium	1.295	53.25	106.50
Calcium	0.416	10.38	20.76
Strontium	0.008	0.09	0.18
Total			605.10
ANIONS			
Chlorine	19.345	545.59	545.59
Bromine	0.066	0.83	0.83
Fluorine	0.0013	0.07	0.07
Sulfate	2.701	28.12	56.23
Bicarbonate	0.145	2.38	—
Boric acid	0.027	0.44	—
Total			602.72

Note: Surplus of cations over strong anions (alkalinity): 2.38.
Source: Kalle 1971.

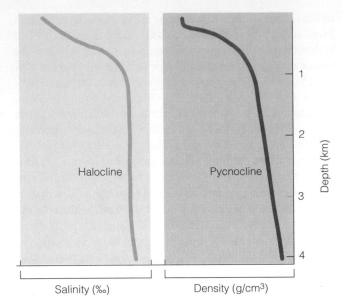

FIGURE 31.3 The halocline, the stratification of salinity in the ocean, and the pycnocline, stable stratification of density. (After Gross 1972.)

1000 atm at its greatest depth. Pressure changes are many times greater in the sea than in terrestrial environments and have a pronounced effect on the distribution of life. Certain organisms are restricted to surface waters, whereas others are adapted to pressure at great depths. Some marine organisms, such as sperm whales (*Physeter macrocephalis*) and certain seals, can dive to great depths and return to the surface without difficulty.

Waves

Waves are generated on the surface of seas (and lakes) by pressures of wind on their surfaces. The frictional drag of the wind on the surface of smooth water ripples it. As the wind continues to blow, it applies more pressure to one side of the ripple, and wave size grows. As the wind becomes stronger, short, choppy waves of all sizes appear; and as they absorb more energy, they continue to grow in size. When the waves reach a point at which the energy supplied by the wind is equal to the energy lost by the breaking waves, they become the familiar whitecaps.

Up to a certain point, the stronger the wind, the higher the waves and the longer the **fetch** (the distance the waves can run without obstruction under the drive of the wind in a constant direction). As waves travel out of the fetch (or if the wind dies), sharp-crested waves change into smooth, long-crested waves or swells, characterized by troughs and ridges. Swells can travel great distances because they lose little energy as they travel.

The waves that break on beaches are generated in and transported from distant seas. Each particle of water remains largely in the same place and follows an elliptical orbit with the passage of the wave form. As the wave moves forward

with a velocity that corresponds to its length, the energy of a group of waves moves with a velocity only half that of individual waves. The wave in front loses energy to the waves behind and disappears, its place taken place by another. Thus the swells that break on the beach are distant descendants of waves generated far out at sea.

As waves approach land, they advance into increasingly shallow water. The height of the waves rises until the wave front grows too steep and topples over. As the waves break on the shore, they dissipate their energy against it, pounding rocky shores or tearing away at sandy beaches at one point and building up new beaches elsewhere.

Langmuir Cells

Wind blowing across the surface may stir up more than surface waves. Winds blowing above 3 m/sec generate circulating vertical cells of water spinning just below the surface. These are called **Langmuir cells** after the physical chemist I. Langmuir (1938), who studied them in Lake George, New York. These cells of water, which may range from a few centimeters to many meters in diameter, are parallel to the water's surface and the direction of the wind (Figure 31.4). Each circulates in a direction opposite to the adjacent one. One circulates clockwise, the adjacent one counterclockwise, so that between one pair there are upwelling or divergent currents and the between the next pair, downwelling or convergent currents. The position of the downwelling currents is marked on the surface by streaks of accumulated materials, such as leaves and strands of submerged aquatic vegetation lifted to the surface. Where upwelling with its divergent currents occurs, no such streaks appear. These internal currents mix heat and nutrients in the surface water during the summer and circulate phytoplankton, accounting for its often patchy and changing distribution in lakes and ponds.

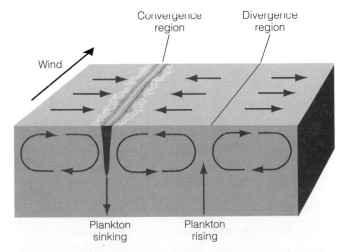

FIGURE 31.4 Langmuir cells in a body of water. Note that these cells are marked by accumulated materials on the surface where the adjacent cells upwell. (After Nybakken 1988:63.)

Ekman Spiral and Upwelling

Also influencing subsurface currents in the oceans is the **Ekman spiral** (Figure 31.5). Ocean currents do not flow parallel to the wind; they are deflected by the Coriolis force. As wind moves the surface water, some of the energy passes down through the water column and sets each successive layer into motion. Energy, however, decreases with depth, and so each layer moves more slowly than the layer above it. Like the surface water, each layer in motion is deflected relative to the one above. Ultimately the water mass at the base of the spiral moves counter to the flow on the surface, although it is at right angles to the wind. In coastal regions, the Ekman spiral can influence upwelling, the bringing of deep water to the surface.

Coupled with horizontal motions of water are other vertical water motions. Wind blowing parallel to continental margins causes surface water to be blown offshore. The water lost is replaced by water moving upward from the deep. Although cold and containing less dissolved oxygen, upwelling water is rich in nutrients that support abundant growth of phytoplankton. For this reason regions of upwelling, such as the west coasts of South America and North America, are highly productive.

Tides

The most familiar feature of marine ecosystems, especially along the coasts, is the tide. Tides are periodic and pre-dictable rising and falling sea levels over a given interval. They result from a combination of the gravitational action of sun and moon on Earth and the centrifugal forces of the Earth-moon system.

Earth and moon form a single system revolving about a common center of gravity. Because Earth is so much larger than the moon, the common center of the mass is within Earth. Gravitational force is a function of the mass of each body and the distance between them. Although the moon is much smaller than the sun, it is also very much closer, and so its gravitational force on Earth is two times that of the sun. The centrifugal force within this revolving system more or less balances the gravitational force acting between these two bodies. The interaction of these two forces pulls the water in the ocean basin into bulges. On the side of Earth facing the moon, the gravitational force exceeds the centrifugal force, and the water is pulled into a bulge—the high tide. On the opposite side of Earth, away from the moon, the centrifugal force exceeds the gravitational force, and it, too, pulls the water into a bulge, creating another high tide (Figure 31.6). As Earth rotates on its axis, the position of the moon relative to Earth changes during the 24-hour day. High tides follow the changing position of the moon around Earth. As Earth rotates eastward on its axis, the tides advance westward. Thus any given place on Earth will in the course of one daily rotation pass through two of the lunar tidal bulges, or high tides, and two of the lows or low tides, at right angles to the high. The moon revolves in a 29.5-day orbit around Earth, and this causes the average period between successive high tides to be approximately 12 hours, 25 minutes.

The sun also causes two tides on opposite sides of Earth, but because they are less than half as high as lunar tides, solar tides are usually partly masked by the lunar tides. Twice during the month, when the moon is full and the sun and moon

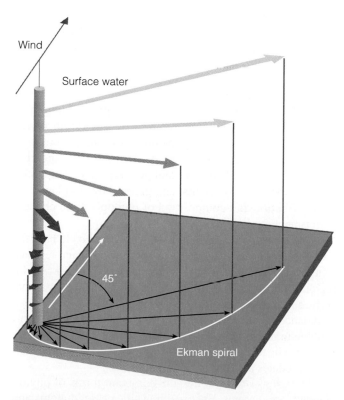

Wind

Surface water

45°

Ekman spiral

FIGURE 31.5 The Ekman spiral. The wind-driven surface current moves at an angle of 45° to the direction of the wind, to the right in the Northern Hemisphere, to the left in the Southern. Successively deeper water layers are deflected even further than those immediately above them and move at slower speeds. Net water movement is at 90° to the wind. (Adapted from Strahler 1971:258.)

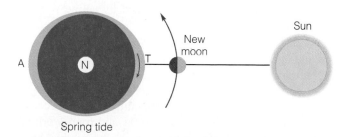

A · N · T · New moon · Sun

Spring tide

FIGURE 31.6 Tides result from the gravitational pull of the moon on Earth. Centrifugal force applied to a kilogram of mass is 3.38 mg. This centrifugal force on a rotating Earth is balanced by gravitational force, except at those (moving) points on Earth's surface that come into direct line with the moon. Thus the centrifugal force at point N, the center of the rotating Earth, is 3.38 mg. Point T on Earth is in direct line with the moon. At this point the gravitational force of the moon is 3.49 mg, a difference of 0.11 mg. Because the moon's gravitational force is greater than the centrifugal force at T, the force is directed away from the Earth and causes a tidal bulge. At point A on the opposite side of the Earth from T, the moon's gravitational force is 3.27 mg, 0.11 mg less than the centrifugal force at N. This difference causes a tidal bulge on the opposite side of Earth. (See text for further details.)

are opposite each other, and when it is new and the sun is behind the moon, Earth, sun, and moon are nearly in line. Then the gravitational pulls of the sun and moon are additive. The intensified force causes what are known as **spring tides,** tides of maximum rise and fall. When the moon is at either quarter, the gravitational pulls of sun and moon interfere with each other, creating **neap tides** of minimum difference between high and low tides.

Tides are not entirely regular, nor are they the same all over Earth. They vary from day to day in the same place, following the waxing and waning of the moon. They may act differently in several localities within the same general area. In the Atlantic, semidaily tides are the rule. In the Gulf of Mexico, flood and ebb tides follow one another at about 24-hour intervals to produce one daily tide. Mixed tides, combinations of semidaily and daily tides, are common in the Pacific and Indian oceans, with different combinations at different places.

Local inconsistencies of tides are due to many variables. The elliptical orbits of Earth about the sun and of the moon about Earth influence the gravitational pull, as does the declination of the moon—the angle of the moon in relation to the axis of Earth. Latitude, barometric pressure, offshore and onshore winds, depth of water, contour of the shore, and internal waves modify tidal movements. Islands in the middle of a tidal basin may experience minimal tides. Geological structures of some basins funnel tidal water into restricted channels, such as the Bay of Fundy in Nova Scotia and Cook Inlet of Alaska. They experience extremely high tides.

THE OPEN SEA

The oceans occupy 70 percent of Earth's surface. All of the seas are interconnected by currents, dominated by waves, influenced by tides, and characterized by saline waters. They are deep, in some places 7 km. The volume of surface waters lit by the sun is small compared with the total volume of water. Dimness and the dilute solution of nutrients limit production.

Compared with terrestrial ecosystems, the pelagic ecosystem lacks well-defined communities, largely because of the lack of the supporting structures and framework of large dominant plant life. Exceptions are the subtidal kelp forests, dominated by brown algae, found throughout the cold temperate regions of the world, and the Sargasso Sea, dominated by floating sargassum weed (*Sargassum*), also a brown algae. However, differences based on physical characteristics and life forms permit a division of the ocean into different regions.

The Arctic Ocean comprises marine waters that lie north of the land masses in the Northern Hemisphere and is open to the Atlantic Ocean throughout the year and to the Pacific only during several short summer months. The Southern or Antarctic Ocean lies about the continent of Antarctica and is open to three oceans, the Atlantic, Pacific, and Indian. Making up the Polar Domain, both have an inner polar zone that is covered by ice for the entire year and an outer zone where drifting ice is encountered during winter and spring. Oceanic waters of the Atlantic and Pacific making up the Temperate Domain vary in the direction of their currents and salinity. They possess deep-mixing processes that in certain regions result in high nutrient concentration and an abundance of plankton. These regions contrast with the blue, nutrient-poor tropical oceans, high in temperature and salinity.

Deep sea benthic ecosystems are distinctively different from the lighted waters. Other important distinctive marine ecosystems include the coral reefs, upwelling systems off the coasts of California, Peru, Northwest Africa, Southwest Africa, India, and Pakistan. Important and distinctive are the sea shelf ecosystems. Shallow, productive, and nutrient-rich, they support a diversity of fish and invertebrate marine life.

Structure

Because the global regions vary in their diversity and nature, ranging from oligotrophic to eutrophic and from cold to tropical, a brief discussion of structure must be general and emphasize the major structural groups. (For a good introduction to the structure and function of marine ecosystems, see Nybakken 1997.)

Phytoplankton   As in lakes, the dominant form of plant life or primary producer in the open sea is phytoplankton. Seawater is so dense that oceanic phytoplankton do not need well-developed supporting structures. In coastal waters and areas of upwelling, phytoplankton 100 μm or more in diameter may be common, but in general plant life is much smaller and widely dispersed.

Because of its requirement for light, phytoplankton is restricted to the upper surface waters. The depth of its occurrence is determined by the depth of light penetration and so may range from tens to hundreds of meters. Because of seasonal, annual, and geographic variations in light, temperature, nutrients, and grazing by zooplankton, the distribution and composition of phytoplankton change with time and place.

Each ocean or region within an ocean appears to have its own dominant forms. Littoral and neritic waters and regions of upwelling are richer in plankton than are midoceans. In regions of downwelling, dinoflagellates concentrate near the surface in areas of low turbulence. They attain their greatest abundance in warmer waters. In summer they may so concentrate in the surface waters that they color it red or brown. Often toxic to other marine life, such concentrations of dinoflagellates are responsible for red tides.

In regions of upwelling, the dominant forms of phytoplankton are diatoms. Enclosed in a silica case, diatoms are particularly abundant in arctic waters.

Smaller than diatoms are the Coccolithophoridae, so small they pass through plankton nets (and so are classified as nanoplankton). Their minute bodies are protected by calcareous plates or spicules embedded in a gelatinous sheath (Figure 31.7). Universally distributed in all waters except the

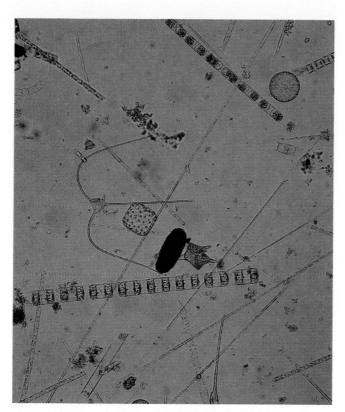

FIGURE 31.7 An array of marine phytoplankton, including diatoms and dinoflagellates.

FIGURE 31.8 Small euphausiid shrimps called krill are eaten by baleen whales.

polar seas, the Coccolithophoridae possess the ability to swim. Droplets of oil aid in buoyancy and storage of food.

In the equatorial currents and in shallow seas, the concentration of phytoplankton is variable. Where both lateral and vertical circulation of water is rapid, the composition reflects in part the ability of the species to grow, reproduce, and survive under local conditions.

Zooplankton Grazing on the phytoplankton is the herbivorous zooplankton, consisting mainly of copepods, planktonic arthropods that are the most numerous animals of the sea, and the shrimplike euphausiids, commonly known as krill (Figure 31.8). Other planktonic forms are the larval stages of such organisms as gastropods, oysters, and cephalopods. Feeding on the herbivorous zooplankton is the carnivorous zooplankton, which includes such organisms as the larval forms of comb jellies (Ctenophora) and arrow-worms (Chaetognatha).

Because its food in the ocean is so small and widely dispersed, zooplankton have evolved ways more efficient and less energy-demanding to harvest phytoplankton than filtering water through pores. They have webs, bristles, rakes, combs, cilia, sticky structures, and even bioluminescence.

The composition of zooplankton, like that of phytoplankton, varies from place to place, season to season, year to year. In general zooplankton fall into two main groups: the larger forms characteristic of shallow coastal waters, and the smaller forms characteristic of the deeper open ocean. Zooplankton forms of the continental shelf contain a large proportion of larvae of fish and benthic organisms. They have a

greater diversity of species, reflecting a greater diversity of environmental and chemical conditions. The open ocean, being more homogeneous and nutrient-poor, supports less diverse zooplankton. In polar waters, zooplankton species spend the winter in a dormant state in the deep water and rise to the surface during short periods of diatom blooms to reproduce. In temperate regions, distribution and abundance depend on temperature conditions. In tropical regions, where temperature is nearly uniform, zooplankton are more abundant, and reproduction occurs throughout the year.

Also like phytoplankton, zooplankton live mainly at the mercy of the currents; but many forms possess sufficient swimming power to exercise some control. Most species of zooplankton undertake a daily vertical migration from as deep as 800 m to 100 m or to the surface. As darkness falls, zooplankton rapidly rise to the surface to feed on phytoplankton. At dawn, the forms move back down to preferred depths.

By feeding in the darkness of night and hiding in the darkened waters by day, zooplankton avoid heavy predation; and by remaining in cooler water by day, they conserve energy during the resting period. Surface currents move zooplankton away from their daytime location during feeding, but they can return home by countercurrents present in the deeper layers. Response to changing light conditions is useful in another way. As clouds of phytoplankton pass over the water above, zooplankton respond to the shadow of food by moving upward. This motion takes them out of the deep current drift and nearer the surface. At night zooplankton can move directly up to the food-rich surface water.

Zooplankton that lack a vertical migration, and even some of those that have it, drift out of the breeding area with surface currents. Survival of a breeding population is assured by a complex cycle of seasonal migration.

Nekton Feeding on zooplankton and passing energy along to higher trophic levels is the **nekton,** swimming organisms that can move at will in the water column. They range in size from small fish to large predatory sharks and whales, seals, and marine birds such as penguins. Some of the predatory fish, such as herring and tuna, are more or less restricted to the photic zone. Others are found in the deeper mesopelagic and bathypelagic zones or can move between them, as does the sperm whale (*Physeter catadon*). Although the ratio in size of predator to prey falls within certain limitations, some of the largest nekton organisms in the sea, the baleen whales (Mysticeti), feed on disproportionately small prey, euphausiids (popularly called krill). The sperm whale, 13 to 18.5 m in length, attacks very large prey, the giant squid (*Architeuthis*), whose overall length (tentacles extended) may be 18 m or more.

Living in a world that lacks any sort of refuge against predation or site for ambush, inhabitants of the pelagic zone have evolved various means of defense and of securing prey. Among them are the stinging cells of jellyfish, the remarkably streamlined shapes that allow speed both for escape and for pursuit, unusual coloration, advanced sonar, a highly developed sense of smell, and a social organization involving schools or packs. Some animals, such as the baleen whales, have specialized structures that permit them to strain krill and other plankton from the water. Others, such as sperm whale and certain seals, have the ability to dive to great depths to secure food. Phytoplankton light up darkened seas, and fish take advantage of that bioluminescence to detect their prey.

The dimly lighted regions of the mesopelagic and the dark regions of the bathypelagic zone depend on a rain of detritus as an energy source. Such food is limited. The rate of descent of organic matter, except for larger items, is so slow that it is consumed, decayed, or dissolved before it reaches the deepest water or the bottom. Other sources include saprophytic plankton, which exist in the darker regions, particulate organic matter, and such material as wastes from the coastal zone, garbage from ships, and dead whales and fish.

Residents of the deep have special adaptations for securing food. Some, like the zooplankton, feed on the upper surface at night. Others remain in the dim waters. Many deep sea fish are darkly pigmented and have weak bodies. They rely on luminescent lures, mimicry of prey, extensible jaws, and expandable abdomens (which enable them to consume large items of food) as means of obtaining sustenance in this zone. Although most fish are small (usually 15 cm or less in length), the region is inhabited by rarely seen large species such as the giant squid.

Bioluminescence reaches its greatest development in the bathypelagic region. Two-thirds of the species produce light. Bioluminescence is not restricted to fish. Squid and euphausiids possess searchlightlike structures complete with lens and iris, and squid and shrimp discharge luminous clouds to escape predators. Fish have rows of luminous organs along their sides and lighted lures that enable them to bait prey and recognize other individuals of the same species.

Benthos   There is a gradual transition of life from the benthos on the rocky and sandy shores to that in the ocean's depths. From the tide line to the abyss, organisms that colonize the bottom are influenced by the nature of the substrate. If the bottom is rocky or hard, the populations consist largely of organisms that live on the surface of the substrate, the epifauna and the epiflora. Where the bottom is largely covered with sediment, most of the inhabitants, chiefly animals, are infauna and live within the deposits. Particle size of the sub-

strate determines the type of burrowing organisms in an area, because the mode of burrowing is often specialized and adapted to a certain type of substrate.

The substrate varies with the depth of the ocean and the relationship of the benthic region to land areas and continental shelves. Near the coast, bottom sediments are derived from the weathering and erosion of land areas along with organic matter from floating marine life. Sediments of deep water are characterized by fine-textured material, which varies with depth and the type of organisms in the overlying waters. Although these sediments are termed organic, they contain little decomposable carbon, consisting largely of skeletal fragments of planktonic organisms. In general, with regional variations, organic deposits down to 4000 m are rich in calcareous matter. Below 4000 m hydrostatic pressure causes some forms of calcium carbonate to dissolve. At 6000 m and lower, sediments contain even less organic matter and consist largely of red clays, rich in aluminum oxides and silica.

Within the sediments are layers that relate to oxidation-reduction reactions. The surface or oxidized layer, yellowish in color, is relatively rich in oxygen, ferric oxides, nitrates, and nitrites. It supports the bulk of the benthic animals, such as polychaete worms and bivalves in shallow water, flatworms, copepods, and others in deeper water, and a rich growth of aerobic bacteria throughout. Below the oxidized layer is a grayish transition zone to the black reduced layer, characterized by a lack of oxygen, iron in the ferrous state, nitrogen in the form of ammonia, and hydrogen sulfide. It is inhabited by anaerobic bacteria, chiefly reducers of sulfates and methane.

In the deep benthic regions, variations in temperature, salinity, and other conditions are negligible. In this world of darkness there is no photosynthesis, and so the bottom community is strictly heterotrophic, depending entirely for its source of energy on what organic matter finally reaches the bottom. Estimates suggest that the quantity of such material amounts to only 0.5 $g/m^2/yr$ (Moore 1958). Bodies of dead whales, seals, and fish may contribute another 2 or 3 g.

Bottom organisms have four feeding strategies: (1) they may filter suspended material from the water, as do stalked coelenterates; (2) they may collect food particles that settle on the surface of the sediment, as do the sea cucumbers; (3) they may be selective or unselective deposit-feeders, such as the polychaetes; or (4) they may be predators, as are the brittle stars and the spiderlike pyenogonids.

Important in the benthic food web are bacteria of the sediments (Figure 31.9). Common where large quantities of organic matter are present, bacteria may reach several tens of grams per square meter in the topmost layer of silt. Bacteria synthesize protein from dissolved nutrients and in turn become a source of protein, fats, and oils for deposit-feeders.

In spite of living in a world of darkness and pressure, the benthic fauna exhibits surprising diversity (Graggle 1991). Faunal composition and diversity increases with depth down to the mid or lower bathyal region and decreases toward the abyssal plain (Rex 1981). Diversity along a gradient of depth appears to be related to the productivity of the waters above, competition among bottom fauna for food, and predation.

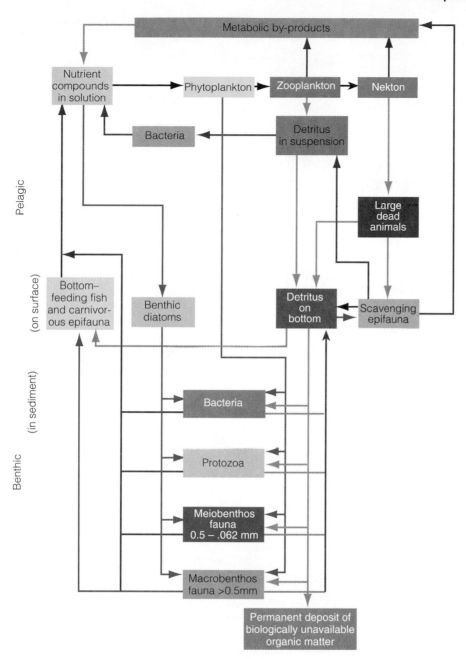

FIGURE 31.9   Simplified marine food web. (After Raymont 1963, from Gross 1972.)

Hydrothermal Vents   In 1977 oceanographers discovered along volcanic ridges in the ocean floor of the Pacific near the Galapagos Islands high-temperature deep sea springs never known to exist before. These springs vent jets of hydrothermal fluids that heat the surrounding water to 8°C to 16°C, considerably higher than the 2°C ambient water. Since then oceanographers have discovered similar vents on volcanic ridges along other fast-spreading centers of ocean floor, particularly in the eastern Pacific (for the geology and chemistry of such vents, see Hayman and McDonald 1983, van Dover 1990, Tunnicliffe 1992).

Vents form when cold seawater flows down through fissures and cracks in the basaltic lava floor deep into the underlying crust, specifically in areas where continental plates are separating from each other. The waters react chemically with the hot ballast, giving up some minerals but becoming enriched with others such as copper, iron, sulfur, and zinc. Heated to a very high temperature, the water reemerges through the floor through mineralized chimneys rising up to 13 m above the floor. Among the types of chimneys are white smokers and black smokers (Figure 31.10). White smoker chimneys rich in zinc sulfides issue a milky fluid under 300°C. Black smokers, narrower chimneys rich in copper sulfides, issue jets of clear water 300°C to over 450°C that is soon blackened by precipitation of fine-grained sulfur-mineral particles.

Associated with these vents is a rich diversity of newly discovered deep sea forms of life confined within a few meters of the vent system. Of the 293 species of vent animals described, 97 percent are endemic to the hydrothermal

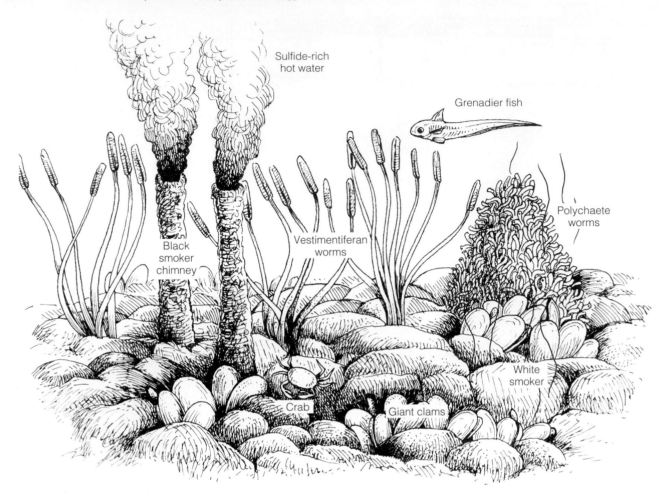

FIGURE 31.10 A typical hydrothermal vent mound resting on flows of black basaltic lava. (Adapted from Hayman and McDonald 1985 and other sources.)

vent habitat. There is one new class, Vestimentifera, the giant tubeworms, which lack a digestive system. There are three new orders and 22 new families (van Dover 1990, Tunnicliffe 1992). These include giant clams, mussels, polychaete worms that encrust the white smokers, and crabs. The primary producers are chemosynthetic bacteria that oxidize the reduced sulfur compounds, such as H_2S, to release energy, which they use to form organic matter from carbon dioxide. Primary consumers, the clams, mussels, and worms, filter bacteria from water and graze on bacterial film on rocks. The giant clam *Calyptogena magnifica* and the large vestimentiferan worm *Riftia pachyptila* contain symbiotic chemosynthetic bacteria in their coelomic tissue. These bacteria need a reduced sulfide source, which is carried to them by the blood of these animals. Aix and Childress (1983) report that *Riftia* has in its blood a sulfide-binding protein that concentrates sulfide from the environment and transports it to the bacteria. Such concentrations would poison normal animals, but the sulfide-bearing protein of the worm and apparently the clam has a high affinity for free sulfides, preventing them from accumulating in the blood and entering the cells (Power and Somber 1983).

Function

The oceans occupy 70 percent of Earth's surface and their average depth is around 4000 m; yet their primary production is considerably less than that of the land. They are less productive because only a superficial illuminated area up to 100 m deep can support plant life; and that plant life, largely phytoplankton, is patchy because most of the open sea is nutrient-poor. A limited to almost nonexistent nutrient reserve can be recirculated. Phytoplankton, zooplankton, and other organisms and their remains sink below the lighted zone into the dark benthic water. While this sinking supplies nutrients to the deep, it robs the upper layers.

The depletion of nutrients is most pronounced in tropical waters. The permanent thermal stratification, with a layer of warmer, less dense water lying on top of colder, denser deep water, prevents an exchange of nutrients between the surface and the deep. Thus in spite of high light intensity and warm temperatures, tropical seas are the lowest in production, about 18 to 50 g $C/m^2/yr$.

The temperate oceans are more productive, largely because a permanent thermocline does not exist. During the spring and to a limited extent during the fall, temperate seas, like temperate lakes, experience a nutrient overturn. The recirculation of phosphorus and nitrogen from the deep stimulates a surge of spring phytoplankton growth. As spring wears on, the temperature of the water becomes stratified and a thermocline develops, preventing a nutrient exchange. The phytoplankton growth depletes the nutrients, and the phytoplankton population suddenly declines. In the fall a similar overturn takes place, but the rise in phytoplankton production is slight because of decreasing light intensity and low winter temperatures. Reduced production in winter holds down annual productivity of temperate seas to a level a little above that of tropical seas, 70 to 110 g $C/m^2/yr$.

Most productive are coastal waters and regions of upwelling, whose annual production may amount to 1000 g $C/m^2/yr$. Major areas of upwelling are largely on the western sides of continents: off the coasts of southern California, Peru, northern and southwestern Africa, and the Antarctic. Upwellings result from the differential heating of polar and equatorial regions that produces the equatorial currents and the winds. As the water is pushed northward or southward toward the equator by winds, it is deflected away from the coasts by the Coriolis force. As the deflected surface water moves away, it is replaced by an upwelling of colder, deeper water that brings a supply of nutrients into the sunlit portions of the sea (see Chapter 2). As a result, regions of upwellings are highly productive and contain an abundance of life. Because of their high productivity, upwellings support (or did support, until overfished) important commercial fisheries such as the tuna fishery off the California coast, the anchoveta fishery off Peru, and the sardine fishery off Portugal.

Other zones of high production are coastal waters and estuaries, where productivity may run to 380 g $C/m^2/yr$. Turbid, nutrient-rich waters are major areas of fish production. Thus between upwellings and coastal waters, the most productive areas of the seas are the fringes of water bordering the continental land masses. A great deal of the measured productivity of the coastal fringes comes from the benthic as well as the surface waters, because these seas are shallow. Benthic production, largely unavailable, is not considered in the productivity of the open sea. An estimate (Koblentz-Mishke et al. 1970) of the total production for marine plankton is 50 Gt dry matter per year; if benthic production is considered, total production may be 55 Gt of dry matter per year.

Carbohydrate production by phytoplankton, largely diatoms, is the base upon which life of the seas exist. Conversion of primary production to animal tissue is accomplished by zooplankton, the most important of which are the copepods. To feed on the minute phytoplankton, most of the grazing herbivores must also be small, measuring between 0.5 and 5.0 mm. Most of the grazing herbivores in the oceans are of the genera *Calanus, Acartia, Temora,* and *Metridia,* probably the most abundant animals in the world. The single most abundant copepod is *Calanus finmarchicus* and its close relative *C. helgolandicus.* In the Antarctic, krill, fed on by the blue whale, are the dominant herbivores.

The copepods then become the link in the classic food chain between the phytoplankton and the second-level consumers, illustrated in the North Sea food web (Figure 31.11). The dominant primary producer is the diatom *Skeletoma,* which is grazed by *Calanus.* The major predator on *Calanus* is the semiplankton sand eel *Ammodytes,* which in turn is food for herring (*Alosa* spp.). The herring, however, can shorten the food chain by bypassing the sand eel and feeding directly on *Calanus.* The herring is also involved in a number of side food chains that add stability to its food supply.

Part of the food chain begins with organisms even smaller. Recent investigations show that bacteria and protists, both heterotrophic and photosynthetic, make up one-half of the biomass of the sea and are responsible for the largest part of energy flow in pelagic systems (Fenchel 1987, 1988). Photosynthetic nanoflagellates (2–20μ) and cyanobacteria (1–2μ) are responsible for a large part of photosynthesis in the sea. These phytoplankton cells excrete a substantial fraction of their photosynthate in the form of dissolved organic material that is utilized by heterotrophic bacteria.

Populations of such bacteria are dense, around 1 million cells per milliliter of seawater. These heterotrophic bacteria account for about 20 percent of primary production. Bacterial growth efficiency does not exceed 50 percent, and so half of phytoplankton primary production in the form of dissolved organic material is consumed by bacteria. Bacterial numbers in the sea remain relatively stable, suggesting predation; but filter-feeding zooplankton cannot retain particles of bacterial size. Therefore, the consumption of bacteria by heterotrophic nanoflagellates, experimentally demonstrated, accounts for the disappearance of bacterial production. This uptake by heterotrophic bacteria of dissolved organic matter produced by the plankton, and the subsequent consumption of bacteria by

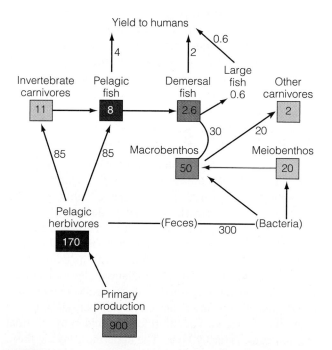

FIGURE 31.11 Food web and energy flow from an area of the North Sea. Note the low energy yield to humans, 6.6 kcal/m²/yr from a primary production of 900 kcal/m²/yr. (From Steele 1971.)

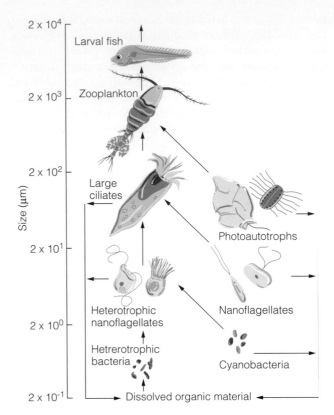

FIGURE 31.12 A microbial loop that feeds into the classic food web. On the right are representative photoautotrophs, nano-flagellates, and cyanobacteria and on the left are associated heterotrophs, bacteria, and heterotrophic nanoflagellates. (Adapted from Fenchel 1988:24.)

FIGURE 31.13 The broad zones of life exposed at low tide on the rocky shore of the Bay of Fundy, Nova Scotia.

nanoplankton introduce a feeding loop, termed the **microbial loop** (Azam et al. 1983) (Figure 31.12). Nanoplankton, a newly discovered group of phytoplankton unicellular cyanobacteria (less than 5 μm in size), and other abundant types of small marine phytoplankton are preyed on by phagotrophic protozoa, adding several more trophic levels to the plankton food chain (Sherr and Sherr 1991).

ROCKY SHORES

All ocean shores have one feature in common. They are alternately exposed and submerged by tides. Roughly, the seashore region is bounded by the extreme high water mark and extreme low water mark. Within this intertidal zone, conditions change from hour to hour with the ebb and flow of tides. At high tide the seashore is a water world; at ebb tide it belongs to the terrestrial environment, with its extremes of temperature, moisture, and solar radiation. Nevertheless the intertidal inhabitants are essentially marine, able to withstand some degree of exposure to air for varying periods of time.

As the sea recedes at ebb tide, life hidden by the tidal water emerges, layer by layer. The uppermost layers, those near the high water mark, are exposed to air, wide temperature fluctuations, intense solar radiation, and desiccation for a considerable period of time, whereas the lowest fringes on the intertidal shore may be exposed only briefly before the tide submerges them again.

Structure

Exposure to varying environmental conditions results in pronounced zonation of life on rocky shores (Figure 31.13). This zonation may be strikingly different from place to place as a result of local and geographic variations in aspect, substrate, wave action, light intensity, shore profile, exposure to prevailing winds, climatic differences, and the like—still, rocky shores possess the same general features. These features are expressed as three basic or universal zones characterized by the dominant organisms occupying them (Figure 31.14).

Where land ends and the seashore begins is difficult to fix. The approach to a rocky shore from the landward side is marked by a gradual transition from lichens and other land plants to marine life, dependent in part on tidal waters (Figure 31.15). The first major change shows up at the **littoral fringe,** whose upper boundary is determined by the upper limits of sea spray. It is marked by a black zone, a patchy or beltlike encrustation of *Varrucaria*-type lichens and crustose algae such as *Calothrix*. Capable of existing under conditions that few other plants could survive, these algae are enclosed in slimy, protective, gelatinous sheaths. They and their associated lichens represent an essentially nonmarine community on which graze basically marine animals, the common periwinkles (*Littorina littora*) (Figure 31.16). On European shores lives the small

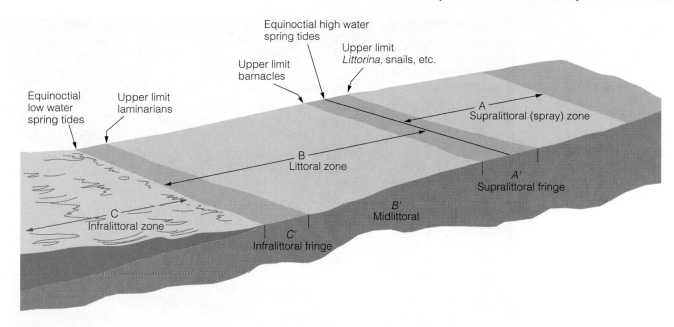

FIGURE 31.14 Basic or universal zonation on a rocky shore. Use this diagram as a guide when studying the drawing in Figure 31.15. (Adapted from Stephenson 1949.)

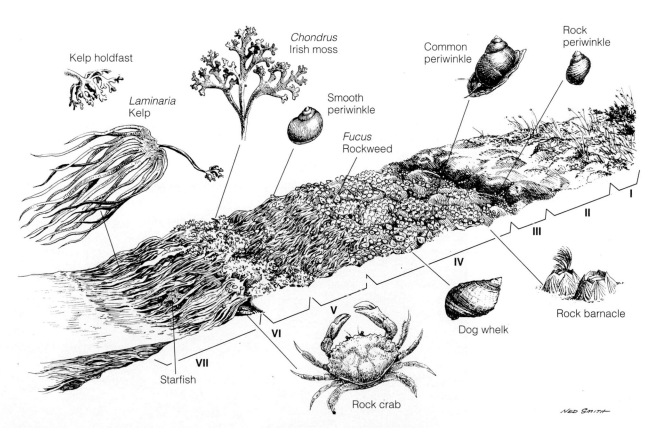

FIGURE 31.15 Zonation on a rocky shore along the North Atlantic. Compare with the diagram in Figure 31.14. I, land: lichens, herbs, grasses; II, bare rock; III, zone of black algae and rock periwinkles; IV, barnacle zone: barnacles, dog whelks, common periwinkles, mussels; V, fucoid zone: rockweed (*Fucus*) and smooth periwinkles; VI, Irish moss (*Chondrus*) zone; VII, kelp (*Laminaria*) zone. (Based on data from Stephenson and Stephenson 1954.)

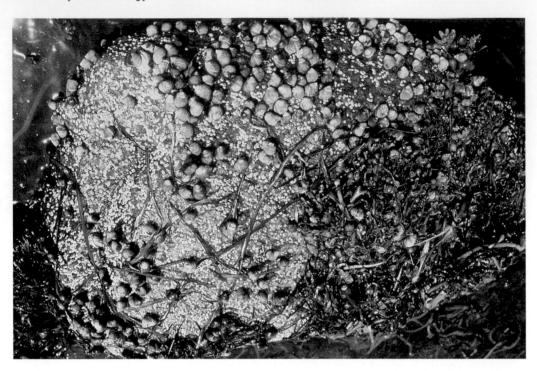

FIGURE 31.16 Low tide reveals common periwinkles clinging to the rocky substrate.

periwinkle (*L. neritoides*), highly adapted to the driest conditions of the black zone. Often living for days, even weeks, without being wetted by salt spray, this periwinkle is the most highly resistant to desiccation of all invertebrate shore animals.

Below the black zone or littoral fringe is the **littoral zone,** marked by the upper limit of barnacles. It is a region covered and uncovered by daily tides. The littoral tends to be divided into subzones. In the upper reaches, barnacles are most abundant. The oyster (*Crassostrea virginica*), blue mussel (*Mytilus edulis*), and limpet (Crucibulum) appear in the middle and lower portions of the littoral, as does the common periwinkle.

Occupying the midlittoral zone of colder climates and in places overlying the barnacles is an ancient group of plants, the brown algae, more commonly known as rockweeds (*Fucus* spp.), and knotted wrack (*Ascophyllum nodosum*) (Figure 31.17). Rockweeds attain their greatest growth on protected shores, where they may grow 2 m long; on wave-whipped shores, they are considerably shorter.

The lower littoral zone may be occupied by blue mussels (*Mytilus edulis*) (Figure 31.18) instead of rockweeds. This is particularly true on shores where hard surfaces have been covered in part by sand and mud. No other shore animal grows

FIGURE 31.17 Knotted wrack (*Ascophyllum nodosum*) exposed at low tide on the Bay of Fundy.

FIGURE 31.18 A tidal flat covered with blue mussels.

in such abundance; the blue-black shells packed closely together may blanket the area. At the lowest reaches of the littoral zone, mussels may grow in association with a red algae, Irish moss (*Chondrus crispus*), which is a low-growing, carpetlike plant. Algae and mussels together often form a tight mat over the rocks. Here, well protected in the dense growth from waves, live infant starfish, sea urchins, brittle stars, and bryzoan sea mats or sea lace (*Membranipora*).

Below the littoral zone is the **sublittoral**, marked by the upper limit of kelp. Uncovered only at the spring tides, the upper sublittoral zone consists of heavy growths of the large brown algae, *Laminaria* spp., one of the kelps, protected by wave action from heavy grazing by urchins. At the intermediate depths, usually greater than 10 m, is the horse mussel (*Modiolus modulus*) and urchin zone. Algal crust dominates the lower zone, maintained by heavy urchin grazing (Witman 1987).

Function

Zonation on the rocky shore results from both physical and biological stress and the amelioration of each. Physical stress results from wave action, the depth and length of time of tidal submergence and exposure, temperature, water loss, and desiccation. For marine intertidal marine organisms, physical stresses increase shoreward with increasing elevation; for ter-

restrial organisms, physical stresses increase seaward. The pattern of life on rocky shores is heavily influenced by biotic interactions of grazing, predation, and competition, and by larval settlement (Underwood et al. 1983). Where wave action is heavy on New England intertidal rocky shores, periwinkles are rare, allowing a more vigorous growth of algae. Absence of grazing favors ephemeral algal species such as *Ulva* and *Enteromorpha,* whereas grazing allows the perennial *Fucus* to become established (Lubchenco 1980, 1983). On the New England coast, the mussel *Mytilus edulis* outcompetes barnacles and algae, either by growing over them (interference competition; see Chapter 14) or by occupying space so densely that other organisms cannot settle there (preemptive competition). On some rocky shores, however, predation by the starfish (*Asterias* spp.) and the snail *Nucella lapillus* prevents dominance by mussels except on the most wave-beaten areas.

A similar situation exists on the Pacific coast. Barnacles of several species tend to outcompete and displace algal species, but in turn the competitively dominant mussel *Mytilus californianus* destroys the barnacles by overgrowing them (Dayton 1971). Where present, the predatory starfish *Pisaster ochraceus* prevents the mussel from completely overgrowing barnacles (Paine 1966).

Although these negative interactions appear to dominate, positive interactions also influence zonation. One is facilitation in which one organism is influenced by the association with another. For example, individual blue mussels growing in dense groups at high intertidal heights buffer one another against an environment in which solitary or a few individuals could not survive (Bertness 1999). At lower intertidal heights, these same individuals living in a group would be competitors. Shore organisms may also modify the environment, making it suitable for other species, a modification that Jones, Lawton, and Shachak (1994) have termed bioengineering. For example, blue mussel trap and bind sediments in the mussel beds, providing habitat and protection for small mobile organisms (Succhanek 1986). The net result of such interactions of the physical and the biotic is a patchy distribution of life across the rocky intertidal shore.

The rocky shore is both autotrophic and heterotrophic. Two sources of primary production are planktonic diatoms and benthic algae ranging in size from microscopic diatoms to large seaweeds and kelps. Grazing on them is a variety of herbivores, such as periwinkles and urchins. Sessile, filter-feeding barnacles and mussels depend on tides to carry plankton to them. Preying on the intertidal herbivores are a variety of predators from dog whelks (*Nucella*) to gulls (Laridae), crows (*Corvus*), and oystercatchers (*Haemotopus*).

Influencing the functioning and productivity of the rocky shores is the indirect energy subsidy they receive from the waves (Leigh et al. 1987, Denny 1988). The force of the waves favorably affects the environmental conditions that promote increased productivity. For one, heavy wave action reduces the activity of such predators of sessile intertidal invertebrates as starfish and sea urchins. Waves bring in a steady supply of nutrients and carry away products of metabolism. They keep in constant motion the fronds of various seaweeds, moving them in and out of shadow and sunlight, allowing for more even distribution of incident light and thus more efficient photosynthesis. By dislodging organisms, both plants and invertebrates, from the rocky substrate, waves open up space for colonization by algae and invertebrates, and reduce strong interspecific competition. In effect, disturbance (see Chapter 23), which influences community structure, is the root of intertidal productivity. Thus, the functioning of the rocky seashore ecosystem involves rather complex interactions between physical and biotic aspects of the ecosystem.

Tide Pools

The ebbing tide leaves behind pools of water in rock crevices, in rocky basins, and in depressions (Figure 31.19). They are tide pools, "microcosms of the sea," as Yonge (1949) describes them. They represent distinct habitats, which differ considerably from the exposed rock and the open sea, and even differ among themselves.

At low tide all the pools are subject to wide and sudden fluctuations in temperature and salinity, but these changes are most marked in shallow pools. Under the summer sun, the temperature may rise above the maximum many organisms can tolerate. As the water evaporates, especially in the smaller and more shallow pools, salinity increases and salt crystals may appear around the edges. When rain or land drainage brings fresh water to the pool, salinity may decrease. In deep pools such fresh water tends to form a layer on the top, developing a strong salinity stratification in which the bottom layer and its inhabitants are little affected. If algal growth is considerable, oxygen content of the water varies through the day. Oxygen will be high during the daylight hours but will be low at night, a situation that rarely occurs at sea. The rise of carbon dioxide at night means a lowering of pH.

Pools near low tide are influenced least by the rise and fall of the tides; those that lie near and above the high-tide line are exposed the longest and undergo the widest fluctuations. Some may be recharged with seawater only by the splash from breaking waves or occasional high spring tides. Regardless of their position on the shore, most pools suddenly return to sea conditions on the rising tide and experience drastic and instantaneous changes in temperature, salinity, and pH. Life in the tidal pools must be able to withstand wide and rapid fluctuations in the environment.

Inhabiting tide pools are green algae (*Enteromorpha* spp. and *Ulva* spp.), filamentous seaweeds (*Cladophera* spp.), small sea anemones, acorn barnacles, and periwinkles. Diversity of life in tide pools is influenced by both the physical environment and interactions among the inhabitants. The density of grazing herbivores influences which algal species are present. Green algae dominate tide pools with few periwinkles, whereas grazer-resistant algae such as Irish moss dominate pools with high periwinkle densities (Lubchenco 1978).

SANDY SHORES AND MUDFLATS

Both the sandy shore and the mudflat at low tide appear barren of life, a sharp contrast to the life-studded rocky shores; but beneath the wet and glistening surface life hides, waiting for the next high tide.

FIGURE 31.19   Tidal pools fill depressions on this piece of rocky coast in Maine.

FIGURE 31.20 A long stretch of sandy beach washed by waves on the southern Australian coast. Although the beach appears barren, life is abundant beneath the sand.

Both in some ways are harsh environments, but the sandy shore is especially so. The very matrix of this seaside environment is a product of the harsh and relentless weathering of rock, both inland and along the shore. Through eons, the ultimate products of rock weathering are carried away by rivers and waves to be deposited as sand along the edge of the sea.

The size of the sand particles deposited influences the nature of the sandy beach, water retention during low tide, and the ability of animals to burrow through it. Beaches with steep slopes usually are made up of larger sand grains and are subject to more wave action.

Beaches exposed to high waves are generally flattened, for much of the material is transported away from the beach to deeper water, and fine sand is left behind (Figure 31.20). Sand grains of all sizes, especially the finer particles in which capillary action is greatest, are more or less cushioned by a film of water about them, reducing further wearing away. Retention of water by the sand at low tide is one of the outstanding environmental features of the tidal flats.

Structure

Life on the sand is very difficult to observe. Sand provides no surface for attachments of seaweed and their associated fauna. The crabs, worms, and snails so characteristic of rocky crevices find no protection here. Most life, then, is forced to live beneath the sand.

Life on sandy and muddy beaches consists of the **epifauna,** organisms living on the surface, and the **infauna,** organisms living within the substrate. Most infauna either occupy permanent or semipermanent tubes within the sand or

mud, like the lugworm (*Arenicola* spp.), or are able to burrow rapidly into the substrate, like the coquina bivalve (*Donax variabilis*). Multicellular infauna obtain oxygen either by gaseous exchange with the water through their outer covering or by breathing through gills and elaborate respiratory siphons.

Within the sand and mud live vast numbers of **meiofauna** with a size range between 0.05 and 0.5 mm, including copepods, ostracods, nematodes, and gastrotrichs. These interstitial fauna are generally elongated forms with setae, spines, or tubercles greatly reduced. The great majority do not have pelagic larval stages. These animals feed mostly on algae, bacteria, and detritus. Interstitial life, best developed on the more sheltered beaches, shows seasonal variations, reaching maximum development in summer months.

Sandy beaches also exhibit zonation related to tides (Figure 31.21), but to discover it requires digging. Sandy beaches and mudflats can be divided into littoral fringe, littoral, and sublittoral zones, based on animal organisms, but a universal pattern similar to that of the rocky shore is lacking. Pale, sand-colored ghost crabs (*Ocypode quadrata*) and beach fleas (not true fleas) (*Talorchestia* and *Orchestia* spp.) occupy the upper beach, the littoral fringe. The intertidal beach, the littoral, is a zone where true marine life appears. Although sandy shores lack the variety found on rocky shores, the populations of individual species of largely burrowing animals often are enormous. An array of animals, among them starfish and the related sand dollar, can be found above and below the low-tide line and in the littoral.

Organisms living within the sand and mud do not experience the same violent fluctuations in temperature as those on the rocky shores. Although the surface temperature of the

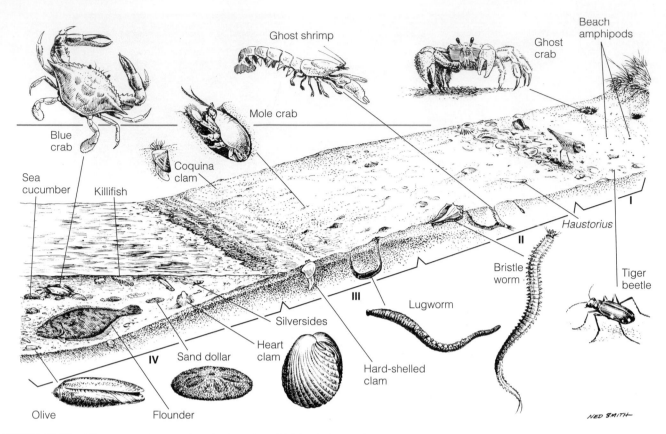

FIGURE 31.21 Life on a sandy ocean beach along the mid-Atlantic Coast. Although strong zonation is absent, organisms still change on a gradient from land to sea. I, supratidal zone: ghost crabs and sand fleas; II, flat beach zone: ghost shrimp, bristle worms, clams; III, intratidal zone: clams, lugworms, mole crabs; IV, subtidal zone. The dashed line indicates high tide.

sand at midday may be 10°C or more higher than the returning seawater, the temperature a few inches below remains almost constant throughout the year. Nor is there a great fluctuation in salinity, even when fresh water runs over the surface of the sand. Below 25 cm, salinity is little affected.

The substrate, however, is somewhat unstable, made more so by the burrowing activities of the infaunal residents. Storms and severe weather move large amounts of sediment, killing infauna. Burrowing activities bring sediments to the surface and loosen the substrate, making it more vulnerable to disturbance.

Function

For life to exist on the sandy shore, some organic matter has to accumulate. Most sandy beaches contain a certain amount of detritus from seaweeds, dead animals, feces, and material blown in. This organic matter accumulates within the sand, especially in sheltered areas. An inverse relationship exists between the turbulence of the water and the amount of organic matter on the beach, with accumulation reaching its maximum on the mudflats. Organic matter clogs the space between the grains of sand and binds them together. As water moves down through the sand, it loses oxygen from both the respiration of bacteria and the oxidation of chemical substances, especially ferrous compounds. The point within the mud or

sand at which water loses all its oxygen is a region of stagnation and oxygen deficiency, characterized by the formation of ferrous sulfides. The iron sulfides cause a zone of black, the depth of which varies with the exposure on the beach. On mudflats, such conditions exist almost to the surface.

The energy base for sandy beach and mudflat fauna is organic matter. Much of it becomes available through bacterial decomposition, which goes on at the greatest rate at low tide. The bacteria are concentrated around the organic matter in the sand, where they escape the diluting effects of water. The products of decomposition are dissolved and washed into the sea at each high tide, which, in turn, brings in more organic matter for decomposition. Thus tidal flats are important sites for biogeochemical cycling, supplying offshore waters with phosphates, nitrogen, and other nutrients.

Energy flow in sandy beaches and mudflats differs from that of terrestrial and aquatic systems because the basic consumers are bacteria. In other systems, bacteria act largely as reducers responsible for conversion of dead organic matter into a form that can be used by producer organisms. In tidal flats, not only do bacteria feed on detrital material and break down organic matter, they are also a major source of food for higher level consumers.

A number of deposit-feeding organisms ingest organic matter largely as a means of obtaining bacteria. Prominent

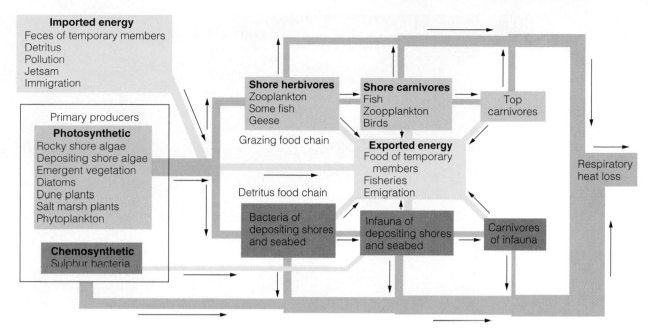

FIGURE 31.22   Diagram of the coastal ecosystem, a supraecosystem consisting of the shore, the fringing terrestrial regions, and the sublittoral zones. It connects two food webs: that of the rocky shore with its algae, herbivores, and zooplankton; and the detrital food webs involving the bacteria of the depositing shore and sublittoral muds and the dependent detritivores and carnivores. Coastal ecosystems are extremely productive; because energy imports exceed exports, the system is continuously gaining energy. (From Eltringham 1971.)

among them on the mudflats are numerous nematodes and copepods (Harpacticoida), the polychaete clam worm (*Nereis*), and the gastropod mollusk (*Hydrobia*). Deposit-feeders on sandy beaches obtain their food by actively burrowing through the sand and ingesting the substrate to obtain the organic matter it contains. The most common among them is the lugworm, which is responsible for the conspicuous coiled and cone-shaped casts on the beach.

Other sandy beach animals are filter-feeders, obtaining their food by sorting particles of organic matter from tidal water. Two of these "surf fishers" who advance and retreat up and down the beach with the ebb and flow of the tides are the mole crab (*Emerita talpoida*) and coquina.

Associated with these essentially herbivorous animals are the predators, always present whether the tide is in or out. Near and below the low-tide line live predatory gastropods, which prey on bivalves beneath the sand. In the same area lurk predatory crabs such as the blue crab (*Callinectes sapidus*) and green crab (*Carcinus maenas*), which feed on mole crabs, clams, and other organisms. They move back and forth with the tides. The incoming tides also bring other predators such as killifish (*Fundulus heteroclitus*) and silversides (*Menidia menidia*). As the tide recedes, gulls and shorebirds scurry across the sand and mudflats to hunt for food.

Because of their dependence on imported organic matter and their essentially heterotrophic nature, sandy beaches and mudflats should be considered not as separate ecosystems, but as part of the whole coastal ecosystem (Figure 31.22). Except in the cleanest of sands, some primary production does take place in the intertidal zone. The major primary producers are the diatoms, confined mainly to fine-grained deposits of sand containing a high proportion of organic matter. Productivity is low. One estimate places productivity of moderately exposed beaches at 5 g C/m²/yr (Orth et al. 1991). Production may be increased temporarily by phytoplankton carried in at high tide and left stranded on the surface. Again, these organisms are mostly diatoms. More important as producers are the sulfur bacteria in the black sulfide or reducing layer. These chemosynthetic bacteria use energy released as ferrous oxide is reduced. Some mudflats are covered with the algae *Enteromorpha* and *Ulva,* whose productivity can be substantial.

In effect, then, sandy shores and mudflats are part of a larger coastal ecosystem involving the salt marsh, the estuary, and coastal waters. They act as sinks for energy and nutrients because the energy they utilize comes not from primary production, but from organic matter that originates outside the area. Many of the nutrient cycles are only partially contained within the borders of the shores.

CORAL REEFS

In subtropical and tropical waters around the world are structures of biological rather than geological origin: coral reefs (Figure 31.23). Coral reefs, wave-resistant accumulations of calcareous skeletal remains, are built by carbonate-secreting organisms, of which coral (Cnidaria, Anthozoa) is the most conspicuous but not always the most important. Also contributing heavily are the coralline red algae (Rhodophyta, Corallinaceae), Foraminifera, and mollusks.

FIGURE 31.23 Coral reef viewed from the air.

Built only underwater at shallow depths, no deeper than 50 to 70 m and preferably 25 m, coral reefs need a stable foundation on which to grow. Such foundations are provided by shallow continental shelves and submerged volcanoes.

Coral reefs are of three types with many gradations among them. (1) **Fringing reefs** project directly seaward from the shore. (2) **Barrier reefs** parallel shorelines and are separated from land by a lagoon. (3) **Atolls** are coral islands that begin as horseshoe-shaped reefs surrounding a lagoon. Such lagoons are about 40 m deep and are usually connected to the open sea by breaks in the reef. Reefs build up to sea level. To become atolls, the reefs have to be exposed by a lowering of the sea level or built up by the action of wind and waves.

Structure

Coral reefs are complex ecosystems involving close relationships between coral organisms and algae. In the tissues of the gastrodermal layer live **zooxanthellae,** symbiotic endozoic dinoflagellate algae (see Chapter 18). On the calcareous skeleton live still other kinds of algae, both the encrusting red and green coralline species and filamentous species, including turf algae. Associated with coral growth are mollusks such as giant clams (*Tridacna, Hippopus*), echinoderms, crustaceans, polychaete worms, sponges, and a diverse array of fishes, both herbivorous and predatory.

Zonation and diversity of coral species are influenced by an interaction of depth, light, grazing, competition, and disturbance (Huston 1985, Jackson 1991). The basic diversity gradient is established by light. Diversity is lowest at the crest, where only species tolerant of intense or frequent disturbance by waves can survive. Diversity increases with depth to a maximum of about 20 m and then decreases as light becomes attenuated, eliminating shade intolerant species (see Wellington 1982a). Imposed on this condition are a variety of biotic and abiotic disturbances that vary in intensity and decrease with depth. Growth rates of photosynthetic coral are highest in shallow depths, and a few species, especially the branching corals, can easily dominate the reef by overgrowing and shading the crustose corals and algae. Disturbances by wave action, storms, and grazing reduce the rate of competitive displacement among corals (Jackson 1991). Heavy grazing of overgrowing algae by sea urchins and fish, such as parrotfish (Scaridae), increases encrusting coralline algae. Light grazing allows rapidly growing filamentous and foliose algal species to eliminate crustose algae (see Wellington 1982b).

Intense disturbances can have pronounced, even disastrous, effects on coral reefs. An outbreak of a major coral predator, the crown-of-thorns starfish (*Acanthaster planci*), destroyed nearly all the coral on certain western Pacific reefs. Overfishing has affected coral reef food webs, especially by removing algal-feeding and other herbivorous fish that keep coral-killing algal growth at a minimum. Dynamiting reefs to collect their colorful fish for the marine aquarium trade destroys both the reefs and its fish population. El Niño–induced warming of ocean surface waters above normal has devastated coral reefs (see Chapter 2). Closely associated with warm surface waters is coral bleaching, which results with the expulsion of zooxanthellae that inhabit coral tissues (for expanded discussion see Nybakken 2000).

Function

Coral organisms are partially photosynthetic and partially heterotrophic. During the day zooxanthellae carry on photosynthesis and directly transfer organic material to coral tissue. At night coral polyps feed on zooplankton, securing phosphates and nitrates and other nutrients needed by the anthozoans and their symbiotic algae. Thus nutrients are recycled in place between the anthozoans and the algae (Pomeroy and Kuenzler 1969, Hatcher 1988). In addition, carbon dioxide concentrations in animal tissue enable the coral to extract the calcium carbonate needed to build the coral skeletons. Adding to the productivity of the coral are crustose coralline

FIGURE 31.24 Colorful fish life in a coral reef.

algae, turf algae, macroalgae, seagrasses, sponges, phytoplankton, and a large bacterial population.

Coral reefs are among the most highly productive ecosystems on Earth. Net productivity ranges from 1500 to 5000 g $C/m^2/yr$, compared with 15 to 50 g $C/m^2/yr$ for the surrounding ocean. Because of the ability of the coralline community to retain nutrients within the system and to act as a nutrient trap, coral reefs are oases of productivity within a nutrient-poor sea.

This high productivity and the diversity of habitats within the reef support a great diversity of life (Figure 31.24). There are thousands of kinds of invertebrates, some of which, such as sea urchins, feed on coral animals and algae, hundreds of kinds of herbivorous fish that graze on algae, and many predatory species (see Yonge 1963, Sale 1980, Jackson 1991). Some of these predators, such as the puffers (Tetraodontidae) and filefishes (Monacanthidae), are corallivores, feeding on coral polyps. Others lie in ambush for prey in coralline caverns. In addition, there is a wide array of symbionts, such as cleaning fish and crustaceans, that pick parasites and detritus from larger fish and invertebrates.

ESTUARIES

Water of all streams and rivers eventually drains into the sea. The place where this fresh water joins and mixes with the salt is called an **estuary.** Estuaries (for example, Chesapeake Bay, and Humboldt Bay of California) are semienclosed parts of the coastal ocean where river water mixes with and measurably dilutes the seawater (Ketchum 1983).

Estuaries differ in size, shape, and volume of water flow, all influenced by the geology of the region in which they occur. As the river reaches the sea, it drops its sediments in quiet water. They accumulate to form deltas in the upper reaches of the mouth and shorten the estuary. When silt and mud accumula-

tions become high enough to be exposed at low tide, tidal flats develop, which divide and braid the original channel of the estuary. At the same time, ocean currents and tides erode the coastline and deposit material on the seaward side of the estuary, also shortening the mouth. If more material is deposited than is carried away, barrier beaches, islands, and brackish lagoons appear.

Structure

Current and salinity, both complex and variable, shape life in the estuary. Estuarine currents result from the interaction of a one-direction stream flow, which varies with the season and rainfall, with oscillating ocean tides and with the wind (Ketchum 1951, 1983, Burt and Queen 1951, Smayda 1983). Because of the complex nature of the currents, generalizations about estuaries are difficult to make (see Lauff 1967).

Salinity varies vertically and horizontally, often within one tidal cycle. Vertical salinity may be the same from top to bottom, or it may be completely stratified, with a layer of fresh water on top and a layer of dense saline water on the bottom. Salinity is homogeneous when currents, particularly eddy currents, are strong enough to mix the water from top to bottom. The salinity in some estuaries is homogeneous at low tide but stratified at high tide. As the tide floods, a surface wedge of seawater moves upstream more rapidly than the bottom water, creating a density inversion of salinity stratification. Seawater on the surface tends to sink as lighter fresh water rises, and mixing takes place from the surface to the bottom. This phenomenon is known as tidal overmixing. Strong winds, too, tend to mix salt water with the fresh in some estuaries (Barlow 1956); but when the winds are still, the river water flows seaward on a shallow surface over an upstream movement of seawater that only gradually mixes with the salt.

Horizontally, the least saline waters are at the river entrance, and the most saline at the mouth of the estuary

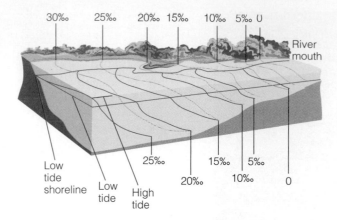

30‰ 25‰ 20‰ 15‰ 10‰ 5‰ 0

River mouth

Low tide shoreline

Low tide

High tide

25‰   15‰   5‰
20‰   10‰   0

FIGURE 31.25 Vertical and horizontal stratification of salinity from the river mouth to the estuary at both low and high tide. At high tide, the incoming seawater increases the salinity toward the river mouth; at low tide, salinity is reduced. Note also how salinity increases with depth, because lighter fresh water flows over denser salt water.

(Figure 31.25). The configuration of the horizontal zonation is determined mainly by the deflection caused by the incoming and outgoing currents (see Officer 1983). In all estuaries of the Northern Hemisphere, outward-flowing fresh water and inward-flowing seawater are deflected to the east because of Earth's rotation. As a result, salinity is higher on the western side.

Salinity also varies with changes in the quantity of fresh water pouring into the estuary through the year. Salinity is highest during the summer and during periods of drought, when less fresh water flows into the estuary. It is lowest during the winter and spring, when rivers and streams are discharging their peak loads. This change in salinity may happen rather rapidly. For example, early in 1957 a heavy rainfall broke the most severe drought in the history of Texas. The resultant heavy river discharge reduced the salinities in Mesquite Bay on the central Texas coast by over 90 percent in a two-month period. At the height of the drought salinities ranged from 35.5 to 50.0 ppt, but after the break in the drought they ranged from 2.3 to 2.9 ppt (Hoese 1960). Such rapid changes have a profound impact on the life of the estuary.

The salinity of seawater is about 35 psv; that of fresh water ranges from 0.065 to 0.30 psv. Because the concentration of metallic ions carried by rivers varies from drainage to drainage, the salinity and chemistry of estuaries differ. The proportion of dissolved salts in the estuarine waters remains about the same as that of seawater, but the concentration varies in a gradient from fresh water to sea.

Exceptions to these conditions exist in regions where evaporation from the estuary may exceed the inflow of fresh water from river discharge and rainfall (a negative estuary). This situation causes the salinity to increase in the upper end of the estuary, and horizontal stratification is reversed.

Temperatures in estuaries fluctuate considerably diurnally and seasonally. Waters are heated by solar radiation and inflowing and tidal currents. High tide on the mudflats may heat or cool the water, depending on the season. The upper

layer of estuarine water may be cooler in winter and warmer in summer than the bottom, a condition that, as in a lake, will result in a spring and autumn overturn.

Mixing waters of different salinities and temperatures acts as a nutrient trap (Officer 1983). Inflowing river waters more often than not impoverish rather than fertilize the estuary, except for phosphorus. Instead, nutrients and oxygen are carried into the estuary by the tides. If vertical mixing takes place, these nutrients are not soon swept back out to sea, but circulate up and down among organisms, water, and bottom sediments (Figure 31.26).

Organisms inhabiting the estuary are faced with two problems—maintenance of position, and adjustment to changing salinity (Vernberg 1983). The bulk of estuarine organisms are benthic and are securely attached to the bottom, buried in the mud, or lodged in crevices and crannies about sessile organisms. Motile inhabitants are chiefly crustaceans and fish, largely young of species that spawn offshore in high-salinity water.

Planktonic organisms are wholly at the mercy of the currents. Because the seaward movement of streamflow and ebb tide transports plankton out to sea, the rate of circulation or flushing time determines the nature of the plankton population. If the circulation is too vigorous, the plankton population may be small. Phytoplankton in summer is most dense near the surface and in low-salinity areas. In winter phytoplankton is more uniformly distributed. For any planktonic growth to become endemic in an estuary, reproduction and recruitment must balance the losses from the physical processes that disperse the population (Barlow 1955).

A gradient of salinity from the point where fresh water enters the estuary, the *head*, to the mouth results in zonation of life (McLusky 1989). The head has limited salt penetration; salinity is less than 5 psu. In the *upper reaches*, fresh and salt water mix. The mixing produces a salinity between 5 and 18 psu, and a minimal current results in turbidity and deposition of mud. The *middle reaches* receiving tidal currents are largely mud, and salinity ranges between 18 and 25 psu. Currents become much faster in the *lower reaches*, bottom deposits become more sandy and the salinity is between 25 and 30 psu. The mouth of the estuary experiences strong tidal currents and the salinity is about 30 psu, close to that of the sea. Within this gradient live the estuarine fauna, low in diversity at the head and maximum in the middle estuary.

The majority of animals living near the head are oligohaline species, ones that prefer low salinities but can tolerate salinities up to 5 psu. In the middle reaches live the euryhaline species, able to live in the sea but largely excluded from it by competition. They are most common at salinities between 5 and 18 psu. Euryhaline species make up the majority of organisms from the sea to the middle reaches. Within this range, each species has its own range of salinity tolerances. At the mouth of the estuary live the stenohaline species. Able to live at a salinity of 25 psu or higher, they cannot tolerate long-term salinity fluctuations.

Two estuarine species illustrate responses to salinity, the clam worm and the scud. Two species of clam worm, *Nereis occidentalis* and *Neanthes succinea*, inhabit the estuaries of the

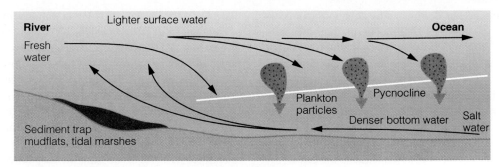

FIGURE 31.26 Circulation of fresh and salt water in an estuary creates a nutrient trap. The trap develops because of a salt wedge of intruding seawater on the bottom, a surface flow of lighter fresh water, and a counterflow of heavier brackish water. This countercurrent serves to trap nutrients, recirculating them toward the tidal marsh. The same countercurrent also sends phytoplankton up the estuary, repopulating the water. When nutrients, especially in the form of particulate matter, are high in the upper estuary, they are taken up rapidly by tidal marshes and mudflats, converted to soluble forms, and exported back to the open waters of the estuary. (From Correll 1978.)

southern coastal plains of North America. *Nereis* is more numerous at high salinities and *Neanthes* at low salinities. In European estuaries the scud *Gammarus* is an important member of the estuarine bottom fauna. Two species, *G. locusta* and *G. marina,* are typically marine species and cannot penetrate far into the estuary. Instead they are replaced by a typical estuarine species, *G. zaddachi.* This species, however, is broken down into three subspecies, separated by salinity tolerance. *G. zaddachi* lives in the lower reaches; *G. z. salinesi* occupies the middle reaches; and *G. z. zaddachi,* which can penetrate fresh water for a short time, lives near the head (Spooner 1947, Segerstrale 1947).

Stage of development influences the range of motile species within the estuarine waters. This influence is particularly pronounced among estuarine fish. Species such as the striped bass spawn near the interface of fresh and low-salinity water (Figure 31.27). The larvae and young fish move downstream to more saline waters as they mature. Thus for the striped bass, the estuary serves both as a nursery and feeding ground for the young. Anadromous species, such as shad (*Alosa sapidissima*), spawn in fresh water; the young fish spend the first summer in the estuary, then move out to the open sea. Other species, such as the croaker (*Micropogon undulatus*), spawn at the mouth of the estuary, and the larvae are transported upstream to feed in the plankton-rich, low-salinity areas. Still others, such as the bluefish (*Pomatomus saltatrix*), move into the estuary to feed. In general, marine species drop out toward fresh water and are not replaced by freshwater forms.

The oyster bed and the oyster reef are the outstanding communities of the estuary. The oyster (*Crassostrea virginica*) is the dominant organism about which estuarine life revolves. Oysters may attach to every hard object in the intertidal zone or form reefs, areas where clusters of living oysters grow cemented to the almost buried shells of past generations. Oyster reefs usually lie at right angles to tidal currents, which bring planktonic food, carry away wastes, and sweep the oysters clean of sediment and debris. Closely associated with oysters are encrusting organisms such as algae, sponges, barnacles, and bryozoans. Beneath and between the oysters live polychaete worms, decapods, pelecypods, and a host of other organisms.

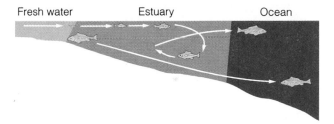

FIGURE 31.27 Relationship of a semianadromous fish, the striped bass, to the estuary. Adults live in the marine environment, but young fish grow up in the estuary. (From Cronin and Mansueti 1971.)

Function

Estuarine systems function on both plankton-based and detrital-based food webs. The producer component, particularly in the middle and lower estuary, consists of dinoflagellates and diatoms. The latter convert some of the carbon intake to high-caloric fats and lipids, rather than the low-energy carbohydrates typical of green plants. This fat provides a high-energy food base for higher trophic levels.

Inflowing water from rivers and coastal marshes carries nutrients, especially nitrogen and phosphorus, into an estuary and can stimulate an increase in phytoplankton production. Phytoplankton production, however, is regulated more by internal nutrient cycling than by external sources. This internal cycling involves excretion of mineralized nutrients by herbivorous zooplankton, release of nutrients remineralized by invertebrates of the bottom sediments, stirring of sediments, and steady-state exchanges between nutrients present in the particulate and dissolved phases (see Smayda 1983).

Nutrient buildup over winter in temperate estuaries stimulates a winter-spring bloom. As the nutrients become depleted and the phytoplankton experience intensive predation by zooplankton, the bloom collapses and falls to the bottom sediments. There it is fed on by bivalves and other filter-feeding invertebrates (see Wolff 1983). In well-mixed estuaries, nutrients remineralized in the benthos are released to the water column, stimulating a summer bloom.

Estuarine zooplankton, dominated by copepods (for ecology and life history, see Miller 1983), undergo their own seasonal fluctuations. Although expansive growth and subsequent declines in zooplankton populations can be associated with phytoplankton blooms, population dynamics are determined by many physical and biotic influences, including flushing rates of the estuary. The rate of increase for many zooplankton populations must balance the rate of loss from river flood and tidal flushing. Because of the unstable physical environment, zooplankton never have evolved species endemic to the estuary.

In shallow estuarine waters, rooted aquatics, such as widgeon grass (*Ruppia maritima*) and eelgrass (*Zostera marina*), assume major importance (for review see Phillips and McRoy 1980, Thayer et al. 1984, Zieman 1982). They support complex systems with a large number of epiphytic and epizootic organisms. Such communities are important to certain vertebrate grazers, such as brant (*Branta bernicla*), Canada geese (*Branta canadensis*), snow geese (*Chen caerulescens*), the black swan (*Cygnus atratus*) in Australia, and sea turtles, and provide a nursery ground for shrimp and bay scallops.

Data for the Pamlico River estuary in North Carolina (Copeland et al. 1974) provide an example of energy flux in an estuary. The estuary is characterized by low salinity, high turbidity, and shallow water. The shallow water supports dense stands of widgeon grass, with associated attached algae and animals such as scuds and grass shrimp (*Palaemonetes pugio*). The benthos is dominated by the clams *Rangia* and *Macoma*. The euphotic zone, the upper 2 m, is dominated by dinoflagellates. Grazing on the phytoplankton are the zooplankters (*Acartia tonsa*) and the harpacticoid copepods that move up to the surface at night. Each day ctenophores crop 30 percent of the zooplankton population. At the same time, considerable detrital material enters the estuarine system as dissolved and particulate organic matter. Because phytoplankton depends heavily on dissolved organic matter, the complex food web can be considered primarily detrital. Both detritus and phytoplankton support a number of trophic levels that lead eventually to fish and humans (Figure 31.28).

Rivers flowing into estuaries carry along with the sediment a load of inland pollutants: domestic wastes, drainage from agri-

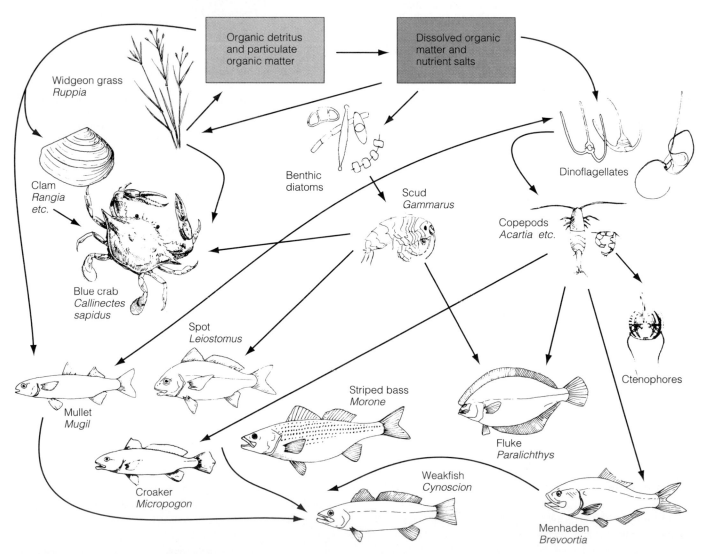

FIGURE 31.28 Simplified estuarine food web based on an estuary in the southeastern United States: The energy base is largely heterotrophic, supported by particulate matter and dissolved organic matter. (Adapted from Copeland et al. 1974.)

FIGURE 31.29 A salt marsh on the Bay of Fundy. Here river and tidal erosion are high because of the great tidal change. The marsh, not well developed, is dominated by *Puccinellia* rather than *Spartina*. Compare this figure with views of a salt marsh on the Virginia coast in Figures 31.32 and 31.33.

cultural lands with nutrients and pesticides, and industrial effluents carrying toxic elements. Industrial, shipping, and housing developments along the estuarine coast add their own load of contaminants, alter flow of tidal water, and increase the anoxic condition of the estuarine bottom sediments. These conditions favor only a few bottom organisms such as polychaete worms and reduce the production of fish, clams, and oysters or make them inedible because of the accumulation of toxic elements and pesticides in their tissues. At the same time, withdrawal of water from coastal aquifers and streams allows salt water to intrude into freshwater aquifers, destroying the water supply.

TIDAL MARSHES

On the alluvial plains mostly in or near the estuary and in the shelter of offshore bars and islands exists a unique community, the salt marsh (Figure 31.29). Although at first glance a salt marsh appears as waving hectares of grass, it is a complex of distinctive and often clearly demarcated plant associations. The nature of the complex is determined by tides and salinity. Tides play a most significant role in plant segregation. Twice each day, at high tide, the outermost tidal flats up to mean high water are submerged in salt water; then at low tide they assume a semiterrestrial existence. These flooded tidal flats make up the low marsh. At mean high water level is a region flooded irregularly at high tides. This area is the high marsh. The low marsh and high marsh form marked

zones and patches of vegetation distinctive in color and texture (Figure 31.30).

Salt marshes develop over time by accretion, subsidence, and rising sea levels. Low marshes develop as sediments build up along the coast or in areas flooded by a rising sea. In eastern North America, these new flats of accumulated sediments are invaded by the seaward encroachment of the dominant plant of the low marsh, smooth cordgrass (*Spartina alterniflora*). High marshes are formed over areas of low marsh where silt deposition raises the level of the marsh, or where high marsh vegetation moves over low upland areas as sea levels rise.

Structure

Plants Salt marshes begin in most cases as tidal sands or mudflats colonized first by algae and, if the water is deep enough, by eelgrass. As organic debris and sediments accumulate, eelgrass is replaced by the first salt marsh colonists— sea poa (*Puccinellia*) on the European coast or smooth cordgrass on the eastern coast of North America. Stiff, leafy, up to 3 m tall, and submerged in salt water at every high tide, *Spartina alterniflora* forms a marginal zone between the open mudflat to the front and the high marsh behind. No litter accumulates in the stand. Strong tidal currents sweep the floor of the low marsh clean, leaving only thick, black mud.

Spartina alterniflora is well adapted to the intertidal flats of which it has sole possession. It has a tolerance for salt water and can live in a semisubmerged state. It can live in the

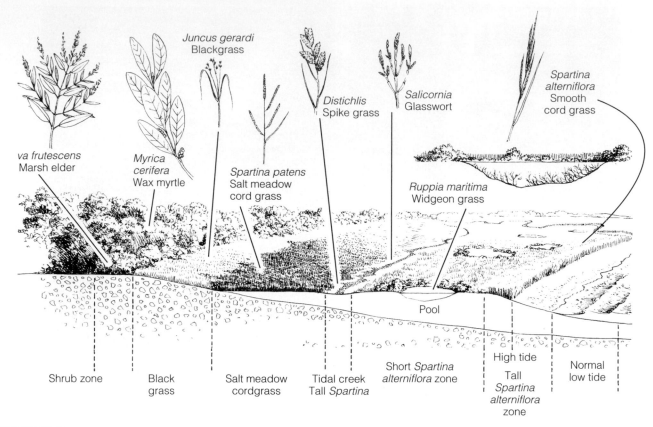

FIGURE 31.30 A stylized transect of part of the salt marsh, showing the relationship of plant distribution to microrelief and tidal submergence.

saline environment by selectively concentrating sodium chloride at a higher level in its cells than in the surrounding seawater, thus maintaining its osmotic integrity. To rid itself of excessive salts, *S. alterniflora* has special salt-secreting cells in its leaves. Water excreted with the salt evaporates, leaving behind sparkling crystals of salt on the leaves to be washed off by tidal water and rain. To get air to its roots buried in anaerobic mud, *Spartina* has hollow tubes (aerenchyma) leading from the leaf to the root, through which oxygen diffuses.

Immediately landward at a higher elevation on the low marsh, the soil is poorly drained and anoxic and belowground plant material accumulates as peat. Here the marsh is marked by one of its most conspicuous plants, a short form of *Spartina alterniflora*. Yellowish, almost chloritic in appearance, short *Spartina* contrasts sharply with the dark green tall form (Figure 31.31). Why such a difference should exist has been the subject of study and debate. Although there may be some genetic difference in the two forms (Gallagher et al. 1988), the short form reflects a deficiency of nitrogen caused by salinity stress and sulfide concentrations that inhibit N uptake (Chalmers 1982, Haines and Dunn 1985).

The short *Spartina* marsh has lower tidal exchange rates, higher salinity, especially in southern salt marshes, a shorter, more open canopy, higher soil temperatures, and higher evaporation rates. This environment provides the opportunity for other highly salt-tolerant plants to grow among the short *Spartina*. Here are the fleshy glassworts (*Salicornia*) that turn

bright red in the fall, sea lavender (*Limonium carolininum*), and spearscale (*Atriplex patula*).

Backing up the low marsh is a slightly elevated area not flooded by daily tides, the high marsh (Figure 31.31). Above mean high water the short *Spartina alterniflora* and its associates are replaced by *Spartina patens,* salt meadow cordgrass, and an associate, spikegrass (*Distichlis spicata*). Unable to deal with low marsh conditions and to adequately oxygenate its roots in an anoxic soil, *S. patens* grows well on the high marsh. So can *S. alterniflora* when transplanted on the high marsh in the absence of neighbors (Bertness 1991b); but it dies out rapidly in the presence of its high marsh competitor salt meadow cordgrass. The lower limits of the monospecific low marsh smooth cordgrass are marked by forming open mudflats subject to subsequent colonization. Competition from salt meadow cordgrass sets its upper limits.

Spartina patens is a fine, small grass that grows so densely and forms such a tight mat that few other plants can grow with it (Figure 31.32). Dead growth of the previous year lies beneath the current growth, shielding the ground from the sun and keeping it moist. Where the soil is more saline or waterlogged, *S. patens* is replaced by or shares the site with spikegrass that invades disturbed or salty areas.

At the uppermost part of the high marsh, flooded only at the highest tides and subject to some intrusion of fresh water, is the zone of black needle rush (*Juncus gerardi* or *J. roeme-*

FIGURE 31.31   The high marsh holds a greater diversity of species than the low marsh.

FIGURE 31.32   The cowlick sweep of salt meadow cordgrass is distinctive next to a stand of salt grass. Bayberry in the background marks the shrub zone.

rianus)—so called because its dark green color becomes almost black in the fall. Possessing thick root mats and underground runners, black needle rush competitively displaces salt meadow cordgrass and spikegrass (Bertness 1991a).

Beyond the rushes are shrubby growths of marsh elder (*Iva frutescens*) and groundsel (*Baccharis halimifolia*). These shrubs tend to invade the high marsh where a slight rise in elevation exists. Intolerant of high salinity, *Iva* depends on the ability of its perennial turf neighbors to ameliorate soil conditions on its the seaward border for the survival of its short, stunted individuals. On the terrestrial border, *Iva* outcompetes the same plants to form a tall marsh elder zone (Bertness and Hacker 1994). On the upland fringe grow other salt marsh shrubs, particularly bayberry (*Myrica* spp.) and sea hollyhock (*Hibiscus palustris*)

Draining the high marsh are tidal creeks, bordered with growths of tall *Spartina*. In estuarine marshes, the river itself forms the main channel, joined by numerous braiding tributaries. The creeks are deepened by scouring and heightened by a steady accumulation of organic debris. At the same time, the heads of the creeks erode backward and more small branch creeks develop. The distribution and pattern of the creeks influence the movement and drainage of water from the high marsh.

The exposed banks of tidal creeks support a dense population of mud algae, the diatoms and dinoflagellates, photosynthetically active all year. Some of the algae are washed out at ebb tide and become part of the estuarine plankton available to such filter-feeders as oysters.

Across the high marsh are many circular and elliptical depressions that hold water, and salt-encrusted flats called salt pans. Water-holding pans (Figure 31.33) come about in several ways. Many are formed as the marsh develops. Bare spots on the marsh become surrounded by vegetation. As the level of the marsh rises, the bare spots lose their water outlet

and become shallow ponds. Others are derived from creeks across which growth of vegetation or organic debris blocks the channel. Death of small patches of vegetation from one cause or another create rotten spots in the peat, which develop into ponds. Such pans support distinctive plant life, which varies with the depth of the water and salt concentration. Pools with a firm bottom and sufficient depth to retain water support dense growths of widgeon grass (*Ruppia maritima*), with long, threadlike leaves and small black, triangular leaves relished by waterfowl. The pools are usually surrounded by forbs such as sea lavender and by spikegrass.

Shallow depressions or pans in which water evaporates are covered with a heavy algal crust and crystallized salt. Most common at intermediate elevations on the southern salt marshes, the salt flats may be invaded about the edges by *Salicornia* species, spikegrass, and even short *Spartina alterniflora* (Figure 31.33). On the centers of these pans, very high salinity precludes any plant growth.

FIGURE 31.33   A salt pan or pool in the high marsh.

The salt marsh described is typical of the North American Atlantic Coast, but many variations exist locally and latitudinally around the world (Chapman 1977). North America has several distinctive types: Arctic salt marshes with few species, East and Gulf Coast marshes dominated by *Spartina* and *Juncus,* and Pacific Coast marshes whose low marshes are dominated by *Spartina foliosa* and high marshes by *Salicornia virginica*. Tidal marshes of Europe, similar to those of the east coast of North America, support *Salicornia angelica*.

Consumers   The three dominant animals of the low marsh are ribbed mussel (*Modiolus dimissus*), fiddler crab (*Uca pugilator* and *U. pugnax*), and marsh periwinkle (*Littorina* spp.). The marsh periwinkle moves up and down the stems of *Spartina* as the tidal cycle changes. At low tide, periwinkles move onto the mud to feed on algae and detritus. At high tide, periwinkles move up the stems of dead cordgrass to forage on wet dead plant material and associated fungi and bacteria (Silliman 1998).

Buried halfway in the mud is the ribbed mussel (Figure 31.34a). At low tide the mussel is closed; at high tide the mussel opens to filter particles from the water, accepting some and rejecting others in a mucous ribbon known as pseudofeces.

Running across the marsh at low tide are fiddler crabs (Figure 31.34b). They are omnivorous feeders, consuming plant and animal remains, algae, and small animals. Fiddler crabs live in burrows, marked by mounds of freshly dug, marble-sized pellets. The burrowing activity of crabs is similar to that of earthworms. In overturning the mud the crabs mix the soil and bring nutrients to the surface (Bertness 1985).

Prominent about the base of *Spartina* stalks and under debris are sandhoppers (*Orchestia*). These detrital-feeding amphipods may be very abundant and are important in the diet of some of the marshland birds. Grazing on tall *Spartina* leaves is the salt marsh grasshopper (*Orchelium*) and sucking the plant juices is the plant hopper (*Prokelsis*). Neither harvests significant portions of the standing crop (Teal 1962).

Two conspicuous vertebrate residents of the intertidal marshes of eastern North America are the diamondback terrapin (*Malaclemys terrapin*) and the clapper rail (*Rallus longirostris*). In the salt marshes of the southern United States, they are joined by the crab-eating marsh rat (*Oryzomys palustris*). The diamondback terrapin feeds on fiddler crabs, small mollusks, and dead fish. The clapper rail finds its diet of fiddler crabs and sandhoppers along the creek banks and in the tall *Spartina* at low tide. Less conspicuous is the seaside sparrow (*Ammospiza maritima*), which eats sandhoppers and other small invertebrates.

On the high marsh, animal life changes almost as suddenly as the vegetation. The pulmonate marsh snail (*Melampus*) replaces the marsh periwinkle of the low marsh. Within the matted growth of *Spartina patens* is a maze of runways made by the meadow mouse (*Microtus pennsylvanicus*), which feeds heavily on the grass. Replacing the clapper rail and seaside sparrow are the willet (*Catoptrophorus semipalmatus*) and the seaside sharp-tailed sparrow (*Ammospiza caudacuta*).

Low tide brings a host of predacious animals into the marsh to feed. Herons, egrets, willets, ibis, raccoons, and others spread over the exposed marsh floor and the muddy banks of tidal creeks to feed. At high tide the food web changes. Such fish as the killifish, silversides, and the four-spined stickleback (*Apeltes*

(a)

(b)

FIGURE 31.34 (a) A ribbed mussel half-buried in the mud at low tide. (b) A fiddler crab emerges from its burrow to forage across the tidal mud at low tide.

quadracus), restricted to creek waters during low tide, spread over the marsh at high tide, as does the blue crab (*Callinectes sapidus*). During the winter migrant waterfowl claim the high marsh, especially snow geese, which graze on *Salicornia.*

Function

The salt marsh is one of the most productive ecosystems. Above-ground production of salt marshes of eastern North America ranges from 650 to 2800 g dry wt/m²/yr; below-ground production is two times greater. Above-ground production of British salt marshes (Ranwell 1961, Jefferies 1972) and California *Salicornia* marshes (Mahall and Park 1976) falls within the same range of values.

The reason for high production is a tidal subsidy. Tidal flushing brings in new nutrients, sweeps out accumulated salts, metabolites, sulfides, and toxic wastes, and replaces anoxic interstitial water with oxygenated water. Added to this is a rather tight internal nutrient cycling. Algal and bacterial populations turn over rapidly, and detrital material is fragmented and decomposed. Up to 47 percent of net primary production is respired by microbes; part is grazed by nematodes and microscopic benthic organisms, both of which are consumed by deposit-feeders. Depending on the nature of the salt marsh, between 11 and 66 percent of decomposed marsh grass is converted to microbial biomass. The rest is lost to the sea. Below-ground production enters an anaerobic food web of fermentation, reducing sulfates and producing methane.

Sulfur, present in tidal water, appears to be significant in energy flow in the salt marsh (Howarth and Teal 1979, Howarth et al. 1983). In the anaerobic environment of the salt marsh, soil bacteria convert the seawater sulfates to sulfites by oxidizing organic compounds. In doing so, bacteria trap a portion of the energy, with the remainder residing in sulfide radicals. These stored sulfides are reoxidized over the year by oxygen diffusing from the roots of salt marsh plants and become available for further growth of sulfur-oxidizing bacteria. An estimated 70 percent of the above-ground net primary production in a New England salt marsh flows through reduced inorganic sulfur compounds, and an equivalent of 20 percent of above-ground net primary production is exported from the marshes as reduced sulfur compounds.

TABLE 31.2 Simulated Annual Carbon Budget for the Duplin River Marshes, Sapelo Island, Georgia

Source or Process	Net Balance (g C/m²/yr)	
Production		
S. alternifolia	1575	
Algae	131	
Total production		1706
Loss		
Respiration ($CO_2 + CH_4$)		
in soil	− 623	
in water	− 222	
in air	− 68	
Tidal movement	− 586	
Total loss		− 1499
Net change		207
Sedimentation	29	
Unexplained	178	

Note: All values prorated over the entire marsh system—soil, creeks, and Duplin River.
Source: Pomeroy and Wiegert 1981:225.

TABLE 31.3 Provisional Nitrogen Budget for the Duplin River Watershed at Sapelo Island, Georgia

Flux Components	Nitrogen Flux (g N/m²/yr)
Inputs	
Rain	0.3
Sedimentation	3.3
Nitrogen fixation	14.8
Tidal exchange	46.6
Losses	
Denitrification	65.0
Internal cycles	
Primary production	70.0
Soil remineralization	70.0

Note: Tidal exchange is depicted as seeking a steady state, offsetting differences in nitrogen fixation and denitrification.
Source: Pomeroy and Wiegert 1981:180.

What happens to excess carbon production in a tidal marsh is not well understood (Table 31.2). Each salt marsh apparently differs. What happens to its excess carbon production is influenced by its route of transformation though a food web, and its route and importance of export. Some salt marshes depend on tidal exchanges and import more than they export (Woodwell et al. 1979), whereas others export more than they import (Valiela et al. 1978, Valiela and Teal 1979). Some of the excess production goes into sediments; some may be transformed microbially in the water in the marsh and tidal creeks. A portion may be exported to the estuary physically as detritus, as bacteria, or as fish, crabs, and intertidal organisms by way of the food web.

The importance of tidal marshes as a nutrient source and sink for the estuary has been a question of long standing. An insight into this relationship is partially provided by phosphorus and nitrogen cycling in the marsh.

The major flux of phosphorus in and out of salt marsh and estuarine waters is dissolved phosphate. Over the year, input equals output and usually phosphorus is surplus to the requirements in both soil and water. Excess phosphorus is mineralized and utilized by autotrophs and moves back through the system by the way of heterotrophs.

Nitrogen is a different story (Table 31.3). Denitrification often exceeds nitrogen fixation, and the marsh depends on inputs to the system. The amount of nitrogen cycled is determined by tidal input, physical and chemical exchanges with air and water, and biological fluxes. In some salt marshes, exemplified by Great Sippewissett Marsh in Massachusetts, groundwater inflow brings in nitrates, some of

which percolate through the peat and are exported to the ocean as organic nitrogen, ammonium, and nitrates. Of the total influx, one-third is exported by denitrification and two-thirds by tides (Valiela et al. 1978). The much larger salt marshes of Sapelo Island, Georgia, depend on inputs of nitrogen from the associated river and tides. Only in summer does the system appear to export any nitrogen. Mostly the marsh is a net sink for nitrogen.

The nitrogen cycle in the salt marsh is not well understood, but a study of nitrogen cycling in *Spartina alterniflora* by Hopkinson and Schubauer (1984) in the Sapelo marshes of Georgia (Figure 31.35) provides some insight into the mechanisms. The above-ground pool of total nitrogen was greater than the below-ground pool. The concentration was highest in young stems, but it decreased with age. Maximum accumulation was above ground in midsummer and below ground in midwinter, when 83 percent of nitrogen was in roots and rhizomes. Uptake of nitrogen was 34.8 g/m²/yr. Total transfer of nitrogen from below-ground to above-ground tissue was 33 g/m²/yr. Forty-six percent of new nitrogen was taken up from the soil. Of N transferred to above-ground biomass, 14.4 g/m²/yr was lost as detritus upon culm death, 0.7 g was leached from the living culm, and 17.9 g was translocated back to rhizomes. The movement of N from above-ground to below-ground biomass during the period of active growth and the dependence of new growth in spring on N stored in the rhizomes suggest strong nutrient conservation and recycling.

Maintenance of this cycling of N within *Spartina* involves an interaction with the plant's animal associate, the ribbed mussel, whose strong byssal threads bind sediments and prevent their erosion. Jordan and Valiela (1982) found that ribbed mussels pump tidal water in excess of the tidal volume of their New England salt marsh and deposit as much particulate matter in the form of feces and pseudofeces on the

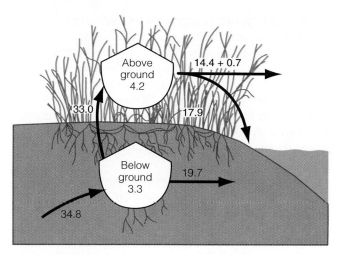

FIGURE 31.35 Nitrogen cycle (g/m²/yr) in the Sapelo salt marshes of Georgia. Note both the degree of internal cycling and the amount exported to the mud and to the estuary. (From Hopkinson and Schubauer 1984:966.)

surface mud as is exported. This nitrogen is immediately available for use by *Spartina*. Investigating further, Bertness (1984), by manipulating density of ribbed mussels, experimentally demonstrated that this invertebrate stimulated production of *Spartina*. Increased height, biomass, and flowering of *Spartina* and soil nitrogen levels were positively associated with mussel density.

MANGROVE SWAMPS

Replacing salt marshes on tidal flats in tropical regions are **mangrove swamps** or **mangal** (Figure 31.36a). The term

mangrove describes a group of halophytic species belonging to 12 genera in eight different families. The term also refers to forest plant communities fringing sheltered tropical shores (Lugo and Snedaker 1974). Mangrove forests develop in tropical coastal areas where wave action is absent, sediments accumulate, and the muds are anoxic. Mangrove forests are dominated by *Rhizophora, Avicennia, Bruguiera, Sonneratia,* and *Laguncularia*. Growing with them are other halophytic species, mostly shrubs.

Mangrove trees are characterized by shallow, widely spreading roots, or by prop roots coming from the trunk and branches (Figure 31.36b). Many species have root extensions or pneumatophores that take in oxygen for the roots. The leaves, although tough, are often succulent and may have salt glands. Red mangroves (*Rhizophora*) have a unique method of reproduction. Seeds germinate on the tree, grow into a seedling with no resting stage between, drop to the water, and float upright until they reach water shallow enough for their roots to penetrate the mud.

Structure

The formation and physiognomy of mangrove swamps is strongly influenced by the range and duration of tidal flooding and surface drainage (Lugo and Snedaker 1974, Lugo 1990). One of the features of this response is zonation, the changes in vegetation from seaward edge to true terrestrial environment. Although often used as an example, the mangrove swamps of the Americas, particularly Florida, have the least pronounced zonation, largely because of the few species involved. The pioneering red mangrove (*Rhizophora mangle*) occupies the seaward edge and experiences the deepest tidal flooding. Red mangroves are backed by pneumatophore-possessing black

(a)

(b)

FIGURE 31.36 (a) Mangrove forests replace tidal marshes in tropical regions. (b) Shallow-rooted mangroves often have prop roots descending from the trunk and branches.

mangroves (*Avicennia germinanas*), shallowly flooded by high tides. The landward edge is dominated by white mangroves (*Laguncularia racemosa*) along with buttonwood (*Conocarpus erectus*), a nonmangrove species that acts as a transition to terrestrial vegetation.

In the Indo-Pacific region, mangrove forests are much better developed, contain up to 30 to 40 species, and have a more pronounced zonation. The seaward fringe is dominated by one or several species of *Avicennia* and perhaps trees of the genus *Sonneratia,* which do not grow well in the shade of other mangrove species. Behind the *Avicennia* is a zone of *Rhizophora,* which grow in areas covered by daily high tide up to the point covered only by the highest spring tides. At and beyond the level of high spring tides is a broad zone of *Bruguiera*. The final and often indefinite mangrove zone is an association of small shrubs, mainly *Ceriops*.

Associated with mangroves are a mix of marine organisms that occupy prop roots and the mud, and those that live in the trees. As in the salt marsh, *Littorina* snails live on the roots and trunks of mangrove trees. Attached to stems and prop roots are barnacles and oysters. Fiddler crabs burrow into the mud during low tide and live on prop roots and high ground during high tide. In the Indo-Pacific mangrove swamps live mudskippers, fish of the genus *Periophthalmus* with modified eyes set high on the head, which live in burrows in the mud, spend time out of the water crawling about the mud, and in many ways act more like amphibians than fish. Closely associated with these mangrove forests are commercially important species of penacid prawns and shrimp that exist on mangrove detritus, phytoplankton, and associated microbial populations (Sasekumar and Ching 1987). Herons and other wetland birds nest in the mangrove trees, and alligators, crocodiles, bears, pumas, and wildcats inhabit the forest interior.

Function

Net productivity of mangrove swamps is variable, ranging in Florida from about 450 to 2700 g C/m²/yr, about the productivity of salt marshes. Productivity is influenced by tidal inflow and flushing, water chemistry, salinity, and soil nutrients, much as in the salt marsh. Highest rates of productivity occur in the mangrove forests that are under the influence of daily tides (Lugo and Snedaker 1974, Careter et al. 1973). For example, gross primary productivity of red mangrove, flooded daily by high tides and receiving some fresh water, is greatest at salinities below that of seawater, whereas gross primary productivity of white and black mangroves is positively correlated with salinity. These two species predominate at higher elevations, zones of increasing salinity (Hicks and Burns 1975). In areas of intermediate salinity, white mangroves have twice the productivity of red mangroves. Zonation of mangrove swamps appears to reflect the optimal productivity niches of the species rather than physical or successional conditions (Lugo 1980).

Nutrient cycling is regulated by hydroperiod, which affects the amount and turnover of litter on the mangrove swamp floor and the ability of plants to translocate and take up nutrients. Data on nutrient cycling in mangrove forests are minimal. Recycling of calcium appears to be low, except on calcium-poor waters. White mangroves appear to have a high recycling efficiency for nitrogen. Australian mangroves seem to be more efficient in recycling phosphorus than the ones in the Malaysian Peninsula (Lugo et al. 1992).

Like the salt marsh, mangrove swamps export a considerable portion, up to 50 percent, of their above-ground primary production to the adjacent estuary. This detrital input is important to the commercial and sport fisheries of the Gulf of Mexico (Odum 1970) as well as those of the mangrove inlets and estuaries of Indo-Pacific regions. In mangrove-fringed peninsular Malaysia, the high commercial harvest of prawns is positively correlated with the presence of mangroves. The prawns assimilate an average of 65 percent of their carbon from the mangroves. The mud substrates are rich in food for benthic meiofauna, which feed on phytoplankton, benthic flora, and mangrove detritus. Prawns in turn feed on mangrove detritus, assimilating its leachable organic matter and microbial populations associated with it (Sasekumar and Ching 1987).

Summary

1. The marine environment is characterized by salinity, waves, tides, depth, and vastness. Salinity is due largely to sodium and chlorine, which make up 86 percent of sea salt. Although sea salt has a constant composition, salinity varies throughout the oceans. It is affected by evaporation, precipitation, and movement and mixing of water masses of different salinities. Because of its salinity, seawater does not reach its greatest density at 4°C but becomes heavier as it cools.

2. Like lakes, the marine environment experiences both zonation of life and stratification of temperature. The open sea can be divided into three main regions. The bathypelagic is the deepest, void of sunlight and inhabited by darkly pigmented, weak-bodied animals characterized by bioluminescence. Above it lies the dimly lit mesopelagic region, inhabited by its own characteristic species such as certain sharks and squid. Both the bathypelagic and mesopelagic region depend on a rain of detrital material from the upper lighted region for their energy source. The sea bottom or benthic region is inhabited by its own unique fauna adapted to life in the total darkness and high pressure of the deep. The species diversity of the deep sea benthos is surprisingly high. The detrital feeding and predaceous fauna are dominated by polychaete worms and crustaceans.

3. Along the volcanic oceanic ridges, especially in the mid-Pacific, are hydrothermal vents inhabited by unique and newly discovered forms of life, including clams, worms, and crabs. The source of primary production for these hydrothermal vent communities are chemosynthetic bacteria that use sulfates as an energy source.

4. The impoverished nutrient status of ocean water results in low productivity. Impoverishment comes about because nutrient reserves in the upper layer of water are limited; phytoplankton and other life sink to the deep water; and a thermocline, permanent in deep water, prevents the circulation of deep water to the upper layer. Most productive are shallow coastal waters and areas of upwelling where nutrient-rich deep water comes to the surface. Copepods are the major herbivores and become the critical link in the food chain between first- and second-level consumers. However, recent studies show that bacteria and protists, both heterotrophic and photosynthetic, make up half of the biomass of the sea and are responsible for the largest part of energy flow in the pelagic system.

5. Sandy shore and rocky coast are places where sea meets land. The drift line marks the furthest advance of tides on the sandy shore. On the rocky shore, the tide line is marked by a zone of black algal growth.

6. The most striking feature of the rocky shore, its zonation of life, results from alternate exposure and submergence of the shore by the tides. The black zone marks the littoral, the upper part of which is flooded only every two weeks by spring tides. Submerged daily by tides is the littoral, characterized by barnacles, periwinkles, mussels, and fucoid algae. Uncovered only at spring tides is the sublittoral, which is dominated by large brown laminarian seaweeds, Irish moss, and starfish. Although the basic pattern of zonation may result from tidal flooding and exposure, distribution and diversity of life across the rocky shore is influenced by disturbance from wave action, competition, herbivory, and predation.

7. Left behind by outgoing tides are tide pools. These are distinct habitats subject over a 24-hour period to wide fluctuation in temperature and salinity and inhabited by varying numbers of organisms, depending on the amount of emergence and exposure, competition, and predation.

8. In contrast, sandy and mud shores appear barren of life at low tide, but beneath the sand and mud conditions are more amenable to life than on the rocky shore. Zonation of life is hidden beneath the surface. The energy base for sandy and muddy shores is organic matter made available by bacterial decomposition. They are important sites for biogeochemical cycling, supplying nutrients for offshore waters. The basic consumers are bacteria, which in turn are a major source of food for both deposit-feeding and filter-feeding organisms. Sandy shores and mudflats are a part of the larger coastal ecosystem including the salt marsh, estuary, and coastal waters.

9. Coral reefs are nutrient-rich oases in nutrient-poor tropical seas. They are complex ecosystems based on anthozoan corals and their symbiotic endozoic dinoflagellate algae and coralline algae. Recycling nutrients within the system and functioning as nutrient sinks, coral reefs are among the most productive ecosystems in the world. Their productive and varied habitats support a high diversity of colorful invertebrate and vertebrate life.

10. Estuaries, where fresh water meets the sea, and their associated tidal marshes and swamps are a unit in which the nature and distribution of life are determined by salinity; as salinity declines from the mouth up through the river to the head, so do estuarine fauna, chiefly marine species. The estuary serves as a nursery for marine organisms, for here the young can develop protected from predators and competing species unable to withstand lower salinity. If vertical mixing between fresh water and salt water occurs in the estuary, the nutrients circulate up and down between organisms and bottom sediments.

11. Salt-tolerant grasses dominate associated salt marshes, flooded daily by tides. The interaction of salinity, tidal height, and flow and competitive interactions among salt marsh plants produce a distinctive zonation of vegetation from the low marsh through the high marsh. Because tidal flooding resupplies nutrients and carries away waste, salt marshes are highly productive. Most of this primary production goes unharvested by herbivores and is converted to microbial biomass. Some production goes into sediments, some accumulates as peat, and another portion is exported to the estuary. Because of their extreme importance to estuarine fisheries and wildlife, the loss of coastal wetlands can have serious ecological and economic implications.

12. Mangrove forests replace salt marshes in tropical regions. Mangroves have shallow, widely spreading roots with additional roots arising from the trunk and branches. Like salt marshes, mangrove forests have a pronounced zonation from the seaward fringe to the level of high spring tides. Associated with and depending on the mangroves is a wide diversity of species from barnacles, snails, and crabs to reptiles and birds. Closely associated with mangroves are commercially important penacid prawns and shrimp.

Review Questions

1. What are the major regions of the oceans?
2. What are the halocline, thermocline, and pycnocline?
3. How are waves generated?
4. What causes tides? What are spring tides? Neap tides?
5. How does the tidal cycle relate to the structure, function, and productivity of intertidal ecosystems?
6. What factors influence the productivity of the open sea?
7. Contrast the productivity of the open sea with that of coastal waters. Explain the difference.
8. Give some possible explanations for the high diversity of the ocean benthos.
9. What is the food source of the nekton of the deep ocean?
10. How are nekton organisms adapted for life in the deep?
11. What feeding strategies are employed by bottom organisms?
12. Why are hydrothermal vents ecologically unique?
13. What is the relationship between photosynthetic nanoflagellates and cyanobacteria and associated heterotrophic bacteria in the marine food web?
14. What are the three major zones of the rocky shore? What are the outstanding features of each?
15. How are barnacles and periwinkles adapted to withstand low tides?
16. Patchy distribution is common among rocky intertidal organisms. What physical and biological interactions give rise to this patchiness?
17. In what ways are tide pools microcosms of the sea?
18. Distinguish between epifauna and infauna.
19. Contrast the energy source of a sandy or muddy shore with that of a rocky shore. What groups of organisms are the basic consumers in each?
20. How does the fact that corals are partially photosynthetic and partially heterotrophic contribute to the high productivity of coral reefs?
21. What are the outstanding features of the coral reef?

22. What is an estuary?
23. Discuss the vertical and horizontal stratification of salinity in the estuary.
24. How is the estuary a nutrient trap?
25. What is the interrelationship between marine fish and an estuary? Oyster reefs and an estuary?
26. What two factors make the estuary an ideal nursery for marine fish?
27. What are salt marshes? Where are they located and how do they develop?
28. What are the three dominant grasses of a salt marsh? What biological and physical interactions determine their distribution within the marsh?
29. In what way is *Spartina alterniflora* adapted to the salt-water environment?
30. What are the dominant animals in the salt marsh? How do they relate to tidal regimes?
31. Why is the salt marsh one of the most productive ecosystems?
32. What is the role of sulfur in the energy flow of salt marshes?
33. Briefly describe the nitrogen cycling in a salt marsh.

34. What are the major features of a mangrove swamp?
35. How do mangrove swamps respond to tidal flooding?
36. What is the ecological and commercial importance of mangrove swamps?
37. How does the reproductive biology of mangroves enable them to colonize shallow water areas of the tropics?
38. What is happening to tidal marshes and swamps? Why? What are the implications of this destruction? (You will have to go to the literature to answer this question. Consult for example Williams 1990, *Wetlands: A Threatened Landscape,* Blackwell.)

Cross References

Coriolis effect, 25; ocean currents, 28; characteristics of water, 51; halophytes, 111; energy flow, 481; nitrogen cycle, 517; phosphorus cycle, 520; plant-herbivore interactions, 291; mutualism, 320; disturbance and community structure on rocky shores, 679.

CHAPTER 32

Global Environmental Change

CONCEPTS

1. Atmospheric concentration of carbon dioxide is rising.

2. Tracking the fate of CO_2 emissions requires an understanding of many processes.

3. Atmospheric CO_2 concentrations affect CO_2 uptake by oceans.

4. Plants respond to increased atmospheric CO_2, influencing both plant growth and net primary productivity.

5. Greenhouse gases may change the global climate.

6. Changes in climate will affect ecosystems at many levels.

7. Changing climate will shift the global distribution of ecosystems.

8. Global warming would raise sea level and affect coastal environments.

9. Climate change will affect agricultural production.

10. Climate change will have both direct and indirect impacts on human health.

11. Understanding global change requires the study of ecology at a global scale.

The term *environmental change* is redundant. Change is an inherent characteristic of Earth's environment. Paleoecology has recorded the response of populations, communities, and ecosystems to changes in climate during periods of glacial expansion and retreat over the past 100,000 years (see Chapter 21). On an even longer time scale, the geological and fossil records recount environmental and evolutionary change since the origin of the planet.

Humans, like other species, are part of Earth's environment. Anthropologists tell us that the first humans were hunters and gatherers with low populations. At that time humans functioned as natural predators and herbivores. With the adoption of agriculture, believed to have begun with the cultivation of grasses in the region of the Middle East some 10,000 years ago, human populations and associated conversion of land to agricultural production began to increase exponentially. That trend has continued into modern times (Figure 32.1). By the mid-1800s, the Industrial Revolution once again changed the nature of human interactions with the global environment. The demand for energy to fuel industrialization and the concentration of populations, or urbanization, brought environmental problems of unprecedented magnitude.

In this chapter we will examine a small subset of the human-induced global environmental changes that are global in scale. The nitrogen and sulfur compounds released in industrial emissions (see Chapter 26), although extensive in their distribution, are largely regional concerns. In contrast, other gaseous emissions, such as carbon dioxide, methane, and chlorofluorocarbons (CFCs), mix readily and circulate globally in the atmosphere. To understand and predict the consequences of these emissions on local and regional environments, scientists must study them at the global scale. The changes in atmospheric chemistry brought about by these emissions have the potential to change Earth's climate, shifting the distribu-

tion and abundance of ecosystems, and directly affecting the health and well-being of the human population.

ATMOSPHERIC CARBON DIOXIDE CONCENTRATIONS

The atmospheric concentration of carbon dioxide (CO_2) has increased by more than 25 percent over the past 100 years. The evidence for this rise comes primarily from continuous observations of atmospheric CO_2 started in 1958 at Mauna Loa, Hawaii, by Charles Keeling (Figure 32.2), and from parallel records around the world (Houghton et al. 1996). Evidence before the direct observations of 1958 comes from a variety of sources, including the analysis of air bubbles trapped in the ice of glaciers in Greenland and Antarctica.

Reconstructing atmospheric CO_2 concentrations over the past 300 years, we see values that fluctuate between 280 and 290 ppmv (parts per million volume) until the mid-1800s (Figure 32.3) (Watson et al. 1996). After 1860, the onset of the Industrial Revolution, the value soars exponentially. The change reflects the combustion of fossil fuels (coal, oil, and gas) as an energy source for industrialized nations (Figure 32.4).

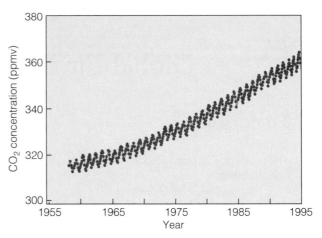

FIGURE 32.2   Concentration of atmospheric CO_2 at Mauna Loa Observatory, Hawaii. The dots depict monthly averages. (Keeling and Wharf 1991.)

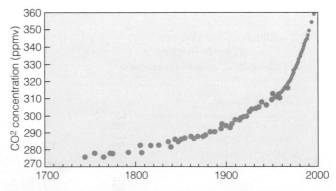

FIGURE 32.3   Historical record of atmospheric CO_2 over the past 300 years. Data prior to direct observation (1958 to present) are estimated from various techniques including analysis of air trapped in Antarctic ice sheets. (Adapted from Watson et al. 1996.)

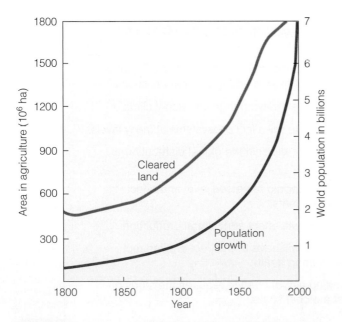

FIGURE 32.1   Changes in global population and land area cleared for agriculture over the past 200 years.

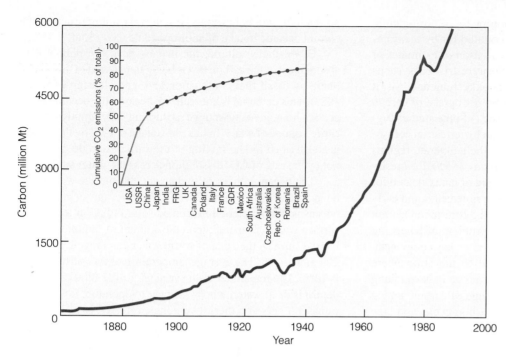

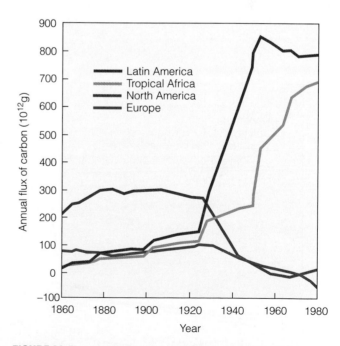

FIGURE 32.4 Historical record of annual input of CO_2 to the atmosphere from the burning of fossil fuels since 1860. The inset graph shows cumulative CO_2 emissions from the 20 largest CO_2-emitting nations in 1989 plotted as a proportion of total emissions. (Adapted from Marland and Boden 1993.)

The burning of fossil fuels is not the only cause of rising atmospheric CO_2 concentration. Deforestation is also a major cause (Figure 32.5) (Schimel et al. 1996). Forested lands are typically cleared and burnt for farming. Although the trees may be harvested for timber or wood pulp, a large part of the biomass, litter layer, and soil organic matter is burnt, releasing the carbon to the atmosphere as CO_2.

Calculations of the contribution of land clearing to atmospheric CO_2 are complex. Following timber harvest on lands managed for forest production, or on lands that have been cultivated and then abandoned, vegetation and soil organic matter become reestablished (see Chapter 22). The net contribution to the atmosphere is the difference between CO_2 released during clearing and burning and CO_2 taken up by photosynthesis and the accumulation of biomass during reestablishment. At one time, scientists used regional estimates of population growth and land use (forestry and agriculture) together with simple models of vegetation and soil succession to estimate the contribution due to changing land use (Houghton et al. 1983). More recent estimates use satellite images to quantify these changes.

The release of CO_2 to the atmosphere through the burning of fossil fuels is currently about 6 Gt C/yr (Marland and Boden 1993). However, only about 56 percent of this amount remains in the atmosphere (Keeling et al. 1995). Using models of ocean circulation and CO_2 exchange in the surface waters of the ocean, it is believed that about 33 percent (2 Gt/yr) of the total input from fossil fuels enters the ocean each year (Siegenthaler and Sarmiento 1993). Taken together, the atmospheric increase and oceanic uptake of CO_2 accounts for 89 percent of the annual emissions from fossil fuels. Given the inputs from terrestrial ecosystems (both vegetation and soils) associated with forest clearing, the global carbon budget does not balance (see Figure 26.4).

A large quantity of carbon that ought to be in the atmosphere is unaccounted for, with estimates ranging from 1.5 to 2.0 Gt C/yr.

It is believed that this "missing carbon" is being taken up by terrestrial ecosystems. Although the processes controlling the exchange of carbon between terrestrial ecosystems and the atmosphere are generally well understood, quantifying these processes at a regional to global scale is

FIGURE 32.5 Historical record of annual input of CO_2 to the atmosphere from the clearing and burning of forest in North America, South America, Central America, Europe, and Africa. (From Houghton 1997.)

extremely difficult. Using a simple process of elimination, carbon emissions that cannot be accounted for by measurements of atmospheric carbon concentration or estimates of oceanic absorption are relegated to the terrestrial ecosystems (Figure 32.6). Prior to this type of bookkeeping analysis, it was generally believed that there is no net uptake of carbon by terrestrial ecosystems at a global scale. Some studies suggest that a possible net uptake of carbon by terrestrial ecosystems may result from reforestation in the temperate regions of the Northern Hemisphere (Melillo et al. 1995). Reforestation follows the large-scale abandonment of lands cleared for agriculture during the latter part of the nineteenth and early part of the twentieth centuries. Although reforestation has not been proven to be the solution to the problem of balancing the global carbon cycle, it is most certainly a key component. Determining the fate of carbon input to the atmosphere through the burning of fossil fuels requires an understanding of the processes controlling the exchange of carbon among the major components of the global carbon cycle, and how the transfers might be influenced by rising atmospheric concentrations of CO_2.

Elevated CO_2 and Oceans

Carbon dioxide diffuses from the atmosphere into the surface waters of the ocean, where it dissolves and undergoes a number of chemical reactions, including the transformation to carbonates and bicarbonates (Chapter 25, Carbon Cycle). As the concentration of CO_2 in the atmosphere rises, the diffusion of CO_2 into the surface waters of the oceans increases. The elevated CO_2 and bicarbonate concentrations represent a potential for increasing rates of net primary productivity in the surface waters; however, this is unlikely. Because of the low nutrient content of seawater, changes in marine net primary

productivity are believed to be relatively unimportant to the current oceanic uptake of anthropogenic CO_2 (Shaffer 1993).

Given their volume, the oceans have the potential to absorb most of the carbon that is being transferred to the atmosphere by fossil fuel combustion and land clearing (Keeling 1983). This potential is not realized, because the oceans do not act as a homogeneous sponge, absorbing CO_2 equally into the entire volume of water. In fact, the uptake of CO_2 by the oceans is constrained by the mixing of the surface and deep waters, not by the rate of CO_2 dissolution across the water surface.

The oceans consist of two layers, the surface waters and deep waters (see Figure 25.10). The average depth of the oceans is 2000 meters. Intercepted solar radiation warms the surface waters. Depending on the amount of radiation reaching the surface, the zone of warm water can range from 75 to 200 m in depth. The average temperature of this surface layer is 18°C. The remainder of the vertical profile (200 to 2000 m depth) is deep water, whose average temperature is 3°C. The transition between these two zones, referred to as the thermocline (see Chapter 25, page 513) is abrupt. In effect, the oceans can be viewed as a thin layer of warm water floating on a much deeper layer of cold water.

The temperature difference between these two layers leads to the separation of many processes. Turbulence caused by winds mixes the surface waters, transferring CO_2 absorbed at the surface into the waters below. Due to the thermocline, however, this mixing does not extend into the deep waters. Mixing between surface and deep waters depends on deep ocean currents caused by the sinking of surface waters as they move toward the poles (see Chapter 2). This process occurs on a time scale of hundreds of years, limiting the short-term uptake of CO_2 by the deep waters. The result is that the amount of CO_2 that can be absorbed by the oceans over the short term is limited, despite the large volume of water.

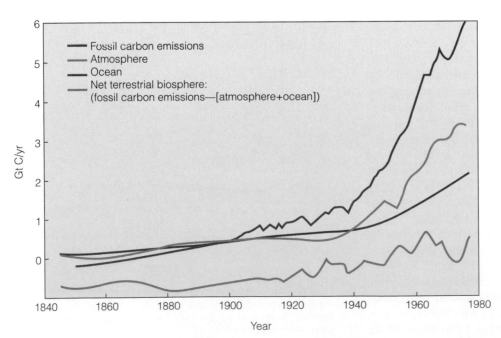

FIGURE 32.6 Emission of carbon from the burning of fossil fuels and changes in the major global reservoirs. Atmospheric values are based on direct observations and estimates from the analysis of ice cores. Ocean uptake is estimated using the Geophysical Fluid Dynamics Laboratory (Princeton University) ocean carbon model, and the net uptake by terrestrial ecosystems is calculated as the remaining difference. (From Houghton 1994.)

If the release of fossil fuels were to cease, nearly all of the CO_2 that has accumulated in the atmosphere from the burning of fossil fuels would eventually dissolve in the oceans, and the global carbon cycle would return to a steady state.

Elevated CO_2 and Terrestrial Ecosystems

The processes of photosynthesis and respiration govern the exchange of carbon dioxide between the atmosphere and terrestrial ecosystems (see Chapter 6). To understand how rising atmospheric CO_2 concentrations influence this exchange, we must understand their influence, both direct and indirect, on a variety of processes operating at a range of temporal and spatial scales.

Photosynthesis Recall that CO_2 diffuses from the air into the leaf through the stomatal openings (see Chapter 6). The higher the CO_2 concentration in the outside air, the greater the rate of diffusion into the leaf. A higher rate of diffusion increases the availability of CO_2 for photosynthesis in the mesophyll cells of the leaf, and so it generally results in a higher rate of photosynthesis. The higher rates of diffusion and photosynthesis under elevated atmospheric concentrations of CO_2 is termed the **CO_2 fertilization effect.** Plant species vary in their response to elevated CO_2, although most show an increase in photosynthetic rates. C_4 plants are less sensitive to elevated CO_2 concentrations than are C_3 plants. The absence of photorespiration in the mesophyll cells and the mechanism for increasing CO_2 concentrations in the bundle sheath cells reduce the influence of elevated external concentrations of CO_2 compared with that on C_3 species (see Chapter 6). In a review of more than 200 experiments that examined the response of tree species (C_3 plants) to a doubling of CO_2 concentration under laboratory conditions, photosynthetic rates increased by an average of 44 percent (Figure 32.7) (Amthor 1995).

Plant Growth Similar patterns to those found for photosynthesis have been observed for plant growth. Poorter (1993) reviewed the results of experiments that examined the growth response of a variety of plant species to elevated CO_2 corresponding to double ambient concentrations. Of the 156 plant

species examined in the review, the average increase in growth was 41 percent for C_3 species and 22 percent for C_4.

The effects of long-term exposure to elevated CO_2 on plant growth and development, however, may be more complicated. In some studies, the enhanced effects of elevated CO_2 levels on plant photosynthesis have been short-lived. With time, some plant species acclimate to the higher CO_2 concentrations and rates of photosynthesis and growth return to levels observed under ambient CO_2 concentrations. This process is referred to as **down-regulation** (Amthor 1995, Curtis 1996, Gunderson and Wullschleger 1994). The mechanisms by which down-regulation occurs are varied. Some plants produce less of the photosynthetic enzyme rubisco under elevated CO_2, reducing photosynthesis to rates comparable to those measured at lower CO_2 concentrations (Arp 1991, Bowes 1991). Other studies reveal that plants grown under increased CO_2 levels allocate less carbon to the production of leaves and more to the belowground processes, including accelerated root turnover (Norby et al. 1992, Zak et al. 1993), mycorrhizal development (O'Neill et al. 1987, Rygiewitz and Anderson 1994, Dhillon et al. 1996), and nitrogen fixation (Soussana and Hartwig 1996). In addition, plants grown under elevated concentrations of CO_2 appear to produce fewer stomata on the leaf surface (Woodward and Kelly 1995).

Transpiration The observed increases in photosynthesis under elevated CO_2 occur even though plants exposed to a doubled CO_2 concentration exhibit a partial closure of the stomata (Field et al. 1995). Although this decrease in stomatal conductance acts to decrease CO_2 diffusion into the leaf, it has a greater impact on reducing water loss through transpiration (see Chapter 6). Thus under elevated CO_2 levels, plants increase their water use efficiency (carbon uptake/water loss).

Although reductions in stomatal conductance under elevated CO_2 have been observed for herbaceous plants and woody seedlings, both in the laboratory and field, mature trees appear not to exhibit partial stomatal closure. In field experiments where the branches of canopy trees are exposed to elevated CO_2, no changes in stomatal conductance have been observed (Jarvis 1995, Teskey 1995, Korner and Warth 1996, Ellsworth et al. 1995).

FIGURE 32.7 The change in photosynthesis for tree species grown under a doubled CO_2 concentration. Response is expressed as the ratio of observed photosynthesis under a doubled CO_2 concentration to that under normal CO_2 level. A value of 1 represents no detected increase in photosynthetic rates under a doubled CO_2 concentration. (From Anthor 1995.)

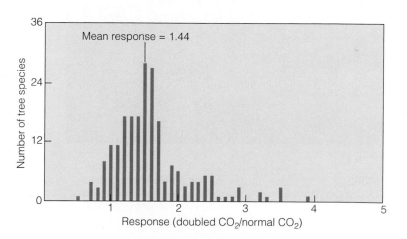

Primary Productivity How the results observed for leaves or single plants translate into changes in the net primary productivity (NPP) of terrestrial ecosystems is largely unknown. A variety of experimental approaches have allowed scientists to examine the response of natural vegetation to elevated CO_2. These methods range from open-top chambers (Figure 32.8a) to large-scale experiments that modify CO_2 concentrations in whole ecosystems—referred to as **FACE experiments** (Free Air Carbon Dioxide Enrichment). Figure 32.8b shows the FACE facility at the Duke University Forest in North Carolina. The exchange of carbon between the atmosphere and ecosystems in these experiments are typically measured in two ways: either direct measures of net CO_2 uptake by the canopy using infrared gas analyzers (IRGA), or by measuring biomass accumulation over the growing season (net primary productivity).

Most ecosystems exposed to double ambient CO_2 show higher net carbon uptake during the peak of the growing sea-

(a)

(b)

FIGURE 32.8 (a) Open-top chamber for growing plants under elevated CO_2. (b) The Free Air Carbon Dioxide Enrichment facility at Duke University.

son (Drake and Leadley 1991). Drake and colleagues (1996) examined the response of a salt marsh in Maryland to continuous exposure of double ambient CO_2. They observed an average 36 percent increase in peak season net CO_2 uptake over a six-year period. Stocker and colleagues (1997) found similar results in a calcareous grassland in England. During two years of CO_2 enrichment, peak season CO_2 uptake increased by 34 percent, while uptake integrated over the entire growing season rose by 22 percent.

Estimates of above-ground net primary productivity under elevated CO_2 reveal similar patterns to those observed for canopy CO_2 exchange. A comparison of field studies in grassland and agricultural ecosystems reveals an average increase in biomass production of 14 percent (Mooney et al. 1999). However, estimates at individual sites ranged from an increase of 85 percent to a decline of almost 20 percent. These results stress the importance of the interactions of elevated CO_2 with other environmental factors, particularly temperature, moisture, and nutrient availability.

Ecosystems characteristic of low-temperature environments tend to show an initial enhancement of productivity following elevated CO_2, followed by down-regulation. A study conducted in Arctic tundra observed an initial increase in productivity, but primary productivity returned to pretreatment levels following three years of continuous exposure to a doubled CO_2 environment (Oechel and Vourlitis 1996). Similar results are reported for an alpine grassland ecosystem (Korner et al. 1997). Productivity increased during the first three years of elevated CO_2, but returned to pretreatment levels by the fourth year of the experiment.

The largest and most persistent responses to elevated CO_2 have been observed in seasonally dry environments, where primary productivity is enhanced during years of below-average rainfall. In a study of a tallgrass prairie ecosystem in Kansas, Owensby and colleagues (1996) found no significant increase in above-ground net primary productivity during wet years (greater than average rainfall) for plots exposed to a doubled CO_2 concentration when compared with control plots receiving ambient CO_2. In contrast, they observed a 40 percent increase in above-ground NPP during years with average rainfall, and an 80 percent increase during years with below-average precipitation. Even though these relative increases in NPP are large, they occur in years of low NPP, so that absolute changes may be quite low.

The enhancement of primary productivity by elevated CO_2 in dry environments arises largely from the influence of reduced stomatal conductance on the water balance of the site rather than the direct effects of increased CO_2 levels on photosynthesis. The largest influence of CO_2 enrichment in periodically dry environments is the small reduction in evapotranspiration resulting from partial closure of the stomata. These small reductions have resulted in measurable changes (increases) in soil moisture in grassland ecosystems, particularly during prolonged dry periods (Field et al. 1996, Zaller and Arnone 1997). Increased soil moisture both extends the growing season and increases soil microbial activity, decomposition, and nitrogen mineralization.

Biogeochemistry In addition to influencing rates of decomposition and nutrient mineralization in periodically dry environments through enhanced soil moisture, elevated CO_2 concentrations may have other indirect effects on nutrient cycling in terrestrial ecosystems. Plants exposed to elevated CO_2 exhibit increased starch production, and consequently their leaves exhibit an increase in the carbon-to-nitrogen (C:N) ratio (Field et al. 1992). Recall from Chapter 9 that the relative abundance of carbon and nitrogen in plant litter has a major influence on the process of immobilization by decomposers and hence an influence on the rate of nitrogen release during decomposition (see Figures 9.11–9.13). It was initially thought that the observed increases in C:N in green leaves under elevated CO_2 would alter patterns of nitrogen mineralization, reducing nutrient availability for plant growth. Recent studies, however, have shown that high starch levels in the leaves disappear during leaf senescence, and the C:N of senescent leaves from plants grown under elevated CO_2 is comparable to values for plants grown in ambient CO_2 (O'Neil 1994, Franck et al. 1997, Hirschel et al. 1997).

Although elevated CO_2 does not appear to influence litter quality directly, a variety of studies examining the response of mixed-species plant communities to elevated CO_2 have observed shifts in relative abundance of species, patterns of species dominance, and diversity (Arnone and Korner 1995, Hattenschwiler and Korner 1996, Korner and Arnone 1992, William et al. 1986, Reekie and Bazzaz 1989). Given the differences among species in leaf characteristics influencing rates of decomposition and nutrient mineralization (see Table 7.1), shifts in species composition under elevated CO_2 could have a significant impact on litter quality, decomposition, and nutrient cycling within ecosystems.

It has been suggested that the increased allocation of carbon to below-ground processes under elevated CO_2 might increase soil carbon; however, small increases in soil carbon are extremely difficult to detect (Hungate et al. 1996). A few studies have shown a trend toward increasing soil carbon storage under elevated CO_2 after periods of treatment ranging from four to six years (Drake et al. 1996, Owensby et al. 1996). Definitive results, however, will no doubt require a much longer period of observation.

Air pollution leading to nitrogen deposition (see Chapter 26) may enhance any increases in primary productivity associated with elevated CO_2; at present, however, this interaction is still speculative.

GREENHOUSE GASES AND THE GLOBAL CLIMATE

Although rising atmospheric concentrations of CO_2 have the potential to influence the productivity and nutrient cycling of both terrestrial and aquatic ecosystems, the larger impact of elevated CO_2 on ecosystems may well be indirect, through its possible influence on the global climate system.

CO_2 in the atmosphere does not readily absorb short-wave radiation from the sun, but it does absorb long-wave or thermal radiation (see Figure 4.1). The Earth's surface emits absorbed solar radiation to the atmosphere as long-wave or thermal energy. Atmospheric CO_2 and water vapor trap this thermal energy, warming the atmosphere. Because it traps heat, carbon dioxide is referred to as a **greenhouse gas.** Were it not for the warming effect of carbon dioxide, water vapor, and a number of other gases present in the atmosphere in much smaller quantities, Earth would be some 20°C cooler than its current average surface temperature of 15°C (Houghton 1997).

As human activities increase the atmospheric concentration of CO_2, will they influence the global climate? Scientists estimate that at current rates of emission, the preindustrial level of 280 ppm of CO_2 in the atmospheric will double by the year 2020 (Houghton 1997). Moreover, CO_2 is not the only greenhouse gas increasing in concentration as a result of human activities. Other greenhouse gases include methane (CH_4), chlorofluorocarbons (CFCs), hydrogenated chlorofluorocarbons (HCFCs), nitrous oxide (N_2O), ozone (O_3), and sulfur dioxide (SO_2) (see Chapter 26). Although much lower in concentration, some of these gases are much more effective at trapping heat than is CO_2. They are a significant component of the total greenhouse effect.

Although the role of greenhouse gases in warming Earth's surface is well established, the specific influence that doubling the CO_2 concentration of the atmosphere will exert on the global climate system is much more uncertain. Atmospheric scientists have developed complex computer models of Earth's climate system—called **general circulation models (GCMs)**—to help determine how increasing concentrations of greenhouse gases may influence large-scale patterns of global climate (for review of GCMs see Gates et al. 1996). Although all use the same basic physical descriptions of climate processes, GCMs at different research institutions differ in their spatial resolution and descriptions of certain features of Earth, including surface and atmosphere. As a result, the models differ in their predictions (Figure 32.9).

Despite these differences, certain patterns emerge consistently. All of the models predict an increase in the average global temperature as well as a corresponding increase in global precipitation (for review of GCM predictions see Kattenberg et al. 1996). Findings published in 1995 by the Intergovernmental Panel on Climate Change (IPCC) suggest an increase in the average global temperature by the year 2100 of 2°C (Houghton et al. 1996). Uncertainty in the estimates of GCMs actually result in a range of predictions between 1.5°C and 3.5°C (Figure 32.9). These changes would not be evenly distributed over Earth's surface. Warming is expected to be greatest during the winter months and in the northern latitudes.

Figure 32.10 shows the spatial variation in changes in mean annual temperature and precipitation for the 48 contiguous states of the United States as predicted by the general circulation model developed by the Geophysical Fluid Dynamics Laboratory (GFDL) at Princeton University (Manabe et al. 1990, Manabe and Stouffer 1993). This model forecasts regional temperature increases ranging from 0.6°C to 7.0°C,

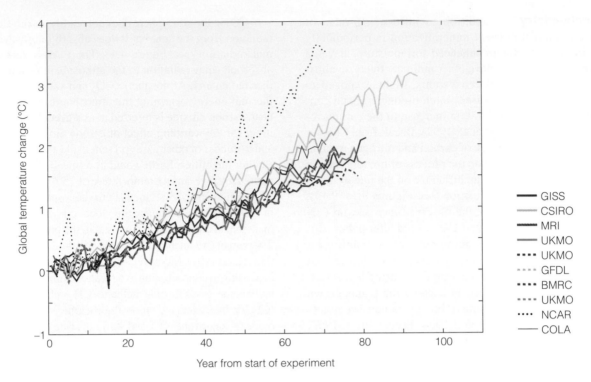

FIGURE 32.9 Comparison of predicted patterns of average global temperature change among a number of general circulation models. Abbreviations refer to different general circulation models. (Adapted from Kattenberg et al. 1996.)

FIGURE 32.10 Changes in (a) precipitation and (b) annual temperature for a doubled CO_2 concentration estimated by the general circulation model developed by the Geophysical Fluid Dynamics Laboratory at Princeton University. Changes in temperature are expressed as absolute increases in °C. Changes in precipitation are expressed as the ratio of current to predicted annual precipitation. A value of 1.0 would imply no change, while a value of 1.5 would imply a 50 percent increase, and a value of 0.8 a 20 percent decrease. (From Vemp 1995.)

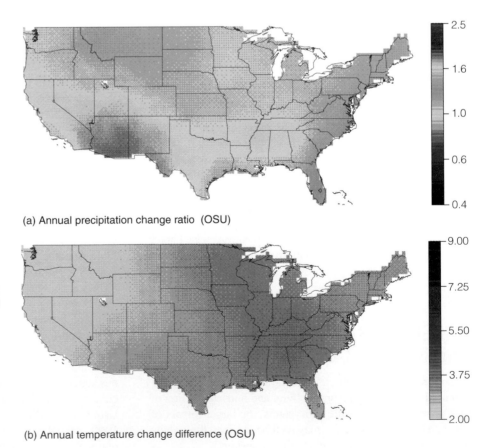

(a) Annual precipitation change ratio (OSU)

(b) Annual temperature change difference (OSU)

with the highest values in the polar regions and the lowest values in the equatorial regions.

Although in popular speech **the greenhouse effect** is synonymous with global warming, the models predict more than just hotter days. One of the most notable predictions is an increased variability of climate, including more storms and hurricanes, greater snowfall (winter precipitation), and increased variability in rainfall, depending on region (Kattenberg et al. 1996).

One recent development that has influenced predicted patterns of climate change is the inclusion of aerosols in calculating Earth's energy balance. Aerosols, or small particles suspended in the atmosphere, both absorb radiation from the sun and scatter it back to space. By scattering solar radiation back to space, they function to reduce the amount of radiation reaching Earth's surface. Aerosols come from a variety of sources. In

desert regions, they originate from winds blowing dust airborne. Over the oceans, aerosols come from sea spray. They also result from the burning of forests and grasslands (referred to as biomass burning). Occasionally, large quantities of particulates are injected into the upper atmosphere through the eruption of volcanoes, as was the case when Mt. Pinatubo erupted in 1991.

A major source of aerosols resulting from human activities is sulfates and soot from the burning of fossil fuels. Sulfate particles are of particular importance. They are formed from sulfur dioxide, a gas produced in large quantities from power stations that burn coal (see Chapter 26). These particles remain in the atmosphere for a very short period (on average, five days), and so their distribution is concentrated in the regions near their source (Figure 32.11a) (Kiehl and Briegleb 1995). In regions of the Northern Hemisphere, their concentration is

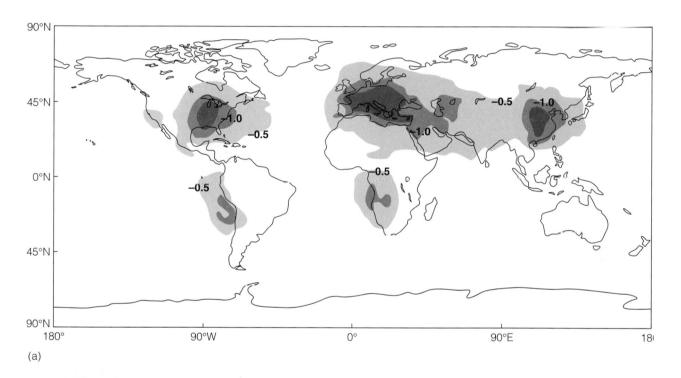

(a)

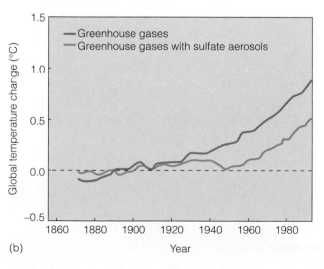

(b)

FIGURE 32.11 (a) Estimates of reduction in radiation (watts per square meter) resulting from anthropogenic sulfate aerosols in the atmosphere. The reduction is largest over regions close to sources of the emissions. (Kiehl and Briegleb 1995.) (b) Predicted changes in mean global temperature for the UKMO general circulation model, with and without the inclusion of sulfate aerosols in the simulation. (Mitchell et al. 1995.)

significant and functions to offset the effects of greenhouse gases, reducing estimates of global warming (Figure 32.11b) (Mitchell et al. 1995a, b).

As the models of global climate improve, there no doubt will be further changes in the patterns and the severity of changes they predict. However, the physics of greenhouse gases and the consistent qualitative predictions of the GCMs lead scientists to believe that rising concentrations of atmospheric CO_2 will have a significant impact on global climate.

Ecosystems and Global Climate Change

Climate influences almost every aspect of the ecosystem: the physiological and behavioral response of organisms (Chapters 5–9); the birth, death, and growth rates of populations (Chapters 10–12); the relative competitive abilities of species

(Chapter 14); community structure (Chapter 20); productivity (Chapter 24); and cycling of nutrients (Chapters 25 and 26). As they change Earth's climate, greenhouse gases will have a major influence on these processes and patterns.

Species Distribution Changes in temperature and water availability will have a direct effect on the distribution and abundance of individual species. For example, the relative abundance of three widely distributed European tree species is plotted as a function of mean annual temperature and rainfall in Figure 32.12 (Miko et al. 1996). These differing environmental responses, the species' realized niches, determine their distribution and abundance over the European landscape. The distribution and abundance of these three important tree species will change as regional patterns of temperature and precipitation change.

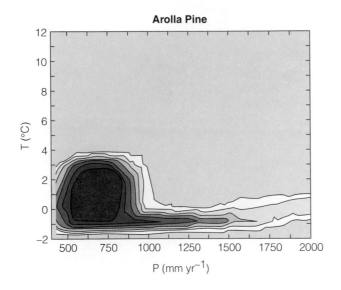

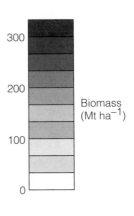

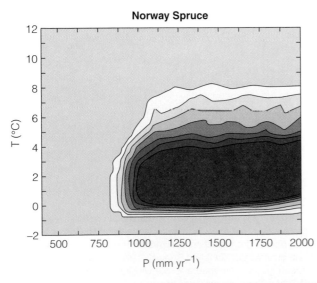

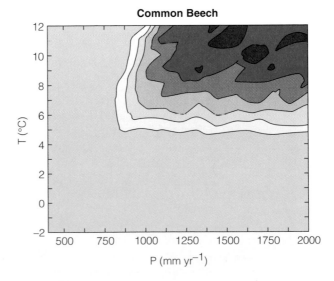

FIGURE 32.12 Abundance (biomass tons/ha) of three common European tree species as a function of mean annual temperature and precipitation. (Adapted from Miko et al. 1996.)

The potential impact of regional climate change on plant species distribution can be more clearly seen from the work of Prasad and Iverson (in press, Iverson et al. 1999). Using data from the Inventory and Analysis program of the U.S. Forest Service, Prasad and Iverson developed statistical models to predict the distribution of 80 different tree species inhabiting the eastern United States. Individual tree species distributions are predicted as a function of variables describing climate, soils, and topography for any location. This framework allows the investigators to predict shifts in the distribution of these tree species based on changes in temperature and precipitation for the region from a variety of GCM predictions under doubled CO_2. The predicted distributions for three major tree species in the eastern United States under current and doubled CO_2 climate using the Princeton GFDL GCM (see Figure 32.10) are presented in Figure 32.13. The predicted changes in temperature and precipitation will have a dramatic impact on the distribution and abundance of tree species that dominate the forest ecosystems of the eastern United States.

The distribution and abundance of animals are also directly related to features of the climate. For example, the northern limit of the winter range of the Eastern phoebe is associated with average minimum January temperatures of −4°C (Root 1988). The phoebe is not found in areas where temperatures drop below this value. Two lines, or isotherms, defining the region of eastern North America where average minimum January temperatures of −4°C occur are plotted in Figure 32.14 (p. 712). Minimum temperatures drop below −4°C in areas to the north and west of the lines, whereas temperatures are above −4°C to the south and east. The two isotherms show the current −4°C average minimum January temperature isotherm and the −4°C isotherm predicted by the GFDL general circulation model for a doubled atmospheric concentration of CO_2. A change in the isotherm would be expected to result in a northern expansion of the Eastern phoebe's winter range.

Community Dynamics Changes in climate will have a direct influence on the growth and reproductive rates of species. These changes in turn will influence their relative competitive abilities, altering patterns of zonation and succession (see Chapters 21 and 22). Given the difficulty of experimentally changing climatic conditions in the field, few studies have examined the effects of changing climate on community dynamics. However, one such experiment was conducted in a meadow community in the Rocky Mountains of Colorado (Harte and Shaw 1995). Using electric heaters suspended 2.6 m above five experimental plots (Figure 32.15, p. 712), Harte and Shaw were able to raise soil temperature and influence soil moisture and the timing of snowmelt. In heated plots, the density of shrubs increased at the expense of grass and forb species. Results suggest that the increased warming expected under an atmosphere with a doubled concentration of CO_2 would shift the dominant vegetation of the widespread mountain meadow habitat. Shrubs would compete better in the altered environment. Such shifts have a major impact not only on plant communities, but on associated animal species as well.

In a similar approach, the International Tundra Experiment (ITEX) was established in late 1990 as a coordinated group of field experiments aimed at understanding the potential impact of warming at high latitudes on tundra ecosystems (Henry and Molau 1997). Investigators from ten countries are applying a range of standard field techniques including passive warming of tundra vegetation using open-top chambers (see Figure 32.8a) and manipulating snow depth to alter growing season length. Studies will examine species-, community-, and ecosystem-level responses to warming in the Arctic region.

Ecosystem Processes The ecosystem processes of net primary productivity, decomposition, and nutrient cycling are directly influenced by climate (see Chapters 24 and 25). Decomposition proceeds faster under warmer, wetter conditions. An ongoing experiment at Harvard Forest in Massachusetts is examining the impact of elevated soil temperatures on rates of decomposition and nutrient cycling in a forested ecosystem. Buried heating cables raise the soil temperature by 5°C. Initial results show a 60 percent increase in rates of soil respiration (CO_2 emissions), a direct result of increased microbial and root respiration (Peterjohn et al. 1993), and a 36 percent decrease in the carbon concentration of the O (organic layer) soil horizon (Melillo et al. 1996). The results are consistent with patterns of soil respiration observed in other forests in warmer regions around the world. They indicate that greenhouse warming will increase rates of decomposition and microbial respiration, leading to a significant rise in emissions of CO_2 from soils to the atmosphere.

Given the difficulty in establishing field studies that can directly examine the impact of warming on plant communities, most of our understanding regarding the potential impacts of shifting climate patterns of ecosystem productivity comes from the use of computer models that simulate basic ecosystem processes (Melillo et al. 1996). One such model, TEM (Terrestrial Ecosystem Model), developed by scientists at Woods Hole Marine Biological Laboratory (McGuire et al. 1995), was used to examine the potential impacts of rising atmospheric CO_2 and associated changes in global climate patterns on net primary productivity of the terrestrial biosphere. The analysis suggests an increase in the net primary productivity of the Earth under elevated CO_2, therefore acting to draw down the rising atmospheric CO_2 concentrations (Melillo et al. 1995).

An analysis of the combined influence of elevated CO_2 and climate change on the net primary productivity of terrestrial ecosystems in the United States was reported by the VEMAP (Vegetation/Ecosystem Modeling Analysis Project) working group (VEMAP 1995). The group used three different ecosystem models to examine the potential impacts of elevated CO_2 and climate change as predicted by three different general circulation models (GCMs). All three ecosystem models predicted an increase in net primary productivity for the United States under the climate change scenarios; however, the magnitude of the increase varied from 2 to 35 percent, depending on the ecosystem model and climate change scenario used. Results of the study point to the problems associated with

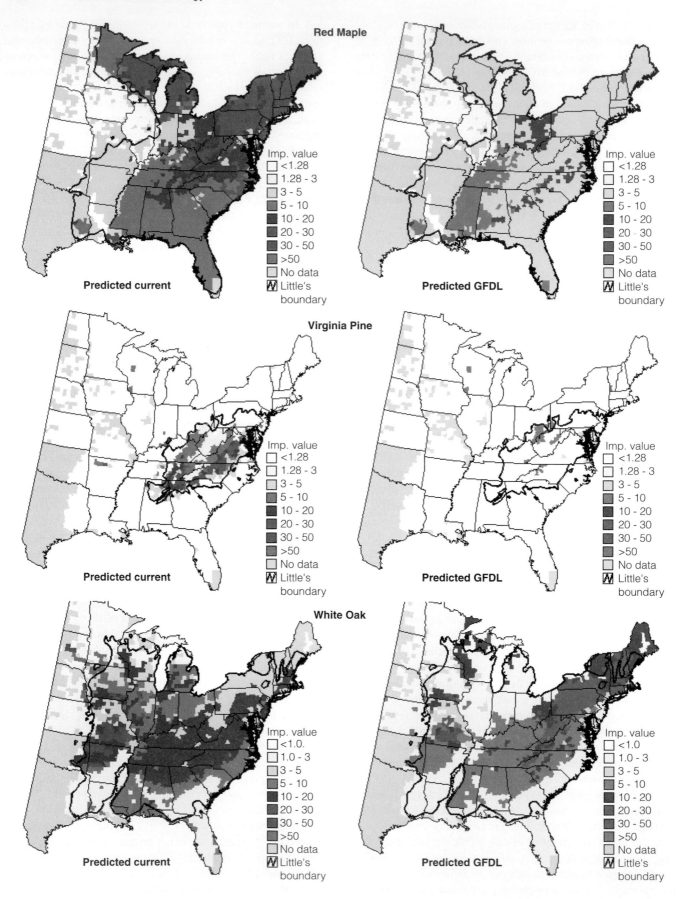

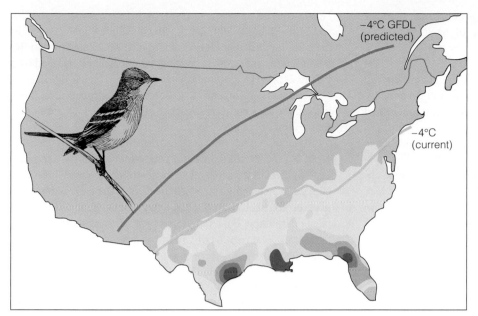

FIGURE 32.14 Map showing the existing distribution of the Eastern phoebe along the current −4°C average minimum January temperature isotherm, as well as the predicted isotherm under a changed climate. The predicted isotherm is based on the changes in temperature due to a doubling of atmospheric CO_2 concentration as predicted by the Geophysical Fluid Dynamics Laboratory general circulation model (shown in Figure 32.10b). (Adapted from Root 1988.)

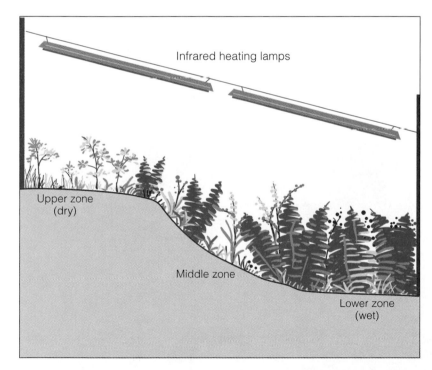

FIGURE 32.15 Experimental design used to elevate temperatures in a Colorado meadow community to examine possible effects of a global climate change. Changes in vegetation composition were monitored through time to track how the community responded to elevated temperatures and drier conditions. See text for discussion of results. (Adapted from Harte and Shaw 1995.)

FIGURE 32.13 The distributions of (a) red maple, (b) Virginia pine and (c) white oak under both current climate and a doubled CO_2 climate as predicted by the GFDL general circulation model. Species abundances expressed in terms of importance value (sum of relative density, basal area and frequency). Little's boundary refers to the observed distribution of the species as reported by Little (1977). See text for description of model used for predicting species distributions based on climate and site factors. (From Iverson et al 1999.)

uncertainties in both the predicted changes in climate and our current understanding of the response of basic plant and ecosystem processes to elevated CO_2.

Distribution of Ecosystems   In addition to influencing ecosystem processes, such as decomposition, and net primary productivity of existing ecosystems, changes in regional and global climate patterns will seriously affect the distribution and abundance of ecosystems on Earth's surface. Ecologists have learned a great deal about the responses of terrestrial ecosystems to changing climate conditions from the study of past climate change. Pollen samples from sediment cores taken in lake beds has allowed paleobotanists to reconstruct the vegetation of many regions during the last 20,000 years. The work of Margaret Davis (1983) in reconstructing the distribution of tree species in eastern North America since the last glacial maximum (see Figure 21.27) is a good example. Tree genera migrated northward at different rates following the retreat of the glaciers. The migration rates depended on how well a species' physiology, dispersal ability, and competitive interactions with other tree species let it respond to changes in climate. Such studies show us that the existing forest ecosystems in eastern

North America are a recent result of different responses of tree species to changing climate. As Earth's climate has changed in the past, the distribution and abundance of organisms and the communities and ecosystems they compose have changed (see Figure 21.26) (Delcourt and Delcourt 1963, Gates 1993).

It is virtually impossible to develop experiments in the field to examine the long-term response of terrestrial ecosystems to a future climate change. Hence, scientists must base predictions on computer models of ecosystem distribution. Perhaps the simplest, but most telling, of these ecosystem models are the biogeographical models that relate the distribution of ecosystems to climate (see Chapter 27). From the days of the early naturalists, plant ecologists have recognized the link between climate and plant distribution. For example, tropical rain forests are found in the wet tropical regions of Central and South America, Africa, Asia, and Australia. According to the biogeographical model developed by L. R. Holdridge (1947, 1967), within these regions the distribution of tropical rain forest is limited to areas where mean annual temperatures are at or above 24°C and annual precipitation is above 2000 mm. The regions of the tropics that meet these climate restrictions are shown in the map in Figure 32.16a. Under the changed temperature and rainfall patterns predicted

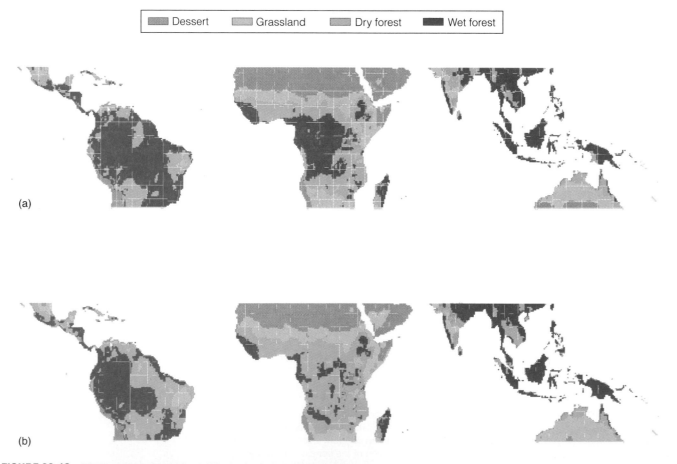

FIGURE 32.16   Maps of the areas in the tropical zone that could possibly support rain (wet) forest ecosystems as predicted by the Holdridge biogeographical model of ecosystem distribution. Map (a) is the area of tropical rain forest under current climatic conditions, and (b) is the predicted area under changed climatic conditions predicted by the UKMO general circulation model for a doubled atmospheric CO_2 concentration. (Smith et al. 1992.)

by the United Kingdom Meteorological Office (UKMO) GCM for a doubled atmospheric CO_2 concentration, this distribution changes dramatically (Figure 32.16b) (Smith et al. 1992). The total area that can support tropical wet forest under this scenario of greenhouse warming shrinks by 25 percent. This decline is a direct result of drying due to higher temperatures. In some areas, the drying is a result of increased temperatures accompanied by decreased precipitation. In other areas, precipitation increases, but the increase is not sufficient to meet the increased demand for water (evaporation and transpiration) resulting from the increased temperatures. Together with the demands of agriculture and forestry, this scenario would devastate both the tropical rain forest ecosystems and the diversity of life they support. Although tropical rain forests cover only 7 percent of the total land area, they are home to over 50 percent of all terrestrial plant and animal species (Wilson 1999, Primack 1998). Currently, deforestation in the tropics is the major single cause of species extinction, with annual rates of extinction in the thousands of species (Wilson 1999). The loss of tropical rain forest predicted by the UKMO climate model would result in a far larger scale extinction of rain forest flora and fauna.

The changes in global climate predicted by GCMs under doubling of atmospheric CO_2 concentrations will affect ecosystems other than tropical rain forest. Smith and colleagues (1992) examined the implications of four climate change scenarios derived from GCMs on the distribution and abundance of terrestrial ecosystems at a global scale. Although the results of the analyses varied for the different climate change scenarios, a number of general patterns emerged. All four climate change scenarios result in a significant (30–60 percent) decline in the global distribution of tundra ecosystems. This decline is a result of warming at the northern latitudes and a northward expansion of the boreal forest zone. The global distribution of desert ecosystems declines under all four scenarios, while the distribution of grassland increases from 20 to 50 percent depending on the climate change scenario. The global distribution of forests either increased or decreased depending on the scenario used. In all four scenarios, there was a poleward shift in forested zones (tropical, temperate, and boreal; see Chapters 27 and 29).

The results of this study and others like it are not predictions, but rather a means of interpreting the climate change scenarios being developed by atmospheric scientists. As the GCM models continue to develop, and our understanding of the impact of elevated greenhouse gases on the global climate system matures, estimates of their impact on the global climate system will no doubt change. The analyses are valuable, however, in showing the sensitivity of terrestrial ecosystems to the magnitude of climate change predicted by the GCM models for a doubled CO_2.

Changes in the global patterns of temperature would also affect the distribution of aquatic ecosystems. For instance, the global distribution of coral reefs is limited to the tropical waters where mean surface temperatures are at or above 20°C. Reef development is not possible where the mean minimum temperature is below 18°C. Optimal reef development occurs in waters where the mean annual temperatures are 23–25°C,

and some corals can tolerate temperatures up to 36–40°C. A warming of the world's oceans would alter the range of waters in which reef development is possible, allowing for reef formation further up the eastern coast of North America.

Ecologists are far from providing a complete analysis of the potential impacts of a global climate change. There is little question, however, that changes in patterns of temperature and precipitation of the magnitude predicted by climate models will have a significant influence on the distribution and functioning of both terrestrial and aquatic ecosystems.

Sea Level Rise and Coastal Environments

During the last glacial maximum, some 18,000 years ago, sea level was 100 meters lower than current levels. The highly productive, shallow coastal waters, such as the continental shelf of eastern North America, were above sea level and covered by terrestrial ecosystems. As the climate warmed and the glaciers melted, sea levels rose. Over the last century, sea level has risen at a rate of 1.8 ($\pm$ 0.1) mm per year (Figure 32.17) (Douglas 1991). This rise is a result of the general pattern of global warming over this period and the associated thermal expansion of ocean waters and melting of glaciers (Gornitz et al. 1982, Warrick and Oerlemans 1990). The 1995 report of the Intergovernmental Panel on Climate Change (IPCC) estimates that under the scenarios of global warming outlined in figure 32.19 (page 706), sea levels will rise from 0.15 to 1.0 m above current levels by the year 2100 (Houghton et al. 1996). A rise in sea level of this magnitude will have serious effects on coastal environments, for both natural ecosystems and human populations.

A large portion of the human population lives in coastal areas—13 of the world's 20 largest cities are located on the coast. Areas that are particularly vulnerable are delta regions, low-lying countries such as The Netherlands, Surinam, and Nigeria, and the smaller low-lying islands of the Pacific and other oceans. Bangladesh, an Asian country of about 120 million inhabitants, is located in the delta region of the Ganges, Brahmaputra, and Meghna Rivers (Figure 32.18). Approximately 25 percent of the country's population lives in areas less than 3 m above sea level, with about 7 percent of the country's habitable land and 6 million people residing in areas less than 1 m above sea level. Estimates of sea level rise in this region due to a combination of land subsidence (a result of land collapsing in response to removal of groundwater) and global warming are 1 m by the year 2050 and 2 m by 2100 (Hue et al. 1995, Nicholls and Leatherman 1995, Milliman et al. 1989). Although there is great uncertainty in these estimates, the impact on Bangladesh would be devastating.

Other coastal regions of Southeast Asia and Africa would be equally affected by the predicted rise in sea level. In Egypt, about 12 percent of the arable land, with a population over 7 million, would be affected by a rise in sea level of 1 m (Broadus 1993). In the coastal areas of eastern China, a sea level rise of just half a meter would inundate an area of approximately

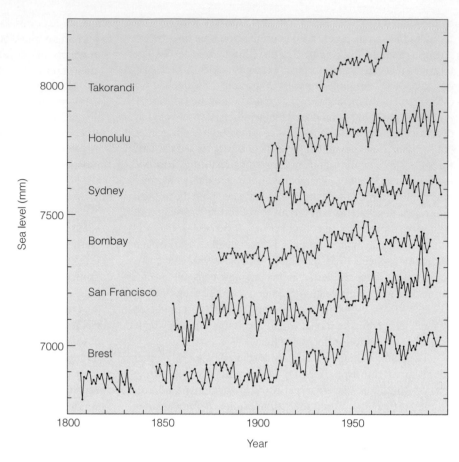

FIGURE 32.17 Long-term sea level records from six coastal regions of the world: Takorandi (Africa), Honolulu (Hawaii), Sydney (Australia), Bombay (Asia), San Francisco (North America), and Brest (Europe). (Adapted from Houghton et al. 1996.)

40,000 square kilometers where more than 30 million people currently live (Worldwide Fund for Nature 1992).

Particularly vulnerable to a sea level rise are small island nations. Over half a million people live in the archipelagos of small islands and coral atolls (Bijlsma 1996). Two examples are the Maldives located in the Indian Ocean and the Marshall Islands of the Pacific Ocean. These island chains lie almost entirely below 3 m elevation above sea level. A half meter or more rise in sea level would dramatically reduce their land area, and in addition it would have a devastating impact on groundwater (fresh water) supply.

A sea level rise will also have a major impact on coastal ecosystems. Among the possible sources of impacts are direct inundation of low-lying wetlands and dryland areas, erosion of shorelines through loss of sediments, increased salinity of estuaries and aquifers, rising coastal water tables, and increased flooding and storm surges. Estuarine and mangrove ecosystems (see Chapter 31) would be highly susceptible to a sea level rise of the magnitude predicted. The coastal salt marshes depend on the twice-daily tidal inundation of salt water mixing with the fresh water provided by streams and rivers. The patterns of water depth, temperature, salinity, and turbidity are critical to maintaining these ecosystems. The invasion of salt water further into the estuary as a result of a rise in sea level would be disastrous, and might also cause salinization of land adjacent to the estuary margins. Estuar-

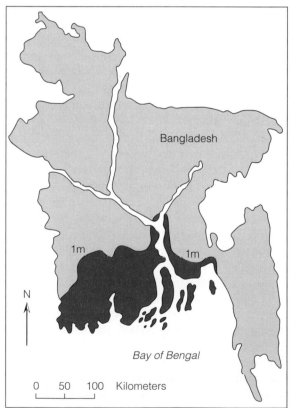

FIGURE 32.18 Land area in Bangladesh that would be submerged (dark green area of map) if sea level rose by 1 m. (From Nicholls and Leatherman 1995.)

ine and mangrove environments are critical to coastal fisheries. Over two-thirds of the fish caught for human consumption, as well as many birds and animals, depend on the coastal marshes and mangroves for part of their life cycle.

Climate Change and Agricultural Production

Despite technological advances in crop varieties and methods of irrigation, climate and weather remain the key factors determining agricultural production. Changes in the global climate patterns will exacerbate an already increasing problem of feeding the world population, which is predicted to double in size over the next half century.

The major cereal crops that feed the world's population, wheat, corn (maize), and rice, are domesticated plant species. Like native plant species, these crops exhibit environmental tolerances to temperature and moisture that control survival, growth, and reproduction. Changes in regional climate conditions will directly influence the suitability and productivity of crop species and, consequently, current patterns of agricultural production. However, these changes will be complex, with economic and social factors interacting to influence patterns of global food production and distribution.

In examining potential impacts of a greenhouse warming on agricultural production, both the increasing concentration of CO_2 and the changes in climate must be considered. Unlike the uncertainties observed in tree species (see Figure 32.7) and terrestrial ecosystems, the results of numerous studies suggest that most crop species (and varieties) will benefit from a rise in CO_2 concentration (Pearcy and Bjorkman 1983, Kimball 1983, Idso and Idso 1994, Pinker et al. 1996, Dijkstra et al. 1996, Baker et al. 1996). For example, in an experiment in Arizona, cotton and spring wheat were grown in field conditions under elevated CO_2 and irrigation. Cotton yield increased by 60 percent and wheat by over 10 percent compared with crops grown under identical field conditions and ambient concentrations of CO_2 (Pinker et al. 1996).

One of the simplest ways to evaluate the potential implications of a climate change on agriculture is to examine shifts in the geographic range of certain crop species as they relate directly to climate. For example, an average daily temperature increase of 1°C during the growing season would shift the "corn belt" (region of highest corn production) of the United States significantly to the north (Figure 32.19a) (Blasing and Solomon 1983). A similar analysis for the shift in suitable regions for irrigated rice production in northern Japan is shown in Figure 32.19b (Yoshino et al. 1988). In both examples, the shifts in agricultural zones imply significant changes in regional land use patterns, with associated economic and social costs. Although analyses of this type can provide an insight into changing patterns of regional agricultural production, to evaluate the actual impacts on global food production and markets requires a more detailed, interdisciplinary approach.

A detailed study of the potential effects of a climate change as predicted by the general circulation models (see Figures 32.9 and 32.10) has been carried out by the Environmental Change Unit at Oxford University (Rosenzweig et al. 1993, 1995a, b). This research group has conducted a collaborative study with agricultural scientists from more than 18 countries to examine the regional and global impacts of a climate change on world agricultural production. Various assumptions were made concerning the ability of farmers to adapt to changing environmental conditions through shifts in the species or varieties of crops grown, or changes in agricultural practices such as irrigation. The analysis also assumes a continuation of current economic growth rates, certain changes in current trade restrictions, and projected estimates of population growth.

One of the major findings of the study is that the negative effects of climate change are to some extent compensated for by increased productivity resulting from elevated atmospheric concentrations of CO_2 (Rosenzweig et al. 1995b). The net effect of a climate change as predicted by the general circulation models, including a doubling of atmospheric CO_2 concentrations, is to reduce the global production of cereal crops by up to 5 percent. An important point is that this reduction in agricultural production is not evenly distributed across the globe or even within a given region or country (Figure 32.20) (Adams et al. 1995a, b).

The predicted changes in agricultural production under a climate change would increase the current disparity in cereal crop production between developed and developing countries (Rosenzweig et al. 1993, 1995a, b). Results of the study tend to show an increase in production in the developed countries, particularly in the mid-latitudes (temperate regions). In contrast, production in the developing nations would, as a group, decline by as much as 10 percent, with an associated increase in the population at risk of hunger. In many of these regions, climatic variability and marginal climatic conditions for agriculture are worsened under the predicted patterns of global climate change.

Climate Change and Human Health

Climatic change will have a variety of direct and indirect effects on human health. Direct effects would include increased heat stress, asthma, and a variety of other cardiovascular and respiratory aliments. Indirect health effects are likely to include increased incidence of communicable diseases, escalating mortality and injury due to increased natural disasters (floods, hurricanes, etc.), and changes in diet and nutrition due to changed agricultural production.

A number of studies have examined the direct relationship between maximum summer temperatures and mortality rates (Figure 32.21) (Kalkstein and Tan 1995). Climate change is expected to change the frequency of very hot days. For example, if average July temperature in Chicago, Illinois, were to rise by 3°C, the probability that the heat index will exceed 35°C (95°F) during the month increases from one in 20 to one in 4 (Kalkstein and Green 1997). In the United States, it is typically warm humid conditions during the summer nights that lead to the highest mortality. The greatest death toll in the United States occurred during the summer of 1936, when 4700 excess deaths were recorded due to heat-related causes. In

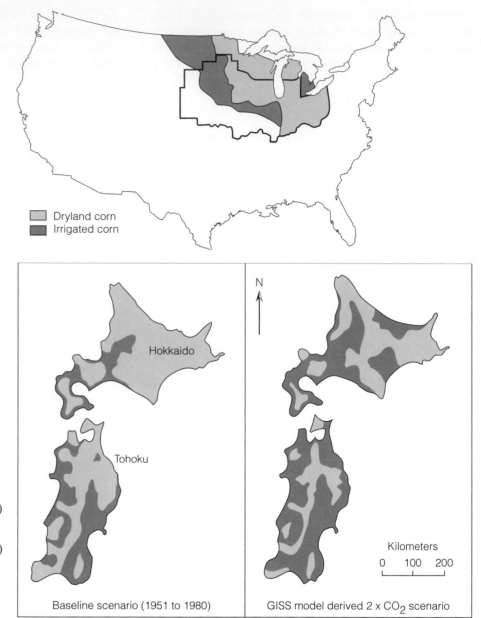

FIGURE 32.19 Regional shifts in areas suitable for crop production under a changed climate as predicted by the Goddard Institute for Space Studies (GISS) GCM: (a) shift in the region suitable for corn production in the United States. (Adapted from Blasing and Solomon 1983.) (b) shift in areas suitable for irrigated rice production in northern Japan. The areas in dark green are suitable for irrigated rice production. (Adapted from Yoshino et al. 1988.)

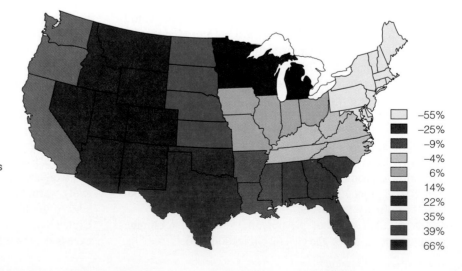

FIGURE 32.20 Changes in regional crop production by the year 2060 for the United States under a climate change as predicted by the Goddard Institute for Space Studies GCM (assuming an average 3°C increase in temperature, 7 percent increase in precipitation, and 530 ppm CO_2). (Adapted from Adams et al. 1995.)

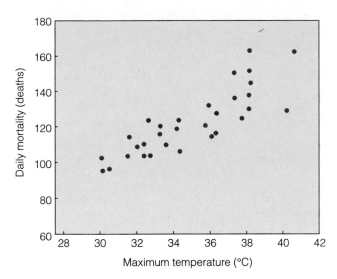

FIGURE 32.21 Relationship between maximum daily temperature and human mortality rate in Cairo, Egypt, during 1982. (Adapted from Kalksten and Tau 1995.)

recent decades, 1200 excess deaths occurred in Dallas, Texas, during the summer of 1980, and 566 in Chicago, Illinois, during July 1995. Analyses of climate change scenarios show a significant rise in heat-related mortality in all regions of the United States over the next several decades (Figure 32.22) (Kalkstein and Green 1997). During heat waves, cardiovascular and respiratory illnesses are the major causes of mortal-

ity. It is the elderly and children who are typically in the greatest danger during these periods.

In addition to direct heat-related mortality, the distribution and rates of transmission for a variety of infectious diseases will be influenced by changes in regional climate patterns. Disease often involves an agent, such as a virus, a bacterium, or a protozoan, and a host organism (a human). Some diseases are transmitted to humans by an intermediate organism, or vector. Insects are a primary vector of human disease. Although acting as carriers, the insects themselves are not affected by the disease agent. Insect-borne viruses (referred to as arboviruses for arthropod-borne viruses) cover a wide variety of diseases. The most common insects involved in the transmission of arboviruses are mosquitoes, ticks, and blood flukes (schistosomes). There are more than 100 known arboviruses that can produce disease in humans (Bryant 1997, Knight 1974). Of this number, about half have been isolated from mosquitoes. The insects that carry these disease agents are adapted to specific ecosystems for survival and reproduction, and they exhibit specific tolerances to features of the climate, such as temperature and humidity. Changes in the climate will affect their distribution and abundance, just as is true of the eastern phoebe (see Figure 32.14).

An example of such an insect-borne disease is malaria. Malaria is a recurring infection produced in humans by protozoan parasites transmitted by the bite of an infected female mosquito of the genus *Anopheles*. The optimal temperature for breeding by the *Anopheles* mosquito lies between

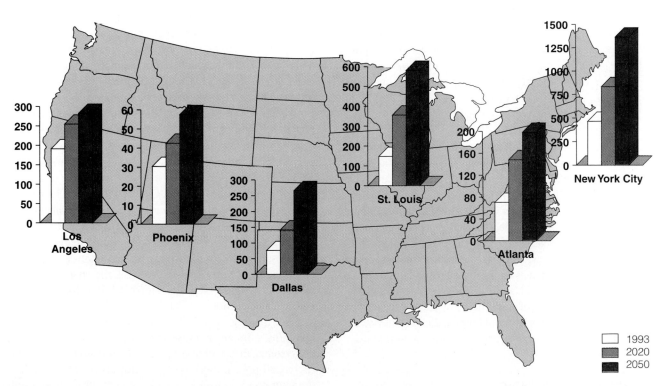

FIGURE 32.22 Average annual excess weather-related mortality for the years 1993, 2020, and 2050 in various cities of the United States. Future projections of weather-related mortality are based on changes in climate predicted by the Geophysical Fluid Dynamics Laboratory GCM. (Adapted from Kalkstein and Green 1997.)

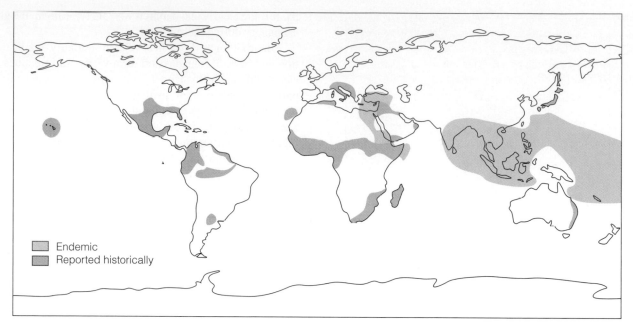

FIGURE 32.23   Global distribution of malaria. (From Bryant 1997.)

20°C and 30°C, with relative humidity over 60 percent. Mosquitoes die at temperatures above 35°C and relative humidity less than 25 percent (Knight 1974). At present, 40 percent of the world's population is at risk, and over 2 million people are killed each year by malaria (Bryant 1997, Knight 1974). The current distribution of malaria (Figure 32.23) will be extensively modified under a climate change. The expansion of the geographic range of the *Anopheles* mosquito into currently more temperate climates is expected to increase the proportion of the worlds population at risk of this infectious disease to over 60 percent by the latter part of the twenty-first century (Bryant 1997, McMichael et al. 1996).

Dengue and yellow fever are related viral diseases that also are transmitted by mosquitoes. In the case of these viruses, the vector is the mosquito *Aedes aegypti,* which is adapted to the urban environment. Colonization by this mosquito is limited to areas with an average daily temperature of 10°C or greater. The virus that causes yellow fever lives in mosquitoes only when temperatures exceed 24°C under high relative humidity (Knight 1974). Epidemics occur when mean annual temperatures exceed 20°C, making this a disease of the tropical forested regions. Yellow fever is currently prevalent across Africa and Latin America, but it has been detected as far north as the mid-latitude ports of Bristol, Philadelphia, and Halifax, where the mosquitoes have survived in the water tanks of ships that have traveled from tropical regions. A climate change would have a direct influence on the distribution of both the virus and its vector, the mosquito.

GLOBAL ECOLOGY

The increasing atmospheric concentrations of CO_2 and other greenhouse gases, and the potential changes in global climate

patterns that may result, present a new class of ecological problems. To understand the impacts of rising CO_2 emissions from fossil fuel burning and land clearing, we have to examine the carbon cycle on a global scale (see Figure 26.4), linking the atmosphere, hydrosphere, biosphere, and lithosphere. Although the discussion in the previous sections focused on the impacts of rising CO_2 concentrations and changes in climate on populations, communities, and ecosystems, the possible impacts are not unidirectional. Ecosystems also influence atmospheric CO_2 and regional climate patterns. For example, if climate changes as shown in Figure 32.9 the global distribution and abundance of tropical rain forest will decline dramatically. Tropical rain forests are the most productive terrestrial ecosystems on the planet. A significant decline in these ecosystems will reduce global primary productivity, the uptake of CO_2 from the atmosphere, and CO_2 storage as organic carbon in biomass (see Smith et al. 1992, Smith and Shugart 1993). In fact, as tropical rain forests shrink, atmospheric CO_2 will increase. The drying of these regions will kill trees, increase fires, and transfer carbon stored in the living biomass to the atmosphere as CO_2 in much the same way as forest clearing does in these regions. The rise in atmospheric CO_2 will increase the greenhouse effect, further exacerbating the problem. In this case, the changes in the terrestrial surface act as a positive feedback loop for rising atmospheric concentrations of CO_2.

On the other hand, if rising atmospheric CO_2 level and changing climate increase the productivity of the world's ecosystems, they will take up more CO_2 from the atmosphere. Increased productivity will function as negative feedback, drawing down atmospheric CO_2 concentrations (see Smith et al. 1992, Smith and Shugart 1993).

In addition to changing climate indirectly through influencing atmospheric concentrations of CO_2, changes in the distribution of rain forests can influence climate directly by altering regional patterns of precipitation. In some regions, such as extensive areas of tropical rain forest, a significant portion of the precipitation is water that has been transpired from the vegetation in the area. In effect, water is being recycled locally through the hydrologic cycle (see Figure 3.13). Removal of the forest (either through deforestation or shifting distribution of ecosystems as in Figure 32.16) reduces transpiration and increases runoff to rivers that transport the water from the area. Experiments using GCMs and regional climate models have examined the potential impacts of large-scale deforestation in the Amazon basin (Dickinson 1989). Findings suggest that the loss of forest cover would result in a significant reduction in annual precipitation, by reducing the internal cycling of water within the forest. This would effectively change the climate of the region, making it unlikely that rain forest would be able to reestablish.

Gordon Bonan (1992) presented another example of a direct influence of terrestrial ecosystems on regional climate. The largest degree of warming under elevated CO_2 is in the northern latitudes. The predicted warming would significantly reduce snow cover in this region and shift the distribution of boreal forests to the north of their current distribution. A major factor influencing the relative absorption and reflection of short-wave radiation (solar radiation) by Earth's surface is its albedo (see Chapter 2). Albedo is an index of the ability of a surface to reflect solar radiation back to space. Snow has a high albedo, or reflectance, while vegetation (with its darker color) has a low albedo. Both the reduced snow cover and the northern movement of boreal forest under a global warming would reduce the regional albedo, therefore increasing the amount of solar radiation absorbed by the Earth's surface. This increase in the absorption of radiation would further increase regional temperatures, functioning as a positive feedback loop.

These are not simple connections. To understand the interactions among the atmosphere, oceans, and terrestrial ecosystems requires that ecologists study Earth as a single, integrated system. It is only through the development of a global ecology that ecologists, working with oceanographers and atmospheric scientists, will come to understand the potential consequences of doubling the concentration of CO_2 in the atmosphere over the next century.

Summary

1. Direct observations beginning in 1958 reveal an exponential increase in the atmospheric concentration of CO_2. The rise is a direct result of fossil fuel combustion and the clearing of land for agriculture.

2. Of the CO_2 released from fossil fuel combustion and land clearing, only about 56 percent remains in the atmosphere. The remainder is taken up by the oceans and terrestrial ecosystems. Calculations of diffusion of CO_2 into surface waters provide an estimate of uptake by the oceans. Carbon uptake by terrestrial ecosystems is calculated as the inputs to the atmosphere less the increase in atmospheric concentration and the uptake by oceans.

3. The oceans have two layers, surface waters (0–200 m depth) and deep (>200 m) waters. Over 85 percent of the ocean volume is deep water. The thermocline prevents mixing of the deep and surface waters. Carbon dioxide diffuses from the atmosphere into the surface waters of the oceans. The rise in atmospheric concentrations is producing an increased uptake of CO_2 into the surface waters. Transfer of dissolved CO_2 from the surface waters into the deep waters of the ocean takes hundreds of years. This fact limits the short-term uptake of CO_2 by the oceans.

4. In general, plants respond to increased atmospheric CO_2 with higher rates of photosynthesis and partial closure of stomata. These responses increase water use efficiency. Responses to long-term exposure vary, including increased allocation of carbon to the production of roots, reduced allocation to the production of leaves, and reduction in stomatal density. The limited number of studies that have examined the response of whole ecosystems to elevated CO_2 show an increase in CO_2 uptake by the canopy and increases in net primary productivity. The observed increases in NPP vary with ecosystem. Ecosystems inhabiting colder environments show the least response, while those in seasonally dry environments exhibit the largest increase.

5. Carbon dioxide is a greenhouse gas. It traps long-wave radiation emitted from Earth's surface, warming the atmosphere. Rising atmospheric concentrations of CO_2 and other greenhouse gases could raise the global mean temperature by 1.5°C to 3.5°C by the year 2100. Warming will not be uniform over Earth. The greatest warming is predicted during the winter months and at northern latitudes. Increased variability in climate is predicted, including changes in precipitation and the frequency of storms. The input of sulfates and other aerosols from human sources acts to reduce the input of solar radiation to Earth's surface, thereby reducing warming.

6. The distribution and abundance of species will shift as temperature and precipitation change. Changes in climate will influence the competitive ability of species and thus change patterns of community zonation and succession. Ecosystem processes such as decomposition, nutrient cycling, and net primary production are sensitive to temperature and moisture, and changing climate will affect them.

7. Changes in climate also will shift the distribution and abundance of terrestrial and aquatic ecosystems. These changes in ecosystem distribution influence global patterns of plant and animal diversity.

8. Sea level is currently rising globally at an average rate of 1.8 mm per year. It is estimated that global warming will cause sea level to rise by 0.15 m to 1.0 m by 2100, as the polar ice caps melt and warmer ocean waters expand. A sea level rise of this magnitude will have major impacts on human populations living in coastal areas. In addition, rising sea level will affect coastal ecosystems such as beaches, estuaries, and mangroves.

9. Climate change will affect global agricultural production. Decreases in crop production from drier conditions will be partly offset by increased rates of photosynthesis under elevated atmospheric CO_2 levels; however, current models project a 5 percent decline in global production of cereal crops. This decline is not distributed evenly. Developed countries in the mid-latitudes will realize a slight increase, while production in the developing countries in the tropics will decline. The result will be increased hunger.

10. Climate change will have both direct and indirect impacts on human health. Mortality rates are expected to rise as a result of heat-related deaths associated with respiratory and cardiovas-

cular ailments. Indirect health effects include greater mortality and injury from increased climate-related natural disasters, as well as changes in diet and nutrition resulting from altered agricultural production. The distribution and transmission rates of a variety of insect-borne infectious diseases that are directly related to climate, such as malaria, will also be affected.

11. To understand the impacts of rising atmospheric concentrations of greenhouse gases and global climate change, we have to study the whole Earth as a single, complex system.

Review Questions

1. What are the major sources of greenhouse gases, especially CO_2?
2. Not all of the CO_2 released to the atmosphere remains there. What happens to the rest?
3. How does elevated CO_2 influence rates of photosynthesis and transpiration?
4. What limits the transfer of CO_2 from the surface waters of the ocean to the deep waters?
5. Why is CO_2 called a greenhouse gas?
6. How do greenhouse gases contribute to global warming?
7. How do forest burning and land clearing affect the global climate?

8. How might changes in climate (temperature and precipitation) influence the distribution of plant and animal species?
9. How might changes in climate influence the distribution and abundance of terrestrial ecosystems?
10. How is sea level currently changing?
11. How will global warming change sea levels?
12. How might rising sea levels influence human populations? Coastal environments?
13. How might changes in climate influence agricultural production? How will rising CO_2 levels influence crop production?
14. What are some direct and indirect influences of climate change on human health?

Cross References

Adaptation, 81; niche, 257; photosynthesis, 86; transpiration, 87; forest ecosystems, 602–612; climate control on NPP, 100; climate control on decomposition, 154; thermocline, 384, 513, 630; ocean circulation, 28; climate, 21–41; ecosystem distribution, 547–552; parasites, 307–319; sulfates, 518–520, 534.

Time Is on Their Side

When a person says that something is about as exciting as watching grass grow, we know that it lacks action and thrills. But for those with lawns to mow every weekend from spring to fall, it is also obvious that however invisible the motion, grass does grow. The seemingly static blade of grass is, in reality, incredibly dynamic: taking up and giving off CO_2 through photosynthesis and respiration, losing water through the stomata and replacing it with soil water taken up through the roots, producing simple sugars and converting them into more complex carbohydrates, transporting these molecules throughout the plant and converting them into energy and new tissues. The apparent lack of action is due to these processes occurring at rates and spatial scales that are beyond our ability to perceive directly.

We humans are not able to directly observe things that are much smaller or larger than we are. Likewise, we cannot directly observe processes that occur over periods greater than that of a human life span or that occur more quickly than our senses can discern. However, human ingenuity has allowed us to break free of some of these constraints. Under the lens of a microscope, an unseen world of microorganisms emerges from a drop of pond water, and from the window of the *Apollo 8* spaceship, humans for the first time viewed our planet against the backdrop of space.

Much of science involves the study of pattern and process outside the realm of normal human perception. Although tools such as microscopes and satellites now regularly allow us to observe phenomena that are either too small or large for unaided human observation, time remains a constraint. Two features of temporal dynamics pose particular problems. First, like watching grass grow, many ecological processes require long periods of observation to detect change. For such processes, long-term research functions in much the same manner as does time-lapse photography when it reveals the almost imperceptible movements that lead to the blooming of a flower. Second, many patterns and processes do not vary through time in a constant fashion, which means that a series of observations must be collected for any trend to emerge.

The study of the Lake Mendota ecosystem in Wisconsin by John Magnuson and Dale Robinson, both from the Center of Limnology of the University of Wisconsin, Madison, is a case in point. An important feature of lakes that influences many ecological processes is the duration of ice cover each winter. A single year's observation of the ice cover on Lake Mendota for the winter of 1997–1998 (Figure IX-A) provides no insight into the long-term behavior of the ecosystem. Only when we examine a series of annual measurements collected over an interval of 10 years, or 50 years, or to the length of the available record—142 years—can we begin to see the importance of data collection over a long period.

A 10-year record of observations reveals that the duration of ice cover in 1998 was 40 days or so shorter than any of the other nine years and far exceeded the typical range of annual variation. Also, we see that the duration of ice cover varies considerably between years. With a 50-year record, it is apparent that all El Niño years, including 1998, had a shorter-than-normal duration of ice cover. With a record of 142 years, a general warming trend is visible that was hidden with only 10- or 50-year records. This general pattern of warming has influenced a variety of processes, including the cycling of nutrients, and primary productivity of the lake ecosystem.

Duration of ice cover at Lake Mendota is but one of a countless number of examples of ecological processes that require extended observation to discern patterns and trends. Environmental processes that extend over decades or even centuries are common, such as the gradual changes associated with community succession, soil development, and populations of large vertebrates. Other processes are rare or episodic, including such disturbances as floods, hurricanes, wildfires, or volcanic eruptions. Long-term studies are essential to understand the response of ecosystems to such phenomena, as the occurrence is so low and the recovery of the ecosystem so slow.

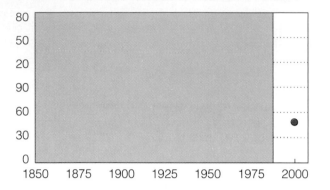

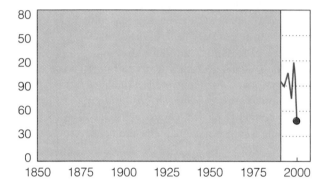

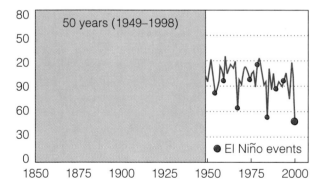

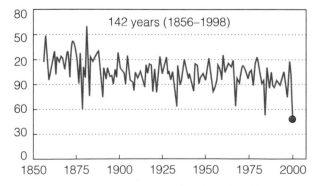

FIGURE IX-A   Variations in the duration of ice cover on Lake Mendota over various observation periods: (a) 1989, (b) 10-year record, (c) 50-year record, (d) complete 142-year record.

In the absence of long-term research, we can make serious misjudgments not only in our attempts to understand and predict change in the world around us, but also in our attempts to manage it. Although serious accidents resulting from an instant of human misjudgment—such as the nuclear disaster at Chernobyl—can have devastating environmental consequences, human-induced destruction of our planet is more likely to occur at a pace hardly noticeable during the course of one's relatively short life. For instance, rising atmospheric concentrations of CO_2 and other greenhouse gases have the potential to change the environment of the planet by increasing sea level, shifting the distribution of ecosystems, affecting global biodiversity, and directly influencing human health (see Chapter 26). These changes will occur over decades to centuries. Because the economic and social costs of reducing emissions of greenhouse gases are enormous, we must reduce the uncertainty of the environmental costs of such emissions. This requires an understanding of ecological and physical processes that can only be gained through long-term research.

Unfortunately, long-term studies are uncommon, despite repeated evidence of the misleading nature of short-term research. Funding lies at the root of the problem. Scientists must obtain results and publish papers quickly enough to satisfy both academic institutions and research funding agencies. Most funding is for one-to-two-year projects, with three-to-five-year projects being very rare, and anything longer rarer still. Unfortunately, the processes controlling promotion and tenure at academic institutions operate on time scales much shorter than most ecological processes.

To address the lack of organized research efforts over long periods, the National Science Foundation initiated a program in Long Term Ecological Research (LTER) in 1980. Researchers at a network of 21 sites (Figure IX-B) across the United States now design their studies on time scales of years, to decades, to a century. They also address a wide range of spatial scales—meters to kilometers to cross-continent comparisons among research sites.

Because of the diverse nature of the North American continent, sampling United States ecosystems is not possible. However, sites in the LTER system currently extend from Puerto Rico to northern Alaska and represent a broad diversity of environments and ecosystems, including temperate and tropical forests, prairie, desert, alpine and arctic tundra, agricultural fields, lakes, rivers, coastal wetlands, and an estuary. All sites are large enough to incorporate moderate to large landscape mosaics, and the majority of these sites include human-manipulated as well as natural ecosystems. In 1997, the LTER network expanded to include two new sites, Baltimore and Phoenix. The objective of adding urban sites is to examine the interactions between ecological and socioeconomic components of the urban environment, a growing area of concern because urbanization continues to increase.

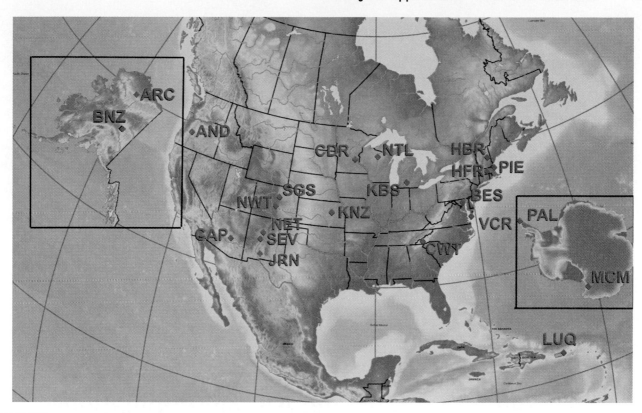

FIGURE IX-B   Location of 21 Long-Term Ecological Research sites.

The LTER sites are as varied in investigative design and objectives as they are in the range of ecosystems they encompass. To provide a focus for the individual projects, core research objectives were established early in the program. All LTER studies must at heart support these core objectives, which are to understand the patterns and controls of primary production, food webs, population abundance and distribution, organic matter accumulation, and biogeochemical cycling, as well as to answer questions related to disturbance frequency and effect. Research approaches include observation, experimentation, comparative analysis, retrospective study, and dynamic modeling.

Our current understanding of how key ecological processes, such as primary productivity and decomposition, vary among ecosystems comes largely from piecing together the results of disparate studies. Consequently, the development of comparable data sets and standardization in analytical methods and equipment were addressed at the beginning of the LTER

program. This permits comparative studies to occur within the LTER network and extend to non-LTER projects. Coordinated and cooperative research across diverse ecosystems takes many forms, including the installation of standard experimental designs across many sites. Such integrated intersite comparisons are essential to addressing questions concerning, for example, climate change and its potential influence on key ecological processes.

As technology advances, our world seems ever smaller as our expectations for timely information grows ever larger. In the era of the Internet, when the collective knowledge of our planet seemingly is available for the price of high-speed modem access, it is too easy to forget that information and understanding are not the same thing. Like wisdom of any sort, a true understanding of ecology often can be gained only by concentrated and intelligent effort over long periods. Thrills and action have their place in our lives, but perhaps there is something to be said after all for watching the grass grow.

APPENDIX A

Sampling Plant and Animal Populations

One of the major field problems in ecology is the determination of population distribution, size, and change in abundance. The problem involves sampling to estimate the true population value of interest, known as the **parameter,** which can be expressed as a number. It may be the total number of plants or animals, population density, average survival rate, proportion of males in a population, seed production per individual plant, and the like. From the sample, taking into account variability within the population, we can make some general inferences about the population as a whole. To be valid the samples must be random; that is, all combinations of sampling units must have an equal probability of being selected.

The object of sampling is to estimate some parameter or a function of some parameter. The mathematical expression that indicates how to calculate an estimation of a parameter from the sample data is the **estimator.** The estimator is usually marked with a "hat" over the parameter to indicate that it is an estimator, not the true parameter. The numerical value obtained by inserting the sample data into the estimator is the **estimate.**

The investigator hopes that the value of the parameter as estimated is **accurate**—that is, close to the true value and free of bias. **Bias** is a systematic distortion due to some flaw in the measurement or in the method of collecting the sample. If the investigator were to repeat the sampling experiment a number of times, each providing a different estimate of the parameter, the average of the estimates should equal the parameter being estimated, but usually it does not. In some statistical procedures bias cannot be avoided, but it is always important to recognize the source of the bias and to take it into account. A biased estimate can never be accurate, although it may be *precise. Precision* is the repeatability of a result, the clustering of sample values about their own mean (Figure A.1). It is measured by the sampling variance and its square root, the **standard error of estimation.**

What the experimenter seeks is a robust estimator. A **robust** estimator has a small bias relative to its standard error, regardless of its ability to detect the true population. An estimator is robust if it is affected very little by the failure of one of the assumptions. A good estimator is (1) robust to critical assumptions; (2) the most precise possible—it exhibits minimal variance; (3) distributed normally; and (4) unbiased (White et al. 1982).

Estimators may be parametric or nonparametric. Parametric estimators involve certain assumptions concerning the specific distribution of the population. Nonparametric estimators involve no specific assumptions about random variables under study. The experimenter hopes to approximate the true parameter. Most field work involving natural populations of plants and animals is nonparametric.

Before plunging into a study or project involving sampling, you should read Chapter 1 in Charles Krebs, 1999,

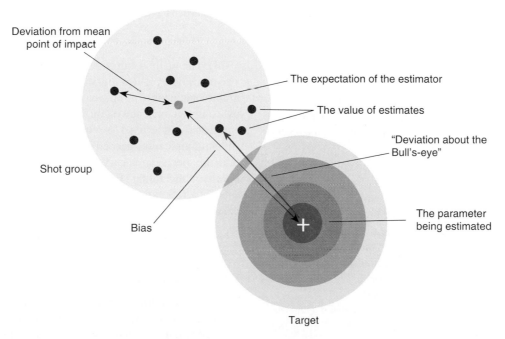

FIGURE A.1   Statistical concepts involved in estimating populations. The solid target represents the parameter being estimated, in this instance population size. The dotted target represents the sampling estimates. The distance of the samples or shots from the true target or bull's-eye is the "deviation." The value of estimates gives the expectation of the estimator. The distances of the shots from the mean point of impact are the deviations. The distance between the mean point of impact or expectation of the estimator and the true value, the bull's-eye (which we may never know) is the bias. A tight shot or clumping of estimates around a mean value indicates a high precision, but the estimates are not necessarily accurate. The shot group close to the parameter being estimated increases accuracy. The variance is the mean squared deviation about the mean point of impact. (From Giles 1969, *Wildlife Management Techniques.*)

Ecological Methods 2nd ed., Benjamin/Cummings, San Francisco, paying particular attention to his ten rules. This book is an excellent advanced text on ecological methodology that expands the basic methods presented in these appendixes.

A number of statistical computer programs are available for solving the computational procedures for the methods presented here. Students, however, should refrain from using them until they understand the basics behind the techniques.

SAMPLING PLANT POPULATIONS

Terrestrial Vegetation

Methods of analyzing the vegetation occupying a given site are numerous, and the literature discussing them, the underlying philosophies, and the statistical treatments extensive. Which method to choose for a specific study is a major decision. The basic references at the end of this appendix should help you decide. This appendix describes selected methods with some comments on their advantages and disadvantages.

Quadrats or Sample Plots

Strictly speaking, the **quadrat** is a square sample unit or plot. It may be a single sample unit or it may be divided into subplots. Quadrats vary in size, shape, number, and arrangement, depending upon the nature of the vegetation and the objectives of the study.

The size of the quadrat must be adapted to the characteristics of the community. The richer the flora, the larger or more numerous the quadrats must be. To sample forest trees, the 100 m^2 plot is a popular size, but it may be too large if trees are numerous or if many species are involved. Smaller 10 m^2 plots can be used to study shrubs and understory trees. For grass and herbaceous plants, 1 m^2 is the usual size.

Quadrats may be square, rectangular, or circular. Circular plots are the easiest to lay out, requiring only a center stake and string of desired length. Rectangular plots appear to furnish a more accurate sampling of vegetation composition.

The number of sample units to be employed always presents a problem. The number will vary with the characteristics of the community, objectives of the investigation, degree of precision, and so on. The final number more often than not is arbitrary. By using statistical methods the reliability of the sample and the number of samples needed for any desired degree of accuracy can be determined, once a normal distribution around a mean has been established.

A second approach to this problem is the use of the **species-area curve** (Figure A.2), obtained by plotting the number of species found in plots of different sizes (vertical axis) against the sample size area (horizontal axis). The curve rises sharply at first because the number of new species found is large. As the sample plot size or number is increased, the quantity of new species added declines to a point of diminishing returns, where there is little to be gained by continuing the sampling. This curve can be plotted on an arithmetic or a logarithmic base (see Mueller-Dumbois and Ellenberg

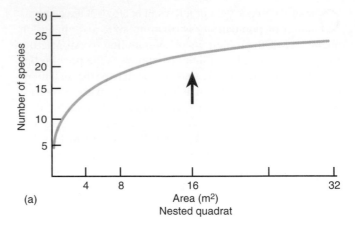

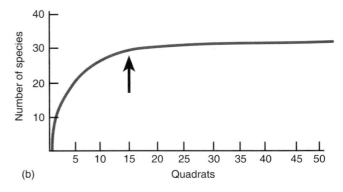

FIGURE A.2 Species-area curves: (a) for minimal area of quadrat; (b) for minimal number of quadrats. Arrows indicate minimums.

1974). The method can be employed to determine the largest size of a single plot (minimal area) needed to survey the community adequately. In this case the sampling should be done by using a geometric system of nested plots (Figure A.3) The curve also can be used to determine the minimum number of small multiple plots needed for a satisfactory sample. In addition, the species-area curve can be used to compare one community with another (Figure A.4).

Quadrats fall into four types, according to the type of data recorded.

1. List quadrat. Organisms found are listed by name. A series of list quadrats gives a floristic analysis of the community and allows an assignment of a frequency index, but nothing else.

2. Count quadrat. Numbers as well as names of species encountered are recorded. Quadrats in browse studies fall into this category, which is widely used in forest survey work. In forest studies additional information, such as height, volume, and basal area, is also taken.

3. Cover quadrat. Actual or relative coverage is recorded, usually as a percentage of the area of the ground surface covered or shaded by vegetation.

4. Chart quadrat. A quadrat is mapped to scale to show the location of individual plants. This tedious job provides

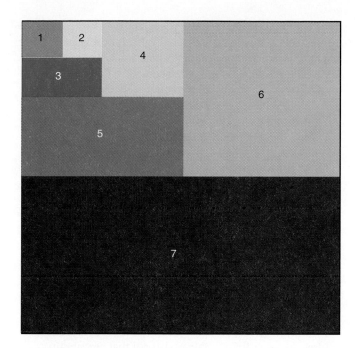

FIGURE A.3 An example of nested quadrats.

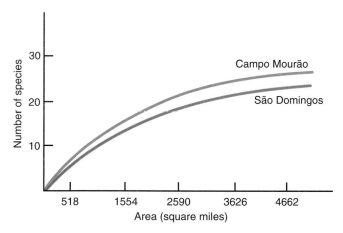

FIGURE A.4 Species-area curves compare two stands of *Araucaria* forest on different sites near Campo Mourão, Parana. (From Cain and Castro 1959.)

an overall view especially useful in long-range studies of herbaceous vegetation, mosses, and lichens.

For statistically reliable estimates, the location of plots must be randomized. This task is rather easy. Numbered grid lines are drawn over an area photo or map of the study area. The numbers of the vertical and horizontal grid lines are written on small squares of paper. To draw random numbers for the two lines, use a table of random numbers, available in statistical tables and statistics texts.

The quadrat method has advantages and disadvantages. It is a popular method, easily employed. If the individual organisms are randomly distributed, then the accuracy of the sample and the estimate of the density depend upon the size

of the sample. However, individuals seldom are randomly dispersed, so the accuracy of quadrat sampling may be low, unless a great number of plots is involved. The quadrat method is tedious and time consuming.

The Belt Transect

A variation of the quadrat method is the belt transect. A **transect** is a cross section of an area used as a sample for recording, mapping, or studying vegetation. Because of its continuity through an area, the transect can be used to relate changes in vegetation along the transect to changes in the environment. As a sample unit the measurements within a transect can be pooled, so that each transect is treated as a single observation. The belts also can be divided into intervals and each interval treated as a plot.

There are three steps in this method:

1. Determine the total area of the site to be sampled; then divide by 5 or 10 to obtain the total sample area.

2. Lay out a series of belt transects of a predetermined width and length, sufficient to embrace the area to be sampled. Then divide the belts into equal-sized segments. These units are sometimes called quadrats or plots, but they differ from true quadrats in that each represents an observational unit rather than a sampling unit.

3. Measure the vegetation in each unit for some attribute, such as abundance, sociability, frequency, or stem count.

A variation of the segmented-belt transect consists of taking observations only on alternate segments. The precision seems to be affected very little (Oosting 1956). For example, ten segments alternately spaced on a 6-m belt are nearly twice as efficient statistically as ten on a 3-m belt.

The belt transect is well adapted to estimate abundance, frequency, and distribution. For estimating the frequency index, it has the disadvantage that frequency by classes is related to the size of the plot. To compare one area with another, the segment size used in sampling must be the same for both areas.

There are several ways to analyze data from quadrats and belt transects. If we have simply recorded the presence of species, it limits analysis to frequency and relative frequency.

If we have recorded the number of individuals of the various species found in the quadrat, these data give a density figure. Because of the variations in growth forms among the various species, numbers mean little. Counts are most useful in certain situations, such as counting the number of stems of shrubby plants available for deer browse or counting the amount of forest reproduction. The samples are broken down into classes, such as 1 m high, 2 to 12 m, or 1 to 3 cm, 4 to 9 cm, and so on.

A third method is that of Braun-Blanquet (1951). It involves a total estimate based on abundance and cover. If the number of individuals in a plant community is estimated but not counted, the data are referred to as **abundance.** Abundance implies a number of individuals, but number does not necessarily reflect dominance or cover. **Cover** is the result of both numbers and massiveness. Although abundance and coverage are separate and distinct, they can be combined in a community description as the total estimate. For many field studies this method works well, but it is subjective and the data are difficult to handle statistically (see Mueller-Dombois and Ellenberg 1974). However, this method does provide a useful general picture of the plant community and a mechanism for classifying vegetation. The scales are given in Table A.1. Along with total estimates an estimate of sociability of each species should be given (as in Table A.2)—whether the plant grows singly, in clumps, mats, and so on.

TABLE A.1 Total Estimate Scale (abundance plus coverage)

+	Individuals of a species sparsely present in the stand; coverage very small
1	Individuals plentiful, but coverage small
2	Individuals very numerous if small; if large, covering at least 5% of area
3	Individuals few or many, collectively covering 6–25% of the area
4	Individuals few or many, collectively covering 26–50% of the area
5	Plants cover 51–75% of the area
6	Plants cover 76–100% of the area

TABLE A.2 Sociability Classes of Braun-Blanquet

Class 1	Shoots growing singly
Class 2	Scattered groups or tufts of plants
Class 3	Small, scattered patches or cushions
Class 4	Large patches or broken mats
Class 5	Very large mats of stands or nearly pure populations that almost blanket the area

The total estimate and the sociability estimate can be expressed together to give a paired value for each species. For example, if a plant species has the value 4.3, the first figure is the total estimate, the second the sociability. Once a number of stands have been surveyed, the community characteristics can be combined in an association table. The plant species usually are listed on the basis of fidelity or presence, the characteristic species of the community often heading the list. A partial example in Table A.3 shows how such tables are constructed. For details see Mueller-Dombois and Ellenberg (1974).

Data so collected describe individual stands. By the use of another attribute, presence, we can compare stands of a community type or of related types. **Presence** is the degree of regularity with which a species recurs in different examples of a community type. It is commonly expressed as a percentage that can be assigned to one of a limited number of presence classes, given in Table A.4. Presence is determined by dividing the total number of stands in which the species is found by the total number of stands investigated. Species that have a high percentage of presence or that fall within presence class 5 often are regarded as characteristic of that community.

Line Intercept

The line intercept is one-dimensional. Most useful for sampling shrub stands and woody understory of the forest, the line intercept or line transect method consists of taking observations on a line or lines laid out randomly or systematically over the study area. The procedure is as follows:

1. Stretch a metric steel tape, steel chain, or a tape between two stakes 50 to 100 m apart.

2. Subdivide the line into predetermined intervals, such as 1 m.

3. Move along the line, and for each interval record the plant species found and the distance they cover along that portion of the line intercept. Consider only those plants touched by the line or lying under or over it. Treat each stratum of vegetation separately, if necessary. (For grasses, rosettes, and dicot herbs, measure the distance along the line at ground level. For shrubs and tall dicot herbs, measure the shadow or distance covered by a downward projection of the foliage above.)

4. Repeat. Usually 20 to 30 such lines are sufficient.

TABLE A.3 Partial Stand Composition, Cumberland Plateau, West Virginia

Herbaceous Species	PLOT NUMBER										Frequency (%)
	1	2	3	4	5	6	7	8	9	10	
Polystichum acrostichoides	2·2	+·1	1·2	1·2	2·2	2·2	2·2	2·2	2·1	1·1	100
Cimicifuga racemosa	3·2	2·2			2·2	3·2	2·2	1·2	2·2	2·2	80
Geranium maculatum		+·1	+·1	+·1	1·2	2·2	2·22	+·2	1·1	+·1	90
Disporum lanuginosum	3·2	3·3		3·3	1·1	2·2	1·1	+·1	+·2	2·2	90
Galium circaezans	+·2	+·2			1·2	+·1	2·2	+·1			60
Thalictrum dioicum		+·2			1·2	1·1	2·1			+·2	50
Sanicula canadensis	+·1	2·2	1·1		+·1						40

TABLE A.4 Presence Classes

Presence Class	Stands in Which Species Occur (%)
1	1–20
2	21–40
3	41–60
4	61–80
5	81–100

The data can be summarized as follows:

1. Number of intervals in which each species occurs along the line.
2. Frequency of occurrence for each species in relation to total intervals sampled.
3. Total linear distance covered by each species along the transect.
4. Total length of line covered by vegetation and total "open" length.
5. Total number of individuals, if they can be so recorded. Because of branching and size variations, it is difficult to count individual plants on a line transect.

This method is rapid, objective, and relatively accurate. The area may be determined directly from recorded observations. The lines can be randomly placed and replicated to obtain the desired precision. The method is well adapted for measuring changes in vegetation if the ends of the lines are well marked. Generally it is more accurate in mixed plant communities than quadrat sampling and is especially suited for measuring low vegetation.

On the debit side, the method is not well adapted for estimating frequency or abundance, because the probability of an individual's being sampled is proportional to its size. Nor is it suitable where vegetation types are intermingled and the boundaries indistinct.

From line-intercept data the following measurements may be calculated. (Calculation of relative density may not be possible if individual plants cannot be identified.)

$$\frac{\text{relative}}{\text{density}} = \frac{\text{total individuals species A}}{\text{total individuals all species}} \times 100$$

$$\frac{\text{dominance}}{\text{(cover)}} = \frac{\text{total intercept length, species A}}{\text{total transect length}} \times 100$$

$$\frac{\text{relative}}{\text{dominance}} = \frac{\text{total intercept length, species A}}{\text{total intercept length, all species}} \times 100$$

$$\text{frequency} = \frac{\text{intervals in which species occurs}}{\text{total number of transect intervals}} \times 100$$

$$\frac{\text{relative}}{\text{frequency}} = \frac{\text{frequency value, species A}}{\text{total frequency value, all species}} \times 100$$

Point-Frequency Intercept

The point-frequency intercept is useful for grassland and herbaceous vegetation. The method involves the use of a point-frequency frame to sample basal and canopy cover of grassland vegetation.

The point-frequency frame, 0.5 m high and 1 m long, is made of wood or aluminum (Figure A.5). Guide holes spaced 5 cm apart are bored perpendicularly through the horizontally fixed laths. Ten sharp-pointed wire pins or steel rods of the same length as the legs are slid through the holes.

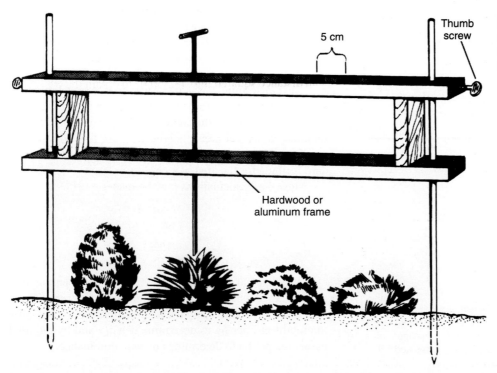

FIGURE A.5 A point-frequency frame sampler. The inside length of the frame is 1 m. The distance between the pins is 5 cm.

Thumb screw

5 cm

Hardwood or aluminum frame

The linear frame is mounted vertically over the herbaceous vegetation to be measured. The pins are lowered vertically one after the other, and the first cover and basal interceptions are recorded by species. This procedure gives a measure of both crown or shoot and basal cover. After reading all points in the frame, move the frame to a new location.

The frame may be set at random points throughout the grassland. The most efficient way is to locate a number of random points in the field. From them establish some present number of transect lines radiating from the center point along randomly selected compass bearings (Figure A.6). The number of frames required depends upon the homogencity of the vegetation. In relatively homogeneous vegetation, 200 points (20 frames) may be sufficient, but usually many more are required.

Although slow, the point-frequency method provides a highly accurate measure of foliar cover of grassland vegetation. However, it will often miss scattered clumps of herbaceous species in the sampling area. The method is most useful in studies of changes in the condition of grassland vegetation over time.

Analysis involves two sampling units—the ten-point frame or "plot" and the ten points within the frame. Both the number of frames and the number of points are utilized.

$$\frac{\text{dominance}}{\text{(coverage)}} = \frac{\text{number of points with species A}}{\text{total number of points}} \times 100$$

$$\frac{\text{relative}}{\text{dominance}} = \frac{\text{number of points with species A}}{\substack{\text{total points all species} \\ \text{(excluding empty points)}}} \times 100$$

$$\text{frequency} = \frac{\text{number of frames with species A}}{\text{total number of frames}} \times 100$$

$$\frac{\text{relative}}{\text{frequency}} = \frac{\text{number of frames with species A}}{\text{number of frames with any species}} \times 100$$

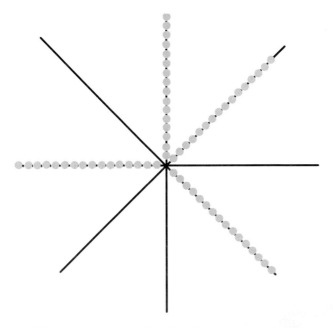

FIGURE A.6 A sampling scheme for use with the point-frequency-frame sampler. From a central stake or point, use random numbers to determine directional and distance coordinates. A minimum of four .lines and 16 frames per line for a total of 64 frames and 640 points should be sampled from each point.

Several variations of the variable plot or "plotless" method have been developed for ecological work. These methods arose from the variable radius method of forest sampling developed in Germany by Bitterlich. He used it to determine timber volume without establishing plot boundaries. The method was introduced into the United States by Grosenbaugh (1952, 1958).

One of the most useful of the plotless methods is the point-quarter method (see Cottam and Curtis 1956, Greig-Smith 1983). It is most useful in sampling communities in which individual plants are widely spaced or in which the dominant plants are large shrubs or trees.

The procedure is as follows:

1. Locate a series of random points within the stand to be sampled, or pick random points along a line transect passing through the stand.

2. At each station mark a point in the ground.

3. Divide the working area into four quarters or quadrants by visualizing a grid line, predetermined by compass bearing, and a line crossing it at right angles, both passing through the point (Figure A.7).

4. Select the tree (or plant) in each quarter that is closest to the point. Record its distance from the point, diameter at breast height, and species. The tally sheet will thus contain data for four trees at each point, one from each quarter.

5. Tally at least 50 such points.

The computations entail several steps. First, add all distances in the samples and divide the total distance by the number of distance to obtain a mean distance of point to plant.

$$\text{mean distance} = \frac{\text{total distance}}{\text{number of distances}}$$

Square the mean distance to obtain the mean area covered on the ground per plant. To obtain the total density of trees, divide the mean area per plant into the unit area on which density is expressed. If the area is in feet, then divide the mean distance squared into 43,560 ft^2 to obtain the total density of trees per acre.

Next, determine basal area for each tree (see Table A.5) from diameter measurements. (Basal area is the area of a plane passed through the stem of a tree at right angles to its longitudinal axis at breast height. Because the cross section approximates a circle, its area can be computed from the standard formula for the area of a circle.)

Now these calculations may be made:

$$N : M :: n : R$$

or

$$N = \frac{nM}{R}$$

Absolute values for the number of trees per unit area of any species and the basal area per unit area of any species are determined by multiplying the relative figures for density by the total trees per ha to determine density and by the total basal area per ha to determine absolute dominance.

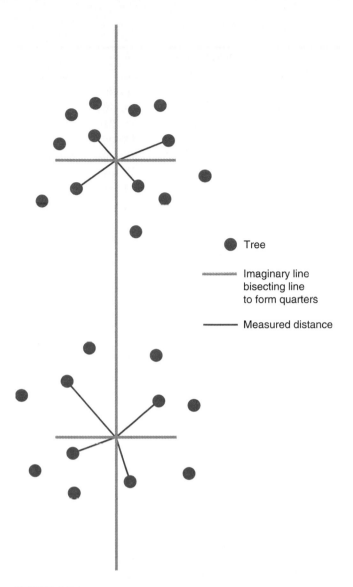

Tree

Imaginary line
bisecting line
to form quarters

Measured distance

FIGURE A.7   The point-quarter method of sampling forest stands.

The point-quarter method is simple, rapid, and effective. The underlying assumption is that individuals of all species together are randomly dispersed. Although this assumption may not be true, it does not seem to produce significant error, except where a deviation from overall randomness is obvious. Relative density and relative dominance are valid even if dispersion is not random. Error would appear in the calculation of absolute density and absolute dominance.

Importance Value

In regions where the plant communities are highly heterogeneous, the classification of communities on the basis of dominants or codominants becomes impractical. Therefore, Curtis and McIntosh (1951) came up with the index of "importance value" to develop a logical arrangement of the stands. This index is based on the fact that most species do not normally reach a high level of importance in the community, but those that do serve as guiding species.

Importance value (IV) is the sum of relative density, relative frequency, and relative dominance for each species. It may be expressed as a range from 0 to 3.00 or 300 percent, or it may be divided by 3 to give importance percentage, which will range from 0 to 1.0 or 100 percent. In situations involving only two values, such as relative frequency and relative density, importance value is expressed in terms of 0 to 2 or 200 percent.

Importance value provides an overall estimate of the influence or importance of a species in a community. Once importance values have been obtained for species within a stand, stands can be grouped by their leading dominants according to importance values, and the groups can be placed in a logical order based on the relationships of several predominant species. In Table A.6, for example, are four species that were the leading dominants in 80 of 95 forest stands in southern Wisconsin. Note that the dominants are arranged in order of decreasing importance value, from stands dominated by black oak to those dominated by sugar maple. Such an arrangement also shows increasing values for sugar maple. Trees intermediate in dominance can be handled in the same way.

Aquatic Vegetation

In aquatic communities algae are the dominant vegetation. Two kinds of growth are involved: the plankton suspended on the water and the periphyton growing attached to some substrate.

Phytoplankton

The phytoplankton can be obtained by drawing water samples from several depths. Cell counts of algae present in each sample, either normal or concentrated, can be made with a Sedgewich-Rafter counting chamber and a Whipple ocular. If necessary, the samples can be concentrated by centrifugation in a Forest plankton centrifuge. The centrifuged samples are then diluted to a suitable volume (100 to 200 ml) in a volumetric flask.

As cells are counted, a separate tally is kept for each species to permit an analysis of community structure at each station. The number of cells for single-celled forms and the number of colonies for colonial forms are recorded. The number of colonies is multiplied by an appropriate factor for each species to convert the colonies into cells. These factors are predetermined by averaging the cell counts from a large number of typical colonies from the area in question.

Another method of handling the phytoplankton is by filtration (see McNabb 1960, Clark and Sigler 1963). The organisms in the sample are first fixed by the addition of 4 parts 40 percent aqueous solution of formaldehyde to each 100 parts of the sample. The analysis is as follows:

1. Thoroughly agitate the sample. Withdraw a fraction with a pipette large enough to hold a sample that will provide an optimum quantity of suspended matter on the filter.

2. Place the sample in the tube of a filter apparatus designed to accommodate a 3 cm-diameter membrane filter. Draw the water through the filter with a vacuum pump.

3. Remove the filter and place it on a glass slide. Put two or three drops of immersion oil over the residue, and store the

TABLE A.5 Area of Circles

Diameter (in./cm)	Circumference (in./cm)	Area (ft²)	(m²)	Diameter (in./cm)	Circumference (in./cm)	Area (ft²)	(m²)
1	3.14	0.005	—	51	160.22	14.186	0.240
2	6.28	0.002	—	52	163.36	14.748	0.212
3	9.42	0.049	0.001	53	166.50	15.321	0.221
4	12.57	0.087	0.001	54	169.65	15.904	0.229
5	15.71	0.136	0.002	55	172.79	16.499	0.238
6	18.85	0.196	0.003	56	175.93	17.104	0.246
7	21.99	0.267	0.004	57	179.07	17.721	0.255
8	25.13	0.349	0.005	58	182.21	18.348	0.264
9	28.27	0.442	0.006	59	185.35	18.986	0.273
10	31.42	0.545	0.008	60	188.50	19.635	0.283
11	34.56	0.660	0.010	61	191.64	20.295	0.292
12	37.70	0.785	0.011	62	194.78	20.966	0.302
13	40.84	0.922	0.013	63	197.92	21.648	0.312
14	43.98	1.069	0.015	64	201.06	22.340	0.322
15	47.12	1.227	0.018	65	204.20	23.044	0.332
16	50.26	1.396	0.020	66	207.34	23.758	0.342
17	53.41	1.576	0.023	67	210.49	24.484	0.352
18	56.55	1.767	0.025	68	213.63	25.220	0.363
19	59.69	1.969	0.028	69	216.77	25.967	0.374
20	62.83	2.182	0.031	70	219.91	26.725	0.385
21	65.97	2.405	0.035	71	223.05	27.494	0.396
22	69.12	2.640	0.038	72	226.19	28.274	0.407
23	72.26	2.885	0.042	73	229.34	29.065	0.418
24	75.40	3.142	0.045	74	232.48	29.867	0.430
25	78.54	3.409	0.049	75	235.62	30.680	0.442
26	81.68	3.687	0.053	76	238.76	31.503	0.454
27	84.82	3.976	0.057	77	241.90	32.338	0.466
28	87.96	4.276	0.062	78	245.04	33.183	0.478
29	91.11	4.587	0.066	79	248.18	34.039	0.490
30	94.25	4.909	0.071	80	251.33	34.907	0.503
31	97.39	5.241	0.075	81	254.47	35.785	0.515
32	100.53	5.585	0.080	82	257.61	36.674	0.528
33	103.67	5.940	0.086	83	260.75	37.574	0.541
34	106.81	6.305	0.091	84	263.89	38.484	0.554
35	109.96	6.681	0.096	85	267.04	39.406	0.567
36	113.10	7.069	0.102	86	270.18	40.339	0.581
37	116.24	7.467	0.108	87	273.32	41.282	0.594
38	119.38	7.876	0.113	88	276.46	42.237	0.608
39	122.52	8.296	0.119	89	279.60	43.202	0.622
40	125.66	8.727	0.126	90	282.74	44.179	0.636
41	128.81	9.168	0.132	91	285.88	45.166	0.650
42	131.95	9.621	0.138	92	289.03	46.164	0.665
43	135.09	10.085	0.145	93	292.17	47.173	0.679
44	138.23	10.559	0.152	94	295.31	48.193	0.694
45	141.37	11.045	0.159	95	298.45	49.224	0.709
46	144.51	11.541	0.166	96	301.59	50.266	0.724
47	147.65	12.048	0.173	97	304.73	51.318	0.739
48	150.80	12.566	0.181	98	307.88	52.382	0.754
49	153.94	13.095	0.189	99	311.02	53.456	0.770
50	157.08	13.635	0.196	100	314.16	54.542	0.785

TABLE A.6 The Average Importance Value Index of Trees in Stands with Four Species as the Leading Dominants

Species	LEADING DOMINANT IN STAND				Ecological Sequence Number
	Quercus velutina	*Quercus alba*	*Quercus rura*	*Acer saccharum*	
Black oak (*Quercus velutina*)	165.1	39.6	13.6	0	2
Shagbark hickory (*Carya ovata*)	0.3	8.8	5.2	5.9	3.5
White oak (*Quercus alba*)	69.9	126.8	52.7	13.7	4
Black walnut (*Juglans nigra*)	1.5	1.2	2.2	1.9	5
Red oak (*Quercus rubra*)	3.6	39.2	152.3	37.2	6
American basswood (*Tilia americana*)	0.3	5.9	19.0	33.0	8
Sugar maple (*Acer saccharum*)	0	0.8	11.7	127.0	10

Source: Adapted from Curtis and McIntosh 1951.

slide in the dark to dry (about 24 hours). The oil replaces the water in the pores of the filter and makes it transparent.

4. Place a cover slip over the transparent filter.

5. Determine the most abundant species by scanning and then choose a quadrat size that will contain individuals of this species approximately 80 percent of the time.

6. Move the mechanical stage so that approximately 30 random quadrats are viewed. Note the presence or absence of individual species. There is no need to count.

7. When 30 quadrats have been surveyed, calculate the percentage frequency.

$$\text{frequency (\%)} = \frac{\text{total number of occurrences of a species}}{\text{total number of quadrats examined}} \times 100$$

Periphyton

The periphyton has not received quite the same attention from ecologists as the phytoplankton, particularly in a quantitative way. Methods for studying the periphyton are given in detail by Sladeckova (1962).

Epiphyton, the periphyton growing on living plants and animals, can be observed in place on the organism if the substrate is thin or transparent enough to allow the transmission of light. If the leaves are thin and transparent, the task is relatively easy, but the growth on one side must be scraped away. If the leaf is opaque, the chlorophyll can be extracted by dipping the leaf in chloral hydrate. Small leaves can be examined over the whole area. Large leaves can be sampled in strips marked by grids on a slide or by an ocular micrometer. With large aquatic plants a square will have to be cut from the leaf or stem. If the leaf is too thick to handle under the microscope, scrape off the periphyton and mount in a counting cell for examination. The results can be related to the total surface area.

Algae growing on such aquatic animals as turtles or mollusks and on stones must be removed for study. Scraping and transfer is difficult, but several techniques are available.

One method employs a simple hollow, square instrument with a sharpened edge, which is pressed closely on or driven into the substrate. It separates out a small area of given size around which the periphyton is washed away. The instrument is then raised and the periphyton remaining in the sample square is scraped into a collecting bottle.

If the stones can be picked up from the bottom, then the periphyton can be removed with an apparatus consisting of a polyethylene bottle with the bottom cut out and a brush with nylon bristles. A section of the stone is delimited by the neck of the bottle held tightly on the surface. The periphyton is scraped loose by the brush and then washed into a collecting bottle with a fine-jet pipette.

The periphyton can be counted in a Sedgewich-Rafter cell recording a predetermined number, usually 100 to 1000, as they appear in the field of view. The results can be expressed as a percentage; or the algae can be checked for frequency in the field of view, using the Braun-Blanquet scale of total estimate (see Table A.1).

Some of the difficulties can be avoided by growing the periphyton on an artificial substrate, usually glass or transparent plastic slides, attached in the water in a variety of ways. In lentic situations they can be placed on sand or stones in the water. In lotic situations they can be placed in saw-cuts on boards, set in holes in bricks, clipped to a rope, attached to a wooden frame, or tucked into rubber corks.

The usual procedure is to use either a Wildco periphyton sampler, which holds glass slides, or a Hester-Dendy sampler (Hester and Dendy 1962), which can be fabricated easily. A Hester-Dendy sampler consists of eight $7\frac{1}{2} \times 7\frac{1}{2}$ cm or $3'' \times 3''$ squares of tempered hardboard (Masonite) with a hole drilled in the center. The boards are held apart by seven spacers, such as 5/8" or 17 mm faucet washers. The unit is assembled using a 2[[1/2]]" or 60 mm stove bolt and tightened with a wing nut. The sampler is attached to a length of stiff wire with a loop on the end to fit around the head of the bolt. The stake carrying the unit is pushed into the stream bottom.

For algae and protozoans the plates should be exposed for one to two weeks; for hydras, sponges, and the like, about one month. At the time of data collection, place each sampler in a plastic bag underwater, then close the bag with the sampler and water inside. Allow the wire stake to protrude from the top of the bag. In the lab, observe the periphyton directly under the microscope if glass slides were used. For hardboard samplers, remove the macroscopic organisms in small bowls. Scrape off algal growth and examine it in a Sedgewich-Rafter cell. For further procedures and data sheets see Brewer and McCann (1982:153).

Stream-Bottom Organisms

Samples of stream-bottom organisms can be taken with a modified Surber bottom-fauna sampler (Figure A.8). The sampler consists of a box made of a brass frame with stainless steel side pieces and a current baffle. To this box are attached on a removable brass frame two cone-shaped nets. A smaller cone, made of coarse net (19 meshes per linear inch, 7 per linear cm), is fitted within a larger cone, made of fine net (74 meshes per linear inch, 29 per linear cm). Flanges on the insert prevent the coarse net from being forced into the fine net.

This modified sampler picks up many small organisms that might otherwise be lost. In fact, collection of virtually all macroorganisms is assured. In addition we obtain two subsamples with respect to size, and the small organisms are associated with fine detritus only.

The sampler encloses a specified area of stream (500 cm²), which is the sample unit. Organisms, detritus, and trash are scrubbed free from the substrate, and the current washes them into the net. The contents can be transferred to a container and taken back to the lab for examination and sorting.

Dendrochronology

Dendrochronology is the science of dating past events by the study of the aging of trees. It is a valuable tool for the ecologist. It has been used in a number of studies—to age trees for management information, and to establish dates of past forest fires, insect outbreaks, ice storms, and periods of suppression and release in the life history of forest trees. It has been involved in hydrological and archaeological studies and even in legal cases involving boundary disputes, in which specimens are taken from fence posts and witness trees. Examples of growth-ring analysis in ecological investigations are Spenser's study (1964) of population fluctuations over the centuries in Mesa Verde National Park and a study by Sinclair and associates (1993) of the relationship among solar cycles, climate, and snowshoe hare cycles.

FIGURE A.8   (a) A modified Surber bottom sampler. (b) Construction details for the Surber bottom sampler. (Redrawn from Withers and Benson 1962.)

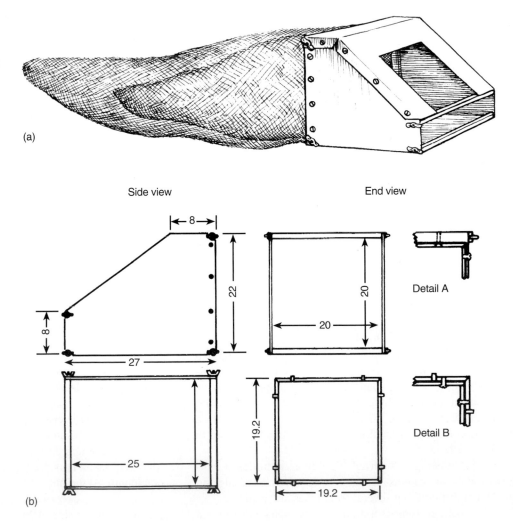

Dendrochronology is based on the variation of growth rings. Growth rings, despite popular belief, are not regular, nor are they all necessarily laid down annually. Because of the failure of the cambium to form a sheath of xylem the entire length of the bole, rings may be omitted, especially near the base. Possible causes are the lack of food manufacture in the crown, drought, fire, extreme cold, insects, and so on. At the other extreme are multiple rings produced by multiple waves of cambial activity during the growing season. They are caused by temporary interruptions in normal growth, such as a late spring frost, or by regrowth after normal seasonal growth has ended. Thus the growth rings reflect the interaction of woody plants and their environment as well as the passage of time.

The fundamental principle of dendrochronology is cross-dating, the correlation of distinctive patterns of growth between trees for a given sequence of years. Because no two plants have exactly the same growing conditions and life history (although the broad features are common to all trees involved), the similarities are relative rather than quantitative. The relative widths of corresponding rings to adjoining rings are the same. By lining up these similarities, the investigator can establish the relative identity of any rings in sequence and aberrant rings in the individual specimen. A great number of specimens must be cross-dated before each ring with a sequence can be dated.

Collection of Material

A recently logged-over area can provide an abundance of material, but new sections must be cut from the stump. Stump sections should be cut at a 30° angle. The cut may be clean enough for examination or may require smoothing with a carpenter's plane or machine sanding. To study shrubs, cut sample stems close to the ground and use the entire cross section.

The usual method of obtaining samples from forest trees is increment boring. The increment borer, available from forestry supply houses, is an instrument designed to bore a core from a tree. It consists of a T handle, a hollow bit, and an extractor. Increment borers are fairly easy to use, but without care they can be damaged or broken. Here are a few hints:

1. For growth and age studies, remove the core as near to the base of the tree as the instrument handle will allow.

2. Coat the screw with heavy-grade oil.

3. To start the borer, use a strong pushing and twisting motion until the borer is engaged in the wood.

4. Line the borer on the radius, keep the borer straight, and attempt to reach the center of the trunk.

5. When the core is drilled, insert the extractor and press firmly to cut the core from the trunk.

6. Remove the borer with reverse rotation.

7. Paint the wound with tree paint.

8. Store the cores in large diameter soda straws or polyethylene tubing. Be sure to label each sample fully, including the directional side of the tree from which it was removed.

To obtain a freshly cut edge for examination, the core is held firmly in a core holder. The groove in a plastic ruler is fine if the ruler is clamped to a table and the end is stopped. With a razor blade a transverse cut is made the length of the top of the core. It can then be brushed with water or kerosene to make the rings stand out better. When the core is ready, clip it to the stage of a microscope for examination. One-hundred-power magnification usually is strong enough. Looking at the whole core under a dissecting scope is often sufficient.

The width of each ring is measured with a graduated mechanical stage, a stage micrometer, or a dial micrometer. The total distance included in the layers observed can be measured and then compared with the accumulated individual measurements. Any error should be distributed over the individual measurements. For serious research a dendrochronometer, a special instrument with a microscope and precise measuring devices, should be used.

Cross-Dating

The methods and problems of cross-dating are complex. The basic procedure is as follows:

1. On graph paper, write down a series of numbers horizontally from left to right to represent growth layers. You can begin with one or with the years, the first number being the season preceding. This setup gives a series of numbers starting with the present and leading backward through the tree's life. A number of such blanks should be made up.

2. Set up a scale on the graph in millimeters so that the largest bars represent the *narrowest* widths.

3. Make a small bar graph for each year of the tree's life.

4. Make such a coded summary of all wood samples available.

5. Compare these graphs visually, two at a time, sliding them along each other. Keep looking for corresponding groups of years with the same pattern of ring sequences. By such a technique, multiple rings can be checked, or extremely narrow growth rings previously missed can be picked up.

A simpler but less precise method is to draw a line under the year that has a ring slightly less than the rings adjacent to it; to draw two lines if the decrease is more pronounced; to draw three lines if very narrow; and to draw two lines above the year for very wide rings.

Statistical Analysis

The data can be reduced to average values and then compared to weather data covering the principal growing season for the species. Comparisons can be made between rainfall and the current year's growth, rainfall and the previous year's growth, monthly evapotranspiration deficits and growth, frost-free periods, and so on.

Data will have to be analyzed by simple or multiple regression, depending upon the variables, using partial correlations and standard errors for tests of significance. For example analysis and interpretations, see Fritts (1962) and Cox (1985:83).

Palynology

Palynology is the study of past plant communities by the analysis of pollen profiles. These studies are especially enlightening if they are coupled with carbon-14 dating.

Peat Collection

Peat cores are bored at one to several stations in a bog. They are taken with a peat borer, available commercially. At each station two separate borings should be drilled, several meters apart. By taking successive samples from alternate borings (for example, first 0.3 m sample from core number one; second 0.3 m sample from core number two), contamination of one sample with another can be prevented. Two 150 cm samples are collected at each boring, one from the lower part of the cylinder, the other from the upper. The samples are placed in glass vials. If the vials are completely filled and tightly sealed, no preservative should be needed (Walker and Hartman 1960).

Treatment of Samples

Back in the laboratory, the samples can be treated as follows:

1. Thoroughly mix each 150 cm sample and remove a pea-size lump for deflocculation.
2. Boil the peat for a few minutes in a dilute solution of NaOH, gently breaking it apart with a wooden cocktail stirrer. Use only one stirrer for each sample to avoid contamination.
3. Add several drops of gentian-violet stain to the boiling mixture.
4. Stir vigorously and strain through fine wire mesh. Then stir again and draw up a 0.5-ml sample into a pipette.
5. Add a very small amount of warm glycerine jelly to the sample and mix.
6. Mount several drops on a slide and add a cover slip.

Examination

The samples, now transferred to slides, should be examined under a microscope equipped with a mechanical stage. Now follow three steps:

1. Tally 100 or 200 pollen grains as they are encountered by systematically moving the slide.
2. Identify each kind of pollen grain, if only by code, and tally the kinds separately. Identification should be made from a reference pollen collection made up beforehand.
3. Record the results from each slide directly as a percentage for each kind of pollen.

Plotting the Pollen Profile

The pollen profile can be constructed by plotting a graph for each species or kind of pollen. The vertical scale is set up for depth in meters; the horizontal scale is percentage, based on counts of 100 or 200 pollen grains for each spectrum level.

SAMPLING ANIMAL POPULATIONS

The study of animals involves considerably more problems than the study of plants. Animals are harder to see and most are not stationary. When it comes to sampling, the animals have something to say about getting caught, and they are more liable to mortality than plants. The following methods of estimating animal numbers, determining age structure, mortality, home range, and so on, enable the field biologist to make some measurements, however rough, of animal populations in the ecosystem.

Trapping and Collecting

The sampling of an animal population involves collecting animals, either alive, for marking and release, or dead. Detailed information on collecting and trapping is available in other publications (see references at the end of this appendix).

Flying Insects

Diurnal insects are collected with aerial nets and heavy-duty sweep nets designed to withstand hard wear when put through grass and woody vegetation. Nocturnal insects may be collected by using traps containing ultraviolet light or a mercury-vapor light or an old sheet fitted on a slant against some support with a strong light above it. Insects can then be picked off the sheet. If the insects are to be killed, they are placed in a killing jar containing a layer, either on the bottom or in a deep lid, of plaster of paris and potassium cyanide. Thin layers of tissue or light cloth in the jar prevent damage to moths and butterflies. Another technique is the use of a Malaise trap. This interceptive device uses a series of baffles to herd insects into a closed chamber that may or may not contain a killing fluid.

Aquatic Organisms

Aquatic organisms may be collected with dip nets, bottom nets for scraping along the bottom of ponds, wire-basket scraper nets, or plankton towing nets. For collecting from the shore, aquatic throw nets are useful. A bottom dredge lowered from a boat can collect bottom organisms in deep water. Fish, tadpoles, and large crustaceans can be collected with seines. A set of assorted widths will be necessary.

Soil Organisms

The most difficult components of soil fauna to study are the soil arthropods. They are the most numerous, the most difficult to identify, and possibly the most difficult to sample accurately. Nematodes, white or pot worms, and protozoans require highly specialized extraction techniques (Murphy 1962).

Soil arthropods can be extracted by means of a Tullgren funnel, an improved version of the Berlese funnel, the construction of which is simple (Figure A.9). Essentially it consists of a heat source, such as a light bulb; a smooth funnel, preferably glass, fitted into a collecting vial; and a screen made of hardware cloth or a sieve meshing inside the funnel.

Light
(heat source)

Household sieve

Shiny paper cone
(cut away to
show sieve)

Forest litter

Wire support

Collecting
vial

(a)

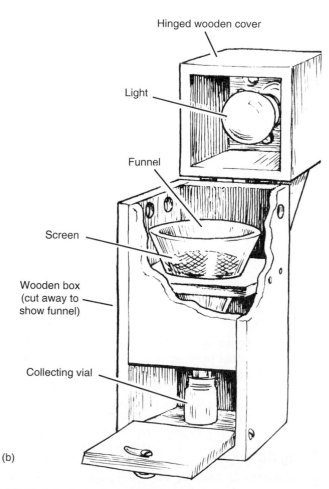

Hinged wooden cover

Light

Funnel

Screen

Wooden box
(cut away to
show funnel)

Collecting vial

(b)

FIGURE A.9 Although the Berlese and other types of funnels can be purchased, they can be constructed easily in the workshop: (a) a simple funnel for introductory work; (b) a more elaborate and efficient design.

The sample is placed on the screen; the heat and then desiccation drive the arthropods downward, until they fall through the funnel into the collecting bottle.

The procedure is simple:

1. Place the sample of litter or soil on the hardware cloth so fitted in the funnel that air space is present between the wire and the wall of the funnel.

2. To begin extraction, open the lid of the funnel 90° and turn on the 100-watt bulb.

3. After about 16 hours, depending on sample size and moisture content, change to a 15-watt bulb and shut the lid. There will be two periods of arthropod exodus, the first wave due to heat, the second due to desiccation. The collecting bottle beneath the funnel may contain alcohol, formalin, or water. Water may be preferable, because it increases the humidity gradient toward which the animals move.

4. Sort and identify the animals under the microscope.

These funnels are adequate for introductory soil biology. For serious studies in soil zoology, a better extractor is required. An extractor for woodland litter has been described by Kempson, Lloyd, and Ghelardi (1963). The funnels are replaced by wide-mouthed bowls filled with an aqueous solution of picric acid. The acid not only preserves the specimens but also produces by evaporation a high humidity in the air just under the sample. The humid air is cooled by conduction from a cold-water bath in which the bowls are immersed.

Another method of extraction is flotation. Procedures, although simple, are too lengthy to include here. Refer to Jackson and Raw (1966) and Andrews (1972).

Larger soil animals, such as spiders and beetles, can be taken in traps made from funnels and cans set in the soil to ground level. Boards placed on the ground may attract millipedes, centipedes, and slugs. Meat bait in small wire traps will attract scavenger insects.

Sampling earthworm populations presents some difficulties, for no really successful method has been devised to extract the animals from lower layers of the soil. One of the better methods is a combination of formalin and a shovel:

1. Apply a dilute solution of formalin (25 ml of 40 percent formalin to 3.75 liters of water) to a quadrat 0.2 m². Within a few minutes worms will come to the surface.

2. After earthworm movement to the surface stops, pour on a second application.

3. When worms cease to come to the surface the second time, dig out the quadrat as deep as necessary.

4. Hand-sort the soil for maximum recovery. Earthworm cocoons can be extracted by the flotation method.

Small Animals in Vegetation

Sweep nets with stout frames to withstand sweeps close to the ground and in woody growth are useful for collecting many types of insects and even some arboreal amphibians and reptiles. Drag nets, consisting of light tubular frames to which are attached canvas bags, are useful on flat ground. Overhead vegetation can be sampled by beating the limbs with sticks to dislodge the animals, which should fall into canvas collecting trays beneath. Other techniques involve traps on which the insects settle on a sticky surface and pit traps in the ground, most useful for spiders and beetles.

Birds and Mammals

Birds can be trapped for banding in specially constructed traps, cannon nets for larger game birds, and mist nets. Both federal and state permits are required for such work. Once a permit is granted, the operator of the banding station will be furnished with plans for suitable traps. For mammals, live traps of wood or wire and snap traps are used. Both are available commercially, but live traps are easily constructed. Traps can be baited with natural foods, dripping water, and so on. For small mammals, a mixture of peanut butter and oatmeal works well. Also useful are grain, apple, meat, and appropriate scents.

Marking Animals

Marking individuals in an animal population is necessary if you wish to distinguish certain members of a population at some future date, to recognize individuals from their neighbors, to study movements, or to estimate populations by the mark-recapture method.

Arthropods and snails are best marked with a quick-drying cellulose paint. It is easily applied with any pointed object. Marking butterflies is a two-person operation. One has to hold the wings together with a pair of forceps, while the other marks the side of the wing exposed at rest. For aquatic mollusks and insects better results are obtained through the use of ship-fouling paint, because acetate paints do not hold up well in water.

Fish are usually marked by tagging in several ways. Strap tags of Monel Metal may be attached to the jaw, the preopercle, or the operculum. Stream or pennant tags attached to various parts of the body, usually at the base of the dorsal fin, are used in some studies. Another method is to insert a plastic tag into the body cavity. The tag is inserted through a narrow incision made in the side of the abdominal wall. Once the incision heals, which it does quickly, the tag is carried by the fish for life and is recovered only when the fish is cleaned. Clipping the fins is still another way to mark fish, but it does not permit the individual recognition of a very large number of fish.

Frogs, toads, salamanders, and most lizards can be marked by some system of toe clipping, which involves the removal of the distal part of one or more toes. One method worked out by Martof (1953) is as follows. The toes on the left hind foot are numbered 1 to 5, the toes on the right foot 10 to 50. The left forefoot toes are numbered 100 to 400, and the toes on the right forefoot 800, 1600, 2400, 3200. Thus one can mark up to 6399 individuals by clipping no more than two toes.

Snakes and lizards can be marked by removing scales or patches or scales in certain combinations.

Birds are usually marked by serially numbered aluminum bands and by cellulose and aluminum colored bands. The colored bands are necessary for individual recognition in the field. In some specialized studies, the plumage is dyed a conspicuous or contrasting color.

Small mammals may be marked by toe clipping in combinations similar to those given for amphibians or by notching the ear.

A number of other methods have been devised for marking mammals. Fur clipping and tattooing may be employed. Bear, deer, elk, moose, rabbits, and hares can be marked with strap tags or plastic discs attached to the ear. Aluminum bands similar to those used on birds can be attached to the forearm of bats. Dyes can be used to mark both large and small mammals. Small amphibians and reptiles can be marked by branding (Clarke 1971).

Some species lend themselves to noninvasive means of identification, such as color patterns on the flukes of humpbacked whales and facial patterns and expressions in gorillas. With patience and observation you can identify individuals of other mammals, such as gray squirrels and chipmunks.

Radioactive tracers are particularly useful for studying animals that are secretive in habits, live in dense cover, spend part or all of their lives underground, or have radically different phases in their life cycle, as the moths and butterflies do. Animals are fed small traces of gamma-emitting radioactive material. The material is metabolically incorporated into the tissue, and the tracer becomes a part of the animal. It is passed along to egg or offspring. Radioactive larvae remain so as they transform to adults. The same is true for birds. This technique is useful for studying dispersal, for identifying specific broods or litters, and for obtaining data on population dynamics and natural selection.

Another method involves the application of a radioactive tracer in or on an animal in such a way that the animal is not seriously injured and behaves in a normal way. Usually a radioactive wire is fastened to the animal or inserted under the skin of the abdomen with a hypodermic needle. The movements of the tagged individual are then followed with a Geiger counter.

Although these techniques have their merits, they also have disadvantages. The greatest is the potential radioactive hazard to the investigator, to other humans, and to the ecosystem. Another disadvantage is the impossibility of separating one animal from another. Most work with radioactive tracers requires a federal license. Specific techniques can be reviewed in Tester (1963), Godfrey (1954), Pendleton (1965), Graham and Ambrose (1967), and T. J. Peterle in Schemnitz (1980).

Aging Animals

Information on the age structure of wild populations is not easily obtained. During the past several decades, a number of aging techniques have been developed, mostly for game and fish.

Fish aging began when Hoffbauer (1898) published his studies on the scale markings of known-age carp. Since then the technique has been refined. It is based on the fact that a fish scale starts as a tiny plate and grows as the fish grows. A number of microscopic ridges, the circuli, are laid down about the center of the scale each year (Figure A.10). When the fish is growing well in summer, the ridges are far apart. During winter, when growth slows down, the ridges are close together. This annual check on growth enables the biologist to determine the age of a fish by counting the number of areas of closed rings, the annuli.

Salmon and some species of trout spend one or two years in streams before migrating out to sea or into lakes. Because stream growth is slower than lake or sea growth, the scales show when the fish migrate. When salmonid fish spawn, reabsorption of scales occurs, eroding the margins of the scales and interrupting the pattern of circular ridges. This erosion leaves a mark that can be detected in later years.

Because the growth of a scale continues throughout the life of a fish, it also provides information on the growth rate. This rate is obtained by measuring the total radius of the scale, the radius to each year's growth ring, and the total body length of the fish. Then by simple proportion, the yearly growth rate can be determined.

Other techniques in aging fish include the length-frequency distribution, vertebral development, and rings or growth layers in the otolith or ear stone (see Jerald 1983).

Because of the large number of year classes (animals born in a population during a particular year) that can be identified, we can determine dominant year classes, learn the age when fish reach sexual maturity, estimate production mortality, and estimate the effects of fish harvest.

Aging techniques for mammals and birds have developed more slowly and are not as refined as those for fish, but a number of methods are in common use. The age of the animal may be indicated by specific characteristics of body parts (Figure A.11). Among birds, plumage development is frequently used. Until molted, the tail feathers of juvenile waterfowl are notched at the tip, in contrast to the normally contoured feather of the winter plumage. The shape of the primary wing feather separates adults from young among many gallinaceous game birds. The presence or depth of the bursa of Fabricus, a blind pouch lying dorsal to the cecum and opening into the cloaca, indicates juvenile birds.

Among mammals the examination of reproductive organs is useful because the majority do not breed until the second year. This method can be used only during the breeding season. The presence of epiphyseal cartilage in rabbits, squirrels, and bats (Figure A.11) identifies juveniles up to 6 or 7 months. Black bars on the pelage of the underside of the tail of juvenile gray squirrels separate the young from the adults. Primeness of pelt on the inside of skins is a good means of aging muskrat during the trapping season. Dark pigmentation on the flesh side of the pelt indicates areas of growing hair. This pigmentation in adults appears in irregular, scattered dark areas, whereas in immature animals it is more or less symmetrical and linear. Skull measurements are useful in beavers and muskrats. Annual growth rings on the roots of canine teeth indicate age for the first few years of life in the fur seal and other pinnipeds and in canids. Growth rings also show up in the horns of mountain sheep. The wear and replacement of teeth in deer and elk permit the determination of different age classes in these mammals (see Schemnitz 1980).

Because the lens of the eye of most mammals (and possibly birds) grows continuously throughout life and because there is only slight variation among individuals in lens size and growth, the measurement of the lens is a feasible method for aging a number of mammals. It has been done successfully for the cottontail rabbit, raccoon, black bear, and fur seal.

The technique involves comparing the weight of the dry lens with the chart of lens weights of known-age individuals (Figure A.12). The investigator may have to develop the chart by rearing young animals in captivity and sacrificing them week by week for their eyes. A table has been prepared for rabbits by Lord (1959, 1963) and for cotton rats by Birney, Jenness, and Baird (1975).

The technique is as follows:

1. Remove eyes as soon as possible after the animal is killed and place in a solution of 10 percent formalin. The formalin will harden the lens so that it can be removed from the vitreous humor.

2. Fix for a minimum of 1 week, but the longer the better.

3. After fixing, remove the lens from the eye and roll it on a paper towel for a few minutes to remove excess moisture.

4. Place the lens in an oven to dry at 80°C.

5. Lenses are considered dry when repeated weighing after intervals of drying results in no additional loss of weight. This step will usually require 24 to 36 hours.

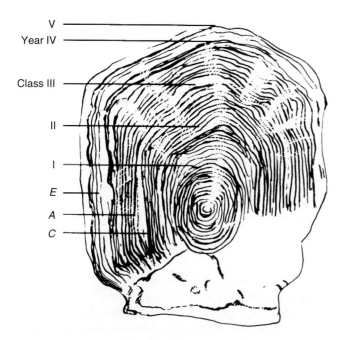

FIGURE A.10 The age of a fish can be determined by the growth rings on the scales. C is the circuli; A, the annuli; E, the erosion of the scale from spawning.

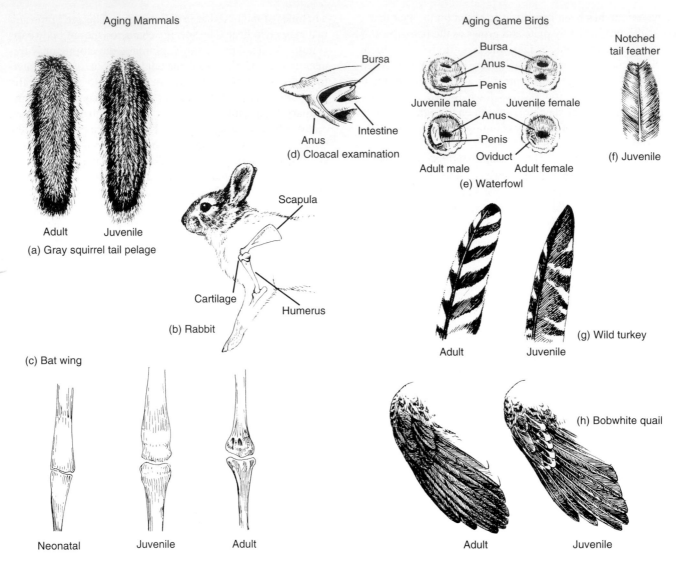

Aging Mammals

(a) Gray squirrel tail pelage

Adult   Juvenile

(b) Rabbit

Scapula

Cartilage

Humerus

(c) Bat wing

Neonatal   Juvenile   Adult

Aging Game Birds

Bursa
Anus
Penis
Juvenile male   Juvenile female

Anus
Penis
Oviduct
Adult male   Adult female
(e) Waterfowl

Bursa
Anus
Intestine
(d) Cloacal examination

Notched tail feather
(f) Juvenile

(g) Wild turkey
Adult   Juvenile

(h) Bobwhite quail
Adult   Juvenile

FIGURE A.11   Age determination in some mammals and birds. (a) Regular barring on the underside of the tail distinguishes the juvenile gray squirrel from the adult. (b) Epiphyseal cartilage on the humerus of juvenile cottontail rabbits separates that age class from the adult. (c) The epiphyseal cartilage of the fourth metacarpal-phalangeal joint separates juvenile from adult living bats. The degree of ossification in the joint can be determined by extending the wing and holding its ventral surface firmly against the glass stage of a dissecting microscope. Also useful in age determination of young bats are forearm lengths between 14.4 mm and 35.5 mm regressed against age (up to 11–12 days). (See Kunz and Anthony 1982.) (d) The bursa of Fabricus (enlarged). Its presence or greater depth indicates a juvenile bird. The depths vary with the species. This method is useful in both waterfowl and some gallinaceous birds. (e) Sexing and aging waterfowl by examining the cloaca. Note the presence of the bursal opening on juvenile waterfowl and its absence on adult waterfowl. (f) The notched tail feather of juvenile waterfowl. (g) The number X (ten) primary in juvenile gallinaceous birds is sharply pointed; in adults it is rounded. The juvenile wild turkey in addition has its outer primary indistinctly barred. (h) The juvenile bobwhite quail, in addition to having a pointed number X primary, possesses buff-tipped primary coverts. (For a more complete discussion on aging see Schemintz 1980, DeBlase and Martin 1981.)

6. Weigh immediately after removal from oven, because the dried lenses are hygroscopic and take on water. Electronic digital scales that read weights rapidly are preferred over other types of balances.

The lens growth curve permits a close approximation of the age of the mammal. For cottontail rabbits, the method permits the determination of the month of birth of young rabbits and the year of birth of young rabbits over one year of age.

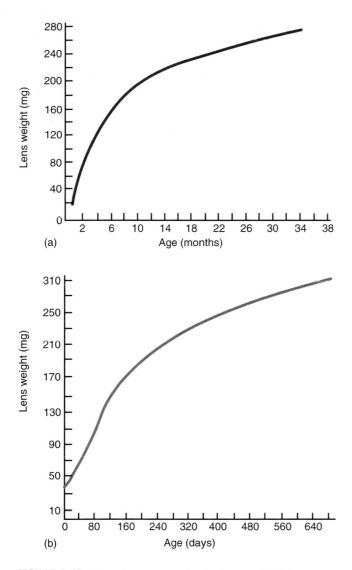

FIGURE A.12 Growth rate curves for the lenses of rabbits. (a) Cottontail rabbit. (From Lord 1961.) (b) Black-tailed jack rabbit. (After Tiemeier and Plenert 1964.)

There are statistical and procedural problems in the use of aging techniques, including the lens-weight technique. See Dapson (1980) for guidelines in the use of statistics in age estimation.

Determining Sex

The sex of mammals in most instances can be determined by examining external genitalia, and the sex of birds by plumage differences (Figure A.13). For example, the male ruffed grouse can be distinguished from the female by the length of the central tail feather; the female prairie chicken can be distinguished from the male by strong mottling or barring of all the tail feathers; the male wild turkey has black-tipped body feathers, whereas the female has brown tipped feathers. (For detailed information, see Schemnitz 1980.)

Determining Home Range and Territory

A number of methods are available for obtaining an approximation of the size of home range. Five methods are offered here.

Home Range Map

On a map, outline and measure the area that includes all the observations made on the movements of individuals. If the observations are obtained by trapping, then assume that the animal could have gone halfway toward an adjacent trap, especially if the traps were set in a regular grid. For techniques and analysis see Anderson (1982), Dixon and Chapman (1980), Samuel and Garton (1985), Samuel et al. (1985).

Center of Activity

Arrange the recaptures on a grid and determine the values on X and Y axes. An average of these locations will give the center of activity (see Hayne 1949). This method has the advantages that the information is easy to summarize and the calculations are not complicated. However, a map of the area must be made and many recaptures of the same individual are required before the extent of the home range can be obtained. As the number of recaptures increases, the area of the known range increases. At least 15 recaptures or more are necessary. This method is unsuited for mammals that follow paths or tunnel underground.

Frequency of Capture

Record the distances between captures in live traps set randomly or in grids. Record the number of captures as a frequency distribution according to the distance between them. Distances between captures are then tallied and proportions calculated for each distance by sex or age categories.

The distances can be measured in two ways. They can be taken from the place where the animal was first marked or observed or from each successive location.

This method has the advantage that the traps can be set out haphazardly, avoiding the labor of setting them out in grids. The recaptures of all individuals can be used. Information can be obtained during a short period of time because the data from animals captured only two or three times can be used. The disadvantages of this method are that short movements are favored and that no definite boundaries of home ranges can be given. Home range is described as a frequency of distances observed.

Territory

Because territorial boundaries are rigidly maintained by birds during the breeding season, territorial boundaries can be mapped by observing the movements of the birds during the day, by plotting singing perches, by observing locations of territorial disputes, and on occasion by chasing the bird (Wakeley 1987c). When the bird arrives at the boundary of its territory, it generally will double back. This technique is most useful in studies of a single species. Do not confuse it with spot mapping (see Verner 1985).

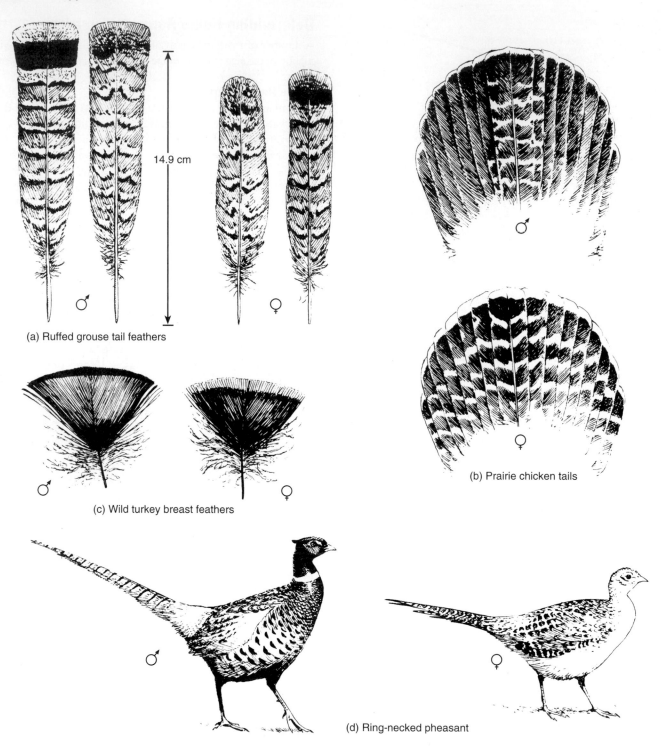

(a) Ruffed grouse tail feathers

(b) Prairie chicken tails

(c) Wild turkey breast feathers

(d) Ring-necked pheasant

FIGURE A.13   Sex determination in game birds. (a) Length of the central tail feather separates male and female ruffed grouse. The mail tail feather is 14.9 cm long or more; the female tail feather is shorter. (b) The female prairie chicken has heavily barred tail feathers; barring is absent in the outer tail feathers of the male. (c) The breast feathers of the male wild turkey are black-tipped; those of the female are brown-tipped. (d) Sexual dimorphism distinguishes males from females in many species of birds.

Radio Tracking

The development of transistors and other miniature electronic devices has made possible the construction of small transmitters that can be attached to animals, usually by a specially designed collar or harness. Mercury cells are the source of power. The transmitter is a transistor-crystal-controlled oscillator with the tank coil for the oscillator acting as the magnetic dipole transmitting antenna. The antenna is constructed of copper or aluminum and has a figure-8 directional pattern. The receiver is a portable battery-powered unit, whose basic components are the receiver, a radio range filter, and two transistorized radio-frequency converters. Positions of stationary animals can be obtained by a single portable direction-finding receiver. Running animals are best located by triangulation, using at least two direction-finding receivers. Radio tracking equipment is available commercially.

Although highly useful in obtaining data that could not be acquired otherwise, radio tracking has its disadvantages for most investigators. The most serious disadvantage is the need for electronic expertise. Unless you are an electronics whiz or have the assistance of an electronics technician, you should not attempt radio tracking. Other disadvantages are cost and the need for an FCC license.

Estimating Numbers

Basic to the study of animal populations is the estimation of their numbers, no small task in wild populations. During the past several decades, much work has gone into the development of techniques and statistical methods to arrive at some estimates of animal populations. Basically the methods of estimating the numbers of animals can be put into three categories: true census, a count of all individuals on a given area; sampling estimate, derived from counts on sample plots; and indices, the use of different types of counts, such as roadside counts, animal signs, and call counts, to determine trends of populations from year to year or from area to area.

True Census

A **true census** is a direct count of all individuals in a given area. It is difficult to do for most wild populations, but there are situations where a total count can be made.

Many territorial species are easily seen and heard and can be located in their specific area. Such a census is regularly used for birds. The spot mapping method is probably the best approach (Wakeley 1987b). A sample plot of at least 10 ha is marked out in a grid with numbered stakes or tree tags placed at intervals of 50 m. Five or more daily counts are made throughout the breeding season. Each time a bird is observed, it is marked on a map of the plot. At the end of the census period all the spots at which a species is observed are placed on one map. The spots should fall into groups, with each group indicating the presence of a breeding pair. The groups for each species can then be counted in order to arrive at the total population for the given area. Results are usually expressed as animals per hectare.

Direct counts can be made in areas of concentration. Deer in open country, herds of elk and caribou, waterfowl on wintering grounds, rookeries, roosts, and breeding colonies of birds and mammals permit direct counting either from the air or from aerial photographs. Coveys of bobwhite quail can be located and counted with the aid of a well-trained bird dog.

Sampling Estimates

A **sampling estimate** of population size involves two basic assumptions: (1) mortality and recruitment during the period the data are being taken are negligible or can be accounted for; and (2) all members have an equal probability of being counted—they are not trap-shy or trap-addicted, they are distributed randomly through the population if marked and released, and they do not group by age, sex, or some other characteristic.

Sampling also involves one major general consideration. The method employed must be adapted to the particular species, time, place, and purpose.

Relatively immobile forms, such as barnacles, mollusks, and cicada emergence holes, can be estimated by the quadrat method, similar to that employed for plants. The data can be analyzed for presence, frequency, and so on, or the results can be converted to a density per hectare. The size and shape of the quadrat will depend upon the density of the population, the diversity of the habitat, and the nature of the organism. A few preliminary surveys are made before settling on a quadrat size.

Foliage arthropods may be sampled by a number of strokes with a standard sweep net over a 10 m^2 area. The number of strokes needed to secure the sample must be predetermined. It will vary with the type of vegetation.

Estimates of zooplankton, obtained by pulling a plankton net through a given distance of water at several depths, can be made by filtering a known volume of sample through a funnel using a filter pump. The filter paper should be marked off in equal squares. With the aid of a hand lens or a binocular microscope, the organisms in each square can be counted. The numbers then can be related back to the total volume of water sampled.

If the organisms are too small to be counted in this manner, a Rafter plankton-counting cell can be used. It consists of a microscope slide base plate ruled into ten 1 cm squares. The slides are made from strips of microscope glass slides cemented to the base with Canada balsam. This cell should hold 1 cc of liquid. After a small volume of water is introduced, the cell is covered with a long cover glass and placed under the microscope. The organisms are counted square by square and the number of each form recorded per square until at least 100 observations have been made. The occurrence of individual species can be recorded as a percentage frequency. (Note: Plankton-counting cells and eyepiece micrometers can be purchased commercially, but they are expensive.)

Capture-Recapture Sampling

Capture-recapture methods are based on trapping, marking, and releasing a known number of marked animals into the population, and then later recapturing individuals from the population after an appropriate interval of time (approximately one week for rabbits and mice). An estimate of the total population is then computed from the ratio of marked to unmarked individuals in the sample, which supposedly reflects the ratio of marked to unmarked animals in the population.

Some points about the capture-recapture method need to be emphasized. The size of the population is estimated by the ratio of marked to unmarked animals in the population, not the number of recaptures. The population estimate is for the time the marked animals are released in the population, not the time of recapture. The estimate is biased because one is examining a sample of the population rather than the whole population, and the precision is dependent on the number of marked animals recaptured.

The capture-recapture method involves a number of assumptions:

1. All individuals in the population have an equal chance of being captured.

2. The ratio of marked to unmarked individuals remains the same from the time of capture to the time of recapture.

3. Marked individuals redistribute themselves homogeneously throughout the population with respect to unmarked ones.

4. Marked individuals do not lose their marks.

5. The population is closed. No emigration or immigration takes place within the sampling period.

Because populations are dynamic, the investigator needs to know some features of the population and adjust accordingly. They include reproductive history, mortality patterns, the effects of marking on behavior, the seasonal patterns of movements, and the bias of age and sex on the individual probability of capture. In effect, what the investigator faces is unequal probabilities of capture.

The *single mark-single recapture method* is known as the *Lincoln index* or *Petersen index* of relative population size. The basic model is

$$N : M :: n : R$$

or

$$N = \frac{nM}{R}$$

where M = number marked in the precensus period, R = number of marked animals trapped in the census period, n = total animals trapped in the census period, and N = the population estimate. The probability of capture on any given occasion is R/n.

This basic model assumes that every animal has the same capture probability p on every capture occasion. It involves only two parameters, N and P, and only two data, the number of captures at each time t.

Capture probability can be influenced by effects of time, behavioral response to capture, and variations in probability of capture among individuals (trap-happy, trap-shy). To account for these factors, White and associates (1982) provide variations on the basic model and a Fortran computer program for them.

As an example, suppose that in a precensus period, biologists tag 39 rabbits. During the census period they capture 15 tagged rabbits and 19 unmarked ones, a total of 34. The following ratio results:

$$N : 39 :: 34 : 15$$

or

$$\frac{nM}{R} = \frac{(34)(39)}{15} = 88$$

The estimated population is 88 rabbits.

The confidence limits at the 95 percent level may be calculated from

$$S.E. = N \sqrt{\frac{(N - M)(N - n)}{Mn(N - 1)}}$$

(S.E. is Standard Error.)

To determine the limits within which the population lies, add and subtract two standard errors from the estimate. A large standard error and rather wide confidence limits are the result of a small number of recaptures and a small sample size. For this example:

$$S.E. = 88 \sqrt{\frac{(88 - 39)(88 - 34)}{(34)(39)(87)}}$$

$$= 88(0.1513) = 13.31$$

Upper limit: $88 + 26 = 114$

Lower limit: $88 - 26 = 62$

The chances are 95 out of 100 that the population of rabbits lies between 62 and 114. Such results are typical in population studies of animals.

Repeated mark and recapture is a variation of the single mark–single recapture method. Investigators capture, mark, and recapture animals over t occasions. The method expands the estimation beyond the $t = 2$ occasions of the Lincoln-Petersen index and provides a series of population estimates, which can be repeated until the investigator is satisfied with the results. The most familiar of these methods is the Schnabel estimate, developed by Zoe Emily Schnabel in 1938.

An example is given in Table A.7 All animals captured are tagged or marked and released daily. A record is kept of the total animals caught each day, the number of recaptures, and the number of animals newly tagged. The assumptions and the calculations are similar to those of the Lincoln-Petersen method. However, in the Schnabel method, M (the number marked in the precensus period) becomes progressively larger. Population estimates can be calculated daily or at intervals during the sampling period, for example weekly. True confidence limits cannot be determined, and the calculation of the standard error becomes complex. (For details, see Ricker 1958.)

Multiple capture–multiple recapture is another variation, in which all animals caught on any particular day are marked and released, including the recaptures. Thus an animal caught on day 1 and again on day 2 will bear the marks of both days. By such a method, one can both keep an account of total marks recaptured each day of trapping and relate recaptures

TABLE A.7 Schnabel Method of Estimating Populations

Period (Date) (P)	Number Trapped (A)	Number Marked	Marked Animals in Area (B)	(A) × (B)	(A) × (B) Sum	Recaptures	Sum of Recaptures (C)	$\dfrac{(A) \times (B)}{(C)}$ Estimated Population
1	4	4	—	00	0	—	—	—
2	4	4	4	16	16	0	0	—
3	2	2	8	16	32	0	—	—
4	6	6	10	60	92	0	—	—
5	10	7	16	160	252	3	3	—
6	4	4	23	92	344	0	3	—
7	8	6	27	216	560	2	5	—
8	4	2	33	132	692	2	7	—
9	5	4	35	175	867	1	8	—
10	7	6	39	273	1140	1	9	—
11	7	6	45	315	1455	1	10	145
12	9	7	51	459	1914	2	12	159
13	6	3	58	348	2262	3	15	150
14	10	6	61	610	2872	4	19	151
15	8	5	67	536	3408	3	22	154
16	6	1	72	432	3840	5	27	142
17	4	2	73	292	4132	2	29	142
18	12	7	75	900	5032	5	34	148
19	8	4	82	656	5688	4	38	149

to the initial day of marking (see Jackson 1939; Fisher and Ford 1947; Blower, Cook, and Bishop 1981; Krebs 1989). The basic equation is

$$\hat{P}_i = \frac{n_i \hat{M}_i}{m_i}$$

where $\hat{P}_i$ is the estimate of population on day i; $\hat{M}_i$ is the number of marked animals or marks in a sample (number of recaptures or number marked previously); n_i is animals captured, marked, and released on day i; and m_i is the number of marked animals or marks in sample on day i. The formula is essentially that of the Lincoln index.

For bookkeeping the trellis diagram is most convenient (Figure A.14). There are two types of trellises. In the Type I trellis, employed here, all marks are counted (Fisher-Ford method). The number of recaptured marks is entered into the cells of the trellis. Because some animals will carry more than one mark, a larger number of marks may be entered than marked animals. In the Type II trellis, marked individuals are entered according to their marks. The Type II trellis is employed in the Jolly-Seber method (Jolly 1965, Seber 1973, Southwood 1978, Blower, Cook, and Bishop 1981).

In the Type I trellis diagram, the marginal column on the left running downward from left to right contains the total animals captured for each day. The marginal column on the right contains the total release each day. In the body of the table are the figures for the recaptures as they relate to the day they

were originally marked. For example, on trapping day 5, 220 captured, 30 had been marked on day 4, 13 on day 3, 8 on day 2, and 2 on day 1. The data can provide two different population estimates. A column starting at any date and running

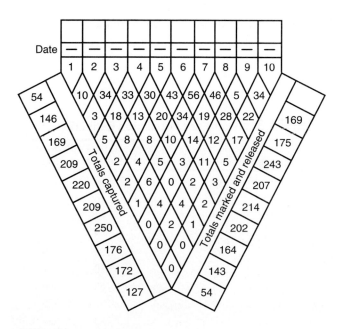

FIGURE A.14   Trellis diagram.

downward to the left gives the necessary information for determining the population on the day of recapture. Columns starting at any date and running downward to the right give the necessary information to determine the population size on the day of release. To use this method, trapping does not have to be done on consecutive days, but only at equally spaced times.

Although new raw data can be used to estimate population, it is more useful if the recapture values are corrected to the number of marked recaptures per 100 marked on day i and per 100 in the recapture sample. This correction must be done for each recapture value in the body of the table. The corrected values are then substituted for the raw values. The formula is

$$y = \frac{X_i}{1} \times \frac{100}{M} \times \frac{100}{n_i}$$

where X_i = number recaptured on day n_i, sample i; M = total number of marked animals initially released (prior to time of sample); n_i = total number of marked and unmarked animals in sample i; and y = corrected recaptures in sample i. For example, on day 5, there were four sets of recaptures: 30, 13, 18, and 2, out of 220 captures. To correct these values:

recaptures from day 4: $y = \frac{30}{1} \times \frac{100}{202} \times \frac{100}{220} = 6.7$

recaptures from day 3: $y = \frac{13}{1} \times \frac{100}{164} \times \frac{100}{202} = 3.6$, and so on

The next step is to calculate a weighted ratio, r, to show the rate of decrease of recapture values. Determine r for each column running to the right, which gives you $r +$ values, size of the population on the day of release. Do the same for each column running to the left, which gives you $r -$ values for the size of the population on the day of recapture. For three or four values for y, the formula is

$$r = \frac{y^2 + y^3 + \ldots y_n}{y_1 + y_2 + \ldots y_{n-1}}$$

For more than four values for y, the formula is

$$r = \frac{y_3 + y_4 + \ldots y_n}{y_1 + y_2 + \ldots y_{n-2}}$$

With this weighted ratio for each method, you can calculate the theoretical number of recaptures that would have been obtained on the day of release:

$$y_0 = \frac{y_1 + y_2 + \ldots y_{n-1}}{r} - (y_1 + y_2 + \ldots + y_{n-2})$$

With the theoretical values for recaptures at the time of release, you can estimate population size for each day (Table A.8) by the following:

$$N \text{ (on day } i) = \frac{100}{1} \times \frac{100}{y_0}$$

For convenience and for comparison of population estimates by the two methods, positive and negative, you may tabulate your calculations in a table headed as follows.

Day	$r +$	y_0	N	$r -$	y_0	N

This method works well where large numbers of animals are involved, such as insect populations.

Removal Sampling

The **removal method** of estimating populations is used in studies of small mammals. The method involves the removal of individuals from the population either permanently or for the duration of the study.

The assumptions underlying this method are: (1) the population is essentially stationary; (2) the probability of capture during the trapping period is the same for each animal exposed to capture; (3) the probability of capture remains constant from trapping to trapping. The last assumption requires that the animals are not trap-shy or trap-prone; and that bait acceptance, weather conditions, and differences in sex and age will not affect the probability of capture. Meeting these assumptions is difficult because dominant mammals, especially males, usually are more trap-prone, and weather can influence mammal movements.

The procedure requires at least two periods of sampling. Sampling requires setting the same number of traps for several days or nights. For small mammals these traps may be snap traps or live traps. If live traps are used, all mammals caught in each sampling period must be marked and not counted if trapped again. The sampling may involve successive removals of animals from the population, the Hayne (1949) method; or it may involve only two periods of sampling, the Zippin (1958) method.

A field procedure is as follows:

1. Set the traps in a grid system, 3 traps to a station; or in two parallel lines 50 meters apart, 20 stations to a line, 3 traps at each station. Space the stations 25 or 50 meters apart, depending upon the vegetation.

2. Prebait for best success.

3. Trap for several successive periods, or for two periods on each of several dates. Depending upon the species, trapping periods may be 24 hours or nighttime only.

TABLE A.8 Fisher-Ford Population Estimates

Day	Captured	Released	Recaptured	Survival	Population Size
2	146	143	10	10	456.3
3	169	164	37	40	460.6
4	209	202	56	84	572.1
5	220	214	53	88	853.5
6	209	207	77	133	659.2
7	250	243	112	210	581.1
8	176	175	86	162	596.2
9	172	169	64	189	725.3
10	127		84	183	384.0

A problem with trapping over a number of successive periods is the possibility of attracting new animals into the sampling area as the population is reduced, particularly if the sampling design is a grid. Traps on the outer stations usually capture more animals than those in the center of the grid, because animals in the border zone react to the sudden removal of animals in the center. These immigrants, picked up by the border traps, contribute significantly to the catch on the outer grid lines and influence population estimates.

In the Hayne method the daily catch is plotted against the number of animals previously caught (Figure A.15). A line can be drawn through the data points to cut the horizontal axis. The point at which the horizontal axis is cut represents the population estimate. A more accurate method is to calculate a simple regression line from the catch data. The slope of the line represents the average proportion of the population removed during each sampling period.

In the Zippin method, N is the population size, n_1 is the number of animals caught and removed in the first sampling period, and n_2 the number caught and removed in the second sampling period. The proportion of the original population captured in the first sampling period is n_1/N; $N - n_1$ animals remain; and the proportion of animals caught in the second sampling period is $n_2/(N - n_1)$. An estimate of N is given by

$$N = \frac{n_1^2}{n_1 - n_2},$$

The two methods give approximately the same answer. In the example given, the Hayne method involving a three-day removal trapline gave an estimated exposed population of 18. The Zippin method, using the catch for the first two days, gives the same answer:

$$N = \frac{36}{6 - 4} = 18$$

The removal method is useful when you desire a relative measure or index figure for small mammal populations to compare one habitat with another. Data will be more useful if details on vegetation and litter are recorded for each station. Often some association can be established between vegetation and trapping success.

Line Transect

The **line transect** is widely used in terrestrial vertebrate ecology. The method involves walking a line established in an area and noting individual animals observed along that line. The results provide an index rather than an absolute measure of density.

The line transect is most useful in situations in which the animals are difficult to see. There are a number of variations of the line transect, which was developed by R. King to census ruffed grouse and became known as the King census.

The method involves the establishment of a straight line or series of straight line segments in the area to be sampled, marking the line or lines well, walking the line, carefully noting any individuals seen or flushed along the line, and measuring the perpendicular distance from the line to the point of sighting or flushing; the sighting distance, the distance of the flushed animal from the observer; or the sighting angle. Preferably all three measurements should be taken (Figure A.16)

The estimator of density from line transect samples is expressed in general as

$$D = \frac{n\,f(0)}{2L}$$

where n is the number of animals sighted, L is the length of the transect in meters, and $f(0)$ is distances or angles.

The following formula suggested by Haynes (1949) is a modification of the original King formula:

$$D = \frac{10^4\,\Sigma\left(1/d_i\right)}{2L}$$

where d_i is distance from the observer to the ith animals sighted, measured in meters to the point where the animal was at the time it was flushed; and 10^4 is a factor for converting m^2 to ha.

This line transect method involves four assumptions:

1. Animals on the line are always seen.

2. Animals do not move before being sighted, and none is counted twice.

3. There are no measurement errors and no rounding errors.

4. Sightings are independent events; that is, flushing one animal does not cause another to flush.

As a sampling scheme the line transect is easy to use, but two problems encountered are defining a straight line and holding to it and obtaining accurate measures of distances and angles. Anderson and associates (1979) offer these suggestions:

1. Keep the center line of the transect straight and well marked.

2. Take care that all animals on the center line of the transect are seen with a probability of 1.

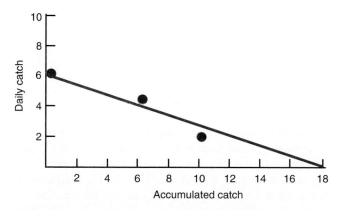

FIGURE A.15   Daily catch on a three-day removal trapline plotted against accumulated catch to estimate the exposed population.

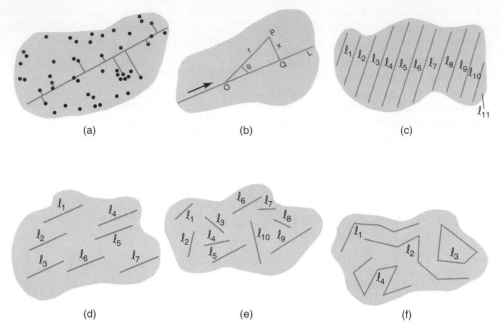

FIGURE A.16 (a) Diagram of a line transect sample. The area, with known boundaries, has points distributed within it. Points detected show the perpendicular distance to the line. Points without lines are not detected. Note that points on the line are always detected, whereas those at a greater distance have a low probability of being detected. (b) Possible measurements for a detected object. The observer is at position O when the object is detected at position P, and Q is that point on the line perpendicular to the object. The sighting distance is r, the sighting angle is θ, and the perpendicular distance from the object to the center line is x. The direction of the observer's travel is from O to Q. (c) Transect lines chosen in a systematic, nonrandom fashion. The initial transverse line is at a random distance from the boundary of the area to be surveyed. (d) Individual transect lines dispersed in a random fashion across the study area. All lines are oriented in the same direction. (e) Transect lines need not parallel each other but can be dispersed across the study area, as long as one line does not meet or cross other lines. (f) In rough terrain transect lines may follow natural features, avoiding boulders, rock outcrops, and other areas difficult to transverse. (Adapted from Burnham, Anderson, and Laake 1980.)

3. Take accurate measurements of distances and angles.

4. Take all three basic measurements: perpendicular distance, sighting (flushing) distance, and sighting (flushing) angle.

5. Record measurements separately for each segment of the total transect length.

6. Record at least 40 animals.

For a successful survey, the observer should take other steps to ensure adequate sampling. Conduct the survey at a time of day appropriate for the species. For songbirds this time would be early morning, for grouse throughout the day, for woodcock in the evening. Conduct surveys under similar weather conditions. Finally, use only well-trained observers, and as few of them as possible to reduce the bias of individual variation in data collecting.

The line transect method may be used with both immobile and mobile populations. Assumptions are most easily met with such immobile populations as dead deer (winter kill), land snails, bird nests, and the like. Most of the problems are encountered with mobile animals—easily flushed animals such as grouse, hares, and rabbits or slow-moving animals such as turtles.

A modification of the transect method is the **strip transect** censusing of songbirds developed by Emlen (1971, 1977; Wakeley 1987a). This method involves tallying birds detected by sight or song on either side of the center transect line. The assumption is that the observer will detect all active birds near the center of the line and for some distance beyond, called the detection distance. Detection distance drops off from the center of the line. The width of the transect line is determined by the full detection distance. For songbirds this distance is about 125 m to either side of the line, assuming the observer's hearing is good. For details consult Emlen (1972, 1977), Cox (1985), and Wakeley (1987a).

Indices

Indices are estimates of animal populations derived from counts of animals signs, calls, roadside counts, and so on. In this type of estimating, all data are relative and must be compared with data from other areas or other time periods. The results do not give estimates of absolute populations, but they do indicate trends of populations from year to year and from habitat to habitat. Often this type of information is all that is needed.

Call counts are used chiefly to obtain population trends of certain game birds, such as the mourning dove, bobwhite quail, woodcock, and pheasant. A predetermined route is established along country roads; it should be no longer than can be covered in one hour's time. Stations are located at quarter- or half-kilometer intervals, depending upon the terrain and the species. The route is run around sunrise for gallinaceous birds and doves and around sunset for woodcock. The exact time to start must be determined for each area by the investigator. The observer stops at each station, listens for a standard period of time (a minute or two), records the calls heard, and goes on to the next station. Routes should be run several times and an average taken. The number of calls divided by the number of stops gives a call-index figure.

The *roadside count* is similar to a call count, with the exception that the number of animals observed along the route is recorded and the results divided by the number of miles. Other variations include counting of animal tracks, browse, signs, active dens and lodges, and so on.

Counting pellets or fecal groups is widely used to estimate big-game populations. This method involves the counting of pellet groups in sample plots or transects located in the study area. It may be used to estimate intensity of use of the range by one or more kinds of animals, to determine trends in animal populations, or in rarer cases to estimate the total population. The last is possible only when an entire herd is known to occupy a given area for a definite period. Intensity of use is usually expressed as the number of pellets or pellet groups per unit area.

The accuracy of estimating populations by this method depends upon some knowledge of the rate of defecation. Herein lies the weakness of the technique, because pellet groups vary with the diet, season, age, sex, rate of decomposition, and type of vegetation (plants can cover the pellet groups). Usually rates vary with the region, so preliminary observations have to be made to arrive at some useful figure. For deer, a pellet-group figure of 15 (per deer) is satisfactory on good range, and 13 for poor range. Rabbits vary too widely in their pellet groups for the technique to have much value with them.

The field procedure is as follows:

1. On randomly located transect lines, establish a number of rectangular plots of 100 m² divided in half longitudinally for ease in counting.

2. Count pellet groups at the most favorable time, when plant growth, leaf fall, and so on are least likely to interfere.

3. Mark the plots permanently and clear or paint the pellet groups at the beginning of the study where age determination of pellets is difficult.

4. Then let

$$t = \frac{1}{na'}y$$

where y is the sum of pellet groups counted over the plots, a' is the area of one plot, n is the number of plots, and t is pellet groups per unit area.

5. Determine the value of t. To translate t to total deer days of use:

6. Assume a defecation rate of 13 pellet groups/deer/day for poor range, 15/day for adequate range.

7. Determine the period, the number of days over which the pellet groups were deposited (for example, since the last count).

8. Divide t by the defecation rate to obtain days of utilization by deer per hectare.

9. Divide the number of days of utilization by the number of days in the period to obtain the number of deer per hectare (assuming a constant population).

10. Multiply the result by 100 ha to obtain the number of deer per square kilometer.

A modification of the pellet-group method for small mammals is the *dropping-board method* (see Emlen et al. 1957).

1. Set out 10 cm squares of weatherproof plywood (in natural color) in lines or grids. Use at least 100 boards. They will cover 0.64 ha if spaced 10 m apart, 1.9 ha with 15 m spacing, and 7.5 ha with 30 m spacing. Be sure the squares are level and placed firmly on the ground.

2. Number each station.

3. Bait the boards or not, depending upon local conditions.

4. Make a series of at least three visits. The time of day the visits are made and their frequency will depend upon local conditions, such as coprophagous insect activity. Daily visits may be necessary.

5. At each station record the presence of droppings by species, and brush the board clean for the next visit. The droppings of small mammals are distinctive and with some experience can be identified (see Murie 1954).

6. Express results as incidence of droppings for each species.

Figures obtained from the record of usage are indices of the population useful in comparative studies of interspecific, interseasonal, and interregional abundance. The dropping-board technique can be used in studies of population trends and fluctuations, local distribution, species association, activity, rhythms, effects of weather and environmental conditions on activity, and movements if the animal is tagged with a radioactive tracer or with dyed bait.

References

Sampling Terrestrial Vegetation

ANDERSON, D. R., ET AL. 1976. *Guidelines for Line Transect Sampling of Biological Populations.* Utah Cooperative Wildlife Research Unit, Logan, UT.

AUBREY, F. T. 1977. Locating random points in the field. *J. Range Management,* 30: 157–158.

BECKER, D. A., AND J. J. CROCKETT. 1973. Evaluation of sampling techniques on tall-grass prairie. *J. Range Management,* 26: 61–65.

BORMANN, F. H. 1953. The statistical efficiency of sample plot size and shape in forest ecology. *Ecology,* 34: 474–487.

BRAY, R., AND J. T. CURTIS. 1957. An ordination of upland forest communities of southern Wisconsin. *Ecol. Monographs,* 27: 325–349.

COTTAM, G., AND J. T. CURTIS. 1956. The use of distance measures in phytosociological sampling. *Ecology,* 37: 451–460.

CURTIS., J. T., AND R. P. MCINTOSH. 1951. The upland forest continuum in the prairie-forest border region of Wisconsin. *Ecology,* 32: 476–496.

GROSENBAUGH, L. R. 1952. Plotless timber estimates—new, fast, easy. *J. Forestry,* 50: 32–37.

GROSENBAUGH, L. R. 1958. *Point-Sampling and Line-Sampling: Probability Theory, Geometric Implications, Synthesis.* Southern Forest Expt. Sta. Occ. Paper 160. U.S. Forest Service Experiment Station, New Orleans, LA.

HAWLEY, T. A. 1978. A comparison of line interception and quadrat estimation methods of determining shrub coverage. *J. Range Management,* 31: 60–62.

HAYS, R. L., C. SUMMERS, AND W. SEITZ. 1981. Estimating wildlife habitat variables. U.S.D.I. Fish and Wildlife Service. FWS/OBS-81/47.

HYDER, D. N., AND F. A. SNEVA. 1960. Bitterlich's plotless method for sampling basal ground cover of bunch grasses. *J. Range Management,* 13: 6–9.

LANG, G. E., D. H. KNIGHT, AND D. A. ANDERSON. 1971. Sampling the density of tree species in a species-rich tropical forest. *Forest Science,* 17: 395–400.

LONG, G. A., P. S. POISSONET, J. A. POISSONET, P. M. DAGET, AND M. P. GORDON. 1972. Improved needle point frame for exact line transects. *J. Range Management,* 25: 228.

POISSONET, P. S., J. A. POISSONET, M. P. GORDON, AND G. A. LONG. 1973. A comparison of sampling methods in dense herbaceous pasture. *J. Range Management,* 26: 65–67.

RICE, E. L. 1967. A statistical method for determining quadrat size and adequacy of sampling. *Ecology,* 48: 1047–1049.

U.S. FOREST SERVICE. 1958. *Techniques and Methods of Measuring Understory Vegetation: A Symposium.* Southern Forest Experiment Station, Washington, DC.

U.S. FOREST SERVICE EXPERIMENT STATION, New Orleans, LA. 1962. *Range Research Methods: A Symposium.* Misc. Publ. 940. U.S. Department of Agriculture, Washington, DC.

Vegetation Analysis

BRAUN-BLANQUET, J. 1951. *Pflanzensoziologie: Grundzuge der Vegetationskunde,* 2nd ed. Springer, Vienna.

BRAUN-BLANQUET, J. 1932. *Plant Sociology.* McGraw-Hill, New York. (An English translation of the 1st ed. of above.)

BREWER, R., AND M. T. MCCANN. 1982. *Laboratory and Field Manual of Ecology.* Saunders, Philadelphia.

BROWER, J. E., AND J. H. ZAR. 1984. *Field and Laboratory Methods for General Ecology,* 2nd ed. Brown, Dubuque, IA.

CAIN, S. A., AND G. M. DE. O. CASTRO. 1959. *Manual of Vegetation Analysis.* Harper & Row, New York.

CHAPMAN, S. B. (ED.). 1976. *Methods in Plant Ecology.* Halsted Press, Wiley, New York.

COX, G. 1985. *Laboratory Manual of General Ecology,* 5th ed. Brown, Dubuque, IA.

GOUNOT, M. 1969. *Méthodes d'étude quantitative de la végétation.* Masson, Paris.

GREIG-SMITH, P. 1983. *Quantitative Plant Ecology,* 3rd ed. Blackwell Scientific Publications, Oxford.

KERSHAW, K. A. 1973. *Quantitative and Dynamic Ecology,* 2nd ed. Edward Arnold, London; Elsevier, New York.

KREBS C. J. 1989. *Ecological Methods.* Harper & Row, New York.

MUELLER-DOMBOIS, D., AND H. ELLENBERG. 1974. *Aims and Methods of Vegetation Ecology.* Wiley, New York.

OOSTING, H. J. 1956. *The Study of Plant Communities.* Freeman, San Francisco.

WHITTAKER, R. (ED.). 1973. *Ordination and Classification of Communities.* Vol. 5, *Handbook of Vegetation Science.* W. Junk, The Hague.

Sampling Aquatic Vegetation

CLARK, W. J., AND W. F. SIGLER. 1963. Method of concentrating phytoplankton samples using membrane filters. *Limnol. Oceanog.,* 8: 127–129.

COWELL, R. R., AND R. Y. MORITA. 1975. *Marine and Estuarine Microbiology Laboratory Manual.* University Park Press, Baltimore.

HARTMAN, R. T. 1958. Studies of plankton centrifuge efficiency. *Ecology,* 39: 374–376.

HESTER, F. E., AND J. S. DANDY. 1962. A multiple-plate sampler for aquatic macroinvertebrates. *Trans. Am. Fish. Soc.* 91: 420–421.

LIND, O. T. 1974. *Handbook of Common Methods in Limnology.* Mosby, St. Louis.

LUND, J. W. G., AND J. F. TALLING. 1957. Botanical limnological methods with special reference to algae. *Botan. Rev.,* 23: 489–583.

MCNABB, C. D. 1960. Enumeration of fresh water phytoplankton concentrated on the membrane filter. *Limnol. Oceanog.,* 5: 57–61.

RODINA, A. G. 1971. *Methods in Aquatic Microbiology.* University Park Press, Baltimore.

SCHWOERBEL, J. 1970. *Methods of Hydrobiology (Freshwater Biology).* Pergamon Press, Elmsford, NY.

SLADECKOVA, ALENA. 1962. Limnological investigation methods for the periphyton (aufwuchs) community. *Botan. Rev.,* 28: 286–350.

WELCH, P. S. 1948. *Limnological Methods.* McGraw-Hill-Blakiston, New York.

WOOD, E. J. F. 1962. A method for phytoplankton study. *Limnol. Oceanog.,* 7: 32–35.

WOOD, R. D. 1975. *Hydrobotanical Methods.* University Park Press, Baltimore.

Dendrochronology

COX, G. W. 1985. *A Laboratory Manual for General Ecology.* Brown, Dubuque, IA.

CREBER, G. T. 1977. Tree rings: A natural data storage system. *Biological Reviews,* 52: 349–383.

FRITTS, H. C. 1960. Multiple regression analysis of radial growth in individual trees. *Forest Sci.,* 6: 344–349.

FRITTS, H. C. 1962. An approach to dendrochronology—screening by multiple regression techniques. *J. Geophys. Res.,* 67: 1413–1420.

FRITTS, H. C. 1971. Dendroclimatology and dendroecology. *Quarternary Res.,* 1: 419–449.

FRITTS, H. C. 1974. Relationship of ring width in arid zone conifers to variations in monthly temperature and precipitation. *Ecol. Monogr.,* 44: 411–440.

FRITTS, H. C. 1976. *Tree Rings and Climate.* Academic Press, New York.

HUGHES, M. K., P. M. KELLY, J. R. PILCHER, AND V. C. LAMARCHE JR. 1982. *Climate from Tree Rings.* Cambridge University Press, London.

KIRTPATRICK, M. 1981. Spatial and age dependent patterns of growth in New England black birch. *Amer. J. Botany,* 68: 535–543.

PILCHER, J. R., AND B. GRAY. 1982. The relationship between oak tree growth in Britain and climate in Britain. *J. Ecology,* 70: 297–304.

ROUGHTON, R. D. 1962. A review of literature on dendrochronology and age determination of woody plants. *Tech. Bull. 15,* Colorado Department of Fish and Game, Denver.

SCHULMAN, E. 1956. *Dendroclimatic Changes in Semiarid America.* University of Arizona Press, Tucson, AR.

SINCLAIR, A. R. E., J. M. GOSLINE, G. HOLDSWORTH, C. T. KREBS, S. BOUTIN, J. N. M. SMITH, R. BOONSTRA, AND M. DALE. 1993. Can the solar cycle and climate synchronize the snowshoe hare cycle in Canada? Evidence of tree rings and ice cores. *Am. Nat.,* 141: 173–198.

SPENSER, D. A. 1964. Porcupine fluctuations in past centuries revealed by dendrochronology. *J. Appl. Ecol.,* 1:127–149.

STOCKTON, C. W., AND H. C. FRITTS. 1971. Conditional probability of occurrence of variations in climate based on width of annual tree rings in Arizona. *Tree-Ring Bull.,* 31: 3–24.

TAYLOR, R. F. 1936. An inexpensive increment borer holder. *J. Forestry,* 34: 814–815.

Palynology

ERDTMANN, G. 1954. *An Introduction to Pollen Analysis.* Ronald Press, New York.

FAEGRI, K., AND J. IVERSEN. 1989. *Textbook of Pollen Analysis,* 4th ed. John Wiley, New York.

FELIX, C. F. 1961. An introduction to palynology. In H. N. Andrews (ed.), *Studies in Paleobotany.* Wiley, New York.

MOORE, P. D., J. A. WEBB, AND M. E. COLLINSON. 1991. *Pollen Analysis,* 2nd ed. Blackwell, Oxford.

WALKER, P. C., AND R. T. HARTMAN. 1960. The forest sequence of the Hartstown bog area in western Pennsylvania. *Ecology,* 41: 461–474.

WODEHOUSE, R. P. 1935. *Pollen Grains.* McGraw-Hill, New York.

Sampling Animal Populations: Comprehensive References

ANDREWARTHA, H. G. 1970. *An Introduction to the Study of Animal Populations,* 2nd ed. University of Chicago Press, Chicago.

ANDREWS, W. A. 1972a. *A Guide to the Study of Freshwater Ecology.* Prentice-Hall, Englewood Cliffs, NJ.

ANDREWS, W. A. 1972b. *A Guide to the Study of Soil Ecology.* Prentice-Hall, Englewood Cliffs, NJ.

BEGON, M. 1979. *Investigating Animal Abundance.* University Park Press, Baltimore, MD.

BLOWER, J. G., L. M. COOK, AND J. A. BISHOP. 1981. *Estimating the Size of Animal Populations.* George Allen & Unwin, London.

BROWER, J. E., AND J. H. ZAR. 1990. *Field and Laboratory Methods for General Ecology,* 3rd ed. Brown, Dubuque, IA.

CAUGHLEY, G. 1977. *Analysis of Vertebrate Populations.* Wiley, New York.

CORMACK, R. M., G. P. PATIL, AND D. S. ROBSON (EDS.). 1979. *Sampling Animal Populations.* International Cooperative Publishing House, Fairland, MD.

COX, G. W. 1985. *Laboratory Manual of General Ecology,* 5th ed. Brown, Dubuque, IA.

DEBLASE, A. F., AND R. E. MARTIN. 1981. *A Manual of Mammalogy.* Brown, Dubuque, IA.

GILES, R. H., JR. (ED.). 1969. *Wildlife Management Techniques.* The Wildlife Society, Washington, DC.

HOLME, N. A., AND A. D. MCINTYRE. 1971. *Methods for the Study of Marine Benthos.* IBP Handbook No. 19. Blackwell, Oxford, England.

JACKSON, R. M., AND F. RAW. 1966. *Life in the Soil.* St. Martin's Press, New York.

KREBS, C. J. 1999. *Ecological Methods.* Benjamin/Cummings, San Francisco.

NIELSEN, L. A., AND D. L. JOHNSON (EDS.). 1983. *Fisheries Techniques.* American Fisheries Society, Bethesda, MD.

PARKINSON, D., T. R. G. GRAY, AND S. T. WILLIAMS. 1971. *Methods for Studying the Ecology of Soil Microorganisms.* IBP Handbook No. 19. Blackwell, Oxford, England.

PHILLIPSON, J. (ED.). 1971. *Methods of Study in Quantitative Soil Ecology.* IBP Handbook No. 18. Blackwell, Oxford, England.

POOLE, R. W. 1974. *An Introduction to Quantitative Ecology.* McGraw-Hill, New York.

RALPH, C. J., AND J. M. SCOTT (EDS.). 1981. Estimating numbers of terrestrial birds. *Studies in Avian Biology,* 6: 1–630.

RICKER, W. E. 1958. Handbook of computations for biological statistics of fish populations. *Fishery Res. Board Can. Bull.,* 119: 1–300.

RICKER, W. E. 1971. *Methods for Assessment of Fish Production in Fresh Waters.* IBP Handbook No. 3. Blackwell, Oxford, England.

SCHEMNITZ, S. D. (ED.) 1980. *Wildlife Management Techniques Manual.* The Wildlife Society, Washington, DC.

SCHWOERBEL, J. 1970. *Methods of Hydrobiology (Freshwater Biology).* Pergamon Press, Elmsford, NY.

SEBER, G. A. F. 1982. *The Estimation of Animal Abundance and Related Parameters.* Macmillan, New York.

SOUTHWOOD, T. R. E. 1978. *Ecological Methods.* Chapman and Hall, London.

VERNER, J., M. L. MORRISON, AND C. J. RALPH (EDS.). 1986. *Wildlife 2000: Modeling Habitat Relationships of Terrestrial Vertebrates.* University of Wisconsin Press, Madison, WI.

WELCH, P. S. 1948. *Limnological Methods.* McGraw-Hill, Blakiston, New York.

Trapping and Collecting

ANDERSON, R. M. 1948. Methods of collecting and preserving vertebrate animals. *National Museum of Canada Bull.,* 69.

BORROR, D. J., AND R. E. WHITE. 1970. *A Field Guide to the Insects* (pp. 4–28). Houghton Mifflin, Boston.

COOPER, R. J., AND R. C. WHITMORE. 1990. Arthropod sampling methods in ornithology. Pp. 29–37 in *Studies in Avian Biology,* Avian Foraging Theory: Modeling and Application. 13.

DEBLASE, A. F., AND R. E. MARTIN. 1981. *A Manual of Mammalogy.* Brown, Dubuque, IA.

KEMPSON, D., M. LLOYD, AND R. GHELARDI. 1963. A new extractor for woodland litter. *Pedobiologia,* 3: 1–21.

KNUDSEN, J. 1966. *Biological Techniques.* Harper & Row, New York.

MURPHY, P. (ED.). 1962. *Progress in Soil Zoology.* Buttersworth, London.

NEEDHAM, J. G. (ED.). 1937. *Culture Methods for Invertebrate Animals* (reprint). Dover, New York.

OMAN, P. W., AND A. D. CUSHMAN. 1948. *Collection and Preservation of Insects.* Mscl. Publ. 60. U.S. Department of Agriculture, Washington, DC.

SOUTHWOOD, T. R. E. 1980. *Ecological Methods with Particular Reference to Insect Populations.* Chapman and Hall, London.

WAGSTAFFE, R. J., AND J. H. FIDLER. 1955. *The Preservation of Natural History Specimens.* Vol. 1, *The Invertebrates.* Philosophical Library, New York.

WILLIAMS, G. E., III. 1974. New technique to facilitate handpicking macrobenthos. *Trans. Amer. Microscop. Soc.,* 93: 220–226.

Marking and Aging Animals

BIRNEY, E. C., R. JENNESS, AND D. D. BAIRD. 1975. Eye lens protein as a criterion of age in cotton rats. *J. Wildlife Management,* 39: 718–728.

CLARK, D. R., JR. 1968. Branding as a marking technique for amphibians and reptiles. *Copeia,* 1971: 148–151.

COCHRAN, W. W., AND R. D. LORD JR. 1963. A radio-tracking system for wild animals. *J. Wildlife Management,* 27: 9–24.

DAPSON, R. W. 1980. Guidelines for statistical usage in age estimation techniques. *J. Wildlife Management,* 44: 541–548.

DAY, G. I., S. D. SCHEMNITZ, AND R. D. TABER. 1980. Capturing and marking wild animals. Pp. 61–88 in S. D. Schemnitz (ed.), *Wildlife Management Techniques Manual.* The Wildlife Society, Washington, DC.

FERNER, J. W. 1979. A review of marking techniques for amphibians and reptiles. *Herptological Circular* No. 9. Society for the Study of Amphibians and Reptiles.

GODFREY, G. K. 1954. Tracing field voles (*Microtus agrestis*) with a Geiger-Muller counter. *Ecology,* 35: 5–10.

GRAHAM, W. J., AND H. W. AMBROSE III. 1967. A technique for continuously locating small mammals in field enclosures. *J. Mammal.,* 48: 639–642.

JERALD, A., JR. 1983. Age determination (fish). Pp. 301–321 in L. A. Nielsen and D. L. Johnson (eds.), *Fisheries Techniques.* American Fisheries Society, Bethesda, MD.

KAYE, S. V. 1960. Gold-198 wires used to study movements of small mammals. *Science,* 13: 824.

KUNZ, T. H., AND E. L. P. ANTHONY. 1982. Age estimation and postnatal growth in the bat *Myotis lucifugus. J. Mamm.,* 63: 23–32.

LARSON, J. S., AND R. D. TAYLOR. 1980. Criteria of sex and age. Pp. 143–202 in S. D. Schemnitz (ed.), *Wildlife Management Techniques Manual.* The Wildlife Society, Washington, DC.

LORD, R. D., JR. 1959. The lens as an indicator of age in cottontail rabbits. *J. Wildlife Management,* 23: 358–360.

LORD, R. D., JR. 1963. The cottontail rabbit in Illinois. *Illinois Dept. Conserv. Tech. Bull.* No. 3.

MADSON, R. M. 1967. *Age Determination of Wildlife: A Bibliography.* Biblio. No. 2. U.S.D.I. Department Library, Washington, DC.

MARION, W. R., AND J. D. SHAMUS. 1977. An annotated bibliography of bird-marking techniques. *Bird-Banding,* 48: 42–61.

MARTOF, B. J. 1953. Territoriality in the green frog *Rana clamitans. Ecology,* 34: 165–174.

NEVILLE, A. C. 1963. Daily growth layers for determining the age of grasshopper populations. *Oikos,* 14: 1–8.

PENDLETON, R. C. 1956. Uses of marking animals in population studies: Labeling animals with radioisotopes. *Ecology,* 37: 686–690.

TABER, R. D. 1956. Marking of mammals: Standard methods and new developments. *Ecology,* 37: 681–685.

TESTER, J. R. 1963. Techniques for studying movements of vertebrates in the field. In *Radioecology.* Van Nostrand Reinhold, New York.

TIEMEIER, O. W., AND M. L. PLENERT. 1964. A comparison of three methods for determining the age of black-tailed jack rabbits. *J. Mammal.,* 45: 409–416.

Estimating Animal Populations

ANDERSON, D. J. 1982. The home range: A new nonparametric estimation technique. *Ecology,* 63: 103–112.

ANDERSON, D. R., J. L. LAAKE, B. R. CRAIN, AND K. P. BURNHAM. 1979. Guidelines for line transect sampling of biological populations. *J. Wildlife Management,* 43: 70–78.

BURNHAM, K. P., D. R. ANDERSON, AND J. L. LAAKE. 1980. Estimation of density from line transect sampling of biological populations. *Wildlife Monograph* No. 44.

CONNER, R. N., AND J. G. DICKSON. 1980. Strip transect sampling and analysis for avian habitat studies. *Wildlife Society Bull.,* 8: 4–10.

DAVIS, D. E. (ED.). 1982. *Handbook of Census Methods for Terrestrial Vertebrates.* CRC Press, Boca Raton, FL.

DESANTE, D. F. 1986. A field test of variable circular plot censusing methods in a Sierran subalpine forest habitat. *Condor,* 88: 129–142.

DIXON, K. R., AND J. A. CHAPMAN. 1980. Harmonic mean measure of animal activity areas. *Ecology,* 61: 1040–1044.

EBERHARDT, L. L. 1978. Transect methods for population studies. *J. Wildlife Management,* 42: 1–31.

EMLEN, J. T. 1971. Population counts of birds derived from transect counts. *Auk,* 88: 323–341.

EMLEN, J. T. 1977. Estimating breeding season bird densities from transect counts. *Auk,* 94: 455–468.

EMLEN, J. T., ET AL. 1957. Dropping boards for population studies of small mammals. *J. Wildlife Management,* 21: 300–414.

FRANZREB, K. E. 1976. Comparison of various transect and spot map methods for censusing avian populations in a mixed coniferous forest. *Condor,* 78: 260–262.

GRODZINSKI, W., Z. PUCEK, AND L. RYSZKOWSKI. 1966. Estimation of rodent numbers by means of prebaiting and intensive removal. *Acta Theriologica,* 11: 297–314.

HAYNE, D. W. 1949. An examination of the strip census method for estimating animal populations. *J. Wildlife Management,* 13: 145–157.

HEYER, W. R., M. A. DONNELLY, R. W. MCDARMID, L. C. HAYEH, AND M. S. FOSTER. 1994. *Measuring and Monitoring Biological Diversity: Standard Methods for Amphibians.* Smithsonian Institute Press, Washington, DC.

JARVINEN, O., AND R. A. VAISANEN. 1975. Estimating relative densities of breeding birds by the line transect method. *Oikos,* 26: 316–322.

MIKOL, S. A. 1980. Field guidelines for using transects to sample nongamebird populations. U.S.D.I. Fish and Wildlife Service. FWS/OBS-80/58.

MURIE, O. J. 1954. *A Field Guide to Animal Tracks.* Houghton Mifflin, Boston.

RALPH, C. J., AND J. M. SCOTT (EDS.). 1981. *Estimating Numbers of Terrestrial Birds. Studies in Avian Biology* 6. Cooper Ornithological Society, Lawrence, KS.

REYNOLDS, R. T., J. M. SCOTT, AND R. A. NUSSBAUM. 1980. A variable circular plot method for estimating bird numbers. *Condor,* 82: 309–313.

ROBINETTE, W. L., R. B. FERGUSON, AND J. S. GASHWEILER. 1958. Problems involved in the use of deer pellet group counts. *Trans. North Amer. Wildlife Conf.,* 23: 411–425.

SAMUEL, M. D., AND E. O. GARTON. 1985. Home range: A weighted normal estimate and tests of underlying assumptions. *J. Wildlife Management,* 49: 513–519.

SAMUEL, M. D., D. J. PIERCE, AND E. O. GARTEN. 1985. Identifying areas of concentrated use within the home range. *J. Anim. Ecol.,* 54: 711–719.

SCHNABEL, Z. E. 1938. The estimation of the total fish population of a lake. *American Mathematics Monthly,* 45: 34–52.

SWIFT, D. M., ET AL. 1976. A technique for estimating small mammal population densities using a grid and assessment lines. *Acta Thierologica,* 21: 471–480.

VERNER, J. 1985. Assessment of counting techniques. Pp. 247–302 in R. J. Johnson (ed.), *Current Ornithology.* Plenum, New York.

WAKELEY, J. S. 1987a. Avian line-transect methods. Section 6.3.2, U.S. Army Corps of Engineers Wildlife Resources Management Manual Tech. Rept. EL-87-5. U.S. Army Engineer Waterways Experiment Station, Vicksburg, MS.

WAKELEY, J. S. 1987b. Avian plot methods. Section 6.3.3, U.S. Army Corps of Engineers Wildlife Resources Management Manual Tech. Rept. EL-87-6. U.S. Army Engineer Waterways Experiment Station, Vicksburg, MS.

WAKELEY, J. S. 1987c. Avian territory mapping. Section 6.3.4, U.S. Army Corps of Engineers Wildlife Resources Management Manual Tech. Rept. EL-87-7. U.S. Army Engineer Waterways Experiment Station, Vicksburg, MS.

WHITE, G. C., D. R. ANDERSON, K. B. BURNHAM, AND D. L. OTIS. 1982. Capture-recapture and removal methods for sampling closed populations. Los Alamos National Laboratory, Los Alamos, NM.

Analyzing Animal Populations

ADAMS, L. 1951. Confidence limits from the Petersen or Lincoln index in animal population studies. *J. Wildlife Management,* 15: 13–19.

BROWNIE, C., D. R. ANDERSON, K. P. BURNHAM, AND D. S. ROBSON. 1978. *Statistical Inference from Band Recovery Data: A Handbook.* U.S.D.I. Fish and Wildlife Service Resource Pub. No. 131.

BURNHAM, K. P., D. R. ANDERSON, AND J. L. LAAKE. 1980. Estimation of density from line transect sampling of biological populations. *Wildlife Monograph* No. 44.

FISHER, R. A., AND E. B. FORD. 1947. The spread of a gene in natural conditions in a colony of the moth *Panaxia dominula. Heredity,* 1: 143–174.

HAYNE, D. W. 1949. Two methods for estimating populations of mammals from trapping records. *J. Mammal.,* 30: 399–411.

JACKSON, C. H. N. 1939. The analysis of an animal population. *J. Animal Ecol.,* 8: 238–246.

JOLLY, G. M. 1965. Explicit estimates from capture-recapture data with both death and immigration-Stochastic model. *Biometrika,* 52: 225–247.

MARTOF, B. J. 1953. Territoriality in the green frog *Rana clamitans. Ecology,* 34: 165–174.

OTIS, D. L., K. P. BURNHAM, G. C. WHITE, AND D. R. ANDERSON. 1978. Statistical inference from capture data on closed animal populations. *Wildlife Monograph* No. 62.

VAN ETTEN, R. C., AND C. L. BENNET JR. 1965. Some sources of error in using pellet group counts for censusing deer. *J. Wildlife Management,* 29: 723–729.

ZIPPIN, C. 1958. The removal method of population estimation. *J. Wildlife Management,* 22: 325–339.

Habitat Analysis

BROWER, J. E., AND J. H. ZAR. 1984. *Field and Laboratory Methods for General Ecology,* 2nd ed. Brown, Dubuque, IA.

COOPERRIDER, A. Y., R. J. BOYD, AND H. R. STUART (EDS.). 1986. *Inventory and Monitoring of Wildlife Habitat.* U.S.D.I. Bur. Land Manage. Service Center, Denver, CO.

FLOOD, B. S., M. E. SANGSTER, R. D. SPARROWS, AND T. S. BASKETT. 1977. *A Handbook for Habitat Evaluation Procedures.* U.S.D.I. Fish and Wildlife Service Resource Publ. No. 132.

MYERS, W. L., AND R. L. SHELTON. 1980. *Survey Methods for Ecosystem Management.* Wiley, New York.

SCHEMNITZ, S. D. (ED.). 1980. *Wildlife Management Techniques Manual,* 4th ed. The Wildlife Society, Washington, DC.

STATES, J. B., P. T. HAUG, T. G. SCHOMAKER, L. W. REED, AND E. B. REED. 1978. *A Systems Approach to Ecological Baseline Studies.* U.S.D.I. Fish and Wildlife Service, Fort Collins, CO.

THOMAS, J. W. (TECH. ED.). 1979. *Wildlife Habitats in Managed Forests: The Blue Mountains of Washington and Oregon.* U.S.D.A. Handbook 553. U.S.D.A. Forest Service, Washington, DC.

U.S. FISH AND WILDLIFE SERVICE. 1980. *Habitat Evaluation Procedures.* U.S.D.I. Fish and Wildlife Service, Washington, DC.

USHER, M. B. (ED.). 1986. *Wildlife Conservation Evaluation.* Chapman and Hall, London.

VERNER, J., M. L. MORRISON, AND C. J. RALPH (EDS.). 1986. *Wildlife 2000: Modeling Habitat Relationships of Terrestrial Vertebrates.* University of Wisconsin Press, Madison, WI.

Measuring Community and Population Structure

MEASURING COMMUNITY STRUCTURE

Population Dispersion

Interspecific Association

Community Similarity

Community Ordination

Species Diversity

MEASURING POPULATION STRUCTURE

Life Table

Fecundity Table

Rate of Increase

Reproductive Value

Age Distribution

REFERENCES

MEASURING COMMUNITY STRUCTURE

Population Dispersion

One of the problems associated with community structure is how the spatial pattern of organisms in the community relates to the interaction of organisms with the environment. Data collected from quadrats, point quadrats, and so on may be used to determine intrapopulation dispersion, as long as we remember that the analysis can be influenced by the size of the sampling unit.

As pointed out in Chapter 10, population dispersion may be uniform, random, or clumped. Where the density of individuals is low for the available surface area or volume, the Poisson method is useful to determine types of dispersion. The Poisson distribution furnishes values expected on the basis of a random dispersion pattern and approximates an extremely asymmetrical distribution. Because the mean of the Poisson is equal to its variance, the Poisson distribution is completely specified by the mean. Thus the theoretical Poisson distribution corresponding to the observed distribution can be constructed from the sample mean μ alone.

To calculate the Poisson we must know the number of sample units, the number of organisms in each sample unit, and the probability that the organism is located in the sample unit or area. The Poisson series is expressed as

$$P_x = e^{\mu}\left(1, x, \frac{\mu^2}{2!}, \ldots, \frac{\mu^4}{i!}\right)$$

The steps for setting up the Poisson distribution are as follows:

1. Determine the sample mean obtained by the equation

$$\mu = \frac{\Sigma f(X)}{N}$$

where f = observed frequencies, X = frequency class, and N = total frequency, individuals, observations. This estimate of the mean density for each sample unit is substituted into the general expression for Poisson probability, $e^{-\mu}$, where e is the base of natural logarithms.

2. Determine from a table of exponential functions (or a scientific calculator) the value of $e^{-\mu}$.

3. Calculate the Poisson probabilities (see example below).

4. Multiply each probability distribution by the total frequency N to convert it to absolute frequency, so that the probabilities are comparable with the observed distribution.

As an example we can use data from Garbutt (1961) on the distribution of fleas on mice. The null hypothesis is that fleas are randomly distributed through a population of mice. The information for *Microtis* is lumped into five classes as follows:

Fleas per mouse (X)	0	1	2	3	4⁺
Mice (f)	44	8	9	3	4

Calculating for the mean,

$$\bar{x} = \frac{\Sigma f(X)}{N} = \frac{(0 \times 44) + (1 \times 8) + (2 \times 9) + (3 \times 3) + (4 \times 4)}{68}$$
$$= 0.75$$

Thus the mean, $\bar{x} = 0.75$ and $e^{-\bar{x}} = e^{-0.75} = 0.472$. To determine the Poisson probability:

$$X(0) = e^{-0.075} = 0.472$$

$$X(1) = \bar{x}e^{-0.75} = 0.75 \times 0.472 = 0.354$$

$$X(2) = \bar{x}^2/2!e^{-0.75} = (0.75)^2/2 \times 0.472 = 0.133$$

$$X(3) = \bar{x}^3/3!e = (0.75)^3/6 \times 0.472 = 0.332$$

$$X(4) = \bar{x}^4/4!e^{-0.75} = (0.75)^4/24 \times 0.472 = 0.0062$$

To obtain the theoretical frequency or distribution, multiply the total number of observations by the Poisson probability for each class (Table B.1). For example, the theoretical distribution for $X(0)$ is $68 \times 0.47 = 31.96$ or 32.

Once the theoretical frequencies have been obtained, the next step is to employ a chi-square goodness-of-fit test to determine how well the data match the Poisson distribution. We set up a chi-square table in which the observed distribution can be compared with the expected distribution (Table B.2).

There are five classes after lumping the data and two constants—$\bar{x}$, the mean, and N, the number of cells. In this example the degrees of freedom are $5 - 2 = 3$. The high value of the chi-square (see statistics books for tabled values of

$$\chi^2 = 34.28$$

$$\text{variance: } s^2 = \frac{f(X - \bar{x})^2}{N - 1} = \frac{\Sigma f(X^2) - N(\bar{x}^2)}{N - 1}$$

$$= \frac{135 - 68(0.56225)}{67} = 1.444$$

$$\frac{s^2}{\bar{x}} = \frac{1.444}{0.75} = 1.925 \qquad P > .005$$

chi-square) with a probability lying well below 0.005 leads to the rejection of the null hypothesis. The fleas are not randomly dispersed among the mice.

TABLE B.1 Poisson Probability and Theoretical Frequency

Number of Fleas per Mouse (X)	Observed Frequency	Poisson Probability	Theoretical Frequency
0	44	0.47	32.0
1	8	0.35	23.8
2	9	0.13	8.9
3	3	0.04	2.7
4+	4	0.01	0.6
Total	68		68.0

TABLE B.2 Comparison of Observed Distribution with Expected Distribution

Number of Fleas per Mouse (X)	Observed Distribution (O or f)	Expected Distribution (E)	O – E	(O – E)²	$\dfrac{(O-E)^2}{E}$
0	44	32.0	12.0	144	4.500
1	8	23.8	−15.8	249.64	10.480
2	9	8.9	0.1	.01	.001
3	3	2.7	0.3	.09	.033
4	4	0.6	3.4	11.56	19.266
Total	68	68.0			$\chi^2 = 34.280$

A further test for randomness is the ratio of the variance, s^2, to the mean, μ. In the Poisson distribution the population mean, μ, is equal to the population variance, s^2. A randomly distributed population would have a ratio of its variance to its mean equal to 1.0. A ratio much less than 1.0 would indicate a uniform distribution, and a ratio much greater than 1.0 would indicate clumped distribution. In the example of fleas on mice, the ratio of 1.925 (see Table B.2) suggests contagious or clumped distribution. Significance of departure from randomness may be assessed by the equation

$$t = \frac{|s^2/\bar{x} - 1.0|}{\sqrt{2/n-1}}$$

Compare the t value to the critical values for t in Student's t table for $n - 1$ degrees of freedom.

For the example of fleas on mice, the t value works out to be

$$t = \frac{1.444/.75}{\sqrt{2/68-1}} - 1.0 = \frac{.921}{.173} = 5.323$$

This value is well above the critical value of 2.660 for 60 DF at the 0.01 level. It indicates that the frequency distribution is significantly different from random.

Another method is to use the chi-square statistic:

$$\text{chi-square} = \text{sum of squares} \frac{(SS)}{\bar{x}}$$

The sum of squares may be computed by

$$SS = (n-1)(s^2)$$

The statistical significance may be obtained by using chi-square tables. The degrees of freedom are $n - 1$. In our example SS = (67) (1.444) = 96.748 and chi-square = 96.748/.75 = 128.99. For 60 DF the chi-square value at the 95 percent level is 79.082 and at the 99 percent level, 88.379. The chi-square value is well above critical values, again indicating significant departure from random.

The relationship between observed and expected Poisson distributions can be compared visually by plotting both values on a graph (Figure B.1). The vertical axis can be either the num-

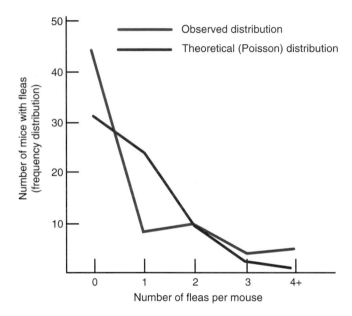

FIGURE B.1 A graph comparing observed distribution with Poisson frequencies for number of fleas per mouse as described in Table B.2.

ber or proportion of samples containing X. The horizontal axis is X. In general, in a clumped distribution the observed number or proportion of empty cells and aggregations would be higher than predicted random probabilities, so there would be a greater chance of finding many individuals in a few plots. In uniform dispersion, there is a higher than random probability of finding only a few individuals in most plots. In this case the observed proportions or numbers would be lower on both the left- and right-hand side of the mean, but higher near the mean.

In the example of fleas on mice, the number of empty cells is much higher than random probability, but the observed distribution at the higher probabilities $P(3)$ and $P(4)$ do not depart much from random. However, the high number of empty cells, mice without fleas, leads to the clumped distribution of fleas on a small portion of the population.

Interspecific Association

Some species in a community may occur together more frequently than by chance. This grouping may result from symbiotic relationships, from food chain coactions, or from similarities in adaptation and response to environmental conditions. Some measurement of this association provides an objective method for recognizing natural groupings of species. Negative associations may indicate interactions detrimental to one or both species, such as interspecific competition, or adaptations to different sets of environmental conditions.

The data are obtained by sampling quadrats, point-centered quadrats, trapping stations, and other means. Presence or absence data for pairs are arranged in a 2×2 or a $2 \times n$ contingency table:

Species B	*Species A* +	−	
+	a	b	$a + b$
−	c	d	$c + d$
+	$a + c$	$b + d$	$a + b + c + d = n$

where

a = samples containing both species A and B

b = samples containing only species B

c = samples containing only species A

d = samples containing neither species

From these data a coefficient of association, C, can be calculated. It will vary from $+ 1.0$ for a maximum positive association to $- 1.0$ for a maximum negative association. A value of 0 suggests that the frequency of association is that expected by chance.

If $bc > ad$ and $d \geq a$, then
$C = ad - bc/(a + b)(a + c)$ (1)

If $bc > ad$ and $a > d$, then
$C = ad - bc/(b + d)(c + d)$ (2)

If $ad \geq bc$ and $c > b$, then
$C = ad - bc/(a + b)(b + d)$ (3)

If $ad > bc$ and $c \geq b$, then
$C = ad - bc/(a + c)(c + d)$ (4)

To learn whether the coefficient of association is significant, we can apply a chi-square test to determine the significant level of the deviations between the observed values of the contingency table and the expected values based on chance association. This test requires a modification of the contingency table as in Table B.3. The number of samples expected in each cell can be determined by

$$a = (a + b)(a + c)/T$$
$$b = (a + b)(b + d)/T$$
$$c = (c + d)(a + c)/T$$
$$d = (c + d)(b + d)/T$$

TABLE B.3 Contingency Table for Determining the Degree of Association

Species B	*Species A* + Observed	Expected	− Observed	Expected
+				
−				

The chi-square value can be calculated on the basis of observed minus expected differences with the formula

$$\text{chi-square} = \sum \frac{(\text{observed} - \text{expected})^2}{\text{expected}}$$

As an example, consider two species of deer mice, *Peromyscus leucopus,* the white-footed mouse, and *P. maniculatus nubiterrae,* the cloudland deer mouse, inhabiting Appalachian hardwood forests. The data are extracted from a three-year study (Violet 1973) on the relations between the two species. For our example, we use the presence of the two species at live-trap stations located on a mesic north-facing slope. The question is: Are the two species mutually exclusive?

Among the 192 stations, *P. maniculatus* occurred exclusively at 133, *P. leucopus,* 14; both species, 26; and none, 19. Because $bc > ad$ and $a > d$, we will use formula 2 to determine the coefficient of association (Table B.4). The two species of mice show a negative association.

$$c = ad - bc/(b + d)(c + d)$$
$$c = 494 - 1862/(33)(152)$$
$$= -1368/5016$$
$$= -0.273$$

TABLE B.4 Coefficient of Association

Peromyscus leucopus	*Peromyscus maniculatus nubiterrae* +	−	
+	a 26	b 14	$a + b$ 40
−	c 133	d 19	$c + d$ 152
Σ	$a + c$ 159	$b + d$ 33	192

To determine whether the coefficient of association is significant, set up a 2×2 contingency table, Table B.5. Because a 2×2 contingency table is the smallest possible and has only one degree of freedom, the absolute value of the difference between observed and expected should be reduced by 0.5, an adjustment known as the ***Yates correction for continuity.***

$$\text{chi-square} = (6.5)^2/33 + (6.5)^2/7 + (6.5)^2/126 + (6.5)^2/26$$
$$= 1.280 + 6.036 + 0.335 + 1.625$$
$$= 9.276$$

The calculated chi-square value of 9.55 exceeds the table value of 3.84, 6.64, and 7.88 for 1 df at the 95, 99, and 99.5 percent levels, respectively. We would reject the hypothesis that the two species are distributed independently and conclude that they are negatively associated.

For a 2×2 contingency table chi-square can be calculated by an alternative method that already has the Yates correction:

$$\text{chi-square} = \frac{(|ad - bc| - 0.5T)^2 (T)}{(a+b)(a+c)(b+d)(c+d)}$$

The chi-square value calculated in this manner is 9.74.

$$\text{chi-square} = \frac{(|1368| - 96)^2 (192)}{(40)(159)(33)(152)}$$
$$= \frac{310,652,928}{31,901,760}$$
$$= 9.74$$

In situations having more than two expressions of each variable, the contingency table will have more than two rows and/or two columns. The expected frequencies for a particular cell are given by the expression

$$\frac{(\text{row total})\,(\text{column total})}{(\text{grand total})}$$

and the chi-square is calculated by the standard formula.

In all chi-square contingency analyses, the degrees of freedom (DF) = (no. of rows $-$ 1) (no. of columns $-$ 1). A 2×2 contingency table has one degree of freedom. A contingency table with three rows and two columns has two degrees of freedom.

Community Similarity

When addressing questions of community structure, ecologists often need to compare the species composition of plants or animals of communities over space or time. The similarity of communities can be measured from data as simple as presence or absence of species or as detailed as density, dominance, frequency, and importance value.

Similarity Coefficients

Similarity coefficients, often called coefficients of community, are the simplest approaches to comparing community structure. They are based solely on presence (indicated with a 1) and absence (indicated with a 0). Among a number of them appearing in the literature are the Jaccard and Sorensen indexes. The values range from 0 when no species are found in both communities to 1, when all species are found in both communities (complete similarity).

The *Jaccard index* is based on the presence-absence relationship between the number of species in each community and the total number of species:

$$SC_j = \frac{c}{A + B - c}$$

where c is the number of common species, A is the total number of species in stand A, and B is the total number of species in stand B. There are variations to this formula (see Mueller-Dombois and Ellenberg 1974). The coefficient expresses the ratio of common species to all species found in the two vegetational groups. The Jaccard coefficient for the stands described in Table B.6 is

$$SC_j = \frac{6}{10 + 7 - 6} = 0.5454 \text{ or } 54.54\%$$

The *Sorensen index* differs from the Jaccard by measuring the ratio of the common to the average number of species in the two samples:

$$SC_s = \frac{c}{\frac{1}{2}A + B}$$

where A is the total number of species in community A, B is the total number of species in community B, and c is the number of species common to both communities. In the Sorensen index, theoretically each species has an equal chance of being present in the two communities. The expression $0.5(A + B)$ represents the sum of the theoretically possible coinciding occurrences, and c is the expression of the actually coinciding occurrences. The Sorensen formula gives greater weight to species common to both areas and less to species unique to either area and results in a greater similarity.

The Sorensen coefficient for the two stands in Table B.6 is

$$SC_s = \frac{6}{\frac{1}{2}(10 + 7)} = 0.7058 \text{ or } 70.58\%$$

The Jaccard and Sorensen indexes do not take into account the relative abundances of species, limiting their usefulness to situations where data on presence and absence of species are sufficient, as in some water pollution studies.

TABLE B.5 Chi-Square Contingency Table

| | | Peromyscus maniculatus nubiterrae | | | |
| | | + | | − | |
		O	E	O	E
Peromyscus	+	26	33	14	7
leucopus	−	133	126	19	26

TABLE B.6 Determination of Percent Similarity

Species	STAND 1		STAND 2	
	Number	% Presence	Number	% Presence
Yellow poplar (*Liriodendron tulipifera*)	83	50.6	0	0*
Red oak (*Quercus rubra*)	25	15.3*	55	34.4
Red maple (*Acer rubrum*)	19	11.6	13	8.1*
Black cherry (*Prunus serotina*)	25	15.3*	27	16.9
Black birch (*Betula lenta*)	2	1.2*	11	6.8
Sugar maple (*Acer saccharum*)	1	0.6*	2	1.3
Sassafras (*Sassafras albidum*)	0	0*	6	3.7
Black locust (*Robinia pseudoacacia*)	1	0.6	0	0*
Black gum (*Nyssa sylvatica*)	3	1.9	0	0*
White ash (*Fraxinus americana*)	1	0.6	0	0*
Chestnut oak (*Quercus prinus*)	4	2.4*	46	28.8
Sum	164		160	

$PS = \Sigma$ (lowest percentage for each species)

    $= 0 + 15.3 + 8.1 + 15.3 + 1.2 + 0.6 + 0 + 0 + 0 + 0 + 2.4$

    $= 42.9$

*Lowest percentage for each species.

Percent Similarity

Another measure of community similarity is the index of percent or proportional similarity, which considers the number of species in each community, the species common to both communities, and the abundance of species. The abundance of a species in a communuity is tabulated as a percentage of the total species presence in that community (Table B.6):

$$\text{percent presence} = \frac{\text{number of individuals of a species}}{\text{total number of individuals in a community}}$$

The lowest percentage for each species in the communities being compared is identified and used to calculate percent similarity using the equation

$$PS = \Sigma(\text{lowest percentage for each species})$$

An example is considered in Table B.6.

Coefficient of Community

A widely used index of similarity between two stands or communities is the coefficient of community. This index ranges in value from 0 to indicate communities with no species in common to 100 to indicate two communities with identical species composition. The proportion of each species can be expressed as density, biomass, frequency, or importance value. The index is calculated using the equation

$$C = \frac{2W}{a + b}(100)$$

where a is the sum of scores for one stand, b the sum of scores for the second stand, and W the sum of lower scores for each species.

An example of the calculations using importance values is given in Table B.8. The two stands involved are Stands 1 and 2, described in Table B.7, which gives stand composition and importance values for canopy trees in eight selected stands. This calculation is the first step in community ordination.

Community Ordination

Ordination is the technique of arranging units (for example, forest stands) in a uni- or multidimensional order in such a manner that the position of each unit along the axis or axes conveys the maximum information about its composition or relationship with the other units.

TABLE B.7 Stand Composition and Importance Values for Canopy Layer

Species	IMPORTANCE VALUES OF STAND							
	1	2	3	4	5	6	7	8
Black oak (*Quercus velutina*)	0	0	0	0	0	0	0	42.90*
Red oak (*Quercus rubra*)	10.25	0	0	0	23.05	0	29.98	7.81
Chestnut oak (*Quercus prinus*)	0	0	0	0	4.68	0	114.46*	0
Sugar maple (*Acer saccharum*)	39.74*	16.94	30.40	27.56	4.46	0	25.56	21.02
Black maple (*Acer nigrum*)	10.16	39.37*	28.95	28.95	0	0	0	0
Slippery elm (*Ulmus rubra*)	20.17	78.87*	0	0	13.60	0	0	0
American elm (*Ulmus americana*)	0	30.77	0	0	0	0	0	11.59
Shagbark hickory (*Carya ovata*)	26.14*	0	0	0	0	0	0	36.67*
Ironwood (*Ostyra virginiana*)	7.02	0	0	0	4.46	0	0	40.49*
American hornbeam (*Carpinus caroliniana*)	0	0	0	0	0	14.01	0	0
Beech (*Fagus grandifolia*)	29.39*	0	46.86*	0	0	161.78*	0	0
Black walnut (*Juglans nigra*)	39.85*	0	17.83	0	0	0	0	0
Yellow poplar (*Liriodendron tulipifera*)	10.44	21.49	51.86*	114.10*	0	0	0	0
White ash (*Fraxinus americana*)	0	0	11.11	0	106.44*	0	0	0
Redbud (*Cercis canadensis*)	6.83	12.98	0	0	0	0	0	32.86*
Flowering dogwood (*Cornus florida*)	0	0	0	0	0	24.21	0	6.66
Black locust (*Robinia pseudoacacia*)	0	12.26	0	0	0	0	0	0
Black cherry (*Prunus serotina*)	0	0	0	17.75	43.31	0	0	0
Sycamore (*Platanus occidentalis*)	0	0	0	4.64	0	0	0	0

*Dominant species.

Ordination is based on the assumption that community composition varies gradually over a continuum of environmental conditions. For this reason communities cannot be classified in discrete units; rather they form a continuum changing in composition and structure over environmental gradients (temperature, elevation, soil, and so on).

Community ordination may be accomplished by two different approaches in deriving the axes. The axes can be based on (1) change in environmental conditions or (2) change in community composition. When the axes represent change in environmental conditions, the position of communities along the axes reflects change in community composition influenced by environmental conditions (gradient analysis). When the axes are based on community composition, the configuration of communities in a geometric space reveals relationship based on similarity in composition.

As an example we will use the Bray-Curtis method, based on the second approach to community ordination. The first step is to determine the degree of similarity among communities or stands using the coefficient of community. The stands

TABLE B.8   Determination of Coefficient of Community

Species	IMPORTANCE VALUE	
	Stand 1	Stand 2
Red oak (*Quercus rubra*)	10.25	0*
Sugar maple (*Acer saccharum*)	39.74	16.94*
Black maple (*Acer nigrum*)	10.16*	39.37
Slippery elm (*Ulmus rubra*)	20.17*	78.87
American elm (*Ulmus americana*)	0*	30.77
Shagbark hickory (*Carya ovata*)	26.14	0*
Ironwood (*Ostyra virginiana*)	7.02	0*
Beech (*Fagus grandiflora*)	29.39	0*
Black walnut (*Juglans nigra*)	39.85	0*
Yellow poplar (*Liriodendron tulipifera*)	10.44*	21.49
Redbud (*Cercis canadensis*)	6.83*	12.98
Black locust (*Robinia pseudoacacia*)	0*	12.26
Sum	199.99	212.68

$$C = \frac{2W}{a+b}(100)$$

a = 199.99

b = 212.68

W = 64.54 (sum of lower scores)

$$C = \frac{2(64.54)}{199.99 + 212.68}(100) = 31.28$$

*Lower score for each species.

involved are described in Table B.7, which gives stand composition and importance values for trees in the canopy layer.

When comparisons are being made among a number of communities, the results are usually presented as a matrix of values representing all pairwise comparisons between communities or stands (see Table B.9).

The next step in the ordination process is to convert the similarity coefficients (Table B.9) to values that express dissimilarity, because the distance between communities in ordination space represents the degree of difference rather than similarity. (Two stands with low values of dissimilarity will appear close together in the ordination arrangement. If the similarity coefficient were used, the greater the similarity, the further apart the two communities would be positioned.) The coefficient of dissimilarity is obtained by subtracting the coefficient of similarity from the highest value of similarity possible. Theoretically this value is 100, but because most replicate samples for a single community show a coefficient of about 85, a more realistic estimate is obtained by using 85 rather than 100 (see Cox 1985). The coefficient of dissimilarity for the stands described in Table B.7 is given in Table B.10.

The position of communities along the ordination axes is determined by calculating values for each stand along the x and y axes (and z axis in the case of three-dimensional ordination). To position the stands along the x axis, terminal points must be determined first. The dissimilarity values between each stand and every other stand are summed (Table B.11). The stand with the highest total dissimilarity values is placed at the 0 point on the x axis and designated as stand A (Table B.11). In our example it happens to be stand 6. The stand with the greatest dissimilarity with A is chosen as the end point along the x axis and designated as stand B (Table B.12) Note that in Table B.12, four stands share the highest dissimilarity with 6, with values of 85.00. They are 2, 4, 5, and 7. Because the four stands are equal, one is chosen arbitrarily as stand B. In the example, 5 is made the end point. The remaining stands are placed along the x axis a given distance, D_x, from stand A using the equation (based on the Pythagorean theorem)

$$D_x = L^2 + (DA^2) - \frac{(DB^2)}{2L}$$

TABLE B.9   Similarity Matrix for Stands in Table B.7

Stand	2	3	4	5	6	7	8
1	31.28	50.75	24.50	16.39	14.70	19.35	31.08
2		35.20	35.20	9.13	0.00	9.26	20.97
3			57.03	8.04	24.21	14.35	10.86
4				11.27	0.00	13.24	10.69
5					0.00	17.40	6.13
6						0.00	3.33
7							15.58

TABLE B.10 Dissimilarity Matrix for Stands in Table B.7

Stand	2	3	4	5	6	7	8
1	55.72	34.25	60.50	68.61	70.30	65.75	53.92
2		49.80	49.80	75.85	85.00	75.74	64.03
3			27.97	76.96	60.79	70.65	74.14
4				73.73	85.00	71.76	74.31
5					85.00	67.60	78.87
6						85.00	81.67
7							69.42

TABLE B.11 Stand Dissimilarity Values

Stand	Sum of Dissimilarity Values (from Table B.10)
1	409.05
2	455.95
3	394.56
4	443.07
5	526.62
6	552.76 stand A
7	518.17
8	496.36

TABLE B.12 Stand Dissimilarity with A

Stand	Dissimilarity with A (Stand 6)
1	70.30
2	85.00
3	60.79
4	85.00
5	85.00 stand B (arbitrary)
7	85.00
8	81.67

For stand 1:

$$D_x = (85)^2 + (70.30)^2 - \frac{(68.61)^2}{170}$$

$$D_x = 43.88$$

where L is the dissimilarity value between A and B, DA the dissimilarity value between A and the stand in question, and DB the dissimilarity value between B and the stand in question.

The calculation of the y coordinate is designed to account for the greatest amount of remaining between-stand variation (Table B.13). First, the stand with the poorest fit along the x axis is determined by calculating a poorness-of-fit value, e, for each stand, using the equation

$$e = \sqrt{DA^2 - x^2}$$

For stand 1:

$$e = \sqrt{(70.3)^2 - (43.88)^2} = 54.92$$

The stand having the largest value of e is designated as stand A′ and given the value of 0 along the y axis (Table B.14). The stand showing the greatest dissimilarity with A′ and located within (0.1) L of A′ along the x axis is chosen as the end point along the y axis and designated as stand B′. The value for stand B′ along the y axis is its dissimilarity value with A′. In the example, stand 4 (stand 6 is excluded because it is the zero point on the x axis) has the greatest dissimilarity value with stand 8 (74.31) and it is within (0.1) L of stand 8 along the x axis, so it becomes the end point along the y axis. The remaining stands are positioned at a given distance, D_y, from A′ (as with the x axis), using the equation

$$D_y = \frac{L^2 + (DA'^2) - (DB'^2)}{2L}$$

where L is the dissimilarity value between A′ and B′, DA' is the dissimilarity value between A′ and the stand in question, and DB' is the dissimilarity value between B′ and the stand in question.

In our example, stand A′ is stand 8 and B′ is stand 4 (Table B.15).

With the points now determined, they can be plotted on an ordination graph, illustrated in Figure B.2. (For an example of construction of a third axis for a three-dimensional ordination, see Mueller-Dombois and Ellenberg 1974).

Species Diversity

Species diversity includes both the number of species and the number of individuals in a community. We also have to consider how the individuals are apportioned among the species. For example, a community consisting of five species and 100 individuals with the individuals equally divided among all

TABLE B.13 Stand Ordination Values, x Axis

Stand	Value along x Axis
1	43.88
2	51.16
3	29.40
4	53.25
5	85.00
6	0.00
7	58.12
8	45.14

TABLE B.14 Stand Poorness-of-Fit Values

Stand	e	
1	54.92	
2	67.88	
3	53.41	
4	66.44	stand B′
5	0.00	
6		NA
7	62.02	
8	68.08	stand A′

TABLE B.15 Stand Ordination Values, y Axis

Stand	Value along y Axis
1	38.07
2	52.04
3	70.23
4	74.31
5	47.11
6	39.24
7	40.56
8	0.00

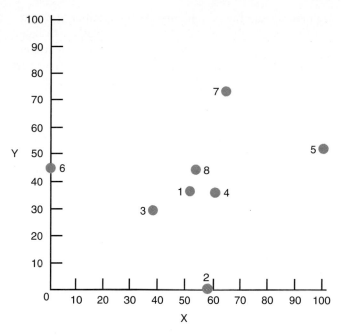

FIGURE B.2 An ordination graph for the eight forest stands described in Tables B.7 to B.15.

five species would be more equitable than a community in which 80 individuals were of one species and the remaining 20 were allotted to the other four species.

Two approaches to species diversity are widely used today: Simpson's index (Simpson 1949) and the Shannon formula (Shannon and Wiener 1963). Both are sensitive to changes in the number of species and to changes in the distribution of individuals among the species. However, the index value of both is influenced by sample size. If the index is to be used to compare diversity among communities, the sample sizes must be equal. If complete census data are used, the areas sampled must be of equal size.

Simpson's index of diversity considers the number of species, the total number of individuals, and the proportion of the total found in each species. It is based on the number of samples of random pairs of individuals that must be drawn from a community to provide at least a 50 percent chance of obtaining a pair with both individuals of the same species. The index is calculated by the formula

$$\lambda = \frac{\Sigma n_i(n_i - 1)}{N(N - 1)}$$

where N is the total number of individuals of all species and n_i the number of individuals of a species. This formula actually measures dominance. A group of species with low dominance will have high diversity. To obtain an index of diversity the formula must be changed:

$$D = 1 - \frac{\Sigma n_i(n_i - 1)}{N(N - 1)}$$

However, Simpson's index is usually inverted to obtain a measure of diversity:

$$D = \frac{N(N - 1)}{\Sigma n_i(n_i - 1)}$$

A community containing only one species would have a value of 1.0. Values would increase to infinity, at which every individual belongs to a different species.

The Shannon formula comes from information theory. In ecological use the function describes the degree of uncertainty of predicting the species of a given individual picked at random from the community. As the number of species increases and as the individuals are more equally distributed among the species present, the more the uncertainty increases. This function has been criticized (see Hurlbert 1971). The Shannon formula in a general form is

$$H' = -\Sigma\, p_i \log p_i$$

where p_i is the decimal fraction of total individuals belonging to the ith species.

In words the formula states that the probability that any one individual belongs to species i is p_i. This value in turn is equal to the ratio n_i/n, where n_i is the number of individuals in the ith species and n is the total number of individuals of all species in the sample. Diversity is greatest if each individual belongs to a different species; the least if all individuals belong to one species. A working formula is

$$H' = \sum_{i=1}^{s}\left(\frac{n_i}{N}\right)\log_2\left(\frac{n_i}{N}\right)$$

where s is the total number of species collected and $\log_2$ is 3.322 $\log_{10}$. For calculation of the index the equation used is

$$H' = 3.322\left[\log_{10}n - (1/n\ \Sigma n_i \log_{10}n_i)\right]$$

To compute:

1. Obtain the appropriate $\log_{10}n$.
2. For each species calculate $\log_{10}n_i$
3. Calculate $n_i\log_{10}n_i$, which means multiplying the number of individuals in each species by $\log_{10}n_i$.
4. Sum the $n_i\log_{10}n_i$ and divide the value by n, the total number of individuals of all the species.
5. Subtract this value from $\log_{10}n$ and multiply by 3.322 to convert the index value to $\log_2$.

Any log base may be used to calculate diversity as long as it is used consistently. Most commonly used logarithmic bases are 10, e, and 2. Because the Shannon formula comes from communication engineering, base 2 is commonly employed and is used here.

If species abundance data are sampled in a nonrandom fashion or if data have been collected for a whole community (such as a total census of organisms), a measure of diversity better suited than Shannon is Brillouin's index:

$$H = \log_{10}\frac{N!}{n_i!}$$

where N is the total number of individuals in all species and n_i the number of individuals in the ith species. For calculation the equation used is

$$H = c(\log_{10}N! - \Sigma \log_{10}n_i)/N$$

where c is a constant for conversion of logarithms from the base 10 to the base chosen for measure. If the base is 2, c is 3.3219; if the base is e, c is 2.3026.

MEASURING POPULATION STRUCTURE

Life Table

The life table has been explained in Chapter 10. Procedures for constructing a life table are considered here.

To construct a life table we must be able to determine the age of the organisms in question and distribute the population members into age classes or age intervals. Age intervals can vary according to the longevity of the organism. For small rodents or lagomorphs the age intervals may be one month, for deer one year, for humans five years. For insects age categories may be instars or life history stages. We need information on survival, mortality, or rate of mortality by age classes for a given population. Data on survivorship in each age class provide the information needed for the survivorship column, l_x. Data on age-specific mortality provide information for the mortality column, d_x. We can use the sum of mortality for each age class over time as the size of the initial population. Thus given information for any one column, we can calculate the others.

To demonstrate the construction of a life table, we will use data for a population of Belding's ground squirrel (*Spermophilus beldingi*) (Sherman and Morton 1984). Data for the life table, Table B.16, were obtained by live-trapping and tagging the entire population of ground squirrels over a period of 11 years. Because the data over the years were pooled, the life table is a dynamic-composite one. The life table considers females only.

These are the steps:

1. Construct a table with the following columns:

    x = age interval or age class

    n_x = number of survivors at start of age interval x (raw field data)

    l_x = proportion of organisms surviving to start age interval x

    d_x = number or proportion dying during age interval x to $x + 1$

    q_x = rate of mortality during the age interval x to $x + 1$

    L_x = number of individuals alive on the average during the age interval x to $x + 1$

    T_x = total years to be lived by individuals of age x in the population

    e_x = mean expectation of life for individuals alive at start of age interval x

2. Tally raw data for survivorship in the n_x column; adjust this survivorship on the basis of 1000 or as a probability of 1.0. The use of 1000 animals makes the table easier to understand, but later the numbers will have to be converted to probabilities to construct fecundity and other tables. If the frequency of each age class in the raw data is not equal to or greater than $x + 1$, the age distribution must be smoothed. Caughley (1977:96) provides a method.

TABLE B.16 Life Table for a Population of Belding's Ground Squirrel

x	n_x	l_x	d_x	q_x	L_x	T_x	e_x
0–1	337	1000	614	0.614	693	1301	1.30
1–2	130	386	190	0.492	291	608	1.58
2–3	66	196	90	0.459	151	317	1.62
3–4	36	106	53	0.500	79.5	166	1.57
4–5	18	53	24	0.452	41	86.5	1.63
5–6	10	29	15	0.517	21.5	45.5	1.57
6–7	5	14	5	0.357	11.5	24	1.71
7–8	3	9	4	0.444	7	12.5	1.38
8–9	2	5	2	0.400	4	5.5	1.10
9–10	1	3	3	1.000	1.5	1.5	0.50

3. Determine mortality by subtracting $l_x + 1$ from l_x. The difference gives d_x.

4. Calculate q_x by dividing d_x by l_x.

5. Calculate additional information:

$$L_x = \frac{l_x + l_{x+1}}{2}$$

Do this work for each age interval.

6. Sum the L_x column cumulatively from the bottom up to obtain T_x.

7. Calculate the life expectancy for each age class by

$$e_x = \frac{T_x}{l_x}$$

Although the construction of a life table is straightforward, given data on survival of various age classes, the life table may be inaccurate or invalid. Especially questionable are life tables based on capture-recapture of marked or banded individuals (Anderson, Wywialowski, and Burnham 1981). Life tables based on such data involve two assumptions. (1) Annual survival rate is age-specific; it varies only by age and not by year. (2) Recovery rates are constant over all ages and all years. Rarely are these assumptions met. In fact, annual survival rates, especially among birds, do vary by year, often influenced more by weather than by age. (For other assumptions in constructing life tables from various sources of data see Caughley 1977 and Krebs 1999).

Fecundity Table

If we know the productivity of each age class of females, m_x, determined by litter counts, brood counts, placental scars, young fledged, and so on, we can construct a fecundity table (Table B.17). The fecundity table includes the age categories, x; age-specific survivorship from the female life table, l_x; age-specific productivity, m_x; and the mean number of female young produced by each female of age x, $l_x m_x$, which is m_x

TABLE B.17 Fecundity Table, Belding's Ground Squirrel

x	l_x	m_x	$l_x m_x$	$x l_x m_x$
0–1	1.000	0.000	0.000	0.000
1–2	0.386	1.040	0.401	0.401
2–3	0.196	2.171	0.426	0.852
3–4	0.106	2.450	0.260	0.780
4–5	0.053	3.113	0.165	0.660
5–6	0.029	1.875	0.054	0.250
6–7	0.014	1.575	0.022	0.132
7–8	0.009	2.000	0.018	0.126
8–9	0.005	1.440	0.007	0.056
9–10	0.003	1.632	0.005	0.045

$$\Sigma l_x m_x = R = 1.358$$
$$\Sigma x l_x m_x = 3.302$$

weighted by survivorship. The sum of the $l_x m_x$ column gives the net reproductive rate, R_o. Added to the fecundity table is another column, $x l_x m_x$, which records the values obtained by multiplying the $l_x m_x$ by the appropriate age. The sum of this column is used to compute the rate of increase.

In Table B.17, age categories have been converted to mean age to permit the correction calculation of $x l_x m_x$ starting with year class 0.

Rate of Increase

Given a life table and a fecundity table, we can determine the rate of increase, r_m, for a population in a particular environment.

An approximation of r can be obtained by the equation

$$r = \frac{\Sigma l_x m_x \log_e \Sigma l_x m_x}{\Sigma x l_x m_x}$$

For the ground squirrel population an approximation is

$$r = \frac{1.358(0.3060)}{3.302} = 0.125$$

Another method requires more calculations but gives the same result. The first step is to determine mean generation time

$$T_c = \frac{\Sigma x l_x m_x}{R_0}$$

For the ground squirrels

$$T_c = \frac{3.302}{1.358} = 2.431$$

To find an approximate value of r_m the following formula applies:

$$r_m = \log_e R_0/T_c = \log_e 1.358/2.431$$
$$= 0.1306/2.431$$
$$= 0.125$$

A more accurate assessment of the rate of increase, also based on life table and fecundity tables at any given density with a stable age distribution, can be obtained by the formula derived by the eighteenth-century French mathematician, Leonard Euler:

$$\Sigma l_x m_x e^{-rx} = 1$$

Because there is no real way to solve for r, it can be obtained only by substituting values for r in the equation by trial and error until the right side balances with the left. The ground squirrel population will serve as an example, although we may erroneously assume it has a stable age distribution (Table B.20).

The first step is to obtain some approximate estimate of r to be used in the Euler equation. The estimate of $r = 0.125$

obtained from $\log_e R_0/T_c$ can be used as the first estimate of r. The procedure follows:

1. Multiply the estimated r value by age to obtain rx. Thus for the first estimated value 0.125 is multiplied by 0, 1, 2, 3, 4, 5, 6, 7, 8, and 9 respectively to give the following values for rx: 0.000, 0.125, 0.250, 0.375, 0.050, 0.625, 0.756, 0.875, 1.000, and 1.125.

2. In a table of functions look up the tabled value for each of the above values of e^{-rx} in the e^{-x} column, or determine the value on a scientific calculator. Record the values as in Table B.18.

3. Multiply the values of e^{-rx} by the appropriate $l_x m_x$ and record. Sum the column. For the r value 0.125, the sum is 1.0158, not close enough to 1.000 to accept.

4. Proceed by substituting a value higher than 0.125 to attempt to bracket $r = 1.000$ between the first estimate and another. A choice of 0.135 results in a sum of 0.9927, too far on the other side of 1.000. A value of 0.132 results in a sum of 0.9995, a very close approximation.

Such iterations can become time consuming. Iteration for the Euler equation can be programmed on a computer or a programmable calculator.

Reproductive Value

The life table and fecundity table provide the data needed to calculate the reproductive values of females age x in the population. The reproductive value is the relative number of female offspring that remain to be born to each female age x. To state it differently, it is the number of offspring that will be produced by a female from age x until the end of her life. It can be estimated relative to the reproductive value of the

TABLE B.18 Determination of r_m, Belding's Ground Squirrel

x	$l_x m_x$	r_x	e^{-rx}	$e^{-rx}l_x m_x$	r_x	e^{-rx}	$e^{-rx}l_x m_x$	r_x	e^{-rx}	$e^{-rx+}l_x m_x$
			r = 0.125			**r = 0.135**			**r = 0.132**	
0	0.000	0.000	0.0000	0.000	0.000	0.0000	0.0000	0.000	0.0000	0.0000
1	.401	0.125	0.8825	0.3539	0.135	0.8737	0.3504	0.132	0.8763	0.3514
2	.426	.250	.7788	.3317	.270	.7634	.3252	.264	.7679	.3271
3	.260	.375	.6872	.1787	.405	.6669	.1734	.369	.6730	.1749
4	.165	.050	.6065	.1000	.540	.5827	.0956	.528	.5898	.0973
5	.054	.625	.5352	.0289	.675	.5091	.0275	.660	.5168	.0279
6	.022	.756	.4695	.0103	.810	.4448	.0098	.792	.4529	.0099
7	.018	.875	.4723	.0085	.945	.3887	.0069	.924	.3969	.0071
8	.007	1.000	.3678	.0026	1.080	.3396	.0024	1.056	.3478	.0024
9	.005	1.125	.3246	.0016	1.215	.2967	.0015	1.188	.3048	.0015
	$\Sigma e^{-rx} l_x m_x =$			1.0158			0.9927			0.9995

female at birth, v_o, which is 1 by the formula given by the geneticist R. A. Fisher:

$$\frac{v_x}{v_x} = \frac{r^{rx}}{l_x} \sum_{y=x}^{\infty} e^{-ry} l_y m_y$$

Because the reproductive value of a female at birth is 1, v_o can be removed from the formula. The formula can also be written as

$$v_x = \frac{e^{rx}}{l_x} \sum_{y=x}^{\infty} e^{-ry} l_y m_y$$

where y is all ages a female passes through from age x on up. Essentially the formula states that the number of female offspring produced at any one moment of time by females aged x and over is divided by the number of females of age x at any one moment.

To calculate reproductive values for females of each age class (Table B.19):

1. Add the $e^{-rx}l_x m_x$ column from the bottom up to obtain a value for $e^{-rx}l_y m_y$ for each age class.
2. Divide this value by $e^{-rx}l_x$ for each age class. The reproductive values for the ground squirrel population are graphed in Figure B.3. Note that the reproductive values rise and fall with age.

Age Distribution

A population growing geometrically with constant age-specific mortality and fecundity rates assumes and maintains a stable age distribution (see Chapter 17). From any set of life tables and fecundity tables for a population the stable age distribution can be determined by the equation

$$C_x = \frac{\lambda^{-x} l_x}{\sum_{i=0}^{\infty} \lambda^{-i} l_i}$$

where C_x is the proportion of organisms in age category x to $x + l$ in a population increasing geometrically, $\lambda = e^{rm}$ finite rate of increase, l_x is the survivorship function from the life table, and x, i subscripts indicate age.

The ground squirrel population again may serve as the example. In this case $\lambda = e^r = e^{0.132} = 1.14$.

Calculating the proportion of squirrels in each age category gives, for example,

$$C_0 = \frac{1.0000}{1.624} = 0.615$$

$$C_5 = \frac{0.0151}{1.624} = 0.009$$

Compare these calculated age distribution values with the actual age structure of the population given in Table B.20.

TABLE B.19 Determination of Reproductive Values, Female Belding's Ground Squirrel

x	l_x	e^{-rx}	$e^{-rx}l_x$	$e^{-rx}l_x m_x$	$e^{-rx}l_y m_y$	v_x
0	1.000	0.000	0.000	0.000	0.9995	1.00
1	.386	.8763	.3383	.3514	.9995	2.95
2	.196	.7679	.1506	.3270	.6481	4.30
3	.106	.6730	.0713	.1749	.3210	4.50
4	.053	.5898	.0312	.0973	.1461	4.68
5	.029	.5168	.0149	.0279	.0488	3.28
6	.014	.4529	.0063	.0099	.0209	3.12
7	.009	.3969	.0036	.0071	.0110	3.05
8	.005	.3478	.0017	.0024	.0039	2.29
9	.003	3048	.0009	.0015	.0015	1.67

TABLE B.20 Stable Age Distribution, Belding's Ground Squirrel

x	l_x	λ^{-x}	$\lambda^{-x}l_x m_x$	C_x
0	1.000	1.000	1.000	0.615
1	.386	.8872	.3425	.210
2	.196	.7695	.1508	.094
3	.106	.6749	.0715	.044
4	.053	.5921	.0314	.019
5	.029	.5194	.0151	.009
6	.014	.4556	.0064	.004
7	.009	.3996	.0036	.003
8	.005	.3505	.0018	.001
9	.003	.3075	.0009	.001

$$\Sigma \lambda^{-x} l_x = 1.6240$$

$$\lambda = e^r = e^{0.132} = 1.14$$

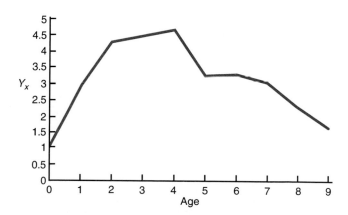

FIGURE B.3   Reproductive values for the female Belding's ground squirrel population described in Tables B.16 to B.19.

References

ANDREWARTHA, H. G. 1970. *Introduction to the Study of Animal Populations.* University of Chicago Press, Chicago.

BEGON, M., AND M. MORTIMER. 1986. *Population Ecology,* 2nd ed. Sinauer Associates, Sunderland, MA.

BRILLOUIN, L. 1962. *Science and Information Theory,* 2nd ed. Academic Press, New York.

BROWER, J. E., AND J. ZAR. 1984. *Field and Laboratory Methods for General Ecology,* 2nd ed. Brown, Dubuque, IA.

CAUGHLEY, G. 1977. *Analysis of Vertebrate Populations.* Wiley, New York.

COX, G. W. 1985. *Laboratory Manual of General Ecology,* 5th ed. Brown, Dubuque, IA.

GARBUTT, P. D. 1961. The distribution of some small mammals and their associated fleas from central Labrador. *Ecology,* 42: 518–525.

GREIG-SMITH, P. 1984. *Quantitative Plant Ecology,* 3rd ed. Blackwell Scientific Publishers, Oxford. Butterworth, London.

HURLBERT, S. H. 1971. The nonconcept of species diversity: A critique and alternative parameters. *Ecology,* 52: 577–586.

KREBS, C. J. 1985. *Ecology: The Experimental Analysis of Distribution and Abundance,* 3rd ed. Harper & Row, New York.

KREBS, C. J. 1999. *Ecological Methodology.* Benjamin/Cummings, San Francisco.

LEVINS, R. 1968. *Evolution in Changing Environments.* Princeton University Press, Princeton, NJ.

LLOYD, M., AND R. J. GHELARDI. 1964. A table for calculating the "equitability" component of species diversity. *J. Animal Ecology,* 33: 217–225.

LUDWIG, J. A., AND J. F. REYNOLDS. 1988. *Statistical Ecology.* Wiley, New York.

MACARTHUR, R. H., AND J. W. MACARTHUR. 1961. On bird species diversity. *Ecology,* 42: 594–598.

MORISITA, M. 1959. Measuring of interspecific association and similarity between communities. *Memoirs of the Faculty of Science, Kyushu University,* Series E (Biology), 3: 66–80.

MUELLER-DOMBOIS, D., AND H. ELLENBERG. 1974. *Aims and Methods of Vegetation Ecology.* Wiley, New York.

PEET, R. K. 1974. The measurement of diversity. *Ann. Rev. Ecology and Systematics,* 5: 285–307.

PIELOU, E. C. 1975. *Ecological Diversity.* Wiley, New York.

POOLE, R. W. 1974. *An Introduction to Quantitative Ecology.* McGraw-Hill, New York.

SHANNON, C. E., AND W. WIENER. 1963. *The Mathematical Theory of Communication.* University of Illinois Press, Urbana.

SHERMAN, P. W., AND M. L. MORTON. 1986. Demography of Belding's ground squirrel. *Ecology,* 65: 1617–1628.

SIMPSON, E. H. 1949. Measurement of diversity. *Nature,* 163: 688.

TRAMER, E. J. 1969. Bird species diversity: Components of Shannon's formula. *Ecology,* 50: 927–929.

VIOLET, R. L. 1973. Microdistribution of small mammals in an Appalachian forest. Unpublished MS thesis, West Virginia University.

ZAR, J. H. 1984. *Biostatistical Analysis,* 2nd ed. Prentice-Hall, Englewood Cliffs, NJ.

APPENDIX C

Distribution of World Vegetation Maps

ECO REGIONS OF THE CONTINENT

HOLDRIDGE LIFE ZONES

Ecoregions of the Continents

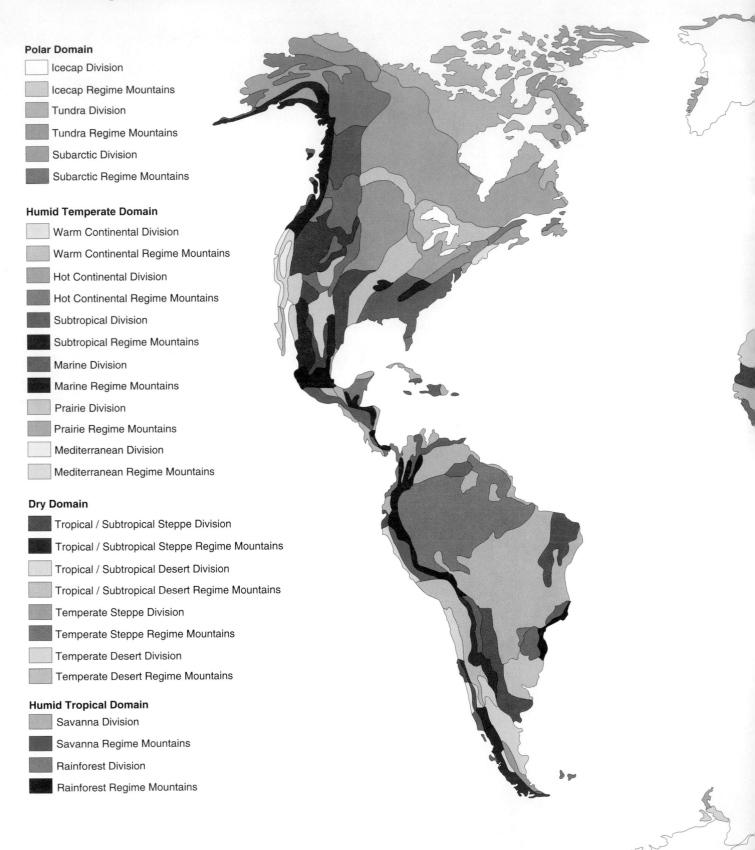

Polar Domain

- Icecap Division
- Icecap Regime Mountains
- Tundra Division
- Tundra Regime Mountains
- Subarctic Division
- Subarctic Regime Mountains

Humid Temperate Domain

- Warm Continental Division
- Warm Continental Regime Mountains
- Hot Continental Division
- Hot Continental Regime Mountains
- Subtropical Division
- Subtropical Regime Mountains
- Marine Division
- Marine Regime Mountains
- Prairie Division
- Prairie Regime Mountains
- Mediterranean Division
- Mediterranean Regime Mountains

Dry Domain

- Tropical / Subtropical Steppe Division
- Tropical / Subtropical Steppe Regime Mountains
- Tropical / Subtropical Desert Division
- Tropical / Subtropical Desert Regime Mountains
- Temperate Steppe Division
- Temperate Steppe Regime Mountains
- Temperate Desert Division
- Temperate Desert Regime Mountains

Humid Tropical Domain

- Savanna Division
- Savanna Regime Mountains
- Rainforest Division
- Rainforest Regime Mountains

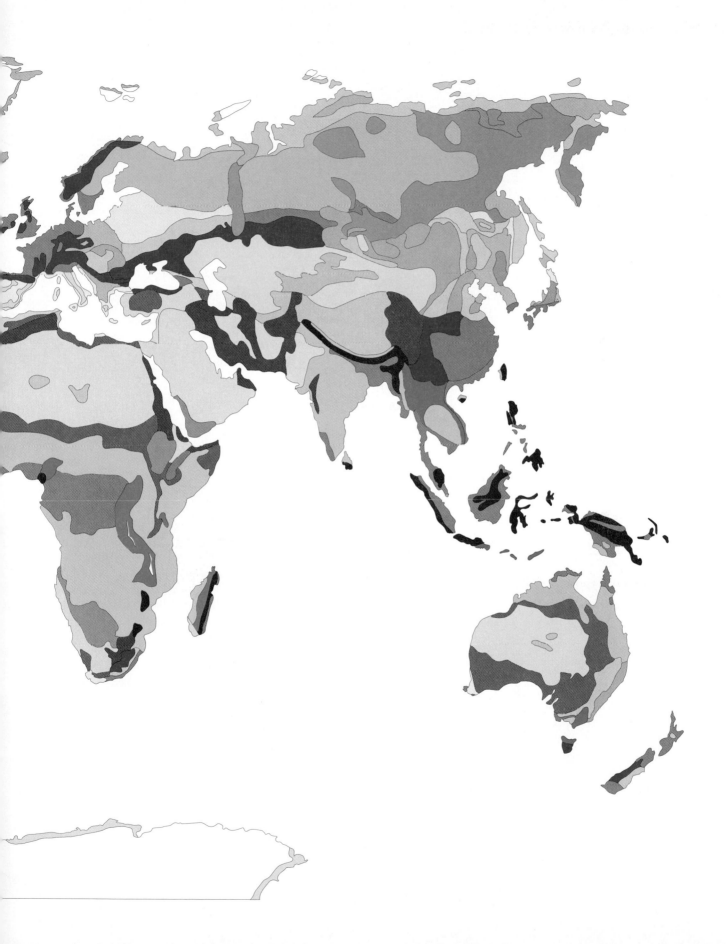

Holdridge Life Zones

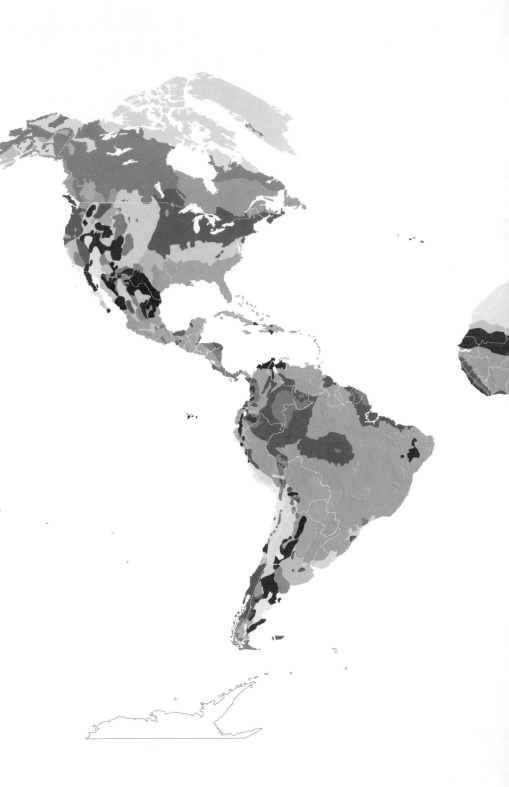

- Tropical / Subtropical Dry Forest
- Tropical / Subtropical Desert Bush
- Tropical / Subtropical Savanna
- Warm Temperate Dry Forest
- Warm Temperate Thorn Steppe
- Temperate Desert Bush
- Boreal Desert / Dry Bush
- Polar Desert / Ice
- Desert
- Tundra
- Steppe
- Boreal Moist Forest
- Boreal Wet / Rain Forest
- Cool Temperate Forest
- Warm Temperate Forest
- Subtropical Moist Forest
- Subtropical Wet / Rain Forest
- Tropical Moist Forest
- Tropical Wet / Rain Forest

GLOSSARY

(Numbers in the parentheses refer to the chapter in which the term appears.)

A **horizon** surface stratum of mineral soil characterized by maximum accumulation of organic matter, maximum biological activity, and loss of such materials as iron, aluminum oxides, and clays (4)

abiotic describes the nonliving component of the environment, including soil, water, air, light, nutrients, and the like 24

abiotic inputs light energy, inorganic substances organic compounds, water, and carbon dioxide that are involved in energy flow and nutrient cycles in ecosystems (24)

abyssal relating to the bottom waters of oceans, usually below 1000 m (31)

acclimation alteration of physiological rate or other capacity to perform a function through long-term exposure to certain conditions (5)

acclimatization changes or differences in a physiological state that appear after exposure to different natural environments (5)

acid deposition wet and dry atmospheric fallout with an extremely low pH. It is brought about by a combination of water vapor in the atmosphere with hydrogen sulfide and nitrous oxide vapors released to the atmosphere from the burning of fossil fuels. The result is sulfuric and nitric acid in rain, fog, and snow, gases, and particulate matter (26)

active surface the surface of any object that receives or is impacted directly by solar radiation (3)

active temperature range (ACT) the range of body temperature over which ectotherms carry out their daily activities (8)

active transport movement of ions and molecules across a cell membrane against a concentration gradient involving an expenditure of energy. The movement of the ion or molecule is in a direction opposite to that it would take under simple diffusion (8)

actual evapotranspiration he amount of evapotranspiration taking place under current ambient conditions (9)

adaptation genetically determined characteristic (behavioral, morphological, physiological) that improves an organism's ability to survive and successfully reproduce under prevailing environmental conditions (5)

adiabatic cooling a decrease in air temperature that results when a rising parcel of warm air cools by expansion (which uses energy) rather than losing heat to the outside surrounding air. The rate of cooling is approximately 1°C/100 m for dry air and 0.6°C/100 m for moist air (3)

adiabatic lapse rate rate at which a parcel of air loses temperature with elevation if no heat is gained from or lost to an external source (2)

adiabatic process one in which heat is neither lost to or gained from the outside (2)

aerenchyma plant tissue with large air-filled intercellular spaces, usually found in roots and stems of aquatic and marsh plants (7)

aerobic living or occurring only in the presence of free uncombined molecular oxygen either as a gas in the atmosphere or dissolved in water (9)

aestivation dormancy in animals through a drought or dry season (See estivation) (8)

age distribution ratio of each age group in a population. See also *stationary age distribution* and *stable age distribution* (10)

age-specific schedule of births the number of offspring produced per unit time by females in different age classes (10)

age structure the number or proportion of individuals in each age group within a population (10)

aggregates soil particles held together in clusters (4)

aggregative response behavior in which consumers spend most of the time in food patches with the greatest density of prey (15)

aggressive mimicry resemblance of a predator or parasite to a harmless species to deceive potential prey (16)

albedo proportion of solar radiation reflected by Earth's surface, by the tops of clouds, and by the atmosphere without heating the receiving surface (2)

Alfisol soil characterized by an accumulation of iron and aluminum in lower or *B* horizon (4)

alkalinity the amount of anions of weak acid in water and of the cations (other than protons) balanced against them; measured by the amount of acid needed to free CO_2 (30)

allele one of two or more alternative forms of a gene that occupies the same relative position or locus on homologous chromosomes (19)

allele frequency the commonness of an allele in a population (19)

allelopathy effect of metabolic products of plants (excluding microorganisms) on the growth and development of other, nearby plants (14)

allogenic succession ecological change or development of species structure and community composition brought about some externally generated force such as fire or storms (22)

allopatric having different areas of geographical distribution; possessing nonoverlapping ranges (19)

allozygous describes two alleles at a locus in question that are not replicates of a single ancestral allele and thus are not identical by descent; see *autozygous* (19)

alluvial soil soil developing from recent alluvium (material deposited by running water); exhibits no horizon development; typical of flood plains (4)

alpha diversity the variety of organisms occupying a given place or habitat (20)

altricial describes the condition among birds and mammals of being hatched or born usually blind and too weak to support their own weight (13)

altruism a form of behavior in which an individual increases the welfare of another at the expense of its own welfare and fitness (19)

ambient refers to surrounding, external, or unconfined conditions (7, 8)

amensalism relationship between two species in which one is inhibited or harmed by the presence of another (14)

ammonification breakdown of proteins and amino acids, especially by fungi and bacteria with ammonia as the excretory by-product (25)

anaerobic adapted to environmental conditions devoid of oxygen (8)

Andosol soils derived from volcanic ejecta; not highly weathered; dark-colored upper layer (4)

anticyclone an area of high atmospheric pressure characterized by subsiding air and horizontal divergence of air near the surface in its central region (2)

applied ecology the application of ecosystem, community, population, and theoretical ecology to resource management (1)

arbuscle the finely bunched network of mycorrhizal hyphae within the cell of the host that serves as the site of exchange between the fungus and the host (17)

area-insensitive species species that are at home in both small and large areas of habitat (23)

area-sensitive species organisms that require large territories or foraging areas, but may or may not be interior species (23)

Aridosol desert soils characterized by little organic matter and high base content (4)

arroyos water-cut canyons in deserts (28)

asexual reproduction any form of reproduction, such as budding, which does not involve the fusion of gametes (12)

assimilation transformation or incorporation of a substance by organisms; absorption and conversions of energy and nutrient uptake into constituents of an organism (24)

assimilation efficiency percentage of energy ingested in food that is assimilated (24)

association a natural unit of vegetation characterized by a relatively uniform species composition and often dominated by a particular species (1, 20, 21, 22)

atoll a ringed-shaped coral reef that encloses or almost encloses a lagoon and is surrounded by open sea (31)

attack-abatement effect antipredator advantage of group living by prey species that reduces individual risk to predation (16)

aufwuchs community of plants and animals attached to or moving about on submerged surfaces; also called *periphyton*, but that term more specifically applies to organisms attached to submerged plant stems and leaves (30)

autogenic self-generated (22)

autosuccession self-replication nature of vegetation (21)

autotrophy ability of an organism to produce organic material from inorganic chemicals and some source of energy (5, 20, 21)

autozygous describes two alleles at a locus in question that have been derived by replication of a single allele in some ancestral population; the alleles are identical by descent (19)

available water capacity supply of water available to plants in a well-drained soil (4)

B **horizon** soil stratum beneath the *A* horizon characterized by an accumulation of silica, clay, and iron and aluminum oxides and possessing blocky or prismatic structure (4)

basal metabolic rate the minimal amount of energy expenditure needed by an animal to maintain vital processes (7)

basin land area contributing to the flow for any one stream (30)

basin wetland wetlands developed in shallow basins or depressions (30)

Batesian mimicry resemblance of a palatable or harmless species, the mimic, to an unpalatable or dangerous species, the model (16)

bathyal pertaining to anything, but especially organisms, in the deep sea, below the photic or lighted zone, and above 4000 m (31)

bathypelagic describes that part of the ocean depths characterized by darkness, low temperature, and great pressure, lying between 100 to 700 and 2000 to 4000 m depending on global location (31)

behavioral ecology the study of the behavior of organisms in their natural habitat; the application of behavioral theories to particular activities (1)

behaviorism the study of behavioral mechanisms, perceptual and physiological (1)

benthic pertaining to the lowermost regions or bottom of aquatic ecosystems (31)

benthic zone the lowermost region of a freshwater or marine profile (21, 30, 31)

benthipelagic province that part of the oceanic province that lies over the major plains of the ocean down to about 6000 m (31)

benthos animals and plants living on the bottom of a lake or sea from high water mark to the deepest depths (30, 31)

beta diversity variety of organisms occupying a number of different habitats over a region; regional diversity, as compared with very local or alpha diversity (20)

biennial plant that requires two years to complete a life cycle, with vegetative growth the first year and reproductive growth (flowers and seeds) the second (12)

biocenosis a term equivalent to the biotic component of an ecosystem (1)

biogeochemical cycle movement of elements or compounds through living organisms and nonliving environment (25)

biogeography the study of the past and present geological distribution of plants and animals at different taxonomic levels, the habitats they occupy, and the ecological relationships involved (27)

biological clock the internal mechanism of an organism that controls circadian rhythms without external time cues (7)

biological magnification process by which pesticides and other substances become more concentrated in each link of the food chain (27)

biological oxygen demand (BOD) a measure of the oxygen needed in a specified volume of water to decompose organic materials; the greater the amount of organic matter in water, the higher the BOD (30)

bioluminescence production of light by living organisms (31)

biomass weight of living material, usually expressed as dry weight per unit area (24, 26)

biome major regional ecological community of plants and animals; usually corresponds to plant ecologists' and European ecologists' classification of plant formations and classification of life zones (27)

biophage organism that feeds on living material (15)

biosequence group of soils in which all soil-forming factors remain constant except vegetation that leaves its imprint on soil formation (4)

biotic community any assemblage of populations living in a prescribed area or physical habitat (20)

biotic inputs influences imposed on an ecosystem by other ecosystems in the landscape and organisms moving into an ecosystem (24)

birth rate number of newborn individuals added to a population in a given time (10)

blanket mire large areas of upland dominated by sphagnum moss and dependent on precipitation for a water supply; a moor (30)

bog wetland ecosystem characterized by an accumulation of peat, acid conditions, and dominance of sphagnum moss (30)

border place where the edge of one patch meets the edge of another (23)

bottleneck an evolutionary term for any stressful situation that greatly reduces a population (19)

bottom-up regulation regulation of populations and community structure through the producers (20)

boundary combination of two edges and the border (23)

breeding dispersal movement of individuals out of a population prior to the initiation of the breeding season (10)

broken-stick model also known as random niche model, views species abundance as a random partitioning of resources distributed along a continuum (21)

brood parasitism the act of an animal laying its eggs in the nest of another species and allows that individual to rear its young (17)

browse part of current leaf and twig growth of shrubs, woody vines, and trees available for animal consumption (8)

browser herbivorous animals that feed on woody growth (8)

buffer a chemical solution that resists or dampens change in pH upon addition of acids or bases (30)

C horizon soil stratum beneath the solum (A and B horizons), relatively little affected by biological activity and soil-forming processes (4)

C$_4$ plant any plant that produces a four-carbon compound, malic or aspartic acid, as its first step in photosynthesis (6)

C$_3$ plant any plant that produces the three-carbon compound phosphoglyceric acid as its first step in photosynthesis (6)

calcicole plants susceptible to aluminum toxicity, acidity, and other factors influenced by the absence of calcium (7)

calcification process of soil formation characterized by accumulation of calcium in lower horizons (4)

calcifuge plants with low calcium requirement that can live in soils with a pH of 4.0 or less (7)

caliche an alkaline, often rocklike salt deposit on the surface of soil in arid regions; it forms at the level where leached Ca salts from the upper soil horizons are precipitated (4)

calorie amount of heat needed to raise 1 gram of water 1°C, usually from 15°C to 16°C (24)

CAM plants (crassulacean acid metabolism) Plants (cactus and other succulents) that separate the process of carbon dioxide uptake and fixation when growing under arid conditions. They take up gaseous carbon dioxide at night when stomata are open and water loss is minimal. During the day, when the stomata are closed, CO$_2$ is released and chemically used (6)

Cambrian earliest period in geological time in the Paleozoic era, from about 570 to 500 million years ago; algae and many marine invertebrates appeared (11, 21)

cannibalism killing and consuming one's own kind; intraspecific predation (15)

canonical distribution particular configuration of the log normal distribution of species abundance (20)

capillary water that portion of water in the soil held by capillary forces between soil particles (4)

Carboniferous second period of the Upper Paleozoic, from about 345 to 280 million years ago; named after extensive coal deposits formed at that time; warm humid period characterized by forests and swamps, appearance of early amphibians, reptiles, and giant ferns (11, 21)

carnivore organism that feeds on animal tissue; taxonomically, a member of the order Carnivora (Mammalia) (8)

carr vegetation dominated by alder and willows and occupying eutrophic peat (30)

carrying capacity (K) number of individual organisms the resources of a given area can support usually through the most unfavorable period of the year; term has acquired so many meanings that it is almost useless (10)

catastrophic extinction a major episode of extinction involving many taxa occurring fairly suddenly in the fossil record (11)

catena a group of related soils (4)

cation part of a dissociated molecule carrying a + electrical charge (4)

cation exchange capacity ability of a soil particle to absorb + charged ions (4)

Cenozoic major division of geological time extending from the Mesozoic era, some 65 million years ago to the present (11, 21)

chamaephyte perennial shoots or buds on the surface of the ground to about 25 cm above the surface (20)

chaparral vegetation consisting of broadleaf evergreen shrubs, found in regions of mediterranean climate of hot, dry summers and mild, wet winters (28)

character displacement divergence of characteristics in two otherwise similar species occupying overlapping ranges; brought about by the selective effects of competition (14)

chasmogamy production of flowers that open to expose reproductive organs, allowing cross-pollination (13)

chemical ecology the study of both chemical substances in the natural world and use by organisms for defense, species recognition, and courtship (1)

chilling tolerance ability of a plant to carry on photosynthesis within a range of +5 to +10°C (11)

chronosequence a sequence of related soils that that differ in the degree of profile development because of differences in age (4, 22)

circadian rhythm endogenous rhythm of physiological or behavioral activity of approximately 24 hours duration (8)

claypan a dense compact layer in the subsoil having a much higher clay content than overlying material; impedes the movement of water and air, and the growth of plant roots (4)

cleistogamy self-pollination within a flower that does not open (13)

climate the average weather conditions experienced at a particular place over a period of time (3)

climate diagram a plot of the monthly variations in temperature and precipitation; the vertical axes are scaled so that the relative positions of the temperature and precipitation graphs reflect the availability of water (3)

climax stable end community of succession that is capable of self-perpetuation under prevailing environmental conditions (21)

climograph a diagram describing a locality based on the annual cycle of temperature and precipitation (2)

cline gradual change in population characteristics over a geographical area, usually associated with changes in environmental conditions (27)

clone a population of genetically identical individuals resulting from asexual reproduction (10)

coevolution joint evolution of two or more noninterbreeding species that have a close ecological relationship; through reciprocal selective pressures, the evolution of one species in the relationship partially depends on the evolution of the other (17)

coexistence two or more species living together in the same habitat, usually with some form of competitive interaction (14)

cohort a group of individuals of the same age (10)

cohort life table see *life table* (10)

cold resistance ability of a plant to resist low temperature stress without injury (7)

collectors aquatic insects that collect and feed on fine detrital particles (30)

colluvium mixed deposits of soil material and rock fragments accumulated near the base of steep slopes through soil creep, landslides, and local surface runoff (4)

commensalism relationship between species that is beneficial to one, but neutral or of no benefit to the other (14)

community a group of interacting plants and animals inhabiting a given area (20, 21, 24)

community ecology that part of ecology that emphasizes the living components of an ecosystem and involves the study of patterns and processes within the community (1)

compensation intensity light intensity at which photosynthesis and respiration balance each other so that net production is 0; in aquatic systems, usually the depth of light penetration at which oxygen utilized in respiration equals oxygen produced by photosynthesis (24)

competition any interaction that is mutually detrimental to both participants; occurs between species that share limited resources (11, 14)

competition coefficient a measure of the degree to which one consumer uses the resources of another, expressed in terms of population interaction (14)

competitive exclusion hypothesis stating that when two or more species coexist using the same resource, one must displace or exclude the other (14)

conduction direct transfer of heat from one substance to another (2, 3)

connectance ratio of potential links or interactions in a food web to those that actually exist (20)

conservation biology an integrated approach to the protection and management of biodiversity based on principles of genetics, ecology, fishery, wildlife, forest, and range management, and social sciences, especially sociology, philosophy, and economics (1)

consumer any organism that lives on other organisms dead or alive (5)

consumption efficiency the ratio of ingestion to production (24)

contest competition competition in which a limited resource is shared only by dominant individuals; this type of competitions results in a relatively constant number of survivors, regardless of initial density (12)

continuum a gradient of environmental characteristics or changes in community composition (20)

continuum index measure of a position of a community on a gradient defined by species composition (20)

control a standard of comparison in scientific experimental work; replicate of the experiment in which a possibly crucial factor is being studied is omitted (1)

convection transfer of heat by the circulation of fluids, liquid or gas (2, 3)

convergent evolution development of similar characteristics in different species living in different areas but under similar environmental conditions (19)

coprophagy feeding on feces (8)

Coriolis effect a physical consequence of the law of conservation of angular momentum—as a result of Earth's rotation, a moving object veers to the right in the Northern Hemisphere and to the left in the Southern Hemisphere relative to the Earth's surface (2)

corridor a narrow connection between two patches of suitable habitat (23)

countercurrent circulation an anatomical and physiological arrangement by which heat exchange takes place between outgoing warm arterial blood and cool venous blood returning to the body core. It is important in maintaining temperature homeostasis in many vertebrates (8)

Cretaceous final period of the Mesozoic, from about 136 to 65 million years ago when much of the present land area was covered by shallow seas; rise of the angiosperms and continued dominance of dinosaurs until mass extinction at the end of the period; evolution of modern birds and fish; appearance of primitive mammals (11)

critical daylength the period of daylight, specific for any given species, that triggers a long-day or a short-day response in organisms (6, 8)

critical temperature during sudden or prolonged cold spells, the point at which insulation is no longer effective and the animal must maintain body heat by increased metabolism (8)

critical thermal maximum temperature at which an animal's capacity to move is so reduced that it cannot escape from thermal conditions that will lead to death (8)

crown fire a forest fire that reaches and spreads across the canopy (23)

crude birth rate number of births per 1000 population (10)

crude density number of individuals per unit area (10)

cryoplanation molding of the tundra landscape by frost action (28)

cryptic coloration coloration of organisms that makes them resemble or blend into their habitat or background (16)

cryptophyte buds buried in the ground on a bulb or rhizome (20, 21)

cultural eutrophication the unintentional overfertilization of freshwater systems by nutrients, notably nitrogen and phosphorus, from anthropogenic sources (26)

cycle recurrent variation in a system that periodically returns to its starting point (10)

cyclic replacement type of succession in which the sequence of seral stages is repeated by imposition of some disturbance so that the sere never arrives at a climax or stable sere (21)

cyclone air spiraling inward around a low-pressure center (2)

dark reaction group of light-independent reactions following the light reactions in photosynthesis; they reduce carbon dioxide to produce glucose and other carbohydrates (6)

day neutral plant a plant that does not require any particular photoperiod to flower (6)

death rate number of individuals in a population dying in a given time interval divided by the number alive at the midpoint of the time interval (10)

decalcification leaching out of carbonates, if present, within a soil body (4)

deciduous (of leaves) shed during a certain season (winter in temperate regions; dry seasons in the tropics); (of trees) having deciduous parts (26)

decomposer organism that obtains energy from the breakdown of dead organic matter to more simple substances; most precisely refers to bacteria and fungi (5, 9)

decomposition breakdown of complex organic substances into simpler ones (9)

deductive method in testing hypotheses, going from the specific to the general (1)

definitive host organism in which parasite becomes an adult and reaches reproductive"? maturity (17)

degradative succession successional process on dead organic matter with maximum energy at the start and a steady decline in energy as succession proceeds, and characterized by a sequence of decomposers (21)

degree days measurement used to relate variations in temperature over a single growing to plant growth; it is calculated as the sum of the departures in temperatures above some base temperature (7)

deletion a chromosomal mutation resulting in the loss of a segment of the genetic material and the genetic information it contains from a chromosome (19)

deme local population or interbreeding group with definable genetic characteristics within a larger population (19)

demographic stochasticity random variation in population birth and death rates due entirely to chance differences experienced by individuals (11)

demography the statistical study of the size and structure of populations and changes within them (10)

denitrification reduction of nitrates and nitrites to nitrogen by microorganisms (9, 25)

density the number of organism per unit area (10)

density dependence regulation of size of a population by mechanisms whose effectiveness increases as population size increases (12)

density independence unaffected by population density; regulation of growth is not tied to population density (12)

dependent variable variable y, that yields the second of two numbers in an ordered pair (x, y); the set of all values taken on by the dependent variable is called the *range* of the function; see *independent variable* (1)

desert an area within which the rate of evaporation exceeds the rate of precipitation most of the time (28)

deterministic extinction extinction that comes about through some force or change from which there is no escape (10)

deterministic model mathematical model in which all relationships are fixed and a given input produces one exact prediction as an output (1)

detritivore organism that feeds on dead organic matter; usually applies to detritus-feeding organisms other than bacteria and fungi (8)

detritus fresh to partly decomposed plant and animal matter (8, 24)

dewpoint temperature at which condensation of water in the atmosphere begins (2)

diameter breast height (dbh) diameter of a tree measured at 1.4 m (4 feet, 6 inches) from ground leve (10)

diapause a period of dormancy, usually seasonal, in the life cycle of an insect in which growth and development cease and metabolism is greatly decreased (8)

diffuse coevolution coevolution involving the interactions of many organisms in contrast to pairwise interactions (17)

diffuse competition type of competition in which a species experiences interference from numerous other species that deplete the same resources (14)

dimorphism existing in two structural forms, two color forms, two sexes, or the like (19)

dioecious describes plants in which male and female reproductive organs are borne on separate individuals (13)

diploid having chromosomes in homologous pairs or twice the haploid number of chromosomes (19)

directional selection selection favoring individuals at one extreme of the phenotype in the population (19)

disease any deviation from normal state of health (17)

dispersal leaving an area of birth or activity for another area (12)

dispersion distribution of organisms within a population over an area (10)

disruptive selection selection in which two extreme phenotypes in the population leave more offspring than the intermediate phenotype, which has lower fitness (19)

disturbance a relatively discrete event that disrupts the structure of an ecosystem, community, or population and changes resource availability or the physical environment (23)

diversity abundance of different species in a given location; species richness (20, 27)

diversity index the mathematical expression of species richness of a given community or area (20)

domain in the ecoregion concept, a subcontinental area of broad climatic similarity (27)

dominance (ecological) control within a community over environmental conditions influencing associated species by one or several species, plant or animal, enforced by number, density, or growth form (20, 21); (social) behavioral, hierarchical order in a population that gives high-ranking individuals priority of access to essential requirements (12); (genetic) ability of an allele to mask the expression of an alternative form of the same gene in a heterozygous condition (19)

dormant state of cessation of growth and suspended biological activity during which life is maintained (7, 8)

drift small organisms and fine particulate matter carried downstream by flowing water and picked up by filter-feeding organisms (30)

drought avoidance ability of a plant to escape dry periods by becoming dormant or surviving the period as a seed (7)

drought resistance sum of drought tolerance and drought avoidance (7)

drought tolerant ability of plants to maintain physiological activity in spite of the lack of water or to survive the drying of tissues (7)

dry adiabatic lapse rate see *adiabatic lapse rate* (3)

dry deposition particulate matter and airborne gases in the atmosphere that return to Earth as pollutants (26)

dryfall nutrients brought to plants by airborne particles and aerosols (25)

duplication (19) the occurrence of extra genes or segments in the sets of chromosomes carried in each cell of an organism (19)

dynamic life table see *life table* (10)

dynamic pool model optimum yield model using growth, recruitment, mortality, and fishing intensity to predict yield (yield)

dynamic-composite life table see *life table* (10)

dystrophic term applied to a body of water with a high content of humic organic matter, often with high littoral productivity and low plankton productivity (30)

E **horizon** zone of maximum leaching or eluviation in a soil (4)

ecesis a term coined by F. Clements to denote initial establishment and growth of vegetation in early succession (27)

ecological density density measured in terms of the number of individuals per area of available living space (10)

ecological efficiency percentage of biomass produced by one trophic level that is incorporated into biomass of the next highest trophic level (24)

ecological release expansion of habitat or increase in food availability resulting from release of a species from interspecific competition (14)

ecophysiology that part of ecological studies concerned with the responses of individual organisms to temperature, moisture, light, nutrients, and other factors of the environment (1)

ecoregion a continuous geographic area across which the interaction of climate, soil, and topography are sufficiently uniform to permit the development of similar types of vegetation (27)

ecosystem the biotic community and its abiotic environment functioning as a system (24)

ecosystem ecology a holistic approach to the study of flow of energy and matter through organisms and their environment (1)

ecosystem management guardianship that emphasizes ecological systems as functional units for land and resource management and emphasizes the long-term sustainability of those systems (1)

ecotone transition zone between two structurally different communities; see also *edge* (23)

ecotype subspecies or race adapted to a particular set of environmental conditions (7)

ectomycorrhizae a type of mycorrhizae in which the fungal hyphae do not penetrate the root, but rather cover the root and grow between the root cells (17)

ectothermy determination of body temperature primarily by external thermal conditions (8)

edaphic relating to soil (4)

edge place where two or more vegetation types meet (23)

edge effect response of organisms, animals in particular, to environmental conditions created by the edge (23)

effective population size the size of an ideal population that would undergo the same amount of random genetic drift as the actual population; sometimes used to measure the amount of inbreeding in a finite, randomly mating population (19)

Ekman spiral a spiraling current pattern from the water's surface to the deeper layers; caused by the Coriolis force; each successive slower moving layer of water is shifted to the right (31)

elaiosome shiny, oil-containing, ant-attracting tissue on the seed coat of many plants (17)

elaiphore specialized oil-secreting organs found in many genera of neotropical plants (17)

eluviation removal of soil materials in suspension or in solution from the upper soil horizon and partial deposition in the lower horizon (4)

emergence theory the idea that natural associations have certain properties not possessed by individual populations and become apparent only at the community level (1)

emigration movement of part of a population permanently out of an area (10)

endemic restricted to a given region (7)

endogenous describes any process that arises within an organism (8)

endomycorrhizae a type of mycorrhizae in which the hyphae of the fungus penetrate the cells of the root (17)

endothermy regulation of body temperature by internal heat production; allows maintenance of appreciable difference between body temperature and external temperature (8, 24)

energy capacity to do work (24)

Entisol embryonic mineral soil whose profile is just beginning to develop; common on recent flood plains and wind deposits, they lack distinct horizons (4)

entrainment synchronization of an organism's activity cycle with environmental cycles (8)

entropy transformation of matter and energy to a more random, more disorganized state (24)

environment total surroundings of an organism including other plants and animals embracing those of its own kind (3)

environmental lapse rate the rate of decrease of air temperature with elevation (2)

environmental stochasticity unpredictable environmental changes, such as adverse weather or food failure, that affect some aspect of population growth (11)

epidemic rapid spread of a bacterial or viral disease in a human population; compare with *epizootic* (17)

epifauna benthic organisms that live on or move over the surface of a substrate (31)

epilimnion warm, oxygen-rich upper layer of water in a lake or other body of water, usually seasonal (25, 30)

epiphyte organism that lives wholly on the surface of plants, deriving support but not nutrients from the plants (20)

epizootic rapid spread of a bacterial or viral disease in a dense population of animals (17)

equilibrium species species whose population exists in equilibrium with resources and at a stable density (23)

equilibrium turnover rate change in species composition per unit time when immigration equals extinction (23)

equitability evenness of distribution of species abundance patterns; maximum equitability is the same number of individuals among all species in the community (20)

estivation dormancy in animals during a period of drought or a dry season (8)

estuary a partially enclosed embayment where fresh water and sea water meet and mix (31)

ethology the study of animal behavior (1)

euphotic zone surface layer of water to the depth of light penetration where photosynthetic production equals respiration (30, 31)

eurythermal able to tolerate a wide range of temperatures (3)

eusocial describes a highly evolved complex social system involving extensive cooperation, communication, and division of labor; found in honey bees, ants, and naked mole rats (19)

eutrophic term applied to a body of water with high nutrient content and high productivity (30)

eutrophication nutrient enrichment of a body of water; called cultural eutrophication when accelerated by introduction of massive amounts of nutrients by human activity (30)

evaporation water lost as vapor from soil, open water, or other surface (2)

evapotranspiration sum of the loss of moisture by evaporation from land and water surfaces and water lost from the surface of a plant mainly through the stomata or transpiration (2)

evenness degree of equitabilty in the distribution of individuals among a group of species; see *equitability* (20, 21)

evolution change in gene frequency through time resulting from natural selection and producing cumulative changes in characteristics of a population (19)

exchangeable cations charged ions that are adsorbed on oppositely charged sites on the surface of clay and humus colloids in the soil; they can replace each other on the surface and are available to plants as nutrients (4)

exothermic describes a chemical reaction that releases heat to the environment (24)

expected energy budget rule applicable to foraging animals, who are expected to be risk prone if their energy budget is negative, risk averse if it is positive; see *risk-sensitive foraging* (15)

exploitative competition competition by a group or groups of organisms that reduces a resource to a point that it adversely affects other organisms (14)

exponential growth instantaneous rate of population growth expressed as proportional increase per unit of time (11)

extinction state of a species no longer represented by living individuals (10)

extinction coefficient point at which the intensity of light reaching certain depth is insufficient for photosynthesis; ratio of intensity of light at a given depth to the intensity at the surface (3)

extra-pair copulation one member of a mated pair cheating on its mate while maintaining his or her relationship with the primary mate and young (13)

F₂ generation offspring produced by selfing or by allowing the F_1 generation to breed among themselves (19)

facilitation model a model of succession in which previous community prepares or "facilitates" the way for a succeeding community (22)

facultative anaerobes bacteria that can living in both the presence of or absence of oxygen (9)

fecundity potential ability of an organism to produce eggs or young; rate of production of young by a female (10, 11)

fellfield area within a tundra characterized by stony debris and sparse vegetation (28)

female choice in lek species the hypothesis that females show a preference for a courtship arena because it is the safest place to mate and enables females to choose a mate from a group of displaying males (12)

fen wetlands, only slightly acidic, dominated by sedges in which peat accumulates (30)

fermentation breakdown of carbohydrates and other organic matter under anaerobic conditions (9)

fetch the distance over water that the wind can blow (31)

field capacity amount of water held by soil against the force of gravity (4)

filter effect the influence of gaps in corridors that allow certain organisms to cross and restricts the movement of others (23)

first law of thermodynamics although energy may change forms, pass from one place to another, or act on matter, it can be neither created or destroyed; regardless of what transfers or transformations take place, no gain or loss in energy occurs (24)

first-level carnivore a carnivore that feeds directly on herbivores (8)

fitness genetic contribution by an individual's descendants to future generations (5, 19)

fixation Process in soil by which certain chemical elements essential for plant growth are converted from a soluble or exchangeable form to a less soluble or nonexchangeable form (25)

fixation index a measure of the reduction of heterozygosity of a subpopulation due to random genetic drift (19)

floating reserve individuals in a population of a territorial species that do not hold territories and remain unmated, but are available to refill territories vacated by death of an owner (11)

foliage height diversity measure of the degree of layering or vertical stratification of foliage in a forest (20)

food chain movement of energy and nutrients from one feeding group of organisms to another in a series that begins with plants and ends with carnivores, detrital feeders, and decomposers (20, 21, 24)

food web interlocking pattern formed by a series of interconnecting food chains (20, 21)

foraging strategy manner in which animals seek food and allocate their time and effort in obtaining it (15)

forb herbaceous plant other than grass, sedge, or rush (28)

formation classification of vegetation based on dominant life forms (27)

founder effect population started by a small number of colonists, which contain only a small and often biased sample of genetic variation of the parent population. It may result in a markedly different new population (19)

fragipan a dense, brittle, compact subsoil horizon associated with acid soil conditions (4)

fragmentation reduction of a large habitat area into small, scattered remnants (23); reduction of leaves and other organic matter into smaller particles (9)

free-running cycle length of a circadian rhythm in the absence of an external time cue (8)

frequency the percentage of quadrats in which a given species occurs (23)

fringe wetlands wetlands that occur along the coastal areas of larger lakes and oceans (30)

frost pocket depression in the landscape into which cold air drains, lowering the temperature relative to the surrounding area. Such pockets often support their own characteristic group of cold tolerant plants (3)

frugivore organism that feeds on fruit (8)

functional response change in rate of exploitation of a prey species by a predator in relation to changing prey density (15)

fundamental niche total range of environmental conditions under which a species can survive (14)

fynbos areas of heathlands in the mediterranaean-type regions of South Africa (28)

gap opening made in a forest canopy by some small disturbance, such as windtrhow or the death of an individual or group of trees, that influences the development of vegetation beneath (20, 21, 23)

gap phase replacement successional development in small disturbed areas within a stable plant community; filling in of a space left by a disturbance, but not necessarily by the species eliminated by the disturbance (21)

garrigue shrub woodland characteristic of limestone areas with low rainfall and thin, poor dry soils; widespread in the Mediterranean countries of southern Europe (28)

gaseous cycle atmospheric cycling of basic elements necessary for life, notably oxygen, nitrogen, carbon, and hydrogen, necessary for life in a gaseous state (25)

Gauses's principle species with the same ecological requirements cannot coexist (14)

Gellisol soil that contains permafrost within 200 cm of the ground surface, characterized by diagnostic perennial coldness rather than by a diagnostic soil horizon (4)

gene unit material of inheritance; more specifically, a small unit of DNA molecule coding for a specific protein to produce one of the many attributes of a species (19)

gene flow exchange of genetic material between populations (19)

gene frequency actually allele frequency; relative abundance of different alleles carried by an individual or a population (19)

gene pool the sum of all the genes of all individuals in a population (10)

genet a genetic individual that arises from a single fertilized egg (10)

genetic drift random fluctuation in allele frequency over time due to chance occurrence alone without any influence by natural selection. Important in small populations (19)

genotype genetic constitution of an organism (19)

genotypic frequency the frequency or percentage of different genotypes within a population (19)

geometric distribution hypothesis also known as the niche preemption hypothesis, it proposes that the most successful or dominant species preempts the most space, the next most successful the next largest share of space, and so on, until little space is left (21)

geometric rate of increase factor by which size of a population increases over a period of time (11)

gigaton a billion tons (31)

gleization process in soil development under poor drainage conditions in which iron in the soils is reduced to ferrous compounds, giving a dull gray or bluish color to the horizon (4)

gley soil soil developed under conditions of poor drainage, resulting in reduction of iron and other elements and in gray colors and mottles (4)

gouger general group of stream invertebrates that live and feed on woody debris (30)

gradient analysis a plot of plant responses to changes along a particular environmental gradient such as soil moisture, temperature, or elevation (22)

granivore organism that feeds on seeds (8, 28)

grazers herbivores that feed principally on leafy vegetation (8)

greenhouse effect selective energy absorption by carbon dioxide in the atmosphere that allows short-wavelength energy to pass through but absorbs longer wavelengths and reflects heat back to Earth (2, 32)

gross production energy fixed per unit area by photosynthetic activity of plants before respiration. Total energy flow at the secondary level is not gross production, but rather assimilation, because consumers use material already produced with respiratory losses (24)

ground fire a forest fire that burns organic matter on the forest floor down to the mineral substrate (23)

group selection elimination of one group of individuals by another group of individuals possessing superior genetic traits; not a widely accepted hypothesis (10, 19)

growth form morphological category of plants, such as tree, shrub, and vine (20)

guild a group of populations that utilizes a gradient of resources in a similar way (20, 21)

gular fluttering a method of evaporative cooling by birds in which they move parts of the gullet, which requires less energy than panting (8)

gully erosion concentration of surface flow of water into channels that cut deep into the soil (4)

gyre circular motion of water in major ocean basins (2)

habitat place where a plant or animal lives (13)

hadal describes that part of the ocean below 6000 m (31)

halocline changes in salinity with depth in the oceans (31)

handicap hypothesis the idea that a secondary male characteristic such as bright plumage that could reduce the male's survival, thus a handicap, attracts females; if the male can carry the handicap and survive, it is proof for the female of a superior genotype (13)

haploid having a single set of unpaired chromosomes in each cell nucleus (19)

Hardy-Weinberg law the proposition that genotypic ratios resulting from random mating remained unchanged from one generation to another, provided natural selection, genetic drift, and mutation are absent (19)

heat form of energy possessed by all substances that results from the random motion of molecules within the substance (3)

heath shrubs that belong to the Ericacaea (28)

heathland shrubland dominated by dwarf shrubs belonging to the Ericacaea (28)

heliothermism acquisition of heat energy by ectotherms through basking in the sun (8)

hemicryptophyte perennial shoots or buds close to the surface of the ground; often covered with litter (20)

hemiparasite a plant parasite that has chlorophyll and carries on photosynthesis but that augments its nutrient supply by feeding on its host or uses its host for mechanical support (17)

herbivore organism that feeds on plant tissue (8)

herbivory feeding on plant material (15)

hermaphrodite organism possessing the reproductive organs of both sexes (12)

heterogeneity state of being mixed in composition; can refer to genetic or environmental conditions (19)

heterotherm organism that during part of its life history becomes either endothermic or ectothermic. Hibernating endotherms become ectothermic, and foraging insects such as bees become endothermic during periods of activity; they are characterized by rapid, drastic, repeated changes in body temperature (8)

heterotroph organism that is unable to manufacture its own food from simple chemical compounds and consumes other organisms for its source of carbon; compare with *autotrophy* (5)

heterotrophic requiring a supply of organic matter or food from the environment (20, 21)

heterotrophic community one that depends on fixed energy, such as organic material from the outside (20)

heterotrophic succession see *degradative succession* (21)

heterozygous containing two different alleles of a gene, one from each parent, at the corresponding loci of a pair of chromosomes (19)

hibernation winter dormancy in animals characterized by a great decrease in metabolism (8)

Histosol soil characterized by high organic matter content (4)

holistic in ecology, describes studies that aim to understand ecosystems as a whole rather than examining their component parts (1)

home range area over which an animal ranges throughout the year (11)

homeostasis maintenance of nearly constant conditions in function of an organism or in interaction among individuals in a population (5)

homeotherm animal with a fairly constant body temperature; also spelled *homoiotherm* or *homotherm* (8)

homeothermy regulation of body temperature by physiological means (8)

homologous chromosomes corresponding chromosomes from male and female parents that pair during meiosis (19)

homozygous containing two identical alleles of a gene at the corresponding loci of a pair of chromosomes (19)

horizon major zones or layers of soil, each with its own particular structure and characteristics (4)

horizontal life table see *life table* (10)

host organism that provides food or other benefit to another organism of a different species; usually refers to an organism exploited by a parasite (17)

hotshot model a hypothesis of lek behavior that proposes a strong hierarchy among males in which the dominant males with most effective displays displace all others, leaving no opportunity for female choice (13)

hotspot model a hypothesis of lek behavior that proposes that males cluster in places where encounters with females are potentially high (13)

humus organic material derived from partial decay of plant and animal matter (4)

hybrid plant or animal resulting from a cross between genetically different parents (19)

hydroperiod defines the duration, frequency depth, and season of flooding in wetlands (30)

hyperosmotic possessing body fluids that are osmotically more concentrated than the medium (8)

hyperthermia rise in body temperature to reduce thermal differences between an animal and a hot environment, thus reducing the rate of heat flow into the body (8)

hypertrophic describes lakes receiving excessive inputs of wastes, raw sewage, or runoff from urban and agricultural areas that result in excessive nutrient load (30)

hypha filament of a fungus thalli or vegetative body (9)

hypolimnion cold, oxygen-poor zone of a lake that lies below the thermocline (20, 29, 30)

hypoosmotic having a lower osmotic concentration than the medium (8)

hypothermia relaxation of body homeostasis that causes the body temperature to drop below normal (8)

hypothesis an idea or concept that can be tested by experimentation (1)

igneous term applied to rocks that have crystallized from molten rock (4)

illuviation the deposition of soil materials either in suspension or solution from an upper horizon into a lower horizon (4)

immigration arrival of new individuals into a habitat or population (10)

immobilization conversion of an element from inorganic to organic form in microbial or plant tissue, rendering the nutrient relatively unavailable to other organisms (9)

importance value sum of relative density, relative dominance, and relative frequency of a species in a community (20, 21)

inbreeding mating among close relatives (19)

inbreeding depression detrimental effects of inbreeding (19)

Inceptisols mineral soils that have one or more horizons in which mineral materials have been weathered or removed; they are only beginning to develop a distinctive soil profile (4)

incipient lethal temperature the temperature at which a stated fraction of a population of poikilothermic animals (usually 50 percent) will die when brought rapidly to it from a different temperature (8)

inclusive fitness sum of the total fitness of an individual and the fitness of its relatives, weighted according to the degree of relationship (19)

independent variable variable *x* that yields the first of two numbers of an ordered pair (*x, y*); the set of all values taken on by the independent variable is called the domain of the function; see *dependent variable* (1)

index species species whose presence is characteristic of a particular community (20, 21)

individualistic concept view, first proposed by H. A. Gleason, that vegetation is a continuous variable in a continuously changing environment; thus no two vegetation communities are identical, and association of species result only from similarities in requirements (1, 22)

induced defense the synthesis and transport by plants of toxic secondary substances to the site of herbivore attack (16)

induced edge edge created and maintained by periodic disturbance (23)

induced resistance the production of relatively large amounts of toxic secondary substances stimulated by herbivore attack that affect herbivore assimilation, reproductive output, or survival (16)

inductive method in testing hypotheses, going from the specific to the general (1)

infauna organisms living within a substrate (31)

infiltration downward movement of water into the soil (3, 4)

inflection point point in the logistic growth curve where population growth is maximal (10)

infralittoral describes the region below the littoral region of the sea (31)

inherent edge edge, usually stable and permanent, brought about by an abrupt environmental change created by long-term natural features (23)

inhibition model model of succession proposing that the dominant vegetation occupying a site prevents colonization of that site by other plants of the next successional community (22)

instar form of insect or other arthropod between successive molts (1)

integrated pest management a program of carefully selected control measures tailored to address a particular pest problem with a minimal application of pesticides (18)

interception retention of precipitation before it reaches the ground with subsequent loss by evaporation (3)

interdemic selection group selection of populations within a species (19)

interference competition competition in which access to a resource is limited by the presence of a competitor (14)

interglacial period period between glaciations during which species recolonize previously occupied areas (22)

interior species organisms, chiefly birds and mammals, whose habitat begins some distance within a forest, shrubland, or grassland (23)

intermediate disturbance hypothesis proposes that species diversity is greatest in habitats that experience moderate amounts of disturbance because of the occurrence of early and late successional species (22)

intermediate host host that harbors developmental phases of a parasite; the infective stage or stages can develop only when the parasite is independent of its definitive host; see *definitive host* (17)

internal cycling tight cycling of nutrients among soil, litter, roots, and above-ground structure of plants (25)

intersexual competition rivalry among members of the same sex for a mate; most common among males and characterized by fighting and display (13)

intersexual selection situation in which selection favors traits in one sex that attract the opposite sex (13)

interspecific between individuals of different species (14)

interspecific competition rivalry between individual of different species (14)

intertropical convergence zone the boundary zone separating the northeast trade winds of the Northern Hemisphere from the southeast trade winds of the Southern Hemisphere (2)

intraguild predation predation among species occupying the same trophic level and using a similar food resource (16)

intrasexual competition competition between individuals of the same sex for a mate (13)

intrasexual selection competition for the opportunity to mate that leads to exaggerated sexual characteristics (13)

intraspecific between individuals of the same species (12)

intrinsic rate of increase the per capita rate of growth of a population that has reached a stable age distribution and is free of competition and other growth restraints (11)

introgression incorporation of genes of one species into the gene pool of another (19)

inversion (genetic) reversal of part of a chromosome so that genes within that part lie in reverse order (19); (meteorological) increase rather than decrease in air temperature with height caused by radiational cooling of Earth (radiational inversion) or by compression and consequent heating of subsiding air masses from high-pressure areas (subsidence inversion) (2)

island biogeography study of distribution of organisms and community structure on islands (23)

isolating mechanism any structural, behavioral, or physiological mechanism that blocks or inhibits gene exchange between two populations (19)

isomorphous substitution replacement of one atom for another without changing the arrangement of the atom or the morphology of lay mineral (4)

isoosmotic describes a solution having the same osmotic pressure or concentration equal to that of anther specified solution, usually taken to be within a cell (8)

iteroparous multiple-brooded over a lifetime (12)

K selection selection under carrying capacity conditions and a high level of competition (12)

k value loss of individuals from a given life history stage as expressed by the difference between common logarithms of the numbers at the beginning and end of the stage (11)

karotype the physical appearance of the chromosome complement of a given species (19)

key factor analysis statistical treatment of population data designed to identify those factors most responsible for change in population size (11)

keystone species a species whose activities have a significant role in determining community structure (20, 21)

kin selection differential reproduction among groups of closely related individuals (19)

kinetic energy energy associated with motion (24)

kleptoparasitism the forcible robbing of food acquired by an individual of one species by an individual or individuals of another (17)

krummholz stunted form of trees characteristic of transition zone between alpine tundra and subalpine coniferous forest (28)

lagg moatlike or shallow water, dominated by sedge, between peat mat and higher ground in a bog (30)

landscape ecology study of the structure, function, and change in a heterogeneous landscape composed of interacting ecosystems (1, 23)

landscape mosaic the quiltlike patches of different types of land cover across a landscape (13)

Langmuir cell a localized area of water a few meters wide and hundreds of meters long created by the wind; involves a local circulation pattern of upwelling water that diverges from the center, converges, and sinks at the center (31)

lapse rate the rate at which the temperature decreases for each unit in height in the atmosphere (2)

late successional species plants characterized by lower rates of dispersal, slower growth rates, large size, and long life (21)

latent heat heat that is released or absorbed by a unit mass of a substance when it undergoes a change of state (3)

latent heat of evaporation heat released during evaporation (3)

latent heat of fusion amount of heat given up when a unit mass of a substance converts from a liquid to a solid state, or the amount of heat absorbed when a substance converts from a solid to a liquid state (3)

laterization soil-forming process in hot, humid climates characterized by intense oxidation resulting in loss of bases and in a deeply weathered soil composed of silica sesquioxides of iron and aluminum, clays, and residual quartz (4)

law of limiting factors maximum quantity of a resource tolerated by an organism that would limit response (5)

law of the minimum refers to limiting effects of availability of resources required by organisms (5)

law of tolerance the presence and success of an organism depends on the extent to which a qualitative or quantitative deficiency or access of any one of several factors may approach the limits of tolerance for that organism (5)

leach to dissolve and remove nutrients by water out of the soil, litter, and organic matter (9)

leaf area index ratio of area of canopy foliage to ground area (3)

lek communal courtship area used by males to attract and mate with females (13)

lentic pertaining to standing water, as lakes and ponds (30)

leptophyll a leaf size up to 25 mm^2 in area (28)

lessivage the washing in suspension of fine clay and fine silt down cracks and other voids in a soil body (4)

life expectancy the number of years on an average to be lived beyond a certain age (10)

life table tabulation of mortality and survivorship schedule of a population; static, time-specific, or vertical life tables are based on a cross section of a population at a given time; dynamic, cohort, or horizontal life tables are calculated on the basis of a cohort of organisms followed throughout life (10)

life zone major area of plant and animal life equivalent to a biome; transcontinental region or belt characterized by particular plants and animals and distinguished by temperature differences; applies best to mountainous regions where temperature changes accompany changes in altitude (27)

light compensation point that value of PAR at which the rate of CO_2 uptake in photosynthesis exactly offsets the loss of CO_2 in respiration (6)

light reaction light-dependent sequences of photosynthetic reactions (6)

light response curve function relating the net exchange of CO_2 (net photosynthesis) for a plant to the received PAR (6)

light saturation point the value of PAR at which any further increase results in no further increase in photosynthesis (6)

limit cycle stable oscillation in the population levels of a species, usually invoking predator and prey interactions (11)

limiting resource resource or environmental condition that limits the abundance and distribution of an organism (12)

limnetic pertaining to or living in the open water of a pond or lake (30)

limnetic zone shallow water zone of lake or sea in which light penetrates to the bottom (30)

limnology the study of freshwater ecosystems, especially lakes (30)

Lithosol soil showing little or no evidence of soil development and consisting mainly of partly weathered rock fragments or nearly barren rock (4)

lithosphere rocky material of Earth's outer crust (4)

littoral shallow water of lake in which light penetrates to the bottom, permitting submerged, floating, and emergent vegetative growth; also shore zone of tidal water between high and low water marks (30, 31)

littoral fringe the upper boundary on a rocky intertidal coast determined by the upper limits of sea spray and growth of crustose lichens (31)

littoral zone rocky coastal zone covered and uncovered daily by tides (31)

locus site on a chromosome occupied by a specific gene (19)

loess soil developed from wind-deposited material (4)

log normal distribution frequency distribution in which the horizontal or x axis is expressed in a logarithmic scale; in frequency distribution of species, the x axis represents the number of individuals and the y axis the number of species (21)

logistic curve S-shaped curve of population growth that slows at first, steepens, and then flattens out at an asymptote, determined by carrying capacity (11)

logistic equation mathematical expression for the population growth curve in which rate of increase decreases linearly as population size increases (11)

long-day organism plant or animal that requires long days—days with more than a certain minimum of daylight—to flower or come into reproductive condition (6, 8)

lotic pertaining to flowing water (30)

macromutation mutation at the level of the chromosome (19)

macronutrients essential nutrients needed in relatively large amounts by plants and animals (2)

macroparasite parasitic worms, lice, fungi, and the like; have comparatively long generation time; spread by direct or indirect transmission, and may involve intermediate hosts or vectors (17)

mallee sclerophyllous shrub community in Australia; most of the species are *Eucalyptus* (28)

mangal a mangrove swamp (31)

maquis sclerophyllous shrub vegetation in the Mediterranean region (28)

marginal value theorem a decision rule model based on economic theory; it holds that a predator should leave a patch of prey when the maximum overall average rate of extraction is the same as that for all other patches in the environment (15)

marine inversion phenomenon in which cool air from the ocean moves in beneath the heated layer above, trapping pollutants in the lower layer (3)

marl unconsolidated deposit formed in freshwater lakes that consists chiefly of calcium carbonate mixed with clay and other impurities (30)

marsh wetland dominated by grassy vegetation such as cattails and sedges (30)

mating system behavioral mechanisms involved in the acquisition of a mate, including the number of mates acquired, the manner in which they are acquired, the nature of the pair bond, and provision of parental care (13)

matric potential tendency of water to adhere to surfaces (7)

matrix in landscape ecology, areas of different species structure or composition in which landscape patches are imbedded (23)

mattoral sclerophyllous shrub vegetation in regions of Chile with mediterranean climate (28)

maximum sustainable yield the maximum rate at which individuals can be harvested from a population without reducing its size; recruitment balances harvesting (18)

mean cohort generation time the average time between the birth of parent and the birth of its offspring (10)

mediterranean-type climate semiarid climate characterized by a hot, dry summer and a wet, mild winter (4, 28)

meiofauna benthic organisms within the size range of 1 to 0.1 mm; interstitial fauna (31)

meiosis two successive divisions by gametic cells, with only one duplication of chromosomes so that the number of chromosomes in daughter cells is one-half the diploid number (19)

melanization changes in color value in soil caused by addition to the organic matter content (4)

mesic describes moderately moist habitat (20)

mesopelagic describes the region of the ocean, ranging from 200 to 1000 m, where little light penetrates and with little seasonal variation in temperature (31)

mesophyll specialized tissue located between the epidermal layers of a leaf; *palisade mesophyll* consists of cylindrical cells at right angles to upper epidermis and containing many chloroplasts; *spongy mesophyll* is next to the lower epidermis and consists of interconnecting, irregularly shaped cells with large intercellular spaces (6)

mesophytic describes plants adapted to environments that are not extremely or extremely dry (21)

mesotrophic term applied to waters having levels of plant nutrients between eutrophic and oligotrophic (30)

Mesozoic middle era in the geological time scale, some 230 to 70 million years; Age of Reptiles; three main periods are Triassic, Jurassic, and Cretaceous (11, 21)

metabolism the chemical reactions that take place in cells responsible for the breaking down of molecules to provide energy (catabolism) and the building up of more complex molecules from simpler molecules (anabolism) (8)

metalimnion transition zone in lake between hypolimnion and epilimnion; region of rapid temperature decline (20, 25, 30)

metamorphic describes rocks characterized by an aggregate of minerals formed by the recrystallization of preexisting rocks in response to a change of pressure and temperature (4)

metapopulation a set of local populations held together by dispersal (10, 19)

micella soil particle of clay and humus carrying positive electrical charges at the surface (4)

microbial loop uptake by bacteria of dissolved organic matter produced by plankton and subsequent consumption of bacteria by nanoplankton, resulting in a feeding loop; adds several more trophic levels to plankton food chain (31)

microbivore organism that feed on microbes, especially in the soil and litter (9)

microclimate climate on a very local scale, which differs from the general climate of the area; influences the presence and distribution of organisms (2)

microflora bacteria and certain fungi inhabiting the soil (9)

microhabitat that part of the general habitat utilized by an organism (13)

micromutation a mutation at the level of the gene; point mutation (19)

micronutrients essential nutrients needed in very small quantities by plants and animals (3)

microparasites viruses, bacteria, and protozoans, characterized by small size, short generation time, and rapid multiplication (17)

migration intentional, directional, usually seasonal movement of animals between two regions or habitats; involves departure and return of the same individual; a round-trip movement (10)

mimicry resemblance of one organism to another or to an object in the environment evolved to deceive predators (18)

mineralization microbial breakdown of humus and other organic matter in soil to inorganic substances (9, 25)

minerotrophic describes peatlands fed by water moving through the mineral soil; contrast with *ombrotrophic* (30)

minimum viable population size of a population that, with a given probability, will ensure the existence of the population for a stated period of time (19)

mire wetland characterized by an accumulation of peat (30)

mitosis cell division involving chromosome duplication, resulting in two daughter cells with full complement of chromosomes, genetically the same as parent cells (19)

model in theoretical and systems ecology, an abstraction or simplification of a natural phenomenon developed to predict a new phenomenon or to provide insights into existing ones; in mimetic association, the organism mimicked by different organism (1)

moder type of forest humus layer in which plant fragments and mineral particles form loose netlike structures held together by a chain of small arthropod droppings (4)

modular organism one that grows by repeated iteration of parts such as branches or shoots of a plant; some may separate and become physically and physiologically independent (10)

moist adiabatic lapse rate see *adiabatic lapse rate* (3)

Mollisol soil formed by calcification characterized by accumulation of calcium carbonate in lower horizons and high organic content in upper horizons (4)

monoecious in plants, describes occurrence of reproduction organs of both sexes on the same individual, either as different flowers (hermaphroditic) or in the same flower (dioceous) (13)

monogamy mating of an animal and maintenance of a pair bond with only one member of the opposite sex at a time (13)

montane related to mountains (7, 29)

moor a blanket bog or peatland (30)

mor type of forest humus layer of unincorporated organic matter usually matted, compacted, or both and distinct from mineral soil; low in bases and acid in reaction (4)

mull humus that contains appreciable amounts of mineral bases and forms a humus-rich layer of forested soil consisting of mixed organic and mineral matter; blends into the upper mineral layer without abrupt changes in soil characteristics (4)

Mullerian mimicry resemblance of two or more conspicuously marked distasteful species that increases predator avoidance (16)

mutation transmissible change in structure of a gene or chromosome (19)

mutualism relationship between two species in which both benefit (7, 14, 17)

mycelium mass of hyphae that make up the vegetative portion of a fungus (9)

mycorrhizae association of fungus with roots of higher plants that improves the plants' uptake of nutrients from the soil (17)

myrmecochores plants that depend on ants to disperse their seeds (17)

myrmecochory dispersal by ants (17)

natal dispersal dispersal of young away from their birth place (10)

natality production of new individuals in a population (10)

natural selection differential reproduction and survival of individuals that results in elimination of maladaptive traits from a population (5)

neap tide (31) tides with little difference between high and low that occur when moon is at either quarter; at that time, the gravitational pull of the sun and moon interfere with each other (31)

negative feedback homeostatic control in which an increase in some substance or activity ultimately inhibits or reverses the direction of the processes leading to the increase (5)

neighborhood size number of individuals in a population found in an area defined by the average dispersal distance of a single individual (19)

nekton aquatic animals that are able to move at will through the water (30, 31)

neritic describes the marine environment embracing the regions where land masses extend outward as a continental shelf (31)

net energy in heterotrophs, the energy left after metabolic losses that is available for maintenance, production, and reproduction (24)

net mineralization rate the difference between the rates of mineralization and immobilization (9)

net production accumulation of total biomass over a given period of time after respiration is deducted from gross production in plants and from assimilated energy in consumer organisms (24)

net reproductive rate number of females produced per females per generation (10)

neutrophilic preferring a habitat that is neither acid or alkaline (7)

niche functional role of a species in the community, including activities and relationships (14)

niche compression contraction of habitat, rather than change in type of food or resources utilized, as a result of competition between two or more species (14)

niche overlap the sharing of niche space by two or more species (14)

niche preemption procurement by a species of a portion of available resources, leaving less for the next (14)

niche shift adoption of changed behavioral and feeding pattern by two or more competing populations to reduce interspecific competition (14)

niche width range of a single niche dimension occupied by a population (14)

nitrification breakdown of nitrogen-containing organic compounds into nitrates and nitrites (25)

nitrogen fixation conversion of atmospheric nitrogen to forms usable by organisms (25)

null hypothesis a statement of no difference between sets of values formulated for statistical testing (1)

numerical response change in size of a population of predators in response to change in density of its prey (15)

nutrient a substance required by organisms for normal growth and activity (3)

nutrient cycle pathway of a element or nutrient through the ecosystems from assimilation by organisms to release by decomposition (25)

nutrient retranslocation transport of a significant percentage of nutrients from the leaves to the perennial parts of the tree prior to leaf-fall (7)

O **horizon** surface layer composed of fresh or partially decomposed organic matter formed or forming above the mineral layer (4)

obligate refers to a response to particular condition or way of life for which there is no alternative (18)

obligate anaerobes bacteria that cannot exist in the presence of oxygen (9)

old-growth forest forest that has not been cut for decades or disturbed by humans for hundreds of years (29)

oligotrophic describes a body of water low in nutrients and in productivity (30)

ombrotrophic describes condition in bogs or mires in which water is highly deficient in mineral salts and has low pH (30)

omnivore an animal that feeds on both plant and animal matter (8)

opportunistic species organisms able to exploit temporary habitats or conditions (21)

optimal foraging tendency of animals to harvest food efficiently—to select food sizes or food patches that will result in maximum food intake for energy expended (15)

optimum sustained yield level of sustained yield that takes into consideration species interaction; amount of material that can be removed from a population that will result in production of maximum amount of biomass on a sustained yield basis (18)

ordination process by which communities are positioned graphically on a gradient of one to several axes so that the distances between them reflect differences in composition (20, 21)

Ordovician second oldest period of the Paleozoic era, some 510 to 440 million years ago; marine invertebrates abundant (11, 21)

organismic concept of the community idea that species, especially plant species, are integrated into an internally interdependent unit; upon maturity and death of the community, another identical plant community will replace it (1, 20)

oscillation regular fluctuation in a fixed cycle above or below some set point (10)

osmosis movement of water molecules across a differentially permeable membrane in response to a concentration or pressure gradient (7)

osmotic potential the pressure needed to stop osmosis; the more concentrated a solution, the higher is its osmotic pressure (7)

osmotic pressure pressure needed to prevent the passage of water or another pure solvent through a semipermeable membrane separating the solvent from the solution (7)

outbreeding production of offspring through preferential mating between nonrelated individuals (19)

outbreeding depression loss of fitness through the contamination of a gene pool with new alleles that produce offspring poorly adapted to the local environment (19)

overdispersion situation in which the distribution of organisms is not random, but clumped so that areas are both empty and heavily overpopulated; contagious distribution (17)

overdominance hypothesis proposes that heterozygotes have higher fitness than either of the homozygotes (19)

overland flow sheet flow of water across the surface of the ground that occurs when the soil can no longer absorb the precipitation (3)

overturn vertical mixing of layers in a body of water brought about by seasonal changes in temperature (30)

Oxisol soil developed under humid semitropical and tropical conditions characterized by silicates and hydrous oxides, clays, residual quartz, deficiency in bases, and low plant nutrients; formed by the process of laterization (4)

paleoecology study of ecology of past communities by means of the fossil record (21)

paludification blanketing of terrestrial ecosystems by an overgrowth of bog vegetation (30)

panting rapid shallow breathing that increases evaporation from the upper respiratory tract to counteract heat stress (8)

parasitism relationship between two species in which one benefits while the other is harmed (although not usually killed directly) (14, 17)

parasitoid insect larva that kills its host by consuming completely the host's soft tissues before pupation or metamorphosis into an adult (14, 15)

partial dominance condition resulting when one allele is not completely dominant to another allele; as a result, the heterozygote has a phenotype between those exhibited by individuals homozygous for either individual allele involved (19)

patch a relatively homogeneous area that differs from its surroundings (23)

peat unconsolidated material consisting of undecomposed and only slightly decomposed organic matter under conditions of excessive moisture (30)

ped soil particles held together in clusters of various sizes (4)

pedalfer podsolic soil possessing a layer of iron accumulation (hardpan) that impedes free circulation of air and water (4)

pedon a three-dimensional sampling unit of soil down to the parent material and wide enough to study all the horizons and their intergradations; smallest unit in the study of soil (4)

pelagic referring to the open sea (31)

per capita rate of increase (see rate of increase)

percent base saturation in the soil, the percentage of exchange sites occupied by ions other than hydrogen (4)

percolation the movement of water downward and radially through the subsurface soil layer, often continuing down to groundwater (2)

perennating tissue vegetative means by which biennial and perennial plants survive periods of unfavorable conditions; aerial parts die back to a minimum during an unfavorable season, such as winter; regrowth during a favorable season is supported by food stored in underground organs—root, bulbs, and rhizomes—or buds on stems of woody plants (20)

periphyton in freshwater ecosystems, organisms that are attached to submerged plant stems and leaves; see *aufwuchs* (30)

permafrost permanently frozen soil (28)

permanent wilting point point at which water potential in the soil and conductivity assume such low values that the plant is unable to extract sufficient water to survive and wilts permanently (7)

Permian most recent period of the Paleozoic, about 280 to 250 million years ago; dominance of reptiles; appearance of modern insects and gymnosperm plants (11)

phanerophyte trees, shrubs, and vines whose perennating buds are borne on aerial shoots (20, 21)

phenology study of the seasonal changes in plant and animal life and the relationship of these changes to weather and climate (6)

phenotype physical expression of a characteristic of an organism as determined by genetic constitution and environment (19)

phenotypic plasticity ability to change form under different environmental conditions (19)

pheromone chemical substance released by an animal that influences behavior of others of the same species (11)

photic zone lighted water column of a lake or ocean inhabited by plankton (30, 31)

photoinhibition the slowing or stopping of a plant process by light (6)

photoperiodism response of plants and animals to changes in relative duration of light and dark (7, 8)

photorespiration respiration that occurs in light in C_3 plants and is not coupled to oxidative phosphorylation and does not generate ATP; a wasteful process decreasing photosynthetic efficiency (6)

photosynthate energy-rich organic molecules produced during photosynthesis (6)

photosynthetically active radiation (PAR) those wavelengths in the radiation spectrum used by plants in photosynthesis (2, 6)

phreatophyte type of plant that habitually obtains its water supply from zone of groundwater (30)

physiological ecology the study of the physiological functioning of organisms in relation to their environment (1)

physiological longevity maximum life span of an individual in a population under given environmental conditions (10)

physiological natality the maximum number of young that a female is capable of producing physiologically (10)

phytoplankton small, floating plant life in aquatic ecosystems; planktonic plants (30, 31)

phytosociology the classification of plant communities based on floristics rather than on life forms or other characteristics (1)

pioneer species plant that are initial invaders of disturbed sites or early seral stages of succession; characterized by rapid growth, short life span, and production of a great number of easily dispersed seeds (21)

plankton small, floating or weakly swimming plants and animals in freshwater and marine ecosystems (30, 31)

playa low basins in the desert that receive water that rushes down from the hills (28)

Pleistocene geological epoch extending from about 2 million to 10,000 years ago, characterized by recurring glaciers; the Ice Age (11, 21)

pneumatophore an erect respiratory root that protrudes above waterlogged soils; typical of bald cypress and mangroves (7, 30)

podzolization soil-forming process resulting from acid leaching of the *A* horizon and accumulation of iron, aluminum, silica, and clays in lower horizon (4)

poikilothermy variation of body temperature with external conditions (8)

polyandry mating of one female with several males (13)

polyclimax theory idea that the endpoint of succession is controlled by one of many local environmental conditions, such as soil and fire, as well as climate (21)

polygamy acquisition by an individual of two or more mates, none of which is mated to other individuals (13)

polygyny mating of one male with several females (13)

polymorphism occurrence of more than one distinct form of individuals in a population (19)

polyploidy having three or more times the haploid number of chromosomes (18, 19)

population a group of individuals of the same species living in a given area at a given time (10)

population cycle oscillations between periods of high and low densities in the number of individuals in a population (11)

population density the number of individuals in a population per unit area (10)

population dynamics study of the factors that influence variations in the number and densities of populations in time and space (10)

population ecology the study of the interaction of a particular species or species with the environment (1)

population genetics study of the heritable changes in population in relation to the underlying individual processes of inheritance and development (1)

population regulation mechanisms or factors within a population that cause it to decrease when density is high and increase when density is low (12)

positive feedback control in a system that reinforces the process in the same direction (5)

potential energy energy available to do work (24)

potential evapotranspiration amount of water that would be transpired under constantly optimal conditions of soil moisture and plant cover (7, 27)

pothole small marshlike ponds found throughout the Dakotas and central Canadian provinces (30)

practical salinity unit (PSU) total amount of dissolved material in seawater expressed as parts per thousand (31)

Precambrian earliest and longest of the geological time periods, some 4600 to 570 million years ago; precedes Paleozoic era; appearance of cyanobacteria and fungi (11, 21)

precipitation all the forms in which water falls to the ground (2)

precocial describes young birds hatched with down, eyes open, and able to move about; also young mammals born with eyes open and able to follow their mother after birth (for example, fawn deer, calves) (13)

predation condition in which one living organism serves as a food source for another (14)

predation risk balancing by a organism of foraging profitability in a foraging area against the risk of being eaten by a predator (15)

preferred temperature range of temperatures within which poikilotherms function most efficiently (8)

presaturation dispersal dispersal of organisms from an area before the population reaches its carrying capacity (11)

primary producers organisms able to produce food from simple inorganic substances; typically photosynthetic plants (5)

primary production production of biomass by photosynthetic and chemosynthetic autotrophs (6)

primary productivity rate at which primary producers produce biomass (6)

primary sex ratio ratio of male to female offspring in a population at birth (10)

primary succession vegetation development starting from a new site never before colonized by life (21)

probability of dying the number that died during a given time period divided by the number alive at the beginning of the time period (10)

probability of surviving the number of survivors remaining during a given time period divided by the number alive at the beginning of the period (10)

producers green plants and certain chemosynthetic bacteria that convert light or chemical energy into organismal tissue (6)

production amount of energy formed by an individual, population, or community per unit time (6, 7, 24)

production efficiency proportion of assimilated energy that goes to production and respiration (24)

productivity rate of energy fixation or storage per unit time; not to be confused with production (24)

profundal describes the deep zone in aquatic ecosystems below the limnetic zone (30, 31)

promiscuity behavior in which males and females copulate with one or many of the opposite sex without forming pair bonds (13)

proportional control temperature regulation by changing body shape and orientation to the sun, a behavior found in many ectotherms (8)

pycnocline layer of ocean water that exhibits a rapid change in density (31)

pyramid of biomass diagrammatic representation of biomass at different trophic levels in an ecosystem (24)

pyramid of energy diagrammatic representation of the flow of energy through different trophic levels (24)

pyramid of numbers diagrammatic representation of the number of individual organisms present at each trophic level in an ecosystem; it is the least useful of the pyramids (24)

quaking bog bog that develops as a floating mat over water (30)

Quanternary geological period from 2 million years ago to the present; includes the Pleistocene and recent epochs (11, 21)

R **horizon** consolidated rock below the *C* horizon (4)

r **selection** selection under low population densities; favors high reproductive rates under conditions of low competition (13)

radiation (radiant energy) energy propagated in the form of electromagnetic waves (2)

radionuclides fission and nonfission products of atomic reaction (26)

rain shadow dry area on lee side of mountains (2)

raised bog a bog in which the accumulation of peat has raised its surface above both the surrounding landscape and the water table; it develops its own perched water table (30)

ramet any individual belonging to a clone (10)

random distribution lacking pattern or order; distribution of individuals is independent of all other individuals (10)

reaction time lag the lag between environmental change and a corresponding change in the rate of population growth (10)

realized natality the amount of successful reproduction that actually occurs over a period of time (10)

realized niche portion of fundamental niche space occupied by a population in face of competition from populations of other species; environmental conditions under which a population survives and reproduces in nature (14)

recessive describes an allele or phenotype that is expressed only in the homozygous state (19)

reciprocal replacement small-scale cyclic succession with a forest (21)

recombination exchange of genetic material resulting from independent assortment of chromosomes and their genes during gamete production followed by a random mix of different sets of genes at fertilization (19)

reductionist describes an approach to the study of ecosystems by examining the function of each part, and in that manner discovering how the whole system works (1)

regolith mantle of unconsolidated material below the soil from which soil develops (4)

regular distribution arrangement of individuals in a pattern that ensures they are more widely separated from each other than would be expected by chance; underdispersion (10)

regulated river a river whose water levels and water flows are controlled by dams, levees, and channels (30)

relative abundance numerical abundance of one species as compared with the total abundance of all species (20, 21)

relative dominance a ratio of basal area occupied by one species to total basal area (20, 21)

relative frequency ratio of the frequency of one species to the total frequency of all species (20, 21)

renosterveld see *strandveld* (28)

replicate individual units receiving the treatment in an experiment (1)

replication in an experiment, the use of a number of individuals or experimental units receiving the treatment to account for uncontrolled variation (1)

reproductive allocation proportion of its available resource input that an organism expends on reproduction over a given period of time (13)

reproductive cost decrease in survivorship or rate of growth experienced by an individual as a result of increasing its current allocation to reproduction; reflected in its decreased potential for future reproduction (13)

reproductive effort proportion of its resources an organism expends on reproduction (13)

reproductive time lag a lag between environmental change and a change in the length of gestation or its equivalent (10)

reproductive value potential reproductive output of an individual at a particular age (*x*) relative to that of a newborn individual at the same time (10, 13)

rescue effect the slowing or halting of extinction by an influx of immigrants (23)

residual reproductive value reproductive value of an individual reduced by its expected present reproduction (13)

resilience ability of a system to absorb changes and return to its original condition (10)

resource allocation action of apportioning the supply of a resource to a specific use (24)

respiration metabolic assimilation of oxygen accompanied by production of carbon dioxide and water, release of energy, and breaking down of organic compounds (6, 7, 24)

restoration ecology study of the application of ecological theory to the ecological restoration of highly disturbed sites (1)

rete a large network or discrete vascular bundle of intermingling small blood vessels carrying arterial and venous blood that acts as a heat exchanger in mammals and certain fishes and sharks (8)

rheotrophic describes wetlands, especially bogs and fens, that obtain much of their nutrient input from groundwater (30)

rhizobia bacteria capable of living mutualistically with higher plants (17)

rhizome a horizontally growing underground stem that through branching gives rise to vegetative structures (20)

rhizoplane root surface (9)

rhizosphere soil region immediately surrounding roots (9, 25)

richness a component of species diversity; the number of species present in an area (20)

rill erosion runoff concentrated in streamlets as it flows across the soil surface (4)

riparian along banks of rivers and streams (30)

riparian woodland seasonally flooded forests along rivers and streams; riverbank forests are often called gallery forests (30)

risk-sensitive foraging decision by a forging animal to revisit a patch that that gives it a constant rate of return or visit a new patch where the return is unknown (15)

riverine wetland wetland found along and associated with a river or stream (30)

root-to-shoot ratio ratio of the weight of roots to that of shoots of a plant (7)

rubisco shortened name for the enzyme ribulose biphosphate carboxylase-oxygenase, used by plant in photosynthesis; the most abundant enzyme on Earth (6)

ruminant an ungulate with a three- or four-chamber stomach. The large first chamber is known as the rumen in which bacterial fermentation of consumed plant matter takes place (8)

salinization enrichment of soil in salts faster than they are leached (4)

saprophage organism that feeds on dead plant and animal matter; they consist mainly of bacteria and fungi and some invertebrates, such as insect larvae (24)

satisficing relative to foraging, strategy in which the decision maker is satisfied after meeting some minimum requirement (15)

saturation dispersal movement of individuals out of a population that has reached or exceeded carrying capacity (11)

saturation vapor pressure the maximum amount of water vapor that the air can hold at a given temperature and pressure (3)

savanna tropical grassland, usually with scattered trees or shrubs (28)

scavenger animal that feeds on other dead animals or on animal products, such as dung (8)

sclerophyll woody plant with hard, leathery, evergreen leaves that prevent moisture loss (28)

scramble competition intraspecific competition in which limited resources are shared to the point that no individual survives (11)

search image mental image formed in predators enabling them to find more quickly and to concentrate on a common type of prey (16)

seasonality describes recurring seasonal responses of organisms influenced by the attendant interactions of light, temperature, and moisture (7)

second law of thermodynamics when energy is transferred or transformed, part of the energy assumes a form that cannot be passed on any further (24)

second-level carnivore animals that feed on other carnivores (8)

secondary production production by consumer organisms (24)

secondary sex ratio ratio of males to females in a population at birth (10)

secondary substances organic compounds produced by plants that are utilized in chemical defense (16)

secondary succession plant succession taking place on sites that have already supported life (21)

sedimentary cycle it involves weathering of existing rock, followed by the erosion of material from it, and subsequent transport largely through the water cycle (salt solution phase), deposition, and burial (rock phase) (25)

sedimentary rocks rocks formed by the deposition and compression of mineral and rock particles (4)

selection coefficient a measure of the fitness of one genotype relative to that of another genotype in the population (19)

selective pressure any force acting on individuals in a population that determines which individuals leave more descendants (contribution to the gene pool) to subsequent generations than others; gives direction to the evolutionary process (5, 19)

self-thinning progressive decline in density of plants associated with the increasing size of individuals in a population of growing individuals (12)

semelparity having only a single reproductive effort in a lifetime over one relatively short period of time (13)

semiarid describes a region of fairly dry climate with precipitation between 25 and 60 cm a year and with an evapotranspiration rate high enough so the potential loss of water to the environment exceeds inputs (28)

sensible heat heat that can be measured and felt in contrast to the actual temperature of the environment that can be measured with a thermometer (3)

seral describes a series of stages that follow one another in succession (21)

seral stage one of the successional stages in a sere (21)

sere the series of successional stages on a given site that leads to a terminal community (21)

serpentine soil soil derived from ultrabasic rocks that are high in iron, magnesium, nickel, chromium, and cobalt and low in calcium, potassium, sodium, and aluminum; supports distinctive communities (7)

sex ratio the relative number of males to females in a population (10)

sex reversal a change in functioning so that a member of one sex behaves as the other (13)

sexual selection selection by one sex for an individual of the other sex based on some specific characteristic or characteristics; usually takes place through courtship behavior (13)

shade intolerant describes plants that cannot grow and reproduce under low light conditions; sun plants (6)

shade tolerant describes plants that can grow and reproduce under low light conditions (6)

sheet erosion more or less even removal of soil over a field by water flowing across its surface (4)

shifting mosaic steady state (21) community composed of a mosaic of patches, each in a phase of successional development; the collection of these patches describes the average state of the forest (21)

shivering uncoordinated, involuntary, high-frequency contraction of skeletal muscles that converts chemical energy to thermal energy (8)

short-day organisms plants and animals that come into reproductive condition under conditions of short days—days with less than a certain maximum length (6, 8)

shredders stream invertebrates that feed on coarse particulate organic matter (30)

shrub a woody plant that branches below or near the ground level into several main stems (28)

sigmoid curve S-shaped curve of logistic growth (11)

Silurian period between the Ordovician and Devonian periods, some 400 to 405 million years ago; characterized by early land plants, invertebrates, and primitive jawless fishes (11, 21)

sink area where local mortality exceeds local reproductive success (12)

skylight diffuse light that reaches Earth's surface resulting from the scattering of solar radiation by dust and water vapor (2)

snag dead or partially dead tree at least 10.2 cm dbh and 1.8 m tall; important for cavity-nesting birds and mammals (29)

social dominance physical dominance of one individual over another, usually maintained by some manifestation of aggressive behavior (13)

social parasite animal that uses other individuals or species to rear its young—for example, the cowbirds (17)

sociobiology integrated study of social behavior with special emphasis on social systems considered as evolutionary ecological adaptations (1)

soil association a group of defined and named soil taxonomic units occurring together in an individual and characteristic pattern over a geographic region (4)

soil erosion carrying away of soil particles by wind and water (4)

soil genesis the changes that have taken place or are taking place in the mineral and organic matter in soil as a natural body in response to climate and organisms (4)

soil horizon developmental layer in the soil with its own characteristics of thickness, color, texture, structure, acidity, nutrient concentration, and the like (4)

soil profile distinctive layering of horizons in the soil (4)

soil series basic unit of soil classification consisting of soils that are essentially alike in all major profile characteristics except texture of the *A* horizon. Soil series are usually named for the locality where the typical soil was first recorded (4)

soil structure arrangement of soil particles and aggregates (4)

soil texture relative proportions of the three particle sizes—sand, silt, and clay—in the soil (4)

soil type lowest unit in the natural system of soil classification, consisting of soils that are alike in all characteristics, including texture of the *A* horizon (4)

solar constant rate at which solar energy is received on a surface just outside of Earth's atmosphere; current value is 0.140 watt/cm^2 (3)

solification slow flow of saturated soil and other unconsolidated materials downslope (4, 28)

species diversity measurement that relates density of organisms of each type present in a habitat to the number of species in a habitat (20, 27)

species richness number of species in a given area (20, 21, 27)

specific birth rate birth rate expressed relative to a criterion, such as age (10)

specific heat amount of energy that must be added or removed to raise or lower temperature of a substance by a specific amount (3)

spiraling mechanism of retention of nutrients in flowing water ecosystems involving the interdependent processes of nutrient recycling and downstream transport (30)

Spodosol soil characterized by the presence of a horizon in which organic matter and amorphous oxides of aluminum and iron have precipitated; includes podzol soils (4)

spring tide tides of maximum rise and fall that occur when Earth, moon, and sun are nearly in line, intensifying the gravitational pull of sun and moon on Earth's water masses (31)

strandveld mediterranean-type shrubland, dominated by broad-sclerophyll woody shrubs found in South Africa (28)

stabilizing selection selection favoring the middle of the phenotype distribution (19)

stable age distribution constant proportion of individuals of various age classes in a population through population changes (10)

stable limit cycle a regular fluctuation in abundance of predator and prey populations, resulting when stabilizing and destabilizing interactions balance (11)

stand unit of vegetation that is essentially homogeneous in all layers and differs from adjacent types qualitatively and quantitatively (23, 29)

standard deviation statistical measure defining the dispersion of values about the mean in a normal distribution (1)

standing crop amount of biomass per unit area at a given time (24)

static life table see *life table* (10)

stationary age distribution special form of stable age distribution in which the population has reached a constant size at which birth rate equals the death rate and age distribution remains fixed (10)

stemflow that portion of precipitation intercepted by trees that flows down their trunks (3)

stenotherms organisms unable to tolerate a wide temperature range (2)

stochastic describes patterns arising from random factors (10)

stochastic extinction extinction caused by some random environmental event (10)

stochastic model mathematical model based on probabilities; predictions of the model are fixed but variable; opposite of *deterministic model* (1)

stoichiometry branch of chemistry dealing with the quantitative relationship of elements in combination (25)

stomata cells on leaves of vascular plants that regulate gas and water exchange (6)

stratification division of an aquatic or terrestrial community into distinguishable layers on the basis of temperature, moisture, light, vegetative structure, and other such factors, creating zones for different plant and animal types (20)

sublimation the direct vaporization of ice (3)

sublittoral describes the lower division of sea from about 40 to 60 m to below 200 m (31)

subsidence inversion atmospheric inversion produced by sinking air movement from aloft (2)

succession replacement of one community by another; often progresses to a stable terminal community (21)

successional sequence see *sere* (21)

sun plant plant able to grow and reproduce only under high light conditions (6)

sunspot relatively dark, sharply defined region on the sun, found mostly in groups of two or more; sunspots are cyclic, with a periodicity of approximately 11 years (16)

supercooling in ectotherms, lowering of body temperature below freezing without freezing body tissue; involves the presence of certain solutes, particularly glycerol (8)

surface fire forest fire that burns surface litter and vegetation but does not reach up into the crown (23)

surface tension phenomenon where the attraction (cohesion) of molecules of a liquid cause its surface to behave like an elastic membrane (3)

survivorship the probability of a representative newborn individual in a cohort surviving to various ages (10)

survivorship curve a graphical description of the survival of a cohort of individuals in a population from birth to the maximum age reached by any one member of the cohort (10)

sustained yield yield per unit time from an exploited population equal to production per unit time (18)

swamp wooded wetland in which water is near or above ground level (30)

switching a predator changing its diet from a less abundant to a more abundant prey species; see also *threshold of security* (15)

symbiosis situation in which two dissimilar organisms live together in close association (17)

system set or collection of interdependent parts or subsystems enclosed within a defined boundary; the outside environment provides inputs and receives attributes transmitted to it by the system (5)

systems ecology the application of general systems theory and methods to ecology (1)

taiga the northern circumpolar boreal forest (29)

temperate rain forest forests in regions characterized by relatively mild climate and heavy rainfall that produce lush vegetative growth; one example is the coniferous forest of the Pacific Northwest of North America (29)

temperature immediate direct measure of the average kinetic energy possessed by individual molecules of a substance (2)

territoriality defense of a defined space by an individual or a social group (11)

territory area defended by an animal; varies among animals according to social behavior, social organization, and resource requirements of different species (11)

Tertiary first period of the Cenozoic cra, about 65 to 2 million years ago; composed of the Paleocene, Eocene, Oligocene, Miocene, and Pliocene epochs; characterized by the emergence of mammals (11, 21)

thermal conductance rate at which heat flows through a substance (3)

thermal neutral zone among homeotherms, the range of temperatures at which metabolic rate does not vary with temperature (8)

thermal tolerance range of temperatures in which an aquatic poikilotherm is most at home (8)

thermocline layer in a thermally stratified body of water in which temperature changes rapidly relative to the remainder of the body (25, 30)

thermogenesis increase in production of metabolic heat to counteract the loss of heat to a colder environment (8)

therophyte life form of plants that survives unfavorable conditions in the form of a seed; annual and ephemeral species (20)

thinning law, 3/2 self-thinning plant populations, sown at sufficiently high densities, approach and follow a thinning line with a slope of roughly −3/2; in a growing population, plant weight increases faster than density decreascs to a point where the slope changes to −1 (12)

threshold of security point in local population density at which the predator turns its attention to other prey (see *switching*) because of harvesting efficiency; the segment of prey population below the threshold is relatively secure from predation (15)

throughfall that part of the precipitation intercepted by vegetation that falls on to the ground (3)

time lag delay in a response to change (12)

time-specific life table see *life table* (10)

tolerance model model of succession that proposes that succession leads to a community composed of those species most efficient in exploiting resources; colonists neither increase or decrease the rate of recruitment or growth of later colonists (22)

top-down regulation regulation of population and community structure through consumers (20)

toposequence a pattern of local soils whose development was controlled by topography of the landscape (4)

torpidity temporary condition of an animal involving a great reduction in respiration; results in loss of power of motion and feeling; usually occurs in response to some unfavorable environmental condition, such as heat or cold, to reduce energy expenditure (8)

trace element element occurring and needed in small quantities; see *micronutrient* (3)

transient polymorphism the occurrence of two or more forms or genes in a population during a time when one form is being replaced by another (19)

translocation transport of materials within a plant; absorption of minerals from soil into roots and their movement throughout the plant (25)

transpiration loss of water vapor by land plants (2, 6)

treatment manipulation of a variable in a predetermined way to monitor the response of the experimental units (1)

Triassic oldest period of the Mesozoic, some 230 to 250 million years ago; increase in primitive amphibians and reptiles (11, 21)

trophic related to feeding (24)

trophic level functional classification of organisms in an ecosystem according to feeding relationships from first-level autotrophs through succeeding levels of herbivores and carnivores (24)

trophic structure organization of a community based on the number of feeding or energy transfer levels (24)

trophogenic zone upper layer of the water column in ponds, lakes, and oceans in which light is sufficient for photosynthesis (30)

tropholytic zone area in lakes and oceans below the compensation point (30)

tundra areas in arctic and alpine (high mountains) regions characterized by bare ground, absence of trees, and growth of mosses, lichens, sedges, forbs, and low shrubs (28)

turgor the state in a plant cell in which the protoplast is exerting pressure on the cell wall because of intake of water by osmosis (7)

turnover rate rate of replacement of a substance or a species when losses to a system are replaced by additions (23)

ultimate incipient lethal temperature upper or lower limit of temperature at which an acclimatized organism will succumb (8)

unitary organism organism, such as an arthropod or vertebrate, whose growth to adult form follows a determinate pathway, unlike modular organisms whose growth involves indeterminate repetition of units of structure (10)

upwellings areas in oceans where currents force water from deep within the ocean into the euphotic zone (2, 31)

vacuole fluid-filled cavity within the cytoplasm (7)

validation an explicit and objective test of the basic hypothesis (1)

vapor pressure the amount of pressure water vapor exerts independent of dry air (3)

vapor pressure deficit the difference between saturation vapor pressure and the actual vapor pressure at any given temperature (3)

vector organism that transmits a pathogen from one organism to another (17)

vegetative reproduction asexual reproduction in which plants propagate themselves by means of specialized multicellular organs such as bulbs, corms, rhizomes, stems, and the like (10)

verification process of testing whether or not a model is a reasonable representation of a real-life system being investigated (1)

vernal pool temporary pond of water filled in the spring; important habitat of many amphibians and aquatic invertebrates (30)

vertical life table see *life table* (2)

Vertisol mineral soil that contains more than 30 percent of swelling clays that expand when wet and contract when dry; associated with seasonally wet and dry environments (4)

vesicular arbuscular mycorrhizae (VAM) a form of mycorrhizae in which the fungus enters and grows within the host's cells and extends widely into the surrounding soil; a type of ectomycorrizae (17)

viscosity property of a fluid that resists the force within the fluid that causes it to flow (3)

wadis an ephemeral river channel in a desert (28)

Wallace's line biogeographic line between the islands of Borneo and the Celebes that marks the eastward boundary of many landlocked Eurasian organisms and the boundary of the Oriental region (27)

water potential measure of energy in an aqueous solution needed to move water molecules across a semipermeable membrane; water tends to move from areas of high or less negative potential to areas of low or more negative potential (3)

water use efficiency ratio of net primary production to transpiration of water by a plant (7)

watershed entire region drained by a waterway that drains into a lake or reservoir; total area above a given point on a stream that contributes water to the flow at that point; the topographic dividing line from which surface streams flow in two different directions (30)

weather air in motion driven by unequal heating (3)

weed a plant species possessing a high rate of dispersal and occurring opportunistically on land or water disturbed by human activity and competing for resources with cultivated plants; a plant growing in the wrong place (21)

wet deposition that component of acid deposition that reaches Earth by the way of some form of precipitation (26)

wetfall nutrients carried to vegetation by precipitation, supplementing nutrients obtained from the soil (25)

wetland areas characterized by presence of water at or near the surface, hydric soils, and vegetation adapted to wet conditions (30)

wilting point moisture content of soil on an oven-dry basis at which plants wilt and fail to recover their turgidity when placed in a dark, humid atmosphere (4)

xeric describes dry conditions, especially relating to soil (21)

zero net growth isocline an isocline along which the population growth rate is zero (14)

zonation characteristic distribution of vegetation along an environmental gradient; this gradient may be latitudinal, altitudinal, or horizontal belts within an ecosystem (21)

zoogeography study of the distribution of animals (27)

zooplankton floating or weakly swimming animals in freshwater and marine ecosystems; planktonic animals (30, 31)

BIBLIOGRAPHY

ABER, J., AND J. MELILLO. 1982. Nitrogen immobilization in decaying hard-wood leaf litter as a function of initial nitrogen and lignin content. *Can. J. Bot.* 60:2263–2269.

———. 1991. *Terrestrial Ecosystems.* Saunders College Publishing, Philadelphia

ABER, J. D., K. J. NADELHOFFER, P. STEUDLER, AND J. M. MELILLO. 1989. Nitrogen saturation in northern forest ecosystems. *Bioscience* 39:378–386.

ACKERT, J. E., G. L. GRAHAM, L. O. NOLF, AND D. A. PORTYER. 1931. Quantitative studies on the administration of variable numbers of nematode eggs (*Ascaridia lineata*) to chickens. *Trans. Am. Microscopical* Soc. 50:206–214

ADAMS, E. S. 1994. Settlement tactics in seasonally territorial animals: Resolving conflicting predictions. *Am. Nat.* 143:939–943.

ADAMS, R. M., R. ALIG, J. M. CALLAWAY, B. A. McCARL, AND S. M. WINNET. 1995. The economic effects of climate change on U.S. agriculture. Final report. EPRI, Climate Change Impacts Program, Palo Alto, CA. 45 pp.

ADAMS, R. M., R. A. FLEMING, C. C. CHANG, AND B. A. McCARL. 1995. A reassesment of the economic effects of global climate change on U.S. agriculture. *Climat. Change* 30:147–167.

ADDICOTT, J. F. 1979. A multispecies aphid-plant association: A comparison of local and metapopulations. *Can. J. Zool.* 56:2554–2564.

———. 1985. On the population consequences of mutualism. In T. Case and J. Diamond (eds.), *Community Ecology.* Harper and Row, New York, pp. 425–436.

———. 1986. Variation in the costs and benefits of mutualism: The interaction between yuccas and yucca moths. *Oecologica* 70:486–494.

ADKISSON, P. L. 1966. Internal clocks and insect diapause. *Science* 154:234–241.

AHMADJIAN, V., AND J. B. JACOBS. 1982. Algal-fungal relationships in lichens: Recognition, synthesis, and development. In L. J. Goff (ed.), *Algal Symbiosis: A Continuum of Interaction Strategies.* Cambridge University Press, Cambridge, England.

AKCAKAYA, H. R., R. ARDITI, AND L. R. GINZBURG. 1995. Ratio-dependent predation: An abstraction that works. *Ecology* 76:995–1004.

AKER, C. L. 1982. Spatial and temporal dispersion patterns of pollinators and their relationship to the flowering strategy of *Yucca whipplei* (*Agavaceae*). *Oecologica* 54:243–252.

ALEXANDER, M. M. 1958. The place of aging in wildlife management. *Am. Sci.* 46:123–131.

ALEXANDER, R. D., AND G. BORGIA. 1978. Group selection, altruism, and levels of organization of life. *Ann. Rev. Ecol. Syst.* 9:449–474.

ALLEE, W. C., A. E. EMERSON, O. PARK, T. PARK, AND K. P. SCHMIDT. 1949. *Principles of Animal Ecology.* Saunders, Philadelphia.

ALLEN, D. L. 1942. *Michigan Fox Squirrel Management.* Michigan Department Conservation, Game Division Publ. 101, Lansing, MI.

ALLEN, T. F. H., AND T. B. STARR. 1982. *Hierarchy: Perspectives for Ecological Complexity.* University of Chicago Press, Chicago.

ALM, G. 1952. Year class fluctuations and span of life of perch. *Rept. Inst. Freshwater Res. Drottningholm* 33:17–38.

ALTMANN, M. 1960. The role of juvenile elk and moose in the social dynamics of their species. *Zoologica* 45:35–40.

AMBUEL, B., AND S. A. TEMPLE. 1983. Area dependent changes in the bird communities and vegetation of southern Wisconsin forests. *Ecology* 64:1057–1068.

AMMAN, G. D. 1992. The role of the mountain pine beetle in lodgepole pine ecosystems: Impact on succession. In W. J. Mattson (ed.), *The Role of Arthropods in Forest Ecosystems.* Springer Verlag, New York.

AMTHOR, J. S. 1995. Terrestrial higher-plant response to increasing atmospheric CO_2 in relation to the global carbon cycle. *Glob. Change Biol.* 1:243–274

AMUNDSON, D. C., AND H. E. WRIGHT, JR. 1979. Forest changes in Minnesota at the end of the Pleistocene. *Ecol. Monogr.* 49:109–127.

ANDERSON, J. M., AND A. MACFADYIN (eds.). 1976. *The Role of Terrestrial and Aqutic Organisms in the Decomposition Process.* Blackwell, Oxford, England.

ANDERSON, M., AND M. O. ERICKSON. 1982. Nest parasitism in goldeneyes, Bucephala clangula: Some evolutionary aspects. *Am. Nat.* 120:1–16.

ANDERSON, R. C. 1963. The incidence, development, and experimental transmission of *Pneumostrongylus tenuis* Dougherty (*Metastrongyloidae: Protostrongyliidae*) of the meninges of the white-tailed deer (*Odocoileus virginianus borealis*) in Ontario. *Can. J. Zool.* 41:775–792.

———. 1965. Cerebospinal nematodiasis (*Pneumostrongylus tenuis*) in North American cervids. Trans. N. Am. Nat. Res. Conf. 13:156–167.

ANDERSON, R. C., AND A. K. PRESTWOOD. 1981. Lungworms. In W. R. Davidson (ed.), *Diseases and Parasites of White-Tailed Deer,* Mscl. Publ. no. 7, Tall Timbers Research Station, Tallahassee, FL, pp. 266–317.

ANDERSON, R. M. 1979. Population biology of infectious diseases. Part I. *Nature* 280:361–367.

———. 1981. Population ecology of infectious diseases. In R. M. May (ed.), *Theoretical Ecology: Principles and Applications,* 2nd ed. Sinauer Associates, Sunderland, MA, pp. 318–355.

ANDERSON, R. M., AND R. M. MAY. 1978. Regulation and stability of host-parasite population interactions. I. Regulatory processes. *J. Anim. Ecol.* 47:219–247.

ANDERSSON, M. 1982. Female choice selects for extreme tail length in a widowbird. *Nature* 299.818–820.

————. The evolution of eusociality. *Ann. Rev. Ecol. Syst.* 15:169–185.

ANDERSSON, M., AND Y. IWASA. 1996. Sexual selection. *TREE* 11:53–58.

ANDREAE, M. O. 1991. Biomass burning: Its history, use and distribution and its impact on environmental quality and global climate. In J. S. Levine (ed.), *Global Biomass Burning.* MIT Press, Cambridge, MA, pp. 3–21.

ANDREN, H. 1994. Effects of habitat fragmentation on birds and mammals in landscapes with different proportions of suitable habitat. A review. *Oikos* 71:35–36.

ANDREWARTHA, H. G., AND L. C. BIRCH. 1954. *The Distribution and Abundance of Animals.* University of Chicago Press, Chicago.

ANDREWS, R. D., D. C. COLEMAN, J. E. ELLIS, AND J. S. SINGH. 1975. Energy flow relationships in a short grass prairie ecosystem. Proc. *1st Inter. Cong. Ecol.* 22–28. W. Junk Publishers, The Hague.

ANDREWS, R., AND A. S. RAND. 1974. Reproductive effort in anoline lizards. *Ecology* 55:1317–1327.

ANDRZEJEWSKA, L., AND G. GYLLENBERG. 1980. Small herbivore subsystem. In A. I. Bretmeyer and G. M. Van Dyne (eds.), *Grasslands, Systems Analysis and Man.* International Biological Programme no. 19, Cambridge University Press, Cambridge, England, pp. 201–268.

ANSTETT, M. C., M. HOSSAERT-MCKEY, AND F. KJELLBERG. 1997. Figs and fig pollinators: Evolutionary conflicts in a coevolved mutualism. *TREE* 12:94–99.

ANTONOVICS, J. A., A. N. BRADSHAW, AND R. G. TURNER. 1971. Heavy metal tolerance in plants. *Adv. Ecol. Res.* 71:1–85.

ANTONOVICS, J. A., AND D. A. LEVIN. 1980. The ecological and genetical consequences of density-dependent regulation in plants. *Ann. Rev. Ecol. Syst.* 11:411–452.

ANTONOVICS, J. A., AND R. B. PRIMACH. 1982. Experimental ecology and genetics in *Plantago.* VI. The demography of seedling transplants of *P. lanceolata. J. Ecol.* 70:55–75.

ARDITI, R., AND A. A. BERRYMAN. 1991. The biological control paradox. *TREE* 6:32.

ARDITI, R., AND L. R. GINSBERG. 1989. Coupling in predator-prey dynamics: Ratio-dependence. *J. Theor. Biol.* 139:311–320.

ARNO, S. F. 1976. The historical role of fire in the Bitterroot National Forest. USDA Forest Service Res. Paper INT 187. Ogden, UT.

ARNONE, J. A., AND CH. KORNER. 1995. Soil and biomass carbon pools in model communities of tropical plants under elevated CO_2. *Oecologia* 104:61–71.

ARP, A. J., AND J. J. CHILDRESS. 1983. Sulfide binding by the blood of the hydrothermal vent tube worm *Riftia pachyptila. Science* 219:295–297.

ARP, W. J. 1991. Effects of source-sink relations on photosynthetic acclimation to elevated CO_2. *Plant Cell Environ.* 14:869–875.

ASCHMANN, H. 1973. Distribution and peculiarity of mediterranean ecosystems. In F. di Castri and H. A. Mooney (eds.), *Mediterranean Type Ecosystems: Origin and Structure.* Springer-Verlag, New York, pp. 11–19.

ASH, J. E., AND J. P. BARKHAM. 1976. Changes and variability in the field layer of a coppiced woodland in Norfolk, England. *J. Ecol.* 64:697–712.

ASHMOLE, N. P. 1963. The regulation of numbers of tropical oceanic birds. *Ibis* 103b:458–473.

ATSATT, P. R., AND D. J. O'DOWD. 1976. Plant defense guilds. *Science* 193:24–29.

ATTIWILL, P. M., AND G. W. LEEPER. 1987. *Forest Soils and Nutrient Cycles.* Melbourne University Press, Melbourne.

AUGSPURGER, C. K. 1982. Light requirements of neotropical tree seedlings: A comparative study of growth and survival. *J. Ecol.* 72:777–795.

AUMANN AND EMLEN, J. 1965. Relationship of population density to sodium availability and sodium selection by microtine rodents. *Nature* 208:198–199.

AUSTIN, M. P., AND T. M. SMITH. 1989. A new model of the continuum concept. *Vegetatio* 83:35–47.

AUSTIN, M. P., R. B. CUNNINGHAM, AND P. M. FLEMING. 1984. New approaches to direct gradient analysis using environmental scalars and statistical curve-fitting procedures. *Vegetatio* 55:11–27.

AUSTIN, M. P., R. B. CUNNINGHAM, AND R. B. GOOD. 1983. Altitudinal distribution in relation to other environmental factors of several Eucalypt species in southern New South Wales. *Aust. J. Ecol.* 8:169–180.

AUSTIN, M. P., R. H. GROVES, L. M. F. FRESCO, AND P. E. KAYE. 1985. Relative growth of six thistle species along a nutrient gradient with multispecies competition. *J. Ecol.* 73:667–684. 1986

AZAM, F., T. FENCHEL, J. D. FIELD, L. A. MEYER-REIL, AND F. THINGSTAD. 1983. The ecological role of water-column microbes in the sea. *Mar. Ecol. Prog. Ser.* 10:257–263.

BACON, P. J. (ed.). 1985. *Population Dynamics of Rabies in Wildlife.* Academic Press, London.

BAER, J. G. 1951. *Ecology of Animal Parasites.* University of Illinois Press, Urbana.

BAES, C. F., JR., H. E. GOELLER, J. S. OLSON, AND R. M. ROTTY. 1977. Carbon dioxide and the climate: The uncontrolled experiment. *Am. Sci.* 65:310–320.

BAILEY, P. C. E. 1986. The feeding behavior of a sit-and-wait predator Ranatra dispar (Heteroptera: Nepidae): Optimal foraging and feeding dynamics. *Oecologica* 68:291–293.

BAILEY, R. G. 1996. *Ecosystem Geography.* Springer, New York.

BAILEY, R. W. 1978. *Description of the Ecoregions of the United States.* U.S.D.A. Forest Service Intermountain Region, Ogden, UT.

BAKER, H. G., K. S. BAWA, G. W. FRANKIE, AND P. A. OPLER. 1983. Reproductive biology of plants in tropical forests. In F. B. Golley (ed.), *Tropical Forest Ecosystems: Structure and Function.* Elsevier, Amsterdam, pp. 183–215.

BAKER, M. C., T. K. BJERKI, H. LAMPE, AND Y. ESPMARK. 1986. Sexual response of female great tits to variation in size of males' song repertoires. *Am. Nat.* 128:491–498.

BAKER, M. C., L. M. MEWALDT, AND R. M. STEWART. 1981. Demography of white-crowned sparrows (*Zonotrichia leucophrys nuttalli*). *Ecology* 62:636–644.

BAKER, S. V., AND J. FRITSCH. 1997. New territory for deer management: Human conflicts on the suburban frontier. *Wildl. Soc. Bull.* 25:404–407

BALDA, R. P. 1975. Vegetation structure and breeding bird diversity. In D. R. Smith (ed.), *Symposium on Management of Forest and Range Habitats for Nongame Birds.* U.S.D.A. Forest Service GTR WB-1.

BALL, D. M., J. F. PEDERSEN, AND G. D. LACEFIELD. 1993. The tall fescue endophyte. *Am. Sci.* 81:370–379.

BALLARD, W. B., J. S. WHITMAN, AND C. L. GARDNER. 1987. Ecology of an exploited wolf population in south-central Alaska. *Wildl. Monogr.* 98. 54 pp

BARBOZA, P., S. FARLEY, AND C. ROBBINS. 1997 Whole body urea cycling and protein turnover during hyperphagia and dormancy in growing bears (Ursus americanus and U. arctos). *Can. J. Zool.* 75:2129–2136.

BARKALOW, F. S., JR., R. B. HAMILTON, AND R. F. SOOTS, JR. 1970. The vital statistics of an unexploited gray squirrel population. *J. Wild. Manage.* 34:489–500.

BARKAN, C. P. L. 1990. A field test of risk-sensitive foraging in black-capped chickadees (Parus atricapillus). *Ecology* 71:391–400.

BARNARD, C. 1984. The evolution of food-scrounging strategies within and between species. In C. J. Barnard (ed.), *Producers and Scroungers: Strategies of Exploitation and Parasitism.* Croom and Helmt, London, pp. 95–126.

BARNARD, C. J., AND D. B. A. THOMPSON. 1985. *Gulls and Plovers: The Ecology and Behaviour of Mixed-Species Feeding Groups.* Columbia University Press, New York.

BARNES, B. M. 1989. Freeze avoidance in a mammal: Body temperature below 0°C in an arctic hibernator. *Science* 244:1593–1595.

BARRETT, J. A. 1983. Plant-fungus symbioses. In D. Futuyma and M. Slakin (eds.), *Coevolution.* Sinauer Associates, Sunderland, MA, pp. 137–160.

BARRETTE, C., AND F. MESSIER. 1980. Scent marking in free-ranging coyotes *Canis latrans. Anim. Behav.* 28:814–819.

BARTHOLOMEW, G. A. 1959. Mother-young relations and the maturation of pup behaviour in the Alaskan fur seal. *Anim. Behav.* 7:163–171.

————. 1970. Bare zone between California shrub and grassland communities: The role of animals. *Science* 170:1210–1212.

————. 1981. A matter of size: An examination of endothermy in insects and terrestrial vertebrates. In B. Heinrich (ed.), *Insect Thermoregulation.* Wiley Interscience, New York, pp. 45–78.

BATJES, N. H. 1996. Total carbon and nitrogen in the soils of the world. *Eur. J. Soil Sci.* 47:151–163.

BATZLI, G. O., AND F. A. PITELKA. 1970. Influence of meadow mouse populations on California grassland. *Ecology* 51:1027–1039.

BATZLI, G. O., B. G. WHITE, S. F. MACLEAN, JR, F. A. PITELKA, AND B. D. COLLIER. 1978. The herbivore-based food chain. In J. Brown, F. L. Bunnell, F. S. MacLean, and L. L. Tieszen, (eds) *An Arctic Ecosystem: The Coastal Tundra of Northern Alaska.* US/IBP Synthesis No. 12. Springer Verlag, New York.

BAUMGARTNER, A. 1968. Ecological significance of the vertical energy distribution in plant stands. In F. E. Eckardt (ed.), *Functioning of Terrestrial Ecosystems at the Primary Production Level.* Proc. Copenhagen

Symposium Natural Resources Research. UNESCO, Paris, pp. 367–374.

BAWA, K. S. 1980. Evolution of dioecy in flowering plants. *Ann. Rev. Ecol. Syst.* 11:15–39.

BAYLEY, P. B. 1995. Understanding large river-floodplain ecosystem. *Bioscience* 45:153–158.

BAZZAZ, F. A. 1975. Plant species diversity in old field successional ecosystems in southern Illinois. *Ecology* 56:485–488.

———. 1979. The physiological ecology of plant succession. *Ann. Rev. Ecol. Syst.* 10:351–371.

———. 1991. Habitat selection in plants. *Am. Nat.* 137:S116–S130.

———. 1996. *Plants in Changing Environments: Linking Physiological, Population, and Community Ecology.* Cambridge University Press, New York

BEACHAM, T. D. 1980. Dispersal during fluctuations of the vole *Microtus townsendii. J. Anim. Ecol.* 49:867–877.

BEALS, E. W. 1968. Spatial pattern of shrubs on a desert plain in Ethiopia. *Ecology* 49:744–746.

———. 1985. Bray-Curtis ordination: An effective strategy for analysis of multivariate ecological data. *Adv. Ecol. Res.* 14:1–55.

BEATLEY, J. C. 1969. Dependence of desert rodents on winter annuals and precipitation. *Ecology* 50:721–724.

BEATTIE, A. J., AND D. C. CULVER. 1981. The guild of myrmecohores in the herbaceous flora of West Virginia forests. *Ecology* 62:107–115.

BECK, S. D. 1980. *Insect Photoperiodism,* 2nd ed. Academic Press, New York.

BEDDINGTON, J. R., M. P. HASSELL, AND J. H. LAWTON. 1976. The components of arthropod predation. II. The predator rate of increase. *J. Anim. Ecol.* 45:165–185.

BEDNARZ, J. C. 1988. Cooperative hunting in Harris' hawk (*Parabuteo unicinctus*). *Science* 239:1525–1527.

BEEHLER, B. M. 1983. Lek behavior of the lesser bird of paradise. *Auk* 100:992–995.

BEEHLER, B. M., AND M. S. FOSTER. 1988. Hotshots, hotspots, and female preference in the organization of lek mating systems. *Am. Nat.* 131:203–209.

BEGON, M., J. L. HARPER, AND C. R. TOWNSEND. 1996. *Ecology: Individuals, Populations, and Communities.* Blackwell Scientific Publishers, New York.

BEKOFF, M. 1977. Mammalian dispersal and the ontogeny of individual behavioral phenotypes. *Am. Nat.* 111:715–732.

BELL, G. 1980. The costs of reproduction and their consequences. *Am. Nat.* 116:45–76.

BELLROSE, F. C., T. G. SCOTT, A. S. HAWKINS, AND J. B. LOW. 1961. Sex ratios and age ratios in North American ducks. *Illinois Nat. His. Surv. Bull.* 27:391–474.

BELOVSKY, G. E. 1981a. Food plant selection by a generalist herbivore: The moose. *Ecology* 62:1020–1030.

———. 1981b. A possible population response of moose to sodium availability. *J. Mamm.* 63:631–633.

BELOVSKY, G. E., and P. F. Jordan. 1981. Sodium dynamics and adaptations of a moose population. *J. Mamm.* 63:613–621.

BELSKY, A. J. 1986. Does herbivory benefit plants? A review of the evidence. *Am. Nat.* 127:870–892.

BELSKY, A. J., AND C. D. CANHAM. 1994. Forest gaps and isolated trees. *Bioscience* 44:77–84.

BEN-ATRI, E. T. 1998. A new wrinkle in wildlife management. *Bioscience* 48:667–673.

BENTLEY, B. L. 1977. Extrafloral nectaries and protection by pugnacious bodyguards. *Ann. Rev. Ecol. Syst.* 8:407–427.

BERG, B., AND C. MCCLAUGHERTY. 1989. Nitrogen and phosphorus release from decomposing litter in relation to the disappearance of lignin. *Can. J. Bot.* 67:1148–1156.

BERG, B., C. MCCLAUGHERTY, AND M. JOHANSSON. 1993. Litter mass-loss rates in late stages of decomposition at some climatically and nutritionally different pine sites. Long-term decomposition in Scots pine forest. VIII. *Can. J. Bot.* 71:680–692.

BERG, R. Y. 1975. Myrmecochorous plants in Australia and their dispersal by ants. *Austral. J. Bot.* 23:475–508.

BERGER, P. J., N. C. NEGUS, E. H. Sanders, and P. D. Gardner. 1981. Chemical triggering of reproduction in *Microtus montanus. Science* 214:69–70.

BERGERUD, A. T. 1971. The population dynamics of Newfoundland caribou. *Wildl. Monogr.* 25.

BERGERUD, A. T., W. WYETH, AND B. SNIDER. 1983. The role of wolf predation in limiting a moose population. *J. Wildl. Manage.* 47:977–988.

BERNARD, J. M., AND E. GORHAM. 1978. Life history aspects of primary production in sedge wetlands. In R. E. Good, D. F. Whigham, and R. L. Simpson (eds.), *Freshwater Wetlands.* Academic Press, New York, pp. 39–51.

BERNAYS, E. A., G. COOPER, DRIVER, AND M. BILGENER. 1989. Herbivores and plant tannins. *Adv. Ecol. Res.* 19:263–302.

BERNHARD-REVERSAT, F. 1982. Biogeochemical cycle of nitrogen in a semiarid savanna. *Oikos* 38:321–332.

BERRY, J. A., AND O. BJORKMAN. 1980. Photosynthetic response and adaptation to temperature in higher plants. *Ann. Rev. Ecol. Syst.* 31:491–453.

BERRYMAN, A. A. 1981. *Population Systems: A General Introduction.* Plenum, New York.

———. 1992. The origins and evolution of predator-prey theory. *Ecology* 73:1539–1535.

BERTNESS, M. D. 1984. Ribbed mussels and *Spartina alterniflora* production on a New England marsh. *Ecology* 65:1794–1807.

———. 1985 Fiddler crab regulation of *Spartina alterniflora* production on a New England salt marsh. *Ecology* 66:1042–1055.

———. 1991a. Interspecific interactions among high marsh perennials in a New England salt marsh. *Ecology* 72:125–137.

———. 1991b. Zonation of *Spartina patens* and *Spartina alterniflora* in a New England salt marsh. *Ecology* 72:138–148.

———. 1999. *The Ecology of Atlantic Shorelines.* Sinauer Associates, Sunderland, MA.

BERTNESS, M. D., AND R. CALLAWAY. 1994. Positive interactions in communities. A Post Cold War perspective. *TREE* 9:191–193.

BERTRAM, B. C. R. 1975. Social factors influencing reproduction in lions. *J. Zool. Lond.* 177:463–482.

BIEL, E. R. 1961. Microclimate, bioclimatology, and notes on comparative dynamic climatology. *Am. Sci.* 49:326–357.

BIERZYCHUDEK, P. 1982a. The demography of Jack-in-the-pulpit, a forest perennial that changes sex. *Ecol. Monogr.* 52:335–351.

———. 1982b. Life history and demography of shade tolerant temperate herbs: A review. New Phytol. 190:757–776.

BIJLSMA, L. 1996. Coastal zones and small islands. In R. T. Watson, M. C. Zinyowera, and R. H. Moss (eds.), *Climate Change 1995: Impacts, Adaptations and Mitigation of Climate Change.* Cambridge University Press, Cambridge, England.

BILBY, R. E. 1981. Role of organic debris dams in regulating the export of dissolved and particulate matter from a forested watershed. *Ecology* 62:1234–243.

BILBY, R. E., AND G. E. LIKENS. 1980. Importance of organic debris dams in the structure and function of stream ecosystems. *Ecology* 1:1107–1113.

BILLINGS, W. D., P. J. GODFREY, B. F. CHABOT, AND D. P. BOURGUE. 1971. Metabolic acclimation to temperature in arctic and alpine ecotypes of *Oxyria digyna. Arctic Alpine Res.* 3:277–289.

BINET, P. 1981. Short-term dynamics of minerals in arid ecosystems. In D. W. Goodall and R. A. Perry (eds.), *Arid Land Ecosystems: Structure, Functioning, and Management,* II. Cambridge University Press, Cambridge, England, pp. 325–356.

BINKLEY, D., AND R. L. GRAHAM. 1981. Biomass, production, and nutrient cycling of mosses in an old-growth Douglas-fir forest. *Ecology* 62:1387–1389.

BIRCH, L. C., AND D. P. CLARK. 1953. Forest soil as an ecological community with special reference to the fauna. *Quart. Rev. Biol.* 28:13–36.

BITMAN, J. 1970. Hormonal and enzymatic activity of DDT. *Agr. Sci. Rev.* 7(4):6–12.

BJORKMAN, O., AND J. BERRY. 1973. High efficiency photosynthesis. *Sci. Amer.* 229(4):80–93.

BLACKBURN, T. H. 1983. The microbial nitrogen cycle. In W. M. Krumbein (ed.), *Microbial Geochemistry.* Blackwell Scientific Publishers, New York, pp. 63–81.

BLACKMAN, F. F. 1905. Optima and limiting factors. *Ann. Bot.* 19:281–298.

BLAIR-WEST, J. R., J. A. Coghlan, D. A. Denton, J. F. Nelson, et al. 1968. Physiological, morphological, and behavioral adaptations to a sodium-deficient environment by wild native Australian and introduced species of animals. *Nature* 217:922–928.

BLAKE, J. G., AND J. R. KARR. 1984. Species composition of bird communities and the conservation benefit of large versus small forests. *Biol. Cons.* 30:193–187.

BLASING, T. J., AND A. M. SOLOMON. 1983. Response of North American Corn Belt to climatic warming. DOE/N88-004, U.S. Department of Energy, Washington, D.C.

BLISS, L. C. 1975. Devon Island, Canada. In T. Rosswall and O. W. Heal (eds.), *Structure and Function of Tundra Ecosystems.* Swedish National Science Research Council, Stockholm, pp. 17–60.

———. 1981. North American and Scandanavian tundras and polar deserts. In L. C. Bliss, O. W. Heal, and J. J. Moore (eds.), *Tundra Ecosystems: A Comparative Analysis.* Cambridge University Press, Cambridge, England, pp. 8–24.

———. 1988. Arctic tundra and polar desert biome. In M. G. Barbour and W. D. Billings (eds.), *North American Terrestrial Vegetation.* Cambridge University Press, New York, pp. 1–32.

BLISS, L. C., G. M. COURTIN, D. L. PATTIE, R. R. WIEWE, D. W. A. WHITFIELD, AND P. WIDDEN. 1973. Arctic tundra ecosystems. *Ann. Rev. Ecol. Syst.* 4:359–399.

BLISS, L. C., O. W. HEAL, AND J. J. MOORE (eds.). 1981. *Tundra Ecosystems: A Comparative Analysis.* International Biological Programme No. 25, Cambridge University Press, Cambridge, England.

BLUM, J. L. 1960. Algal populations in flowing waters. *In Ecology of Algae.* Spec. Pub. No. 2, Pymatuning Lab. of Field Biology, pp. 11–21.

BOAG, P. T., AND P. GRANT. 1986. Intense natural selection in a population of Darwin's finches. *Science* 241:83.

BOCK, C. E., AND J. H. BOCK. 1998. Factors controlling the structure and function of desert grassland: A case study from southeastern Arizona. In B. Tellman, D. M. Finch, C. Edminster, and R. Hamre (eds.) *The Future of Arid Grasslands: Identifying Issues, Seeking Solutions.* 1986. October 9–13; Tucson, AZ. Proceedings RMRS-P-3. U.S. Department of Agriculture, Forest Service Rocky Mountain Research Station, Fort Collins, CO, pp. 33–44.

BOERZYCHUDEK, P. 1984. Assessing optimal life histories in a fluctuating environment: The evolution of sex-changing by jack-in-the-pulpit. *Am. Nat.* 123:829–840.

BONAN, G. B., F. S. CHAPIN, AND S. L. THOMPSON. 1995. Boreal forest and tundra ecosystems as components of the climate system. *Climat. Change* 29:145–168.

BOND, R. R. 1957. Ecological distribution of breeding birds in the upland forests of southern Wisconsin. *Ecol. Monogr.* 27:351–384.

BONNELL, M. C., AND R. K. SELANDER. 1974. Elephant seals: Genetic variation and near extinction. *Science* 184:908–909.

BORCHERT, M. I., AND S. K. JAIN. 1978. The effect of rodent seed predation on four species of California annual grasses. *Oecologica* 33:101–113.

BORGIA, G., AND K. COLLIS. 1989. Female choice for parasite-free male satin bowerbirds and the evolution of bright male plumage. *Behav. Ecol. Sociobiol.* 25:445–454.

BORMANN, F. H., AND G. E. LIKENS. 1979. *Pattern and Process in a Forested Ecosystem.* Springer-Verlag, New York.

BORNEBUSCH, C. H. 1930. *The Fauna of Forest Soils.* Nielsen and Lydiche, Copenhagen.

BOTKIN, D. B. 1977. Forest, lakes, and the anthropogenic production of carbon dioxide. *Bioscience* 27:325–331.

BOTKIN, D. B., J. F. JANAK, AND J. R. WALLIS. 1972. Some ecological consequences of a computer model of forest growth. *J. ECOL.* 60:948–972.

BOTT, T. L., AND T. D. BROCK. 1968. Bacterial growth rates above 90°C in Yellowstone hot springs. *Science* 164:1411–1412.

BOUCHER, D. H., S. JAMES, AND H. D. KELLY. 1982. The ecology of mutualism. *Ann. Rev. Ecol. Syst.* 13:315–347.

BOUCHER, C., AND E. J. MOLL. 1981. South African Mediterranean shrublands. In F. di Castri, D. W. Goodall, and R. L. Specht (eds.), *Mediterranean-Type Ecosystems* (Ecosystems of the World, Vol. 11). Elsevier, Amsterdam, pp. 233–248.

BOUL, S. W., F. D. HOLE, R. J. MCCRACKEN, AND R. J. SOUTHWARD. 1997. *Soil Genesis and Classification.* Iowa State University Press, Ames.

BOURGERON, P. S. 1983. Spatial aspects of vegetation structure. In F. B. Golley (ed.), *Tropical Rain Forest Ecosystems* (Ecosystems of the World, Vol. 10). Elsevier, Amsterdam, pp. 29–47.

BOURLIERE, F. (ed.). 1983. *Tropical Savannas* (Ecosystems of the World, Vol. 13). Elsevier, Amsterdam.

BOURLIERE, F., AND M. HADLEY. 1983. Present-day savannas: An overview. In F. Bourliere (ed.), *Tropical Savannas* (Ecosystems of the World, Vol. 13). Elsevier, Amsterdam, pp. 1–18.

BOWDEN, W. B. 1986a. Gaseous nitrogen emission from undisturbed terrestrial ecosystems: An assessment of their impact on local and global nitrogen budgets. *Biogeochemistry* 2:249–279.

———. 1986b. Nitrification, nitrate reduction, and nitrogen immobilization in a tidal freshwater marsh sediment. *Ecology* 67:88–99.

BOWES, G. 1991. Growth at elevated CO_2: Photosynthetic responses mediated through Rubisco. *Plant Cell Environ.* 14:795–806.

BOX, T. W., J. POWELL, AND D. L. DRAWE. 1967. Influence of fire on south Texas chaparral. *Ecology* 48:955–961.

BOYER, B. B., AND B. M. BARNES. 1999. Molecular and metabolic aspects of mammalian hibernation. *Bioscience* 49:713–724.

BRADBURY, J. 1981. The evolution of leks. In R. D. Alexander and D. W. Tinkle (eds.), *Natural Selection and Social Behavior: Research and New Theory.* Chiron Press, New York, pp. 138–169.

BRADBURY, J. W., AND R. GIBSON. 1983. Leks and male choice. In D. Bateson (ed.), *Mate Choice.* Cambridge University Press, Cambridge, England, pp. 109–138.

BRADBURY, J., R. M. GIBSON, C. E. MCCARTHY, AND S. I. VEHRENCAM. 1989. Dispersion of displaying male sage grouse: The role of female dispersion. *Behav. Ecol. Sociobiol.* 24:15–24.

BRADLEY, A. F., N. V. NOSTE, AND W. C. FISCHER. 1992. Fire ecology of forest and woodland in Utah. Gen. Tech. Rept. INT-287 USDA Forest Service. Intermountain Research Station, Ogden, UT.

BRADLEY, W. G., AND R. A. MAUER. 1971. Reproduction and food habits of Merriam's kangaroo rat, *Dipodomys merriami. J. Mamm.* 52:497–507.

BRADSHAW, A. D., R. W. LODGE, D. JOWETT, AND R. W. SNAYDON. 1964. Experimental investigations into the mineral nutrition of several grass species. IV. Nitrogen level. *J. ECOL.* 52:665–676.

BRATTON, S. P. 1994. Logging and fragmentation of broadleaf deciduous forests: Are we asking the right questions? *Cons. Biol.* 8:295–297.

BRAY, J. R., AND E. GORHAM. 1964. Litter production in the forests of the world. Adv. Ecol. Rec. 2:101–157.

BRAY, J. R., AND J. T. CURTIS. 1957. An ordination of the upland forest communities of southern Wisconsin. *Ecol. Monogr.* 27:325–349.

BREYMEYER, A. I., AND G. M. VAN DYNE (eds.). 1980. *Grasslands, Systems Analysis, and Man.* International Biological Programme No. 19. Cambridge University Press, Cambridge, England.

BRIAND, F., AND J. E. COHEN. 1984. Community food webs have scale-invariant structure. *Nature* 307:264–266.

BRICKLEMYER, E. C., JR., S. LUDICELLO, AND H. J. HARTMANN. 1989. Discarded catch in U.S. commercial marine fisheries. *In Audubon Wildlife Report* 1989/1990. Academic Press, San Diego, pp. 259–295.

BRIGHT, C. 1998. *Life out of Bounds: Bioinvasion in a Borderless World.* W. W. Norton, New York.

BRISKE, D. D. 1996. Strategies of plant survival in grazed systems: A functional interpretation. In J. Hodgson and A. W. Illius (eds.), *The Ecology and Management of Grazing Systems.* Wallingford Oxon, UK CAAB International, pp. 37–68.

BROADUS, J., J. MILLIMAN, S. EDWARDS, D. AUBREY, AND F. GABLE. 1986. Rising sea level and damming of rivers: Possible effects in Egypt and Bangladesh. In J. G. Titus (ed.), *Effects of Stratospheric Ozone and Global Change.* EPA, Washington, DC.

BROCK, T. R. 1979. *Biology of Microorganisms.* Prentice-Hall, Englewood Cliffs, NJ.

BRODSKY, L. M., AND P. J. WEATHERHEAD. 1984. Behavioral and ecological factors contributing to American black-mallard hybridization. *J. Wildl. Manage.* 48:846–852.

BROOKS, M. B. 1951. Effect of black walnut trees and their products on vegetation. *W. Va. Univ. Agr. Exp. Stat. Bull.* 347:1–31.

BROWER, J. V. Z. 1958. Experimental studies of mimicry in some North American butterflies: 1. The monarch, *Danaus plexippus,* and *viceroy, Limenitis archippus;* 2. *Battus philenor* and *Papilio troilus, P. polyxenes* and *P. glaucus;* 3. *Danaus glippus berenice* and *Limenitis archippus floridensis. Evolution* 12:32–47, 123–136, 273–285.

BROWER, L. P. 1984. Chemical defense in butterflies. *Symp. Roy. Entomol. Soc. London* 11:109–134.

———. 1988. Avian predation on the monarch butterfly and its implication for mimicry theory. *Am. Nat.* 131:54–56.

BROWER, L. P., AND L. S. FINK. 1983. A natural toxic defense system in butterflies versus birds. *Ann. N.Y. Acad. Sci.* 443:171–186.

BROWN, C. R. AND M. B. BROWN. 1989. Behavioral dynamics of interspecific brood parasitism in colonial cliff swallows. *Anim. Behav.* 37:777–796.

BROWN, J. H., AND A. KODRICH-BROWN. 1977. Turnover rates in insular biogeography: Effect of immigration on extinction. *Ecology* 58:445–449.

BROWN, J. L. 1987. *Helping and Communal Breeding in Birds.* Princeton University Press, Princeton, NJ.

BROWN, J. R., AND E. J. HESKE. 1990. Control of a desert-grassland transition by keystone rodent guild. *Science* 250:1705–1707.

BROWN, L. 1976. *British Birds of Prey.* Collins, London.

BROWN, R. J. E. 1970. *Permafrost in Canada.* University of Toronto Press, Toronto.

BROWN, R. J. E., AND G. H. JOHNSON. 1964. Permafrost and related engineering problems. *Endeavour* 23:66–73.

BRUNIG, E. F. 1983. Vegetation structure and growth. In F. Golley (ed.), *Tropical Rain Forest Ecosystems: Structure and Function.* Elsevier, Amsterdam, pp. 49–75.

BRYANT, E. 1997. *Climate Process and Change.* Cambridge University Press.

BRYANT, J. A., AND P. J. KUROPAT. 1980. Selection of winter forage by subarctic browsing vertebrates: The role of plant chemistry. *Ann. Rev. Ecol. Syst.* 11:261–285.

BRYANT, J. P., G. D. WIELAND, T. CLAUSEN, AND P. J. KUROPAT. 1985. Interactions of snowshoe hares and feltleaf willow (Salix alaxensis) in Alaska. *Ecology* 66:1564–1573.

BRYLINSKY, M. 1980. Estimating the productivity of lakes and reservoirs. In E. D. Le Cren and R. H. Lowe-McConnell (eds.), *The Functioning of Freshwater Ecosystems.* International Biological Programme No. 22. Cambridge University Press, Cambridge, England, pp. 411–418.

BUCHMANN, S. L. 1987. The ecology of oil flowers and their bees. *Ann. Rev. Ecol. Syst.* 18:343 369.

BUELL, M. F., AND R. E. WILBUR. 1948. Life form spectra of the hardwood forests of the Itaska Park region, Minnesota. *Ecology* 29:352 359.

BULL, J., AND E. CHARNOV. 1989. Energetic reptilian sex ratios. Evolution 43:1561–1566. (13)

BULLOCK, J. M. 1996. Plant populations and population dynamics. In J. Hodgson and A. W. Illius (eds.), *The Ecology and Management of Grazing Systems.* Wallingford Oxon, UK CAB International, pp. 69–100.

BUNNELL, F. L., S. F. MCCLEAN, JR., AND J. BROWN. 1975. Barrow, Alaska, U.S.A. In T. Rosswall and O. W. Heal (eds.), *Structure and Function of Tundra Ecosystems.* Swedish Natural Science Institute, Stockholm, pp. 73–124.

BUNNELL, F. L., O. K. MILLER, P. W. FLANAGAN, AND R. E. BENOIT. 1980. The microflora: Composition, biomass, and environmental relations. In J. Brown, P. C. Miller, L. L. Tieszen, and F. L. Bunnell (eds.), *An Arctic Ecosystem: The Coastal Tundra at Barrow, Alaska.* Dowden, Hutchinson, and Ross, Stroudsburg, PA, pp. 255–290.

BUNNING, E. 1964, *The Physiological Clock,* 2nd ed. Academic, New York.

BURBANCK, M. P., AND R. B. PLATT. 1964. Granite outcrop communities of the Piedmont Plateau in Georgia. *Ecology* 45:292–306.

BURNS, R. C., AND R. W. F. HARDY. 1975. *Nitrogen Fixation in Bacteria and Higher Plants.* Springer-Verlag, New York.

BURT, W. V., AND J. QUEEN. 1957. Tidal overmixing in estuaries. *Science* 126:973–974.

BUSS, D. M. 1994. The strategies of human mating. *Amer. Sci.* 82:238–249.

BUTCHER, S. S., R. J. CHARLSON, G. H. ORIANS, AND G. V. WOLFE (eds.). 1992. Global Biogeochemical Cycles. Academic Press, New York.

CALDWELL, M. M. 1971. Solar ultraviolet radiation as an ecological factor for alpine plants. *Ecol. Monogr.* 38:243–268.

———. 1985. Cold desert. In B. F. Chabot and H. A. Mooney (eds.), *Physiological Ecology of North American Plant Communities.* Chapman and Hall, New York, pp. 198–212.

CALDWELL, M. M., R. ROBBERRECHT, AND W. D. BILLINGS. 1980. A steep latitudinal gradient of solar ultraviolet-B radiation in the arctic-alpine life zone. *Ecology* 61:600–611.

CALDWELL, M. M., A. H. TERAMURA, AND M. TEVINI. 1989. The changing solar ultraviolet climate and the ecological consequence for higher plants. *TREE* 4:363–367.

CARACO, T. S., S. MARTINDALE, AND T. W. WHITHAM. 1980. An empirical demonstration of risk-sensitive foraging preferences. *Anim. Behav.* 28:820–830.

CAREY, F. C. 1982. A brain heater in the swordfish. *Science* 216:1327–1329.

CARLSON, J. D., JR., W. R. CLARK, AND E. E. KLAAS. 1993. A model of the productivity of the northern pintail. U.S. Fish and Wildlife Service Biological Report 7. 20 pp.

CARPENTER, E. J., AND K. ROMANS. 1991. Major role of the cyanobacterium Trichodesmium in nutrient cycling in the North Atlantic ocean. *Science* 254:1356–1358.

CARPENTER, F. L. 1987a. Food abundance and territoriality: To defend or not to defend? *Am. Zool.* 27:387–399.

———. 1987b. The study of territoriality: Complexities and future directions. *Am. Zool.* 27:401–409.

CARPENTER, S. R. 1980. Enrichment of Lake Wingra, Winconsin, by submerged macrophyte decay. *Ecology* 61:1145–1155.

CARPENTER, S. R., AND J. F. KITCHELL. 1984. Plankton community structure and limnetic primary production. *Am. Nat.* 124:159–172.

CARRICK, R. 1963. Ecological significance of territory size in the Australian magpie *Gymnorhina tibiten. Proc. Int. Orn. Cong.* 13:740–753.

CARROLL, C. R., AND C. A. HOFFMAN. 1980. Chemical feeding deterrent mobilized in response to insect herbivory and counter adaptation by by *Epilachnia tridecimnoata. Science* 209:414–416.

CARROLL, G. C. 1979. Forest canopies: Complex and independent subsystems. In R. H. Waring (ed.), *Forests: Fresh Perspectives from Ecosystem Analysis.* Proc. Ann. Biol. Coll., Oregon State University Press, Corvallis, pp. 87–107.

CARROLL, J. F., AND J. D. N. NICHOLS. 1986. Parasitization of meadow *voles Microtus pennsylvanicus* (Ord) by American dog ticks *Dermacenter variabilis* (Say) and adult tick movement during high host density. *J. Entomol. Sci.* 21:102–113.

CARSON, R. 1962. *Silent Spring.* Houghton Mifflin, Boston.

CARTER, R. V. 1991. A test of risk sensitive foraging in wild bumblebees. *Ecology* 72:887–895.

CASTRI, F. DI, AND H. A. MOONEY (eds). 1973. *Mediterranean Type Ecosystems: Origin and Structure.* Springer-Verlag, New York.

CASTRI, F. DI. 1981. Mediterranean-type shrublands of the World. In F. Di Castri, D. W. Goodall, and R. L. Specht (eds.), *Mediterranean-Type Shrublands* (Ecosystems of the World, Vol. 11). Elsevier, Amsterdam.

CATCHPOLE, C. K. 1987. Bird song, sexual selection, and female choice. *TREE* 2:94–97.

CAUGHLEY, G. 1976a. Wildlife management and the dynamics of ungulate populations. *App. Biol.* 1:183–246.

———. Plant and herbivore systems. In R. B. May (ed.), *Theoretical Ecology: Principles and Applications.* Saunders, Philadelphia, pp. 94–113.

———. 1977. *Analysis of Vertebrate Populations.* Wiley, New York.

CERNUSA, A. 1976. Energy exchange within individual layers of a meadow. Oecologia 23:141–149.

CHAFFEE, R. R. J., AND J. C. ROBERTS. 1971. Temperature acclimation in birds and animals. *Ann. Rev. Physiol.* 33:155–202.

CHALMERS, A. G. 1982. Soil dynamics and the productivity of Spartina alternifolia. In V. S. Kennedy (ed.), IAcademic, New York, pp. 231–243.

CHANGNON, S. A. 1968. La Porte weather anomaly: Fact or fiction? *Bull. Am. Meteorol. Soc.* 49:4–11.

CHAPIN, F. S. The mineral nutrition of wild plants. *Ann. Rev. Ecol. Syst.* 11:233–260.

CHAPIN, F. S., P. C. MILLER, W. D. BILLINGS, AND R. I. COYNE. 1980. Carbon and nutrient budgets and their control in coastal tundra. In J. Brown, P. C. Miller, L. L. Tieszen, and F. L. Bunnell (eds.), *An Arctic Ecosystem: The Coastal Tundra at Barrow, Alaska.* Dowden, Hutchinson, and Ross, Stroudsburg, PA, pp. 458–482.

CHAPIN, F. S., III, E. D. SCHULZE, AND H. A. MOONEY. 1990. The ecology and economics of storage in plants. *Ann. Rev. Ecol. Syst.* 21:846–852.

CHAPIN, F. S., L. L. TIESZEN, M. C. LEWIS, P. C. MILLER, AND B. H. MCCOWEN. 1980. Control of tundra plant allocation patterns and growth. In J. Brown, P. C. Miller, L. L. Tieszen, and F. L. Bunnell (eds.), *An Arctic Ecosystem: The Coastal Tundra at Barrow, Alaska.* Dowden, Hutchinson, and Ross, Stroudsburg, PA, pp. 140–185.

CHAPMAN, V. J. 1976. *Coastal Vegetation,* 2nd ed. Pergamon Press, Oxford, England.

CHARNOV, E. L. 1976. Optimal foraging: The marginal value theorem. *Theor. Pop. Biol.* 9:129–136.

———. 1982. *The Theory of Sex Allocation.* Princeton University Press, Princeton, NJ.

CHARNOV, E. L., AND J. J. BULL. 1977. When is sex environmentally determined? *Nature* 266:828–830.

CHARNOV, E. L., AND W. M. SCHAFFER. 1973. Life history consequences of natural selection: Cole's results revisited. *Am. Nat.* 107:791–793.

CHASKO, G. G., AND M. R. CONOVER. 1988. Too much of a good thing? *Living Bird Quart.* 7:8–13.

Chasko, G. G., and J. E. Gates. 1982. Avian habitat suitability along a transmission line corridor in an oak-hickory forest region. *Wildl. Monogr.* 82. The Wildlife Society.

CHAZDON, R. L. 1988. Sunflecks and their importance to forest understory plants. *Adv. Ecol. Res.* 18:1–63.

CHAZDON, R. L., AND R. W. PEARCY. 1986. Photosynthetic responses to light variation in rainforest species. II. Carbon gain and photosynthetic efficiencies during light flecks. *Oecologica* 69:524–531.

———. 1991. The importance of sunflecks for forest understory plants. *Bioscience* 41:760–765.

CHEATUM, E. L., AND C. W. SEVERINGHAUS. 1950. Variations in fertility of white-tailed deer related to range conditions. *Trans. North Am. Wildl. Conf.* 15:170–189.

CHESSER, R. K. 1983. Isolation by distance: Relationship to the management of genetic resources. In C. Schonewald-Cox, S. Chambers, B. MacBryde, and W. Thomas (eds.), *Genetics and Conservation: A Reference for Managing Wild Animal and Plant Populations.* Benjamin/Cummings, Menlo Park, CA, pp. 51–65.

CHESSON, P. L. 1986. Environmental variation and coexistence of species. In J. Diamond and T. Case (eds.), *Community Ecology.* Harper and Row, New York, pp. 240–256.

CHESSON, P. L., AND T. CASE. 1986. Overview: Nonequilibrium community theories: Chance, variability, history, and coexistence. In J. Diamond and T. Case (eds.), *Community Ecology.* Harper and Row, New York, pp. 229–239.

CHEW, R. M., AND A. E. CHEW. 1970. Energy relationships of the mammals of a desert shrub (*Larrae tridentatia*). *Ecol. Monogr.* 40:1–21

CHRISTENSEN, N. L. 1977. Fire and soil-plant nutrient relations in a pine-wiregrass savanna on the coastal plain of North Carolina. *Oecologia* 31:27–44.

CHRISTIAN, J. J. 1963. Endocrine adaptative mechanisms and the physiologic regulation of population growth. In W. V. Mayer and R. G. Van Gelder (eds.), *Physiological Mammalogy,* Vol. 1, Mammaliam Populations. Academic Press, New York, pp. 18–35.

———. 1971. Fighting, maturity, and population density in Microtus pennsylvanicus. *J. Mamm.* 52:556–567.

———. 1978. Neurobehavioral endocrine regulation of small mammal populations. In D. P. Snyder (ed.), *Populations of Small Mammals Under Natural Conditions* (Pymatuning Symposia in Ecology, Vol. 5). University of Pittsburgh Press, Pittsburgh, PA, pp. 143–158.

CICERONE, R. J. 1987. Changes in stratosphere ozone. *Science* 237:35–42.

CLARKE, J. F. 1969. Nocturnal urban boundary layer over Cincinnati, Ohio. *Monthly Weather Rev.* 97:582–589.

CLAY, K. 1988. Fungal endophytes of grasses: A defensive mutualism between plants and fungi. *Ecology* 69:10–16.

———. 1990. Fungal endophytes of grasses. *Ann. Rev. Ecol. Syst.* 21:275–297.

CLAY, K., S. MARKS, AND G. P. CHEPLICK. 1993. Effects of insect herbivory and fungal endophyte infection on competitive interactions among grasses. *Ecology* 74:1767–1777.

CLEMENTS, F. E. 1916. *Plant Succession: An Analysis of the Development of Vegetation.* Carnegie Inst. Wash. Publ. 242.

CLEMENTS, F. E., AND V. E. SHELFORD. 1939. *Bio-ecology.* McGraw-Hill, New York.

CLOUDSLEY-THOMPSON, J. L. 1956. Studies in diurnal rhythms: VII. Humidity responses and nocturnal activity in woodlice (*Isopoda*). *J. Exp. Biol.* 33:576–582.

CLUTTON-BROCK, T. H. 1984. Reproductive effort and terminal investment in iteroparous animals. *Am. Nat.* 123:212–229.

CLUTTON-BROCK, T. H., S. D. ALBON, AND F. E. Guinness. 1986. Great expectations: Dominance, breeding success, and offspring sex ratios in red deer. *Anim. Behav.* 34:460–471.

CLUTTON-BROCK, T. H., D. GREEN, M. HIRAIWA-HASEGAWA, AND S. D. ALBON. 1988. Passing the buck: Resource defense, lekking, and mate choice in the fallow deer. *Behav. Ecol. Sociobiol.* 23:281–296.

CLUTTON-BROCK, T. H., M. HIRAIWA-HASEGAWA, AND A. ROBINSON. 1989. Mate choice in fallow deer leks. *Nature* 340:463–465.

CLUTTON-BROCK, T. H., F. E. GUINNESS, AND S. D. ALBON. 1982. *Red Deer: Behavior and Ecology of the Two Sexes.* University of Chicago Press, Chicago.

CLUTTON-BROCK, T. H., O. F. PRICE, S. D. ALBON, A. ROBERTSON, AND P. A. JEWELL. 1991. Population regulation in Soay sheep. *J. Anim. Ecol.* 60:593–608.

CODY, M. L. 1966. A general theory of clutch size. Evolution 20:174–184.

———. 1986. Structural niches in plant communities. In J. Diamond and T. Case (eds.), *Community Ecology.* Harper and Row, New York, pp. 381–405.

COE, M. J., D. H. CUMMINGS, AND J. PHILLIPSON. 1976. Biomass and production of large African herbivores in relation to rainfall and primary production. *Oecologia* 22:341–354.

COKER, R. E. 1947. *This Great and Wide Sea.* University of North Carolina Press, Chapel Hill.

COLE, D. W., S. D. GESSEL, AND S. F. DICE. 1968. Distribution and cycling of nitrogen, phosphorus, potassium, and calcium in second growth Douglas-fir ecosystem. *In Symposium on Primary Productivity and Mineral Cycling in Natural Ecosystems.* University of Maine Press, Orono, pp. 197–232.

COLE, D. W., AND M. RAPP. 1981. Elemental cycling in forest ecosystems. In D. E. Reichle (ed.), *Dynamic Properties of Forest Ecosystems.* Cambridge University Press, New York, pp. 431–409.

COLE, L. C. 1951. Population cycles and random oscillations. *J. Wildl. Manage.* 15:233–252.

———. 1954. The population consequences of life history phenomena. *Quart. Rev. Biol.* 29:103–137.

CONFER, J. L., AND K. KNAPP. 1981. Golden-winged warblers and blue-winged warblers: The relative success of a habitat specialist and a habitat generalist. *Auk* 98:108–114.

CONNELL, J. H. 1961. The effects of competition, predation by *Thais lapillus,* and other factors on the distribution of the barnacle Balanus balanoides. *Ecol. Monogr.* 31:61–104.

———. 1978. Diversity in tropical rain forests and coral reefs. *Science* 199:1302–1310.

———. 1983. On the prevalence and relative importance of interspecific competition: Evidence from field experiments. *Am. Nat.* 122:661–696.

CONNELL, J. H., AND G. ORIAS. 1964. The ecological regulation of species diversity. *Am. Nat.* 98:399–414.

CONNELL, J. H., AND R. O. SLATYER. 1977. Mechanisms of succession in natural communities and their role in community stability and organization. *Am. Nat.* 111:1119–1144.

CONOVER, M. R., AND G. G. CHASKO. 1985. Nuisance Canada Goose problems in the eastern United States. *Wildl. Soc. Bull.* 13:228–233.

COOPER, S. M., AND N. OWEN-SMITH. 1986. Effects of plant spinescence on large mammalian herbivores. *Oecologica* 68:446–455.

CORNELISSEN, J. 1996. An experimental comparison of leaf decomposition rates in a wide variety of temperate plant species and types. *J. Ecol.* 84:573–582.

CORNFORTH, I. S. 1970. Leaf-fall in a tropical rain forests. *J. App. Ecol.* 7:603–608.

CORRELL, D. L. 1978. Estuarine productivity. *Bioscience* 28:646–650.

COUGHENOUR, M. B. 1985. A mechanistic simulation analysis of water use, leaf angles, and grazing in East African graminoids. *Ecol. Modeling* 6:203–230.

COUGHENOUR, M. B., S. T. MCNAUGHTON, AND L. L. WALLACE. 1984. Modeling primary production of perennial graminoids: Uniting physiological processes and morphometric traits. Ecol. Modeling 23:101–134.

COUPLAND, R. T. 1958. The effects of fluctuations in weather upon the grassland of the Great Plains. *Bot. Rev.* 24:273–317.

COVICH, A. P., M. A. PALMER, AND T. A. CROWL. 1999. The role of benthic invertebrate species in freshwater ecosystems. *Bioscience* 49:119–128.

COWAN, I., AND V. GEIST. 1961. Aggressive behavior in deer of the genus *Odocoileus. J. Mamm.* 42:522–526.

COWAN, R. L. 1962. Physiology of nutrition as related to deer. Proc. 1st Natl. White-Tailed Deer Disease Symp., pp. 1–8.

COWLES, H. C. 1899. The ecological relations of the vegetation on the sand dunes of Lake Michigan. *Bot. Gaz.* 27:95–117, 167–202, 281–308, 361–391.

COWLES, R. B., AND C. M. BOGERT. 1944. A preliminary study of the thermal requirements of desert reptiles. *Bull. Am. Mus. Natur. Hist.* 83:265–269.

COX, C. R., AND B. J. LEBOEUF. 1977. Female incitation of male competition: A mechanism of mate selection. *Am. Nat.* 111:317–355.

COX, G. W., AND R. E. RICKLEFS. 1977. Species diversity and ecological release in Caribbean land bird fauna. *Oikos* 28:113–122

CRAWLEY, M. J. 1989. Insect herbivore and plant population dynamics. *Ann. Rev. Entomol.* 34:531–564.

CRITCHFIELD, W. B. 1971. *Profiles of California Vegetation.* USDA Forest Serv. Res. Paper PSW-76, Pacific Southwest Forest and Range Exp. Sta., Berkeley, CA.

CROCKER, R. L., AND J. MAJOR. 1955. Soil development in relation to vegetation and surface age at Glacier Bay, Alaska. *J. Ecol.* 43:427–488. (210

CROLL, N. A. 1966. *Ecology of Parasites.* Harvard University Press, Cambridge, MA.

CROMACK, K. 1973. Litter production and litter decomposition in a mixed-hardwood watershed and in a white pine watershed at Coweeta Hydrologic Station, North Carolina. Dissertation. University of Georgia, Athens.

CROMBIE, A. C. 1947. Interspecific competition. *J. Anim. Ecol.* 16:44–73.

CROW, J. F., AND M. KIMURA. 1970. *An Introduction to Genetic Theory.* Harper and Row, New York.

CROWLEY, J. 1967. Biogeography. *Can. Geog.* 11:312–326.

CUMMINS, K. W. 1974. Structure and function of stream ecosystems. *Bioscience* 24:631–641.

———. 1988. The study of stream ecosystems: A functional review. In L. R. Pomeroy and J. J. Alberts (eds.), *Concepts of Ecosystem Ecology.* Springer-Verlag, New York, pp. 247–262.

CUMMINS, K. W., AND M. J. KLUG. 1979. Feeding ecology of stream invertebrates. *Ann. Rev. Ecol. Syst.* 10:147–172.

CURRIE, D. J. 1991. Energy and large-scale biogeographical patterns of animal and plant species richness. *Am. Nat.* 137:27–49.

CURRIE, D. J., AND V. PAQUIN. 1987. Large-scale biogeographical patterns of species richness in trees. *Nature (London)* 329:326–327.

CURTIS, J. T. 1959. *The Vegetation of Wisconsin.* University of Wisconsin, Madison.

CURTIS, J. T., AND R. P. MCINTOSH. 1951. An upland forest continuum in the prairie-forest border region of Wisconsin. *Ecology* 32:476–496.

CURTIS, P. S. 1996. A meta-analysis of leaf gas exchange and nitrogen in trees grown under elevated carbon dioxide. *Plant Cell Environ.* 19:127–137.

D'ANGELO, D. J., J. R. WEBSTER, AND E. F. BENFIELD. 1991. Mechanisms of stream phosphorus retention: An experimental study. *J. N. Am. Benthol. Soc.* 10:225–237.

DARBY, H. C. 1956. The clearing of woodland in Europe. In W. L. Thomas, Jr. (ed.), *Man's Role in Changing the Face of the Earth.* University of Chicago Press, Chicago. pp. 183–216.

DARLINGTON, P. J., JR. 1957. *Zoogeography: The Geographical Distribution of Animals.* Wiley, New York.

DARWIN, C. 1859. *The Origin of Species.* Murray, London.

DASH, M. C., AND A. K. HOTA. 1980. Density effects on the survival, growth rate, and metamorphosis of *Rana tigrina* tadpoles. *Ecology* 61:1025–1028.

DAUBENMIRE, R. 1992. Palouse prairie. In R. T. Coupland (ed.), *Natural Grasslands. Introduction and Western Hemisphere.* (Ecosystems of the World, Vol. 8A). Elsevier, Amsterdam, pp. 291–312.

DAVIDSON, C. I., J. R. HARRINGTON, M. J. STEVENSON, M. C. MONAGHAN, J. PUDYKEIWICZ, AND W. R. SCHNELL. 1987. Radioactive cesium from the Chernobyl accident in the Greenland ice sheet. *Science* 237:633–634.

DAVIDSON, D. W. 1985. An experimental study of diffuse competition in a desert ant community. *Am. Nat.* 125:500–506.

DAVIDSON, W. R. 1981. Diseases and parasites of white-tailed deer. *Southeastern Cooperative Wildlife Disease Study.* Mscl. Pub. No. 7. Tall Timbers Research Station, Tallahassee, FL.

DAVIES, N. B. 1977. Prey selection and social behavior in wagtails (*Aves Monticillidae*). *J. Anim. Ecol.* 46:37–57.

———. 1978. Ecological questions about territorial behaviour. In J. R. Krebs and N. B. Davies (eds.), *Behavioural Ecology: An Evolutionary Approach.* Blackwell, Oxford, England, pp. 317–350.

———. 1991. Mating systems. In J. R. Krebs and N. B. Davies (eds.), *Behavioural Ecology: An Evolutionary Approach,* 3rd ed. Blackwell, London, pp. 263–294.

DAVIES, N. B., AND A. I. HOUSTON. 1984. Territory economics. In J. R. Krebs and N. B. Davies (eds.), *Behavioural Ecology: An Evolutionary Approach,* 2nd ed. Blackwell, Oxford, England, pp. 148–169.

DAVIS, D. E. 1978. Physiological and behavioral responses to the social environment. In D. P. Snyder (ed.), *Populations of Small Mammals Under Natural Conditions,* Vol. 5. Special Publication Ser., Pymatuning Laboratory of Ecology, University of Pittsburgh Press, Pittsburgh, PA, pp. 84–91.

DAVIS, M. B. 1981. Quaternary history and the stability of forest communities. In D. C. West, H. H. Shugart, and D. B. Botkins (eds.), *Forest Succession: Concepts and Application.* Springer-Verlag, New York, pp. 132–153.

———. 1983. Holocene vegetational history of the eastern United States. In H. E. Wright, Jr. (ed.), *Late Quaternary Environments of the United States.* Vol. II, The Holocene. University of Minnesota Press, Minneapolis, pp. 166–188.

DAY, F. P., JR., AND D. T. MCGINTY. 1975. Mineral cycling strategies of two deciduous and two evergreen tree species on a southern Appalachian watershed. In F. C. Howell, J. B. Gentry, and M. H. Smith (eds.), *Mineral Cycling in Southeastern Ecosystems.* National Technical Information Service, U.S. Department of Commerce, Washington, DC, pp. 736–743.

DAY, T. A., T. C. VOGELMANN, AND E. H. DELUCIA. 1992. Are some plant life forms more effective than others in screening out ultraviolet radiation? *Oecologica* 92:513–516.

DAYTON, P. 1971. Competition, disturbance, and community organization: The provision and subsequent utilization of space in a rocky intertidal community. *Ecol. Monogr.* 41:351–389.

———. 1975. Experimental evaluation of ecological dominance in a rocky intertidal algal community. *Ecol. Monogr.* 45:137–159.

DEAN, R., L. E. ELLIS, R. W. WHITE, AND R. E. BEMERET. 1975. Nutrient removal by cattle from a short-grass prairie. *J. App. Ecol.* 12:25–29.

DEANGELIS, D. L., R. H. GARDNER, AND H. H. SHUGART. 1980. Productivity of forest ecosystems studies during the IBP: The woodlands data set. In D. E. Reichle (ed.), *Dynamic Properties of Forest Ecosystems.* International Biological Programme 23. Cambridge University Press, Cambridge, England.

DEBACH, P. 1974. *Biological Control by Natural Enemies.* Cambridge University Press, Cambridge, England.

DECOURSEY, P. J. 1960. Phase control of activity in a rodent. Cold Spring Harbor Symp. Quant. Biol. 25:49–54.

DEEVEY, E. S. 1947. Life tables for natural population of animals. *Quart. Rev. Biol.* 22:283–314.

DELCOURT, H. R. 1987. The impact of prehistoric agriculture and land occupation on natural vegetation. *TREE* 2:39–44.

DELCOURT, H. R., AND P. A. DELCOURT. 1985. Quaternary palynology and vegetational history of the southeastern United States. In W. M. Bryant, Jr., and R. G. Holloway (eds.), *Pollen Records of Late Quaternary North American Sediments.* American Association of Stratigraphic Palynologists Foundation, Washington, DC, pp. 1–37.

DELCOURT, P. A., AND H. R. DELCOURT. 1981. Vegetation maps for eastern North America, 40,000 yr BP to present. In R. Romans (ed.), *Geobotany.* Plenum Press, New York, pp. 123–166.

DEMPSTER, J. P. 1975. *Animal Population Ecology.* Academic Press, London.

———. 1984. Late Quaternary paleoclimates and biotic responses in eastern North America and the western North Atlantic Ocean. *Paleogeography, Paleoclimatology, Paleoecology* 48:263–284.

DERRICKSON, E. M. 1992. Comparative reproduction strategies of altricial and precocial eutherian mammals. *Funct. Ecol.* 6:57–65.

DHILLION, S. S., J. ROY, AND M. ABRAMS. 1996. Assessing the impact of elevated CO_2 on microbial activity in a Mediterranean model ecosystem. *Plant Soil* 187:333–342.

DHYSTERHUIS, E. J., AND E. M. SCHMUTZ. 1947. Natural mulches or "litter of grasslands," with kinds and amounts on a southern prairie. *Ecology* 28:163–179.

DIAMOND, J. 1975. The island dilemma: Lessons of modern biogeographic studies for the design of natural preserves. *Biol. Cons.* 7:129–146.

———. 1986. Overview: Laboratory experiments, field experiments, and natural experiments. In J. Diamond and T. Case (eds.), *Community Ecology.* Harper and Row, New York, pp. 3–22.

DIAMOND, J. D., AND T. J. CASE (eds.). 1986. *Community Ecology.* Harper and Row, New York.

DICKERMAN, J. A., AND R. G. WETZEL. 1985. Clonal growth in Typha latifolia: Population dynamics and demography of the ramets. *J. Ecol.* 73:535–552.

DILGER, W. C. 1956. Hostile behavior and isolating mechanisms in the avian genera *Catharus* and *Hylocichla. Auk* 73:313–353.

———. 1960. Agonistic and social behavior of captive redpolls. *Wilson Bull.* 72:115–132.

DINERSTEIN, E., AND G. F. MCCRACKEN. 1990. Endangered greater one-horned rhinoceros carry high levels of genetic variation. *Cons. Biol.* 4:417–423.

DIX, R. L. 1960. The effects of burning on the mulch structure and species composition of grassland in western North Dakota. *Ecology* 41:49–56.

Dobkin, D. S. 1985. Heterogeneity of tropical floral microclimates and the response of hummingbird flower mites. *Ecology* 66:536–543.

Dobkin, D. S., I. Olivieri, and P. R. Ehrlich. 1987. Rainfall and the interaction of microclimate with larval resources in the population dynamics of checkerspot butterflies (*Euphydryas editha*) inhabiting serpentine grassland. *Oecologica* 71:161–166.

Dobson, A. P., and P. J. Hudson. 1986. Parasites, disease, and structure of ecological communities. *TREE* 1:11–15.

Dobson, F. S., and W. T. Stone. 1985. Multiple causes of dispersal. *Am. Nat.* 126:855–858.

Dobzhansky, T. 1951. *Genetics and the Origin of Species,* 3rd ed. Columbia University Press, New York.

Dobzhansky, T., and S. Wright. 1941. Genetics of natural populations. V. Relations between mutation rates and accumulation of lethals in populations of *Drosophila pseudoobscura*. *Genetics* 26:23–51.

Dolbeer, R. A., and W. R. Clark. 1975. Population ecology of snowshoe hares in the central Rocky Mountains. *J. Wildl. Manage.* 39:535–549.

Dorst, R. 1958. Uber die Ansiedlung von jung ins Binnerien verfrachteten Silbernowesn (*Larus argentatus*). *Vogelwarte* 17:169–173.

Doty, M. S. 1956. Rocky intertidal surfaces. In J. W. Hedgpeth (ed.), *Treatise in Marine Ecology and Paleoecology:* 1. Ecology. Memoir 67, Geological Soc. Am., pp. 535–585.

Downs, A. A., and W. E. McQuilkin. 1944. Seed production of southern Appalachian oaks. *J. For.* 42:913–920.

Doyle, T. W. 1981. The role of disturbance on the gap dynamics of a montane rain forest: An application of a tropical forest succession model. In D. C. West, H. H. Shugart, and D. B. Botkin (eds.), *Forest Succession: Concepts and Application.* Springer-Verlag, New York, pp. 56–73.

Drake, B. G., and P. W. Leadley. 1991. Canopy photosynthesis of crops and native plant communities exposed to long-term elevated carbon dioxide. *Plant Cell Environ.* 14:853–860.

Drake, B. G., G. Peresta, E. Beugeling, and R. Matamala. 1996. Long-term elevated CO_2 exposure in a Chesapeake Bay wetland: Ecosystem gas exchange, primary productivity, and tissue nitrogen. In G. Koch and H. A. Mooney (eds.), *Carbon Dioxide and Terrestrial Ecosystems.* Academic Press, San Diego.

Drake, J. A. 1990. Communities as assembled structures: Do rules govern pattern? *TREE* 5:159–164.

Drew, M. C. 1979. Root development and activities. In D. W. Goodall and R. A. Perry (eds.), *Arid Land Ecosystems: Structure, Functioning, and Management,* Vol. 1. Cambridge University Press, Cambridge, England, pp. 573–608.

Drew, M. C., M. B. Jackson, and S. Gifford. 1979. Ethylene-promoted adventitious roots and development of cortical air spaces (*aerenchyma*) in roots may be adaptive responses to flooding in *Zea mays. C. Planta* 147:83–88.

Drift, J. Van Der. 1951. Analysis of the animal community in a beech forest floor. *Tijdschrift voor Entomologie* 94:1–168.

———. 1971. Production and decomposition of organic matter in an oak wood in the Netherlands. In P. Duvigneaud (ed.), *Productivity of Forest Ecosystems* (Proc. Brussels Symposium 1969). UNESCO, Paris, pp. 631–634.

Duce, R. A., P. A. Liss, J. T. Merrill, E. L. Atlas, P. Buat-Menard, B. B. Hicks, J. M. Miller, J. M. Prospero, R. Arimoto, T. M. Church, W. Ellis, J. N. Galloway, L. Hansen, T. D. Jickells, A. H. Knap, K. H. Reinhardt, B. Schneider, A. Soudine, J. J. Tokos, S. Tsunogai, R. Wollast, and M. Zhou. 1991. The atmospheric input of trace species to the world ocean. *Glob. Biogeochem. Cycl.* 5:193–259.

Duffey, E. 1974. *Grassland Ecology and Wildlife Management.* Chapman and Hall, London.

Duffy, D. C., and A. J. Meier. 1992. Do Appalachian herbaceous understories ever recover from clearcutting? *Cons. Biol.* 6:196–201.

Dugdale, R. C., and J. J. Goering. 1967. Uptake of new and regenerated forms of nitrogen in primary productivity. *Limno. Oceanog.* 12:196–206.

Duncan, J. S., H. W. Reed, R. Moss, J. P. P. Phillips, and A. Watson. 1978. Ticks, louping ill, and red grouse on moors in Speyside, Scotland. *J. Wildl. Manage.* 42:500–505.

Dussourd, D. E., and R. F. Denno. 1991. Deactivation of plant defense: Correspondence between insect behavior and secretory canal architecture. *Ecology* 72:1383–1396.

Dussourd, D. E., and T. Eisner. 1987. Vein-cutting behaviour: Insect counterploy to the latex defense of plants. *Science* 237:898–901.

Duvigneaud, P., and S. Denaeyer-Desmet. 1970. Biological cycling of minerals in a temperate deciduous forest. In D. Reichle (ed.), *Analysis of Temperate Forest Ecosystems.* Springer-Verlag, New York, pp. 109–115.

Dwyer, P. D., J. Kikkawa, and G. J. Ingram. 1979. Habitat relations of vertebrates in subtropical heathlands of coastal southeastern Queensland. In R. L. Specht (ed.), *Heathlands and Related Shrublands* (Ecosystems of the World, Vol. 9A). Elsevier, Amsterdam, pp. 281–300.

Dyer, M. I., C. L. Turner, and T. R. Seastedt. 1993. Herbivory and its consequences. *Ecol. App.* 3:10–16.

Eadie, J., and H. G. Lumsden. 1985. Is nest parasitism always deleterious to goldeneyes? *Am. Nat.* 126:859–866.

Eaton, J. S., G. E. Likens, and F. H. Bormann. 1973. Throughfall and stemflow chemistry in a northern hardwood forest. *J. Ecol.* 61:495–508.

Edwards, C. A., and G. W. Heath. 1963. The role of soil animals in the breakdown of leaf materials. In J. Doeksen and J. van der Drift (eds.), *Soil Organisms.* North Holland, Amsterdam, pp. 76–84.

Ehleringer, J. R. 1980. Leaf morphology and reflectance in relation to water and temperature stress. In N. C. Turner and P. J. Kramer (eds.), *Adaptations of Plants to Water and Temperature Stress.* Wiley Interscience, New York, pp. 295–308.

———. 1983. Ecology and leaf physiology in North American desert plants. In E. Rodriguez, P. Heley, and I. Mehta (eds.), *Biology and Chemistry of Plant Trichomes.* Plenum Press, New York, pp. 113–132.

———. 1985. Annuals and perennials of warm deserts. In B. F. Chabot and H. A. Mooney (eds.), *Physiological Ecology of North American Plant Communities.* Chapman and Hall, New York, pp. 166–190.

Ehleringer, J. R., and I. Forseth. 1980. Solar tracking by plants. *Science* 210:1094–1098.

Ehleringer, J. R., and R. K. Monson. 1993. Evolutionary and ecological aspects of photosynthetic pathway variation. *Ann. Rev. Ecol. Syst.* 24:411–440.

Ehleringer, J. R., R. F. Sage, L. B. Flanagan, and R. W. Pearcy. 1991. Climate change and the evolution of C_4 photosynthesis. *TREE* 6:95–99.

Ehrenfeld, J. G. 1980. Understory response to canopy gaps of varying size in a mature oak forest. *Bull. Torrey Bot. Club* 107:29–41.

Ehrlich, P. R., and A. H. Ehrlich. 1981. *Extinction: The Causes and Consequences of the Disappearance of Species.* Random House, New York.

Ehrlich, P. R., and P. H. Raven. 1965. Butterflies and plants: A study in coevolution. Evolution 18:586–608.

Ehrlich, P. R., and J. Roughgarden. 1987. *The Science of Ecology.* Macmillan, New York.

Eisner, T. 1970. Chemical defense against predation in arthropods. In E. Sondheimer and J. B. Simeone (eds.), *Chemical Ecology.* Academic Press, New York, pp. 157–217.

Eisner, T., and J. Meinwald. 1966. Defensive secretions of arthropods. *Science* 153:1341–1350.

Elgar, M. A., and B. L. Crespi. 1992. *Cannibalism: Ecology and Evolution Among Diverse Taxa.* Oxford University Press, Oxford, England.

Ellis, D. H., J. C. Bednarez, D. G. Smith, and S. P. Fleming. 1993. Social foraging in raptorial birds. *Bioscience* 43:14–20.

Ellison, L. 1954. Subalpine vegetations of the Wasatch Plateau, *Utah. Ecol. Monogr.* 24:89–184.

Ellsworth, D. S., R. Oren, C. Huang, N. Phillips, and G. R. Hendrey. 1995. Leaf and canopy responses to elevated CO_2 in a pine forest under free-air CO_2 enrichment. *Oecologia* 104:139–146.

Elner, R. W., and R. N. Hughes. 1978. Energy maximization in the diet of the shore crab Carcinus maenas. *J. Anim. Ecol.* 47:103–116.

Elton, C. S. 1927. *Animal Ecology.* Sidgwich and Jackson, London.

Elton, C., and M. Nicholson. 1942. The ten-year cycle in numbers of lynx in Canada. *J. Anim. Ecol.* 11:215–244.

Eltringham, S. W. 1971. *Life in Mud and Sand.* Russak, New York.

Emanuel, W. R., H. H. Shugart, and M. L. Stevenson. 1985. Climate change and the broad scale distribution of terrestrial ecosystem complexes. *Climat. Change* 7:29–43.

Emlen, J. M. 1973. *Ecology: An Evolutionary Approach.* Addison-Wesley, Reading, MA.

Emlen, S. T. 1991. Evolution of cooperative breeding in birds and mammals. In J. R. Krebs and N. B. Davies (eds.), *Behavioural Ecology,* 3rd ed. Blackwell, Oxford, England, pp. 301–337.

Emlen, S. T., and L. W. Oring. 1977. Ecology, sexual selection, and the evolution of mating systems. *Science* 197:215–223.

ENGLE, L. G. 1960. Yellow-poplar seedfall pattern. *Central States Forest. Expt. Stat.* Note 143.

ENRIGHT, J. T. 1975. Orientation in time: Endogenous clocks. In O. Kinne (ed.), *Marine Ecology.* Vol. 2, *Physiological Mechanisms,* Part 2. Wiley, New York, pp. 917–944.

EPSTEIN, H. E., I. C. BURKE, AND A. R. MOSIER. 1998. Plant effects on spatial and temporal patterns of nitrogen cycling in short-grass steppe. *Ecosystems* 1:374–385.

EPSTEIN, H. E., W. K. LAUENROTH, AND I. C. BURKE. 1997. Effects of temperature and soil texture on ANPP in the U.S. Great Plains. *Ecology* 78:2682–2631.

EPSTEIN, H. E., W. K. LAUENROTH, I. C. BURKE, AND D. P. COFFIN. 1997. Productivity patterns of C_3 and C_4 functionl types in the U.S. great Plains. *Ecology* 78:722–731.

ERIKSSON, E. 1960. The yearly circulation of chloride and sulfur in nature; meteorological, geochemical and pedological implications. Part II. *Tellus* 12:63–109.

———. 1963. The yearly circulation of sulfur in nature. J. Geophys. Res. 68:4001–4008.

ERKERT, H. G., AND S. KRACHT. 1978. Evidence for ecological adaptation of circadian systems: Circadian activity rhythms of neo-tropical bats and their re-entrainment after phase shifts of the Zeitgeber-LD. *Oecologica* 32:71–78.

ERLINGE, S., G. GORANSSON, G. HOGSTEDT, G. JANSSON, O. LIBERG, J. LORMAN, I. N. NILSSON, T. VON SCHANTZ, AND M. SYLVEN. 1984. Can vertebrate predators regulate their prey? *Am. Nat.* 12:125–133.

ERRINGTON, P. L. 1943. Analysis of mink predation upon muskrats in the north-central U.S. *Iowa Agr. Exp. Sta. Res. Bull.* 320:797–924.

———. 1945. Some contributions of a fifteen-year local study of the northern bobwhite to a knowledge of population phenomena. *Ecol. Monogr.* 15:1–34.

———. 1946. Predation and vertebrate populations. *Quart. Rev. Biol.* 21:144–177, 221–245.

———. 1963. *Muskrat Populations.* Iowa State University Press, Ames.

ETHERINGTON, J. R. 1976. *Environmental and Plant Ecology.* Wiley, New York.

EVANS, F. C., AND S. A. CAIN. 1952. Preliminary studies on the vegetation of an old-field community in southeastern Michigan. *Contrib. Lab. Vert. Biol. U. Michigan* 51:1–17.

EVANS, J. R. 1989. Photosynthesis and nitrogen relations in leaves of C_3 plants. *Oecologia* 78:9–19.

EVENARI, M. 1985. The desert environment. In M. Evenari, I. Noy-Meir, and D. W. Goodall (eds.), *Hot Deserts and Arid Shrublands A* (Ecosystems of the World, Vol. 12A). Elsevier, Amsterdam, pp. 1–22.

———. 1986. *Hot Deserts and Arid Shrublands* (Ecosystems of the World, Vol. 12B). Elsevier, Amsterdam.

EWALD, P. W. 1983. Host-parasite relations, vector and the evolution of disease severity. *Ann. Rev. Ecol. Syst.* 14:465–485.

EYRE, S. R. 1963. *Vegetation and Soils: A World Picture.* Aldine, Chicago.

FAEGRI, K., AND L. VAN DER PIJL. 1979. *The Principles of Pollination Ecology,* 3rd ed. Pergamon, Oxford, England.

FARINA, A. 1998. *Principles and Methods in Landscape Ecology.* New York: Chapman and Hall.

FEINSINGER, P. 1983. Coevolution and pollination. In D. Futuyma and M. Slatkin (eds.), *Coevolution.* Sinauer Associates, Sunderland, MA, pp. 282–310.

FENCHEL, T. 1987. *Ecology—Potentials and Limitations* (Excellence in Ecology 1). Ecology Institute, Oldendorf/Luhe, Federal Republic of Germany.

———. 1988. Marine plankton food chains. *Ann. Rev. Ecol. Syst.* 19:19–38.

FENCHEL, T., AND T. H. BLACKBURN. 1979. *Bacteria and Mineral Cycling.* Academic Press, London.

FENCHEL, T., AND P. HARRISON. 1976. The significance of bacterial grazing and mineral cycling for the decomposition of particulate detritus. In J. M. Anderson and A. MacFayden (eds.), *The Role of Terrestrial and Aquatic Organisms in the Decomposition Process.* Blackwell, New York, pp. 285–321.

FENNER, F., AND F. N. RATCLIFFE. 1965. *Myxamatosis.* Cambridge University Press, Cambridge, England.

FERGUSEN, D. E., AND R. J. BOYD. 1988. Bracken fern inhibition of conifer regeneration in northern Idaho. USDA Forest Service Intermountain Research Station Res. Pap. INT-388, pp. 1–11.

FIELD, C. B., F. S. CHAPIN, N. R. CHIARIELLO, E. A. HOLLAND, AND H. A. MOONEY. 1996. The Jasper Ridge CO_2 experiment: Design and motivation. In G. Koch and H. A. Mooney (eds.), *Carbon Dioxide and Terrestrial Ecosystems.* Academic Press, San Diego, pp. 212–245.

FIELD, C. B., F. S. CHAPIN, P. A. MATSON, AND H. A. MOONEY. 1992. Responses of terrestrial ecosystem to the changing atmosphere: A resource based approach. *Ann. Rev. Ecol. Syst.* 23:201–235.

FIELD, C. B., R. B. JACKSON, AND H. A. MOONEY. 1995. Stomatal responses to CO_2: Implications from the plant to global scale. *Plant Cell Environ.* 18:1214–1225.

FIELD, C., AND MOONEY, H. 1986. The photosynthesis-nitrogen relationship in wild plants. In T. J. Givinish (ed.), *On the Economy of Plant Form and Function.* Cambridge University Press, New York, pp. 25–55.

FISCHER, A. G. 1960. Latitudinal variation in organic diversity. *Evolution* 14:64–81.

FISCHER, E. A. 1981. Sexual allocation in a simultaneously hermaphroditic coral. *Am. Nat.* 117:64–82.

FISCHER, S. G., AND G. E. LIKENS. 1973. Energy flow in Bear Brook, New Hampshire: An integrative approach to stream ecosystem metabolism. *Ecol. Monogr.* 43:421–439.

FISH AND WILDLIFE SERVICE. 1980. *Habitat as a Basis for Environmental Assessment.* Ecological Services Manual 101, USDI Fish and Wildlife Service, Division of Ecological Services, Washington, DC.

FISHER, R. A. 1930. The Genetical Theory of Natural Selection. Clarendon, Oxford, England.

FITTER, A. H. 1986. Acquisition and utilization of resources. In M. J. Crawley (ed.), *Plant Ecology.* Blackwell, Oxford, England, pp. 375–406.

FITZPATRICK, S. 1998. Bird tails as signaling devices: Markings, shape, length, and feather quality. *Am. Nat.* 151:157–173.

FLANAGAN, P. W., AND F. L. BUNNELL. 1980. Microflora activities and decomposition. In J. Brown, P. C. Miller, L. L. Tieszen, and F. L. Bunnell (eds.), *An Arctic Ecosystem: The Coastal Tundra at Barrow, Alaska.* Dowden, Hutchinson, and Ross, Stroudsburg, PA, pp. 291–334.

FLINT, R. F. 1970. *Glacial and Quaternary Geology.* Wiley, New York.

FLOOK, D. R. 1970. Causes and implications of an observed sex differential in the survival of wapiti. *Can. Wildl. Serv. Rept. Ser. No.* 11.

FOGARTY, M. J., AND S. A. MURAWSKI. 1998. Large-scale disturbance and the structure of marine systems: Fishery impacts on Georges Bank. *Ecol. App.* 8(1) Supplement:S6–S22.

FOLTZ, D. W., AND J. L. HOOGLAND. 1983. Genetic evidence of outbreeding in the black-tailed prairie dog (*Cynomys ludovicianus*). *Evolution* 37:273–281.

FOOSE, T. J. 1983. The relevance of captive populations to the conservation of biotic diversity. In C. M. Schonewald-Cox, S. M. Chambers, B. Macbryde, and W. L. Thomas (eds.), *Genetics and Conservation.* Benjamin/Cummings, Menlo Park, CA, pp. 374–401.

FOOSE, T. J., AND E. FOOSE. 1984. Demographic and genetic status and management. In B. Beck and C. Wemmer (eds.), *Pere David's Deer: The Biology and Conservation of an Extinct Species.* Noyes, Park Ridge, NJ.

FORBES, S. A. 1887. The lake as a microcosm. Bull. Peoria Sci. Assoc. 87; reprinted 1925, *Ill. Nat. History Surv. Bull.* 15:537–550.

FORCIER, L. K. 1975. Reproductive strategies and co-occurrence of climax tree species. *Science* 189:808–810.

FOREL, F. A. 1901. *Handbuch der Seenkunde,* Allgemeine Limnologie. J. Engelhorn, Stuttgart.

FORMAN, R. T. T. 1986. Landscape Ecology. John Wiley, New York.

———. 1995. *Land Mosaics: The Ecology of Landscapes and Regions.* Cambridge University Press, Cambridge, England.

FORMAN, R. T. T., AND L. E. ALEXANDER. 1998. Roads and their major eological effects. *Ann. Rev. Ecol. Syst.* 29:207–31.

FORMAN, R. T. T., AND M. GODRON. 1981. Patches and structural components for a landscape ecology. *Bioscience* 31:733–740.

FORSETH, I. N., AND J. R. EHLERINGER. 1983. Ecophysiology of two solar tracking desert annuals. IV. Effects of leaf orientation on calculated water gain and water use efficiency. *Oecologica* 58:10–18.

FORTNEY, J. L. 1974. Interactions between yellow perch abundance, walleye predation, and survival of alternate prey in Oneida Lake, New York. *Trans. Am. Fish Soc.* 103:15–24.

FOWLER, C. W. 1981. Density dependence as related to life history strategy. *Ecology* 62:602–610.

FOWLER, N. L., AND J. ANTONOVICS. 1981. Small scale variability in the demography of transplants of two herbaceous species. *Ecology* 62:1450–1457.

FOWLER, T. D., AND C. W. SMITH. 1981. *Dynamics of Large Animal Populations.* Wiley, New York.

Fox, F. M., and M. M. Caldwell. 1978. Competitive interaction in plant populations exposed to supplementary ultraviolet-B radiation. *Oecologica* 36:173–190.

Fox, L. R. 1975a. Some demographic consequences of food shortage for the predator *Notoneta hoffmanni*. *Ecology* 56:868–880.

———. 1975b. Factors influencing cannibalism, a mechanism of population limitation in the predator Notoneta hoffmanni. *Ecology* 56:933–941.

———. 1975c. Cannibalism in natural populations. *Ann. Rev. Ecol. Syst.* 6:87–106.

———. 1981. Defense and dynamics in plant-herbivore systems. *Am. Zool.* 21:853–864.

Francis, W. J. 1970. The influence of weather on population fluctuations in California quail. *J. Wildl. Manage.* 34:249–266.

Franck, V. M., B. A. Hungate, F. S. Chapin, and C. B. Field. 1997. Decomposition of litter produced under elevated CO_2: Dependence on plant species and nutrient supply. *Biogeochemistry* 36:223–237.

Frankel, O. H., and M. E. Soule. 1981. *Conservation and Evolution.* Cambridge University Press, Cambridge, England.

Franklin, I. R. 1980. Evolutionary change in small populations. In M. E. Soule and B. A. Wilcox (eds.), *Conservation Biology: An Evolutionary-Ecological Approach.* Sinauer, Sunderland, MA, pp. 135–149.

Franklin, J. F., and C. T. Dyrness. 1973. Natural Vegetation of Oregon and Washington. U.S.D.A. Forest Service Gen. Tech. Rept. PNW 8. USDA Forest Service, Corvallis, OR.

Franklin, J. F., and T. A. Speis. 1991. Composition, function, and structure of old-growth Douglas-fir forests. In Ruggiero, L. F., K. B. Aubry, A. B. Carey, and M. H. Huff (tech. coord.), *Wildlife and Vegetation of Unmanaged Douglas-fir Forests.* Gen. Tech Rept. PNW-GTR-285. USDA Forest Service; Pacific Northwest Research Station, Portland, OR, pp, 71–80.

Franklin, J. F., and R. H. Waring. 1979. Distinctive features of the northwestern coniferous forest: Development, structure, and function. In R. H. Waring (ed.), *Forests: Fresh Perspectives from Ecosystem Analysis.* Proc. 40th Ann. Biol. Coll., Oregon State University Press, Corvallis.

Franz, H. 1958. *Bodenzoologie als Grundlage der Bodenpflege.* Akademie-Verlag, Berlin.

Freeland, W. J. 1983. Parasites and the coexistence of animal host species. *Am. Nat.* 121:223–236.

Freeman, C. L., K. T. Harper, and E. L. Charnov. 1980. Sex change in plants: Old and new observations and new hypotheses. *Oecologica* 47:222–232.

Freemark, K. E., and H. G. Merriam. 1986. Importance of area and habitat heterogeneity to bird assemblages in temperate forest fragments. *Biol. Cons.* 36:115–141.

French, A. R. 1988. The patterns of mammalian hibernation. *Am. Sci.* 76:569–575.

French, C. E., L. C. McEwen, N. C. Magruder, R. H. Ingram, and R. W. Swift. 1955. Nutritional requirements of white-tailed deer for growth and antler development. *Penn. State Univ. Agr. Exp. Sta. Bull.* No. 600.

French, N. R., R. K. Steinhorst, and D. M. Swift. 1979. Grassland biome trophic pyramids. In N. R. French (ed.), *Perspectives in Grassland Ecology.* Springer Verlag, New York, pp. 59–87.

Fretwell, S. D., and H. L. Lucas. 1969. On territorial behavior and other factors influencing habitat distribution in birds. *Acta Biotheoretica* 19:16–36.

Friche, H., and S. Friche. 1977. Monogamy and sex change by aggressive dominance in coral reef fish. *Nature* 266:830–832.

Fritts, S. H., and L. D. Mech. 1981. Dynamics, movements, and feeding ecology of a newly-protected wolf population in northwestern Minnesota. *Wildl. Monogr.* 80.

Fritz, A. W. 1987. Commercial fishing for carp. In E. L. Cooper (ed.), *Carp in North America.* American Fishery Society, Bethesda, MD, pp.17–30

Fryxell, J. M., and C. M. Doucet. 1993. Diet choice and functional response of beavers. *Ecology* 74:1297–1306.

Futuyma, D. J. 1983. Evolutionary interactions among herbivorous insects and plants. In D. J. Futuyma and M. Slaktin (eds.), *Coevolution.* Sinauer Associates, Sunderland, MA, pp. 207–231.

Gaines, M. M., and M. L. Johnson. 1987. Phenotypic and genotypic mechanisms for dispersal in Microtus populations and the role of dispersal in population regulation. In B. B. Chepko-Sade and Z. T. Halpen (eds.), *Mammalian Dispersal Patterns.* University of Chicago Press, Chicago, pp. 162–179.

Gaines, M. S. and L. .R. McClenaghan, Jr.1980. Dispersal in small mammals. *Ann. Rev. Ecol. Syst.* 11:163–198.

Galli, A. E., C. F. Leck, and R. T. T. Forman. 1976. Avian distribution patterns in forest islands of different sizes in central New Jersey. *Auk* 93:356–364.

Galloway, J. N., G. E. Likens, and M. E. Hawley. 1984. Acid precipitation: Natural versus anthropogenic components. *Science* 226:829–831.

Gallun, R. L. 1977. The genetic basis of Hessian fly epidemics. *Ann. N.Y. Acad. Sci.* 287:223–229.

Gasaway, W. C., R. O. Stevensn, J. L. Davis, P. E. K. Shepherd, and O. E. Burns. 1983. Interrelationships of wolves, prey, and man in interior Alaska. *Wildl. Monogr.* 84.

Gates, D. 1962. *Energy Exchange in the Biosphere.* Harper and Row, New York.

———. 1965. Radiant energy: Its receipt and disposal. *Meteorol. Monogr.* 6:1–26.

———. 1966. Special distribution of solar radiation at the earth's surface. *Science* 151:523–528.

———. 1968. Energy exchange between organisms and environment. In W. P. Lowry (ed.), *Biometeorology.* Proc. 28th Ann. Biol. Colloq. Oregon State University Press, Corvallis.

———. 1972. *Man and His Environment: Climate.* Harper and Row, New York.

———. *Climate Change and Its Biological Consequences.* Sinauer Associates, Sunderland, MA.

Gause, G. F. 1934. *The Struggle for Existance.* Williams and Wilkins, Baltimore.

Geiger, R. 1965. *Climate near the Ground.* Harvard University Press, Cambridge, MA.

Geis, A. D., R. I. Smith, and J. P. Rogers. 1971. Black duck distribution, harvest characteristics, and survival. U.S. Fish and Wildl. Serv. Spec. Sci. Rept., Wildl. No. 139.

Gemmel, R. P., and G. T. Goodman. 1980. The maintenance of grassland on smelter wastes in the lower Swansea Valley. III. Zinc smelter wastes. *J. App. Ecol.* 17:461–468.

Gentry, A. H. 1986. Endemism in tropical versus temperate plant communities. In M. E. Soule (ed.), *Conservation Biology: The Science of Scarcity and Diversity.* Sinauer Associates, Sunderland, MA, pp. 153–181.

Getty, T. 1981. Competitive collusion: The preemption of competition during sequential establishment of territories. *Am. Nat.* 118:426–431.

Ghiselin, M. T. 1974. *The Economy of Nature and the Evolution of Sex.* University of California Press, Berkeley.

Gholz, H. L. 1982. Environmental limits on aboveground net primary production, leaf area, and biomass in vegetation zones of the Pacific northwest. *Ecology* 63:469–481.

Gilbert, L. E. 1975. Ecological consequences of a coevolved mutualism between butterflies and plants. In L. E. Gilbert and P. H. Raven (eds.), *Coevolution of Animals and Plants.* University of Texas Press, Austin, pp. 210–240.

Gill, D. E. 1975. Spatial patterning of pines and oaks in the New Jersey pine barrens. *J. Ecol.* 63:291–298.

Gill, F. B., and L. L. Wolf. 1975. Economics of feeding territoriality in the golden-winged sunbird. *Ecology* 56:333–345.

Gillon, Y. 1983. The invertebrates of the grass layer. In F. Bourliere (ed.), *Tropical Savannas* (Ecosystems of the World, Vol. 13). Elsevier, Amsterdam, pp. 289–311.

Gilpin, M. E., and M. E. Soule. 1986. Minimum viable populations: The processes of species extinction. In M. E. Soule (ed.), *Conservation Biology: The Science of Scarcity and Diversity.* Sinauer Associates, Sunderland, MA

Gimingham, C. H., S. B. Chapman, and N. R. Webb. 1981. European heathlands. In R. L. Specht (ed.), *Heathlands and Related Shrublands* (Ecosystems of the World, Vol. 9A). Elsevier, Amsterdam, pp. 365–419.

Ginsborne, H. T. 1941. How the wind blows in the forest of northern Idaho. Northern Rocky Mt. Forest Range Expt. Sta.

Gist, C. S., and D. A. Crossley, Jr. 1975. A model of mineral-element cycling for an invertebrate food web in a southeastern hardwood forest litter community. In F. Howell, J. B. Gentry, and M. H. Smith (eds.), *Mineral Cycling in Southeast Ecosystems.* National Technical Information Service, U.S. Dept. Commerce, pp. 84–106.

Gittleman, J. L., and S. D. Thompson. 1988. Energy allocation in mammalian reproduction. *Am. Zool.* 28:863–877.

GIVNISH, T. J. 1984. Leaf and canopy adaptations in tropical forests. In E. Medina, H. A. Mooney, and C. Vásquez-Yánes (eds.), *Physiogical Ecology of Plants of the Wet Tropics*. Dr. Junk, The Hague, pp. 51–84.

GLEASON, H. A. 1917. The structure and development of the plant association. *Bull. Torrey Bot. Club* 53:7–26.

———. 1926. The individualistic concept of the plant association. *Bull. Torrey Bot. Club* 62:1–20.

———. 1927. Further views on the succession concept. *Ecology* 8:299–326.

GODFRAY, H. C. J., AND A. B. HARPER. 1990. The evolution of brood reduction by siblicide in birds. *J. Theor. Biol.* 145:163–175.

GOLDBERG, D. E., AND A. M. BARTON. 1992. Patterns and consequences of interspecific competition in natural communities: A review of field experiments with plants. *Am. Nat.* 139:771–801.

GOLDBERG, D. E., AND P. A. WERNER. 1983. Equivalence of competitors in plant communities: A null hypothesis and a field experimental approach. *Am. J. Bot.* 70:1098–1104.

GOLDWASSER, S. D., D. GAINES, AND S. R. WILBUR. 1980. The Least Bell's Vireo in California. A de facto endangered race. *Am. Birds* 34:742–745.

GOLLEY, F. B. 1960. Energy dynamics of a food chain of an old field community. *Ecol. Monogr.* 30:187–206.

———. 1993. *A History of the Ecosystem Concept in Ecology*. Yale University Press, New Haven, CT.

GOLLEY, F. B., AND H. LEITH. 1972. Basis of organic production in the tropics. In P. M. Golley and F. B. Golley (eds.), *Tropical Ecology with an Emphasis on Organic Production*. University of Georgia Press, Athens, pp. 1–26.

GOLTERMAN. H. L., AND F. A. KOUWE. 1980. Chemical budgets and nutrient pathways. In E. D. Cren and R. H. Lowe-McConnell (eds.), *The Functioning of Freshwater Ecosystems* (International Biological Programmes No. 22). Cambridge University Press, Cambridge, England, pp. 88–140.

GORDON, M. S. 1972. *Animal Physiology*, 2nd ed. Macmillan, New York.

GORE, A. J. P. (ed.). 1983. *Mires: Swamp, Bog, Fen, Moor* (Ecosystems of the World, Vol. 13). Elsevier, Amsterdam.

GORE, J. A., AND E. D. SHIELDS, JR. 1995. Can large rivers be restored? *Bioscience* 45:142–152.

GOSS-CUSTARD, J. D. 1977a. The energetics of prey selection by redshank Tringa totanus (L) and a preferred prey *Corophium volutator* (Pallas). *J. Anim. Ecol.* 46:21–35.

———. 1977b. The energetics of prey selection by redshank *Tringa totanus* (L) in relation to prey density. *J. Anim. Ecol.* 46:1–19.

GOSZ, J. R., E. LIKENS, AND F. H. BORMANN. 1976. Organic matter and nutrient dynamics of the forest and forest floor in the Hubbard Brook Forest. *Oecologica* 22:305–320.

GOTELLI, N. J. 1995. *A Primer of Ecology*. Sinauer Associates, Sunderland, MA.

GRACE, J. B., AND R. G. WETZEL. 1981. Habitat partitioning and competitive displacement in cattails (Typha): Experimental field studies. *Am. Nat.* 118:463–474.

GRAHAM, W. F., AND R. A. DUCE. 1979. Atmospheric pathways of the phosphorus cycle. *Geochimica et Cosmochimica Acta* 43:1195–1208.

GRANHALL, R., AND V. LID-TORSVIK. 1975. Nutrition fixation by bacteria and free-living blue-green algae in tundra ecosystem. In F. E. Wielgolaski (ed.), *Fennoscandian Tundra Ecosystems*. Part I, *Plants and Microorganisms*. Springer Verlag, New York, pp. 305–315.

GRANT, P. R. Interspecific competition in fluctuating environments. In J. Diamond and T. J. Case (eds.), *Community Ecology*. Harper and Row, New York, pp. 173–191.

GRANT, P. R., AND B. R. GRANT. 1992. Demography and the genetically effective sizes of two populations of Darwin's finches. *Ecology* 73:766–784.

GRANT, R., AND P. R. GRANT. 1987. Mate choice in Darwin's finches. *Biol. J. Linn. Soc.* 32:247–270.

GREENWOOD, P. I. 1980. Mating Systems, philopatry, and dispersa, in birds and mammals. *Anim. Behav.* 28:1140–1162

GREENWOOD, P. I., AND P. H. HARVEY. 1982. The natal and breeding dispersal of birds. *Ann. Rev. Ecol. Syst.* 13:1–21.

GREGORY, S. V., F. J. SWANSON, W. A. McKEE, AND K. W. CUMMINS. 1991. An ecosystem perspective of riparian zones. *Bioscience* 41:541–551.

GREIG, J. C. 1979. Principles of genetic conservation in relation to wildlife management in Southern Africa. *S. Afr. J. Wildl. Res.* 9:57–78.

GRENFELL, B. T. 1988. Gastrointestinal nematode parasites and the stability and productivity of intensive ruminal grazing systems. *Phil. Trans. Royal Soc. London, Br. Biol. Sci.* 321:541–563.

———. 1992. Parasitism and the dynamics of ungulate grazing systems. *Am. Nat.* 139:907–929.

GRIER, C. C., AND R. S. LOGAN. 1977. Organic matter distribution and net production of plant communities of a 400-year old Douglas-fir ecosystems. *Ecol. Monogr.* 47:373–400.

GRIER, C. C., AND S. W. RUNNING. 1977. Leaf area of mature northwest conifer forests: Relation to a site water balance. *Ecology* 58:893–899.

GRIME, J. P. 1966. Shade avoidance and shade tolerance in flowering plants. In F. Bainbridge (ed.), *Light as an Ecological Factor*. Blackwell, Oxford, England, pp. 187–207.

———. 1977. Evidence for the existence of three primary strategies in plants and its relevance to ecological and evolutionary theory. *Am. Nat.* 111:1169–1194.

GRIME, J. P., AND D. W. JEFFREY. 1965. Seedling establishment in vertical gradients of sunlight. *J. Ecol.* 53:621–642.

GRIMM, N. 1987. Nitrogen dynamics during succession in a desert stream. *Ecology* 68:1157–1170.

GRINNELL. J. 1904. The origin and distribution of the Chestnut-backed Chickadee. *Auk* 21:364–382

———. 1917. The niche relationships of the California thrasher. *Auk* 34:427–433.

———. 1924. Geography and evolution. *Ecology* 5:225–229.

———. 1928. Presence and absence of animals. *U. Calif.* Chronicle 30:429–450.

GROSS, M. 1972. *Oceanography*. Prentice-Hall, Englewood Cliffs, NJ.

GUILFORD, T. 1988. The evolution of conspicuous coloration. *Am. Nat.* 131:S7–S21.

GUNDERSON, C. A., AND S. C. WULLSCHLEGER. 1994. Photosynthetic acclimation in trees to rising atmospheric CO_2: A broader perspective. *Photosynth. Res.* 39:369–388.

GUO, Q., AND P. W. RUNDEL. 1998. Self-thinning in postfire chaparral succession: Mechanisms, implications, and a combined approach. *Ecology* 79:579–586.

GUREVITCH, J., L. L. MORROW, A. WALLACE, AND J. J. WALSH. 1992. Meta-analysis of competition in field experiments. *Am. Nat.* 140:539–572.

GUSTAFSSON, L., AND W. J. SUTERLAND. 1988. The costs of reproduction in the collared flycatcher Ficedula albicollis. *Nature* 335:813–815.

GUTHERY, F. S., and R. L. BINGHAM. 1992. On Leopold's principle of edge. *Wildl. Soc. Bull.* 20:340–344.

HADLEY, E. B., AND B. J. KIECKHEFER. 1963. Productivity of two prairie grasses in relation to fire frequency. *Ecology* 44:389–395.

HAECKEL, E. 1869. Über Entwichelunge Gang 4. *Aufgabe de Zoologie Jemaische z.* 5:353–370.

HAINES, B. L., AND E. L. DUNN. 1985. Coastal marshes. In B. F. Chabot and H. A. Mooney (eds.), *Physiological Ecology of North American Plant Communities*. Chapman and Hall, New York, pp. 323–347.

HAIRSTON, N. G., F. E. SMITH, AND S. L. SLOBODKIN. 1960. Community structure, population control, and competition. *Am. Nat.* 94:421–425.

HALDANE, J. B. S. 1932. *The Causes of Evolution*. Longman, Green, London.

———. 1954. The measurement of natural selection. *Proc. 9th Int. Congr. Genetics*, pp. 421–425.

HALL, C. A. S., C. J. CLEVELAND, AND R. KAUFMAN. 1986. *Energy and Resource Quality: The Ecology of the Economic Process*. Wiley, New York.

HALLE, F. R., A. A. OLDMEMAN, AND P. B. TOMLINSON. 1978. *Tropical Trees and Forests: An Architectural Analysis*. Springer-Verlag, New York.

HAMILTON, W. D. 1964. The genetical evolution of social behavior I, II. *J. Theoret. Biol.* 82:1–16, 17–52.

———. 1972. Altruism and related phenomena, mainly in insects. *Ann. Rev. Ecol. Syst.* 3:193–283.

HAMILTON, W. D., AND M. ZUK. 1982. Heritable true fitness and bright birds: A role for parasites? *Science* 218:384–387.

HAMILTON, W. J., III. 1959. Aggressive behavior in migrant pectoral sandpipers. *Condor* 61:161–179.

HANDEL, S. N. 1978. The competitive relationship of three woodland sedges and its bearing on the evolution of ant-dispersal of Carex penduculata. *Evolution* 32:151–163.

HANES, T. L. 1981. California chapparal. In F. di Castri and H. A. Mooney (eds.), *Mediterranean-Type Shrublands: Origin and Structure*. Springer-Verlag, New York, pp. 139–174.

HANSKI, I. A. 1987. Populations of small mammal cycles—unless they don't. *TREE* 2:55–56.

HANSKI, I. A., AND M. E. GILPIN. 1997. *Metapopulation Biology: Ecology, Genetics, and Evolution*. Academic Press, San Diego.

HANSKI, I. A., AND D. SIMBERLOFF. 1997. The metapopulation approach, its history, conceptual domain, and application to conservation. In I. Hanski and M. Gilpin (eds.), *Metapopulation Biology*. Academic Press, San Diego, pp. 5–26.

HANSON, W. C. 1971. Seasonal patterns in native residents of three contrasting Alaskan villages. *Health Phys.* 20:585–591.

HANSON, W. C., D. G. WATSON, AND R. W. PERKINS. 1967. Concentration and retention of fallout radionuclide in Alaska arctic ecosystems. In B. Aberg and F. P. Hungate (eds.), *Radioecological Concentration Processes*. Pergamon Press, London, pp. 233–245.

HANZAWA, F. M., A. J. BEATTIE, AND D. C. CULVAR. 1988. Directed dispersal: Demographic analysis of ant-seed mutualism. *Am. Nat.* 131:1–13.

HARDIN, G. 1960. The competitive exclusion principle. *Science* 131:1292–1297.

HARDY, I. C. W., N. T. GRIFFITHS, AND H. C. J. GODFRAY. 1992. Clutch size in a parasitoid wasp: A manipulation experiment. *J. Anim. Ecol.* 61:121–129.

HARESTAD, A. S., AND E. L. BUNNELL. 1979. Home range and body weight—a reevaluation. *Ecology* 60:389–402.

HARLEY, J. L., AND S. E. SMITH. 1969. *The Biology of Mycorrhiza*. Academic Press, London.

HARMEL, D. E. 1983. Effects of genetics on antler quality and body size in white-tailed deer. In R. D. Brown (ed.), *Antler Development in Cervidae*. Caeser Kleberg Wildlife Research Institute, Kingsville, TX.

HARMON, M. E., J. F. FRANKLIN, F. S. SWANSON, et al. 1986. Ecology of coarse woody debris in temperate ecosystems. *Adv. Ecol. Res.* 15:133–302.

HARPER, J. L. 1977. *Population Biology of Plants*. Academic Press, New York.

HARPER, J., AND A. D. BELL. 1979. The population dynamics of growth form in organisms with modular construction. In R. M. Anderson (ed.), *Population Dynamics*, 20th Symposium, British Ecological Society. Blackwell, Oxford, England, pp. 29–52.

HARRIS, G. G., A. W. EBLING, D. R. LAUR, AND R. J. ROWLEY. 1984. Community recovery after storm damage: A case of facilitation in primary succession. *Science* 224:1136–1138.

HARRIS, L. D. 1984. *The Fragmented Forest*. University of Chicago Press, Chicago.

———. 1988. Edge effects and the conservation of biological diversity. *Cons. Biol.* 2:212–215.

———. 1989. The faunal significance of fragmentation of southeastern bottomland forests. In D. D. Hook and L. Ross (eds.), *Proc. Symp. the Forested Wetlands of Southern United States*. USDA Forest Service Gen. Tech. Rept. SE-50. Southeastern Forest Experiment Station, Asheville, NC, pp. 126–134.

HARRIS, L. D., AND J. MCELVEEN. 1981. *Effect of Forest Edge on North Florida Breeding Birds*. IMPAC Rept. 6. University of Florida School of Forest Resources and Conservation, Gainesville, FL.

HARRIS, L. D., AND G. SILVA-LOPEZ. 1992. Forest fragmentation and the conservation of biological diversity. In P. L. Feidler and S. K. Jain (eds.), *Conservation Biology: The Theory and Practice of Nature Conservation, Preservation, and Management*. Chapman and Hall, New York, pp. 197–237.

HARRISON, J. L. 1962. Distribution of feeding habits among animals in a tropical rain forest. *J. Anim. Ecol.* 31:53–63.

HARRISON, R. G. 1980. Dispersal polymorphisms in insects. *Ann. Rev. Ecol. Syst.* 11:95–118.

HARRISON, S., AND A. D. TAYLOR. 1997. Empirical evidence for metapopulation dynamics. In I. A. Hanski and M. Gilpin (eds.), *Metapopulation Biology*. Academic Press, San Diego, pp. 27–42.

HARTE, J., AND R. SHAW. 1995. Shifting dominance with a montane vegetation community: Results of a climate-warming experiment. *Science* 267:876–880.

HARTENSTEIN, R. 1986. Earthworm biotechnology and biogeochemistry. *Adv. Ecol. Res.* 15:179–407.

HARTL, D. L. 1988. *A Primer of Population Genetics*. Sinauer Associates, Sunderland, MA.

HARTLEY, J. L., AND S. E. SMITH. 1983. *Mycorrhizal Symbiosis*. Academic Press, London.

HARTSHORN, G. S. 1978. Tree falls and tropical forest dynamics. In P. B. Tomlinson and M. H. Zimmerman, (eds.), *Tropical Trees as Living Systems*. Cambridge University Press, Cambridge, England, pp. 617–638.

HARVEY, P. H., AND R. M. ZAMMUTO. 1985. Patterns of mortality and age at first reproduction in natural populations of mammals. *Nature* 315:319–320.

HASLER, A. D. 1969. Cultural eutrophication is reversible. *Bioscience* 19:425–431.

HASSELL, M. P. 1966. Evaluation of parasite or predator response. *J. Anim. Ecol.* 35:65–75.

HASSELL, M. P., J. L. LAWTON, AND J. R. BEDDINGTON. 1976. The components of arthropod predation. *J. Anim. Ecol.* 45:135–164.

———. 1977. Sigmoid functional responses by invertebrate predators and parasitoids. *J. Anim. Ecol.* 46:249–262.

HASSELL, M. P., AND R. M. MAY. 1973. Stability in insect host-parasite models. *J. Anim. Ecol.* 42:693–726.

———. 1974. Aggregation in predators and insect parasites and its effect on stability. *J. Anim. Ecol.* 43:567–597.

HASTINGS, A., AND S. HARRISON. 1994. Metapopulation dynamics and genetics. *Ann. Rev. Ecol. Syst.* 25:167–188.

HATTENSCHWILER, S., MIGLIETTA, F., RASCHI, A. AND KORNER, CH. 1997. Thirty years of in situ tree growth under elevated CO_2: a model for future forest responses. *Glob. Change Biol.* 3:463–471.

HAWKINS, B. A., AND M. HOLYOAK. 1998. Transcontinental crashes of insect populations. *Am. Nat.* 152:480–484.

HAWKINS, C. P., M. L. MURPHY, AND N. H. ANDERSON. 1982. Effects of canopy, substrate composition, and gradient on the structure of macroinvertebrate communities in the Cascade Range streams of Oregon. *Ecology* 63:1840–1856.

HAYMON, R. M., AND K. C. MCDONALD. 1985. The geology of deep sea hot springs. *Am. Sci.* 73:441–449.

HEADY, H. F., J. W. BARTOLOME, M. D. PITT, G. D. SAVELLE, AND M. C. STROUD. 1992. California prairie. In R. T. Coupland (ed.), *Natural Grasslands. Introduction and Western Hemisphere* (Ecosystems of the World, Vol. 8A). Elsevier, Amsterdam, pp. 313–335.

HEAL, O. W., H. E. JONES, AND J. B. WHITTAKER. 1975. Moore House, U.K. In T. Rosswall and O. W. Heal (eds.), *Structure and Function of Tundra Ecosystems*. Swedish Natural Science Institute, Stockholm, pp. 295–320.

HEARD, D. C. 1992. The effect of wolf predation and snow cover on muskox group size. *Am. Nat.* 139:190–204.

HEATH, J. E. 1965. Temperature regulation and diurnal activity in horned lizards. *Univ. Cal. Pub. Zool.* 64:97–136.

HEDRICK, P. W., AND P. S. MILLER. 1992. Conservation genetics: Techniques and fundamentals. *Ecol. App.* 2:30–46.

HEICHEL, G. H., AND N. C. TURNER. 1976. Phenology and leaf growth of defoliated hardwood trees. In J. F. Anderson and H. K. Karpus (eds.), *Perspectives in Forest Entomology*. Academic Press, New York, pp. 31–40.

HEIMANN, M. (ed.). 1993. *The Global Carbon Cycle*. Springer-Verlag, Berlin.

HEINRICH, B. 1976. Heat exchange in relation to blood flow between thorax and abdomen in bumblebees. *J. Exp. Biol.* 64:567–585.

———. 1979. *Bumblebee Economics*. Harvard University Press, Cambridge, MA.

HEITHAUS, E. R. 1981. Seed predation by rodents on three ant-dispersed plants. *Ecology* 63:136–145.

HEITHAUS, E. R., D. C. CULVER, AND A. J. BEATTIE. 1980. Models of some ant-plant mutualisms. *Am. Nat.* 116:347–361.

HENNTTONEN, H. 1985. Predation causing extended low densities in microtine cycles: Further evidence from shrew dynamics. *Oikos* 45:156–157.

HERKERT, J. R. 1994. The effect of habitat fragmentation on midwestern grassland bird communities. *Ecol. App.* 4:461–471.

HERRERA, C. M. 1982. Defense of ripe fruit from pests: Its significance in relation to plant disperser interactions. *Am. Nat.* 120:218–241.

———. 1985. Determinants of plant-animal coevolution: The case of mutualistic dispersal of seeds by vertebrates. *Oikos* 44:132–141.

HETT, J., AND O. L. LOUCKS. 1976. Age structure models of balsam fir and eastern hemlock. *J. Ecol.* 64:1029–1044.

HICKS, R. R., JR., AND D. E. FOSBROKE. 1987. Stand vulnerability: Can gypsy moth damage be predicted? In S. Fosbroke and R. Hicks (eds.), *Proc. Cooperative Workshop. Coping with the Gypsy Moth in the New Frontier*. West Virginia Office of Publications, Morgantown, pp. 73–80.

HIGGS, A. T., AND M. B. USHER. 1980. Should nature reserves be large or small? *Nature (London)* 285:568–569.

HILBERT, D. W., AND W. C. OECHEL. 1987. Response of tussock tundra to elevated carbon dioxide regimes: Analysis of ecosystem CO_2 flux through nonlinear modeling. *Oecologica* 72:466–472.

HILDEN, O. 1965. Habitat selection in birds: A review. *Ann. Zool. Fennica* 2:53–75.

HILL, E. P., III. 1972. Litter size in Alabama cottontails as influenced by soil fertility. *J. Wildl. Manage.* 36:1199–1209.

HILL, R. W. 1976. *Comparative Physiology of Animals: An Environmental Approach.* Harper and Row, New York.

HILL, R. W., AND G. A. WYSE. 1989. *Animal Physiology,* 2nd ed. Harper and Row, New York.

HILLBRICHT-ILKOWSKA, A. 1974. Secondary productivity in freshwaters, its value and efficiencies in plankton food chain. In Proc. *1st Inter. Congr. Ecol.,* pp. 164–167.

HILLS, T. L. 1965. Savannas: A review of a major research problem in tropical geography. *Can. Geog.* 9:216–228.

HIRSCHEL, G., CH. KORNER, AND J. A. ARNONE. 1997. Will rising atmospheric CO_2 affect litter quality and in situ decomposition in native plant communities? *Oecologia* 110:387–392.

HOBBS, R. J., AND D. A. NORTON. 1996. Toward a conceptual framework for restoration ecology. *Restor. Ecol.* 4:93–110.

HOCKING, B. 1975. Ant-plant mutualism: Evolution and energy. In L. E. Gilbert and P. H. Raven (eds.), *Coevolution of Animals and Plants.* University of Texas Press, Austin, pp. 78–90.

HOFMANN, D. J., AND T. DESHLER. 1991. Evidence from balloon measurements for chemical depletion of stratospheric ozone in the Arctic winter of 1989–1990. *Nature* 349:300–305.

HOFMANN, R. R. 1989. Evolutionary steps of ecological adaptation and diversification of ruminants: A comparative view of their digestive system. *Oecologica* 78:443–457.

HOLDRIDGE, L. R. 1947. Determination of wild plant formations from simple climatic data. *Science* 105:367–368.

———. 1967. Determination of world plant formation from simple climatic data. *Science* 130:572.

HOLDRIDGE, L. R., W. C. GRENKE, W. H. HATHEWAY, T. LIANG, AND J. A. TOSI, JR. 1971. *Forest Environments in Tropical Life Zones: A Pilot Study.* Pergamon Press, New York.

HOLLING, C. S. 1959. The components of predation as revealed by a study of small mammal predation of the European pine sawfly. *Can. Entomol.* 91:293–320.

———. 1965. The functional response of predators to prey density and its role in mimicry and population regulation. *Mem. Entomol. Soc. Can.* No. 45.

———. 1966. The functional response of invertebrate predators to prey density. *Mem. Entomol. Soc. Can.* No. 48.

HOLMES, J. 1983. Evolutionary relationships between parasitic helminths and their hosts. In D. J. Futuyma and M. Slatkin (eds.), *Coevolution.* Sinauer Associates, Sunderland, MA.

HOLMES, R. T., AND S. K. ROBINSON. 1981. Tree species preferences of foraging insectivorous birds in a northern hardwoods forest. *Oecologica* 48:31–35.

HOLMES, W. G. 1984. Sibling recognition in thirteen-lined ground squirrels: Effects of genetic relatedness, rearing association, and olfaction. *Behav. Ecol. Sociobiol.* 14:225–233.

HOLMES, W. T., AND P. W. SHERMAN. 1982. The ontogeny of kin recognition in two species of ground squirrels. *Am. Zool.* 22:491–517.

———. 1983. Kin recognition in animals. *Am. Sci.* 7:46–55.

HOOGLAND, J. C. 1982. Prairie dogs avoid extreme inbreeding. *Science* 215:1639–1641.

———. 1985. Infanticide in prairie dogs: Lactating females kill offspring of close kin. *Science* 230:1037–1040.

HOOK, D. D. 1984. Adaptation to flooding with fresh water. In T. Kozlowski (ed.), *Flooding and Plant Growth.* Academic Press, Orlando, FL, pp. 265–294.

HOPKINSON, C. S., AND J. P. SCHUBAUER. 1984. Static and dynamic aspects of nitrogen cycling in the salt marsh graminoid *Spartina alternifolia. Ecology* 65:961–969.

HORNBECK, J. W., G. E. LIKENS, AND J. S. EATON. 1976. Seasonal patterns in acidity of precipitation and implications for forest-stream ecosystems. In L. S. Dochinger and T. A. Seligar (eds.), *Proc. 1st International Symp. on Acid Precipitation and the Forest Ecosystem.* USDA Forest Service Gen. Tech. Rept. NE-23.

HORSLEY, S. B. 1977. Allelopathic inhibition of black cherry by ferns, grass, goldenrod, and aster. *Can. J. For. Res.* 7:205–216.

HOUGEN-EITZMAN, D., AND M. D. RAUSHER. 1944. Interactions between herbivorous insects and plant-insect coevolution. *Am. Nat.* 143:677–697.

HOUGH, A. F., AND R. D. FORBES. 1943. The ecology and silvics of forests in the high plateaus of Pennsylvania. *Ecol. Monogr.* 13:299–320.

HOUGHTON, J. T. 1997. *Global Warming: The Complete Briefing.* Cambridge University Press, Cambridge, England.

HOUGHTON, J. T., L. G. MEIRA FILHO, J. BRUCE, H. LEE, B. A. CALLANDER, E. HAITES, N. HARRIS, AND K. MASKELL (eds). 1995. *Climate Change 1994.* Cambridge University Press, Cambridge, England.

HOUGHTON, J. T., L. G. MEIRA FILHO, B. A. CALLANDER, N. HARRIS, A. KATTENBERG, AND K. MASKELL (eds.). 1996. *Climate Change 1995: The Science of Climate Change.* Intergovernmental Panel on Climate Change. Cambridge University Press, Cambridge, England.

HOUGHTON, R. A. 1995. Land-use change and the carbon cycle. *Glob. Change Biol.* 1:275–287.

HOUSTON, D. B. 1982. *The Northern Yellowstone Elk: Ecology and Management.* Macmillan, New York.

HOWARD, D. V. 1974. Urban robins: A population study. In J. Noyes and D. R. Progulske (eds.), *Wildlife in an Urbanizing Environment.* Massachussetts Cooperative Extension Service, Amherst, MA, pp. 67–75.

HOWARD, W. E. 1960. Innate and environmental dispersal of individual vertebrates. *Amer. Mid. Natur.* 63:152–161.

HOWARD-WILLIAMS, C. 1970. The ecology of *Becium homblei* in Central Africa with special reference to metaliferous soils. *J. Ecol.* 58:745–763.

HOWARTH, R. W., AND A. GIBLIN. 1983. Sulfate reduction in the salt marshes at Sapelo Island, Georgia. *Limno. Oceanogr.* 28:70–82.

HOWARTH, R. W., AND J. M. TEAL. 1979. Sulfate reduction in a New England salt marsh. *Limno. Oceanogr.* 24: 999–1013.

HOWE, H. F. 1980. Monkey dispersal and waste of a neotropical fruit. *Ecology* 61:944–959.

———. 1985. Gomphothere fruits: A critique. *Am. Nat.* 125:853–865.

———. 1986. Seed dispersal by fruit-eating birds and mammals. In D. R. Murra (ed.), *Seed Dispersal.* Academic Press, Sydney, pp. 123–190.

HOWE, H. F., AND J. SMALLWOOD. 1982. Ecology of seed dispersal. *Ann. Rev. Ecol. Syst.* 13:201–238.

HOWE, H. F., AND L. C. WESTLEY. 1988. *Ecological Relationships of Plants and Animals.* Oxford University Press, New York.

HUBBARD, S. F., AND R. M. COOK. 1978. Optimal foraging by parasitoid wasps. *J. Anim. Ecol.* 47:593–604.

HUDSON, J. W. 1973. Evolution of thermal regulation. In G. W. Whittow (ed.), *Comparative Physiology of Thermoregulation,* Vol. 3. Academic Press, New York, pp. 97–165.

HUEY, R. B. 1991. Physiological consequences of habitat selection. *Am. Nat.* 137:S91 S115.

HUEY, R. B., C. R. PETERSON, S. J. ARNOLD, AND W. P. PORTER. 1989. Hot rocks and not-so-hot rocks: Retreat site selection by garter snakes and its thermal consequences. *Ecology* 70:931–944.

HUFFAKER, C. B. 1958. Experimental studies on predation: Dispersion factors and predator-prey oscillations. *Hilgardia* 27:343–383.

HUHEEY, J. E. 1988. Mathematical models of mimicry. *Am. Nat.* 131: S22–S41.

HUNGATE, B. A., R. B. JACKSON, C. B. FIELD, AND F. S. CHAPIN. 1996. Field CO_2 enrichment experiments lack statistical power to detect changes in soil carbon. *Plant Soil* 187:135–145.

HUNGATE, R. E. 1975. The rumen microbial ecosystem. *Ann. Rev. Ecol. Syst.* 6:39–66.

HUNTER, G. W., III, AND W. S. HUNTER. 1934. Further studies on bird and fish parasites. Supp. 24th Ann. Rept. N.Y. State Dept. Cons., No. 9, *Rept. Biol. Surv. Mohawk-Hudson Watershed,* pp. 267–283.

HUNTER, J. E., F. L. SCHMIDT, AND G. B. JACKSON. 1982. *Meta-analysis: Cumulating Research Findings Across Studies.* Sage, Beverly Hills, CA.

HUNTER, M. D., AND J. C. SCHULTZ. 1995. Fertilization mitigates chemical induction and herbivore response within damaged oak trees. *Ecology* 76:1226–1232.

HUNTER, M. L., AND YONZON, P. 1992. Altitudinal distributions of birds, mammals, people, forests and parks in Nepal. *Cons. Biol.* 7:420–423.

HUNTLEY, B. 1991. How plants respond to climate change: Migration rates, individualism, and the consequences for plant communities. *Am. Bot.* 67 (Supp. 1):15–22.

HUNTLEY, B. J., AND B. H. WALKER (eds.). 1982. *Ecology of Tropical Savanna.* Springer Verlag, New York.

HUSTON, M. 1979. A general hypothesis of species diversity. *Am. Nat.* 113:81–101.

———. 1985. Patterns of species diversity on coral reefs. *Ann. Rev. Ecol. Syst.* 16:149–177.

———. 1994. *Biological Diversity: The Coexistence of Species on Changing Landscape.* Cambridge University Press, New York.

———. 1997. Hidden treatments in ecological experiments: Reevaluating the ecosystem function of biodiversity. *Oecologica* 110:449–460.

HUSTON, M., AND T. M. SMITH. 1987. Plant succession: Life history and competition. *Am. Nat.* 130:168–198.

HUTCHINS, H. E., AND R. M. LANNER. 1982. The central role of Clark's nutcracker in the dispersal and establishment of white-bark pine. *Oecologica* 55:192–201.

HUTCHINSON, G. E. 1957. A Treatise on Limnology, Vol. 1. *Geography, Physics, Chemistry.* Wiley, New York.

———. 1978. *An Introduction to Population Ecology.* Yale University Press, New Haven, CT.

———. 1957. Concluding remarks. *In Population Studies: Animal Ecology and Demography.* Cold Springs Harbor Symposium on Quantitive Biology 22:415–427. Long Island Biological Association, NY.

HUTCHINSON, T. C. 1967. Comparitive study of the ability of species to withstand prolonged periods of darkness. *J. Ecol.* 55:291–299.

HUTTO, R. L. 1985. Habitat selection by nonbreeding migratory land birds. In M. Cody (ed.), *Habitat Selection in Birds.* Academic Press, Orlando, FL, pp. 455–476.

HUXLEY, J. S. 1934. A natural experiment on the territorial instinct. *Brit. Birds* 27:270–277.

HYNES, H. 1970. *Biology of Running Water.* University of Toronto Press, Ontario.

IDSO, K. E., AND S. B. IDSO. 1994. Plant response to atmospheric CO_2 enrichment in the face of environmental constraints: A review of the past ten years research. *Agri. For. Meteorol.* 69:153–203.

IKUSIMA, I. 1965. Ecological studies on the productivity of aquatic plant communities: Measurement of photosynthetic activity. *Bot. Mag. Tokyo* 78:202–211.

IMS, R. A. 1990. On the adaptive value of reproductive synchrony as a predator swamping strategy. *Am. Nat.* 136:485–498.

IVERSON, L.R., PRASAD, A.M., HALE, B.J. AND SUTHERLAND, E.K. 1999. Atlas of current and potential future distributions of common trees of the eastern United States. US Department of Agriculture/Forest Service General Technical Report NE 265.

IWOA, K., AND M. D. RAUSHER. 1997. Evolution of plant resistance to multiple herbivores: Quantifying diffuse coevolution. *Am. Nat.* 149:316–335.

JACKSON, M. B. 1982. Ethylene as a growth promoting hormone under flooded conditions. In P. F. Waring (ed.), *Plant Growth Substances.* Academic Press, London, pp. 291–301.

———. 1985. Ethylene and responses of plants to soil waterlogging and submergence. *Ann. Rev. Plant Physiol.* 36:145–174.

JACKSON, M. B., AND M. C. DREW. 1984. Effects of flooding on growth and metabolism of herbaceous plants. In T. T. Kozlowski (ed.), *Flooding and Plant Growth.* Academic Press, Orlando, FL, pp. 47–128.

JACKSON, R. R. 1992. Eight-legged tricksters. *Bioscience* 42:590–598.

JACKSON, S. T., R. P. FUTUYMA, AND D. A. WILCOX. 1988. A paleoecological test of a classical hydrosere in the Lake Michigan dunes. *Ecology* 69:928–936.

JAENIKE, J. 1992. Mycophagous Drosphila and their nematode parasites. *Am. Nat.* 139:893–906.

———. 1998. On the capacity of macroparasites to control insect populations. *Am. Nat.* 154:84–96.

JAMES, F. 1971. Ordination of habitat relationships among birds. *Wilson Bull.* 83:215–236.

JAMES, S. W. 1991. Soil, nitrogen, phosphorus, and organic matter processing by earthworms in tallgrass prairie. *Ecology* 72:2101–2109.

JANZEN, D. H. 1966. Coevolution of mutualism between ants and acacias in Central America. *Evolution* 20:249–275.

———. 1967. Synchronization of sexual reproduction of trees within the dry season in Central America. *Evolution* 21:620–637.

———. 1971. Seed predation by animals. *Ann. Rev. Ecol. Syst.* 2:465–492.

———. 1980. When is it coevolution? *Evolution* 34:611–612.

———. 1986. Chihuahuan desert nopaleras: Defaunated big mammal vegetation. *Ann. Rev. Ecol. Syst.* 17:595–636.

JARVINEN, A. 1985. Predation causing extended low densities in microtine cycles: Implications from predation on hole-nesting passerines. *Oikos* 45:157–158.

JARVIS, P. G. 1995. The role of temperate trees and forests in CO_2 fixation. *Vegetatio* 121:157–174.

JEFFERIES, R. L. 1972. Aspects of salt marsh ecology with particular reference to inorganic plant nutrients. In R. S. K. Barnes and J. Green (eds.), *The Estuarine Environment.* Applied Science, London, pp. 61–85.

———. 1988. Pattern and process in arctic coastal vegetation in response to foraging by leser snow geese. In M. J. A. Werger (ed.), Plant Form and Vegetation Structure, SPB Academic Publishing, pp. 281–300.

JEFFERS, J. N. R. 1988. *Practitioner's Handbook on the Modelling of Dynamic Change in Ecosystems* (Scope 34). Wiley, New York.

JENNY, H. 1941. *Factors of Soil Formation: A System of Quantitative Pedology.* McGraw-Hill, New York.

———. 1980. *The Soil Resource.* Springer-Verlag, New York.

JOHNSGARD, P. A. 1965. *Handbook of Waterfowl Behavior.* Cornell University Press, Ithaca, NY.

JOHNSON, B. L., W. B. RICHARDSON, AND T. J. NAIMO. 1995. Past, present, and future concepts in large river ecology. *Bioscience* 45:134–152.

JOHNSON, C. N. 1996. Interactions between mammals and ectomycorrhizal fungi. *TREE* 11:503–506.

JOHNSON, D. W., D. W. COLE, S. BLEDSOE, K. CROMACK, R. L. EDMONDS, S. P. BESSEL, C. C. GRIER, B. N. RICHARDS, AND K. A. VOGT. 1982. Nutrient cycling in forests of the Pacific Northwest. In R. L. Edmonds (ed.), *Analysis of Coniferous Forest Ecosystems in the Western United States.* Hutchinson, Ross, Stroudsburg, PA, pp. 186–232.

JOHNSON, D. W., D. O. RICHTER, H. MIEGROET, AND D. W. COLE. 1983. Contribution of acid deposition and natural processes to cation leaching from forest soils: A review. *J. Air Pollution Control Assoc.* 33:1036–1041.

JOHNSON, J. A. 1984. Small woodlot management by single-tree selection: 21-year results. *Northern J. App. For.* 1:69–71.

JOHNSON, M. C., AND M. S. GAINES. 1987. The selective basis for dispersal of the prairie vole *Microtus ochrogaster. Ecology* 68:684–694.

JOHNSON, R. W., AND J. C. TOTHILL. 1985. Definition and broad geographic outline of savanna lands. In J. C. Tothill and J. J. Mott (eds.), *Ecology and Management of the World's Savannas.* Commonwealth Agricultural Bureau, Australian Academy of Science, Canberra, pp. 1–13.

JOHNSTON, J. W. 1936. The macrofauna of soils as affected by certain coniferous and hardwood types in the Harvard forest. PhD Dissertation, Harvard University Library, Cambridge, MA.

JOHNSTON, R. F. 1956. Predation by short-eared owls in a Salicornia salt marsh. *Wilson Bull.* 68:91–102.

JONES, H. G. 1983. *Plants and Microclimate.* Cambridge University Press, Cambridge, England.

———. 1992. Plants and Microclimate, 2nd ed. Cambridge University Press, Cambridge, England.

JONES, R. L., AND H. C. HANSON. 1985. *Biogeochemistry, Mineral Licks, and Geophagy of North American Ungulates.* Iowa State University Press, Ames.

JONES, R. L., AND H. P. WEEKS. 1985. Ca, Mg, and P in the annual diet of deer in south-central Indiana. *J. Wildl. Manage.* 49:129–133.

JORDAN, C. F. 1982. Amazon rain forest. *Am. Sci.* 70:394–401.

———. 1986. Local effects of tropical deforestation. In M. Soule (ed.), *Conservation Biology: The Science of Scarcity and Diversity.* Sinauer, Sunderland, MA, pp. 410–446.

JORDAN, C. F., J. R. KLINE, AND D. S. SASSIER. 1972. Relative stability of mineral cycles in forest ecosystems. *Am. Nat.* 106:237–253.

JORDAN, W. R., III, M. E. GILPIN, AND J. D. ABER (eds.). 1987. Restoration Ecology: A Synthetic Approach to Ecological Research. Cambridge University Press, Cambridge, England.

JORDEN, T. E., AND I. VALIELA. 1982. The nitrogen cycle of the ribbed mussel Geukinsea demissa and its significance in nitrogen flow in a New England salt marsh. Limno. *Oceanogr.* 27:75–90.

JORGENSEN, J. R., AND C. G. WELLS. 1986. Forester's Primer in Nutrient Cycling. USDA Forest Service Gen. Tech. Rept. SE-37.

JUDAY, C. 1940. The annual energy budget of an inland lake. *Ecology* 21:438–450.

JUNK, W. J., P. B. BAYLEY, AND R. E. SPARKS. 1989. The flood pulse concept in river floodplain-systems. *Can. Spec. Publ. Fish. Aquatic Sci.* 106:110–127.

KABAT, C., AND D. R. THOMPSON. 1963. *Wisconsin Quail, 1834–1962: Population Dynamics and Habitat Management.* Tech. Bull. No. 30. Wisconsin Conserv. Dept., Madison.

KALKSTEIN, L. S., AND J. S. GREEN. 1997. An evaluation of climate/mortality relationships in large U.S. cities and possible impacts of a climate change. *Environ. Health Persp.* 105:84–93.

KALKSTEIN, L. S., AND G. TAN. 1995. Human health. In K. Strzepek and J. Smith (eds.), *As Climate Changes: International Impacts and Implications.* Cambridge University Press. Cambridge, UK

KALLIO, S., AND P. KALLIO. 1975. Nitrogen fixation in lichens at Kevo, North Finland. In F. E. Wiegolaski (ed.), *Fennoscandian Tundra Ecosystems*. Part 1, *Plants and Microorganisms*. Springer Verlag, New York, pp. 292–304.

KAMIL, A. C., J. R. KREBS, AND H. R. PULLIAN (eds.). 1987. *Foraging Behavior*. Plenum, New York.

KARR, J. R. 1975. Production and energy pathways, and community diversity in forest birds. In F. Golley and E. Medina (eds.), *Tropical Ecological Systems: Trends in Terrestrial and Aquatic Research*. Springer-Verlag, New York, pp. 161–176.

KARR, J. R., AND R. R. ROTH. 1971. Vegetation structure and avian diversity in several new world areas. *Am. Nat*. 105:423–435.

KASTINGS, J. F. 1993. Earth's early atmosphere. *Science* 259:921–926.

KASTINGS, J. F., AND T. P. ACKERMAN. 1986. Climatic consequences of very high carbon dioxide levels in the earth's early atmosphere. *Science* 234:1383–1385.

KATTENBERG, A., F. GIORGI, H. GRASSL, G. A. MEEHL, J. F. B. MITCHELL, R. J. STOUFFER, T. TOKIOKA, A. J. WEAVER, AND T. M. L. WIGLEY. 1996. *Climate Models—Projections of Future Climate*. Pp. 285–357. in J. T. Houghton et al. (eds.) Climate Change 1995. The Science of Climate Change. Intergovernmental Panel on Climate Change. Cambridge University Press, Cambridge, UK.

KAYS, S., AND J. L. HARPER. 1974. The regulation of plant and tiller density in a grass sward. *J. Ecol*. 62:97–105.

KEELER, K. 1981. A model of selection for facultative mutualism. *Am. Nat*. 18:488–498.

KEELING, C. D., T. P. WHORF, M. WAHLEN, AND J. VAN DER PLICHT. 1995. Interannual extremes in the rate of rise of atmospheric carbon dioxide since 1980. *Nature* 375:666–670.

KEEVER, C. 1950. Causes of succession in old fields of the Piedmont, North Carolina. *Ecol. Monogr*. 20:229–250.

KEITH, L. B. 1974. Some features of population dynamics in mammals. *Proc. Inter. Congr. Game Biol. Stockholm* 11:17–58.

———. 1983. Role of food in hare population cycles. *Oikos* 40:385–395.

KEITH, L. B, J. R. CARY, O. J. RONGSTAD, AND M. C. BRITTINGHAM. 1984. Demography and ecology of declining snowshoe hare population. *J. Wildl. Manage*. 90:1–43.

KEITH, L. B., AND L. A. WINDBERG. 1978. A demographic analysis of the snowshoe hare cycle. *Wildl. Monogr*. 58.

KELLER, L. F., AND P. ARCESE. 1998. No evidence for inbreeding avoidance in a natural population of song sparrows (*Melospiza melodia*). *Am. Nat*. 152:380–392.

KELLER, L. F., P. ARCESE, J. N. M. SMITH, W. M. HOCHACHA, AND C. S. STERNS. 1994. Selection against inbred song sparrows during a natural population botleneck. *Nature (London)* 372:356–357.

KERBES, R. H., P. M. KOTANEN, AND R. L. JEFFRIES. 1990. Destruction of wetland habitats by lesser snow geese: A keystone species on the west coast of Hudson Bay. *J. App. Ecol*. 27:242–258.

KERNER, A. 1904. *The Natural History of Plants, Their Forms, Growth, Reproduction, and Distribution*. Greshma, London.

KETCHUM, B. H. (ed.). 1983. *Estuaries and Enclosed Seas* (Ecosystems of the World, Vol. 26). Elsevier, Amsterdam.

KETTLEWELL, H. B. D. 1961. The phenomenon of industrial melanism in *Lepidoptera*. *Ann. Rev. Entomol*. 6:245–262.

———. 1965. Insect survival and selection for pattern. *Science* 148:1290–1296.

KIESTER, A. R. 1971. Species density of North American amphibians and reptiles. *Syst. Zool*. 20:127–137.

KIKKAWA, J., AND W. T. WILLIAMS. 1971. Altitudinal distribution of land birds in New Guinea. *Search* 2:64–69.

KIMBALL, B. A. 1983. Carbon dioxide and agricultural yield. An assemblage and analysis of 430 prior observations. *Agron. J*. 75:779–788.

KINGSLAND, S. 1985. *Modelling Nature*. University of Chicago Press, Chicago.

KINGSTON, T. J., AND M. J. COE. 1977. The biology of the giant dung beetle (*Heliocopris dilloni*) (*Coleoptera: Scarabaeidae*). *J. Zool. (London)* 181:243–263.

KIRKPATRICK, M. 1986. The handicap mechanism does not work. *Am. Nat*. 127:222–240.

———. 1987. Sexual selection by female choice in polygynous animals. *Ann. Rev. Ecol. Syst*. 18:43–70.

KITTREDGE, J. 1948. *Forest Influences*. McGraw-Hill, New York.

KLEMOW, K. M., AND D. J. RAYNAL. 1983. Population biology of an annual plant in a temporally variable habitat. *J. Ecol*. 71:691–703.

———. 1985. Demography of two facultative biennial plant species in an unproductive habitat. *J. Ecol*. 73:147–167.

KLOPFER, P. 1963. Behavioral aspects of habitat selection: The role of early experience. *Wilson Bull*. 75:15–22.

KLOPFER, P. H., AND J. U. GANZHORN. 1985. Habitat selection: Behavioral aspects. In M. Cody (ed.), *Habitat Selection in Birds*. Academic, Orlando, FL.

KNAPP, A. K., AND T. R. SEASTEDT. 1986. Detritus accumulation limits productivity of tall grass prairie. *Bioscience* 36:662–668.

KNAPP, R. A., AND R. C. SARGENT. 1989. Egg-mimicry as a mating strategy in the fantail darter *Etherostoma flabellare:* Females prefer males with eggs. *Behav. Ecol. Sociobiol*. 25:321–326.

KNIGHT, C. G. 1974. The geography of vectored diseases. In J. M. Hunter (ed.), *The Geography of Health and Disease*. U. North Carolina, Dept. of Geography Studies in Geography No. 6.

KNOX, E. G. 1952. *Jefferson County (NY) Soils and Soil* Map. NY State College Agr., Cornell University, Ithaca, NY.

KOBLENTZ-MISHKE, J. J., V. V. VOLKOVINSKY, AND J. G. KABANOVA. 1970. Plankton primary production of the world's oceans. In W. S. Wooster (ed.), *Scientific Exploration of the South Pacific*. National Academy of Science, Washington, DC, pp. 183–193.

KODRICH-BROWN, A. K., AND J. H. BROWN. 1984. Truth in advertising: The kinds of traits favored by sexual selection. *Am. Nat*. 124:309–3223.

KOENIG, W. D., J. HAYDOCK, AND M. T. STANBACK. 1988. Reproductive rolesin the cooperatively breeding acorn woodpecker: Incest avoidance versus reproductive competition. *Am. Nat*. 151:243–255.

KONISHI, M. 1973. How the owl tracks its prey. *Am. Sci*. 61:414–424.

KOPEC, R. J. 1970. Further observations of the urban heat island of a small city. *Bull. Am. Meteorol. Soc*. 51:602–606.

KORNER, CH., AND J. A. ARNONE. 1992. Responses to elevated carbon dioxide in artificial tropical ecosystems. *Science* 257:1672–1675.

KORNER, CH., M. DIEMER, B. SCHAPPI, P. NIKLAUS, AND J. ARNONE. 1997. The response of alpine grassland to four seasons of CO_2 enrichment: A synthesis. *Acta Oecologia* 18:165–175.

KORNER, CH., AND M. WURTH. 1996. A simple method for testing leaf responses of tall tropical trees to elevated CO_2. *Oecologia* 107:421–425.

KORPIMAKI, E., AND C. J. KREBS. 1996. Predation and population cycles of small mammals. *Bioscience* 46:754–764.

KORPIMAKI, E., AND K. NORRDAHL. 1991. Numerical and functional responses of kestrels, short-eared owls, and long-eared owls to vole densities. *Ecology* 72:814–826.

KOWAL, N. E. 1966. Shifting agriculture, fire, and pine forests in the Cordillera, Central Luzon, Philippines. *Ecol. Monogr*. 56:389–419.

KOZLOWSKI, T. T. 1984a. Plant responses to flooding of soil. *Bioscience* 34:162–168.

———. 1984b. Response of woody plants to flooding. In T. T. Kozlowski (ed.), *Flooding and Plant Growth*. Academic, Orlando, FL, pp. 129–163.

KOZLOWSKI, T. T., AND S. G. PALLARDY. 1984. Effects of flooding on water, carbohydrate, and mineral relations. In T. T. Kozlowski (ed.), *Flooding and Plant Growth*. Academic, Orlando, FL, pp. 165–194.

KRAMER, P. J. 1983. *Water Relations of Plants*. Academic, Orlando, FL.

KRATZ, T. K., AND C. D. DEWITT. 1986. Internal factors controlling peatland-lake ecosystem development. *Ecology* 67:100–107.

KREBS, C. J. 1985. Ecology: *The Experimental Analysis of Distribution and Abundance*, 3rd ed. Harper and Row, New York.

———. 1989. *Ecological Methodology*. Harper and Row, New York.

KREBS, C. J., S. BOUTIN, R. BOONSTRA, A. R. E. SINCLAIR, J. N. M. SMITH, M. R. T. DALE, K. MARTIN, AND R. TURKINGTON. 1995. Impact of food and predation on the snowshoe hare cycle. *Science* 269:1112–1115.

KREBS, C. J., B. L. KELLER, AND R. H. TAMARIN. 1969. Microtus population biology: Demographic changes in fluctuating populations of M. ochrogaster and M. pennsylvanicus in southern Indiana. *Ecology* 75:214–223.

KREBS, C. J., I. WINGATE, J. LEDUC, J. A. REDFIELD, M. TAITT, AND R. HILBORN. 1976. Microtus population biology: Dispersal in fluctuating populations of M. townsendii. *Can. J. Zool*. 54:79–95.

KREBS, J. R. 1971. Territory and breeding density in the great tit Parus major. *Ecology* 52:2–22.

KREBS, J. R., AND N. B. DAVIES (eds.). 1978. *Behavioral Ecology: An Evolutionary Approach*. Blackwell, Oxford, England.

———. 1984. *Behavioral Ecology: An Evolutionary Approach*, 2nd ed. Blackwell, Oxford, England.

KREBS, J. R., A. KACELNIK, AND P. TAYLOR. 1978. Optimal sampling by foraging birds: An experiment with great tits (*Parus major*). *Nature* 275:27–31.

KREMER, J. N., AND S. W. NIXON. 1978. *A Coastal Marine Ecosystem.* Springer-Verlag, New York.

KRUCKEBERG, A. R. 1954. The ecology of serpentine soils. III. Plant species in relation to serpentine soils. *Ecology* 35:267–274.

KRUGER, F. J. 1979. South African heathlands. In R. L. Specht (ed.), *Heathlands and Related Shrublands Descriptive Studies* (Ecosystems of the World, Vol. 9A), Elsevier, Amsterdam.

KRUUK, H. 1972. *The Spotted Hyena.* University of Chicago Press, Chicago.

KUCERA, C. L., R. C. DAHLMAN, AND M. R. KOELLING. 1967. Total net productivity and turnover on an energy basis for a tallgrassprairie. *Ecology* 48:536–541.

KUCERA, C. L. 1992. Tall-grass prairie. In R. T. Coupland (ed.), *Natural Grasslands. Introduction and Western Hemisphere* (Ecosystems of the World, Vol. 8A). Elsevier, Amsterdam, pp. 227–268.

KUCHLER, A. W. 1964. *Potential Natural Vegetation of the United States, U. S. Geological Survey National Atlas.* U.S. Geological Survey, Washington, DC.

KUENZLER, E. J. 1958. Niche relations of three species of Lycosid spiders. *Ecology* 39:494–500.

KURLANSKY, M. 1997. *Cod: The Biography of a Fish that Changed the World.* Walker and Company, New York.

LACK, D. L. 1954. *The Natural Regulation of Animal Numbers.* Clarendon Press, Oxford, England.

———. 1966. *Population Studies of Birds.* Clarendon Press, Oxford, England.

———. 1971. *Ecological Isolation in Birds.* Harvard University Press, Cambridge, MA.

LACK, D. L., AND L. S. V. VENABLES. 1939. The habitat distribution of British woodland birds. *J. Anim. Ecol.* 8:39–71.

LAFFERTY, K. D., AND A. KIMO MORRIS. 1996. Altered behavior of parasitized killifish increases susceptibility to predation by bird final hosts. *Ecology* 77:1390–1397.

LAMOTTE, M. 1975. The structure and function of a tropical savannah ecosystem. In F. B. Holley and E. Medina (eds.), *Tropical Ecological Systems: Trends in Terrestrial and Aquatic Research.* Springer-Verlag, New York, pp. 179–222.

LAMPSON, N. I. 1987. D.b.h./crown diameter relationships in mixed Appalachian hardwood stands. Res. Pap. NE-610, USDA Forest Service, Northeastern Forest Experiment Station.

LANCINANI, C. A. 1975. Parasite-induced alternations in host reproduction and survival. *Ecology* 56:689–695.

LANDAHL, J., AND R. B. ROOT. 1969. Differences in the life tables of tropical and temperate milkweed bugs, genus *Oncopeltus* (Hemiptera: Lygaeidae). *Ecology* 50:734–737.

LANDE, R., AND G. F. BARROWCLOUGH. 1987. Effective population size, genetic variation, and their use in population management. In M. E. Soule and B. A. Wilcox (eds.), *Conservation Biology: An Evolutionary-Ecological Approach.* Sinauer Associates, Sunderland, MA, pp. 135–149.

LANDSBERG, H. E. 1970. Man-made climatic changes. *Science* 170:1265–1274.

LANE, P. A. 1985. A food web approach to mutualism in lake communities. In D. H. Boucher (ed.), *The Biology of Mutualism.* Oxford University Press, New York, pp. 344–374.

LANG, G. E. 1985. Forest turnover and the dynamics of bole wood litter in subalpine balsam fir forest. *Can. J. For. Res.* 15:262–288.

LANG G. E., AND R. T. FORMAN. 1978. Detrital dynamics in a mature forest: Hutchinson Memorial Forest. *Ecology* 59:580–595.

LANG, G. E., W. A. REINERS, AND R. R. HEIER. 1976. Potential alterations of precipitation chemistry by epiphytic lichens. *Oecologica* 25:229–241.

LARCHER, W. 1980. *Physiological Plant Ecology.* Springer-Verlag, New York.

———. 1995. *Physiological Plant Ecology,* 3rd ed. Springer, Berlin.

LARSEN, K. W., AND S. BOUTIN. 1994. Movements, survival, and settlement of red squirrel (*Tamiasciurus hudsonicus*) offspring. *Ecology* 75:214–223.

LASSOIE, J. P., T. M. HINCKLEY, AND C. C. GRIER. 1985. Coniferous forests of the Pacific Northwest. In B. F. Chabot and H. A. Mooney (eds.), *Physiological Ecology of North American Plant Communities.* Chapman and Hall, New York, pp. 127–161.

LAURHAN, M. 1992. A technique for estimating seed production of common moist-soil plants. U.S. Fish and Wildlife Service Fish and Wildlife Leaflet 13.4.5. 8 pp.

LAUCK, T., C. W. CLARK, M. MANGEL, AND G. R. MUNRO. 1998. Implementing the precautionary principle in fisheries management through marine reserves. *Ecol. App.* 8(1, Suppl.): S72–S78.

LAUENROTH, W. K. 1979. Grassland primary production: North American grasslands in perspective. In N. F. French (ed.), *Perspectives in Grassland Ecology.* Springer-Verlag, New York, pp. 3–24.

LAUENROTH, W. K., AND D. J. MILCHUNAS. 1992 Short-grass steppe. In R. T. Coupland (ed.), *Natural Grasslands. Introduction and Western Hemisphere* (Ecosystems of the World, Vol. 8A). Elsevier, Amsterdam, pp. 183–226.

LAUFF, G. (ed.). 1967. *Estuaries.* American Association for the Advancement of Science, Washington, DC.

LAW, R., A. D. BRADSHAW, AND P. W. PUTWAIN. 1977. Life history variation in Poa annua. *Evolution* 3:233–246.

LAWRENCE, W. T., AND OECHEL, W. 1983a. Effects of soil temperature on the carbon exchange of taiga seedlings. I. Root respiration. *Can. J. For. Res.* 13:840–849

———. 1983b. Effects of soil temperature on the carbon exchange of taiga seedlings. II. Photosynthesis, respiration and conductance. *Can. J. For. Res.* 13:850–859.

LAWTON, J. H., M. P. HASSELL, AND J. R. BEDDINGTON. 1975. Prey death rates and rate of increase of arthropod predator populations. *Nature* 255:60–62.

LEEMANS, R., AND W. CRAMER. 1990. The IIASA climate database for land area on a grid of 0.50 resolution. WP-41, International Institute for Applied Systems Analysis, Laxenburg.

LEES, D. R., AND E. R. CREED. 1975. Industrial melanism in *Biston betularia:* The role of selective predation. *J. Anim. Ecol.* 44:67–83.

LEIBIG, J. 1840. *Organic Chemistry* and Its Application to Vegetable Physiology and Agriculture. T. B. Peterson, Philadelphia.

LEITH, H. (ed.). 1974. *Phenology and Seasonality Modeling.* Springer-Verlag, New York.

LEITH, H. 1975a. Modeling primary productivity of the world. In H. Leith and R. Whittaker (eds.), *Primary Production of the Biosphere.* Springer Verlag, New York, pp. 237–264.

———. 1975b. Primary productivity in ecosystems: Comparative analysis of global patterns. In W. H. van Dobben and R. H. Lowe-McConnell (eds.), *Unifying Concepts in Ecology.* Junk, The Hague, pp. 67–88.

LENINGTON, S. 1980. Female choice and polygyny in redwinged blackbirds. *Behaviour* 28:347–361.

LEOPOLD, A. 1933. *Game Management.* Scribner, New York.

LESSELS, C. M. 1986. Brood size in Canada geese: A manipulation experiment. *J. Anim. Ecol.* 55:669–689.

———. 1991. The evolution of life histories. In J. R. Krebs and N. B. Davies (eds.), *Behavioural Ecology,* 3rd ed. Blackwell, London, pp. 32–68.

LETT, P. F., R. K. MOHN, AND D. F. GRAY. 1981. Density-dependent processes and management strategy for the northwest Atlantic harp seal populations. In C. W. Fowler and T. D. Smith (eds.), *Dynamics of Large Mammal Populations.* Wiley, New York, pp. 135–158.

LEVENSON, J. B. 1981. Woodlots as biogeographic islands in southeastern Wisconsin. In R. L. Burgess and D. M. Sharpe (eds.), *Forest Island Dynamics in Man-Dominated Landscapes.* Springer-Verlag, New York, pp. 13–39.

LEVERLICH, W. J., AND D. A. LEVIN. 1979. Age specific survivorship and reproduction in *Phlox drummondii. Am. Nat.* 113:881–903.

LEVIN, D. A. 1976. The chemical defenses of plants to pathogens and herbivores. *Ann. Rev. Ecol. Syst.* 7:121–159.

LEWONTIN, R. C. 1970. The units of selection. *Ann. Rev. Ecol. Syst.* 1:1–18.

LIDICKER, W. Z. 1975. The role of dispersal in the demography of small mammals. In F. B. Golley, E. Petrusewicz, and L. Ryszkowski (eds.), *Small Mammals, Their Productivity and Population Dynamics.* Cambridge University Press, Cambridge, England, pp. 103–128.

———. 1985. Dispersal. In R. H. Tamarin (ed.), *Biology of New World Microtus,* Sp. Pub. 8, American Society of Mammalogists.

LIGON, F. K., W. E. DIETRICH, AND W. J. TRUSH. 1995. Downstream ecological effects of dams. *Bioscience* 45:183–192.

LIKENS, G. E., AND F. H. BORMANN. 1974. Linkages between terrestrial and aquatic ecosystems. *Bioscience* 24(8):447–456.

———. 1975. Nutrient-hydrologic interactions (eastern United States). In A. D. Hasler (ed.), *Coupling of Land and Water Systems.* Springer-Verlag, New York, pp. 1–5.

LIKENS, G. E., F. H. BORMANN, M. N. JOHNSON, AND R. S. PIERCE. 1967. The calcium, magnesium, potassium, and sodium budgets for a small forested ecosystem. *Ecology* 38:46–49.

LILLYWHITE, H. B. 1970. Behavioral temperature regulation in the bullfrog, *Rana catesbeiana. Copeia* 1970:158–168.

LIMA, S. L., AND L. M. DILL. 1990. Behavioural decisions make under the risk of predation: A review and prospectus. *Can. J. Zool.* 68:619–640.

LINDBERG, S. E., G. M. LOVETT, D. D. RICHTER, and D. W. JOHNSON. 1986. Atmospheric deposition and canopy interactions of major ions in a forest. *Science* 231:141–145.

LINDEMAN, R. 1942. Trophic-dynamic aspects of ecology. *Ecology* 23:399–418.

LINDQUIST, B. 1942. Experimentelle Untersuchingen uber die Bedeutung einiger Landmollusken fur die zersetgung der Waldstreu. *Kgl. Fysiograf. Sallskap. Lund. Forh.* 11:144–156.

LINDSAY, J. H. 1985. A food web approach to mutualism in lake communities. In D. H. Boucher (ed.), *The Biology of Mutualism.* Oxford University, Oxford, England, pp. 344–374.

LINDSTROM, E. R., H. ANDREN, P. ANGELSTAM, AND G. CEDERLUND. 1994. Disease reveals the predator: Sarcoptic mange, red fox predation, and prey population. *Ecology* 75:1042–1049.

LITTLE, C., AND J. A. KITCHING. 1996. *The Biology of Rocky Shores.* Oxford University Press, Oxford, England.

LITVIATIS, J. A., J. A. SHERBURNE, AND J. A. BISSONETTE. 1985. Influence of understory characteristics on a snowshoe hare habitat use and density. *J. Wildl. Manage.* 49:866–873.

LLOYD, D. G. 1987. Selection of offspring size at independence and other size vs. number strategies. *Am. Nat.* 129:800–817.

LLOYD, J. E. 1951. Mimicry in the sexual signal of fireflies. *Sci. Amer.* 245:110–117.

LODGE, D. J., F. N. SCATENA, C. E. ASBURY, AND M. J. SANCHEZ. 1991. Fine litterfall and related nutrient inputs resulting from Hurricane Hugo in subtropical wet and lower montane rain forests of Puerto Rico. *Biotropica* 23:36–342.

LOERY, G., AND J. D. NICHOLS. 1985. Dynamics of a black-capped chickadee population, 1958–1983. *Ecology* 66:1195–1203.

LORD, R. D. 1960. Litter size and latitude in North American mammals. *Am. Midl. Nat.* 64:488–499.

LORIMER, C. G. 1989. Relative effects of small and large disturbances on temperate hardwood forest structure. *Ecology* 70:565–567.

LOSSAINT, P. 1973. Soil-vegetation relationships in Mediterranean ecosystems of southern France. In F. di Castri and H. A. Mooney (eds.), *Mediterranean Type Ecosystems: Origin and Structure.* Springer-Verlag, New York, pp. 199–210.

LOTKA, A. J. 1925. *Elements of Physical Biology.* Williams and Wilkens, Baltimore.

LOUMA, J. R. 1997. Whittling Dixie. Audubon 99:38–45, 97–100.

LOVEJOY, T. E., R. O. BIERREGAARD, JR., H. B. RYLANDS, J. R. MALCOLM, C. E. QUINTELA, L. H. HARPER, K. S. BROWN, JR., A. H. POWELL, G. V. N. POWELL, H. O. R. SCHUBERT, AND M. B. HAYS. 1986. Edge and other effects of isolation on Amazonian forest fragments. In M. Soule (ed.), *Conservation Biology.* Sinauer Associates, Sunderland, MA, pp. 257–285.

LOVETT, G. M., W. A. REINERS, AND R. K. OLSON. 1982. Cloud droplet deposition in subalpine balsam fir forests: Hydrological and chemical input. *Science* 218:1303–1304.

LOVETT DOUST, J., AND P. B. CAVERS. 1982. Sex and gender dynamics in jack-in-the-pulpit *Arisaema triphyllum* (Araceae). *Ecology* 63:797–808.

LOWE, V. P. Q. 1969. Population dynamics of red deer (*Cervus elaphus L.*) on Rhum. *J. Anim. Ecol.* 38:425–457.

LUBCHENCO, J. 1978. Plant species diversity in a marine intertidal community: Importance of herbivore food preferences and algal competitive abilities. *Am. Nat.* 112:23–29.

———. 1980. Algal zonation in the New England rocky intertidal community: An experimental analysis. *Ecology* 61:333–344.

———. 1983. *Littorina* and *Fucus:* Effects of herbivores, substratum heterogeneity, and plant escapes during succession. *Ecology* 64:1116–1123.

———. 1986. Relative importance of competition and predation: Early colonization by seaweeds in New England. In J. Diamond and T. Case (eds.), *Community Ecology.* Harper and Row, New York, pp. 537–555.

LUBCHENCO, J. A., M. OLSON, L. B. BRUBAKER, et al. 1992. The sustainable biosphere initiative: An ecological research agenda. *Ecology* 72:371–412.

LUGO, A. E. 1980. Mangrove ecosystems: Successional or steady state? *Biotropica* 12:65–72.

LUGO, A. E. (ed.). 1990. *The Forested Wetland.* Elsevier, Amsterdam.

LUGO, A. E., AND S. BROWN. 1986. Steady state terrestrial ecosystem and the global carbon cycle. *Vegetatio* 68:83–90.

———. 1991. Comparing tropical and temperate forests. In J. Cole, G. Lovett, and S. Findlay (eds.), *Comparative Analysis of Ecosystems.* Springer Verlag, New York, pp. 319–330.

LUGO, A. E., AND S. C. SNEDAKER. 1974. The ecology of mangroves. *Ann. Rev. Ecol. Syst.* 5:39–64.

LUGO, A. E., M. BRINSON, AND S. BROWN (eds.). 1990. Forested Wetlands. Ecosystems of the World 15. Amsterdam: Elsevi 1995. Tropical Forests: Management and Ecology, New York.

LULL, H. W., AND W. E. SOPPER. 1969. Hydrologic effects from urbanization of forested watersheds in the northeast. USDA Forest Service Res. Paper, NE-146.

LUNDBERG, P. 1988. Functional response of a small mammalian herbivore: The disk equation revisited. *J. Anim. Ecol.* 57:999–1006.

LUSSENHOP, J. 1992. Mechanism of microarthropod-microbial interactions in soil. *Adv. Ecol. Res.* 23:1–33.

LUTZ, H. J. 1956. Ecological effects of forest fires in the interior of Alaska. USDA Tech. Bull. No. 1133.

LUTZ, H., AND R. F. CHANDLER. 1954. Forest Soils. Wiley, New York.

LUXTON, M. 1982. Quantitative utilization of energy by soil fauna. *Oikos* 39:342–354.

LYMAN, C. P., J. S. WILLIS, A. MALAN, AND L. C. H. WANG. 1982. *Hibernation and Torpor in Mammals and Birds.* Academic Press, New York.

LYNCH, J. F., AND R. F. WHITCOMB. 1977. Effects of insularization of the eastern deciduous forest and avifaunal diversity and turnover. In *Classification, Inventory, and Analysis of Fish and Wildlife Habitat.* FWS/OBS-78/76, U.S. Fish and Wildlife Service, Washington, DC, pp. 461–489.

MABBERLEY, D. J. 1983. *Tropical Rain Forest Ecology.* Blackie, London.

MACARTHUR, R. H. 1958. Population ecology of some warblers of northeastern coniferous forests. *Ecology* 39:599–619.

———. 1960. On the relative abundance of species. *Am. Nat.* 94:25–36.

———. 1972. *Geographical Ecology.* Harper and Row, New York.

MACARTHUR, R. H., AND R. LEVINS. 1967. The limiting similarity, convergence, and divergence of coexisting species. *Am. Nat.* 101:377–385.

MACARTHUR, R. H., AND J. W. MACARTHUR. 1961. On bird species diversity. *Ecology* 42:594–598.

MACARTHUR, R. H., AND E. R. PIANKA. 1966. On optimal use of a patchy environment. *Am. Nat.* 100:603–609.

MACARTHUR, R. H., AND E. O. WILSON. 1963. An equilibrium theory of insular zoogeography. *Evolution* 17:373–387.

———. 1967. *The Theory of Island Biogeography.* Princeton University Press, Princeton, NJ.

MACKENZIE, J. J., AND M. T. EL-ASBRY (eds.). 1989. *Air Pollution's Toll on Forests and Crops.* Yale University Press, New Haven, CT.

MACLEAN, S. F., JR. 1980. The detritus-based trophic system. In J. Brown, P. C. Miller, L. L. Tieszen, and F. L. Bunnell (eds.), *An Arctic Ecosystem: The Coastal Tundra at Barrow, Alaska.* Dowden, Hutchinson and Ross, Stroudsburg, PA, pp. 411–457.

MACLINTOCK, L., R. F. WHITCOMB, AND B. L. WHITCOMB. 1977. Island biogeography and the "habitat islands" of eastern forest. II. Evidence for the value of corridors and minimization of isolation in preservation of biotic diversity. *Am. Birds* 31:6–12.

MACLULICH, D. A. 1937. Fluctuations in the numbers of varying hare (Lepus americanus). *Univ. Toronto Biol.* Ser. No. 43.

MACMAHON, J. A., AND F. H. WAGNER. 1985. The Mojave, Sonoran, and Chihuahuan deserts of North America. In M. Evenardi, I. Noy-Meir, and D. W. Goodall (eds.), *Hot Deserts and Arid Shrublands* (Ecosystems of the World, Vol. 12A). Elsevier, Amsterdam, pp. 105–202.

MADGWICK, H. A. I., AND J. D. OVINGTON. 1959. The chemical composition of precipitation in adjacent forest and open plots. *Forestry* 32:14–22.

MAGUIRE, D. A., AND R. T. T. FORMAN. 1983. Herb cover effects on tree seedling patterns in a mature hemlock-hardwood forest. *Ecology* 64:1367–1380.

MAHALL, B. E., AND R. B. PORK. 1976. The ecotone between Spartina foliosa. Trin. and Salicornia virginica L. in salt marshes of northern San Francisco Bay. 1. Biomass and production. *J. Ecol.* 64:421–433.

MAIN, A. R. 1981. Fire tolerance of heathland animals. In R. Specht (ed.), *Heathlands and Related Shrublands* (Ecosystems of the World, Vol. 9B). Elsevier, Amsterdam, pp. 85–90.

MAISUROW, D. K. 1941. The role of fire in the perpetuation of virgin forests of northern Wisconsin. *J. Fores.* 39:201–207.

MALCOLM, S. B. 1990. Mimicry: Status of a classical evolutionary paradigm. *TREE* 5:57–62.

MALTHUS, T. R. 1798. *An Essay on Principles of Population.* Johnson, London (numerous reprints).

MANABE, S., AND R. J. STOUFFER. 1993. Century-scale effects of increased atmospheric carbon dioxide on ocean-atmosphere system. *Nature* 364:215–218.

———. 1994. Multiple century response of a coupled ocean-atmosphere model to an increase of atmospheric carbon dioxide. *J. Clim.* 7:5–23.

MANABE, S., R. J. STOUFFER, M. J. SPELMAN, AND K. BRYAN. 1991. Transient response of a coupled ocean-atmosphere model to a gradual change in atmospheric carbon dioxide. Part I: Annual mean response. *J. Clim.* 4:785–818.

MANUAT, J. 1983. The vegetation of African savannas. In F. Bourliere (ed.), *Tropical Savannas* (Ecosystems of the World, Vol. 13). Elsevier, Amsterdam, pp. 109–149.

MARKS, P. L. 1974. The role of pin cherry (*Prunus pensylvanica L.*) in the maintenance of stability in northern hardwood ecosystems. *Ecol. Monogr.* 44:73–88.

MARLAND, G., AND T. BODEN. 1993. The magnitude and distribution of fossil-fuel related carbon releases. In M. Heimann (ed.), *The Global Carbon Cycle.* Springer-Verlag, New York, pp. 117–138

MARQUIS, D. A. 1974. The impact of deer browsing on Allegheny hardwood regeneration. USDA. Forest Service Res. Paper NE-308.

———. 1981. Effect of deer browsing on timber production in Allegheny hardwood forests of northwestern Pennsylvania. USDA Forest Service Res. Paper NE-475.

MARQUIS, D. A., AND T. J. GRISEZ. 1978. The effect of deer exclosures on the recovery of vegetation in failed clearcuts on the Allegheny plateau. USDA Forest Service Res. Note NE-270.

MARTIN, M. M. 1970. The biochemical basis of the fungus-attine ant symbiosis. *Science* 169:16–20.

MARX, D. H. 1971. Ectomycorrhizae as biological deterrents to pathogenic root infections. In E. Hacskaylo (ed.), *Mycorrhizae.* USDA Misc. Pub. 1189, pp. 81–96.

MASER, C., R. G. ANDERSON, K. CROMAC, J. T. WILLIAMS, AND R. E. MARTIN. 1979. Dead and down woody material. In J. W. Thomas (ed.), *Wildlife Habitats in Managed Forests: The Blue Mountains of Washington and Oregon.* Agr. Handbook 533, USDA, Washington, DC, pp. 78–95.

MASER, C., AND J. M. TRAPPE (eds.). 1984. *The Seen and the Unseen World of the Fallen Tree.* USDA Forest Service Gen. Tech. Rept. PNW-164.

MASER, C., J. M. TRAPPE, AND R. A. NUSSBAUM. 1978. *Fungal-small mammal interrelationship with emphasis on Oregon coniferous forests.* *Ecology* 59:799–809.

MASSEY, A., AND J. D. VANDENBERG. 1980. Puberty delay by a urinary cue from female house mice in feral populations. *Science* 209:821–822.

MATLACK, G. 1994. Plant demography, land use history, and the commercial use of forests. *Cons. Biol.* 8:298–299.

MATTES, H. 1994. Coevolutional aspects of stone pines and nutcracker. In W. C. Schmidt and F.-K. Holtmeier (comps.), *Proceedings—International workshop on subalpine stone pines and their environment: The status of our knowledge; 1992 September 5–11; St. Moritz, Switzerland.* Gen Tech. Rep. INT-grt-309. Ogden Utah USDA Forest Service, Intermountain Research Station, Ogden, UT, pp. 31–35.

MATTHIESSEN, P. 1985. Contamination of wildlife with DDT insecticides in relation to tsetse fly control operations. *Env. Pollution* (B) 10:189–211.

MATTSON, W. J., JR. 1980. Herbivory in relation to plant nitrogen content. *Ann. Rev. Ecol. Syst.* 11:119–161.

MATTSON, W. J., AND R. H. HOACH. 1987a. The role of drought in outbreaks of plant-eating insects. *Bioscience* 37:110–118.

———. 1987b. The role of drought stress in provoking outbreaks of phytophagous insects. In P. Barbosa and J. Schultz (eds.), *Insect Outbreaks: Ecological and Evolutionary Perspectives.* Academic Press, Orlando, FL.

MAY, R. M. 1973. *Stability and Complexity in Model Ecosystems.* Princeton University Press, Princeton, NJ.

———. 1976. Models for single populations. In R. M. May (ed.), *Theoretical Ecology: Principles and Applications.* Saunders, Philadelphia, pp. 4–29.

———. 1981. Models for two interacting populations. In R. M. May (ed.), *Theoretical Ecology,* 2nd ed. Sinauer Associates, Sunderland, MA, pp. 78–104.

———. 1983. Parasitic infections as regulators of animal populations. *Am. Sci.* 71:36–45.

MAY, R. M., AND R. M. ANDERSON. 1978. Regulation and stability of host-parasite population interactions. II. Destabilizing processes. *J. Anim. Ecol.* 47:249–267.

———. 1979. Population biology of infectious diseases: Part II. *Nature* 280:455–461.

———. 1983. Parasite-host coevolution. In D. J. Futuyma and M. Slatkin (ed.), *Coevolution.* Sinauer Associates, Sunderland, MA, pp. 186–206.

MAYBECK, M. 1993. C, N, P and S in rivers: From sources to global inputs. In R. Wollast, F. T. MacKenzie, and L. Chou (eds.), *Physical and Chemical Weathering in Geochemical Cycles.* Kluwer Academic Publishers, Dordrecht, The Netherlands, pp. 163–193

MAYNARD SMITH, J. 1956. Fertility, mating behavior, and sexual selection in *Drosophila subobscura. J. Genet.* 54:261–279.

———. 1971. The origin and maintenance of sex. In G. C. Williams (ed.), *Group Selection.* Aldine, Chicago, pp. 163–175.

———. 1976. A comment on the Red Queen. *Am. Nat.* 110:325–330.

———. 1991. Theories of sexual selection. *TREE* 6:146–151.

MAYR, E. 1963. *Animal Species and Evolution.* Harvard University Press, Cambridge, MA.

MCARDLE, R. E., W. H. MEYER, AND D. BRUCE. 1949. The yield of Douglas-fir in the Pacific Northwest. USDA Tech. Bull. No. 201 (rev.).

MCARTHUR, R. H., AND E. O. WILSON. 1967. *The Theory of Island Biogeography.* Princeton University Press, Princeton, NJ.

MCBEE, R. H. 1971. Significance of intestinal microflora in herbivory. *Ann. Rev. Ecol. Syst.* 2:165–176.

MCCLAUGHERTY, C., J. PASTOR, J. ABER, AND J. MELILLO. 1985. Forest litter decomposition in relation to soil nitrogen dynamics and litter quality. *Ecology* 66:266–275.

MCDOWELL, D. M., AND R. J. MAIMAN. 1986. Structure and function of a benthic invertebrate stream community as influenced by beaver (Castor canadensis). *Oecologica* 68:481–489.

MCELROY, M. B., AND R. J. SALAWITH. 1989. Changing composition of the global stratosphere. *Science* 243:763–770.

MCGEE, C. E. 1984. Heavy mortality and succession in a virgin mixed mesophytic forest. USDA Southern For. Exp. Stat. Res. Paper SO-209.

MCGINNES, W. G. 1972. North America. In C. M. McKella, J. P. Blaisdell, and J. R. Goodwin (eds.), *Wildland Shrubs: Their Biology and Utilization.* USDA Forest Service Gen. Tech. Rept. INT-1, pp. 55–66.

MCGRAW, J. B. 1985a. Experimental ecology of Dryas actopetela ecotypes: Relative response to competition. *New Phytol.* 100:23–241.

———. 1985b. Experimental ecology of *Dryas actopetela* ecotypes. III. Environmental factors and plant growth. *Arct. Alp. Res.* 17:229–239.

———. 1989. Effects of age and size on life histories and population growth of *Rhododendron* maximum shoots. *Am. J. Bot.* 76:113–123.

MCGRAW, J. B., AND J. ANTONOVICS. 1983. Experimental ecology of Dryas actopetela ecotypes. I. Ecotypic differentiation and life cycles stages of selection. *J. Ecol.* 71:879–897.

MCGRAW, J. B., AND F. S. CHAPIN. 1989. Competitive ability and adaptation to fertile and infertile soils in two *Eriophorum* species. *Ecology* 70:736–749.

MCINTOSH, R. P. 1976. Ecology since 1900. In B. J. White and T. J. White (eds.), *Issues and Ideas in America.* University of Oklahoma Press, Norman.

———. 1980. The background of some current problems of theoretical ecology. *Synthese* 43:195–255.

———. 1985. *The Background of Ecology: Concept and Theory.* Cambridge University Press, Cambridge, England.

———. 1987. Pluralism in ecology. *Ann. Rev. Ecol. Syst.* 18:2321–341.

MCKELL, C. M., J. P. BLAISDELL, AND J. R. GOODWIN (eds). 1972. *Wildland Shrubs: Their Biology and Utilization.* USDA Forest Service Gen. Tech. Rept. INT-1.

MCLELLAN, C. H., A. D. DOBSON, D. S. WILCOVE, AND J. F. LYNCH. 1986. Effects of forest fragmentation on new- and old-world bird communities: Empiricial observations and theoretical implications. In J. Verner, M. L.. Morrison, and C. J. Ralph (eds.), *Wildlife 2000: Modeling Habitat Relationships of Terrestrial Vertebrates.* University of Wisconsin Press, Madison, pp. 305–313.

McNab, B. K. 1963. Bioenergetics and the determination of home range size. *Am. Nat.* 97:133–140.

McNaughton, S. J. 1975. r and K selection in *Typhya. Am. Nat.* 109:215–261.

————. 1979. Grazing as an optimization process: Grass-ungulate relationships in the Serengeti. *Am. Nat.* 113:691–703.

————. 1983. Serengeti grassland ecology: The role of composite environmental factors and contingency in community organization. *Ecol. Monogr.* 53:291–320.

————. 1984. Grazing lawns: Animals in herds, plant forms, and coevolution. *Am. Nat.* 124:863–886.

————. 1985. Ecology of a grazing ecosystem: The Serengeti. *Ecol. Monogr.* 55:259–294.

————. 1993. Grasses and grazers, science and management. *Ecol. App.* 3:17–20.

McNaughton, S. J., and N. J. Georgiadis. 1988. Ecology of African grazing and browsing mammals. *Am. Rev. Ecol. Syst.* 17:39–65.

McNaughton, S. J., M. Osterheld, D. A. Frank, and K. J. Williams. 1989. Ecosystem-level patterns of primary productivity and herbivory in terrestrial habitats. *Nature* 341:142–144.

McNaughton, S. J., R. W. Ruess, and S. W. Seagle. 1988. Large mammals and process dynamics in African ecosystems. *Bioscience* 38:794–800.

McNaughton, S. J., and L. L. Wolf. 1979. *General Ecology.* Holt, Rinehart and Winston, New York.

McPherson, J. K., and C. H. Muller. 1969. Allelopathic effects of *Adenostoma fasciculatum* "chamise" in the California chaparral. *Ecol. Monogr.* 39:177–179.

Mech, L. D. 1970. *The Wolf: The Ecology and Behavior of an Endangered Species.* Doubleday, Garden City, NY.

Mech, L. D., R. E. McRoberts, R. O. Peterson, and R. E. Page. 1987. Relationships of deer and moose populations to previous winter's snow. *J. Anim. Ecol.* 56:615–627.

Meentenmeyer, V. 1978. Macroclimate and lignin control on litter decomposition rates. *Ecology* 59:465–472.

Meetintemeyer, V., E. O. Box, and R. Thompson. 1982. World patterns and amounts of terrestrial plant litter production. *Bioscience* 32:108–113.

Meffe, G. K., and C. R. Carroll. 1997. Genetics: Conservation of diversity within species. In G. K. Meffe, C. R. Carroll and contributors, *Principles of Conservation Biology,* 2nd ed. Sinauer Associates, Sunderland, MA, pp. 161–201.

Melillo, J. M., J. D. Aber, and J. F. Muratore. 1982. Nitrogen and lignin control on hardwood leaf decomposition dynamics. *Ecology* 63:621–626.

Melillo, J. M., D. W. Kicklighter, A. D. McGuire, B. Moore, C. J. Vorosmarty, and A. L. Schloss. 1995. The effects of CO_2 fertilization on the storage of carbon in terrestrial ecosystems: A global modeling study. *Glob. Biogeochem. Cycl.*

Melillo, J. M., A. D. McGuire, D. W. Kicklighter, B. Moore, C. J. Vorosmarty, and A. L. Schloss. 1993. Global climate change and terrestrial net primary production. *Nature* 363:234–240.

Melillo, J. M., K. M. Newkirk, C. E. Catricala, P. A. Steudler, J. B. Aber, K. J. Nadelhoffer, and R. D. Boone. 1996. The soil warming experiment at Harvard Forest 1991–1996. *Bull. Ecol. Soc. Amer.* 77:300.

Menzie, C. M. 1969. Metabolism of pesticides. U.S. Fish and Wildlife Service Sp. Sci. Rept. Wildl. No. 127.

Merendino, M. T., C. O. Ankney, and D. G. Dennis. 1993. Increasing mallards, decreasing black ducks: More evidence for cause and effect. *J. Wildl. Manage.* 57:199–208.

Mettler, L. E., and T. G. Gregg. 1969. *Population Genetics and Evolution.* Prentice-Hall, Englewood Cliffs, NJ.

Meybeck, M. 1982. Crabon, nitrogen and phosphorus transport by world rivers. *Am. J. Sci.* 282:401–450.

Michener, W. K., and R. A. Haeuber. 1998. Flooding: Natural and managed disturbances. *Bioscience* 48:677–680.

Michod, R. E. 1982. The theory of kin selection. *Ann. Rev. Ecol. Syst.* 13:23–56.

Miko, U. F. 1996. Climate change impacts on forests. In R. T. Watson, M. C. Zinyowera, and R. H. Moss (eds.), *Climate Change 1995: Impacts, Adaptations and Mitigation of Climate Change.* Cambridge University Press, Cambridge, England.

Miller, A. H. 1942. Habitat selection among higher vertebrates and its relation to intraspecific variation. *Am. Nat.* 76:25–35.

Milliman et al 1989. Environmental and economic implications of rising sea-level and subsiding deltas: The Nile and Bangladesh examples. *Ambio* 18:340–345.

Mills, L. S., and P. E. Smouse. 1994. Demographic consequences of inbreeding on remnant populations. *Am. Nat.* 144:412–431.

Milne, A. 1957. Theories of natural control of insect populations. Cold *Spring Harbor Symp.* Quant. Biol. 22:253–271.

Minchella, D. J., and M. E. Scott. 1991. Parasitism: A cryptic determinant of animal community structure. *TREE* 6:250–254.

Minshall, G. W., K. W. Cummins, R. C. Petersen, C. E. Cushing, D. A. Bruns, J. R. Sedell, and R. L. Vannote. 1985. Developments in stream ecology. *Can. J. Fish. Aquat. Sci.* 42:1045–1055.

Minshall, G. W., R. C. Petersen, K. W. Cummins, T. L. Bott, J. R. Sedell, C. E. Cushing, and R. L. Vannote. 1983. Interbiome comparison of stream ecosystem dynamics. *Ecol. Monogr.* 53:1–25.

Mitchell, J. F. B., R. A. Davis, W. J. Ingram, and C. A. Senior. 1995. On surface temperature, greenhouse gases and aerosols: Models and observations. *J. Clim.* 10:2364–2386.

Mitchell, J. F. B., T. J. Johns, J. M. Gregory, and S. B. F. Tett. 1995. Climate response to increasing levels of greenhouse gases and sulfate aerosols. *Nature* 376:501–504.

Mobius, K. 1877. An oyster bank is a bioconose, or a social community. Transl. by H. J. Rice from Die Auster und die Austerwirtschaft. Wiegundt, Hemfel, and Parey, Berlin. In *Report of U.S. Commission of Fisheries 1889,* pp. 683–675.

Mock, D. W., and M. Fujioka. 1990. Monogamy and long-term pair bonding in vertebrates. *TREE* 5:39–43.

Moehlman, P. D. 1979. Jackal helpers and pup survival. *Nature* 277:382–383.

————. 1983. Socioecology of silverbacked and golden jackals, Canis mesomelas and C. aureus. In J. F. Eisenberg and D. G. Kleinman (eds.), *Recent Advances in the Study of Mammalian Behavior.* Sp. Pub. 7, American Society of Mammalogists, pp. 423–453.

————. 1986. Ecology of cooperation in canids. In D. I. Rubenstein and R. W. Wrangman (eds.), *Ecological Aspects of Social Evolution: Birds and Mammals.* Princeton University Press, Princeton, NJ, pp. 64–85.

Mohler, C. I., P. I. Marks, and D. G. Sprugel. 1978. Stand structure and allometry of trees during self-thinning of pure stands. *J. Ecol.* 66:599–614.

Møller, A. 1991a. Parasites, sexual ornamentation, and male choice in barn swallows. In J. E. Loye and M. Zuk (eds.), *Bird-Parasite Interactions.* Oxford University Press, Oxford, England, pp. 328–343.

————. 1991b. Parasite load reduces song output in a passerine bird. *Anim. Behav.* 41:723–730.

Monro, J. 1967. The exploitation and conservation of resources by populations of insects. *J. Anim. Ecol.* 36:531–547.

Monsi, M. 1968. Mathematical models of plant communities. In F. E. Eckardt (ed.), *Functioning of Terrestrial Ecosystems at the Primary Production Level.* UNESCO, Paris, pp. 131–149.

Mook, L. J. 1963. Birds and spruce budworm. In R. Morris (ed.), *Entomol. Soc. Can. Mem.* 31, pp. 244–248.

Mooney, H. A. (ed.). 1977. *Convergent Evolution in Chile and California Mediterranean Climate Ecosystems.* Academic Press, New York.

Mooney, H. A. 1981. Primary production in mediterranean-type shrubland. In F. di Castri, D. Goodall, and R. Specht (eds.), *Mediterranean-Type Shrublands* (Ecosystems of the World, Vol. 11). Elsevier, Amsterdam, pp. 249–256.

Mooney, H. A., and W. D. Billings. 1961. Comparative physiological ecology of arctic and alpine populations of *Oxyria digyna. Ecol. Monogr.* 31:1–29.

Mooney, H. A., O. Bjorkman, J. Ehleringer, and J. Berry. 1976. Photosynthetic capacity of in situ Death Valley plants. *Carnegie Inst. Yearbook* 75:410–413.

Mooney, H. A., and C. E. Conrad. 1977. *Proc. Symp. Environmental Consequences of Fire and Fuel Management in Mediterranean Ecosystems.* USDA Forest Service Gen. Tech. Rept. WO-3, USDA, Washington, DC.

MOONEY, H. A., J. EHLERINGER, AND O. BJORKMAN. 1977. The energy balance of leaves of the evergreen desert shrub *Atriplex hymenelytra*. *Oecologica* 29:301–310.

MOONEY, H. A., C. FIELD, W. E. WILLIAMS, J. A. BERRY, AND O. BJORKMAN. 1983. Photosynthetic characteristics of plants of a California cool coastal environment. *Oecologica Journal*, 57:38–42.

MOONEY, H. A., AND S. L. GULMON. 1979. Constraints on the leaf structure and function in reference to herbivory. *Bioscience* 32:198–206.

MOONEY, H. A., AND P. C. MILLER. 1985. Chaparral. In B. F. Chabot and H. A. Mooney (eds.), *Physiological Ecology of North American Plant Communities*. Chapman and Hall, New York, pp. 213–231.

MOONEY, H. A., AND D. J. PARSONS. 1973. Structure and function of the California chaparral: An example from San Dimas. In F. di Castri and H. A. Mooney (eds.), *Mediterranean-Type Ecosystems: Origin and Structure*. Springer-Verlag, New York, pp. 83–112.

MOORE, H. B. 1958. Marine *Ecology*. Wiley, New York.

MOORE, J. J., P. DOUDING, AND B. HEALY. 1975. Glenamoy, Ireland. In T. Rosswall and O. W. Heal (eds.), *Structure and Function of Tundra Ecosystems*. Swedish Natural Science Research Council, Stockholm, pp. 321–343.

MOORE, N. W., AND M. D. HOOPER. 1975. On the number of bird species in British woods. *Biol. Cons.* 8:239–250.

MOORE, P. D., AND D. J. BELLAMY. 1974. Peatlands. Springer-Verlag, New York.

MOORE, W. S., AND R. A. DOLBEER. 1989. The use of banding recovery data to estimate dispersal rates and gene flow in avian species: Case studies in the red-winged blackbird and common grackle. *Condor* 91:242–253.

MORRIS, R. F., W. F. CHESHIRE, C. A. MILLER, AND D. G. MOTT. 1958. The numerical response of avian and mammalian predators during a gradation of the spruce budworm. *Ecology* 39:487–494.

MOSS, R., A. WATSON, AND R. PARR. 1996. Experimental prevention of a population cycle in red grouse. *Ecology* 77:1512–1530.

MOSSE, B., D. P. STRIBLEY, AND F. LETACON. 1981. Ecology of mycorrhizae and mycorrhizal fungi. *Adv. Microb. Ecol.* 5:137–210.

MOULDER, B. C., AND D. E. REICHLE. 1974. Significance of spider predation in the energy dynamics of forest floor arthropod communities. *Ecol. Monogr.* 42:473–498.

MOULDER, B. C., D. E. REICHLE, AND S. I. AUERBACH. 1970. Significance of spider predation in the energy dynamics of forest floor arthropod communities. Oak Ridge National Laboratory Report ORNL 4452.

MOULTON, M. P., AND S. L. PIMM. 1986. The extent of competition in shaping an introduced avifauna. In J. Diamond and T. J. Case (eds.), *Community Ecology*. Harper and Row, New York, pp. 80–97.

MUDRICK, D., M. HOOSEIN, R. HICKS, AND E. TOWNSEND. 1994. Decomposition of leaf litter in an Appalachian forest: Effects of leaf species, aspect, slope position and time. *For. Ecol. Manag.* 68:231–250.

MUELLER-DOMBOIS, D., AND H. ELLENBERG. 1974. *Aims and Methods of Vegetation Ecology*. Wiley, New York.

MULLER, C. H., R. B. HANAWALT, AND J. K. MCPHERSON. 1968. Allelopathic control of herb growth in the fire cycle of California chaparral. *Bull. Torrey Bot. Club* 95:225–231.

MULROY, T. W., AND P. W. RUNDEL. 1977. Annual plants: Adaptations to desert environments. *Bioscience* 27:109–114.

MURDOCH, W. W. 1969. Switching in general predators: Experiments on predator specificity and stability of prey populations. *Ecol. Monogr.* 39:335–354.

MURDOCH, W. W., AND A. OATEN. 1975. Predation and population stability. *Adv. Ecol. Res.* 9:1–131.

MURPHY, G. I. 1966. Population biology on the Pacific sardine. *Proc. Calif. Acad. Sci.* 4th Ser. 34:1–84.

———. 1967. Vital statistics of the Pacific sardine and the population consequences. *Ecology* 48:731–736.

MURPHY, P. G., AND A. E. LUGO. 1986. Ecology of tropical dry forests. *Ann. Rev. Ecol. Syst.* 17:67–85.

MURRAY, B. G., JR. 1967. Dispersal in vertebrates. *Ecology* 48:975–978.

MUSCATINE, L., AND J. W. PORTER. 1977. Reef corals: Mutualistic symbioses adapted to nutrient-poor environments. *Bioscience* 27:454–460.

MYERS, J. H. 1988. Can a general hypothesis explain population cycles of forest Lepidoptera? *Adv. Ecol. Res.* 18:179–284.

———. 1990. Population cycles of western tent caterpillars: Experimental introduction and synchrony of fluctuations. *Ecology* 71:986–995.

———. 1993. Population outbreaks in forest *Lepidoptera*. *Am. Sci.* 81:240–251.

MYERS, K., C. S. HALE, R. MYKYTOWYCZ, AND R. L. HUGHS. 1971. The effects of varying density and space on sociality and health in animals. In A. H. Esser (ed.), *Behavior and Environment: The Use of Space by Animals and Men*. Plenum, New York, pp. 148–187.

MYERS, R. A., J. A. HUTCHINGS, AND N. J. BARROWMEIN. 1997. Why do fish stocks collapse? An example of cod in Atlantic Canada. *Ecol. App.* 7:91–106.

NAIMAN, R. J., J. M. MELILLO, AND J. E. HOBBIE. 1986. Ecosystem alteration of boreal forest streams by beaver (*Castor canadensis*). *Ecology* 67:1254–1269.

NATIONAL RESEARCH COUNCIL. 1983. *Acid Deposition: Atmospheric Processes in Eastern North America*. National Academy Press, Washington, DC.

NATIONAL RESEARCH COUNCIL OF CANADA. 1981. *Acidification in the Canadian Aquatic Environment: Scientific Criteria for Assessing the Effects of Acid Deposition on Aquatic Ecosystems*. NRCC No. 18475. National Research Council of Canada, Ottawa.

NEGUS, N. C., P. J. BERGER, AND L. G. FORSLUND. 1977. Reproductive strategy of Microtus montanus. *J. Mamm.* 58:347–353.

NELSON, R. A. 1980. Protein and fat metabolism in hibernating bears. *Fed. Proc.* 39:2955–2958.

NELSON, R. A., AND T. D. I. BECK. 1984. Hibernation adaptation in the black bear: Implications for management. *Proc. East. Workshop Black Bear Manage. Res.* 7:48–53.

NELSON, R. A., T. D. I. BECK, AND D. L. STENGER. 1983. Ratio of serum urea to serum creatinine in wild black bears. *Science* 226:841–842.

NELSON, R. A., G. E. FOLK, JR., E. W. PFEIFFER, J. J. CRAIGHEAD, C. J. JONKEL, AND D. L. STEIGER. 1983. Behavior, biochemistry, and hibernation in black, grizzly, and polar bears. *Int. Conf. Bear Res. Manage.* 5:284–290.

NEWBOLD, J. D., J. W. ELWOOD, R. V. O'NEILL, AND A. L. SHELDON. 1983. Phosphorus dynamics in a woodland stream: A study of nutrient spiraling. *Ecology* 65:1249–1265.

———. 1984. Phosphorus dynamics in a woodland stream ecosystem. *Bioscience* 34:43–44.

NEWBOLD, J. D., R. V. O'NEILL, J. W. ELWOOD, AND W. VAN WINKLE. 1982. Nutrient spiraling in streams: Implications for nutrient and invertebrate activity. *Am. Nat.* 20:628–652.

NEWMAN, E. I. 1988. Mycorrhizal links between plants: Their functioning and significance. *Adv. Ecol. Res.* 18:243–270.

NEWMAN, J. A., AND M. A. ELGAR. 1991. Sexual cannibalism in orb-weaving spiders: An economic model. *Am. Nat.* 138:1372–1395.

NEWSOME, A. E. 1990. The control of vertebrate pests by vertebrate predators. *TREE* 5:187–191.

NEWTON, I. 1998. *Population Limitation in Birds*. Academic Press, New York.

NICE, M. M. 1943. Studies in the life history of the song sparrow: 2. *Trans. Linn. Soc. New York* 6:1–329.

———. 1962. Development of behavior in precocial birds. *Trans. Linn. Soc. New York* 8.

NICHOLLS, R. J., AND S. P. LEATHERMAN. 1995. Global sea-level rise. In K. Strzepek and J. B. Smith (eds.), *As Climate Changes: International Impacts and Implications*. Cambridge University Press, Cambridge, England.

NICHOLSON, A. J. 1954. An outline of the dynamics of animal populations. *Aust. J. Zool.* 2:9–65.

———. 1957. The self-adjustment of populations to change. *Cold Spring Harbor Symp. Quant. Biol.* 22:153–173.

NICHOLSON, A. J., AND V. A. BAILEY. 1935. The balance of animal populations: Part 1. *Proc. Zool. Soc. London* 3:551–598.

NOBLE, I. R., AND R. O. SLATYER. 1980. The use of vital attributes to predict successional changes in plant communities subject to recurrent disturbances. *Vegetatio* 43:5–21.

NOBLE, P. S. 1978. Surface temperature of cacti: Influence of environmental and morphological factors. *Ecology* 59:986–996.

———. 1985. Desert succulents. In B. F. Chabot and H. A. Mooney (eds.), *Physiological Ecology of North American Plant Communities*. Chapman and Hall, New York, pp. 131–197.

NORBY, R. J., N. A. GUNDERSON, S. D. WULLSCHLEGER, E. G. O'NEILL, AND M. K. MCCRACKEN. 1992. Productivity and compensatory responses of yellow-poplar trees to elevated CO_2. *Nature* 357:322–324.

NOY-MEIR, I. 1973. Desert ecosystems: Environment and producers. *Ann. Rev. Ecol. Syst.* 4:25–51.

————. 1974. Desert ecosystems: Higher trophic levels. *Ann. Rev. Ecol. Syst.* 5:195–214.

————. 1975. Stability of grazing systems: An application of predator-prey graphs. *J. Ecol.* 63:459–481.

————. 1985. Desert ecosystem structure and function. In M. I. Evenardi, I. Noy-Meir, and D. W. Goodall (eds.), *Hot Deserts and Arid Shrublands* (Ecosystems of the World, Vol. 12A). Elsevier, Amsterdam, pp. 93–103.

NYBAKKEN, J. W. 1988. *Marine Biology: An Ecological Approach,* 2nd ed. Harper and Row, New York.

O'BRIEN, S. J., AND J. F. EVERMANN. 1988. Interactive influence of infectious disease and genetic diversity in natural populations. *TREE* 3:254–259.

O'BRIEN, S. J., D. E. WILDT, D. GOLDMAN, D. R. MERREL, AND M. BASH. 1983. The cheetah is depauperate in genetic variation. *Science* 221:459–462.

O'DOWD, D. J., AND M. E. HAY. 1980. Mutualism between harvester ants and a desert ephemeral: Seed escape from rodents. *Ecology* 61:531–540.

O'NEILL, E. G. 1994. Response of soil biota to elevated atmospheric carbon dioxide. *Plant Soil* 165:55–65.

O'NEILL, E. G., R. J. LUXMORE, AND R. J. NORBY. 1987. Elevated atmospheric CO_2 effects on seedling growth, nutrient uptake, and rhizosphere bacterial populations of *Liriodendron tulipifera L. Plant Soil* 104:3–11.

O'NEILL, R. V. 1976. Ecosystem persistence and heterotrophic regulation. *Ecology* 57:1244–1253.

O'NEILL, R. V., W. F. HARRIS, B. S. AUSMUS, AND D. E. REICHLE. 1975. A theoretical basis for ecosystem analysis with particular reference to element cycling. In F. G. Howell, J. B. Gentry, and M. H. Smith (eds.), *Mineral Cycling in Southeastern Ecosystems.* ERDA Symposium Series, National Technical Information Service, U.S. Department of Commerce, pp. 28–40.

ODUM, E. P. 1964. The new ecology. *Bioscience* 14:14–16.

————. 1971. *Fundamentals of Ecology,* 3rd ed. Saunders, Philadelphia.

————. 1983. *Basic Ecology.* Saunders, Philadelphia.

ODUM, E. P., AND L. J. BIEVER. 1984. Resource policy, mutualism, and energy partitioning in food chains. *Am. Nat.* 100:65–75.

ODUM, H. T. 1970. Summary: An emerging view of the ecological system at El Verde. In H. T. Odum and R. F. Pigeon (eds.), *A Tropical Rain Forest.* U.S. Atomic Energy Commission, Washington, DC. pp. I191–I218.

————. 1983. Systems Ecology: An Introduction. Wiley, New York.

ODUM, W. E., AND M. A. HEYWOOD. 1978. Decomposition of intertidal freshwater marsh plants. In R. E. Good, D. F. Whigham, and R. L. Simpson (eds.), *Freshwater Wetlands.* Academic Press, New York, pp. 89–97.

ODUM, W. E., T. J. SMITH III, J. K. HOOVER, AND C. C. McIVOR. 1984. *The Ecology of Tidal Freshwater Marshes of the United East Coast: A Community Profile.* U.S. Fish and Wildlife Service FWS/OBS-87/17, Washington, DC, 177 pp.

OECHEL, W. C., AND W. T. LAWRENCE. 1985. Taiga. In B. F. Chabot and H. A. Mooney (eds.), *Physiological Ecology of North American Plant Communities.* Chapman and Hall, New York, pp. 66–94.

OECHEL, W. C., AND G. L. VOURLITIS. 1996. Direct effects of elevated CO_2 on Arctic plant and ecosystem function. In G. Koch and H. A. Mooney (eds.), *Carbon Dioxide and Terrestrial Ecosystems.* Academic Press, San Diego, pp. 163–176.

OFFICER, C. B. 1983. Physics of estuarine circulation. In B. Ketchum (ed.), *Estuaries and Enclosed Seas* (Ecosystems of the World, Vol. 26). Elsevier, Amsterdam, pp. 15–42.

OKE, T. R., AND C. EAST. 1971. The urban boundary layer in Montreal. *Boundary-Layer Meteorol.* 1:411.

OLIVIERI, I., Y. MICALAKIS, AND P. H. GOUYON. 1995. Metapopulation genetics and the evolution of dispersal. *Am. Nat.* 146:202–228.

OLSON, J. S. 1958. Rates of succession and soil changes on southern Lake Michigan sand dunes. *Bot. Gazette* 119:125–170.

————. 1970. Carbon cycles and temperate woodlands. In D. E. Reichle (ed.), *Analysis of Temperate Forest Ecosystems.* Springer-Verlag, New York, pp. 226–241.

ORIANS, G. H. 1969. On the evolution of mating systems in birds and mammals. *Am. Nat.* 103:589–603.

OSAWA, A., AND R. B. ALLEN. 1993. Allometric theory explains self-thinning relationships of mountain beech and red pine. *Ecology* 74: 1010–1032.

OSBORN, F. 1949. *Our Plundered Planet.* Little, Brown, Boston.

OUSEY, A., AND M. WOLF. 1996. Inbreeding avoidance in animals. *TREE* 11:201–206.

OVERGAARD, C. 1949. Studies on the soil microfauna: II. The soil-inhabiting nematodes. *Nat. Jutland* 2:131–150.

OWEN, D. F. 1980. How plants may benefit from animals that eat them. *Oikos* 35:230–235.

OWEN, D. F., AND R. G. WIEGERT. 1981. Mutualism between grasses and grazers: An evolutionary hypothesis. *Oikos* 36:376–378.

OWENSBY, C. E., J. M. HAM, A. KNAPP, C. W. RICE, P. I. COYNE, AND L. M. AUEN. 1996. Ecosystem-level responses of tallgrass prairie to elevated CO_2. In G. Koch and H. A. Mooney (eds.), *Carbon Dioxide and Terrestrial Ecosystems.* Academic Press, San Diego, pp. 147–162.

PACKER, C. A., A. E. PUSEY, H. ROWLEY, D. A. GILBERT, J. MARTENSON, AND S. J. O'BRIAN. 1991. Case study of a population bottleneck: Lions of Ngorngoro Crater. *Cons. Biol.* 5:219–230.

PACKER, C. A., L. HERBST, A. E. PUSEY, J. P. BYGOTT, J. P. HANBY, S. J. CAIRNS, AND M. B. MULDER. 1988. Reproductive success in lions. In T. H. Clutton-Brock (ed.), *Reproductive Success: Studies of Individual Variation in Contrasting Breeding Systems.* University of Chicago Press, Chicago, pp. 363–383.

PACKER, C., AND L. RUTTAN. 1988. The evolution of cooperative hunting. *Am. Nat.* 132:159–198.

PAGE, K. N., AND T. G. WHITHAM. 1987. Overcompensation in response to mammalian herbivory: The advantage of being eaten. *Am. Nat.* 129:407–416.

PAINE, R. T. 1966. Food web complexity and species diversity. *Am. Nat.* 100:65–75.

————. 1969. The Pisaster-Tegula interaction: Prey patches, predator food preference and intertidal community structure. *Ecology* 50:950–961.

PALACA, S., AND J. ROUGHGARDEN. 1984. Control of arthropod abundance by Anolis lizards on St. Eustatius (Nethl. Antilles). *Oecologica* 64:160–162.

PALMER, H. E., W. C. HANSON, B. I. GRIFFIN, AND W. C. ROESCH. 1963. Cesium-137 in Alaskan Eskimos. *Science* 142(3588):64–65.

PALMER, J. D. 1976. *An Introduction to Biological Rhythms.* Academic Press, New York.

————. 1990. The rhythmic lives of crabs. *Bioscience* 40:352–358.

PALO, R. T., A. PEHRSON, AND P. KNUTSSON. 1983. Can birch phennolics be of importance in the defense against browsing vertebrates? *Finn. Game Res.* 41:75–80.

PARK, T. 1948. Expeimental studies of interspecies competition: 1. Competition between poulations of flour beetles., *Trilobium confusum* Duval and *Trilobium castaneum* Herbst. *Ecol. Monogr.* 18:265–308.

————. 1954. Experimental studies of interspecies competition: 2. Temperature, humidity and competition in two species of *Trilobium. Physiol. Zool.* 27:177–238.

PARKER, G. A., AND R. A. STUART. 1976. Animal behavior as a strategy optimizer: Evolution of resource assessment strata and optimal emigration thresholds. *Am. Nat.* 110:1055–1076.

PARRY, M. 1990. *Climate Change and World Agriculture.* Earthscan Publications, London.

————. 1992. The potential effect of climate changes in agriculture and land use. *Adv. Ecol. Res.* 22:63–91.

PARSONS, J. 1971. Cannibalism in herring gulls. *Brit. Birds* 64:528–537.

PASTEUR, G. 1982. A classificatory review of mimicry systems. *Ann. Rev. Ecol. Syst.* 13:169–199.

PASTOR, J., J. D. ABER, C. A. McCLAUGHERTY, AND J. M. MELILLO. 1984. Aboveground production and N and P cycling along a nitrogen mineralization gradient on Blackhawk Island, Wisconsin. *Ecology* 65:256–268.

PASTOR, J., R. J. NAIMAN, B. DEWEY, AND P. McINNES. 1988. Moose, microbes, and boreal forest. *Bioscience* 88:770–777.

PATTERSON, D. T. 1975. Nutrient return in stemflow and throughfall of individual trees in the Piedmont deciduous forest. In F. G. Howell et al. (eds.), *Mineral Cycling in Southeastern Ecosystems.* National Technical Information Service, U.S. Dept. Commerce, pp. 800–812.

PATTON, D. R. 1975. A diversity index for quantifying habitat "edge." *Wildl. Soc. Bull.* 3:171–173.

PAYNE, R. B. 1968. Among wild whales. N.Y. Zool. *Soc. Newsletter,* November 1968.

————. 1977. The ecology of brood parasitism in birds. *Ann. Rev. Ecol. Syst.* pp. 1–28.

PEARCY, R. W. 1976. Temperature effects on growth and CO_2 exchange rates of *Atriplex leniformis. Oecologica* 26:245–255.

————. 1977. Acclimation of photosynthetic and respiratory CO_2 to growth temperature in *Atriplex lentiformis* (Torr.). *Wats. Plant Physiol.* 61:484–486.

PEARL, R. 1927. The growth of populations. *Quart. Rev. Biol.* 2:532–548.

PEARL, R., and L. J. Reed. 1920. On the rate of growth of the population of the United States since 1790 and its mathematical representation. *Proc. Nat. Acad. Sci.* 6:275–288.

PEARSON, D. L. 1971. Vertical stratification of birds in a tropical dry forest. *Condor* 73:46–55.

PEASE, J. L., R. H. VOWLES, AND L. B. KEITH. 1979. Interaction of snowshoe hares and woody vegetation. *J. Wildl. Manage.* 43:43–60.

PECKARSKY, B. L. 1982. Aquatic insect predator-prey relations. *Bioscience* 32:261–266.

PEET, R. K. 1974. The measurement of species diversity. *Ann. Rev. Ecol. Syst.* 5:285–307.

———. 1981. Changes in biomass and production during secondary forest succession. In D. C. West, H. H. Shugart, and D. B. Botkin (eds.), *Forest Succession: Concepts and Applications.* Springer-Verlag, New York, pp. 324–338.

PEET, R. K., AND N. L. CHRISTENSEN. 1980. Succession: A population process. *Vegetatio* 43:131–140.

———. 1987. Competition and tree death. *Bioscience* 37:586–594.

PEMBERTON, R. W. 1995. Cactoblastis cactorum in the United States: An immmigrant biological control agent or an introduction of the nursery industry? *Amer. Entomol.* 41:230–232.

PERLIN, J. 1991. *A Forest Journey: The Role of Wood in the Development of Civilization.* Harvard University Press, Cambridge, MA.

PETERJOHN, W. T., J. M. MELILLO, F. P. BOWLES, AND P. A. STEUDLER. 1993. Soil warming and trace gas fluxes: Experimental design and preliminary flux results. *Oecologia* 93:18–24.

PETERKEN, G. F. 1996. *Natural Woodland: Ecology and Conservation in Northrn Temperate Regions.* Cambridge University Press, New York.

PETERS, R. P., AND L. D. MECH. 1975. Scent-marking in wolves. *Am. Sci.* 63:628–637.

PETERS, R. L., AND T. E. LOVEJOY. 1992. *Global Warming and Biological Diversity.* Yale University Press, New Haven, CT.

PETERSON, R. O. 1977. Wolf ecology and prey relationships in Isle Royale. U.S. Nat. Park Serv. Sci. Monogr. Ser. 11. 210 pp.

PETERSON, R. O., AND R. E. Page. 1983. Wolf-moose fluctuations at Isle Royale National Park, Michigan U.S.A. *Acta Zool. Fenn.* 174:251–253.

PETERSON, R. O., R. E. PAGE, AND K. M. DODGE. 1984. Wolves, moose, and the allometry of populatin cycles. *Science* 244:1350–1352.

PETRIE, M. 1986. Reproductive strategies of male and female moorhens (*Gallinula chloropus*). In D. I. Rubenstein and R. W. Wrangham (eds.), *Ecologial Aspects if Social Evolution.* Princeton University Press, Princeton, NJ, pp. 43–63.

PETRIE, M., AND A. P. MOLLER. 1991. Laying eggs in others' nests: Intraspecific brood parasitism in birds. *TREE* 6:315–320.

PETRINVICH, L., AND T. L. PATTERSON. 1982. The white-crowned sparrow: Stability, recruitment, and population structure in the Nuttal subspecies (1975–1980). *Auk* 99:1–14.

PFEIFFER, W. 1962. The fright reaction of fish. *Biol. Rev.* 37:495–511.

PFENNING, D. W. 1997. Kinship and cannibalism. *Bioscience* 47:667–675.

PFISTER, R. D., B. L. KOVALCHIK, S. E. ARNO, AND P. C. PRESBY. 1977. *Forest Habitat Types of Montana.* USDA Forest Servics Gen. Tech. Rept. INT-34 Ogden UT.

PHILANDER, S. G. H. 1983. El Nino southern oscillation phenomenon. *Nature* 302:295–301.

PIANKA, E. 1967. On lizard species diversity, North American flatlands desert. *Ecology* 48:333–351.

———. 1972. r and k selection or b and d selection? *Am. Nat.* 100:65–75.

———. 1975. Niche relations of desert lizards. In M. Cody and J. Diamond (eds.), *Ecology and Evolution of Communities.* Harvard University Press, Cambridge, MA, pp. 292–314.

———. 1978. *Evolutionary Ecology,* 3rd ed. Harper and Row, New York.

———. 1980. On r and K selection. *Am. Nat.* 102:592–597.

———. 1981. Competition and niche theory. In R. M. May (ed.), *Theoretical Ecology: Principles and Application.* Blackwell, Oxford, England, pp. 167–196.

———. 1994. *Evolutionary Ecology,* 5th ed. HarperCollins, New York.

PICKETT, S. T. A. 1982. Population patterns through twenty years of old field succession. *Vegetatio* 49:15–59.

PICKETT, S. T. A., AND P. WHITE (eds.). 1985. *The Ecology of Natural Disturbances and Patch Dynamics.* Academic Press, Orlando, FL.

PIEHLER, K. G. 1987. *Habitat relationships of three grassland sparrow species on reclaimed surface mines in Pennsylvania.* Unpublished M.S. thesis, West Virginia University, Morgantown.

PIELOU, E. C. 1972. Niche width and niche overlap: A method for measuring them. *Ecology* 53:687–692.

———. 1974. *Population and Community Ecology.* Gordon and Breach, New York.

———. 1975. *Ecological Diversity.* Wiley, New York.

———. 1981. The usefulness of ecological models: A stock-taking. *Quart. Rev. Biol.* 56:1423–1437.

———. 1994. *After the Ice Age: The Return of Life to Glaciated North America.* University of Chicago Press, Chicago.

PIERCE, B. A. 1985. Acid tolerance in amphibians. *Bioscience* 35:239–243.

PIJL, L. VAN DER, AND C. L. DOTSON. 1966. *Orchid Flowers: Their Pollination and Evolution.* University of Miami Press, Miami.

PILSON, D., AND M. D. RAUSHER. 1988. Clutch size adjustment by a swallowtail butterfly. *Nature* 333:361–363.

PIMENTEL, D., J. E. DEWEY, AND H. H. SCHWARDT. 1951. Space-time structure of the environment and the survival of the parasite-host system. *Am. Nat.* 97:141–167.

PIMM, S. L. 1980. Food web design and the effect of species deletion. *Oikos* 35:139–149.

———. 1982. *Food Webs.* Chapman and Hall, London.

———. 1987. Determining the effects of introduced species. *TREE* 2:106–107.

———. 1991. *The Balance of Nature: Ecological Issues in the Conservation of Species and Communities.* University of Chicago Press, Chicago.

PINTER, P. J., B. A. KIMBALL, R. L. GARCIA, G. W. WALL, D. J. HUNSAKER, AND R. L. LaMORTE. 1996. Free-air CO_2 enrichment: Response of cotton and wheat crops. In G. Koch and H. A. Mooney (eds.), *Carbon Dioxide and Terrestrial Ecosystems.* Academic Press, San Diego, pp. 215–249.

PITELKA, F. A. 1973. Cyclic patterns in lemming populations near Barrow, Alaska. In M. E. Butler (ed.), *Alaskan Arctic Tundra.* Arctic Institute of America Tech. Paper No. 25, pp. 199–216.

PITELKA, L. F., AND E. D. SCHULZE. 1999. Ecosystem physiology responses to global change. In B. Walker, W. Steffen, J. Canadell, and J. Ingram (eds.), *The Terrestrial Biosphere and Global Change.* Cambridge University Press, Cambridge, England, pp. 141–189.

PIVNICK, K. A., AND J. N. MCNEIL. 1986. Sexual differences in the thermoregulation of *Thymelicus lineola* adults (*Lepidoptera: Hesperiidae*). *Ecology* 67:1024–1035.

PLATTS, W. J., AND D. R. STRONG. 1989. Tree fall gaps and forest dynamics. *Ecology* 70:535–576.

PLESCZYNSKA, W. K., AND R. HANSELL. 1980. Polygyny and decision theory: Testing of a model in lark buntings (*Calamospiza melanocorys*). *Am. Nat.* 116:821–830.

POLICANSKY, D. 1982. Sex change in plants and animals. *Ann. Rev. Ecol. Syst.* 13:471–495.

POLIS, G. 1981. The evolution of intraspecific predation. *Ann. Rev. Ecol. Syst.* 12:225–251.

POLIS, G. A., AND R. D. FARLEY. 1980. Population biology of a desert scorpion: Survivorship, microhabitat, and the evolution of life history strategy. *Ecology* 61:620–629.

POLIS, G., AND R. D. HOLT. 1992. Intraguild predation: The dynamic of complex trophic interactions. *TREE* 7:151–154.

POLIS, G. A., C. A. MYERS, AND R. D. HOLT. 1989. The ecology and evolution of intraguild predation: Potential competitors that eat each other. *Ann. Rev. Ecol. Syst.* 20:297–330.

POLLARD, E., M. D. HOOPER, AND N. W. MOORE. 1974. *Hedges.* W. Collins, London.

POMEROY, L. R., H. M. MATHEWS, AND H. SHIKMIN. 1963. Excretion of phosphate and soluble organic phosphorus compounds by zooplankton. *Limno. Oceanogr.* 4:50–55.

POMEROY, L. R., AND R. G. WIEGERT (eds.). 1981. *The Ecology of a Salt Marsh.* Springer-Verlag, New York.

POORE, M. E. D. 1968. Studies in Malaysian rain forests. 1. The forest on Triassic sediments in the Jenka forest reserve. *J. Ecol.* 56:143–196.

POORTER, H. 1993. Interspecific variation in the growth response of plants to elevated ambient CO_2 concentration. *Vegetatio* 104/105:77–97.

POST, W. M., W. R. EMANUEL, P. J. ZINKE, AND A. G. STANGENBERGER. 1982. Soil carbon pools and world life zones. *Nature* 298:156–159.

POST, W. M., J. PASTOR, P. J. ZINKE, AND A. G. STANGENBERGER. 1985. Global patterns of soil nitrogen storage. *Nature* 317:613–616.

POST, W. M., C. C. TRAVIS, AND D. L. DEANGELIS. 1980. Evolution of mutualism between species. In C. L. Cooke and S. Brisenberg (eds.), *Dif-*

ferential Equations and Applications in Ecology, Epidemics, and Population Problems. Academic Press, New York, pp. 183–201.

———. 1985. Mutualism, limited competition, and positive feedback. In D. H. Boucher (ed.), *The Biology of Mutualism.* Oxford University Press, New York, pp. 305–325.

POUGH, F. H. 1988. Mimicry of vertebrates: Are the rules different? *Am. Nat.* 131:S67–S192.

PRESTON, F. W. 1960. Time and space and the variation of species. *Ecology* 41:611–627.

———. 1962. The canonical distribution of commonness and rarity: Parts 1 and 2. *Ecology* 43:185–215, 410–432.

PRICE, P. W. 1975. Reproductive strategies of parasitoids. In P. W. Price (ed.), *Evolutionary Strategies of Parasitic Insects and Mites.* Plenum, New York, pp. 87–111.

———. 1980. *Evolutionary Ecology of Parasites.* Princeton University Press, Princeton, NJ.

PRIMACK, R. B. 1979. Reproductive effort in annual and perennial species of *Plantago (Plantaginaceae). Am. Nat.* 114:51–62.

PROCTOR, J., AND S. R. J. WOODWELL. 1975. The ecology of serpentine soils. *Adv. Ecol. Res.* 9:256–366.

PRUITT, W. O., JR. 1970. Some aspects of interrelationships of permafrost and tundra biotic communities. In *Productivity and Conservation in Northern Circumpolar Lands,* IUCN Publ. 10:33–41, Geneva, Switzerland.

PULLIAM, H. R. 1988. Sources, sinks, and population regulation. *Am. Nat.* 132:652–661.

PULLIAM, H. R., AND B. J. DANIELSON. 1991. Sources, sinks, and habitat selections: A landscape perspective on population dynamics. *Am. Nat.* 137:S50–S66.

PUSEY, A., AND M. WOLF. 1996. Inbreeding avoidance in animals. *TREE* 11:201–206.

PUTMAN, R. J. 1978a. Patterns of carbon dioxide evolution from decaying carrion: Decomposition of small mammal carrion in temperate systems. *Oikos* 31:49–57.

———. 1978b. Flow of energy and organic matter from a carcass during decomposition: Decomposition of small mammal carrion in temperate systems. 2. *Oikos* 31:58–68.

———. 1983. *Carrion and Dung: The Decomposition of Animal Wastes.* Edward Arnold, London.

PUTMAN, R. J., AND S. D. WRATTEN. 1984. *Principles of Ecology.* University of California Press, Berkeley.

PUTWAIN, P. D., AND J. L. HARPER. 1970. Studies of dynamics of plant populations: 3. The influence of associated species on populations of Rumex acetosa L. and R. acetosella L. in grassland. *J. Ecol.* 58:251–264.

RAFFAELLI, D., AND S. HAWKINS. 1996. *Intertidal Ecology.* Chapman and Hall, London.

RALLS, K., P. H. HARVEY, AND M. A. LYLES. 1986. Inbreeding in natural populations of birds and mammals. In M. Soule (ed.), *Conservation Biology: The Science of Diversity.* Sinauer Associates, Sunderland, MA, pp. 35–56.

RANDOLPH, S. E. 1975. Patterns of distribution of the tick *Ioxodes trianguliceps Birula* on its host. *J. Anim. Ecol.* 44:451–474.

RANNEY, J. W. 1977. Forest island edges: Their structure, development, and importance to regional forest ecosystem dynamics. EDFB/IBP Cont. No. 77/1, Oak Ridge National Laboratory, Oak Ridge, TN.

RANNEY, J. W., M. C. BRUNNER, AND J. B. LEVENSON. 1981. The importance of edge in the structure and dynamics of forest islands. In R. L. Burgess and D. M. Sharpe (eds.), *Forest Island Dynamics in Man-Ddominated Landscapes* (Ecological Studies No. 41). Springer-Verlag, New York, pp. 67–95.

RANTA, E., J. LINDSRTROM, V. KAITALA, H. KOKKO, H. LINDEN, AND E. HELLE. 1997. Solar activity and hare dynamics: A cross-continental comparison. *Am. Nat.* 149:765–775.

RANWELL, D. S. 1961. *Spartina* salt marshes in southern England: 1. The effects of sheep grazing at the upper limits of *Spartina* marsh in Bridgewater Bay. *J. Ecol.* 49:325–340.

———. 1974. The salt marsh to tidal woodland transition. *Hydrobiol. Bull.* 8: 139–151.

RAPP, M., AND P. LOSSAINT. 1981. Some aspects of mineral cycling in the garrigue of southern France. In F. di Castri, D. Goodall, and R. L. Specht (eds.), *Mediterranean-Type Shrublands* (Ecosystems of the World, Vol. 11). Elsevier, Amsterdam, pp. 289–302.

RASMUSSEN, R. A., AND M. A. K. KAHLIL. 1986. Atmospheric trace gases: Trends and distribution over the last decade. *Science* 232:1623–1624.

RAVEN, P. H. 1973. The evolution of Mediterranean flora. In F. di Castri and H. A. Mooney (eds.), *Mediterranean-Type Ecosystems: Origin and Structure.* Springer-Verlag, New York, pp. 213–224.

RAWLINS, J. E. 1980. Thermoregulation by the black swallowtail butterfly *Papilio polypenes. Ecology* 61:345–357.

REAL, L. 1977. The kinetics of functional response. *Am. Nat.* 111:289–300.

———. 1983. *Pollination Ecology.* Academic Press, Orlando, FL.

REAL, L., AND T. CARACO. 1986. Risk and foraging in stochastic environments. *Ann. Rev. Ecol. Syst.* 17:371–390.

REED, D. C., AND M. S. FOSTER. 1984. The effects of canopy shading on algal recruitment and growth in a giant kelp forest. *Ecology* 65:937–948.

REEKIE, E. G., AND F. A. BAZZAZ. 1987. Reproductive efforts in plants. 3. Effect of reproduction on vegetative activity. *Am. Nat.* 29:907–919.

———. 1989. Competition and patterns of resource use among seedlings of five tropical trees grown under ambient and elevated carbon dioxide. *Oecologia* 79:212–222.

REEM, C. H. 1976. Loon productivity, human disturbance, and pesticide residues in northern Minnesota. *Wilson Bull.* 88:427–431.

REGIER, H. A., AND K. H. LOFTUS. 1972. Effects of fisheries exploitation on salmonid communities in oligotrophic lakes. *J. Fish. Res. Board Can.* 29:959–968.

REICH, P. B., M. B. WALTERS, AND D. S. ELLSWORTH. 1992. Leaf life-span in relation to leaf, plant, and stand characteristics among diverse ecosystems. *Ecol. Monogr.* 62:365–392.

REICHARDT, P. B., J. P. BRYANT, T. P. CLAUSEN, AND G. WIELAND. 1984. Defense of winter-dormant Alaska paper birch against snowshoe hare. *Oecologica* 68:58–59.

REICHLE, D. E. (ed.). 1981. *Dynamic Properties of Forest Ecosystems.* Cambridge University Press, Cambridge, England.

REICHLE, D. E., B. E. DINGER, N. T. EDWARDS, W. F. HARRIS, AND P. SOLLINS, et al. 1973. Carbon flow and storage in a forest ecosystem. In G. M. Woodwell and E. V. Pecan (eds.), *Carbon and the Biosphere.* National Technical Information Service, Springfield, VA, pp. 345–365.

REIFSNYDER, W. E., AND H. W. LULL. 1965. Radiant energy in relation to forests. USDA Tech. Bull. No. 1344.

REX, M. A. 1981. Community structure in the deep-sea benthos. *Ann. Rev. Ecol. Syst.* 12:331–354.

REY, J. R. 1981. Ecological biogeography of arthropods on *Spartina* islands in northwest Florida. *Ecol. Monogr.* 51:237–265.

RICE, E. L. 1972. Allelopathic effects of *Andropogon virginicus* and its persistence in old fields. *Am. J. Bot.* 59:752–755.

RICE, W. R. 1982. Acoustical location of prey by the marsh hawk: Adaptation to concealed prey. *Auk* 99:403–413.

RICH, P. H., AND R. G. WETZEL. 1978. Detritus in the lake ecosystem. *Am. Nat.* 112:57–71.

RICHARDS, P. W. 1996. *The Tropical Rain Forest: An Ecological Study,* 2nd ed. Cambridge University Press, New York.

RICHTER, D. D., AND L. I. BABBAR. 1991. Soil diversity in the tropics. *Adv. Ecol. Res.* 21:315–389.

RICKLEFS, R. 1979. *Ecology,* 2nd ed. Chiron Press, New York.

———. 1980. Geographical variation in clutch size in passerine birds: Ashmoles's hypothesis. *Auk* 97:38–49.

———. 1987. Community diversity: Relative roles of local and regional processes. *Science* 235:167–171.

———. 2000. Density dependence, evolutionary optimization, and diversification of avian life histories. *Condor* 102:9–22.

RIDLEY, B. A., J. E. DYE, J. G. WALEGA, J. ZHENG, F. E. GRAHEK, AND W. RISON. 1996. On the production of active nitrogen by thunderstorms over New Mexico. *J. Geophys. Res.* 101:20985–21005.

RIECHERT, S. E. 1981. The consequences of being territorial: Spiders, a case study. *Am. Nat.* 117:871–892.

RIEDMAN, M. L. 1982. The evolution of alloparental care and adoption in mammals and birds. *Quart. Rev. Biol.* 57:405–435.

RILEY, G. A. 1973. Particulate and dissolved organic carbon in the oceans. In G. M. Woodwell and E. V. Pecan (eds.), *Carbon and the Biosphere.* National Technical Information Service, Springfield, VA, pp. 204–220.

RINGLER, N. 1979. Selective predation by drift-feeding brown trout (*Salmo trutta*). J. Fish. Res. Bd. Can. 26:392–403.

RISEBROUGH, R. W., W. WALKER, T. T. SCHMIDT, B. W. DELAPPE, AND C. W. CONNERS. 1976. Transfer of chlorinated biphenyls to Antarctica. *Nature (London)* 264:738–739.

RISSER, P. G. 1985. Grasslands. In B. F. Chabot and H. A. Mooney (eds.), *Physiological Ecology of North American Plant Communities.* Chapman and Hall, New York, pp. 236–256.

RISSER, P. G., E. C. BIRNEY, H. D. BLOCKER, S. W. MAY, W. J. PARTON, AND J. A. WIENS. 1981. *The True Prairie Ecosystem* (US/IBP Synthesis Series 16). Hutchinson, Ross, Stroudsburg, PA.

RITLAND, D. B. 1994. Variation in palatability of queen butterflies (*Danaus gilippus*) and implications regarding mimicry. *Ecology* 75:732–746.

RITLAND, D. B., AND L. P. BROWER. 1991. The viceroy butterfly is not a batesian mimic. *Nature* 350:497–498.

ROBBINS, C. S., D. K. DAWSON, AND B. A. DOWELL. 1989. Habitat requirements of breeding forest birds of the Middle Atlantic States. *Wildl. Monogr.* 103. The Wildlife Society.

ROBBINS, C. T., S. MOLE, A. E. HAGERMAN, AND T. A. HANLEY. 1987. Role of tannins in defending plants against ruminants: Reduction dry matter digestion? *Ecology* 68:1606–1615.

ROBINSON, J. 1990. Lignin, land plants, and fungi: biological evolution affecting Phanerozoic oxygen balance. *Geology* 15:607–610.

ROBINSON, S. K., AND R. T. HOLMES. 1984. Effects of plant species and foliage structure on the foraging behavior of forest birds. *Auk* 101:672–684.

RODDA, G. H., T. H. FRITTS, AND D. CHISZAR. 1977. The disappearance of Guam's wildlife. *Bioscience* 47:565–574.

RODIN, L. Y., AND N. I. BAZILEVIC. 1967. *Production and Mineral Cycling in Terrestrial Vegetation* (translated from Russian by Scripta Technica). G. E. Fogg (ed.). Oliver and Boyd, Edinburgh.

———. 1968. World distribution of biomass. In F. E. Eckhard (ed.), *Functioning of Terrestrial Ecosystems at the Primary Production Level*. UNESCO, Paris, pp. 45–52.

ROFF, D. A. 1997. *Evolutionary Quantitative Genetics*. Chapman and Hall, New York.

ROGERS, L. L. 1987. Factors influencing dispersal in black bears. In B. D. Chepko-Sade and Z. T. Halpin (eds.), *Mammalian Dispersal Patterns*. University of Chicago Press, Chicago, pp. 75–84.

ROHRIG, E., AND B. ULRICH (eds). 1991. *Temperate Deciduous Forests* (Ecosystems of the World, Vol. 7). Amsterdam: Elsevier.

ROHWER, F. C., AND S. FREEMAN. 1989. The distribution of conspecific nest parasitism in birds. *Can. J. Zool.* 67:239–253.

ROMME, W. H., AND D. H. KNIGHT. 1982. Landscape diversity: The concept applied to Yellowstone Park. *Bioscience* 32:664–670.

ROOT, R. B. 1967. The niche exploitation pattern of the blue-gray gnatcatcher. *Ecol. Monogr.* 37:317–350.

ROOT, T. 1988. Energy constraints on avian distributions and abundances. *Ecology* 69:330–339.

ROSEBERRY, J. L., AND W. D. KLIMSTRA. 1984. *Population Ecology of the Bobwhite*. Southern Illinois University Press, Carbondale.

ROSENZWEIG, C., AND M. L. PARRY. 1994. Potential impact of climate change on world food supply. *Nature* 367:133–138.

ROSENZWEIG, C., M. L. PARRY, G. FISCHER, AND K. FROHBERG. 1993. *Climate Change and World Food Supply*. Research Report No. 3, Environmental Change Unit, University of Oxford., Oxford UK

ROSENZWEIG, M. 1968. Net primary productivity of terrestrial communities: Prediction from climatological data. *Am. Nat.* 102:67–74.

ROSENZWEIG, M. L., AND R. H. MACARTHUR. 1963. Graphical representation and stability conditions of predator-prey interactions. *Am. Nat.* 97:209–223.

ROSSWALL, T., J. G. K. FLOWER-ELLIS, L. G. JOHANSSON, S. JOHANSSON, B. E. RYDÉN, AND M. SONESSON. 1975. Stordalen (Abisko), Sweden. In T. Rosswall and O. W. Heal (eds.), *Structure and Function of Tundra Ecosystems*. Swedish Natural Science Research Council, Stockholm, pp. 265–294.

ROSSWALL, T., AND U. GRANHALL. 1980. Nitrogen cycling in a subarctic ombrotrophic mire. In M. Sonesson (ed.), *Ecology of a Subarctic Mire*. Ecol. Bull. 30. Swedish Natural Science Research Council, Stockholm.

ROUGHGARDEN, J. 1974. Species packing and the competition function with illustrations from coral reef fish. *Theor. Pop. Biol.* 5:163–186.

———. 1986. A comparison of food-limited and space-limited animal competition communities. In J. Diamond and T. J. Case (eds.), *Community Ecology*. Harper and Row, New York, pp. 492–516.

ROWLEY, I., E. RUSSELL, AND M. BROOKER. 1997. Inbreeding in birds. In N. W. Thornhill (ed.), *Natural History of Inbreeding and Outbreeding: Theoretical and Empirical Perspectives*. University of Chicago Press, Chicago, pp. 304–328

ROYAMA, T. 1970. Factors governing the hunting behavior and selection of food by the great tit. *J. Anim. Ecol.* 39:619–668.

RUESINK, J. L., I. M. PARKER, M. J. GROOM, AND P. M. KAREIVA. 1995. Reducing the risks of nonindigenous species introductions. *Bioscience* 45:465–477.

RUINEN, J. 1962. The phyllosphere: An ecologically neglected region. *Plant Soil* 15:81–109.

RUNDEL, P. W. 1980. The ecological distribution of C_3 and C_4 grasses in the Hawaiian Islands. *Oecologica* 45:354–359.

———. 1981. The mattoral zone of central Chile. In F. di Castri, D. Goodall, and R. Specht (eds.), *Mediterranean-Type Ecosystems* (Ecosystems of the World, Vol. 11). Elsevier, Amsterdam, pp. 175–210.

RYAN, M. G., D. BINKLEY, AND J. H. FOWNES. 1997. Age-related decline in forest productivity: pattern and process. *Adv. Ecol. Res.* 27:213–262.

RYAN, M. G., AND R. H. WARING. 1992. Maintenance respiration and stand development in a subalpine lodgepole pine forest. *Ecology* 73:2100–2108.

RYGIEWICZ, P. T., AND C. P. ANDERSON. 1994. Mycorrhizae alter quality and quantity of carbon allocation belowground. *Nature* 369:58–60.

RYTHER, J. H., AND C. S. YENTSCH. 1957. The estimation of phytoplankton production in the ocean from the chlorophyll and light data. *Limno. Oceanogr.* 2:381–386..

SAIKKONEN, K., S. H. FAETH, M. HELANDER, AND T. J. SULLIVAN. 1998. Fungal endophytes.: A continuum of interactions with host plants. *Ann. Rev. Ecol. Syst.* 29: 319–344.

SAINO, N., A. M. BOLZERN, AND A. P. MOLLER. 1997. Immunocopentence, ornamentation, and viability of male barn swallows (*Hirundo rustica*). *Proc. Nat. Acad. Sci.* 97:579–585.

SAINO, N., R. STRADI, P. NINNI, E. PINI, AND A. P. MOLLER. 1999. Carotenoid plasma concentrations, immune profile, and plumage ornamentation of male barn swallows (*Hirundo rustica*). *Am. Nat.* 154:41–448.

SANDON, H. 1927. *The Composition and Distribution of Protozoan Fauna of the Soil*. Oliver and Boyd, London.

SANTOS, P. F., N. Z. ELKINS, Y. STEINBERGER, AND W. G. WHITFORD. 1984. A comparison of surface and buried *Larrea tridentata* leaf litter decomposition in North American hot deserts. *Ecology* 65:278–284.

SARGENT, R. C. 1989. Allopaternal care in the fathead minnow, *Pimiphales promelas:* Stepfathers discriminate against their adopted eggs. *Behav. Ecol. Sociobiol.* 25:379–385.

SARGENT, R. C., AND M. R. GROSS. 1985. Parental investment decision rules and the Concorde fallacy. *Behav. Ecol. Sociobiol.* 17:43–45.

SARGENT, R. C., P. D. TAYLOR, AND M. P. GROSS. 1987. Parental care and the evolution of egg size in fish. *Am. Nat.* 129:32–46.

SARMIENTO, G. 1984. *The Ecology of Neotropical Savannas*. Harvard University Press, Cambridge, MA.

SARMIENTO, G., AND M. MONASTERRIO. 1975. A critical consideration of environmental conditions associated with the occurrence of savanna ecosystems in tropical America. In F. B. Golly and E. Medina (eds.), *Tropical Ecological Systems*. Springer-Verlag, Berlin, pp. 223–250.

SARUKHAN, J., M. MARTINEZ-RAMOS, AND D. PINERO. 1984. The analysis of demographic variability at the individual level and its population consequences. In R. Drizo and J. Sarukhan (eds.), *Perspectives in Plant Population Ecology*. Sinauer Associates, Sunderland, MA, pp. 83–106.

SASEKUMAR, A., AND C. V. CHING. 1987. Mangroves and praws: Further perspectives. Proc. *10th Annual Seminar of the Malaysian Society of Marine Sciences*, pp. 10–22.

SCHAEFER, R. 1973. Microbial activity under seasonal conditions of drought in Mediterranean climates. In F. di Castri and H. A. Mooney (eds.), *Mediterranean-Type Ecosystems: Origin and Structure*. Springer-Verlag, New York, pp. 191–198.

SCHAFFER, W. M. 1974. Optimal reproductive effort in fluctuating environments. *Am. Nat.* 108:783–790.

———. 1981. Ecological abstraction: The consequences of reduced dimensionality in ecological models. *Ecol. Monogr.* 51:383–401.

SCHALL, B. A. 1984. Life history variation, natural selection, and maternal effects in plant populations. In R. Drizo and J. Sarukhan (eds.), *Perspectives in Plant Population Ecology*. Sinauer, Sunderland, MA, pp. 188–211.

SCHALLER, G. B. 1972. *Serengeti: A Kingdom of Predators*. Knopf, New York.

SCHEEL, D., AND C. PACKER. 1991. Group hunting behavior of lions: A search for cooperation. *Anim. Behav.* 41:697–709.

SCHEFFER, V. C. 1951. The rise and fall of a reindeer herd. *Sci. Month.* 73:356–362.

SCHIMEL, D. S., I. G. ENTING, M. HEIMANN, T. M. L. WIGLEY, D. RAYNAUD, D. ALVES, AND U. SIEGENTHALER. 1995. CO_2 and the carbon cycle. In J. T. Houghton, L. G. Meira Filho, J. Bruce, H. Lee, B. A. Callander, E. Haites, N. Harris, and K. Maskell (eds.), *Climate Change* 1994. Cambridge University Press, Cambridge, England.

SCHINDLER, D. W. 1977. Evolution of phosphorus limitation in lakes. *Science* 195:260–262.

———. 1978. Factors regulating phytoplankton production and standing crop in the world's freshwaters. *Limno. Oceanog.* 23:478–486.

SCHLESINGER, W. H. 1997. *Biogeochemistry: An Analysis of Global Change,* 2nd ed. Academic Press, London.

SCHMIDT-NIELSEN, K. 1977. *Animal Physiology: Adaptation and Environment,* 5th ed. Cambridge University Press, New York.

———. 1960. The salt secreting gland of marine birds. *Circulation* 21:955–967.

SCHOENER, T. 1983. Simple models of optimal feeding-territory size: A reconciliation. *Am. Nat.* 121:608–629.

SCHOENER, T. 1983. Field experiments on interspecific competition. *Am. Nat.* 122:240–285.

SCHOENER, T. W., AND D. A. SPILLER. 1987. Effect of lizards on spider populations: Manipulative reconstruction of a natural experiment. *Science* 236:949–953.

SCHOENER, T. W., AND C. A. TOFT. 1983. Spider populations: Extraordinarily high densities on islands without top predators. *Science* 21:1353–1355.

SCHOFIELD, C. L., AND J. R. TROJNAR. 1980. Aluminum toxicity to brook trout (*Salvelinus fontinali*) in acidified waters. In T. Y. Toribara, M. W. Miller, and P. E. Morrow (eds.), *Polluted Rain.* Plenum, New York, pp. 341–365.

SCHOLANDER, P. F., R. HOCK, V. WALTHERS, F. JOHNSON, AND L. IRVING. 1950. Heat regulation in some arctic and tropical birds and mammals. *Biol. Bull.* 99:237–258.

SCHOLANDER, P. F., V. WALTERS, R. HOCK, L. IRVING, AND F. JOHNSON. 1950. Body insulation of some arctic and tropical mammals and birds. *Biol. Bull.* 99:225–236.

SCHROEDER, M. J., AND C. C. BUCK. 1970. *Fire Weather.* USDA Ag. Handbook 360, USDA Forest Service, Washington, DC.

SCRIBER, J. M. A., AND F. SLANSKY, JR. 1981. The nutritional ecology of immature insects. *Ann. Rev. Entomol.* 26:182–211.

SEARCY, W. A. 1979. Male characteristics and pairing success in red-winged blackbird. *Auk* 96:353–363.

SEARCY, W. A., AND M. ANDERSSON. 1986. Sexual selection and the evolution of song. *Ann. Rev. Ecol. Syst.* 17:507–534.

SEARCY, W. A., P. D. MCARTHUR, AND K. YASUKAWA. 1985. Song repertoire size and male quality in song sparrows. *Condor* 87:222–228.

SEARS, P. B. 1935. *Deserts on the March.* University of Oklahoma Press, Norman. Segerstrale, S. G. 1947. New observations on the distribution and morphology of the amphipod Gammarus zaddachi Sexon, with notes on related species. J. Marine Biol. Assoc. U.K. 27:219–244.

SEMLER, D. E. 1971. Some aspects of adaptation in a polymorphism for breeding in the three-spined stickleback (*Gasterostermus aculeatus*). J. Zool. (London) 165:291–302.

SETTLE, W. H., H. ARIAWAN, E. T. ASTUTU, et al. 1996. Managing tropical rice pests through conservation of generalist natural enemies and alternative prey. *Ecology* 77:1975–1988.

SHAPIRO, D. Y. 1979. Social behavior, group structure, and the control of sex reversal in hermaphroditic fish. Adv. Study Behav. 10:43–102.

———. 1980. Serial female sex changes after simultaneous removal of males from social groups of a coral reef fish. *Science* 209:1136–1137.

———. 1987. Differentiation and evolution of sex change in fishes. *Bioscience* 37:490–497.

SHAW, S. P., AND C. G. Fredine. 1956. *Wetlands of the United States.* U.S. Fish and Wildl. Circ. 39.

SHELDON, W. G. 1967. *The Book of the American Woodcock.* University of Massachusetts Press, Amherst.

SHELFORD, V. E. 1911. Physiological animal geography. *J. Morph.* 22:551–618.

———. 1913. Animal communities in temperate America. *Bull. Geog. Soc.* Chicago 5:1–368.

SHERMAN, P. W. 1977. Nepotism and the evolution of alarm calls. *Science* 197:1246–1253.

———. 1981. Kinship, demography and Belding's ground squirrel nepotism. *Behav. Ecol. Sociobiol.* 8:251–259.

SHERMAN, P. W., J. V. M. JARVIS, AND R. D. ALEXANDER. 1991. *The Biology of the Naked Mole Rat.* Princeton University Press, Princeton, NJ.

SHERR, E. B., AND B. F. SHERR. 1991. Planktonic microbes: Tiny cells at the base of the ocean's food webs. *TREE* 6:50–54.

SHIELDS, W. M. 1987. Dispersal and mating systems: Investigating their causal connections. In B. D. Chepho-Sade and Z. T. Halpi (eds.), *Mammalian Dispersal Patterns.* University of Chicago Press, Chicago, pp. 3–24.

———. 1993. The natural and unnatural history of inbreeding and outbreeding. In N. W. Thornhill (ed.), *Natural History of Inbreeding and Outbreeding.* University of Chicago Press, Chicago, pp. 143–169.

SHIPLEY, L. A., J. E. GROSS, D. E. SPALINGER, N. T. HOBBS, AND B. A. WUNDER. 1994. The scaling intake rate in mammalian herbivores. *Am. Nat.* 143:1055–1082.

SHUGART, H. H. 1984. *A Theory of Forest Dynamics: The Ecological Implications of Forest Succession.* Springer-Verlag, New York.

SHUGART, H. H., AND D. C. WEST. 1977. Development of an Appalachian deciduous forest model and its application to assessment of the impact of the chestnut blight. *J. Environ. Manage.* 5:161–169.

SHURE, D. J., AND H. S. RAGSDALE. 1977. Patterns of primary succession on granite outcrop surfaces. *Ecology* 58:993–1006.

SIEGENTHALER, U., AND J. L. SARMIENTO. 1993. Atmospheric carbon dioxide and the ocean. *Nature* 365:119–125.

SIMBERLOFF, D. S. 1974. Equilibrium theory of island biogeography and ecology. *Ann. Rev. Ecol. Syst.* 5:161–182.

SIMBERLOFF, D., AND N. GOTELLI. 1984. Effects of insularization on plant species richness in the prairie-forest ecotone. *Biol. Cons.* 29:27–46.

SIMBERLOFF, D., AND P. STILING. 1996. How risky is biological control? *Ecology* 77:1965–1974.

SIMBERLOFF, D. S., AND E. O. WILSON. 1969. Experimental zoogeography of islands: The colonization of empty islands. *Ecology* 50:278–296.

———. 1970. Experimental zoogeography of islands: A two-year record of colonization. *Ecology* 50:278–296.

SIMPSON, G. G. 1964. Species density of North American recent mammals. *Syst. Zool.* 13:57–73.

SIMS, P. L. 1988. Grasslands. In M. G. Barbour and W. D. Billings (eds.), *North American Terrestrial Vegetation.* Cambridge University Press, New York, pp. 266–286.

SIMS, P. L., AND J. S. SINGH. 1971. Herbage dynamics and net primary production in certain grazed and ungrazed grasslands in North America. In N. R. French (ed.), *Preliminary Analysis of Structure and Function in Grasslands.* Range Sci. Dept. Sci. Ser. No. 10. Colorado State University, Fort Collins, pp. 59–124.

SINCLAIR, A. R. E. 1977. The African Buffalo: A Study of Resource Limitation of Populations. University of Chicago Press, Chicago.

———. 1979. Dynamics of the Serengeti ecosystem. In A. R. E. Sinclair and M. Norton-Griffiths (eds.), *Serengeti: Dynamics of an Ecosystem.* University of Chicago Press, Chicago, pp. 1–30.

SINCLAIR, A. R. E., AND P. ARCESE (eds.). 1995. *Serengeti II: Dynamic, Management, and Conservation of an Ecosystem.* University of Chicago Press, Chicago.

SINCLAIR, A. R. E., J. M. GISLINE (GOSLINE), G. HOLDSWORTH, C. T. KREBS, S. BOUTIN, J. N. M. SMITH, R. BOONSTRA, AND M. DALE. 1993. Can the solar cycle and climate synchronize the snowshoe hare cycle in Canada? Evidence of tree rings and ice cores. *Am. Nat.* 141:173–198.

SINCLAIR, A. R. E., AND J. M. GOSLINE. 1997. Solar activity and mammal cycles in the northern hemisphere. *Am. Nat.* 1149:776–784.

SINCLAIR, A. R. E., C. J. KREBS, J. N. M. SMITH, AND S. BOUTIN. 1988. Population biology of snowshoe hares. III. Nutrition, plant secondary compounds, and food limitation. *J. Anim. Ecol.* 57:787–806.

SINCLAIR, A. R. E., AND J. N. M. SMITH. 1984. Do secondary compounds determine feeding preferences of snowshoe hares? *Oecologica* 61:403–410.

SISK, T. D., N. M. HADDAD, AND P. R. EHRLICH. 1997. Assemblages in patchy woodlands: Modeling the effects of edge and matrix habitats. *Ecology* 78:1770–1180.

SKUTCH, A. 1986. *Helpers at Birds' Nests: A World-wide Survey of Cooperative Breeding and Related Behavior.* University of Iowa Press, Iowa City.

SLATKIN, M. 1987. Gene flow and the geographic structure of natural populations. *Science* 236:787–792.

SLOBODKIN, L. B. 1962. *Growth and Regulation of Animal Population.* Holt, Rinehart and Winston, New York.

SMALLEY, A. E. 1960. Energy flow of a salt marsh grasshopper population. *Ecology* 41:672–677.

SMAYDA, T. J. 1983. The phytoplankon of estuaries. In B. Ketchum (ed.), *Estuaries and Enclosed Seas* (Ecosystems of the World, Vol. 26). Elsevier, Amsterdam.

SMITH, A. P., AND T. P. YOUNG. 1987. Tropical alpine plant ecology. *Ann. Rev. Ecol. Syst.* 18:137–158.

SMITH, B. D. 1978. *Mississippi Settlement Patterns.* Academic Press, New York.

SMITH, C. C., AND S. D. FRETWELL. 1974. The optimal balance between size and number of offspring. *Am. Nat.* 108:499–506.

SMITH, D. W. 1986. *Principles of Silviculture.* McGraw-Hill, New York.

SMITH, H. J. 1978. Parasites of red foxes in New Brunswick and Nova Scotia. *J. Wildl. Dis.* 14:366–370.

SMITH, M. H., R. K. CHESSER, E. C. COTHRAN, AND P. E. JOHNS. 1983. Genetic variability and antler growth in a natural population of white-tailed deer. In R. D. Brown (ed.), *Antler Development in Cervidae.* Caesar Kleberg Wildl. Res. Inst., Kingsville, TX.

SMITH, R. A. H., AND A. D. BRADSHAW. 1979. The use of heavy metal tolerant plant populations for the reclamation of metalliferous wastes. *J. App. Ecol.* 16:595–612.

SMITH, R. E., AND B. A. HORWITZ. 1969. Brown fat and thermogenesis. *Physiol. Rev.* 49:330–425.

SMITH, R. L. 1956. An evaluation of conifer plantations as wildlife habitat. Ph.D. Dissertation, Cornell University, Ithaca, NY.

———. 1959. Conifer plantations as wildlife habitat. *N.Y. Fish Game J.* 5:101–132.

———. 1962. Acorn consumption by white-footed mice (*Peromyscus leucopus*). Bull. 482T, WV Univ. Agr. Expt. Sta.

———. 1963. Some ecological notes on the grasshopper sparrow. *Wilson Bull.* 75:159–165.

SMITH, S. M. 1978. The "underworld" in a territorial adaptive strategy for floaters. *Am. Nat.* 112:570–582.

SMITH, T. M. 2000. Patterns of variation in litterfall, chemistry, and decomposition in a diverse temperate forest. Ecology (submitted).

SMITH, T. M., AND P. GOODMAN. 1987. The effect of competition on the structure and dynamics of Acacia savannas in southern Africa. *J. Ecol.* 75:1013–1044.

SMITH, T. M., AND K. GRANT. 1986. The role of competition in the spacing of trees in a *Burkea africana-Terminalia sericea savanna. Biotropica* 18:219–223.

SMITH, T. M., P. N. HALPIN, H. H. SHUGART, AND C. SECRETT. 1995. Global forests. In K. Strzpeck and J. Smith (eds.) *As Climate Changes:International Impacts and Implications.* Cambridge University Press, Cambridge UK.

SMITH, T. M., AND M. HUSTON. 1987. A theory of spatial and temporal dynamic of plant communities. *Vegetatio* 83:49–69.

SMITH, T. M., R. LEEMANS, AND H. H. SHUGART. 1992. Sensitivity of terrestrial carbon storage to CO_2 induced climate change: Comparison of five scenarios based on general circulation models. *Climat. Change* 21:367–384.

SMITH, T. M., AND H. H. SHUGART. 1987. Territory size variation in the ovenbird: The role of habitat structure. *Ecology* 68:695–704.

———. 1993. The transient response of terrestrial carbon storage to a perturbed climate. *Nature* 361:523–526.

SMITH, T. M., H. H. SHUGART, G. B. BONAN, AND H. B. SMITH. 1992. Modeling the potential response of vegetation to global climate change. *Adv. Ecol. Res.* 22:93–116.

SMITH, T. M., H. H. SHUGART, AND J. B. SMITH. 1992. Modeling the response of tropical forests to climate change: Integrating regional and site specific studies. In J. G. Goldammer (ed.), *Tropical Forests in Transition: Ecology of Natural and Anthropogenic Disturbance Processes in Tropical Forest Biomes.* Birkhauser-Verlag, Basel-Boston.

SMITH, T. M., AND D. L. URBAN. 1988. Scale and resolution of forest structural pattern. *Vegetatio* 74:143–150.

SMITH, T. M., AND B. H. WALKER. 1983. The role of competition in the spacing of savanna trees. *Proc. Grassland Soc. S. Africa* 18:159–164.

SMITH, T. M., A. B. WELLINGTON, AND M. P. AUSTIN. 2000 Performance of four *Eucalyptus* spp. across an experimental water gradient: Linking individual response to competitive outcomes. *J. Ecol.* (in review)

SMITH, W. H. 1981. *Air Pollution and Forests: Interactions Between Air Contamination and Forest Ecosystems.* Springer-Verlag, New York.

———. 1991. Ontogeny and adaptiveness of tail-flagging behavior in white-tailed deer. *Am. Nat.* 138:190–200.

SOLLINS, P. 1982. Input and decay of coarse woody debris in coniferous forest stands in western Oregon and Washington. *Can. J. For.* 12:18–28.

SOLLINS, P., C. C. GRIER, F. M. McCORSIN, K. CROMACK, JR., R. FOGEL, AND R. L. FREDRIKSEN. 1980. The internal element cycles of an old-growth Douglas-fir ecosystem in western Washington. *Ecol. Monogr.* 50:275–282.

SOLOMON, A. M. 1986. Transient response of forests to CO_2 induced climate change: Simulation modeling experiments in eastern North America. *Oecologica* 68:567–579.

SOLOMON, M. E. 1949. The natural control of animal populations. *J. Anim. Ecol.* 18:1–32.

———. 1957. Dynamics of insect populations. *Ann. Rev. Entomol.* 2:121–142.

SOLOMON, P. M., R. DE ZAFRA, A. PARRISH, AND J. W. BARRET. 1984. Diurnal variation of stratospheric chlorine monoxide: Critical test of chlorine chemistry in the ozone layer. *Science* 224:1210–1214.

SORENSON, M. D. 1993. Parasitic egg laying in canvasbacks: Frequency, success, and individual behavior. *Auk* 110:57–69.

SORIANO, A. 1972. South America. In C. M. McKell et al. (eds.), *Wildland Shrubs: Their Biology and Utilization.* USDA Forest Service Gen. Tech. Rept. INT-1, pp. 31–54.

SOULE, M. E. 1986. *Conservation Biology: The Science of Scarcity and Diversity.* Sinauer Associates, Sunderland, MA.

SOUSA, W. P. 1979. Disturbance in marine intertidal boulder fields: The non-equilibrium maintenance of species diversity. *Ecology* 60:1225–1239.

———. 1984. Intertidal mosaics: Patch size, propagule availability, and spatially variable patterns of succession. *Ecology* 65:1918–1935.

SOWLS, L. K. 1960. Results of a banding study of Gambel's quail in southern Arizona. *J. Wildl. Manage.* 24:185–190.

SPAETH, J. N, AND C. H. DIEBOLD. 1938. Some interrelations between soil characteristics, water tables, soil temperature, and snow cover in the forest and adjacent open areas in south central New York. *Cornell Univ. Agr. Exp. Stat. Mem.* 213.

SPALINGER, D. E., AND N. T. HOBBS. 1992. Mechanisms of foraging in mammalian herbivores: New models of functional response. *Am. Nat.* 140:325–348.

SPECHT, R. L. 1979. Heathlands and related shrublands of the world. In R. L. Specht (ed.), *Heathlands and Related Shrublands: Descriptive Studies* (Ecosystems of the World, Vol. 9A). Elsevier, Amsterdam, pp. 1–18.

———. 1981. Mallee ecosystems in southern Australia. In F. di Castri, D. W. Goodall, and R. L. Specht (eds.), *Mediterranean-Type Shrublands* (Ecosystems of the World, Vol. 11). Elsevier, Amsterdam, pp. 203–231.

SPENSER, C. N., B. McCLELLAND, AND J. A. STANFORD. 1991. Shrimp stocking, salmon collapse, and eagle displacement. *Bioscience* 41:14–21.

SPIES, T. A., AND J. F. FRANKLIN. 1991. The structure of natural young, mature, and old-growth Douglas-fir forests in Oregon and Washington. In L. F. Ruggiero, K. B. Aubry, A. B. Carey, and M. H.Huff. (tech coord.), *Wildlife and Vegetation of Unmanaged Douglas-fir forests.* Gen. Tech Rept. PNW-GTR-285, USDA Dept. Agriculture Forest Service; Pacific Northwest Research Station, Portland, OR, pp. 91–121.

SPOONER, G. M. 1947. The distribution of *Gammarus* species in estuaries: Part I. *J. Marine Biol. Assoc.* U.K. 27:1–52.

SPRUGEL, D. G. 1976. Dynamic structure of wave generated Abies balsamea forests in northeastern United States. *J. Ecol.* 64:889–911.

SPURR, S. H. 1957. Local climate in the Harvard Forest. *Ecology* 38:37–56.

STACEY, P. B., AND M. TAPER. 1992. Environmental variation and the persistence of small populations. *Ecol. App.* 2:18–29.

STAFF, H., AND BERG, B. 1982. Accumulation and release of plant nutrients in decomposing Scots pine needle litter: Long-term decomposition in a Scots pine forest. II. *Can. J. Bot.* 60:1561–1568.

STAMPS, J. A., AND V. V. KRISHNAN. 1990. The effect of settlement tactics on territory sizes. *Am. Nat.* 135:527–546.

STANLEY, E. H., S. T. FISHER, AND N. B. GRIMM. 1997. Ecosystem expansion and contraction in streams. *Bioscience* 47:427–436.

STANTON, N. L. 1988. The underground in grasslands. *Ann. Rev. Ecol. Syst.* 19:573–589.

STEBBINS, G. L. 1972. Evolution and diversity of arid-land shrubs. In C. M. McKell et al. (eds.), *Wildland Shrubs: Their Biology and Utilization.* USDA Forest Service Gen. Tech. Rept. INT-1

STECK, F. 1982. Rabies in wildlife. *Symp. Zool. Soc. Lond.* 50:57–75.

STEELE, J. H. 1974. *The Structure of Marine Ecosystems.* Harvard University Press, Cambridge, MA.

STENGER, J., AND J. B. FALLS. 1959. The utilized territory of the ovenbird. *Wilson Bull.* 71:125–140.

STEPHENS, D. W. 1981. The logic of risk-sensitive foraging preferences. *Anim. Behav.* 29:628–629.

STEPHENSON, T. A., AND A. STEPHENSON, 1949. The universal features of zonation between the tide-marks on rocky coasts. *J. Ecology* 37:289–305.

————. 1954. Life between the tide-marks in North America: 3A. Nova Scotia and Prince Edward Island: The geographical features of the region. *J. Ecol.* 42:14–45, 46–70.

STEPONKUS, P. L. 1981. Responses to extreme temperatures: Cellular and subcellular bases. In O. Lange, P. S. Nobel, C. B. Osmund, and H. Zeigler (eds.), *Physiological Plant Ecology.* I. Vol 12A. *Encyclopedia of Plant Physiology.* Springer Verlag, New York. pp. 371–402.

STERN, W. L., AND M. F. BUELL. 1951. Life-form spectra in a New Jersey pine barren forest and Minnesota jack pine forest. *Bull. Torrey Bot. Club* 78:61–65.

STEVENS, D. W., AND J. R. KREBS. 1986. *Foraging Theory.* Princeton University Press, Princeton, NJ.

STEVENS, G. C. 1989. The latitudinal gradient in geographical range: How so many species can coexist in the tropics. *Am. Nat.* 133:240–256.

STEVENS, G. C., and J. F. Fox. 1991. The causes of treelines. *Ann. Rev. Ecol. Syst.* 22:177–192.

STILES, E. W. 1980. Patterns of fruit presentation and seed dispersal in bird disseminated woody plants in the eastern deciduous forest. *Am. Nat.* 116:670–688.

————. 1982. Fruit flags: Two hypotheses. *Am. Nat.* 120:500–509.

STILES, F. G. 1975. Ecology, flowering phenology, and hummingbird pollinatio of some Costa Rican *Heliconia* species. *Ecology* 56:285–301.

STOCKER, R., P. W. LEADLEY, AND CH. KORNER. 1997. Carbon and water fluxes in a calcareous grassland under elevated CO_2. *Funct. Ecol.* 11:222–230.

STODDARD, H. 1932. *The Bobwhite Quail: Its Habits, Preservation, and Increase.* Charles Scribner, New York.

STOUT, R. J., B. A. KNUTH, AND P. D. CURTIS. 1997. Preferences of suburban landowners for deer management techniques: A step toward better communication. *Wildl. Soc. Bull.* 25:348–359.

STOWE, L. G., AND J. A. TEERI. 1978. The geographic distribution of C_4 species of the Dicotyledonae in relation to climate. *Am. Nat.* 112:609–623.

STRAHLER, A. 1971. *The Earth Sciences.* Harper and Row, New York.

STRAIN, B. R., AND J. D. CURE (eds.). 1985. *Direct Effects of Increasing Carbon Dioxide on Vegetation.* U.S. Department of Energy Publication DOE/ER-0238. Washington, DC.

STRAYER, D. L., N. F. CARACO, J. J. COLE, S. FINDLAY, AND M. L. PACE. 1999. Transformation of freshwater ecosystems by bivalves. *Bioscience* 49:19–27.

STRONG, D. R., J. H. LAWTON, AND R. SOUTHWOOD. 1984. *Insects and Plants: Community Patterns and Mechanisms.* Harvard University Press, Cambridge, MA.

STRONG, D. R., JR., AND J. R. REY. 1982. Testing for MacArthur-Wilson equilibrium with the arthropods of the miniature *Spartina* archipelago at Oyster Bay, Florida. *Am. Zool.* 22:350–360.

STRONG, D. R., D. SIMBERLOFF, L. G. ABELE, AND A. B. THISTLE (eds.). 1984. *Ecological Communities: Conceptual Issues and the Evidence.* Princeton University Press, Princeton, NJ.

STRZEPED, K. M., J. SMITH (eds.) *As Climate Changes: International Impacts and Implications.* Cambridge University Press, Cambridge, England.

SUGG, D. W., R. K. CHESSER, F. S. DOBSON , AND J. L. HOOGLAND. 1996. Population genetics meets behavioral ecology. *TREE* 11:338–342.

SUHONEN, J. 1993. Predation risk influences the use of foraging sites by tits. *Ecology* 74:1197–1203.

SWANK, W. T., J. W. FITZGERALD, AND J. T. ASH. 1983. Microbial transformation of sulfate in forest soils. *Science* 223:182–184.

SWIFT, M. J., O. W. HEAL, AND J. M. ANDERSON. 1979. *Decomposition in Terrestrial Ecosystems.* Blackwell, Oxford, England.

SYDEMAN, W. J., AND S. D. EMSLIE. 1992. Effect of parental age on hatching asynchrony, egg size, and third chick disadvantage in western gulls. *Auk* 109:242–248.

TAHVANAINEN, J., E. HELLE, R. JULKUNEN-TITTO, AND A. LAVOLA. 1985. Phenolic compounds of willow bark as deterrents against feeding by mountain hare. *Oecologica* 65:319–323.

TAIT, R. V. 1968. *Elements of Marine Ecology.* Plenum, New York.

TAJCHMAN, S. J., M. H. HARRIS, AND E. C. TOWNSEND. 1988. Variability of the radiative index of dryness in an Appalachian watershed. *Agric. For. Meteorol.* 42:199–207.

TAMARIN, R. H. 1978. Dispersal, population regulation, and K-selection in field mice. *Am. Nat.* 112:545–555.

TAMM, C. O. 1951. Removal of plant nutrients from tree crowns by rain. *Physiol. Plant* 4:184–188.

TANNER, J. T. 1975. The stability and intrinsic growth rates of prey and predator populations. *Ecology* 56:855–867.

TANSLEY, A. G. 1935. The use and abuse of vegetational concepts and terms. *Ecology* 16:284–307.

TAYLOR, C. R. 1969. The eland and the *oryx. Sci. Am.* 220(1):88–95.

TAYLOR, C. R., AND C. P. LYMAN. 1972. Heat storage in running antelopes: Independence of brain and body temperatures. *Am. J. Physiol.* 222:114–117.

TAYLOR, R. J. 1984. *Predation.* Chapman and Hall, New York.

TEAL, J. M. 1962. Energy flow in the salt marsh ecosystem of Georgia. *Ecology* 43:614–624.

TEMPLE, S. A. 1977. The dodo and the tambalacoque tree. *Science* 203:1364.

————. 1986. Predicting impacts of habitat fragmentation on forest birds: A comparison of two models. In J. Verner, M. L. Morrison, and C. T. Ralph (eds.), *Wildlife 2000: Modeling Habitat Relations of Terrestrial Vertebrates.* University of Wisconsin Press, Madison, pp. 301–304.

TEMPLETON, A. R. 1986. Coadaptation and outbreeding depression. In M. E. Soule (ed.), *Conservation Biology.* Sinauer Associates, Sunderland, MA, pp. 105–121.

TEMPLETON, J. W., R. M. SHARP, J. WILLIAMS, D. DAVIS, D. HARMEL, B. ARMSTRONG, AND S. WARDROUP. 1983. Single dominant gene effect on the expression of antler point numbers in the white-tailed deer. In R. D. Brown (ed.), *Antler Development in Cervidae.* Caesar Kleberg Wildlife Research Institute, Kingsville, TX, pp. 469–470.

TERAMURA, A. H. 1990. Implication of stratospheric ozone depletion upon plant production. *Hort. Sci.* 25:1557–1559.

TERBROUGH, J. 1992. *Diversity and the Tropical Rainforest.* Scientific America Library, William Freeman, New York.

————. 1995. Wildlife in managed tropical forests. In A. E. Lugo and C. Lowe (eds.), *Tropical Forests: Management and Ecology.* Springer-Verlag, New York, pp. 331–330.

TERRI, J. A. 1979. The climatology of the C_4 photosynthetic pathway. In O. T. Solbrig, S. Jain, G. B. Johnson, and P. H. Raven (eds.), Topics in *Plant Population Biology.* Columbia University Press, New York, pp. 356–374.

TERRI, J. A., AND L. STOWE. 1976. Climate patterns and distribution of C_4 grasses in North America. *Oecologia* 23:1–12.

TESKEY, R. O. 1995. A field study of the effects of elevated CO_2 on carbon assimilation, stomatal conductance and leaf and branch growth of Pinus taeda trees. *Plant Cell Environ.* 18:565–573.

TESTER, J. R., AND W. H. MARSHALL. 1961. A study of certain plant and animal interrelations on a native prairie in Northwestern Minnesota. *Minn. Mus. Nat. Hist.* Occasional Paper No. 8.

TEVINI, M., AND A. H. TERAMURA. 1989. UV-B effects on terrestrial plants. Photochem. *Photobiol.* 50:479–487.

THIENEMANN, A. 1927. Der Bau des Seebeckens in seiner Bedeutung fur den Ablouf des Lebens im See. *Zool. Bot Ges. Vienna Verhandl.* 77:87–91.

Thomas, J. W., R. G. Anderson, C. Maser, and E. L. Bull. 1979. Snags. In J. W. Thomas (ed.), *Wildlife Habitats in Managed Forests* (The Blue Mountains of Oregon and Washington). USDA Forest Service Ag. Handb. No. 553, pp. 60–77.

THOMAS, S. C., C. B. HALPERN, D. A. FALK, D. A. LIGUORI, AND K. A. AUSTIN. 1999. Plant diversity in managd forests: Understory responses to thinning and fertilization. *Ecol. App.* 9:864–879.

THOMPSON, C. W., N. HILLGARTH, M. LEU, AND H. E. MCCLURE. 1977. High parasite loads in house finches (*Carpodacus mexicanus*) is correlated with reduced expression of a sexually selected trait. *Am. Nat.* 149:270–294.

THOREAU, H. D. 1860. Succession of forest trees. *Mass. Board Agric. Rept.* VIII.

THORNHILL, N. W. (ed.). 1993. *The Natural History of Inbreeding and Outbreeding.* University of Chicago Press, Chicago.

THORNTHWAITE, C. W. 1931. Climates of North America according to a new classification. *Geog. Rev.* 21:633–655.

————. 1948. An approach to a rational classification of climate. *Geog. Rev.* 38:55–94.

THURMAN, N. C., AND J. C. SENCINDIVER. 1986. Properties and classification of mine soils at two sites in West Virginia. *Soil. Sci. Soc. Amer. J.* 36:181–185.

TILMAN, D. 1980. Resources: A graphical-mechanistic approach to competition and predation. *Am. Nat.* 116:362–393.

————. 1982. Resources: *Competition and Community Structure.* Princeton University Press, Princeton, NJ.

———. 1985. The resource ratio hypothesis of succession. *Am. Nat.* 125:827–852.

———. 1986. Evolution and differentiation in terrestrial plant communities: The importance of the soil resource-light gradient. In J. Diamond and T. Case (eds.), *Community Ecology.* Harper and Row, New York, pp. 359–380.

———. 1987. The importance of the mechanisms of the interspecific interaction. *Am. Nat.* 129:769–774.

———. 1988. *Plant Strategies and the Dynamics and Structure of Plant Communities.* Princeton University Press, Princeton, NJ.

TILMAN, D., M. MATTSON, AND S. LANGER. 1981. Competition and nutrient kinetics along a temperature gradient: An experimental est of a mechanistic approach to niche theory. *Limno. Oceanogr.* 26:1020–1033.

TINER, R. W. 1991. The concept of a hydrophyte for wetland identification. *Bioscience* 41:236–237.

TINKLE, D. W. 1969. The concept of reproductive effort and its relation to the evolution of life histories of lizards. *Am. Nat.* 103:501–516.

TINKLE, D. W., AND R. E. BALLINGER. 1972. Sceloporus undulatus, a study of the intraspecific comparative demography of a lizard. *Ecology* 53:570–585.

TINLEY, K. L. 1982. The influence of soil moisture balance on ecosystem patterns in southern Africa. In B. J. Huntley and B. H. Walker (eds.), *Ecology of Tropical Savannas.* Springer-Verlag, New York, pp. 175–192.

TOBEY, R. C. 1981. Saving the Prairies: *The Life Cycle of the Founding School of American Plant Ecology,* 1895–1955. University of California Press, Berkeley.

TOMASELLE, R. 1981a. Main physiognomic types and geographic distribution of shrub systems related to Mediterranean climates. In F. di Castri, D. W. Goodall, and R. L. Specht (eds.), *Mediterranean-Type Shrublands* (Ecosystems of the World, Vol. 11). Elsevier, Amsterdam, pp. 95–106.

———. 1981b. Relations with other ecosystems: Temperate evergreen forests, coniferous forests, savannas, steppes, and desert shrubland. In F. di Castri, D. W. Goodall, and R. L. Specht (eds.), *Mediterranean-Type Shrublands* (Ecosystems of the World, Vol. 11). Elsevier, Amsterdam, pp. 123–136.

TOMBACH, D. F. 1982. Dispersal of whitebark pine seeds by Clark's nutcracker: A mutualism hypothesis. *J. Anim. Ecol.* 51:451–467.

———. 1994. Ecological relationship between Clark's nutcracker and four wingless-seed *Strobus* pines of western North America. In W. C. Schmidt, F.-K. Holtmeier (comps.), *Proceedings*—International workshop on subalpine stone pines and their environment: the status of our knowledge; 1992 September 5–11; St. Moritz, Switzerland. Gen Tech. Rep. INT-grt-309. USDA Forest Service, Intermountain Research Station, Ogden, UT, pp. 221–224.

TOMLINSON, P. B. 1983. Structural elements of the rain forest. In F. B. Golley (ed.), *Tropical Rain Forest Ecosystems: Structure and Function* (Ecosystems of the World, Vol. 14A). Elsevier, Amsterdam, pp. 9–28.

TRABALKA, J. R. (ed.). 1985. *Atmospheric Carbon Dioxide and the Global Crabon Cycle.* U.S. Department of Energy Report DOE/ER-0239, Washington, DC.

TRABALKA, J. R., AND D. E. REICHLE (eds.). 1994. *The Changing Carbon Cycle: A Global Analysis.* Springer-Verlag, New York.

TRABAUD, D. L. 1981. Man and fire: Impacts on Mediterranean vegetation. In F. di Castri, D. W. Goodall, and R. L. Specht (eds.), *Mediterranean-Type Shrublands* (Ecosystems of the World, Vol. 11). Elsevier, Amsterdam, pp. 523–538.

TRACY, C. R. 1976. A model of the dynamic exchanges of water and energy between a terrestrial amphibian and its environment. *Ecol. Monogr.* 46:293–326.

TRANSEAU, E. N. 1926. The accumulation of energy by plants. *Ohio J. Sci.* 26:1–10.

TRIMBLE, G., JR. 1973. The regeneration of central Appalachian hardwoods with emphasis on the effects of site quality and harvesting. USDA Forest Service Res. Paper NE 282.

TRIMBLE, G., JR., AND E. H. TYRON. 1966. Crown encroachment into openings cut into Appalachian hardwood stands. *J. For.* 64:104–108.

TRISKA, F. J., AND K. CROMACH, JR. 1980. The role of wood debris in forests and streams. In R. H. Waring (ed.), *Forests: Fresh Perspectives from Ecosystem Analysis.* Oregon State University, Corvallis, pp. 171–190.

TRISKA, F. J., J. R. SEDELL, K. CROMACH, JR., S. V. GREGORY, AND F. M. MCCOUSON. 1984. Nitrogen budget for a small coniferous forest stream. *Ecol. Monogr.* 54:119–140.

TRISKA, F. J., J. R. SEDELL, AND S. V. GREGORY. 1982. Coniferous forest streams. In R. L. Edmonds (ed.), *Analysis of Coniferous Forest Ecosystems in Western United States.* US/IBP Synthesis Ser. No. 14. Dowden, Hutchinson and Ross, Stroudsburg, PA, pp. 292–332.

TRYON, E. H., AND G. R. TRIMBLE, JR. 1969. Effect of distance from stand border on height of hardwood reproduction in openings. *WV Acad. Sci. Proc.* 41:125–132.

TULLAR, B. F. 1979. The management of foxes in New York State. *Conservationist* 34(3):33–36.

TUNNICLIFFE, V. 1992. Hydrothermal vent communities of the deep sea. *Amer. Sci.* 80:336–349.

TURNER, G. F., AND T. J. PITCHER. 1986. Attack abatement: A model for group protection by combined avoidance and dilution. *Am. Nat.* 128:228–240.

TURNER, M. G. 1989. Landscape ecology: The effect of pattern on proces. Ann. Rev. Ecol. Syst., 20:171–197.

———. 1998. Landscape Ecology: Livingin a Mosaic. In S. I. Dotson, T. F. H. Allen, S. R. Carpenter, et al. (eds.), Ecology. Oxford University Press, New York, pp. 78–122.

TUXILL, J. 1998. Losing strands in the web of life: vertebrate declines and the conservation of biological diversity. Worldwatch Paper 131. Worldwatch Institute, Washington, DC.

TYRTIKOV, A. P. 1959. Perennially frozen ground. In Principles of Geocryology: Part I, *General Geocryology* (trans. from Russian by R. E. Brown). *Nat. Res. Cun. Canada Tech. Trans.* 1163(1964):399–421.

UDVARDY, M. D. F. 1958. Ecological and distributional analysis of North American birds. *Condor* 60:50–66.

UHL, C., AND C. F. JORDAN. 1984. Succession and nutrient dynamics following forest cutting and burning in Amazonia. *Ecology* 65:1467–1492.

UNDERWOOD, A. J. 1986. The analysis of competition by field experiments. In J. Kikkawa and D. J. Anderson (eds.), *Community Ecology: Pattern and Process.* Blackwell, Melbourne, pp. 240–268.

UNDERWOOD, A. J, E. J. DENLEY, AND M. J. MORAN. 1983. Experimental analyses of the structure and dynamics of mid-shore rocky intertidal communities in New South Wales. *Oecologica* 56:202–219.

URBAN, D. L., AND T. M. SMITH. 1989. Microhabitat pattern and the structure of forest bird communities. *Am. Nat.* 133:811–829.

USDA SOIL CONSERVATION SERVICE SOIL SURVEY STAFF. 1975. *Soil Taxonomy: A Basic System of Soil Classification for Making and Interpreting Soil Surveys.* Agr. Handbook 436. USDA, Washington, DC.

VALIELA, I., AND J. M. TEAL. 1979. The nitrogen budget of a salt marsh ecosystem. *Nature* 20:652–656.

VALIELA, I., J. M. TEAL, AND W. G. DENSER. 1978. The nature of growth forms in salt marsh grass *Spartina alterniflora. Am. Nat.* 112:461–470.

VALLENTYNE, J. R. 1974. *The Algal Bowl-Lakes and Man.* Misc. Spec. Publ. 22. Department of Environment, Ottawa, Canada.

VAN DER HAMMER, T., T. A. WIJMSTRA, AND W. H. ZAGWIGN. 1971. The floral record of late Cenozoic of Europe. In K. K. Turekian (ed.), *The Late Cenozoic Glacial Ages.* Yale University, New Haven, CT, pp. 391–424.

VAN DER VALK, A. G. 1981. Succession in wetlands: A Gleasonian approach. *Ecology* 62:68–696.

VAN DER VALK, A. G., AND C. B. DAVIS. 1978. The role of seed banks in the vegetation dynamics of prairie glacial marshes. *Ecology* 59:322–335.

VAN DOVER, C. L., C. J. BERG, AND R. D. TURNER. 1988. Recruitment of marine invertebrates to hard substrates at deep-sea hydrothermal vents on the East Pacific Rise and Galapagos spreading center. *Deep-Sea Res.* 35:1833–1849.

VAN HOOKE, R. I. 1971. Energy and nutrient dynamics of spider and orthopteran populations in a grassland ecosystem. *Ecol. Monogr.* 41:1–26.

VANDERMEER, J. H. 1980. Indirect mutualism: Variations on a theme by Stephen Levine. *Am. Nat.* 116:441–448.

VANDERMEER, J. H., AND D. H. BOUCHER. 1978. Varieties of mutualistic interaction in population models. *J. Theor. Biol.* 74:594–558.

VANNOTE, R. L., G. W. MINSHALL, K. W. CUMMINS, J. R. SCHELL, AND C. E. CUSHING. 1980. The river continuum concept. *Can. J. Fish. Aq. Sci.* 37:130–137.

VAUGHAN, T. A. 1978. *Mammalogy,* 2nd ed. Saunders, Philadelphia.

VEALE, P. T., AND H. L. WASCHER. 1956. Henderson County soils. Illinois Univ. Agr. Expt. Sta. Soil Rept. No. 77.

VEMAP PARTICIPANTS. 1995. Vegetation/ecosystem modeling and analysis project: Comparing biogeography and biogeochemistry models in a continental-scale study of terrestrial ecosystem responses to climate change and CO_2 doubling. *Glob. Biogeochem.* 9:407–437.

VERME, L. J., AND D. I. ULTRY. 1984. Physiology and nutrition. In L. Halls (ed.), *White-Tailed Deer Ecology and Management.* Stackpole, Harrisburg, PA, pp. 91–118.

VERNER, J., M. L. MORRISON, AND C. J. RALPH. 1986. *Wildlife 2000: Modeling Habitat Relationships of Terrestrial Vertebrates.* University of Wisconsin Press, Madison.

VEZINA, P., AND D. W. K. BOULTER. 1966. The spectral composition of near ultraviolet and visible radiation beneath forest canopies. *Can. J. Bot.* 44:1267–1283.

VITOUSEK, P. M., S. W. ANDARIESE, P. A. MATSON, L. MORRIS, AND R. L. SANFORD. 1992. Effects of harvest intensity, site preparation, and herbicide use on soil nitrogen transformations in a young loblolly pine plantation. *For. Ecol. Manag.* 49:277–292.

VITOUSAK, P. M., AND R. L. SANFORD, JR. 1986. Nutrient cycling in moist tropical forests. *Ann. Rev. Ecol. Syst.* 17:131–167.

VITT, L. J. 1992. Lizard mimics millipede. *Nat. Geog. Res. Explor.* 8:76–95.

VOGEL, S. 1969. Flowers offering fatty oil instead of nectar. *XI Proc. Intl. Bot. Congress,* Seattle, WA. p. 229.

VOIGHT, W. 1948. *The Road to Survival.* William Sloan, New York.

VOIGT, G. K. 1971. Mycorrhizae and nutrient mobilization. (pp. 122 - 131) In E. Hacskaylo (ed.), *Mycorrhizae.* USDA Forest Service Misc. Pub. No. 1189, pp. 122–131.

VOLTERRA, V. 1926. Variation and fluctuations of the numbers of individuals in animal species living together. Reprinted in R. M. Chapman (1931), *Animal Ecology.* McGraw-Hill, New York, pp. 409–448.

WALBOT, V., AND C. A. CULLIS. 1985. Rapid genomic changes in higher plants. *Ann. Rev. Plant Physiol.* 36:367–396.

WALDMAN, B., AND J. S. MCKINNON. 1993. Inbreeding and outbreeding in fishes, amphibians, and reptiles. In N. W. Thornhill (ed.), *Natural History of Inbreeding and Outbreeding: Theoretical and Empirical Perspectives.* University of Chicago Press, Chicago, pp. 250–283.

WALDMAN, J. M., J. W. MUNGER, D. J. JACOB, R. C. GLAGAN, J. J. MORGAN, AND M. R. HOFFMAN. 1982. Chemical composition of acid fog. *Science* 218:677–680.

WALKER, J. C. 1980. The oxygen cycle. In O. Hutzinger (ed.), *The Natural Environment and the Biogeochemical Cycles.* Springer-Verlag, New York, pp. 87–104.

———. 1984. How life affects the atmosphere. *Bioscience* 43:486–491.

WALKER, J. M., A. GARGER, R. H. J. BERGER, AND H. C. HELLER. 1979. Sleep and aestivation (shallow torpor): Continuous process of energy conservation. *Science* 204:1098–1100.

WALKER, P. C., AND R. T. HARTMAN. 1960. Forest sequence of the Hartstown bog area in western Pennsylvania. *Ecology* 41:461–474.

WALKER, R. B. 1954. Ecology of serpentine soils. II. Factors affecting plant growth on serpentine soils. *Ecology* 35:259–274.

WALKINGSHAW, L. 1983. *Kirtland's Warbler.* Cranbrook Institute of Science, Bloomfield Hills, MI.

WALLACE, B. 1968. *Topics in Population Genetics.* Norton, New York.

WALLACE, J. B., J. R. WEBSTER, AND T. F. COFFNEY. 1982. Stream detritus dynamics regulation by invertebrate consumers. *Oecologica* 53:197–200.

WALLER, D. M. 1982. Jewelweed's sexual skills. *Nat. History* 91(5):32–39.

WALLER, D. M., AND W. S. ALVERSON. 1997. The white-tailed deer: A keystone herbivore. *Wildl. Soc. Bull.* 25:217–226.

WALTER, H. 1977. *Ecology of Tropical and Subtropical Vegetation.* Oliver and Boyd, Edinburgh.

———. 1979. *Vegetation of the Earth and Ecological Systems of the Geosphere,* 2nd ed. Springer-Verlag, New York.

WALTER, H., E. HARNICKELL, AND D. MUELLER-DOMBOIS. 1975. *Climate-Diagram Maps of the Individual Continents and the Ecological Climatic Regions of the Earth.* Springer Verlag, New York.

WALTERS, C. J. 1986. *Adaptive Management of Natural Resources.* Macmillan, New York.

WARNER, R. E. 1968. The role of introduced diseases in the extinction of the endemic Hawaiian avifauna. *Condor* 70:101–120.

WARRICK, R. A., C. LEPROVOST, M. F. MEIER, J. OERLEMANS, AND P. L. WOODWORTH. 1996. Changes in sea level. In J. T. Houghton, L. G. Meira Filho, B. A. Callander, et al. (eds.), *Climate Change 1995: The Science of Climate Change.* Intergovernmental Panel on Climate Change. Cambridge Univ. Press, Cambridge, England.

WASER, N. M., AND L. A. REAL. 1979. Effective mutualism between sequentially flowering plant species. *Nature* 281:670–672.

WASER, P. M. 1985. Does competition drive dispersal? *Ecology* 66:1170–1175.

WATERHOUSE, F. L. 1955. Microclimatological profiles in grass cover in relation to biological problems. *Quart. J. Roy. Meteorol. Soc.* 81:63–71.

WATERS, T. F. 1972. The drift of stream insects. *Ann. Rev. Entomol.* 17:253–212.

WATSON, R. T., M. C. ZINYOWERA, R. H. MOSS, AND D. J. DOKKEN (eds.). 1998. *The Regional Impacts of Climate Change* (A Special Report of IPCC Working Group II). Cambridge University Press, Cambridge UK.

WATT, A. S. 1947. Pattern and process in the plant community. *J. Ecol.* 35:1–22.

———. 1955. Bracken versus heather: A study in plant sociology. *J. Ecol.* 43:490–406.

WAUTERS, L., AND A. A. DOHONDT. 1989. Body weight, longevity, and reproductive success in red squirrels (*Sciurus vulgaris*). *J. Anim. Ecol.* 58:637–651.

WEAVER, J. E. 1954. North American Prairie. Johnson, Lincoln, NE.

WEAVER, J. E., AND F. W. ALBERTSON. 1956. *Grasslands of the Great Plains: Their Nature and Use.* Johnson, Lincoln, NE.

WEAVER, M., AND M. KELLMAN. 1981. The effects of forest fragmentation on woodlot tree biotas in Southern Ontario. *J. Biogeogr.* 8:199–210.

WEAVER, P. L. 1995. The colorado and dwarf forests of Puerto Rico's Luquillo Mountains. In A. E. Lugo and C. Lowe (eds.), *Tropical Forests: Management and Ecology.* Springer Verlag, New York, pp. 109–141 .

WEBSTER, J. 1956–1957. Succession of fungi on decaying cocksfoot culms: Parts 1 and 2. *J. Ecol.* 44:517–544; 45:1–30.

WEBSTER, J. R., D. J. D'ANGELO, AND G. T. PETERS. 1991. Nitrate and phosphate uptake in streams at Coweeta Hydrological Laboratory. *Vehn. Internat. Verein Limnol.* 24:1681–1686.

WEEKS, H. P., JR., AND C. M. KIRKPATRICK. 1976. Adaptations of white-tailed deer to naturally occurring sodium deficiencies. *J. Wildl. Manage.* 40:610–625.

———. 1978. Salt preferences and sodium drive phenology in fox squirrels and woodchuck. *J. Mamm.* 59:531–542.

WEIR, J. S. 1972. Spatial distribution of elephants in an African national park in relation to environmental sodium. *Oikos* 23:1–13.

WEISE, C. M. 1974. Seasonality in birds. In H. Leith (ed.), *Phenology and Seasonality Modeling.* Springer-Verlag, New York, pp. 139–147.

WERNER, E. E., AND D. J. HALL. 1976. Niche shift in sunfishes: Experimental evidence and significance. *Science* 191:404–406.

———. 1977. Competition and habitat shift in two sunfishes (*Centrarchidae*). *Ecology* 58:867–876.

———. 1979. Foraging efficiency and habitat switching in sunfishes. *Ecology* 60:256–264.

WEST, N. E. 1979. Formation, distribution, and function of plant litter in desert ecosystems. In D. W. Goodall and R. A. Perry (eds.), Arid-Land Ecosystems: Structure, Functioning and Management, Vol. 1. Cambridge University Press, London, pp. 647–659.

———. 1981. Nutrient cycling in desert ecosystems. In D. W. Goodall, R. A. Perry, and K. M. W. Howes (eds.), *Arid-Land Ecosystems: Structure, Functioning and Management,* Vol. 2. Cambridge University Press, Cambridge, England, pp. 301–324.

WEST, N. E., AND J. J. SKUJINS (eds.). 1978. *Nitrogen in Desert Ecosystems,* US/IBP Synthesis Series 9. Dowden, Hutchinson and Ross, Stroudsburg, PA.

WEST, N. E., AND P. T. TUELLER. 1972. Special approaches to studies of competition and succession in shrub communities. In C. M. McKell, J. P. Blaisdell, and J. R. Goodin, *Wildland Shrubs; Their Biology and Utilization.* USDA Forest Service Gen. Tech. Rept INT-1, pp. 172–181.

WESTOBY, M. 1984. The self-thinning rule. *Adv. Ecol. Res.* 14:167–225.

WETZEL, R. 1975. *Limnology.* Saunders, Philadelphia.

WETZEL, R. G., AND H. L. ALLEN. 1970. Function and interactions of dissolved organic matter and the littoral zone in lake metabolism and eutrophication. In Z. Kabap and A. Hillbricht-Ilkowska (eds.), *Productivity Problems of Freshwater.* PWN Polish Sci. Publ., Warsaw, pp. 333–347.

WETZEL, R. G., P. H. RICH, M. C. MILLER, AND H. L. ALLEN. 1972. Metabolism of dissolved and particulate detrital carbon in a temperate hardwater lake. *Mem. 1st. Ital. Idrobiol.* 29(Supp.):185–243.

WHICKER, A. D., AND J. K. DETLING. 1988. Ecological consequences of prairie dog disturbances. *Bioscience* 38:778–785.

WHIGHAM, D. F., J. MCCORMICK, R. E. GOOD, AND R. L. SIMPSON. 1978. Biomass and primary production of freshwater tidal marshes. In R. E. Good, D. F. Whigham, and R. L. Simpson (eds.), *Freshwater Wetlands: Ecological Processes and Management Potentials.* Academic, New York, pp. 243–257.

WHITCOMB, R. F., J. F. LYNCH, P. A. OPLER, AND C. S. ROBBINS. 1976. Island biogeography and conservation: Strategy and limitations. *Science* 193:1030–1032.

WHITE, J. 1979. The plant as a metapopulation. *Ann. Rev. Ecol. Syst.* 10:109–145.

WHITE, J., AND J. L. HARPER. 1970 Correlated changes in plant size and number in plant populations. *J. Ecol.* 58:467–485.

WHITFORD, W. G., D. W. FRECHMAN, N. Z. ELKINS, L. W. PARKER, R. PARMALEO, J. PHILLIPS, AND S. TUCKER. 1981. Diurnal migration and response to simulated rainfall in desert soil microarthropods and nematodes. *Soil Biol. Biogeochem.* 13:17–425.

WHITHAM, T. G. 1980. The theory of habitat selection: Examined and extended using Pemphigus aphids. *Am. Nat.* 115:449–466.

WHITMORE, T. C. 1984. *Tropical Rain Forest of the Far East*, 2nd ed. Oxford, London.

———. 1990. *An Introduction to Tropical Rain Forests.* Oxford University Press, New York.

WHITTAKER, R. H. 1953. A consideration of the climax theory: The climax as a population and pattern. *Ecol. Monogr.* 23:41–78.

———. 1954. The ecology of serpentine soils. IV. The vegetational response to serpentine soils. *Ecology* 35:275–288.

———. 1956. Vegetation of the Great Smoky Mountains. *Ecol. Monogr.* 26:1–80.

———. 1960. Vegetation of the Siskiyou Mountains, Oregon and California. *Ecol. Monogr.* 30:279–338.

———. 1962. Classification of natural communities. *Bot. Rev.* 28:1–239.

———. 1963. Net production of heath balds and forest heaths in the Great Smoky Mountains. *Ecology* 44:176–182.

———. 1965. Dominance and diversity in land plant communities. *Science* 147:250–260.

———. 1967. Gradient analysis of vegetation. *Biol. Rev.* 42:207–264.

———. 1972. Evolution and the measurement of species diversity. *Taxon.* 21:213–251.

———. 1974. Climax concepts and recognition. In R. Knapp (ed.), *Vegetation Dynamics.* Junk, The Hague, pp. 137–154.

———. 1975. *Communities and Ecosystems,* 2nd ed. Macmillan, New York.

———. 1977. Evolution of species diversity in land communities. *Evol. Biol.* 10:1–67.

WHITTAKER, R. H., F. H. BORMANN, G. E. LIKENS, AND T. G. SICCAMA. 1974. The Hubbard Brook ecosystem study: Forest biomass and production. *Ecol. Monogr.* 44:233–252.

WHITTAKER, R. H., AND P. R. FEENEY. 1971. Allelochemics: Chemical interactions between species. *Science* 171:757–770.

WHITTAKER, R. H., S. A. LEVIN, AND R. B. ROOT. 1973. Niche, habitat, and ecotope. *Am. Nat.* 107:321–338.

WHITTAKER, R. H., AND G. M. WOODWELL. 1968. Dimension and production relations of trees and shrubs in the Brookhaven forest, New York. *Ecology* 56:1–25.

———. 1969. Structure, production, and diversity of the oak-pine forest at Brookhaven, New York. *J. Ecol.* 57:155–174.

WIELAND, N. K., AND F. A. BAZZAZ. 1975. Physiological ecology of three codominant successional annuals. *Ecology* 56:681–688.

WIELGOLASKI, F. E. 1975. Productivity of tundra ecosystems. In F. E. Wielgolaski (ed.), *Fennoscandian Tundra Ecosystems.* Part 1, *Plants and Microorganisms.* Springer-Verlag, New York, pp. 1–12.

WIELGOLASKI, F. E., L. C. BLISS, J. SUOBODA, AND G. DOYLE. 1981. Primary production of tundra. In L. C. Bliss, O. W. Heal, and J. J. Moore (eds.), *Tundra Ecosystems: A Comparative Analysis.* Cambridge University Press, Cambridge, England, pp. 187–225.

WIELGOLASKI, F. E., AND S. KJELVIK. 1975. Energy content and use of solar radiation of Fennoscandian tundra plants. In F. E. Wielgolaski (ed.), *Fennoscandian Tundra Ecosystems.* Part 1, *Plants and Microorganisms.* Springer-Verlag, New York, pp. 201–207.

WIENS, J. A. 1973. Pattern and process in grassland bird communities. *Ecol. Monogr.* 43:237–270.

———. 1976. Population responses to patchy environments. *Ann. Rev. Ecol. Syst.* 7:81–120.

———. 1977. On competition and variable environments. *Am. Sci.* 65:590–597.

———. 1985. Habitat selection in variable environments: Shrub-steppe birds. In M. Cody (ed.), *Habitat Selection in Birds.* Academic Press, Orlando, FL, pp. 227–251.

———. 1997. Metapopulation dynamics and landscape ecology. In I. Hanski and M. Gilpin (eds.), *Metapopulation Biology.* Academic Press, San Deigo, pp. 43–62.

WIENS, J. A., J. F. ADDICOTT, T. J. CASE, AND J. DIAMOND. 1986. Overview of the importance of spatial and temporal scale in ecological investigations. In J. Diamond and T. Case (eds.), *Community Ecology.* Harper and Row, New York, pp. 229–239.

WIENS, J. A., AND J. T. ROTENBERRY. 1981. Habitat associations and community structure of birds in shrub-steppe environments. *Ecol. Monogr.* 51:21–41.

WILBUR, H. M., AND J. E. FAUTH. 1990. Experimental aquatic food webs: Interactions between two predators and two prey. *Am. Nat.* 135:176–204.

WILCOVE, D. S. 1986. Nest predation in forest tracts and the decline of migratory songbirds. *Ecology* 66:1211–1214.

WILLIAMS, C. B. 1966. *Adaptation and Natural Selection.* Princeton University Press, Princeton, NJ.

WILLIAMS, C. E. 1965. Soil fertility and cottontail body weights: A reexamination. *J. Wildl. Manage.* 28:329–337.

WILLIAMS, C. E., AND A. I. CASHEY. 1965. Soil fertility and cottontail fecundity in southeastern Missouri. *Am. Midl. Nat.* 74:211–224.

WILLIAMS, G. C. 1966. *Adaptation and Natural Selection.* Princeton University Press, Princeton, NJ.

———. 1975. *Sex and Evolution.* Princeton University Press, Princeton, NJ.

WILLIAMS, K. S., AND L. E. GILBERT. 1981. Insects as selective agents on plant vegetative morphology: Egg mimicry reduces egg laying by butterflies. *Science* 212:467–469.

WILLIAMS, K., K. G. SMITH, AND F. M. STEVEN. 1993. Emergence of 13-year periodical cicadas (*Cicadidae: Magicicada*): Phenology, mortality, and predator satiation. *Ecology* 74:1143–1152.

WILLIAMS, M. 1989. *Americans and Their Forests: A Historical Geography.* Cambridge University Press, Cambridge, England.

WILLIAMS, W. E., K. GARBUTT, F. A. BAZZAZ, AND P. M. VITOUSEK. 1986. The response of plants to elevated CO_2. IV. Two deciduous forest tree communities. *Oecologia* 69:454–459.

WILLIS, A. J. 1963. Braunton burrows: The effects on vegetation of the addition of mineral nutrients to the dune soils. *J. Ecol.* 51:353–374.

WILLIS, E. O. 1963. Is the zone-tailed hawk a mimic of the turkey vulture? *Condor* 65:313–317.

———. 1984. Conservation, subdivision of reserves, and antidismemberment hypothesis. *Oikos* 42:396–398.

WILLOUGHBY, L. C. 1974. Decomposition of litter in freshwater. In C. H. Dickinson and G. J. F. Pugh (eds.), *Biology of Plant Litter Decomposition,* Vol. II. Academic, London, pp. 659–661.

WILSON, D. 1975. A theory of group selection. *Proc. Nat. Acad. Sci.* 72:143–146.

———. 1977. Structured demes and the evolution of group-advantageous traits. *Am. Nat.* 111:157–185.

———. 1979. Structured demes and trait-group variation. *Am. Nat.* 113:606–610.

———. 1980. *The Natural Selection of Populations and Communities.* Benjamin/Cummings, Menlo Park, CA.

———. 1983. The group selection controversy: History and current status. *Ann. Rev. Ecol. Syst.* 14:159–187.

WILSON, D. E., AND G. L. GRAHAM (eds.). 1992. *Pacific Island Flying Foxes: Proceedings of an International Conservation Conference.* U.S. Fish and Wildlife Service Biological Report 90(23). 176 pp.

WILSON, E. O. 1971. Competitive and aggressive behavior. In J. Eisenberg and W. Dillim (eds.), *Man and Beast: Comparative Social Behavior.* Smithsonian Institute Press, Washington, DC, pp. 522–533.

———. 1975. *Sociobiology: The New Synthesis.* Harvard University Press, Cambridge, MA.

———. 1999. *The Diversity of Life.* Norton. New York.

WILSON, E. O., AND W. H. BOSSERT. 1971. *A Primer of Population Biology.* Sinauer Associates, Sunderland, MA.

WILSON, E. O., AND D. S. SIMBERLOFF. 1969. Experimental zoogeography of islands: Defaunation monitoring techniques. *Ecology* 50:267–278.

WING, L. D., AND I. D. BUSS. 1970. Elephants and forests. *Wildl. Monogr.* 19. The Wildlife Society, Washington, DC.

WITHERSPOON, J. P., JR. 1964. Cycling of cesium-134 in white oak trees. *Ecol. Monogr.* 34:403–420.

WITHERSPOON, J. P., S. I. AVERBACH, AND J. S. OLSON. 1962. Cycling of cesium-134 in white oak trees on sites of contrasting soil type and moisture. *Oak Ridge Nat. Lab.* 3328:1–143.

WITKAMP, M., AND D. A. CROSSLEY. 1966. The role of arthropods and microflora on the breakdown of white oak litter. *Pedobiologia* 6:293–303.

WOLFF, J. O. 1980. The role of habitat patchiness in the population dynamics of snowshoe hares. *Ecol. Monogr.* 50:111–130.

———. 1988. Maternal investment and sex ratio adjustment in American bison calves. *Behav. Ecol. Sociobiol.* 23:127–133.

WOLIN, M. J. 1979. The rumen fermentation: A model for microbial interactions in anaerobic systems. *Adv. Microb. Ecol.* 3:49–77.

WOODWARD, F. I. 1987. *Climate and Plant Distribution.* Cambridge University Press, New York.

WOODWARD, F. I., AND C. K. KELLY. 1995. The influence of CO_2 concentration on stomatal density. *New Phytol.* 131:311–327.

WOODWARD, F. I., AND T. M. SMITH. 1993. Predictions and measurements of the maximum photosynthetic rate, A_{max}, at a global scale. In E. D. Schulze M. M. and Caldwell (eds.), *Ecophysiology of Photosynthesis,* Vol. 100. Springer, Berlin. pp. 491–508.

———. 1994. Global photosynthesis and stomatal conductance: Modeling the controls of soil and climate. Adv. Bot. Res. 20:367–384.

WOODWARD, I. 1987. *Climate and Plant Distribution.* Cambridge University Press, Cambridge, England.

WOODWELL, G. M., R. A. HOUGHTON, C. A. S. HALL, D. E. WHITNEY, R. A. MOLL, AND D. W. JUERS. 1979. The Flax Pond ecosystem study: The annual metabolism and nutrient budgets of a salt marsh. In R. L. Jeffries and A. J. Davies (eds.), *Ecological Processes in Coastal Environments.* Blackwell, Oxford, England, pp. 491–511.

WOOLFENDEN, G. E. 1975. Florida scrub jay helpers at the nest. *Auk* 92:1–15.

WOOLFENDEN, G. E., and J. W. Fitzpatrick. 1984. *The Florida Scrub Jay: Demography of a Cooperatively Breeding Bird.* Princeton University, Princeton, NJ.

WRIGHT, H. A., AND R. W. BAILEY. 1982. *Fire Ecology: United States and Southern Canada.* Wiley, New York.

WRIGHT, W. G. 1988. Sex change in Mollusca. *TREE* 3:137–140.

WUNDERLE, J. M., JR., AND Z. COTTO-NAVARRO. 1988. Constant vs. variable risk-aversion in foraging bananaquits. *Ecology* 69:1434–1438.

WYMAN, R. L. 1998. Experimental assessment of salamanders as predators of detrital food webs: Effects on invertebrates, decomposition, and the carbon cycle. *Biodiv. Conserv.* 7:641–650.

YAHNER, R. H. 1988. Changes in wildlife communities near edges. *Cons. Biol.* 2:333–339.

YODA, K., T. KIRA, H. OGAWA, AND K. HOZUMI. 1963. Self-thinning in overcrowded pure stands under cultivated and natural conditions. *J. Biol. Osaka Univ.* 14:107–129.

YOM-TOV, Y. 1980. Intraspecific nest parasitism in birds. *Biol. Rev. Cont. Phil. Soc.* 55:93–108.

YONGE, C. M. 1963. The biology of coral reefs. *Adv. Mar. Biol.* 1:209–260.

YOSHINO, M., T. HORIE, H. SEINO, H. TSUJII, T. UCHIJIMA, AND Z. UCHIJIMA. 1988. The effects of climate variations on agriculture in Japan. In M. Parry, T. R. Carter, and N. T. Konijn (eds.), *The Impacts of Climate Variation on Agriculture,* Volume 1, *Assessments in Cool Temperate and Cold Regions.* Kluwer, Dordrecht, The Netherlands.

ZACK, R., AND J. B. FALLS. 1976a. Ovenbird hunting behavior in a patchy environment: An experimental study. *Can. J. Zool.* 54:1863–1879.

———. 1976b. Foraging behavior, learning, and exploration by captive ovenbirds. *Can. J. Zool.* 54:1880–1893.

———. 1976c. Do ovenbirds hunt by expectation? *Can. J. Zool.* 54:1894–1903.

ZAHAVI, A. 1975. Mate selection for a handicap. *J. Theor. Biol.* 53:205–214.

———. 1977. The cost of honesty (further remarks on the handicap principle). *J. Theor. Biol.* 67:603–605.

ZALLER, H., AND J. A. ARNONE. 1997. Activity of surface-casting earthworms in a calcereous grassland under elevated atmospheric CO_2. *Oecologia* 111:249 254.

ZEDLER, P. H., C. R. GAUTIER, AND G. S. MCMASTER. 1983. Vegetation change in response to extreme events: The effect of a short interval between fires in California chaparral and coastal shrub. *Ecology* 64:809–818.

ZEEVALKING, H. S., AND L. F. M. FRESCO. 1977. Rabbit grazing and diversity in a dune area. *Vegetatio* 35:193–196.

ZIEMAN, J. C. 1982. The ecology of the seagrasses of South Florida: A community profile. U.S. Fish and Wildlife Service program FWS/OBS-82/25.

ZIMEN, E. 1978. *The Wolf: A Species in Danger* (1981 translation). Dell, New York.

ZIMMERMAN, J. L. 1971. The territory and its density dependent effect in *Spiza americana. Auk* 88:591–612.

ZOELLICK, B. W., AND N. S. SMITH. 1992. Size and spatial organization of home ranges of kit foxes in Arizona. *J. Mamm.* 73:83–88.

ZUK, M. 1991. Parasites and bright birds: New data and new predictions. In J. E. Loge and M. Zuk (eds.), *Bird-Parasite Interactions.* Oxford University Press, Oxford, England, pp. 317–327.

ACKNOWLEDGEMENTS

PHOTO ACKNOWLEDGEMENTS

Unless otherwise acknowledged, all photographs are the property of Benjamin Cummings. Page abbreviations are as follows: (T) top, (B) bottom, (C) center, (L) left, (R) right, (INS) inset.

PART OPENING PHOTOS: Page 1: Corbis/Kevin R. Morris *Page 19:* FPG International/Ron Thomas *Page 79:* Corbis/Galen Rowel *Page 159:* Corbis/Robert Yin *Page 241:* Corbis/Michael and Patricia Foyden *Page 355:* Allstock Inc. *Page 381:* Photo Researchers, Inc./M. Harvey *Page 477:* R. L. Smith *Page 561:* R. L. Smith

Page 2: Corbis/Kevin R. Morris *Page 4L:* Culver Pictures *Page 4C&R:* Courtesy Hunt Inst. For Botanical Documentation/Carnegie Mellon Unversity, Pittsburgh, PA *Page 6:* Courtesy of Mr. G. P. Darwin/By permission of the Darwin Museum, Down House *Page 10:* Erich Hartmann/Magnum Photos *Page 14L:* U.S. Forestry Service *Page 14R:* R. L. Smith *Page 20:* Corbis/Eisele Reinhard *Page 42:* Corbis/Wolfgang Kaehler page 57:* Corbis/Charles O'Rear *Page 80:* The Image Bank/Joseph Van Os *Page 85:* PhotoDisc/Robert Glusic Part 90: Corbis/Steve Kaufman *Page 98:* Corbis/Eric Crichton *Page 110:* Walter Chandoha *Page 111:* John Eastcott/YVA Momatiuk/DRK Photo *Page 112:* R. L. Smith *Page 120:* Corbis/K. M. Westerman *Page 125:* R. L. Smith *Page 127T:* Corbis/Tom Brakefield *Page 127C&B:* R. L. Smith *Page 145:* R. L. Smith *Page 149:* William E. Ferguson Photography *Page 151:* R. L. Smith *Page 160:* Corbis/David Muench *Page 161:* Robert and Linda Mitchell *Page 181:* Corbis/Wolfgang Kaehler *Page 194:* Corbis/Layne Kennedy *Page 214:* The Image Bank/Paul McCormick *Page 218:* C. C. Lockwood/DRK Photo *Page 220:* M. Anderson/VIREO *Page 222:* Tom Bledsoe/Allstock Inc./Tony Stone Images *Page 224:* James Randkler/Allstock Inc./Tony Stone Images *Page 232:* Doug Wechsler/VIREO *Page 242:* Kennan Ward Photography *Page 252:* Ken Brate/Photo Researchers *Page 264:* Corbis/Michael and Patricia Fogden *Page 284:* Minden Picture/Frans Lanting *Page 285:* John Bora/Photo Researchers *Page 286:* S. J. Krasemann/Peter Arnold, Inc. *Page 288:* D. Cavagnaro/DRK Photo *Page 289:* John Serrao/Photo Researchers *Page 290:* R. L. Smith *Page 294:* Corbis/Gary W. Carter Page 295T John Cancalosi/Peter Arnold, Inc. *Page 285B:* Corbis/Joe McDonald *Page 296L:* Gunter Ziesler/Peter Arnold Inc. *Page 296R:* S. J. Krasemann/Peter Arnold Inc. *Page 306:* Corbis/Malcom Kilto *Page 308:* Robert and Linda Mitchell *Page 310:* R. L. Smith *Page 315:* Gary R. Zahm/DRK Photo *Page 320:* Richard R. Hansen/Photo Researchers *Page 321:* Robert and Linda Mitchell *Page 322:* Gary Braasch *Page 323:* Bruce Iverson *Page 325:* Courtesy of D. M. Ball/Auburn University Page 326L Darrell Gulin/Allstock Inc./Tony Stone Images *Page 326R:* Robert and Linda Mitchell *Page 327:* Dr. Morley Read/SPL/Photo Researchers *Page 328:* Kim Taylor/Bruce Coleman Inc. *Page 329ALL:* R. L. Smith *Page 330:* Gregory K. Scott/Photo Researchers *Page 331L:* Ted Wood/Tony Stone Images *Page 331R:* Norbert Wu/Allstock Inc./Tony Stone Images *Page 335:* AP Wide World Photos/Al Grillo *Page 343:* Jeffrey David/The Image Bank *Page 345:* Ron Austing Frank Lane Picture Agency/Corbis *Page 347:* R. L. Smith *Page 355:* Art Wolfe All Stock Inc./Tony Stone Images *Page 356:* Super Stock *Page 360:* Photo Researchers, Inc. *Page 361:* The Photo Library of Australia/Index Stock Photography *Page 368T:* A&E Morris/VIREO *Page 368B:* Nigel Dennis/Photo Researchers, Inc. *Page 381M:* Harvey/Photo Researchers, Inc. *Page 382:* Corbis/Adam Woolfitt *Page 403:* Craig Brewer/PhotoDisc, Inc. *Page 406:* R. L. Smith *Page 408:* Alan Pitcairn/Grant Heilman Photography *Page 409T:* © Glen M. Oliver/Visuals Unlimited *Page 410:* Corbis/Jeffery L. Rotman *Page 411:* R. L. Smith *Page 411INS:* R. L. Smith *Page 415:* Roger Tidman/Corbis *Page 427:* Don Bryan/AP Wide World Photos *Page 446:* R. L. Smith *Page 449:* Emma Lee/Lifefile/Photodisc *Page 450:* R. L. Smith *Page 465ALL:* Photos Courtesy of NIFC/Photography by Tianna Glenn *Page 466:* Tony Aruzza/Bruce Coleman Inc. *Page 467:* Corbis/Kevin R. Norris *Page 489T:* R. L. Smith *Page 489B:* Jay Brown/Courtesy Robert Clark Photography *Page 469:* Stephen J. Krasemann/Photo Researchers *Page 470T:* Peter D. Pickford/DRK Photo *Page 470B:* The Image Bank/Chuck Kuhn Photography *Page 471L:* Tom Bean/DRK Photo *Page 471R:* T. A. Wiewandt/DRK Photo *Page 474T:* APWW/Rhonda Simpson *Page 474B:* Academy of Natural Sciences Philadelphia/VIREO *Page 478:* Corbis/David A. Northcott *Page 488:* Stephen J. Krasemann/DRK Photo *Page 499T:* R. L. Smith *Page 499M:* Joseph R. Pearce/DRK Photo *Page 504:* Wolfgang Kaehler *Page 518L:* Breck P. Kent/Earth Scenes *Page 518R:* R. Degginger/Earth Scenes *Page 524:* Corbis/JS 001447 *Page 535:* R. L. Smith *Page 543:* Corbis/Hubert Stadler *Page 549ALL:* R. L. Smith *Page 562:* Diana Lo Stratton/Tom Stack & Associates *Page 565L:* Michael Black/Bruce Coleman Inc. *Page 565R:* Breck P. Kent/Earth Scenes 566TL: Tom Bean/DKR Photo *Page 566TR:* Keven Mageel/Tom Stack & Associates *Page 566B:* Stpehen J. Krasemann/DRK Photo *Page 567L:* Walt Anderson *Page 567R:* Neil Cooper/Panos Pictures *Page 570:* Dave Watts/Tom Stack & Associates *Page 575TL:* Jim

Holmes/Panos Pictures *Page 575TR:* Bill Roth/Bruce Coleman Inc. *Page 575BL:* Dale & Marian Zimmerman/Bruce Coleman Inc. *Page 575RT:* Eli A. Mittermeier/Brice Coleman Inc. *Page 579:* R. L. Smith Pge 580: Dan Suzio *Page 581B:* Matthew Kneale/Panos Pictures *Page 582:* The Photo Library-Sydney/Michael James *Page 583:* Michael P. Gradowski/Bruce Coleman Inc. *Page 586T:* Brian Parker/Tom Stack & Associates *Page 586B:* Adris Apse/Bruce Coleman Inc. *Page 587:* Stephen J. Krasemann/DRK Photo *Page 588:* Des Bartlett/Bruce Coleman Inc. *Page 589:* John Eastcott/YVA Momatiuk/DRK *Page 592T:* R. L. Smith *Page 592BL:* Steve McCutcheon *Page 592BR:* R. L. Smith *Page 593ALL:* R. L. Smith *Page 594:* Fred Bruemmer/ DRK Photo *Page 595BL:* Steve McCutcheon *Page 595BR:* Robin Smith/The Photo Library-Sydney *Page 596:* Walt Anderson *Page 601:* Corbis/Steve Kaufman *Page 603:* Rick Buzzeli/Tom Stack & Associates *Page 604T:* Tom Bean/DRK Photo *Page 604B:* Ken Lewis/Earth Sciences *Page 605TL:* Patti Murray/Earth Sciences *Page 605TR:* Tom Bean/DRK Photo *Page 605B:* Larry Ulrich/DRK Photo *Page 606TL:* Heather Angel/Biophoto Associates *Page 606TR:* Hans Reinhard/Bruce Coleman Inc. *Page 607:* Jeremy Hartley/Panos Pictures *Page 608TL:* Larry Ulrich/DRK Photo *Page 608TR:* Tom Coker/Yesteryear Photography *Page 608B:* Jack Wilburn/Earth Scenes *Page 609L:* David, C. Fritts/Earth Scenes *Page 609R:* Doug Wechsler/Earth Scenes *Page 610TL:* Dr. Nigel Smith/Earth Scenes *Page 610TR:* James P. Rowan/DRK Photo *Page 610BL:* Belinda Wright/DRK Photo *Page 610BR:* Peter French/Bruce Coleman *Page 611:* Michael Fogden/Bruce Coleman *Page 612T:* Corbis/Kevin Schafer *Page 612B:* Corbis/Richard Hamilton *Page 613:* N. H. Cheatham/DRK Photo *Page 615:* R L Smith *Page 617:* Corbis/Gary Braasch *Page 619:* Michael Fogden/Bruce Coleman *Page 627:* R L Smith *Page 628:* Dr. Nigel Smith/Earth Scenes *Page 629:* Corbis/Paul A. Souders *Page 631T:* Bob and Ira Spring *Page 631CL:* Bob and Ira Spring *Page 631CR:* Alex Maclean/Landslides *Page 631BL:* Landslides *Page 631BR:* Rich Buzzelli/Tom Stack & Associates *Page 634BL:* Gary Braasch *Page 635TL:* Brian Parker/Tom Stack & Associates *Page 635TR:* David Wrobel/Biological Photo Service *Page 639T:* Gary Braasch page *639B:* Brian Parker/Tom Stack & Associates *Page 641T:* Dale Jorgenson/Tom Stack & Associates *Page 641B:* TSA/Tom Stack & Associates *Page 642TR:* B. "Moose" Peterson/Wildlife Research Photography *Page 644:* Bob and Ira Spring *Page 646:* Kirkendale/Spring Photography, Edmunds, WA *Page 653:* Gary Braasch *Page 657TR:* B. "Moose" Peterson/Wildlife Research Photography *Page 657BR:* Gary Braasch *Page 658TL:* John Gerlach/DRK Photo Page *658TR:* Mark Rollo/Photo Researchers *Page 665:* Joel W. Rogers/Corbis *Page 676TR:* Stephen J. Krasemann/Allstock Inc./Tony Stone Images page *679L:* Tom Bean /Allstock Inc./Tony Stone Images *Page 680:* Tom Bean/Allstock Inc./Tony Stone Images *Page 681:* R. L. Smith *Page 684:* Marty Snyderman *Page 685:* Marty Snyderman *Page 689:* R. L. Smith *Page 691ALL:* R. L. Smith *Page 692:* R. L. Smith *Page 693L:* Gary Braasch *Page 693R:* Bill Curtsinger *Page 695L:* Gary Braasch *Page 695R:* Bill Curtsinger *Page 699:* Courtesy of NASA

ECOLOGICAL APPLICATIONS: Page II-A: Corbis/Betteman *Page II-B(L):* Michael Fogden. Animals Animals *Page II-B(R):* Joe McDonald/Animals Animals *Page III-B:* Ralph A. Clevenger *Page V-B:* Corbis/Ralph A. Clevenger *Page V-C:* Corbis/Paul A. Souders *Page VIIA:* © Stouffer Prod./Animals Animals *PageVII-B:* Corbis/NASA *Page VIII-Aa:* Corbis/W. Cody *Page VIII-A(R):* R. L. Smith *Page VIII-B:* Courtesy Janet Vail *Page VIII-C:* Corbis/Michael Boys *Page IX-C:* Courtesy National Science Foundation LTER website.

ILLUSTRATION ACKNOWLEDGEMENTS

The author wishes to acknowledge the following for kind permission to adapt, reprint or redraw from their publications the figures listed below. Full information on source and citation can be found in the captions and bibliography.

AAAS
Figure 2.23 © 1970

Academic Press
Figures 7.6 © 1994, 17.12 © 1969, 30.31 © 1978, 30.32 © 1978, 30.33 © 1978

Academy of Natural Sciences of Philadelphia
Figure 28.14 © 1978

Addison Wesley Longman
Figures 3.2 © 1972, 6.1 © 1998, 8.9 © 1989, 8.15 © 1989, 14.11 © 1986, 14.13 © 1986, 14.14 © 1986, 14.19 © 1978, 14.21 © 1978

Allen Press Inc.
Figure 6.13 © 1967

American Institute of Biological Sciences
Figures 12.8 © 1987, 30.24 © 1991

American Scientist
Figure 10.10 © 1958

American Society of Mammalogists
Figures 1212 © 1961, 12.18 © 1992

American Society of Limnology & Oceanography
Figure 14.5 © 1981

American Society of Plant Physiologists
Figure 7.7 © 1977

Annuals Reviews Inc.
Figures 19.11 © 1983, 31.12 © 1988

Austrial Academy of Science
Figure 28.17 © 1985

Benjamin Cummings
Figures 19.20 © 1983, 19.21 © 1983, 19.22 © 1983

Blackwell Science
Figures 7.26 © 1986, 7.27 © 1964, 9.2 © 1979, 9.3 © 1979, 10.3 © 1969, 10.12 © 1978, 10.15 © 1969, 10.18 © 1983, 10.20 © 1969, 10.21 © 1969, 12.21 © 1987, 13.18 © 1989, 14.18 © 1970, 15.9a © 1974, 15.9e © 1979, 15.12 © 1974, 15.13 © 1974m 15.15 © 1964, 17.7 © 1975, 21.17 © 1976, 23.28 © 1976, 25.15 © 1983

Botanical Society of America
Figure 10.6 © 1989

Bulletin of the American Meteorological Society
Figure 2.24 © 1970

Cambridge University Press
Figures 6.5 © 1992, 7.11 © 1987, 7.25 © 1986, 8.10 © 1997, 8.17 © 1979, 19.19 © 1981, 28.3 © 1988, 28.13 © 1980, 28.31 © 1981, 28.34 © 1981

Central States Forest Expt.
Figure 10.6 © 1960

Cold Spring Harbor
Figures 8.22 © 1960, 12.1 © 1957

Dr. W. Junk Publishers
Figure 7.19 © 1984

Ecological Monographs
Figure 23.27 ©1974

Ecological Society of America
Figures 3.9 © 1976, 7.28 © 1992, 10.7 © 1994, 10.8 © 1994, 10.11 © 1992, 12.2 © 1980, 12.6 © 1981, 12.9 © 1994, 12.13 © 1986, 12.17 © 1971, 12.19 © 1979, 13.3 © 1992, 14.17 © 1974, 15.8 © 1994, 15.9d © 1993, 16.17 © 1993, 16.23 © 1994, 17.8 © 1975, 18.7 © 1988, 19.1 © 1962, 21.20 © 1960, 23.8 © 1962, 24.1 © 1976, 24.15 © 1974, 28.35 © 1962, 31.35 © 1984

Entomology Society of Canada
Figures 15.7 © 1959, 15.14 © 1963

Elsevier Science
Figures 1.9 © 1985, 13.4 © 1987, 15.11 © 1991, 16.10 © 1990, 16.18 © 1992, 16.21 © 1987, 20.12 © 1990

Eric Hobbie
Figure 21.8 © 1994

Harcourt
Figure 9.14 © 1991

Harper Collins Ltd.
Figure 10.5

Harvard University Press
Figure 28.19 © 1984

Hutchinson & Ross
Figures 28.44 © 1980, 29.26 © 1982

IBM
Figure 7.10 © 1972

Iowa State University Press
Figure 4.10 © 1977

John Wiley & Sons
Figures 8.18 © 1981, 12.3 © 1981, 12.4 © 1981, 21.22 © 1971, 21.24 © 1971, 24.4 © 1966

Kluwer Academic
Figure 12.12 © 1969

Macmillan Magazines
Figures 13.6 © 1982, 19.14 © 1979, 27.14 © 1987, 27.15 © 1987

McGraw Hill
Figure 12.7 © 1982

National Research Council of Canada
Figure 15.19 © 1978

New York State Conservation Department
Figure 17.4© 1934

Noyes
Figure 19.18 © 1984

Oak Ridge National Laboratory
Figures 23.6 © 1977, 25.3 © 1962, 26.6 © 1991, 26.16, 26.17 © 1971

Oliver & Boyd
Figure 2.7 © 1977

Plenum
Figures 10.17 © 1975, 21.25 © 1984, 21.26 © 1984, 31.2 © 1968

Prentice Hall
Figures 19.17 © 1997, 31.3 © 1972, 31.5 © 1942, 31.9 © 1972

Proceedings of the First National White-Tailed Deer Disease Symposium
Figure 24.17 © 1962

Saunders
Figures 16.8 © 1976, 18.1 © 1976

Science
Figure 27.6 © 1967

Sinauer Associates
Figures 11.3 © 1996, 11.5 © 1996, 11.8 © 1996, 19.3 © 1971, 26.2 © 1993, 26.3 © 1987

Southern Illinois University
Figure 12.5 © 1994

Springer-Verlag
Figures 3.3 © 1980, 3.10 © 1976, 6.12 © 1974, 13.12 © 1985, 21.5 © 1979, 21.27 © 1981, 23.5 © 1981, 23.16a&c © 1981, 24.3 © 1974, 24.29 © 1975, 28.11 © 1979, 28.12 © 1979, 28.31 © 1973

Thomson Learning
Figure 2.5 © 1991

Transactions of the American. Microscopical Society
Figure 17.11 © 1391

Transactions of the Linn. Society.
Figure 13.2 © 1962

University of California
Figure 15.6 © 1958

University of Chicago Press
Figures 10.5 © 1979, 10.14 © 1979, 12.1 © 1987. 12.16 © 1978, 13.3 © 1986, 13.11 © 1964, 13.14 © 1964, 13.19 © 1977, 16.16 © 1992, 16.22, 17.28 © 1988, 27.16 © 1991, 30.20 © 1982

University of Georgia Press
Figures 17.3 © 1981, 24.8 © 1972

University of North Carolina Press
Figure 2.10 © 1977

University of Texas Press
Figures 8.11 © 1970, 16.5 © 1980

University of Toronto
Figure 16.19 © 1937

University of Wisconsin Press
Figures 23.14 © 1986, 23.16b © 1986

U.S. Department of Agriculture
Figures 2.2 © 1965, 23.4 © 1980, 23.17 © 1986, 27.7 © 1978, 27.9 © 1978

U. S. Department of Commerce
Figures 24.23 © 1975, 25.2 © 1975

Vernon Meentemeyer
Figure 9.16 © 1978

W.H. Freeman
Figure 2.6 © 1982

West Virginia University Office of Publications
Figure 25.1 © 1976

The Wildlife Society
Figures 1.7 © 1985, 8.3 © 1985, 8.6 © 1972, 12.22 © 1960, 23.15 © 1989

Wilson Bulletin
Figures 12.11 © 1963, 13.21 © 1971

Yale University Press
Figure 21.23 © 1971

Zoological Association of London
Figure 17.6 © 1982

INDEX